McGRAW-HILL
YEARBOOK
OF SCIENCE &
TECHNOLOGY

1992

McGRAW-HILL YEARBOOK OF SCIENCE & TECHNOLOGY

1992

COMPREHENSIVE COVERAGE OF RECENT EVENTS
AND RESEARCH AS COMPILED BY THE STAFF OF THE
McGRAW-HILL ENCYCLOPEDIA OF SCIENCE & TECHNOLOGY

McGRAW-HILL, INC.

New York   St. Louis   San Francisco

Auckland   Bogotá   Caracas
Lisbon   London   Madrid   Mexico   Milan
Montreal   New Delhi   Paris   San Juan
São Paulo   Singapore   Sydney   Tokyo   Toronto

1234567890 DOW/DOW 987654321

Library of Congress Cataloging in Publication data

McGraw-Hill yearbook of science and technology.
1962–   . New York, McGraw-Hill Book Co.

    v. illus. 26 cm.
    Vols. for 1962–   compiled by the staff of the
McGraw-Hill encyclopedia of science and
technology.
    1. Science—Yearbooks. 2. Technology—
Yearbooks. 1. McGraw-Hill encyclopedia of
science and technology.
Q1.M13   505.8   62-12028

ISBN 0-07-046707-2
ISSN 0076-2016

INTERNATIONAL EDITORIAL ADVISORY BOARD

041903

EDITORIAL STAFF

Sybil P. Parker, Editor in Chief

Arthur Biderman, Senior Editor
Jonathan Weil, Editor
Betty Richman, Editor
Ginger Berman, Editor
Patricia W. Albers, Editorial Administrator

Ron Lane, Art Director
Vincent Piazza, Assistant Art Director

Joe Faulk, Editing Manager
Ruth W. Mannino, Editing Supervisor

Thomas G. Kowalczyk, Production Manager

Art supplier: North Market Street Graphics, Lancaster, Pennsylvania

This book was set in Times Roman and Helvetica Bold Condensed by McGraw-Hill's Professional Publishing composition unit.

It was printed and bound by R. R. Donnelley & Sons Company, The Lakeside Press at Willard, Ohio.

CONSULTING EDITORS

CONTRIBUTORS

A list of contributors, their affiliations, and the titles of the articles they wrote appears in the back of this volume.

PREFACE

The 1992 *McGraw-Hill Yearbook of Science & Technology*, continuing in the tradition of its predecessors, presents the outstanding recent achievements in science and technology. Thus it serves as an annual review and also as a supplement to the *McGraw-Hill Encyclopedia of Science & Technology*, updating the basic information in the sixth edition (1987) of the Encyclopedia.

The Yearbook contains articles reporting on those topics that were judged by the consulting editors and the editorial staff as being among the most significant recent developments. Each article is written by one or more authorities who are actively pursuing research or are specialists on the subject being discussed.

The *McGraw-Hill Yearbook of Science & Technology* provides librarians, students, teachers, the scientific community, and the general public with information needed to keep pace with scientific and technological progress throughout the world. The Yearbook has long served this need through the ideas and efforts of the consulting editors and the contributions of eminent international specialists.

SYBIL P. PARKER
EDITOR IN CHIEF

McGRAW-HILL
YEARBOOK
OF SCIENCE &
TECHNOLOGY

1992

A-Z

1992

Acid rain

Acid rain has been acidified by sulfuric and nitric acids because of chemical reactions in the atmosphere between the two gaseous pollutants sulfur dioxide and nitrogen dioxide and tiny droplets of cloud moisture. The polluted rain carries hydrogen ions, sulfate compounds, and nitrate compounds. The effect of these three components on plant life has been the subject of much investigation.

Acid rain is not the only or most important form of atmospheric pollution. It is the most important form in which sulfur dioxide interacts with living systems, but ozone is the pollutant that is most damaging to plant life. Ozone is produced in the atmosphere by the interaction of excessive quantities of precursor pollutant gases (nitrogen oxides and hydrocarbons) with oxygen in the presence of intense sunlight. Research continues concerning the effects of acid rain and ozone on plants, both separately and in combination, and with various biological and climatological stress factors.

Research focuses on long-lived, or perennial, trees and on short-lived, or annual, agricultural crops. Both areas of investigation present difficult problems, and both have yielded important findings; although investigators have learned much about the impact of acid rain on natural as well as agricultural plant communities, critical uncertainties remain.

Forest studies. Studies that involve the effects of acid rain on forests must deal with the long life-spans and large sizes of trees. Experiments on mature trees attempt to control the environmental variables that influence tree growth, as well as to account for the years it may take for response to stressful conditions. These factors make controlled experiments difficult or impossible in forest ecosystems.

Experiments in a controlled laboratory or field environment are usually limited to seedling trees that can fit inside plant growth chambers and greenhouses. Responses of seedlings to simulated acid rain may not reflect those of mature trees, and this information

cannot be used to estimate the long-term, cumulative damage for long-lived species.

Often, forest studies do not include carefully controlled experiments but involve surveys of tree health along gradients of pollutant exposure. Such studies cannot readily pinpoint the cause of forest decline or damage to individual trees, but they can correlate tree injury with the occurrence of specific pollutants. Surveys can be coordinated with controlled laboratory experiments on seedlings to implicate further particular pollutants as the cause of observed tree injury or death. *See Forest and forestry*.

Effects on forests. Striking declines of forests in Europe in the 1980s initiated widespread concern among scientists that acid rain was damaging trees. Subsequent observations of tree decline in high-elevation forests in the eastern United States brought about several collaborations between European and North American research teams. Studies on both sides of the Atlantic have not yet come to firm conclusions, though findings suggest that air pollution may significantly contribute to tree decline.

Studies in the eastern United States suggest that acid rain, or more accurately, acid cloud moisture, may be involved in the decline of red spruce on mountain tops. Cloud particles deposit more moisture in these high-elevation forests than rainfall, and recent chemical analysis found pollutant concentrations more than an order of magnitude greater in cloud droplets than in rainfall.

Numerous seedling experiments support the field survey observations that acid rain or cloud moisture damages red spruce. Seedlings exposed to simulated acid rain or mist at pH 3.0 or lower (the lower the pH, the greater the hydrogen ion concentration or acidity) exhibited visible needle injury. Simulated acid rain treatments also altered the ratio of root mass to shoot mass; this increased the likelihood of injury to red spruce seedlings from winter frost.

Although the evidence suggests that acid rain and other atmospheric pollutants, such as ozone, may be a factor in red spruce decline in the eastern United States, other environmental factors are almost certainly involved. A key factor is the widespread decline of a close companion of red spruce, the fir tree. Fir trees are attacked by a particular species of aphid, a tiny insect pest. As the firs are wiped out of an area by the aphids, openings are created in the forest. These openings change the temperature and moisture conditions of the forest, which may be a factor in the observed decline of the red spruce that are left when the firs die out.

There is little additional evidence that points to damaging direct effects of acid rain on any other trees in North America. However, acid rain may indirectly affect forest ecosystems so as to lead to tree injury or even death. A body of evidence from large-scale surveys suggests that acid rain has caused a long-term accumulation of sulfur in forest soils in certain areas of North America. This accumulation may be causing nutrient imbalances in sensitive soils that could eventually affect forest health.

Uncertainties in the forest. Although there is suggestive evidence that acid rain, in certain situations, may contribute to tree injury and death and thus forest decline, the controlled experiments required to establish a cause–effect relationship are nearly impossible to perform. Incremental steps toward demonstrating a certain role for acid rain in forest decline include relating short-term effects on seedlings to long-term effects on mature trees, and determining the influence of insect pests and climatic factors, including temperature and moisture, on tree growth in forests.

A poorly understood yet potentially crucial role of acid rain may involve soils. This indirect effect of acid rain could be a factor in current forest declines in Europe and the United States, though the evidence is even weaker than for the direct effects. More disturbing are two facts concerning soil responses. First, soil responses may be even more difficult to study than tree responses because of the extreme diversity and complexity of soils. Second, soils may store up subtle, cumulative effects of acid rain over long periods of time that could at some point reach a critical threshold and damage forest trees.

Agricultural research. Unlike forest trees, agricultural crops are generally small, herbaceous plants with short life-spans. Experiments require only one growing season, and thus all stages of the life cycle can be examined. As a result, well-controlled experiments on crops usually can determine whether or not simulated acid rain had any deleterious effects.

However, the tremendous variety of crop species and cultivars, the array of pests and diseases that attack agricultural crops, and crop sensitivity to a variety of environmental factors often make it difficult to pinpoint the effects of acid rain in realistic, field experiments. Furthermore, carefully controlled greenhouse or laboratory experiments cannot reflect a crop's response to pollutants and other stress factors in the field.

Effects on crops. More than a decade of research in the United States coordinated by the National Acid Precipitation Assessment Program failed to demonstrate any consistent effects of simulated acid rain (with tests including levels of acidity from pH 5.5 to 3.0) on most major crops. Although the evidence indicates that acid rain is probably not a threat to agricultural production, in certain limited situations acid rain or mist might influence the growth or yield of sensitive cultivars.

Most of the large agricultural field experiments examined one crop, soybean. Results varied with location, environmental conditions, and soybean cultivar. Overall, there were no consistent effects of simulated acid rain on soybean growth or yield in these field experiments. Similar but less extensive field experiments with corn, potatoes, snap bean, alfalfa, and radish disclosed no significant effects.

Recent experiments in California suggest that localized and highly acidic fogs could damage alfalfa, spinach, strawberry, and other crops, although no such damage has yet been observed in the field. Of all locations in the United States, the combination of the topography, climate, and large sources of pollution in southern California have the potential for producing the most occurrences of acidic low-elevation fog.

Greenhouse and laboratory experiments have demonstrated that crop sensitivity to simulated acid rain varies with species and cultivar, and that under very limited circumstances simulated acid rain inhibits the reproductive processes that lead to seed and fruit production in corn and cotton. In similar studies of reproduction, acid rain had no demonstrable effects on seed production in wheat or soybean, and in no case has it been found to influence crop plant reproduction in the field.

Agricultural soils are modified readily and frequently in response to routine soil tests. These modifications, accomplished by additions of fertilizer or lime, can easily swamp any potential effects of acid rain. As a result, scientists do not believe that acid rain could have any damaging effects on agricultural soils.

Agricultural uncertainties. There are thousands of cultivars of agricultural crops, of which relatively few have been investigated for their response to acid rain. However, representative selections of the most important species have been tested and found to be highly tolerant of acid rain. There is widespread agreement that acid rain is not a significant threat to agricultural production.

Only under highly specific circumstances can simulated acid rain injure certain crops, so the potential impact is small.

The most significant uncertainty concerning the effect of acid rain on crops is its interaction with other pollutants and environmental stress factors. Ozone, drought, pests, and diseases are the factors most likely to interact with acid rain, although initial studies of such interactions show no cause for immediate alarm.

For background information SEE ACID RAIN; AIR POLLUTION; FOREST ECOLOGY; FOREST SOIL in the McGraw-Hill Encyclopedia of Science & Technology.

Denis DuBay

Acoustic levitation

The use of high-intensity acoustic waves to suspend an object in a fluid medium without obvious mechanical support is commonly referred to as acoustic levitation. This technique is used to overcome gravity in Earth-based applications as well as to position materials for studies in the microgravity environment of space. The National Aeronautics and Space Administration (NASA) has been supporting the development of several acoustic positioning approaches for high-temperature containerless processing of materials. Several acoustic positioning modules are being evaluated on space shuttle flights in preparation for more elaborate materials studies on the proposed space station *Freedom*.

Technologies and applications. The concept of levitation implies the use of a noncontacting force to support an object against the force of gravity. Many types of force fields have been proposed for levitation applications, such as acoustic, electromagnetic, electrostatic, light-beam, and air-jet. The ability to levitate an object opens up a new approach for the study of materials since interactions with the surrounding container are significantly reduced or eliminated. Under these noncontaminating containerless conditions, it is possible to isolate corrosive materials, cool materials below their melting point without solidification, and develop higher-purity materials and possibly new high-temperature materials. This new containerless capability is enhanced by performing these studies on orbital space flights where the force of gravity is essentially eliminated. In the space environment, the magnitude of the forces required to stably position and isolate a sample is significantly reduced. Furthermore, the microgravity environment significantly reduces gravity-induced convection and hydrostatic pressure in a fluid, buoyancy forces on an object in a fluid, and sedimentation of particles within a fluid.

Acoustic forces. Acoustic levitation has an advantage over most other levitation techniques because acoustic forces can position an object independent of the material properties of that object. This is in contrast to electromagnetic and electrostatic forces, which require the material to be conducting or to have free surface charges, respectively. In 1866 A. Kundt was one of the first to observe the effect of acoustic forces through the motion of dust particles in resonant tubes, and in 1878 Lord Rayleigh gave the first theoretical explanation for these forces. The forces associated with a sound field are produced by the acoustic radiation pressure acting on the object. The acoustic force is essentially zero for low-intensity sound fields where the fluid particle oscillations approximately obey simple harmonic motion. However, at higher intensity levels the resultant acoustic force is finite due to second-order effects associated with higher-order terms in the equations of motion that lead to more complicated fluid particle motion.

The time average of the acoustic radiation pressure can be thought of as a constant acoustic pressure component that is added to the existing ambient gas pressure (usually 1 atm). In general, the magnitude of

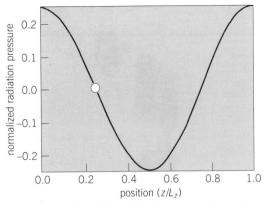

Fig. 1. Position dependence of the time-averaged acoustic radiation pressure for the fundamental plane-wave resonance. An object located anywhere along the tube will be forced to the center plane, $Z/L_z = 0.5$.

this time-averaged acoustic pressure will depend on position in the sound field. For example, a vibrating piston can be attached to the end of a tube of length L_z with a rigid end plate. The end plate reflects the sound waves to the source. When these reflected waves reach the source piston, they are again reflected. This reflection process continues until the sound-wave amplitude vanishes due to losses in the gas and the walls of the tube. The tube is said to be in resonance if the frequency of the vibrating piston is adjusted so that the reflected waves are exactly in phase with the new generated wave. When this condition occurs, the pressure amplitude of the sound field in the tube is significantly increased due to the superposition of all the in-phase reflected waves of varying amplitudes.

The dependence of the time-averaged acoustic pressure on position z along the tube is shown in **Fig. 1** for the case of the lowest (fundamental) plane-wave resonance of the tube. For this case, a sample (the circle in Fig. 1) located anywhere within the tube will experience a larger time-averaged acoustic pressure on the side of the sample away from the tube center, and the net force on the sample (obtained by integrating the time-averaged pressure over the surface of the sample) will always be directed toward the center of the tube. Thus, the sample will be acoustically positioned in the center plane of the tube. The sample will not be uniquely positioned by this fundamental plane wave alone since it can be located anywhere within the midplane of the tube. To position the sample uniquely, additional acoustic resonances that produce forces within the center plane would be needed. The acoustic forces associated with high-intensity sound fields are weak, requiring an acoustic pressure level of approximately 160 dB for this fundamental plane-wave example to produce a force of approximately 8×10^{-4} newton (1.8×10^{-4} lbf).

Acoustic positioning techniques. The Jet Propulsion Laboratory is developing several levitation techniques for NASA. One of the most versatile positioners is called the triple-axis levitator. The method used to position objects in this device, which has already flown on several space shuttle missions, is based on the plane-wave example given above. The device

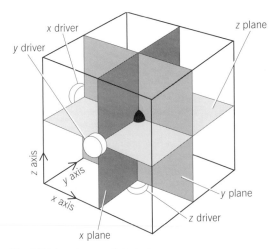

Fig. 2. Triple-axis acoustic levitation concept. The object is positioned at the chamber center by the excitation of three orthogonal plane-wave resonances. *(After G. E. Rindone, ed., Materials Processing in the Reduced Gravity Environment of Space, Elsevier, 1982)*

(**Fig. 2**) consists of a rectangular chamber with three orthogonal drivers, each of which generates a fundamental plane wave along the x, y, or z axis. These waves produce forces that position a sample in the midplane along the given axis. The intersection of the three orthogonal midplanes is a unique levitation position at the center of the chamber. Besides positioning an object, the triple-axis levitator can produce acoustic torques that rotate the suspended sample. These acoustic torques are generated by having a square cross section in the xy plane so that the x and y plane waves have identical (degenerate) resonant frequencies. In this case, the two identical orthogonal plane waves can interact with each other, and an acoustic torque of varying magnitude may be generated by varying the phase relationship between the x and y drivers.

A further advantage of the triple-axis levitator is the ability to modulate a suspended liquid sample by appropriate modulation of one or more of the excited plane waves. Modulation of the driver voltage leads to a modulation of the acoustic force applied to the sample. Detectable sample oscillations may be excited by tuning the modulation frequency to an oscillational resonance of the liquid drop. Since the oscillational frequency depends on the surface tension of the drop, the surface tension may be determined in a noncontact manner by experimentally measuring the modulation frequency that produces the maximum drop oscillation.

More recent advances in acoustic technology at the Jet Propulsion Laboratory have led to the development of single-mode acoustic positioners. These positioners use only one chamber resonance to suspend an object uniquely, in contrast to the triple-axis levitator concept, which requires the simultaneous excitation of three chamber resonances. These single-mode levitators provide greater sample stability than the triple-axis levitator and should simplify the development of advanced containerless processing space modules.

Another levitation approach uses the interference method, in which sound waves from a source such as a high-powered loudspeaker or a piezoelectric driver are directed to a rigid reflector. The reflected waves interfere with the incoming waves leading to one or more regions where the spatial dependence of the time-averaged acoustic pressure reaches a unique minimum similar to that shown in Fig. 1. The main advantage of the interference method is that an enclosed resonant chamber is not required and thus the positioned sample is much more accessible. The disadvantage of this method is that the acoustic forces generated are weaker than those produced in the resonant technique.

Containerless processing. In a typical containerless processing experiment, a material would be released in the presence of acoustic positioning fields that would suspend the sample within a furnace. The furnace temperature would then be increased sufficiently that the sample would be heated above its melting point, and after a sufficient amount of time the furnace would be cooled to resolidify the sample. The sample may be processed in a variety of ways during this thermal cycle and several of its material properties measured in a noncontact manner. More advanced processing techniques use local heating methods, such as laser beams, arc lamps, and microwave or electromagnetic fields, to heat and process the sample while it is being levitated or positioned. With these advanced approaches, samples can be rapidly heated and cooled.

The application of acoustic levitators for containerless materials studies is still in the developmental stage. It is anticipated that significant advancements and verification of these positioning techniques will take place with the availability of future long-duration space shuttle and space station *Freedom* flights. The combination of advanced acoustic levitation and noncontact measurement techniques should lead to unique studies of the thermophysical properties of materials, particularly at high temperatures.

For background information *see* ACOUSTIC LEVITATION; SPACE PROCESSING in the McGraw-Hill Encyclopedia of Science & Technology.

Martin Barmatz

Bibliography. L. V. King, On the acoustic radiation pressure on spheres, *Proc. Roy. Soc. London Ser. A*, 147:212–240, 1934; B. R. McAvoy (ed.), *Ultrasonic Symposium Proceedings*, IEEE, 1984; G. E. Rindone (ed.), *Materials Processing in the Reduced Gravity Environment of Space*, 1982; T. G. Wang et al., Shapes of rotating free drops: Spacelab experimental results, *Phys. Rev. Lett.*, 56:452–455, 1986.

Acoustics

The human head is an acoustic structure with some interesting features. The contours of the head, the skull, and the soft tissues on it act as acoustic reflectors and diffractors. The sound that enters the two ear canals and impinges on the eardrums is a modified version of sound in free space, and the ears along with the brain are capable of performing a remarkable amount of signal processing to determine not only the content of acoustic signals but also the direction of their origin. Speech production is a process involving the very efficient generation by the vocal cords of sound that is shaped by the rapidly variable geometry of the mouth and lips. Both speech production and hearing involve some tissues that are acoustically soft and some that are hard.

Because the acoustics of speech and of hearing are such complex phenomena, it is often very difficult to predict accurately the performance of human communication. Such assessments are important in designing headsets for telephones and flight helmets, studying how respirators muffle speech, determining the sound level under a communications or personal stereo headset, and studying room acoustics. Over the past few years, sophisticated head simulators have been developed that mimic the acoustics of the head very closely.

Acoustics of the head. Speech production and hearing are separate, yet closely related, functions. In the development of the species, the two evolved together, and the capacity of the ear to detect sounds closely matches the ability of the articulatory system to produce them. The ear is also to some degree capable of ignoring the effects of nearby reflecting surfaces and interfering noise sources. In the human head, there is a substantial direct acoustic coupling of sound from the mouth and nose to the ear. This coupling takes two separate paths: through the air outside the head, and through the tissues of the head, between the mouth and the inner ear.

Speech is a combination of voiced and unvoiced sounds, pops, hisses, and clicks of different parts of the articulatory tract. The sounds are influenced by the shape of the mouth, which is responsible for the differences between vowels. The articulatory tract can change extraordinarily rapidly and precisely, allowing a person to speak very fast. The acoustic energy comes from the release of compressed air in the lungs. Normally, people speak into relatively unobstructed air. The closest objects are typically more than 1 ft (30 cm) away from the mouth. The acoustic near-field is therefore not affected by those objects. The situation is quite different if there is an acoustically important object, such as a close-talking microphone or a gas mask, near the mouth. Such changed sound can easily be verified by placing a hand very close to the mouth while speaking. The interaction between the mouth and the object is again complex, and it is sometimes important to assess the quality of the speech sounds. For example, such an assessment might be important in the choice of a respirator for use by fire fighters, who need to communicate with each other. *See* SPEECH TECHNOLOGY.

Simulation of acoustic properties. The **illustration** shows an anthropometric head simulator, in which the ear and its surroundings have been accurately modeled. The external ear consists of the pinna (the visible

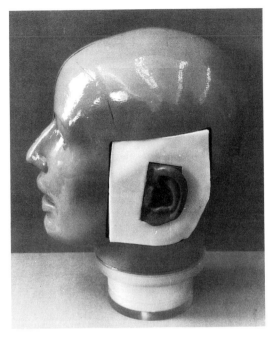

Anthropometric head with simulated soft tissues in the ear canal as well as around the external ear. There is a microphone at the end of the ear canal, and the measured sound levels accurately reflect those in the normal human ear under a wide range of circumstances.

part) and the ear canal. The pinna is a funnel for sound, but also shapes the spectrum of the incoming signal according to the direction of the source. Thus, the same acoustic field sounds different for different head orientations. This can be modeled by an average human pinna. The ear canal is about 1 in. (2.5 cm) long, tapered, oval in cross section, and has slight twists and turns. At frequencies below a few kilohertz, the wavelength of the sound is much larger than the dimensions of the ear canal. Acoustically, the ear canal can then be represented by a cylindrical cavity. Provided the cavity has the same length and volume as the ear canal, it will display resonances in the 2–5-kHz range, just like the real ear canal.

In any acoustic structure, including the ear, when sound impinges on an interface to another medium, such as the eardrum, some energy is transmitted while some is reflected. The amount that is reflected depends on the difference in acoustic impedance of the two media. For example, if a room is divided by a very thin plastic film, practically all the acoustic energy is transmitted through the film. If the film is replaced by a solid wall, most of the energy is reflected. The eardrum is connected to the bones of the middle ear, whose purpose is to convey the vibrations to the inner ear and to the acoustic nerve. The eardrum therefore represents an acoustic impedance that is due in part to the membrane itself and in part to the mechanical structures behind it. Effectively, it represents an acoustic impedance at the end of the ear canal.

In modeling the ear, a microphone is usually placed at the position of the eardrum. Thus, head simulators measure the sound pressures at that location, and not in locations that may in fact be more interesting from a physiological point of view, such as the inner ear. In order to ensure that the sound level is modeled correctly at the eardrum, it is necessary to verify the correctness of the acoustic impedance at the microphone site. Most microphones are very stiff; that is, the displacement of the membrane is extremely small. Thus, artificial elements must be introduced in an ear simulator to obtain the appropriate impedance. This is typically done by using a system of tubes and volumes near the microphone membrane.

A model of the human voice production system is a more complex task in practice, for the accurate representation of the many moving, flexible parts of the articulatory tract (jaw, tongue, and lips) is extremely complicated. In practical head simulators, a loudspeaker is coupled through a fixed tube to the mouth opening. Thus, the articulation is taken care of by the signal from the speaker, and not by the variable geometry of the acoustic pathway. By means of appropriate electronic filtering of the signal to the speaker, it is possible to create a realistic sound field from the artificial mouth, but only when there are no objects near the head. Some applications, such as respirators, effectively block the mouth, and speech intelligibility is severely affected. In order to model the speech signal properly in such cases, the mouth simulator should take into account the load impedance of the occluding object. This can be achieved through electronic feedback: A very small microphone is embedded in the lips of the head simulator. The output from the microphone in response to a known test signal is compared to the response when the head is radiating sound into an unobstructed space. The difference between the two spectra makes it possible to design a filter for the electronic signal to the speaker to compensate for the load impedance. The design and implementation of the filter can be performed automatically by digital signal processing techniques. Without this correction, the speech signal would be modified in an unrealistic way by the mask.

The shape and texture of the head form are important. The sound field near the head experiences propagation delay, reflection, and diffraction due to the shape of the head. Thus, the sound at the two eardrums is different in time of arrival and in spectrum. The mechanical properties of the head surface are important acoustically as well. For instance, a hearing protector of the plug type will, especially at low frequencies, exhibit vibrations that depend critically on the mechanics of the soft tissues that it contacts in the ear canal. Similarly, gas masks and other face coverings that contact the skin of the face will acoustically combine the properties of the soft tissues and the mask. Thus, if a head simulator is to be used for sound transmission measurements, the soft tissues of the face are important.

In practical head models (see illus.), great care is taken to ensure that all important anthropometric data are reproduced accurately. Mechanical impedance measurements of normal human skin, including the underlying soft tissues, can be made with a small vibrator equipped with an accelerometer and a force transducer. Artificial soft tissue can be fabricated from specially formulated silicone rubber. The differences in mechanical impedance over the face are obtained by varying the thickness of the rubber layer. The artificial soft tissue is mounted on a hard plastic artificial skull, and the total mass of the head simulator should ideally be the same as the dynamic mass of the normal human head.

Many practical head simulators lack some of these properties, such as soft tissue simulation or simultaneous voice production and ear function. There are several commercial units available, and a number of one-of-a-kind head simulators are produced by scientific laboratories.

Applications. The applications of acoustic head simulators are numerous. One important use is in the evaluation of the objective performance of hearing aids. Head simulators can be used for all kinds of aids and make possible the study of the effect of the design on the frequency response of the unit. Another important application is the testing of hearing protectors. Here head simulators are particularly useful, because there is no need to expose a human ear to loud sounds while testing, and also because it is possible to obtain more accurate data from the measurements. Finally, it is possible to test for sounds that are at dangerous levels, or for sounds that human ears are not well equipped to evaluate. Impulsive noise such as the sound from hammering is a good example.

The acoustics of telephone headsets can be evaluated with head simulators, as can those of other devices that are close to the head and important acoustically, such as respirators, gas masks, and flight helmets. When testing such devices, it is often desirable to use a simplified form of synthetic speech, consisting of narrow-band noise modulated by a low-frequency sine wave. The effect of the head-worn gear will be to distort the signal, and the degree of distortion represents the decrease in speech intelligibility.

An important problem in hearing conservation is the increasing use of personal stereos. It is not easy to measure the sound levels under a headphone or communications headset. The artificial head allows for an accurate measurement of the sound levels at the eardrum. If necessary, these levels can subsequently be referred to the corresponding levels in the free field that are normally measured with sound level meters, and for which there are regulations and practical experience.

For background information SEE ACOUSTIC IMPEDANCE; ACOUSTIC NOISE; ARTIFICIAL VOICE; EAR; EAR PROTECTORS; EARPHONES; HEARING (HUMAN); HEARING AID; SOUND; SPEECH in the McGraw-Hill Encyclopedia of Science & Technology.

Hans Kunov

Bibliography. H. Kunov and C. Gigure, An acoustic head simulator for hearing protector evaluation, I. Design and construction, *J. Acous. Soc. Amer.*, 85(3):1191–1196, 1989; J. Schroeter and C. Poesselt, The use of acoustical test fixtures for the measurement of hearing protector attenuation, Part II. Modelling the external ear, simulating bone conduction, and comparing test fixture and real-ear data, *J. Acous. Soc. Amer.*, 80(2):505–527, 1986.

Anemia

Anemia is a condition characterized by a hemoglobin concentration below normal (13.5–18 grams of hemoglobin per deciliter of blood in men, 11.5–16.4 g/dl in women, 11.2–12.9 g/dl in children, and 13.6–19.6 g/dl in newborns) or a decrease in the number of circulating red blood cells. It can be caused by an insufficient production of red blood cells, an increase in the destruction of red blood cells, or a loss of red blood cells through bleeding. The hemoglobin molecule inside the red blood cell captures and transports oxygen to various parts of the body for cellular metabolism, including the generation of adenosine-triphosphate (ATP). The blood of anemic patients is unable to provide an adequate amount of oxygen to the body's tissues. As a result, the patients experience lower energy levels and often suffer from fatigue or exhaustion, and they cannot live a normal, active life.

Anemia is a major and predictable complication of chronic and acute renal failure because the red blood cell–producing hormone, erythropoietin, is produced mainly by the kidney. Patients with renal failure cannot produce adequate erythropoietin to support the formation of blood cells. Approximately 400,000 patients worldwide have end-stage renal disease and require hemodialysis or peritoneal dialysis. About 90% of them are anemic, and many of them require regular blood transfusions or androgen therapy to treat the anemia. These traditional treatments have undesired side effects. For example, the efficacy of androgen therapy is limited and often causes unacceptable adverse reactions, including liver dysfunction, and masculinization in females.

A significant recent therapeutic advance for treating anemia in patients with chronic renal failure has been achieved with the production of recombinant human erythropoietin through genetic engineering technology. Recombinant human erythropoietin therapy has proven to be safe and effective in correcting the anemia of chronic renal failure in patients on dialysis. This therapy eliminates the transfusion dependence of virtually all patients. In so doing, it eliminates the risks attendant to blood transfusion, such as acquired immune deficiency syndrome (AIDS), hepatitis, and the toxicity of iron overload. Recombinant human erythropoietin therapy may increase the chance of successful kidney transplantation because of the reduction or elimination of transfusion-induced foreign cytotoxic antibody reactions. The therapy has been approved for treating anemia in AIDS patients. It is also being evaluated for use in several other anemic conditions, such as in cancer and in rheumatoid arthritis, and results are promising.

Production of red blood cells. The level of red blood cells in the human body is tightly regulated by the carbohydrate-containing protein hormone erythropoietin. The hormone exists in the circulation in minute quantities, in the range of picograms per milliliter of serum. In the fetus, erythropoietin is primarily produced by the liver; its production switches to the kidney after birth. In healthy individuals, the kidney produces approximately 90% of erythropoietin; the remainder is produced by the liver and a subpopulation of macrophages in the bone marrow. Under normal physiological conditions, the kidney senses the oxygen supply to tissues and adjusts its level of erythropoietin synthesis accordingly. Erythropoietin is produced in the kidney by specific cells that employ a heme-containing protein as an oxygen-sensing molecule. It is secreted as an active molecule into the circulation. It activates the red cell progenitors in the bone marrow to proliferate and differentiate into mature red blood cells, erythrocytes. When kidney function becomes impaired, as in chronic and acute renal failure, erythropoietin production is insufficient and anemia arises.

Development of recombinant human erythropoietin. The human gene that codes for the synthesis of erythropoietin has been isolated from genomic deoxyribonucleic acid (DNA) and spliced into a DNA vector. This vector was introduced into Chinese hamster ovary cells and incorporated into the genome of host cells. The DNA vector carries a marker that allows for amplification of the erythropoietin gene copy number, thereby achieving high levels of recombinant human erythropoietin production in the host cells. The cells are cultured in a defined serum-free medium in a sterile environment. The recombinant human erythropoietin secreted into the medium by the cells is then isolated and formulated for clinical uses. Recombinant human erythropoietin has biological and immunological properties similar to those of the natural erythropoietin that was isolated in very minute amounts from the urine of patients with aplastic anemia. These patients excrete an elevated level of erythropoietin into urine.

Therapy for anemia of chronic renal failure. Recombinant human erythropoietin was approved for the treatment of anemia associated with chronic renal failure in Europe in July 1988, in the United States in June 1989, and in Japan in April 1990. As of January 1991, more than 150,000 patients with chronic renal failure have been successfully treated worldwide with recombinant human erythropoietin and no longer require blood transfusions.

The recombinant human erythropoietin therapy enhances significantly their quality of life. It improves sleep–wake cycles and increases energy and activity levels, exercise capacity, body warmth, appetite, and libido.

Recombinant human erythropoietin treatment increases hematocrit (the volume percentage of red cells in whole blood) and hemoglobin levels proportionally to the doses given (see **illus.**). Virtually all treated patients become free of transfusions and achieve the target hematocrit range of 30–33% within 12 weeks of therapy, 50–300 units of erythropoietin per kilogram of body weight administered intravenously three times weekly. Recombinant human erythropoietin treatment does not seem to suppress endogenous production of erythropoietin and has no deleterious effects on kidney and liver functions.

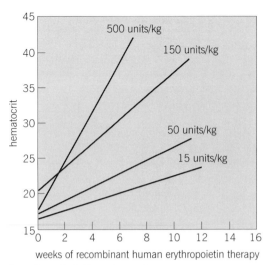

Dose-dependent effect of recombinant human erythropoietin on hematocrit. (After J. W. Eschbach et al., Correction of the anemia of end-stage renal disease with recombinant human erythropoietin: Results of a combined phase I and II clinical trial, N. Engl. J. Med., 316:73–78, 1987)

The therapy has no effect on the total leukocyte counts. It increases platelet counts by about 11%, which is within the normal platelet range, improves platelet function, and reduces bleeding because the blood clots faster. The increase in platelet counts could be due both to the direct action of recombinant human erythropoietin on megakaryocytes (platelet progenitor cells) and to the iron deficiency developed transiently as a result of accelerated red blood cell formation during the course of the therapy.

Side effects related to erythropoietin therapy are minor. No patient developed antibodies to recombinant human erythropoietin following long-term intravenous administration. While erythropoietin has no direct pressor effects, blood pressure may rise in some chronic renal failure patients during the early phase of therapy when hematocrit is increasing. Blood pressure should be closely monitored and controlled with antihypertensive medications. The increase in the blood pressure in dialysis patients may be related to an increase in the peripheral vascular resistance.

The hypertensive reaction is not observed in individuals with normal renal function who are undergoing recombinant human erythropoietin therapy, such as healthy volunteers, autologous blood donors, rheumatoid arthritis patients, and multiple myeloma patients. Therefore, the hypertensive reaction is not caused directly by recombinant human erythropoietin, but may be related to the underlying renal disease. A small percentage of patients may have a slightly higher rate of seizures during the first 90 days of therapy than at later times. Since seizures occur at a similar rate in untreated chronic renal failure patients, it is difficult to determine if the therapy affects the seizure rate.

Pharmacokinetics of the therapy. Recombinant human erythropoietin can be effectively administered intravenously or subcutaneously. For patients with chronic renal failure, the circulating elimination half-life is about 4–13 h following intravenous administration. Recombinant human erythropoietin is probably eliminated mainly by the liver, with less than 10% excreted unchanged by the kidney into the urine.

Poor responders to the therapy. Some patients demonstrate relative resistance to recombinant human erythropoietin and require a higher dose to achieve target hemoglobin and hematocrit levels. Delayed or diminished response has been observed in patients with iron deficiency; deficiencies of vitamin B_{12} and folic acid; occult blood loss; hemolysis; aluminum excess; osteitis fibrosa cystica; underlying hematologic diseases, for example, thalassemia, refractory anemia, myelodysplastic disorders, AIDS, and aplastic anemia; infections; inflammatory or malignant processes; and certain genetic defects, for example, pyruvate kinase deficiency.

Therapy for anemia of other etiologies. A number of clinical trials are in progress to evaluate the efficacy of recombinant human erythropoietin in correcting anemia of other etiologies. Recombinant human erythropoietin has been approved for treatment of anemia-related therapy with Zidovudine (AZT) in AIDS patients in the United States beginning in January 1991. Preliminary results indicate that recombinant human erythropoietin may be useful for anemia resulting from diseases such as rheumatoid arthritis, multiple myeloma, and Gaucher's disease, and anemia in premature infants.

Therapy for autologous blood transfusion. Autologous blood donations before elective surgery are now widely popular, particularly in view of the risk of transmission of AIDS and hepatitis virus from the transfusion of blood or blood products. Preliminary data indicate that administration of recombinant human erythropoietin can increase the amount of blood that can be safely collected during the preoperative period for elective orthopedic surgery.

For background information SEE ANEMIA; BLOOD; GENETIC ENGINEERING; HEMATOPOIESIS; HEMOGLOBIN in the McGraw-Hill Encyclopedia of Science & Technology.

Fu-Kuen Lin

Bibliography. J. W. Eschbach et al., Recombinant human erythropoietin in anemic patients with end-stage renal disease: Results of a phase III multicenter clinical trial, *Ann. Intern. Med.*, 111:992–1000, 1989; F.-K. Lin et al., Cloning and expression of the human erythropoietin gene, *Proc. Nat. Acad. Sci. USA*, 82:7580–7584, 1985; A. P. Lundin, Quality of life: Subjective and objective improvements with recombinant human erythropoietin therapy, *Semin. Nephrol.*, 9(suppl. 1):22–29, 1989; W. J. Williams et al. (eds.), *Hematology*, 4th ed., 1990.

Animal evolution

The sea floor of the world's oceans exhibits a remarkable range of environments, which are perhaps best characterized by their depth below the ocean surface.

Thus, coastal areas vary from the very shallow near-shore zones, which are continually disturbed by the turbulent activity of breaking waves, to somewhat deeper settings on the continental shelves, which are relatively calm except when storm waves touch bottom. Near the outer edges of the continental shelf lies an area that is too deep for storm waves to reach, and beyond that is the continental slope and the deep sea where the effects of waves and currents on the sea floor are also typically lacking.

Sedimentary rocks preserve the fossil remains of invertebrate organisms that have lived on and in the sea floor for approximately the last 600 million years (m.y.). Evolutionary changes in invertebrate marine life have long been studied by paleontologists so that details in the oceanwide history of sea-floor life are now known. Until recently the chance for evolutionary change was considered to be about equal in any of the wide array of sea-floor environments. New studies, however, show that the processes of evolution may act differently and occur at varying rates in different sea-floor environments.

Time–environment diagrams. The new studies have relied on the production of time–environment diagrams (**Figs. 1–3**). These diagrams represent an attempt to map the history of invertebrates in the different sea-floor environments through geological time. Three types of information are needed to produce such diagrams: the identity of fossils from specific deposits of sedimentary rock; the geological age of each of these deposits of rock; and the type of sea-floor environment in which each deposit of rock, with its fossils, was deposited. Such information has been collected for hundreds of fossil faunas from hundreds of individual sedimentary deposits.

This information was first collected and systematically analyzed for fossils and sedimentary rocks of Paleozoic age, ranging approximately 550–250 m.y. old (Fig. 1). These studies showed a remarkable pattern of change linked to environment. Cambrian sea-floor environments were dominated by assemblages characterized by trilobites (a common Paleozoic arthropod), along with representatives of the brachiopods, echinoderms, and mollusks. Beginning in the Ordovician, however, a new type of assemblage appeared, characterized by brachiopods and accompanied by cnidarians, echinoderms, and bryozoans. This new assemblage first appeared in nearshore settings, and during the Ordovician steadily replaced the trilobite-dominated assemblage, from onshore (in general, subtidal settings near the shoreline) to offshore environments. By the end of the Ordovician the trilobite-dominated assemblage that characterized the Cambrian no longer existed, although trilobites themselves persisted until the end of the Paleozoic. Shortly after the appearance of the brachiopod-dominated assemblage in the Ordovician, another new assemblage appeared, characterized by mollusks; this assemblage also proceeded to expand from onshore to offshore environments and, while doing this, to replace the brachiopod-dominated assemblage. Thus, the 300-m.y. history of the Paleozoic is char-

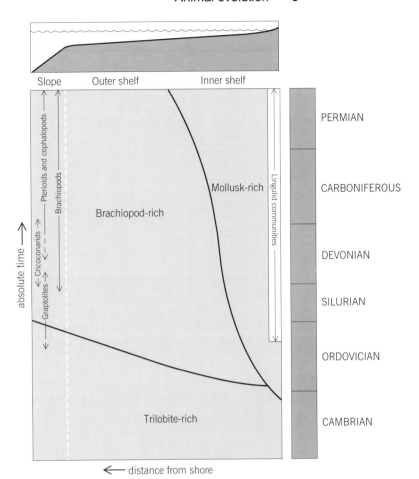

Fig. 1. Summary time–environment diagram of the results of statistical analysis of several hundred Paleozoic data points. The fields labeled trilobite-rich, brachiopod-rich, and mollusk-rich delineate the distributions of the three major assemblage types indicated by the analysis. The boundaries between these fields display an intergradation of the assemblage types. The taxa listed at left and right margins are the principal members of assemblages in very shallow and deep-water environments that do not fit easily into any of the major assemblage types; the arrows show the approximate geologic ranges of these taxa within these environments in the analyzed data. (*After J. J. Sepkoski and A. I. Miller, Evolutionary faunas and the distribution of Paleozoic benthic communities in space and time, in J. W. Valentine, ed., Phanerozoic Diversity Patterns, Princeton University Press, 1985*)

acterized by the appearance of three assemblage types, which show expansion from their nearshore environmental origins to more offshore settings as well as retreat offshore with loss of onshore representatives (Fig. 1).

Trace fossils. The patterns described above were revealed by analysis of fossils that represent the hard mineralized remains of the skeletons of once-living invertebrates. However, paleontologists must rely on different kinds of fossils for analysis of invertebrates that, in life, did not have mineralized skeletons. The activity of marine invertebrates on and in the sea floor is commonly preserved in sedimentary rocks, and the products of such activity are termed trace fossils. Trace fossils include tracks, trails, and burrows made by animals. They provide valuable evidence on the history of life because they are commonly produced by organisms that, since they do not have a mineralized skeleton, are only rarely preserved as fossils. Several distinctive trace fossils show onshore–offshore patterns resembling

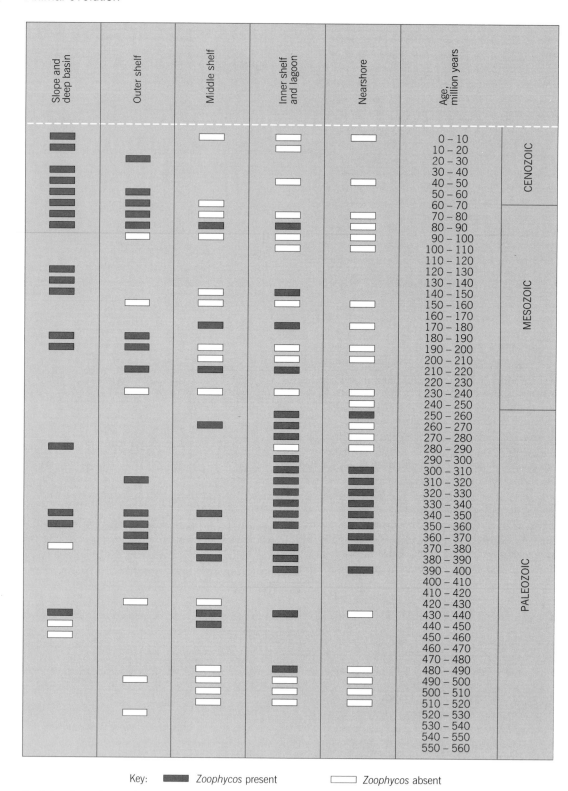

Key: ▮ *Zoophycos* present   ▯ *Zoophycos* absent

Fig. 2. *Zoophycos* time–environment diagram. (*After D. J. Bottjer, M. L. Droser, and D. Jablonski, Palaeoenvironmental trends in the history of trace fossils, Nature, 333:252–255, 1988*)

those documented for Paleozoic assemblages. For example, the trace fossil genus *Zoophycos* (Fig. 2), generally interpreted to represent a feeding burrow, first appeared in onshore environments in the Ordovician, then expanded both onshore and offshore so that it

occurred in all environments by the Devonian. *Zoophycos* was then commonly produced throughout ocean sea-floor environments for at least 150 m.y. A slow retreat, however, began in the Triassic (the beginning of the Mesozoic), which has resulted in this trace fossil being formed only in modern offshore environments from the margin of the continental shelf to the deep sea. Paleontologists are uncertain as to what type of organism has been producing this trace fossil for the past 500 m.y. However, it is known that the soft-bodied organism that makes *Zoophycos* exhibits an onshore–offshore pattern akin to that shown by the fossils of organisms with hard, mineralized skeletons (Fig. 2).

Crinoids. The analysis of Paleozoic onshore–offshore patterns has been made by examining fossil assemblages that lived in sea-floor environments, while the analysis of *Zoophycos* was for just one type of fossil. Most studies of post-Paleozoic (Mesozoic and Cenozoic) onshore–offshore patterns have also been of single fossil groups. For example, a major group of echinoderms is the crinoids, which typically attach by a stalk to the sea floor and feed by filtering organic material from seawater with a series of arms (although some fossil groups were planktonic and some modern crinoids lack a stalk and can detach from the sea floor and swim). Crinoids were decimated by the huge mass extinction at the end of the Paleozoic, with only a few survivors lasting into the Mesozoic. These gave rise to all the following crinoids, called articulates. A common and characteristic group is the isocrinids, which originated in onshore environments in the Early Triassic (Fig. 3). The isocrinids expanded into environments of greater water depth until they occurred at all ocean depths in the Middle Jurassic, and for almost 100 m.y. they could be found in this great range of environments. However, beginning in the Late Cretaceous the isocrinids commenced a retreat that has resulted in a present-day distribution limited to offshore environments, from the edge of the continental shelf to the deep sea. Other major groups of post-Paleozoic invertebrates that live on and in the sea floor, such as the cheilostome bryozoans and the tellinacean bivalves, also show onshore–offshore patterns.

Thus, much work to date has shown that onshore environments appear to preferentially foster evolutionary change when compared with other oceanic sea-floor settings. Early studies on the Paleozoic assemblage-level data concluded that they represent community evolution, with assembly of new communities in onshore environments, which then expanded across the shelf, displacing other communities. However, the finer-scale analyses done for post-Paleozoic fossil groups, illustrated here with the time–environment diagram for the isocrinid crinoids, indicate that each invertebrate group may have its own onshore–offshore history which, if viewed on the coarse scale of geological time and summed for all the invertebrate components of an assemblage, gives the appearance of community-wide patterns. Thus, paleontologists are currently debating the role of several evolutionary processes that could produce a pattern where novel morphological types, typically classified as new higher taxa (such as new orders), could preferentially appear in onshore environments.

For background information SEE ANIMAL EVOLUTION;

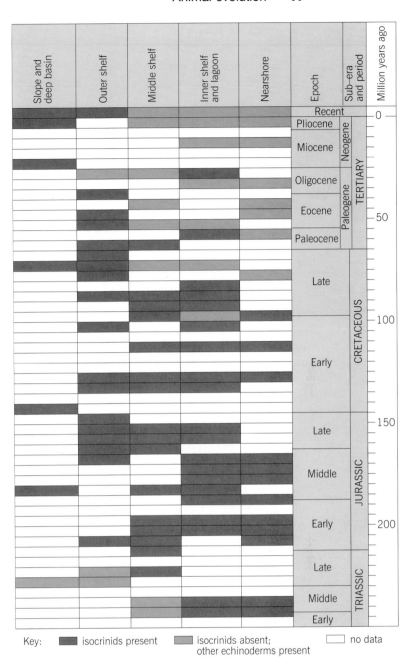

Fig. 3. Isocrinid crinoid time–environment diagram. (*After D. J. Bottjer and D. Jablonski, Paleoenvironmental patterns in the evolution of post-Paleozoic benthic marine invertebrates, Palaios, 3:540–560, 1988*)

FOSSIL; MARINE ECOSYSTEM; PALEOECOLOGY; PALEONTOLOGY; TRACE FOSSILS in the McGraw-Hill Encyclopedia of Science & Technology.

David J. Bottjer; David Jablonski

Bibliography. D. J. Bottjer and D. Jablonski, Paleoenvironmental patterns in the evolution of post-Paleozoic benthic marine invertebrates, *Palaios*, 3:540–560, 1988; D. J. Bottjer, M. L. Droser, and D. Jablonski, Palaeoenvironmental trends in the history of trace fossils, *Nature*, 333:252–255, 1988; D. Jablonski et al., Onshore–offshore patterns in the evolution of Phanerozoic shelf communities, *Science*, 222:1123–1125, 1983; J. W. Valentine (ed.), *Phanerozoic Diversity Patterns*, 1985.

Animal virus

Recent research on animal viruses has dealt with the identification of specific viral receptors and with the viral envelope protein of human immunodeficiency virus (HIV).

VIRUS RECEPTORS

Advances in the techniques of molecular biology have made possible the identification of the receptors for several common viruses, including human immunodeficiency virus, influenza virus, and the common cold virus (see **table**). These advances have led both to new insights into the evolution of viruses and to the promise of new strategies to combat virus infections.

Viruses are intracellular parasites; they can multiply and mature only inside a host cell, using the cell's biochemical machinery. The first essential step in a virus infection is the attachment of the virus to a cell. Animal viruses attach one of their surface proteins to a molecule on the host-cell surface, which is usually called the virus receptor. The receptor is often a protein, but other molecules, such as carbohydrates, are used by some viruses.

The normal function of the receptor is not, of course, to allow a virus infection, which usually damages or destroys the cell; rather, the virus has evolved so as to make use of a molecule that the cell cannot do without. Some viruses can multiply efficiently only inside certain cells, and so they use as a receptor a molecule that is present only on those cells. For example, HIV, the virus that causes acquired immune deficiency syndrome (AIDS), mainly infects the subset of T lymphocytes known as helper T cells. These T cells carry the CD4 molecule, which is the HIV receptor, on their surface.

Apart from the interest to virologists studying the biology and in particular the evolution of viruses, the receptors are of great practical importance. If the attachment of a virus to the cell can be blocked, the beginning or spreading of an infection can be prevented. Since most existing antiviral agents, such as alpha-interferon,

acyclovir, azidothymidine (AZT), and ribavirin, act only after the virus has entered its host cell, they cannot prevent all the damage done by a virus.

Influenza virus receptor. The first well-characterized virus receptor was the one for influenza virus. The receptor, sialic acid, is a carbohydrate structure present on several different surface molecules. Sialic acid was suspected to be the influenza virus receptor early on, because treatment of cells with neuraminidase, an enzyme that destroys sialic acid, also destroyed the cell's ability to bind influenza virus. The identity of the receptor was confirmed as follows: (1) reintroducing sialic acid allowed the virus to bind again; (2) addition of other molecules containing sialic acid residues competed for virus binding; and (3) x-ray crystallography was used to determine the structure of the molecule that binds to the receptor, the influenza hemagglutinin (HA) protein, in a complex with sialic acid. The sialic acid neatly fills the pocket in the hemagglutinin molecule, forming a complex.

This three-dimensional structure has also thrown light on the evolution of influenza virus. If an antibody binds to the hemagglutinin molecule near the receptor-binding pocket, the antibody may obstruct the attachment of the virus to a cell. By comparing their protein sequences, it became clear that newly evolved strains of flu virus, which may cause epidemics, often differ from previous strains in the structure of the hemagglutinin molecule surrounding this pocket. Because of this structural difference, antibodies against an earlier flu strain fail to bind to the new strain. The new strain is therefore able to attach to the receptor, sialic acid, and infect the cell.

Interestingly, flu is not the only virus to use sialic acid as a receptor. Reoviruses also bind to sialic acid, but they are less particular than flu about its exact chemical form. As a result they probably have a wider range of possible host cells, since many different proteins carry one or the other form of sialic acid.

AIDS virus receptor. An important feature of AIDS is the destruction of many T lymphocytes carrying the CD4 surface molecule. These lymphocytes are the helper T cells, necessary for an effective immune response to most foreign proteins. The normal function of the CD4 molecule is to bind proteins of the major histocompatibility complex (MHC). These proteins play a central part in allowing the immune system to distinguish between foreign proteins and the body's own tissues. Since HIV infects lymphocytes carrying the CD4 molecule so efficiently, it was suspected that the CD4 molecule itself might be the receptor for HIV. This was confirmed in the mid-1980s by the demonstration that anti-CD4 antibodies block HIV infection of T lymphocytes carrying the CD4 molecule, and that introducing the gene for CD4 into cells that lack this protein makes the cells susceptible to HIV infection.

The CD4 molecule may not be the only receptor for HIV. Some cells in brain and muscle that lack CD4 can be infected in cultures with this virus, albeit less efficiently than cells containing the CD4 molecule. The unidentified receptors are of great impor-

Known receptors for animal viruses	
Virus	Receptor
Influenza	Sialic acid
Epstein-Barr	CD21*
Human immunodeficiency virus (HIV)	CD4
Simian immunodeficiency virus (SIV†)	CD4
Rhinovirus (common cold)	Intercellular adhesion molecule 1 (ICAM-1)
Poliovirus	Immunoglobulin superfamily member (three Ig domains)
Murine leukemia virus (ecotropic‡)	Novel protein with many transmembrane domains
Reovirus	Sialic acid

*Also known as CR2 or the receptor for the C3d component of complement.
†The monkey homolog of HIV.
‡Having the ability to infect only mouse cells.

tance because HIV frequently infects many parts of the central nervous system and can cause severe dementia.

Receptors for rhinovirus and poliovirus. The CD4 molecule is a member of the immunoglobulin (Ig) superfamily, comprising related proteins that include immunoglobulin molecules. Members of this family share certain structural features called immunoglobulin domains, which are loops of protein about 90 amino acids long, each protein having a similar three-dimensional shape.

In 1989 two more virus receptors, those for rhinovirus (the common cold virus) and for poliovirus, were identified. Surprisingly, these receptors are both members, like CD4, of the Ig superfamily. The CD4 molecule has four immunoglobulin domains, the rhinovirus receptor has five domains, and the poliovirus receptor has three domains. The normal function of the poliovirus receptor protein is not known. However, just like the CD4 molecule, the molecule that turned out to be the rhinovirus receptor, intercellular adhesion molecule 1 (ICAM-1), was already known to play an important role in cell–cell adhesion. Since the Ig superfamily is now known to include a substantial proportion of the molecules on the surface of mammalian cells, and since it is likely that these molecules have changed little over the millennia (because they often have a fundamental biological role), it may simply be chance that three of the recently discovered proteinaceous virus receptors are Ig superfamily members. Two other recently cloned virus receptors, those for the Epstein-Barr virus, which causes infectious mononucleosis, and a mouse leukemia virus, are indeed quite different in structure from the Ig superfamily.

As previously mentioned, the antibodies that bind to flu hemagglutinin can prevent virus-receptor binding. However, the actual receptor-binding pocket or canyon on the rhinovirus seems to be too narrow to be penetrated by an antibody and so is protected from the immune system. The same may be true of the CD4-binding pocket on the HIV envelope protein. In the case of rhinoviruses, this probably explains why antibodies against one strain almost never protect against another strain.

Receptor analogs as potential antiviral agents. There is a highly effective vaccine available against poliovirus, but no vaccine against HIV or rhinovirus. The identification of the receptors for these latter two has raised the possibility of using soluble versions of these molecules to smother the virus's receptor-binding proteins and so block attachment to cells. Experiments in cell cultures with soluble CD4-derived proteins have been very efficient in blocking HIV infection of cells. Similarly, soluble ICAM-1 has successfully blocked rhinovirus infection of cells.

The use of artificial receptor analogs to block virus attachment could be a particularly attractive strategy in viral infections. For example, once the rabies virus has entered a nerve cell, it is beyond the reach of the immune system, and nothing can be done to prevent the advance of the disease. However, there are a number of problems to be overcome before receptor analogs can be used as effective antiviral agents in humans. In particular, they must be produced in a form that is stable; reaches the appropriate tissue in high concentrations and stays there; and does not elicit an allergic reaction, even after repeated doses. Lastly, since these receptor analogs have a similar structure to normal surface molecules with important functions, the analogs must not interfere with these normal functions.

The next few years will undoubtedly see the identification of more virus receptors. This will lead to a better understanding not only of the biology of viruses but of the normal role of the receptor molecules. It may also lead to effective treatments for virus infections for which there has been no specific therapy.

Charles R. M. Bangham

HIV Envelope Protein

HIV, the causative agent of AIDS, infects primarily the helper subset of T cells and cells of the macrophage lineage. The first steps in infection are the binding of the envelope of the virus with a cellular protein called CD4 and the subsequent fusion of the viral and cellular membranes. However, each isolate of HIV differs in its ability to infect different types of cells. This difference in virus tropism is due to mutations within the HIV envelope protein that do not affect the ability of the virus to bind to its receptor but do affect the ability of the virus to enter cells.

Role and production of envelope protein. The outer core of HIV is a lipid bilayer containing two viral proteins called the envelope proteins. Because these proteins are on the surface of the virus, they are primary targets of the immune system. These envelope proteins are made as a precursor of about 860 amino acids that is cleaved in the cell, giving rise to two proteins called the TM and SU proteins. The TM protein spans the viral membrane and anchors the SU protein to the surface. The SU protein is responsible for allowing the virus to bind to particular cells that contain the CD4 protein on their surface, while the TM protein allows the virus to fuse with the cells to which the virus has bound. After initial synthesis the envelope proteins are highly modified by the addition of complex sugars to the amino acid backbone. About half of the molecular mass of the resulting envelope glycoprotein is made of carbohydrates. It is perhaps this high amount of modification by the addition of carbohydrate that accounts for the fact that antibodies made by people infected with HIV do not protect them from disease.

The production of the envelope proteins is highly regulated by the virus. One HIV protein, called the Tat protein, controls the amount of viral ribonucleic acid (RNA) made in the cell. This small protein increases the amount of full-length RNA that initiates at one end of the virus. This type of protein is unusual for a transcriptional regulator because its target sequence is downstream of the start site of transcription. A second regulatory protein, the Rev protein, determines whether or not the RNA will be able to leave the cell nucleus to be translated into protein. The Rev protein binds to a sequence that is present in the viral RNA species that will encode the structural proteins (including envelope) of the virus. It is not yet known how the Rev protein brings about the transport of these RNAs into the cell cytoplasm. Both Rev and Tat are made early in infection. It is only at later stages of infection that envelope protein begins to be produced. The virus has probably evolved these strategies of regulated control of its envelope proteins because it can thus avoid the host's immune system and because expression of the envelope proteins on the surface of infected cells can cause the cell to fuse with itself.

HIV, like other RNA viruses, has a high mutation rate; HIV isolated from every individual is unique, although each isolate is clearly related. Even HIV isolates from the same person at different times or from different tissues are different from one another. When the sequences of different HIV isolates are compared, the envelope gene is the most variable.

Virus tropism. HIV infects all cells that display the CD4 receptor. This includes the T lymphocytes that are responsible for regulating immune responses (helper T cells), and macrophages. In addition, microglial cells, which are cells in the brain related to macrophages, can be infected. It is the infection of the CD4-bearing lymphocytes that impairs the immune system in persons suffering from AIDS. Surprisingly, not every isolate can infect every cell that has the CD4 receptor. Some isolates will infect T cells but will not infect macrophages. In contrast, some isolates infect macrophages much better than they infect T cells. Several different T-cell lines and monocyte cell lines (monocytes are the precursors to macrophages) have been established in cell culture from patients with leukemias. These cell lines have been used to characterize the different tropism of each HIV isolate.

The finding that every HIV isolate infects a particular subset of cells that display the CD4 receptor suggests that the binding of the virus to the CD4 protein is not sufficient to allow virus infection. In fact, when the human CD4 protein is expressed on the surface of mouse cells, HIV will bind to the surface of the mouse cells but will not enter. This shows that infection involves additional steps after the virus has attached to its receptor that are specific to primate cells.

Mutations in the envelope that change HIV tropism. When the nucleotide sequence of many different HIV isolates was determined, it could be shown that despite their divergence the envelope sequences contained six regions that were unusually conserved. Mutations in each of these regions had an effect on virus function. In particular, mutations of a single amino acid, a point mutation, in one region affected the ability of the envelope protein to bind to CD4. Viruses with this mutation were not infectious.

Other point mutations in the same region did not affect the ability of the envelope to bind to CD4. Surprisingly, these mutations restricted the virus tropism. While the original virus isolate could productively infect both T-cell lines and monocyte cell lines, the point mutations in the envelope abolished the ability of the virus to infect monocytes but did not at all affect the ability to infect T cells. The mutant envelope retained the ability to specifically bind to the surfaces of both the T cells and the monocyte cells. These results show that the tropism of the virus isolate is determined by sequences in the envelope gene. The tropism, however, is not determined by the ability of the virus to bind to the surface of the CD4-containing cells, but rather by the ability of the virus to enter the cells once it has bound.

There are several important implications of these new findings. Because the virus has such a high mutation rate, the envelope sequences can evolve rapidly. During the course of the disease, individuals may develop more virulent strains of HIV that will more efficiently infect their microglial cells or macrophages. This might explain the widely different time course of disease in persons with AIDS. The prevention of virus spread by treatment with antiviral drugs would be expected to decrease the possibility of acquiring a more virulent strain of HIV after the initial infection.

Additionally, the finding that HIV variants can bind to cells but not enter them suggests that a second step is needed before the virus can fuse with the cell. This step might be the binding of HIV to a secondary receptor on the surface of macrophages. Identification of the second, or entry, receptor would be very useful in the design of vaccines to prevent HIV infection.

For background information SEE ACQUIRED IMMUNE DEFICIENCY SYNDROME (AIDS); ANIMAL VIRUS; VIRUS in the McGraw-Hill Encyclopedia of Science & Technology.

Michael Emerman

Bibliography. C. R. M. Bangham and A. J. McMichael, Nosing ahead in the cold war, *Nature*, 344:16, 1990; A. Cordonnier, L. Montagnier, and M. Emerman, Single amino acid changes in HIV envelope affect viral tropism and receptor binding, *Nature*, 340:571–574, 1989; B. Cullen and W. C. Greene, Regulatory pathways governing HIV-1 replication, *Cell*, 58:423–426, 1989; A. Fauci, The human immunodeficiency virus: Infectivity and mechanisms of pathogenesis, *Science*, 239:617–622, 1988; W. Weis et al., Structure of the influenza virus haemagglutinin complexed with its receptor, sialic acid, *Nature*, 333:426–431, 1988; J. M. White and D. R. Littman, Viral receptors of the immunoglobulin superfamily, *Cell*, 56:725–728, 1989.

Antibiotic

Recent research in antibiotics has included the development of new quinolone compounds and the elucidation of the molecular genetics involved in the biosynthetic pathways to penicillin and cephalosporin.

Ciprofloxacin, a new 4-quinolone. Among the more exciting developments in antimicrobial therapy in the 1980s was the introduction of the new quinolone compounds. Foremost among these is ciprofloxacin (I). The parent compound of this group, nalidixic acid

(I)

(II), was first synthesized in 1962. Its major limita-

(II)

tions were its narrow bacterial spectrum, its weak activity relative to the amount of drug absorbed, and its ability to encourage the selection of bacterial strains during therapy. Therefore, nalidixic acid has been limited to treating urinary tract infections. In contrast, ciprofloxacin is well absorbed and is highly active against a wide variety of important pathogens. These characteristics enable the physician to treat a variety of serious infections orally that heretofore would have required intravenous therapy and prolonged hospitalization. The selection of resistant bacterial strains on therapy, although not completely overcome, appears to be less of a phenomenon during therapy with ciprofloxacin than with nalidixic acid.

All quinolones with antimicrobial activity contain an oxo group at position 4 [see structure (II)] and hence are known as 4-quinolones. The important structural changes in the new quinolones are the introduction of a fluorine atom at position 6, which is responsible for improved activity against gram-positive bacteria such as the staphylococci, and a piperazine group at position 7, which is responsible for improved activity against gram-negative species, in particular *Pseudomonas aeruginosa*.

Mechanism of action. The quinolones act on bacteria by inhibiting deoxyribonucleic acid (DNA) gyrase, a critical enzyme in DNA synthesis. The length of the bacterial DNA molecule is about 1100 micrometers, whereas the length of the bacterial cell is only 1–2 μm. Thus, in order to fit into the cell, the DNA has to be very tightly coiled and compressed. The DNA molecule, despite its extraordinarily compressed state, must still be able to replicate and segregate into daughter chromosomes without lethally entangling itself. DNA gyrase plays an essential role in this process. It is a tetramer, composed of functionally and structurally distinct subunits, two α and two β subunits. During DNA replication, gyrase, through the action of the α subunits, nicks the double-stranded DNA. Through the twisting action of the β subunits, gyrase introduces negative supercoils to permit the

DNA's tight packaging into the tiny bacterial cell. Finally the α subunits reseal the nicked ends of the DNA (**Fig. 1**). The 4-quinolones appear to bind to both α and β subunits and to the DNA, but their main mode of action appears to be to inhibit the sealing of the nick by the α subunits. Eukaryotic cells, including those of humans, contain topoisomerase, which is similar in function to bacterial DNA gyrase, but topoisomerase is not inhibited by the 4-quinolones due to the structural and functional differences between the prokaryotic and the eukaryotic enzymes.

Antimicrobial activity. Ciprofloxacin possesses potent activity against a wide variety of important gram-negative pathogens. The Enterobacteriaceae, such as *Escherichia coli*, *Klebsiella*, *Proteus*, *Salmonella*, and *Shigella*, are inhibited by less than 0.1 μg/ml of ciprofloxacin. These bacteria are important causes of urinary tract and enteric infections. *Pseudomonas aeruginosa*, a common pathogen in hospital-acquired infections, burns, and complex wound and bone infections, is inhibited by 0.5 μg/ml. The bacteria that cause sexually transmitted diseases such as gonorrhea and chancroid are also extremely susceptible to ciprofloxacin. Ciprofloxacin is also active against many gram-positive bacteria, such as the staphylococci and the streptococci. Staphylococci, which are the cause of soft tissue infections and septicemia, are inhibited by 0.5 μg/ml of ciprofloxacin. The streptococci, however, are only moderately susceptible, being inhibited at 2 μg/ml. The anaerobic flora of the mouth and intestine are for the most part resistant to ciprofloxacin.

Pharmacology. The main features of the pharmacology of ciprofloxacin are (1) its good oral absorption such that levels of 2–3 μg/ml in serum are attained after a 500-mg dose (levels similar to those achieved by a standard intravenous dose of 300 mg); (2) its high volume of distribution, reflecting good tissue and intracellular penetration of the drug; (3) its relatively long half-life in serum, 4 h, enabling twice-daily dosage in most situations; and (4) its partial metabolism in the liver, with the main route of excretion being the kidneys.

Clinical use. Ciprofloxacin has been extensively studied and has been shown to be an effective therapy for urinary tract infections, soft tissue infections, osteomyelitis, hospital- and community-acquired pneumonia, enteric infections such as traveler's diarrhea, and a variety of sexually transmitted diseases. However, many of these infections, particularly the respiratory, urinary, and soft tissue infections, can be treated just as well by using the more traditional therapies, such as penicillins, cephalosporins, or sul-

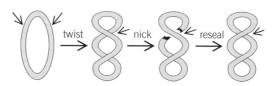

Fig. 1. Action of DNA gyrase on DNA supercoiling formation. The short arrows show the site of action of DNA gyrase.

fonamides. Moreover, ciprofloxacin is relatively costly compared to traditional therapies, and its indiscriminate usage may promote the selection of resistant strains. For these reasons, most authorities feel the use of ciprofloxacin should be restricted to problematic cases. Thus its major areas of use are in the therapy of osteomyelitis, pneumonia due to gram-negative bacteria, urinary tract infections due to resistant bacteria, prostatitis, and certain enteric infections.

Toxicity. In general, adverse effects resulting from ciprofloxacin therapy are mild. Gastrointestinal effects, such as nausea, are the most frequent and occur in about 5–10% of patients. Skin rash, headache, or insomnia are seen in 1–2% of patients. Severe renal, hepatic, or neurological toxicity is rare. There are two important drug interactions, one with a magnesium- or calcium-containing antacid drug that blocks almost completely the absorption of ciprofloxacin, and the other with theophylline, a drug for asthma, whose clearance from the body is slowed by ciprofloxacin.

Nai Xun Chin; Brian E. Scully

Molecular genetics of penicillin and cephalosporin biosynthesis. Two natural penicillins, several semisynthetic penicillins, and many cephalosporins are important in treatment of bacterial infections in humans. Their antibacterial activity depends on a β-lactam ring in their structures. The most extensively studied penicillin-producing microbes are *Penicillium chrysogenum* and *Aspergillus nidulans*, both filamentous fungi. The best-studied cephalosporin producers are *Cephalosporium acremonium*, a filamentous fungus, and *Streptomyces clavuligerus*, a gram-positive bacterium.

Prior to 1985, isolation of biosynthetic intermediates and test-tube stepwise conversions of these intermediates to β-lactam antibiotics elucidated the biosynthetic pathways to penicillin G in *P. chrysogenum*, cephalosporin C in *C. acremonium*, and cephamycin C in *S. clavuligerus*. Since early intermediates in these pathways are the same, the three pathways can be depicted in the format of a single branched pathway, as in **Fig. 2**. Genetic manipulation of penicillin and cephalosporin biosynthesis in these organisms was empirical, relying on random mutagenesis, natural selections, and extensive screening. These procedures significantly increased productivity, but—prior to 1985—specific alterations of β-lactam biosynthetic genes were not feasible.

Since 1985, molecular genetics has significantly advanced understanding of the genes and enzymes that are responsible for synthesis of penicillins and cephalosporins. Cloned biosynthetic genes have been used to increase antibiotic yield and to explore new approaches to discovering novel β-lactam antibiotics.

Application of molecular genetics to β-lactam antibiotics began with development of efficient genetic transformation systems for organisms that produce penicillins and cephalosporins. Transformation allowed the systematic introduction of chosen genes into the organisms. Genes involved in penicillin and cephalosporin biosynthesis were then cloned. Many of these genes were identified and cloned via the follow-

ing process, known as reverse genetics. A short amino acid sequence from a purified β-lactam biosynthetic enzyme was determined. From this information, corresponding DNA sequences were deduced, synthesized as oligonucleotides, and made radioactive. These radioactive probes had the capacity to locate the corresponding full-length gene in a library of DNA prepared from the host antibiotic-producing organism. Once located, the gene was subcloned, its DNA sequence determined, and its protein coding region expressed in a bacterium. This proved the identity of the gene because without the added gene the bacterium could not produce the β-lactam biosynthetic enzyme.

pcbAB gene. Two genes (*pcb*AB and *pcb*C) are involved in the synthesis of all naturally occurring penicillin and cephalosporin antibiotics. The compound L-α-aminoadipyl-L-cysteinyl-D-valine (LLD-ACV) is the first intermediate in biosynthesis of all penicillins and cephalosporins. The *pcb*AB gene encodes LLD-ACV synthetase, which catalyzes formation of LLD-ACV from three amino acid substrates: L-α-aminoadipic acid, L-cysteine, and L-valine. The *pcb*AB gene was recently identified and cloned from several different organisms. DNA sequence analysis revealed the presence of three highly related domains within this large multidomain enzyme. It has been proposed that each domain is essential for incorporation of one of the amino acid substrates. LLD-ACV synthetase is structurally related to other peptide antibiotic synthetases such as gramicidin S synthetase and tyrocidin synthetase, which operate via a thiotemplate mechanism. Thus it is likely that formation of LLD-ACV occurs by a similar mechanism.

pcbC gene. The *pcb*C gene codes for isopenicillin N (IPN) synthetase; *pcb*C was the first β-lactam biosynthetic gene to be cloned. Site-specific mutagenesis of the cloned *pcb*C gene clarified the functional significance of two highly conserved cysteine residues in IPN synthetase. Provision of novel substrates to IPN synthetase under catalytic conditions has led to the formation of many novel β-lactam structures. By 1990, seven *pcb*C genes had been cloned from divergent organisms. A surprisingly high degree of identity of the nucleotide sequences and deduced amino acid sequences led to a proposal that penicillin and cephalosporin biosynthetic genes had been transferred horizontally from a prokaryote to a eukaryotic ancestor of the filamentous fungi. Data from six other genes that are involved in penicillin and cephalosporin biosynthesis supported this evolutionary hypothesis.

penDE gene. In the last step of penicillin G biosynthesis, the L-α-aminoadipoyl side chain of isopenicillin N is replaced with a phenylacetyl moiety. This moiety can be derived from phenylacetylCoA. The reaction is catalyzed by acylCoA:IPN acyltransferase, an enzyme encoded by the *pen*DE gene. The gene from both *P. chrysogenum* and *A. nidulans* has been cloned and sequenced. The *pen*DE gene is the only β-lactam biosynthetic gene cloned to date that has introns, that is, regions of DNA within a protein coding region that do not direct polypeptide synthesis.

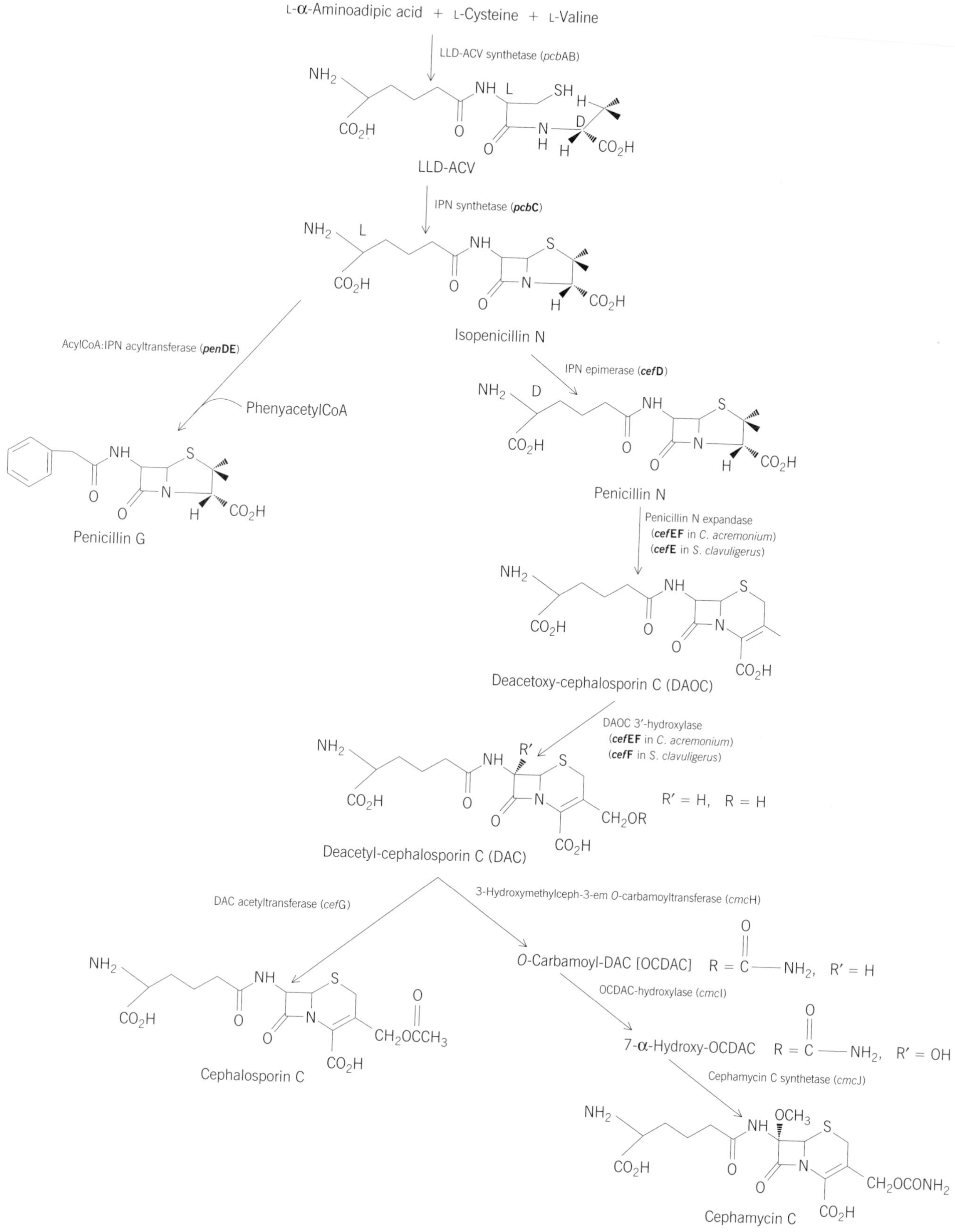

Fig. 2. Biosynthetic pathways for the sulfur-containing β-lactam antibiotics penicillin G, cephalosporin C, and cephamycin C. When gene designations are shown in bold type, the gene has been cloned and expressed in *Escherichia coli*; *pcb* denotes a gene that functions in both penicillin and cephalosporin biosynthesis; *pen* denotes a gene that functions in the synthesis of penicillin G; *cef* denotes a gene that functions in the synthesis of cephalosporins; *cmc* denotes a gene specific to the synthesis of cephamycins. Stereochemistry of isopenicillin N (IPN) and of penicillin N is designated by L and D. (*After S. W. Queener, Molecular biology of penicillin and cephalosporin biosynthesis, Antimicrob. Agents Chemother., 34:943–948, 1990*)

The gene encodes both of the two nonidentical subunits of IPN acyltransferase, a 30-kilodalton subunit and a 10-kD subunit. Adding copies of the *pen*DE gene to low-producing strains of *P. chrysogenum* increased penicillin production. Improvement was not observed in industrial strains that already had the entire pathway amplified several fold.

cefD gene. Cephalosporin C and cephamycin C are each biosynthesized from deacetylcephalosporin C (DAC). In both pathways, isopenicillin N must be epimerized to penicillin N by IPN epimerase in a pyridoxal-dependent reaction. The enzyme is encoded by *cef*D in *S. clavuligerus* and *C. acremonium*. The cloned gene from *S. clavuligerus* indicated a size of 55 kD for IPN epimerase. The *cef*D gene differs markedly from other cloned penicillin/cephalosporin genes. This was expected since the epimerase differs mechanistically from the synthetases, oxido-reductases, and transferases in the biosynthetic pathway.

cefEF gene. The five-membered thiazolidine ring of penicillin N is ring-expanded to the six-membered dihydrothiazine ring of deacetoxycephalosporin C (DAOC). DAOC is hydroxylated at the 3′-carbon to form DAC. In *C. acremonium*, both ring expansion and hydroxylation are catalyzed by a bifunctional enzyme encoded by the *cef*EF gene. Examination of large-scale cephalosporin fermentations revealed accumulation of the intermediate penicillin N in the broth. Therefore, it was reasoned that adding extra copies of the *cef*EF gene to this *C. acremonium* production strain should improve antibiotic production. The *cef*EF gene from *C. acremonium* was cloned via reverse genetics. One extra copy of this gene was added to the production strain by transformation. When compared to the untransformed strain, the transformant exhibited twice the expandase/hydroxylase activity, dramatically reduced accumulation of penicillin N, and increased cephalosporin C production. Transverse alternating-field gel electrophoresis was used to characterize the strain and transformant. Eight chromosomes were detected in each, ranging in size from 1.7 to 4.0 megabases. The naturally occurring copy of the *cef*EF gene was found on chromosome II. The extra copy of *cef*EF in the transformant had integrated ectopically into chromosome III. Regulatory approval at both national and local levels have been obtained to use this improved organism in large-scale fermentations.

cefE and cefF genes. *Streptomyces clavuligerus* possesses two monofunctional genes, *cef*E and *cef*F, coding for expandase and hydroxylase, respectively. Both genes have been cloned and characterized. They share extensive homology with each other and with the bifunctional *cef*EF. Expandase, hydroxylase, and bifunctional expandase/hydroxylase each require α-ketoglutarate, ferrous iron, and molecular oxygen for activity. These data are consistent with horizontal transfer of a bifunctional *cef*EF gene from a prokaryote to a eukaryotic ancestor of the fungi. Later, a gene duplication may have occurred in the prokaryotic ancestor, followed by functional drift that formed two virtually monofunctional enzymes.

Several oral cephalosporins are manufactured from 7-ADCA, which is made by chemical ring expansion of penicillin G. 7-ADCA can be formed by removal of the D-α-aminoadipoyl side chain from DAOC by a new enzymatic process. Thus, biosynthetic DAOC might allow development of a simple aqueous-based process for making 7-ADCA. This would be compatible with efforts to reduce use of organic solvents in industry in order to lessen environmental pollution. The monofunctional nature of expandase produced by *cef*E from *S. clavuligerus* is particularly attractive for this effort because its product is DAOC. This gene has been expressed in *P. chrysogenum*. To produce a strain suitable for DAOC production at industrial scale, expression of *cef*D and removal of acyltransferase activity from this transformant would be necessary.

cefG gene. The gene encoding DAC acetyltransferase, the enzyme catalyzing the last step in the cephalosporin C pathway, has very recently been cloned. It was located near the *pcb*C gene in *C. acremonium*. Therefore, all genes involved in the natural production of penicillin G and cephalosporin C have been cloned and characterized.

Clustering of genes. All cloned *pcb* and *cef* biosynthetic genes are found clustered together in the bacterium *S. clavuligerus*. The *pcb*AB, *pcb*C, and *pen*DE genes are found as a cluster in *P. chrysogenum*. The *pcb*AB, *pcb*C, and *cef*G genes are clustered on chromosome II in *C. acremonium*. Genes specific for cephamycin C production (*cmc* genes) remain to be cloned.

For background information SEE ANTIBIOTIC; MOLECULAR BIOLOGY in the McGraw-Hill Encyclopedia of Science & Technology.

Stephen W. Queener; Paul L. Skatrud

Bibliography. V. T. Andriole, *The Quinolones*, 1988; T. D. Ingolia and S. W. Queener, Beta-lactam biosynthetic genes, *Med. Res. Rev.*, 9:245–264, 1989; H. C. Neu, Ciprofloxacin: An overview and prospective appraisal, *Amer. J. Med.*, 82(suppl. 4A):395–404, 1987; H. C. Neu, New oral and parenteral quinolones, *Amer. J. Med.*, 87(suppl. 5A):283–287, 1989; J. V. Piddock and R. Wise, The antibacterial action of the 4-quinolones, *Antimicrob. Newsl.*, 2(1):1–4, 1985; S. W. Queener, Molecular biology of penicillin and cephalosporin biosynthesis, *Antimicrob. Agents Chemother.*, 34:943–948, 1990; S. M. Samson et al., Isolation, sequence determination and expression in *E. coli* of the isopenicillin N synthetase gene from *Cephalosporium acremonium*, *Nature* (London), 318:191–194, 1985; P. L. Skatrud et al., Use of recombinant DNA to improve production of cephalosporin C by *Cephalosporium acremonium*, *Bio/Technology*, 7:477–485, 1989.

Antibody

The availability of hybridoma antibodies since 1975 has revolutionized the science of immunology. A more recent advance in the production of antibodies is

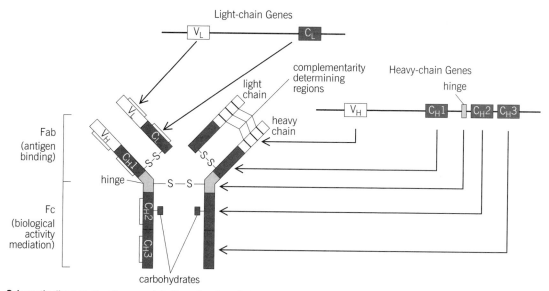

Schematic diagram of an immunoglobulin molecule and the genes that encode it. Arrows indicate the correspondence between the DNA segments and the different domains of the immunoglobulin polypeptide chain that they encode. The light chain consists of a variable region (V_L) and a constant region (C_L) domain. The heavy chain consists of a variable region (V_H) and three constant region domains (C_H1, C_H2, C_H3). Fab fragments contain the antigen binding sites, and the Fc fragment contains most of the sequences determining effector functions. Within the V_H and V_L are the complementarity-determining regions (CDRs) that form the majority of the contacts with antigen. S—S = interchain disulfide bonds.

the ability to express genetically engineered antibody genes. Antibody genes can be manipulated by using standard techniques of molecular biology to encode antibodies with altered combining specificities or effector function. Antibody genes can be assembled by using segments derived from different species to produce chimeric antibodies; antibody gene segments can be joined to nonimmunoglobulin sequences to produce molecules with novel functions. The resulting engineered antibodies can be produced in mammalian cells, yeast, bacteria, or insect cells. Recent innovative methods may lead ultimately to antibodies that can be produced completely in the laboratory without using an immunized animal.

Structure. Antibody molecules are composed of two different polypeptide chains, a heavy chain and a light chain, that are held together by disulfide bonds (see **illus.**). Each chain folds into discrete domains, each of which consists of about 110 amino acids. There are two domains in the light chain and four or five in the heavy chain. Each domain is encoded by a discrete genetic element, an exon.

The variable domains of the heavy chain (V_H) and the light chain (V_L) form the antibody combining site. These variable regions can be divided into complementarity-determining regions (CDRs) and framework regions. It is the complementarity-determining regions that interact primarily with the antigen and determine the combining specificity; the framework residues serve as scaffolds to support and position the complementarity-determining regions. The constant region domains (C_L and C_H) determine the biologic properties and effector functions of the antibody, such as the ability to activate complement. Some antibodies also contain a hinge region that provides segmental flexibility. Human antibodies can

have any of nine different constant regions and, depending on the associated constant region, will express different effector functions.

Chimeric antibodies. One limitation inherent in the initial hybridoma antibodies was that they were primarily of rodent origin; it is still difficult to produce human monoclonal antibodies. Rodent antibodies are adequate for use in laboratory culture, but their foreignness makes it difficult to use them for therapy in the organism. One solution to this problem has been to produce chimeric antibodies in which the variable regions from the rodent antibodies are joined to human constant regions. The resulting antibodies are mostly human and possess human constant regions, but have rodent variable regions. Chimeric antibodies have now been produced with many binding specificities, including those with antitumor specificities, and exhibit the expected biologic properties; for example, they continue to bind the antigen recognized by the original rodent hybridoma, and they exhibit the effector functions characteristic of the human constant regions to which they are joined. The limited clinical data indicate that chimeric antibodies exhibit decreased immunogenicity and have improved biologic properties compared to their totally rodent counterparts.

Although the chimeric antibodies are a decided improvement, the murine variable regions elicit immune responses in some individuals. A technique termed CDR grafting has been developed to attempt to reduce this immunogenicity. For example, this technique has been used to graft the complementarity-determining regions from mouse antibodies onto human framework regions, resulting in variable regions with mostly human sequences. Frequently these humanized variable regions retain their specificity for antigen, although often with reduced affinity. However, CDR grafting is not always successful, and the humanized variable regions sometimes no longer recognize antigen.

Fusion proteins with novel functions. One of the major advantages of producing antibodies by genetic engineering is that investigators are no longer limited to using antibodies as they occur in nature. In particular, it is now possible to make fusion proteins in which antibody combining specificities are joined to nonantibody proteins. Antibody combining specificities have been joined to toxins, enzymes, growth factors, and oncogenes. Antibody specificities joined to toxins potentially provide specific cytotoxic agents. Antibody combining specificities joined to an enzyme provide reagents for use in immunoassays. In an exciting example of this approach to producing immunotherapeutics, an anti–fibrin combining specificity was joined to the deoxyribonucleic acid (DNA) sequences encoding the catalytic beta chain of the tissue-type plasminogen activator. When this heavy chain was expressed with a specific light chain, the fusion protein bound fibrin, although with reduced affinity, and exhibited the peptidolytic activity characteristic of the native tissue-type plasminogen activator. Antibody specificities can be joined to genes that encode proteins whose purification is difficult, for example, the products of some oncogenes such as *c-myc*. The resulting fusion proteins can be purified on an antigen column. Another approach is to join antibody combining specificities to growth factors. These fusion proteins potentially provide more effective reagents for localizing tumor cells since they can be targeted to tumor-specific antigens through their variable regions and to receptors on the surface of the malignant cell through the growth factor.

Although the emphasis generally has been to join antibody combining specificities with other molecules, it has also been possible to join antibody constant region to foreign proteins in an attempt to alter their biologic properties. One example is soluble CD4 joined to immunoglobulin constant regions. Soluble CD4 has been proposed as an acquired immune deficiency syndrome (AIDS) therapeutic because it binds to the envelope of the human immunodeficiency virus (HIV), the causative agent of AIDS, and may slow or halt the spread of infection by preventing HIV from binding to cellular receptors (CD4 on helper T cells). *See Animal virus.*

Fusion proteins represent vastly different structures from the initial antibody molecule. Laboratory mutagenesis of cloned immunoglobulin genes provides a method for making more subtle changes in the antibody molecule. Laboratory mutagenesis of the variable region can be used to alter combining specificity, affinity, and idiotype. Laboratory mutagenesis of the constant region can be used to alter effector functions such as receptor binding or complement activation, or biologic properties such as serum half-life. The net result of these changes would be the production of an antibody better suited for the desired purpose.

Expression systems. Several expression systems are available for producing antibodies. Mammalian expression systems have great appeal, for they provide a biosynthetic environment similar to that in which the antibody is naturally produced. In particular, cultured myeloma cells that are frequently used for antibody expression are known to be able to properly assemble, glycosylate, and secrete antibody molecules. Limitations of mammalian expression systems are that it is difficult to achieve a high production level and that they are expensive. Yeast provides an alternative nonmammalian eukaryotic expression system. An antibody can be properly assembled and secreted by yeast. A shortcoming of the yeast expression system is that the carbohydrate added by yeast differs in structure from that of mammalian cells and antibodies containing the yeast carbohydrate do not possess the full range of effector functions.

Production of antibodies in bacteria has great appeal because of the potential for producing large quantities at low cost. However, attempts to produce complete antibodies in bacteria have met with failure. Production of antibody fragments in bacteria has been much more successful. Fc-like fragments of immunoglobulin E (IgE) have been produced that continue to bind to the Fc receptor on cultured human basophils. Fab fragments have been produced that are able to bind to their specific antigen. However, a persisting problem is that bacteria do not glycosylate proteins, so antibody products made in bacteria will not contain any carbohydrate and therefore will not express any of the functional properties dependent on its presence.

Recently, a mouse immunoglobulin was synthesized in insect cells by using a baculovirus expression system. The polypeptide chains were correctly processed at their amino terminus, glycosylated, and assembled. The resulting antibodies were able to bind antigen and express the expected idiotypes.

Expression of immunological repertoire in E. coli. To date, the major emphasis has been to express antibodies whose variable regions were cloned from cells synthesizing antibodies of a known specificity. It is, however, possible that the immunoglobulin repertoire can be completely constructed in the laboratory and then screened for molecules of the desired specificity. Two basic approaches to this problem have been taken. One approach was motivated by the observation that the variable domains of the heavy chain of an antilysozyme antibody expressed in *Escherichia coli* was sufficient to bind antigen. A library of amplified genomic DNA made from a mouse immunized with lysozyme was expressed in *E. coli*, and about 2000 colonies were screened for antilysozyme activity. It was possible to isolate variable domains of the heavy chain with antilysozyme activity from this library and from a similar library made from a mouse immunized with an irrelevant antigen.

The second approach has been to produce a combinatorial library of Fab fragments in *E. coli* by using a bacteriophage lambda expression system generated by amplification of messenger ribonucleic acid (mRNA) from immunized mice. Since the library is screened by using phage plaque lift techniques, large numbers of recombinants can be screened. In this case, antigen binding colonies could be detected only when Fab was expressed; neither light chains nor heavy chain fragments exhibited antigen binding activity.

For background information SEE ANTIBODY; ANTIGEN-ANTIBODY REACTION; IMMUNOGLOBULIN in the McGraw-Hill Encyclopedia of Science & Technology.

S. L. Morrison

Bibliography. C. A. Hasemann and J. D. Capra, High-level production of a functional immunoglobulin heterodimer in a baculovirus expression system, *Proc. Nat. Acad. Sci. USA*, 87:3942–3946, 1990; W. D. Huse et al., Generation of a large combinational library of the immunoglobulin repertoire in phage lambda, *Science*, 246:1275–1281, 1989; S. L. Morrison and V. T. Oi, Genetically engineered antibody molecules, *Adv. Immunol.*, 44:65–92, 1989; S. E. Ward et al., Binding activities of a repertoire of single immunoglobulin variable domains secreted from *Escherichia coli, Nature*, 341:544–546, 1989.

Antifertility agent

The work on steroid receptors and antihormones, the progress in understanding progesterone action in the human menstrual cycle and pregnancy, the concern regarding insufficient available fertility control methods, and the expertise of chemists and pharmacologists have merged to produce RU486 (generic name, mifepristone), a steroid with high affinity for the progesterone receptor. RU486 is the first efficient, safe noninstrumental method of early pregnancy interruption. In 1989–1990, more than 60,000 women in France chose RU486 over surgery for abortion. The evidence for the safety and efficacy of this method is compelling. RU486 also has other potential applications in medicine.

Interruption of pregnancy by RU486. A short interruption of progesterone activity is sufficient to stop early pregnancy and the luteal phase of a nonfertile cycle. RU486 works by blocking the normal action of the hormone progesterone during pregnancy. While the success rate for pregnancy interruption with a single dose of 600 milligrams of RU486 is 80% in pregnancies of less than 42 days of amenorrhea, this percentage decreases significantly in longer pregnancies. However, a small dose of prostaglandin given 36–48 h after RU486 raises the success rate to at least 96% for pregnancies of up to 49 days of amenorrhea. Failures of 3% are due to incomplete abortions or bleeding, necessitating instrumental intervention.

In 1% of women tested, RU486 exerted no effect. Although it is not known why RU486 is ineffective in those women, one possibility may be that there is a defect in the steroid-binding domain in a small percentage of women. This hypothesis is supported by studies with chicks where it was found that RU486 does not act as an antiprogesterone. The progesterone receptor does not bind RU486 in this species, apparently because of a single amino acid difference in the steroid-binding domain.

Although the method of combining a 600-mg dose of RU486 with a small dose of prostaglandin results in a very high success rate, it may be possible to improve the method. Doses of RU486 lower than 600 mg may yield the same results in the combined treatment with prostaglandin, but may necessitate higher doses of prostaglandin, which could lead to increased discomfort. The search for an active prostaglandin that would cause less severe uterine cramps should have high priority in ongoing research. Furthermore, the discovery of an oral prostaglandin could make voluntary pregnancy interruption less intimidating to patients; currently, in France, prostaglandin is administered as an injection or vaginal pessary. If a prostagladin derivative could exert a delayed effect and be administered at the same time as RU486, then the separate medical visit that is required in France for the administration of prostaglandin could be eliminated. Medical supervision of the RU486-plus-prostaglandin treatment is necessary for detection of ectopic pregnancy, which is not terminated by RU486, and for prevention and treatment of complications such as hemorrhages, incomplete evacuation, and cardiovascular effects of prostaglandin.

In all other European countries (except Ireland), although different laws regulating abortion exist, RU486 should become available to women in the next few years. In the United States, RU486 lacks approval from the Food and Drug Administration. Moreover, the current uncertainty over the legal status of abortion has discouraged the large drug companies from even applying for licensing of RU486.

Contraception by RU486. Besides the successful clinical use of RU486 for early abortion, animal studies indicate that this compound inhibits follicle maturation, ovulation, and egg implantation.

The use of RU486 late in the cycle, specifically with a dose of 100–400 mg/day on the last 2 days, induces earlier menses in the nonpregnant woman and no change in the next cycle. Statistically, a sexually active woman has a 20% chance of becoming pregnant if engaging in unprotected sex. The use of RU486 late in the cycle as a contraceptive method has a 20% failure rate. Thus, the combined failure rate for unprotected sex followed by late luteal contraception is the product of the two separate failure rates ($0.20 \times 0.20 = 0.04$). A failure rate of 4% would lead to a pregnancy on the average of once every 2 years, which is clearly unacceptable for a contraceptive method. The failure rate may be reduced by administering a small dose of prostaglandin or antigonadotropin-releasing hormone; however, trials have not yet been conducted.

Periovulatory (more precisely, in the 2–3 days after the luteinizing hormone peak) administration of RU486 at a very low dose does not modify the luteal function and luteinizing hormone output. It provokes an alteration in the endometrium, which may then prevent implantation at midcycle.

Much work has been done on interruption of the

menstrual cycle at the midluteal phase. A dose of 50–200 mg RU486/day, taken 2–4 days at the midluteal phase, is very effective in blocking ovulation. However, this method often causes a change in the time of the next ovulation. Therefore this method is not yet a viable form of contraception for more than 1 month.

Conventional contraception by prevention of fertilization could also be obtained, because RU486 administered during the follicular phase delays and eventually suppresses ovulation. A successive administration of RU486, then progestin, and then again RU486 to ensure anovulation and cyclic menses has been proposed. However, even if this appealing estrogen-free contraception proves to have high efficacy, it probably will not be available in the near future, since it requires that a drug company carry out safety testing and development of a new contraceptive pill when several such pills already exist.

Other uses of RU486 as an antiprogestin. During pregnancy, RU486 can facilitate intrauterine maneuvers that require opening and softening of the cervix, changes that accompany the normal onset of labor. Such a use of RU486 would be helpful in pregnancy interruptions due to pathological states in the mother or fetus. At delivery time, RU486 can trigger labor in cases of undue delay. Experiments in monkeys have indicated its effectiveness. Incidentally, the antiprogesterone effect of RU486 also facilitates early milk secretion.

Beyond pregnancy management, RU486 may have a role in the treatment of endometriosis and breast cancer. Some breast tumors require a combination of estrogen and progesterone for growth. The value of RU486 in arresting these tumors is being investigated. Clinical trials are also being performed on patients with progesterone receptor–containing neurological tumors, for example, meningiomas. In addition, the ability of RU486 to decrease the output of luteotropic hormone and follicle-stimulating hormone may be of therapeutic importance during menopause. In any case, for long-term administration of RU486, evaluation of the safety of the drug will be necessary.

RU486 as an antiglucocorticosteroid. Originally, RU486 was found to be an anticorticosteroid, and indeed it binds with high affinity to the glucocorticosteroid receptor. The counteraction of RU486 on the negative-feedback activity of corticosteroids provokes an increase of adrenocorticotropic hormone (ACTH), endorphin, and cortisol in humans. Thus, large doses of RU486 are needed to overcome this reaction in order to obtain a state of hypocorticism.

Four categories of use of RU486 as an antiglucocorticosteroid may be envisaged: (1) basic research on the hypothalamus-pituitary-adrenals system, thus providing a new way to test this important axis in stressful conditions and immunological reactions; (2) therapeutic intervention in the regulation of the hypothalamus-pituitary-adrenals system in some states of depression, immunological abnormalities, and stressful conditions; (3) intervention in certain conditions where

excess cortisol is produced independently of the central nervous system–pituitary regulatory mechanism, as in inoperable adrenal cancers of ACTH-producing ectopic tumors; in conjunction with RU486 surgical removal of a tumor has occasionally become possible; (4) local administration of RU486, permitting a localized antiglucocorticosteroid effect, for example, in treating glaucoma or accelerating the healing of wounds and burns.

For background information SEE ANTIFERTILITY AGENT; MENSTRUATION; OVARY; PREGNANCY; PREGNANCY DISORDERS; PROGESTERONE in the McGraw-Hill Encyclopedia of Science & Technology.

Etienne-Emile Baulieu

Bibliography. E. E. Baulieu, Contragestion and other clinical applications of RU486, an antiprogesterone at the receptor, *Science*, 234:1351–1357, 1989; E. E. Baulieu and S. J. Segal (eds.), *The Antiprogestin Steroid RU486 and Human Fertility Control*, 1985; W. Herrmann et al., Effet d'un stéroïde anti-progestérone chez la femme: Interruption du cycle menstruel et de la grossesse au début, *C. R. Acad. Sci.*, 294:933–938, 1982; S. R. Milligan (ed.), *Oxford Reviews of Reproductive Biology*, 1989.

Apical meristem

All seed plants elongate by means of shoot apical meristems, which are regions at the shoot tip where small, densely cytoplasmic cells divide frequently. Shoot apical meristems have three primary functions in plant morphogenesis: (1) They produce new cells. (2) They establish tissue patterns of the shoots. (3) They remain relatively constant genetically, thereby minimizing the occurrence of mutations and the damage caused by them. Studies that exploit the unique types of growth and development in cacti have elucidated many aspects of the metabolism of shoot apical meristems for all plants.

Producing new cells. The first role of shoot apical meristems is well understood in terms of cell biology. Newly formed vascular tissues (xylem and phloem) transport sugars, minerals, and water to the shoot apical meristem, where they are incorporated into protoplasm as carbohydrates, proteins, lipids, and nucleic acids. This causes cells to grow, with cell division resulting in new cells. When averaged over a long period of time, half the cells flow out of the meristem and half remain, so that the meristem stays constant despite the inflow of nutrients and the export of cells.

Producing tissues in an orderly pattern. In many seed plants, the apex develops a particular structure at germination, then maintains this structure until the time of flowering, perhaps years later. The structure of the meristem is so constant that it appears to be essential for establishing the proper tissue patterns in the stem. The structure of the apex consists of a dome-shaped mass of cells called the corpus, which

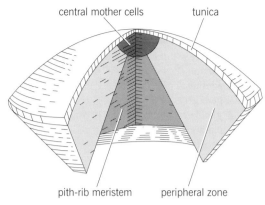

Fig. 1. Zones of shoot apical meristems.

includes three zones (**Fig. 1**): a central columnar pith rib meristem, located below an oval group of central mother cells, both of which are surrounded by a skirt of cells termed the peripheral zone. Cells in the corpus divide in all directions, so that the tissues produced by each zone can enlarge in all dimensions. Overlying the corpus is a fourth zone, the tunica, whose cells divide only with new walls perpendicular to the surface of the shoot apical meristem. Thus, the tunica grows in two dimensions as a sheet, covering the tissues produced by the corpus but never becoming thicker. Quantitative transmission electron microscopy has proven that each zone is made of cells that differ from those of the other zones in many aspects of cytology.

In cacti, this organization of the shoot apical meristem is not necessary for the orderly production of leaves or stem tissues. The seedlings of many cactus species do not have fully zoned shoot apical meristems at germination; frequently the corpus is an irregular mass of cells. However, the shoot apical meristem still functions normally and produces stem tissues in the proper spatial arrangement. As the seedlings age, full zonation develops gradually, with the central mother

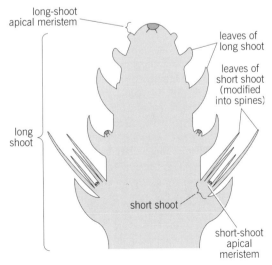

Fig. 2. Main stem of a cactus.

cell zone becoming distinct earliest, followed by the pith rib meristem and the peripheral zone. Even with partial zonation, the meristem functions normally. From this observation it was concluded that the metabolic mechanisms that control tissue morphogenesis are not linked to or part of the mechanisms that control zonation. The constant association of these mechanisms in other species is probably coincidental.

Further understanding of tissue morphogenesis was made possible because the bodies of cacti consist of at least two types of shoots and leaves. The bulk of the body, such as the prickly pear pad or the barrel of barrel cacti, is a shoot that bears microscopic leaves (**Fig. 2**). It is known as a long shoot and is equivalent to the shoots of all other seed plants. Just as in other plants, the leaves bear axillary buds; but whereas in most plants these would typically grow out as a branch (another long shoot), in cacti they develop into shoots only 0.08–0.12 in. (2–3 mm) long, known as short shoots. The leaves of short shoots are modified into spines. In cacti, the shoot apical meristems of long shoots and short shoots have identical structure and zonation, but those of long shoots are much larger, up to 1500 micrometers in diameter, than those of short shoots, less than 70 μm in diameter.

Studies of dormant shoot apical meristems in tissue culture have shown that applications of the plant hormone benzylaminopurine (a cytokinin) causes shoot apical meristems to become active mitotically, to enlarge, and to begin acting as long-shoot apical meristems. These meristems produce a shoot with long internodes and ordinary, small cactus leaves. Alternatively, applications of the hormone gibberellic acid cause dormant shoot apical meristems to become mitotically active and to act as short-shoot apical meristems, producing a short shoot with spines. Differences in shoot structure correlate not with shoot apical meristem zonation structure (which is identical in both types of morphogenesis) but with mitotic activity within zones. With applications of cytokinin, all zones are very active in cell division, and a long shoot results. With application of gibberellic acid, the two central zones—the central mother cells and the pith rib meristem—remain dormant and a short shoot results. Only the tunica and peripheral zone become active, resulting in the production of spine primordia. Consequently, shoot morphogenesis is controlled by the type of hormone acting on the shoot apical meristem and involves the differential activation of particular zones, not the alteration of shoot apical meristem structure.

Maintaining genetic health. To produce the new cells of a plant, meristems must undergo cell division, which is preceded by deoxyribonucleic acid (DNA) replication. Paradoxically, this necessary step is a dangerous one, since the replicating enzymes may cause mutations that damage the DNA. There are several elaborate proofreading and error-correcting mechanisms that keep the mutation rate down to less than 1 error in 10^7 nucleotides replicated, but because each nucleus has up to 3×10^9 nucleotides, only rarely does an error-free replication occur prior to cell division.

Many mutations are trivial, occurring in junk DNA, spacer DNA, introns, and other regions where they have no effect. Also, a mutation in a shoot apical meristem may affect a gene that is active only in roots; because that gene is not required or activated in the shoot, it is not detrimental. However, if a shoot apical meristem functions for several years, mutations accumulate and the cells become increasingly defective. This is an important problem because some of the shoot apical meristem cells will be involved in the formation of flowers or cones. Since flowers and cones produce the sex cells, if the shoot apical meristem cells are defective, then the sperms and egg cells will also be defective.

This problem can be minimized in several ways. In an annual plant, the meristem functions for less than 1 year, so the accumulation time for mutations is short. Also, since annual plants tend to be small, fewer cell divisions are needed to produce the small number of cells in the plant body. In perennial plants, meristems must function longer than they do in annual plants. If the perennial plant is highly branched, each branch has its own shoot apical meristem so that the work of producing new cells is divided among many shoot apical meristems. Since the number of cell divisions required of each shoot apical meristem cell is much lower than if a single shoot apical meristem had to produce the whole plant, the number of mutations is minimized.

Neither of these methods to minimize mutations has evolved in cacti; instead, the body of many species consists of a single unbranched long shoot that grows from a single shoot apical meristem during a period of hundreds of years. If these plants had typical shoot apical meristems, the number of cell divisions per meristem cell would be enormous and the shoot apical meristems would probably become nonfunctional before the plants were old. However, these species have the largest shoot apical meristems in the plant kingdom, up to 1500 μm in diameter, and each contains thousands of cells. In contrast, plants such as beans and sunflowers have meristems less than 100 μm in diameter with only a few cells. Consequently, because there are so many meristem cells in cacti, the number of cell divisions per cell is low and the number of dangerous DNA replication cycles is kept minimized.

For background information SEE APICAL MERISTEM in the McGraw-Hill Encyclopedia of Science & Technology.

James Mauseth

Bibliography. E. J. Klekowski, Jr., *Mutation, Development, Selection, and Plant Evolution*, 1988; J. D. Mauseth, A morphometric study of the ultrastructure of *Echinocereus engelmannii* (Cactaceae), *Amer. J. Bot.*, 69:1524–1526, 1982; J. D. Mauseth, Effect of growth rate, morphogenetic activity, and phylogeny on shoot apical ultrastructure in *Opuntia polyacantha* (Cactaceae), *Amer. J. Bot.*, 71:1283–1292, 1984; J. D. Mauseth, *Plant Anatomy*, 1988.

Arctic Ocean

Recent research in the Arctic Ocean has involved studies of basin-scale circulation and of the basin plate tectonics and sedimentology of the northern Norwegian–Greenland Sea.

Basin-scale circulation. Oceanographers can determine circulation by measurements of flow or free drift or from calculations based upon the distribution of density. Sometimes the sense of flow can be inferred from variations in space or time of physical or chemical characteristics. However, the Arctic Ocean is sometimes referred to as the moon of oceanography, because data of any of these types are mostly sparse or even nonexistent there; little is known about the circulation. The principal exception is the surface waters, where circulation has been increasingly well understood from studies based upon ice drift. The dynamics relating ice drift to wind speed and ocean circulation are generally understood, and observations from instruments on drifting ice floes have provided a steady and, at times, coordinated stream of data.

Partly because the primary vertical variation of density in the Arctic Ocean is very strong and relatively shallow, occurring mostly through the upper 300 m (1000 ft), it is difficult to deduce the energetics or even direction of the circulation of the deeper layers from the shallow field. Direct measurement of the subsurface field of motion has been restricted to boundary regions accessible to Western countries. The interior circulation of the Arctic Ocean basins and the flow adjacent to most boundaries are virtually unknown, especially in quantitative terms.

In this general absence of direct measurements and basin-crossing oceanographic sections, some of the most fruitful studies of the general basin-scale circulation of the Arctic Ocean have considered the unique interactions of ocean and climate representative of the Arctic region. The Arctic Ocean is a mediterranean sea, that is, almost surrounded by land; its periphery consists of broad, shallow continental-shelf seas constituting about 30% of the total surface area. Waters of the shelf seas begin relatively fresh due to river input, and freezing in winter and melting in summer further rectifies the vertical salinity structure; that is, a layer of relatively fresh water is left on the top. Hence the shelf seas feed low-salinity water into the basins in patterns that vary greatly in space and time, and this input helps maintain the permanent ice pack. Oceanographers now hypothesize that the deeper layers of the interior are also fed from the shelves; this annual freezing of surface seawater leaves behind dense brines, which may sink to depressions on the shelf, accumulate in winter, and spill into the basin interior, potentially contributing their condensed characteristics to all levels (**Fig. 1**).

There are also external sources of supply: through the eastern side of the Greenland–Spitsbergen Passage (Fram Strait), which is the only deep interconnection to the other oceans, the Arctic Ocean receives at intermediate depth relatively warm highly saline water derived from Atlantic Ocean sources, and cold young deep waters from the Greenland and Norwegian seas. These spread through the basins under the upper layers from the shelves. Little is known about the details of the middepth and deeper circulation other than that it is expected to follow boundaries, crossing submarine ridges where permitted, and interacting with the shelf-derived dense waters before eventually exiting through western Fram Strait.

Field programs. Confidence in this hypothetical scheme comes from recent field programs, which for the first time have provided oceanographic data to a common high standard from nearly all the principal deep-water domains of the Arctic. Contributing field programs during 1979–1987 included winter Greenland Sea coverage, a Fram Strait expedition, measurements from air-supported temporary ice camps staged at two Canadian sector sites and at a critical site spanning the ridge separating the Canadian and Eurasian sectors of the Arctic Ocean (**Fig. 2**), and the first shipborne oceanographic section across a major deep basin of the Arctic Ocean (the Nansen Basin, which is the basin north of the Barents Sea).

Scientists have learned from these data that topography plays an even stronger role than was previously supposed in determining the distributions and circulation of the water masses, with evidence of separate circulation patterns and separate water masses in each of the deep basins. Their sections crossing boundaries reveal narrow features that the scientists interpret to be cores of boundary currents at intermediate depths and intrusions of shelf waters into the interior. Also it has been found that some anthropogenic substances—products of civilization—are mostly absent in the deep Canada basin, though found throughout the water column in the Eurasian section of the Arctic Ocean (east of the Lomonosov Ridge). However, the most recently introduced substances (chlorofluorocarbons and modern radionuclides) are absent in the bottom waters. The interpretation is that the deep waters of the Canadian sector, which are the most physically isolated, are in addition the most isolated from deep-water renewal processes, and that the time scale for deep-water renewal in the Eurasian sector, though shorter, is still more than 50 years. Obviously there must be a transition, but it is not yet known how this transition manifests itself spatially in the various layers, to what extent the boundary regimes hinted at indirectly in data really do exist, and what role they have in determining the characteristics and sensitivity of the waters that are found over the deep basins.

Relationship to global ocean climate. This new understanding of the interior of the Arctic Ocean provides insight into its relationship to global ocean climate. Oceanographers have learned that the Arctic Ocean and its peripheral seas act as a large machine that receives upper layers of warm salty water from the North Atlantic Ocean, carries this water underneath cold air in the Norwegian and Greenland seas, and transforms it into various colder and denser water masses. At the same time, through different mecha-

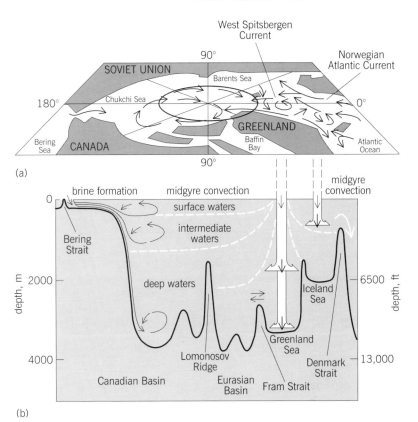

Fig. 1. Schematic diagrams of the Arctic Basin and its adjacent seas showing (a) circulation and (b) oceanic structure. Relatively warm waters carried northward with the Norwegian Atlantic Current and West Spitsbergen Current are cooled and thus sink and spread at intermediate depths both into the Arctic Basin and into the Greenland and Iceland gyres. Midgyre convection produces dense water masses, which then spread throughout the region and into the remainder of the world ocean. (*After K. Aagaard, J. H. Swift, and E. C. Carmack, Thermohaline circulation of the Arctic mediterranean seas, J. Geophys. Res., 90(C5):4833–4846, 1985*)

nisms, the Arctic Ocean produces and exports sea ice. Where this very cold and relatively fresh surface regime intrudes upon the warm salty regime, stratification is increased (as in the central Arctic). Over the shelf seas the water column is shallow, and the brines released during ice formation can significantly raise bottom salinities there. The resulting dense waters can slide off the shelves and feed the basin interiors; through the outflows through the Greenland–Spitsbergen Passage (Fram Strait), the middepth and deep Arctic Ocean then influences the other oceans.

The primary gap in knowledge and understanding of the large-scale circulation in the Arctic Ocean lies in the extremely sparse data now available. Hence a top priority for new research is for high-quality basin-scale sections of full-depth density-related and tracer measurements. Shipborne sections remain the single most productive tool available to physical oceanographers and are within the capabilities of modern icebreakers. Accompanying these must be direct measurement of the flow field of the basin interior over periods of months to years, and computer-based Arctic modeling that can reproduce the primary oceanographic features with realistic simulated-climate forcing. This combination can provide most of the needed

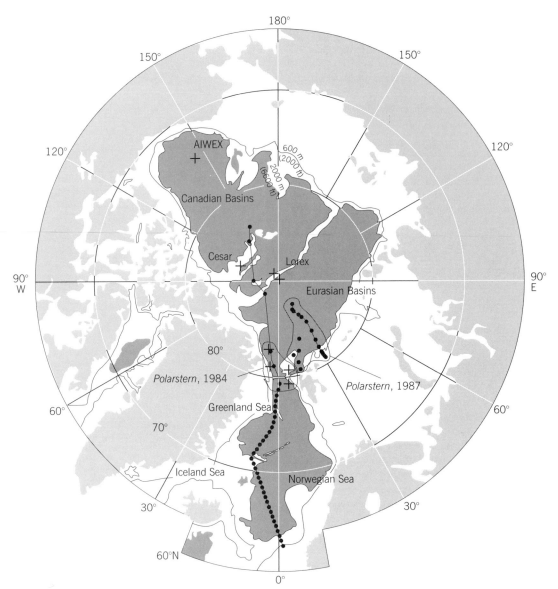

Fig. 2. Map showing bathymetry (600 and 2000 m, or 2000 and 6500 ft), nomenclature, and positions of recent hydrographic measurements of the Arctic Ocean.

evaluation of the large-scale circulation at a lower cost than any other feasible method.

James H. Swift

Basin plate tectonics and sedimentology. The high-latitude Norwegian–Greenland Sea is a natural laboratory to study the growth of a narrow, young ocean basin dominated by very slow opening across one of the world's longest fracture zones (**Fig. 3**). Because of its proximity to the Arctic ice pack, this sea has remained relatively unexplored. It is a remote region that has been of great interest to the international scientific community because within it lies the only deep-water entrance into the Arctic Ocean (known as the Fram Strait). It is thought that the strait was created by the opening across the lengthy Spitsbergen Fracture Zone, which was one of the largest continuous faults on the surface of the Earth.

Fault-controlled passages between ocean basins exercise important influence on the hydrography of

the world ocean, because their morphology and bathymetry control exchange of water masses. The Fram Strait is one of the most important of these deep passages, because it plays a pivotal role in the Earth's recent ocean hydrography and paleoenvironmental evolution. Heat exchange via transport of water and sea ice through this passage influences global climate and oceanography. However, this region is very complex, and nearly all geophysical parameters, which are used to reconstruct the geological history of any area, are abnormal relative to the rest of the world ocean basins. Therefore, deciphering when and at what rate the Norwegian–Greenland Sea and the Fram Strait formed is a formidable task, but nonetheless it is a necessary goal if the paleoclimatic history is to be unraveled.

The complexity of the plate boundary in the Norwegian–Greenland Sea reflects the complex opening history in this area. Sea-floor spreading in the

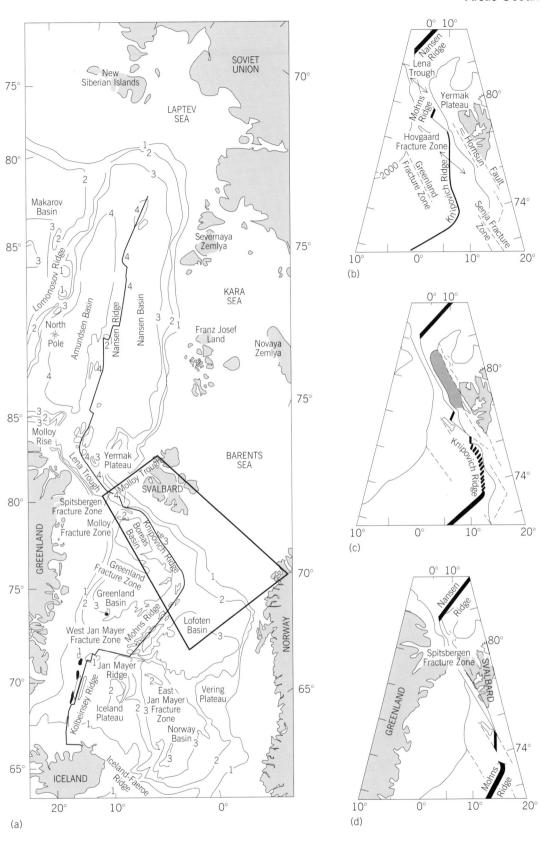

(a)

(b)

(c)

(d)

Fig. 3. Norwegian–Greenland Sea and the eastern Arctic Ocean. (a) Geomorphology. The heavy lines indicate plate boundaries, and the trapezoid indicates the area mapped with the SeaMARC II system. Contours are in kilometers of depth; 1 km = 0.6 mi. (b) Plate reconstructions at the present, (c) 9.5 m.y.a. (darker-shaded region represents the area of overlap between the continental shelf when the Eurasian and the North American plates are rotated back to 36 m.y.a.), and (d) 36–50 m.y.a.

Norwegian–Greenland Sea and the Arctic Ocean started at approximately 66–57 million years (m.y.) before the present. The relative motion between Svalbard (Spitsbergen) and Greenland was approximately northeast–southwest from the Mohns Ridge, with no crustal extension in the northern Norwegian–Greenland Sea. The Spitsbergen Fracture Zone acted as the plate boundary between the incipient southern

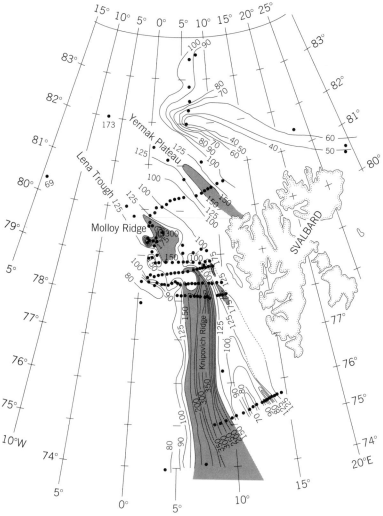

Fig. 4. Observed heat flow on the western Svalbard margin and the Yermak Plateau. Contours are in milliwatts/m². Shaded areas indicate bands of relatively high heat flow (>150 mW/m²). Dots represent heat-flow stations.

Norwegian–Greenland Sea and the Arctic Ocean (the Spitsbergen Fracture Zone; **Fig. 4**).

About 36 m.y.a. the east–west component of opening increased, allowing the mid-oceanic ridge to move northward into the Spitsbergen Fracture Zone. This started sea-floor spreading in the northern Norwegian–Greenland Sea along the Knipovich Ridge. Heat-flow data also confirm that spreading began earlier in the south than in the north. The rate of opening along the Mohns Ridge in the south is presently 0.91 cm (0.36 in.) per year. The component of opening along the more northerly Knipovich Ridge is even slower. By using heat-flow data, it has been estimated that the North American Plate in this region is growing 1.5 times as fast as the Eurasian Plate.

Evidence for rift propagation. Analysis of heat-flow data suggests that the Knipovich Ridge propagated northward at a rate of 1° per 10 m.y. commencing 60 m.y.a. at a latitude of 75°N.

Evidence for further northward rift propagation exists on the northerly Yermak Plateau adjacent to Svalbard. Additional heat-flow data (Fig. 4) suggest that faults along the plateau (which were a part of the

more extensive and ancient Spitsbergen Fracture Zone system) are serving as present-day channels along which heat from a subjacent body of magma escapes to the sea floor.

SeaMARC II survey. To resolve the evolution of the northern Norwegian–Greenland Sea, a SeaMARC II side-looking sonar and bathymetric swath-mapping expeditions aboard the Norwegian ship *Haakon Mosby* took place in the fall of 1989 and 1990. The SeaMARC II system (operated out of the Hawaii Institute of Geophysics) is a shallow-tow vehicle (50–100 m or 160–320 ft) with port–starboard coverage and a tow-speed capability of up to 10 knots (5 m/s). It operates at frequencies of 11 and 12 kHz with short transmission pulses. The qualitative backscattering measurements from the imagery are useful for determination of bottom slope and texture and large-scale regional trends.

The program, a cooperative effort involving the University of Bergen, Norway, the U.S. Naval Research Laboratory, and the Lamont-Doherty Geological Observatory, investigated the Knipovich Ridge, Molloy Ridge, Molloy Transform Fault, southwestern Yermak Plateau, the northern Spitsbergen Margin, and the Spitsbergen–Barents Sea continental margin.

Navigation was by transit, Global Positioning System satellites, and range-range-mode Loran C. Sea ice placed a northern limit of 80°05′N on the first expedition and 82°N on the second expedition. About 11,000 line-kilometers (6600 line-miles) of geophysical data were collected, including SeaMARC II sidescan and bathymetry, 3.5-kHz bathymetry, gravity, and magnetics. High-resolution 38-kHz single-beam bathymetry was obtained at water depths less than 2500 m (8200 ft). About 80,000 km² (31,000 mi²) of the sea floor was mapped by side-looking sonar, and roughly half that area was bathymetrically swath-mapped.

Sedimentary processes. Volcanic-tectonic structures and their modification by mass wasting dominate the active plate boundary. Sedimentary features include pockmarks, iceberg plow marks, channels, gullies, slumps, and bottom-current effects.

Roughly 100 pockmarks were imaged, scattered across thickly sedimented parts of the area. The pockmarks appear as small dark speckles, barely resolvable in the side-looking sonar images and unresolvable in the swath bathymetry. At least half of the pockmarks occur in a narrow crestal belt along Vestnesa Ridge, a 1300–1500-m-deep (4300–4900-ft) spur extending from the west Svalbard margin almost to the Molloy Ridge. Where crossed by single-channel echosounders, the pockmarks are bathymetric dimples around 100–200 m (330–660 ft) across and 10 m (33 ft) deep. The features are thought to have been derived by methane venting from a shallow clathrate (solid, icelike gas hydrates) layer. High heat flow from the nearby volcanically active spreading could cause the transformation from the frozen to the gaseous phase. Gas trapped below the clathrate would migrate up and collect under the Vestnesa "anticline," escaping through available cracks.

A reconnaissance made across the southwest Yermak Plateau revealed unambiguous iceberg plow-marks to depths of 600 m (2000 ft). Broader, more subtle lineations extend to 900 m (3000 ft) deep, even for large Pleistocene icebergs. The depression between the Yermak Plateau and the Vestnesa Ridge is occupied by a narrow, isolated channel extending from the Kongsfjord continental margin to the Molloy Ridge valley.

One track from Svalbard along the 1000-m (3300-ft) contour south to the Bear Island Fan reveals the absence of large canyons along the western Svalbard margin (79°N–75°N), but canyons are prominent on the steep margin between 74°15′N and 74°30′N (west of Bear Island), with channels of lesser relief occurring intermittently from there southward. Three east–west transects down the Bear Island Fan suggest that the fan comprises long (several hundred kilometers), complex fingers 5–10 km (3–6 mi) wide and incised by channels.

Adjacent to the Norwegian margin the sea floor is subjected to constant mass wasting. Large erosion

channels and debris fields mark the imagery. Only a few slump scars were imaged, one at 74°55′N, 15°E and several on steep, thickly sedimented slopes in the Molloy Ridge/Transform area. Features that have possibly been generated by bottom currents appear in the side-looking sonar near the Molloy Transform. However, their proximity to this active fault means that a tectonic origin cannot be ruled out.

Volcanic-tectonic structure. The SeaMARC II imaged the accreting Mohns and Knipovich ridges, the Molloy Transform Fault, and the Molloy Ridge. Diagonal contiguous sidescan swaths mapped the outcropping rift mountain and ridge flank basement topography along the northern Knipovich Ridge (**Fig. 5**). The rift valleys are characterized by scalloped shaped faults, which mark their boundaries. Within the valleys, faults and fissures trend obliquely to the more regional rift valley trend. Young lava flows occur at 74°35′N, 75°30′N, 76°05′N, and 76°50′N (65–100 km or 40–60 mi apart). Oval-shaped basement highs that have been emplaced on the flanks of the rift are aligned parallel to the trend of the Molloy Transform Fault. These are inferred to be once-active volcanic centers that were rafted off the spreading center with plate motion.

Off-axial faults are not parallel to the overall trend of the spreading center. Instead, the farther north the fault, the closer its position to the axis of spreading. Conjugate pairs of flank faults make a V-shaped pattern about the rise axis, with the point of the V intersecting with the Molloy Transform Fault (Fig. 4).

The neotectonic expression of the Molloy Transform is dramatic because sediments have everywhere buried older basement topography. Two parallel transcurrent fault traces make up the transform itself, but a broader belt of deformed lineations striking 30–45° with respect to the transform trend extends up to 10–30 km (6–8 mi) southwest from the transform axis. These structures curve gradually into the northern Knipovich Ridge.

The Molloy Ridge, a pull-apart basin within the Spitsbergen Fracture Zone, has little conspicuous neovolcanic expression, as its dramatic relief (5500–1500 m or 18,000–4900 ft) is dominated by mass wasting.

The topographic and bathymetric maps produced by the SeaMARC II system provide additional evidence that the Knipovich Ridge is propagating northward. If this model is correct for the opening of the Norwegian–Greenland Sea, and if the analysis of the heat-flow data is reasonable, then it can be suggested that the Fram Strait began to separate approximately 40 m.y.a., allowing exchange of water masses between the Arctic and the Atlantic oceans. Propagation of the ridge suggests that faulting and volcanic activity will continue to occur with greater magnitude on and north of the Svalbard for millions of years to come.

For background information SEE ARCTIC OCEAN; BASIN; EARTH, HEAT FLOW IN; LORAN; PLATE TECTONICS; SATELLITE NAVIGATION SYSTEMS; SONAR in the McGraw-Hill Encyclopedia of Science & Technology.

Kathleen Crane; Peter Vogt; Eirik Sundvor

Bibliography. K. Aagaard, J. H. Swift, and E. C.

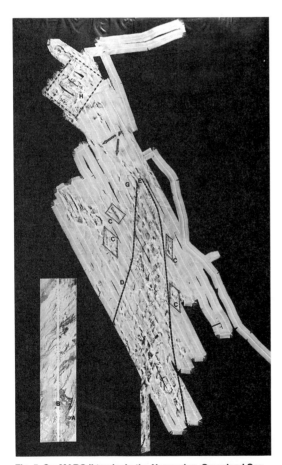

Fig. 5. SeaMARC II tracks in the Norwegian–Greenland Sea. A: curvilinear faults marking the boundary of the rift valley of the Knipovich Ridge. B: Recent lava flow, 10 × 10 km (6 × 6 mi). C: off-axial elongate highs suggesting regions of ancient excess volcanism. D: V-shaped region within which faulting related to the ridge axis extension occurs; this V shape suggests that the ridge is propagating northward. E: the intersection between the Knipovich Ridge and the Molloy Transform Fault. F: double fault traces that make up the Molloy Transform Fault. G: off-axial faults that bend into the Molloy Transform Fault. H: the Molloy Ridge, which is a pull-apart basin formed by oblique extension between the Molloy and Spitsbergen transform faults.

Carmack, Thermohaline circulation in the Arctic mediterranean seas, *J. Geophys. Res.*, 90C5:4833–4846, 1985; L. G. Anderson et al., The first oceanographic section across the Nansen Basin in the Arctic Ocean, *Deep-Sea Res.*, 36:475–482, 1989; K. Crane et al., Thermal evolution of the western Svalbard margin, *Mar. Geophys. Res.*, 9:165–194, 1988; B. G. Hurdle (ed.), *The Nordic Seas*, 1986; A. M. Myhre and O. Eldholm, The western Svalbard margin (74°N–80°N), *Mar. Petrol. Geol.*, 5:134–156, 1987; M. Talwani and O. Eldholm, Evolution of the Norwegian–Greenland Sea, *Geol. Soc. Amer. Bull.*, 88:969–999, 1977; D. W. R. Wallace and M. Krysell, Arctic Ocean ventilation studied with a suite of anthropogenic halocarbon tracers, *Science*, 242:746–749, 1988.

Art conservation chemistry

Wall paintings and stained glass are two major categories of artistic endeavor usually encountered in architectural settings. This circumstance places these artifacts at greater risk than works of art kept in the controlled environments of museums and libraries. They are exposed to great fluctuations in temperature and humidity as well as to air pollution and harmful agents introduced by public use. In recent years, there have been major campaigns to conserve such world-renowned Renaissance wall paintings as Michelangelo's Sistine Chapel frescoes in Rome, Leonardo da Vinci's *Last Supper* in Milan, and the frescoes of the Brancacci Chapel in Florence painted by Masaccio and Filippino Lippi. Other significant projects involving conservation of wall paintings include the Lescaux cave paintings in France, the Tomb of Nefertari in Egypt, and the Ajanta and Ellora caves in India.

Though medieval stained glass is primarily found in western Europe, with the major proportion, some 80,000 m² (900,000 ft²), in France, the importance of these works in the architectural context of the great

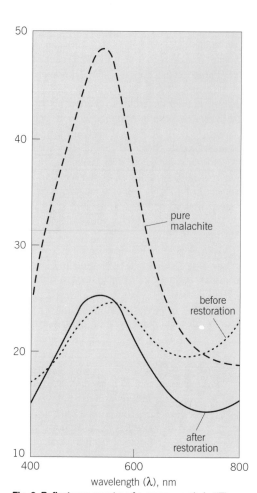

Fig. 2. Reflectance spectra of a green mantle in "The Resurrection of the Son of Teofilo" from the Brancacci Chapel frescoes before and after restoration, compared to the reflectance spectrum of pure malachite in a fresco sample. (*After M. Bacci et al., A colour analysis on the Brancacci Chapel frescoes, in N. S. Baer, C. Sabbioni, and A. I. Sors, eds., Science, Technology and European Cultural Heritage, Butterworth-Heinemann, 1991*)

Gothic cathedrals of Europe has made their conservation an area of intense scientific investigation.

Wall paintings. Wall paintings fall into two general categories: those painted on a dry wall (*secco*) and those painted on a setting lime plaster wall (*fresco*) so that the pigment is bound to the plaster. The ancient paintings at Lescaux were created directly on the wall of the cave. The *Last Supper* was painted on a prepared wall. *Secco* paintings require some binding material, such as drying oils, gums, glues, or waxes, to effect adhesion of the pigment particles to the wall. In true fresco, the major portion of the painted surface requires no adhesive binder. Only areas to be corrected or employing pigments incompatible with the setting plaster (such as gold or cinnabar) are painted *a secco*. The setting reaction (**Fig. 1**) follows a simple path: the evaporation of water (H_2O) and the reaction of the slaked lime plaster [$Ca(OH)_2$] with atmospheric carbon dioxide (CO_2) lead to the formation of calcium carbonate [$CaCO_3$; reaction (1)].

$$Ca(OH)_2 + CO_2 \rightarrow CaCO_3 + H_2O \qquad (1)$$

The remarkable stability of fresco paintings is offset

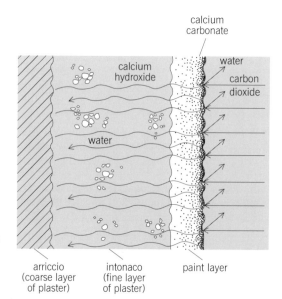

Fig. 1. Diagram showing the setting reaction for a fresco; the image is painted on the *intonaco*. (*After P. Mora, L. Mora, and P. Philippot, Conservation of Wall Paintings, Butterworth, 1984*)

Table 1. Glass compositions, in wt %

Component	Modern soda-lime-silica glass	Average Roman glass	Excavated* Crusted	Excavated* Uncrusted
Silica (SiO_2)	73.6	67.0	55.2–55.7	57.5–59.9
Soda (Na_2O)	16.0	18.0	2.7–3.1	2.5–3.3
Potash (K_2O)	0.6	1.0	10.8–11.4	9.0–11.3
Lime (CaO)	5.2	8.0	12.8–14.8	8.3–13.0
Magnesia (MgO)	3.6	1.0	7.2–9.1	7.3–8.7
Alumina (Al_2O_3)	1.0	2.5	—	—
Iron oxide (Fe_2O_3)	—	0.5	—	—
Antimony oxide (Sb_2O_3)	—	1.5	—	—
Manganese dioxide (MnO_2)	—	0.5	—	—
Lead oxide (PbO)	—	0.01	—	—

*From Bagot's Park, Staffordshire, England, excavated in 1967 and dated to 1523.
Source: After R. Newton, The weathering of medieval window glass, *J. Glass Stud.*, 17:161–168, 1975.

by the limitations of the types of pigment permitted the artist and the severe time constraint that the painting be done in sections of drying plaster equal to a day's work (*giornata*). In contrast, the far less stable *secco* wall paintings can employ the entire palette with no time constraints. The reliance on a binder, however, introduces a significant source of instability, for the painting will suffer substantial damage with any deterioration of this adhesive.

Degradation processes. The direct contact of paintings with their supporting walls leaves them vulnerable to a broad range of deterioration processes, including evaporation of water from the surface with the consequent crystallization of soluble salts, condensation of moisture on the walls, deposition of dust and other particulate matter, surface encrustation, embrittlement of glue and gum media, biological attack on the binder, mechanical damage, and the alteration of chemical composition of certain pigments with time. In some cases, the presence of calcium sulfate ($CaSO_4$) on the surface is evidence of the direct attack of sulfur dioxide (SO_2) from atmospheric pollution [reaction (2)].

$$CaCO_3 + \tfrac{1}{2}O_2 + SO_2 + 2H_2O \rightarrow$$
$$CaSO_4 \cdot 2H_2O + CO_2 \quad (2)$$

Conservation. A wide range of conservation procedures have been applied to the problems associated with wall paintings. The Lescaux caves had 1 million visitors between 1948 and 1965. Air conditioning maintaining a temperature of 14°C (57°F) and 98% relative humidity and providing four air changes per hour did not prevent the growth of algae that damaged the paintings. Antibiotics, dispersed as aerosols, made possible a 96% reduction in algae growth. Then, in 1965, it was observed that a thin veil of calcite was forming, obscuring the paintings. This phenomenon was associated with the reaction of emissions of CO_2 from visitors with the moisture on the walls that was saturated with calcium carbonate from the limestone of the cave. In 1983, a same-size copy was constructed nearby for visitors, while access to the actual cave was limited to five persons per day, five days per week.

The *Last Supper*, painted by Leonardo da Vinci in an experimental medium, began to suffer damage soon after its creation. Attempts at restoration included substantial repainting. In a painstaking campaign, where daily progress is measured in square

centimeters (1 cm^2 = 0.155 $in.^2$) restored under the binocular microscope, these overpainted layers are being removed, leaving behind the remnants of the original. Paint sections and chemical analyses assist the conservator in determining which portions of the painting are original.

The Sistine Chapel frescoes had become obscured by layers of soot deposited from the burning of candles and oil lamps. In previous centuries, restorers applied layers of glue to brighten the appearance of the paintings. This glue was contracting and so was leading to the detachment of small portions of the pigment layer. In addition, a roof leak in the past had caused the deposition of salts on the surface. These factors led to a projected 12-year campaign to clean and conserve all of the Michelangelo frescoes in the Sistine Chapel. The cleaning of the lunettes and the ceiling frescoes was completed in 1990, when work was begun on the final section, *The Last Judgment.* The standard cleaning mixture included water; a quaternary ammonium salt as surfactant; carboxymethyl cellulose to hold the solution on the vertical surfaces; ammonium and sodium bicarbonates; and, in some cases where salts were to be removed, the chelating agent ethylenediaminetetracetic acid (EDTA).

The Brancacci Chapel frescoes in Florence, painted in the fifteenth century, were also obscured by particulate matter from candle smoke and from a fire that caused considerable damage to the church in 1771. The effect of this particulate matter was not only to darken the paintings but also to cause a color shift

Table 2. Composition of excavated glass*

Component	Crusted, mole %	Uncrusted, mole %
Silica (SiO_2)	57.0–57.8	60.6–61.5
Lime (CaO)	14.1–16.5	9.4–14.4
Magnesia (MgO)	11.2–14.0	11.2–13.7
Soda (Na_2O)	2.7–3.1	2.5–3.3
Potash (K_2O)	7.2–7.5	5.9–7.6
All others	4.3–4.7	3.4–6.1

*From Bagot's Park, Staffordshire, England, excavated in 1967 and dated to 1523.
Source: After R. Newton, The weathering of medieval window glass, *J. Glass Stud.*, 17:161–168, 1975.

(**Fig. 2**). After cleaning, the reflectance spectral distribution of a section painted with malachite approaches that of a reference specimen. At the Sistine Chapel, similar color shifts, documented by reflectance spectroscopy, confirmed the observation that deposited particulate matter has a definite hue, leading not only to loss of color saturation but also to changes in the colors observed. Further conservation measures include microclimate monitoring, visitor limitation, and illumination selected for color fidelity.

The frescoes by Domenico Ghirlandaio in the Cappella Maggiore of S. Maria Novella in Florence were completed in 1490. The paintings suffered significant damage due to the action of soluble salts introduced by rainwater. In addition, substantial amounts of gypsum ($CaSO_4 \cdot 2H_2O$) were found, revealing the effects of atmospheric pollution. The gypsum crusts were detaching the painting from the support. The multistep conservation procedure involved removal of particulate matter with deionized water, application of ammonium carbonate [$(NH_4)_2$-CO_3] in a cellulose pulp compress to cause swelling of accumulated organic substances, and final consolidation achieved by diffusion to the affected areas of barium hydroxide [$Ba(OH)_2$] applied in a cellulose pulp compress.

Though general principles are applied in the conservation of all wall paintings, each work of art by its nature is a unique act of creation in a singular physical environment. It requires a careful program of environmental monitoring and chemical analysis to determine the types of damage, the causative factors, the appropriate conservation sequence, and the conditions under which the restored painting can best be preserved for future generations.

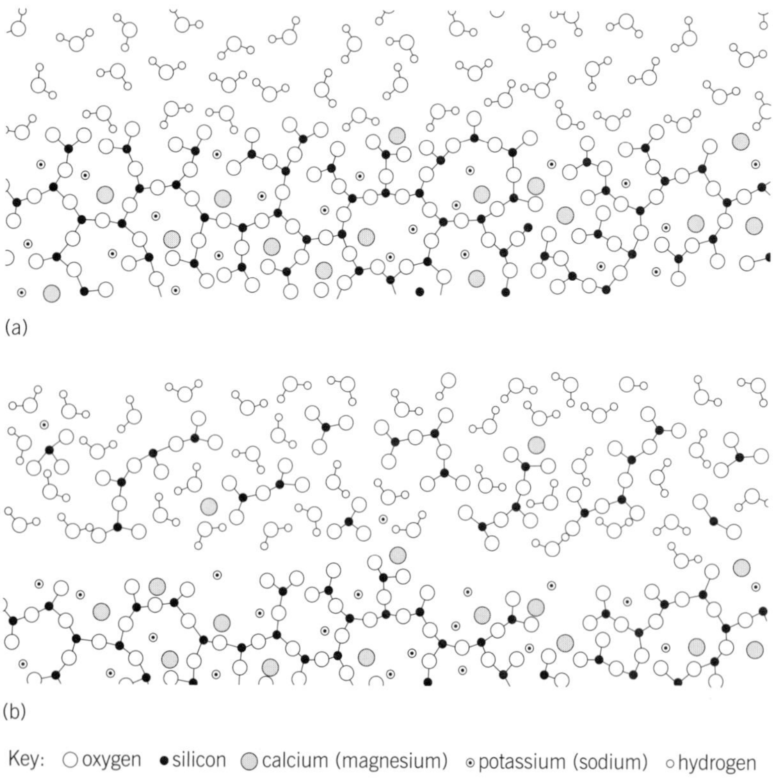

Stained glass. Traditional glasses are the product of the solidification without crystallization of fused inorganic materials. Their primary component is the network-former silica (SiO_2), which melts at 1720°C (3128°F). Common glasses have various oxides added to act as stabilizers, network modifiers, and colorants. The network modifiers break up the silica network, changing such properties as viscosity and durability. Oxides of sodium (Na_2O) and potassium (K_2O) reduce the viscosity, while those of calcium (CaO) and magnesium (MgO) improve stability. The composition of a typical modern soda-lime-silica glass is given in **Table 1**.

Composition of medieval glasses. Fully preserved specimens of ancient Egyptian and Roman glasses are quite common. The stability of these glasses, even under the harsh conditions of long-term burial, reflects the skill of ancient craft workers and their care in the selection of raw materials. The composition of an "average" Roman glass is given in Table 1 together with the results of analyses for crusted and uncrusted glass excavated in Bagot's Park, Staffordshire, England, in 1967. The date of the glass is placed at 1523, based on archeological evidence. No major differences are apparent when only the composition as weight percentages is considered. More revealing is the composition expressed in mole percent (**Table 2**). The combined high total of some 30% MgO and CaO plus 10% combined Na_2O and K_2O reduced the silica content to a critical level for the crusted glass. Once the silica content falls below 66.7 mole %, each silicon atom is associated with a modifier ion as a second neighbor, so that mobile ions are readily leached from aqueous solution. Another factor is the greater mobility of potassium ions compared to those of sodium, further reducing the stability of those glasses where potassium is in excess. Though some glasses from the medieval period demonstrate admirable stability, in general their durability is considerably less than would be expected for an inherently stable material. Though there is increased understanding of the compositional variables that lead to reduced glass stability, the detailed mechanism of corrosion is not fully understood.

Weathering. It is generally agreed that the rate of deterioration of medieval stained glass windows has increased in the twentieth century. Among the factors cited are increased air pollution, and poor conditions of storage during World War II, when the windows were removed for safe keeping and because of lack of maintenance. The fundamental process in the weathering of glass windows is the leaching out of soluble species in a hydrous gel layer: sodium and potassium ions, and to a lesser degree those of magnesium and calcium (**Fig. 3***a*). As they are removed, a broken network remains, often covered with a crust of sulfates formed by the reaction of sulfur-bearing air pollutants— SO_2, sulfuric acid (H_2SO_4), ammonium

(a)

(b)

Key: ○ oxygen ● silicon ◯ calcium (magnesium) ⊙ potassium (sodium) ○ hydrogen

Fig. 3. Decomposition of medieval stained glass. (a) Initial wetting of original surface. (b) Weathering with leaching and depolymerization. (*After G. Frenzel, The restoration of medieval stained glass, Sci. Amer., pp. 126–135, May 1985*)

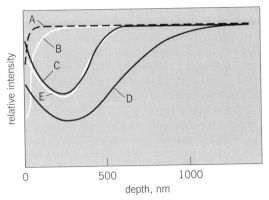

Fig. 4. X-ray photoelectron spectroscopy profiles demonstrating the reactive effect of sulfur dioxide (SO_2) on the leaching of calcium from a sample of reactive glass. A = blank (reference curve). B = 1.5 h in climate chamber. C = 1.5 h in climate chamber + 5 ppm sulfur dioxide (SO_2). D = 3 h in climate chamber + 5 ppm SO_2. E = 50 days outdoors, ~0.05 ppm SO_2. (*After D. R. Fuchs et al., Preservation of stained glass windows: New materials and techniques, in N. S. Baer, C. Sabbioni, and A. I. Sors, eds., Science, Technology and European Cultural Heritage, Butterworth-Heinemann, 1991*)

sulfate [$(NH_4)_2SO_4$], and others—with the alkaline-enriched wet surface layer (Fig. 3*b*). What has not been resolved is whether the actions of sulfur-bearing species are causative or passive, that is, simply forming a crust with the products of the leaching reaction. Recent evidence obtained in a series of chamber studies (**Fig. 4**) supports the hypothesis that SO_2, especially in the presence of oxidizing gases, may play an active role in the deterioration process, though the mechanism is not yet understood. Studying the depth of reaction of a highly reactive test glass, a group at the Fraunhofer Institut für Silikatforschung in Würzburg, Germany, demonstrated a substantially enhanced depth of reaction in the presence of SO_2. The profile of a chamber specimen (curve C, Fig. 4) exposed for 1.5 h at 5.0 ppm SO_2 matched that of a similar specimen exposed for 50 days to the ambient environment at approximately 1/1000 the concentration of SO_2, that is, 0.05 ppm (curve E).

Conservation. Two basic approaches are taken to the conservation of stained glass. After removal of the encrustation by mechanical or chemical means and the reattachment of broken shards, the window is most commonly protected from the atmosphere by application of a polymeric coating, or is isolated from the atmosphere by isothermal glazing—the rehanging of the entire window several centimeters inward, with ordinary glass installed in its original place, keeping the front and rear surfaces of the window at approximately the same temperature. The system is equivalent to setting the stained glass window in a controlled museum environment. Depending on heat transfer from incident sunlight, central heating, and visitors, the airflow between the two windows may change in direction and speed. Since this is of concern regarding possible thermal stresses, condensation, corrosion, and particulate deposition, the airflow, temperature, and corrosivity of special test-glass surrogates are

studied by means of sensors placed on the glazed surfaces (**Fig. 5**). Empirical evidence suggests that this type of protection, for which there is some 40 years of experience, has been effective in eliminating further deterioration of medieval glass windows. The primary disadvantages are esthetic in that the historic glass is no longer in its original position and a protective glass is in its place. This protective glass can substantially alter the architectural appearance of the exterior.

Among the protective layers used in the past were polyurethanes and an epoxy-bonded polymethyl methacrylate–glass sandwich. A promising recent development is a coating system based on a heteropoly-siloxane (I) synthesized by a hydrolysis and conden-

$$\left[\begin{array}{c} CH_3 \\ | \\ O-Si- \\ | \\ CH \\ || \\ CH_2 \end{array} \right. \left. \begin{array}{c} OH \\ | \\ O-Si- \\ | \\ OC_2H_5 \end{array} \right. \left. \begin{array}{c} C_6H_5 \\ | \\ O-Si- \\ | \\ C_6H_5 \end{array} \right]_n$$

(I)

sation process. The polymer is combined with solvents and acrylates to form a lacquer. To increase the diffusion path through the coating, platelike inorganic pigments are added. In tests of natural weathering on a sensitive glass, the protective coating was quite successful. It is now being tested at various sites where there are substantial quantities of medieval stained glass.

When the paint, that is, the fired enamel or stain

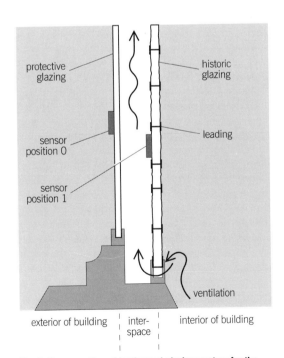

Fig. 5. Cross section of isothermal glazing system for the protection of historic glass windows. (*After D. R. Fuchs et al., Preservation of stained glass windows: New materials and techniques, in N. S. Baer, C. Sabbioni, and A. I. Sors, eds., Science, Technology and European Cultural Heritage, Butterworth-Heinemann, 1991*)

used to define images on the clear glass, is unstable, consolidation is required. Organic polymers, such as epoxy resins or cyanoacrylates, that were used for this purpose in the past have demonstrated inadequate aging properties. Current experiments involve the use of sol-gel techniques leading to a network of silicon-oxygen-zirconium [Si-O-Zr; structure (II)] that adds

(II)

glassy material, filling pores and reattaching the endangered surface layer.

For background information SEE AIR POLLUTION; ANALYTICAL CHEMISTRY; PLASTER; STAINED GLASS in the McGraw-Hill Encyclopedia of Science & Technology.

Norbert S. Baer

Bibliography. N. S. Brommelle and P. Smith (eds.), *Case Studies in the Conservation of Stone and Wall Paintings*, 1986; G. Frenzel, The restoration of medieval stained glass, *Sci. Amer.*, pp. 126–135, May 1985; P. Mora, L. Mora, and P. Philippot, *Conservation of Wall Paintings*, 1984; R. Newton and S. Davison, *Conservation of Glass*, 1989.

Asymmetric synthesis

Asymmetric epoxidation of olefins is one of the most successful and promising strategies for the synthesis of optically active organic compounds. In particular, the epoxidation of allylic alcohols by titanium-tartrate complexes has emerged as the most generally applicable reaction in asymmetric synthesis. Although there is as yet no general method for the asymmetric epoxidation of olefins that lack other functional groups, highly promising methods including the direct epoxidation of unfunctionalized olefins catalyzed by Schiff base or porphyrin complexes have been reported recently. Also, chiral catalysts based on osmium tetroxide that oxidize simple olefins enantioselectively have been found; this provides an indirect route to optically active cyclic sulfates, which are equivalent synthetically to epoxides in many respects.

Epoxides are valuable intermediates for the synthesis of a wide assortment of organic molecules. Steric and electronic factors associated with the strained three-membered ring make it possible to induce nucleophilic attack at either of the two adjacent electrophilic carbons in unsymmetrical epoxides regioselectively (forming a specific structural isomer) and without competing elimination reactions. Ring opening of epoxides also occurs stereoselectively (producing one stereoisomer more rapidly than another), with inversion of configuration at the carbon undergoing attack, and retention at the other carbon. Enantiomer-

ically pure epoxides can thus be converted to enantiomerically pure alcohols, providing entry to a variety of natural products. Because of nature's tendency to employ chiral materials in optically active form, living systems often discriminate strongly between enantiomers. Enantioselective routes to chiral epoxides are therefore desirable for the ultimate preparation of pharmaceuticals, agrochemicals, pheromones, fragrances, and flavors in optically pure form.

Titanium-tartrate–catalyzed epoxidation. While various methods for the synthesis of epoxides are known, reactions involving transfer of oxygen atoms to olefins by using metal catalysts have proven to be the most amenable to asymmetric synthesis. After concerted efforts by several research groups, a procedure for the catalytic asymmetric epoxidation of allylic alcohols was reported in 1980 and was thoroughly developed throughout the ensuing decade. Allylic alcohols make up a small but important class of olefins in which a hydroxy group (OH) is bonded to a carbon atom adjacent to a doubly bonded carbon atom. This landmark achievement represented the first (and is still the most) practical nonenzymatic asymmetric oxidative transformation of olefins. The utility of the titanium-tartrate–catalyzed epoxidation procedure (**Fig. 1**) can be appreciated by the numerous total syntheses in which it has been employed. These include the preparation of disparlure (the Gypsy moth sex attractant), erythromycin, propanolol, fluoxetine, members of the leukotriene and amphotericin families, and a series of carbohydrates and macrolides.

The titanium-tartrate–catalyzed epoxidation procedure embodies several important features. The chiral ligands are simple esters of tartaric acid, and both enantiomers of this acid are readily available. The L enantiomer, often referred to as natural tartaric acid, is obtained as a by-product of the wine industry and is one of the most inexpensive optically active compounds available. The D form is moderate in price and is prepared synthetically, although it too can be obtained from natural sources. Epoxy alcohols of either configuration are thus accessible from allylic alcohols, depending on which enantiomer of the tartrate ester is chosen for the epoxidation reaction. The other two components of the catalyst system are also available at low cost and are convenient to use. Titanium is nontoxic, and *t*-butylhydroperoxide is nonhazardous as compared with other common sources of oxygen atoms, such as peracids and hydrogen peroxide.

The titanium-tartrate catalysts display remarkable generality with respect to substituents on the double bond undergoing epoxidation. The composition of a nonracemic mixture is usually described in terms of the percentage of enantiomeric excess (% ee), corresponding to the difference in the relative percentages of the two enantiomers. Thus, a 95:5 ratio of enantiomers is reported as 90% ee. Over 300 optically active epoxy alcohols have been prepared with the titanium-tartrate catalysts, and in each case the major enantiomer can be predicted with the simple model shown in Fig. 1, with nearly all allylic alcohols giving 90% ee or better.

Substrate requirements. Although simple reagents are used in the titanium-tartrate epoxidation system, the catalytically active species appears to have a complex and highly fluxional structure. As a result, the mechanism of the reaction remains incompletely understood and very controversial; it is the focus of continued study.

From a synthetic standpoint, the only limiting requirement in the titanium-tartrate epoxidation method is that the substrate be an allylic alcohol (see Fig. 1). The OH functional group coordinates to the titanium prior to oxidation of the double bond, leading to a highly ordered transition state in the asymmetry-inducing epoxidation step. Substrate precoordination is ubiquitous in enzymatic processes, but the titanium-tartrate catalysts are not limited by the substrate specificity that is typical of enzymes. Nonetheless, replacing the OH group in allylic alcohols with any other type of functional group results in either very low selectivity in the epoxidation or no reaction at all. Hydroxy groups bind strongly and exchange rapidly to the titanium(IV) center. Either other functional groups do not bind well to the metal center, or they induce severe changes in the catalyst structure that result in poor selectivity. Unfunctionalized olefins do not react.

While hydroxy groups are uniquely effective as directing groups, the separation between the double bond and the OH group is also crucial, with allylic alcohols giving by far the best selectivity. Some success has been achieved in the asymmetric epoxidation of homoallylic alcohols, olefins in which the double bond is separated by two carbons from a hydroxy group. Up to 77% ee has been reported for these substrates by using zirconium-tartrate catalysts. No generally useful methods exist for the asymmetric epoxidation of olefins in which the double bond and the OH group are separated by more than two carbons. Thus, the titanium-tartrate catalytic epoxidation and related methods are enormously successful, but they are restricted to a relatively narrow substrate pool.

General method. Substantial effort is currently devoted to finding a completely general method for the asymmetric epoxidation of unsaturated substrates that do not contain any other functional groups. Osmium-catalyzed asymmetric dihydroxylation of alkenes was reported in 1988, representing the first synthetically viable example of asymmetric catalytic oxidation of simple olefins (**Fig. 2**). The chiral auxiliaries in these catalysts are simple derivatives of either quinine or quinidine, two readily available alkaloids. While the products of these reactions are vicinal diols (dialcohols in which the two OH groups are bonded to neighboring carbon atoms) rather than epoxides, diols may be converted in two steps to cyclic sulfates, which have been shown to behave much like epoxides in their reactions with nucleophiles. Thus, the dihydroxylation reaction provides an indirect route to intermediates that display epoxidelike reactivity. The osmium-catalyzed process is most effective with trans-disubstituted olefins, with certain substrates of this type being oxidized with better than 90% ee. Olefins with other substitution patterns undergo dihydroxylation with substantially lower selectivity, and this appears to be a limitation of this method. The high expense and toxicity of osmium are largely compensated for by the extremely low concentrations of the metal that can be used. Concerns that trace amounts of osmium may nonetheless contaminate the organic products have given rise to attempts to immobilize the catalysts on solid supports; preliminary results in this area are extremely promising.

A direct method for asymmetric epoxidation of alkenes was reported in 1990, involving catalysis by manganese-imine complexes (**Fig. 3**). Enantioselectivities are the highest reported to date for epoxidation of unfunctionalized olefins by nonenzymatic catalysts, with up to 85% ee obtained with cis-disubstituted alkenes. Iron and manganese derivatives of certain synthetic chiral porphyrins also catalyze the asymmetric epoxidation of simple olefins, with cis-disubstituted alkenes again giving best results but with lower enantioselectivity than the manganese-imine systems. Iodosylarenes are typically used as the stoichiometric oxygen atom source in these processes, although sodium hypochlorite (commercial bleach) has been shown in several cases to be an effective and far more practical substitute. Chiral catalysts based on imine and porphyrin ligands that give synthetically useful selectivities for a broad range of substrates are

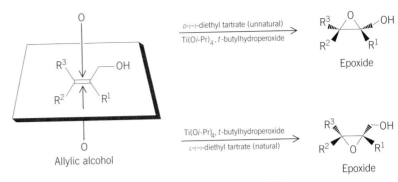

Fig. 1. The titanium-tartrate–catalyzed epoxidation reaction. Ti(O*i*-Pr)$_4$ = titanium isopropoxide. *t*-BuOOh = *tert*-butylhydroperoxide. O = oxygen. R^1, R^2, R^3 = alkyl or aryl groups. OH = hydroxy group.

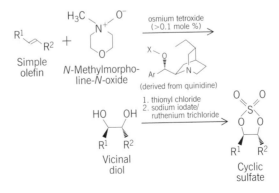

Fig. 2. Osmium-catalyzed asymmetric oxidation of a simple olefin. R^1 and R^2 = functional groups. Ar = aryl group (C$_6$H$_5$). X = aryl group or ester group. Numbers at the second arrow indicate the order in which the reagents are added.

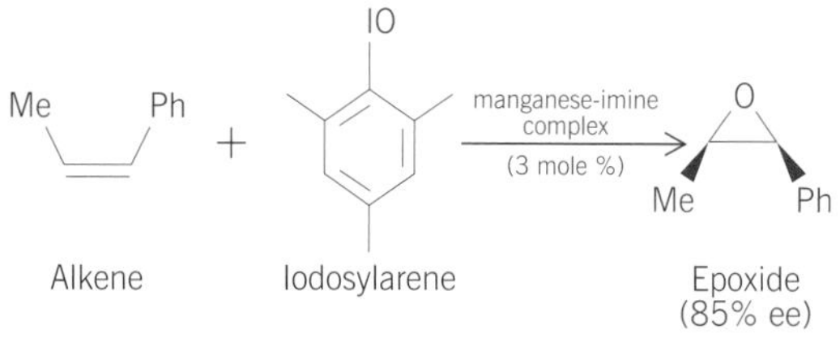

Fig. 3. A direct method for asymmetric epoxidation of simple alkenes catalyzed by a manganese-imine complex. Me = methyl group. Ph = phenyl group.

likely to emerge as these systems continue to be explored and optimized.

For background information *SEE ASYMMETRIC SYNTHESIS; CATALYSIS; ENANTIOMER; EPOXIDATION; ORGANIC CHEMICAL SYNTHESIS; STEREOCHEMISTRY* in the McGraw-Hill Encyclopedia of Science & Technology.

Eric N. Jacobsen

Bibliography. E. N. Jacobsen et al., Enantioselective epoxidation of unfunctionalized olefins catalyzed by (salen)manganese complexes, *J. Amer. Chem. Soc.*, 112:2801–2803, 1990; J. D. Morrison (ed.), *Asymmetric Synthesis*, vol. 5, 1985; K. B. Sharpless et al., Catalytic asymmetric epoxidation and kinetic resolution: Modified procedures including in situ derivatization, *J. Amer. Chem. Soc.*, 109:5765–5780, 1987; K. B. Sharpless et al., Vicinal diol cyclic sulfates: Like epoxides only more reactive, *J. Amer. Chem. Soc.*, 110:7538–7539, 1988.

Atom

The control of neutral atoms with laser light has dramatically improved in the last several years. The atoms have been cooled to temperatures as low as 2.4 μK above absolute zero, the lowest temperature achieved to date, and stored in a variety of optical, magneto-optic, and magnetic traps. Devices analogous to optical components such as mirrors, diffraction gratings, and lenses have been fashioned to control atoms. Even more significantly, devices that have no optical counterparts, such as atomic fountains, funnels, and trampolines, have been demonstrated. Some of these experimental advances have led to a more subtle understanding of light–atom interactions, which have then become the basis of further experimental advances.

Laser cooling. Laser cooling of atoms, proposed in 1975, is accomplished by generating a force on the atoms that always opposes their motion. Every time an atom scatters a photon with momentum $p = h/\lambda$, where λ is the wavelength of the laser light and h is Planck's constant, it experiences a small change in velocity. Since the photons are not scattered into a preferred direction, the net average velocity change is $\Delta v = p/m$, where m is the mass of the atom. By tuning the laser to the low-frequency side of the resonance, an atom moving against the direction of a laser beam will interact with the light Doppler-shifted into resonance, while the beam copropagating with the atom will be

Doppler-shifted out of resonance. Thus, the atom will preferentially scatter photons from the beam opposing the direction of motion. For Doppler shifts much less than the width of the atomic absorption line, a viscous drag force results that is linearly proportional to the velocity of the atom. This sea of photons was called an optical molasses by the group that first demonstrated the cooling effect in 1985.

The first experiments demonstrating optical molasses seemed to confirm the theoretical expectations. However, the tidy picture soon began to unravel as it was realized that the molasses worked far better than predicted by theory in virtually all respects. The greatest surprise was the finding in 1988 that the atoms could be cooled to temperatures more than an order of magnitude lower than the Doppler limit if the magnetic field on the atoms was reduced below about 50 milligauss and the laser was tuned several linewidths below resonance.

The current theoretical understanding of laser cooling gives an explanation of the colder temperatures based on a new cooling mechanism. The mechanism depends on effects due to optical pumping, Stark shifts in time-varying fields, and the motion of atoms in light fields with polarization gradients. While many of these properties have been exploited in earlier laser cooling techniques, the effects due to a combination of these properties had to await experimental discovery. A prediction of the theory is that atoms can be cooled to average velocities corresponding to three times the recoil of a single photon. In the case of cesium, this corresponds to a temperature of 2.4 μK.

Neutral-atom manipulation. In addition to cooling atoms to very low temperatures, there has been rapid progress in developing techniques to trap and otherwise manipulate neutral atoms. Magnetic traps, optical traps, and magneto-optic traps have been demonstrated. Atomic densities of 10^{11} atoms/cm^3 have been achieved. Laser-cooled atoms stored in magnetic traps offer the possibility of creating a quantum gas of atoms where the interatomic spacing is comparable to the de Broglie wavelength of the atoms. Atoms with integer angular momentum would then be expected to undergo Bose condensation, while atoms with half-integer angular momentum would form a degenerate Fermi gas. High-density, spin-aligned samples of hydrogen and cesium atoms have been stored in magnetic traps in efforts to reach this regime.

Atomic fountains and funnels. All neutral-atom traps work by strongly perturbing the energy levels of the atoms in a spatially dependent manner. In applications such as high-resolution spectroscopy, it is desirable to avoid the trapping perturbations, and one solution is to work with freely falling atoms in slow atomic beams. An extreme limit of a slow beam is a so-called atomic fountain in which the atoms directed upward will return because of gravity. The first attempt to build an atomic fountain was in the early 1950s, but the first successful demonstration had to await the advent of slow-atom technology. An atomic fountain was constructed by first trapping atoms from a thermal beam in a magneto-optic trap and then pushing the atoms upward with a pulse of light from a continuous-wave laser (**Fig. 1**). Near the top of the atomic trajectory, transitions between the ground-state hyperfine levels were induced by using the Ramsey method of separated oscillatory fields. A measurement time of 0.25 s (the time between the two microwave pulses used to induce the transition) yielded a linewidth of 2 Hz, and after 15 min of integration the center of the line could be resolved to 10 mHz. Straightforward engineering should make possible a cesium microwave time standard based on this device that could be two orders of magnitude better than the current atomic clock time standards.

Since the precision of a spectroscopic measurement depends on both the high Q (Q is the quality factor of the resonance, defined by $Q = \nu/\Delta\nu$, where ν is the frequency and $\Delta\nu$ is the width of the resonance) and the signal-to-noise ratio of the signal, it is important to create as high a flux of cold atoms as possible. Also, while the first atomic fountain worked in a pulsed mode, many applications would benefit from a continuous beam of atoms. A high-flux beam of slow atoms that can be used in a continuous atomic fountain has been generated by creating a so-called atomic funnel (**Fig. 2**). The funnel employs a current wire to generate a magnetic quadrupole field, and trapping and cooling laser beams. Molasses-trapping beams, which flood the central volume with light, both confine the atoms in the transverse directions and cool them, whereas molasses beams along the axial direction of the current wire only cool the atoms. Atoms enter the funnel from a slowed atomic beam and exit along the center axis of the current wire. The wire loop generates the weak magnetic quadrupole field needed for the two-dimensional version of a magneto-optic trap; this field increases linearly from 0 at the center of the funnel. The funnel presently accepts atoms with velocities as large as 20 m/s (65 ft/s), and cools them into a localized and collimated beam. Because of the dissipative properties of optical molasses, the funnel has no optical analog and has produced a beam of atoms with a peak phase-space density [(number of atoms)/$(\Delta v_x \Delta v_y \Delta v_z \Delta x \Delta y \Delta z)$, where Δx, Δy, and Δz are ranges of spatial coordinates, and Δv_x, Δv_y, and Δv_z are ranges of velocity coordinates] over 10^4 times higher than that of the original beam.

Other techniques have been developed for neutral-atom manipulation. Grazing-incidence atom mirrors have been constructed by using an evanescent wave outside a glass prism from a laser beam that is totally internally reflected. When the atoms are moving slowly enough, normal-incidence reflection in the form of an atomic trampoline has also been demonstrated, wherein atoms are dropped and then bounce up from the repulsive potential created by the light field. Atom diffraction gratings formed with both standing-wave patterns of light and matter gratings have also been demonstrated. These devices will be useful as analogs to beam splitters and mirrors in the construction of atom interferometers.

Applications of atom manipulation. An application that is receiving considerable attention is the construc-

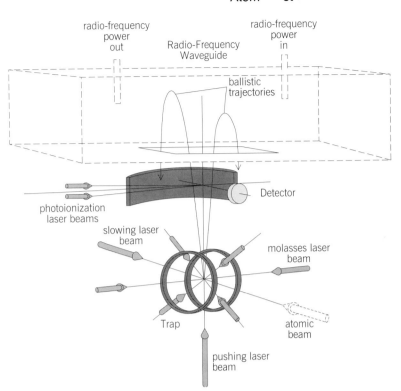

Fig. 1. Atomic fountain. The loading of the magneto-optic trap from a conventional atomic beam, the ballistic trajectory of the cooled atoms through the radio-frequency waveguide, and the resonant photoionization detection region are shown. (*After M. Kasevich et al., RF spectroscopy in an atomic fountain, Phys. Rev. Lett., 63:612–615, 1989*)

tion of better time standards in the microwave or optical domain. Many of the most troublesome systematic shifts associated with a time standard will be reduced as the velocity of the atoms is reduced. Equally important, the trapping technology has been simplified to the point where practical clocks using laser cooling are possible. Cesium atoms have been laser-cooled and trapped with small, inexpensive di-

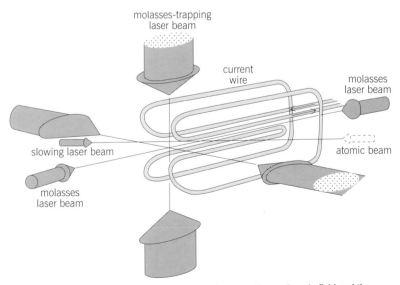

Fig. 2. Atomic funnel. The wire used to generate the magnetic quadrupole field and the trapping and cooling laser beams are shown. (*After E. Riis et al., Atom funnel for the production of a slow, high-density atomic beam, Phys. Rev. Lett., 64:1658–1661, 1990*)

ode lasers in a sealed vapor cell with the use of the magneto-optic trap. A time standard based on an optical transition of either stored ions or neutral-atom fountains is also receiving considerable attention.

The ability to laser-cool and trap atoms in a sealed cell also means that experiments that require the optical manipulation of rare species of atoms (for example, rare isotopes) can be carried out, since the sealed cell needs an atomic density of roughly 3×10^8 atoms/cm^3. Thus, as few as 10^{11} atoms of the rare species (corresponding to 2×10^{-11} grams in the case of cesium) are sufficient to do an experiment. Measurements of parity nonconservation in a large number of isotopes of cesium or the creation of spin-polarized samples for beta-decay studies will be possible.

Optical tweezers. The first optical trap based on a single focused laser beam has been used to trap not only atoms but also particles ranging in size from 10 micrometers to ~20 nanometers. This trap works by using the strong electric field of the laser beam to induce a dipole moment on the object being trapped. As long as frequency of the laser field is below the natural resonances of the particle being trapped (for example, the atomic transition of an atom or the absorption edge of a polystyrene sphere), the energy of the induced dipole $\mathbf{p}$ in the laser field $\mathbf{E}$, given by $W = -\mathbf{p} \cdot \mathbf{E}$, is less than zero. Hence, the particle achieves a lower energy state by moving into the focal spot of the laser beam. Since the focal spot of a laser beam can be easily moved with mirrors or lenses, this type of laser trap has been named an optical tweezers. In the case of micrometer-sized particles, the laser beam is focused with a microscope objective of high numerical aperture, and the object being held can be simultaneously viewed through the same objective.

An application of the optical tweezers of importance for biological research was the discovery that single cells can be moved about in a water solution without apparent damage to the organism. Objects inside a living cell such as organelles or filaments of cytoplasm can also be manipulated without puncturing the cell wall.

Single macroscopic molecules have also been manipulated with optical tweezers. Even though biological molecules are too small to be trapped at room temperature, the molecule may be held by trapping a micrometer-sized sphere that has been attached to the molecule. A single molecule of deoxyribonucleic acid (DNA) has been attached to a polystyrene sphere via a biotin-avidin-biotin bridge. The manipulation of the molecules in aqueous solution has been seen in real time by staining the DNA with a dye and observing the fluorescence with an image-intensified video camera. Optical tweezers may allow the manipulation and observation of biochemical processes on a single-molecule basis with the same degree of control that is now beginning to be possible with trapped ions and atoms.

For background information SEE ATOMIC CLOCK; BOSE-EINSTEIN STATISTICS; CELL (BIOLOGY); DEOXYRIBO-NUCLEIC ACID (DNA); LASER SPECTROSCOPY; MOLECULAR BEAMS in the McGraw-Hill Encyclopedia of Science & Technology.

Steven Chu

Bibliography. S. Chu and C. Wieman (eds.), Laser cooling and trapping (special issue), *J. Opt. Soc. Amer.*, vol. B6, no. 11, 1989; M. Kasevich et al., RF spectroscopy in an atomic fountain, *Phys. Rev. Lett.*, 63:612–615, 1989; E. Riis et al., Atom funnel for the production of a slow, high-density atomic beam, *Phys. Rev. Lett.*, 64:1658–1661, 1990.

Atomic force microscope

Since its invention in 1986, the atomic force microscope has rapidly grown in popularity as a tool for studying the properties of materials. Initially, atomic force microscopy (also called simply force microscopy) was conceived as a technique for determining the topography of materials. Unlike other imaging instruments such as the scanning tunneling microscope, which can image only conducting materials, the atomic force microscope can image both conducting and insulating materials. The atomic force microscope is a powerful technique for determining topography because of its ultrahigh resolution, to the point where it can detect individual atoms on the surface of a material.

Recently, the technique of force microscopy has been used to understand the physics of the different types of forces that exist between materials, such as van der Waals, frictional, magnetic, and electrostatic forces. With the force microscope, it is often possible to map these forces over the surface of a material with a much better resolution than was previously available.

Operating principles. In atomic force microscopy the goal is to measure the force acting on the last few atoms at the end of a sharp tip as the tip is gently moved over a surface. The force varies slightly as the tip moves from one atom to the next on the surface.

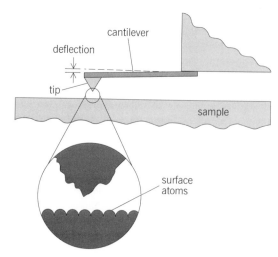

Fig. 1. Diagram of atomic force microscope.

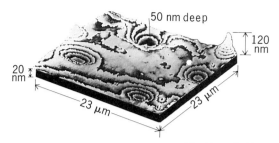

Fig. 2. Atomic-force-microscope image of liquid menisci in small pore openings. (*From A. M. Homola, C. M. Mate, and G. B. Street, Overcoats and lubrication for thin film disks, MRS Bull., 15:45–52, Materials Research Society, 1990*)

Consequently, if the force microscope has a sufficiently high degree of sensitivity, it is possible to detect the individual atoms on a surface. This is accomplished by mounting the microscope tip at the end of a small cantilever with a known spring constant (**Fig. 1**). A force acting on the end of the tip causes the cantilever to deflect by an amount proportional to the force. The key to making a successful atomic force microscope is to be able to measure the very small deflections of the cantilever caused by the atomic-scale forces acting on the end of the tip. To be able to sense the force from individual atoms, it is necessary to measure deflections much smaller than the size of an atom, that is, less than $\sim 10^{-10}$ m ($\sim 4 \times 10^{-9}$ in.). The detection schemes developed for measuring the very small cantilever deflections include electron tunneling methods, optical interference, optical deflection, and capacitance.

For determining the topography of surfaces, the atomic force microscope can use either the repulsive forces from the tip in contact with the surface or an attractive force with the tip separated a small distance from the sample. In the repulsive mode, the tip is scanned across a hard surface while touching it with constant force. The tip then moves up and down over the features of the surface in much the same way as a stylus moves in the groove of a phonograph record. In this mode, very high resolution topographs can be obtained, occasionally even with atomic resolution. In the attractive mode, the tip is held 2 to 10 nanometers (8×10^{-8} to 4×10^{-7} in.) away from the sample, so that it senses only an attractive van der Waals force. Since the attractive van der Waals force is much weaker than the repulsive force, a more sensitive detection method is used. The force on the tip slightly changes the natural frequency of the cantilever. Therefore, a sensitive way of measuring the attractive force is to vibrate the cantilever a small amount at a frequency near its natural vibration frequency and to monitor the amplitude of vibration as the force shifts the natural frequency on and off resonance with the applied frequency. By using a feedback loop to maintain a constant shift in frequency by adjusting the height of the tip over the surface as the tip is scanned over the surface, the topography of the surface can be determined without the tip ever contacting the sample. This is particularly useful for studying fragile materials where a tip would disturb the material on the surface if it came into contact. **Figure 2** shows an

example of an attractive-force topographic image of liquid menisci in several very small pore openings. If the tip were to touch the liquid surface, the liquid would wick up around the tip, disturbing its topography.

Friction force microscopy. Once highly sensitive atomic force microscopes had been constructed, it was realized that these versatile instruments not only were capable of topographic measurements but also were excellent tools for learning about the nature of the forces acting between two materials. An early example was the use of the atomic force microscope to measure the frictional force acting on a sharp tungsten tip as it was dragged across a graphite surface under very low loads (less than 10^{-4} newton). As the tip was dragged across the surface, the frictional force was observed to increase and decrease at regular intervals of 0.25 nm (**Fig. 3**). The 0.25-nm spacing observed in the frictional force is the same as the periodicity of graphite atoms on the graphite basal plane. Therefore, even though the contact area can be several hundred nanometers across for these loads, the atomic force microscope has the ability to observe the atomic variations in the frictional force. Future work should use these measurements of atomic-scale friction to develop an understanding of how friction, lubrication, and wear occur at the atomic level. Such atomic-scale understanding should be invaluable in improving the mechanical durability of material surfaces.

Magnetic force microscopy. Another form that is finding widespread use is magnetic force microscopy. Here, the force measured comes from the gradient of a magnetic field acting on a tip made up of a magnetic material. The variation of magnetic field above a sample is determined by monitoring the shift of the natural frequency of the cantilever due to the magnetic force as the tip is scanned over the sample. By using this method, maps of the variation of magnetic field over a magnetic material have been obtained with

Fig. 3. Frictional force, indicated by brightness, on a sharp tungsten tip dragged from left to right across a graphite surface as a function of sample position. (*From C. M. Mate et al., Atomic-scale friction of a tungsten tip on a graphite surface, Phys. Rev. Lett., 59:1942–1945, 1987*)

lateral resolutions as high as 25 nm (10^{-6} in.). Such maps not only are useful for obtaining a fundamental understanding of the magnetic materials but also are beginning to play an important role in developing magnetic materials for technological applications such as magnetic recording.

Electrostatic force microscopy. As the tip of a force microscope is brought near an electric charge on a surface, it will sense an electrostatic force from the electric charge. Even though this is a weak force, it can be measured in the same manner as the weak van der Waals force by vibrating the cantilever near its natural frequency. Electrostatic force microscopy has been used to measure the variations in the dielectric properties of the near-surface regions of several types of materials, including photoresists and *pn* junctions on silicon substrates that are used in the manufacture of semiconductor integrated circuits. As these intricate electronic circuits are continually made smaller, electrostatic force microscopy should become a useful tool for understanding their performance.

Another recent application of electrostatic force microscopy is to study contact electrification. Here, the tip is brought into contact with a surface and pulled away. During contact, a small amount of electric charge is transferred between the tip and the sample. The tip is then scanned over the surface while measuring the electrostatic force due to the localized charges on the sample. From the force image, the distribution of charge on the sample after contact is determined. Currently, the electrostatic force microscope has the sensitivity to measure as little as three electric charges, and in the future it may be possible to image the charge from a single electron. Such an ability should be extremely helpful in understanding the physics of contact electrification, a fairly common phenomenon that results in static electricity. Technologically, contact electrification is important to such areas as xerography, where it is used to charge the toner particles in photocopying machines.

Dipstick microscopy. In dipstick microscopy, the tip of a force microscope is lowered into a thin liquid film covering a surface. When the tip touches the top of the liquid surface, it experiences a strong attractive force from the surface tension of the liquid meniscus that forms around the tip. After penetrating the film, the tip touches the hard surface of the underlying sample. The thickness of the liquid film can be calculated from the distance between these two events with an accuracy of 0.5 nm (2×10^{-8} in.). By repeatedly dipping the tip into the liquid film at different locations, the variation of thickness of the film can be mapped over a small area. Dipstick microscopy has already found use in measuring the thickness of lubricant films used to protect computer hard disks and for studying the wetting and spreading of molecular liquid films that are only a few molecules thick.

For background information SEE FRICTION; INTEGRATED CIRCUITS; INTERMOLECULAR FORCES; PHOTOCOPYING PROCESSES; STATIC ELECTRICITY in the McGraw-Hill Encyclopedia of Science & Technology.

C. Mathew Mate

Bibliography. G. Binnig, C. F. Quate, and C. Gerber, Atomic force microscope, *Phys. Rev. Lett.*, 56:930–933, 1986; C. M. Mate et al., Atomic-scale friction of a tungsten tip on a graphite surface, *Phys. Rev. Lett.*, 59:1942–1945, 1987; C. M. Mate, M. R. Lorenz, and V. J. Novotny, Atomic force microscopy of polymeric liquid films, *J. Chem. Phys.*, 90:7550–7555, 1989; R. Pool, Children of STM, *Science*, 247:634–636, 1990.

Auxin

Auxins, a group of plant hormones, regulate various growth and developmental processes, including cell elongation, cell division, and cell differentiation in plants. However, the mechanisms by which auxins regulate diverse developmental processes are poorly understood. Auxins are being intensively studied over a wide range of applications by using a variety of techniques. This article discusses auxin-regulated gene expression and auxin-binding proteins.

AUXIN-REGULATED GENE EXPRESSION

Auxin-induced cell elongation has been used widely to investigate the biochemical and molecular events involved in auxin action. Two theories, the gene expression hypothesis and the acid growth theory, have been proposed to explain auxin-induced cell elongation. The gene expression hypothesis, proposed in the 1960s, suggested that auxin regulates the expression of specific genes whose products are involved in auxin-induced response. The observation that the treatment of tissue with auxin for several hours increases ribonucleic acid (RNA) and protein synthesis supported the gene expression hypothesis. Furthermore, the inability of auxin to induce cell elongation in the presence of inhibitors of RNA and protein synthesis suggested that auxin-induced cell elongation is dependent on RNA and protein synthesis. However, at that time, auxin-regulated changes in specific messenger RNAs (mRNAs) or proteins prior to auxin-induced cell elongation could not be detected owing to the lack of sufficiently sensitive techniques. Hence, it was argued that the primary action of auxin is not at the level of gene expression. In the 1970s the acid growth theory was proposed, suggesting that auxin induces proton secretion into the cell wall, resulting in loosening of the cell wall and cell elongation. Although auxin-induced proton extrusion is sufficient to account for a rapid stimulation of cell elongation, it is inadequate to explain auxin-induced sustained cell elongation as well as prolonged effects of auxin, where cell division, cell elongation, or cell differentiation is involved. As a result of developments in molecular biology in the 1980s, there was renewed interest in investigating the gene expression hypothesis. Recombinant deoxyribonucleic acid (DNA) techniques have enabled the isolation, characterization, and analysis of expression of several auxin-regulated genes.

Auxin-induced gene expression. Most molecular studies on auxin-induced cell elongation have been performed by using excised soybean hypocotyl and pea epicotyl sections. This is due to the rapid and dramatic effect of auxin on cell elongation and the ease of handling and experimental manipulation. Electrophoretic analysis of labeled proteins from auxin-treated and control tissues reveals that auxin treatment induces the synthesis of specific polypeptides while repressing the synthesis of others. Translation of poly(A)$^+$ RNA from auxin-treated and control tissues in cell cultures reveals that up to 40 mRNAs undergo either upward or downward shift in relative concentration in response to auxin treatment. Several auxin-regulated complementary DNA (cDNA) clones have been isolated by differential hybridization screening. In brief, poly(A)$^+$ RNA from auxin-treated tissue is used to synthesize cDNA, which is inserted into a vector. These recombinant molecules are transferred into bacteria by transformation. The transformed bacteria containing the recombinant molecules represent a cDNA library. Duplicate filters containing the DNA from each of the recombinant molecules of the cDNA library are made. One set of filters is hybridized with labeled first-strand cDNA prepared from poly(A)$^+$ RNA of auxin-treated tissue, and another set is hybridized with labeled first-strand cDNA from poly(A)$^+$ RNA of control tissue. Clones that hybridize exclusively to probes made from auxin-treated tissue represent auxin-induced cDNAs.

Nine independent cDNA clones to auxin-induced mRNAs have been isolated from soybean hypocotyls and two from pea epicotyl. The mRNAs corresponding to these clones are induced in 2.5–30 min by auxin. Small auxin up RNAs (SAUR), which are small mRNAs of about 550 nucleotides, are induced within 2.5 min following auxin treatment. This is the fastest response to a plant hormone at the gene level. The induction of all but two of these mRNAs is specific to auxins. Nonauxin analogs, other plant hormones, and stress conditions are ineffective in inducing these mRNAs. Protein synthesis inhibitors do not interfere with auxin induction of some of the mRNAs, indicating that the auxin effect is direct and does not require protein synthesis. Some of the auxin-induced mRNAs are induced by protein synthesis inhibitors in the absence of auxin, suggesting that these auxin-induced mRNAs are under the control of rapidly turning-over protein.

It has been demonstrated that the expression of SAUR genes correlates strongly with cell elongation and growth during gravitropic bending. The gravitropic bending is caused by differential growth on the upper and lower sides of the responding organ, with the differential growth being attributed to differential distribution of auxin. In vertically growing hypocotyls the SAUR genes are evenly expressed. However, within 20 min after the hypocotyls are placed in a horizontal position, expression of SAUR genes becomes asymmetrical with very high expression on the lower side, the side that grows rapidly by cell elongation in response to gravity. After 180 min, the apex of the hypocotyl regains vertical position, and the SAUR mRNAs are symmetrically distributed once again (see **illus.**).

Seven auxin-induced cDNAs have been isolated from tobacco cells in suspension culture. The mRNA corresponding to these clones is induced as early as 15–60 min after auxin addition. Studies with these clones indicate that the expression of these genes may be essential for cell division. A cDNA (*par*) to an auxin-induced mRNA has been isolated from the mesophyll protoplasts. Nucleotide analysis shows that the *par* codes for a hydrophilic protein of 24.5 kilodaltons. Accumulation of *par* mRNA prior to DNA synthesis indicates a role for *par* gene product in the initiation of meristematic activity in differentiated mesophyll cells. Two cDNA clones corresponding to auxin-induced mRNAs have been isolated from strawberry fruit. Application of auxin induced the mRNA level corresponding to these clones. A positive correlation between accumulation of auxin-induced mRNAs and fruit growth is observed.

Auxin-repressed gene expression. Twelve auxin-repressed cDNAs have been isolated from soybean; they hybridize to just three different mRNA species,

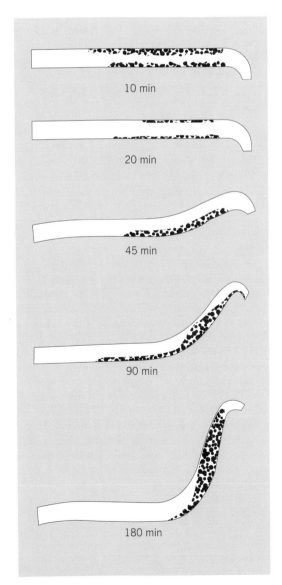

Schematic diagram of the expression of SAURs (small auxin up RNAs; black areas) in soybean seedlings in response to gravity. Soybean hypocotyls were gravistimulated by being placed horizontally, and tissue prints were made after the indicated times and probed with labeled SAUR cDNA. (*After B. A. McClure and T. Guilfoyle, Rapid redistribution of auxin-regulated RNAs during gravitropism, Science, 243:91–93, 1989*)

10 min

20 min

45 min

90 min

180 min

and based on this these clones are divided into three groups. The mRNA corresponding to these clones is reduced in concentration after 4 h of auxin application, and disappears within 20 h of auxin treatment. A cDNA (λSAR5) to an auxin-repressed mRNA has been isolated from strawberry fruit and found to accumulate within 1 h of auxin depletion. A very high level of this mRNA was observed in unpollinated fruit (fruit without auxin), and the level reduced to low level following auxin application. A strong positive correlation is found between accumulation of the SAR5 mRNA and cessation of fruit growth.

Transcriptional regulation. The increase or decrease in auxin-regulated mRNAs could be due to an altered rate of transcription or regulation at posttranscriptional events such as processing of precursor transcript, transport of mRNA from the nucleus to cytoplasm, or stabilization of the mature mRNA. Auxin could control one or more of these events to regulate gene expression. Transcriptional rates of seven auxin-induced clones from soybean and six auxin-regulated sequences from tobacco suspension culture have been shown to be higher in nuclei isolated from auxin-treated tissue. These studies indicate that the auxin-induced increase in mRNAs is controlled, at least in part, at the level of transcription. Studies on auxin-repressed mRNAs in soybean hypocotyl indicate that auxin regulation is at the posttranscriptional level. Preliminary studies with strawberry auxin-repressed mRNA indicate auxin regulation at both the transcriptional and posttranscriptional levels. Regulation of gene expression at the posttranscriptional level by auxin has not been assessed thoroughly. It is likely that auxin regulates gene expression at multiple steps.

Genomic clones and their characterization. Genomic clones for six auxin-responsive soybean cDNAs have been isolated and sequenced. Genomic sequences for two of them are present in one to two copies per haploid genome and contain introns. They code for highly hydrophilic proteins of 26.8 kDa and 21.5 kDa. Five regions of high amino acid homology constituting one-third of the coding region are found. Two conserved sequences have been identified, and these sequences are found at similar distances upstream of the transcription initiation site in each gene. The genes for SAUR cDNAs are clustered within about 5 kilobases, contain no introns, and code for small proteins (9–10.5 kDa) with a high degree of amino acid homology. Alignment of sequences upstream of the initiation codon revealed two conserved regions. Analysis of DNA sequences upstream of the transcription initiation site of several auxin-inducible genes showed a conserved sequence (see **table**). Whether this or other conserved elements within the gene family are involved in auxin regulation is yet to be determined.

Prospects. Research on auxin-regulated gene expression has raised two important issues. The first is the elucidation of how auxin regulates the transcription of specific genes and the sequence of events involved in this process. This requires the identification of regulatory DNA sequences involved in the

Conserved DNA sequence elements of various auxin-inducible promoters*		
Sequence[†]	Promoter	Position[‡]
GCA- -CATACGT	Nos	−114
GCAC-CATGCGT	Aux28-B′	−210
GCAG-CATGCAC	Aux28-B	−285
GCAG-CATGCAT	Aux22-B	−207
GCATTCATACGC	Gmhsp26	−178
CCA- -TATGCCC	SAURs	−130
GCGT-CACGCGC	mas	−67
GCAN-CATRCRY	Consensus	

*From Gynheung An, M. A. Costa, and Sam-Bong Ha, Nopaline synthase promoter is wound inducible and auxin inducible, *Plant Cell*, 2:225–233, 1990.
[†]G = guanine; C = cytosine; A = adenine; T = thymine; R = purine; Y = pyrimidine; N = any nucleotide; - indicates gap.
[‡]The position is given with respect to the transcription start site, with a negative number indicating an upstream position.

auxin regulation and proteins that interact with these sequences. The second one is the determination of the identity and function of the products of auxin-regulated genes and how they are involved in eliciting the auxin-induced response. Studies have shown a strong correlation between the expression of auxin-regulated genes and auxin-induced responses, indicating that the proteins coded by auxin-regulated genes could be important in the physiology of auxin-regulated processes. However, there is no direct evidence that the proteins coded by auxin-regulated genes mediate the auxin-induced responses. Localization of the auxin-responsive gene products in the cell using the antibodies may provide some insight into the function of these polypeptides. Over- and underexpression of the auxin-regulated gene products in transgenic plants might provide information on the function of the gene products.
A. S. N. Reddy

AUXIN-BINDING PROTEINS

It is generally believed that plant cells communicate via intrinsic signals such as hormones in the same way as other eukaryotic cells communicate. The perception of a hormone begins when the specific hormone is bound to a receptor, thereby causing a hormone-induced conformational change in the polypeptide structure of the appropriate hormone receptor. Perception ends with a specific interaction of the hormone–hormone receptor complex with the next member in the signal transduction chain. Thus, a receptor has two functions: it provides the selectivity in signal perception by binding the hormone; and it initiates a specific chain of events in the hormone transduction path that leads to a response.

Auxins have a number of effects, but promotion of cell elongation is the operational definition of this class of hormones. Members of the auxin family have a range of structures that include indole-, naphthalene-, phenyl-, and phenoxy-substituted carboxylic acids. The most studied endogenous auxin is indole-3-acetic acid, and the most widely used synthetic auxin is naphthalene-1-acetic acid.

Auxins are important in many developmental processes in plants. There are many enzymes that are involved in the auxin economy of the cell, ranging from synthesis to degradation. Also, auxins are transported and stored by specific enzymes. Therefore, besides auxin receptors, it is expected that many other proteins that bind auxins in a specific manner will be found. This expectation is being realized since, by a variety of techniques, several different auxin-binding proteins in plants have been identified. Currently, the function of each is unknown, but progress is being made to elucidate their functions.

Although auxin receptors in plants have not yet been identified, several proteins that bind auxin have been identified and partially characterized. While it has not been possible to demonstrate unequivocally that any of these auxin-binding proteins initiate the cell-elongation response, at least one is a good candidate for the auxin receptor that mediates cell elongation.

Auxin-binding assays were developed in the 1970s to identify auxin receptors and to characterize the binding of auxins to their putative receptors. The objectives in studying auxin binding are to characterize the auxin binding site and to identify the auxin receptor that mediates growth. The rationale behind these studies is that the binding-site structure will direct the design of new growth regulators, while the identification of the auxin receptor will open the possibility of manipulating growth through genetic engineering of this regulatory molecule.

Auxin-binding proteins in maize. Maize has been the preferred organism to use in auxin-binding studies, probably because it has the highest level of auxin binding sites among all the plants surveyed and the nature of the isolated maize microsomal fraction is suited for the binding assays. Studies using maize have found at least three detectable auxin binding sites: site I, located on the endoplasmic reticulum; site II, probably located on the tonoplast or Golgi apparatus; and site III, located on the plasma membrane. Site I has been the most intensively studied because it meets the requirements for the auxin receptor that mediates cell elongation. Site I binds auxins and auxin analogs optimally at pH 5.5 and with the affinities expected based upon the growth-promoting efficacies of each of the tested auxins and analogs. Site I behaves like 40-kDa protein and is located within the growing regions of the maize shoot; in addition, the abundance of site I decreases after irradiation by red light, a treatment that causes the growth rate of the shoot tissue to also decrease. These observations are consistent with the hypothesis that site I auxin represents the binding activity of the receptor that mediates growth.

In 1985, an auxin-binding protein was purified that had all the properties of site I, that is, pH optimum, binding specificities, and molecular mass. It has been demonstrated that this auxin-binding protein is distributed in the growing regions of the maize shoot and that its abundance is down-regulated in a parallel fashion as site I. Down-regulation means that the abundance of the auxin-binding protein decreases in a pattern parallel to the decrease in auxin-binding and growth. The cDNA for this auxin-binding protein has been identified, and several features of the encoded protein have been deduced. For example, this auxin-binding protein is probably a dimer of two 22-kDa subunits, each subunit containing 201 amino acids, of which the first 38 constitute a signal sequence. The auxin-binding protein has a single glycosylation site to which is N-linked a high-mannose-type sugar moiety, and there are no predicted alpha-helical secondary structures. Finally, it has been shown that this auxin-binding protein contains an endoplasmic reticulum retention signal at the carboxy terminus, consistent with the observation that the auxin-binding protein is located in the endoplasmic reticulum. It is this last feature that is unique, since, if this auxin-binding protein is shown to be the auxin receptor, it will be the only known growth hormone receptor located in the endoplasmic reticulum.

The main task facing researchers today is to determine if this protein is the auxin receptor. There are several approaches to solve this problem. One is to produce transgenic plants that overexpress or inappropriately express the auxin-binding protein. This altered expression would be expected to cause altered growth rates or patterns and may provide the clue for its function. Another approach is to add antibodies against the maize auxin-binding protein to protoplasts and determine if these antibodies block an auxin response. Both approaches are under way.

If receptor function is unequivocally demonstrated, then the task of characterization remains. One important character of the receptor is the nature of the binding site, since this knowledge will be useful in designing new growth regulators and understanding the mechanism of hormone perception. It will also be necessary to determine how the auxin receptor interacts with the next member of the hormone signal transduction path.

Auxin-binding proteins in other plants. Auxin-binding proteins have been identified in plants other than maize by using a variety of methods. One extremely useful method is photoaffinity labeling, a technique to covalently and specifically tag auxin-binding proteins in crude samples. This technique has been used to identify not only the auxin-binding protein in maize discussed above but also the auxin-binding proteins in zucchini, tomato, soybean, and henbane tissue cultures. In most cases, a dimeric protein having 40-kDa subunits has been identified, in contrast to the 22-kDa-subunit auxin-binding protein in maize. Furthermore, this auxin-binding protein is located at the plasma membrane rather than the endoplasmic reticulum. While it has not been demonstrated that this auxin-binding protein is the auxin receptor, the observation that an auxin-insensitive tomato mutant lacks the photolabeled auxin-binding protein in the shoot tissue is interesting and is consistent with the postulated receptor function of this auxin-binding protein.

In mung bean, a protein complex containing 15- and 37-kDa subunits has been shown to bind auxins. Very little is known of this protein, and further purification and characterization are necessary.

Recently, anti-idiotypic antibodies have been prepared against an auxin-binding site, and these antibodies have been shown to recognize a single protein within a large protein complex found in the cytoplasm of a wide variety of plants, including maize. The protein containing the binding site is 65 kDa. Also, this auxin-binding complex probably contains nucleic acids. Therefore, in many ways, this auxin-binding protein is different than any other discovered so far. As with the other auxin-binding proteins, the function is not known yet.

For background information *SEE AUXIN; GENETIC CODE; PLANT MOVEMENTS* in the McGraw-Hill Encyclopedia of Science & Technology.

Alan Jones

Bibliography. T. J. Guilfoyle, Auxin-regulated gene expression in higher plants, *CRC Crit. Rev. Plant Sci.*, 4:247–276, 1986; G. Hagen, Molecular approaches to understanding auxin action, *New Biologist*, 1:19–23, 1989; A. M. Jones, P. L. Lamerson, and M. A. Venis, Comparison of site I auxin binding and a 22-kilodalton auxin-binding protein in maize, *Planta*, 179:409–413, 1989; J. L. Key, Modulation of gene expression by auxin, *BioEssays*, 11:52–58, 1989; D. Klambt, A view about the function of auxin-binding proteins at plasma membranes, *Plant Mol. Biol.*, 14:1045–1050, 1990; R. M. Napier and M. A. Venis, Receptors for plant growth regulators: Recent advances, *J. Plant Growth Regul.*, 9:113–126, 1990; A. Theologis, Rapid gene regulation by auxin, *Annu. Rev. Plant Physiol.*, 37:407–438, 1986.

Biohydrometallurgy

Biohydrometallurgy is the practice of using microorganisms to recover metals from ores. Its most widespread application is for the recovery of copper from dumps of low-grade ore. The dumps are generally formed from overburden rock stripped away to gain access to underlying high-grade ores. The ores typically contain less than 0.4% copper, too low to permit economic recovery by conventional processes. Recent promising results indicate that biohydrometallurgy can also be used to recover uranium from underground mines as well as gold from refractory sulfide ores. Laboratory studies suggest that other metals, including silver, manganese, zinc, nickel, and lead, can be recovered by biohydrometallurgy. As reserves of high-grade ores are depleted around the world, biohydrometallurgy is likely to play an increasingly important role in recovering many different types of metals from waste ores and residual low-grade ores.

Copper. Copper can be recovered from dumps by sprinkling dilute sulfuric acid onto the surface (**Fig. 1***a*). The acid percolates down through the dump, lowering the pH and promoting the growth of acidophilic microorganisms. These microorganisms oxidize the copper from the ore, putting it into solution in the acid. The microorganisms also produce their own sulfuric acid. The acid, often blue with dissolved copper, is collected at the bottom of the dump, where it is pumped to a recovery station. Copper is recovered

Fig. 1. Recovery of copper from dumps. (*a*) Dilute sulfuric acid being sprinkled onto a copper dump. (*b*) A solvent-extraction plant where dissolved copper is recovered from acid solution.

from the acid by solvent extraction and electrowinning (electroplating on cathodes; Fig. 1*b*) or by precipitation onto scrap iron, and subsequent smelting. The precipitation process, known as cementation, has been in use for over 500 years. Residual acid is recycled back to the dump to preserve water, reduce acid costs, and lower environmental pollution.

Current estimates place the cost of biologically produced copper at about half the price of conventionally recovered copper. One reason for the low cost of biologically produced copper is that the cost of making the dump is not figured into the price, because the waste rock has to be dumped in any case to provide access to the higher-grade ore.

A copper dump represents a complex and heterogeneous microbiological habitat. It contains solids ranging in size from boulders to fine sand and includes material of heterogeneous mineralogy. The interior of a dump is frequently hot (104–158°F or 40–70°C), and it is often anaerobic or microaerophilic. The exterior of the dump is at ambient temperature, undergoing changes in temperature that reflect seasonal and diurnal fluctuations.

Many different microorganisms have been isolated from copper dumps, some of which have been studied in the laboratory. These include a variety of mesophilic, aerobic iron and sulfur oxidizers; thermophilic iron and sulfur oxidizers; and anaerobic microorganisms, including sulfate-reducing bacteria. Also found are heterotrophic bacteria and protozoa that indirectly affect metal solubilization by affecting the growth and activity of metal-solubilizing bacteria.

The biological solubilization of copper in dumps is known to involve both a direct, enzymatic oxidation of the copper and an indirect, chemical oxidation of the copper by biologically produced ferric irons. The most studied microorganism involved in these processes is *Thiobacillus ferrooxidans*.

Thiobacillus ferrooxidans can directly, enzymatically oxidize copper (Cu) by a process that is still poorly understood [reaction (1)]. It can also oxidize

$$CuFe_2 + 4O_2 \xrightarrow{bacteria}$$

Chalcopyrite   Oxygen

$$CuSO_4 + FeSO_4 \qquad (1)$$

Copper(II) sulfate   Iron(II) sulfate

ferrous (Fe II) iron and reduced sulfur found in sulfide ores such as pyrite (FeS_2), as shown in reaction (2).

$$4FeS_2 + 15O_2 + 2H_2O \xrightarrow{bacteria}$$

$$2Fe_2(SO_4)_3 + 2H_2SO_4 \qquad (2)$$

Iron(III) sulfate   Sulfuric acid

Thiobacillus ferrooxidans obtains energy from the process, and the resulting ferric iron [Fe(III)] is a powerful chemical oxidant, oxidizing Cu(I) to Cu(II), as shown in reaction (3), in which bacteria do not

$$CuFeS_2 + 2Fe_2(SO_4)_3 + 2H_2O + 3O_2 \longrightarrow$$

$$CuSO_4 + 5FeSO_4 + 2H_2SO_4 \quad (3)$$

participate directly. Ferrous iron is regenerated in the process. This again serves as an energy source for *T. ferrooxidans*, which converts ferrous to ferric iron [reaction (4)], thus completing a ferrous iron-ferric

$$4FeSO_4 + O_2 + 2H_2SO_4 \xrightarrow{bacteria}$$

$$2Fe_2(SO_4)_3 + 2H_2O \quad (4)$$

iron cycle (**Fig. 2**). The microorganisms also create sulfuric acid [reaction (2)]; this promotes the solubilization of the Cu(II) and reduces the amount of exogeneous acid that must be added to the dump. Direct oxidation of copper [reaction (1)] or pyrite [reaction (2)] requires attachment of *T. ferrooxidans* to the ore, whereas the reoxidation of ferrous iron can be accomplished by unattached bacteria in solution. The relative importance of attached versus unattached microorganisms to the process of copper solubilization is not fully understood.

Although superior strains of bioleaching microorganisms have been developed in the laboratory, there are no examples of their use in commercial dump bioleaching operations, nor has an evaluation of the activity and survivability of such strains in a dump been undertaken. A recent advance is to improve the design of the dump to promote bacterial activity, for example, by the construction of low, narrow, so-called finger dumps to increase the ratio of surface area to volume for better aeration.

Uranium. Bioleaching has been used successfully to obtain uranium (U) from waste ore at Buffelsfontein in South Africa and from the walls and stopes of several underground uranium mines in the Elliot Lake and Agnew Lakes regions of Canada.

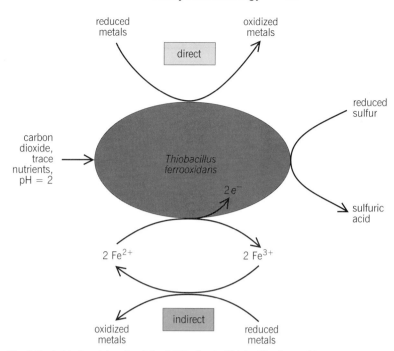

Fig. 2. Basic biochemistry of metal solubilization by *Thiobacillus ferrooxidans*.

Although microorganisms such as *T. ferrooxidans* are capable of directly oxidizing tetravalent uranium into the soluble hexavalent form, it is thought that the main contribution of microorganisms to uranium bioleaching is in the production of ferric iron according to reaction (2). The ferric iron can then oxidize the uranium chemically [reaction (5)], resulting in soluble

$$UO_2 + Fe_2(SO_4)_3 \longrightarrow$$

Uranium oxide

$$UO_2SO_4 + 2FeSO_4 \quad (5)$$

Uranyl sulfate

uranium sulfate and ferrous iron.

Gold. Existing procedures for the recovery of gold from ore generally incorporate a step where the gold is complexed with cyanide. Ores that do not respond well to direct cyanidation require some method of pretreatment to render them amenable to cyanidation. It has recently been discovered that oxidation by microorganisms can successfully accomplish this pretreatment, and interest has been heightened by the successful initiation of the first commercial-scale biooxidation plants in Zimbabwe and South Africa, and several successful pilot-scale plants in North America.

Biooxidation of refractory gold ores generally takes place in tanks or bioreactors. Conditions such as temperature, pH, and oxygen and nutrient availability can be manipulated in these vessels to promote bacterial activity. This has led to a renewed interest in the development of genetically engineered strains of microorganisms. However, as yet, only naturally occurring strains are being used.

Future applications. It is anticipated that biohydrometallurgy will have applications in recovery of copper from heaps and concentrates; in addition, the technique will be used for recovery of copper ore using in-place techniques.

Heaps. Copper can be recovered by bioleaching from specially prepared heaps of ore in a manner similar to its recovery from dumps. However, compared to dumps, heaps generally contain ore that has been ground to a finer and more uniform particle size, and they are usually built by deposition from conveyor belts so that surface compaction by haulage trucks is not a problem. Also, heaps are generally smaller than dumps and have adequate aeration. All of this leads to much faster and more efficient copper solubilization, and copper recovery is accomplished in days or weeks compared to years in a dump. In heap bioleaching, mine development and comminution processes are attributed to the costs of copper recovery.

Concentrates. Copper is conventionally recovered from high-grade ores by smelting copper concentrates. Because of the high capital and operating costs and environmental pollution problems associated with smelting, biological means for recovering copper from concentrates have long been considered as an alternative. The technical feasibility of such a process has been demonstrated, but no plant has ever been developed. The reluctance to adopt bioleaching may exist because the process is only marginally more attractive economically. However, increasingly stricter environmental laws may cause renewed interest in biohydrometallurgical alternatives to smelting of copper concentrates.

In-place process. Ore bodies that are too low-grade or too small to be mined by conventional methods have potential for recovery by in-place biohydrometallurgy. In this approach, the ore body is shattered underground and copper recovered by biohydrometallurgical procedures. Although the cost of rock breakage appears prohibitive at present for virgin ore bodies, recovery of copper from shattered rocks in underground abandoned mine workings has proved an economic method for recovering additional copper when conventional minable reserves have been depleted. A successful example was the recovery of copper by in-place leaching at Magma Copper's Miami mine for nearly 30 years.

Metal concentration and pollution control. Problems due to the release of metals from industrial and urban wastes constitute a significant proportion of total environmental pollution. Many organisms can concentrate or trap metal ions from solution, and technologies are being developed to use such organisms for the remediation of toxic metal wastes. It has been suggested that this approach may permit the metals to be recovered for recycling, enabling wastes to be regarded as a resource rather than a liability. If successful, this technology might also be used to recover metals from bioleaching operations; in this case, some mining companies might become biotechnology companies, using microorganisms both to solubilize and to recover metals.

For background information *see* Copper metallurgy; Electrochemistry; Gold metallurgy; Solvent extraction; Uranium metallurgy in the McGraw-Hill Encyclopedia of Science & Technology.

David S. Holmes

Bibliography. H. L. Ehrlich and C. L. Brierley (eds.), *Microbial Mineral Recovery*, 1990; G. Rossi, *Biohydrometallurgy*, 1990; J. Salley, R. G. L. McCready, and P. L. Wichlacz (eds.), *Biohydrometallurgy*, 1989.

Bioinorganic chemistry

Many metal ions are required for metabolic functions in organisms. For the physiologically important metal ions such as copper, iron, and zinc, both stimulatory and inhibitory effects can be observed. Normal physiology exists only near an optimal concentration of a given metal ion. High external concentrations result in enhanced metal-ion uptake in eukaryotic cells (cells with a defined nucleus) due to poorly regulated membrane transport systems. Intracellular concentrations in excess of the optimal level may result in a myriad of toxic effects. Metal ions with no beneficial effects, for example, cadmium and mercury, can also be transported inside cells and can evoke toxic effects. Cells must be competent to respond to changes in these fluxes and minimize toxic effects arising from elevated concentrations of essential as well as nonessential metal ions.

Sequestration molecules. Cells have evolved a variety of mechanisms to regulate the intracellular concentration of unbound metal ions. Sequestration is a major mechanism by which eukaryotic cells resist toxic effects of metals such as copper (Cu) and cadmium (Cd). Sequestration of metal ions typically occurs with formation of intracellular metal:protein complexes. Protein ligation of metal ions effectively buffers the free-metal-ion concentration and also minimizes formation of insoluble metal hydroxides and oxides.

Two classes of eukaryotic molecules have been described to function in the complexation of copper and cadmium ions. One class is a group of polypeptides known as metallothioneins. The second class is a series of short peptides. Metallothioneins are unique in that 20–30% of the amino acid content in the polypeptides from different species is the sulfur-containing amino acid, cysteine (Cys). A common structural motif in metallothioneins is repeating Cys-X-Cys sequences (X refers to any other amino acid) in which the cysteinyl sulfurs function as metal-ion ligands. The structure of mammalian metallothionein consists of a monolayer of polypeptide that is folded around two metal:sulfur polynuclear clusters.

The second class of molecules is the peptides composed of only glutamic acid (Glu), cysteine (Cys), and glycine (Gly) with the general structure (γGlu-Cys)$_n$Gly, in which *n* designates the number of dipeptide repeats. The linkage between the repeating glutamylcysteine dipeptide is an isopeptide linkage through the γ-carboxyl function of glutamate (**Fig. 1**). The (γGlu-Cys)$_n$Gly peptides are derived from the abundant cellular tripeptide, glutathione (γGlu-Cys-Gly), involved in maintaining the oxidation-reduction state of the cell. The biosynthesis of the (γGlu-

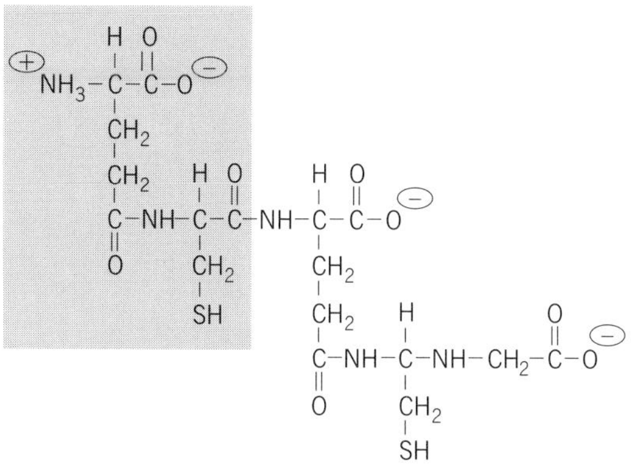

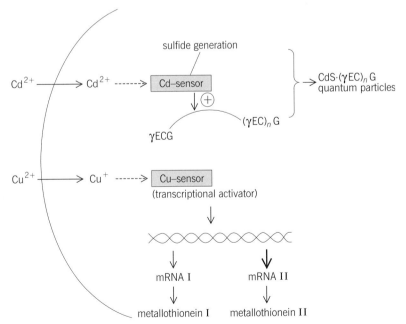

Fig. 1. Structure of the (γ-glutamyl-cysteinyl)₂glycine peptide. The shaded area indicates the dipeptide repeat.

Cys)₂Gly peptide in plants appears to consist of an enzymatic reaction in which 2 moles of glutathione are condensed, with the liberation of 1 mole of glycine. Peptides with a number of dipeptide repeats in excess of 2 are then formed by an analogous reaction of glutathione and $(\gamma Glu\text{-}Cys)_n Gly$ to yield $(\gamma Glu\text{-}Cys)_{n+1} Gly$. Typically, a heterogeneous mixture of $(\gamma Glu\text{-}Cys)_n Gly$ peptides with n varying from 2 to 5 are synthesized in cells in response to metal-ion stress. The sulfur atoms of the $(\gamma Glu\text{-}Cys)_n Gly$ peptides bind metal ions in a metal:sulfur cluster involving multiple $(\gamma Glu\text{-}Cys)_n Gly$ peptides.

Intracellular sensors for metal ions. Metallothioneins and $(\gamma Glu\text{-}Cys)_n Gly$ peptides are effective in providing cellular resistance to metal toxicity in that the concentrations of these molecules are regulated by the intracellular concentration of certain metal ions. Intracellular sensors detect the metal-ion concentration and signal the biosynthesis of the metal-sequestering molecules.

The sensor for metallothionein synthesis in the yeast *Saccharomyces cerevisiae* is a protein that activates the expression of the metallothionein gene in a metal-specific manner. This regulation occurs by the binding of Cu(I) ions to the sensor protein. Copper binding to the sensor stabilizes a structure of the sensor protein that enables the molecule to bind specifically to deoxyribonucleic acid (DNA) that is encoding the metallothionein gene. This binding activates expression of the metallothionein gene. Similar sensor proteins are presumed to mediate the induction of mammalian metallothionein genes by a variety of metal ions.

The sensor for $(\gamma Glu\text{-}Cys)_n Gly$ peptide synthesis in plant cells is the enzyme γ-glutamylcysteine dipeptidyl transpeptidase. The activity of this enzyme appears to be activated by a variety of metal ions. Whether this is the general mechanism in all species containing the peptides has not been determined.

Most species studied to date elaborate a single class of sequestration molecules in response to metal ions. Most organisms within the phylum Fungi synthesize either metallothioneins or $(\gamma Glu\text{-}Cys)_n Gly$ peptides but not both. It has been demonstrated that the pathogenic yeast *Candida glabrata* synthesizes both

metallothioneins and $(\gamma Glu\text{-}Cys)_n Gly$ peptides in a metal-specific manner (**Fig. 2**). *Candida glabrata* synthesizes a family of metallothionein polypeptides in response to copper stress and of $(\gamma Glu\text{-}Cys)_n Gly$ peptides in response to cadmium stress.

Cadmium:sulfide crystallites. The cadmium-$(\gamma Glu\text{-}Cys)_n Gly$ clusters from Fungi, Plantae, and Protista contain sulfide ions as an additional component. Sulfide ions arise from a Cd-mediated enhancement of cellular sulfide production, although the source of the sulfide ions is unresolved. The incorporation of sulfide ions into Cd:peptide complexes yields nanometer-sized crystallites coated with $(\gamma Glu\text{-}Cys)_n Gly$ peptides. The crystallites consist of cadmium and sulfide ions in a lattice. In addition, several yeasts use sulfide ions to precipitate metal ions as metal:sulfide particles on the cell surface or in the growth medium. These reactions in effect decrease the concentration of bioavailable metal ions, and therefore are part of the repertoire of mechanisms available to eukaryotic cells to resist the adverse effects of metal ions. The production of sulfide is expected to be highly regulated, as unregulated sulfide production may precipitate essential metal ions and impede cell growth.

The cadmium:sulfide (CdS) crystallites that form in the $(\gamma Glu\text{-}Cys)_n Gly$-coated particles from yeast range up to 2 nm in diameter as determined by transmission electron microscopy and powder x-ray analysis. Heterogeneity exists in the range of particle sizes, and this heterogeneity is related to the magnitude of the metal-mediated sulfide response within the cell. The greater the stimulation in sulfide production, the more uniform the crystallites will be and the greater the proportion of particles with a diameter of 2 nm. The heterogeneity in particle size is readily apparent from the ultraviolet absorption spectrum, as the optical

Fig. 2. Metal detoxification pathways in *Candida glabrata*. Cd = cadmium; Cu = copper; γECG = one-letter code for the tripeptide glutathione; (γEC)ₙG = one-letter amino acid code for the $(\gamma Glu\text{-}Cys)_n Gly$ peptide; mRNA = messenger ribonucleic acid.

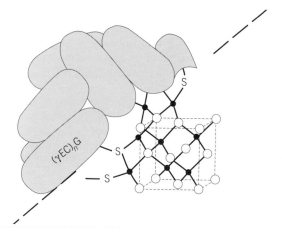

Fig. 3. Model of the lattice structure of the cadmium:sulfide crystallite coated with (γGlu-Cys)$_n$Gly peptides.

properties of the crystallites are size-related. This property makes the particles analogous to quantum, semiconductor clusters. Synthetic CdS clusters with particle diameters less than 10 nm are too small to exhibit bulk electronic properties. Rather, the energy necessary to excite an electron varies with particle size. The energy barrier increases with diminishing particle diameter, and this translates into a spectral transition in the near-ultraviolet/visible region whose wavelength maxima correlate with the excitation energy and therefore particle size.

Crystallites of CdS from *C. glabrata* (lattice diameter 2 nm) exhibit an energy of excitation of nearly 1.5 eV above that observed for bulk cadmium:sulfide minerals. As the sulfide content of these CdS particles decreases, the optical transition is shifted to higher energies. The normal CdS crystallites from *C. glabrata* are calculated to contain 85 Cd-S pairs in a lattice coated with the peptides. Identification of actual lattice structure has not been possible due to deviations from pure crystallinity. Such deviation is also common in small synthetic CdS particles. Approximately 30 peptides are expected to coat the lattice structure with cysteinyl sulfurs of the peptides providing ligands to the core (**Fig. 3**).

The stability of the CdS crystallites is a function of the coating peptides. Uncoated CdS clusters readily aggregate and precipitate in aqueous solutions. The (γGlu-Cys)$_n$Gly peptides stabilize the CdS core particle, yet conditions that favor dissociation of (γGlu-Cys)$_n$Gly peptides such as low pH or elevated temperatures lead to particle accretion.

Another property of yeast CdS crystallites analogous to semiconductor clusters is the high redox potential of excited-state electrons. The CdS crystallites of *C. glabrata* are capable of mediating the photoreduction of electron-accepting dyes (methyl viologen) and oxygen. Decay of photoexcited electrons to the ground state can occur by a radiative process, with the energy of photon emission being related to the lattice diameter.

Discrete sizes of CdS crystallites form in laboratory ware when (γGlu-Cys)$_n$Gly peptides are available for coating. Titration of sulfide ions into a Cd–peptide complex readily leads to crystallite formation in the laboratory. The CdS particles thus formed exhibit properties analogous to those of native CdS clusters. The Cd–(γGlu-Cys)$_n$Gly peptide complex appears to be the matrix for the biomineralization process. A variety of peptides containing either α- or γ-peptide linkages with two to four cysteinyl residues in the sequence are effective in terminating CdS crystallite growth near the 2-nm-diameter size. Peptides containing four or more cysteinyl residues form reasonably stable Cd(II) complexes in the absence of sulfide ions, and sulfide incorporation to generate CdS crystallites is impeded. Polycysteinyl peptides such as Cd(II)–metallothioneins do not serve as a matrix for CdS formation.

The intracellular formation of CdS quantum particles in fungi is regulated by cells only in that Cd(II) ions stimulate the synthesis of (γGlu-Cys)$_n$Gly peptides and the enhancement of production of sulfide anions. The mineralization process that follows probably proceeds spontaneously.

Biomineralization processes have been divided into two types, termed matrix-mediated and biologically induced. Matrix-mediated biomineralization occurs from a preformed structural framework, and such minerals have a well-defined crystal orientation and a narrow size distribution. The biologically induced process yields random crystal orientations of diverse sizes. The mineralization process in yeasts is typical of a matrix-mediated mineralization. The size range of the crystallites is controlled primarily by the magnitude of the cellular sulfide response.

For background information SEE BIOINORGANIC CHEMISTRY; COMPLEX COMPOUNDS; ENZYME; EUKARYOTE; MOLECULAR BIOLOGY; PEPTIDE; SEMICONDUCTOR in the McGraw-Hill Encyclopedia of Science & Technology.

Dennis R. Winge; Charles T. Dameron

Bibliography. L. Brus, Electronic wave functions in semiconductor clusters: Experiment and theory, *J. Phys. Chem.*, 90:2555–2560, 1986; C. T. Dameron et al., Biosynthesis of cadmium sulfide quantum semiconductor crystallites, *Nature*, 338:596–598, 1989; C. T. Dameron and D. R. Winge, Characterization of peptide-coated cadmium sulfide crystallites, *Inorg. Chem.*, 29:1343–1348, 1990; C. T. Dameron and D. R. Winge, Peptide-mediated formation of quantum semiconductors, *Trends Biotech.*, 8:3–6, 1990.

Biosphere

The Earth is experiencing environmental changes that are of global concern. These changes will affect the lives of all people, yet there is only a limited understanding of their causes and effects, and of the role of human activity in inducing them. Trying to actively modify these changes without a comprehensive understanding of the intricate couplings between oceans, atmosphere, land, and biota could have severe repercussions upon the global economy and might compound the existing environmental situation.

Mission to Planet Earth will generate the knowledge needed to understand changes in the Earth's environment. It will be a program of space-based and in-situ measurements, interdisciplinary research, and modeling, coordinated among many federal agencies and other nations as part of a worldwide research effort on global change. The program is designed to offer a more complete understanding of the processes by which the Earth's various systems interact. Such understanding will give policy makers the scientific information needed to make sound environmental policy.

The changing Earth. There is irrefutable evidence that human activity has altered Earth's nature by changing its landscape and the composition of its atmosphere. This evidence suggests that changes such as global warming, sea levels, upper-atmosphere ozone levels, and massive deforestation can affect lives and global economic well-being. Although the potential for these changes has been recognized, no consensus has been reached on their causes or long-term consequences.

While research and data on some aspects of Earth science are extensive, other aspects remain virtually unexplored. The greenhouse warming problem illustrates the complexity of the current situation. Increasing carbon dioxide emissions may increase global temperatures through greenhouse warming; however, some of these emissions are absorbed by the Earth's biota. The extent or timing of these processes remains unclear. Similar uncertainties exist concerning the role of clouds in regulating the temperature of the Earth, the effect on cloud formation of increasing concentrations of greenhouse gases in the atmosphere, and the effect of the global circulation of heat through the oceans.

The Earth is a dynamic planet whose components are entwined and interconnected over time. The response of the Earth to human activities depends on the interactions that occur among these components; the feedback mechanisms generated will either enhance or mitigate the effects of global change. Understanding which feedback mechanisms are dominant, predicting how the Earth will respond, and separating human-induced change from natural phenomena require long-term observations and present a key challenge to Earth-science research.

Global change research. The United States and other nations have devised a broad, interdisciplinary research effort to gain comprehensive understanding of global change and to respond to it. In the United States, the multiagency Committee on Earth and Environmental Sciences (CEES) in the Office of Science and Technology Policy coordinates the planning and execution of the national effort. The CEES devised the U.S. Global Change Research Program (USGCRP) as an integrated, government-wide program. The planning for the program has had significant scientific guidance from the National Academy of Sciences and other advisory groups. Individual agencies are responsible for implementing specific activities, coordinating their programs through the

CEES. In turn, United States efforts contribute to the international global change research program coordinated through the World Climate Research Program and the International Geosphere Biosphere Program.

The goal of the USGCRP and international research programs is to provide a sound scientific basis for effective and rational national and international decision making on global change issues. These programs are dedicated to a predictive understanding of the interactive physical, geological, chemical, biological, and social processes that regulate the total Earth system.

Mission to Planet Earth. The National Aeronautics and Space Administration (NASA) will contribute to the global change research effort through its Mission to Planet Earth program. Applying its expertise in space-based remote sensing, NASA will use the global perspective available only from space. Other agencies will contribute ground, in-situ, and satellite observations, conduct research, and design models. Other nations will conduct their own research and satellite missions, participate in bilateral and multilateral cooperative efforts, and coordinate their activities through international scientific and government organizations. This worldwide coordination will enable a comprehensive research and observation program and the establishment of standards for the exchange of data, models, and technical support.

NASA's Mission to Planet Earth consists of several interrelated elements: (1) 19 flight missions, such as the Upper Atmosphere Research Satellite, to expand knowledge of various aspects of global change; (2) an integrated data system available by 1994 to draw maximum information from existing and upcoming missions; (3) an expanded scientific research program; (4) a series of small, focused missions known as Earth Probes; and (5) the Earth Observing System (EOS), which will provide comprehensive, simultaneous, long-term observations of the Earth.

Earth Observing System. While the near-term missions and Earth Probes will advance knowledge of the Earth, they will not provide a comprehensive understanding of the couplings within the Earth system. The Earth Observing System will obtain a 15-year set of continuous, simultaneous global observations of a number of key variables in order to characterize the processes that control the global-scale cycles of energy, water, and biogeochemical activity and to completely understand the Earth as a coupled system.

The Earth Observing System program consists of two series of polar orbiting spacecraft with complementary payloads launched on expendable launch vehicles. The spacecraft in each series will be replaced every 5 years to achieve the goal of a 15-year mission lifetime. Observations from European and Japanese spacecraft, in orbit at the same time as the Earth Observing System, will complement the measurements acquired by the Earth Observing System spacecraft, and may provide flight opportunities for additional United States instruments. Instruments on some of the spacecraft will make simultaneous measurements of the often subtle interactions between the elements of the Earth system to serve as the basis for predictive models of the future of the planet.

The scientific rationale for simultaneity is based on the need for measurements that can be compared to and correlated with each other with a minimum of error and uncertainty. Simultaneous measurements allow for extensive cross-calibration of instruments and avoid the impact of rapidly occurring atmospheric and illumination changes. Moreover, flying companion sensors capable of measuring atmospheric effects are required so that primary measurements can be better analyzed or corrected for the effects of the intervening atmosphere. The spacecraft in the first series are sized to accommodate those instruments that need to observe the Earth simultaneously. This will require 3000–3500 kg (6600–7700 lb) of payload to be flown on each spacecraft.

Each of the Earth Observing System spacecraft series is designed to investigate a different aspect of the Earth system. The first spacecraft (EOS-A) series is designed to achieve simultaneous observations of the surface and characterization of the atmospheric and oceanic conditions that influence both surface conditions and their measurement through the intervening atmosphere. Earth Observing System imagers will measure biological and physical productivity of the land, study surface mineralogy, and make critical measurements of the clouds, oceans, and other elements of the hydrological cycle. Other sensors will provide more accurate profiles of atmospheric temperature and water vapor. The second spacecraft (EOS-B) series, which will be launched approximately 2½ years later, will be devoted to measurements of the chemistry of the upper atmosphere, the circulation of the oceans, and the behavior of the solid Earth. The series will include a radiometer to measure the trace constituents of the stratosphere and the temperature of the upper atmosphere, an altimeter to observe sea-surface height and ocean topography, as well as spectral, imaging, and sounding instruments.

EOS Data and Information System. The data from these two EOS spacecraft series, and other elements of Mission to Planet Earth, may produce in excess of 5×10^{16} bytes of data, creating one of the largest data sets in history. The EOS Data and Information System (EOSDIS) is planned as an advanced system to process and archive this 15-year global data set, to maximize its utility for scientific purposes, and to facilitate its easy access by the research community.

Development of EOSDIS is expected to proceed immediately by building on the existing infrastructure within the research community. It will implement an open and distributed architecture to be able to evolve with advances in computing and networking technology. Near-term operations will support the research and analysis of existing data and data from the early elements of Mission to Planet Earth, along with complementary in-situ and satellite data from CEES agencies other than NASA. Researchers will also be able to gain valuable early experience with the data system and to provide input on its design and capabilities.

EOSDIS data policy specifies that all data and derived products be available to all researchers. The system will generate standard data products, while individual investigators will generate specialized data products and make them available to EOSDIS for archiving and distribution as appropriate. Research users will pay only the nominal cost of data reproduction and delivery; they will be required to agree to publish their results and to make available supporting information. International researchers may propose cooperative projects and contributions in exchange for access to Earth Observing System data on similar terms to United States researchers. NASA will provide, according to existing law, for commercial distribution of data on a nondiscriminatory basis to all other users.

For background information SEE APPLICATIONS SATELLITES; ATMOSPHERE; ATMOSPHERIC OZONE; BIOGEOCHEMISTRY; BIOSPHERE; CLIMATIC CHANGE; GREENHOUSE EFFECT; OCEAN-ATMOSPHERE RELATIONS; REMOTE SENSING in the McGraw-Hill Encyclopedia of Science & Technology.

Lennard A. Fisk

Bibliography. Committee on Earth Sciences, *Our Changing Planet: The FY 1991 U.S. Global Change Research Program*, 1990; Goddard Space Flight Center, NASA, *Earth Observing System: 1990 Reference Handbook*, 1990; G. S. Wilson and W. T. Huntress, *Mission to Planet Earth*, NASA, 1990.

Biotechnology

The relationship between biotechnology and inorganic chemistry has recently grown in importance. Although the term inorganic biotechnology is not yet in common usage, studies merging these two separate fields are of expanding fundamental interest and practical value. Metals have unique characteristics (catalytic activity, multiple unpaired electrons, desirable and versatile radioactivity) and can be conjugated with larger, more complex biomaterials to produce new species called bioconjugates with greatly enhanced utility. The conjugating agent is usually a bifunctional species with a reactive group to attach to the biomolecule and a chelating group to bind to the metal. In turn, biotechnological products are useful to the inorganic chemist in understanding the roles of metals in biology and medicine and in the production of new materials.

The overall goals of inorganic biotechnology fall into two complementary classifications: using biotechnology developments to further both fundamental and practical goals in the fields of inorganic biochemistry and inorganic materials, and exploiting properties of inorganic metal centers to enhance the effectiveness or utility of biotechnological applications.

Site-directed mutagenesis. Inorganic biochemistry generally encompasses the study and applications of metal species in biology and medicine (although nonmetallic inorganic species are also important). Enzymes are a special type of catalyst; they enhance the rates of the reactions they govern. Understanding the details of this process is potentially useful for designing energy-saving, less polluting chemical processes. Enzymes are usually composed mainly of proteins. The majority of atoms in proteins are those found in the constituent amino acids and include carbon, hydrogen, nitrogen, and oxygen and, less frequently, sulfur, phosphorus, and selenium; but metalloenzymes also contain metal centers such as zinc, copper, manganese, molybdenum, nitrogen, and cobalt. About one-third of enzymes contain metals at the active site, that is, the catalytic center. Each metal is typically attached to combinations of four to six atoms of nitrogen, oxygen, or sulfur in amino acid side chains, although in some cases other ligating atoms (for example, the four nitrogen atoms of a porphyrin bound to iron in heme-containing enzymes) are involved in metal binding.

Biotechnology has begun to play an important role in elucidating details of the chemical steps leading to catalysis and the central role of the inorganic metal in the processes. In particular, recombinant deoxyribonucleic acid (DNA) technology allows the investigator to specifically change an individual amino acid into a different amino acid by altering the DNA sequence. Events in nature in which changes in DNA sequence alter protein sequence are known as mutations. In the biotechnological approach, the scientist is in control, and the process is known as site-directed mutagenesis. By altering specific amino acids that participate in the catalysis, either by binding the metal center or by influencing the substrate (the starting reagent being converted to a product by the enzyme), scientists have begun to unravel how nature promotes chemical reactions such as the oxidation of hydrocarbons by the enzyme cytochrome $P450_{cam}$.

Catalytic antibodies. Antibodies specifically recognize foreign molecules or cells and bind to them tightly; the foreign substances are inactivated and eventually eliminated by the organism. One of the most important areas of biotechnology is the controlled expression and rapid evaluation of antibodies. The hybrid cells originally used to produce large quantities of a particular antibody were selected from one colony, leading to the term monoclonal antibodies; each antibody is only one specific protein. However, in a normal immune response polyclonal antibodies, which consist of numerous different antibodies (each a different protein) against the same foreign substance, are produced. Research is directed at developing antibodies against chemical species formed during the chemical reaction. This binding to the antibody stabilizes these species, which exist in the transition-state phase of the reaction. Such stabilization makes the reaction more rapid. The transient species formed in the transition state are too short-lived to induce antibody formation. Therefore, chemists have employed stable molecules that are analogs of these transient chemical species. Such transition-state analogs have been used successfully to induce antibodies that are catalytic, and these artificial antibody enzymes are known as abzymes. Antibodies do not contain metals, and the special properties of metal ions are required in many catalytic processes. Recently, antibodies against metal-containing molecules have been obtained to develop artificial-antibody-based metalloenzymes.

One natural role of antibodies is to attach to foreign cells, such as are found in tumors. Considerable success has been achieved in attaching metal centers to such antibodies without destroying immunogenicity, the ability of the antibody to identify the tumor cell. These bioconjugates (the chemically modified biological molecule) can have radioactive metals attached, for example, technetium-99m and indium-111, which emit gamma rays readily detected by gamma counters. This technology allows nuclear-medicine physicians to search for and image tumors. In a related valuable therapeutic approach, it is also possible to administer antibodies conjugated with rhenium-188, copper-67, bismuth-212, yttrium-90, and other metalloradionuclides that emit damaging radiation. Selective binding of the conjugated antibody to the tumor concentrates the radiation damage in the tumor, selectively killing tumor cells in preference to normal cells.

Applications of antibodies to imaging are not limited to tumors and hold promise for therapy for blood clots or heart disease. In addition to possessing radioactivity, metal centers often have unpaired electrons. An electron has a magnetic effect roughly a thousand times that of a typical atomic nucleus. Gadolinium(III) has seven unpaired electrons, the most that can exist in any stable atom or ion. Such paramagnetic metals influence the nuclear properties of surrounding protons in water, specifically the rate of interchange of the two spin states, $+\frac{1}{2}$ and $-\frac{1}{2}$, of the hydrogen nucleus. Differences in this rate for water in different regions of the body can be used to create images when an organism is placed in a magnetic field gradient. This magnetic resonance imaging (MRI) has the advantage of avoiding x-rays or gamma rays, using instead radio-frequency radiation that is believed to be harmless. The image obtained depends on the compound and its method of administration. Paramagnetic gadolinium(III) compounds decrease the time needed to create an image. Such methods may eventually permit better diagnosis, although the potential toxicity of the metals must be considered. *SEE BIOINORGANIC CHEMISTRY.*

Proteins. Proteins are involved with detoxification of metal species. The same types of amino acids useful for holding metals in catalytic sites are effective in detoxifying metal species by selective binding. The metal ions exhibiting the greatest toxicity are "soft," that is, polarizable species that have loosely held outer electrons. Such soft metals prefer binding to the softest atom found in a common amino acid side chain, namely sulfur. The most important such amino acid is cysteine. The cysteine-rich protein metallothionein can bind up to seven cadmium atoms. Some organisms package cadmium in defined cadmium sulfide crystallites. An important development in biotechnology is the advent of automated synthesizers, which use solid-state chemical methods to prepare synthetic peptides and proteins. Synthetic peptides may prove valuable in preparing well-defined crystalline inorganic materials needed for industrial applications.

Natural proteins expressed in organisms as a response to challenge by toxic metals have very high metal-binding constants. These high constants suggest that such proteins have potential for direct determination of metals and also for developing analytical methods to detect low levels of toxic metals. Such biosensors would have to be expressed in large quantities by biotechnological approaches in order to permit widespread use in testing for toxic metals in lakes, rivers, and oceans.

The first step in the expression of genes as proteins is isolation of the gene from the DNA. Ideally, only DNA from the gene will be cut from the organisms bearing the gene and transformed to a plasmid (circular DNA), which can be incorporated into a cell that can produce the protein. This recombinant DNA technology, as usually practiced, requires enzymes (restriction enzymes) that cut DNA at specific sequences. Available enzymes characteristically recognize six-base-pair sequences of DNA, which occur too frequently to allow sufficiently precise cleavage of DNA. This lack of selectivity is also an impediment in the use of these enzymes for mapping chromosomes and sequencing DNA. Artificial restriction enzymes can be constructed with oligonucleotides (short sequences of DNA). Oligonucleotides with up to 20 bases recognize a particular DNA sequence by triple-helix formation. A metal attached at the end of the oligonucleotide [for example, an iron complex with an ethylenediamine tetraacetic acid (EDTA) type of linker] then cleaves the DNA. Such artificial restriction enzymes are other examples of bioconjugates.

Therapeutic agents. DNA and ribonucleic acid (RNA) cleavage can be used to develop therapeutic agents. Cleavage of these biomolecules in bacterial or cancer cells or viruses will destroy the cells or viruses. For example, a sequence directed toward a nucleic acid related to the virus causing acquired immune deficiency syndrome (AIDS) or important in a cancer cell could have obvious value. Indeed, metal species in conjunction with specific RNA or RNA/DNA hybrids have been shown to adopt conformations that cleave RNA and DNA. The recent discovery of these ribozymes established that enzymes are not necessarily proteins. Ribozymes cleave the RNA of human immunodeficiency virus 1 (HIV-1), which causes AIDS, and thus may find use as therapeutic agents.

Metal oligonucleotide species are also important for understanding mechanisms of action of metal anticancer drugs and are useful in assessing interactions of metal reagents that recognize and cleave specific sequences of DNA by either chemical or photochemical means. Nonselective cleaving species are important for probing fundamental processes essential in biotechnology since, for example, proteins or enzymes bound to DNA protect that region from cleavage. In gel electrophoresis methods, it is possible to identify every point of cleavage; the absence of cleavage leaves a so-called footprint, indicating the protein- or enzyme-binding site. Indeed, many of these proteins or enzymes contain metal species such as the grouping known as a zinc finger.

For background information SEE AMINO ACIDS; ANTIBODY; BIOINORGANIC CHEMISTRY; ENZYME; MEDICAL IMAGING; MONOCLONAL ANTIBODIES; MUTATION in the McGraw-Hill Encyclopedia of Science & Technology.

Luigi G. Marzilli

Bibliography. G. L. Eichhorn and L. G. Marzilli (eds.), *Metal-Ion Induced Regulation of Gene Expression*, 1990; B. L. Iverson and R. A. Lerner, Sequence-specific peptide cleavage catalyzed by an antibody, *Science*, 243:1184–1188, 1989; L. G. Marzilli, Tutorial on biotechnology: Opportunities for inorganic chemistry, *Abstracts of the 199th American Chemical Society National Meeting*, 1–25, 1990; S. S. Roberts, Catalytic antibodies: Reinventing the enzyme, *J. NIH Res.*, 2:77–82, 1990; N. Sarver et al., Ribozymes as potential anti-HIV-1 therapeutic agents, *Science*, 247:1222–1225, 1990; S. A. Strobel and P. B. Dervan, Site-specific cleavage of a yeast chromosome by oligonucleotide-directed triple-helix formation, *Science*, 249:73–75, 1990.

Blood

Crustaceans have an open circulatory system where fluid known as hemolymph pumped from the heart circulates and bathes internal organs. Cells carried in the hemolymph are known as hemocytes. None of the hemocytes are equivalent to vertebrate red blood cells because the respiratory pigment (hemocyanin) is free in the plasma. In contrast, hemocytes are considered analogous to vertebrate white blood cells since recent studies have shown that hemocytes are involved in blood coagulation, wound healing, phagocytosis, and encapsulation of foreign material. In addition, hemocytes are involved in hardening of the exoskeleton following molting in crustaceans.

A workable scheme to identify specific types of hemocytes involved with particular physiologic functions has not previously been available. Many schemes have been proposed, but rarely have they been useful beyond the species from which they were developed, because the schemes relied excessively on cell morphology. Most schemes focus attention on the absence or presence of granules and, if granules are present, on their sizes. In some species, hemocytes lack cytoplasmic granules; these are known as hyaline cells, referring to the clear nature of the cytoplasm, to distinguish them from the granulocytes. However, in other species, all hemocytes contain conspicuous granules. Research is being conducted to develop a classification scheme for crustacean hemocytes that will correlate their morphology and function. Realization of this goal will allow for increased understanding of the immune responses of crustaceans, and the detection and characterization of disease in crustaceans in natural and aquaculture settings in much the same way it can be performed on vertebrates.

The classification scheme being described recognizes two main categories of hemocytes, hyaline cells and granulocytes, but distinguishes between them by using a different set of morphological features, cytochemical reactions, and cellular functions.

Morphology. Hyaline cells (**illus.** *a*) are the smallest type of hemocyte (usually 15 micrometers or less in diameter) and have a high nucleocytoplasmic ratio (greater than 35%). While in most species these hemocytes lack granules, in a few species, such as the California spiny lobster (*Panulirus interruptus*), these cells contain a small number of large conspicuous granules, and in the Maine lobster (*Homarus americanus*) there are abundant small cytoplasmic granules. Observations of any hyaline cells must be made on samples fixed immediately after removal from the body because these cells rapidly lyse as the hemolymph clots.

Granulocytes are larger cells (18 μm or more in diameter), with an increased amount of cytoplasm and a nucleocytoplasmic ratio of less than 30%. They may be subdivided into two categories: small-granule (illus. *b*) and large-granule (illus. *c*) hemocytes. In the former, the nucleus is readily observed, and the cytoplasm contains a variable number of small (0.1-μm-diameter) dark granules. Large-granule hemocytes are highly refractile when viewed with phase optics and contain so many large (0.2-μm-diameter) granules that the nucleus is obscured.

By using morphological criteria, the distinction between hyaline cells and granulocytes is obvious in some species but extremely subtle in others. Therefore, other factors must be considered to accurately classify hemocytes, such as cytochemical reactions and functional activities.

Cytochemistry. The categories of hemocytes may be distinguished by three cytochemical tests using Sudan black B, acid phosphatase, or phenoloxidase. Hyaline cells are stained by Sudan black B, possibly because of a reaction with the abundant small lipoprotein deposits in their cytoplasm. In contrast, granulocytes show cytoplasmic reaction sites for acid phosphatase and phenoloxidase. Acid phosphatase is a lysosomal enzyme that should be abundant in cells that phagocytize and degrade foreign material. Phenoloxidase is an enzyme involved in the formation of melanin, the black layer encapsulating large foreign bodies such as parasites. For each cytochemical test, differential counts correspond well to those based on morphology alone.

Functional studies. Three functional tests have also been developed to determine the role of particular types of hemocytes. Coagulation of the hemolymph is an important physiologic process that prevents excessive blood loss at a site of injury and spreading of pathogens throughout the body. When hemolymph is mixed with seawater to simulate a break in the exoskeleton, the hemolymph clots rapidly. Examination of the hemocytes shows that the hyaline cells lyse during coagulation, whereas there is no change in the granulocytes. The lysing hyaline cells are surrounded

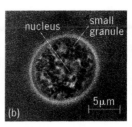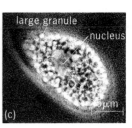

Phase-contrast photograph of the three types of hemocytes in the crab *Loxorhynchus grandis*: (*a*) agranular, (*b*) small-granule, and (*c*) large-granule.

by circular halos or multiple long strands of clotted hemolymph. This suggests that a factor released from the lysing hyaline cells triggers coagulation of the hemolymph. The released factor may be a serine proteinase or perhaps a transglutaminase that crosslinks the plasma protein coagulogen to form an insoluble gel. In species where hyaline cells comprise more than half of all circulating hemocytes, large expanses of coagulated hemolymph can be produced. In other species where fewer hyaline cells are circulating, coagulation is accomplished by aggregation of granulocytes, with little clot material formed.

When hemocytes are cultured in the presence of bacteria, both types of granulocytes, but not the hyaline cells, engulf bacteria by means of phagocytosis. More small-granule hemocytes ingest bacteria than large-granule cells, and individual small-granule cells contain more bacteria. Phagocytosis is carried out faster and in greater numbers by both types of granulocytes when the bacteria have been treated by preincubation in cell-free hemolymph.

When particles too large to be phagocytosed are cultured with hemocytes, they are encapsulated by the adhesion of both types of granulocytes, but with no involvement of the hyaline cells. This observation correlates with the situation in the organism, where foreign objects are covered by layers of melanin-producing cells.

Categories of hemocytes. By using the morphological, cytochemical, and functional criteria described above, the two major types of decapod crustacean hemocytes, hyaline cells and granulocytes, have been shown to be distinct both morphologically and functionally.

This is the first classification scheme that links hemocyte structure and function. It recognizes that morphology alone can be a potentially confusing feature in classifying hemocytes. This understanding helps explain the confusion as to whether the hyaline cells or the granulocytes initiate coagulation. In some decapods, there are no truly agranular cells, yet there are two main categories of granulated cells. One group lyses and initiates clot formation, stains with Sudan black B, and lacks lysosomal enzymes and phenoloxidase. These are considered to be the hyaline cells in those species. The second group comprises small- and large-granule hemocytes that do not lyse during coagulation of the hemolymph but instead contain lysosomal enzymes and phenoloxidase and are phagocytic and encapsulate foreign material.

Hematopoiesis. The recognition of two main categories of hemocytes is also supported by preliminary studies on the formation and maturation of hemocytes. In decapods, blood-cell-producing tissue lies on the dorsolateral surface of the foregut. Mitosis, maturation, and release of blood cells seem to be correlated with the molt cycle. Within the hematopoietic tissue, mitosis has been observed in both hyaline cells and small-granule hemocytes. This suggests that there are two distinct lines of development but that the large-granule hemocytes may develop from the small-granule cells.

Prospects. This hemocyte classification scheme has now been tested and found effective on a variety of decapods. It remains to be determined whether these categories are valid in nondecapod crustaceans as well. The development of a unified classification scheme for crustacean hemocytes could facilitate further research in the fields of physiology and comparative immunology and also provide the tools to assess crustacean health and disease by using hematology.

For background information SEE CYTOCHEMISTRY; DECAPODA (CRUSTACEA) in the McGraw-Hill Encyclopedia of Science & Technology.

Gary G. Martin; Jo Ellen Hose

Bibliography. J. E. Hose et al., Cytochemical features of shrimp hemocytes, *Biol. Bull.*, 173:178–187, 1987; J. E. Hose et al., A decapod hemocyte classification scheme integrating morphology, cytochemistry and function, *Biol. Bull.*, 178:33–45, 1990; J. E. Hose and G. G. Martin, Defense functions of granulocytes in the ridgeback prawn, *Sicyonia ingentis* Burkenroad 1938, *J. Invert. Pathol.*, 53:335–346, 1989; N. A. Ratcliffe and A. F. Rowley (eds.), *Invertebrate Blood Cells*, vol. 2, 1981.

Carboniferous

The recent discovery of an early community of terrestrial animals and plants in the Lower Carboniferous rocks of Scotland is providing the first evidence of the successful colonization of land by arthropods and vertebrates. The fossils collected so far include the earliest harvestman spiders, the earliest air-breathing scorpions, and the largest terrestrial arthropods ever recorded, together with almost complete individuals of the earliest temnospondyls, a group of salamander-like forms ancestral to all living amphibians, and of the earliest reptiles, the antecedents of the dinosaurs, birds, and mammals. This remarkable fauna was preserved in the mud and ash of a small fresh-water lake into which drained the hot springs associated with nearby volcanic activity. Over time, this mud and ash was converted into a hard siliceous rock, the East Kirkton Limestone, which now contains a unique record of life on land 3.4×10^8 years ago. The fossils found at East Kirkton are preserved in limestone laid down in water, so additional evidence is needed to confirm that they are the remains of members of a terrestrial community. Evidence comes from two main sources: the geology of the rocks in which the fossils have been found, and the anatomy of the fossils themselves.

Invertebrate fossils. This limestone is now exposed in a small, disused quarry in the Bathgate Hills, 17 mi (27 km) west of Edinburgh. The quarry was originally opened in the 1820s to provide walling stones for local farmers, and the first fossils found at the site were described in 1836. Although the peculiar lithology of the rocks has been studied by generations of geologists, it was not until 1984 that the importance of the quarry to paleontology was fully recognized. In that year, a wealth of previously unknown fossil animals were discovered by S. Wood in the slabs originally removed from the quarry to make up the local walls. In the following year, work began in the quarry itself; since then a large number of specimens have been discovered in situ.

Among the first fossils discovered by Wood were scorpions, myriapods (millipedes), and the earliest record of the arachnid group Opiliones, represented by a moderately large harvestman with limbs exceeding 1.6 in. (4 cm) in length. Also found were incomplete specimens of gigantic eurypterids, a group of water scorpions that became extinct in the Permian Period, and some of the earliest known tetrapods, the major group of vertebrates, which includes amphibians, reptiles, birds, and mammals.

Tetrapod fossils. Temnospondyls are the most common tetrapod found at the quarry: more than 20 articulated skeletons have been collected, ranging from 6 to 32 in. (15 to 80 cm) long (**Fig. 1**). Skeletons of the earliest anthracosaurs, a group of amphibians closely related to reptiles, are much less common, and the aïstopods, a group of snakelike amphibians lacking limbs, are represented by three incomplete skeletons. In addition, a large number of isolated bones have been collected, including limb elements from animals that must have exceeded 6.5 ft (2 m) in length. But perhaps the most remarkable discovery was an articulated skeleton of the earliest known reptile.

Fossil reptiles are well known in the Upper Carboniferous of Europe and North America, but their discovery in the Lower Carboniferous of Scotland predates the earliest of these records by almost

Fig. 1. Almost complete 12-in.-long (30-cm) skeleton of a temnospondyl amphibian from the East Kirkton Limestone. (*National Museums of Scotland specimen G. 1985.4.1; photograph courtesy of National Museums of Scotland*)

Fig. 2. Almost complete 7-in.-long (18-cm) skeleton of the earliest known reptile from the East Kirkton Limestone. (*National Museums of Scotland specimen G. 1990.72.1; photograph courtesy of S. P. Wood*)

4×10^7 years. The new specimen is represented by an articulated skeleton with an overall length of approximately 7 in. (18 cm). In general body form, it closely resembles a small lizard, with well-developed if slightly short limbs and a long tail (**Fig. 2**). As in other reptiles, it has two special bones in the ankle, the astragalus and calcaneum, but it lacks the otic notch and associated eardrum present in most living reptiles.

One of the principal differences that distinguish extant amphibians and reptiles is their mode of reproduction. Amphibians lay spawn in water, whereas reptiles lay eggs on land. Reptiles have achieved this independence from water by developing a special membrane called the amnion that encloses the growing embryo in its own reservoir of water within the egg. This development was fundamental to the successful colonization of land by vertebrates, and is one of the principal reasons why amniotes (reptiles, birds, and mammals) have been the dominant tetrapods for the last 3×10^8 years. East Kirkton provides the earliest indirect evidence of the development of the amniote condition.

Terrestrial community. The fossiliferous rock is a peculiar laminated cherty limestone of alternating light and dark bands varying in thickness from 0.2 to 0.8 in. (0.5 to 2.0 cm). This lamination may have arisen as a result of periodic changes in temperature and pH following the entry into the lake of hot water springs associated with local volcanic activity. This

Fig. 3. Almost complete 16-in.-long (40-cm) scorpion from the East Kirkton Limestone. (*National Museums of Scotland specimen G. 1987.7.136; photograph courtesy of National Museums of Scotland*)

then led to alternate precipitation of silica and calcium carbonate, which covered wide areas of the lake floor and the fossils lying on it. In comparable modern environments, high temperatures and fluctuating pH produce conditions that are unable to support a macrobiota. It is not surprising therefore that at East Kirkton there is no shelly fauna or any trace of fossil fishes. In addition, the occurrence of numerous articulated skeletons of both vertebrates and arthropods suggests rapid burial with no benthic organisms disturbing the carcasses.

Both the myriapods and the scorpions show features which indicate that they were able to breathe air. A new chilognathan millipede is the oldest known form with spiracles or respiratory pores. The scorpions had book lungs, as do modern scorpions, and not the book gills of their aquatic ancestors. Some of the East Kirkton scorpions were quite large terrestrial predators, with one specimen more than 16 in. (40 cm) long (**Fig. 3**). Fragments of other, larger individuals indicate that they grew to a length of at least 28 in. (70 cm). The tetrapod skeletons are all well ossified and belonged to animals that were capable of successful terrestrial locomotion. In many specimens the bones of the wrist and ankle are well ossified, a feature of Palaeozoic tetrapods normally associated with terrestriality. Furthermore, none show evidence of a lateral-line system, the aquatic pressure-sensing system present in fishes and many early amphibians; whereas members of one group, the temnospondyls, do have a well-developed otic notch, which probably housed a tympanum used for atmospheric hearing. All of these features indicate that the fossils found at East Kirkton represent members of a terrestrial community that had lived around the margins of the small lake.

Significance of finds. The discoveries at East Kirkton provide remarkable evidence of a flourishing and diverse community of early terrestrial animals. It includes the oldest known members of a number of arthropod and vertebrate taxa, among which the discovery of the earliest reptile was perhaps the most unexpected. This discovery suggests that tetrapods must have evolved much earlier than is currently inferred from the presence of amphibian fossils in the Upper Devonian of Greenland and the Soviet Union. It is quite likely that tetrapod fossils will eventually be found in Middle Devonian rocks, and it is even possible that they may be present in those from the Lower Devonian, which were laid down more than 4×10^8 years ago.

For background information SEE ANIMAL EVOLUTION; CARBONIFEROUS; REPTILIA in the McGraw-Hill Encyclopedia of Science & Technology.

T. R. Smithson

Bibliography. A. Jeram, When scorpions ruled the world, *New Scient.*, 126:52–55, 1990; A. R. Milner et al., The search for early tetrapods, *Mod. Geol.*, 10:1–28, 1986; W. D. I. Rolfe et al., An early terrestrial biota preserved by Visean vulcanicity, *Geol. Soc. Amer. Spec. Pap.*, 244:13–24, 1990; T. R. Smithson, The earliest known reptile, *Nature*, 342:676–678, 1989.

Cartography

Digital mapping is the digital representation of natural and cultural features in a standard format on a medium such as magnetic tape, floppy disk, or hard disk. The features may have various descriptors attached to them in the digital data file, such as geographic position, height, length, width, and surface composition. The resultant digital database may be displayed on a computer monitor or printed as hard copy; it may be similar in appearance to a paper map if the proper database is used. There are two principal categories into which most of the various digital mapping databases can be placed: raster (matrix) and vector.

Initially, aerial photographs and paper maps were used as the source for generating the digital mapping databases. Satellite photography from *Landsat* (United States) and *SPOT* (France) are currently used to complement the previously referenced sources. The satellite data are also available directly in a digital format. The development of emerging weapon systems that require tremendous quantities of map information has led to a rapid evolution of digital mapping technology. Commercial applications and local governmental agencies employing electronic map displays are also taking advantage of this ever-increasing database. Digital mapping databases include Digital Terrain Elevation Data (DTED), Digital Feature Analysis Data (DFAD), World Vector Shoreline (WVS) Data, and ARC Digitized Raster Graphics (ADRG) Data. The Defense Mapping Agency (DMA) and the U.S. Geological Survey (USGS) are the major producers of digital mapping data in the United States. The Defense Mapping Agency works closely in the collection of data with map production agencies in other countries. This agency also is charged with the responsibility for developing both national and international digital map standards. The National Committee for Digital Cartographic Data Standards and the Federal Interagency Coordinating Committee on Digital Cartography in 1989 published a set of digital mapping standards that are similar to the standards developed by the Defense Mapping Agency. New distribution media for the digital mapping databases have been developed that enable large amounts of digital mapping data to be explored on personal computers as well as on the highly sophisticated weapon systems.

History. The demand for digital mapping information only recently increased greatly. The subject was initially driven by a requirement that arose in 1970. Digital Terrain Elevation Data and Digital Feature Analysis Data were needed to provide simulated radar returns in aircraft trainers. Aircraft crews could fly simulated missions over virtually any place in the world if they had the appropriate information from these databases. However, the existing database was incomplete over large areas of the globe. To produce the Digital Terrain Elevation Data and Digital Feature Analysis Data, photographs or similar digital data are required over the area to be digitally mapped. As of January 1990, Digital Terrain Elevation Data was available over approximately 60% of the world and Digital Feature Analysis Data was available over approximately 35%. The Defense Mapping Agency is attempting to complete worldwide coverage by the end of the 1990s.

Digital Terrain Elevation Data. This database is a matrix of terrain elevation values. Basic quantitative data are provided to systems that require terrain elevation, slope, line-of-sight, and surface roughness information. The data are stored in cells, approximately 1° square, with the southwest corner coordinates (latitude, longitude) being the primary geographic identifier for the cell. The information content is approximately equivalent to the contour information found on a 1:250,000-scale map. (On a 1:250,000-scale map, 1 in. on the paper map represents 250,000 in. on the Earth's surface.) Since most of the 1:250,000 maps used are on a transverse Mercator projection, the matrices will be smaller in longitude at the higher northern latitude zones (in the Southern Hemisphere the reverse is true). As noted in **Table 1**, a cell of Digital Terrain Elevation Data between 0° and 50° North or South latitude will contain 1201 by 1201 elevation values, since there are 1201 3-arc-second increments in 1°. This number of data points (1,442,401) is not astronomical, but when large quantities of cells are employed, the data storage requirement becomes a problem for small computers.

To solve this problem, the Defense Mapping Agency began using the compact disk–read only memory (CD-ROM) as a storage device for Digital Terrain Elevation Data in 1989. The CD-ROM holds over 200 cells of Digital Terrain Elevation Data, the equivalent of 20 magnetic tapes or 1500 floppy disks. The technology used to produce the CD-ROM is identical to that for the compact disks used in the music industry. The Digital Terrain Elevation Data control datum for the horizontal coordinate is the World Geodetic System 84 (WGS 84). Thus all the elevation points are referenced horizontally to the World Geodetic System 84, a system used to establish a very precise worldwide geographic (latitude, longitude) reference system. The World Geodetic System 84 also contains worldwide gravity values. The vertical control datum is mean sea level (MSL). The absolute horizontal accuracy objective is 130 m (426 ft), and the absolute vertical accuracy objective is 30 m (98 ft) for each elevation point. **Figure 1** illustrates the appearance of Digital Terrain Elevation Data when plotted as a graphical representation.

Table 1. Elevation data spacing for Digital Terrain Elevation Data		
Zone	Latitude N/S	Spacing, latitude/ longitude, arc-seconds
I	0 to 50°	3 by 3
II	50 to 70°	3 by 6
III	70 to 75°	3 by 9
IV	75 to 80°	3 by 12
V	80 to 90°	3 by 18

New system requirements indicate the need for a higher resolution in Digital Terrain Elevation Data. The Defense Mapping Agency has produced very small quantities of this type of data at 1-arc-second spacing. Thus one cell of this higher-resolution product will have 3601 by 3601 (12,967,201) elevations. This product is referred to as Level 2 Digital Terrain Elevation Data, and it is extremely useful for exploiting terrain roughness, slope activity, or line-of-sight calculations.

Digital Feature Analysis Data. This vector-oriented digital mapping database consists of selected natural and human-made features. The features are classified as point (such as towers, wells, or lighthouses), line (such as roads, railroads, or fences), or area (such as buildings, forests, or towns). Several hundred different feature categories are in use in the various digital map products derived from the Digital Feature Analysis Data. As noted in a few examples given in **Table 2**, each feature is assigned a feature identification code along with attributes giving height, length, width, orientation, and surface composition. The orientation is the angular displacement between true north and the major axis of the feature. The geographic coordinates, referred to the World Geodetic System 84, are also provided in the header information for each feature.

The first application of this type of digital map data was the simulation of radar displays. The composition descriptor gives an indication of the material of which the feature is made and consequently its radar reflectivity value. The lower the number, the better the reflectivity. Today databases composed of Digital Feature Analysis Data are used in electronic map displays, navigation and mission planning systems, planning of commercial truck routes, and other national and state government functions. The U.S. Department of Defense, the Census Bureau, and the Department of Agriculture are among the principal users of such databases.

Digital Feature Analysis Data is available in various levels of feature density. Level 1 comprises the major features found on a 1:250,000-scale map and, like Digital Terrain Elevation Data, is produced in 1°-square cells. A typical Level 1 cell will have approximately 3500 features. Level 2 contains most of the features found on a 1:50,000-scale map. Level 2 is generally produced in patches of 2 by 2 nautical miles around a major item of interest, since it is expensive to produce maps of large areas with current technology.

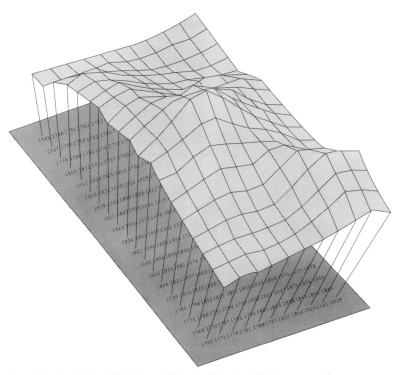

Fig. 1. Graphical plot of Digital Terrain Elevation Data. Level 1: 3-arc-second increments. Level 2: 1-arc-second increments. The numbers represent elevation values in meters. 1 m = 3.28 ft.

The Defense Mapping Agency is in the process of modernizing its production processes in order to produce this digital map database much more economically. **Figure 2** illustrates a typical graphical plot of Digital Feature Analysis Data.

Another new product of Digital Feature Analysis Data, Level 1C (Second Edition), contains every feature found on a Joint Operational Graphic Chart (JOG); this has a 1:250,000 scale. It is very useful in a wide variety of planning applications. All Digital Feature Analysis Data products are referenced to the World Geodetic System 84 datum. Levels 1 and 2 have an absolute accuracy of approximately 100 m (330 ft), while the accuracy of 1C is dependent on the Joint Operational Graphic Chart used to produce it; the accuracy typically may be 120 m (396 ft). Digital Feature Analysis Data is currently available on magnetic tape only, but in the near future the CD-ROM will be used as a distribution medium. Two CD-ROMs will contain all of the Digital Feature Analysis Data coverage in the United States.

Table 2. Sample identification codes and attributes used in Digital Feature Analysis Data

Feature identification code number	Surface composition number	Orientation*	Minimum size selection criteria		
			Length, m (ft)	Width, m (ft)	Height, m (ft)
102 (quarry)	3	001°	300 (990)	300 (990)	−10 (−33)
104 (offshore platform)	2	360°	30 (99)	Any	3 (10)
240 (roads)	9	003°	300 (990)	Any	Any
260 (bridges)	3	002°	150 (495)	Any	Any
908 (marsh)	11	—	300 (990)	300 (990)	0

*Angular displacement between true north and the major axis of the feature.

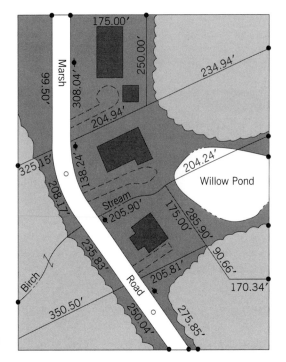

Fig. 2. Typical representation of Digital Feature Analysis Data.

ARC Digitized Raster Graphics. The digital map from this database was initially developed in 1989 to assist aircraft in-cockpit electronic map display systems. ARC Digitized Raster Graphics is a database produced by scanning existing maps and charts in a raster format. The scanning resolution is 100 micrometers per pixel. The digital data are recorded at 24 bits per pixel, that is, 8 bits of red, 8 bits of green, and 8 bits of blue. The resolution was chosen in order to be able to scan the very fine feature line widths on 1:50,000 maps. The ARC Digitized Raster Graphics database gives an excellent rendition of the original paper map or chart when displayed on a high-resolution monitor. After the scanning process, the digital data are transformed to an equal arc-second reference frame known as ARC to minimize the distortion in any of the system's 18 worldwide latitude zones. The ARC system is also a worldwide seamless database; that is, there is no overlap between adjacent map sheets. This is an important consideration for electronic map display systems. Since hundreds of megabytes of data are required for each map or chart, the CD-ROM media is used for distribution. Data compression techniques should enable as many as 15 or 20 maps to be placed on one CD-ROM. The accuracy of ARC Digitized Raster Graphics is determined by the accuracy of the paper map scanned, since the scanning process introduces no significant additional error. The U.S. Department of Defense, as well as local, state, and national governments, and international organizations are the primary users of this database.

World Vector Shoreline. This is a digital data file containing the shorelines, international boundaries, and country names of the world. Boundaries and mean-high-water shorelines are referenced to the World Geodetic System 84. The absolute horizontal accuracy requires that all shoreline features be located within a 500-m (1650-ft) circular error of their true geographic position. The entire database is contained on a single CD-ROM. The United States merchant marine, the Navy, and industry are among the primary users of World Vector Shoreline.

Miscellaneous digital mapping databases. There are many other digital mapping databases, and some have important applications. The World Data Base II is a worldwide digital mapping system produced by the U.S. Central Intelligence Agency (CIA) at an approximate scale of 1:1,000,000; it was revised in January 1990. A digital map database under development for the Defense Mapping Agency, the Digital Chart of the World (DCW), should be available in 1991. Nautical digital mapping databases currently in production or under development are the Digital Bathymetric Data Base and the High Speed Digital Chart. Another mapping product required by the U.S. Army is Tactical Terrain Data (TTD); this will contain all features found on a 1:50,000 map plus Digital Terrain Elevation Data and other terrain analysis information. The Census Bureau has a digital mapping file known as DIME. The DIME database contains state and county boundaries taken from 1:24,000 topographic maps where available.

It is anticipated that the demand for digital mapping databases, in a wide variety of formats and content, will increase significantly. Geographic Information Systems (GIS), aircraft simulators, and new weapon systems are among the major users of these databases. The challenge of the future will be to develop media that can, in a reasonable size, contain the digital mapping database. The CD-ROM is currently the preferred medium, but with the anticipated requirement for terabytes (10^{12} bytes) of data, the CD-ROM is not a practical media choice.

For background information SEE CARTOGRAPHY; COMPUTER GRAPHICS; ELECTRONIC DISPLAY; GEODESY; MAP PROJECTIONS in the McGraw-Hill Encyclopedia of Science & Technology.

B. L. Klock

Bibliography. J. R. Carter, *Computer Mapping*, 1984; Defense Mapping Agency, *Department of Defense World Geodetic System 1984*, DMA Tech. Rep. 8350.2, 1987; Defense Mapping Agency, *Digitizing the Future*, DMA Stock No. DDIPDIG-ITALPAC, 1987.

Cell cycle

Major transitions subdivide the cell cycle. Mitosis, or M phase, the stage in which chromosomes are segregated and cell division occurs, is morphologically distinct from interphase, during which growth occurs. When biochemical probes allowed the detection of deoxyribonucleic acid (DNA) synthesis, interphase was divided. A cell normally progresses from mitosis to a first-gap (G1) phase, defined only by the absence of mitotic morphology or DNA synthesis activity, to a period of DNA synthesis, or S phase, to a second-gap (G2) phase, and then back to mitosis. There are two major control points in this cycle where a cell can rest, one prior to DNA synthesis and the other prior to mitosis.

The transition from interphase to mitosis is tightly regulated. New advances have identified many of the molecules involved in this regulation. These molecules are assembled into a molecular trigger that induces an abrupt transition to the mitotic state. Many components of this trigger have been found to be highly conserved in evolution. For example, molecular components of the human mitotic trigger can function in yeast to substitute for the analogous yeast components. The operation of this mitotic trigger and its control by the many factors that govern the progression of the cell cycle are beginning to be understood.

Mitotic protein kinase. The key component in controlling the transition to mitosis is a protein kinase, which is an enzyme that attaches phosphates to other proteins. There are many protein kinases in a cell, each of them attaching phosphates to a specific group of proteins. The protein kinase that activates mitosis attaches phosphates on the serines and threonines of select proteins. The functions of these proteins are altered by this phosphorylation, and it is the changes in their activities that bring about the events of mitosis.

The mitotic protein kinase was first characterized in yeasts. Mutations in yeast that arrest the cell cycle occur in cell division control (*cdc*) genes. Mutations in the *cdc*2 gene of the fission yeast *Schizosaccharomyces pombe* inactivate the key cell-cycle gene and block progression of the cell cycle at the two control points. Recombinant DNA techniques uncovered a human gene that plays a similar role. When introduced into yeast, this gene substitutes for the *cdc*2 gene of *S. pombe*. The human and the yeast genes were isolated and their DNA sequences determined. Both the human and yeast genes encode proteins with molecular weights of 34,000 whose amino acid sequence is 62% identical. A similar protein has also been identified in organisms representing many major branches of eukaryotic evolution: starfish (echinoderm), clam (mollusk), fruit fly (arthropod), and plants. Based on the molecular weight and the homology to the yeast gene, these proteins are referred to as p34^{cdc2} kinases.

Tests were developed to detect the activity that induces mitosis. For example, the large unfertilized eggs of frogs are arrested in a particular stage of the cell cycle but can be experimentally activated by injection of cytoplasm from dividing cells. The mitosis-promoting factor was purified by fractionating active cytoplasm and then following the activity in a simplified test. The frog version of mitosis-promoting factor is a complex of two or three proteins, one of which is the frog p34^{cdc2} kinase. This result suggested that yeast genetics had correctly identified the role of this protein kinase as an activator of the mitotic state.

Consistent with its proposed role, the p34^{cdc2} kinase is only transiently active before and during mitosis. The next research problem became the search for the regulators of the p34^{cdc2} activity.

Regulation of p34^{cdc2} by cyclins. One of the regulators of p34^{cdc2} activity was identified in a remarkable series of experiments using the eggs of clams, sea urchins, and frogs. Analysis of the proteins synthesized in synchronously dividing clam eggs uncovered two proteins with unusual behaviors. These two proteins are synthesized throughout the cell cycle but are abruptly destroyed at the completion of mitosis. Because of their cyclical patterns of accumulation and destruction, these proteins were named cyclin A and cyclin B. The amino acid sequences of the two cyclins are very similar, and it is thought that they play similar but subtly distinct roles. Related proteins were found in sea urchin eggs, and ultimately, in a wide variety of eukaryotes, including yeasts.

In *S. pombe* and in fruit flies, mutations in the genes that encode for cyclin cause the cell cycle to arrest prior to mitosis. Thus, cyclins appear to be required for mitosis, and their pattern of accumulation and destruction suggests that they could serve as an intracellular clock; that is, time would be measured by the gradual accumulation of cyclins to a threshold at which mitosis would be activated. The abrupt destruction of the cyclin proteins could reset the clock.

The mitosis-promoting factor purified from frog contained not only p34^{cdc2} but also cyclin. Indeed, p34^{cdc2} is active only when it is bound to cyclin. Consequently, the periodic accumulation and destruction of cyclin can control the p34^{cdc2} kinase activity.

Most cell cycles are controlled by many factors such as those that sense whether a cell has duplicated its mass and its important constituents. However, in the embryos of many organisms, the earliest cell cycles are unusually simple. Division appears to be timed directly by the oscillating levels of cyclins. The enormous egg cell of a frog (about 1.2 mm or 0.05 in. in diameter) is subdivided by synchronous divisions every 30 min. These divisions require protein synthesis. Ruptured frog eggs yield an acellular extract that continues to go through cell-cycle oscillations such as periodic pulses of p34^{cdc2} kinase activity. Each new pulse of kinase activity requires either protein synthesis or addition of new cyclin. Thus, the only reason that protein synthesis is needed is to produce cyclin. The frequency of pulses of kinase activity depends on the rate of cyclin synthesis. Thus, in the early frog embryo the oscillating levels of cyclin drive a simple program of periodic division.

But relentless periodic division as seen in early embryos is not the usual pattern of cell-cycle control. Embryonic production of an animal with coherent shape and detailed structure requires intricate and orderly control of cell proliferation. Indeed, after about 12 rapid divisions, when the enormous frog egg has been subdivided into much smaller cells, the cell cycle slows dramatically and adopts a more elaborate and typical program of control. This more elaborate program involves additional modes of regulating

$p34^{cdc2}$ kinase activity. These modes of control were identified in yeast.

Additional modes of regulating p34^{cdc2} kinase. Changes in nutritional conditions can dramatically alter the rate of growth of yeast, but the distribution of cell sizes remains roughly constant. This relative size constancy is achieved by controls acting at mitosis. The activation of $p34^{cdc2}$ kinase occurs when cell size reaches a particular set point. Mutants having smaller cells were isolated, and the affected gene was called *wee*1. The product of the normal *wee*1 gene inhibits the $p34^{cdc2}$ kinase. In the mutant, absence of *wee*1 function leads to hyperactivity of the $p34^{cdc2}$ kinase, which causes precocious cell division and hence creates small cells. In contrast, if the *wee*1 product is overproduced, $p34^{cdc2}$ is more strongly inhibited, mitosis is delayed, and cells are large.

The $p34^{cdc2}$ protein occurs with and without attached phosphate groups. Addition of phosphate occurs on several amino acids (including threonine 14 and tyrosine 15) that are within the active site of the $p34^{cdc2}$. Phosphate in these positions inhibits $p34^{cdc2}$ kinase activity. During the cell cycle these phosphates are abruptly removed, and the $p34^{cdc2}$ kinase is activated just prior to mitosis. The *wee*1 gene is required to put on the inhibitory phosphates, and another gene, *cdc*25, is required to take them off. While it is still uncertain how *wee*1 promotes phosphorylation of $p34^{cdc2}$ or how *cdc*25 promotes its dephosphorylation, it is clear that *wee*1 inhibits and *cdc*25 activates the kinase activity and mitosis. Together with the earlier studies of cyclin, these analyses lead to the conclusion that, to be active, $p34^{cdc2}$ needs to be both bound to cyclin and dephosphorylated.

Control of cell division in fruit fly embryos. As in other organisms, early embryos of the fruit fly *Drosophila melanogaster* have a simple rapid cell cycle. But in this organism it has been possible to examine the processes that control the later divisions. In later divisions, cells in different parts of the embryo divide according to very different schedules. These schedules of division times are precise and reproducible from embryo to embryo. It can be deduced that the timing of these divisions might be controlled by either the time of accumulation of cyclin or the time of dephosphorylation.

Cyclin is found to accumulate at uniform rates in all the cells despite their different schedules of division. Thus, cyclin levels cannot control the timing of these divisions. Rather, the key controlling role is played by a gene called *string*, which encodes a protein required for mitosis. Because both the messenger ribonucleic acid (mRNA) encoding *string* and the *string* protein itself are very unstable, *string* protein levels can be maintained only by continuous synthesis of the mRNA and the protein. Thus, mitosis requires concurrent *string* gene expression. Transcription of the *string* gene occurs in pulses. About 15 min after the start of a pulse of *string* transcription, the mRNA and hence the protein product accumulate to a level that induces mitosis. All transcription stops during mitosis, and *string* protein declines as cells proceed to the

interphase of the next cell cycle. But the pulses of *string* transcription do not occur at the same time in all the cells of the embryo. Cells in different parts of the embryo have different schedules for expression of *string*. Examination of an embryo reveals an intricate pattern of *string* gene expression. This pattern of expression changes continuously as some cells stop expressing the gene and other cells start. Since the cells enter mitosis toward the end of a pulse of *string* expression, the cells of the embryo divide in a pattern that follows the pattern of *string* expression. To show that *string* transcription controls mitosis, recombinant DNA techniques were used to induce *string* transcription at unusual times and in unusual places. This unscheduled expression provoked cell division 15 min later. Thus, the time of *string* transcription controls mitosis.

The *string* gene of the fruit fly is homologous to the *cdc*25 gene of the fission yeast, and it will function in the fission yeast to substitute for the yeast homolog. Thus, like its yeast homolog, the *string* gene promotes the dephosphorylation of $p34^{cdc2}$ and consequently induces its kinase activity and mitosis.

Multiple controls of the cell cycle. The relentless periodic divisions of early frog eggs are controlled by oscillating cyclin levels, whereas the highly programmed divisions of a fly embryo are controlled by intricate patterns of transcription of *string*. It is likely that additional modes of control will be found. For example, more recent work has offered a preliminary view of the control prior to DNA synthesis. It appears that entry into the DNA synthesis phase is controlled by a second trigger related to the one controlling initiation of mitosis. The $p34^{cdc2}$ kinase is an essential component of both triggers, but it interacts with a different set of proteins to trigger the two distinct regulatory transitions.

Cancer and the cell cycle. At maturity, the growth and division of cells are strictly regulated by various growth factors and inhibitors. Cancer arises when a cell escapes its constraints and begins to grow and divide without heeding the normal regulators. It is hoped that identification of the molecules and mechanisms that control progression through the cell cycle will identify the normal restraints that prevent unscheduled proliferation, and will reveal how these constraints are circumvented in cancerous cells.

For background information SEE CELL CYCLE; CELL DIVISION; MITOSIS in the McGraw-Hill Encyclopedia of Science & Technology.

Patrick H. O'Farrell

Bibliography. K. L. Gould and P. Nurse, Tyrosine phosphorylation of the fission yeast $cdc2^+$ protein kinase regulates entry into mitosis, *Nature*, 342:39–45, 1989; M. G. Lee and P. Nurse, Complementation used to clone a human homologue of the fission yeast cell cycle control gene *cdc2*, *Nature*, 327:31–35, 1987; A. W. Murray and M. W. Kirschner, Dominoes and clocks: The union of two views of the cell cycle, *Science*, 246:614–621, 1989; P. H. O'Farrell et al., Directing cell division during development, *Science*, 246:635–640, 1989.

Cell senescence

Normal human and rodent cells in culture exhibit a finite life-span and eventually cease proliferating and die, a process termed cell senescence or cell aging. It is possible, but not yet proven, that aging at the cell level contributes to the aging of the individual. The risk of cancer greatly increases with age, yet many cancer cells differ from normal cells in that they do not senesce, and have an indefinite life-span in culture. This suggests that alterations relating to cell senescence are involved in the neoplastic evolution of tumor cells. Recent experimental results strongly support a genetic basis for cell senescence. Defects in the senescence program in cancer cells can be corrected by introduction of a specific chromosome from normal cells into the abnormal cells. By using this approach, possible senescence genes can be mapped to specific chromosomes. An understanding of cell senescence at the genetic and molecular levels may provide new insights into both the cancer and aging processes.

Cell senescence and aging. In 1961, it was reported that normal human embryo fibroblasts are able to undergo a limited, fixed number of cell divisions and then cease proliferation. The key determinant in the life-span of cells in culture is the number of cell doublings, not the length of time in culture.

Three lines of evidence suggest that the aging of cells in culture is related to the aging of the organism: (1) The doubling potential of cells in culture is inversely proportional to the age of the donor. Cells from embryonic tissue exhibit the longest life-span, with approximately 50–60 cell doublings. Cells from adult tissue can be grown for only 14–29 doublings, with a general decrease in the life-span of cells in culture with increasing age of the donor tissue. (2) Cells from individuals who exhibit premature aging have a decreased life-span in culture. Those with progeria (Hutchinson-Gilford syndrome) manifest signs of aging at the end of their first decade of life that are typical of normal individuals in the seventh decade. Werner syndrome individuals also have accelerated aging but in later years; the mean age at death is 47. The main causes of death for individuals with progeria and Werner syndrome are cancer and cardiovascular disease, which are also leading causes of death for normal individuals. Fibroblasts derived from individuals with the premature aging characteristics senesce prematurely in culture, after only 2–10 population doublings. (3) The life-span of cells in culture is generally correlated with the maximum life-span of the species, although the variability in the experimental data is great. Cells from humans (maximum life-span = 100–120 years) can be grown for 50–60 population doublings, whereas cells from rodents (3–5-year life-span) can be grown for only 20–40 population doublings. Cells from the Galápagos tortoise (175–200-year life-span) undergo 90–125 population doublings.

These lines of evidence, although not conclusive, support the hypothesis that aging of cells is related to the aging process of the organism.

Cell senescence and cancer. Escape from cell senescence is an important step in neoplastic progression of human and rodent cancers. Many tumor cells can be grown indefinitely in culture and therefore have escaped senescence. They are termed immortal. That treatment of normal cells with diverse carcinogenic agents, including chemical carcinogens, viruses, and oncogenes, allows cells to escape senescence indicates that this change is important in cancer induction. Immortality is not sufficient for neoplastic transformation, but most immortal cells have an increased propensity for spontaneous, carcinogen- or oncogene-induced neoplastic progression. Therefore, escape from senescence is a change that predisposes a cell to neoplastic conversion. It is clear that immortal cells are further along the multistep pathway to neoplasia than normal cells.

Genetic basis for cell senescence. Two major theories of cell senescence have been proposed. One is the error catastrophe, or damage model, which proposes that random accumulation of damage or mutations in deoxyribonucleic acid (DNA), ribonucleic acid (RNA), or protein leads to the loss of proliferative capacity. A second hypothesis is that senescence is a genetically programmed process. The experimental evidence supporting the error accumulation hypothesis has been criticized, whereas support for a genetic basis for senescence has recently been provided. It is possible to fuse cells of different origins and then to select for the hybrid cells by using biochemical markers for drug sensitivity or resistance that differ in the parental cells. When cells with a finite life-span are fused to immortal cells with an indefinite life-span, the majority of these hybrids senesce, indicating that senescence is dominant over immortality. Even hybridization of two different immortal human cell lines can result in senescence, indicating that different complementation groups exist for the senescence function lost in these cells. Four complementation groups have been established, suggesting that multiple genes are lost or inactivated during escape from senescence. If this hypothesis is correct, it should be possible to map the genes involved in cell senescence to specific chromosomes. Recent findings with hamster and human interspecies hybrids mapped a putative senescence gene to human chromosome 1.

Mapping genes for cell senescence. When normal human cells with a finite life-span are fused to immortal hamster cells, the hybrids that form exhibit a finite life-span characteristic of the normal human cells. At the end of this life-span, the cells display signs of cellular senescence characteristic of the parental human cells at the end of their life-span. Furthermore, the senescence of the hybrids is an active process dictated by the senescence program of the human cells.

Although the majority of the human–hamster hybrids senesce, some of the hybrids ultimately escape senescence. Senescent cells appear in all of the hybrid clones after two to three passages, but in some of the clones a few nonsenescent cells persist and continue to proliferate indefinitely. It was observed that, without exception, all of the human–hamster hybrid clones that escaped senescence had lost both copies of human chromosome 1. One or two copies of all other human chromosomes were present in at least one of the immortal hybrids. Furthermore, the introduction of a single copy of human chromosome 1 into the hamster cells by microcell fusion caused typical signs of cellular senescence. Transfer of chromosome 11 had no effect on the growth of the cells. These findings indicate that human chromosome 1 may participate in the control of cellular senescence and add further support to a genetic basis for cell senescence.

The gene or genes involved in the senescence of hamster cells are located on the long arm of human chromosome 1. Since alterations of chromosome 1 have been reported in a variety of human tumors, including intestinal, mammary, ovarian, uterine, and colonic tumors, as well as tumors in the myeloproliferative disorders, it is possible that a unifying biological mechanism is responsible for these chromosomal changes. SEE TUMOR.

For background information SEE CANCER (MEDICINE); CELL SENESCENCE AND DEATH; MUTATION; ONCOLOGY; SENESCENCE; TISSUE CULTURE in the McGraw-Hill Encyclopedia of Science & Technology.

J. Carl Barrett

Bibliography. L. Hayflick, The cell biology of human aging, *New Engl. J. Med.*, 295:1302–1308, 1976; A. Macieira-Coelho, *Biology of Normal Proliferating Cells In Vitro: Relevance for In Vivo Aging*, Interdisciplinary Topics in Gerontology, vol. 23, 1988; O. M. Pereira-Smith and J. R. Smith, Genetic analysis of indefinite division in human cells: Identification of four complementation groups, *Proc. Nat. Acad. Sci. USA*, 85:6042–6046, 1988; O. Sugawara et al., Induction of cellular senescence in immortalized cells by human chromosome 1, *Science*, 247:707–710, 1990.

Cellular immunology

Recent advances in immunology research have suggested that T cells bearing gamma/delta antigen receptors function as a first line of defense against infection. This hypothesis is based on the finding of large numbers of gamma/delta T cells in tissues that interface with the exterior environment, such as skin, gut, and lungs in healthy mice. Furthermore, the early murine (pertaining to mice) immune response to *Mycobacterium tuberculosis* infection is characterized by a tremendous increase in the number of gamma/delta T cells. Lymphocytes that are formed in the thymus gland are known as thymocytes. A large fraction of neonatal thymocytes, cells that have never been exposed to foreign pathogens, have been shown to be gamma/delta T cells that do respond to mycobacteria. Also, human gamma/delta T cells from the peripheral blood of normal donors have been shown to proliferate in response to mycobacteria. Finally, extraordinary numbers of gamma/delta T cells are found to accumulate in infectious skin lesions of patients with leprosy and leishmaniasis.

T-cell receptors. Antigen recognition by T cells is accomplished via a set of cell-surface proteins called T-cell receptors. These receptors recognize antigen in the context of antigen-presenting cells. Two such classes of T-cell receptors, termed alpha/beta and gamma/delta, have been isolated. In general, these receptors are composed of segments, called constant-region domains, that are invariant to each class, and segments, termed variable-region domains, that differ greatly within a class of receptors. The ability of the T-cell repertoire to recognize a great diversity of foreign antigens is made possible by the genes that encode the variable region of the T-cell receptor.

Actually, there is a difference in the mechanism by which the alpha/beta and gamma/delta T-cell receptors generate diversity in antigen recognition. For the alpha/beta receptor, diversity is encoded by 50–100 variable-region gene segments. In thymic alpha/beta T cells, these variable gene segments undergo genetic recombination with a small number of adjacent gene segments termed joining and constant. Such alpha/beta T cells constitute the majority of peripheral blood T cells. Unlike the alpha/beta T cells, the diversity of gamma/delta T cells is due, not to variable gene regions, of which there are less than 20 in these cells, but rather to incredible events of genetic recombination. In gamma/delta T cells that reside in the thymus, nucleotides are deleted and randomly inserted at the junction where the variable and joining gene segments splice. Gamma/delta T cells make up less than 10% of murine peripheral blood cells, but are more common in the murine epidermis and gut.

The role of gamma/delta T cells in the immune response is a fundamental question of current immunology research. Analysis of the function and recognition patterns of these cells has received heightened interest due to the reported reactivity of gamma/delta T cells to mycobacteria and the finding of increased gamma/delta T cells in the lesions of several human diseases.

Probably, gamma/delta T cells function in the early process, until the alpha/beta T-cell response can eradicate the unwanted pathogen. The gamma/delta and alpha/beta T-cell responses may cooperate, perhaps by mechanisms including granuloma formation, to wall off and eliminate the infectious pathogen.

Response of gamma/delta T cells to mycobacteria. Mycobacteria are important human pathogens. *Mycobacterium tuberculosis* is the cause of human tuberculosis, a disease that is recently on the increase. Leprosy is due to infection with *M. leprae* and currently affects 15 million people worldwide. It is a significant health and economic burden in third world countries. There are other mycobacteria that are opportunistic pathogens, causing disease in immunocompromised individuals such as those with HIV infection.

A number of studies in the mouse have pointed toward stress-induced proteins (often termed heat shock proteins) as potential antigens for gamma/delta T cells. Stress proteins represent a family of proteins that are conserved over many species, including mycobacteria and humans. Stressful stimuli, such as heat or deprivation of oxygen or nutrient, induce the organisms to synthesize increased amounts of these proteins. The conservation of amino acid sequence of these stress proteins among organisms would cause gamma/delta T cells to respond to a wide variety of pathogens. In addition, gamma/delta T cells would respond to infected host cells that have been stressed to express autologous proteins.

Initially, it was found that gamma/delta T cells have a propensity to react with mycobacterial antigens. For example, there is a 25-fold increase in the absolute numbers of gamma/delta T cells in lymph nodes that drain regions of primary immunization with heat-killed *M. tuberculosis*. These isolated gamma/delta T cells proliferated and produced lymphokines in response to *M. tuberculosis* in cell culture. Similarly, it was found that resident pulmonary gamma/delta T cells could be stimulated by exposure of the mice to aerosolized *M. tuberculosis* antigens. A significant proportion of gamma/delta T cells derived from neonatal thymus displayed reactivity to *M. tuberculosis* antigens.

A large percentage of these gamma/delta T cells from the thymus were found to react with a mycobacterial stress protein of molecular mass 65 kilodaltons (kD), thus providing the first evidence for gamma/delta T-cell reactivity to stress proteins. There was some reactivity for the murine stress protein as well, supporting the hypothesis that gamma/delta T cells could respond to autologous cells that have been stressed, perhaps by intracellular infection. Another piece of supporting evidence for the idea that murine gamma/delta T cells may cross-react with autologous stress proteins was the observation that the resident pulmonary gamma/delta T-cell subset began to proliferate in response to heating above the physiologic temperature of 98.6°F (37°C). Incubating at 107.6°F (42°C) will induce autologous stress protein expression.

Gamma/delta T cells and human disease. Although gamma/delta T cells have not been found to constitute a large resident population in normal human tissues, they have been found to accumulate in the lesions of several disease states. The increase in gamma/delta T cells in leprosy skin lesions has provided an opportunity to determine the role of these cells at the site of immunopathologic reaction. Experimental evidence suggests that gamma/delta T cells contribute to immune defense against infection by inducing the formation of granulomas (an organized collection of lymphocytes and macrophages) in response to infectious pathogens. First, gamma/delta T cells constitute a strikingly high percentage of the T-cell population in infectious disease lesions that contain recently formed granulomas. These lesions include the sites of skin tests that are the result of experimental challenge with injection of *M. leprae*. In addition, gamma/delta T cells are found to accumulate in reversal reactions in leprosy, which are due to the sudden increase in host resistance against *M. leprae*. Gamma/delta T cells are also prominent in the skin lesions of localized

American cutaneous leishmaniasis. These lesions are a granulomatous response to infection with the intracellular parasite *Leishmania brasiliensis*. Second, the gamma/delta T cells derived from these leprosy lesions proliferate upon stimulation with mycobacterial antigens in cell culture. The gamma/delta T cells derived from lesions appear to release a lymphokine that may contribute to the formation of tissue granulomas by inducing critical cellular events such as macrophage adhesion and aggregation.

The finding of gamma/delta T cells that are reactive to mycobacteria in the blood of humans has further sparked interest in the nature of the antigen recognized by these T cells. It is generally thought that these cells do not exist solely to defend against mycobacteria, but rather that mycobacteria share highly conserved antigens shared by other organisms. As in the murine studies, investigation has centered on the stress proteins that are highly conserved in many different organisms, including humans. Several research groups have reported human gamma/delta T-cell clones that are reactive to mycobacteria, and specifically to the 65-kD heat shock protein. It appears that a low percentage of the human gamma/delta T-cell repertoire recognizes the 65-kD antigen. However, an aberrant response to self stress proteins would result in autoimmune disease. Indeed, human gamma/delta T cells that are reactive to the 65-kD protein have been reported in the synovial fluid of a patient with the autoimmune disease of rheumatoid arthritis. Gamma/delta T cells may contribute to other diseases in humans as well. For example, in the blood of malaria patients, there is an increase in the percentage of gamma/delta T cells. Gamma-delta T cells have also been reported to accumulate in the intestinal epithelium of patients with celiac disease, a gastrointestinal disease that is due to sensitivity to gluten. Gamma/delta T cells are increased in the blood and lungs of patients with sarcoidosis, a systemic disorder in which granulomas are prominent and which has an enigmatic etiology. Finally, there is an alteration in the gamma/delta T-cell population in patients with the immunodeficiency disorder ataxia telangiectasia.

For background information SEE AUTOIMMUNITY; CELLULAR IMMUNOLOGY; HYPERSENSITIVITY; IMMUNOLOGICAL ONTOGENY; LEISHMANIASIS; LEPROSY; MYCOBACTERIAL DISEASES; TUBERCULOSIS in the McGraw-Hill Encyclopedia of Science & Technology.

Drew M. Pardoll; Robert L. Modlin

Bibliography. E. M. Janis et al., Activation of gamma delta T cells in the primary immune response to *Mycobacterium tuberculosis*, *Science*, 244:713–716, 1989; D. Kabelitz et al., A large fraction of human peripheral blood gamma/delta[+] T cells is activated by *Mycobacterium tuberculosis* but not by its 65-kD heat shock protein, *J. Exp. Med.*, 171:667–679, 1990; R. L. Modlin et al., Lymphocytes bearing antigen-specific gamma/delta T-cell receptors in human infectious disease lesions, *Nature*, 339:544–548, 1989; R. L. OBrien et al.,

Stimulation of a major subset of lymphocytes expressing T cell receptor gamma delta by an antigen derived from *Mycobacterium tuberculosis, Cell,* 57:667–674, 1989.

Chaos

The motion of fluid particles can be quite complicated even when the flow carrying (advecting) the particles is simple. This somewhat surprising regime of fluid motion has been termed chaotic advection and provides a physical example of a simple deterministic system that possesses chaotic dynamics. The recent recognition of chaotic advection has led to new insights into the stirring and mixing of fluids. An important practical consequence of this chaotic behavior is that it provides an efficient means of stirring very viscous liquids.

Flow kinematics. The description of fluid motion can be achieved by two alternative means. Most commonly a flow is described by giving the velocity **u** of the fluid at each point in space **x** as a function of time *t*. This field representation, known as the eulerian representation, is usually the most convenient way of expressing the equations of fluid motion. The equations of motion written in eulerian coordinates are a set of partial differential equations (such as the Navier-Stokes equations) from which the velocity field **u**(**x**,*t*) can be determined. However, in some circumstances it is more natural to follow the trajectories of labeled fluid particles, the lagrangian representation. For example, in oceanography, the positions of floats are routinely tracked to infer ocean currents, and in experimental fluid mechanics smoke or dye is frequently used to visualize flows. Although both representations fully describe the fluid motion, it is not always a simple matter to transform lagrangian quantities into their eulerian counterparts. However, if the eulerian field is known, then the trajectories of lagrangian particles can be computed easily. *See* Fluid flow.

Chaotic particle motion. Until recently the common assumption was that the complexity of particle trajectories reflected the complexity of the eulerian velocity field; if the eulerian flow were simple (laminar flow), then the motion of fluid particles would be regular. Conversely, if the flow were complicated (turbulent flow), then the motion of particles would be highly irregular. Now it is known that even some simple flows can generate particle trajectories that are complicated. This complicated behavior is chaotic in the technical sense of the word: the system possesses sensitive dependence on its initial conditions. Thus, the trajectories of particles that are initially quite close diverge rapidly. The rate of divergence is, on average, exponential in time. In contrast, regular trajectories diverge only at a rate that is linear in time.

The motion of a passive particle is described by the set of equations (1). These equations relate the motion

$$\frac{dx}{dt} = u(x,y,z,t) \qquad (1a)$$

$$\frac{dy}{dt} = v(x,y,z,t) \qquad (1b)$$

$$\frac{dz}{dt} = w(x,y,z,t) \qquad (1c)$$

of a particle to the instantaneous eulerian velocity field **u** = (*u,v,w*). The advection equations (1) are a set of ordinary differential equations (a dynamical system). It is now well known that simple dynamical systems can possess extremely complicated behavior. For two-dimensional steady (time-independent) flow, Eqs. (1) reduce to Eqs. (2). Dividing Eq. (2a) by

$$\frac{dx}{dt} = u(x,y) \qquad (2a)$$

$$\frac{dy}{dt} = v(x,y) \qquad (2b)$$

Eq. (2b) and rearranging gives Eq. (3). This equation

$$\frac{dx}{u} = \frac{dy}{v} \qquad (3)$$

simply expresses the fact that in a steady flow the particle trajectories are identical to the streamlines of the flow. Streamlines are closed curves or curves that terminate at the boundaries of the flow (which can be at infinity). Consequently, the motion of particles in two-dimensional steady flows is always simple and predictable. If the flow is not steady, namely, if it is given by Eqs. (4), then an expression such as Eq. (3),

$$\frac{dx}{dt} = u(x,y,t) \qquad (4a)$$

$$\frac{dy}{dt} = v(x,y,t) \qquad (4b)$$

in general, cannot be written. In such a case, Eqs. (4) are said to be nonintegrable. Chaotic behavior is possible only when the advection equations are non-

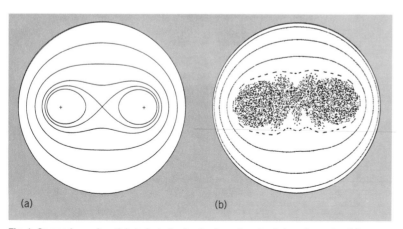

Fig. 1. Comparison of particle trajectories in steady and unsteady two-dimensional flow.
(a) Streamlines of steady flow; the crosses indicate the positions of the vortex agitators.
(b) Stroboscopic portrait of unsteady flow.

(a)   (b)

integrable. In contrast to the trajectories generated by steady two-dimensional flows, these trajectories are irregular and essentially unpredictable.

Figure 1 illustrates the difference between the particle trajectories in a steady two-dimensional flow and an unsteady two-dimensional flow. The flow is an idealized model of stirring in a tank by a pair of vortex agitators. In one case (Fig. 1a), both of the stirring agitators are on simultaneously. The flow is steady and the particle trajectories coincide with streamlines. The motion is quite regular. In the second case, the agitators operate alternately for a fixed period of time. Figure 1b is a stroboscopic portrait of the flow. It is constructed by plotting the positions of a set of particles after each cycle of the two agitators being turned off and on. After many such cycles, some particles trace out smooth curves, but the trajectories of particles placed near the agitators have no apparent regularity. This mixture of regular regions (sometimes called islands) and chaotic regions (the chaotic sea) is characteristic of these kinds of flows. The differences between Fig. 1a and b are a consequence of flow kinematics; the balance of forces is the same in both flows. The only difference between them is that one flow is steady and the other is time-dependent.

In three spatial dimensions, even steady flows can produce chaotic advection. In this case, Eqs. (1) reduce to Eqs. (5). Dividing Eqs. (5a) and (5b) by

$$\frac{dx}{dt} = u(x,y,z) \tag{5a}$$

$$\frac{dy}{dt} = v(x,y,z) \tag{5b}$$

$$\frac{dz}{dt} = w(x,y,z) \tag{5c}$$

Eq. (5c) gives Eqs. (6), where $U = u/w$ and

$$\frac{dx}{dz} = U(x,y,z) \tag{6a}$$

$$\frac{dy}{dz} = V(x,y,z) \tag{6b}$$

$V = v/w$. Equations (6) are identical in form to Eqs. (4) and, in general, are capable of generating chaotic trajectories.

An example of steady three-dimensional chaotic advection is provided by the so-called ABC flow. This flow is a solution of the momentum equation (ignoring the effects of viscosity) for flow in a periodic cube with sides of length 2π. The advection equations for the ABC flow are Eqs. (7), where A, B, and C are

$$\left.\begin{array}{l} \dfrac{dx}{dt} = A \sin z + C \cos y \\[2ex] \dfrac{dy}{dt} = B \sin x + A \cos z \\[2ex] \dfrac{dz}{dt} = C \sin y + B \cos x \end{array}\right\} \bmod 2\pi \tag{7}$$

adjustable parameters. **Figure 2**a and b show particle trajectories, Fig. 2a for a particle initially placed

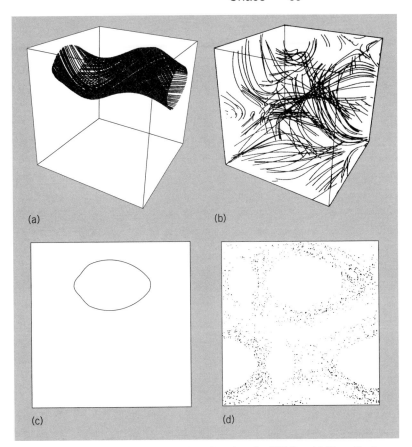

Fig. 2. Streamlines and corresponding Poincaré sections for ABC flow. (a) Streamline for particle initially placed inside a regular region. **(b)** Streamline for particle initially placed within a chaotic region. **(c)** Poincaré section for regular trajectory. **(d)** Poincaré section for chaotic trajectory.

inside a regular region, and Fig. 2b for a particle initially placed within a chaotic region. Figure 2c and d show how the particle trajectory repeatedly intersects one of the faces of the cube. For the regular trajectory, the intersection points again lie on a smooth curve (Fig. 2c). For the chaotic trajectory, the intersection points of Fig. 2d are spread over a broad region. In the language of dynamical systems theory, a figure constructed in this manner is called a Poincaré section. *See Fractals.*

The values used in Fig. 2 are $A = 1$, $B = \frac{1}{2}$, and $C = \frac{1}{2}$. For different values of A, B, and C, the size of islands and the chaotic sea will vary. For example, if the product $ABC = 0$ (for example, because $C = 0$), then the advection equations (7) are completely integrable (the solution can be written in terms of elliptic functions) and there are no chaotic trajectories. For $ABC \neq 0$, chaotic trajectories and regular trajectories will coexist. If both A and B are held constant and C is increased, the size of the chaotic sea will grow. However, although the islands might become quite small, they never completely disappear. In fact, there is always an infinite hierarchy of smaller islands around larger islands.

Stirring and mixing. The stirring and mixing of fluids is the area of applied science in which an understanding of chaotic advection can have its most profound impact. Stirring is defined as the mechanical

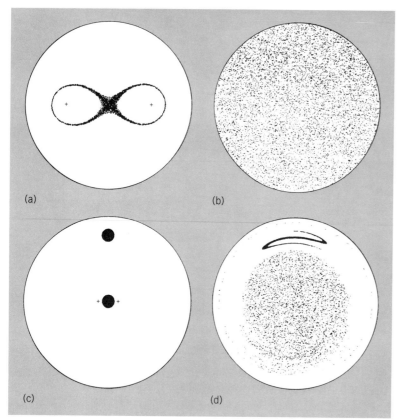

Fig. 3. Stirring of a small circular disk for (a) steady flow and (b) unsteady flow. (c) Initial positions of two small disks. (d) Stirring of the two disks after 10 cycles of unsteady flow.

stretching of a material interface. Mixing is the diffusion of substance across this interface. Mixing depends on the physical properties of the fluids, whereas stirring is governed by the nature of the flow. In general, stirring enhances mixing by producing more interface for the substance to diffuse across, although it is possible to stir fluids that do not mix at all. Only stirring will be discussed.

The standard procedure for efficiently stirring two fluids is to make the flow turbulent. This is intuitively obvious from such everyday experience as stirring cream in coffee. However, there are occasions when it is too costly, or otherwise undesirable, to make a flow turbulent. For a flow to be turbulent, the Reynolds number must be large. The Reynolds number is the ratio of inertial forces to viscous forces in the flow. Consequently, for very viscous fluids it might not be practical to generate a turbulent flow.

The features of turbulent flow that lead to efficient stirring are that (1) the material interface stretches at an exponential rate and (2) the stirring occurs over a wide range of length scales. Fortunately, these are two of the properties associated with chaotic advection. Moreover, chaotic advection is a consequence of flow kinematics and can occur at any Reynolds number. Thus, chaotic advection provides the ideal strategy for stirring viscous fluids. **Figure 3***a* and *b* contrast two examples of the stirring of 10,000 particles by the two-vortex flow discussed above. The particles are initially

located inside a small disk whose center is at the center of the tank. Both Fig. 3*a* and *b* show the stirring of the disk after the same amount of time. In Fig. 3*a* the flow is steady and all trajectories are regular. The stirring is poor and particles are confined to a figure-eight-shaped region. Figure 3*b* shows the stirring of the disk when the chaotic sea is large. Clearly the disk has been stirred much better by the time-dependent flow. In this case, particles are distributed throughout the tank.

An additional factor to consider is the size and distribution of islands. To ensure good stirring, it is necessary to minimize the fraction of the flow domain that is occupied by islands. Figure 3*c* shows the initial positions of two disks of fluid each containing 5000 particles. Figure 3*d* shows the positions of the particles after both disks are stirred by the same flow for 10 cycles of the flow. The upper disk is located inside an island and is not stirred very well. The lower disk is in the chaotic sea and obviously is stirred much better. The minimization of regular islands usually can be achieved by a judicious selection of the parameters governing the flow.

For background information SEE FLUID FLOW; FLUID-FLOW PRINCIPLES; REYNOLDS NUMBER; TURBULENT FLOW; VISCOSITY in the McGraw-Hill Encyclopedia of Science & Technology.

Scott Jones

Bibliography. H. Aref, Stirring by chaotic advection, *J. Fluid Mech.*, 143:1–21, 1984; J. C. Chaiken et al., Experimental study of Lagrangian turbulence in a Stokes flow, *Proc. Roy. Soc. Lond.*, A408:165–174, 1986; T. Dombre et al., Chaotic streamlines and Lagrangian turbulence in the ABC flows, *J. Fluid Mech.*, 167:353–391, 1986; J. M. Ottino, The mixing of fluids, *Sci. Amer.*, 260:56–67, 1989.

Chemoreception

Cnidarians are the simplest metazoan animals that have a nervous system. Since image-forming vision is absent in most cnidarians, they rely primarily on chemical and mechanical stimuli in order to detect and capture prey. Consequently, chemoreception, the ability to sense chemicals in the environment, is highly developed in this group.

Characteristic of all cnidarians is the presence of tiny organelles called cnidae that are contained within specialized, secretory cells called cnidocytes. The cnidae are intracellular, eversible, secretory products primarily used to capture prey or to defend against predators. When suitably triggered by combined chemical and mechanical stimuli, the cnidae, also known as nematocysts, discharge within 3 milliseconds by everting a hollow tubule along with the extruded soluble contents. The response to the chemical stimulus, which predisposes the cnidocyte to respond to a triggering mechanical stimulus, is referred to as chemosensitization.

Cnidocytes. Functionally mature cnidocytes possess a cilium surrounded by a parallel array of stereocilia. The structure as a whole projects into the ambient seawater. In anthozoans this region is referred to as the ciliary cone and consists of a typically structured but nonmotile cilium plus an array of stereocilia that originate from the neighboring supporting cells. In hydrozoans and scyphozoans the region is referred to as the cnidocil apparatus or "trigger," which consists of a modified cilium, termed the cnidocil, plus an array of stereocilia that originate from the cnidocytes. The ciliary cone and the cnidocil apparatus are structurally similar to the mechanoreceptors of hair cells of the acousticolateralis system in vertebrates and are presumed to be sensitive to the mechanical stimuli that trigger discharge of cnidae.

Accessory cells. Accessory cells are ectodermal cells with functional significance in the control of the discharge of cnidae. These cells are characterized by their close association with cnidocytes, of which there are three types: (1) In hydra, several kinds of cnidocytes occur together within a single battery cell such that the only contact the cnidocytes have with the tentacle is via the battery cell, which in turn is attached to the mesoglea. Morphological evidence suggests that the battery cell may also interact with elements of the nerve net. (2) In tentacles of sea anemones, each cnidocyte is surrounded by two or more supporting cells that provide the stereocilia to the sensory cone of the cnidocytes. (3) Each cnidocyte in the Portuguese man-of-war (*Physalia* sp.) is anchored directly to the underlying mesoglea and is surrounded only at its apical region by a single neighbor cell.

Discharge processes. The discharge of cnidae may be divided into four sequential and hierarchical component processes: (1) The mechanics of discharge entails changes in the structural relationships among the various parts of the cnida during eversion of the tubule. (2) The cause of discharge includes changes within or to the cnida that generate or release the force causing the tubule to evert. (3) The activation of discharge refers to the action of the cnidocyte or other cells on the cnida to initiate tubule eversion. (4) The control of discharge comprises those cellular functions that determine whether or not the cnida is activated to discharge. Of these four processes, control of discharge is the least understood and undoubtedly the most complicated.

Control of discharge. The control of discharge consists of two integrated processes, chemoreceptor-mediated sensitization and mechanoreceptor-mediated triggering. Both processes involve specific external stimuli and their respective receptors, as well as transduction and signal-processing mechanisms associated with these receptors, along with any subsequent biochemical or ultrastructural effects resulting from stimulating the receptors. In addition, control of discharge involves the integration of intracellular signals originating from the sensory receptors with respect to the threshold and mechanism of activation. The control of discharge also includes modulation from the nervous system, especially with respect to signals originating from remote sensory receptors.

There are two widely accepted views regarding the control of discharge of cnidae, both of which are being challenged in light of recent findings. In the first view, concomitant mechanical and chemical stimulation is required to activate discharge; that is, neither alone is sufficient to elicit discharge. In the second view, known as the independent effector hypothesis, the cnidocyte is an independent effector in which causally related sensory and effector systems are both located and controlled without an intervening nervous system.

Chemoreceptor-mediated sensitization. At least two classes of naturally occurring molecules, including a variety of low-molecular-weight amino compounds and free and conjugated *N*-acetylated sugars, sensitize tentacle cnidocytes in sea anemones to discharge cnidae in response to tactile triggering. Thus, amino acids commonly found in the body fluids of prey, as well as mucins and mucopolysaccharides containing conjugated *N*-acetylated sugars commonly found on the surfaces of prey, are capable of chemosensitizing cnidocytes.

Dose-response curves to all tested chemosensitizers using nonvibrating test probes are biphasic in shape, having a region of sensitization at low doses of sensitizer, a region of maximum response at optimal concentrations, and a region of desensitization at higher concentrations. It is not yet known what causes the apparent desensitization observed at higher concentrations, but one result is that the desensitization prevents excessive discharge of cnidae against already-captured prey, thereby conserving cnidae for future use.

Since the dose responses to chemosensitizers are saturable, and since the sensitizing effect is reversible and specific, it is concluded that the chemosensitization of cnida discharge to tactile triggering is mediated by surface receptors. Furthermore, since the response to amino agonists is competitively inhibited by certain antihistamines, such as diphenhydramine, whereas the response to *N*-acetylated sugars and mucins is unaffected, it can be concluded that there exist at least two classes of chemoreceptors that are involved in sensitizing cnidae to discharge.

Contrary to the independent effector hypothesis, in which it is proposed that the sensory receptors involved in triggering discharge of cnidae are located on the cnidocyte, it has been shown by using mucin-labeled colloidal gold that the "sugar" chemoreceptors are located only on the supporting cells adjacent to cnidocytes.

Mechanoreceptor-mediated triggering. It has recently been shown that cnidocytes in fishing tentacles of naive (that is, nonchemosensitized) sea anemones (*Haliplanella luciae*) discharge nematocysts preferentially into targets vibrating at 30, 55, 65, and 75 Hz. However, in the presence of micromolar doses of soluble *N*-acetylated sugars or mucin the frequency response shifts to lower frequencies of 5, 15, 30, and 40 Hz. These lower frequencies match those recorded from swimming prey, such as brine shrimp (*Artemia salina*) nauplii. Thus, chemoreceptors for *N*-acetylated sugars tune cnidocyte mechanoreceptors to frequencies matching the movements of swimming prey. This is the only known case of a frequency-tunable mechanoreceptor, and the first report that cnidae preferentially discharge in response to vibrating targets.

All cnidocyte mechanoreceptors consist of a hair cell–like central kinocilium surrounded by a row of stereocilia. However, unlike hair cells of the acousticolateralis system of vertebrates, which are both structurally and functionally static, the stereocilia associated with the cnidocyte mechanoreceptors in *H. luciae* lengthen by as much as 70% upon exposure to sensitizing doses of mucin or *N*-acetylated sugars, in addition to tuning to lower frequencies. Furthermore, the stereocilia on the Portuguese man-of-war cnidocyte form extracellular bridges with the cnidocil in the presence of fish mucin. Thus, chemoreceptor-mediated morphological changes have been reported to occur in cnidocyte mechanoreceptors, and in sea anemones such morphological changes are associated with changes in frequency responsiveness.

Cnidocyte-supporting cell complexes. Classically, the independent effector hypothesis as it applies to cnidocytes has held that all cnidocytes act as both sensory receptors and effectors of discharge. Recent findings are critical of this view. This evidence includes the localization of specific chemosensitizing receptors on supporting cells adjacent to cnidocytes and the correlation of surface binding of ligand on supporting cells to cnidocyte responsiveness, plus the existence of functionally dynamic vibration-sensitive mechanoreceptors whose morphodynamic stereocilia originate on the supporting cells. It is a convincing indication that in sea anemones, at least, the functional and structural unit of discharge is not the cnidocyte but the cnidocyte-supporting cell complex.

With respect to the discharge of nematocysts from sea anemone fishing tentacles, three types of cnidocyte-supporting cell complexes can be identified, denoted as types A, B, and C. Type A complexes represent those that are triggered by targets vibrating at specific frequencies and that possess chemoreceptors that shift the frequency-response of discharge toward lower frequencies. Such chemoreceptors, however, are not obligatory for discharge. Thus, selectivity of discharge is provided by a combination of the frequency and amplitude of the vibrational stimulus plus the chemical stimulus, which functions to shift the optimal frequencies for discharge. Type B complexes are triggered only by stationary probes in the presence of a chemical sensitizer. Selectivity of discharge is, therefore, provided solely by the nature and dosage of the chemical stimulus. Type C complexes are triggered by stationary probes in the absence of any sensitizer, presumably do not possess chemoreceptors, and are not selective in their requirements for discharge.

Dose–response curves of type A and type B complexes to *N*-acetylated sugars, such as *N*-acetylneuraminic acid, differ in two respects. (1) The doses of *N*-acetylneuraminic acid that half-maximally sensitize are significantly lower at each of the optimal vibrational frequencies tested than for nonvibrating targets. Specifically, the doses that cause half-maximal sensitization at 40, 30, 15, 5, and 0 Hz (no vibrations) are approximately 10^{-15}, 10^{-13}, 10^{-12}, 10^{-10}, and 10^{-8} molar *N*-acetylneuraminic acid, respectively. Thus, the sensitivities of type A complexes to *N*-acetylneuraminic acid are frequency-dependent and, in all cases, higher than the sensitivities of type B complexes; that is, the doses that cause half-maximal sensitization are lower. The higher sensitivity of type A complexes to *N*-acetylneuraminic acid may enable type A complexes to detect *N*-acetylneuraminic acid earlier than do type B complexes and, therefore, to elongate the stereocilia before arrival of the prey. (2) The dose–response curves of type A complexes are sigmoidal rather than biphasic, as is typical of type B complexes. Since desensitization occurs at high agonist concentrations in type B complexes, the response of type B complexes appears to be self-limiting, whereas desensitization appears to be lacking in type A complexes. Since captured prey will eventually stop moving because of envenomation by penetrant cnidae, that is, nematocysts, there may be no need for type A complexes to desensitize since their discharge will cease once the prey is fully subdued.

For background information *see* CHEMORECEPTION; COELENTERATA in the McGraw-Hill Encyclopedia of Science & Technology.

David A. Hessinger

Bibliography. D. A. Hessinger and H. M. Lenhoff (eds.), *The Biology of Nematocysts*, 1988; G. U. Thorington and D. A. Hessinger, Control of cnida discharge: III. Spirocysts are regulated by three classes of chemoreceptors, *Biol. Bull.*, 178:74–83, 1990; G. M. Watson and D. A. Hessinger, Cnidocyte mechanoreceptors are tuned to the movements of swimming prey by chemoreceptors, *Science*, 243:1589–1591, 1989; G. M. Watson and D. A. Hessinger, Cnidocytes and adjacent supporting cells from receptor-effector complexes in anemone tentacles, *Tiss. Cell*, 21:17–21, 1989.

Chlorophyll

Chlorophylls are the green pigments that serve as the photoreceptors of light energy in photosynthesis. Recent studies have elucidated the role of light regulation in pigment synthesis in plants and have characterized the light-harvesting chlorophyll complexes.

LIGHT REGULATION IN PIGMENT SYNTHESIS

Light plays a critical role in the coordinated regulation of the complex photosynthetic machinery of a phototrophic cell, one that utilizes light as its primary source of energy. In addition to participating in the photochemical reaction of photosynthesis, light has been shown to regulate the enzyme activities that control the synthesis of the chlorophyll that acts as the photoreceptor molecule in that reaction. Recent advances in the understanding of the role of light in controlling pigment synthesis have centered on the involvement of light in regulating the biosynthesis of the basic building block of chlorophyll, a five-carbon molecule known as δ-aminolevulinic acid (ALA). Investigators have revealed a previously unknown mechanism for the formation of ALA itself, the involvement of transfer ribonucleic acid (tRNA) in this enzymatic conversion, and the role of light in determining the level of activity of the ALA-forming enzymes.

Aminolevulinic acid. It is well established that ALA is the first committed biosynthetic precursor of chlorophylls, hemes, and bilins, and that the enzymatic pathway from ALA to all three of these tetrapyrrole pigments via protoporphyrin IX is a single branched pathway (**Fig. 1**). At protoporphyrin IX, a critical division occurs. If magnesium is inserted into the ring structure, a distinct branch of the pathway begins, which leads to the formation of chlorophylls and bacteriochlorophylls. If the porphyrin is chelated with iron, hemes and bilins will be formed. The pathway from ALA to the tetrapyrroles is common to plants and animals.

In contrast, it is now clear that there are two distinct mechanisms for the formation of ALA, the precursor to this tetrapyrrole pathway, one essentially a "plant" route and the other essentially an "animal" route. The condensation of glycine and succinyl coenzyme A (CoA), catalyzed by the enzyme ALA synthase, has long been known to exist, and it was assumed to be the universal pathway for formation of ALA. This mechanism occurs in the mitochondria of animal cells, yeast, fungi, and some bacterial species. A second pathway, occurring in an unexpectedly large number of species, specifically in higher plants, red and green algae, cyanobacteria, and some other bacterial groups (including *Escherichia coli*), utilizes the carbon skeleton of glutamate to form ALA, and ultimately chlorophylls, bilins, and hemes. Although evidence for this 5-carbon pathway was presented more than 15 years ago, its existence and wide distribution in nature are not generally known. An interesting discovery about this pathway is the requirement for glutamate-specific tRNA in what appears to be the first step of this enzymatic conversion (**Fig. 2**). In all species studied so far, glutamate-specific tRNA is required for ALA formation from glutamate. In addition to its glutamate specificity, the tRNA, which often carries the chemical genetic code for glutamate (UUC) in a highly modified state, is organelle-specific in eukaryotic organisms, being of chloroplast origin.

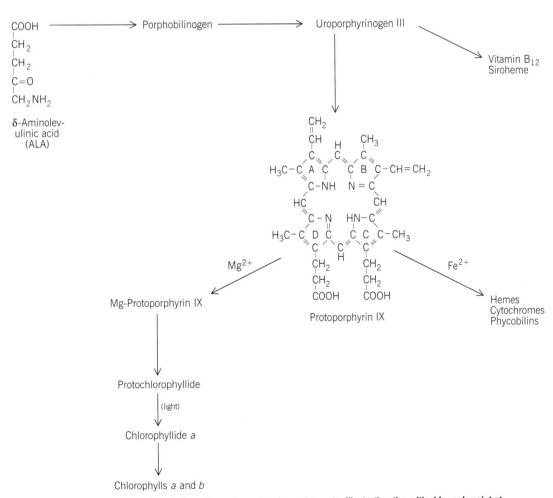

Fig. 1. Proposed pathway for chlorophyll and heme formation from glutamate, illustrating the critical branch point at protoporphyrin IX.

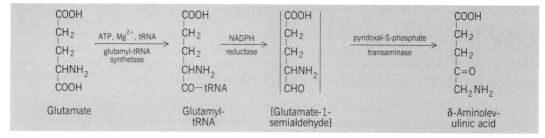

Fig. 2. Conversion of glutamate to δ-aminolevulinic acid (ALA). In the first step, in which adenosinetriphosphate (ATP) and magnesium ion (Mg^{2+}) are required, glutamate is ligated to tRNA^{Glu} by glutamyl-tRNA synthetase. The activated glutamate is reduced by the NADPH-linked reductase, and the reduction product is rearranged by an aminotransferase to form ALA. Glutamate-1-semialdehyde is in brackets because the exact structure of the intermediate has not been determined; it may be cyclized.

Enzyme activity. Levels of activity of the enzymes and, in some cases, the tRNA involved in formation of ALA from glutamate are profoundly affected by the presence or absence of light, the intensity of light, the spectral quality of the light, and the amount of time of light exposure. A convenient model for the study of these effects on the 5-carbon pathway is *Euglena gracilis*, a unicellular green photosynthetic phytoflagellate. *Euglena* has the apparently unique characteristic of possessing both of the ALA-forming pathways. Mitochondrial heme *a* is formed from ALA synthesized from glycine and succinyl CoA, and chloroplast tetrapyrrole pigments are synthesized from ALA formed from glutamate. *Euglena* is useful because the mechanism of the glutamate-to-ALA reaction appears to be nearly identical in algae and higher plants, viable plastidless mutants of *Euglena* are readily available for control experiments, light is required for chlorophyll formation in *Euglena*, and cell material is much more quickly obtained from *Euglena* grown in liquid culture than from plants. *Euglena* also shares with plants the subcellular compartmentalization of the chloroplast, which is the site of ALA synthesis from glutamate.

The unfractionated enzyme system that converts glutamate to ALA has been shown to be dependent on light for induction in *Euglena*, barley, maize, cucumber, some strains of *Chlorella*, as well as other species. Organisms grown in the dark have only a basal amount of ALA-synthesizing activity—in some cases, undetectable amounts. When dark-germinated plants or dark-adapted algae are exposed to white light, there is a lag phase, which may last several hours, after which rapid synthesis of chlorophyll occurs; this is known as the greening period. It has been proposed that it is during the lag phase of greening that the ALA-synthesizing enzymes are formed. Results of experiments involving exposure of algae or leaves to cycloheximide, an inhibitor of protein synthesis, support that proposal.

Studies of the greening period in wild-type *Euglena* cells have shown that at 12, 24, and even 48 h after the beginning of illumination the chlorophyll content of the cells is still not equal to that of cells that have been constantly maintained in the light. The dark-adapted cells require 72 h of illumination to reach chlorophyll content levels comparable to that of the light-adapted *Euglena*, although 24–36 h seems to be sufficient for some higher plants. The same

study was done with a white *Euglena* mutant that appears to have no chloroplast elements, although it is not certain that no remnants exist. This mutant does not form chlorophyll or ALA via glutamate in the cell because the required chloroplast tRNA is absent, but the ALA-forming activity of cell extracts with the tRNA added exhibits the same time-dependent pattern of light induction as the chlorophyll content of the wild-type cells.

Photoinduction. The intensity of light required for the photoinduction of enzyme activity in wild-type *Euglena* is quite low. ALA-forming activity essentially equal to that obtained from cells grown under bright light can be induced even at a much lower illumination. Even when the intensity of illumination is reduced further, to a small fraction of bright light, enzyme activity is still induced but not to the same level. Chlorophyll accumulation follows the same pattern of dependence on light intensity as the enzyme activity, with the response being sensitive to light of the same low intensity.

Manipulation of the spectral quality of light under which the *Euglena* are grown has yielded results that give some information about the photoreceptors involved in light induction of ALA-forming enzymes. Wild-type cells and white mutant cells that have been dark-adapted and subsequently grown in white, blue, red, or blue + red light have very different levels of ALA-forming enzymes. In wild-type cells, blue light alone and, to a lesser extent, red light can set in motion the events that lead to ALA formation. However, white light or blue + red light is required to achieve full induction of the enzymes, and they are equal in their ability to achieve this effect. In contrast, ALA-forming enzymes are present in extracts from the mutant only if the cells have been grown in white or blue light; red light has no inducing effect.

Euglena has two photoreceptor systems, the red/blue light system based on a protochlorophyll pigment, and the blue light system, which is incompletely described and based on a yet-uncharacterized receptor. It is possible that the blue photoreceptor system, which is cytoplasmic and would be operative in both types of *Euglena*, controls induction of the enzymes, which are encoded in the nucleus. It must be noted, however, that red light can induce the system to some degree in wild-type cells—this may be a direct or indirect effect. The observation that cycloheximide treatment prevents the synthesis of ALA-forming enzymes lends support to the theory that the genes are localized in the nucleus, because cycloheximide specifically inhibits protein synthesis on cytoplasmic ribosomes from nuclear-encoded genes. It can also be proposed that the red/blue photoreceptor, which is chloroplast-localized and would not be present in a plastidless mutant, induces chloroplast-encoded components. A likely component for the chloroplast-localized regulation and synthesis is the glutamyl tRNA involved in the ALA-forming reaction, since it has been shown to be of chloroplast origin and has not been found in extracts from white mutants without detectable chloroplasts.

The actual regulatory mechanism is likely to be more complex. Although some biological events can be identified as being controlled by one or the other photoreceptor system, coordinate control by both systems is probably very common. Syntheses occurring within the chloroplast are necessarily coordinated with those taking place in the cytoplasm, especially during chloroplast development in the early stages of light exposure.

Light regulation of reaction components. The three known enzyme components (synthetase, reductase, and transaminase) and the tRNA can be separated and isolated from cell extracts. Serial affinity chromatography based on the known cofactor requirements of the particular enzyme (Fig. 2) and ion-exchange chromatography based on the ionic properties of the molecules are used to separate the activities. The activity levels of the RNA and the separated enzymes are quite different for organisms maintained in the dark, maintained in the dark and then grown in the light, or maintained in the light.

The level of the specific tRNA required for ALA formation from glutamate has been shown to be down-regulated by about two-thirds in dark-grown *Euglena* cells versus light-grown cells. This is in contrast to the results from studies in *Chlorella* and barley enzyme systems, in which the required tRNA does not increase during the greening period. In addition, tRNA from a white *Euglena* apparently has no trace of ability to participate in the enzymatic reaction. This lends further support to the likelihood that the tRNA is chloroplast-encoded.

The level of activity of glutamyl-tRNA synthetase that effects the first step of the reaction (Fig. 2), ligation of glutamate and tRNA, is also affected by the presence or absence of light during growth. Aminoacyl-tRNA synthetases are very specific for both the tRNA and the amino acid that they ligate. In plant cells with subcellular compartments such as mitochondria and chloroplasts, there exist different species of a specific aminoacyl-tRNA synthetase, dependent on the subcellular location, including the cytoplasm. Total glutamyl-tRNA synthetase activity is twice as high in extracts of light-grown cells as in dark-grown cells. However, the RNA charged by either enzyme is equally suitable as the substrate for ALA synthesis; this suggests that light does not affect the ability of the enzyme to ligate, but only the total activity present in the cell.

The second step of the reaction, reduction of glutamate to glutamate-1-semialdehyde (GSA), is facilitated by a reductase dependent on nicotinamide adenine dinucleotide phosphate (reduced form; NADPH); this reductase increases at least fourfold when dark-grown cells are transferred to the light.

The final step of this enzymatic conversion, transamination of GSA to form the product ALA, is the step most profoundly affected by the presence or absence of light during growth. In extracts of *Euglena* cells grown in the dark, this enzyme is undetectable, whereas growth in the light results in extracts with extremely high transaminase activity levels. Isolated barley chloroplasts exhibit the same light response specifically for induction of the transaminase, as well as for the total enzyme system. Experiments with barley leaves using an inhibitor that blocks the aminotransferase show that steps prior to the transamination are also light-regulated, but the regulated enzymes have not been identified.

Photoregulation models. Very little information exists about the precise mechanisms of light regulation of ALA synthesis. A number of models exist to explain photoregulation, based on light influences on gene transcription and/or translation; chloroplast biogenesis; posttranslational modification; transport of proteins (and other elements) across the chloroplast membrane; feedback inhibition; and substrate limitation, where the limiting substrate is glutamyl-tRNA.

Light influences on the expression of plant genes may be exerted directly or through intermediate molecules, and the effectors for photocontrol of expression of nuclear genes may be in either the chloroplast or the cytoplasm, or both. The same is true for those of chloroplast genes. Many plant genes have been determined to be under light control, and the particular gene sequences that respond to light have been identified and located in the regulatory regions of the genes. None of these genes has been identified as coding for components of the 5-carbon ALA-synthesizing pathway, however.

Because heme has been shown to be a potent inhibitor of glutamate-to-ALA conversion, an attractive model for light regulation has been proposed that involves feedback inhibition by heme. In this scheme, heme builds up in the dark as an indirect result of the lack of reduction of protochlorophyllide by light. Heme then prevents ligation of tRNA and glutamate, and possibly reduction of glutamyl-tRNA, and ALA synthesis stops. When light exposure takes place, protochlorophyllide is reduced, the heme supply is lowered, and inhibition of the ALA pathway is relieved.   *Sandra M. Mayer*

LIGHT-HARVESTING COMPLEXES

The characteristic green color of higher plants and algae is due to chlorophyll molecules in the internal membranes of subcellular organelles (chloroplasts). All the chlorophyll molecules are attached to proteins in the membrane. This concept is a change from the earlier view that since isolated chlorophyll is a hydrophobic molecule (that is, it preferentially dissolves in nonpolar solvents rather than in polar solvents such as water) it could be located only in the lipid bilayer

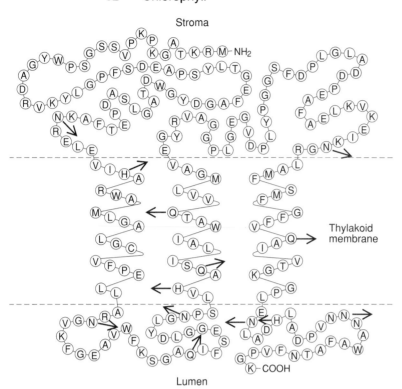

Fig. 3. A proposed model for the folding of one of the polypeptides of the light-harvesting complex. The chlorophyll molecules are proposed to be coordinated to the polypeptide chain via liganding to the conserved histidine (H), glutamine (Q), and asparagine (N) residues, indicated by arrows. The other letters represent other amino acid residues as identified by the standard one-letter code. In this model, relatively hydrophobic sequences form transmembrane helices, connected by relatively hydrophilic stretches on the outside of the photosynthetic membrane (stromal side) and the inside (lumenal side). (*After S. E. Stevens, Jr., and D. A. Bryant, eds., Light-energy transduction in photosynthesis: Higher plant and bacterial models, Proceedings of the 3d Annual Penn State Symposium in Plant Physiology, American Society of Plant Physiologists, 1989*)

portion of the membrane. This view did not take into account that transmembrane proteins located in the bilayer have areas that are relatively nonpolar because of a clustering of amino acids bearing nonpolar side chains, and that chlorophyll could, in fact, be associated with proteins.

Roles of chlorophyll–protein complexes. Since the microenvironment of a molecule affects its chemical properties, the function and properties of a chlorophyll molecule depend, at least partly, on the nature of the apoprotein portion of a chlorophyll–protein complex. Differences in the absorption spectrum of chlorophyll caused by the apoprotein help to identify different chlorophyll–protein complexes.

In the chloroplast membrane of green plants and algae are two different types of photosynthetic particles, photosystem I (PS I) and photosystem II (PS II), which play different roles in the whole photosynthetic process in these organisms. Both types of particle have a two-part structure: a reaction center where the primary events of photosynthesis occur, and antennal complexes that gather light energy and transfer it to the reaction center. Chlorophyll molecules attached to their apoproteins are central to both functions.

Photochemical function in green plants. In reaction centers, absorption of light energy by a chlorophyll molecule causes its oxidation (loss of an electron), which initiates the further steps of photochemistry.

Chemically, these molecules are chlorophyll a, and while they are central to the process of photosynthesis, they represent only a small fraction of the total chlorophyll in the plant.

Antennal function in green plants. Most of the chlorophyll in the green plant functions as an energy-gathering apparatus for chlorophylls of the reaction center, in so-called antennal complexes. Some antennal complexes have chlorophyll a, and different complexes have both chlorophylls a and b. Chemically, the two types of chlorophyll differ only in the exchange of a methyl (CH_3) side chain in chlorophyll a for a formyl ($HC=O$) side chain in chlorophyll b. The protein environment of both kinds of chlorophyll molecules in antennal complexes is such that after light absorption by either molecule the molecules lose excitation energy, not by oxidation (as in the case of reaction-center chlorophylls) but by a process known as Förster resonance transfer. In this process, the excitation energy is transferred to another chlorophyll in the environment, the efficiency of the transfer depending on the closeness and orientation of the two chlorophyll molecules. The transfer can be repeated, and the excitation is said to migrate around the pigment bed. The energy-gathering function is important mainly in dim light, where it is possible that not every reaction-center chlorophyll will absorb a photon. However, since each reaction center is surrounded by a large number of chlorophyll molecules located in antennal complexes, statistically there is a good chance that the energy absorbed by any of the chlorophyll molecules in the surrounding complex will ultimately be transferred to the reaction-center chlorophylls. In this case, the reaction-center chlorophyll will be in the same state as if it had absorbed a photon, and it can lose an electron to start the photochemical reactions.

Structure of the antennal complex. In order to characterize the structure of the antennal complex, it is necessary to isolate and study the chlorophyll–protein complexes.

Isolation and structure of chlorophyll–protein complexes. A large number of chlorophyll-binding protein complexes have been isolated from the photosynthetic membranes of plants and green algae by the judicious use of mild detergents to remove the proteins from surrounding lipids, without totally removing the chlorophyll from the apoprotein. A solubilized preparation can be further purified by various chromatographic and electrophoretic techniques. The carefully isolated proteins are green because of attached chlorophyll. The carotenoids in the photosynthetic membranes are also associated with the chlorophyll–protein complexes.

No antennal chlorophyll–protein from a plant membrane has yet been completely structurally characterized by x-ray crystallography, and current models of the shape of the protein molecules are based mostly on amino acid sequences. In the apoproteins of chlorophyll–protein complexes, as in membrane proteins in general, there are stretches of amino acids that are rich in hydrophobic amino acids; these seem to be located mostly in the lipid bilayer. It is believed that the pigment molecules, both chlorophyll and carotenoids, are located here. Little information about how the pigment molecules are actually bound by the protein is available, although it is suggested that there is liganding to residues of histidine, glutamine, and asparagine (**Fig. 3**).

Classes of chlorophyll–protein complexes. Most of the chlorophyll in the plant (more than 50% of total chlorophyll *a* plus chlorophyll *b*, depending on species and growth conditions) and all the chlorophyll *b* is bound to a structurally related group of proteins known as chlorophyll *a + b*–binding (Cab) proteins. The Cab proteins are the products of a nuclear gene family (structurally related genes), and they are synthesized on ribosomes in the cytoplasm as precursor proteins somewhat larger than their final size. The extra length of protein interacts with a binding site on the outer chloroplast membrane. Once bound, the protein is simultaneously inserted into the chloroplast across the outer membranes and trimmed to its final size. It appears to insert itself into the photosynthetic membrane spontaneously.

A second class of chlorophyll-binding antennal proteins differs from the Cab proteins in that only chlorophyll *a*, not *b*, is bound and that the apoproteins are encoded on chloroplast genes. There are no current exceptions to the observation made in higher plants that chlorophyll-binding proteins encoded in the nucleus bind both chlorophyll *a* and chlorophyll *b*, and chloroplast-coded chlorophyll-binding proteins bind only chlorophyll *a*.

Organization of chlorophyll–proteins into antennal complexes. Photosystem I is associated with several Cab-type proteins that are collectively referred to as the light-harvesting complex I (LHC I). LHC I has 4 polypeptides whose molecular masses are between 20,000 and 24,000 daltons, each of which bind about 10 chlorophyll *a* and *b* molecules, although the exact number of chlorophyll molecules per protein is not certain. It is not known how the the individual chlorophyll–protein complexes are assembled into the antenna in the native membrane. These complexes are easily stripped from the reaction-center core by using mild detergents. The reaction-center complex, or core of PS I, is composed of two large (molecular masses 82,000 and 83,000 daltons) polypeptides that between them appear to bind some of the prosthetic groups that carry on electron transport of PS I, including the special chlorophyll molecules that lose one electron each. However, both of these large proteins also bind carotenoids and about 50 antennal molecules of chlorophyll *a* each. Thus, a portion of these large proteins serves as the reaction-center core complex, while other parts function as internal antennal complexes.

The antennal complexes of PS II are more complex than those of PS I. Just as in the case of PS I, PS II is associated with Cab proteins known as the light-harvesting complex of PS II (LHC II), the most abundant type of chlorophyll–protein complex in the chloroplasts of higher plants. The major proteins have molecular masses of about 25,000–28,000 daltons, each bearing about 13 or 14 chlorophyll molecules and about 3 xanthophyll molecules. In the membrane the Cab proteins of LHC II appear in electron micro-

graphs as particles 8 nanometers in diameter, although it is not known how many individual chlorophyll–proteins assemble with each other in each particle. These particles can dissociate from the rest of the PS II particle, thus decreasing the efficiency of the antenna. The peripheral complex LHC II is considered to play a role in regulating the flow of energy to the PS II reaction center.

Biochemical studies on isolated membranes have shown that it is possible to strip off the peripheral antenna LHC II without affecting electron transport functions (**Fig. 4**). Further detergent treatment of the particle removes at least two more classes of Cab proteins that are not part of the peripheral mobile antenna, and leaves behind a particle consisting of the reaction-center complex plus two inner antennal complexes, CP47 and CP43, which have chloroplast-encoded proteins (molecular masses 51,000 and 44,000 daltons, respectively). These latter complexes appear to be functional antennae, since mutants that lack chlorophyll *b* (and thus have no functional LHC II) photosynthesize well with only CP47 and CP43 as antennal complexes.

These chloroplast-encoded antennal complexes can also be removed from the reaction center proper by further treatment, but in nature they appear to stabilize the reaction-center structure. This is based on evidence from the Cyanobacteria, which lack antennae based on Cab chlorophyll–proteins like LHC I or LHC

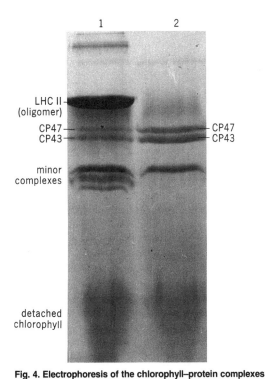

Fig. 4. Electrophoresis of the chlorophyll–protein complexes in sequentially extracted photosystem II (PS II) particles. Lane 1 shows an array of chlorophyll *a + b*–proteins as well as the two chlorophyll *a* complexes CP47 and CP43. In lane 2, treatment with a detergent results in removal of the chlorophyll–proteins of the LHC II peripheral antenna. The chlorophyll complex of the reaction center of PS II has lost its chlorophyll during electrophoresis and so is not visible in this unstained gel. The dark area at the bottom of the gel is chlorophyll that became detached from its apoprotein during electrophoresis.

II; instead they have a complex antennal system based on phycobiliproteins. However, the PS II core complex from these prokaryotes is very similar in structure and function to the core complexes in green plants, and it is often used as a model. Cyanobacterial mutants that lack the genes corresponding to CP47 or CP43 are unable to accumulate stable forms of the reaction center, suggesting that the chloroplast-coded chlorophyll antennal complexes have a structural as well as light-harvesting role.

In the PS II particles of higher plants, the reaction-center proteins bind the prosthetic groups needed for photosynthetic activity; however, in contrast to PS I, they probably do not contain any antennal chlorophyll molecules. Both PS I and PS II can be seen to have a basically common pattern of reaction-center chlorophyll molecules served by a chlorophyll *a* chloroplast-encoded internal antenna complex, surrounded by a more peripheral nuclear-encoded chlorophyll $a + b$ antennal complex.

Other photosynthetic organisms. Photosynthesis based on chlorophyll occurs in a large variety of organisms, and while the exact chemical structure of the chlorophyll molecule varies among photosynthetic organisms (including chlorophylls *a* and *b* in green plants and green algae, chlorophylls *a* and *c* in chromophyte algae, bacteriochlorophylls *a* and *b* in purple photosynthetic bacteria, and bacteriochlorophylls *c, d*, and *e* in green photosynthetic bacteria), in all cases examined so far photosynthetic chlorophyll molecules are attached to special chlorophyll-binding protein complexes. Further, all known photosynthetic organisms have the two-stage system of reaction centers and antennal complexes. In photosynthetic eukaryotes, the chloroplast-encoded components of reaction-center polypeptides and antennal complexes containing chlorophyll *a* are generally similar in amino acid sequence. However, the more peripheral antennal complexes vary somewhat among different evolutionary groups. For example, the proteins of the antennal complexes of chromophyte algae are only slightly related to LHC I and LHC II, and the phycobiliproteins of the rhodophyte algae are quite dissimilar.

Photosynthetic prokaryotes are also very diverse with regard to antennal structure. Other than in the case of the Cyanobacteria, which share similar chlorophyll *a* antennae with green plants, there are great contrasts in the arrangement of pigments and proteins with antennal complexes of green plants. Nonetheless, all follow the general pattern of reaction-center and antennal complexes, in which proteins complexed with the various kinds of chlorophyll play major roles.

For background information *SEE* CELL PLASTIDS; CHLOROPHYLL; CYTOCHROME; ENZYME; PHOTOSYNTHESIS in the McGraw-Hill Encyclopedia of Science & Technology.

Edith Camm

Bibliography. S. I. Beale and P. A. Castelfranco, The biosynthesis of δ-aminolevulinic acid in higher plants, II. Formation of ^{14}C-δ-aminolevulinic acid from labeled precursors in greening plant tissues, *Plant Physiol.*, 53:297–303 1974; S. I. Beale and J. D. Weinstein, Tetrapyrrole metabolism in photosynthetic organisms, in H. A. Dailey (ed.), *Biosynthesis of Heme and Chlorophylls*, 1989; B. R. Green, The chlorophyll–protein complexes of higher plant photosynthetic membranes or Just what green band is that?, *Photosyn. Res.*, 15:3–22, 1988; S. M. Mayer and S. I. Beale, Light regulation of δ-aminolevulinic acid biosynthetic enzymes and tRNA in *Euglena gracilis, Plant Physiol.*, 94:1365–1375, 1990; P. B. Moses and N.-H. Chua, Light switches for plant genes, *Sci. Amer.*, pp. 88–93, April 1988; M. A. Schneegurt, S. Rieble, and S. I. Beale, $tRNA^{Glu(UUC)}$ of plants and algae: HPLC separation into two subfractions, and determination of their relative effectiveness in homologous in vitro translation and ALA-forming systems, *Plant Physiol.*, 86S:61. 1988; S. E. Stevens, Jr., and D. A. Bryant (eds.), Light-energy transduction in photosynthesis: Higher plant and bacterial models, *Proceedings of the 3d Annual Penn State Symposium in Plant Physiology*, American Society of Plant Physiologists, 1989.

Circuit (electricity)

The understanding of the phenomenon now known as chaos has developed rapidly since the mid-1970s, chiefly as a result of the advent of widespread improved computing resources. Chaos is a universal phenomenon that can appear in systems (of which electronic and electrical systems form a very small subset) that are sufficiently nonlinear and are described by continuous variables.

In the case of electrical and electronic circuits, nonlinearity may be due to the following factors: the finite nature of power supplies; nonlinear circuit elements such as diodes, varicap diodes, saturating inductors, and solid-state and vacuum-tube active devices; arbitration circuits (comparators that drive switches, and arrival time arbiters); phase ambiguity in phase detectors; and the finite memory capacity of computer registers and memory. The other condition for chaos is fulfilled because all electrical and electronic circuits operate with continuous variables of time, current, and voltage. This is true even for so-called digital systems, in which arbitration is usually inherent.

Simple chaotic system. A good way to introduce the phenomenon is to describe a simple system that has been built to demonstrate chaotic behavior. The system consists of two parts.

The first part is an amplifier with the gain characteristic shown in **Fig. 1.** For sufficiently positive input voltages, the gain (the slope of the curve) is negative with absolute value greater than unity. At a lower positive input threshold, the gain switches suddenly but continuously to a positive value greater than unity, and then remains at this level until the input and output both fall to near zero. Then the negative gain characteristic is the inverse of the positive characteristic.

The second part of the system consists of a clocked two-stage shift-register feedback circuit, isolating the input from the output, which captures the amplifier

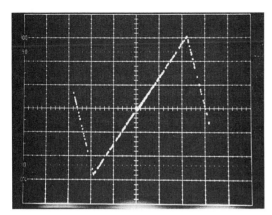

Fig. 1. Chaotic outline of the transfer characteristic described in the text. Horizontal axis is input voltage; vertical axis is output voltage (same scale); center of graticule is the point (0,0).

output voltage as an analog value, and then transfers it to the amplifier input at the next clock cycle, at the same time acquiring the new output voltage.

To understand the behavior of the system, it is useful to follow it after an initial input voltage has been selected and the clock has been started. It will be assumed that the positive gain is set to 1.3, the negative gain slopes are set to -8, the thresholds for inversion of gain are at input signals of ± 5 V, and the speed of the clock is such that dynamic effects due to the amplifier frequency response and shift-register droop can be neglected. Starting at an input voltage of 1 V, the output will be 1.3 V, which is transferred to the input after the next clock edge, and gives rise to an output of 1.69 V. The following sequence of output voltages results on the first seven successive clock edges: 1.3, 1.69, 2.197, 2.8561, 3.71293, 4.826809, 6.2748517. The last value is bigger than the gain threshold of 5 V, and so the next output voltage is $(5 \cdot 1.3) - (6.2748517 - 5) \cdot 8$, which at -3.6988136 is again within the range of positive gain. The next output voltage values are -4.80845768 and -6.250994984, which causes an overflow at the negative threshold, giving the next value at $+3.507959872$ V. The output then sequences chaotically forever, and most probably never repeats previous values. The continuous nature of the voltage variable is important; any slight error introduced, for example, by noise, will grow exponentially with increasing numbers of clock cycles since the gain modulus is everywhere greater than unity.

Properties of chaotic systems. There are two fundamentally important properties of this sequence. First, output values depend critically on the initial voltage; for example, given an error of 1 mV in the initial voltage, the final error produced after 11 clock cycles is $1.3^7 \cdot 8 \cdot 1.3^2 \cdot 8$, or about 678 mV. This is an exponential increase in error (with so-called Liapunov exponent). Second, it is not possible to deduce the starting input voltage from a knowledge of the output voltage, for although the transfer characteristic is unique in that an input in the range defines an output uniquely, the converse is not true, since for each output in the range ± 6.5 V, there are three possible

input values to choose from and no way of making the decision. These two properties are loosely referred to as stretching (sensitivity to initial conditions) and folding (noninvertibility of the so-called return map). They cause chaotic mixing, analogous to the process of kneading dough.

Time sequences of voltages in such a system built experimentally are shown in **Fig. 2.** This system has the additional feature that the gain is reduced (over a very small region near zero) to a value less than one, resulting in a trap in the mapping. When the chaotic sequence falls into the trapping region, successive output voltages decay exponentially to zero, resulting in trapping at zero, as is seen in Fig. 2a and b. If set off repeatedly, the system takes a random time to trap because the starting values cannot be set to exactly the same voltage with indefinite precision. Figure 1 is an oscilloscope photograph of a particular chaotic sequence of points (which can never be exactly repeated) plotted as output against input voltage. This kind of analysis provides the basis for an experimental instrument known as a return map plotter, which can be used to investigate apparently random or noiselike sequences of voltages for short-range correlation.

In the example described above, as in most chaotic systems, it is possible to show that however long the system is left to run, there will be many values of output voltage that are never visited. An infinite sequence of voltages may be selected from the range 0 to 1, for example, by choosing the values 1, ½, ¼, ⅛, . . . ; and if this set formed a chaotic sequence, there would be many other nonoverlapping infinite sets that

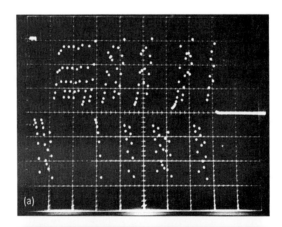

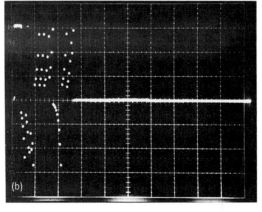

Fig. 2. Two experimental time sequences of chaotic output voltages corresponding to Fig. 1. Horizontal axis is time; vertical axis is output voltage. The starting voltage is the same in each case. Random times to trapping: (*a*) 175 clock cycles, (*b*) 54 clock cycles.

could be chosen from the range. It is a property of chaotic systems that the sequences need never repeat, yet approach so-called attracting sets that are mathematically said to be of measure zero within the variable range. Many such chaotic attractors have the structure of a mathematical construction known as a Cantor set. Such sets are termed attracting because a system that is perturbed away from the sparse infinite chaotic set will fall back onto it as time progresses.

This property has application in engineering. While information is lost about where the system is within its attracting set, it is known that wherever the system starts from, it will approach this set as time progresses. Even if the system is perturbed by other stimuli on its way to the attracting set, it will resume its course provided such stimuli do not drive the system beyond its so-called basin of attraction. The attracting set of values may be a property of use in engineering when precise knowledge of the location within that set is unnecessary.

Important electrical chaotic systems. Systems of interest in electrical engineering that are known to behave chaotically include diode-inductor circuits, class C and E power amplifiers, multivibrator circuits, fluorescent lamps, ferroresonant circuits, phase-locked loops, digital filters, neural networks, communications protocols, and nonlinear piezoelectric acoustic devices. In general, when sufficient nonlinearity is combined with feedback, chaotic behavior is a possibility. It is commonplace in systems that are modeled as nonlinear differential equations of second or higher order, driven by periodic forcing functions.

Applications of chaotic systems. Such properties may possibly be put to use in pattern recognition systems because of the robustness of the chaotic attractor in the presence of noise and the simplicity of the generating algorithm. It is known that the mammalian heartbeat forms a chaotic sequence in time. It may further be conjectured that there are important applications of chaotic dynamics to control engineering, since a controlled system that is exhibiting chaos may utilize the robustness of the chaotic attractor as a means of coping with nonrandom noiselike stimuli, as the mammalian heart is believed to do.

Sometimes a sequence of values does repeat. In many chaotic systems the chaotic sequences are approached by successive doubling in the lengths of the closed cycles as system parameters are varied, a phenomenon known as period doubling. By using this phenomenon, it is possible to construct an inductor-resistor-diode circuit that divides the frequency of a modulated carrier while keeping the modulation information substantially intact.

Chaotic processes can give rise to problems in radar systems on ships where the naturally occurring diodes and inductances due to the superstructure can lead to chaotic frequency down-conversion (period multiplication). This can cause interference with communication channels from single carriers, even though intermodulation effects, which are due to the interaction of multiple carriers with nonlinearities, are not expected.

The science of chaotic dynamics is still in its infancy. It may take some time for the important engineering applications to become apparent, similar to the state of affairs shortly after the invention of the laser.

For background information *SEE PERIOD DOUBLING* in the McGraw-Hill Encyclopedia of Science & Technology.

David J. Jefferies

Bibliography. P. Cvitanovic (ed.), *Universality in Chaos*, 1984; J. H. B. Deane and D. J. Jefferies, Periodic, intermittent and chaotic states of clocked autonomous digital systems, *Int. J. Circ. Theory Appl.*, 18:175–187, 1990; D. J. Jefferies, J. H. B. Deane, and G. G. Johnstone, An introduction to chaos, *Electr. Commun. Eng. J.*, 1(3):115–123, 1989; Special issue on chaos in electronic circuits, *IEEE Trans. Circ. Sys.*, vol. 75, no. 8, August 1987.

Cold hardiness (plant)

Low-temperature tolerance of plants is lowest during active growth, but it increases in many species in response to certain environmental conditions. The most common acclimating conditions are shortening day length, low temperature, and lack of water, all of which are also effective at slowing growth. In many temperate perennials, cold acclimation begins in late summer or early fall as day length and ambient temperatures decline, coinciding with the onset of dormancy. During acclimation, changes can occur at the whole-plant level, the cellular level, and the molecular level. New proteins are produced, while synthesis of others declines or stops. The plant hormone abscisic acid (ABA), applied under nonacclimating conditions, can increase freezing tolerance, and causes some of the same changes in protein synthesis.

Translocatable factors in cold acclimation. If a red-osier dogwood plant is exposed to shortened day length (8 h of light and 16 h of darkness, for example), it will begin to cold-acclimate. If it is defoliated before 4 weeks of the short-day regime have passed, it will not reach its maximum hardiness. When one branch is defoliated and the rest of the plant is exposed to acclimating conditions, the defoliated branch will harden (increase in freezing tolerance) to the same extent as the rest of the plant. Similar observations have been made on several other plant species, demonstrating the active synthesis of a promoter of cold acclimation that can move to sites remote from where it is made. Analogous experiments in which plants and plant parts are exposed to artificial long days have demonstrated the production of an inhibitor to cold acclimation, which also moves within plants.

Paper chromatography of extracts from leaves of cold-acclimating plants yields fractions that inhibit the growth of other plants in bioassays. These fractions behave chromatographically the same as abscisic acid in several different solvent systems. An extract from leaves of cold-acclimated box elder when applied to leaves of a box elder grown under long days (non-cold-acclimated) increases its freezing tolerance. Applying abscisic acid has the same effect as the extract.

Exogenous application of abscisic acid. In addition to box elder, other plants that increase in freezing tolerance in response to abscisic acid treatment include alfalfa, mulberry, *Solanum commersonii* (a wild potato species), barley, wheat, and *Arabidopsis thaliana*. Abscisic acid has also been applied to plant cells in culture. Bromegrass, tobacco, *Brassica napus*, alfalfa, winter wheat, winter rye, and *S. commersonii* suspension culture cells increase their freezing tolerance when grown at temperatures less than about 41°F (5°C) or when grown at room temperature with abscisic acid added to the growth medium. The effective abscisic acid concentrations are low, in the 10–100 micromolar range, so the effect cannot be simply an increase in solute concentration; the abscisic acid must be initiating a response in the plant cells that causes them to become more tolerant of freezing. When abscisic acid is applied to cells in suspension culture of plants that do not have the ability to cold-acclimate, such as carrot, datura, periwinkle, soybean, and vetch, freezing tolerance is not enhanced.

Dependence of endogenous abscisic acid concentration on cold acclimation. The advent of analytical methods for quantification of plant hormones has led to a deeper understanding of the dynamics of synthesis, degradation, and movement of plant hormones in general and abscisic acid in particular. The concentration of abscisic acid in leaves of *S. commersonii* plantlets remains constant at first after the growth temperature is reduced from 68 to 36°F (20 to 2°C). Then it increases threefold, and finally falls back to the initial value. Freezing tolerance of *S. commersonii* begins to increase only after this spike in abscisic acid concentration. Spinach, another plant that can acclimate, also shows this transient increase in abscisic acid concentration.

Protein synthesis during cold acclimation. If protein synthesis in *S. commersonii* is blocked by applying cycloheximide at the beginning of cold acclimation, induced either by low temperature or by exogenous application of abscisic acid, cold acclimation does not occur. If applied 5 days after induction, however, cycloheximide does not block cold acclimation. The ability of inhibitors of protein synthesis to block cold acclimation has been observed in several other plant species as well. Thus, it appears that protein synthesis in the early stages of induction is an essential part of cold acclimation.

New protein synthesis has been observed in response to lowered growth temperature, which also induces cold acclimation, in a number of plants, including spinach, mulberry, *B. napus*, bromegrass, winter rye, *S. commersonii*, alfalfa, and *A. thaliana*. Cessation of synthesis of certain other proteins is also observed in these plants in response to lowered growth temperature. In several of these species, new messenger ribonucleic acids (mRNAs) with translation products that are very similar to the new

proteins are produced at the lower temperature. Whether all or any of these new proteins are essential to cold acclimation is not known. They could be required for growth at the lower temperature, for vernalization, or for dormancy, as well as for cold acclimation. The identities and functions of these proteins are being actively sought. Calling these substances proteins is an oversimplification, since the electrophoresis method that is used separates oligomeric proteins into their individual subunits so that the spots on the gels actually indicate polypeptides.

Protein synthesis in response to abscisic acid application. Because of the growing evidence that abscisic acid is a central factor in cold acclimation, several of the recent studies of new protein synthesis in response to cold-acclimating conditions have included abscisic acid treatment. In suspension cultures of a number of plants, including bromegrass, *B. napus*, *S. commersonii*, *A. thaliana*, spinach, rice, and alfalfa plants, new proteins are also synthesized in response to treatment with abscisic acid. There are new mRNAs for these proteins, suggesting that they are not post-translational modifications of proteins that are also synthesized at room temperature, but that abscisic acid treatment initiates new gene expression. The constellation of new polypeptides produced in response to abscisic acid treatment is different than that produced in response to lowered temperature, but some of the new polypeptides are common to both treatments. These polypeptides are considered prime candidates for involvement in cold acclimation.

Genes responsive to abscisic acid and to water stress. Water stress causes an increase in freezing tolerance in plants capable of cold acclimating, and the increase is usually more rapid than that which occurs in response to shortened day length or lowered temperature. For example, 24 h of exposure to 40% relative humidity causes plumules of winter rye seedlings to acclimate to the same level that they achieve after 14 days at 37°F (3°C). Water stress can also increase the resistance to chilling damage of chilling-sensitive plants.

Although it may seem strange that a connection exists between these environmental stresses, there is sound reason for the relation. When a plant freezes under normal circumstances, ice is initiated on the outer surface and works its way through the tissues, remaining in spaces outside individual cells. Water within the cells is slightly supercooled but still liquid. The vapor pressure of supercooled, liquid water within the cells is higher than the vapor pressure of the ice outside, resulting in an efflux of water from within cells to the surrounding ice. During a freeze, then, plant tissue is subjected to dehydration. The relationship between chilling stress and water stress is more obscure, but there is good evidence that one of the manifestations of chilling stress is desiccation.

A large body of evidence supports a role for abscisic acid in the water-stress response, raising the possibility that the responses to both water stress and low-temperature stress share elements of the same pathways. The progress made in understanding how abscisic acid is involved in water stress has been more rapid than that for how abscisic acid is involved in cold acclimation. Abscisic acid stimulates osmotic adjustment, a process in which the osmotic concentration increases within cells of a water-stressed or salt-stressed plant. The identities have been established of several proteins that are known to be induced by water stress and by abscisic acid, and specific genes have been cloned for others, making possible the determination of the amino acid sequences of the gene products, although the function of the gene products have not yet been established. It is well to keep in mind that there are factors other than abscisic acid involved in the induction of the water-stress response.

Modulation of gene expression by abscisic acid. The molecular basis for the ability of certain genes to respond to changes in concentrations of plant hormones is being actively examined. As an example, a small piece of an abscisic acid–responsive gene from wheat is spliced to the β-glucuronidase reporter gene and inserted into tobacco, where it is expressed in response to exogenous abscisic acid application. It is expressed in the same tissue (embryos) and at the same developmental stage as is the intact gene in wheat. From analysis of various gene constructs, the abscisic acid–responsive section of the regulatory part of the gene has been identified. Whether this regulatory sequence operates in vegetative tissue as it does in embryos, and whether it is the sequence that is involved in regulating genes that are responsive to water stress, are subjects of current research.

For background information SEE ABSCISIC ACID; COLD HARDINESS (PLANT); PHOTOPERIODISM; PLANT-WATER RELATIONS in the McGraw-Hill Encyclopedia of Science & Technology.

John V. Carter

Bibliography. C. L. Guy, Cold acclimation and freezing stress tolerance: Role of protein metabolism, *Annu. Rev. Plant Physiol.*, 41:187–223, 1990; V. Lang, P. Heino, and E. T. Palva, Low temperature acclimation and treatment with exogenous abscisic acid induce common polypeptides in *Arabidopsis thaliana* (L.) Heynh, *Theor. Appl. Genet.*, 77:729–734, 1989; W. R. Marcotte, Jr., S. H. Russell, and R. S. Quatrano, Abscisic acid–responsive sequences from the Em gene of wheat, *Plant Cell*, 1:969–976, 1989; K. Skriver and J. Mundy, Gene expression in response to abscisic acid and osmotic stress, *Plant Cell*, 2:503–512, 1990.

Communications satellite

Recent advances in the area of communications satellites include the development of inexpensive, lightweight satellites (LIGHTSATs) and the application of satellite communications to civil aviation.

LIGHTSATS

Lightweight satellites are a new initiative involving the building of less costly single-purpose, small satel-

Fig. 1. One of the two digital, store-and-forward *Multiple Access Communications Satellites* (*MACSATs*) of the LIGHTSAT program.

lites that are inexpensive to launch. LIGHTSATs are intended to serve functions not adequately served by large satellites; they will create new satellite applications and will make these affordable to tactical military and scientific users. Commercial applications in remote sensor readout and worldwide paging are beginning to emerge.

Development. In 1987 the Defense Advanced Research Project Agency (DARPA) initiated a program to develop inexpensive satellites for tactical military and scientific purposes. The program had two objectives: to make satellites for certain missions more affordable, and to develop satellites to support tactical military operations not adequately supported by large satellites. Since the cost of a satellite is closely related to its weight, low cost implies small size and light weight, or new technology to provide more capability within the same weight limits. Satellites built as part of the DARPA program became known as LIGHTSATs. This name was extended later to encompass other satellites of up to several hundred pounds.

The first DARPA-sponsored small satellite, a precursor of the LIGHTSAT program, was the *Global*

Message Relay (*GLOMR*), a digital store-and-forward data relay satellite built by Defense Systems, Inc. (DSI), and launched in October 1985 from the space shuttle *Challenger*. The first satellites built in the LIGHTSAT program were the two *Multiple Access Communications Satellites* (*MACSATs*), built by DSI and launched on May 9, 1990, into a 400-nautical-mile (740-km) polar orbit by a Scout launch vehicle. After insertion into orbit, the satellites separated from one another and from the launch vehicle to become free-flyers.

Description. Each *MACSAT* (**Fig. 1**) weighs about 150 lb (68 kg), is a 16-sided cylinder of 24-in. (0.6-m) diameter, and has a 20-ft (6-m) gravity gradient boom for Earth-pointing stabilization, with damping provided by hysteresis rods. A three-axis magnetometer provides attitude knowledge, and an electromagnet is used to invert the satellite if it should stabilize in the inverted position. The *MACSAT* communications mission payload and spacecraft command and control share the use of the transmitter and receiver. *MACSAT* has digitally tuned dual receivers and transmitters operating in the ultrahigh-frequency (UHF) band. It has dual computers, mass memory, a redundant electric power system, and sophisticated on-board software to enable the commanding and scheduling of communications service, provide communications protocols, and collect and process on-board telemetry to be sent to multiple ground stations and many user terminals.

Store-and-forward satellites function like electronic mail systems. As the satellite comes into view of a ground user terminal, that terminal can uplink messages with sender and destination code headers. It can also receive messages or data that carry the satellite's code as the destination.

When the satellite reaches the message destination position, it can be accessed from the ground to read out the message, or it can transmit the message to the ground automatically. On board the satellite, messages are stored in so-called mailboxes for each user destination code. In other words, the ground user puts a message into an on-board mailbox, and the satellite delivers the contents of the mailbox to the message destination.

Communications capability. Store-and-forward digital communications satellites provide global coverage. A typical satellite circles the Earth about 15 times per day, has an orbital period of about 100 min, and is visible from any point on Earth only 5–15 times per day, depending on the latitude of the user. It is seen 5 times a day at the Equator and up to 15 times a day at high latitudes.

Each visibility window is typically only 8–14 min long. Herein lies the limitation of low-Earth-orbiting communications satellites. Satellite access to uplink or downlink messages may cause a delay in message delivery of many hours, if only one satellite is available. For this reason, applications of store-and-forward communications satellites are primarily in monitoring unattended devices to collect data automatically, and for communications to or from low-powered transceivers. There are numerous such applications that even one or two satellites can satisfy. Key among these is the monitoring of oceanographic buoys and environmental sensors, where even a single satellite can provide a global sensor data collection service at low cost. Including satellites built for others than DARPA, six LIGHTSATs were launched in 1990.

Clusters of LIGHTSATs. To provide continuous communications coverage of regions of about 1000-nmi (1850-km) radius anywhere on Earth, the LIGHTSAT program will launch a cluster of seven MICROSATs (**Fig. 2a**). These satellites are different from the store-and-forward digital communications

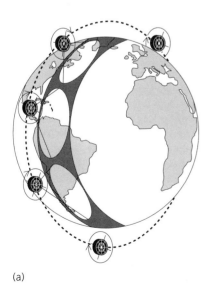

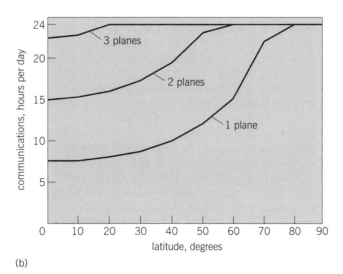

(a)   (b)

Fig. 2. MICROSAT cluster of satellites, providing continuous coverage within a 1000-nmi (1850-km) radius. (a) Configuration. Footprints of individual satellites and a swath formed by a composite of the footprints are shown. (b) Number of hours of service per day provided by one, two, or three planes of satellites within 1000 nmi (1850 km) of any point on Earth as a function of the latitude of the ground communication terminal.

satellites. Each MICROSAT weighs only 50 lb (23 kg). A set of seven will be launched by the Pegasus air-launched missile, dropped from a B-52 aircraft, and the satellites will be placed into a 400-nmi (740-km) polar orbit. Once in orbit, the satellites will spin up to provide gyroscopic stabilization through the interaction of the Earth's magnetic field and the on-board, computer-controlled electromagnetic torque coils.

The on-board cold-gas propulsion system will be used to move the satellites around in the orbit plane so that they are evenly spaced around the Earth. Each satellite will have a 1000-nmi (1850-km) instantaneous footprint, within which it is visible from the ground and communications can be conducted without time delay. The composite of these footprints will provide a swath on the ground (Fig. 2a). The propulsion systems will also be used to provide station keeping for 3 years. From any point within the swath, at least one satellite will be visible at all times. As the Earth rotates, the point on the ground will move out from under the swath, and there will be a waiting period while the Earth rotates before the swath includes the point and the satellites become visible again. Figure 2b shows the number of hours of service per day that one, two, or three planes of satellites can provide within 1000 nmi (1850 km) of any point on Earth as a function of the latitude of the ground communication terminal.

The MICROSATs will contain so-called bent-pipe transponders that will permit analog voice, digital voice, and digital data communications to be relayed between ground terminals. Three planes of seven MICROSATs each can provide worldwide tactical (regional) communications service.

Ground control. An integral part of lightweight, inexpensive satellites is the need to keep operating costs low. Each of the LIGHTSATs described above can be operated by personal-computer-based ground stations containing comprehensive software for mission planning, satellite control, and communications and message handling. Unattended ground stations can operate the satellite system for long periods of time.

Prospects. While no longer LIGHTSATs, the 77 satellites of the Iridium system, announced by Motorola, will function in a manner analogous to the MICROSAT system. Iridium satellites in multiple planes of polar orbits are intended to provide cellular telephone coverage worldwide.

The demand for small communications satellites, and for small satellites in general, is growing rapidly. These satellites are not replacements for the existing classes of large satellites; they fill a new market in which the need and the applications are just beginning to emerge.   *George Sebestyen*

SATELLITE COMMUNICATIONS FOR AVIATION

Currently, civil aviation communications are provided through terrestrial facilities operating in the high-frequency (HF) and very high frequency (VHF) radio-frequency bands. (The high-frequency band spans 2–30 MHz, and the aeronautical very high frequency band is 118–137 MHz.) Very high frequency radio is the preferred medium, but is restricted to line-of-sight range of a ground station. The maximum range for an en-route aircraft at an altitude of, say, 30,000 ft (9000 m) is about 250 mi (400 km). The radio range decreases at lower altitudes to a strictly localized coverage when the aircraft is on the ground.

High-frequency radio provides communication beyond the line-of-sight limitations of very high frequency radio. It is used for aeronautical communication in oceanic regions and other areas where very high frequency coverage is not available, perhaps 85% of the Earth's surface. However, very few frequencies in the high-frequency band are available for dedicated aviation use, and its propagation characteristics make high-frequency radio subject to poor communications quality due to noise, fading, and interference.

Nature of communication. A single satellite in geostationary orbit (**Fig. 3**) can provide high-quality communications to one-third of the Earth's surface (and its airspace). Three or more such satellites situated as in **Fig. 4a** can provide complete coverage of the world's airspace between latitudes of about 70° North and South. **Figure 5** depicts the coverage available from the existing constellation of INMARSAT satellites. (Polar-orbiting satellites, such as those used by GLONASS and the Global Positioning System, are needed to cover the regions above 70°.) *SEE SATELLITE NAVIGATION SYSTEMS.*

Aeronautical mobile satellite services (AMSS) occupy a portion of the radio-frequency spectrum encompassing 1530–1670 MHz for the satellite–aircraft links. This part of the spectrum, referred to as the L band, is lower in frequency than those parts commonly used for fixed, point-to-point satellite communication (4000–6000 and 11,000–14,000 MHz). Use of these lower frequencies affords an insensitivity to atmospheric effects, such as losses due to precipitation.

Applications. The continuous availability to aviation of high-quality en-route communications almost anywhere in the world is opening a new era of enhanced safety and efficiency of flight. One obvious example is air-traffic control, which is primarily concerned with the maintenance of safe separation of aircraft. Other examples are communications regarding weather conditions; communications regarding operational parameters of flight; performance monitoring of the aircraft's avionics, flight control, and propulsion systems; and flight status reports. Passengers will also enjoy direct benefits from aeronautical mobile satellite communication. Personal telephone service is now available, soon to be followed by facsimile, teletext, and data connections for personal computers. Display of information services in the passenger cabin, including news and financial market reports, is in the planning stage. The cabin crew will be able to offer ticketing and reticketing, ground transportation, lodging arrangements, and other passenger services.

System architecture. Because aircraft rapidly travel over large areas of the globe, it is essential that the standards for aeronautical satellite communications be uniform on a global basis. International agreement on standards has been achieved through such organizations as the International Civil Aviation Organization (ICAO), the Airlines Electronic Engineering Committee (AEEC), and the Radio Technical Commission for Aeronautics (RTCA). These organizations, comprising representatives from all sectors of aviation, have developed standards for the architecture of the system and requirements for the individual pieces of equipment needed for operations of the aeronautical mobile satellite services.

Until later in the 1990s, the only satellites providing aeronautical mobile satellite services will be those of the International Maritime Satellite Organization (INMARSAT), which also has made major contributions to the design of these services. INMARSAT satellites of the first and second generations utilize global beam transponders, meaning that the satellite antennas are designed to have a beamwidth approximating the solid angle (17.5°) subtended by the Earth (Fig. 3). By using global beams for both uplink and downlink directions, a link can be established between any ground Earth station and any mobile terminal from which the same satellite is visible.

Discrete channels are formed by using the single-channel-per-carrier (SCPC) technique of frequency-division multiple access (FDMA). The transponders operate as bent pipes; that is, a signal is received by the satellite from the direction of the Earth, translated in frequency, suitably amplified, and then is retransmitted in the direction of the Earth without modification other than frequency translation and linear amplification.

Communication channels. The system architecture provides for a variety of channel types and corresponding applications, all based on digital modulation. All data communications are packetized. In the forward (ground-satellite-aircraft) direction, one or more packet channels (P channels) carry packetized messages addressed to the aircraft tuned to those channels. As this is a single-point to multipoint, common-channel network, the packets are synchronously multiplexed in time to achieve high efficiency. In each satellite coverage area, at least one of these channels, designated as the P_{smc} channel, carries the information necessary for system management and control.

In the return direction, two types of packet-mode channels are employed. A random-access channel (R channel) allows aircraft to initiate communication on an unscheduled basis. Such a multipoint to single-point channel necessarily is subject to collisions among packets that may be transmitted during the same time intervals by multiple aircraft. If this occurs, each aircraft retransmits its packet after a random wait interval. In order to keep the number of retransmissions and the concomitant communication delay reasonably low, an R channel must be operated at a relatively low throughput efficiency, about 15% maximum.

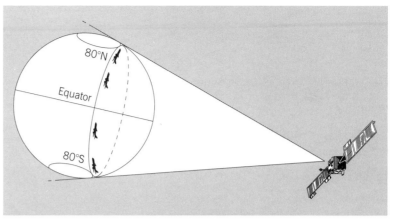

Fig. 3. Aeronautical mobile communications coverage from a single geostationary satellite. Coverage = 160° in latitude and longitude. (*INMARSAT*)

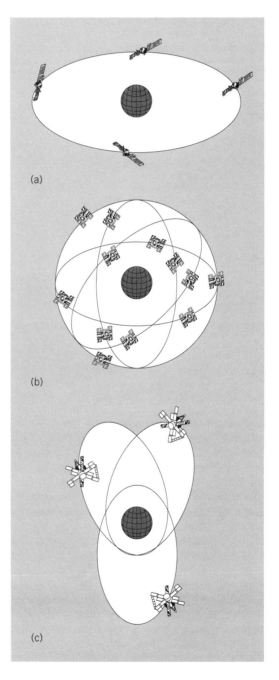

Fig. 4. Satellite constellations for mobile services. (*a*) Geostationary orbit. (*b*) Circular inclined orbits. (c) Highly elliptical inclined orbits. (*INMARSAT*)

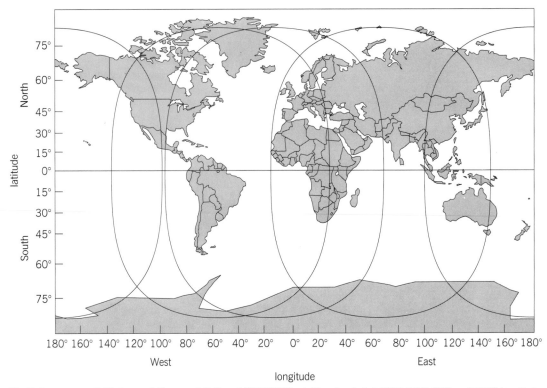

Fig. 5. Coverage available from existing constellation of INMARSAT satellites, located at 55°W, 18°W, 63°E, and 180°E longitude. Contours are shown, along which one of the satellites is at the horizon, as viewed from the ground. (*Av Com*)

A second type of return-direction channel is a synchronous time-division multiple-access (TDMA) channel (T channel). While the efficiency of a T channel can be high, its synchronous nature requires that unique time-slot assignments be made by the controlling ground station for a particular aircraft. Normally, requests are initiated by an aircraft over an R channel and are confirmed by the ground station over a P channel. Thus, efficiency is gained at the expense of channel setup delay. T channels are best suited for longer messages and for communications that have predictable periodicity such as position reports.

Communications requiring longer-term, continuous connectivity, such as voice calls and transfers of large amounts of data, are provided by circuit-mode C channels. A C channel operates simultaneously in both directions, and can be initiated on demand by either a ground station or an aircraft. The call setup procedure is similar to that for a T channel.

Data rates. P, R, and T channels are defined for modulation rates ranging from 600 to 10,500 bits per second, including half-rate error-correction encoding. C channels operate at 21,000 bits per second, also with half-rate coding, a rate that supports digitized voice at 9600 bits per second and a subband channel for signaling purposes.

Aircraft satellite terminals. The internationally agreed terminology for a satellite communication terminal (an Earth station) located on an aircraft is aircraft Earth station (AES). Aside from the seeming contradiction in terms, the design of an aircraft Earth station presents difficult issues due to the constraints of an aircraft as a platform.

If, as many envisage, satellites are to provide communications services for all aviation, then it must be possible to fit an aircraft Earth station to any aircraft, including the small aircraft in the general-aviation fleet. Even for the largest commercial transport aircraft, there are significant constraints on the cost, size, weight, power consumption, cooling, and maintenance requirements of the equipment constituting an aircraft Earth station. Environmental factors, such as vibration, temperature, and pressure extremes, also contribute to the problem.

Perhaps the greatest challenge lies in the design of the antenna for the aircraft Earth station. The Airlines Electronic Engineering Committee has defined two general types of antennas: a low-gain antenna, having an essentially omnidirectional characteristic; and a high-gain, directional antenna. Smaller aircraft can accommodate only the low-gain antenna, which resembles a very high frequency blade antenna.

Using current satellites, aircraft equipped with a low-gain antenna are limited to operation at the lowest data rates and cannot support voice. Equipage with a high-gain antenna affords a full range of data and voice services through simultaneous multiple channels.

With one exception, all current high-gain antenna designs are electronically steerable phased arrays having frontal areas of about 2 ft² (0.2 m²). Some array designs are single podlike assemblies of several inches height, mounted on the top of aircraft. Others are thin (fraction of an inch) assemblies designed to mount essentially flush with the sides of an aircraft; two such assemblies, placed at approximately 45° from the vertical, are required for nearly hemispherical coverage. High-gain antennas also require an interior beam steering unit. The choice of the high-gain antenna type for a particular installation involves a set of trade-offs considering cost, route structure, satellite locations, both antenna and aircraft performance characteristics, and maintenance issues.

Prospects. Potential new developers of satellites for the aeronautical mobile satellite service, as well as INMARSAT, are planning higher-power, spot-beam satellites that may be in service as early as 1994. Coupled with advances in digital voice encoding, these satellites can provide higher data rates and single-channel voice services to aircraft equipped with a low-gain antenna. High-gain antennas will support six or more simultaneous voice and high-speed data channels. If these satellites are advantageously placed in the geostationary orbital plane, many of the high-gain-antenna coverage problems will be ameliorated.

In the more distant future, new system architectures may be available. For example, constellations based on low-Earth-orbit satellites (Fig. 4b) and highly elliptical orbits (Fig. 4c) are being investigated for potential application to mobile satellite communications, and may have characteristics attractive for use by aviation.

For background information SEE ANTENNA (ELECTRO-MAGNETISM); COMMUNICATIONS SATELLITE; DATA COMMUNICATIONS; MULTIPLEXING; PACKET SWITCHING; RADIO SPECTRUM ALLOCATIONS; RADIO-WAVE PROPAGATION in the McGraw-Hill Encyclopedia of Science & Technology.

R. Andrew Pickens

Computer

All microelectronic devices produce heat that must be removed so that the device temperature does not become too high. Reliable operation of computer chips, for example, typically requires that the semiconductor junction temperatures be kept below 85°C (185°F). The usual modes of heat transfer are applicable: conduction, convection, and radiation.

Although the heat generation of individual devices is very small, the number of devices in a silicon chip has increased rapidly, and exceeds 10^6 for ultralarge-scale integration. Because of signal delays, even at the speed of light, the device packing density is very high. The result is a significant power that must be removed from a small surface area. This means a high heat flux, possibly exceeding 100,000 W/m^2 [31,700 Btu/(h · ft^2)] at the device. Depending on the packaging scheme for the chip and its carrier (module), the heat flux may be reduced by a process called spreading, in which an increased area is provided by conductive material. In any case, the energy must eventually be carried away by a moving fluid; therefore, convective cooling is necessary for the thermal control of all computer chips.

Techniques for convective cooling of chip arrays take two basic forms. Numerous single-chip modules are installed on a circuit board, or many chips are bonded to a common ceramic substrate with embedded wiring to connect chips and to communicate externally. The multichip-module approach reduces the signal delays among chips, for example, compressing the capabilities of four 0.3-m-square (1-ft-square) printed circuit board assemblies into a single 0.13-m-square (0.42-ft-square) module. Novel approaches have been proposed to package the chip arrays so that the electrical and fluid flow paths are efficient.

Air cooling. Air is the traditional cooling medium, and forced convection is generally required to accommodate the high power levels.

Cooling of circuit boards. The single-chip modules (**Fig. 1**) are capped for protection, with good thermal contact between the chip and the cap assured with thermal grease. Chip carriers are soldered into cards which in turn are plugged into a board. Blowers push, pull, or both push and pull the air through the gaps between the cards.

Heat removal is thus to the air through the module caps, and to the card via the pins and then to the air. For high-power chips, the caps may be finned. The heat-transfer situation is typically described in terms of thermal resistances (the R's in Fig. 1).

The thermal-fluid performance of the forced-air cooling is mathematically simulated with the aid of data from sophisticated experiments. The main problem is to assure adequate airflow distribution among passages and over each component within a card. The technology is not new, but the predictive capability has increased recently to the point where trial-and-error methods of arranging chips on a board can be avoided.

Cooling of multichip modules. An example of the latest air-cooling technology for multichip modules is depicted in **Fig. 2**. The silicon integrated circuits are located in cut-out areas of the signal-carrier substrate and are electrically connected to the substrate. The chips are thermally connected to the airstream by an adhesive (epoxy filled with diamond particles), baseplate (chromium copper), and pin-finned heat sink

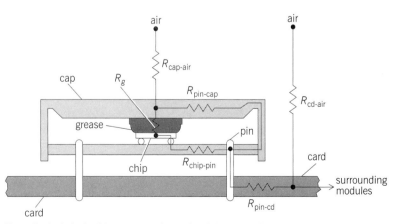

Fig. 1. Typical single-chip module and associated thermal paths.

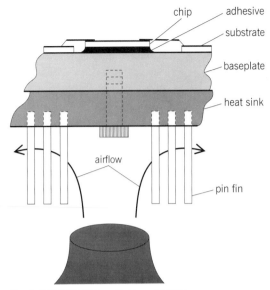

Fig. 2. DEC VAX 9000 multichip unit (MCU).

(aluminum). The materials are chosen for their thermal and mechanical properties so that they and the associated interfaces have a low thermal resistance. For instance, the baseplate and the heat sink have smooth joining surfaces and an evenly distributed bolt pattern.

An assembly of 16 multichip units receives air from a blower directed through a converging nozzle for each unit, and the impinging air flows across the pin fins (Fig. 2). The 600 pins in each unit provide an increase in surface area of about 8:1. The combination of large area and high heat-transfer coefficient, resulting from the small pin diameter, assures efficient cooling; however, this part of the thermal path still represents 60% of the total thermal resistance. The temperature of the semiconductor junction was found to be about 65°C (149°F) with a 25°C (77°F) airstream.

Indirect water cooling. Although improvements have been made in air cooling, air cannot support the packaging density and heat-flux levels of the multichip modules required in more powerful machines. Water-cooled heat sinks or cold plates are utilized in combination with innovative conduction coupling of the chips to the cold plate. The overall thermal resistance is substantially reduced through the use of water as a coolant so that high module power can be accommodated.

The thermal conduction module (**Fig. 3**) was introduced in the 308X series of IBM processors and has been continuously upgraded to handle cooling requirements of the more powerful 3090 series. Over 100 chips are electrically connected to a substrate. A spring-loaded, crowned piston is in contact with each chip. The module is filled with helium because it has a thermal conductivity six times greater than air. The heat is transferred from the chip to the coolant through the various materials and interfaces, including several gas gaps. The cold plate itself is internally finned to improve convective heat transfer to water by both extending the surface and enhancing the heat-transfer coefficient.

In the thermal conduction module, as well as somewhat similar schemes, the key to accommodating higher module power is proper selection of geometry and materials as well as thermal interface management. Ideally, the latter is accomplished by elimination of interfaces; however, conducting pastes are often utilized.

Direct liquid cooling. The rather obvious solution to materials and interface problems is to place the liquid coolant in direct contact with chips or chip packages, as is done with air. Water is ruled out, however, because the coolant must be dielectric and inert. Fluorocarbon liquids satisfy these requirements, but they have poor heat-transfer characteristics for single-phase convection, and, with a phase change, the maximum heat flux associated with nucleate boiling is relatively low.

The packaging may be arranged for natural circulation or forced circulation. One arrangement uses the boiling of a subcooled pool (open bath tester) to provide heat removal during testing of the chip–substrate assembly normally cooled by the thermal conduction module. This experiment has provided much confidence for the use of boiling liquid in actual computers.

With natural circulation and boiling, provision must be made for heat removal at the package. This is generally done by cooling one or more of the side walls, or the top, with air or water, or by immersing a water-cooled coil in the liquid itself above the module. An example is the early prototype liquid-encapsulated module (**Fig. 4**). Fins extend from the air- or water-cooled wall into the hermetically sealed chamber filled with liquid. Advances continue to be made in the prediction of the performance of such systems, especially the condensation limit of the generated vapor,

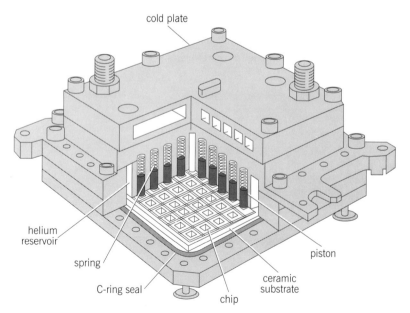

Fig. 3. Thermal conduction module (TCM) with water-cooled cold plate.

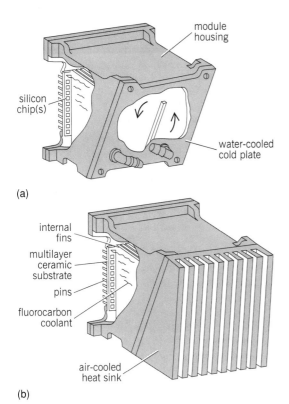

(a)

(b)

Fig. 4. Liquid-encapsulated modules. (a) Water-cooled module. (b) Air-cooled module.

which determines one upper limit of module power. The other limit is given by the maximum nucleate-boiling heat flux at the chip. The surface of the chip itself may be enhanced by laser drilling or channel etching; however, care must be taken not to disturb the devices and circuits in the silicon. Current research is directed instead toward enhancing this maximum flux by area extension through use of finned heat sinks secured to the chip by an adhesive. Alternatively, the heat sinks may be prepared with one of the contemporary nucleate-boiling enhancements, such as a sintered metallic layer, tunnel and pore structuring, or T-shaped fins.

With forced circulation, single-phase heat transfer may be adequate, especially if the heat flux is low. This is the case in the Cray 2 supercomputer, where a low-velocity flow of fluorocarbon liquid is circulated through horizontal cards containing single-chip modules. Basically, the liquid coolant replaces the air depicted in Fig. 1. It appears that a similar scheme will be used in the Cray 3 supercomputer. In a more advanced concept, the chips are packaged on a multichip module forming one wide wall of a rectangular channel. If high heat fluxes are to be accommodated, the chips must be filled with extended surfaces. With extensive finning, the thermal resistance from coolant to semiconductor junction can be reduced to a very low value.

Also in the exploratory stage is the strategy of providing cooling channels directly in the chip itself, thereby avoiding the problem of multiple thermal resistances. Channels, typically 50 micrometers wide

and 300 μm deep, are chemically etched into the silicon. With water, chip powers approaching 1 kW have been accommodated with very low thermal resistance in a 1-cm-square (0.4-in.-square) chip. The heat removal capability is considerably less with fluorocarbon liquid. While this approach is promising, the mechanical and hydraulic problems are formidable, especially when coolant flow at rather high pressure must be evenly supplied to an array of chips.

Other technologies under investigation include direct liquid cooling by jet impingement and transfer of the heat from the chips to air or water by means of a heat pipe.

While the experiences with direct liquid cooling systems have been very good, the commercial adoption of such systems has been slow. This is due to the increased complexity of liquid packaging, the cost of the coolant, poor heat-transfer characteristics, and concern about possible long-term fouling and corrosion.

For background information SEE CONDUCTION (HEAT); CONVECTION (HEAT); HEAT TRANSFER; PACKAGING OF EQUIPMENT; PRINTED CIRCUIT in the McGraw-Hill Encyclopedia of Science & Technology.

Arthur E. Bergles

Bibliography. A. Bar-Cohen and A. D. Kraus (eds.), *Advances in Thermal Modeling of Electronic Components and Systems*, vol. 1, 1988, vol. 2, 1990; A. E. Bergles (ed.), *Heat Transfer in Electronic and Microelectronic Equipment*, 1990; A. D. Kraus and A. Bar-Cohen, *Thermal Analysis and Control of Electronic Equipment*, 1983; D. P. Seraphim, R. Lasky, and C.-Y. Li (eds.), *Principles of Electronic Packaging*, 1989; R. R. Tummala and E. J. Rymaszewski (eds.), *Microelectronics Packaging Handbook*, 1989.

Consciousness

In science, the comparison of the presence and absence of a phenomenon is needed in order to collect evidence and develop theories about it. The presence and absence of consciousness is quite clear for the state of consciousness; there are massive neurological and behavioral differences between an individual who is in deep sleep or coma and one who is awake and alert. The neuroanatomical structures that are necessary for the state of consciousness are well known and include the brainstem reticular formation, the outer shell of the thalamus (its reticular nucleus), and the fountain of neuronal fibers that projects widely from the thalamus to all parts of the cortex. In the absence of any of these anatomical structures, human beings become comatose. On the other hand, any part of the cerebral cortex can be damaged or removed without producing coma.

Any adequate theory concerning consciousness must also deal with the contents of consciousness. Obviously, it is possible to be awake and experience an indefinitely large number of distinct conscious contents. Psychologically, accurate comparisons can be made between mental representations that are clearly conscious, that is, those that can be talked about, distinguished, remembered, and acted upon, and those very similar representations that are not conscious, such as the processing of stimuli that have faded from consciousness. Sometimes, when a highly predictable train of events, such as the repetitive sound of a noisy refrigerator pump, suddenly stops, the stimulus becomes conscious; it becomes apparent to one that something has changed, even though one was not aware of the train of events while it was going on. Research on such habituation phenomena suggests that the nervous system must maintain a rather accurate, unconscious representation of the repetitive stimulus, since a change in any parameter of the stimulus may bring it to consciousness. Similarly, when one is asked a question while reading a book, a first impulse may be to ask what was said. However, the memory of what was said suddenly comes to mind, indicating that the words must have been stored unconsciously— or else they would not be remembered a few seconds later. This is one aspect of selective attention, the phenomenon of an individual's being absorbed in a dense stream of conscious events while certain unconscious stimuli are being processed and represented. Both of the above examples raise the question of what is the difference between the same stimulus representation when it is conscious versus unconscious. There are dozens of examples in which there is a clear contrast between two mental representations of the same event, one of which is conscious and the other not. A complete theory of conscious experience must be able to account for all such contrasts and is therefore highly constrained.

Models of consciousness. The rise of an information-processing approach to scientific psychology has seen a concomitant development of increasingly sophisticated theoretical models. Until recently, most of these models did not refer to consciousness explicitly but used terms and mechanisms that are very closely associated with conscious experience. One of the first models was of short-term memory, which is roughly the domain of rehearsing telephone numbers while walking from the phone book to the telephone, or somewhat more broadly, the domain of people mentally talking to themselves. It is true, of course, that not all items in short-term memory are conscious; of the seven digits of a new telephone number that can be mentally rehearsed, only one or two are conscious at any single moment. Thus, short-term memory is not identical to consciousness, but the two seem to be closely related. The domain of visual imagery, sometimes called the mind's eye, has a great overlap with consciousness. Visual imagery may be one of several domains of conscious access and experience. Availability has been used as a pseudonym for the most consciously accessible item in a problem-solving task, and strategic control is used to refer to conscious, voluntary control of a skilled task that has become automatic and faded from consciousness with practice. Such pseudonyms have obvious utility in defining concepts with precision in particular, experimental tasks; unfortunately, the multiplicity of pseudonyms tends to obscure the fact that all these phenomena are intimately involved with the fundamental issue of conscious experience.

All phenomena associated with conscious experience share a common feature: they show remarkably limited capacity. When a baseball game is watched on television, a football game that is overlaid on the same screen cannot be experienced simultaneously, nor can a normal conversation be followed while any other unpredictable, skilled task is being performed. Short-term memory is characterized by a limit of 7 ± 2 items of information. Voluntary control, which is also intimately involved with conscious experience, seems similarly limited. Finally, the flow of conscious information is relatively serial, as suggested by the expression "the stream of consciousness." Limited-capacity mechanisms of some kind have been incorporated into many psychological theories, and in that sense, theoretical psychologists have been concerned with the cluster of phenomena associated with consciousness since the 1950s.

Models of unconscious processes. There is a curious dichotomy between limited-capacity models, which attempt to explain phenomena such as immediate memory, selective attention, and voluntary control, and the great body of evidence that suggests that the nervous system as a whole has enormous capacity. The obvious illustration is the sheer size and capacity of the brain. The cerebral cortex alone has on the order of 50 billion neurons, with similar numbers for the large subcortical mass, the spinal cord, and the multitude of peripheral neurons. Each neuron fires, or sends an electrochemical signal to its neighbors, on an average of 40 times per second. The system is thoroughly interconnected: neurons can have up to 10,000 connections with their neighbors, and a path can be traced from any single neuron in the brain to any other in only seven steps. All neurons seem to be processing information all of the time, certainly during the waking state. This is a picture of an enormously complex organ, one that operates in parallel rather than serially, with many things going on at once, and that performs highly complex and intelligent tasks, such as language analysis and visual perception, with minimal conscious involvement. *SEE NEURAL NETWORK.*

It is only recently that psychological models have begun to approach this complexity. The new perspective, which is sometimes called a society model of the nervous system, suggests that the great bulk of information processing is performed by local networks that are unconscious, though they are complex, symbolic, and intelligent, and can operate in parallel. With such sophisticated, distributed processors, it would be of interest to determine the role of the limited-capacity system that is so closely tied to consciousness. There is no consensus at present, with some suggesting that conscious experience corresponds to a domain of interaction between the many unconscious processors.

For background information *see* Brain; Conscious-ness; Psychology, physiological and experimental in the McGraw-Hill Encyclopedia of Science & Technology.

Bernard J. Baars

Bibliography. J. R. Anderson, *Cognitive Psychology and Its Implications*, 2d ed., 1985; B. J. Baars, *A Cognitive Theory of Consciousness*, 1988; A. Newell and H. A. Simon, *Human Problem Solving*, 1974; D. E. Rumelhart, J. E. McClelland, and the PDP Group, *Parallel Distributed Processing: Explorations in the Microstructure of Cognition*, vol. 1: *Foundations*, 1986.

Control systems

The field of intelligent control systems is multidisciplinary, drawing upon computer science, operations research, control theory, and so on. Recent advances have involved hierarchical control, communications, architecture, intelligent machines, multiple sensors, and learning capabilities.

Hierarchical control. Hierarchies consist of decision-making units arranged in a multilevel structure where at any one level decision-making units operate in parallel and have an overall common goal, with information being passed up and down the hierarchy. The existence of hierarchies emerges from the fact that systems have become so large and complex that any one decision unit typically cannot perform all required functions in a uniform and coordinated manner. Decomposition of the command structure leads to parallel processing capability at any level such that more tasks can be performed in a specified time period, with coordination alleviating or eliminating information flow among the decision units at a given level.

The concept of hierarchical control structures is being developed by many organizations and research institutes. One such example is the structure developed by the National Institute of Standards and Technology (NIST; formerly the National Bureau of Standards) that was initially designed for robot control systems with extensions to a totally automated factory. The hierarchical reference model used by NIST in their Automated Manufacturing Research Facility (AMRF) is shown in **Fig. 1**. In general, a reference model serves as an abstract model showing a structure of cooperating decision-making units. Such models have been proposed within recent years in an attempt to provide a conceptual framework for identifying generic manufacturing tasks. The reference model shown in Fig. 1 is partitioned into vertical levels of control, where levels pertain to a time decomposition; that is, higher levels have a long time horizon whereas lower levels approach real-time computational needs. Goals and plans generated at the top level are decomposed into sequences of subgoals (commands) at lower levels.

The NIST model is segmented into three separate but interacting hierarchies: a goal (task decomposi-

tion) hierarchy, a sensory interactive (feedback) hierarchy, and a world model hierarchy. The sensory hierarchy has the functions of determining the state of the environment at any level, providing feedback for corrective action, and providing status reports up through the hierarchy. The world model has the function of generating predictions of the state of the environment through a knowledge base that contains all information relevant to the task. Such a knowledge base may consist of an expert system, analytical models, or a combination of both. Predictions are compared to sensory observations for measurement of degrees of difference (or correlation) to determine if the model used is accurate. Much current research is focused on obtaining improved world models of the

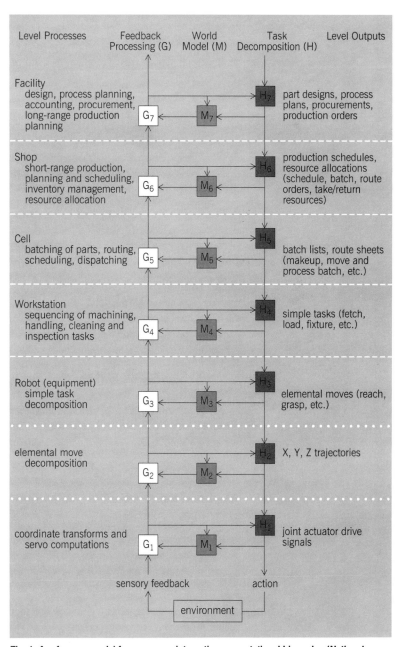

Fig. 1. A reference model for a sensory-interactive computational hierarchy. (*National Institute of Standards and Technology*)

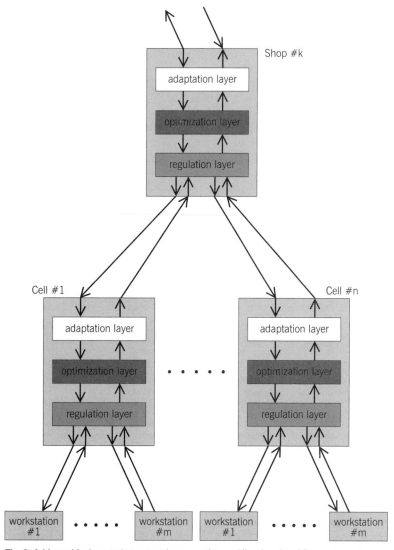

Fig. 2. A hierarchical control structure incorporating multilevel and multilayer concepts.

need for development of expert systems, current and future efforts will involve, among others, the formulation of complex analytical models, interactive computer graphics with natural language interfaces or voice communication, programming language development, hierarchical database management systems, and control strategies and algorithms. *See Database management systems*.

Communications. Issues pertaining to reliable communications are being facilitated at the equipment level by the development of local-area networks (LANs). On a factory floor, there typically is a disparate group of robots, machine tools, automated materials-handling devices, and automated storage and retrieval systems that may differ not only in their controllers but also in communication capability. Incompatibilities in control languages, data structure, and operating systems among various components at the equipment level coupled with competing mainframe computer architecture of different manufacturers have led to considerable activity by organizations involved with generation of standards.

Architecture. Aside from the issues of computer communications, much of the thrust in hierarchical control is in the development of mathematical models at the various levels of the hierarchy or for the total hierarchy. These mathematical models, commonly known as an architecture, pertain to purely algorithmic or heuristic efforts that include expert system approaches, modern control theory approaches, and the concepts of fuzzy control.

One such approach for a hierarchical control structure incorporating both multilevel and multilayer concepts in its design is given in **Fig. 2** for the shop, cell, and workstation levels of a hierarchy. The multilevel approach characterizes the division of control into local control units with local goals. Local control units are coordinated by a higher-level unit, for example, a shop level coordinating multiple cells. Three layers of control are incorporated into the shop and cell levels with the function of further subdividing control into algorithms, each of which acts at different time intervals. These layers are termed adaptation, optimization, and regulation.

An example of the function of these layers can be seen by considering the shop level, which (as noted in Fig. 1) has as level outputs the production schedules and resource allocations (workers, machines, materials, or tooling). The optimization layer at the shop level establishes the optimal open-loop production plan of activities by determining the routing and rates of parts flow among cells in addition to tool and fixture allocations, buffer (storage) allocation, and materials-handling considerations. The function of the adaptation layer is to make any structural (model) change or parameter adjustments of algorithms used in the optimization layer that may arise, for example, by a change in the mix of parts or new technology for handling material. The regulation layer provides the real-time closed-loop coordination of flow among cells to compensate for disturbances (such as machine breakdowns) and maintain the open-loop production

processes and on developing sensors to enhance measurements.

The NIST reference model consists of five control levels: facility, shop, cell, workstation, and equipment (individual devices such as robots, machine tools, and fixtures); each level is a separate process with a limited scope of responsibility. At the facility level, goals are formulated and strategies are determined. Decisions at this highest planning horizon are decomposed throughout the structure with the intent of providing unified and coordinated actions for given goals. The issues of real-time control become more pronounced descending from one level to another through the hierarchy, and they are particularly dominant at the workstation and equipment levels. Implementation of this model at NIST is based primarily on state-transition tables, that is, expert systems with IF-THEN production rules. The advantages of such an approach is that smaller, more tractable software can be developed and can be easily extended. In addition, since system states are well defined, diagnostics and debugging are enhanced. In addition to the continuing plan. The overall outcome of such a multilayer, multilevel decision-making approach is a real-time adaptive control scheme capable of handling the dynamics of a flexible manufacturing shop with the ability to adjust and compensate for planned or unplanned events.

Intelligent machines. There is a continuing effort by researchers to detail and develop control algorithms for reference model hierarchies. Even more intensive efforts have been applied to the development of intelligent controls at the lower levels of the hierarchy, particularly at the equipment level. This has led to the concepts of intelligent machines or robots. Further decomposition of the control structure has been pursued actively by NIST and others in the development of an intelligent robot, that is, a robot that can perform autonomously in an unstructured environment.

As with overall factory models and architecture, the development of intelligent machines is multidisciplinary. It is a difficult task since a machine must be given a capacity for reasoning and understanding based on input information. Attainment of intelligent control means that the machine must reach its goal autonomously without any intervention by a human operator. Fundamental research issues include not only the architecture or structure of control but also sensor strategies, multisensor integration or fusion, learning and decision making, programming, mobility and navigation, and overall systems integration. *SEE TELEROBOTICS.*

Multiple sensors. To increase the capabilities of intelligent systems and enhance intelligent control of such systems and machines, the first step is to select the most suitable sensor or combination of sensors for the task. Sensors can range from simple proximity ones, to tactile/force sensors, to sophisticated three-dimensional vision systems for use in guidance and navigation of a robotic arm. ROBART II (**Fig. 3**), a robot developed by the U.S. Navy for use in a security environment, has more than 128 sensors; these include tactile sensors, ultrasonic ranging sensors, proximity sensors, motion detectors, and a stereoscopic vision system. The computer hardware architecture for ROBART II consists of a two-level nine-microprocessor hierarchy with a 16-bit planner at the top level. The planner functions are path planning, obstacle avoidance, position estimation, automated map making, sonar plotting, and security assessment. Second-level microprocessors are dedicated to functions such as steering, vision, and speech processing.

To enhance the understanding of the environment for an autonomous machine, it is effective to have more than one sensor providing information upon which to act. A single sensor is typically not sufficient in an intelligent system because effectiveness for one measurement does not imply effectiveness for other measurements. Sensor fusion refers to the combining of multiple sources of sensory information into one representational format. This has been distinguished from multisensor integration, which refers to the synergistic use of multiple sensor information to assist in the accomplishment

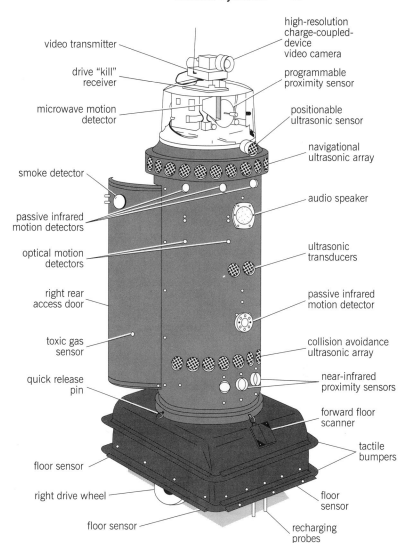

Fig. 3. A diagram of ROBART II, a U.S. Navy mobile security robotic testbed incorporating over 128 sensors.

of a task. Various approaches to solving the multisensor integration and fusion problem are being studied. These include various paradigms and frameworks for integration, control structures, and world models to enable a multiple sensor system to store and reason with prior sensory information. In general, the majority of fusion methods has been statistically based and falls within the context of statistical pattern recognition. Other techniques being investigated include fuzzy logic, neural network approaches, evidential reasoning, logical sensors, and production rule-based systems. Research is needed for multisensor systems to function in unstructured and dynamic environments, and for consideration of their implementation on parallel computer architecture. Concurrent with such research must be the development of interface standards and solid-state chips with multisensor capability.

Learning. To have a truly autonomous system, there is a need to develop a learning capability that enables adaptive behavior. This has led to considerable interest in neural network research. A neural network is a computational structure that is modeled on biological processes, specifically the nervous system. A neural network consists of many simple processors operating in parallel, with the network function determined by the connection topology. The advantages of neural networks over other computational structures is that such networks have extremely high speed due to the inherent parallel structure, and they are adaptable in that there is capability for learning. The adaptive or training feature, typically generated by algorithms, falls into three general categories, namely, supervised, unsupervised, or self-supervised training. In each case, whether there is an external teacher or not, data are input to the system, and any incorrect response generates an error signal that serves as a feedback to the system. Iterations of this process are continued until a correct response is attained. Learning is accomplished in that the algorithms allow the responses of the processing elements to input signals to change over time. Typically, researchers have focused on issues pertaining to classification problems in vision system applications or speech processing, sensory data processing and multisensor fusion, and various nonlinear computational problems. The applications of neural networks in an industrial environment have been directed toward vision systems and intelligent robot control. It is anticipated that heightened activity in both the theoretical and practical developments of neural networks will continue into the twenty-first century. SEE NEURAL NETWORK.

For background information SEE CONTROL SYSTEMS; DATA COMMUNICATIONS; EXPERT SYSTEMS; INTELLIGENT MACHINE; LOCAL-AREA NETWORKS; OPTIMIZATION in the McGraw-Hill Encyclopedia of Science & Technology.

Nicholas G. Odrey

Bibliography. *DARPA Neural Network Study*, AFCEA, 1988; H. R. Everett and G. A. Gilbreath, ROBART II: A robotic security testbed, Naval Ocean Systems Center, San Diego, Tech. Doc. 1450, 1989; R. C. Luo and M. G. Kay, Multisensor integration and fusion in intelligent systems, *IEEE Trans. Syst., Man, Cyber.*, 19(5):901–931, 1989; A. Meystel, Intelligent control in robotics, *J. Robot. Syst.*, 5(4):269–308, 1988; N. G. Odrey and G. R. Wilson, Hierarchical planning and control for a flexible manufacturing shop, *Proceedings of NSF Design and Manufacturing Systems Conference*, Arizona State University, January 8–12, 1990; G. N. Saridis, Analytical formulation of the principle of increasing precision with decreasing intelligence for intelligent machines, *Automatica*, 25(3):461–467, 1989.

Coordination chemistry

Nonlinear optics is a discipline concerned with how the electromagnetic field of a light wave interacts with the electromagnetic fields of matter and of other light waves. As light travels through nonlinear optical material, strong interactions can exist among the various fields. These interactions may change the frequency, phase, polarization, or path of incident light. The magnitude of these nonlinear optical effects depends on many variables. One set of variables that chemists are attempting to understand, and thus perhaps rationally control, is that of the molecular electronic hyperpolarizabilities. These hyperpolarizabilities are an intrinsic measure of the ability of a molecule to induce various nonlinear optical effects. Since the properties of light can be manipulated (modulated) in a nonlinear optical material, in a predictable manner, light can be used instead of electricity for communication and storage of information. The desire to use nonlinear optical materials for a variety of applications has motivated chemists to search for new approaches for optimizing nonlinear optical hyperpolarizabilities.

Organometallic and coordination compounds provide unexplored variables for engineering nonlinear optical (NLO) hyperpolarizabilities that are not found in other organic compounds. It is possible to change the metal element, its oxidation state, and the number of *d* electrons; it is also possible to examine the differences between diamagnetic and paramagnetic complexes and the effect of novel bonding geometries and coordination patterns.

Organometallic and coordination compounds possess a number of interesting properties that have led to their being the subject of active study as nonlinear optical materials. For example, these compounds can have metal-to-ligand or ligand-to-metal charge transfer bands in the ultraviolet-to-visible region of the spectrum. These optical absorption bands are often associated with large second-order optical nonlinearities. Another interesting property is that chromophores containing metals are among the most intensely colored materials known. The strength of the optical absorption band (that is related to its transition dipole moment) is also associated with large optical nonlinearities.

Organometallic and coordination compounds are often strong oxidizing or reducing agents, since metal centers may be rich or poor in electrons, depending on their oxidation state and ligand environment. Thus, the metal center may be either an extremely strong donor or an extremely strong acceptor. Metals can also stabilize unusual or unstable organic fragments such as carbenes, carbynes, or cyclobutadienes; therefore it becomes possible to investigate new classes of materials. In addition, metals can be used to tune the electronic properties of organic fragments.

Although the studies of organometallic and coordination compounds for optical nonlinearity are in their infancy, a variety of interesting effects that are usually very weak or nonexistent in organic or inorganic materials have been observed. These results suggest that research on organometallic and coordination compounds is fruitful from both scientific and technological perspectives.

Second-order materials. There are several manifestations of second-order nonlinear optical effects. One important effect is second-harmonic generation, in which the frequency of a light beam is doubled by its interaction with a second-order nonlinear optical medium. Second-harmonic generation is relevant to laser technology and optical information storage. For example, by frequency-doubling the 820-nanometer light from an inexpensive semiconductor laser, light at 410 nm is generated. The density of information stored on an optical disk is inversely related to the spot size of the light beam that is used to read the information, and this is in turn related to the square of the wavelength of the light beam. Thus, by using frequency-doubled light (the wavelength halved), information density enhancement with a factor of 4 can be realized. Another second-order effect, the electrooptic effect, can be used for optical modulation or optical switching. This manifestation has applications in telecommunications and in integrated optics. In this manner, electrooptical switches can be used to convert electrical signals (from a microphone, for example) to optical signals that can be transmitted through optical fibers. Electrooptic switches can also be used to switch light from one optical fiber to an adjacent fiber and can thus be used to control the route of the light beam.

Large second-order optical nonlinearities are associated with structures that are very polarizable (that is, structures in which the electrons are free to move easily) in an asymmetric manner (that is, the ease of polarization in one direction is different from that in the opposite direction). The polarization is a nonlinear function of the applied field, and the efficacy of a molecule to be polarized in this nonlinear manner is known as its first hyperpolarizability. Molecules with donor–acceptor interactions (resulting from charge transfer between electron donating and withdrawing groups) are promising candidates to fulfill the above requirements.

Theoretical and experimental studies have demonstrated significant enhancement of hyperpolarizability by extending the conjugation length between the donor and the acceptor (by increasing the number of double bonds, for example). Since second-order nonlinear optical effects result from asymmetric polarization, they can be induced only in molecules lacking a center of symmetry. If asymmetric molecules are aligned in the same direction within a material, the molecular nonlinearities will build up in the material; however, if the asymmetric molecules are oriented in a centrosymmetric manner, the molecular nonlinearities will cancel and no effect will be observed.

The original studies of metal-organic materials relied on a method known as the Kurtz powder technique. In this technique a powdered sample is illuminated with laser light and, if the material is second-harmonic-generation active, light at the second harmonic frequency is collected; its intensity is then compared to some reference. Relatively large powder second-harmonic-generation efficiencies of several metallocene complexes [structure (I), where

(I)

X = hydrogen (H) or methyl group (CH_3); $n = 1$ or 2; Y = groups CN, CHO, or NO_2 (C = carbon, O = oxygen, N = nitrogen); and M = iron (Fe) or ruthenium (Ru)] and square planar platinum and palladium aryl complexes [structure (II), where

(II)

X = bromine (Br) or iodine (I); M = palladium (Pd) or platinum (Pt); Y = CHO or NO_2; L = $P(CH_2CH_3)_3$ (P = phosphorus)] provided the impetus for systematic studies of structure–property relations for metal-organic complexes. The effects on the hyperpolarizability arising from variation of the metal center, bonding patterns, conjugation, and ligands on the metal for various classes of organometallic and coordination complexes were of interest.

The magnitude of the second-harmonic-generation signal obtained from the Kurtz powder test is determined not only by the molecular hyperpolarizability but also by crystallographic (orientational) factors. Therefore, little insight into molecular structure–property relationships can be inferred. Another technique known as solution-phase dc electric-field-induced second-harmonic generation can be used to obtain meaningful information about the molecular first hyperpolarizability.

Metallocene derivatives, iron group metallocenes (ferrocenes) in particular, represented a logical starting point for the examination of organometallic nonlinear optical compounds. The metal center in these complexes is known to be an excellent donor; substituted metallocenes are often stable in air, and the organic chemistry of ferrocenes is well understood. Compounds of general structure (I) have been examined by the technique of electric-field-induced second-harmonic generation. Ferrocene derivatives were found to have significant optical nonlinearities, approaching that of 4,4′-dimethylaminonitrostilbene, which is an organic molecule with a large hyperpolarizability. Variation of the organic fragment and the metal center resulted in dramatic changes in the magnitude of the observed nonlinearity. The results indicated that metal-to-ligand charge transfer plays an important role in the nonlinear optical responses of these molecules.

Also examined were platinum and palladium aryl compounds [structure (II)] of the form $(aryl)MX(P(CH_2CH_3)_3)_2$, where aryl = C_6H_5; M = nickel (Ni), Pd, or Pt; and X = I, Br, chlorine (Cl), or others. The interest in these compounds arose from the knowledge that the metal centers are good electron donors. Platinum compounds of the general form shown in structure (II) were more nonlinear than the analogous palladium compounds, reflecting the superior inductive donating strength of platinum as compared to palladium. Square-planar platinum and palladium benzenes are moderately nonlinear, with hyperpolarizabilities comparable to methoxy substituted benzenes.

Derivatives of tungsten pentacarbonyl [$W(CO)_5$] pyridine [structure (III), where Y = H, CHO,

(III)

$COCH_3$, C_6H_5, or NH_2] are known to have strong metal-to-ligand [the antibonding orbitals (π^*) of the pyridine] charge transfer bands. These compounds show respectable nonlinearities that depend on the substituent in the 4-position of the pyridine ring, with acceptor substitution leading to enhanced nonlinearities. The hyperpolarizability values of metal carbonyl derivatives are always negative, indicating reversal of the charge transfer direction in the ground and excited states. This observation is consistent with simple molecular orbital pictures for these compounds.

These preliminary studies indicate that strategies involving donor-bridge-acceptor molecules with organometallic end groups are useful for engineering molecular hyperpolarizability, resulting in species with nonlinearities comparable to those found in non-metal-containing organic compounds. Many of the same chemical factors that influence the magnitude of hyperpolarizabilities of organic materials, that is, the nature of the π-electron bridge and donor/acceptor strength, are operative in determining the hyperpolarizabilities of metal-containing systems. In addition, interesting behavior not commonly seen in organic compounds is observed with organometallic species. It is currently believed that absorption bands in the visible region of the electronic spectrum will limit the utility of many organometallic and coordination compounds for harmonic generation of ultraviolet or visible light. In contrast, these compounds may have some utility for electrooptic applications using near-infrared light.

Third-order materials. Third-order nonlinear optical effects include frequency tripling, optical phase conjugation (that may be useful for restoring distorted optical images), and optical limiting. In optical limiting, light of low-incident intensity is transmitted through a sample as a linear function of the input intensity, whereas at input intensities above a threshold the intensity of the transmitted light is lower than would be predicted by using a linear extrapolation. At very high input intensities (but below the damage threshold intensity for the medium), the output intensity saturates at a value determined by nonlinearity of the medium. This process can be used to protect detectors (or eyes) from very bright light.

In contrast to the relatively well-developed models that have been used to guide the synthesis of second-order nonlinear optical molecules, the models for molecular third-order optical nonlinearities are less well developed. Frequently molecules with strong,

low-energy optical absorption bands have large second hyperpolarizabilities. Such strong transitions are often seen in materials with highly delocalized electrons, like conjugated polymers and semiconductors. By using metal systems, it may be possible to develop completely new design criteria for materials with large third-order nonlinearities that do not invoke the use of large delocalized electronic systems. To date, investigations of metal-containing materials with third-order optical nonlinearities have been relatively limited. Most research has focused on resonant or near-resonant nonlinearities, that is, those which occur in a region of the spectrum where the molecule absorbs light. Two classes of molecules that have been examined are metal dithiolenes [structure (IV), where R = C_6H_5 or CH_3] and phthalocyanines [structure (V), where ML_n = $Si(OSi(C_6H_{13})_3)_2$, Si = silicon;

(IV)

(V)

or AlCl (Al = aluminum)], both of which have low-energy absorption bands (between 650 and 1200 nm). It has recently been suggested that metal dithiolene complexes may have sufficiently large nonlinearities and sufficiently low absorption losses to be used as all-optical switches for certain applications. Phythalocyanine compounds of the generic form shown in structure (V) have been shown to have large resonant optical nonlinearities and, in particular, to be effective as optical limiters in the visible region of the spectrum. As a result, these materials either alone or in combination with semiconductors are being investigated for eye and sensor protection from laser light.

For background information *see* COORDINATION CHEMISTRY; ELECTROOPTICS; FERROCENE; METALLOCENE;

Nonlinear optics; Organometallic compound in the McGraw-Hill Encyclopedia of Science & Technology.

Seth R. Marder

Bibliography. D. S. Chemla and J. Zyss (eds.), *Nonlinear Optical Properties of Organic Molecules and Crystals*, 1987; D. R. Coulter et al., Optical limiting in solutions of metall-aphthalocyanines and naphthalocyanines, *Proc. Soc. Photo-optic Instrum. Eng.*, 1105:42–51, 1989; M. L. H. Green et al., The synthesis and structure of (*cis*)-[1-ferrocenyl-2-(4-nitrophenyl)ethylene]: An organometallic compound with a large second-order optical nonlinearity, *Nature*, 330:360–362, 1987; S. R. Marder et al. (eds.), *Materials for Nonlinear Optics: Chemical Perspectives*, ACS Symp. Ser., vol. 455, 1991.

Cosmic background radiation

The cosmic microwave background radiation is the only phenomenon that currently allows for direct quantitative measurements of the physical conditions in the early universe. The universe is believed to be about 10^{10} years old, and the cosmic microwave background radiation is a relic from when the universe was only 4×10^5 years old (which, in terms of a human life, corresponds to a person a few hours old). The primary feature of the cosmic microwave background radiation is that it has an energy distribution as a function of wavelength that is very close to that of a Planck blackbody at a temperature of 2.7 K. In addition, this radiation is extremely uniform over the entire sky. Clearly, any model that seeks to explain the physics of the early universe must accommodate these striking properties of the cosmic microwave background radiation.

Because of its importance, many attempts have been made to measure the cosmic microwave background radiation. A variety of techniques and strategies have been used to characterize both the distribution of its energy as a function of wavelength (spectrum) and its uniformity over the entire sky (isotropy). This experience has culminated in the development of NASA/Goddard Space Flight Center's *Cosmic Background Explorer* (*COBE*), launched on November 18, 1989. This satellite, dedicated primarily to cosmological studies, is providing extremely accurate measurements of the nature of the cosmic microwave background radiation.

Origin. The current models of the early universe call for an initial very hot and dense phase (the so-called big bang) in which photons (light) and matter interacted so strongly that the universe was opaque. This makes it impossible to observe the photons that existed during still earlier epochs, since the physical processes that were occurring intercepted and changed the nature of these photons.

These models also call for an expanding universe, so that at an age of roughly 4×10^5 years the universe had expanded (and thus had cooled) so that there was insufficient thermal energy to keep the atoms ionized. This era is commonly referred to as the decoupling (of photons from matter), and the places observed on the cosmic microwave background radiation are on the surface of last scattering. The photons that last interacted during this era have traveled through the transparent universe until they are now observed.

Because the early universe is believed to have been in approximate thermal equilibrium, the spectrum of the cosmic microwave background radiation at decoupling was very nearly that of a Planck blackbody with a characteristic temperature of roughly 4000 K. As the universe expanded, this radiation was modified in a very specific way (adiabatically), so that the spectrum is still that of a Planck blackbody but at lower and lower temperatures. The observed temperature now is roughly 2.7 K.

The universe may not have been perfectly transparent after decoupling. Energetic processes related to galaxy formation may have added their imprint and distorted the spectrum of the cosmic microwave back-

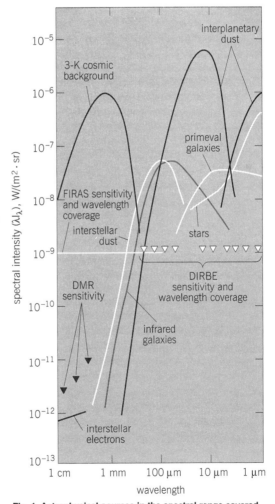

Fig. 1. Astrophysical sources in the spectral range covered by the *Cosmic Background Explorer*. The projected sensitivity for each of the instruments for each spatial resolution element and 1 year of integration is also shown. Of the two curves labeled interplanetary dust, the one on the left is for infrared emission, and the one on the right is for scattering of sunlight.

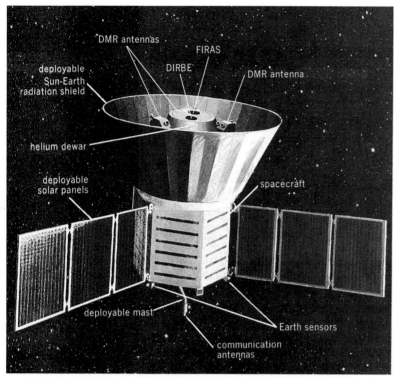

Fig. 2. The *Cosmic Background Explorer* showing the spacecraft, instrument apertures, and Sun–Earth radiation (radio-frequency and thermal) screen. Omnidirectional antennas communicate with the Tracking and Data Relay Satellite System (TDRSS).

ground radiation. However, it is unlikely that such processes had total energies close to that at the decoupling, so any such spectral distortions are expected to be small.

Measurement difficulties. Since the spectrum of the cosmic microwave background radiation is very close to that of a 2.7 K Planck blackbody, the peak intensity

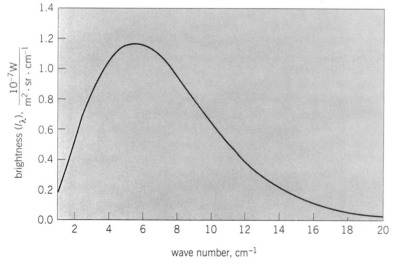

Fig. 3. Preliminary cosmic microwave background radiation spectrum from the Far Infrared Absolute Spectrophotometer (FIRAS) on the *Cosmic Background Explorer*. The best-fit temperature curve is 2.735 K. (*After J. C. Mather et al., A preliminary measurement of the cosmic microwave background spectrum by the Cosmic Background Explorer (COBE) satellite, Astrophys. J. Lett., 354:L37–L40, 1990*)

is at a wavelength of roughly 2 mm (in the microwave–far-infrared region of the electromagnetic spectrum). Unfortunately, there are many competing sources of emission in this spectral range that contaminate high-accuracy measurements of the cosmic microwave background radiation.

The foremost contaminant is emission from the Earth's atmosphere. In the far-infrared, the atmosphere is far from transparent and is extremely hot (300 K) compared to the cosmic microwave background radiation. For ground-based observations, atmospheric emission easily dominates the cosmic microwave background radiation by up to a few orders of magnitude. This problem has motivated many observers to use balloon-borne instruments, which observe at altitudes of up to 140,000 ft (43 km) but for a limited amount of time (typically one night). Sounding rockets can reach higher altitudes (200 mi or 320 km) but at the expense of even shorter observing times (several minutes total).

Even if the atmosphere were not a problem, there are local astrophysical sources that emit in the spectral range of interest (**Fig. 1**). At wavelengths longer than 1 cm, galactic radio emission from synchrotron and bremsstrahlung processes begins to interfere. At shorter wavelengths, emission from interstellar dust becomes important. Fortunately, these foreground emission sources have spectral signatures that are different from that of the cosmic microwave background radiation, so that with sufficient spectral coverage their effect can be modeled and subtracted. In addition, the combined emission from these sources is expected to have a minimum at around the 0.3–1-cm portion of the spectrum. Measurements in this wavelength range will therefore require the least correction for galactic contamination.

COBE instrumentation. The *Cosmic Background Explorer* (**Fig. 2**) contains three science instruments: the Diffuse Infrared Background Experiment (DIRBE), the Far Infrared Absolute Spectrophotometer (FIRAS), and the Differential Microwave Radiometers (DMR).

The DIRBE is a photometer that measures absolute flux in 10 spectral bands from 1 to 300 μm with a 0.7° spatial resolution. Its primary scientific goal is to detect the light emitted during the formation of the first pregalactic or galactic objects in the universe (the cosmic infrared background). It is also providing data of unprecedented quality for understanding interplanetary dust scattering (zodiacal light) and emission, and interstellar dust emission, all of which must be modeled and subtracted to find the cosmic infrared background.

The FIRAS is a scanning Michelson interferometer that measures the spectral energy distribution of the sky between 100 μm and 1 cm with a 7° spatial resolution. It has four channels, two sensitive between 500 μm and 1 cm (long-wavelength channels) and two between 100 and 500 μm (short-wavelength channels). The primary scientific goal of the long-wavelength channels is to measure the spectrum of the cosmic microwave background radiation over the entire sky. The short-wavelength channels measure the emission from interstellar dust, which can then be modeled and distinguished from the cosmic microwave background radiation emission.

The DMR is a set of three pairs of microwave receivers that measure the difference in intensity between two 7° patches in the sky separated by 60°. These radiometers operate at wavelengths of 0.92, 0.57, and 0.33 cm and are used to measure the isotropy of the cosmic microwave background radiation to an accuracy of better than 1 part in 10^5.

The orbit and attitude of the *Cosmic Background Explorer* are designed to allow all three instruments to observe the entire sky over a 6-month period. The DIRBE and FIRAS are cryogenic instruments that are cooled by a 160-gal (600-liter) liquid helium cryostat on the spacecraft. This cryogen was designed to last approximately 1 year, allowing these two instruments to observe the entire sky twice. The DMR can be operated for as long as the data are useful, and will continue operations until at least December 1991.

The legacy of the mission will be a set of calibrated full-sky maps that cover four decades in wavelength (1 μm to 1 cm). The relative stability of the space environment, careful instrument design, and long life (allowing for systematic performance tests) ensure that these maps will be free from instrument effects, and will be true maps of the sky intensities. The vast spectral range of the data will provide an unprecedented tool for sorting out the various foreground components in the sky in order to study the cosmic microwave background radiation.

Preliminary results. The most striking preliminary result from the *Cosmic Background Explorer* is the cosmic microwave background radiation spectrum measured by the FIRAS instrument. The results (**Fig. 3**) show that there are no deviations from the simple theoretical planckian shape greater than 1% of the peak cosmic microwave background radiation intensity. The sensitivity of this result is expected to improve by at least a factor of 10 as further data are analyzed.

The preliminary DMR results show that there are no variations in the intensity of the cosmic microwave background radiation at the level of 1 part in 10,000 for all angular scales greater than 10° and less than 180°. This result is also expected to improve by at least a factor of 10.

Significance. On the surface, these results lend remarkable support to the big bang theory of the universe. However, there are some very troubling consequences. The major problem is that the present universe is far from smooth and homogeneous. The universe contains planets, stars, galaxies, clusters of galaxies, regions with no galaxies, and regions with an enhanced number of galaxies; in short, it contains a rich assortment of structures extending to very large scales. The *Cosmic Background Explorer* isotropy results indicate that at the last scattering surface the universe was very smooth, and the spectrum results imply that nothing extraordinary happened between the surface of last scattering and the current epoch. Assuming that there are seed fluctuations at the surface of last scattering that are consistent

with the observed smoothness, theoretical calculations show that there is no way that gravity, working by itself on the known kinds of matter, could have formed the structures observed now. Much work is under way to try to create these structures without distorting the shape of the cosmic microwave background radiation spectrum, or to find mechanisms that may help gravity to form the required structures in a shorter amount of time.

Data processing. The preliminary results were based on ground-analysis computer programs intended primarily for verifying proper instrument operation. With the end of cryogenic operations in late 1990, data processing was begun using more sophisticated programs to produce the final sky intensity maps. This effort is expected to last for several years and to result in a set of comprehensive, high-quality sky maps that will play a crucial role in solving the cosmic evolution puzzle.

For background information *SEE* B*IG BANG THEORY*; C*OSMIC BACKGROUND RADIATION*; C*OSMOLOGY* in the McGraw-Hill Encyclopedia of Science & Technology.

Edward S. Cheng

Bibliography. S. Gulkis et al., The *Cosmic Background Explorer, Sci. Amer.*, 262(1):132–139, January 1990; J. C. Mather et al., A preliminary measurement of the cosmic microwave background spectrum by the *Cosmic Background Explorer* (*COBE*) satellite, *Astrophys. J. Lett.*, 354:L37–L40, 1990; J. Silk, *The Big Bang*, 1989; S. Weinberg, *The First Three Minutes*, 1977; R. Weiss, Measurement of the cosmic background radiation, *Annu. Rev. Astron. Astrophys.*, 18:489–535, 1980.

Crosstalk

Optical-fiber cable has already replaced much coaxial cable in telecommunication systems, reducing though not totally eliminating crosstalk problems. However, coaxial cables are still commonplace and are likely to be employed for some time in applications involving high power, short jumper connections, and so forth. Recent investigations have highlighted methods for eliminating electric-field leakage through cable screens, designing braided cable screens, and planning multiple-cable layouts.

Electromagnetic interference problems can be characterized by calculating the crosstalk between the interference generator and the cable that is subject to interference (victim). When a screened cable (either coaxial or twisted-pair) is either the generator or victim, a major factor in determining the interference is the surface transfer impedance of the screen, Z_T. This quantity accounts for the resistive and the inductive crosstalk through the screen and is specific to the cable. It is given by the equation below, where $R(f)$

$$Z_T = R(f) + j\omega L \qquad \Omega/\text{m}$$

is the longitudinal resistance of the cable screen at the

frequency of operation f, in ohms per meter, and L is the net screen inductance, in henrys per meter.

The surface transfer admittance, Y_F, accounts for the conductive and capacitive crosstalk through the braid. However, its value depends on the cable's physical environment. For most radio-frequency cable screens the quantity Y_F is not significant, and inductive leakage dominates. By placing a semiconductive (for example, high-carbon-content) layer over the braid, the capacitive part of Y_F can be eliminated.

The crosstalk is directly related to the leakage through the screen of the cable concerned. The leakage in turn is directly related to Z_T and Y_F.

Where high levels of screening are required, a coaxial cable that is badly installed (that is, with connectors inadequately attached) or subsequently damaged can undo all the good achieved by the screen. Also, if pigtail connections from coaxial cables to other circuits are used, care must be taken to screen the pigtail sections from external systems.

The following discussion is primarily relevant to the frequency range 100 kHz–1 GHz. The term cable is used throughout instead of coaxial cable for brevity and because much of what is said is relevant to any screened type of cable.

Control through choice of cable type. The main types of cable screen, in order from lowest to highest screening, are (1) single-braided (of wires or tapes); (2) double-braided (most often with both braids made of wires, sometimes separated by dielectric); (3) superscreened (as with double-braided, but with braids separated by mumetal tape); and (4) solid screen.

Figure 1 shows typical $|Z_T|$–frequency characteristics for the first three of these screens on a coaxial cable. The inductive component of the leakage causes the linear variation of Z_T at higher frequencies. The frequency at which the inductive leakage becomes dominant depends on the particular cable. A solid screen is best at the higher frequencies because the skin effect traps the signal within the cable. Other considerations such as mechanical flexibility, weight,

ease of installation, and cost will influence the eventual choice of cable.

In the frequency range where the inductive leakage is dominant, replacing single-braided screen cable with double-braided screen cable will reduce crosstalk typically by 25 dB; superscreened cable will reduce crosstalk typically by 53 dB; and solid-screen cable ($L = 0$) will give still lower crosstalk. Further reduction in leakage is possible by using a cable type that has a screen radius of larger size, although the benefits are less than those obtained by changing the type of screen used.

The single-braided screen design can be optimized (as can each of the individual braids in the double-braided and superscreened designs). The inductance in the equation above for Z_T is the net inductance because in general braids have two inductances involved in the leakage process. The first is due to the braid weave and the second to the holes in the braid. Both are dependent on the braid geometry. In braid designs with braid angles (that is, angles between crossing wires or tapes) of less than 45°, the two inductances cause currents that oppose each other, thus giving rise to the possibility of optimized braid designs (Z_T = minimum). Optimized braids can easily lose their optimization (for example, because of twisting or flexing), and therefore can have widely different leakages in practice from those expected. However, optimized designs tend to have an intrinsically lower leakage than nonoptimized ones.

The double-braided screen uses the fact that two screens will be better than one, but does not provide a squaring of Z_T for two braids with the same value of Z_T. Indeed, double-braided screens with dielectric spacing may have only marginally better leakage than single-braided screens at some frequencies because the two screens and the dielectric spacer form a transmission line around the inner transmission line (or lines) of the screened cable, and resonances can occur unless the transmission line formed by the screens and spacer is correctly terminated in its characteristic impedance.

Superscreened cables essentially are double-braided cables with a mumetal spacer. The effect of the mumetal is to make the characteristic impedance and longitudinal attenuation of the outer transmission line high. The mumetal will also eliminate most of the electric leakage. Due to increased longitudinal resistance in the frequency range 100 kHz–1 MHz, superscreened cables sometimes give less screening than single-braided ones.

Control by variation of cable run layout. Two problems will be discussed: crosstalk between screened cables and other electronic systems (such as printed circuit boards or antennas), and crosstalk between cables within a particular run. The first problem is complex since it depends on many factors, for example, the angle of incidence and frequency of the interfering signal, and whether shielding is being attempted against a single frequency (or a narrow band) or a broad range of frequencies. Obvious methods to consider for reducing crosstalk are to

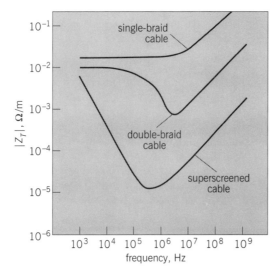

Fig. 1. Graphs of the absolute value of surface transfer impedance ($|Z_T|$) versus frequency for various types of coaxial cable.

surround the cable run with either a metallic duct or high-tangent δ material (where δ is the dielectric loss angle), or reorientation of the cable run. There are applications where all or some of these methods would be unsatisfactory. Assuming that the cable-run cross section is electrically small (less than λ/10, where λ is the wavelength), and that for most of the run the cables are parallel, the major crosstalk contribution into or out of the cable will probably be via TEM (transverse electric and magnetic) waves propagating along the multiconductor transmission line formed by the cables in the run and any sufficiently close (that is, <λ/10) return path. If this is the case, the methods given below for the reduction of cable-to-cable crosstalk may be considered as applicable since the crosstalk occurs through the same multiconductor transmission line.

Wide-band shielding. An important factor in the problem of screening against a wide band of frequencies is the ability of an unmatched multiconductor transmission line to resonate, regardless of whether the cable screens are short-circuited together periodically, only at their ends, or not at all. High levels of crosstalk occur at the resonant frequencies. Experimental evidence shows that cable bundles have more resonances in a given frequency range than a flat-pack arrangement where neighboring cables are more than 2.5 times the cable diameter apart (assuming the cables to be all the same size). Resonant crosstalk levels can be reduced only by matching or by increasing the losses of the multiconductor transmission line. Matching is difficult but may be attempted. Increased losses can be achieved by using higher-resistance conductors (normally undesirable), through the choice of the dielectric, or by enhancing the radiation from the cable. Increased radiation is clearly undesirable if the interference problem is crosstalk with an external system. Radiation from the multiconductor transmission line can be increased by adding discontinuities, and, conversely, it can be reduced by removing them. If none of these is an option, then only a cable with a better screen can solve the problem. If cable screens in a run are shorted together at exactly regular intervals, each section will resonate at the same frequency, resulting in higher crosstalk than for a single section. If the intervals are not exactly equal, then many resonances will occur but will not add to give higher resonant crosstalk.

Narrow-band shielding. To eliminate interference for a single frequency (or a narrow band of frequencies) in circumstances where the physical environment of the multiconductor transmission line is under control, advantage may be taken of the classic cable crosstalk characteristic shown in **Fig. 2**. A crosstalk curve for one end of the open-circuit multiconductor transmission line would be very similar except that the resonances would occur at frequencies where the run length was equal to odd multiples of λ/4. For cable runs that have an unmatched multiconductor transmission line, crosstalk to remote and noncoaxial systems will have a similar crosstalk characteristic, although the levels will be different. Low levels of crosstalk can be achieved by tuning the short-circuiting intervals in

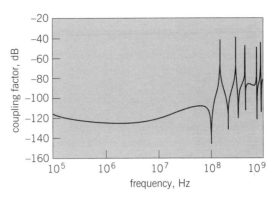

Fig. 2. Typical crosstalk–frequency characteristic (for RG 59–size cables, 1 m or 3.28 ft in length with screens shorted together at both ends).

the cable screen so that the crosstalk is at a minimum on the curve. Where the environment of the multiconductor transmission line is not sufficiently under control, a run length (or any section of it) that is a multiple of λ/4 should be avoided. The crosstalk may then be minimized by placing all the cables in the run as close to a ground plane or duct as the cable jackets permit. To reduce crosstalk between the cables in the run, they should also be spaced as far apart as possible. Conversely, to reduce crosstalk to circuits too remote to be part of the multiconductor transmission line, the cables should be kept as close to each other (still on the plane) as possible.

For background information SEE COAXIAL CABLE; COMMUNICATIONS CABLE; CROSSTALK; ELECTRICAL INTERFERENCE; ELECTROMAGNETIC COMPATIBILITY; TRANSMISSION LINES in the McGraw-Hill Encyclopedia of Science & Technology.

Frank A. Benson; Peter A. Cudd

Bibliography. F. A. Benson et al., Coupling between a pair of cables in a multicable environment with various ground-plane configurations, *IEE Proceedings, Part A*, 129(6):377–380, 1982; P. A. Cudd et al., Calculation of coupling between coaxial cables, *Proceedings of International Conference on EMC*, IERE-56:141–149, 1982; P. A. Cudd et al., Prediction of leakage from single braid screened cable, *IEE Proceedings, Part A*, 133(3):144–151, 1986; J. M. Tealby et al., Coupling between jacketed braided coaxial cables, *IEE Proceedings, Part A*, 134(10):793–798, 1987.

Cystic fibrosis

Cystic fibrosis is the most common, life-threatening, autosomal recessive disease in the Caucasian population. Approximately 1 in 2500 live births is affected by this genetic disorder. Obstructive lung disease, pancreatic enzyme insufficiency, and elevated secretion of sodium and chloride in sweat are the hallmarks for cystic fibrosis, but the severity of these symptoms varies from patient to patient. Patients with cystic fibrosis die at an early age (often, in childhood) due to lung infection. With advances in clinical treatments, which are directed against the symptoms, the mean survival age for patients has increased to 26 years.

Despite intensive research efforts since the 1940s, the basic defect in cystic fibrosis remains speculative. It is generally believed that the heavy mucus found in the respiratory tracts and the blockage of exocrine secretion from the pancreas are due to reduced water secretion. This imbalance is believed to be the consequence of a defect in the regulation of ion transport in the epithelial cells. This article discusses the identification of the gene responsible for the development of cystic fibrosis as well as the implications of such a discovery.

Chromosome localization. With the development of gene mapping and recombinant deoxyribonucleic acid (DNA) cloning techniques in the 1980s, it became possible to tackle the cystic fibrosis problem through a molecular genetic approach. The first step was to localize the disease gene to a subchromosomal region so that molecular techniques could subsequently be employed to detect candidate gene sequences.

The basic strategy in cystic fibrosis linkage analysis was to try to detect an association between the inheritance of the disease locus and a test marker with the use of a large number of families, each with two or more children affected with cystic fibrosis. In 1985, a series of markers were found linked to the disease gene, with the first being the gene locus for the liver enzyme paraoxonase (*PON*), then a DNA marker (*D7S15*) that allowed the assignment of the linkage group to chromosome 7, and subsequently two additional DNA marker segments (*MET* and *D7S8*) that tightly flank both sides of the cystic fibrosis locus (**Fig. 1**). The search was thus narrowed to a specific region less than 0.1% of the human genome.

Gene identification. The precise localization of the cystic fibrosis locus on a particular region of the long arm of chromosome 7 facilitated various molecular strategies to isolate the gene; many more DNA segments were isolated from the region. Two DNA markers (*D7S122* and *D7S340*) were found to be much closer to the cystic fibrosis region than *MET* and *D7S8*; their physical location was confirmed by long-range restriction-enzyme mapping studies. The two markers were then used as the starting points to isolate neighboring chromosome DNA. This effort resulted in the isolation of a series of overlapping DNA segments where candidate genes might be identified. Upon careful examination of the entire region, several gene sequences were found, and the one responsible for cystic fibrosis was detected toward the end of the stretch (**Fig. 1**).

The presence of the cystic fibrosis gene sequence was initially indicated by the ability of a DNA fragment to hybridize with DNA from other animal species. It was then confirmed through the isolation of a matching complementary DNA (cDNA) from a normal sweat-gland cell culture. Isolation of the remainder of the cystic fibrosis gene was accomplished by repeated screening of different cDNA libraries and alignment with genomic DNA sequence. *See Human genetics.*

As there were no functional assays for the cystic fibrosis gene product, the only way to ensure that the

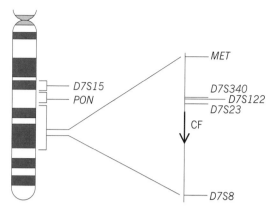

Fig. 1. Localization of the cystic fibrosis gene to the long arm of human chromosome 7. The relative positions of the genetically linked DNA markers and gene loci are shown, with the direction of transcription of the cystic fibrosis gene indicated by an arrow.

gene involved was the cystic fibrosis gene was to identify a mutation in it. Upon sequence comparison between cDNA clones from normal individuals and from individuals with cystic fibrosis, a 3-base-pair (bp) deletion was discovered among the cystic fibrosis cDNA clones. This deletion was shown to have a strict correlation with the disease (that is, the sequence alteration was found only in the mutant chromosomes, not in any of the normal ones). This result provided the strongest argument for the gene being responsible for cystic fibrosis. Its identity is now beyond doubt, as an overwhelming number of other mutant alleles have been discovered.

Gene product. The cystic fibrosis gene encodes a messenger ribonucleic acid (mRNA) of about 6500 nucleotides in length. Transcription of this gene was observed in various tissues that are affected in cystic fibrosis patients, such as lung, pancreas, liver, sweat gland, and nasal epithelia. The gene product, a putative protein deduced to consist of 1480 amino acids, is called cystic fibrosis transmembrane conductance regulator (CFTR) to reflect its possible role in cells (**Fig. 2**).

Based on sequence alignment with other proteins of known functions, CFTR is thought to be a membrane-spanning protein requiring adenosinetriphosphate (ATP) for its activity; the internal sequence identity between the first and second half of CFTR resembles that of the other prokaryotic and eukaryotic transport proteins, notably, the mammalian P glycoprotein. More recent studies show that CFTR may be a pore structure that controls ion transport in epithelial cells.

Mutations. The most frequent mutant allele of the cystic fibrosis gene involves a 3-base-pair deletion that results in the deletion of a single amino acid residue (phenylalanine) at position 508, within the first ATP-binding domain of the predicted polypeptide (Fig. 2). Although this mutation, termed ΔF508, accounts for about 70% of all cystic fibrosis chromosomes, there is marked difference in its proportion among different populations, from 30% in the Ashkenazic Jewish population to 85% in Denmark. A consortium has been formed among 80 laboratories around the world to study the remaining 30% of cystic fibrosis mutations and, thus far, more than 75 different mutations have been detected. The nature of these mutations are heterogeneous, including amino acid substitutions (missense) or various nucleotide alterations, deletions, or insertions that lead to premature termination of CFTR synthesis. Most of these mutant alleles are rare, however, with some represented by only single examples.

The large number of disease-causing mutations in the CFTR gene provides a rich source of information about the structure and function of the protein. For example, the mutation screening study confirms that the ATP-binding domains detected by sequence alignment are important for CFTR function; some of the mutations have been found at several of the highly conserved amino acid residues in these regions. Among the other mutations, there is a second 3-base-pair deletion resulting in the deletion of an isoleucine residue at position 506 or 507 of the putative protein, while amino acid substitutions at positions 506 or 508 are apparently benign. These observations suggest that the number of amino acid residues in this region is more important than the amino acid composition. Further, certain mutations predict the synthesis of the grossly nonfunctional CFTR proteins, suggesting that the total absence of CFTR protein is not lethal.

Genotype and phenotype. The varied symptoms among different cystic fibrosis patients suggest that disease severity is at least in part related to the mutations in the cystic fibrosis gene. Such association is expected to be concordant among patients within the same family, since they should have the same genotype at the cystic fibrosis locus. This is, in fact, observed for pancreatic function. Approximately 85% of cystic fibrosis patients are severely deficient in pancreatic enzyme secretion and are thus diagnosed as pancreatic-insufficient, and the other 15% have sufficient enzyme and are diagnosed as pancreatic-sufficient. Studies have shown that there is almost complete concordance of the pancreatic status among patients within the same family, leading to the suggestion that pancreatic insufficiency or sufficiency is predisposed by the patients' genotypes. Subsequent studies have shown that patients homozygous for the ΔF508 mutation described earlier are almost exclusively pancreatic-insufficient.

There are other mutations that would also be classified in the same group as ΔF508. These are the so-called *severe* mutant alleles with respect to pancreatic function. In contrast, patients with one or two copies of the so-called *mild* mutant alleles are expected to be pancreatic-sufficient. Meconium ileus (intestinal obstruction in the newborn with cystic fibrosis due to trypsin deficiency), which is observed in about 30% of cystic fibrosis patients, appears to be a clinical variation of pancreatic insufficiency and not directly determined by the cystic fibrosis genotype. Other clinical manifestations are more complex, and no apparent associations with a particular genotype have yet been detected.

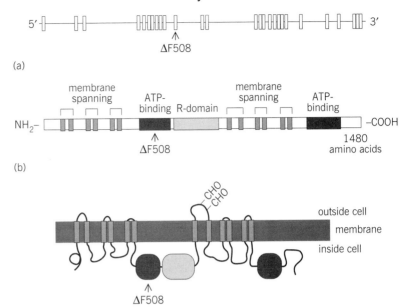

Fig. 2. Schematic diagram of the cystic fibrosis gene and its protein product. The mutation ΔF508 results in the deletion of phenylalanine at the first ATP-binding domain. (*a*) Gene structure with exons (open boxes), which contain coding portions; (*b*) computer-predicted primary structure of cystic fibrosis transmembrane conductance regulator (CFTR); and (*c*) model of CFTR showing possible relationship with cellular membrane and potential glycosylation sites (—CHO). The R-domain is thought to have a regulatory role. It contains a large proportion of charged amino acid residues and many potential sites for phosphorylation. The elements in part *c* correspond to those in *b*.

DNA diagnosis. An immediate application using the knowledge of the various cystic fibrosis mutations at the DNA sequence level is in the area of diagnosis. Previous genetic testing, based on closely linked DNA markers, was limited to prenatal diagnosis for families that already had children with cystic fibrosis or to detection of carriers in immediate family members of the patients. The ability to now detect the various cystic fibrosis mutations directly offers the opportunity to perform the test in a general population.

Prospects. With the identification of the cystic fibrosis gene, a better understanding of the basic defect and physiology of the disease can now be attained. In addition to the detection of additional cystic fibrosis mutations and correlation of genotype and phenotype, current experiments are directed toward the development of effective assay systems for testing the function of CFTR. Advances are also being made in studies of the regulatory mechanisms governing the expression of this gene, and in the biosynthesis and subcellular localization of the protein. Suitable animal models, such as those generated by transgenic mice and other lower organisms, are being explored as experimental systems to increase understanding of the basic function of CFTR. Development of rational therapies, including gene therapy, is also under way.

For background information SEE BRONCHIAL DISORDERS; GENE ACTION; GENETIC MAPPING, LINKAGE (GENETICS); MUTATION; PANCREAS DISORDERS in the McGraw-Hill Encyclopedia of Science & Technology.

Lap-Chee Tsui; Johanna M. Rommens
Bibliography. N. Kartner et al., Expression of the cystic fibrosis gene in non-epithelial invertebrate cells

produces a regulated anion conductance, *Cell*, 64:681–691, 1991; B. Kerem et al., Identification of the cystic fibrosis gene: Genetic analysis, *Science*, 245:1073–1080, 1989; J. R. Riordan et al., Identification of the cystic fibrosis gene: Cloning and characterization of complementary DNA, *Science*, 245:1066–1073, 1989; J. M. Rommens et al., Identification of the cystic fibrosis gene: Chromosome walking and jumping, *Science*, 245:1059–1065, 1989.

Dating methods

The objective of obsidian hydration dating is to provide an absolute age for prehistoric artifacts by determining the rate of formation for the surface hydration layer. The occurrence of datable obsidian artifacts at archeological sites means that these prehistoric tools and, by association, the prehistoric occupation surfaces, middens, and structures where they have been found may all be placed within a temporal framework. These data are often critical to the documentation of changes in material culture, subsistence practices, and sociopolitical organization.

The process of hydration is controlled by the diffusion of atmospheric moisture into the core of the obsidian artifact. When a freshly fractured surface on a piece of obsidian is created, ambient water is attracted to the surface and diffuses into the glass, depolymerizing the structure of the silica network. This disruption of the glass network forms a rim, known as a hydration rim, which exhibits optical birefringence when viewed in transmission with polarized light. The hydration rim increases in thickness as the obsidian artifact ages. New experimental data indicate that the rate of molecular water diffusion, or hydration rim formation, is controlled by the effective hydration temperature, atmospheric relative humidity, and glass composition. Current research efforts are directed toward developing an equation whereby hydration rates can be predicted from these controlling variables. Advances made toward this goal are discussed below.

The successful application of this dating method involves four basic steps: (1) measurement of the archeological hydration rim; (2) chemical characterization of the obsidian; (3) measurement of the ground temperature at the archeological site; and (4) development of a laboratory hydration rate.

Hydration rim measurement. The dating process begins with an optical measurement of the thickness of the hydration rim which normally ranges between 1.0 and 8.0 micrometers. A standard petrographic thin section of 30 μm is prepared. The thin section is viewed in fully polarized light while utilizing a first-order red gypsum plate. This method allows maximum transmittance of light through the sample while providing an enhanced blue–red or blue–yellow contrast, thereby increasing the accuracy of the measurement process. Measurement of the rim thickness using an image shearing eyepiece is the preferred method

since it is an interferometry technique that relies on phase contrast, thereby improving image resolution.

Chemical characterization. Obsidian fragments at an archeological site may originate from a variety of distantly located geological obsidian sources, each of which has a different chemical composition and, as a result, a different rate of hydration. Identification of the chemically distinct obsidians permits the application of the hydration rate constants developed for each geological flow.

Archeological specimens are matched with the geological parent flow by comparing their minor-element and trace-element abundances. A variety of instrumental techniques that include x-ray fluorescence analysis, flame emission spectrophotometry, and neutron activation have routinely provided useful data. In addition to contributing to the hydration analysis, the successful correlation of artifacts to specific sources permits the study of prehistoric trade and exchange patterns, and the documentation of hunter-gatherer territorial size.

Effective hydration temperature. The rate of hydration increases exponentially as a function of temperature, demonstrating classical Arrhenius behavior. As a result of this exponential dependency, small errors in estimation of the soil temperature can affect profoundly the hydration rate. Accurate soil temperatures are determined routinely by using a thermal cell that is planted in the ground for a period of 1 year. At the end of the burial period, the cell is weighed and the increase in weight is converted into an integrated soil temperature. This temperature is used to adjust the laboratory hydration rate to reflect the temperature of the archeological site.

The first thermal cell to be routinely used consisted of a dry desiccant encased in polycarbonate. While accurate and reliable, the cell had to be replaced every 6 months if the upper temperature reached around 68°F (20°C). Recently, a salt-based, wet-desiccant cell was developed that can measure temperatures for as long as a year at up to 86°F (30°C) without replacement. As a result, this cell design may be used in most environments throughout the world, including deserts.

Laboratory hydration rate. Early attempts at developing hydration rates for a particular obsidian involved correlating archeological hydration rims with other forms of chronometric data such as radiocarbon dates. This approach resulted in the development of over a dozen hydration rates for certain widely used glasses and suggested that hydration rates are specific to the environmental context at each archeological site. In contrast, rate development within the laboratory offers a flexible strategy whereby hydration rates may be adjusted to reflect the environmental conditions at specific locations, thereby improving the accuracy of the technique and alleviating the need for painstaking rate development using traditional methods.

The development of hydration rates in the laboratory involves the hydration of freshly fractured glass fragments at temperature increments between 305 and 359°F (150 and 180°C) for durations of up to 24 days. The optically measured hydration rims are used to calculate the hydration rate at 320°F (160°C), A, and the activation energy, E_a. With these data, the hydration rate at high temperature may be extrapolated to ambient conditions by use of the Arrhenius equation below, where K = archeological hydration rate

$$K = Ae^{E_a/RT}$$

(μm^2/1000 years), A = hydration rate (μm^2/day) at 320°F (160°C), E_a = activation energy (kJ/mol), R = universal gas constant (J/mol), T = temperature (kelvin), and e = base of the natural logarithms (2.718...).

In developing hydration rates within the laboratory, previous experimental programs have reacted samples under a variety of conditions that include saturated steam, distilled deionized water, silica-saturated distilled water, and 100% relative humidity. The last reaction environment is considered the preferred condition since it is a more representative model of field conditions at many archeological sites where the artifacts are buried beneath the surface at depths in excess of 16 in. (50 cm). Once the laboratory rate has been developed and adjusted for the ground temperature at the prehistoric site, the measured hydration rim is converted to an absolute date by using the formula below, where Y = years before present, x = thick-

$$Y = \left(\frac{x^2}{r}\right) \times 1000$$

ness of the hydration rim (μm), and r = archeological hydration rate.

New developments. Experimental studies on other glass compositions have found that parameters such as intrinsic (chemically bonded) water content of the glass and atmospheric relative humidity affect the reactivity of a glass. These factors also appear to hold for obsidians.

Obsidians typically exhibit a relatively narrow range of water contents (0.05–0.4 wt %), yet recent experiments demonstrate that a strong relationship exists between the rate of hydration at high temperature (320°F or 160°C) and the intrinsic water content of the glass (**Fig. 1**). The greater the intrinsic water

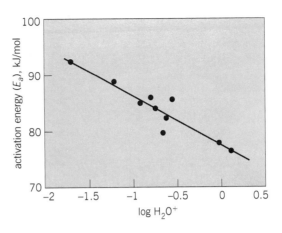

Fig. 2. Relationship between the logarithm of the intrinsic water content of obsidian and the activation energy, E_a.

content of the obsidian, the faster the hydration rate. Several other compositional parameters were examined, but the strongest correlation to rate of hydration was found to be the intrinsic water content. A linear regression fit to these data resulted in a correlation of 0.9, and was superior to those found when the same data set was analyzed by using other compositional dependence models. These improved results are attributed primarily to improved methods for measuring obsidian water contents using infrared spectroscopic techniques. A similar positive correlation was found to hold between the intrinsic water content and the activation energy (**Fig. 2**). When these two relationships are firmly established, it will be possible to calculate hydration rates for a glass once the intrinsic water content and environmental conditions of the archeological site are known.

To examine the relationship between the rate of hydration and atmospheric relative humidity, obsidians were reacted at 60, 90, and 100% relative humidity at temperatures of 305, 320, and 347°F (150, 160, and 175°C). This range of relative humidities encompasses most archeological field conditions experienced in temperate and desert environments. The hydration rims produced in the laboratory at elevated temperature and varying relative humidities were then compared. Preliminary data indicate that the rate of hydration decreases with declining relative humidity. The relationship is nonlinear and appears to be most pronounced when the relative humidity varies between 90 and 100%. These data reveal that hydration rates may need to be reduced substantially when the relative humidity of a soil is less than 100%.

New laboratory experimentation has shed some light on the fundamental principles and controlling parameters behind the dating method. These revelations have been exciting, yet they require refinement, validation, integration into a mathematical model, and practical application to the archeological record.

For background information see *Archeological chemistry; Archeological chronology; Birefringence; Dating methods; Hydration; Obsidian; X-ray fluorescence analysis* in the McGraw-Hill Encyclopedia of Science & Technology.

C. M. Stevenson; J. J. Mazer

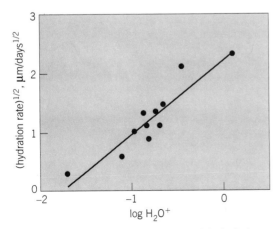

Fig. 1. Relationship between the logarithm of the intrinsic water content of obsidian and the square root of the hydration rate for obsidian hydrated at 320°C (160°C), *A*.

Bibliography. S. Newman, E. M. Stolper, and S. Epstein, Measurement of water in rhyolitic glasses: Calibration of an infrared spectroscopic technique, *Amer. Mineralog.*, 71:1527–1541, 1986; F. Trembour, F. Smith, and I. Friedman, Diffusion cells for integrating temperature and relative humidity over long periods of time, *Materials Issues in Art and Archaeology: Material Research Society Symposium Proceedings*, 123:245–251, 1988; C. M. Stevenson, J. Carpenter, and B. E. Scheetz, Obsidian dating: Recent advances in the experimental determination and application of hydration rates, *Archaeometry*, 31:193–206, 1989; C. K. Wu, Nature of incorporated water in hydrated silicate glasses, *J. Amer. Ceram. Soc.*, 63:453–457, 1980.

Deep-sea fauna

Recent research on deep-sea fauna has included investigations of hydrothermal vent communities and of a group of oceanic jellyfishes, the narcomedusae.

HYDROTHERMAL VENT COMMUNITIES

The dense populations of animals living at the hydrothermal vents of oceanic spreading centers are in strong contrast to typical deep-sea communities. In most parts of the deep sea, life is sparse because there is little food. For most deep-sea communities, the food is organic matter that has settled from above and was originally the product of plant photosynthesis in the ocean's surface waters or on land, using sunlight as the energy source. Very little of this food ever reaches the ocean floor thousands of meters below. At hydrothermal vents, the energy source is hydrogen sulfide, which is produced by geochemical reactions as the water percolates through the hot basalt. Chemoautotrophic bacteria oxidize the sulfide to obtain energy for synthesis of organic material, principally from water, carbon dioxide, oxygen, and nitrate that is taken from the surroundings. Except for the energy source, this process of chemosynthesis is much the same process as plant photosynthesis. This bacterial productivity constitutes a rich source of food for the vent animals, with the result that animal biomass is among the highest in the oceans.

The opportunity to partake of this large food supply is limited by the fact that hydrogen sulfide is toxic to animals because it poisons one of their fundamental metabolic pathways. However, vent animals have evolved mechanisms for detoxifying the sulfide. Some species immobilize it by binding it to another molecule; others alter it to thiosulfate, a sulfide derivative that does not harm animal metabolism. The animals that have developed these abilities are very different from those in the surrounding deep sea. Only some of the more conspicuous species will be mentioned here.

The most unusual animals are among those that live directly in the vent opening. In the eastern Pacific these animals include giant tubeworms with brilliant red gills (Vestimentifera), white clams (Vesicomyidae), and mussels (Mytilidae); in the western Pacific they include a coiled snail (*Alvinoconcha*). These vent animals have carried their utilization of bacteria to an extreme by incorporating the living microbes into their digestive organs or gills. They supply the bacteria with sulfide and other inorganic nutrients, and the multiplying bacteria act as an internal food supply. This is an excellent example of mutualistic symbiosis; the animal provides an optimal environment for the bacteria, which in turn provide their host with food.

Limpets (Archaeogastropoda), barnacles (Balanomorpha), crabs (Bythograeidae), shrimp (Bresiliidae), and fish (Zoarcidae or Bythitidae) also live at the vent opening. The limpets graze on bacteria that grow on rocks, worm tubes, and shells; barnacles filter bacteria from the passing water; crabs and shrimp scavenge bits of dead food or nip at living animals; zoarcids are carnivores; the food source of bythitids remains a mystery.

The rock farther away from the vent openings is dominated by filter feeders, such as anemones, serpulid worms, and scallops, that feed on bacteria or plankton settling from the plume of vent water as it drifts from the openings. Here too live scavengers and carnivores, such as galatheid crabs, polynoid worms, and whelks.

Vent fields are small; the zone occupied by vent-endemic animals has a radius of only tens of meters, presumably because dilution quickly reduces hydrogen sulfide to levels too low for chemosynthesis. Beyond this zone live the typical deep-sea fauna, although at the edge of the vent field they may occur in greater numbers. Here they are close enough to take advantage of vent productivity, but not so close as to be poisoned by the toxic waters.

The temperature of emerging vent water varies from nearly 400°C (750°F) to only a little above that of the surrounding bottom water (2°C or 36°F), depending on the amount of subsurface dilution. Most vent animals are limited to water no hotter than 25°C (80°F). However, alvinellid worms, whose tubes encrust the precipitated chimneys that form when high-temperature water emerges from the rock, are known to tolerate 50°C (120°F).

Hydrothermal vents do not last forever; some persist for only a decade before they cease to flow. Then the vent fauna dies. Of course, as old vents cease, new ones arise. For a vent species to survive, it must have dispersal mechanisms to find these new vents. While members of a few species, such as crabs, might walk to a new vent if the distance is short, most must rely on planktonic stages for dispersal. Little is known about this facet of vent natural history. The mussel has planktotrophic larvae, which feed as the currents carry them about. Clams have lecithotrophic larvae, which use stored yolk for nutrition while they are up in the water. Some species, such as amphipods, have no larval stage, but these can swim as adults. It is not known to what extent dispersal stages can sense new vents and direct themselves toward them, as opposed to accidentally drifting upon them.

Because the dispersal between vents decreases with

increasing distance, vent faunas become more dissimilar the farther apart they are. The closest vent fields, such as along a single segment of the spreading mid-ocean ridge, share the same species. The largest gaps, for example between the East Pacific Rise and the Juan de Fuca/Explorer ridges, have only a few species in common. Yet even the most distant vents show a general similarity at a high taxonomic level. Vestimentiferans, vesicomyids, mytilids, archaeogastropods, bythograeids, galatheids, bresiliids, and other types are characteristic components of vent fauna from many different vents, although the species or genus may vary.

There are some exceptions to this distribution. The coiled, symbiotic snail and sessile barnacles are known only from western Pacific vents, though they live there with many members of the more cosmopolitan fauna. This shows that there have been regional invasions of hydrothermal vents by important new groups that have not had the opportunity to become widespread. Vents in many parts of the oceans have still not been explored; exciting discoveries will undoubtedly accompany future exploration of different oceanic spreading centers.   *Robert R. Hessler*

NARCOMEDUSAE

The Narcomedusae are a group of worldwide, oceanic jellyfishes, generally rather small, with diameters of 0.4–4 in. (1–10 cm), constituting an order of the class Hydrozoa. Although narcomedusae were among the first midwater animals to be photographed in place, aspects of their life history have eluded biologists until recently. Narcomedusae feed on large gelatinous zooplankton, including other hydromedusae, ctenophores, salps, doliolids, pteropods, chaetognaths, and veliger larvae of mollusks. Most narcomedusae live at specific depths, although a few species have broad depth ranges. One species undergoes a vertical migration in which it travels approximately 1650 ft (500 m) up and down through the water column daily.

Distribution. All narcomedusae are holoplanktonic, meaning their entire life cycles take place in the plankton, never attached to solid substrates. A few species live in coastal waters (for instance, some members of the genus *Solmaris*), but most occur in the less turbulent waters of the open seas. Such oceanic species can be roughly divided into two groups: those living in the upper mixed layer of the water column (the upper 650 ft or 200 m), which are called epipelagic and include some species in the genera *Pegantha* and *Cunina*; and those living in the deep water layers, which are called mesopelagic or bathypelagic, such as *Aeginura grimaldii* (**Fig. 1**) and *Solmissus incisa*. Only a few species of narcomedusae occupy really broad depth ranges spanning most of the water column. Such species are termed eurybathic and include *Solmissus marshalli* (**Fig. 2**) and *Aegina citrea*. Most species are restricted to specific depths, but the factors that make these depths favorable to each species are not yet known. Most species of narcomedusae are cosmopolitan, occurring

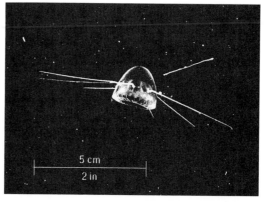

Fig. 1. The narcomedusa *Aeginura grimaldii*. Photographed in the North Atlantic Ocean off Massachusetts at 2350 ft (716 m) from the *Johnson-Sea-Link I* crewed submersible. (*Photograph by C. Mills*)

in most oceans worldwide. A few species have restricted geographic distributions, such as *Solmissus albescens*, which is confined to the Mediterranean Sea, where it is endemic.

In deep water, narcomedusae are sometimes the most numerous jellyfishes. They may attain numerical densities as high as one medusa per cubic meter (35 ft^3) of water, as seen in the *Solmissus albescens* population in the Mediterranean Sea, but more typical densities are on the order of one narcomedusa per 100–1000 m^3 (3500–35,000 ft^3) of water. Although numerical density may be high, the species diversity of narcomedusae in any one region of the deep sea seems to be much lower than that of other groups of hydromedusae, such as the trachymedusae or anthomedusae.

Morphology. Narcomedusae can be distinguished from hydrozoan medusae of other orders by their tentacles, which originate from the umbrella some distance above the margin rather than at the margin itself, although there may be secondary tentacles attached at the margin. The umbrella is divided at its base by grooves leading up to the primary tentacles, giving the umbrella margin a lobed appearance; the stomach is broad with either an entire circular periph-

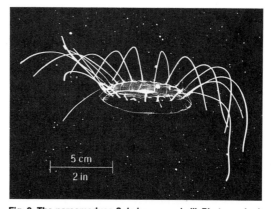

Fig. 2. The narcomedusa *Solmissus marshalli*. Photographed in the North Atlantic Ocean in the Bahamas at 2000 ft (610 m) from the *Johnson-Sea-Link II* crewed submersible. (*Photograph by C. Mills*)

ery or with rectangular peripheral pouches; the gonads are on the stomach walls; there are no radial canals, although a peripheral canal system is sometimes present; the sense organs are in the form of free sensory clubs at the bell margin.

Some species of narcomedusae are colorless and extremely transparent (such as *Solmissus marshalli*, Fig. 2), whereas others are pigmented. Most of the darkly pigmented species occur in deep water; they are usually red (such as *Aeginura grimaldii*, Fig. 1), black, or brown. The variation in color and intensity suggests that several different pigments may be involved. In some species the entire umbrella is colored, whereas others are basically transparent except for the stomach. It has been suggested that such pigmentation can serve to mask any bioluminescent light emitted by ingested prey. Some narcomedusae are themselves able to produce bioluminescent light.

Of approximately 800 described species of hydromedusae, there are presently about 60 described species of narcomedusae. There may really be only about 25 valid species of narcomedusae, with the others not being recognizable from their brief formal descriptions. Because they are fragile, narcomedusae are usually heavily damaged when collected in nets, but through recent access to deep water by crewed submersibles, more is being learned about the life habits of these jellyfishes.

Swimming and feeding behavior. Narcomedusae were originally assumed to be relatively inactive. Recent observations in the wild indicate that all narcomedusae are good swimmers and that some species are very active, pulsating nearly all of the time. Most narcomedusae have thick, stiff jelly in the central region of their bodies, so that swimming contractions generally flex only the thin side walls of the umbrella. Each bell contraction propels the medusa approximately one bell diameter's distance. In the ocean, narcomedusae can swim in any direction, including upside down or sideways.

Some narcomedusae undergo daily vertical migrations, moving toward the surface every evening and back down to their preferred depth in the early morning. Such behavior is typical of many species of medusae, but *Solmissus albescens*, a narcomedusa that is very abundant in mesopelagic waters throughout the Mediterranean Sea, has the longest known daily vertical migration. The *S. albescens* population usually occurs as a more or less continuous layer that is about 330–660 ft (100–200 m) thick, located at a depth of 1300–2300 ft (400–700 m) during the day. In the mid to late afternoon, presumably as the light intensity falls in these deep waters, the entire population begins to swim toward the surface. Moving upward at a rate of 150–500 ft/h (50–150 m/h), the population migrates in about 4–6 h, with the swimming being circuitous. They remain in the upper 330 ft (100 m) for a few hours before beginning to swim downward with the onset of light. It is believed that *S. albescens* follows an isolume, or a particular low value of light intensity, up and down through the water column. This behavior implies that the jellyfish

are able to perceive light. In fact, *S. albescens* is one species among many that seem to respond to light, but in which no specific light-sensitive cells have yet been identified. Narcomedusae have neither ocelli nor other known light receptors.

The number of tentacles in narcomedusae varies from 2 to 40 by species. These are usually held upward over the bell or projected outward, although during escape swimming, many species hold their tentacles stiffly down below the bell. In general, narcomedusae forage by slow continuous swimming with tentacles outstretched (Figs. 1 and 2). Unlike the tentacles of most other hydromedusae, those of narcomedusae are not contractile, but the grooves in the bell beneath each tentacle allow the tentacles to bend or coil inward across the lower portion of the bell in order to make contact with the mouth.

Narcomedusae, like all hydromedusae, are carnivores. Although data on their diets are sparse, it appears that they prey primarily on other gelatinous zooplankton and on pteropods (free-swimming snails). Prey items that have been found in the digestive cavity of narcomedusae include other hydromedusae, ctenophores, salps, doliolids, pteropods, chaetognaths, and veliger larvae of mollusks. Narcomedusae in the laboratory are known to ingest hydromedusae, ctenophores, and salps. *Solmissus albescens* in the Mediterranean feed on pteropods living in the upper water layers; the medusae gain access to these prey during the nocturnal vertical migration. They may also feed on a deep population of ctenophores (*Bathocyroe*) with which they coexist during the daylight hours.

Medusae capture and narcotize their prey by using microscopic intracellular stinging structures called nematocysts, which are densely arrayed on their tentacles and lips. The nematocysts of narcomedusae are highly specialized, having tubules up to 0.3 in. (8.5 mm) long. It is known that they penetrate. These unusually long, barbed nematocysts penetrate the prey and perhaps also inject a toxin, enabling narcomedusae to capture and retain their thick-bodied gelatinous prey. The nematocysts are concentrated along the upper side of the tentacle surfaces of narcomedusae. Prey are caught on the extended tentacles, which are then bent inward to bring the prey to the mouth. The normally flat mouth and inconspicuous lips bulge outward to meet the prey, and nematocysts around the lips help to grasp it. The lips slowly creep over the prey, pulling it into the stomach, presumably by activity of the cilia lining the lips and stomach. The mouth of a narcomedusa does not seal very tightly, so most specimens caught by nets drop their prey and appear to have empty guts when later examined. In their natural environment, however, these medusae rarely experience great turbulence, so prey can be retained easily, usually at the perimeter of the circular stomach, during digestion.

Life cycle and reproduction. Although the life-span of narcomedusae is not known, it is likely that these jellyfish live much less than 1 year. Narcomedusae carry out their entire life cycle in the water column.

Individual medusae are of separate sexes. Although they lack the benthic hydroid stage typical of many hydromedusae, the life cycle of most species of narcomedusae is complicated, often involving a stage that is parasitic on other planktonic medusae. This parasitic stage may be either a pelagic hydroid that is able to bud new medusae, or an asexually produced bit of tissue called a stolon that is released from its parent medusa and then subsequently buds more medusae, or single juvenile medusae that simply attach to host medusae and grow for a while. These parasites typically attach in the subumbrellar cavity or even within the gastric cavity of their host medusae. A minority of narcomedusae have direct development from the fertilized egg through a free-living planula larva that metamorphoses directly into a free-living medusa. In other directly developing species, embryos are brooded by the adult jellyfish and later released as tiny, fully formed medusae.

For background information SEE COELENTERATA; DEEP-SEA BACTERIA; DEEP-SEA FAUNA; MARINE BIOLOGICAL SAMPLING; MARINE MICROBIOLOGY in the McGraw-Hill Encyclopedia of Science & Technology.

Claudia E. Mills

Bibliography. J. Bouillon, Considérations sur le développement des Narcoméduses et sur leur position phylogénétique, *Indo-Malayan Zool.*, 4:189–278, 1987; J. J. Childress, H. Felbeck, and G. N. Somero, Symbiosis in the deep sea, *Sci. Amer.*, 255:114–120, 1987; J. F. Grassle, The ecology of deep-sea hydrothermal vent communities, in J. H. S. Blaxter and A. J. Southward (eds.), *Advances in Marine Biology*, vol. 23, pp. 301–362, 1986; R. R. Hessler et al., Temporal change in megafauna at the Rose Garden hydrothermal vent (Galápagos Rift, eastern tropical Pacific), *Deep-Sea Res.*, 35:1681–1709, 1988; R. J. Larson, C. E. Mills, and G. R. Harbison, In situ foraging and feeding behaviour of Narcomedusae (Cnidaria: Hydrozoa), *J. Mar. Biol. Ass. U.K.*, 69:785–794, 1989; C. E. Mills and J. Goy, In situ observations of the behavior of mesopelagic *Solmissus* Narcomedusae (Cnidaria, Hydrozoa), *Bull. Mar. Sci.*, 43:739–751, 1988; F. S. Russell, *The Medusae of the British Isles: Anthomedusae, Leptomedusae, Limnomedusae, Trachymedusae, and Narcomedusae*, 1953; V. Tunnicliffe, Biogeography and evolution of hydrothermal-vent fauna in the eastern Pacific Ocean, *Proc. Roy. Soc. Lond.*, B233:347–366, 1988.

Deoxyribonucleic acid (DNA)

Reproduction is the fundamental event of living systems, and much recent effort has been devoted to the question of replication at the molecular level. In biochemical systems, the enzyme-catalyzed replication of deoxyribonucleic acid (DNA) or the transcription of DNA into messenger ribonucleic acid (mRNA) is reasonably well understood; one strand of the nucleic acid acts as a template for the synthesis of the other, complementary strand. Unraveling the myster-

ies of the enzyme reverse transcriptase, which uses RNA as a template for DNA synthesis in retrovirus replication, has seen intensive activity, particularly in the context of research involving acquired immune deficiency syndrome (AIDS). Inhibition of this enzyme is one of the goals of AIDS therapies.

A number of nonenzymatic models of self-replicating systems have also been developed. In part, these studies are spurred by the discovery of ribozymes, or catalytic RNA molecules. The self-splicing intron of the protozoon *Tetrahymena* catalyzes a number of phosphoester transfer reactions, and it has recently been demonstrated that this intron is capable of splicing multiple oligonucleotides aligned on a template strand to yield a fully complementary product strand. Such self-replicating RNA or related polynucleotide structures are believed to be the key intermediates in the evolution of living systems from prebiotic chemicals. This article discusses the use of short, nucleotide sequences as templates for self-replication and also the recent work in self-assembly of molecules, especially those using base pairing as a vehicle for amide synthesis and other reactions. SEE RIBONUCLEIC ACID (RNA); RIBOZYME.

Templates for self-replication. The initial studies used a self-complementary hexanucleotide DNA template to couple two trinucleotides, but template effects have recently been observed with even smaller molecules. These usually involve dehydrating agents that form phosphodiesters in the key step. Autocatalysis has been observed in such reactions; that is, the dimeric product dissociates and reenters the cycle as a catalyst. Attempts to polymerize single nucleotides on a single strand were less successful, because individual base pairs are unstable in aqueous solutions. Higher nucleotides or their analogs are required in order to observe template effects, but the minimum number needed for self-replicating systems has not been established.

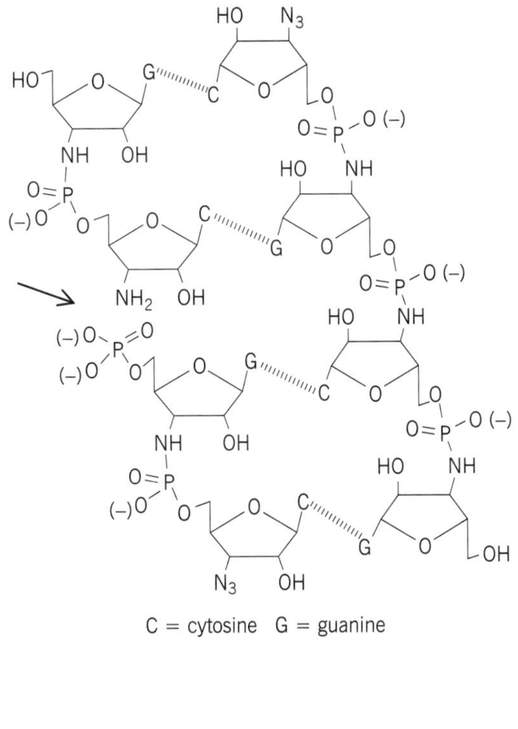

Fig. 1. A tetranucleotide analog acts as a replication template for the coupling of two dinucleotides. The arrow indicates the site of covalent N—P bond formation.

C = cytosine   G = guanine

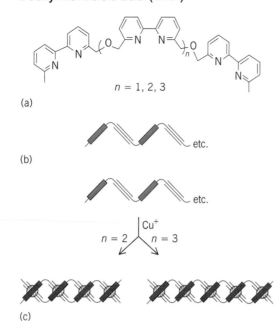

n = 1, 2, 3

(a)

etc.

(b)

etc.

Cu^+

n = 2   n = 3

(c)

Fig. 2. Assembly of polybipyridyl helicates by copper(I) ions in a process involving self-recognition. (a) Structure of the polybipyridyl ligand. (b) Symbolic representation of the polybipyridyl ligand. (c) Formation of double-stranded helicates.

plate molecules as a miniduplex. A number of similar studies using nucleotides or closely related compounds have appeared.

Molecular assemblies. Research in bioorganic chemistry has been involved with model systems for the nucleotide studies described above. The emerging research trend involves molecular assemblies in which self-recognition plays a key role. An example is given in **Fig. 2**. A series of metal ion–chelating agents was prepared from alternating segments of bipyridyl and ether subunits. Treatment of the polybipyridyl ligands with copper(I) (Cu^+) salts causes the spontaneous formation of double-stranded species known as helicates. Significantly, no mismatching occurs between helices of different lengths. The resulting strings of metal ions can reach up to 2.7 nanometers in length. The metal ion reads out the information of the ligating sites and arranges them on its surface in a tetrahedral geometry. The central bipyridyl core has been modified with chiral auxiliaries to induce asymmetric helix formation. Most recently, nucleic acid components have been added to the periphery of this system. The new molecules feature positive charges at their cores, stacking between the pyrimidine groups, and deoxynucleosides on the outside of the double helix. These compounds are reminiscent of the structures originally proposed by Linus Pauling for DNA.

Other examples of how the specific geometry requirements of metal ions can be used to assemble unusual shapes exist in the synthesis of catenanes, structures that are characterized by two interlocked rings. They can be formed from a copper(I) ion center to which two appropriately substituted *ortho*-phenanthrolines are bound. After the ring-forming reactions, which take place on the periphery of the structure, are completed, the metal ion can be removed, leaving the catenane. An even more complex structure, a molecular trefoil knot, has been synthesized through this strategy.

Modules. This research also exemplifies the advantages offered by modules, that is, subunits used repeatedly to assemble structures of considerable size with greater efficiency. Rather than continuing the development of amino acids or nucleotides as subunits, many researchers in bioorganic chemistry (a branch of organic chemistry), especially those whose work involves molecular recognition, are exploring a variety of structures such as crown ethers, bipyridyls, and binaphthyls. Combinations of these units lead to a rich assortment of rapidly synthesized, midsized molecules that can express a number of the characteristics of biological macromolecules. Moreover, the use of nonaqueous media permits the observation of phenomena involving weak intermolecular forces. For example, in solvents such as acetonitrile or chloroform, hydrogen bonding and aryl stacking effects are magnified to an extent that allows their characterization by nuclear magnetic resonance methods.

Model systems have been developed to show behavior such as transport of amino acids and nucleosides across liquid membranes; allosteric effects in which binding interactions at one site induce confor-

A specific case involving the synthesis of a tetranucleotide is shown in **Fig. 1**. This was shown to catalyze the formation of itself (self-replication) in the presence of water-soluble carbodiimide as a condensing agent. In a reaction mixture where the initial concentration of the tetramer was 0.25 millimole, each such molecule acted as a template for the synthesis of about 0.4 new tetramer during the course of the reaction. The observed dependence of the reaction rate followed the square root of the total concentration of the template. This is the anticipated result, given the extensive dimerization of two tem-

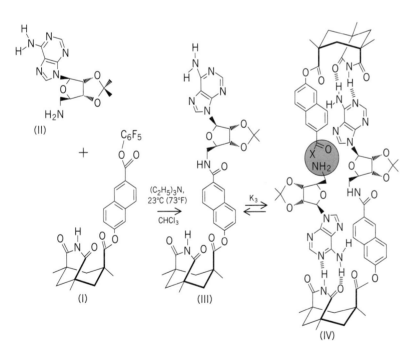

Fig. 3. Autocatalysis in a self-replicating system that uses base pairing to catalyze amide formation. X = OC_5F_5. K_3 = equilibrium constant: [IV]/[III][II][I]. The shaded area indicates the site of covalent N—C bond formation.

mational changes that regulate binding at a remote site; catalysis of biorelevant chemical reactions, such as phosphate and acyl transfers; and highly specific molecular recognition of neurotransmitters and barbiturates. These phenomena were once believed to require structures comparable in size to biological macromolecules such as proteins.

Self-replicating systems. A self-replicating system involving the formation of amide bonds has been developed recently. Specifically, the imide pentafluorophenyl ester (I) shown in **Fig. 3** provides hydrogen bonding and aromatic stacking interactions for base pairing with adenine derivatives. An amino derivative of adenosine (II) was shown to base-pair and couple to give the template product (III). The latter was shown to catalyze its own formation by gathering the two components from which it is composed on its surface and positioning them for efficient acyl transfer reactions. The termolecular complex (IV) is much more reactive toward amide formation than are the separated amine and ester components in the bulk solution. The rate of formation of the coupling product could be increased by more than 40% by the addition of the template to the components in CHCl$_3$ solution, even though only about 2% of the components were in the form of the termolecular complex under the reaction conditions.

This system presents the possibilities of developing a code in which the information involved in the hydrogen bonds of base pairing is translated into the formation of a specific amide bond. This could be of significance in prebiotic chemistry. Many researchers believe that nucleic acids taught proteins how to reproduce, a process that requires molecules that can act as adaptors between the two very different structures. This function is currently performed by the enzymes that recognize both the amino acids and the specific transfer RNA (tRNA) molecules to which the amino acids become covalently attached for protein synthesis. The search for an ancestral adaptor system and the origins of the genetic code continues to generate much speculation.

Another system capable of catalyzing formation of products through a termolecular complex has been developed. An artificial enzyme provides the surface on which an amine nucleophile and an activated bromide are arranged in space for an intramolecular substitution reaction. The rate enhancement observed was six times that of the background reaction. The system was subsequently modified to provide distinct recognition sites for nucleophile and electrophile components. As a result, the rate increase has been improved to greater than a factor of 20.

Significance. The recurrent theme in all of the systems described above involves the promotion of otherwise bimolecular reactions to unimolecular ones on a template surface. This strategy for formation of bonds provides a means by which the entropy of the system can be temporarily reduced to stabilize the transition state for the reaction. The enhanced coupling rates observed augur well for the development of wholly synthetic enzyme mimics. Such molecules could provide catalysts for processes involved in chemical manufacture for which no biological methods are available.

For background information SEE BIOINORGANIC CHEMISTRY; DEOXYRIBONUCLEIC ACID (DNA); ENZYME; LIFE, ORIGIN OF; OLIGONUCLEOTIDE in the McGraw-Hill Encyclopedia of Science & Technology.

Julius R. Rebek

Bibliography. T. R. Kelly, C. Zhao, and G. J. Bridger, A bisubstrate reaction template, *J. Amer. Chem. Soc.*, 111:3744–3745, 1989; J.-M. Lehn and A. Rigault, Helicates: Tetra and pentanuclear double helix complexes of Cu^I and poly(bipyridine) strands, *Angew. Chem. Int. Ed. Engl.*, 27:1095–1096, 1988; T. Tjivikua, P. Ballester, and J. Rebek, Jr., A self-replicating system, *J. Amer. Chem. Soc.*, 112:1249–1250, 1990; W. S. Zielinski and L. Orgel, Autocatalytic synthesis of a tetranucleotide analogue, *Nature*, 327:346–347, 1987.

Developmental biology

This article concentrates on the role of extracellular matrix in development and in maintenance of tissue-specific gene expression. Most cells in higher animals are in contact with each other and with an extensive network of glycoproteins and proteoglycans referred

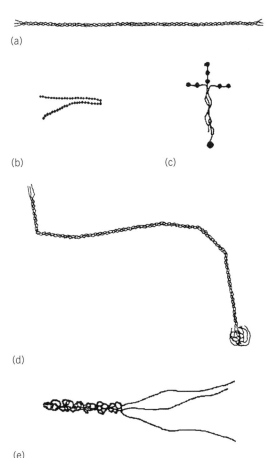

Fig. 1. Major components of the extracellular matrix drawn to the same scale. (*a*) Types I, II, and III collagen. (*b*) Fibronectin. (*c*) Laminin. (*d*) Type IV collagen. (*e*) Perlecan, a proteoglycan with three long heparan sulfate chains. (*Reproduced with permission of R. L. Trelsad, Matrix molecules: Spatial relationship in two dimensions, Ann. N.Y. Acad. Sci., 580:391–420, 1990*)

to as extracellular matrix. Extracellular matrices are instrumental in guiding cell migration, adhesion, differentiation, morphogenesis, and tissue-specific gene expression. The biochemistry of extracellular matrix molecules has been well studied; most of the genes have been cloned and the proteins are well characterized (**Fig. 1**). In connective tissues the extracellular matrix is composed mainly of type I and type III collagen (Fig. 1*a*), proteoglycans, and glycoproteins such as fibronectin (Fig. 1*b*). The collagens are a family of large proteins, rich in glycine and proline, and characterized by a triple-helical structure. When aggregated, collagens produce fibrils of exceptional tensile strength. There are 13 known distinct collagens. Dimers of fibronectin are important cell adhesion molecules, linking many cell types to collagen and proteoglycan matrices. Fibronectin is one of the key molecules that guides cell migration during development.

In organs such as exocrine and endocrine glands or skin, sheets of epithelial cells, joined together by cell–cell adhesion molecules, are separated from the underlying stroma (composed of stromal cells and a form of extracellular matrix called interstitial matrix) by a basement membrane. Basement membranes are thin sheets of another type of extracellular matrix that contain laminin (Fig. 1*c*), type IV collagen (Fig. 1*d*), heparan sulfate proteoglycan (now referred to as perlecan; Fig. 1*e*), and entactin. Laminin, a potent regulator of cell behavior, is a molecule formed from three glycoprotein chains that are assembled in a cruciform structure.

In such glandular tissues, both the epithelial and stromal cell types interact with components of the extracellular matrix via cell surface receptors (**Fig. 2**). Cytokines and hormones bind to their own specific cell surface receptors to influence aspects of cell behavior such as growth and differentiation; some of these factors bind to extracellular matrix, which then presents the factors to cells. Certain developmental processes and cell migrations depend upon local remodeling of the extracellular matrix, which is accomplished by protease-dependent matrix degradation and the synthesis of new extracellular matrix components.

Historically, extracellular matrix molecules were thought to provide only structural and physical support. However, a rich literature identified the stroma as a key regulator of morphogenesis and function. With that in mind, it was suggested that extracellular matrix molecules are involved in directing tissue-specific function. It was proposed that the extracellular matrix is linked to the nucleus via transmembrane receptors and the cytoskeleton, and hence has a direct effect on gene expression. This model implies that the unit of function in the higher organisms is the cell plus its extracellular matrix, and that the functional unit in the organism is the tissue itself (epithelium, mesenchyme, and the extracellular matrix). Much evidence has accumulated to support this concept. Extracellular matrix proteins, as well as hormones and growth factors, have emerged as important regulatory elements in determining tissue-specific form and function. The following is a review of some of the evidence for the involvement of extracellular matrix molecules in development and in expression of tissue-specific functions.

Role in development. Three recent experimental approaches have been useful in unraveling the role of extracellular matrix in development. The first approach uses natural mutants, where the phenotype is manifested but the genes responsible have to be tracked down. The second approach uses transgenic animals, where a gene is introduced and the resultant phenotype is dissected. The third approach disrupts developmental processes by using specific inhibitors or antibodies to extracellular matrix molecules.

Three developmental mutants, the lethal spotted mutant mouse, the white axolotl mutant, and the *Steel* mouse, have aberrations in extracellular matrix structure and function. The lethal spotted (*ls/ls*) mutant mouse has been shown to be deficient in ganglion cells (derived from neural crest cells) in the terminal segment of the colon. The animals develop a congenital megacolon in which functional motility is absent in the aganglionic zone. The neural crest cells, however, appear to be normal in coculture experiments, suggesting that non-neural-crest-derived cells of the *ls/ls* terminal bowel are responsible for the failure of neural crest cells to migrate. Recent studies have shown that basement membrane components in the terminal bowel are abnormal, which argues that the defect in *ls/ls* mutant is related to faulty extracellular matrix

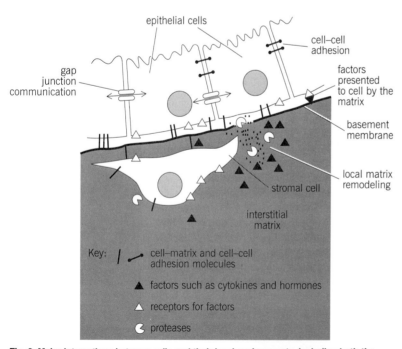

Fig. 2. Major interactions between cells and their local environments, including both the extracellular matrix (basement membrane and interstitial matrix) and adjacent cells. (*After A. W. Stoker et al., Designer microenvironments for the analysis of cell and tissue function, Curr. Opin. Cell Biol., 2:864–874, 1990*)

accumulation. In white axolotl mutants, trunk pigmentation is restricted because the epidermis does not support the subepidermal migration of pigment cells from the neural crest. By using membrane microcarriers to absorb extracellular matrix molecules in the organism and by transplanting them into embryos, recent studies indicate that it is the extracellular matrix from mutant skin that fails to promote migration at the correct developmental juncture. Finally, in *Steel* mice, where a mutation adversely affects the differentiation of three migratory stem-cell populations (including melanoblasts derived from the neural crest), it has been shown that melanogenesis is enhanced on normal extracellular matrix but not on mutant-derived extracellular matrix.

Studies with transgenic animals are not yet refined. However, the importance of intact collagen I can be seen in transgenic animals in which introduction of a mutant gene leads to an embryonic lethal phenotype. In addition, many human diseases are known to be related to mutations in collagen genes.

Finally, studies with antagonists indicate that an intact extracellular matrix is required for many developmental processes, including neural tube morphogenesis. Thus, extracellular matrix is becoming firmly established as an important component of inductive processes in development.

Maintenance of tissue-specific function in differentiated cells. Extracellular matrix is also necessary for maintenance of tissue-specific function once the cells have differentiated. The use of cell culture is necessary in order to study the regulation of gene expression under strictly defined conditions. However, as soon as the cells are removed and placed under conventional culture conditions (that is, in plastic dishes in the presence of serum), the characteristic traits to be studied are lost. This occurs because conditions that were defined for growing cells in culture are not necessarily conducive for keeping them differentiated.

The importance of cell–cell and cell–extracellular matrix interactions in functional differentiation was realized when epithelial cells (notably hepatocytes and mammary epithelial cells) were cultured on exogenous physiological matrices. In many cases the cells, which on tissue-culture plastic had lost both their epithelial morphology and tissue-specific traits, could now resume their "correct" form and some specialized functions. With the recognition that a functional basement membrane can provide an even more physiologically relevant substratum for epithelial cells, reflecting their close apposition in the organism (see Fig. 2), and with the availability of a preparation of basement membrane from a tumor (EHS matrix), high levels of tissue-specific gene expression can now be achieved in a wide variety of cell types. For example, hepatocytes (epithelial cells that constitute the major cell type in the liver) in culture require such a matrix for sustained expression of liver function.

Biological effects of the extracellular matrix apply

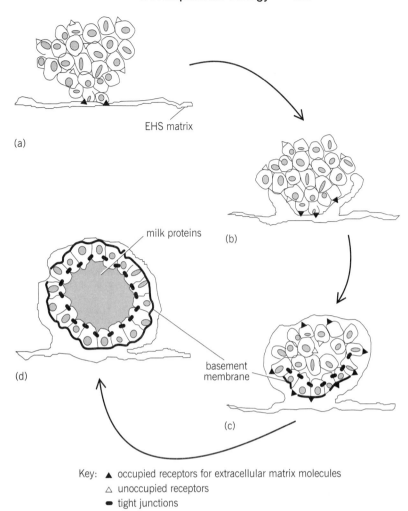

Key: ▲ occupied receptors for extracellular matrix molecules
  △ unoccupied receptors
  ● tight junctions

Fig. 3. Model of alveolarlike morphogenesis by mammary epithelial cells cultured on Engelbreth-Holm-Swarm (EHS) matrix. (*a*) Clusters of cells interact with components in the EHS matrix via extracellular matrix receptors and (*b*) roll themselves up in the matrix by a mechanism that is not understood. (*c*) The cells then polarize, forming a basal surface at the site of contact with the EHS matrix. (*d*) This results in the establishment of full polarity with tight junctions forming at the apical surface of the cells. Subsequent vectorial secretion of milk products in an apical direction results in the swelling of cell clusters into alveolarlike structures. (*From M. H. Barcellos-Hoff and M. Bissell, in L. C. Krey, B. J. Gulyas, and J. A. McCracken, eds., Autocrine and Paracrine Mechanisms in Reproductive Endocrinology, Plenum Press, 1989*)

to mesenchymal as well as epithelial cells. For example, liver-derived lipocytes express one kind of extracellular matrix on tissue-culture plastic (type I collagen), but on basement membrane matrix they switch exclusively to type III collagen synthesis, a phenotype resembling the situation in the organism.

Vascular endothelial cells form tubular networks on EHS matrix in a laminin-dependent process; peptides that inhibit interaction of these cells with laminin inhibit tube formation. Angiogenesis is regulated by the extent of interaction of endothelial cells with extracellular matrix molecules. In addition to directly influencing cellular function, such matrices sequester growth factors and thus affect functional differentiation and growth (see Fig. 2). The subendothelial matrix, for example, binds basic fibroblast growth factor, which directly stimulates endothelial cell growth. Tubule formation in the kidney also depends on both extracellular matrix and sequestered transforming growth factor alpha.

Mammary epithelial cells from pregnant mice present one of the more dramatic examples of cell–extracellular matrix interactions. Epithelial cells from the glands of pregnant animals, plated on type I collagen gels that are then floated, contract the gel and establish intimate cell–cell interactions (a three-dimensional culture model closer to the actual structure of the gland). Under these conditions the cells change shape, become polar, develop extensive microvilli, and secrete a number of milk proteins. All messenger ribonucleic acids (mRNAs) for milk proteins are upregulated (some as much as 50–70-fold), and the degree of response to lactogenic hormones is considerably elevated; these processes are regulated at both transcriptional and posttranscriptional levels. On an exogenously supplied basement membrane (EHS matrix), the cells undergo complex morphological reorganization leading to formation of functional ''alveoli'' (**Fig. 3**). In these alveoli, most milk proteins are secreted vectorially from the apical surface of the polarized epithelial cells into an enclosed luminal space, whereas other proteins such as extracellular matrix–degrading enzymes are secreted basally. In addition, cells on this type of basement membrane matrix are capable of producing whey acidic protein, a major mouse milk protein that is absent in cells on tissue-culture plastic and is greatly reduced on other substrata. The inability to synthesize whey acidic protein under the latter conditions appears to be related to the production of a secreted inhibitor that, by autocrine action, selectively turns it off. Thus, the formation of the inhibitor and the ability of cells to make whey acidic protein are dependent on the nature of the extracellular matrix on which the cells are cultured.

For background information *see* CELL DIFFERENTIA-TION; CELL-SURFACE DIFFERENTIATIONS; CELLULAR ADHE-SION; CONNECTIVE TISSUE; CYTOSKELETON; DEVELOPMEN-TAL BIOLOGY in the McGraw-Hill Encyclopedia of Science & Technology.

M. J. Bissell; A. R. Howlett; C. H. Streuli

Bibliography. M. J. Bissell, H. G. Hall, and G. Parry, How does the extracellular matrix direct gene expression?, *J. Theoret. Biol.*, 99:31–68, 1982; K. Morrison-Graham and J. A. Weston, Mouse mutants provide new insights into the role of extracellular matrix in cell migration and differentiation, *Trends Genet.*, 5:116–121, 1989; E. J. Sanders, The roles of epithelial-mesenchymal cell interactions in developmental processes, *Biochem. Cell Biol.*, 66:530–540, 1988; A. W. Stoker et al., Designer microenvironments for the analysis of cell and tissue function, *Cur. Opin. Cell Biol.*, 2:864–874, 1990.

Electric power systems

The basic function of a modern electric power system is to satisfy the system load requirement as economically as possible and with reasonable assurance of continuity and quality. The concern regarding the ability of the system to provide an adequate supply of electrical energy is usually designated by the term reliability. The concept of power system reliability is extremely broad, covering all aspects of the ability of the system to satisfy customer requirements. Reliability has a wide range of meanings and cannot be limited to a single specific definition. It is therefore often used to indicate the overall ability of the power system to perform.

The concern designated as system reliability can be subdivided by considering reliability evaluation in terms of adequacy and security assessment. In general, adequacy assessment involves steady-state analysis, while security evaluation includes consideration of dynamic conditions. A power system can be divided into three hierarchical frameworks for the purpose of conducting reliability assessments. Hierarchical level I (HLI) is concerned only with the generation facilities. Hierarchical level II (HLII) includes both generation and transmission, while hierarchical level III (HLIII) includes all three functional zones of generation, transmission, and distribution in an assessment of reliability at the customer load point.

Hierarchical level I. Most utilities now use probabilistic techniques in adequacy assessment at hierarchical level I. These methods have largely replaced the conventional deterministic approaches, such as the percent-reserve method. Security evaluation at hierarchical level I is usually confined to evaluation of operating risk, and in this area most utilities use a deterministic approach.

The most recent innovation in assessment at hierarchical level I is the use of a reliability cost–reliability worth approach to determine the optimum reserve level. The basic concept is illustrated in **Fig. 1**. Electric power utility costs increase as consumers are provided with increased reliability. Costs to electrical energy consumers that are associated with power failures decrease as system reliability increases. The total cost is the sum of the two curves and exhibits a low point at what is designated an optimum level of reliability. The major difficulty with this approach is the determination of the customer costs associated with power failures. This is a major area of research activity.

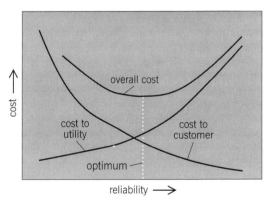

Fig. 1. System reliability cost-worth curve.

The most convenient adequacy index to use in combining the predicted adequacy and the interruption costs at hierarchical level I is the expected energy not supplied. This index has been available for some time but has only recently been used as a primary adequacy criterion. Customer interruption costs can be expressed in the form of customer damage functions (CDF) and can be presented for a single customer or for a combined group of customers in the form of a composite customer damage function (CCDF).

Hierarchical level II. Reliability evaluation at hierarchical level II is usually referred to as composite system assessment because it involves a combined evaluation of the generating and major transmission facilities. Very few utilities now utilize quantitative reliability techniques at hierarchical level II, but a large number of groups are engaged in development and program-testing activities. The bulk of the work has been performed in the adequacy domain, and very little progress has been made in security assessment at hierarchical level II.

The basic objective in adequacy evaluation at hierarchical level II is to quantitatively assess the ability of the electric power system to satisfy the energy requirements at the major load points in the system. **Figure 2** shows the single-line diagram of a small test system.

The basic indices of load-point inadequacy are the probability of failure and the expected frequency of failure. It is also possible to produce additional indices involving the extent of load and energy curtailment at each bus, namely, expected number of voltage violations, expected number of load curtailments, expected load curtailed, expected energy not supplied, and expected duration of load curtailment.

The major requirement in adequacy assessment at hierarchical level II is the determination of what constitutes a system failure. The ability to do this depends on the solution technique used. The fastest and least discriminating network-analysis technique is the network-flow approach. The most widely used approach in studies at hierarchical level II is the dc-load-flow method, which permits the recognition of system overloads and transmission limitations but not the consideration of voltage and volt-ampere-reactive effects. The most comprehensive technique and the slowest in terms of computer solution time is the ac-load-flow method, which permits the calculation of all the load-point inadequacy indices listed above.

The load-curtailment indices at each load point in the system shown in Fig. 2 are quite different and are influenced by the priorities placed on supplying these loads. The reliability cost–reliability worth considerations discussed above can be used to recognize individual-consumer reliability requirements at each load point by creating composite customer damage functions at the individual load points. This is a new area of reliability research and development and promises to be a significant advancement in comprehensive system planning.

The individual load-point indices can be combined to produce an overall system-risk index that provides a useful management and global planning criterion. Possible system-inadequacy indices are the bulk-power interruption index, the bulk-power supply average megawatt curtailment per disturbance, the bulk-power energy-curtailment index, the modified bulk-power energy-curtailment index, and the system-severity index.

The individual load-point inadequacy indices and the system inadequacy indices do not replace each other but serve quite different functions in composite system adequacy evaluation. The addition of a line between any two load points improves the system indices but has maximum impact on the load points in the immediate proximity of the addition. The actual magnitude of this impact can be determined only from the individual load-point indices.

The most useful indices at hierarchical levels I and II are the expected-energy-not-supplied values, since these can be related to the customer interruption costs through factors known as interrupted energy assessment rates (IEAR). These factors are obtained from

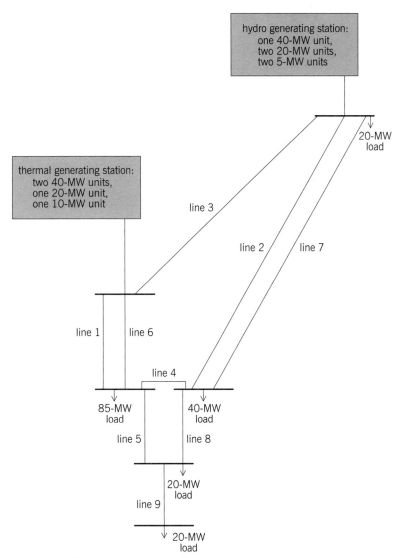

Fig. 2. Composite generating and transmission system.

the composite customer damage function profiles of the individual load points and the overall system. Reliability cost-worth evaluation is an immature technology that promises to be useful in future system decision making.

Techniques. Adequacy assessment indices at both hierarchical levels I and II can be calculated by two fundamentally different techniques. Analytical techniques represent the system by a mathematical model and evaluate the indices from this model by using mathematical solutions. Monte Carlo simulation methods estimate the indices by simulating the actual process and random behavior of the system. These methods treat the problem as a series of real experiments. Both methods have advantages and disadvantages. Monte Carlo simulation usually requires a large amount of computing time and should not be used if alternative analytical techniques are available. However, Monte Carlo simulation can, at least in theory, include system effects that may have to be severely approximated in an analytical approach. It can also produce statistical distributions of the adequacy indices that cannot be obtained by an analytical approach.

Prospects. The availability of high-speed dedicated digital computers has permitted the development of new solution techniques and digital computer programs for system assessment. Composite generation and transmission facilities are important segments of an overall power system; therefore, this is an important area of reliability research and development. The ability to predict these indices and the availability of representative customer outage cost data will permit quantitative reliability cost–reliability worth assessment to become an important activity in electric power utility planning and operation.

For background information *see* Electric power systems; Electric power systems engineering; Monte Carlo method; Reliability of equipment in the McGraw-Hill Encyclopedia of Science & Technology.

Roy Billinton

Bibliography. R. Billinton and R. N. Allan, *Reliability Assessment of Large Electric Power Systems*, 1988; R. Billinton and R. N. Allan, *Reliability Evaluation of Power Systems*, 1984; *Reliability Assessment of Composite Generation and Transmission Systems*, IEEE Tutorial Course Text 90EH0311-1-PWR, December 1989.

Electrical utility industry

The year 1990 was unique in the history of the electrical utility industry in that it witnessed the first hostile takeover attempts by one utility against another. Pacificorp, the parent company of Pacific Power & Light Company, tendered an unsolicited bid for the Arizona Public Service Corporation, a fully owned subsidiary of the Pinnacle Corporation. In a second incident, the Kansas City Power & Light Company pursued a hostile takeover of the Kansas

Gas & Electric Company. Pacific Power & Light Company, based in the winter-peaking Northwest, was seeking additional winter capacity available on the summer-peaking Arizona Public Service system, and seeking to sell energy available in the summer to the Arizona Public Service region. Kansas City Power & Light is short of capacity to meet its summer demand, while the contiguous Kansas Gas & Electric system has a surplus of capacity available. Both takeover attempts ultimately failed when the parties negotiated other mutually acceptable agreements. The fear of hostile takeovers has led to adoption of poison-pill takeover measures, that is, measures that would make it economically unattractive to take control of the company through hostile action, by 21 different investor-owned utilities, 7 in 1990 alone.

In another milestone case, Connecticut-based Northeast Utilities, the largest of the New England electrical utilities, made a friendly offer to take over Public Service Company of New Hampshire, which had become insolvent due to costs associated with the

Table 1. United States electric power industry statistics for 1989*

Parameter	Amount	Change compared to 1988, %
Generating capacity, MW		
Hydroelectric	86,980 (11.9%)	0.85
Fossil-fueled steam	475,563 (65.1%)	1.15
Nuclear steam	106,722 (14.6%)	4.73
Combustion turbines, internal combustion	61,245 (8.4%)	6.17
TOTAL†	730,517	1.84
Noncoincident demand,‡ MW	523,410	−1.1
Energy production, TWh§	2,780.8	2.83
Energy sales, TWh		
Residential	896.2	2.90
Commercial	718.5	4.69
Industrial	903.3	3.31
Miscellaneous	91.0	1.89
TOTAL	2,609.0	2.34
Revenues, total, 10⁹ dollars	168.0	3.61
Capital expenditures, total, 10⁶ dollars	26,850	0.85
Customers, 10³		
Residential	96,265	2.55
TOTAL	107,997	1.54
Residential usage, (kWh/customer)/ year	9,479	0.22
Residential bill, cents/kWh (average)	7.61	1.87

*After 41st annual electric utility industry forcast, *Elec. World*, 204(10):61–68, October 1990; 1990 annual statistical report, *Elec. World*, 204(4):11–24, April 1990; and advanced release of data from Edison Electric Institute, *Statistical Yearbook of the Electric Utility Industry*, 1989.
†Includes 7 MW of wind and solar capacity. Does not include nonutility capacity available to utilities.
‡Noncoincident demand is the sum of the peak demands of all individual utilities, regardless of the day and time at which they occurred.
§TWh (terawatt-hour) = 10¹² Wh.

failure to obtain the necessary permission to bring on line the Seabrook nuclear unit or the financial relief to pay for its construction. The offer has been accepted, but the terms are still under consideration by the legislatures and regulatory agencies involved.

Regulatory decisions. There were two especially notable regulatory decisions made during 1990. The first was the decision by the Federal Energy Regulatory Commission to defer immediate action on the question of requiring utilities to grant to third parties the right of open access to their transmission networks. Technical questions of reliability and security advanced vigorously by utilities and success in moving toward this goal through case-by-case rulemaking reduced the urgency with which the former chairperson had approached this matter. The issue has been deferred for further consideration.

The second decision, made by the Environmental Protection Agency, has potentially far-reaching implications for the industry's plans for rebuilding older generating stations rather than constructing new ones to meet forecast demands. The Wisconsin Electric Power Company had planned to rebuild the units at its Port Washington power station. The Environmental Protection Agency ruled that, even though it was an existing station, the revisions being made to upgrade the generating units were extensive enough to subject the exhaust from the plant to the same emission standards as would be applied to new units. Until this ruling, such plant improvements were considered exempt from new source standards. Should this ruling stand, any plant that was renovated to any major degree would probably have to be retrofitted with exhaust-gas cleanup that employed the best available technology. This would generally mean that flue-gas scrubbers would have to be retrofitted to any coal-fired unit at a cost that would make such upgrading economically impractical. The Environmental Protection Agency had based its rulings on a comparison of emissions during actual operation of the units in the past with potential future emissions, assuming that the renovated units would operate at maximum capacity 24 h a day for 365 days a year. This specific interpretation was struck down in the courts, but the issue remains open.

Capacity additions. Capacity additions continue to lag increase in system demand nationwide. During 1988, electrical utilities in the United States added a total of 7625 MW to their systems. This was followed by 6665 MW in 1989 and an estimated increment of 3597 MW in 1990. The 1990 figure is the lowest in the last 41 years.

Reasons for low growth. The historically low figure for annual capacity additions can be attributed to several factors. First, there was the abrupt shift from long-range growth forecasts in the 7–8% range that obtained prior to the energy crisis of 1973 to the current consensus long-term growth forecast of about 2.5–3%. This caused utilities to cancel construction of those plants that, considering a normal construction lead time of roughly 8–10 years, would have entered service in the 1981–1988 period. Second, regulatory

discouragement of new construction through the disallowance of any return or only partial return on capital expenditures for new capacity ruled to be in excess of present need, coupled with intense environmental opposition, has caused utilities to sustain the hiatus in ordering new units. Consequently, annual capacity additions may continue to decline until the 1995–1997 period. There is rising concern on the part of utilities and regulators that insufficient capacity will be available to meet the demands of the 1990s, and a renewal of orders over the next few years should drive capacity additions up from present levels beginning in the late 1990s. This concern was confirmed in a heat wave during the summer of 1990 by the forced imposition of rolling brownouts and blackouts in Maryland and Florida because of inability of local utilities to meet the extraordinary demands imposed on their systems.

Aging of generating base. Because of the long hiatus in construction, the generating base in the United States is aging. In decades past, utilities generally estimated 30 years as the economic life of a generating unit, after which the accelerating cost of maintenance and low availability to serve load made it more economical to replace it with a new unit. An analysis of existing plant shows that by the year 2000 about 70,000 MW of existing fossil-fired and nuclear generating capacity will be more than 40 years of age. This will be roughly 10% of total installed capacity at that time. Further, more than 100,000 MW of capacity will be between 30 and 40 years of age by that date. This could pose serious problems in reliability.

Division by plant type and utility type. The division of generating capacity by plant type in 1989, along with other United States electric power industry statistics for that year, is given in **Table 1**. The evolution of this breakdown in past years and a forecast for future years are shown in the **illustration**. Capacity additions in

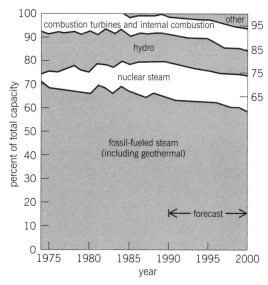

Probable mix of net generating capacity. "Other" includes capacity of nonutility generators available to utilities on a firm basis. (*After 41st annual electric utility industry forecast, Elec. World, 204(10):61–68, October 1990*)

Table 2. Generating capacity added in 1990, and comparison of capacity in 1989 and 1990

Plant type	Capacity added in 1990		Capacity, MW	
	Power, MW	Number of units	1989	1990*
Fossil-fueled steam	578	4	475,563	476,139 (64.9%)
Nuclear steam	2,205	2	106,722	108,927 (14.8%)
Combustion turbines	783	11	54,215	54,998 (7.5%)
Hydroelectric	25	2	86,980	87,005 (11.9%)
Diesel generators	6	1	7,030	7,036 (1.0%)
Solar			3	3
Wind-driven generators			4	4
TOTAL	3,597	20	730,517	734,114

*Percentages of total capacity are indicated.

1990 and a comparison of capacity in 1989 and 1990 are given in **Table 2**.

The electrical utility industry in the United States is pluralistic, divided among investor-owned utilities, cooperatives, municipal utilities, federal organizations such as the Tennessee Valley Authority, and state or public power districts. The division of capacity among these entities at the end of 1990 was as follows: investor-owned, 564,777 MW (76.9%); cooperatives, 26,381 MW (3.6%); federal, 63,963 MW (8.7%); municipals, 40,586 MW (5.5%); and state and public districts, 35,393 MW (4.8%).

Alternative measures of capacity. All capacity figures above represent the nameplate rating of the units, that is, the rating assigned by the manufacturer at the time of installation. However, some units may not be capable of achieving their nameplate rating for various reasons such as low water levels in hydroelectric reservoirs, boiler-tube leaks, component degradation, or partial derating caused by operating problems. Therefore, the North American Electric Reliability Council prefers to assess the ability of individual units or entire utility systems in terms of capability, that is, the ability to perform to a given level at periods of annual maximum demand, rather than by capacity. The actual capability of the United States systems during the seasonal summer peak demand period in August 1990 was only 652,492 MW, considerably below the aggregate nameplate capacity of those same systems.

The aggregate nameplate capacity or capability of the United States understates to some degree the capacity available to serve the maximum demand. The reluctance of utilities to construct new base-load generating units in the present regulatory climate, together with the provisions of the Public Utility Regulatory Policy Act and state regulations, both of which provide incentives for nonutility entities to build generating facilities and sell power to utilities, has spawned a vigorous sector of nonutility generators of electricity. Addition of the capacity of these entities that were available to utilities at the time of annual peak demand raises the industry's aggregate capability to 677,443 MW. Estimates by the North American Electric Reliability Council show nonutility capacity additions amounting to 18,120 MW over the 10-year period 1989–1999. This will raise the total of non-utility capacity available to utilities to 34,800 MW.

Canadian imports. A portion of the total annual peak demand of the United States is met by importation of electricity from Canada. Although of great importance to the New England region, where such imports constitute slightly more than 10% of total consumption, on a national level, imports make up only about 2% of the total.

Plant retirement. Utilities expect to retire 6540 MW during the 10-year period 1989–1999. This figure, however, could increase sharply, depending upon the final provisions of the acid rain legislation under final debate in the Congress. Older plants, which might otherwise be upgraded through plant rebuilding programs to provide capacity in lieu of building new units, could be retired should the strictures of the new clean-air laws require the installation of flue-gas scrubbers.

Demand. The demand in summer is, on a national basis, the critical measure of load on the national utility system. The high temperature of the air creates less efficient cooling of electrical equipment, and warmer water in the steam condensers of steam-turbine units reduces the efficiency of the thermal cycle in which such units operate. Preliminary estimates are that the total national summer demand in 1990 was 541 GW, an increase of 3.34% over the comparable figure of 523.4 GW in 1989. This peak demand is on a noncoincident basis; that is, it is the sum of the peak demands reached by individual utilities but not necessarily at the same exact time.

One major measure of the reliability of the electric system is the reserve margin, that is, the surplus of capacity installed over maximum peak demand. On a national basis, a margin of 25% is generally accepted as adequate. Reserve margin in 1989 was 37.6% and in 1990, 34.9%. This figure may be deceptive because it does not fully account for capacity that may be unavailable for various reasons at the time of maximum annual peak. The North American Electric Reliability Council prefers a slightly different measure of adequacy, the capability margin. This is a measure of the surplus of capacity actually available at the time of peak over actual maximum demand. Capability margin was 27.1% in 1989 and only 23.4% in 1990. The danger in using national figures, however, is that while an adequate reserve margin or capability margin may exist at the national level, some individual utilities or regions may actually have severe shortages of capacity or capability while others have a substantial surplus. In 1990, for example, although national reserves were more than adequate, Baltimore Gas & Electric Co. had to resort to rolling blackouts at the time of peak demand to avoid a system shutdown. This happened because two nuclear units on their system were shut down for repairs, modifications, and inspection by the Nuclear Regulatory Commission, and because transmission capacity into the service area was inadequate to import the equivalent capacity.

Usage. In 1989, the last year for which figures are available, energy sales to the ultimate customer rose 2.3% over those in 1988, reaching a total of 2609×10^9 kWh. For 1990, preliminary figures indicate a further rise of about 2%, pushing sales to 2661×10^9 kWh.

Residential sales, which constitute about 34% of the total, in 1989 rose by 2.9% over 1988 to $896,156 \times 10^6$ kWh. Commercial sales, which constitute about 28% of the total, rose sharply higher, jumping 4.7% to $718,523 \times 10^6$ kWh. Industrial sales, which are roughly 35% of the total, ended up by 3.3% to $903,276 \times 10^6$ kWh. The remaining percentage is made up of street lighting, railroads, public authorities, and interdepartmental usage.

Use per customer. Annual residential use per customer in 1989 was 9479 kWh, which represents only a slight increase from the previous all-time high of 9456 kWh in 1988. This small gain came despite an unusually hot summer in the southern sections of the country, and reflects low use in an extremely mild winter. Individual residential customers paid an average of 7.61 cents/kWh, which translates to an average annual bill to the residential customer of approximately $722. This is an increase of about 2% over the 1988 figure of $708 and, in real terms, reflects a continuation of the steady decline in real costs to the customer that has been maintained since 1983. In those regions where oil makes a major contribution to total generation, specifically New England and Florida, the rapid run-up in oil prices since the Iraqi invasion of Kuwait in 1990 will exert upward pressure on these per-kilowatt-hour costs, since regulators permit fuel-cost adjustments in rates.

On the basis of total ultimate customers rather than just residential customers, the annual average energy use was 24,347 kWh in 1989, a rise of only 1.5% over 1988. Average cost per kilowatt-hour on this same basis was 6.44 cents/kWh, or an annual average bill of $1567 per customer of all classes.

Number of customers. The number of ultimate customers served by electrical utilities rose 1.5% in 1989 to 107,997,082. These were divided as follows: residential, 95,265,110 (82%); commercial, 11,861,168 (9.8%); industrial, 502,004 (4.6%); street and highway lighting, 176,231 (1.6%); other public authorities, 190,743 (1.8%); and railroads and interdepartmental use, 1826 (0.02%).

Revenues. Total revenues of the electrical utility industry in 1989 amounted to $167.96 billion. This was an increase of 3.6% over 1988, roughly matching inflation. Residential revenues made up 40.6%, or

$68.23 billion, a 2.9% increase over 1988; commercial customers contributed 30.6%, or $51.29 billion, a 4.7% increase over 1988; and industrial customers paid 25.6% of the total, or $42.98 billion, a 3.3% jump from the previous year. The remaining 3.2% was paid by street lighting, railroads, and interdepartmental customers. The relatively low increases in residential and industrial revenues is a measure of the slowdown in housing completions during 1989 and the general softness of the economy in the industrial sector.

Fuels. Fuel, along with the cost of capital, is one of the two largest expenses that utilities incur. The relatively low price of natural gas is exerting pressure for switching to that fuel where possible. Additional pressure in that direction is growing from the increasing use of combustion turbines fired with gas to which utilities are turning increasingly to avoid construction of coal-fired, base-load capacity. Gas consumption in 1989 was 2767×10^{12} ft^3 (78.3×10^{12} m^3), compared to 2634×10^{12} ft^3 (74.5×10^{12} m^3) in 1988, a rise of 5%. Total energy generated by gas as a fuel in 1989 amounted to 265×10^9 kWh, or 9.5% of the total, slightly higher than the percentage in 1988. Oil, as might be expected, dropped from 247.8×10^6 barrels (39.4×10^6 m^3) burned in 1988 to only 158.2×10^6 barrels (25.2×10^6 m^3) in 1989, a decrease of 36%. Oil-generated energy represented just 5.7% of total generation during 1989. Coal continues its dominance as the fuel of choice. In 1989, utilities fired 765.8×10^6 short tons (692.7×10^6 metric tons). This compares to 1988 quantities of 757.5×10^6 tons (685.2×10^6 metric tons). Coal, in 1989, supplied 55.8% of the total energy generated, slightly down from the 56.9% of total in the previous year. Nuclear fuel contributed 529.4×10^9 kWh in 1989, 19% of the total. This compares to the 1988 total of 526.9×10^9 kWh, which was 19.9% of the total generated by all fuels and sources. Hydroelectric generators produced 265.1×10^9 kWh during 1989, which was 9.5% of total generation, compared to 222.9×10^9 kWh in the previous year, which was 8.3% of total national generation. All other sources, including wind, solar, geothermal, and other nonconventional sources, contributed only 0.4% of the total, about the same as during 1988.

Expenditures. Total expenditures for capital accounts during 1989 were $26.8 billion, of which $23.8 billion was by the investor-owned utilities. Expenditures for generating equipment constituted roughly half of the capital expenditures, about $10 billion for the investor-owned sector. This amount, however, included $2.3 billion in allowance for funds used during construction. This is a noncash item of current income that is added to the capital plant accounts for future collection when facilities currently under construction are completed and enter the rate base. Nuclear fuel made up $1.7 billion of the total. Total capital budgets for 1990 should rise to $28.6 billion.

Capital expenditures for new and reinforced distribution facilities for the investor-owned sector of the industry amounted to $8.2 billion in 1989. Rural cooperatives and municipals spent an additional $3.1 billion for distribution lines and substations during that year. In 1990, capital spending by the investor-owned utilities was slightly lower at $8.1 billion, and the rural cooperatives and public power entities also decreased spending to $2.8 billion.

Capital expenditures for transmission facilities in 1989 were $2.0 billion in the investor-owned sector and $1.1 billion for the public-power and cooperative utilities. During 1990, investor-owned utilities raised their expenditures substantially to $3.1 billion, while the cooperatives and public power sector saw only a mild increase to $1.3 billion.

Electric operating expenses of the investor-owned sector rose 4.3% in 1989, reaching $111.7 billion, with the cost of fuel representing 29% of that total. The oil increases of 1990 were not yet in effect, so fuel costs as a percent of total operating expense actually decreased from 29.3% in 1988. Comparable figures were not available for the other types of utilities.

Total assets of the investor-owned utilities at the end of 1989 were $516.6 billion.

For background information SEE ELECTRIC POWER GENERATION; ELECTRIC POWER SYSTEMS; ENERGY SOURCES in the McGraw-Hill Encyclopedia of Science & Technology.

William C. Hayes

Bibliography. Edison Electric Institute, *1989 Post-Summer Electric Power Survey*, 1989; Edison Electric Institute, *1989 Statistical Yearbook of the Electric Utility Industry*, 1991; 41st annual electric utility industry forecast, *Elec. World*, 204(10): 61–68, 1990; 1990 annual statistical report, *Elec. World*, 204(4):11–24, April 1990; North American Electric Reliability Council, *1990 Electricity Supply & Demand*, 1990.

Electromagnetism

The contributions of J. C. Maxwell to the mathematical framework of electromagnetism in the second half of the nineteenth century were so far-reaching and are of such permanent value that it is customary to use the term Maxwell's equations as a description of the entire content of the subject. This may give the impression that nothing now remains to be done in the field apart from the routine application of these equations to particular electromagnetic systems. This, however, is a mistaken view. Both electromagnetic theory and applied electromagnetics are still developing and changing, partly in response to technological needs such as arise in communication systems and power devices, partly in response to new mathematical and computational techniques, and partly as a result of an improved physical understanding of the electromagnetic phenomena themselves.

Physical understanding. Progress in the subject in the past has been due to clearer physical insight, and this is likely to be equally important in the future. Maxwell's central contribution lay not in his equations but in his description of the phenomena in terms of transverse waves and his discovery that these waves propagate at a constant velocity equal to the velocity of light. At first the propagation velocity was attributed to an elastic ether, which supported these waves, although Maxwell did not commit himself totally to this view. A deeper significance of the velocity emerged from A. Einstein's studies of the electrodynamics of moving bodies. Einstein was guided in his investigations by the search for invariant properties of physical systems, and he grasped the fact that the velocity of electromagnetic waves is an inherent property of space-time, so that there is no need for an additional substance called ether. This explained why it had been impossible to determine the ether velocity relative to a particular reference frame. It also indicated that electric and magnetic propagation effects are aspects of the single phenomenon of energy propagation and are therefore inseparable from each other. The change of view also clarified the role of the potentials as measures of energy.

Maxwell's investigations dealt with energy propagation in space but not with the related problem of the interaction of electromagnetic energy and matter. This problem was settled by quantum theory, which showed that the invariant property of this interaction is not the energy itself but the energy divided by the frequency of the wave. The frequency converts the space-time behavior of electromagnetic waves into purely spatial energy packets. The notion of frequency involves the substitution of time by the idea of phase associated with a steady succession of waves of constant length.

Although these ideas have been generally accepted in particle physics, their influence has been less felt in applied electromagnetics, where the essential link between space and time in propagation effects is often obscured by treating time as an additional independent coordinate. Moreover, the standard textbook treatment seldom explains the difference between time and phase effects. The consequences of Einstein's theory are generally discussed in terms of small corrections to measurements, and the implications of quantum theory are similarly confined to phenomena involving subatomic particles. In these respects the present state of electromagnetic theory is still unsatisfactory.

Numerical methods. Maxwell's partial differential equations are simple to write down, yet extremely difficult to solve. Before the advent of computers, solutions were available only for systems having a high degree of symmetry such as cylinders and spheres. These could be solved in terms of special functions that were tabulated. Such restrictions were removed when it became possible to handle large amounts of numerical information directly. One of the most influential methods for doing so is the method of finite elements, which enables electromagnetic systems to be divided into regions of arbitrary shape that are then interconnected. The differential equations are replaced by a set of simultaneous algebraic equations, which can be solved numerically by standard methods. The accuracy of the solution can be improved by increasing the number of the finite elements into which the system is divided. The method is simple in concept and extremely flexible and has been brought to a high degree of perfection so that it can deal with systems containing regions of different kinds of materials of arbitrary shape, which may have nonlinear characteristics. It requires considerable computing power, but this presents few obstacles in view of the continuing rapid progress in computer hardware.

The chief difficulty lies in the preparation of software. The construction of finite element programs requires a rare combination of skills, including physical insight as well as numerical and mathematical ability of a very high order. The time involved in the development of programs is always considerable. The requirements of users vary greatly. Some need a general suite of programs to deal with a whole class of systems, whereas others require extremely accurate programs dealing with particular systems or devices. The subject matter is so varied and the possibilities are so enormous that further rapid development of computer software for electromagnetic calculations is extremely probable.

There are also a number of variants of the finite element method such as boundary elements, integral-equation methods, and finite difference methods that can be used with advantage in certain applications. Recently the novel method of tubes and slices has been introduced, which uses the vector properties of electromagnetic fields graphically and avoids the need for simultaneous algebraic equations.

Mathematical methods. The mathematical basis for the numerical methods in electromagnetic calculations deserves close attention. These calculations involve information about the physical phenomena described in mathematical terms. It is vital that appropriate mathematics be used to handle the information economically. Moreover, the mathematics must control the numerical processes and limit the spurious information generated by the approximations, which is an inevitable feature of all numerical methods.

This can be illustrated by the mathematical basis of finite element calculations. The reason for the success of the method lies in the fact that it calculates a so-called functional, which in general is the system energy rather than the vector field itself. Thus, instead of attempting to solve Maxwell's equations, it seeks the energy distribution that gives rise to the equations. There is an interesting parallel here with the methods of variational mechanics, in which the equations of motion are derived from an invariant system property called the action.

The relationship between the energy and the vector field has only recently been clarified by the methods of differential geometry and by the notation of differential forms. These forms associate the field vectors with the space in which they act. They show that the so-called constitutive equations relating flux densities to field strengths are really geometrical transformations that associate an area element with a line element and so define a product that is the energy in a small

volume. This energy density is always positive (or zero), and any arbitrary displacement from equilibrium will provide an upper bound for the system energy. Moreover, the transformations are reciprocal, and this makes it possible to obtain the lower bound as well as the upper bound, thus giving confidence limits for the unknown system parameters. The full advantages of this insight have yet to be developed. Another clarification achieved by differential geometry is that it gives guidance in the choice of suitable functionals for economical computation. Often these functionals are equivalent to the circuit parameters used in engineering. Further developments are likely from the application of topology to electromagnetic systems. The mathematical methods of differential geometry and topology are not new in themselves, but their application to electromagnetic calculation is still in its infancy.

Technological needs. All this progress in computation is of course driven by technological needs. The subject range is immense. For example, in the area of antenna development, there has been rapid progress in satellite communication systems, radar, radio astronomy, mobile communication systems for airplanes and ships, and microwave links. Progress is being made by the use of both large mainframe computers and inexpensive personal computers. The former have the advantage of being able to handle very large amounts of data, but the disadvantage that the physical insight into the problem is reduced. Small personal computers are well adapted to relatively simple design calculations. In electrical power applications, computers have made possible the accurate design of insulation systems in high-voltage apparatus, including transmission lines, transformers, rotating machines, and ancillary apparatus. Electrostatic systems involving powder deposition, protection of electronic circuits, and avoidance of explosive hazards have recently attracted attention. Magnetic fields in power

Plot of the magnetic field of a switched reluctance motor, based on a finite element calculation using 4232 elements. (*Vector Fields Ltd.*)

devices need to be calculated in order to use materials more economically while maintaining reliability. For example, the successful design of switched reluctance motors, which are very robust mechanically and are suitable for domestic appliances and lawn mowers, depends on knowledge of the magnetic fields everywhere in the motor (see **illus.**). Dielectric heating is important as an industrial process and depends on calculating the energy distribution. Nondestructive testing is of great importance and often involves the use of electromagnetic energy. Induction heating by means of eddy currents is used in furnaces and involves the optimization of the energy distribution. In medical applications there is the design of magnetic-resonance-imaging magnets, and in physics there are enormous problems in the design of magnets for fusion technology, as well as particle accelerators and electron optical systems. There is a specialist literature for all these fields and for many others. It is manifestly impossible for any individual to master more than a few specialties and keep abreast of present-day developments. However, a knowledge of the physical and mathematical structure of electromagnetism is a necessary foundation for all the varied applications, and this knowledge must be linked to an appreciation of modern computing methods. SEE BRAIN; COMMUNICATIONS SATELLITE; LASER.

For background information SEE ANTENNA (ELECTROMAGNETISM); DIFFERENTIAL GEOMETRY; ELECTROSTATICS; FINITE ELEMENT METHOD; LIGHT; MAXWELL'S EQUATIONS; QUANTUM MECHANICS; RELATIVITY in the McGraw-Hill Encyclopedia of Science & Technology.

P. Hammond

Bibliography. D. Baldomir and P. Hammond, Geometrical formulation for variational electromagnetics, *IEE Proc. A*, 137:321–330, November 1990; *5th International Conference on Antennas and Propagation*, 1987; B. Schutz, *Geometrical Methods of Mathematical Physics*, 1980; P. P. Silvester and R. L. Ferrari, *Finite Elements for Electrical Engineers*, 2d ed., 1989; J. K. Sykulski, Computer package for calculating electric and magnetic fields exploiting dual energy bounds, *IEE Proc. A*, 135(3):145–150, 1988.

Electronic power supply

Resonant-mode is a technique of sine-wave shaping that can generate high-frequency, ultracompact power supplies far beyond what is possible with present techniques. The largest category, called quasi-resonant-mode, is regarded as the major future power supply technology by both researchers and power supply designers. It is expected that, although few quasi-resonant-mode power supplies are manufactured now, in 1995, 55% of all power supplies manufactured will be of this type. The advantages of quasi-resonant-mode techniques include the creation of smaller, lighter, more efficient power supplies. How a quasi-resonant-mode power supply operates, and its benefits, will be described, along with an overview of the variety of power supplies that can be designed with quasi-resonant-mode techniques. In this article, resonant-mode refers to the subcategory quasi-resonant-mode. True resonant power supplies are rare and have distinctly different characteristics.

Advantages of resonant-mode supplies. With the growing use of very large scale integration (VLSI), hybrid assemblies, and surface-mount packaging, more functionality is being packed into less space. To prevent the power supply from taking up excessive space in an electronic system, it must be reduced in size. Quasi-resonant-mode as well as true resonant-mode techniques have been used for many years in military and aerospace power supplies because of the resulting reduced size and weight, and the low electromagnetic and radio-frequency interference. Recently these benefits have become critical requirements for commercial and industrial applications, including portable computers, high-definition computer monitors, commercial avionics, and telecommunications. With the recent introduction of new high-frequency components, the choice of resonant-mode techniques is becoming cost-competitive with the present technique of pulse-width modulation. **Figure 1** shows a prototype of a resonant-mode power supply.

Pulse-width modulation is characterized by its square-wave switching waveforms of voltage and current that are applied to the power transformer. In pulse-width modulation, high voltages and currents exist simultaneously during turn-off and turn-on of the power switches, resulting in high switching losses. Higher frequencies are necessary to reduce the size of power supplies, but switching losses in pulse-width modulation contribute significantly to the poor efficiency as frequencies are pushed into the megahertz range. Resonant-mode techniques, with minimized switching losses, are better suited for high-frequency operation.

The central difference between pulse-width modulation and resonant-mode techniques is the insertion of a small inductor and capacitor in the current path from the input to the power transformer. This inductor-capacitor (*LC*) network is called a resonant tank since it takes the voltage and current from the input and

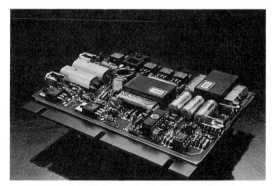

Fig. 1. Prototype 100-W dc-dc resonant-mode power supply, using high-frequency planar magnetics and a monolithic resonant controller. (Gennum Corp.)

stores energy alternately in the inductor and capacitor. Current and voltage are shaped to sinusoids that drive the power transformer, allowing the power switches [usually power metal-oxide-semiconductor field-effect transistors (MOSFETs)] to commutate at zero crossing points in the resonating tank waveforms.

Resonant-mode techniques, because of their sinusoidal current and voltage, have very low electromagnetic interference and radio-frequency interference. The sinusoids also reduce the component stresses, which in turn reduces the chances of noise problems and reliability problems. The zero-current or zero-voltage switching makes possible high-frequency operation with high efficiency by reducing the switching losses. The high frequency allows the inductor of the output filter to be very small. One prototype design operates at frequencies up to 15 MHz with a power density of 30 W/in.[3] (1.8 W/cm[3]), as compared to a frequency of 100 kHz with a power density of 5 W/in.[3] (0.3 W/cm[3]) for a typical pulse-width-modulation design. The ripple and transient response of a power supply are also improved by increasing the switching frequency. Because of the added complexity, resonant-mode power supplies will replace pulse-width-modulation power supplies only in those high-performance applications where these benefits are required, leaving unaffected a large market for pulse-width-modulation power supplies.

Diverse topologies. Resonant-mode refers to a large variety of power supplies. Presently there are about 24 basic categories with many different topologies in each of them. The most general classifications are based on the connection of the load to the resonant tank, the mode of switching, and the number of power switches.

The resonant tank can be either series-loaded or parallel-loaded, with the load being directly connected to the tank or indirectly connected through a power transformer. The mode of switching is either zero-voltage-switched (ZVS) or zero-current-switched (ZCS). In the zero-voltage-switched mode, the power switch turns on and off at the specific time when the voltage across the switch is zero. In the zero-current-switched mode (the dual of the zero-voltage-switched mode), the power switch turns on and off at the specific time when the current through the switch is zero. Each of these basic categories has many different topologies. Some of the topologies, such as the push-pull and half-bridge topologies, require two power switches. The most diverse area is that of single-switch topologies that include the buck, boost, flyback, and forward. These topologies represent a large variety of actual power supplies that can be designed to meet the needs of every conceivable application.

Circuit operation. The circuit shown in **Fig. 2** is a good example to illustrate the various electrical waveforms of a typical resonant-mode design. This is a quasi-resonant dc-dc converter that produces a 5-V output. The series resonant tank is parallel-loaded by the isolation transformer T. The topology is a resonant forward converter with zero-current switching. The waveforms in **Fig. 3** are the resonant tank voltage and current (V_C and I_L), and the corresponding waveforms for the power switch gate-source and drain-source voltages (V_{gs} and V_{ds}).

At the beginning of interval A, the power FET Q_1 is off and the voltage on both sides of the resonant capacitor C is $+V_{in}$. There is no current flowing through either the transformer or the resonant inductor L. The output inductor L_o and output capacitor C_o are large so that the load can be viewed as a constant direct current equal to the load current. As Q_1 turns on, a voltage of $+V_{in}$ is applied across the tank,

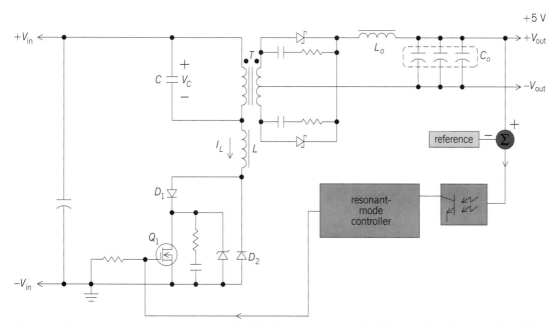

Fig. 2. Circuit diagram of resonant-mode power supply. This dc-dc converter uses the _LC_ resonant tank to apply a sinusoidal voltage to transformer _T_. The controller uses variable frequency to regulate the output voltage to +5 V.

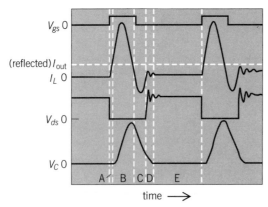

Fig. 3. Waveforms for the circuit of Fig. 2. They are the gate-source voltage V_{gs} across Q_1, the current I_L in the resonant inductor L, the drain-source voltage V_{ds} across Q_1, and the voltage V_C across the resonant capacitor C.

causing the current in the inductor I_L to increase. During interval A, it is linearly increasing until it reaches the dc load current I_{out}. At this point (the beginning of interval B) the capacitor starts charging. The inductor current becomes sinusoidal, as does the capacitor voltage V_C. The effect of the dc load is to shift the dc level of the inductor current waveform. Interval C begins when the inductor current crosses zero. Since this is a resonant cycle, this is always at a fixed time from the turn-on of Q_1. Current now is flowing back from the inductor into the capacitor and back to $+V_{in}$. The transformer is still providing a constant direct current to the load. The MOSFET can only carry forward current because of the series blocking diode D_1. The reverse current instead flows through the antiparallel fast-recovery diode D_2. The resonant cycle ends when the current returns to zero at the beginning of interval D. Sometime during interval C, Q_1 had been turned off by having its gate pulled low. With no path for forward current, the resonance is stopped. This still leaves a voltage on the resonant capacitor C that will linearly discharge through the power transformer at a rate determined by the load direct current. During the interval E, there is no voltage across C and no current in L. This time interval has no energy transfer. The next cycle will start the next time Q_1 is turned on.

By varying the duration of interval E, it is possible to control the average amount of energy that is transferred to the load. This modulation is usually done by an integrated controller chip. The output voltage is sensed and compared to a precise reference. The error voltage is transmitted through an optocoupler to the controller to set the frequency of pulses that turn on the FET.

Advanced components. The viability of resonant-mode depends strongly on what components are available. The transformer is the primary limiting component in resonant power supplies. This is due to the many different losses that reduce efficiency. Assuming the core is not saturated, high-frequency core losses result from hysteresis, residual, and eddy-current losses. Special new materials that maximize the ac core resistance have greatly reduced eddy-current losses for cores operating below 2 MHz. Also, a new technique called planar magnetics produces low-profile, low-leakage inductance transformers.

The requirements of a controller for resonant-mode are distinctly different from those of standard pulse-width-modulation controllers. Discontinuous resonant converters still depend on the on-time–to–off-time ratio to control the input-to-output conversion ratio, which is exactly the same as in pulse-width modulation. The difference is that resonant-mode controllers control this conversion ratio by varying the frequency, whereas pulse-width modulation varies the duty cycle. To avoid destroying the FET during a shutdown, the controller must synchronize any shutdown with the dead time of the switching cycle.

Status and prospects. The present commercial designs that use resonant-mode techniques operate at frequencies up to a megahertz. Prototype designs operating from 5 to 22 MHz have been demonstrated. The improvements in density with increasing frequency slow down in these high ranges, mainly because of the limitations of the present magnetic materials. Even without further increases in the frequency, there is still potential for improvement. Space utilization can be optimized by using hybrid techniques. Parasitic losses in interconnects can be reduced by using planar magnetic transformers and radio-frequency packaging techniques. When resonant-mode techniques are used in volume production, they will be able to benefit from high levels of integration in customized integrated circuits and smart power devices.

For background information SEE ELECTRICAL INTERFERENCE; ELECTRONIC POWER SUPPLY; INTEGRATED CIRCUITS; PULSE MODULATION; TRANSFORMER in the McGraw-Hill Encyclopedia of Science & Technology.

Frederick E. Sykes

Bibliography. K. K. Sum, *Recent Developments in Resonant Power Conversion*, 1988; F. Sykes, Resonant-mode power supplies: A primer, *IEEE Spectrum*, 25(5):36–39, 1989; P. Todd, Practical resonant power converters: Theory and application, *PowerTechnics*, 2(4):30–34, 2(5):29–35, and 2(6): 32–37, 1986.

Electronics

Electronic devices that are based on the movement of electrons through vacuum but employ electron field emission and standard microelectronics technology have been developed and show great promise.

Vacuum versus solid-state electronics. All electrical and electronic devices are based on the fact that electrons can be moved from one point in space to another. For example, electrons can be transported through metal wires such as copper or aluminum, through semiconductors such as silicon or gallium arsenide, or through vacuum (where the density of atoms or molecules is low enough that the elastic and inelastic scattering of those electrons by atoms or molecules is insignificant). In electronic devices, voltages can be applied to electrodes inside the device to modulate (or control) this charge flow.

What is known as the first electronics technology, developed just after 1900, was based on electron transport in vacuum using vacuum tubes. The chief limitation of vacuum tubes was not the vacuum charge-transport process itself; rather, it was the fact that very few electrons could be boiled off the thermionic cathodes. For example, when vacuum-tube electronics was the dominant technology (before 1950), thermionic cathodes could provide a current density of only about 1 A/cm^2 with reasonable lifetimes. In order to produce sufficiently high total currents to make the devices work adequately (that is, to change the voltage on grids and capacitors quickly enough), a large-area cathode was needed. This physically large cathode led to large vacuum tubes, which resulted in large and electronically simple devices.

The solid-state transistor (developed after 1950) provided device current densities in excess of 1000 A/cm^2, which resulted in devices that were three orders of magnitude smaller than vacuum tubes. Solid-state transistors led to integrated circuits composed of thousands of fundamental electronic devices on a single chip. That number is now greater than a million. The electronics industry based on solid-state transistors is the second electronics technology.

Solid-state devices superseded vacuum devices due to the creation of the fundamentally smaller structures that dissipated less power and could be batch-processed, thereby yielding a smaller unit cost. This changeover was not due to the superiority of the solid as a charge transport medium compared to the vacuum. In fact, the vacuum is vastly superior to the solid with respect to saturation drift velocity and associated transit times. For example, the saturation drift velocity is limited to less than 5×10^5 m/s (1.5×10^6 ft/s) in all solids, whereas the electron velocity in vacuum has a practical value of about 6–9×10^6 m/s (2–3×10^7 ft/s). This means that vacuum transistors should be an order of magnitude faster than solid-state transistors for a device of the same size. Vacuum electronics is also vastly superior to solid-state electronics with respect to transient and permanent radiation damage, voltage isolation, and operation in extremely harsh temperature environments. The fundamental problem with vacuum electronics has been the lack of significant current density. The fabrication technology, although based on mechanical assembly, forming, and hand assembly, is of secondary importance.

Electron field emission. Electron emitters with higher current density have been sought for almost a century. The electron field emitter might be the ultimate answer. Electron field emission can produce current densities up to 10^8 A/cm^2 without heat. Only an electrostatic voltage is required. Unfortunately, the required electric field is greater than about 5×10^7 V/cm, or about two orders of magnitude greater than that needed for the voltage breakdown of most materials. To avoid

voltage breakdown, a high-curvature surface is created, for example, a sharp point, to provide the necessary field enhancement. Sharp wires and razor-blade structures, in both single and multiple bed-of-nails configurations, have been developed. But these structures required extraction voltages greater than 1000 V and were frequently destroyed by ion bombardment.

The original microminiature field emitter with an integral low-voltage extraction gate was called the thin-film field-emitter cathode (TFFEC). This device was to be a replacement for thermionic cathodes used in linear beam tubes such as the traveling-wave tube. Macroscopic current densities exceeding 1000 A/cm^2 have been reported from TFFECs, comparable to those obtained in the early solid-state transistors, and less than 0.1% of the emitted current is intercepted by the extraction gate. Unfortunately, TFFECs are fabricated by using nonstandard multiple electron-beam evaporation sources to form sharply pointed field emitters inside self-aligned extraction apertures. Nonetheless, the TFFEC pointed the way to a new type of high-input-impedance, high-current-density

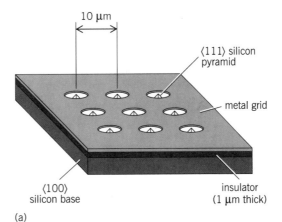

(a)

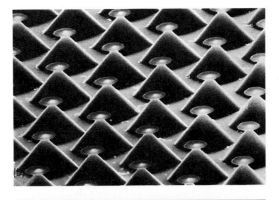

Fig. 1. First silicon field-emitter array (FEA). (*a*) Schematic diagram. (*b*) Scanning electron microscope picture of the array without the extraction gate. The spacing between pyramids is 10 micrometers. The disks on top of the pyramids are silicon masking dots. (*c*) Scanning electron microscope picture of one cell of the array. The diameter of the extraction gate aperture is about 1.5–2.0 μm. (*Naval Research Laboratory*)

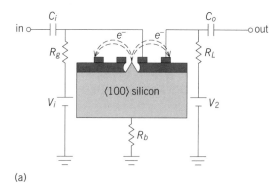

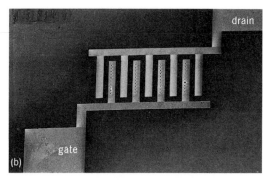

Fig. 2. First silicon planar field-emitter array (FEA) vacuum field-effect transistor (FET). (*a*) Schematic diagram. C_i = input coupling capacitance; R_g = gate resistance; V_i = gate voltage; R_b = emitter-base resistance; C_o = output coupling capacitance; R_L = load resistance; V_2 = collector voltage. (*b*) Scanning electron microscope picture. The substrate serves as the source electrode. The interdigitated design (with 10 FEA cells in each gate finger) increases total device current. (*Naval Research Laboratory*)

device, analogous to the transistor, that would be very fast, radiation-hard, and temperature-insensitive.

Vacuum transistors. In the 1970s, noting that the TFFEC was based on metallic field emitters that had a tendency to catastrophically "blow up" if too much current was emitted, and that it used nonstandard electron-beam technology, H. F. Gray designed a variety of new structures based on standard microelectronics technology. These field-emitter arrays (FEAs; **Fig. 1**) are the basis of vacuum transistors. The breakthrough came with the planar vacuum transistor (**Fig. 2**). This vacuum device, similar to a silicon solid-state field-effect transistor (FET), had the following features.

1. The use of silicon provided protection from tip blow-up, such as that observed in the TFFEC. This protection is due to velocity saturation of the electrons inside the silicon field tip. In essence, the tip is electronically blunted if there is an attempt to emit too much current. This blunting causes a decrease in the extraction field, which limits the emitted current, thereby preventing tip blow-up.

2. This electronic blunting phenomenon created two separate operating regions. One region exhibited a highly nonlinear current–voltage characteristic, similar to solid-state transistors. The second operating region exhibited a very linear current–voltage characteristic (which would minimize harmonic distortion).

3. Essentially no electrons were intercepted by the gate, which meant that the structure had a very high input impedance, similar to a solid-state field-effect transistor.

4. Complete planar devices could be fabricated on-chip, thereby eliminating the separately fabricated anode required by the previous vertical structures. Electronic properties of the device could be altered by implantation, sputtering, or depositing a low-work-function material onto the surface.

This first vacuum transistor exhibited both voltage and power gain. Calculations indicated that it was an inherently fast transistor, with a total transit time of less than 5 picoseconds (corresponding to a frequency of 200 GHz), even though the electrode spacings were an order of magnitude larger than state-of-the-art solid-state transistors. Smaller structures promised the possibility of subpicosecond switching and terahertz amplification.

Vacuum microelectronics. The demonstration of this vacuum transistor resulted in the development of numerous device concepts, processing techniques, materials, and applications. The term vacuum microelectronics was introduced to distinguish this field from solid-state microelectronics and conventional vacuum tubes.

Since 1987, many novel vertical vacuum microelectronic devices have been developed, including those fabricated with a silicon mold technology. A variety of horizontal structures based on thin-film technology have also been reported. Single-crystal silicon field emitters have been made atomically sharp, and silicon surface atoms have been replaced atom for atom with metal. It has now been amply demonstrated that the fabrication and processing of vacuum microelectronics structures are extremely flexible. The technology is not restricted to single-crystal materials or extremely pure materials. Material defects are probably irrelevant. The use of organic materials and high-temperature superconductors has been reported, showing that any conducting material can be used. Stable and long-lifetime operation even at atmospheric pressure has been reported, thereby decreasing the requirement for a hard vacuum. With the creative use of vias, truly three-dimensional vacuum microelectronic circuits might be possible. Furthermore, because these devices appear temperature insensitive, the well-known heat dissipation problem experienced with solid-state devices might be essentially eliminated. Consequently, vacuum microelectronics might be the third electronics technology. *See* Computer.

Applications. With this new vacuum microelectronics technology, it might be possible to address applications that are difficult or impossible to address with devices that are completely solid-state, or to go beyond their limits. Possible applications include the following.

1. A superior full-color flat-panel display that has all the advantages inherent to both cathode-ray tubes and liquid-crystal displays.

2. Efficient, small, and lightweight radio-frequency amplifiers that operate at frequencies greater than 100 GHz with output powers in the 1–1000-W region.

These amplifiers could be used for supercellular telephones, direct Earth–satellite communication, and a new class of phased-array radars.

3. Extremely radiation-hard digital and analog electronics that could be put inside nuclear reactors to eliminate the fragile metal–ceramic feedthroughs in the reactor walls. These ultra-radiation-hard electronics could also be used next to lightweight unshielded nuclear reactor power supplies for deep-space probes.

4. Temperature-insensitive electronics for non-fuel-cooled electronics inside jet and rocket engines, ceramic automobile engines, devices in oil wells, and space probes that penetrate the hot atmospheres of planets such as Venus.

5. Superfast computers and memory that can be truly three-dimensional in integrated circuit form in order to minimize the distance and transit time between interacting elements and to provide a smaller, more compact instrument.

6. A new electron source for high-power microwave, millimeter-wave, and submillimeter-wave tubes. This electron source could be either gate- or laser-modulated.

7. A new class of multi-electron-beam lithographic machines in which each beam is individually gated on and off inside a common deflection system, or in which microdeflectors are integrated on-chip at each emitting site. These could be coupled with large parallel computers and massive system-cell libraries to fabricate advanced application-specific integrated circuits, without the need for the batch processing associated with optical and x-ray lithographies.

8. A new class of optical detectors with the possibility of noiseless energy gain, a property lacking in solid-state detectors.

9. High-brightness, high-spatial-resolution, low-energy-dispersion electron sources for a variety of scientific instruments such as scanning and transmission electron microscopes, scanning tunneling microscopes, and field emitter array appearance potential spectrometers.

This technology is still in its initial stages of development, and it might be several decades before a significant number of these ideas can be demonstrated and commercialized.

For background information SEE ELECTRONICS; FIELD EMISSION; INTEGRATED CIRCUITS; TRANSISTOR; VACUUM TUBE in the McGraw-Hill Encyclopedia of Science & Technology.

Henry F. Gray

Bibliography. F. L. Alt and M. Rubinoff (eds.), *Advances in Computers*, vol. 2, 1961; H.-O. Andrén and H. Nordén (eds.), *Proceedings of the 29th International Field Emission Symposium*, 1982; R. Davies and A. Hartman (eds.), *IEDM Technical Digest 1986*, 1986; H. F. Gray and G. J. Campisi, A silicon field emitter array planar vacuum FET fabricated with microfabrication technology, *Mater. Res. Soc. Proc.*, 76:25–30, 1987; C. A. Spindt, A thin-film field-emission cathode, *J. Appl. Phys.*, 39:3504–3505, 1968; R. E. Turner (ed.), *Vacuum Microelectronics 1989*, Inst. Phys. Conf. Ser. 99, 1989.

Electrosensory systems (biology)

The South American gymnotiform fish and the African mormyriform fish generate weak electric signals by discharging electric organs within their tails. They sense the associated currents with electroreceptors distributed over their body surfaces. An electric fish is able to "electrolocate" an object in its environment by evaluating distortions of its own electric field that are caused by the object's presence. Electric fish also sense the electric organ discharges of conspecific neighbors, and they communicate socially through particular modulations of their electric discharges.

A close integration of behavioral, neurophysiological, and neuroanatomical approaches has guided research on the neural basis of electrosensation and the generation of behaviors associated with this sensory modality. By postulating neuronal implementations of specific computations in sensory information processing, behavioral studies have been crucial in focusing studies at the neuronal level onto behaviorally relevant structural and functional aspects. Physiological and anatomical studies have analyzed (1) neural networks underlying the distributed processing of sensory information; (2) the role of descending recurrent pathways and efference copy mechanisms for the filtering of incoming information; (3) the significance of multiple topographic representations for sensory information processing; and (4) the modulation of sensory and motor structures through various transmitters and receptor subtypes.

Electrosensory systems offer particular experimental advantages by being simpler than more highly evolved systems, such as vision and audition in birds and mammals. Yet, the basic neuronal designs found in electrosensation, appearing old and conservative in the evolutionary sense, are largely the same as those in higher and more derived systems. Most significantly, some behavioral responses of electric fish are so robust that they remain intact in physiological preparations, thus allowing simultaneous studies at the behavioral and cellular level.

Neuronal networks. One of the most extensively studied behaviors is the jamming avoidance response in gymnotiform electric fish of the genus *Eigenmannia*. These fish produce nearly sinusoidal electric organ discharges of highly stable, though individually different, frequencies. When exposed to an interfering sinusoidal signal of a frequency close to that of its own electric discharge, *Eigenmannia* lowers its frequency in response to a slightly higher interfering frequency, and raises its frequency in response to a slightly lower frequency. Frequency differences of a magnitude between 2 and 6 Hz cause the strongest responses.

Behavioral experiments on the jamming avoidance response have shown that these fish are able to assess modulations in the amplitude and in the phase (timing) of zero crossings of sinusoidal signals resembling electric organ discharges. Such modulations result from the interference of the fish's own electric discharge with that of a neighbor of similar frequency, and the rate of these modulations is equal to the magnitude of the frequency difference between the two interfering signals. By evaluating the spatial pattern of amplitude and phase modulations distributed over its body surface, the fish can determine the sign of the frequency difference and then shift its own frequency in the correct direction, that is, away from that of its neighbor.

Information concerning amplitude and phase is coded by different types of electroreceptors and processed in separate structures of the hindbrain and midbrain. These structures include the electrosensory lateral-line lobe of the hindbrain and the torus semicircularis of the midbrain, which are organized somatotopically (according to the spatial order on the body surface as preserved within a layer of neurons in the brain); thus they preserve the spatial pattern of the recruitment of electroreceptors. A convergence of amplitude and phase information within the torus yields higher-order neurons suitable for the control of the jamming avoidance response. A projection from the torus to the nucleus electrosensorius of the diencephalon abandons the somatotopic order of electrosensory information and, instead, generates a motor map consisting of two subnuclei. Whereas excitation of one subnucleus raises the frequency of the fish's own electric organ discharge, excitation of the second subnucleus lowers the frequency.

Neuronal networks involved in the processing of electrosensory information have been analyzed by intracellular labeling of physiologically identified cells. In several instances, their synaptic morphology was studied at the ultrastructural level, and immunohistochemical studies led to the identification of their transmitters. The role of specific transmitters can be tested by local application of receptor blockers and simultaneous monitoring of the responses of individual neurons.

Through several levels of sensory processing, the neuronal substrate of the jamming avoidance response is distributed and lacks neurons that appear to be individually essential for this behavior. Neurons up to the midbrain show very general response properties, and responses of single neurons reveal little about the nature of stimulus patterns in the environment. It is apparent, however, that such patterns could be identified unambiguously by monitoring large numbers of these neurons simultaneously. The obvious advantage of this form of organization, apart from computational speed, is a relative immunity to physical damage and loss.

A comparison of neurons from lower to higher hierarchical levels reveals a gradual sharpening of their sensitivity to stimulus features relevant to the jamming avoidance response as well as growing immunity to irrelevant features. At the highest level, neurons are encountered that recognize jamming signals so selectively and reflect dynamic properties of the jamming avoidance response so faithfully that recordings from a single neuron yield rather accurate predictions of the behavioral performance.

Descending recurrent inputs. Through descending recurrent inputs, sensory structures of higher order can modulate the function of lower-order structures. Although such connections are known in all sensory systems, little has been learned about their functional significance. It is assumed that descending inputs adapt the filter properties of networks to the processing of particular stimulus features. The electrosensory lateral-line lobe of gymnotiform fish receives at least two kinds of recurrent, descending input. Intracellular studies have shown that these two forms of inputs originate from different types of cells in the nucleus praeeminentialis of the midbrain. One descending pathway can be interrupted reversibly by local application of anesthetics. This input appears to affect the electrosensory lateral-line lobe in its entirety by rendering the responses of pyramidal cells more phasic, that is, more rapidly adaptable to sustained inputs. This form of general inhibition functions as a gain control by enabling the pyramidal cells of the electrosensory lateral-line lobe to operate over a wide range of mean amplitude levels, much as a visual system can operate under vastly different light levels.

A second path of descending input to the electrosensory lateral-line lobe forms a topographically reciprocal projection to it and appears to generate more localized effects. The function of this input is still unknown, but on the basis of its topographic restriction it can be speculated that it forms the basis of an attention mechanism.

Efference copy mechanism. Sensations caused by the animal's own activity can be suppressed by filtering the flow of sensory information by means of an efference copy or corollary discharge reflecting the animal's intended activity. Although this mechanism has attracted considerable theoretical interest, little has been learned about its neuronal implementation, a single exception being the case of efferent control of electrosensory information flow in mormyrid fish.

Mormyrid fish generate discrete, pulselike electric organ discharges. Within a subsystem of particular electroreceptors and higher-order neurons adapted to social communication, these fish discriminate the sensation of the discharges generated by neighbors by selectively blocking the sensation of their own discharges. This exclusion of self-induced excitation is achieved by inhibiting inputs from electroreceptors at each moment when feedback from the fish's own electric organ discharges is expected to arrive, and this inhibition is triggered by the neuronal pacemaker system that generates the fish's discharges.

Another class of electroreceptors is adapted to the perception of low-frequency signals. Although these receptors also respond to the fish's own electric organ discharges, higher-order neurons respond only to novel inputs and ignore the predictable feedback from the fish's own discharges. Recent research has revealed three effects: (1) the electrosensory system nulls predictable inputs of this form at a central level by generating a negative copy of the pattern of expected input activity; (2) this copy is updated continually, with a time constant on the order of a few minutes; and (3) this updating is controlled by a corollary discharge of the pacemaker. Mormyrids are thus able to suppress predictable inputs at a central level to focus on novelties. This modifiable form of an efference copy should provide a superb model system for the study of neuronal plasticity in general.

Multiple neuronal representations. The electrosensory lateral-line lobe contains four electrosensory maps of the body surface, three of which receive identical sensory information. This suggests an evolutionary multiplication of a single structure, with different copies available for adaptations to different processing functions.

Recent physiological and neuroanatomical studies have shown that the receptive fields of neurons differ between these three maps with regard to their temporal and spatial response properties. Whereas the neurons in the most lateral map show the greatest temporal resolution and poorest spatial resolution, neurons in the most medial map show the opposite specialization.

Modulation of motor system and sensory processors. Gymnotiform electric fish can modulate the firing rate of their pacemaker in various ways, and particular modulations serve as signals in social communication. Modulations of the pacemaker activity are induced by inputs descending from a diencephalic prepacemaker nucleus. Different classes of prepacemaker neurons induce different forms of modulations; and these modulations, in turn, are mediated by different transmitters or receptor subtypes. Sensory information processing in the electrosensory lateral-line lobe is modulated by descending inputs that affect the temporal and spatial properties of receptive fields.

For background information SEE ELECTRIC ORGAN (BIOLOGY); INFORMATION PROCESSING in the McGraw-Hill Encyclopedia of Science & Technology.

Walter Heiligenberg

Bibliography. T. H. Bullock and W. Heiligenberg (eds.), *Electroreception*, 1986; W. Heiligenberg, *Neural Nets in Electric Fish*, Computational Neuroscience Series, 1991; S. B. Laughlin (ed.), Principles of sensory coding and processing, *J. Exp. Biol.*, vol. 146, 1989; A. Popper, J. Atema, and R. Fay (eds.), *Sensory Biology of Aquatic Animals*, 1987.

Embryonic induction

Cells and tissues, as they develop during the early stages of embryogenesis along specific pathways, receive signals, stimuli, and cues from at least two general sources. The first source, known as cytoplasmic localizations, is the endowment of information that cells inherit from their precursor cells; the second source, known as embryonic induction, is the coaxing that cells and tissues receive from direct contact with neighboring cells. The endowment of information usually consists of cytoplasmic organization systems, such as cytoskeletal elements, proteins, or messenger ribonucleic acids (mRNAs) that code for cell-type specific proteins. The cytoplasmic localizations are present in eggs at fertilization and are usually inherited in an asymmetric fashion such that different cell lineages contain different sets of cytoplasmic localizations.

Recent discoveries have provided substantial insight into the molecular nature of such cytoplasmic localizations. It is generally believed that cytoplasmic localizations, such as mRNAs, enzyme proteins, or their derivative products, migrate from the cell's cytoplasm into the nucleus and act there to stimulate the expression of specific sets of genes. Hence, in some embryos, one source of the signals or cues that guide cells toward specific pathways, for example, prospective muscle, is the egg cytoplasm. Its components are inherited by specific regions of the early embryo and eventually interact with specific genes to direct gene expression patterns. These cytoplasmic localizations are often referred to as regulatory molecules, since they appear to act directly by regulating gene expression. The most useful information has been collected from the analyses of genetic mutations that yield abnormal embryos in simple invertebrates such as the fruit fly (*Drosophila*) and nematode (*Caenorhabditis*).

Conceptual framework. The other major source of stimuli or cues that establish or determine cell fates or specialization pathways during early embryogenesis, embryonic induction, represents a process whereby two groups of cells or tissues interact with one another, usually by direct cell-to-cell contact. That such close apposition between the interacting cells is required has historically been demonstrated with microsurgical techniques. The technique involves either inserting a cellophane barrier between the presumed interacting groups of cells or physically removing one of the prospective embryological tissues, thereby erasing the developmental pathway that the responding tissue would normally yield.

Embryonic induction is probably more complex than the cytoplasmic localization scenario, due to the intricacies involved in a situation in which two or more independently developing tissues must interact at the appropriate place and time. Hence, until recently, embryonic induction schemes have been less understood than cytoplasmic localization phenomena. That lack of understanding does not, however, reflect a secondary role for embryonic induction in establishing developmental pathways or determining cell fates. Indeed, the list of tissues or organs that require embryonic induction at one or another stage of their development is long and includes the neural tube (spinal column), heart, eye, pancreas, ear, and kidney.

The lack of progress in understanding the molecular mechanisms that comprise specific embryonic inductions can be attributed to several factors, some of which were overcome in the past several years. These include the following three factors: (1) There has been a lack of appropriate genetic systems. The best-documented examples of embryonic induction have emerged from studies of vertebrate embryos, especially early amphibian and avian embryos (see **illus.**). Genetic manipulation in vertebrate embryos is extraordinarily difficult, compared to the simple invertebrate embryo types that are used for analyzing cytoplasmic localizations. The analysis of genetic mutations gives

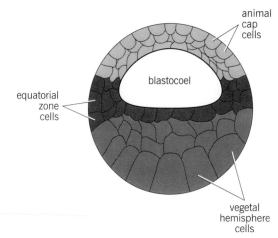

Cross section of the blastula stage of an amphibian embryo.

insights that are often counterintuitive and would go unnoticed if they were not recognized by their exaggeration in abnormal (mutant) embryos. (2) Bioassays that are carried out with cell cultures, so that purified cellular components can be used to replace the inducing tissue, frequently lack precise quantifiable indices of development; that is, the extent to which the responding tissue (for example, prospective optic cup) develops often requires the use of histological examination, rather than accurate enzyme assays, in order to estimate the effectiveness of the inducer. (3) Some responding tissues (for example, prospective neural-tube tissue) can be stimulated to develop by treatment with inducer preparations obtained from heterologous tissues. Guinea pig bone marrow extracts and liver extracts can substitute for natural tissue sources in amphibian organ cultures. That observation baffled scientists, because embryonic induction was conceptualized, from the earliest days of its research, as involving inducing molecules that were tissue- and species-specific.

Molecular mechanism. Against that backdrop of potential limitations, tremendous progress has been made in the past few years toward understanding the molecular mechanism of embryonic induction. For example, it has been found that mesoderm induction, perhaps the first embryonic induction event in most vertebrates, is guided by the action of peptide growth factors. The equatorial zone cells of the mesoderm, or middle layer of the embryo (see illus.), are induced by vegetal hemisphere cells immediately beneath them. The induced, equatorial zone cells develop mostly into skeletal muscle and notochord, as well as several minor components such as blood-forming tissue. When vegetal hemisphere tissue is experimentally combined with animal cap cells of the ectoderm, or outer layer of the embryo, those cap cells are induced to develop into muscle. In normal embryogenesis, most of those cap cells would develop into nervous system components.

The discovery that peptide-growth-factor action represents the molecular mechanism of embryonic induction was made by culturing amphibian animal-cap ectoderm in a solution containing minute amounts of a particular type of peptide growth factor. A survey of the various types of peptide growth factors, which have been known for a decade or longer to be active in promoting the growth of normal as well as tumor-forming cells, revealed two groups of active molecules. One group is related to the β type of transforming growth factor; the other group is related to acidic and basic fibroblast growth factor.

The actual identity of the type of mesodermal tissue that is induced varies with the type of peptide growth factor employed in the organ culture system. The β type of transforming growth factor induces the broadest range of mesodermal tissue types, including muscle, notochord, and kidney. Fibroblast growth factor, however, yields a less complete range of induced tissues. Notochord is conspicuously absent, for example, in fibroblast-growth-factor-induced cultures.

Searches for evidence that such peptide growth factors are actually present in the early embryo, as would be expected if they represent the endogenous inducer, have been rewarding. Evidence that the mRNAs for both transforming growth factor and fibroblast growth factor are present in the egg has been accumulated. Also, some of the proteins have been found in the egg (for example, fibroblast growth factor).

The reasons for the existence and presumed function of more than one growth factor in mesoderm induction are not fully understood. It may be that during the phylogenetic history of the induction mechanism, more than one molecule emerged as an effective inducer and all were maintained during subsequent evolution. Alternatively, local concentrations of different peptide growth factors may act in combination to induce a particular cell specialization pathway. For example, one ratio of two or three growth factors might bias the responding tissue to develop into skeletal muscle, whereas another ratio of those same growth factors might favor notochord development. In this regard, the cytoplasmic localization and embryonic induction concepts converge. For the earliest embryonic inductions, such as those involved in mesoderm formation, which may occur as soon as the 64-cell stage, cytoplasmic endowments (for example, mRNA) may serve to generate the molecular mechanism (for example, peptide growth factors) that guides embryonic induction. Later inductions, such as those involved in the development of the pancreas or lung, are so far removed from the egg stage and its cytoplasmic endowment that induction involves a strategy that no longer employs components of the egg cytoplasm.

Evidence that different peptide growth factors do indeed specify different mesodermal cell fates has been accumulated in the last few years. The expression of the homeobox gene, one of the genes that is widely believed to be directly involved in establishing the fate of cells, has been monitored in amphibian animal-cap cultures that were treated with different peptide growth factors. The relative amount of homeobox gene expression varied according to the type of growth factor used in the culture medium and the concentration of the growth factor.

Taking together the evidence that peptide-growth-factor action can mimic natural induction and that specific presumed regulatory gene expression is elevated by growth-factor treatment, a strong case can be made that one of the major mechanisms for establishing cell specialization pathways, embryonic induction, is finally being understood in molecular terms.

Prospects. The amphibian embryo has been exploited in embryonic induction studies for four reasons: (1) its external development facilitates microsurgical manipulation; (2) its yolk stores permit organ rudiments to be cultured in a simple salt solution; (3) the classical embryological literature provides substantial background information for conceptualizing; and (4) molecular markers, such as muscle-specific actin gene products, myosin gene products, and homeobox mRNA, can be conveniently quantitated. A future challenge is to extend the blueprint of information collected from amphibian embryos to more complex vertebrates, such as mammals. The higher-level animals will likely prove to be formidable experimental systems. Further research on the amphibian embryo is also likely to encounter technical obstacles. The establishment of precise cause-and-effect relationships between inducer and target genes will require deleting the inducer molecules from whole embryos. In simpler systems, such as the fruit fly or nematode, that can be accomplished by straightforward genetic manipulation. But for higher vertebrates such as amphibia, the prospects for eliminating a specific function, such as the action of a peptide growth factor, through genetic mutation are remote. Yet the single most important piece of remaining corroborating evidence to fully legitimize the growth-factor induction scheme is observation of mesoderm development in whole embryos from which growth factors have been genetically eliminated. Such embryos should exhibit extreme defects in mesoderm formation.

One alternative avenue for analysis could involve transferring the conceptual and technical approaches that have been successful in understanding mesoderm induction to the study of other early embryonic amphibian induction systems. Neural induction, in which the mesoderm induces overlying ectoderm to form a neural tube, is one obvious candidate for research.

For background information *SEE* CELL DIFFERENTIATION; CYTOPLASMIC INHERITANCE; DEVELOPMENTAL BIOLOGY; DEVELOPMENTAL GENETICS; EMBRYOLOGY; EMBRYONIC DIFFERENTIATION in the McGraw-Hill Encyclopedia of Science & Technology.

George Malacinski

Bibliography. S. F. Gilbert, *Developmental Biology*, 1988; O. Nakamura and S. Toivonen, *Organizer: A Milestone of a Half Century from Spemann*, 1978; A. Ruiz i Altaba and D. A. Melton, Axial patterning and the establishment of polarity in the frog embryo, *Trends Genet.*, 6:57–64, 1990; J. C. Smith, Mesoderm induction and mesoderm-inducing factors in early amphibian development, *Development*, 105:665–677, 1989.

Enzyme

The heme prosthetic group consists of an iron (Fe) atom coordinated by the four nitrogen atoms of the porphyrin macrocycle. When this cofactor binds to a protein, one or two additional ligands to the iron are supplied by amino acid side chains to anchor the macrocycle and form the heme enzyme. Recently, there has been considerable success in understanding the means by which these heme enzymes use dioxygen (O_2) or hydrogen peroxide (H_2O_2) to effect the oxidation of a variety of substrates. Techniques have been developed for preparing model compound analogs for reactive intermediates in the catalytic process so that their chemical and spectroscopic properties can be studied. In both peroxidases (a class of heme enzyme that uses hydrogen peroxide as a substrate) and oxidases (a class of heme enzyme that uses dioxygen as a substrate), time-resolved spectroscopic techniques, protein crystallography, molecular genetics, and trapping techniques have been used to provide insight into the sequence of events that occur between substrate binding and product release.

Model compounds. In terms of elucidating heme enzyme catalysis, synthesis of model compounds for potential reaction intermediates has been of key importance. The initial attempts to study oxygen binding, activation, and reactions of model hemes in solution were frustrated by the formation of the μ-oxo dimer, as shown in reaction (1), where the μ-oxo

$$4 \text{ heme-Fe}^{2+} + O_2 \rightarrow 2 \text{ heme-Fe}^{3+}\text{-O-Fe}^{3+}\text{-heme} \quad (1)$$
$$\text{(I)}$$

product (I) contains an oxo dianion bridging two heme groups. Two strategies have evolved to overcome formation of the μ-oxo dimer. The first involves synthesizing sterically hindered heme macrocycles. These constructions are designed to impede the close physical approach that is required for dimer formation, thereby stabilizing precursors to the μ-oxo state. In essence, this approach mimics one of the functions of protein in heme enzymes by controlling access to the iron site. The second strategy involves using a lower temperature to control the kinetics of the formation of the μ-oxo dimer.

Both approaches have been successful, and a variety of unstable heme/oxygen adducts have been prepared and characterized. Dioxygen binds to the heme iron to form an oxy adduct. Structure (II) is a ferrous

(II)

oxy heme complex in which imidazole is present as the trans ligand to the bound dioxygen. In the model complexes, O_2 binds in an end-on fashion but assumes a bent geometry to maximize orbital overlap with the iron. The specificity of the binding site for O_2 in a variety of heme proteins appears to be due in part to its three-dimensional structure, which favors bent, end-on ligand binding. Reduction of the oxy adduct by two electrons produces a peroxy species, which has often been postulated as a reaction intermediate in heme enzyme catalysis. Whereas the oxy complex has been relatively easy to trap and characterize, the peroxy species, unfortunately, has been more elusive. Five coordinate, μ-peroxy structures have been proposed in the autooxidation of ferrous hemes according to reaction (2). In addition, a side-on peroxide binding

$$2 \text{ heme-Fe}^{2+} + O_2 \rightarrow$$
$$\text{heme-Fe}^{3+}\text{-O}^-\text{-O}^-\text{-Fe}^{3+}\text{-heme} \quad (2)$$

to a single heme Fe to form a monomeric peroxide heme complex has spectroscopic support; reaction (3)

$$\text{heme–Fe}^{2+} + O_2^{\ominus \bullet} \rightarrow \text{heme–Fe}^{3+} \quad (3)$$

shows one route by which this complex has been synthesized.

Heterolytic cleavage of the oxygen-oxygen bond to produce a ferryl-oxo species is a common feature of heme protein oxidases and peroxidases. A variety of synthetic means have been developed by which to generate these complexes; structure (III) is an example

(III)

of a six-coordinate ferryl-oxo complex. The iron-oxygen bond is short (~ 0.17 nanometer) in this class of compound, and the spectroscopic properties of the iron are typical of the $+4$ valence state. Hence, the iron-oxygen bond is generally formulated as $\text{Fe}^{IV}\text{=O}$. Like oxy model compounds, ferryl oxo species are reasonably stable, and a variety of spectroscopic and chemical reactivity studies have been carried out. Recently, spurred by new results from heme protein peroxidases, interest has developed in generating ferryl-oxo heme models in which the porphyrin ring is in an oxidized, cation radical state. The initial results indicate that, in this class of compound, the properties of the ferryl-oxo bond are similar to those in which the ring is in its neutral form.

Peroxidases. Heme protein peroxidases use hydrogen peroxide to effect the two-electron oxidation of substrate. These enzymes are ubiquitous and carry out a range of physiologically important reactions. Myeloperoxidase, for example, uses H_2O_2 to oxidize chloride ion to hypochlorous acid in the leukocyte antimicrobial reaction, and prostaglandin H synthase catalyzes key reactions in the conversion of arachidonate to prostaglandins and thromboxanes. The active form of the enzyme contains Fe^{3+}-heme, which reacts in a three-step process with H_2O_2 and the reducing substrate, here designated as AH_2 [reactions (4)–(6), where P = peroxidase protein].

$$\text{P-heme-Fe}^{3+} + H_2O_2 \rightarrow$$
$$(\text{P-heme})^{+}\text{Fe}^{4+}\text{=O} + H_2O \quad (4)$$

$$(\text{P-heme})^{+}\text{Fe}^{4+}\text{=O} + AH_2 \rightarrow$$
$$\text{P-heme-Fe}^{4+}\text{=O} + AH\cdot + H^+ \quad (5)$$

$$\text{P-heme-Fe}^{4+}\text{=O} + AH\cdot \rightarrow$$
$$\text{P-heme-Fe}^{3+} + A + OH^- \quad (6)$$

The first intermediate, $(\text{P-heme})^+\text{Fe}^{4+}\text{=O}$, is usually called compound I or compound ES; its formal valence state is two electrons more oxidized than the resting enzyme. One of the oxidizing equivalents is stored at the iron as a ferryl species; the second oxidizing equivalent is localized either on the heme ring as a heme cation radical or on an amino acid in the protein. In the plant enzyme horseradish peroxidase, for example, the heme ring is oxidized, whereas in cytochrome c peroxidase the heme ring is neutral and the second oxidizing equivalent is stored at tryptophan 191. In prostaglandin H synthase, the stable form of compound I involves a protein tyrosine radical, but the heme cation radical can be detected as a transient intermediate. In the peroxidases, compound II (P-heme-$\text{Fe}^{4+}\text{=O}$) formed as substrate is oxidized according to reaction (5); it is one oxidizing equivalent more oxidized than the resting enzyme. The heme or protein radical of compound I is reduced, and the ferryl species is retained in compound II. In several peroxidases, compound II is surprisingly stable, and both compounds I and II have been characterized by spectroscopic techniques in a number of different proteins.

With the crystallization and determination of the three-dimensional structure of cytochrome c peroxidase, new insights into the mechanism by which this protein binds and metabolizes H_2O_2 have emerged. **Figure 1** shows one proposed mechanism that is generally consistent with a variety of spectroscopic and kinetic data. Key elements in the substrate binding pocket include the heme, its histidine axial ligand, and arginine, histidine, and tryptophan residues. The peroxide binds to the resting enzyme (Fig. 1a) and undergoes heterolytic oxygen-oxygen bond cleavage in a process catalyzed by the histidine and arginine groups in the pocket (Fig. 1b and c). The structures shown in Fig. 1d–f represent postulated forms for the compound ES intermediate; recent data favor the tryptophan radical formulation shown in Fig. 1f as

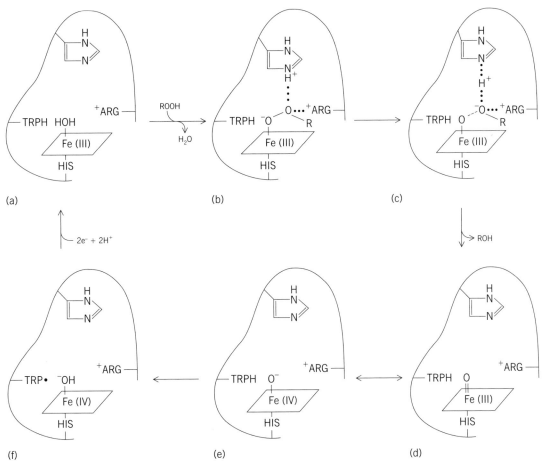

Fig. 1. A proposed mechanism for the reaction between peroxide (ROOH) and the active site in cytochrome c peroxidase.
(a) Resting enzyme. **(b–c)** Heterolytic oxygen-oxygen bond cleavage. **(d–f)** Postulated forms for compound ES intermediate.
R = organic functional group. ARG = arginine group. HIS = histidine group. TRP = tryptophan group. TRPH = tryptophan
group (reduced form). **(After T. L. Poulos and J. Kraut, The stereochemistry of peroxidase catalysis, J. Biol. Chem.,**
255:8199–8205, 1980)

the most probable. In this peroxidase, reduction to resting enzyme occurs without producing a significant compound II population.

Cytochromes P450 represent an important class of oxygen-metabolizing heme enzymes that catalyze a variety of metabolically significant reactions. There are analogies between the cytochromes P450 and peroxidases in terms of mechanism, although the former is more complex, in that O_2, not H_2O_2, is the substrate, and the redox activity of the cytochromes P450 is more complex. Recently a crystal structure of the enzyme has been determined that has provided important insight into the active site structure and O_2-reducing, substrate-metabolizing function of the enzyme.

Cytochrome oxidases. As opposed to peroxidases and cytochromes P450, which usually are important for producing metabolic products, cytochrome oxidase is critical for sustaining the electron transfer reactions involved in cellular respiration. In respiration, highly reducing and thus energetic electrons are stripped from foodstuffs in a series of preliminary reactions in the Krebs cycle. In eukaryotes, these electrons move down the mitochondrial electron-transport chain, which couples spontaneous electron transfer to synthesis of adenosinetriphosphate (ATP). For sustained electron transfer, the electrons must be cleared from the cell. Cytochrome oxidase catalyzes this activity and uses O_2 as a vehicle by which to rid the cell of spent, low-energy electrons. Cytochrome oxidase achieves this function in a remarkably rapid fashion and without releasing toxic, partially metabolized intermediates, such as O_2^- and OH·, into the cell. The overall reaction can be written as shown in reaction (7), where cyt c represents mitochondrial

$$4 \text{ cyt } c^{2+} + O_2 + 4H^+ \rightarrow 2H_2O + 4 \text{ cyt } c^{3+} \qquad (7)$$

cytochrome c. Recent work has shown that cytochrome oxidase functions as a proton pump, thereby contributing directly to the electrochemical gradient that is used for ATP synthesis.

To achieve these functions, oxidase uses four metal centers—two copper atoms and two heme-bound iron atoms. One copper and one iron are used in oxidizing cytochrome c, and the other two metals make up a binuclear center where O_2 is bound and reduced. Since 1980, key insights have been obtained as to the individual steps that occur during reduction of O_2 at this site, and a reasonably detailed mechanism (**Fig. 2**) has emerged. The first step involves binding of O_2 at

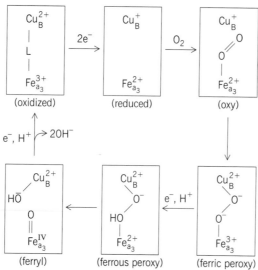

Fig. 2. A proposed mechanism for the cytochrome oxidase–catalyzed reduction of O_2. Cu_B and Fe_{a_3} represent the copper and iron ions, respectively, at the O_2 reducing site. L represents an unidentified bridging ligand between Cu_B and Fe_{a_3}. (After D. F. Blair, S. N. Witt, and S. I. Chan, Mechanism of cytochrome c oxidase-catalyzed dioxygen reduction at low temperatures: Evidence for two intermediates at the three-electron level and entropic promotion of the bond-breaking step, J. Amer. Chem. Soc., 107:7389–7399, 1985)

the heme iron to form an oxy intermediate, which has been observed directly in time-resolved spectroscopic studies. Subsequently, participation of the copper in the site to form a peroxy intermediate is postulated—although, thus far, such an intermediate has escaped detection. Electron transfer from the cytochrome *c* oxidizing heme iron and copper leads to heterolytic cleavage of the oxygen-oxygen bond to produce a ferryl intermediate, somewhat analogous to compound II in the peroxidases. Finally, the ferryl is reduced and the second water molecule is released. It appears that the reactions that occur in the iron/copper binuclear center are intimately involved in the proton-pumping function mentioned above; considerable current effort is directed at establishing the mechanism of this linkage.

For background information SEE CATALYSIS; CYTOCHROME; ENZYME; KREBS CYCLE; PROTEIN; REACTIVE INTERMEDIATES in the McGraw-Hill Encyclopedia of Science & Technology.

Gerald T. Babcock

Bibliography. R. J. Kulmacz et al., Prostaglandin H synthase: Spectroscopic studies of the interaction with hydroperoxides and with indomethacin, *Biochemistry*, 29:8760–8771, 1990; W. A. Oertling et al., Factors affecting the iron-oxygen vibrations of ferrous-oxy and ferryl-oxo heme proteins and model compounds, *Inorg. Chem.*, 29:2633–2645, 1990; T. L. Poulos et al., The 2.6-Å crystal structure of *Pseudomonas putida* cytochrome P-450, *J. Biol. Chem.*, 260:16122–16130, 1985; M. Wilkstrom, Identification of the electron transfers in cytochrome oxidase that are coupled to proton-pumping, *Nature*, 338:776–778, 1989.

Ergonomics

Since the 1940s, the term ergonomics has come to mean the study of humans at work, with the intent of understanding the complex interaction of people with various types of hardware (and now software) required to perform specific tasks.

Scope. It is now recognized by industrial managers and engineers that to develop world-class manufacturing systems they must consider the worker as a vital part of the system. In other words, by considering explicitly known human attributes in the design of a system, not only will the probability of human error be reduced, but an improvement in performance of the system will be realized; that is, productivity and quality of finished goods and services will be improved. Since the early 1970s this need to better understand worker–hardware interactions in industry has been reinforced by the growing cost of occupational injuries. Many of the injuries are due to physical interactions between workers and hardware that result in impact or overstress trauma. The annual costs of such trauma are extremely high in monetary terms and in the number of workers who are disabled. In recent years in the United States, ergonomics research has been increased within the National Institute for Occupational Safety and Health, and some initial ergonomics-based regulations have been proposed by the Occupational Safety and Health Administration within the Department of Labor.

Though ergonomics has been used traditionally to improve the design and use of consumer products, such as a remote control for a television or video cassette recorder, or the shape of a new camera, a major new application has been in the occupational setting. To accomplish this, programs of interdisciplinary research and education have been developed since the 1960s in several large universities. These programs combine knowledge about people from the life and behavioral sciences with knowledge about production systems obtained from engineering and applied mathematical studies. These eclectic efforts have begun to produce graduates with specialized expertise in a number of subdisciplines, as illustrated in **Fig. 1**.

A number of advances have been achieved by this interdisciplinary approach. Production system surveillance methods have been developed that are able to identify production, quality, and injury or illness statistical trends. Computer-assisted work and time study methods allow a trained job analyst to classify and evaluate the variety of techniques that a worker can use in performing certain jobs, particularly if highly repetitive motions are required. Occupational biomechanical methods are used to predict the mechanical stresses produced when a person performs a manual task, which in turn predict the type and extent of trauma that may result. Occupational kinesiological methods provide the basis for determining how people learn to produce complex body movements and what job conditions affect such movements. Work physiology methods can predict how long a specific manual

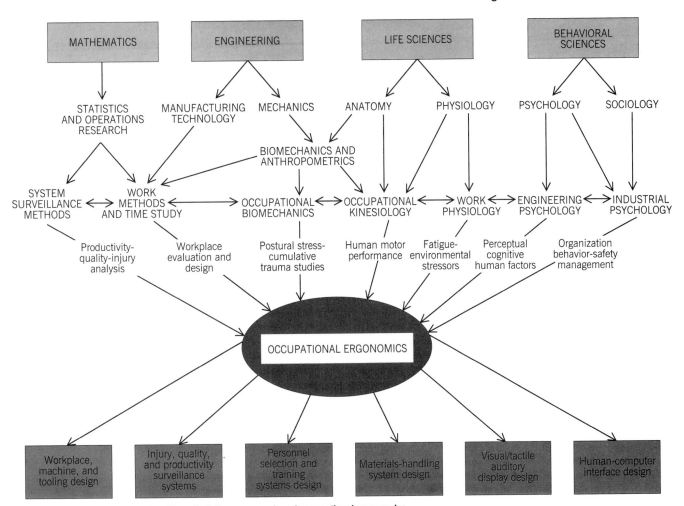

Fig. 1. Chart showing relationship of interdisciplinary research and occupational ergonomics.

task can be performed without fatigue, and how changes in the work requirements can affect the type and degree of fatigue-related performance decrements that will occur during a specific period. Methods based on engineering psychology are used to predict the performance of individuals on tasks that have a high information content, for example, monitoring complex control panels and radar screens, as well as using computers to make financial transactions and plane reservations or to process written text. Industrial psychology is used to establish methods for determining how individuals and organizations respond to the introduction of new manufacturing processes, tools, materials, and methods. The combined knowledge of these specialties has been integrated into the discipline of occupational ergonomics. The goal of this discipline is improvement of the design of workplace, machines, and tooling; surveillance systems for statistically evaluating injury or illness trends on specific jobs; selection tests and training programs; materials-handling systems; information displays; and user-friendly computer systems.

Anthropometric methods. The design of a workplace should ensure that controls, tools, and other frequently used objects will be within easy reach of most workers. Yet, the wide variations in the capabil-

ity of worker reach that exist today, particularly for women of short stature, older employees, or physically limited workers, has greatly increased the complexity of the problem facing the job analyst or job designer.

Not long ago a job designer would assume that if an overhead reach to a control was required, a male would be given such a job. Today, that is not the case; over 46% of the labor market in the United States is composed of women. Where an easy overhead grasp-type reach distance was prescribed to be about 77 in. (195 cm) in some earlier design books, now approximately 73 in. (185 cm) is recommended to accommodate 95% of women. Such a change certainly makes it easy to reach the object of concern, but there are complications. For instance, a control that is 73 in. (185 cm) high, if not carefully located in the workspace, would hit the heads of over 20% of men walking under it.

The large anthropometric variations in the present work force require great concern to assure that the majority of people can physically reach all the objects necessary to perform a job. Overhead, side, and forward reach data must be scrutinized carefully before a final workplace is specified.

If an object to be moved requires an extended arm

posture overhead, to the side, or forward, combined with a forceful exertion requirement, then even greater care is necessary in the layout of the workplace. Strength is greatly limited when the arm is extended and lifting vertically an object in front of the body. An average male can lift for a few seconds about 100 lb (45 kg) when the hands are close to the shoulders. But this capability diminishes to less than 40 lb (18 kg) when the hands are 20 in. (49 cm) in front of the shoulders.

The traditional approach to solve such problems has been to have a job designer consult tables of anthropometric data. Static anthropometric data describe a person's size and shape either at rest (sitting or standing) or while holding an extreme posture. Most of the data depict a specific body segment (for example, length, circumference, or shape) or joint mobility. Thus, the user must often combine the data to form a description of the population attribute of interest.

For instance, to estimate overhead reach while seated, the lengths of the upper extremity and torso can be combined to provide the necessary prediction for the population. However, the prediction may have considerable error, since the amount of upward rotation of the shoulder and torso erection (flattening) of the spine would not be depicted, and thus the estimate might be too short. Such an error may be quite acceptable, however, especially in the early stages of a job design process when fairly close approximations will suffice. Fortunately, varied sources of static anthropometric data are available in tabular form.

In the last decade, the need to combine these data into a more useful form than a tabulation has resulted in the development of drawing-board manikins. These are clear plastic scale models of the human body articulated at various major joints. The most recent development in design manikins allows the torso to flex and extend in a realistic fashion.

The use of drawing-board manikins in combining

static data to predict a functional (dynamic) anthropometric approach does not reduce the error by much. However, manikins are useful. Because the error may be unacceptable in many final design situations, various functional anthropometric data have been developed for typical work situations (for example, overhead reach, seated reach forward, or reaches to each side). One method of depicting these types of functional data is to regress each variable onto the stature (standing height) of the person, to assist in scaling the data for a particular segment of the population.

Since the mid-1980s, computerized human form models have been developed to provide more generality in the use of population size data for widely varying design situations. Though these are still mostly research tools, some limited applications have been reported for advanced designs of aircraft cockpits and general workstations. A typical human form model is shown in **Fig. 2**. As better techniques are developed to predict the postures that are preferred by a person when performing a reach task, these models will provide the means for matching either a person's or a population's size and shape with a specific workspace requirement.

Biomechanical job design guidelines. Recent in-plant statistical studies of injuries from physical exertions have been used to determine the potential for adverse effects on a worker's health when performing general types of physical acts. As an example, one study indicated that lifting of compact (tote box) loads of over 100 lb (45 kg) resulted in about eight times more frequent complaints of low-back pain per worker-hour than did lighter (<35 lb or 16 kg) lifts. Unfortunately, such field studies are difficult to control, and the results are often too general to provide design guidance for a specific task. Rather, these data indicate that there is some type of a problem related to the physical exertions required on a job and that more detailed evaluation by other means is needed.

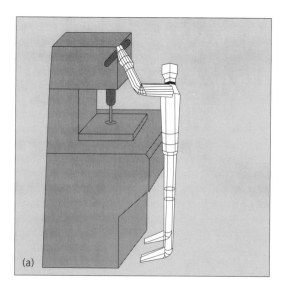 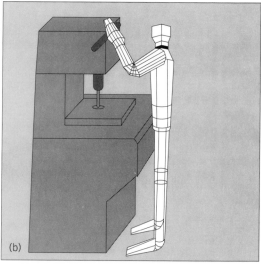

Fig. 2. Human form graphic model for evaluating reach postures for persons operating a drill press. (a) Small person, 5% anthropometry (only 5% of the population would be smaller). (b) Large person, 95% anthropometry (dimensions exceed 95% of the population's dimensions). (*University of Michigan*)

Such detailed evaluations may rely on a second approach, that of creating a biomechanical model of the human musculoskeletal system, which has been shown to be effective in comparing the effects on a person of performing various industrial manual materials-handling tasks. In this approach, the body is considered to be a set of solid links articulated at the major body joints. Statistical studies of injuries have indicated that low-back pain develops when handling heavy loads; thus, one major joint of concern in the kinematic linkage is the lumbosacral joint (a disc at the base of the lumbar spine).

Biomechanical studies have shown that it is not unusual to expect 800 lb (360 kg) of compressive force to develop at the lumbosacral disc while lifting 35-lb (16-kg) loads. If the load is heavier than 35 lb (16 kg) or if it cannot be held close to the body, compressive forces on the lumbosacral spine have been shown to exceed 2000 lb (900 kg). Studies on cadaver spines have shown disc failure when compression forces exceed only 600 lb (270 kg).

Today, biomechanical models of a person's musculoskeletal system are programmed for personal computers, so job designers can simulate manual activities required on a job and predict when spinal disc compression forces and other strength requirements may be excessive. By this means, alternative job requirements can be evaluated before a job is actually performed.

In addition, the National Institute for Occupational Safety and Health has published a guideline for simple manual lifting of loads. This allows a job analyst to predict the risk of injury depending on the magnitude of the load, how close it can be held while lifting, where it is located at the beginning and end of the lift, and how often it is lifted. Similar guidelines are being proposed by the Occupational Safety and Health Administration for jobs that require a great deal of grip force or repetitive hand or finger exertions, which have been shown to cause excessive cumulative trauma disorders to the upper extremities.

Future directions. Ergonomics is a rapidly growing field with several major disciplines that are divided into two arbitrary groups. One discipline deals with the stress placed on personnel involved in information processing originating from poor design of displays, controls, and software that workers must use, often without adequate training. Such stress has been well defined over the years for traditional meters, dials, and similar displays, particularly within the military, aerospace, and transportation industries. Since the mid-1970s the results of this work were gradually applied to more general job design problems in industry. The widespread use of computers in both the office and production environments has led to situations in which workers can be overwhelmed with the information to which they must respond, sometimes very quickly; examples are airline controllers or reservation clerks. Studies are under way to understand how best to use the computer to present information in a way that will achieve maximum human performance with minimum mental stress. This area of investigation will remain a challenge to the psychology-oriented ergonomists for many years.

Experience has shown that where physical work requirements exceed normal worker performance capabilities, the result is numerous injuries that require costly treatments. Thus, the study of how people interact physically with the machines and hardware provided in a job has become an area of intense concern in ergonomics. Sometimes known as physical ergonomics, this has brought engineers, physiologists, anatomists, and kinesiologists together to form a scientific discipline at various universities and industrial laboratories. Their work has provided computerized biomechanical models that help explain why certain injuries occur when a specific manual task is performed.

Guidelines are now being proposed and reviewed. These provide more objective means to determine the relative safety of certain types of manual activity in industry. At present, the ergonomics knowledge is sufficient to understand what happens to a person's musculoskeletal system during relatively simple manual acts. Because the human body can perform manual work with a variety of postures and motions, however, many common work situations will need to be carefully studied in the future. These studies are being facilitated by use of new miniaturized motion-and-force measuring systems and powerful small computers. Their goal is improvement of work environments.

For background information SEE ANTHROPOMETRY; HUMAN FACTORS ENGINEERING; HUMAN-MACHINE SYSTEMS; INDUSTRIAL HEALTH AND SAFETY in the McGraw-Hill Encyclopedia of Science & Technology.

Don B. Chaffin

Bibliography. D. B. Chaffin and G. B. J. Andersson, *Occupational Biomechanics*, 2d ed., 1991; V. Putz-Anderson, *Cumulative Trauma Disorders*, 1988; S. H. Rodgers, *Ergonomic Design for People at Work*, vols. 1 and 2, 1983.

Ethylene

Ethylene, a gas that is naturally produced by plants, controls many plant responses, including growth, abscission, ripening, seed germination, and in some cases cell division. Ethylene, whether endogenous or exogenous, acts in very low concentrations. Externally applied, it is used as a ripening agent and is of considerable commercial importance. Preventing endogenously produced ethylene from inducing a ripening response before it is wanted is also commercially important. Special storage atmospheres, including low-oxygen and high–carbon dioxide mixtures, are used to reduce the effects of ethylene. However, it is not known exactly why atmospheres that are low in oxygen and high in carbon dioxide are effective. Knowledge of ethylene action is incomplete, but an increasing understanding should lead to ways of using this regulator for human benefit.

Action by metabolism. Early researchers thought that ethylene might act by being metabolized since some ethylene is metabolized by many plants. The first metabolic product appears to be ethylene oxide; subsequently some of this is converted to carbon dioxide and some is incorporated into tissues. It was thought that ethylene oxide might act in the same manner as the product of a receptor might act. However, one of the criteria for a hormone receptor is that it saturate at low levels; and while most ethylene responses saturate at 10 microliters/liter, ethylene oxidation does not saturate, even at much higher levels. Also, since not all plant tissues metabolize ethylene, and reagents that block ethylene metabolism do not correspondingly block ethylene action, it is questionable whether ethylene acts in this way. However, ethylene oxidation may have a modulating effect. There does not appear to be an exchange of hydrogens on ethylene during ethylene action, so it is likely that the molecule remains intact during action.

Action by a receptor. Ethylene is not the only compound that causes the array of effects known as the ethylene effect. It is by far the most effective compound, but other compounds, such as propylene, 1-butene, acetylene, carbon monoxide, and isocyanides, also induce the effect. These compounds follow kinetics similar to those involved in enzymatic reactions and hormone binding. For this reason it is thought that ethylene acts by first binding to a receptor. Since the order of action and binding of the compounds that induce an ethylene response is the same as the order in which they bind to silver, it is thought that a metal similar in properties to silver is involved. Because silver is not known to be an essential element for plant growth, it is probable that binding is to some other element.

Action versus inaction. Not all compounds that bind to the ethylene receptor induce an ethylene response. For example, 2,5-norbornadiene prevents ethylene from inducing leaf abscission. By competing with ethylene for the receptor, this compound as well as several others are effective in blocking the action of ethylene. 2,5-Norbornadiene was the first competitive compound discovered, and it is now widely used in experimentation. Recently it has been shown that *trans*-cyclooctene and some *trans*-cyclooctadienes are even more effective in blocking the action of ethylene. The order of binding to silver among compounds that block ethylene action seems to correspond to their order of effectiveness in blocking ethylene. Those that bind to silver at a low concentration are effective at blocking ethylene responses at a low concentration, indicating that binding alone is not sufficient to induce an ethylene response. It has been suggested that back-acceptance of electrons from the metal to ethylene or another compound is necessary to induce a response.

Silver ions, usually applied as silver thiosulfate, overcome the effect of ethylene in low concentrations and are widely used for this purpose. It is not known exactly how silver acts, but it does not appear to be by exerting a competitive effect. Somehow it interferes with ethylene binding; it may also inactivate part of the ethylene response mechanism.

Ethylene binding. Plant tissues have been shown to bind ethylene in a reversible manner. Ethylene binding can be measured by exposing plant material to low amounts of ethylene labeled with carbon-14. The labeled ethylene binds, and after removal of the plant material from it the labeled ethylene can be trapped as a mercury perchlorate complex whose radioactivity can be measured. The amount of nonbound dissolved ethylene can be determined by treating the plant material with the same amount of labeled ethylene that also contains a large amount of unlabeled ethylene to displace the labeled ethylene from the binding sites. The difference between these two values (bound ethylene plus dissolved ethylene and dissolved ethylene) is the amount of specifically bound ethylene.

More than one type of binding site has been identified. By using labeled ethylene, it can be shown that bound ethylene diffuses from the binding site in distinct phases. In intact mung bean sprouts, three distinct phases have been identified. One has a half-life of about 6 min, another a half-life of about 2 h, and a third a half-life of 50 h or more. Ground mung beans do not show all three distinct sites. It is not known whether one or all of these binding components are the physiological receptor.

Purification. Efforts have been made to purify the binding components. A component purified from bean cotyledons has been reported to contain copper. This component binds ethylene for a long time. No function of ethylene in bean cotyledons has ever been shown, and it is not known if this component is a receptor.

Another binding component has been partially purified from mung bean sprouts. This component has somewhat different properties than the component from bean cotyledons. Again, it is not known whether this component is a receptor. This component binds ethylene for a short time in intact tissue, but does not seem to bind when the tissue is ground. New methods must be developed to purify this component; then it must be determined if this component is a receptor and if there are additional binding components and receptors.

Mutants. Recently, by using mutagenic agents, a mutant of *Arabidopsis* has been produced that does not respond to ethylene. Measurements with labeled ethylene indicate that the mutant has a much reduced level of ethylene binding. The receptor may either be missing or altered so that it does not bind ethylene. This mutant plant should be useful in deciphering the mechanism of ethylene action. Since the plant is capable of completing its life cycle, it illustrates the regulatory role rather than the essential role of ethylene.

Photoaffinity labels. Recently, the compound diazocyclopentadiene has been shown to be an effective photoaffinity label that could be used to tag the receptors. Research using photoaffinity labeling may soon identify which components are physiological receptors and where they are located. Since a photoaffinity label appears to permanently block the ethylene receptor, it may be of great practical value in prolonging the life of fruits and vegetables.

Action by induction of new enzymes. After plants are exposed to exogenous ethylene, new enzymes are induced. The type of enzyme induced depends on the plant tissue. In many fruits, the enzymes that are associated with ripening are induced or increased. In abscission zones, enzymes such as cellulase and polygalacturonase, both of which are associated with the breakdown of cell walls, are induced. In other plants, there appears to be a role for ethylene in the induction of components that are associated with disease resistance.

Action by other modes. Ethylene plays a role in growth responses, in seed germination, and in many other plant responses. For example, by using ethylene blocking agents, it can be shown that dandelion flowers require ethylene to close. It is not known whether responses such as this require new protein synthesis. It remains a challenge to show how ethylene acts in such cases. It is possible that ethylene acts through interaction with other plant hormones. For example, ethylene action and indole-3-acetic acid (IAA) action are closely related. Ethylene inhibits IAA transport, and IAA stimulates ethylene biosynthesis. Because of this close relationship, it is often difficult to distinguish between ethylene action and IAA action.

Research at the genome level. Work at the genome level is being undertaken to determine how the ethylene response is controlled, particularly as it relates to the ripening response. At present, nothing is known about how the ethylene signal is transferred, from the receptor level to the genome level.

For background information SEE ABSCISSION; DORMANCY; ETHYLENE; PLANT GROWTH; PLANT HORMONES in the McGraw-Hill Encyclopedia of Science & Technology.

Edward C. Sisler

Bibliography. K. E. Broglie et al., Functional analysis of DNA sequences responsible for ethylene regulation of a bean chitinase gene in transgenic tobacco plants, *Plant Cell*, 1:599–607, 1989; A. K. Matoo and J. C. Suttle (eds.), *The Plant Hormone Ethylene*, 1990; R. P. Pharis and S. Rood (eds.), *Plant Growth Substances 1988*, 1990; E. C. Sisler and S. F. Yang, Ethylene, the gaseous plant hormone, *Bioscience*, 34:234–238, 1984.

Event-related potentials

Event-related potentials (ERPs) are electrical responses in the brain that are produced by discrete occurrences (events) that provide information to the senses and nervous system. Possibly, study of ERPs will help increase understanding of the mind by describing how the brain works.

Electroencephalogram and event-related potentials. There are two broad classes of electrical brain activity. The electroencephalogram (EEG) consists of intrinsic rhythms or waves that are integral to the brain's functioning independent of any observable changes (events) in the environment. The intrinsic rhythms are continuously present, even during sleep. EEGs have been useful in the study of organic changes in the brain, such as in the diagnosis of epilepsy and tumors, and in the study of sleep stages.

The trace in **illus.** *a* shows that the EEG can occur in the absence of observable stimulation. This activity alternates rapidly between negative and positive voltage values, in this case 10 Hz. Hence, two characteristics of the EEG are its polarity (positive or negative voltage) and its frequency (waves per second). Wave height (peak to trough) yields a third descriptor, amplitude, which is expressed in microvolts.

A second type of electrical brain activity is the event-related potential, which is derived from the EEG and which occurs when the brain responds to stimulation. Illustration *b* shows that an electrical potential with a downward (positive) deflection can be seen following the occurrence of a stimulus, which was in this case a dim light flash. This type of inflection point in the EEG trace permits the detection of a component, which is an identifiable wave that regularly occurs after a stimulus. Two important characteristics of a component are the location on the scalp where it appears most clearly, and its relationship to stimulus properties such as latency (the time required for the peak of a component to occur after a stimulus) and amplitude. The value of the peak voltage (whether positive or negative) relative to the

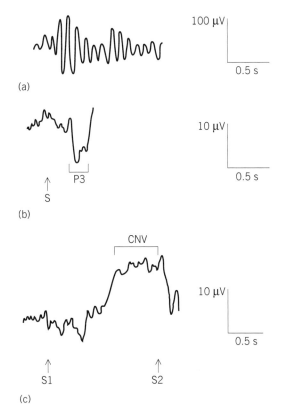

Three types of electrical brain activity based on the presence or absence of events (stimuli; S), which are indicated by arrows. (a) Electroencephalogram. (b) Event-related potential (single stimulus). (c) Event-related potential (paired stimuli). An upward deflection indicates negativity. Terms are explained in text.

neutral EEG baseline that existed before the event is its amplitude, in microvolts. As with the EEG, amplitude can also be expressed by measuring peak-to-trough values.

The amplitude of an ERP component is usually much smaller than the background EEG from which it is derived. Nevertheless, the ERP can be extracted from the EEG, since components have positive and negative voltage waves that appear at approximately the same time following a stimulus, whereas the background EEG waves are not as synchronized. Consequently, with repeated presentations of the same stimulus, the random occurrences of positive and negative waves of the background EEG will tend to cancel one another and permit ERPs to stand out more clearly. This averaging process is done by computers. The enhancement of potentials may also be achieved by removing unwanted, selected frequencies through filtering, thus permitting only designated EEG signals to enter computer analyses. ERPs reveal information on how the brain responds to events.

Two classes of ERPs. There are two broad classes of ERPs. One type is called exogenous potentials, since they are determined primarily by stimuli from outside the organism. Since exogenous potentials are caused or evoked by the stimulus, they are also called evoked potentials. Their appearance is determined by how the physical energy in the stimulus is processed by the nervous system, especially the brain. Hence, they reflect the integrity of primary sensory systems in the brain. A second type of ERP is called endogenous, since these potentials are primarily a function of internal states that determine an individual's interpretation of the stimulus. For example, events considered important, such as hearing one's own name, will elicit a larger endogenous ERP than a relatively unimportant event, such as hearing the name of a chemical element. Exogenous ERPs do not show this type of relationship. Endogenous ERPs are considered to be evoked potentials when they are responses to physically present stimuli. However, an important feature of endogenous components is that they can occur in the absence of an external event, such as when a regularly occurring event is unexpectedly withheld. In this case, they are called emitted potentials. The distinction between exogenous and endogenous ERPs is not always a mutually exclusive one, and some potentials can be classified in both groups.

Exogenous ERPs. Exogenous ERPs have been studied primarily in three modalities: auditory, visual, and tactile. These ERPs typically occur soon after the stimulating event. For example, in the auditory modality, clicks given at the rate of 10 per second will evoke five positive waves within 10 milliseconds after each stimulus. These components originate from successive points in the auditory pathway. They are called brainstem auditory evoked potentials or far-field potentials, since the structures generating them are deep in the brain and are far from the recording site on the scalp. Because of the recording distance involved, their amplitudes are low (0.5–1.5 μV), and thousands of clicks are needed for these waves to be measured reliably. Amplitude measurements permit the assessment of hearing thresholds in premature babies who are at risk for hearing impairments, in individuals for whom behavioral testing is not feasible, and in the diagnosis of brain death, multiple sclerosis, and other neurological disorders. These potentials are insensitive to changes in psychological states, such as lack of alertness, and are resistant to effects from centrally acting drugs, such as general anesthesia; consequently, they are valuable in the detection of selective organic hearing problems with a degree of specificity lacking in most other ERPs.

In the visual modality, patterned light, such as checkerboard designs with black and white squares reversed in alternate presentations, has been found clinically useful in the assessment of visual problems. The P100 component (positive in polarity and approximately 100 ms in latency) can detect silent lesions, that is, damage to the optic nerve, before visual problems become clinically apparent, and thus can assist in the early detection of multiple sclerosis (a demyelinating disease that can produce lesions in the visual system and eventual problems in vision). Abnormalities in P100 can also indicate retinal deterioration, degeneration of the optic nerve, and damage to other visual pathways in the brain. A loss of normal visual acuity, as in blurring, can result in an abnormal P100.

The third modality of stimulation involves the presentation of small, painless electric shocks or other tactile stimuli to the skin in order to stimulate nerves and cause brain activation. These nerve pathways in the somatosensory system are unusually long and run from peripheral nerves to various parts of the brain. These ERPs have short latencies, typically within 50 ms, although later components occur 200 ms or more after stimulation. The clinical application of short-latency potentials includes diagnosis of multiple sclerosis and other demyelinating diseases as well as the assessment of nerves and evaluation of both the spinal cord and brainstem itself. These potentials have also been useful in assisting in the determination of brain death.

Endogenous potentials. Endogenous potentials typically have longer latencies than exogenous potentials, ranging from 100 ms to several seconds, and can be elicited by any sense modality. A late positive complex of waves has a latency of approximately 300 ms, and is referred to as P300 or P3. An example of a P3 wave appears in illus. *b*. It is based on an average of 12 responses. The late positive complex is readily elicited in a task where successive, identical stimuli are occasionally and unpredictably interrupted by different stimuli. For example, if a low tone occurs randomly one out of five times in a series of high tones, it will produce the P3 wave. The occasional omission of the high tone will also emit a P3. These potentials have been described as reflecting a variety of factors: information transmission, stimulus meaning and evaluation, orienting, memory revision, attentiveness, and surprise. Latencies of P3 appear delayed in clinical populations having dementia, depression, and schizophrenia, suggesting its sensitivity, but not selectivity, in the assessment of clinical status.

There are several negative endogenous ERPs. When tones are presented randomly to one ear or the other, a component (latency of 100 ms) exhibits a heightened amplitude to tones appearing in the ear to which the individual is attending. This wave, the Nd wave, contains a prolonged negative wave that lasts for several hundred milliseconds and appears to be reflecting processing of the stimuli beyond selective attention to a channel (ear) that yields the information. Similarly, P3 components represent a more fine grained processing of the stimulus after the early selection represented in the Nd wave. Attention-related components similar to the Nd wave have been reported for visual and somatosensory modalities.

A negative potential with a latency of about 200 ms appears to reflect the brain's recognition of a mismatch between the evoking stimulus and memories from previous stimulation. This N200 or N2 wave is called mismatch negativity. A later negative wave (N400) seems to reflect a similar type of incongruity for semantic processing. For example, an N400 wave can occur in response to the word ketchup in the sentence, ''I took off my shoes and ketchup.''

In a reaction time task, where a preparatory stimulus (S1) is followed by a second stimulus (S2) to which a motor response is made, there appears in the S1-S2 interval a negative shift in the EEG baseline (illus. c). This wave is called the contingent negative variation or CNV, since its occurrence depends upon the association or contingency between the two stimuli. Since a voluntary motor response alone generates a slow negative potential prior to its occurrence, the contingent negative variation complex contains this motor component as well as both early and late cognitive components. Late contingent negative variation amplitude (that part of the wave just prior to S2) is the most widely studied component and has been related to a variety of functions, such as attention, arousal, expectancy, preparation, and motivation. It is a particularly sensitive measure of arousal and attention impairment caused by distraction. For example, when letters are presented in the S1-S2 interval for memorization, contingent negative variation amplitude is reduced (CNV distraction effect). When the letters are unexpectedly omitted, contingent negative variation amplitude is increased to above-normal values (CNV rebound effect). Contingent negative variation amplitude is also reduced by psychoactive drugs that interfere with attention in both normal elderly groups and clinical populations, such as patients with Alzheimer's disease and schizophrenia, who characteristically have impaired attention functions. The delay in the return of contingent negative variation to baseline after S2 (considered an imperative stimulus because of the required response) is called postimperative negative variation and has been useful in the evaluation of psychiatric populations.

A new technique for recording ERPs as well as the EEG itself is the magnetoencephalograph (MEG), which records the magnetic fields of the brain. An-other is a computerized method called imaging, which permits pictorial displays of amplitudes and latencies of components against a graphic template of the brain. This procedure gives an immediate visual impression of the extent of component presence for different scalp areas and, by inference, different brain areas.

For background information SEE BRAIN; ELECTROEN-CEPHALOGRAPHY in the McGraw-Hill Encyclopedia of Science & Technology.

J. J. Tecce

Bibliography. K. H. Chiappa, *Evoked Potentials in Clinical Medicine*, 1990; E. Niedermeyer and F. Lopes da Silva (eds.), *Electroencephalography*, 1987; T. W. Picton (ed.), *Human Event-Related Potentials*, 1988.

Extinction (biology)

Extinctions occurring over geological ages can have significant effects on the composition of subsequent biotas. This article discusses the biotic transitions of the Late Precambrian to Cambrian and the Triassic to Jurassic.

LATE PRECAMBRIAN-CAMBRIAN BIOTIC TRANSITION

One of the major episodes in life's history occurred during the transition from the Proterozoic to the Cambrian Period of the Paleozoic Eon, between about 580 and 540 million years ago (m.y.a.). During this period, phytoplankton in the seas became extinct and reradiated, the first known animals evolved and then disappeared, other animals left trackways on the sea floor and later began to burrow, the first tiny animals with skeletons appeared, and finally larger skeletonized animals appeared and radiated into the biota of the Paleozoic (**Fig. 1**). Extraordinary tectonic, paleoclimatic, glacial, and paleoceanographic changes took place during this time that may have been connected with the biological events. It was a period of complex, dynamic evolution that involved organisms at all grades of organization from single cells to multicellular animals. At present, many workers in all parts of the world are struggling to understand three basic questions: (1) what is the sequence of fossil occurrences; (2) what is the relationship between the fossil occurrences and the physical events known in the geologic record; and (3) what caused the apparent radiation of so many kinds of organisms in this relatively short period of geologic time?

Ages. The ages of rocks containing these Precambrian and Cambrian fossils are currently undergoing revision. For years, the extinction of the first soft-bodied animals and phytoplankton was thought to be as long ago as 650 million years (m.y.) and the first skeletonized fossils were thought to be about 570 m.y. old. Recent determinations of radiometric ages and stratigraphic relationships between rock units indicate that both of these dates are too old. The soft-bodied fauna now appears to be 580 or 570 m.y. and the first skeletons are about 550 m.y., thus reducing the time between these appearances from nearly 100 m.y. to

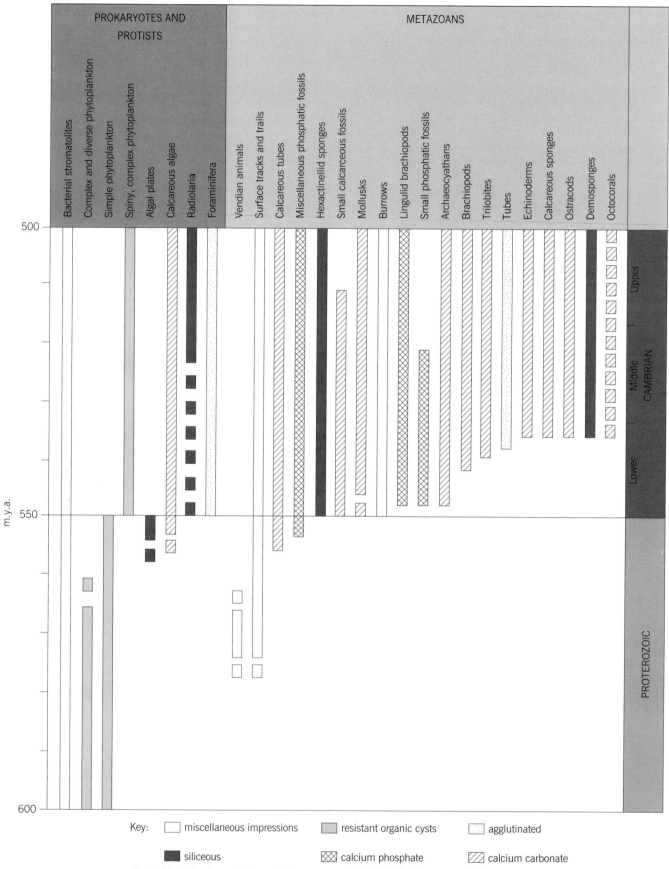

Fig. 1. Chart showing extinctions and appearances of fossils in the Late Proterozoic-Cambrian, 600–500 million years ago. Both single-celled organisms (bacteria and protists) and multicellular animals (metazoans) have apparent extinctions in the latest Proterozoic. Radiations in single-celled organisms and metazoans occurred near the base of the Cambrian and involved fossils with different body walls (soft or skeletonized with a variety of materials) as indicated by key. (*Skeletal fossil data modified from S. Bengtson and S. Conway Morris*)

Fig. 2. Vendian frond animal, *Dickinsonia brachina*, from the Ediacaran fauna of South Australia. Specimen is 182 mm or 7.15 in. long. (*Holotype, South Australian Museum, Adelaide, P. 17998; photograph by C. S. Hickman*)

perhaps only 20 m.y. The actual Precambrian-Cambrian boundary is thus close to 550 m.y. old.

Single-celled fossils. About 1800 m.y.a., the first eukaryotic cells had joined the preexisting bacteria that dominated the world for the previous 2000 m.y. These cells were simple, smooth microscopic cysts made of resistant organic material; the cysts are known as acritarchs and may represent several groups of planktonic algae. An onshore and an offshore assemblage can be detected in many parts of the world. About 850 m.y.a., the acritarchs became morphologically more complex and their diversity began to increase. There were some large acritarchs with spines, sculptured vesicles with double walls, tiny vase-shaped microfossils, and later, microscopic algal scales. Algal protists were probably much more diverse in the Late Proterozoic, but many simply were not preserved. A few kinds of macroscopic algae also existed then. All these fossils indicate that the rate of evolution in the oceans had increased significantly between 850 and 550 m.y.a., compared to the previous 2500 m.y. for which a fossil record is known.

All known single-celled fossil organisms from the Precambrian were photoautotrophic or chemoautotrophic; none seems to have captured or eaten other organisms. Because so many kinds of phytoplankton and bacteria were present, heterotrophic protists may

well have been present too, but are simply not preserved for lack of hard parts. Some time near the end of the Proterozoic when the first metazoans appeared, the phytoplankton underwent a major extinction, leaving only rather simple, spherical cysts. Later, at the base of the Cambrian, the phytoplankton underwent a radiation that increased diversity and morphologic complexity.

First animals. The first fossil animals appeared about 580 m.y.a. during this period of relatively rapid evolutionary activity in the single-celled organisms. They are large soft-bodied animals, known as the Ediacaran or Vendian fauna. Vendian animals are found on all continents except Antarctica, but they number fewer than 100 identifiable species. Because of their unique appearance, they have been placed in a number of different systematic categories. At first, they were placed in known phyla of animals, but increasingly, their differences are being recognized and most of them are considered to be unrelated to known phyla. The various forms can be divided into several groups: (1) thin, bilateral, frondlike forms (**Fig. 2**), some with holdfasts on one end, others with equally tapering ends, and some kinds growing up to a meter in length (*Dickinsonia*); (2) medusoids (jellyfish) of various sizes, shapes, and symmetries (**Fig. 3**), some with thick tentacles, but all differing from modern forms in that they lack the characters, particularly fourfold symmetry, of later scyphozoans; (3) an unusual group of fossils with triradial symmetry, including *Tribrachidium*; (4) tracks and trails (up to a centimeter wide) on but not in the sea floor made by unknown, probably wormlike, soft-bodied animals; (5) fossils questionably assigned to the arthropods or in one case the echinoderms; and (6) fossils of unknown affinities.

The Vendian soft-bodied fauna disappeared more or less forever well before the start of the Cambrian. Most Vendian animals have no descendants easily attributed to them, although some vestiges of this

Fig. 3. Vendian medusoid animal, *Mawsonites spriggi*, from the Ediacaran fauna of South Australia. Specimen is 96 mm or 3.8 in. top to bottom. (*Holotype, South Australian Museum, Adelaide, F. 17009; photograph by C. S. Hickman*)

biota can be seen in later rocks. The various tracks and trails, for example, do not change their character across the Proterozoic-Cambrian boundary, but new burrowing kinds did appear; the unusual trilateral symmetry occurs in a few new Early Cambrian forms. Whether the disappearance of the Vendian animals marked a significant extinction or not remains to be determined. They were few in number, preserved only in certain kinds of rocks, and lived prior to the evolution of infaunal animals that could disrupt the traces of their bodies left on the soft sea floor. Thus, they may have been slowly replaced just as easily by the Cambrian fauna without leaving a good record. Contrarily, the phytoplankton did experience an extinction close to this time, which might indicate that extinctions also took place in other parts of the marine ecosystem.

Cambrian radiation. The base of the Cambrian is marked by the appearance of the first skeletonized animals and protists, burrowing animals, and a reradiation of phytoplankton. The phytoplankton became complex, with generally smaller cysts carrying spines and wall patterns of various kinds that were not present previously. The animal skeletons are generally small, calcareous, phosphatic, or agglutinated shells or skeletal parts of various shapes. Many are cap- or tube-shaped; others are spicules, plates, spines, or bivalved shells. Most of the modern phyla of invertebrate animals were present in this assemblage, but so were other ones that are placed in their own phyla. Foraminiferans and radiolarians may be represented among them as well. These early Cambrian assemblages gave way to a fauna dominated by trilobites, brachiopods, bivalves, and other fossils more typical of the rest of geologic time. Likewise, these biotas were the part of a typical marine ecosystem structured by longer food chains based on the phytoplankton, detrital feeding invertebrates, herbivores, and predators. The nature of marine ecosystems underwent the biggest change in the history of the world, and this change led to the the kinds of ecosystems that exist in the present.

Causes. One of the most difficult questions in paleobiology is determination of the reason for the appearance and radiation of animals and protists in this time interval. Over 25 hypotheses have been proposed to account for these events, but none is generally accepted. The hypotheses can be divided into those that assume a longer, unrecorded history of metazoans, and those that link the biological events to some environmental factors.

The fossil record shows no evidence that animals existed prior to the appearance of the Vendian fauna, but that does not exclude a possible long history. The Vendian fauna is meager, unusual in morphology, and preserved in unique circumstances. It probably cannot be considered to represent the very first animals. Hypotheses that assume the preexistence of unknown metazoans then attribute their appearance in the fossil record to intrinsic factors such as the attainment of large size or sedentary habit, the evolution of sex, the invasion of preservable habitats, the development of

predators, or the attainment of regulatory genes or metazoan developmental patterns. These hypotheses are unacceptable because all causes affect only metazoans, yet planktonic and benthonic protists also show similar patterns of evolution.

Very many physical events occurred during the Precambrian-Cambrian interval that have been implicated in the observed evolutionary patterns. These include the breakup of a supercontinent, sea-level fluctuations, or the emergence of continents. Such events all provided increased ecological opportunities for radiations; for changes in the oxygen levels; for the carbonate or the phosphate in the oceans to provide the necessary chemical components for the construction of muscles or skeletons; or for the glaciation of the Earth near 600 m.y.a., just prior to the radiations, to provide new circulation patterns and intensities in the world's oceans, perhaps affecting nutrient supplies and trophic relationships. Many hypotheses in this class make ecological and evolutionary sense, but the models are so complex that they cannot be tested in the fossil record. Each model, to one degree or another, can account easily for the observed patterns, but no way to test the models exists.

The marine biota near the Proterozoic-Phanerozoic boundary shows rapid diversification in nearly all groups and at different grades of organization, different ecologies, different skeletal types, and different trophic levels. These simultaneous histories seem to require a process acting across ecosystems and affecting each biota. Reasonable causes that might be involved include paleoceanographic or paleoclimatic changes influencing trophic resources that then cascaded through the various food chains. At present, no resolution to this question is possible.   *Jere H. Lipps*

TRIASSIC-JURASSIC BIOTIC CHANGE

The transition from the Triassic to the Jurassic Period, currently dated as 201 ± 2 m.y. before present, marks one of the most profound of the 13 or so major episodes of extinction that punctuate the Phanerozoic history of life. Paleontologists first found unequivocal evidence for a mass extinction at the end of the Triassic (Norian stage) among marine invertebrates, which severely affected bivalves, brachiopods, cephalopods, and gastropods. This event apparently more or less coincided with a major extinction among terrestrial vertebrates. Some uncertainty in relating these biotic changes to each other is due to a lack of precision in stratigraphic correlation. Geologists originally established the currently recognized stages of the Triassic and Jurassic periods in sequences of fossiliferous marine sedimentary rocks, especially in Europe; thus, direct correlation to continental strata proved, for the most part, impossible. Correlation using fossil pollen and spores and radiometric dating of volcanic rocks interbedded with the sedimentary rocks both have provided effective indirect means of linking marine and continental strata.

Pattern of biotic change on land. The Newark Supergroup of eastern North America provides an ideal setting for studies of biotic change across the Triassic-

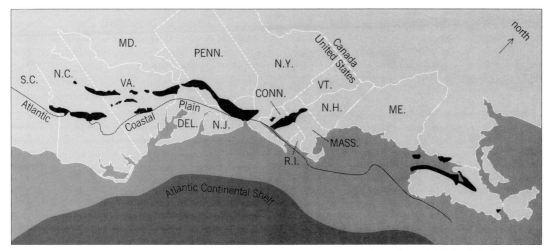

Fig. 4. Exposures of the Newark Supergroup in eastern North America. (*After Proceedings of the 2d U.S. Geological Survey Workshop on Early Mesozoic Basins of the Eastern United States, USGS Circ. 946, 1984*)

Jurassic boundary (**Fig. 4**). It represents the remnants of thousands of meters of sedimentary and volcanic rocks deposited in 13 or so major rift basins that were created during a 45-m.y. episode of crustal extension and thinning preceding the Jurassic breakup of the supercontinent Pangaea and the opening of the northern Atlantic Ocean. Many of the Newark Supergroup basins were occupied by large fresh-water lakes, which waxed and waned in accordance with cycles induced by changes in the Earth's orbit (Milankovitch cycles). These changes in lake levels produced a cyclical sedimentary record that permits rather precise inferences regarding the timing of the biological changes across the Triassic-Jurassic boundary.

The sedimentary rocks of the Newark Supergroup contain several well-dated occurrences of fossil tetrapods ranging in age from the Middle Triassic to the Early Jurassic. The recent discovery of diversified vertebrate assemblages of earliest Jurassic age in the lower part of the McCoy Brook Formation of Nova Scotia (Canada) is of particular interest for discussions of faunal change across the Triassic-Jurassic boundary, because the assemblages appear to postdate the boundary by less than 1 m.y. Although these assemblages represent a number of different paleoenvironments, they all share the absence of certain dominant groups of reptiles and amphibians that characterized Late Triassic communities in North America and elsewhere. The groups include several lineages of archosaurian reptiles related to crocodilians, such as the crocodilelike phytosaurs, the heavily armored, herbivorous aetosaurs, and the carnivorous rauisuchians. A number of smaller forms also disappeared at the end of the Triassic Period, including the vaguely lizardlike procolophonids (**Fig. 5**). On the other hand, most of the vertebrate groups found in the McCoy Brook Formation already occur in Late Triassic assemblages and apparently were unaffected by the extinction event. This abrupt disappearance of many groups of terrestrial vertebrates suggests a major extinction event at or near the boundary. Only a few Early Jurassic tetrapod occurrences are known from

other parts of the world (Europe, southern Africa, and China), but they all reflect the same pattern of change in faunal composition.

Similarly, the fossil record of pollen and spores from the sedimentary rocks of the Newark Supergroup and elsewhere indicates a major turnover among land plants at or near the Triassic-Jurassic boundary. Late Triassic assemblages of pollen and spores, which indicate high levels of plant diversity, are replaced by Early Jurassic ones that are almost exclusively made up of pollen of cheirolepidiaceous conifers.

Pattern of biotic change in seas. In the richly fossiliferous sequences of marine sedimentary rocks in

Fig. 5. Vertebrates from Late Triassic assemblages. (*a*) Skeleton of a prosauropod dinosaur, *Plateosaurus*; these herbivorous dinosaurs survived well into the Early Jurassic (*after D. B. Weishampel and F. Westphal, Die Plateosaurier von Trossingen, ATTEMPTO Verlag GmbH, 1986*). (*b*) Skull (front view) of the procolophonid reptile *Hypsognathus*; procolophonids became extinct at the end of the Triassic Period (*after W. Manspeizer, ed., Triassic-Jurassic Rifting: Continental Breakup and the Origin of the Atlantic Ocean and Passive Margins, pt. A, Elsevier, 1988*).

Europe, some 92% of bivalve species became extinct between the latest Triassic (Sevatian substage) and earliest Jurassic (Hettangian stage), and 80% of the latest Triassic species of brachiopods disappeared before the Jurassic. Ammonites, a group of shelled cephalopods, almost completely vanished, with perhaps only a single genus surviving into the earliest Jurassic. The large communities of reef-building scleractinian corals and calcareous sponges in the Alpine region suddenly collapsed and disappeared. The structure of marine invertebrate assemblages was severely affected by these extinctions, which removed many ecologically dominant taxa and represented far more than a mere elimination of rare or highly endemic forms. The Triassic-Jurassic boundary also marks the demise of the enigmatic conodonts, known only from minute, toothlike elements with possible chordate affinities. Among vertebrates, a number of lineages of marine reptiles disappeared at the end of the Triassic Period.

Two possible mass-extinction events. Several recent compilations of Triassic biotic diversity have suggested the possibility that the mass extinctions among marine organisms and land vertebrates, respectively, at the end of the Triassic Period were preceded by comparable drops in diversity during the Late Triassic, specifically at the transition from the Carnian to the Norian stage. This suggestion has been based on literature surveys of the familial and generic diversity of marine invertebrates and of vertebrates. Preliminary data from pollen and spores and tetrapods from the Newark Supergroup, however, do not support this idea. Rather, they indicate the occurrence of a considerable turnover among tetrapods within the Carnian portion of the Newark Supergroup. Apparent major losses in diversity among marine organisms at the Carnian-Norian transition may simply reflect very high evolutionary rates among ammonites as well as inadequate sampling of early and middle Norian sedimentary rocks.

Possible causes of extinction. A number of possible causes have been advanced to explain the mass-extinction event at the end of the Triassic Period: (1) climatic changes (possibly related to changes in sea level); (2) competitive superiority of newly evolved animal and plant groups; (3) very large volcanic eruptions; and (4) impacts of extraterrestrial objects such as asteroids or comets. Early Jurassic sediments appear to indicate climatic changes in many parts of the world, some of which may be related to changes in sea level and the onset of the breakup of the supercontinent Pangaea. These climatic perturbations, however, do not appear to coincide with each other on a global scale or with the major faunal and floral changes. Changes in sea level did occur in many parts of the world and are a possible cause of at least some of the extinctions among marine organisms. Extinctions as the result of competitive replacement are unlikely in view of the fact that the survivor groups originated during the Triassic Period and coexisted for millions of years with those that became extinct at the end of the Triassic.

Widespread massive extrusions of basaltic lavas in eastern North America and in western and southern Africa, respectively, appear to postdate the Triassic-Jurassic boundary by perhaps less than 100,000 years, based on recent estimates using sedimentary cycles in the lake deposits of the Newark Supergroup. While quite close in geological time to the extinctions, this increased volcanic activity could then not account for the observed biological changes. Most recently, several scientists have suggested the impact of an extraterrestrial object as a causative agent. Speculations focus on the large Manicouagan impact crater (>70 km or 42 mi diameter) in Quebec. Radiometric dates for the impact that formed this structure, however, range from 206 ± 6 to 215 ± 4 m.y.a.; this scattering of dates may be due to excess argon released into the impact melt from the underlying Precambrian basement rocks. Attempts at more refined redating of the Manicouagan impact are currently under way, along with a search for possible geochemical abnormalities and layers of impact ejecta such as shocked quartz in sedimentary rocks straddling the Triassic-Jurassic boundary in eastern North America and in Europe. In absence of such unequivocal evidence, it is premature to hypothesize any causal connection between the Manicouagan impact event and the faunal and floral changes at the end of the Triassic Period.

Effects of mass extinction. The Triassic Period witnessed the origination and initial diversification of most of the major groups of vertebrates that subsequently came to dominate the world's ecosystems. Mammals, teleostean fishes, and dinosaurs all first appeared during the latter part of the Triassic, along with precursors or representatives of most major groups of present-day vertebrates. Toward the end of the Triassic, the dinosaurs show a rapid and dramatic rise to dominance on land, which they maintained for the next 145 m.y. They appear to have been unaffected by the events at the Triassic-Jurassic boundary. Initially, the newly originated vertebrate groups coexisted with surviving lineages from the Permian Period and with some indigenous Triassic forms, but most of the latter were eliminated at the end of the Triassic, leaving terrestrial vertebrate communities of essentially modern appearance.

For background information SEE CAMBRIAN; CONTINENTAL DRIFT; EXTINCTION (BIOLOGY); GEOLOGICAL TIME SCALE; JURASSIC; ORGANIC EVOLUTION; PRECAMBRIAN; ROCK AGE DETERMINATION; TRIASSIC in the McGraw-Hill Encyclopedia of Science & Technology.

Hans-Dieter Sues

Bibliography. M. D. Brasier, Evolutionary and geological events across the Precambrian-Cambrian boundary, *Geol. Today*, pp. 141–146, September–October 1985; S. Conway Morris, The search for the Precambrian-Cambrian boundary, *Amer. Sci.*, 75: 157–167, 1987; M. F. Glaessner, *The Dawn of Animal Life*, 1984; *Global Catastrophes in Earth History: An Interdisciplinary Conference on Impacts, Volcanism, and Mass Mortality*, Lunar and Planetary Institute and National Academy of Sciences, 1988; M. A. S. McMenamin and D. L. S. McMenamin, *The Emergence of Animals: The*

Cambrian Breakthrough, 1990; K. Padian (ed.), *The Beginning of the Age of Dinosaurs: Faunal Change across the Triassic-Jurassic Boundary*, 1986.

Flower

Flowers, the reproductive structures of angiosperms, generally consist of a short length of stem with modified leaves attached to it. In a typical flower, these modified leaves occur in four sets, or whorls. In the course of evolution, the members of any whorl may have become fused with one another or with those of another whorl, resulting in a diversity of floral forms. These fusions are usually connected with precise pollination systems. For example, nectar in flowers visited by hummingbirds is often held within strong tubes where it can be obtained only by the bird's beak and not, in general, by insects. In many such flowers, these tubes are formed by fusion of the margins of neighboring petals. Other fusions occur between neighboring carpels (the female reproductive structures) to yield many-chambered fruits; tomatoes and bell peppers are examples.

Plant development. Several processes are unique to the development of multicellular plants. Surrounding each plant cell is a rigid wall that usually prevents cell migration early in development. This lack of cell migration means that there is usually no new contact between cells that have not been formed following cell division; neighboring cells are truly sister cells. Also, during flower development, organs are sometimes initiated independently and then fused together to form the final structure.

The formation of regular and predictable patterns during the development of multicellular plants requires that positional information be available to cells, which then differentiate accordingly. The mechanisms by which these patterns of discrete cell types are formed in response to position is a long-standing problem in developmental biology. A cell that reacts in a special way in consequence of its association with another can only do so if it acquires information from that other cell, information that must be conveyed through chemical or physical signals. The interacting cells then behave in their characteristic ways because they are programmed to transmit and receive their particular signals. The nature of the signals, their origin, and how they are transmitted are unknown. However, the essence of the regulation of differentiation must be the selective expression of one type of cellular specialization while leaving other types unexpressed.

Plant organ fusions usually involve a profound change in the outer, or epidermal, cells of the fusing organs. Although fusions occur in the development of many flowers, the fusion of vegetative plant parts is relatively rare because epidermal cells usually display a resistance to change or redifferentiation. In fact, these cells show no redifferentiation response even after months of intimate contact with either wounded or nonwounded surfaces. Fusing floral organs are naturally occurring exceptions to this rule. Studying how and why the epidermal cells in fusing floral organs redifferentiate may lead to a better understanding of plant development in general.

Plant organ fusions. The fusion of floral organs is an important event in the development of many flowers in several families of higher plants. The actual course of the fusion does not vary much in the different families. The parts are forced together by growth, the epidermal cells interlock, and the enclosed cuticle (if present) disappears. In the fused epidermal layers, cell divisions usually occur parallel to the plane of fusion, disturbing the original cell pattern and obscuring the exact place of fusion. In the region of coalescence, the identity of the originally separate epidermal layers is lost. These fusions involve a change in the developmental fate of the contacting epidermal cells. A well-described example

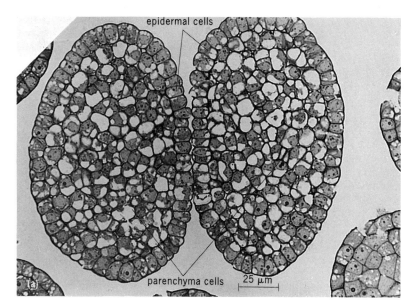

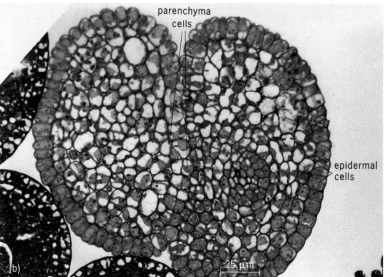

Fig. 1. Light micrographs of transverse sections of (*a*) prefusion carpels and (*b*) postfusion carpels of *Catharanthus roseus.*

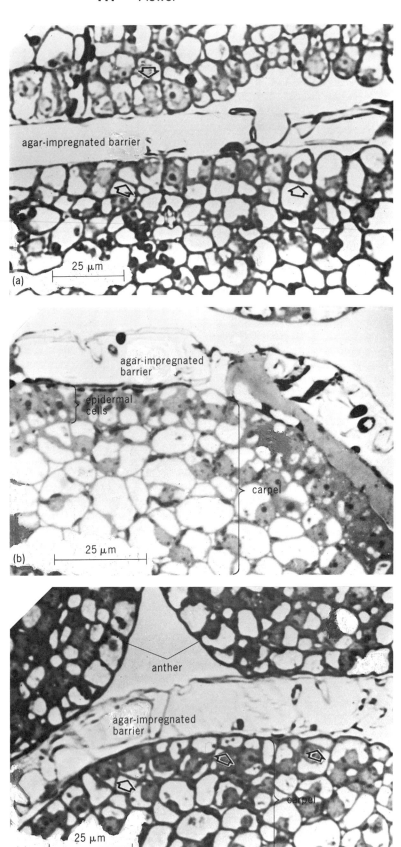

Fig. 2. Light micrographs of transverse sections of *Catharanthus roseus*.
(*a*) Redifferentiating epidermal cells (arrows) in the carpels. (*b*) Normally nonredifferentiating (that is, abaxial) carpel surfaces with agar-impregnated polycarbonate barrier placed against a single carpel, and (*c*) with barrier transferred to this carpel from another one; arrows indicate redifferentiating epidermal cells.

of this type of fusion occurs in the developing gynoecium of *Catharanthus roseus* (Madagascar periwinkle). During normal floral ontogeny in *C. roseus*, the surfaces of the two carpels, which are originally separate, touch. Within 9 h, approximately 400 contacting epidermal cells then redifferentiate into parenchyma cells. The changes in these fusing cells are dramatic (**Fig. 1**). Prefusion epidermal cells have a thin cuticle, are rectangular in section and densely cytoplasmic, and divide strictly in a plane that is perpendicular to the surface of the organ. Following contact, these cells still have the remnants of the cuticle but become almost spherical in section and highly vacuolated; they divide predominantly in a plane that is parallel to the plane of fusion.

Cell-to-cell communication. Cell-to-cell communication has been implicated in the distinctive response seen in the fusing carpels of *C. roseus*. Preventing cell contact either by surgically removing one of the carpels or by inserting an impermeable barrier between the two carpels before contact prevents the change in cellular fate, and the cells remain epidermal. When porous barriers separate the two carpels, epidermal cells redifferentiate regardless of the barrier's composition or pore size, suggesting that water-soluble diffusible agents induce the specific response of epidermal cell redifferentiation. To further test the hypothesis that diffusible factors control these responses, polycarbonate barriers impregnated with agar were placed between two carpels prior to fusion. These agar-impregnated barriers do not block redifferentiation (**Fig. 2***a*); the factors diffuse through the agar from one carpel to the other. Additionally, the factors can be trapped in agar; agar-impregnated barriers are placed between the carpels (so that the factors diffuse into the agar) and then transferred to the outer (abaxial) epidermal surface of one of the carpels. The cells at the abaxial surface normally remain epidermal throughout development, but in these experiments the cells always redifferentiate. Also, epidermal cell redifferentiation factors can be collected at the abaxial surface. If factors are collected in a barrier at the abaxial surface of only one carpel, the cells touching that barrier remain epidermal. However, if that barrier is then transferred to the abaxial surface of the opposing carpel, redifferentiation occurs in 10–18 h (Fig. 2*b* and *c*). This suggests that the carpels are not identical. If this is true, a barrier placed at the abaxial surface of one carpel and transferred at random to an abaxial carpel surface in another bud should cause redifferentiation only half the time. This is precisely the result of that type of experiment. All these experiments lend credence to the hypothesis that the carpels in this system are not identical and that factors from each are essential for redifferentiation.

For background information *SEE* CELL DIFFERENTIATION; FLOWER in the McGraw-Hill Encyclopedia of Science & Technology.

Judith A. Verbeke

Bibliography. R. Moore, Cellular interactions during the formation of approach grafts in *Sedum telephoides* (Crassulaceae), *Can. J. Bot.*, 62:2476–

2484, 1984; B. A. Siegel and J. A. Verbeke, Diffusible factors essential for epidermal cell redifferentiation in *Catharanthus roseus, Science,* 244: 580–582, 1989; J. A. Verbeke and D. B. Walker, Morphogenetic factors controlling differentiation and dedifferentiation of epidermal cells in the gynoecium of *Catharanthus roseus*, II. Diffusible morphogens, *Planta*, 168:43–49, 1986.

Fluid dynamics

The flow of a real fluid where viscosity effects lead to the conversion of kinetic energy of the fluid into heat is termed viscous flow. Hence, viscous flow is dissipative in nature. The equations governing the motion of a viscous fluid are the unsteady Navier-Stokes equations. Because these equations are nonlinear, they are not amenable to closed-form solution for general geometry. Under certain conditions, a flow exhibits the property of self-similarity such that the solution profile at various streamwise locations can be made congruent by the use of appropriate scaling or normalization. Self-similarity leads to the elimination of the streamwise direction from the scaled mathematical problem and, for two-dimensional flows, it leads to the reduction of partial differential equations to ordinary differential equations. The numerical solution of the self-similar problem can be obtained with great accuracy and constitutes an "exact" solution, serving as a benchmark for evaluating the corresponding numerical solution of the partial differential equations. The development of high-speed digital computers made it possible to numerically solve the complete nonlinear partial differential equations governing more general fluid flows efficiently and accurately. Thus a new methodology evolved in the 1960s to complement the earlier classical theoretical and experimental approaches for understanding fluid-flow phenomena. This is the field of computational fluid dynamics (CFD). Just as nonlinear problems in fluid dynamics provided a major impetus to the development of many areas of applied mathematics, the emergence of computational fluid dynamics provided significant impetus to development of the high-speed computers of the 1980s.

Steps of the method. Computational fluid dynamics deals with the study of fluid flows by numerically computing the solution of the nonlinear partial differential equations describing fluid flow. The analysis of a flow problem using computational fluid dynamics involves four basic steps, as follows.

Formulation of suitable mathematical model. This step requires knowledge of the principles of fluid mechanics analysis and leads to a well-posed mathematical problem consisting of the differential equations describing fluid motion, together with appropriate boundary and initial conditions specifying the flow.

Preparation of computational model. This is the key step of grid generation, that is, the process of defining a network of elements and nodes to define the given geometrical configuration. This step invokes the fun-

damentals of geometry, mappings, and transformations, and its significance increases with the geometrical complexity of the application. It leads to a model for a "digital tunnel" analogous to a physical model for a wind tunnel.

Generation of solution. This third step relies significantly on a sound understanding of numerical analysis and numerical methods and on techniques of computer programming and code development. For computing efficiency, programming strategies must reflect the architecture of the computer to be used for performing the computation.

Postprocessing of solution and documentation. Graphical display of the computed solutions, which is the fourth step, constitutes an efficient means for analysis of the results and understanding fluid-flow physics, as well as for detecting errors and discrepancies committed in the analysis or computations. Documentation is crucial to communicating the knowledge gained and enhancing the state of computational fluid dynamics. The inadequacy of the print media for presenting three-dimensional and two-dimensional unsteady flow results has led to the development and use of video-animated representation of these flows.

The computer-resource intensity of a computation depends largely on the numerical method employed. Finite-difference and finite-volume methods of second- and higher-order accuracy are well developed for both low- and high-speed flows, and are used widely for scientific and engineering calculations in computational fluid dynamics. On the other hand, spectral methods with their superalgebraic accuracy are generally used in conjunction with large-eddy simulation and direct numerical simulation techniques for scientific calculations to provide understanding of fundamental phenomena such as transition and turbulence. Random-vortex methods are being developed to track the evolution of vortices; these methods do not require the use of the grid-generation step. Finally, finite element methods have also been developed, especially for low-speed flows.

While a number of major analyses and codes are still in a research mode for computational fluid dy-

Fig. 1. Pressure contours over the Boeing 747–200 calculated by using transonic flow code AIRPLANE. In the original computer-generated image, colors (rather than the shades of gray shown here) clearly distinguish regions having different pressures and velocities. (*From S. Lekoudis, Aerospace '89: Fluid dynamics, Aerosp. Amer., December 1989*)

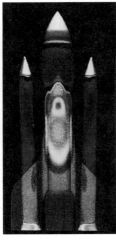

Fig. 2. Pressure coefficient distribution for the shuttle solid-rocket-booster separation sequence. (*From T. Baker and A. Jameson, Computational methods and aerodynamics, CRAY Channels, CRAY Research, Inc., Summer 1986*)

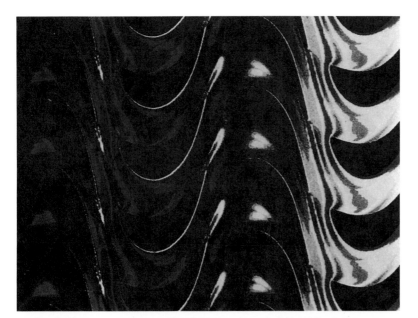

Fig. 3. Results of six-stage computational fluid dynamics calculation of flow in low-pressure oxidizer turbopump of the space shuttle main engine. Entropy contours are shown in fourth and fifth stages.

namics, several analyses and codes are now available with well-documented users' manuals. In general, codes are compute-intensive as well as memory-intensive and were traditionally developed for mainframes and supercomputers. However, with the significant improvements achieved, many of these codes are regularly accommodated on personal computers and workstations.

Viscous-flow modeling on small computers. Viscous flows of practical interest exhibit physical features that span a wide range of spatial and temporal scales. Solutions of engineering quality may be obtained within practical computer resources by either approximate representations of the effects of some of the smaller scales or neglecting them altogether, with the expectation that their influence on the overall flow is minimal. This permits flows to be simulated satisfac-

torily on modern personal computers, workstations, and small computers. The codes are often derived from the corresponding large-computer codes by simply reducing the grid size.

Not only is computational fluid dynamics being used increasingly in engineering design analysis, but software for it is being designed, developed, and marketed. Some of the commonly used software is capable of solving two-dimensional steady incompressible/compressible, laminar and turbulent flows with and without heat and mass transfer by using finite-difference techniques. Some of these also consider unsteady, three-dimensional, multiphase and chemically reacting flows.

Codes based on finite element methods are also available. These too are designed to analyze steady and transient two-dimensional, axisymmetric, and three-dimensional flows; the complex geometries analyzed include carburetors and artificial heart valves.

Modeling of complex three-dimensional flows on supercomputers. The extensive computational resources required for complex viscous-flow analysis make researchers continue to pursue inviscid analyses, especially when interest is in aerodynamic flows at high Reynolds number (Re) for which the viscous effects are often confined within a thin boundary layer near the body surface. The inviscid-flow equations, namely, the Euler equations, coupled with boundary-layer analyses, are computationally less demanding and, hence, widely used as design tools in the aerospace industry. Inviscid transonic flow past a complete Boeing 747–200 aircraft was completed in 1985 (**Fig. 1**). Transonic viscous flow past an F-16A fighter configuration was simulated in 1990. More recently, the flow field for the shuttle solid-rocket booster separation sequence (**Fig. 2**) and the complex flow through the space shuttle main engine (SSME) components (**Fig. 3**) can be successfully computed. The X-30, the National Aerospace Plane (NASP), and its propulsion system being designed to reach Mach number M = 25 is relying heavily on computational fluid dynamics, since wind-tunnel tests cannot currently be performed beyond M = 8. Flow simulation for submarine configurations and entire automobiles is also ongoing. Fundamental mechanisms responsible for observed flow phenomena, such as dynamic stall, are also being investigated by using the method.

Supercomputers have enabled pursuit of large-eddy simulation and direct numerical simulation and, hence, have advanced the understanding of transition and turbulence structure. Direct numerical simulation has been used successfully in simulation of laminar flow breakdown on a flat plate as well as on a swept wing. It is the dominant tool for research on turbulence structure in boundary-layer flows, mixing layers, development of phenomenological turbulence models, compressible flows, noise generation, chemically reacting flows, and theory of dynamical systems and chaos. However, for flows involving complex geometries, today's supercomputers can support direct numerical simulation only for low Reynolds numbers; for complex flows of engineering interest, the method generally used is large-eddy simulation.

Fluid flow   **147**

Prospects. Together with powerful computers, computational fluid dynamics represents a sophisticated tool for analyzing complex fluid-flow problems and important fluid mechanics phenomena. Numerical simulations provide far more detailed flow information than is possibly obtainable from experimental tests. The responsibility of deducing conclusions correctly from this information lies with the researcher. Care must be exercised to ensure that a computed discrete numerical solution represents a valid approximation to the true solution of the original continuous problem. Causes of inaccuracy in numerical solutions include approximations incurred in discretized representation of the problem geometry and the nonlinear partial differential equations (leading to truncation error), round-off error in the machine computation, and, for iterative solution algorithms, the tolerance in solution convergence. Used cautiously, computational fluid dynamics can significantly reduce the cost of experimental testing as well as the time needed to advance from a concept to a product.

The success and further potential of computational fluid dynamics have led to establishment of state and national supercomputer centers; these, in turn, have had an immense impact on the maturation of the technology. The supercomputer centers encourage the training of students and professionals for supercomputer/supergraphics usage.

The maturity achieved by basic computational fluid dynamics during the 1980s has generated enthusiasm for pursuing increasingly challenging simulations of engineering and scientific interest. These include maneuvering-body flows involving fluid-structure and fluid-control interactions, chemically reacting hypersonic vortical flow, transition for more complex flows, and turbulence of reactive flows. In addition, research will continue via multidisciplinary studies of fluid dynamics together with one or more of the following fields: aeroelasticity, controls, electromagnetics, artificial intelligence, theory of nonlinear dynamics and chaos, bioengineering, and manufacturing. These are the promising areas where the method can make a significant impact. It will require continual improvements in numerical algorithms through rigorous analyses, especially for very stiff equations, adaptive grids, convergence acceleration techniques, and full exploitation of the vector-parallel architecture of supercomputers.

For background information SEE FLUID FLOW; FLUID-FLOW PRINCIPLES; NUMERICAL ANALYSIS; SUPERCOMPUTER in the McGraw-Hill Encyclopedia of Science & Technology.

Kirti N. Ghia; Urmila Ghia

Bibliography. D. A. Anderson, J. C. Tannehill, and R. H. Pletcher, *Computational Fluid Mechanics and Heat Transfer*, 1984; T. Baker and A. Jameson, Computational methods and aerodynamics, *CRAY Channels*, CRAY Research, Inc., Summer 1986; C. A. J. Fletcher, *Computational Techniques for Fluid Dynamics*, vols. 1 and 2, 1988; I. P. Jones, S. Simcox, and N. S. Wilkes, Modeling the King's Cross Station fire, *CRAY Channels*, CRAY Research, Inc., Fall 1989; S. Lekoudis, Aerospace '89: Fluid dynamics, *Aerosp. Amer.*, December 1989; J. P. McCarty, J. M. Murphy, and H. McConnaughey, U.S. liquid propulsion technology, *Aerosp. Amer.*, July 1990.

Fluid flow

Information on the physical state of fluid flow can be obtained by using either the flow measurement or flow visualization method. The information obtained by flow measurement is local at the position of a monitoring probe but is quantitative and continuous with time. Flow visualization provides information about the entire flow field under observation, but of a qualitative nature. However, with the development of more powerful digital computers and image-processing techniques, the visual flow pattern can be evaluated and interpreted to generate quantitative information. Three methods of flow visualization will be discussed: surface-flow visualization, tracer methods, and optical techniques.

Surface-flow visualization. The visualization of a flow pattern close to a solid surface is useful in estimating the rates of momentum, heat, and mass transfer, and in understanding the interaction of the flow with the surface. There are three distinct interaction processes (**Table 1**), as follows.

Mechanical interaction process. The surface oil film technique, a standard practice in wind-tunnel experimentation, involves coating a solid surface with a paint consisting of a mixture of an oil and a powdered pigment. The airstream under study carries the oil away, leaving a streaky deposit of the pigment, which provides information on the direction of flow, the location of the transition from laminar to turbulent flow inside the boundary layer, and the positions of flow separation and reattachment (except in higher-pressure flows). Fluorescent pigments illuminated with ultraviolet light provide very high contrast and visibility.

In the oil dot method, oil is administered on a model surface in the form of a small drop along a straight line normal to the main flow direction. A thin oil film, coating the surface, results, and its thickness is monitored by interferometric means. The technique is applied in skin-friction measurements.

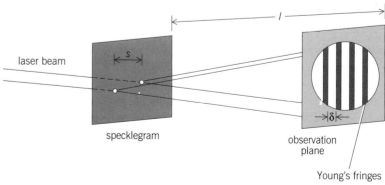

Fig. 1. Specklegram and Young's fringes.

Table 1. Surface-flow visualization methods

Interaction	Method	Phenomenon	Visualization and measurement
Mechanical	Surface oil film	Coating of (oil/pigment) paint → pigment trace	Visualize surface flow pattern, boundary-layer transition and separation
		Coating of oil → variation in oil thickness	Measure skin friction
	Tuft	Flexible string or yarn or flow cone → local flow direction	Visualize surface flow pattern
Thermal	Phase-change coating (single-component paint)	Irreversible color change	Visualize one isotherm
	Temperature-sensitive paint (multi-component paint)	Irreversible color change	Visualize multiple isotherms over a wide range of temperature
	Thermography	Surface radiation	Visualize multiple isotherms over a medium range of temperature
	Liquid crystal	Reversible color change	Visualize multiple isotherms over a narrow range of temperature
Chemical or mass transfer	Naphthalene sublimation	Change of surface structure	Visualize a surface mass transfer rate, measure local transfer coefficient
	Chemical reaction	Color change	Visualize local concentration difference, measure local mass-transfer rate
		pH value change → color change	Visualize boundary-layer separation line, measure local mass-transfer rate

Thermal interaction process. In this process, aerodynamic heating loads are visualized by means of surface coatings whose pattern or color reacts to changes in the surface temperature. Thermal-sensitive paints and liquid crystals have been used in wind-tunnel testing. The former consist of single or multiple components, each of which undergoes a visible change in its phase or internal structure at a specific temperature. Such a color change is an irreversible process. On the other hand, the color change in a liquid-crystal coating with a wall-temperature variation is reversible. The application of liquid crystals is limited to a narrow range of temperature, while thermal-sensitive paints can be used over wide temperature ranges.

Infrared thermography is a noninvasive thermal surface mapping technique that can survey large surface areas. A television-type infrared camera can produce a real-time display of the temperature history of model surfaces in a wind tunnel.

Chemical interaction or mass transfer. Mass transfer of a vapor from a solid surface can be utilized to visualize the boundary-layer flow. The vapor is generated through the sublimation of naphthalene or azobenzene that has been coated on the surface. The flow pattern, either laminar or turbulent, affects the rates of subli-

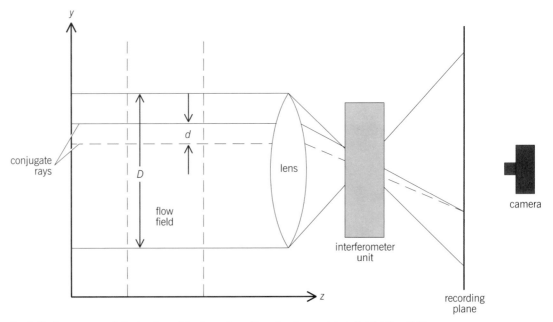

Fig. 2. Basic arrangement of a two-beam interferometer with conjugate rays traversing the flow field.

mation and local mass transfer, as evidenced in the change of surface profile.

Some coated materials may transfer into the flow and induce a chemical reaction, resulting in a color change on the solid surface. Another technique is to coat a layer of a pH indicator on the wall surface. The color of the layer changes, depending on the rate of local mass transfer as it reacts with ammonia vapor in the airstream.

The tuft method is one of the earliest techniques of surface flow visualization. Tufts are point indicators of the local flow direction and are most often used to distinguish between attached and separated flows. Fluorescent minitufts and flow cones have recently been adapted for use.

Tracer methods. The use of a foreign substance, called a tracer, to visualize fluid flow is a traditional method. The tracers must be neutrally buoyant (with the same density as the host fluid), slowly mixing, and highly visible. Tracer methods can be divided into direct-injection and internal-reaction methods (**Table 2**).

Direct-injection methods. In these methods, dyes are commonly employed in water flow (typically in water tunnels and towing tanks), and smokes in airflow (typically in wind tunnels). Smokes include mists, vapors, steam, and aerosols, while dyes comprise food colorings, milk, inks, fluorescent dyes, rhodamine, and potassium permanganate. Tracers are introduced directly into the flow stream from small ejector tubes.

Internal-reaction methods. In these methods, tracers are produced within the fluid by means of electrolytic reactions, photochromatic reactions, or electric ionization processes. A line of tracers, called a time line, is produced by imposing an applied voltage or irradiating laser beam over a short time interval, and it deforms, following the contour of a local velocity profile. By pulsating the applied voltage or the irradiating light beam at a certain frequency, a series of time lines with constant intervals can be generated for visualizing the flow field.

An electrolytic reaction can result in the generation of hydrogen bubbles (hydrogen bubble method); production of a cloud of tellurium (tellurium method); or change of the pH value of a pH indicator, commonly thymol blue, resulting in a color change (thymol blue method). All three electrolytic reaction methods use a similar electrode system, but with different applied voltages. Hydrogen bubbles are generated through electrolysis of water using wire electrodes. The tellurium and hydrogen bubble methods use almost identical apparatus.

The spark-tracer method for mapping high-velocity fields and the smoke-wire technique for low-velocity cases are both in the category of electric control methods. In the spark-tracer method, application of a high voltage between two electrodes results in a series of spark discharges that form a system of time lines. The technique is used in studying boundary-layer flow, especially swirling flow in rotating machinery. The smoke-wire method (a counterpart of the hydro-

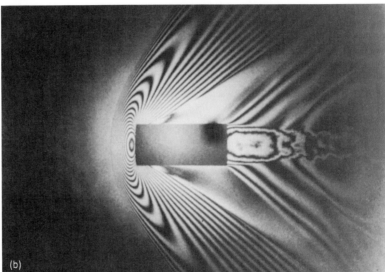

Fig. 3. Flows visualized by interferometry. (*a*) Flow around rotating blades inside the low-pressure stage of a steam turbine, visualized by means of Mach-Zehnder interferometry. (*b*) Flow around a circular cylinder flying at a supersonic speed in air, visualized by means of double-exposure-type, laser holographic interferometry. (*From Flow Visualization Society of Japan, ed., Handbook of Flow Visualization (in Japanese), Asakura Publishing, 1986*)

gen bubble method in airflow) uses a thin stainless-steel wire coated with a thin film of kerosine to generate smoke through electrical heating. This device can introduce smoke into an airstream with greater precision than an ejector tube.

Photochromism and the laser-induced fluorescence method are optical control methods of dye production. In the former, a dye is produced in a photochromatic solution by irradiation of appropriate light. This photochromatic dye, initially transparent, changes to an opaque state under photostimulation. The laser-induced fluorescence method uses fluorescent tracers that emit a characteristic, fluorescent light upon excitation by a light of proper wavelength. The exciting laser beam can be introduced as a light sheet to observe a two-dimensional surface of the flow field.

Optical flow visualization. Optical methods are based on the interaction of a light wave with a fluid

Table 2. Flow visualization tracer methods

Induced by	Means	Fluid	Method	Assisting agent	Tracer	Remarks
Direct injection	Dye	Liquid	Dye	Ejector tube	Dye*	Nearly ideal
	Smoke	Gas	Smoke	Ejector tube	Smoke	Popular
Internal reaction	Chemical reaction	Liquid	Tellurium	1 electrode	Tellurium cloud	Electrolytic reaction, irreversible
		Liquid	Thymol blue	1 electrode	pH indicator*	Electrolytic reaction, reversible
	Electrical control	Liquid	Hydrogen bubble	1 electrode	Hydrogen bubbles	Electrolysis
		Gas	Smoke-wire	1 wire	Smoke	Electrical pulse heating
		Gas	Spark-tracer	2 electrodes	Plasma (ionized) column	Pulses of electric spark discharge
	Photo-chemical reaction	Gas, liquid	Laser-induced fluorescence	Laser beam	Photofluorescent particle	Inelastic scattering, reversible
		Liquid (organic)	Photo-chromism	Ultraviolet laser beam	Photochromatic dye*	Insoluble in water, photochromic reaction, reversible

*Neutrally buoyant.

flow. They are preferred because of their nondisturbing nature, in contrast to the introduction of mechanical probes. Information on the interaction process can be obtained in two different ways. One is to record the light transmitted through the fluid, and the other is to record the light scattered from a certain location in the fluid in a specified direction. In the former case, the information is integrated along the entire path of the light in the fluid, while the recorded information in the scattered case is local.

Light-sheet method. The light sheet is an illumination system in the form of a planar, thin sheet for visualizing tracer particles within the slice. It can be produced by either a conventional lamp or a laser. A light beam is expanded by means of a cylindrical lens onto a plane sheet, which is applied normally to the main flow direction. The light scattered from tracer particles in the illuminated plane exhibits the flow pattern. The method is applied in conjunction with direct-injection tracer methods, speckle velocimetry, and laser-induced fluorescence.

Speckle photography. This method was originally developed to measure surface deformations in solid mechanics. Later, its use was extended to measure flow velocities in fluids seeded with tracer particles. In most cases, a light sheet is used to illuminate the flow field. The velocity components of particles in the sheet can be measured and a pattern of velocity contours produced for viewing.

The technique takes a double exposure of the flow field. The resulting photographic record is called a specklegram. By measuring the displacement of a tracer particle s during the time interval of the double exposures t on the specklegram, the particle velocity can be determined as s/t.

When the specklegram is illuminated by a thin laser beam, the two images of a tracer particle, separated by the distance s, form Young's fringes in an observation plane (**Fig. 1**). The fringes are perpendicular to the direction of particle displacement, and the fringe spacing δ is related to s by the equation below, where

$$\delta = \frac{\lambda l}{s}$$

λ is the laser wavelength and l is the distance between the specklegram and the observation plane. Hence, the particle velocity and flow direction can be determined. This method of velocity measurements is called speckle velocimetry. Recently, a system was developed for a fully automated evaluation of a specklegram.

Methods utilizing refractive index changes. There are three optical methods of flow visualization that are based on changes in the fluid's refractive index: shadowgraphy, schlieren photography, and interferometry (**Table 3**). Both shadowgraphy and schlieren photography are based on light deflection in the flow field, while interferometry is dependent upon the change of optical phase in the fluid.

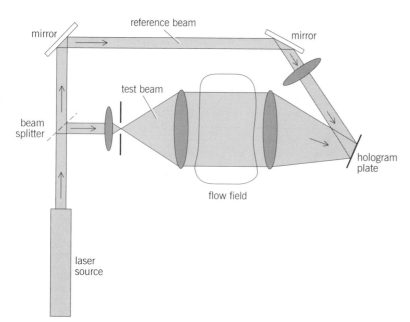

Fig. 4. Basic arrangement of a holographic interferometer.

Table 3. Three principal optical flow visualization methods

Method	Quantity measured	Sensitive to change*	Recorded feature
Shadowgraphy	Deflection length	$\dfrac{\partial^2 n}{\partial y^2}$	Shades or grayness in flow pictures
Schlieren system	Deflection angle	$\dfrac{\partial n}{\partial y}$	Shades or grayness in flow pictures
Interferometry Shearing type	Change of optical phase	$\dfrac{\partial n}{\partial y}$	Fringe pattern in interferograms
Reference-beam type	Change of optical phase	n	Fringe pattern in interferograms

*n is the refractive index of the fluid; y is the coordinate perpendicular to the direction of light propagation.

In a two-beam interferometer (**Fig. 2**), two parallel conjugate light rays separated by distance d traverse the flow field of diameter D. They will coincide after passing through an interferometer unit. If the condition for optical coherence is met, the conjugate pair will interfere and produce an interference pattern in the recording plane. Bright interference fringes appear at the locations where the difference in optical path lengths is an integral multiple of the wavelength. Two extreme cases correspond to two kinds of interferometers: the reference-beam interferometer, in which d is equal to or greater than D; and the shearing interferometer, in which d is much less than D.

In a reference-beam interferometer, a light is divided into two beams by a light splitter. One beam (called the test beam) travels through the flow field with refractive index n, while the other beam (the reference beam) travels outside the test field and remains undisturbed, with a constant refractive index n_0. The interference fringes are lines of constant n or density. When a mirror or beam splitter is tilted or when a transparent plane wedge is inserted into the path of one light beam, parallel, equidistant fringes appear on the recording plane. The best-known reference-beam interferometer is the Mach-Zehnder interferometer (**Fig. 3a**).

In a shearing interferometer, both the test beam and the reference beam pass through the flow field, where they are sheared or separated by the small distance d. Interference fringes are sensitive to the refractive-index (or density) gradient, as in the schlieren system; therefore, shearing interferometers are also referred to as schlieren interferometers. In the most common arrangement, the interferometer unit consists of a Wollaston prism (which functions as the shearing element) and two crossed polarizers. This method has been used widely in visualizing fluid flows and convective heat transfer.

Conventional photography records information on only the amplitude of a light wave. It provides the recording of a two-dimensional flow field. In contrast, holography is a technique to freeze all the information contained in a light wave, including both the amplitude and the phase, and to later reconstruct the wave. It provides the recording of a three-dimensional flow field. The principle can be employed in the interference of two (or more) light waves, which is called holographic interferometry.

In a holographic interferometer (**Fig. 4**), laser light is divided into a test beam and a reference beam by a beam splitter. After traversing the flow field, the test beam coincides with the reference beam at the plane of the holographic plate. The holographic interferometer is another kind of reference-beam interferometer, with the spatial separation d between the test and reference beams replaced by a separation time. The replacement of spatial separation by temporal separation is achieved by means of a double exposure of a holographic plate (called a holographic interferogram; Fig. 3b). The two exposures are taken on the flow field without flow and with flow, respectively, equivalent to the reference beam and the test beam in a conventional beam interferometer. When a holographic interferogram is illuminated by the reconstruction light of the holographic system, a superposition of the two recording light waves appears.

For background information SEE ACID-BASE INDICATOR; BOUNDARY-LAYER FLOW; DYE; ELECTRIC SPARK; FLOW MEASUREMENT; FLUID FLOW; FLUORESCENCE; HOLOGRAPHY; INTERFEROMETRY; LIQUID CRYSTALS; SCHLIEREN PHOTOGRAPHY; SHADOWGRAPH OF FLUID FLOW; SPECKLE; THERMOGRAPHY in the McGraw-Hill Encyclopedia of Science & Technology.

Wen-Jei Yang

Bibliography. W. Merzkirch, *Flow Visualization*, 2d ed., 1987; W. Merzkirch, *Techniques of Flow Visualization*, AGARDograph 302, NATO, 1987; J. D. Trolinger, *Laser Applications in Flow Diagnosis*, AGARDograph 296, NATO, 1988; W. J. Yang (ed.), *Handbook of Flow Visualization*, 1989.

Fog

Fog, cloud, mist, and steam consist of water droplets a few micrometers in diameter. Fog is cloud in contact with the ground or sea and limits visibility to less than 1 km (0.6 mi). Mist is thin fog in which visibility ranges 1–4 km (0.6–2.4 mi). A rough relationship between visibility V, droplet radius r, and liquid water content w is given by the equation below, where c is

$$V = \frac{cr}{w}$$

a coefficient depending to some extent on background lighting. This equation shows that the lowest visibilities occur when a given liquid water content is composed of very small droplets.

Basic physics. Fog droplets form on dust particles (condensation nuclei) in air that is supersaturated. Supersaturation is induced in the atmosphere directly by cooling (due to adiabatic ascent of air or to radiative loss of heat) or indirectly by the mixing of two nearly saturated air masses initially at different temperatures (**Fig. 1**). Fog droplets may be evaporated by warming (due to adiabatic descent or radiation) or by mixing with drier air; or they may be removed by gravitational settling to the ground, or by scavenging on impact with trees, buildings, and other objects.

There is little dispute about the basic mechanisms of fog formation and dispersal. However, understanding of the interaction of these mechanisms in various situations is less sure, and recent research has been devoted to understanding these interactions.

Research studies. Field research studies of fog are relatively few, although some have been made over land and sea in the United States and the United Kingdom since 1970. The small-scale variability of land fog in time and space makes the interpretation of observations at a given site very difficult, even when assisted by numerical models of fog, which usually assume the fog structure to be horizontally stratified. Over the sea, investigations have to be made by research aircraft, whose flights are expensive and difficult to mount logistically. Recent work on radiation fog and sea fog are summarized in this article; the conclusions stated are provisional.

Radiation fog. This is the type of fog most commonly experienced. Radiation fog forms on calm, cloudless nights over land, and it is the type mainly responsible for disruption of land and air transport. The formation of radiation fog is induced by direct radiative cooling and inhibited by turbulence. A specific sequence of events is thought to be characteristic of radiation fog formation.

In the absence of solar radiation, the emission of infrared radiation from the ground to space cools the ground at a rate dependent upon its thermal conductivity. Ground cooling induces cooling of the lowest layers of air mainly by radiative transfer in the strongly absorbing bands of water vapor and carbon dioxide, and it generates a temperature inversion in the lowest few meters of atmosphere. This cooling of the air is offset to some extent by weak turbulent diffusion of heat from warmer air above, but the increasing thermal stability of the air near the ground may eventually quench turbulence.

When the air temperature near the ground falls to dew point, air in contact with the ground begins to deposit its excess water vapor as dew. This begins to dry out the lower layers of air and to create a water-vapor gradient down which weak turbulent diffusion continues to carry water vapor to be deposited as dew. Fog does not form at this stage.

Field research has determined that when the wind speed at 2 m (6.6 ft) above the ground falls below about 0.5–1 m/s (1.1–2.2 mi/h), turbulence ceases. This removes the mechanism for further deposition of dew, but as radiative cooling continues, excess water vapor continues to condense in the air onto available condensation nuclei as fog droplets. Local fog development was found to proceed most rapidly when the wind speed dropped below about 0.5 m/s (1.1 mi/h).

Fog formation usually begins near the ground and spreads upward. As soon as fog forms, the fog droplets begin to cool radiatively themselves and to shield the ground from radiative loss. The ground temperature begins to rise, thereby eroding the surface radiation inversion as the fog deepens. Eventually, the radiation inversion detaches from the ground and migrates upward with the fog top. The upward heat flux from the ground into the atmosphere is now removed by weak convection as ground temperature rises slightly above air temperature. The heat introduced into the lower layers of fog by this convection may lead to some thinning and even lifting of the fog. The liquid water content of the fog depends upon a balance among radiative cooling, gravitational settling, and turbulence. Fog droplets settle out at about 1 cm (0.4 in.), so that a fog 50 m (165 ft) deep replaces itself in about 1 h.

Further upward development of the fog may be limited by the wind field above the fog. Nocturnal jets of 5–10 m/s (11–22 mi/h) at 50–100 m (165–330 ft) above the fog top were often found in some field studies in the United Kingdom. These jets fluctuate and cause fluctuations in the height of the fog top. The weak turbulent mixing within the developing radiation inversion normally associated with radiation fog inhibits fog formation. However, the breakdown of the whole radiation inversion by rising wind—which may occur after a clear radiation night with no fog formation—may lead to the sudden formation of thick fog in the early morning caused by the mixing mechanism in air initially cooled by radiation.

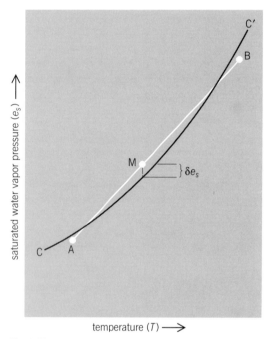

Fig. 1. Diagram showing induction of supersaturation. The (schematic) curve CC′ of saturated water vapor pressure e_s is a function of temperature. If two parcels of nearly saturated air initially at the vapor pressures and temperatures indicated by points A and B are mixed, the vapor pressure and temperature of the resultant mixture will lie somewhere along the line AB. If this point M lies above CC′, the mixture will be saturated and fog will form. However, the initial temperatures must be separated by several degrees Celsius for the initial supersaturation δe_s to be sufficient to produce a dense fog.

The local variability of terrain structure and type ensure (as is commonly observed) local variability in the balance of radiation and turbulence, and therefore in the time of fog formation over even a small area. This results in patchy fog, which drifts around and is particularly dangerous to transport. The local aviation forecaster may be able to forecast the general formation of fog in a specific area to within 1–2 h, but has to rely on real-time observations (for example, visibility meters on an airfield) to provide short-term warning.

Sea fog. This type of fog may form on a local scale (~10 km or 6 mi) or a large scale (100–300 km or 60–180 mi). Local sea fog may form in air moving over a cold patch of sea (such as occurs typically around coasts due to local upwelling) if the sea surface temperature drops below the dew point of the air. The cooling of the air takes place radiatively in part, but probably the main mechanism is turbulent mixing of the cooled air near the surface with air at progressively higher levels. Wind speeds are typically much stronger than in the radiation fog situation on land.

The mixing mechanism is also predominant in the formation of so-called sea smoke, where very cold air is carried over warm sea. The difference here is that most of the vapor condensing to form sea smoke has been evaporated from the sea surface in the immediate vicinity, whereas sea fog is condensed mainly from vapor already contained by the air for some time.

On the large (synoptic) scale, areas of order 10^5 km^2 (10^4 mi^2) of sea fog (**Fig. 2**) often form in poleward flows of warm, moist air, and they may migrate 1000–2000 km (600–1200 mi), affecting temperate latitude coasts for several days. It is not known whether these fogs develop from the surface upward or result from the gradual lowering of the base of existing stratus or stratocumulus cloud to the sea surface, but a recent field study of an established area of sea fog that was interpreted by using a numerical model of marine stratocumulus suggested the following sequence of events.

Air moving poleward is initially cooled by turbulent transfer of its heat to the relatively cold sea surface. When fog forms, radiative cooling of the fog soon becomes significant, although it is partly offset by solar heating during the day. Eventually, radiative cooling becomes predominant, and the fog temperature eventually drops 1–2°C (1.8–3.6°F) below the sea temperature. The initially "warm" advection fog becomes a diurnally modulated, self-maintaining "cold" radiation fog, held at a temperature 1–2°C (1.8–3.6°F) below sea temperature. Radiative cooling is balanced by convection of sensible heat from the sea surface, entrainment of warm air into fog top, and solar heating.

The loss of liquid water from the fog by drizzle is balanced by condensation of water vapor evaporated into the fog from the relatively warm sea surface. Air entrained into fog top is sometimes drier (lower vapor mixing ratio) than air in the fog; it may or may not evaporate some fog. An established sea fog could persist almost indefinitely in the absence of major weather disturbances. Persistent fog and stratus clouds off the Californian coast in spring and summer may

Fig. 2. Distribution of sea fog at 1448 GMT on April 27, 1984, off the Scottish coast as shown by *NOAA* 7 satellite. The circulation set up by a heat low forming over the Scottish mainland has cleared the fog from a patch of sea in the Moray Firth. (*University of Dundee*)

demonstrate this persistence in a climate less subject to weather disturbance than in temperate latitudes.

The movement of areas of established sea fog may be strongly influenced by local circulations induced by contrasts in land–sea temperatures. The forecaster at coastal stations must therefore have satellite imagery and aircraft reports in order to monitor the movement and development of sea fog, but such information is usually sporadic (depending on times of local satellite passes) or even nonexistent. Maps of sea-temperature distribution are now available from satellite data, and forecasts of local wind fields output by fine-scale numerical models are becoming available in the United Kingdom and show promise.

For background information SEE ATMOSPHERE; DEW; DEW POINT; DIFFUSION IN GASES AND LIQUIDS; SATELLITE METEOROLOGY; TEMPERATURE INVERSION; WEATHER FORECASTING AND PREDICTION in the McGraw-Hill Encyclopedia of Science & Technology.

W. T. Roach

Bibliography. J. Findlater et al., The haar of northeast Scotland, *QJRMS*, 115:581–608, 1989; S. Nicholls, The dynamics of stratocumulus: Aircraft observations and comparisons with a mixed layer model, *QJRMS*, 110:783–820, 1984; R. J. Pilie et al., The formation of marine fog and the development of fog-stratus systems along the Californian coast, *J. Appl. Meteorol.*, 18:1275–1286, 1979; W. T. Roach et al., The physics of radiation fog, *QJRMS*, 102:313–354, 1976.

Food engineering

Recent advances in food engineering involve microwave heating, freeze concentration, solid–liquid extraction, spray drying, fouling and cleaning, and membrane technology.

Microwave heating. The use of domestic microwave ovens has demonstrated the advantages in speed and convenience of microwave heating. However, application of microwave heating in the food industry is limited, even though it can provide high power utilization and preferential heating of water in foods. The limited use has been attributed to high investment costs in many applications, and problems of controlling heating uniformity, which also occur in microwave heating in domestic ovens.

In some recent industrial installations for microwave pasteurization and sterilization, process economics have been improved by combining microwave heating with conventional heating methods, for example, by preheating to 122°F (50°C) and using the microwave units only for temperature increases to pasteurization temperatures of 167°F (75°C).

Problems of nonuniform heating have been well documented over the years; and interactions between food and its placement in a microwave field, and between food and package geometry and composition have been demonstrated both in experimental studies and theoretical analysis.

Understanding of these problems has been helped by the development of improved mathematical modeling methods. These models use solutions of Maxwell's equation for electromagnetic field distribution inside microwave ovens containing food loads. Analytical solutions in one dimension have been used for some time to calculate energy distributions in loads with simple geometries such as slabs, cylinders, and spheres. Some of these calculations show that the interface between the food and the microwave oven field affects the distribution of microwaves inside the food. Center heating phenomena in spherical and cylindrical samples have been analyzed by such models.

For more complicated situations involving two- and three-dimensional fields, including both foods and containers of various shapes, numerical mathematical models must be used. Among numerical methods, the transmission-line matrix method and the method of moments require large computer memory capacity and use of mainframe computers. Finite element methods (FEM), available as commercial software packages, require the construction of a network of small elements, representing the oven and food. This can require a great deal of effort. However, easy combination of calculations of electromagnetic field distribution and heat transfer is a major advantage of the finite element method. The finite element method, as well as the finite difference time-domain method, can be implemented on powerful workstations. Finite difference models are easy to construct. **Figure 1** shows an example of the results of such calculations; it illustrates the electromagnetic field distribution at a sharp corner of a rectangular food inside a simplified oven cavity.

These improved numerical methods will improve understanding of the interactions between microwave oven fields and foods. This will help not only in analyzing problems of heating uniformity but also in synthesizing microwave oven fields, and food geometries and compositions, thus leading to improvement in the usefulness of microwave heating in food engineering applications.

Thomas Ohlsson

Freeze concentration. In food freeze concentration systems, ice production and removal is used to concentrate fruit juices, beverage extracts, vinegar, milk, wine, and beer. Multistage countercurrent systems for freeze concentration (**Fig. 2**) have largely replaced single-stage systems. Their use has substantially reduced energy expenditure and capital costs per unit of concentrate produced.

In the multistage countercurrent systems, ice crystals with diameters of 5–10 micrometers are produced in scraped freezers, a type of freezer in which a flowing solution or slush comes in contact with a chilled surface that is periodically scraped. These crystals are too small and too branched in shape to separate cleanly from concentrate. They are transferred to ripening tanks and mixed with concentrate and larger crystals. Most of the small crystals melt, removing heat, which causes large crystals and surviving small crystals to round off and grow until they ultimately become large enough to separate cleanly. This combination of melting and growth is known as ripening.

In countercurrent freeze concentration systems, crystals are initially produced in the stage containing the most concentrated liquid. Concentrate is with-

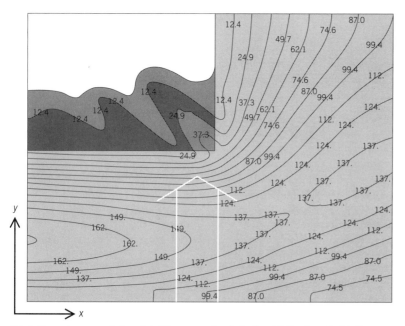

Fig. 1. Computer-generated model of the power density in the vicinity of and inside the rectangular corner of a food being heated in a microwave oven. White arrow indicates direction traveled by the microwaves. Numbers represent electrical field density.

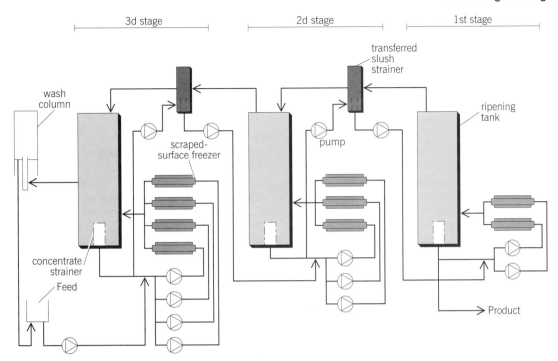

Fig. 2. Schematic diagram of three-stage freeze concentration system.

drawn from that stage as product, and the surviving crystals of intermediate size are progressively transferred through a sequence of ripening tanks containing less and less concentrated liquid. Liquid flows in the opposite direction, increasing in concentration as it does. The transferred crystals grow because of ripening caused by continual introduction of freshly produced small ice crystals. Crystals with mean effective diameters of 230–260 μm are ultimately produced. **Figure 3**, a flow diagram for countercurrent freeze

concentration, shows that most growth occurs at low concentrations where temperatures are less depressed, viscosities are lower, and ripening is rapid.

Ice crystals are occasionally separated from concentrate in screening centrifuges. More frequently, wash columns are used. There, uniformly deposited beds of crystals, transferred from the most dilute ripening tank, are propelled upward through a stationary layer of melt water. If propulsion is slow enough, the water cleanly displaces concentrate surrounding the crystals,

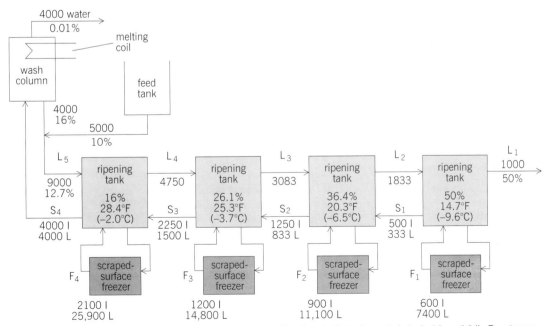

Fig. 3. Flow diagram for four-stage freeze concentration system. Numbers indicate flow rate in kg/h. 1 kg = 2.2 lb. F = streams of slush discharged from scraped-surface freezer. S = slush. I = ice. L = liquid. Weight percents of solute concentrations in the unfrozen portions of the streams or mixtures are indicated.

and solute losses are less than 0.01%. If propulsion is too rapid, large amounts of concentrate are discharged with the ice. Production of large crystals minimizes solute losses and needed wash column area, because allowable propulsion rates are proportional to the square of the diameters of the crystals.

Large-bore freezers scraped by rotating blades are used to produce ice in food freeze concentration systems. These freezers are expensive. The scraping produces frictional heat, which has to be removed by refrigeration, and this significantly reduces efficiency. Multitube scraped freezers in which shuttling balls remove ice from tube walls, and the use of plastic coatings to reduce ice adhesion and friction and facilitate flow- or gravity-induced ice removal are being investigated in attempts to reduce freezer cost and frictional energy expenditure.

Modular components, for example, ripening tanks that are equal in volume and freezers that are equal in

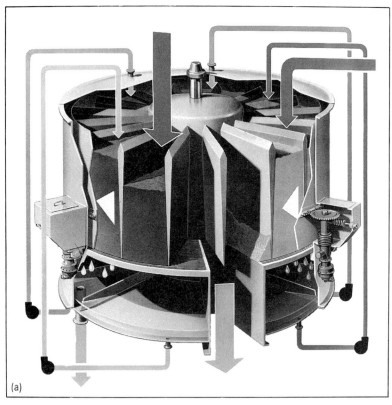

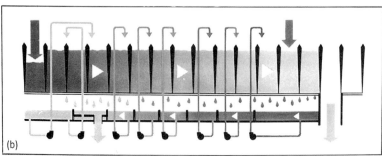

Fig. 4. Solid–liquid extraction in the carrousel extractor. (a) Configuration. (b) Diagram showing material flowing through the unit. (*Extraktionstechnik Gesellschaft für Anlagenbau M.B.H.*)

size, are usually used in freeze concentration systems. Three or four scraped-surface freezers are usually used in each ripening tank, and four or five parallel wash columns of equal size are used to process discharged ice.

Ice crystal nucleation, growth, and ripening slow down markedly as the concentration of the solute increases. In some cases, eutectics form at high solute concentrations, preventing useful separation. Therefore, maximum concentrations achieved by freeze concentration are usually in the 40–55% range.

Henry G. Schwartzberg

Solid–liquid extraction. Solid–liquid extraction, or leaching, is a process by which solutes are removed from solids by a liquid solvent. Typical food applications are extraction of vegetable oils from flaked oilseeds, sugar from sugarbeets, coffee from roasted ground coffee beans, and caffeine from green coffee beans. Extracting solvents include water, aqueous solutions, organic liquids, and supercritical fluids.

Commercial extraction equipment ranges from batch single-stage tanks to continuous multistage countercurrent extractors. Countercurrent extraction is generally the most economical alternative, since it requires less solvent per unit of solid. In the carrousel extractor (**Fig. 4**), solids are contacted with percolating solvent as they are conveyed over a stationary slitted plate; fresh solvent is fed near the solids discharge point for countercurrent operation.

Recent advances in solid–liquid extraction include development of improved mathematical solutions for most types of commercial extractors, and evaluation of technological alternatives aimed at increasing yields and rates of extraction.

Mathematical solutions. Reliable methods for the design, scale-up, and optimization of solvent extraction equipment require the simultaneous use of relationships involving appropriate equilibrium, mass balance, and mass transfer. If the solute is held freely in open pores, internal solute transport can be described as diffusion in porous media; thus the mean solute concentration in the solid, X, can be determined from the equation below, where X_0 and X_∞ are the initial

$$\frac{X - X_\infty}{X_0 - X_\infty} = \sum_{n=1}^{\infty} C_n \exp\left(-\frac{q_n^2 D_s t}{a^2}\right)$$

and equilibrium values of X, D_s is the apparent diffusivity of solute in the solid, a is the particle half-thickness (radius for sphere and cylinder), and t is the extraction time. The constants q_n and C_n depend on particle geometry, solid–liquid ratio, and Biot number. This equation is directly applicable to single-stage batch and semibatch extractions, as well as to continuous cocurrent extraction.

Solutions for multistage extractors are obtained from this equation by using superposition (Duhamel's theorem) to account for the nonuniform concentration in the solids entering the second and subsequent stages. For systems that do not obey this equation, for example, solvent extraction of oilseeds, superposition-based solutions based on empirical, single-stage extraction behavior can sometimes be developed. Analytical solutions have also been developed for continuous countercurrent extractors, fixed-bed extractors, and extraction accompanied by a first-order solubilization reaction.

Improving extraction performance. In most commercial extractions, the rate-limiting step is mass transfer within the solid. Cellular solids sometimes exhibit increased mass-transfer resistance due to unbroken cell walls, which may have to be disrupted to achieve economic yields. According to the equation above, extraction rate can be increased by reducing particle thickness or increasing apparent diffusivity. This is normally accomplished by grinding, flaking, and by other solid-conditioning methods which include heating, swelling, puffing, and chemical or enzymatic treatments.

Innovative extraction techniques tested at bench scale include the use of pulsating flows, ultrasonic energy, and electric fields. These may enhance rates of mass transfer by inducing microconvective currents within the solid.

Roberto R. Segado

Spray drying. Developments and advances in established technologies such as spray drying normally evolve gradually; in many instances they are conceived long before they become operational. Advances in the application of spray drying to the food industry have followed this trend. Recent developments have been related to hygiene, design and control, and energy utilization.

Hygiene. Stringent food-product requirements of a physical, chemical, and bacteriological nature necessitate strict cleanliness of the drying medium. Mere filtration of air prior to either indirect heating or burning of a fossil fuel for use as the drying medium may no longer be adequate. Newly developed, high-efficiency, heat-resistant filters allow filtration of the hot drying medium just before it enters the drying chamber, and reduce contamination levels to a minimum.

The common use of permanently applied insulation and cladding to drying chambers inhibits discovery of cracks that have developed in the chamber wall. Therefore adjacent insulation can become wet, which promotes bacteriological growth and produces cold spots that facilitate adhesion and buildup of the product. Readily removable insulation panels that are separated from the chamber wall by a controlled air gap have been developed, permitting easy inspection of the walls and preventing insulation wetting, local product buildup, and bacterial growth.

Dry-powder deposition on hot, dry surfaces can be a major source of product contamination; it is minimized by elimination of all protuberances, such as duct intakes, and instrument sensors, inside the drying chamber. To reduce further the risk of such powder deposition, retractable nozzles for clean-in-place (CIP) installations have been developed. Previously surface-mounted, these devices have been designed to withdraw to form a seal flush with the inner surface of the dryer when not in use.

Design and control. In food processing, multistage drying is most commonly used, with a spray dryer as the

first stage and fluidized-bed, rotary, moving-belt dryers as the subsequent stages. This arrangement usually requires that the product be reasonably dry as it leaves the spray dryer. In some instances the height of the spray dryer has been reduced considerably by locating a short, usually axially diverging spray-drying zone directly over a moving-belt dryer in an integrated manner with the only partially spray-dried product falling directly on the belt. A relatively new development is the replacement of the moving-belt dryer with a fluidized-bed dryer (**Fig. 5**). Successful application of this integrated two-stage dryer, which requires the short spray-drying zone to be axially convergent to accommodate the smaller diameter of the fluidization zone, has been made possible by proper reintroduction of recycled fines separated from the spent drying medium and use of vigorous fluidization. This dual-dryer combination gives considerable flexibility to the process, while the operation of the nonintegrated spray dryer is generally quite inflexible.

New improvements in the design of drying-medium dispersers include adjustable rather than stationary vanes. This allows on-site variation of gas-flow patterns to improve mixing of the drying medium and the spray, to minimize secondary flow zones in the drying chamber, and to reduce wall deposition.

Previous control strategies for dryer operation were based exclusively on temperature sensing. More realistic control strategies based on both humidity and temperature sensing have been introduced recently. This has been facilitated by development of robust and rapidly responding humidity sensors. Consequently, control of operating conditions in response to rather rapid changes in ambient humidity as well as temperature is now possible.

Energy utilization. Considerable progress has been made during the past few years in the improvement of overall dryer efficiency by recovery of thermal energy from the dust-containing exhaust gases. Heat-exchanger systems have been developed that can

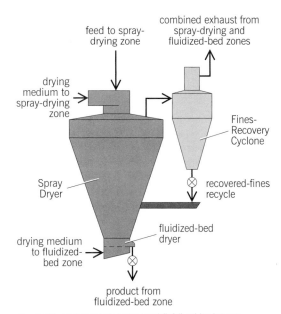

Fig. 5. Spray dryer with integrated fluidized-bed dryer.

handle large quantities of gases; they are relatively compact and can be readily freed of dust deposits and cleaned in place. Systems that exchange energy with incoming air, with a dehumidified drying medium, or with a heat-transfer medium are used. Efforts also have been made to increase drying-chamber efficiency by raising the inlet temperature and lowering the outlet temperature of the drying medium; however, the degree of improvement is limited by the type of feedstock being dried. Further progress in the reduction of energy consumption has been made indirectly by recent improvements in the preconcentration of feedstocks.

E. Johansen Crosby

Fouling and cleaning. One of the most important unit operations in food engineering is cleaning. Cleaning is the removal from surfaces of materials that decrease heat-transfer efficiency, increase hydraulic load, or contribute to microbiological contamination of products. Cleaning is applied to both batch and continuous process systems; it may involve use of chemicals as well as mechanical and thermal energy. Although there has been a great deal of research on cleaning processes, there has been much less emphasis on the cause of the problem: the unwanted deposition of material on surfaces, or fouling. Fluids of biological origin are particularly aggressive fouling agents, because many have constituents that reduce interfacial tension. Thus, these materials migrate to interfaces, where they may be subjected to oxidative and thermally driven reactions that result in chemical reactions and polymerization, with subsequent deposition at the surface.

For heat exchangers, fouling causes progressive declines in efficiency and performance. Unwanted deposits can lead to increased capital costs, energy loss, maintenance costs, and loss of production. In addition, when the fouled material is removed, a polluting stream is generated by the cleaning solution. Eventually material deposited on the surfaces interferes with transfer of heat or momentum. When there is no effect of the deposit on practical heat- and momentum-transfer operations, the surface is said to be thermally and hydraulically cleaned even though it

may not be microbiologically or chemically clean.

Most of the work on fouling and cleaning of food equipment has focused on milk. Milk has several fouling systems, including undenatured whey proteins, fats and lipids, carmelization reaction products, casein particulate deposition, and calcium phosphate deposition. Most of the recent research has been directed at understanding interaction forces, velocities, concentrations, and temperatures in boundary layers and how they affect deposition mass flux. In addition, there is a removal mass flux due to erosion, dissolution, and spalling, which occurs under the hydrodynamic situation at the solid–liquid interface.

Factors influencing fouling in both heat exchange and isothermal operations include concentration of chemical species, velocity, temperature of the surface, temperature of the bulk solution, pretreatment of the fluid, and pretreatment of the surface. It has been shown that fouling rate is insensitive to surface finish and surface free energy as measured by contact angle. In addition, studies have been completed on the use of chemical treatments to minimize fouling of food fluids. These treatments include pH control, agents for blocking sulfhydryl groups, and special fatty acids. However, use of chemical treatments may not be feasible, since they can alter the flavor of the fluid foods. The most promising opportunity for minimizing fouling due to milk is in a design of the heat exchanger that improves mechanical erosion.

Cleaning is the mirror image of fouling, in the sense that chemicals, temperature, and mechanical energy are used to remove material from a fouled surface to render it hydraulically, thermally, and chemically clean. Factors influencing the rate of cleaning are the same as those that influence the rate of fouling. Fouling and cleaning cycles can be visualized as shown in **Fig. 6**, in which a clean surface is fouled and a fouled surface is cleaned.

An important consequence of fouling is the deposition of materials on sensors that are used to control processes. This deposition of unwanted materials on sensors can interrupt and distort the entire process. When severe fouling occurs, stand-by sensors can be put into the system circuit alternately to ensure presence of relatively clean sensors during the entire process.

Daryl B. Lund

Membrane technology. Membrane separation processes are based on the ability of semipermeable membranes with the appropriate physical and chemical properties to discriminate between molecules primarily on the basis of size and, to a lesser extent, on shape and chemical composition. The chemical and physical characteristics of the membrane control which of the components are retained and which of them permeate through the membrane. Depending on the properties of the membrane (primarily pore-size distribution), separation can be achieved by concentration or dewatering by reverse osmosis, fractionation of components in solution by nanofiltration or ultrafiltration, and clarification of slurries or removal of suspended matter by microfiltration, as shown in **Fig. 7.**

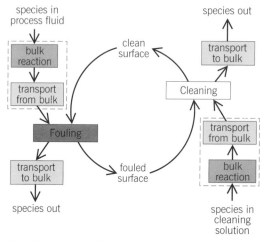

Fig. 6. Flow diagram of the fouling–cleaning cycle.

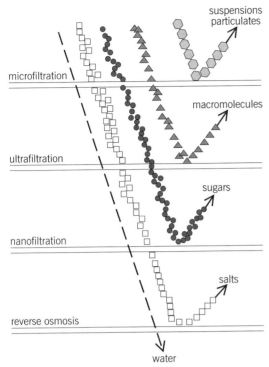

Fig. 7. Diagram showing types of membrane separations and the components separated.

In all three processes, hydraulic pressure (through a pump) is used to provide the driving force for permeation. In the case of reverse osmosis, pressure is used to overcome the difference in chemical potential between the concentrate and the permeate, expressed in terms of the osmotic pressure. Pressures in reverse osmosis are frequently of the order of 2–5 megapascals (300–750 lb/in.^{2}). Since ultrafiltration is designed to retain macromolecules that exert little osmotic pressure, the pressure required is much less (1–7 MPa or 15–100 lb/in.^{2}), primarily to overcome hydraulic resistance of the polarized macromolecular layer on the membrane surface, a phenomenon known as concentration polarization. The hydraulic pressure applied in microfiltration (about 1–5 MPa or 15–75 lb/in.^{2}) is also used primarily to overcome resistance of the filter cake. If the polarization phenomenon is serious enough, mass transfer can also be limiting, in which case high cross-flow velocities will also be required, especially for ultrafiltration and microfiltration processes. Another membrane process, electrodialysis, uses voltage as the driving force to separate ionic solutes.

Advantages. Reverse osmosis and ultrafiltration are perhaps the first continuous molecular separation processes that do not involve a phase change or interphase mass transfer. The removal of solvent or water is accomplished without a change in its phase from liquid to vapor (as in evaporation) or liquid to solid (as in freeze concentration).

Thus, membrane processes can also be operated at ambient temperatures if necessary. This avoids the problems of product degradation associated with thermal processes, resulting in products with better func-

tional and nutritional properties. Higher temperatures are frequently used, however, to lower viscosity, reduce pumping costs, and minimize microbial growth during processing.

For these reasons, energy requirements are low compared to other dewatering processes. Typically, while open-pan evaporation may need over 600 kilowatt-hours per 1000 kg water removed, and while a 5-7 effect evaporator with mechanical vapor recompression requires 37–53 kWh/1000 kg water removed, reverse osmosis for desalination requires 5–20 kWh/1000 kg water removed. Energy needed for concentration of whole milk to 31% solids by membrane processes requires only 6–7 kilocalories/kg milk, compared to 70–90 kcal/kg using evaporators and 330 kcal/kg by conventional open-pan boiling.

No complicated heat-transfer or heat-generating equipment is needed, and the membrane operation requires only electrical energy to drive the pump motor. Since small molecules normally pass freely through a membrane for ultrafiltration or microfiltration, their concentration on both sides of the membrane should be equal during processing, thus minimizing changes in the microenvironment such as pH and ionic strength.

Limitations. Membrane processes cannot take a liquid stream to dryness. In fact, they are quite limited in their upper boundaries for solids. In the case of reverse osmosis, it is the osmotic pressure of the concentrated solutes that limits the process. For example, milk and whey exert an osmotic pressure of about 700 kPa (100 lb/in.^{2}) at room temperature. Since reverse osmosis is commonly conducted at pressures of 2–3 MPa (300–400 lb/in.^{2}), there is a maximum concentration 3 to 4 times that of milk and whey. In ultrafiltration, it is rarely osmotic pressure but rather low mass transfer rates and high viscosity of the concentrate that limit the process. In microfiltration, the limiting factor is the resistance of the filter cake built up on the membrane.

Until the late 1970s, the major limitation was the availability of appropriate membranes in well-engineered systems that could meet the rigorous sanitation and safety demands of the food industry at a reasonable cost. Other problems that plagued early membrane applications were fouling, poor cleanability, and narrow operating limits, but many improvements have been made.

Membrane materials. Superior membrane materials and vastly improved module and system configurations have become available (see **table**), and it is rarely the cost of the membrane itself that is the limiting factor. For food-processing applications, membrane selection should be based on the following factors: (1) stability at a wide range of pH and temperature, especially under cleaning and sanitizing conditions, (2) minimum membrane–solute interactions, which affect the rate of fouling, cleaning, yields, and rejection of individual feed components, and (3) acceptability of the membrane as a food contact material, which essentially implies using materials that are inert and do not leach out of the membrane into the food.

Commercially available membranes for specific applications*

Material	Micro-filtra-tion	Ultra-filtra-tion	Reverse osmosis
Cellulose esters (mixed)	A	N	N
Cellulose nitrate	A	N	N
Gelatin	A	N	N
Polycarbonate (track-etch)	A	N	N
Polyester (track-etch)	A	N	N
Polypropylene	A	N	N
Polytetrafluoroethylene (PTFE)	A	N	N
Polyvinyl alcohol (PVA)	A	N	N
Polyvinylchloride (PVC)	A	N	N
Alumina	A	A	N
Carbon-carbon composites	A	A	N
Cellulose (regenerated)	A	A	N
Ceramic composites (Zr on C,Al)†	N	A	N
Polyvinylchloride copolymer	A	A	N
Polysulfone (PS)	A	A	N
Polyvinylidenefluoride (VF)	A	A	N
Polyacrylonitrile (PAN)	N	A	N
Polyelectrolyte complex	N	A	N
Cellulose acetate (CA)	A	A	A
Cellulose triacetate	A	A	A
Polyamide (aromatic)	A	A	A
Polyimide	N	A	A
Polyacrylic acid + zirconia	N	A	A
CA/triacetate blend	N	N	A
Polybenzimidazole (PBI)	N	N	A
"TFC" (polyamide or polyether urea on PS)	N	N	A

*A = available; N = not available.
†Zr = zirconium; C = carbon; Al = aluminum.

Applications. Since the birth of modern membrane separations technology in 1960, a vast array of applications have been developed. In the 1980s, major advances of this technology occurred in the food industry. In the 1990s, the emphasis will be on biotechnology, where membrane technology is finding increasing use as a gentle and efficient way of fractionating, concentrating, and clarifying a variety of products.

On a worldwide basis, the dairy industry is probably the largest user of membrane technology in food processing, and it will continue to be a significant user of this technology. Application of ultrafiltration and reverse osmosis for reducing costs of hauling bulk milk (on-farm membrane processing) is also gaining attention, especially in countries where there are large distances between areas of milk production and consumption.

Processing of fruit juices and other beverages will probably be the next large application in the food industry. Ultrafiltration is ideal for clarification of clear juices, where the goal is to replace the conventional holding, filtration, and decanting steps, and perhaps the final pasteurization step. Higher yields, improved juice quality due to removal of pectins, polyphenol oxidase and tannin–protein complexes, and elimination of filter-aid and fining agents are its features.

Other food applications with major potential include the following: production of vegetable protein concentrates and isolates from soybeans, sunflowers, cottonseeds, peanuts, and other plant materials; removal of glucose and partial concentration prior to dehydration in egg processing; extraction of blood and gelatin as animal products; purification and preconcentration in sugar refining; removal of yeast from beer or alcohol distilleries by microfiltration or ultrafiltration instead of by centrifuge or filter; and concentration of fruit juices by reverse osmosis. In biotechnology, major applications are separation and harvesting of enzymes and microbial cells, use of continuous high-performance bioreactors for enzymatic and microbial conversion processes, and use of tissue-culture systems utilizing plant and animal cells as biocatalysts or for production of monoclonal antibodies.

For background information SEE DRYING; FILTRATION; FOOD ENGINEERING; MEMBRANE SEPARATIONS; ULTRAFILTRATION in the McGraw-Hill Encyclopedia of Science & Technology.

Munir Cheryan

Bibliography. J. M. Aguilera and D. W. Stanley, *Microstructural Principles of Food Processing and Engineering*, 1990; S. Bruin (ed.), *Preconcentration and Drying of Food Materials*, 1988; M. Cheryan and M. A. Mehaia, Membrane bioreactors, *CHEMTECH*, 16:676–681, 1986; M. Cheryan, *Ultrafiltration Handbook*, 1986; S. A. Goldblith, L. Rey, and W. W. Rothmayr (eds.), *Freeze Drying and Advanced Food Technology*, 1975; H. G. Kessler and D. B. Lund (eds.), *Fouling and Cleaning in Food Processing*, 1989; R. W. Rousseau (ed.), *Handbook of Separation Process Technology*, 1987; H. Schwartzberg and M. A. Rao (eds.), *Biotechnology and Food Process Engineering*, 1990; W. E. L. Spiess and H. Schubert (eds.), *Proceedings of the 5th International Congress on Engineering and Food*, 1990; W. Van Pelt and J. Roodenrijs, *Multi-Stage Countercurrent Concentration System and Method and Separator*, U. S. Patent 4,430,104, February 7, 1984.

Food manufacturing

Recent advances in food manufacturing have been made in emulsification and homogenization, aseptic packaging, crystallization, powder conditioning, and baking technology.

Emulsification and homogenization. The production of emulsions generally involves two essential mechanisms: the comminution of the disperse phase and the stabilization of the obtained disperse state. In recent years there have been major modifications and improvements with respect to industrially applied emulsification machines and in the agents used for the stabilization of the emulsions. Extensive progress has been made in establishing and quantifying the interrelation between the emulsification machinery and the specific properties of the emulsifiers, stabilizers, and emulsions. Putting this knowledge into practice has led to improved processes and products.

Typical emulsification machines. Stirrers are the oldest type of emulsification apparatus. The disperse phase is formed mainly by shearing within the turbulent flow. The energy input per unit volume is very low compared with other devices. Stirrers, therefore, are usually suitable only for coarse dispersions or emulsions; further processing by other means is necessary if a fine emulsion is desired for technical or functional reasons.

In rotor-stator systems the geometry of the rotor and stator is a decisive factor in controlling the droplet size. As the mechanical energy is dissipated within a small volume, rotor-stator systems are particularly suitable for the production of finely dispersed emulsions.

Generally speaking, there are two types of dispersing machines based on the rotor-stator principle: colloid mills and toothed-disk machines. Colloid mills are used to produce medium- or high-viscosity emulsions. The minimum viscosity of the continuous phase should be about 20 millipascal-seconds (mP · s). The rotor wings accelerate the fluid tangentially, which has the effect of both mixing and coarsely comminuting the disperse phase. Then the fluid passes through the dispersing gap between rotor and stator, where the droplet size of the disperse phase is further reduced, primarily by shear stress. Residence time and shear intensity can be controlled by adjusting the relative position between rotor and stator, that is, the gap width. Toothed-disk dispersing machines contain a slitted rotor and stator, which consists of several toothed disks arranged concentrically. While passing the rotor-stator arrangement, the emulsion is repeatedly subjected to stress due to tangential acceleration and deceleration.

In high-pressure homogenizers the disperse phase is comminuted by special valves consisting of a plunger and valve seat. The gap width, usually in the range of 10–200 meters, is adjusted by the force applied to the plunger. In commercial homogenizers, three-to-five-piston pumps provide pressures of 100–1000 bars (10^4–10^5 kilopascals) and force the fluid through the valves. Thus, very high specific-energy input and extremely short residence time in the valve are achieved. According to current theories, the droplet disruption in the homogenizing valve is mainly caused by cavitation.

Stabilization of disperse state. Instability is manifested by sedimentation or creaming, aggregation, and coalescence of droplets. Two kinds of additives are used to facilitate the formation of an emulsion and to stabilize it thereafter.

Emulsifiers are surface-active compounds whose molecules contain hydrophilic and lipophilic groups. This enables them to reduce interfacial tension, thus facilitating droplet disruption. They also charge the droplet interface, thus inducing repulsive forces and reducing coalescence probability. The determining factor in the emulsification result obtained is the velocity at which the emulsifier stabilizes the interphase newly formed by droplet comminution. Generally, a distinction is made between fast and slow emulsifiers; the latter are predominantly used in food processing. When using slow emulsifiers, it would be a waste of energy to introduce high-shear intensity in the dispersing gap; this is because the emulsifier is the determining factor for the stabilization and because the intermediately obtained disperse state is partially reversed because of coalescence. In this case, a recycle mode is advantageous; at the beginning of the process, low specific-energy input is more than sufficient.

Stabilizers are compounds that increase the viscosity of the continuous phase and thus considerably reduce the rate of coalescence, sedimentation, or creaming. If, in addition, the stabilizer produces a yield stress of sufficiently high magnitude, the mobility of the droplets is practically eliminated and a long-term stability can be achieved.

Harald Armbruster; Manfred Kerner

Aseptic packaging. Aseptic packaging is a food preservation technique that has the potential for producing shelf-stable products with so-called kitchen-fresh quality. Consumer demands for convenience and quality in preserved foods have intensified commercial interest in this technology. Aseptic packaging means packaging in an environment free of microorganisms. The process involves filling a sterile product into sterile containers, which are sealed within a sterile environment.

Aseptic packaging was formerly known as aseptic canning, since the technology is similar in principle to the preservation of canned foods. The first commercially successful aseptic packaging system utilized cans or glass jars. Current technology permits the use of metal, glass, and plastic containers in sizes ranging from single-serving portions, to institutional one-gallon packages, to bulk-storage tanks of several thousand gallons. The technology has two subareas: product sterilization and sterile packaging.

Processes for product sterilization involve heating fluid foods rapidly to 266°F (130°C) or higher for low-acid foods and to 203–221°F (95–105°C) for acid foods and holding at those temperatures for a few seconds, followed by rapid cooling. Since exposure to the high temperatures for sterilization is very short, very little quality degradation occurs during heating. Current technology is suitable for fluid foods that do not contain particles, such as milk, yogurt, mixes for soft-serve ice cream, purees and concentrates of fruit juice, beverages, tomato soup, and clear soups. Methods for effectively sterilizing foods (for example, vegetable soup and stews) that contain large particles are being investigated; however, no commercial system is currently used for packaging such foods.

A packaging machine contains a system for container forming and sterilization, for filling product into containers, for sealing the containers, and for maintaining the integrity of the sterile packaging environment. Some aseptic packaging systems use preformed containers that are sterilized after loading into the unit. Some units use flexible film bags presterilized by ethylene oxide gas or ionizing radiation, followed by special handling to prevent recontamination prior to filling. Features designed into packaging machines permit breaking seals for filling of product and resealing without recontamination.

The most widely used aseptic packaging machine is of the type known as form, fill, and seal. A sheet of packaging material in a roll is formed into pouches or cups, filled, and sealed. Some systems form, fill, and seal brick-shaped packages continuously. In-line sterilization of the packaging material is accomplished by dipping in a 35% solution of hydrogen peroxide at 86–140°F (30–60°C), followed by heating with hot sterile air; radiant heat from an electrical heating element vaporizes the hydrogen peroxide prior to filling. Flushing with sterile air is necessary to remove hydrogen peroxide vapors that can remain trapped in the package headspace or can dissolve in the product, and also to maintain a positive pressure within the sterile chamber to prevent entry of microorganisms. Hydrogen peroxide residues in packages are limited to 0.1 part per million. The packaging system is presterilized by spraying with 35% hydrogen peroxide at ambient temperature, and drying with sterile hot air prior to the start of the filling operation. The filler and lines leading from the product sterilization system to the filler are presterilized by heating.

Aseptic packaging technology allows packaging in plastic, thereby providing products that can be microwaved directly in the package prior to serving. For this application, cheaper packaging and better product quality can be expected compared to current, plastic-canned products that are sterilized by conventional retorting.

Romeo T. Toledo

Crystallization. Crystallization of various food components during processing plays a major role in determining overall product quality. Controlling the formation of crystalline structure in foods is important in developing textural properties of many products. Recent research on crystallization phenomena of importance in foods has focused on the control of crystallization processes (crystal nucleation and growth) in order to enhance final product quality or to control processing steps more efficiently. Control of ice crystallization during freezing of foods, sugar crystallization during confectionery manufacture or sugar (sucrose, lactose, and fructose) refining, and fat crystallization for production of fat fractions for ingredient use are typical applications of these techniques.

Ice crystallization. Ice crystallization in foods is the main physical process involved in food preservation and processing by freezing. Recent research has focused on methods of controlling this ice crystal formation in order to maximize the quality of frozen foods or frozen concentrates. Processes involving control of ice crystal nucleation and subsequent crystal growth through the use of thermal tempering processes prior to freezing, rapid freezing processes, and the addition of bacterial ice nucleators found in nature are under development. These processes result in the formation of large numbers of small ice crystals within the food, allowing preservation in essentially the native state with little damage to the cell structure of the food, yielding products of excellent quality. The

use of naturally occurring ice-nucleating bacteria, such as *Pseudomonas syringae*, can also enhance ice crystal formation, resulting in reduction of the amount of energy required for the freezing process. Another active area of research involves the partial freezing of fluid foods with subsequent removal of the ice crystals to manufacture concentrates that have not undergone thermal treatment (see **illus.**). Heat-induced changes that degrade product quality are thereby avoided. The development of efficient and economical processes for freeze concentration by optimal control of ice crystal nucleation and growth may lead to more efficient concentration with increased quality for such products as fruit juices, vinegar, and fluid dairy products (milk and whey).

Sugar crystallization. One recent advance in sugar crystallization has been the development of a crystalline fructose product refined from corn syrups. This process involves the selective crystallization of fructose from sugar mixtures that result from corn wet milling. Crystalline fructose may be used in instant drink mixes and other products requiring moisture control and also as a sweetener for diabetics.

Fat crystallization. Another area of current commercial importance is fat crystallization. Here, a crystallization process is used to fractionate readily available

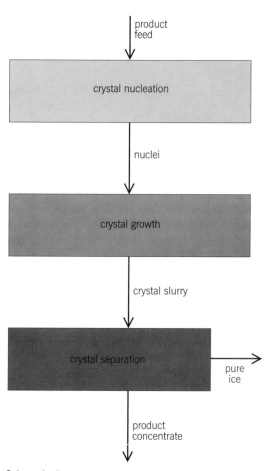

Schematic diagram of a freeze concentration process involving production of a concentrated product by separation of pure ice crystals from a partially frozen product slurry.

fats into components with selected properties. Fractions separated at various conditions have different textural and flavor properties when used as food ingredients. For example, butterfat may be separated by a crystallization process into fractions for use in such products as milk chocolates and baked goods. Development of crystallization technologies is under way to enhance use of these fat fractions for specific applications.

Richard W. Hartel

Powder conditioning. Powder conditioning is a treatment in which a finely divided powder is admixed with a host powder of a larger size to improve the latter's flowability and to inhibit caking. The added powder may be called an anticaking agent, flow conditioner, free-flowing agent, glidant, or antiagglomerant. Its effective concentration rarely exceeds 2% by weight, and with few exceptions it has no nutritional value. Conditioners are considered food additives; and their use in foods is governed in the United States by the Code of Federal Regulations. Most are considered ''generally regarded as safe'' (GRAS); in some cases, such as phosphates or zinc oxide, they are classified as nutrients.

Types. Numerous commercial conditioners and anticaking agents have been developed. They differ in chemical composition, particle size, density, and sorptive capacity. The main types are silica and silicates, phosphates, stearates, polysaccharides, and iron salts.

Silica and the silicates include silicon dioxide, calcium silicate, aluminum calcium silicate, magnesium silicate, sodium alumino silicate, and sodium calcium aluminum silicate. These are extremely fine powders with surface area on the order of hundreds of square meters per gram. They have very low bulk density and are chemically inert, but they usually have very high sorptive capacity.

The phosphates are mainly varieties of tricalcium phosphate having a particle size on the order of 50–100 micrometers and medium bulk density.

The stearates are mainly calcium and magnesium stearates, but can also be aluminum and zinc stearate. They have a mean particle size on the order of 40–80 μm and a low bulk density.

The polysaccharides are mainly starch or starch derivatives.

The iron salts are ferric ammonium citrate and sodium ferrocyanide, for use in salt (sodium chloride) only at up to 25 and 13 ppm, respectively.

Activity. In order to be effective, the agent must be compatible with the host powder; that is, its particles must adhere to the surface of the host-powder particles. Otherwise, the agent's particles will aggregate with its own kind. This not only will render the agent ineffective but also can be a potential source of segregation or dust problems. Similar situations may arise when excessive amounts of conditioner are used, because once all the available absorption sites on the host particles have been occupied, unabsorbed conditioner particles can move freely. Absorption patterns vary considerably, from complete coverage of the host-particle surface to sparse adherence. In general,

the former causes more dramatic effects on the powder bulk properties.

Once absorbed, the conditioner particles can act in one of the following manners or a combination. They serve as a physical barrier between the cohesive host particles. They create and expose new chemically inert surfaces replacing originally active surfaces of the host particle. They disrupt the formation or continuity of liquid bridges between particles, thus reducing the powder's cohesion. They can serve as a lubricant, thus reducing interparticle friction (this is particularly characteristic of stearates). Potentially, when they have a large sorptive capacity, for example, silicates, they can compete with the host particles for absorbed moisture. If, however, moisture sorption is undisturbed, as in the case of storage under high-relative-humidity conditions, then the conditioner presence offers no protection at all. Conditioners are known to affect the electrostatic charging pattern of powders and therefore can inhibit or reduce caking by changing the surface properties of the host particles. In the presence of iron salts the sodium chloride crystal changes its structure from cubic to dendritic. Thus, even if caking does occur, the resulting agglomerates are soft and fragile.

Other effects. Because, in general, conditioners reduce the cohesion of powders, they also affect the bulk properties. The most notable effect is a significantly increased bulk density. (Crystalline powders such as ground sugar conditioned with 0.5% agent can have a density increase on the order of 10–15%.) The density increase is accompanied by reduced compressibility under small loads. Since most conditioners are insoluble in water, dissolving a conditioned powder, such as a dry beverage or soup, causes sedimentation, suspension, or flotation. This may affect the reconstituted product appearance, but since the conditioner concentration is usually very low, the effect is considered minor when compared with the benefits of flowability improvement and caking inhibition.

Micha Peleg

Baking technology. Recent advances in baking technology include development of dough dividers using extrusion methods, chamber ovens with spiral conveyors, and controlled-atmosphere packaging for increased shelf life.

Dough dividers. The conventional type of dough divider used in large commercial bread and roll plants separates bulk dough into pieces of relatively uniform size by compressing a portion of dough into a pocket of fixed volume and then cutting off the excess. These machines work satisfactorily for the most part but have a few disadvantages. Chief among these is the constant gassing that occurs in the bulk dough as it awaits processing; compression in the divider does not fully eliminate this gas, so that density of the dough varies continually, and while the pieces have uniform volume they are not uniform in weight. These dividers must also be very strong. They have many moving parts with close clearances.

A new type of dough divider utilizes a totally different method of dividing a mass of dough into loaf- or roll-size pieces. These machines, commercial models of which have become available only since 1989, pressurize a large mass of dough by mechanical means, causing it to extrude from a circular orifice in the form of a cylinder a few inches in diameter. Because the dough has been worked to eliminate gas pockets, the density of the extruded dough is relatively uniform. Pieces of uniform length are cut from the cylinder by a blade that rotates past the orifice. The speed of rotation can be varied over a wide range, and it is selected to cut pieces of a length that will have the desired weight.

Ovens. Virtually all types of high-production bakery ovens have, until recently, depended on straight-line front-to-back or back-to-front conveying of the product during the baking process. This often required very long baking chambers—300 ft (90 m) is not an unusual length for cookie ovens. Although on the whole these ovens were reliable and permitted satisfactory production rates, their high initial cost and large dimensions were disadvantages. Furthermore, the conveying mechanism in such equipment was often cumbersome, expensive, and hard to maintain and adjust.

A new design utilizes a flexible conveying means with detachable pan assemblies. This permits the utilization of an approximately spiral track within the baking chamber, so that shorter and higher ovens become practical, yielding a more efficient use of space. Because the oven walls and conveying mechanism are much less massive than in previous types of ovens, the temperature can be brought up or down much more quickly.

One major difficulty with these new types of ovens is that it is considerably more difficult to obtain zone heating effects, that is, to pass the product through regions having different temperature and humidity conditions. With straight-line flow, it is a simple matter to divide the baking chamber into sections where the heating elements are adjusted differently and the air flow is varied.

Packaging. Some early attempts at controlled-atmosphere packaging of bakery products have indicated that this process has considerable potential for improving the quality of foods reaching the consumer and for decreasing the cost of distribution (but increasing the cost of packaging) for the manufacturer. New developments in plastic containers offer the possibility of designing new methods and materials for improving the shelf life of nonfrozen bakery products by packaging them in environments having controlled combinations of gases.

Packaging of bakery products (bread, sweet rolls, and cakes) in hermetically sealed cans is a process that is at least 50 years old. Though quite uneconomical for distribution of common bakery products by ordinary routes of distribution, it has some value for military rations, dietetic foods used in amounts too small to justify distribution by ordinary means, and so-called gourmet or ethnic foods not otherwise available outside major metropolitan areas. Since around 1987, military procurement agencies have provided an impetus for commercial production of bakery products in hermeti-

cally sealed metal trays, about half the size of the trays used in cafeteria steam tables. Satisfactory preservation of canned bakery products depends upon maintaining a very low oxygen tension within the container (to prevent growth of mold and other obligate aerobic organisms). This was achieved in the previous generation of products by sealing the container when it was just out of the oven, relying on cooling to draw a considerable vacuum (upward of 25 in. Hg or 85 kilopascals) in the can. However, external air pressure will collapse economically feasible structures, since the large surface area leads to a need to support walls against several hundred pounds of force. This problem can be overcome by sealing the tray in a chamber that is evacuated after the just-baked product is inserted, and then normalizing the pressure by injecting nitrogen into the chamber. The sealed tray has an internal pressure approximating atmospheric air pressure but contains very little oxygen. However, equipment for hermetically sealing (double-seaming) a rectangular tray in a chamber provided with means for pulling a vacuum and introducing a gas is in very limited supply, is very expensive, and lacks versatility.

For background information *SEE* CAVITATION; FOOD MANUFACTURING in the McGraw-Hill Encyclopedia of Science & Technology.

Samuel A. Matz

Bibliography. P. Becher, *Encyclopedia of Emulsion Technology*, vol. 1: *Basic Theory*, 1983, vol. 2: *Applications*, 1985, vol. 3: *Basic Theory, Measurement, Applications*, 1988; S. A. Matz, *Bakery Technology*, 1988, *Equipment for Bakers*, 1988, *The Materials of Baking*, 1989; J. H. Nash, G. G. Leiter, and A. P. Johnson, Effect of antiagglomerant agents on physical properties of finely divided solids, *Ind. Eng. Chem. (Product R&D)*, 4:140–145, 1965; M. Peleg and A. M. Hollenback, Flow conditioners and anticaking agents, *Food Technol.*, 38:93–102, 1984; L. Phoenix, How trace additives inhibit the caking of inorganic salts, *Chem. Eng. (Brit.)*, 11:34–38, 1966; H. G. Schwartzberg, D. Lund, and J. L. Bomben (eds.), Food process engineering, *AIChE Symp. Ser.*, 78(218):31, 1982; R. T. Toledo, Strategies for overcoming shortcomings in technologies for achieving and maintaining aseptic sterility, *Proceedings of Aseptipak 88*, pp. 47–62, 1988; R. T. Toledo and S. Y. Chang, Advantages of aseptic processing fruits and vegetables, *Food Technol.*, 44(2):72, 1990.

Forest and forestry

Recent research in the field of forestry has focused on, among other things, the interactions between forests and global climate change, and the effects of forest gaps on forest ecology.

FORESTS AND GLOBAL CLIMATE CHANGE

There is a strong consensus among scientists that rising concentrations of heat-trapping greenhouse gases in the Earth's atmosphere will increase average global temperatures by 2.7–8.1°F (1.5–4.5°C) over the next half-century. This seemingly small change in temperature has the potential to drastically alter climatic patterns that govern the large-scale distribution and composition of forest ecosystems. Carbon dioxide (CO_2) and other greenhouse gases have been increasing in the atmosphere because of a variety of human activities.

Historical records of atmospheric CO_2 concentrations depict a rise from 292 parts per million (ppm) in the late 1800s to about 345 ppm in 1988, with over half the increase occurring since 1958. Roughly two-thirds of the increase in the past 40 years has been the result of burning fossil fuels; the remainder of the increase is mostly from deforestation worldwide.

Simulation of greenhouse effects. General-circulation models are used to simulate greenhouse effects on the Earth's climate; they all attempt to model the Earth's climate system, but vary in certain assumptions. A recent comparison of three general-circulation models by the Environmental Protection Agency (EPA) showed that average summer temperatures in the United States could increase over the next 50 years to the equivalent of a 4–19° shift southward in latitude, with the largest potential warming occurring in the west (**Fig. 1**). This means, for example, that the conifer forests of northern California might experience summer temperatures like those of the current chapparal and deserts of southern California, and the northern hardwood forests of central New York might have a summer temperature like that of pine forests in present-day southern Georgia. In mountains, summer warming could equal a 2145–2970-ft (650–900-m) shift in elevation, which means that tree species that presently grow at high elevation, such as spruces and firs, could find their elevation pushed higher and higher until they are essentially shoved off the top of mountains (**Fig. 2**). Changes will probably not be limited to simple shifts in current climate zones, but may involve entirely new climatic patterns.

Winter temperatures are predicted to rise less than summer temperatures, and precipitation patterns are likely to change, although there is much uncertainty regarding greenhouse effects on precipitation (a crucial factor in determining how forests will respond). Moreover, greenhouse effects are predicted to increase the probability of insect infestations, wildfires, and severe windstorms; these factors may be even greater determinants in the future of forests than the warmer temperatures.

Forest health depends on a complex of interactions within the ecosystem, and between the ecosystem, atmosphere, and parts of the globe that may be far removed. Forest response to changing climate emerges from three factors: the direct effects of climatic changes on plant physiology, the ability of forests to resist increased levels of disturbance, and the level of maintenance of soil nutrients and microbes during the change from one vegetation type to another. Because of feedbacks between terrestrial vegetation and the atmosphere, the greenhouse effect could

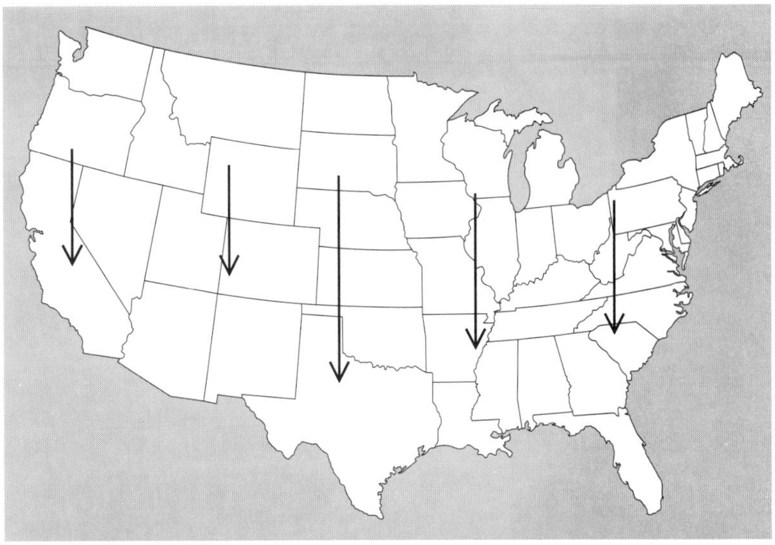

Fig. 1. Estimated greenhouse-related temperature changes expressed as southward shifts in latitude. The magnitude of the shift (indicated by the length of the arrow) varies by location. These are average predictions of three different models. (*After R. P. Neilson et al., Sensitivity of ecological landscapes and regions to global climatic change, Environmental Protection Agency, EPA 600/3-89-073, 1989*)

be either dampened or exacerbated by the response of plant communities, especially forests, which are the largest sinks for carbon on land.

Effects on plant physiology. Three factors related to the greenhouse scenarios under consideration affect the performance of plants in ecosystems: higher temperature, higher CO_2 concentrations, and shifts in water availability.

Temperature. The rate of carbon fixation in photosynthesis is not nearly as sensitive to increasing temperature as is the rate of carbon loss in respiration. On this basis alone, higher temperatures would lead to a decrease in the net amount of carbon fixed by trees and, consequently, would lead to a decrease in their growth. But in many regions, higher temperatures would also extend the length of the growing season, perhaps compensating for lower growth rates. This picture is complicated by the fact that certain trees require a minimum winter temperature for proper growth and development during the ensuing growing season. Normal growth may not be possible if winters become so warm that this chilling requirement is not met.

CO_2 concentration. Plant species are not consistent in how they respond to higher atmospheric CO_2 concentrations, but most grow better. One study with tree seedlings demonstrated that CO_2 fertilization increased photosynthesis of several species, but this did not always translate into better growth. A major benefit of high CO_2 concentrations is more efficient water use by plants. Stomatal openings control the plant–atmosphere exchange of both CO_2 and water vapor. During moisture deficits, stomates close to conserve water, thus restricting the diffusion of CO_2 into the leaf. However, with higher atmospheric CO_2 concentrations, plants can take up more CO_2 during periods when stomates are open; hence, the rate of carbon fixation per unit of water increases. The result

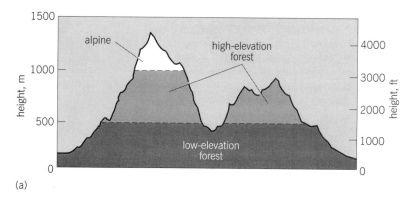

(a)

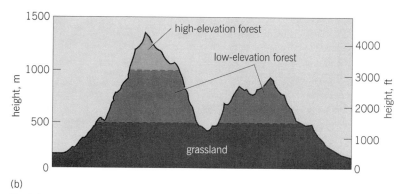

(b)

Fig. 2. Hypothetical change in elevational range of alpine, high-elevation forest, low-elevation forest, and grassland. (a) Current elevational range. (b) Elevational range with a 5.4°F (3°C) change in average temperature.

might be to overcome drought stress, but only if overall water demand is decreased.

Water availability. Even with no decrease in precipitation, forest ecosystems will become somewhat drier because of greater evaporation associated with higher temperatures. Plant responses to drier conditions are species-specific and complicated by interactions with increased temperature and CO_2 concentration. Whether a site becomes more prone to drought will depend on whether precipitation increases sufficiently to compensate for greater water losses brought about by higher temperatures.

The net effect of increased temperature, higher CO_2 concentrations, and altered precipitation patterns is very difficult to predict. It will vary among plant communities and their environments. For example, higher temperatures and CO_2 concentrations might benefit trees growing at high elevations and latitudes, but those in warmer environments might experience reduced growth rates because of increased respiration. In semiarid forests, increased CO_2 concentrations might benefit plants by improving water-use efficiency, unless the increase in CO_2 is accompanied by significant decreases in precipitation.

The response of any one species cannot be isolated from that of other species in the forest community. Competition among species may exacerbate the disadvantages that certain species have in adapting to a changing climate. But interactions among some plants may be mutually beneficial rather than competitive, in

which case some species may sustain others that are less favored by the altered climate.

Climate change and disturbance in ecosystems. Disturbances are common in all forests and are usually beneficial to the health of the ecosystem. However, normal disturbances are likely to be intensified by changing climatic conditions; hence, understanding the role of disturbance is essential to predicting how forest ecosystems may respond.

Biotic disturbances. In healthy forests, invertebrate pests and other disease organisms are generally found in small numbers, but outbreaks can be precipitated by plant stress factors, for example, drought, nutrient deficiency, and pollution. A warmer, drier climate could stress forests sufficiently to cause outbreaks of insects or pathogens. Recent studies indicate that elevated atmospheric CO_2 concentrations can increase plant carbon-to-nitrogen ratios. This would prompt insects to increase their consumption of leaves in order to acquire sufficient nitrogen. However, enhanced plant growth or production of defensive biochemicals in a CO_2-enriched atmosphere may offset greater consumption.

Abiotic disturbances. Events such as temperature and precipitation fluctuations, ice, wind, flood, and fire periodically alter the structure and composition of forests. In many forest types, especially those that are seasonally dry, the most rapid and extensive changes are those wrought by fire. In these areas, the species composition, biomass accumulation, and overall distribution on the landscape reflect in large part the historical influence of fire as it interacts with climate. Generally, forests in warmer, drier climates tend to have less biomass and more frequent but less severe fires than forests in cooler, wetter climates. This relationship suggests that if climatic conditions rapidly become warmer and drier, overall fire frequency will increase. However, it is possible that fire severity will also be exacerbated since large amounts of biomass (that is, fuel) from the previous cooler, wetter climate would be carried over into the new climatic regime.

Disturbances of human origin. Arguably, the major agents of disturbance in forest ecosystems are human. Forest clearing, air pollution, and other human activities have wholly or partially altered large areas of the Earth's surface. A more subtle example of human interference is the twentieth-century history of fire control in the forests of western North America. These control measures have altered the species composition and structure of many forest types, and allowed accumulations of biomass that would have burned in a more natural fire regime. Such a condition compounds the risk of more frequent, severe fires that will result from climate change.

Acid deposition has been implicated in the decline of forests in central Europe and of certain species in eastern North America. Although debate on the cause and mechanism of forest decline is ongoing, forest ecosystems already stressed by acid deposition are likely to have little resilience in the face of rapid climate shifts and increased frequency of disturbance.

Plant migrations and ecosystem stability. Plant species will shift northward and upward in response to predicted warming trends. Models predict that temperature changes will be equivalent to a yearly latitude shift of several thousand meters southward. However, at the end of the last glaciation, maximum tree migration rates were much less—on the order of a few hundred meters per year. Even if humans were to facilitate the process of tree migration, by the time seedlings planted along expected migration corridors reached maturity the environment might no longer be favorable for seed production and dispersal, or even for survival. In the absence of viable populations of tree species, soil fertility may be reduced and more opportunistic species such as weeds favored. Degradation of forest ecosystems over sufficiently large areas will probably augment the amount of CO_2 added to the atmosphere from soil and biomass.

Feedback processes and ecosystems. The least understood aspects of global climate change involve the many complex interactions of aquatic and terrestrial ecosystems with the climate system. A simplified example of interactions between systems involves soil carbon storage, temperature, and atmospheric CO_2 concentrations. In this scenario, rising temperatures increase the decomposition rate of soil organic matter, a carbon store about twice the size of that in the Earth's atmosphere or vegetation. Decomposition releases CO_2 to the atmosphere, elevating global temperatures that in turn increase the rates of decomposition and CO_2 evolution. This is an example of positive feedback. Other greenhouse gases may be involved in similar positive feedback processes. For example, methane and nitrous oxide are mostly associated with organic matter decomposition in anaerobic environments. Methane is particularly important, because it is perhaps 20 times as effective as CO_2 in its greenhouse effect. Feedback processes remain one of the most daunting obstacles to obtaining accurate model simulations of future climates.

Jeffrey G. Borchers; David A. Perry

FOREST GAPS

Forests are mosaics of different sizes, types, and densities of vegetation. One of the most distinctive aspects of this mosaic is the gap, an opening created by disturbance in an otherwise dense canopy of foliage. The process of gap formation and canopy closure is common to all forests and has important effects on forest composition, successional change, and ecosystem function.

Disturbances and gap dynamics. Disturbances to canopy trees create canopy gaps. Gaps can assume a wide range of sizes, from small openings created by the death of an individual branch or a tree to extensive areas of hundreds of hectares created by windstorms or wildfire. Two major types of gap-forming disturbances can be recognized: those that have no direct effect on the forest floor and soil, such as insect attacks that leave dead trees standing, and wind that breaks tree boles; and those that directly affect the forest floor and soil, such as fire that burns understory

vegetation and the organic layers, and wind that uproots trees. These two types of gap formation produce distinctly different environments and opportunities for seedling establishment and regeneration.

The amount and number of gaps in a forest is a function of gap dynamics, that is, the rate of gap formation and the rate of gap closure. Gaps less than about 0.04 acre (0.1 hectare) in mature temperate and tropical forests form at an average rate of 0.2–2.0% of the forest area per year. However, episodic disturbances can create a much larger proportional area of gaps in a single year. Turnover time has been used to characterize the gap dynamics of a forest by accounting for both creating and filling gaps. Turnover is defined as the mean time between disturbances. This definition incorporates both the time required for vegetation to fill a gap and the mean period of time before another gap forms at that place. Turnover rate is often calculated as the inverse of the frequency of gap events, assuming the area of gaps created per unit time is balanced by the area of gaps that closed during the same period. For many forests this assumption may not be valid because disturbance rates and regeneration vary with time and forest stand development. Estimates of turnover times for both tropical and temperate forests range between approximately 50 and 500 years.

Characteristics of gaps. Uniform definitions or standards of gap measurement do not exist. Gaps have been typically characterized by an opening in the canopy that extends down to a particular height above the forest floor. The area of a gap is determined either by measuring the distance between the horizontally projected edges of the canopy surrounding the gap or by measuring the distances between the boles of the trees surrounding the gap. In forests with relatively open canopies, gaps may form a connected network that cannot be described as a well-defined hole.

Gap characteristics such as size, shape, and environment are important in vegetative response to gaps. Gap sizes can range from several square yards to thousands of acres. However, the term forest gap most often refers to smaller openings created by the death of one to dozens of canopy trees. In mature forests these smaller gap events typically create openings of 300–11,000 ft^2 (25–1000 m^2), with most single-tree gaps covering less than 1100 ft^2 (100 m^2). In many forests, gaps typically cover at least 3–25% of the forest area. Gap shapes can vary, depending on the disturbance. For example, disturbances such as root pathogens can create roughly circular gaps as they radially spread through a uniform forest soil. In contrast, gaps formed by windstorms or avalanches can create very narrow, elongated gaps. The shape and orientation of the gap can affect the amount of light that penetrates to the forest floor, thereby influencing tree regeneration and forest development.

The environment of a gap is distinct from the environment underneath canopies. Total incident sunlight is typically higher in gaps than under closed canopies. Temperature patterns frequently consist of higher maximums and lower minimums under gaps.

In some forests, gaps may become areas of frost or snow accumulation. Soil moisture is often higher under gaps than under canopies. Initial research on small gaps does not show large differences between soil nutrients in gaps and soil nutrients under canopies. This lack of differences may be a consequence of nutrient uptake by the remaining understory vegetation. It is expected that in gaps where soil is disturbed, or where gap size and vegetation loss are large, more nutrients will become available, although some leaching of nutrients out of the gaps will occur. In coniferous forests in cool wet climates, such as southeastern Alaska, uprooting of trees by wind can improve soil fertility by churning the soil, mixing the deep organic layers of the forest floor into the mineral soil below.

Gaps are heterogeneous in terms of both their microclimate and their substrates. Regular patterns of physical properties associated with light and soil are observed from the gap center to the edge. At low latitudes and high incidence angles of the sun, the center of the gap is the location of the highest light intensity. With increasing latitude and lower incidence angle of the sun, the zone of highest light intensity shifts from the center of the gap toward its edge. In high latitudes, this zone may extend several meters outside the projected area of the gap. Soil moisture is typically highest in the center of the gap, where root density is presumably lowest. Substrates in gaps can include tipped-up root wads, mineral soil, tree boles, and piles of branches and leaves from fallen tree crowns. The internal heterogeneity of gaps makes it difficult to generalize about gap effects based solely on characteristics such as gap size and shape.

Ecological function of gaps. Gaps play two major roles in vegetation dynamics: providing growth opportunities for understory trees, shrubs, and herbs that survive the disturbance; and allowing new individuals and species to become established in a forest. Gap formation most commonly benefits tree seedlings and saplings already present in the understory. Growth rate for seedlings and saplings increases with gap size primarily as a function of light, although the relative importance of above- and belowground resources in controlling vegetative response to gaps is not well known for many forests.

For most seedlings and saplings, gap formation may provide their only opportunity to grow into the canopy. In many cases, an understory tree may experience several cycles of gap formation and canopy closure during its lifetime. While the gap is open, trees in the understory can grow more rapidly toward the canopy. Tree seedlings and saplings that do not experience a sufficient number of small gaps or a larger, more persistent gap will die before they reach the canopy. Most gap events benefit the growth of shade-tolerant tree species. Shade-intolerant trees typically have few if any seedlings and saplings under intact canopies and cannot compete with shade-tolerant trees in small, single-tree gaps.

The best opportunities for establishment of new individuals occur in larger gaps and in gaps in which preexisting vegetation is killed by fire or soil disturbance. For many plant species, removal of understory vegetation and exposure of mineral soil provide sites for seedling establishment. When large, multiple-tree gaps form, shade-intolerant species may become established, and if light levels are high enough these species can frequently outgrow regeneration of shade-tolerant species to reach the canopy. The tree seedlings that first become established in a new gap may be the ones that eventually fill the canopy opening because they can capture available resources and preclude further establishment.

Canopy gaps may close in one of two ways: trees grow from below to fill the opening, or crowns of canopy trees surrounding the gap extend laterally into the gap. For very small gaps, lateral closure is probably most common. In some forests, gaps may close very slowly or never close because understory shrubs and herbs occupy the area beneath the gap, precluding the establishment of canopy tree species.

In many types of old-growth forest the composition of canopy species can be maintained solely by tree regeneration occurring in small canopy gaps. Even shade-intolerant species can maintain a small number of canopy trees in older forests through the infrequent occurrences of large gaps. In tropical forests characterized by high species richness, it has been hypothesized that many tree species can be differentiated based on their competitive abilities across a range of gap sizes and environments. This suggests that gaps may have a direct control over species richness and community composition.

Effects of gaps on other components of the biota are relatively unknown. For some animals, gaps may provide food resources and physical habitat that does not occur in areas of closed canopies. Bird species richness in some tropical forests has been found to be greater in gaps than in undisturbed areas of forest. Dead trees that occur in gaps are important habitat features for a large number of cavity-nesting birds and for small mammals and amphibians that use decayed tree boles for perches and cover. The production of flowers and fruits in many forest shrub and herb species is higher in gaps, and some species may be dependent on gaps to persist in the understories of dense forests. Consequently, in many forests, gaps function as biological hot spots that provide a shifting mosaic of available resources, maintaining the overall diversity and productivity of the entire forest ecosystem.

For background information SEE CLIMATOLOGY; FOREST AND FORESTRY; FOREST ECOLOGY; GREENHOUSE EFFECT in the McGraw-Hill Encyclopedia of Science & Technology.

Thomas A. Spies

Bibliography. J. S. Denslow and T. A. Spies (organizers), Canopy Gaps in Forest ecosystems (symposium), *Canad. J. For. Res.*, 20(5):619–667, 1990; D. A. Lashof, The dynamic greenhouse: Feedback processes that may influence future concentrations of atmospheric trace gases and climatic change, *Climat. Change*, 14:213–242, 1989; R. P. Neilson et al.,

Sensitivity of ecological landscapes and regions to global climatic change, Environmental Protection Agency, EPA 600/3-89-073, 1989; D. A. Perry et al., Species migration and ecosystem stability during climate change: The belowground connection, *Conserv. Biol.*, 4:266–274, 1990; S. T. A. Pickett and P. S. White (eds.), *The Ecology of Natural Disturbance and Patch Dynamics*, 1985: W. J. Platt and D. R. Strong (eds.), Special feature: Treefall gaps and forest dynamics, *Ecology*, 70(3):535–576, 1989; S. H. Schneider, The greenhouse effect: Science and policy, *Science*, 243:771–781, 1989; W. E. Shands and J. S. Hoffman, *The Greenhouse Effect, Climate Change, and U. S. Forests*, 1987; C. Uhl et al., Vegetation dynamics in Amazonian treefall gaps, *Ecology*, 69(3):751–763, 1988.

Forest ecology

The forest canopy is the layer of the forest ecosystem that is photosynthetically active, and is an important interface between the atmosphere and the biosphere. The structure of the canopy determines the diversity of habitat and food resources for associated plants and animals. Canopy interactions with the atmosphere contribute to forest health, air quality, and climate.

Canopy structure. Canopy structure develops as forests age, often culminating in dense, multilayered canopies that may exceed 225 ft (70 m) in height in temperate and tropical rainforests. Forest canopies in less favorable environments are more open, as a result of wider tree spacing, and often reach only a few meters in height. Canopy structure is determined primarily by resource availability and disturbances.

Effect of resources. Canopy structure reflects tree adaptations to acquire adequate water and light. Water availability and snow accumulation influence tree crown height, shape, and spacing. The conical shape of boreal conifers is an adaptation for shedding snow and reducing winter breakage; the funnel shape of crowns in xeric forests is an adaptation for channeling water to roots; the arching form of rainforest trees is an adaptation for shedding water.

Forests with closed canopies or foliage characteristics that limit light penetration have shallower canopies than do forests with open canopies. Similarly, forests at low latitudes with high sun angles (tropical forests) maximize interception of light from above with shallowly domed crowns (**Fig. 1**). Subcanopies form at heights where cones of light penetrating shallow upper canopies intersect to form a zone of adequate light intensity. By contrast, trees at high latitudes and low sun angles (boreal forests) maximize light interception from the side with narrow crowns (boreal conifers). Light penetration under these conditions is insufficient for subcanopy development.

Effect of disturbance. All forests periodically experience disturbances that alter canopy structure. Catastrophic disturbances, such as wildfire, landslides, and hurricanes, remove the canopy over relatively large areas every 50–500 years and initiate canopy development. Small-scale disturbances that injure or kill individual trees or small groups of trees are more frequent; such disturbances include wind and storm damage, drought, breakage under the weight of snow or accumulated epiphytes, factors that increase competition for light and other resources, and infestations of insects and pathogens.

All tree species are adapted to survive some adverse conditions that kill or suppress other tree species. For example, shade-adapted tree species tolerant of low light intensities will be favored in gaps that are small relative to sun angle or canopy height; shade-intolerant tree species requiring full sunlight will be favored in large gaps. In the absence of further disturbances or other limiting factors, shade-intolerant species are eventually replaced, through the process of succession, by shade-tolerant species that can germinate and grow under the canopy. Therefore, disturbances tend to maximize forest diversity by maintaining a patchwork of varied canopy structures.

Canopy composition. Half of all known organisms may inhabit forest canopies. Older, more complex canopies with moderate temperature and humidity are especially diverse. Epiphytic plants intercept water and nutrients, and provide additional food resources and, in many cases, aquatic habitats. Small pools in tree cavities and epiphytic bromeliads are protected habitats for many amphibians and invertebrates. A variety of vertebrate and invertebrate herbivores, predators, and decomposers also make their homes in forest canopies and influence forest processes. Insects are especially well represented, with perhaps several million species in tropical canopies.

Most canopy organisms are highly specialized and rarely, if ever, are seen living near the forest floor. Many are restricted to particular tree species or canopy layers (**Fig. 2**). In turn, the importance of a tree species to canopy structure and function often reflects the presence of associated species that affect pollination, seed dispersal, growth rate, or branching pattern.

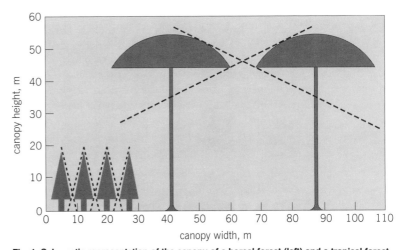

Fig. 1. Schematic representation of the canopy of a boreal forest (left) and a tropical forest (right) drawn to scale. The limiting angle for light penetration through the boreal canopy is 15° from the vertical, and for the tropical canopy 60°. 1 m = 3.3 ft. (*From J. Terborgh, The vertical component of plant species diversity in temperate and tropical forests, Amer. Natural., 126:760–776, 1985*)

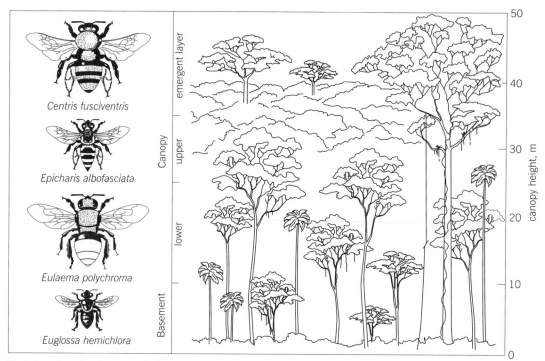

Fig. 2. Layering of environments in the rainforest. Each level of the forest has its own array of plants and animals, including pollinating insects. The two bee species at the top left were found only at the upper margin of the canopy, and the lower two species only in the forest below. 1 m = 3.3 ft. (*From D. R. Perry, The canopy of the tropical rain forest, Sci. Amer., 251(5):138–147, 1984*)

Nutritional quality and density of foliage and other plant tissues, size and orientation of branches, depth of bark crevices, and size of sunlit openings vary widely, especially in older forests. The variety of these resources determines the diversity of associated plant and animal species. For example, large epiphytes with high moisture requirements are restricted to large branches low in the canopy; insects that feed on young, succulent foliage occur on the outer margins of the canopy.

Canopy resources vary in availability, suitability, and visibility as food or habitat for associated organisms. Availability varies spatially over the patchwork of canopy structures and often varies seasonally in trees that shed foliage during unfavorable periods. Suitability as food is determined by chemical composition. Nutritional quality varies with tree species and with the age and condition of the foliage. Healthy trees of all species produce a characteristic array of defensive chemicals (such as phenols, terpenes, alkaloids, and animal hormone analogs) to protect sensitive tissues from ultraviolet radiation, and to provide protection against animal and pathogen species that have not adapted to detoxify or avoid these chemicals. Some of these defensive compounds are small, highly volatile molecules. When carried on the airstream, these chemicals become powerful signals that advertise plant location to adapted animal species that track host aerosols.

Feeding pressure by organisms adapted to plant defenses favors plants that produce new compounds. The biochemical diversity found in canopy plants and other organisms is gaining increased attention. Animals produce their own array of biochemicals, used to detoxify host chemicals, attract mates, and deter predators. Although the potential wealth of canopy compounds has barely been tapped, chemical screening, especially of tropical species, is yielding promising new pharmaceutical chemicals and natural pesticides. Natural products may regain attention as petroleum-derived products become more expensive.

Canopy function. Forest canopies are recognized for their importance to photosynthesis, evapotranspiration, and reduced soil disturbance and erosion. Research on canopy function has been limited by the difficulty of canopy access. However, development of canopy access and remote sensing techniques in the 1980s has stimulated canopy research. Elaborate rope-and-pulley systems suspended among neighboring trees have increased canopy access from the ground, permitting the first detailed observation of canopy communities. Sky cranes (tall structures with revolvable booms) are being used to access larger canopy areas, to map canopy structure, and to install monitoring equipment. Remote sensing via satellite and laser imagery now permits studies of canopy productivity, evapotranspiration, thermal flux, and foliar and atmospheric chemistry over extensive areas. As a result of these new techniques, canopy processes contributing to forest health, air quality, and global climate are becoming better understood. **Figure 3** summarizes these canopy processes.

Forest health. The diversity of canopy species functions to maintain forest health and minimize disruption of ecological processes. Ground cover by the canopy reduces precipitation impact and erosion, and maintains the moderate environment essential for survival of many key species. The vast surface area of tree foliage and epiphytes serves to intercept water, nutrients, and other particulate materials from the atmosphere. Some epiphytic lichens, algae, and bacteria capture and convert atmospheric nitrogen, a critical and often limiting nutrient, into usable organic nitrogen through the process of nitrogen fixation. Decomposition of dead plant and animal tissues within the canopy makes nutrients available for canopy processes. Nutrients eventually reaching the forest floor increase soil fertility.

Even herbivorous insects, which are often viewed as pests, contribute to long-term forest health by pruning plant tissues that are photosynthetically inefficient or poorly defended, and releasing nutrients from these tissues. Nutrient availability and forest productivity can be enhanced for decades following insect population outbreaks.

Diverse and healthy canopies tend to prevent outbreaks of particular species. Because many canopy organisms are restricted to a particular tree species, a diverse canopy limits species ability to increase population size and exhaust resources. Necessary resources are widely scattered, chemically defended, and hidden by mixing of host and nonhost aerosols in the airstream. In contrast, forest canopies composed of one or a few tree species provide abundant resources for adapted organisms. Adverse conditions stress most trees simultaneously in such forests, increasing the likelihood of pest outbreaks and forest decline.

Canopy processes are essential to the long-term health and stability of forest ecosystems. Disruption of these processes through crop species selection, reduced biodiversity, canopy removal, and other management practices can reduce soil fertility and forest productivity, exacerbating adverse conditions and leading to forest decline.

Air quality. The filtering capacity of the canopy is instrumental in removing particles, including pollutants, from the air. Deposition of atmospheric particles on canopy surfaces augments other sources of essential nutrients. Filtering of fog and rain removes additional materials from the atmosphere; conifer forest canopies may intercept and filter as much as 30% of bulk precipitation, and broadleaved canopies 20%.

Unfortunately, this filtering capacity also aggravates canopy vulnerability to atmospheric pollutants. Acid deposition and toxic chemicals affect a substantial portion of the canopy during air passage through the forest. Toxic aerosols and acid precipitation deposited on foliage or absorbed through stomata destroy plant tissues and disrupt photosynthesis. Air pollutants have been implicated in forest declines in industrial areas of Europe and eastern North America.

Climate. Forest canopies have a substantial effect on local and global climate. Shading provided by the canopy reduces soil temperature and convective radiation, stabilizing diurnal and regional temperatures. Evapotranspiration further cools the forest and contributes to cloud formation and regional rainfall and

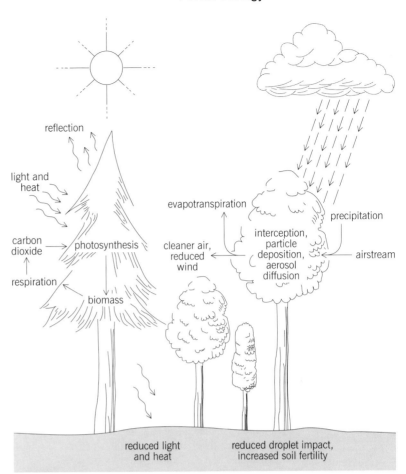

Fig. 3. Canopy processes influencing interactions between the biosphere and the atmosphere. The net result of the process of photosynthesis and respiration is reduced atmospheric carbon dioxide.

humidity. Deforestation can raise regional temperatures and reduce precipitation.

Incorporation of atmospheric carbon dioxide, a greenhouse gas, into forest biomass through photosynthesis in the canopy regulates the absorption of infrared radiation by the atmosphere. Forest biomass accounts for at least 60% of global organic carbon storage. Young trees are more efficient photosynthetically than are older trees, leading some resource managers to contend that conversion of older forests to younger forests will reduce atmospheric carbon dioxide. However, young forests harvested every 50–100 years cannot accumulate as much carbon dioxide as is released by harvest, burning, or conversion of older forests. Hence, harvest likely will increase the carbon dioxide content of the atmosphere. Increased atmospheric carbon dioxide, in the absence of cooling factors, is projected to increase global temperatures.

For background information SEE EVAPOTRANSPIRA-TION; FOREST AND FORESTRY; FOREST ECOLOGY in the McGraw-Hill Encyclopedia of Science & Technology.

Timothy D. Schowalter

Bibliography. R. E. Dickinson (ed.), *The Geophysiology of Amazonia*, 1987; A. W. Mitchell, *The*

Enchanted Canopy: A Journey of Discovery to the Last Unexplored Frontier, the Roof of the World's Rainforests, 1986; E. A. Norris, *Ancient Forests of the Pacific Northwest*, 1990; T. D. Schowalter, W. W. Hargrove, and D. A. Crossley, Jr., Herbivory in forested ecosystems, *Annu. Rev. Entomol.*, 31:177–196, 1986.

Forest management and organization

Conflicts between environmental and commodity interests over management of forest lands have increased dramatically during the last several decades. Increased scientific knowledge of the detrimental ecological impacts of traditional forest practices, such as use of even-aged forest monocultures based upon periodic clear-cutting, has contributed to the conflict. Issues being debated include the potential impacts of these traditional practices on nongame wildlife species, watershed values (including levels of sediment production and flooding), fish production, and long-term site productivity. Another issue has been the increased societal concern for protection of biological diversity, as reflected in the National Forest Management and Endangered Species Acts.

In the United States, debate over disposal of the remaining unreserved old-growth forests and preservation of the northern spotted owl in the Pacific Northwest epitomizes the conflict between environmental and commodity interests. Conservationists favor preservation of most of the remaining old-growth forests for their environmental and recreational values. One major environmental concern is maintenance of old-growth forest habitat for species such as the northern spotted owl, which was recently listed by the U.S. Fish and Wildlife Service under the Endangered Species Act as a threatened species. The forest products industry, on the other hand, favors logging of the remaining unreserved old-growth forests, arguing that sufficient acreage of old-growth forests is already preserved in National Parks, National Wilderness, and other reserved areas.

Traditionally, such conflicts have been resolved by allocating lands to primary or single uses, such as intensive timber production or preservation. New approaches to forest management that better integrate maintenance of ecological values with commodity production, sometimes called New Forestry, have emerged as an alternative to allocation. Although many of the specific practices advocated under New Forestry have their historical roots in traditional forestry, there are many differences. New Forestry incorporates an ecosystem view of the forest and provides tools that allow a more balanced weighting of ecological values with commodity values.

Scientific basis for New Forestry. New Forestry draws heavily on modern ecological concepts to design forest practices that retain ecological values while providing for production of wood products. The ecosystem paradigm, which recognizes the forest as a biological system with many essential and highly integrated parts, is one critical concept. For example, natural forests typically have high levels of spatial heterogeneity and structural diversity (including trees varied in size, species, vigor, and soundness). This richness in structural characteristics is one of the major reasons that natural forests provide habitat for a wide variety of organisms and exert great influence on ecological processes, such as those involved in regulation of nutrient and hydrologic cycling. For example, the immense surface areas of the multilayered canopies of the old-growth forests provide condensing surfaces for moisture and other atmospheric materials.

A second concept, biological legacies, relates to the recovery of natural forests following catastrophic disturbances, such as wildfire and windstorm. There are typically large legacies of living organisms in forms ranging from mature and seedling trees to spores and seeds found in the forest floor and fossorial animals. Extensive legacies of dead organic materials, including large structures, such as logs on the forest floor and standing dead trees (snags), also carry over from the ecosystem before the disturbance to the recovering ecosystem after the disturbance. For example, wildfire typically kills trees but does not consume much of the wood. Because of these legacies, natural young forests are typically diverse ecosystems with high levels of structural and compositional diversity rather than simply communities of young trees. *See* FOREST AND FORESTRY.

Consideration of large spatial scales, or landscapes, and longer temporal scales is a third concept underpinning New Forestry. This concept addresses the critical importance of issues such as patch sizes and arrangements, edge phenomena, corridors to wildlife, and the size of watersheds. Increasing evidence of ecological dysfunction at the landscape level, such as cumulative negative impacts on water quality from excessive cutting or fragmentation of forest cover, has spurred the interests of scientists and managers in landscape ecology.

Application of concepts at stand level. A basic principle of New Forestry at the level of the forest stand is maintenance of structural and compositional diversity within stands managed primarily for wood production. This contrasts with traditional forest practices that have emphasized simplification of forests, for example, creation of even-aged monocultures.

Many techniques that will promote greater species richness and structural diversity can be applied to young managed stands. An aggressive effort to establish mixtures of tree species, rather than monocultures, is one approach. For example, hardwood trees can be included within a stand dominated by conifers; or soil-building coniferous species, such as members of the Cupressaceae, can be added to plantations. Early successional plant and animal species can also be retained for longer periods by using wide spacings to delay tree canopy closure.

Maintaining a continuous supply of coarse woody debris (large standing dead trees and downed logs) is one technique increasingly used to provide structural diversity within managed forest stands and associated streams (**Fig.** 1*a*). This reflects the recent understanding that dead trees have as many ecological roles as live trees. Coarse woody debris provides habitat critical to many vertebrate and invertebrate animal species. Woody debris also contributes to many other essential ecological processes. For example, rotting wood is a long-term source of energy and nutrients, improves soil structure, and provides habitat for nitrogen-fixing microorganisms. Woody debris in streams provides for a greater diversity of habitats for aquatic organisms and increases a stream's ability to retain allochthonous inputs, sediments, and water.

Harvest cutting systems that retain significant numbers of green trees on cutover areas, variously referred to as green tree retention, partial retention, or partial cutting, are another important variation from traditional practices. **Figure 2** contrasts traditional clear-cutting and partial cutting. Partial cutting involves retention of a significant number of green trees at the time of harvest with the intention of keeping them until the next harvest, perhaps 50–100 years in the future. Such cuttings lead to the creation of mixed-structure stands that retain many ecological values, such as habitat for animal species more typical of old-growth forests. Partial cutting also provides for moderated environments on cutover areas, opportunities to grow larger and higher-quality trees, and development of managed stands with high levels of structural diversity. Logging is more difficult and expensive when using a partial-cutting approach, however.

Partial cutting takes many forms depending upon forest and environmental conditions and management objectives (Fig. 1). Residual timber volumes in partial cuttings typically vary from 10 to 40% of the original stand and include at least some of the vigorous dominant trees. Retained trees may be dispersed evenly over the logged area or clustered in small patches or both.

Potential applications of partial cutting for wildlife purposes are diverse. For example, one requirement of the threatened red-cockaded woodpecker found in the southeastern United States is large longleaf pine trees for nesting; partial cutting could be used to create and maintain stands with such trees while providing for forest harvest. Northern spotted owls sometimes occur in forests that are mixed in age and structure; partial cutting could be used to create such stands, thereby re-creating suitable habitat for the owl less than 100 years following harvest.

The partial-cutting approach is different than either traditional selection or shelterwood cutting systems. Selection cutting involves removal of either small groups of trees or selected individuals and leaves most of the stand in place; entries are made at frequent intervals, such as every 10 years. In contrast, there would typically be only one entry per rotation under partial cutting. In traditional shelterwood cutting,

Fig. 1. Two types of partial cutting in mature Douglas-fir forests on the Willamette National Forest, Oregon. (*a*) Harvest cutting where an average of six green trees and three snags are retained on each acre with the objective of providing for a continuous supply of the latter category. (*b*) Harvest cutting with retention of twelve green trees per acre with the objective of creating a mixed-age and mixed-structure stand during the next rotation. (*Courtesy of J. F. Franklin*)

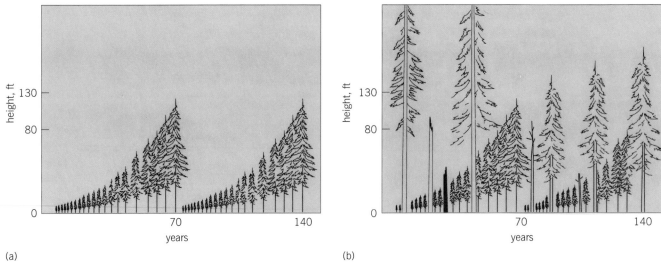

Fig. 2. Contrast between forest structure developed by using (*a*) traditional clear-cutting and (*b*) partial cutting. 1 ft =0.3 m.

residual green trees are removed after tree regeneration is established.

Application of concepts at landscape level. New Forestry at the level of the forest landscape is based on analysis of immediate and long-term effects of forest activities using large spatial scales (hundreds and thousands of acres) and over long time intervals. Foresters have, however, typically considered individual stands or patches in isolation. This approach has often produced unforeseen and unexpected cumulative negative impacts on larger-scale watershed or wildlife values.

One critical landscape concern is the issue of appropriate patch size. Results of some of the research have been counterintuitive. For example, a common practice has been to disperse small clear-cut areas through a forest matrix. A result is fragmentation of the remaining forest into small patches. Because of edge effects, these patches do not provide suitable habitat for some forest species, notably those identified as interior forest species. For example, when a patch of mature forest is adjacent to a clear-cut area, microenvironmental and habitat characteristics along the margin of the forested patch will typically be influenced by the climatic conditions on the clear-cut. As a specific example, in northwestern Douglas-fir forests wind and relative humidity may be modified for 650–1300 ft (200–400 m) from a boundary with a clear-cut; a forest patch surrounded by clear-cuts must be 37–62 acres (15–25 ha) in size before any location within it is free of external influences. Hence, in some parts of the Pacific Northwest and the Great Lakes region, it may be preferable to select larger patch sizes for both cut and uncut areas, thereby reducing the amount of high-contrast edge and edge-influenced environment in the landscape.

There are many other important landscape issues that are considered in New Forestry. For example, integration of reserved areas into managed landscapes is a high priority; geographical distribution and connectivity between such areas are critical consider-ations. Long-term landscape planning is necessary to avoid undesirable cumulative effects on hydrologic conditions, such as increased flood or sediment levels that may be associated with recent clear-cuts. The relative merits of concentrating or dispersing forest cutting within the landscape are another consideration.

For background information SEE FOREST AND FORESTRY; FOREST ECOLOGY; FOREST MANAGEMENT AND ORGANIZATION; SILVICULTURE in the McGraw-Hill Encyclopedia of Science & Technology.

Jerry F. Franklin

Bibliography. J. F. Franklin, Toward a new forestry, *Amer. For.*, 95(11–12):37–44, November/December 1989; J. F. Franklin and R. T. T. Forman, Creating landscape patterns by cutting: Ecological consequences and principles, *Landsc. Ecol.*, 1(1):15–18, 1987; E. A. Norse, *Ancient Forests of the Pacific Northwest*, 1990; D. A. Perry (ed.), *Maintaining the Long-Term Productivity of Pacific Northwest Forest Ecosystems*, pp. 82–97, 1990.

Fractals

Fractal descriptions of fluid motions have been recently put forward in several areas. Most notably, fractals have been used to quantify some simple flows that are evolving toward turbulence and also to describe the interface that exists between the turbulent and nonturbulent zones in flows such as jets or boundary layers.

Chaotic flows and strange attractors. The length of a classical curve can be measured by covering it with ever-smaller yardsticks until a yardstick is found that is small enough that the length of the curve remains the same for any measurement with smaller yardsticks. However, the situation is different for fractal curves. Such lines have an ever-increasing length as the length of the yardstick is decreased; a classical example is that of a coastline, discussed below. The manner in which the length of a fractal curve increases with a decreasing yardstick depends on the particular curve. The higher the convolution of the line on small scales, the faster its length increases with a decreasing yardstick. A measure characterizing this dependence is given by an intrinsic parameter called the fractal dimension. A fractal curve, for instance, would have a fractal dimension of any number greater than 1 and up to and including 2 (that of a classical surface). In this case, the curve would be space-filling. The same argument can be made for fractal surfaces, which would therefore have fractal dimensions greater than 2 and up to 3 (that of a volume). *See Fractons.*

Physical systems such as turbulent flows can be represented in a so-called phase space by two systems of variables, one representing, for example, the position vectors and the other the velocity vectors of the fluid particles. Such systems may be chaotic, and in such cases it is known that the orbits (trajectories of points having these variables for coordinates) lie on a geometrical surface called an attractor. For chaotic systems, the attractor has a fractal dimension, and is termed a strange attractor. The notion of the attractor is indissolubly linked to that of energy dissipation. It also can be seen that dissipation leads to a contraction of areas (or volumes) in phase space and that the dimension of the attractor is always lower than that of the phase space in which the flow is represented. In many instances, instead of studying the trajectories in the phase space, their intersections with a plane are investigated. Such a representation is obviously of lower dimension than that of the original phase space (by one) and is referred to as a Poincaré section (or surface of section). *See Chaos; Circuit (electricity).*

In his attempt to study numerically convective motions in the atmosphere, E. N. Lorenz had to solve a system of three nonlinear partial differential equations describing the Rayleigh-Bénard problem of fluid heated from below in a rectangular box. He reduced this system to that of three coupled, nonlinear, first-order differential equations, (1)–(3), in the three

$$x = \sigma y - \sigma x \qquad (1)$$
$$y = -xz + rx - y \qquad (2)$$
$$z = xy - bz \qquad (3)$$

variables *x, y,* and *z*. The parameters σ and r are well known in heat-transfer studies (where they correspond to the Prandtl and Rayleigh numbers), whereas the parameter *b* is a geometric factor. Among the many solutions provided, as these parameters are varied, Lorenz chose to study those corresponding to $\sigma = 10$ and $b = 8/3$ as the control parameter r takes on positive values. The well-known solution corresponding to this particular system is shown in the phase portrait of **Fig. 1**.

The structure of this attractor is highly layered, and its dimension is 2.06. Although the fractal dimension is not a measure of irregularity, it indicates how smooth curves pile upon each other. In this case, the fact that the dimension is slightly greater than 2 (but much smaller than 3) clearly shows that the attractor

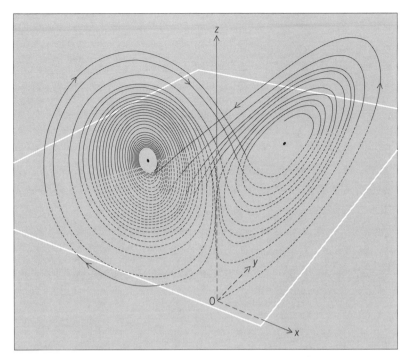

Fig. 1. Lorenz attractor, an example of a strange attractor that describes a complicated flow. *(After W. K. George and R. Arndt, eds., Advances in Turbulence, Hemisphere, 1989)*

more closely resembles a surface than a volume. The orbits (trajectories representing the solutions) are located on this geometrical object, which is layered and ''fuzzy'' but not far from being a standard surface.

All trajectories spiral around one of the two attracting points for some time and then switch to the other in a seemingly haphazard yet deterministic manner. Lorenz noted that such behavior has philosophical and practical consequences for long-range weather forecasting. Indeed, one of the attributes of deterministic chaos is the fact that two trajectories that are initially very close separate exponentially on the average; the difficulty of forecasting the weather is a good illustration of this property. The Lorenz attractor is an example of a strange attractor describing a very complicated although extremely important flow. Many other such examples are known, and more are being discovered and studied. Perhaps this will lead eventually to a unified, nonstochastic description of turbulence, but this stage has not been reached yet.

Turbulent–nonturbulent flow interfaces. High-Reynolds-number free flows (such as jets) or semi-constrained flows (such as boundary layers) present an interface separating the turbulent from the nonturbulent regions. The interface itself, a sharp and highly convoluted surface, is quite thin, of the order of the smallest scales of turbulence; its random motion has been described by many, following the pioneering work of S. Corrsin. This interface separates clearly the two flow regimes, the irrotational flow field on one side (where the turbulent shear stresses are zero) and the fully turbulent one on the other. The interface (called the superlayer by Corrsin) is obviously not a material surface, as can be seen from photographs such as those taken of the eruption of Mount Saint

Helens, Washington (**Fig. 2**); it propagates throughout the flow region. This propagation, by a process akin to nibbling, is partially responsible for the irrotational flow outside the interface becoming turbulent. The rate at which the flow crosses the interface and becomes turbulent is termed entrainment; it is an essential feature of free shear flows, for without it no growth in turbulence would be observed with downstream distance and most of the salient features of such flows would therefore not exist.

Another example of such an interface and the corresponding entrainment is that of diffusion flames (such as the flame of a Bunsen burner), so called because one of the reactants, typically air, diffuses through the interface to react with the injected species (that is, a gas). The role of the convolutions of such interfaces is very important in enhancing the entrainment process.

A section through the interface of this and similar flows would look like the picture of a coastline presented in **Fig. 3**. As mentioned above, the length of the coastline is a function of the divider opening ϵ, and a good model of it is provided by the snowflakelike Koch curve. The fractal dimension of a coastline is easily calculated from this model to be 1.26. Measured values of the dimensions of flow interfaces vary from close to 1 to 1.37. The lower bound has been observed for cumulus clouds that are highly strained by strong vertical winds, and the higher values, more typical of interfaces, have been observed both for large clouds in nature and for laboratory flows.

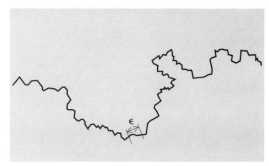

Fig. 3. Typical portion of a coastline, resembling a section of a turbulent–nonturbulent interface.

It is interesting that most flow interfaces have similar fractal dimensions. Perhaps this indicates that, while the large-scale indentations of the interface are mainly responsible for the engulfment of the flow, it is the small scale that is eventually responsible for mixing and subsequent chemical reactions. Although the exact mechanism is not established, models have been postulated that take into account the nonlinear processes involved in turbulence to explain these phenomena. Fractal models have been proposed to understand the physical processes involved at turbulent–nonturbulent interfaces and to reconstruct flow patterns such as clouds. Research in this direction, although promising, is still at a very early stage.

For background information SEE BOUNDARY-LAYER FLOW; DIFFUSION IN GASES AND LIQUIDS; FLAME; FRACTALS; JET FLOW; REYNOLDS NUMBER; TURBULENT FLOW in the McGraw-Hill Encyclopedia of Science & Technology.

Rene Chevray

Bibliography. S. Corrsin and A. L. Kistler, *Free Stream Boundaries of Turbulent Flows*, NACA Rep. 1244, 1954; E. N. Lorenz, Deterministic nonperiodic flow, *J. Atmos. Sci.*, 20:130–139, 1963; S. Lovejoy, Area-perimeter relation for rain and cloud areas, *Science*, 216:185–187, 1982; K. R. Sreenivasan and C. Meneveau, The fractal facets of turbulence, *J. Fluid Mech.*, 173:357–386, 1986.

Fractons

Sound is an excitation of matter that propagates in the form of waves of well-defined frequencies. The corresponding quanta of excitation are called acoustic phonons. These phonons also occur in disordered media, such as gases, liquids, and amorphous solids, provided these materials are effectively translationally invariant at the scale of the phonon wavelength. However, the situation rapidly becomes complicated for media that are appreciably inhomogeneous. There is increased interest in shorter length scales, or in media that are not translationally invariant, such as rocks, compacted powders, sinters of all kinds, and many other porous substances. Sometimes these materials have the same appearance when looked at with widely different magnifications. That property, which is called self-similarity, allows the use of fractal

Fig. 2. Mount Saint Helens eruption, providing an example of an interface separating turbulent and nonturbulent regions of flow. (*Gary Rosenquist, Earth Images*)

geometry and scaling laws for the description of their physical properties. *See* Fractals.

The vibrational excitations of random fractals cannot be propagating waves, but should be highly localized packets. The corresponding quantum is called fracton, a term coined by S. Alexander and R. Orbach, who first worked out the theory of these excitations. By extension, the term has been used to describe the vibrational quanta of regular fractals, such as Sierpinski gaskets, in which case the excitations might be delocalized in view of the high degeneracy of many vibrational modes. By further extension, the term has sometimes been employed to describe localized excitations in disordered systems that are not fractal, or whose fractal nature has not been established. In the following discussion, the properties of fractons in the first, narrower meaning of the term are described.

Numerical simulations. Computer simulations can help to develop a mental picture of the complex processes that occur in disordered systems, and, in particular, of their vibrations. The most important models of disordered systems are the so-called percolation networks. These are randomly decimated nets of sites or bonds such that connections between different regions of a net occur only along sparse and tortuous paths. They have fractal properties. An example of site percolation is shown in **Fig. 1**. In this case, each square represents a mass connected to the other masses by springs. To simplify, a two-dimensional plane net is used here, and it is assumed that the masses can move only perpendicularly to the plane. The model has been thereby reduced to the minimum complexity for the vibration of a percolating network. The number of distinct vibrational modes is very large, equal to the number of occupied sites, 2740 in the case of Fig. 1. The figure shows one of the vibrational modes. This is a motion of the masses that persists in time and space with a well-defined frequency. In other words, it is a particular fracton. To obtain average properties, it is necessary not only to sum over many modes but also to average over many realizations of the percolation network.

The vibration illustrated in Fig. 1 is strongly localized, meaning that it covers a limited spatial extent, of linear size l, without any propagation. A study of all modes indicates that, on the average, modes with higher frequency ω have a smaller spatial extent l. However, in any given region of the net, fractons of all frequencies and sizes can exist. The simulation can in fact be used to derive statistical information, for example, to count the number of modes per unit frequency interval.

Complicated as it is, the simplified model of Fig. 1 remains very different from real experimental systems. The latter are generally three-dimensional. Further, only rarely can the forces be described just by the stretching of springs, a case that is called scalar elasticity. Usually, the forces also include bending and torsion, and elasticity is usually tensorial. Recent simulations have treated some of these more complex situations. So far they have been mostly restricted to

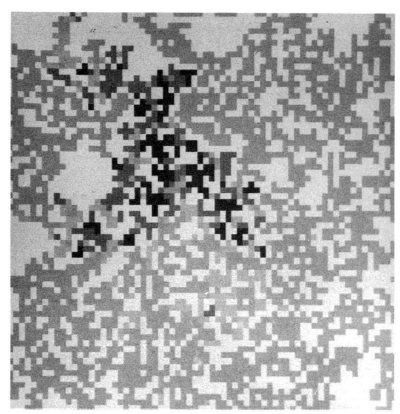

Fig. 1. A model of a percolation cluster, in which the massive points are shown in gray. A single fracton eigenmode is illustrated by the darker shades. The darker the shade, the higher is the corresponding vibrational amplitude of the masses at that point. (*Simulations by E. Stoll*)

percolation networks, although the real fractals that occur in nature are, of course, much more varied.

Scaling laws. The self-similarity of fractals makes it possible to write simple power laws that connect the important physical quantities such as the macroscopic density, the sound velocity, and the low-temperature specific heat. These relations are called scaling laws. They are both remarkable and very useful, given the great complexity of random fractals.

In a solid with a fractal material distribution, the average mass M within a sphere of linear size l, centered on a solid particle, is proportional to a power of l, as given by Eq. (1). Here, D is the fractal

$$M \propto l^D \qquad (1)$$

dimension, which is smaller than 3. Alexander and Orbach have shown that the average frequency ω of fractons of size l is also given by a power law, Eq. (2).

$$\omega^{\bar{\bar{d}}} \propto l^{-D} \qquad (2)$$

The new dimension $\bar{\bar{d}}$ is called the spectral, or fracton, dimension. The number of fracton states in the frequency interval $d\omega$ is also related to $\bar{\bar{d}}$ by Eq. (3). The quantity $\mathcal{N}(\omega)$ is called the density of vibrational

$$\mathcal{N}(\omega)d\omega \propto \omega^{\bar{\bar{d}}-1}d\omega \qquad (3)$$

states. The question of diffusion on the fractal, either diffusion of a particle, such as a charge carrier, or diffusion of heat, is also characterized by an anomalous power law. The diffusion distance l scales with time as in Eq. (4). In the nonfractal case, $\theta = 0$,

$$l \propto t^{1/(2+\theta)} \tag{4}$$

whereas, for a fractal, θ is related to the two dimensions D and $\bar{\bar{d}}$ by Eq. (5). Here, the appropriate value

$$2 + \theta = \frac{2D}{\bar{\bar{d}}} \tag{5}$$

of $\bar{\bar{d}}$ is the one characterizing scalar elasticity, since it is the scalar elastic problem that maps onto the diffusion problem.

Observation. Real materials can be mass fractals only over a restricted range of length scales. At sufficiently large lengths, larger than the fractal persistence length, the materials must become homogeneous to ensure their connectivity. Otherwise, they would collapse. Consequently, at sufficiently low frequencies, phonons always exist in bulk matter. A phonon–fracton crossover occurs at intermediate frequencies where the phonon wavelength becomes of the order of the fractal persistence length. Measurements of wave propagation in porous substances, at wavelengths long enough for the materials to appear homogeneous, have revealed the end of the phonon regime. These experiments do not extend into the fracton regime, since fractons do not propagate.

Information on localized excitations can be obtained by observing how they scatter a beam of light or neutrons. Extensive measurements have been performed, in particular, on silica aerogels, highly porous, monolithic, silica-based materials that probably are the most perfect fractal solids now known.

Both light- and neutron-scattering measurements have been performed on silica aerogels, and have now provided a very detailed understanding of their vibrational properties. A large part of the effort has been devoted to the measurement of the fracton dimension $\bar{\bar{d}}$. It was found that the vibrations should be characterized by at least two different values of $\bar{\bar{d}}$, one associated with bending, $\bar{\bar{d}}_b$, and another with stretching, $\bar{\bar{d}}_s$. In tenuous solids, the elasticity can be dominated by bending motions at low frequencies, and by stretching at higher ones. For the density of vibrational states, this leads to the situation sketched in **Fig. 2**. At low frequencies, up to the phonon–fracton crossover, the usual law of acoustic phonons, Eq. (6), holds.

$$\mathcal{N}(\omega) \propto \omega^2 \tag{6}$$

Just above the crossover, Eq. (7) holds. Here, the

$$\mathcal{N}(\omega) \propto \omega^{\bar{\bar{d}}_b - 1} \tag{7}$$

dimension $\bar{\bar{d}}_b$ can be smaller than 1, in which case the density of states decreases as ω increases. Then an elastic crossover toward a law given by Eq. (8)

$$\mathcal{N}(\omega) \propto \omega^{\bar{\bar{d}}_s - 1} \tag{8}$$

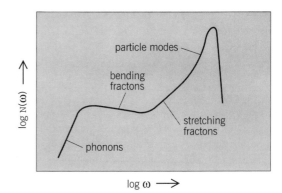

Fig. 2. The density of vibrational states $\mathcal{N}(\omega)$ of a fractal material versus frequency ω. Both variables are plotted on logarithmic scales. Four distinct regimes are shown, as revealed by neutron scattering spectroscopy.

occurs, where $\bar{\bar{d}}_s > \bar{\bar{d}}_b$. Finally, the end of the fracton regime is reached, and is followed by the domain of particle excitations, where $\mathcal{N}(\omega)$ is a sum of contributions in ω and ω^2. The experimental results clearly indicate that $\bar{\bar{d}}$, just like D, is a dimension that characterizes the particular fractal, and that its value does not have a universal character.

For background information SEE DIFFUSION IN GASES AND LIQUIDS; FRACTALS; LATTICE VIBRATIONS; PHONON; SCATTERING OF ELECTROMAGNETIC RADIATION; SLOW NEUTRON SPECTROSCOPY; SOUND in the McGraw-Hill Encyclopedia of Science & Technology.

Eric Courtens; René Vacher

Bibliography. S. Alexander and R. Orbach, Density of states on fractals: "Fractons," *J. Phys. (Paris)*, 43:L625–L631, 1982; E. Courtens et al., Fractons observed, *Phys. D*, 38:41–55, 1989; S. Feng, Crossover in spectral dimensionality of elastic percolation systems, *Phys. Rev. B*, 32:5793–5797, 1985; R. Vacher et al., Crossover in the density of states of fractal aerogels, *Phys. Rev. Lett.*, 65:1008–1011, 1990.

Fungi

Protoplast fusion is a method for obtaining hybrids between strains and species of plants, yeasts, filamentous fungi, and bacteria by circumventing the mating process. The methods for fusion of protoplasts of any sort are relatively simple and are usually less expensive than those methods used in construction of plasmids or artificial chromosomes for transformation of yeast and fungal cells. In protoplast fusion the cell walls are removed enzymatically, and the protoplasts are mixed and fused by either chemical means (polyethylene glycol) or physical means (electrofusion). After fusion and regeneration of the cell wall, the genome of the resulting hybrids may be of two types: in intraspecific hybrids, both partners will probably contribute equally to the hybrid genome (see **illus.**); while in interspecific hybrids, one or the other nucleus will become dominant and the other, subordinate partner will contribute only a small fraction of the genome. The latter type of hybrid will resemble the dominant partner, although it will have a few, possibly very useful, genes from the other. For example, in interspecific fusions of plant protoplasts the only demonstrable genes from the minor partner in the hybrid encode a few isozymes.

Yeasts. The first reports of protoplast fusion in yeasts (*Saccharomyces cerevisiae*) appeared in the mid-1970s. The method was tried for construction of new strains of *S. cerevisiae* for the brewing, wine-making, and bakers' yeast industries. The breeding of new brewing yeasts was not an immediate success since the yeast genome proved to be more complex than was at first thought. However, in Japan it was first used to transfer killer character to sake and winery yeasts in order to immunize the production strain against invasion of the fermentation by wild yeasts. Killer yeasts secrete a toxin to which they are immune but which is lethal to sensitive cells. Protoplast fusion has since been used extensively for construction of new strains of winery and bakers' yeasts, and industrial strains of filamentous fungi. For instance, strains of *Torulaspora delbrueckii*, used in Japan for raising sweet doughs, were fused to yield large polyploid cells that improved recovery during production. Protoplast fusion is also used in Europe and North America for obtaining new strains of brewing yeasts. It has been used widely in research for transferring subcellular organelles, for example, killer viruslike particles or mitochondria, between strains, and for determining the nature of chromosomal rearrangements in hybrid genomes.

Interspecific fusion of yeast species has also been investigated, though the first hybrids obtained were often unstable. Stable fusion products were eventually obtained by maintaining the hybrids on selective media for a longer period of time.

The nature of the events occurring immediately after fusion are still not well understood. At first it was thought that after breakdown of the membranes the cytoplasms fused and then the nuclei fused. The genome of the hybrid was thought to be composed of approximately equal proportions of the genomes of the two original strains. This picture was probably reasonably accurate for intraspecific fusions, but it has since been shown that in interspecific fusion the minor partner contributes only a small part of its genome to that of the hybrid. The genome of the hybrid, therefore, was essentially that of the dominant partner. This has its advantages in construction of industrial yeast strains, since the desirable features of the production strain are retained while a few useful characteristics from the minor strain are incorporated into the genome. The remaining, possibly undesirable genes are eliminated shortly after fusion. Unfortunately, in some circumstances the dominant nucleus is not the desired one.

However, it is possible to select for desirable characteristics in hybrids obtained by interspecific fusions. *Saccharomyces cerevisiae* was successfully fused with *Zygosaccharomyces rouxii*, and

Schematic illustration of the main events in protoplast fusion, including formation of a dikaryotic cell, nuclear fusion (diploid formation), haploid segregation, and transfer of mitochondria.

strains with increased osmotic tolerance were obtained. *Saccharomyces cerevisiae* was also successfully fused with *Hansenula capsulata*, yielding a hybrid with enhanced ability to ferment starch. The hybrids were stable in both cases.

Hybrids of strains of *Candida albicans*, an important pathogen of humans that has no sexual state, have been made by fusion, and genetic analysis has been done by inducing mitotic recombination using low doses of ultraviolet light.

The location of deoxyribonucleic acid (DNA) sequences, integrated into the genome of the recipient strain after fusion, is assumed to be random. Determination of chromosome mobility by pulsed-field gel electrophoresis has shown two types of events: alterations of chromosome mobility due to integration of a sequence into an existing chromosome, and transfer of one or more chromosome-sized DNA sequences.

Mitochondria. Recombination between mitochondrial DNAs is also probable in hybrids obtained by fusion. For example, in the hybrid obtained by fusion of *S. diastaticus* and *H. capsulata*, the numbers and development of cristae in the mitochondria were intermediate between that in the parental strains. In hybrids from fusion of *Aspergillus nidulans, A. nidulans* var. *echinulatus, A. nidulans* var. *latus,* and *A. rugulosus*, considerable recombination was observed among the mitrochondrial DNAs. This was probably because sequence homology among the respective mitochondrial DNAs was high, with slight differences in molecular sizes being due mostly to inserts of additional DNA sequences.

Filamentous fungi. The major difference in techniques of protoplast fusion for filamentous fungi and yeasts is in the enzymes used for removal of the cell wall. The older regions of the fungal cell wall are more resistant to enzymatic dissolution than are the younger regions of growth. To circumvent this problem, young mycelium has been used for production of protoplasts.

Numerous fungal species have been induced to form hybrids by fusion of protoplasts. These hybrids have been used extensively in genetic investigations, as well as for strain improvement. For example, crosses between *Penicillium chrysogenum* and numerous other *Penicillium* species yielded diploids, polyploids, and altered colony phenotypes; crosses of *Cephalosporium acremonium* strains gave hybrids producing higher titers of cephalosporin C; crosses of production strains of *P. chrysogenum* by protoplast fusion gave hybrids with improved characteristics; and mycoviruses have been successfully transferred from infected to uninfected strains of fungi via protoplast fusion.

For background information *SEE CHEMOTAXONOMY; FUNGI; YEAST* in the McGraw-Hill Encyclopedia of Science & Technology.

J.F.T. Spencer; D.M. Spencer

Bibliography. K. N. Kao and M. R. Michayluk, A method for high-frequency intergeneric fusion of plant protoplasts, *Planta*, 115:355–367, 1974; J. F. Peberdy, Fungi without coats: Protoplasts as tools for mycological research, *Mycol. Res.*, 93:1–20, 1989; J. F. T. Spencer et al., The use of mitochondrial mutants in hybridization of industrial yeasts, V. Relative parental contribution to the genomes of interspecific and intergeneric yeast hybrids obtained by protoplast fusion, as determined by DNA reassociation, *Curr. Genet.*, 9:623–635, 1985; J. F. T. Spencer et al., The use of mitochondrial mutants in hybridization of industrial yeast strains, VI. Characterization of the hybrid, *Saccharomyces diastaticus* × *Saccharomyces rouxii*, obtained by protoplast fusion, and its behavior in simulated dough-raising tests, *Curr. Genet.*, 9:649–652, 1985.

Fuzzy sets and systems

Fuzzy sets were introduced in 1965 by L. Zadeh as a new way to represent vagueness in everyday life. They are a generalization of conventional set theory, one of the basic structures underlying computational

mathematics and models. Fuzzy control theory has recently found a wide variety of system applications. This article describes the basic structure of fuzzy sets, gives an example of a simple control system for a pacemaker to illustrate the basic ideas, and describes a few of the latest products that utilize this new technology.

Everyday language is one example of the ways that vagueness is propagated. For example, an instructor who must advise a driving student approaching a red light when to apply the brakes would not say "Begin braking 74 feet from the crosswalk," since this instruction is too precise to be implemented, but rather would say something like "Apply the brakes pretty soon." Imprecision in data and information gathered from and about the environment is either statistical (for example, in a coin toss), wherein the outcome is a matter of chance, or nonstatistical (for example, in the instruction "Apply the brakes pretty soon"). Nonstatistical uncertainty is called fuzziness.

Humans quickly learn to assimilate and use (act on) fuzzy data and instructions (such as "Go to bed about 10"). Accordingly, computational models of real systems should also be able to recognize, represent, manipulate, interpret, and use (act on) both fuzzy and statistical uncertainties. Statistical models deal with random events and outcomes; fuzzy models attempt to capture and quantify nonrandom imprecision.

Sets and membership functions. Conventional (crisp) sets contain objects that satisfy precise properties required for membership. The set of numbers H from 6 to 8 is crisp; it can be written as in Eq. (1),

$$H = \{r \in \mathfrak{R} : 6 \leq r \leq 8\} \qquad (1)$$

where $\mathfrak{R}$ represents the real numbers. Equivalently, H is described by its membership function, m_H, specified by Eq. (2). The set H is shown in **Fig. 1***a*,

$$m_H(r) = \begin{cases} 1 \; ; 6 \leq r \leq 8 \\ 0 \; ; \text{otherwise} \end{cases} \qquad (2)$$

and the graph of m_H in Fig. 1*b*. Every real number r either is in H or is not. Since m_H maps all numbers onto the two points $\{0,1\}$, crisp sets correspond to two-valued logic: is or is not, on or off, black or white, 0 or 1. In logic, values of m_H are called truth values with reference to the question "Is r in H?"

Fuzzy sets, on the other hand, contain objects that satisfy imprecise properties to varying degrees, for example, the set of numbers F that are "close to 7." In 1965 Zadeh proposed representing F by a membership function, say m_F, that maps numbers into the entire unit interval [0,1]. This membership function is the basic idea in fuzzy set theory. The value $m_F(r)$ measures the degree to which r satisfies some imprecisely defined property, and is called the grade of membership of r in F. This construct corresponds to continuous logic; all shades of gray between black and white can be described. In this context, $m_F(r)$ is sometimes regarded as the truth or belief in the proposition that r does indeed satisfy the imprecise property (or properties) defining F.

Since the property "close to 7" is fuzzy, there is no unique membership function for F. Rather, it is left to the modeler to decide, based on the potential application and properties desired for F, what m_F should be like. Properties that seem plausible for this fuzzy set include the following:

1. Normality: $m_F(7) = 1$.

2. Monotonicity: the closer r is to 7, the closer $m_F(r)$ is to 1, and conversely.

3. Symmetry: numbers equally far left and right of 7 have equal memberships.

Given these intuitive constraints, any of the functions shown in Fig. 1c might be useful representatives of F. The function m_{F1} is discrete (the staircase graph); m_{F2} is continuous but not smooth (the triangle graph); and m_{F3} is very smooth (the bell graph). Every number may have some membership in F, but numbers "far from 7," for example, 20,000,987, should not have much. These fuzzy sets may be more useful than the conventional set H. For example, the numbers in H and F might be the heights of basketball players. The information that $m_H(p) = 1$ indicates only that player p's height lies between 6 and 8 ft. On the other hand, the knowledge that $m_F(q) = 0.98$ leads to the inference that q's height is pretty close to 7 ft, and thus this seems to be a more useful piece of information.

Every crisp set is fuzzy, but not conversely. Thus, fuzzy sets generalize (but do not replace) conventional ones.

Fuzzy logic and control. Commercial applications of fuzzy sets center on controllers, devices that automatically adjust the performance of some aspect of a piece of machinery when environmental conditions dictate the need for change. The basic ideas underlying fuzzy logic for control will be illustrated by following the steps in the design of a simplified cardiac pacemaker as the device being controlled (**Fig. 2**).

1. Given a physical system (the heart and pacemaker) with two input variables, T = temperature and B = blood pH, and one output variable, P = pacing rate, the goal is to devise a fuzzy logic controller that adjusts the pacing rate automatically. The idea is to measure the values t of T and b of B at some instant, compute an optimal pacing rate P^* by using fuzzy logic, and then adjust the current pacing rate P by ΔP so that the new pacing rate is given by Eq. (3). The new optimal pacing rate P^* is then fed

$$P^* = P + \Delta P \qquad (3)$$

back into the heart (and also the fuzzy logic unit, for the next comparison cycle), thereby completing compensation for the changes sensed in the input control variables at this instant in time (Fig. 2a). There are two reasons to use fuzzy logic: first, to accommodate measurement errors, but much more importantly, to enable the designer to spread uncertainties about the unknown effect each variable would have on the output (and on other inputs) across several measurement intervals in a fuzzy way.

2. A set of three to seven easily understood linguistic terms that describe each input and output variable

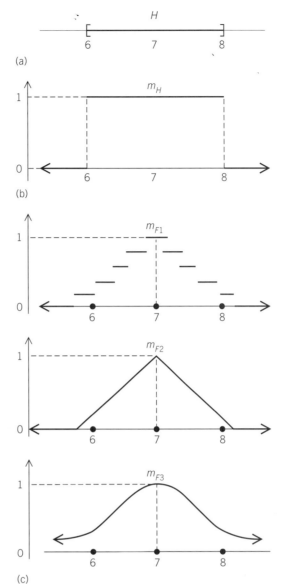

Fig. 1. Comparison of crisp and fuzzy sets. (a) The crisp set H of numbers from 6 to 8. (b) Graph of the membership function of H. (c) Graphs of membership functions of fuzzy sets of numbers "close to 7."

is chosen. Then, membership functions are chosen to represent each linguistic term. The fuzzy membership functions typically (1) peak at central tendencies of domain variables; (2) disperse systemic uncertainties by spreading them across numerical measurement intervals about their peaks; and (3) provide, through different shapes, different ways to weight the effects of imprecisely known domain relationships. Usual choices look like the sets and functions shown in Fig. 2b, c, and d. Different kinds of linguistic term sets are used for different variable domains, for example, {low, medium, high}, or {not alike, somewhat alike, alike, very alike, identical}. The membership functions need not be linear, symmetric, normal, and so forth, but triangular and trapezoidal shapes such as those shown in Fig. 2 are used most often because they are easy to work with and seem to have enough representational power and flexibility for most controller applications.

3. Mathematical operators are chosen for the union

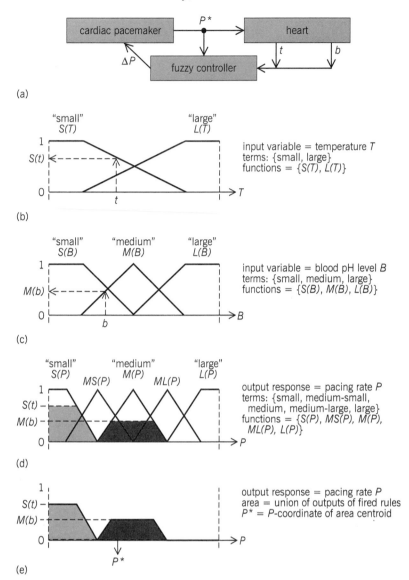

(a)

(b)

(c)

(d)

(e)

Fig. 2. Application of fuzzy logic, approximate reasoning, and control to the operation of a cardiac pacemaker. (a) Diagram of system. P^* = new optimal rate of pacemaker. (b) Term sets and membership functions for temperature, (c) for blood pH level, and (d) for pacing rate, showing truncation of functions. (e) Overall function calculated for pacing rate, and calculation of new optimal rate.

5. At some instant in time, actual values (t and b) are measured for each input variable (Fig. 2b and c). The rules are searched, and every instance is found where a set of conditions (the IF portion of the rule) matches the descriptors at this instant. When a match is found, the rule is "fired" (the THEN portion of the rule is executed) by truncating the appropriate membership function for each output variable (Fig. 2d).

By using the rules and the OR operator, the individual truncations are combined to produce an overall function on the output domain such as the one shown in Fig. 2e. Finally, an output P^* is found by defuzzifying the output area. This is usually done by finding the area centroid and taking the horizontal coordinate P^* of the centroid to be the new optimal pacing rate. P^* is compared to the current pacing rate P, and $(P - P^*) = \pm \Delta P$ is the control signal sent to the pacemaker.

Recent applications. Beginning in 1985 with the subway system in Sendai, the Japanese have introduced a remarkably diverse and growing line of products that use fuzzy technology. There are roughly 2000 Japanese patents on various devices based on fuzzy sets. Japanese products that have some form of fuzzy control include washing machines, vacuum cleaners, cameras, heat pumps, air conditioners, and palm top computers. In the United States, some refrigerators use fuzzy control for defrosting. The application of fuzzy technology to videocassette recorders, brake and transmission controls in cars, and many factory automation techniques is in prospect.

For background information *SEE CONTROL SYSTEMS; LOGIC; SET THEORY* in the McGraw-Hill Encyclopedia of Science & Technology.

James C. Bezdek

Bibliography. J. C. Bezdek, *Pattern Recognition with Fuzzy Objective Function Algorithms*, 1981; G. Klir and T. Folger, *Fuzzy Sets, Uncertainty and Information*, 1988; *Proceedings of the International Conference on Fuzzy Logic and Neural Networks*, Iizuka, Japan, 1990; L. A. Zadeh, Fuzzy sets, *Info. Control*, 8:338–353, 1965.

(disjunction, OR, $\cup$) and intersection (conjunction, AND, $\cap$) of pairs of fuzzy sets. The usual choices for fuzzy sets F and G are given by Eqs. (4) and (5).

$$m_{F \cup G}(r) = \text{maximum } \{m_F(r), m_G(r)\} \quad (4)$$

$$m_{F \cap G}(r) = \text{minimum } \{m_F(r), m_G(r)\} \quad (5)$$

4. A knowledge base of fuzzy rules about relationships between inputs and outputs is defined or acquired. Some typical fuzzy rules for the pacemaker example are as follows:

Rule R1 : IF [T = Small (S)] AND [B is Large (L)]

THEN [P = Medium (M)]

Rule R2 : IF [T = Small (S)] AND [B is Medium (M)]

THEN [P = Medium-Large (ML)]

Galaxy, external

Recent surveys of large numbers of galaxies have revealed inhomogeneities in the distributions of both the underlying mass and the galaxies themselves at the largest scales that have been surveyed. The first section of this article discusses the large-scale distribution of mass and the identification of a very large structure, known as the Great Attractor. The second section discusses the distribution of galaxies in space and, in particular, the discovery of the largest known structure in the universe, named the Great Wall.

LARGE-SCALE DISTRIBUTION OF MASS

Redshifts in the optical spectra of galaxies are interpreted as Doppler shifts caused by the galaxies' motions in an expanding universe; specifically, redshifts are roughly proportional to distance. This provides a relatively easy way to go from the two-dimensional projection of galaxy positions in the night sky to true space dimensions. From redshift studies, as discussed in more detail in the second section of this article, a tapestrylike distribution has been found for the galaxies, including filamentary and sheetlike collections of galaxies surrounding large, nearly empty voids.

Origins of large-scale structures. A widely accepted picture is that such large-scale clumpiness of the universe is the result of gravitational instability. The extreme smoothness of the 3-K cosmic background radiation, which emanates from the epoch of decoupling of matter and radiation about 300,000 years after the big bang, suggests that the universe was much more uniform in its youth. The small density variations that did exist grew in cosmic time as mutual gravitation drew more and more matter into overdense regions while emptying underdense regions. Although the origin of these perturbations is unknown, it has been suggested that quantum fluctuations in the early big bang could have been amplified sufficiently in an epoch of rapid inflation of the size of the universe by some 40 orders of magnitude. This inflationary growth could have been driven by a first-order phase transition involving symmetry-breaking in the primordial material. *See Cosmic background radiation.*

A natural choice for the spectrum of fluctuations in this and other models is the scale-invariant form, so named because the amplitude of a characteristic fluctuation is the same for each mass scale as it first comes within the event horizon. With this input spectrum and the relatively simple physics of gravity, computers are used to create model universes that can be compared with the real universe to study the validity of the physical model. These simulations are most accurate on large scales where density contrasts are of order unity or smaller. Furthermore, on large scales gravity should be the only important mover; the energy released by star formation in young galaxies or the effects of gaseous dissipation should be negligible.

Comparing these numerical simulations with the real universe is complicated by the fact that the models predict not the distribution of galaxies but rather the distribution of the mass of the universe, most of which is now believed to be in a nonluminous, nonbaryonic form. Fortunately, nature has provided a way to map the mass distribution directly; measurements can be made not only of the positions of galaxies in space but also of their motions. Although the major component of a galaxy's motion toward or away from the Earth is from the expansion of the universe, a nearby overdensity of dark matter will pull galaxies, imparting additional motions. Thus, galaxies will stream toward overdense regions and away from underdense regions, revealing the distribution of underlying dark matter directly.

Observations of peculiar motions. The first indication that such peculiar motions might be large came in 1976 with the measurement of an anisotropy in the temperature of the cosmic background radiation, interpreted as a peculiar motion (the technical term for

an additional motion to that of the expansion) of 600 km/s (370 mi/s) for the Milky Way Galaxy. To find the source of this motion and relate it to the distribution of mass in the local universe, it is necessary to make a map of the peculiar motions of galaxies over a large volume of space and look for regions of convergence.

The main component of a galaxy's motion is its participation in the expansion of the universe, called the Hubble flow. Thus, to determine the amplitude of any additional peculiar motion requires a distance measurement (one independent of the redshift distance discussed in the second part of this article) to the galaxy so that the expected expansion velocity in a pure Hubble flow can be removed. Methods currently employed rely on empirical relationships found separately for spiral and elliptical galaxies between the characteristic stellar or gas velocity in the galaxy and its total luminosity. Once the luminosity of a galaxy has been predicted, the distance to the galaxy follows from a comparison with the apparent brightness made with a photometric measurement with a telescope. Spectroscopic techniques are used to measure the characteristic speed: for spirals, the speed of circular rotation of gas in the galaxy (the Tully-Fisher relation); for ellipticals, the virial velocity of the stellar system (the D_n-σ relation).

In 1982 the Tully-Fisher relation was applied to a sample of spiral galaxies in a fundamental study of the

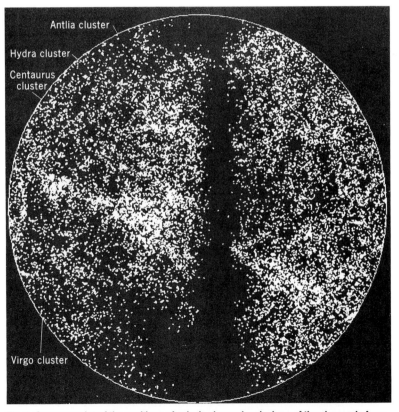

Fig. 1. Computer plot of the positions of galaxies in one hemisphere of the sky, made from galaxy catalogs. The picture is centered on the direction of local streaming. The dark band where no galaxies are recorded is the result of the absorption of light by dust in the Milky Way Galaxy. The large, diffuse feature just below the Centaurus cluster is the overdensity of galaxies associated with the Great Attractor. (*O. Lahav, Cambridge University*)

peculiar motions of galaxies in the Local Supercluster, the region of high galaxy density, roughly 100×10^6 light-years across, in which the Milky Way Galaxy is located. Distances to several hundred spiral galaxies were measured, and the expansion-field component was subtracted from the observed recession velocity, leaving a field of peculiar motions. This study showed a clear pattern of infall of galaxies toward the region of highest galaxy density. (Actually, the Hubble expansion velocity is larger than the infall, so it is more correct to say that expansion of the Local Supercluster is retarded relative to the average expansion rate for the universe.) This pattern confirmed that the higher local density of galaxies does indeed outline a region of higher mass density, and that the effect of gravity over the denser region could be observed. These first measurements of large-scale gravity confirmed that there is at least 10 times as much dark matter as can be observed in the visible galaxies.

Great Attractor. This pioneering study did not resolve the question of the source of the Milky Way's motion of 600 km/s (370 mi/s) with respect to the cosmic background radiation. The component that could be attributed to the relative contraction of the Local Supercluster is only about 250 km/s (150 mi/s), and is in a direction quite different from that implied by the microwave dipole anisotropy. The 1982 measurements suggested that the major component was yet to be found and that the Local Supercluster was moving as a whole toward some more distant point.

A 1988 study of the peculiar motions of more than 400 elliptical galaxies, covering more than 10 times the volume of the 1982 study, found that the motion of the Milky Way Galaxy is shared by other galaxies over a region about 70×10^6 light-years across. However, a gradient in the direction of the Centaurus region was also observed in the peculiar velocities of galaxies in this region, indicating that a mass concen-

tration about 150×10^6 light-years distant is the source of the peculiar motion of the Galaxy and its neighbors. It was suggested that a great supercluster of galaxies and dark matter, named the Great Attractor, is the location of the excess mass density, and the overdensity was identified in galaxy maps of the Hydra-Centaurus to Pavo-Indus region of the southern sky. For example, **Fig. 1** is a computer plot of the positions of galaxies, based on galaxy catalogs, in one hemisphere of the sky centered on the direction of local streaming. The Virgo, Centaurus, Hydra, and Antlia clusters are indicated, but the greatest concentration of light is a larger, more diffuse feature just below the Centaurus cluster, and this overdensity of galaxies is associated with the center of the Great Attractor. The amount of extra mass in the Great Attractor region, over and above what would be found in an average piece of the universe, is roughly equivalent to the mass of 10,000 galaxies, including the dark matter that is thought to surround them.

The extent and amplitude of the flow pattern, which includes galaxies with peculiar motions in excess of 1000 km/s (600 mi/s), was a surprise to theoretical astrophysicists who had made numerical simulations of expected mass distributions in the universe. Fluctuations of this size indicated hills and valleys in the mass distribution that could not be reproduced with popular models (particularly the cold-dark-matter model) of the properties of dark matter and the physical processes of the big bang.

Because of the controversial nature of the result and the strong constraints provided by large-scale flows, more extensive observations have been made and are in progress. A redshift survey confirmed that the galaxies seen in the galaxy maps were between 100 million and 200 million light-years from the Milky Way Galaxy, as predicted in the Great Attractor model. More important, however, are new measure-

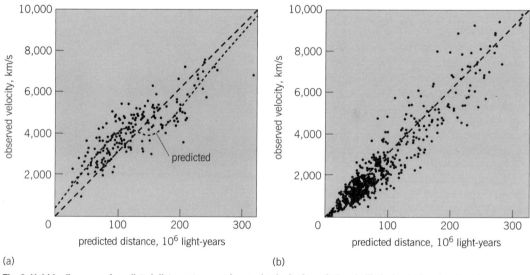

(a)   (b)

Fig. 2. Hubble diagrams of predicted distance versus observed velocity for spiral and elliptical galaxies. A value of 31 km per second per 10^6 light-years or 100 km per second per megaparsec is assumed for the Hubble constant. 1 km = 0.62 mi. (a) Galaxies inside the Great Attractor region. The curve is a prediction based on the Great Attractor model, made before the data were collected. (b) Galaxies outside the Great Attractor region.

ments of peculiar motions. New data for the peculiar motions of 134 elliptical galaxies and 117 spiral galaxies in the Great Attractor region show the same pattern found in the 1988 study, but with a much larger, more extensive sample. The new observations reach the center of the Great Attractor and beyond, so that it is now possible to confirm the point of convergence of the peculiar motions. These remain large and positive out to over 100×10^6 light-years, but are then observed to decline, becoming approximately zero at 150×10^6 light-years distance. Beyond this the peculiar motions are negative: galaxies have a net motion toward the Milky Way. This pattern of peculiar motions, for which the velocities go from positive to negative over a region, is exactly what is expected from the gravitational contraction (or slowed expansion) caused by mass overdensity.

All this is illustrated in **Fig. 2,** showing Hubble diagrams of predicted distance versus observed velocity, for galaxies both inside (Fig. 2a) and outside (Fig. 2b) the Great Attractor region. The S-shaped curve in Fig. 2a is a prediction based on the Great Attractor model before the pictured data were collected. The agreement is good, especially considering that the weight of the original data on which this prediction was based came from galaxies on only one side of the Great Attractor at distances of 70×10^6 to 180×10^6 light-years from the Great Attractor center.

The influence of the Great Attractor is further illustrated in **Fig. 3,** a map of peculiar motions of galaxies in a volume of space generally aligned to the direction of local flow. The Milky Way Galaxy is located at the center of the figure, and peculiar motions toward and away from it are distinguished. Galaxies to the left of the Milky Way have peculiar motions away from the Milky Way and toward the Great Attractor center, which is directly to the left of the Milky Way at a distance of roughly 150×10^6 light-years. Infall to the Great Attractor from the back side can be seen as vectors farther to the left, which point back in toward the Great Attractor center (and also toward the Milky Way). In the shaded triangular regions above and below the Milky Way Galaxy, it is impossible to detect motion toward the Great Attractor because this motion would be across the line of sight and would not produce a Doppler shift. The dominance in these regions of galaxies moving toward the Milky Way shows the tides raised by the Great Attractor, and also the contraction of the Local Supercluster. This contraction is particularly obvious as the abutting of open and solid circles immediately above the Milky Way's position.

In summary, the observations largely confirm the Great Attractor model for the large-scale flow: one large overdensity dominates the velocity field over a region of space roughly 400×10^6 light-years across. This is still a small volume of the universe, much less than 1%, but it does represent a significant reach to the largest structures yet discovered. Though of similar scale to the Great Wall and other prominent superclusters, the Great Attractor is unique in that it is the first large-scale feature mapped in the mass distribution, the more fundamental quantity of large-scale structure.

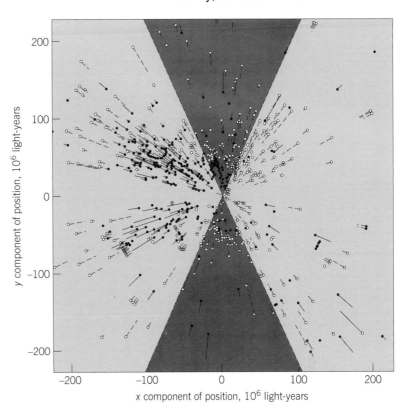

Fig. 3. Map of peculiar motions of galaxies in a volume of space generally aligned to the direction of local flow. Each line represents a galaxy (or the median motion of a group of galaxies), with the circle showing its location in space and the line showing the amplitude of its motion [more precisely, the component of its peculiar motion toward (open circles) or away from (solid circles) the Milky Way Galaxy]. The Milky Way Galaxy is located at the origin at the center of the figure, and peculiar motions toward and away from it are distinguished as indicated in the key. A value of 31 km per second per 10^6 light-years or 100 km per second per megaparsec is assumed for the Hubble constant. 1 km = 0.62 mi.

Because present techniques limit the accuracy of distance measurements to 20% per galaxy, it is difficult to determine peculiar velocities in the more distant Great Wall in order to see if a similar underlying mass structure is present. No doubt there are many other great attractors that are just beyond the reach of present techniques.

Comparison of galaxy and mass distributions. In addition to programs to increase the numbers of galaxies for which distances, and thus peculiar motions, have been measured, astronomers are trying to assemble a better map of the distribution of all galaxies in local universe. A comparison of a map of the galaxy distribution with the map of the mass distribution made from the measurements of peculiar motion will provide crucial information about whether galaxy positions alone can be used to measure the large-scale structure of the universe. The extent to which galaxies do or do not follow the underlying mass distribution will provide important clues to how galaxies formed in the early universe.   *Alan Dressler*

DISTRIBUTION OF GALAXIES IN SPACE

One of the frontiers of astronomy is the exploring and mapping of the universe over distances of hundreds of millions, and even billions, of light-years. These maps can be used to learn about the formation and evolution of the very large patterns in the distribution of galaxies such as the Milky Way.

The enormous size of the universe presents challenges to cosmic mapmakers. Galaxies can be observed only from afar, and only indirect measures of the distances of galaxies from Earth are available. Furthermore, even though 90% (perhaps even 99%) of the matter in the universe is dark, only objects like galaxies that emit light can be used as markers of the structure of the universe. A map of the universe is actually a chart of the positions of galaxies in a three-dimensional space called redshift space.

Redshift surveys. The first step in making a map is the construction of a catalog of the positions of galaxies on the sky. During the 1960s, astronomical plates were examined by eye to catalog the positions of more than 30,000 galaxies in the northern celestial hemisphere brighter than a limiting apparent blue magnitude of 15.7, approximately 10,000 times fainter than the limit reached with the unaided eye. Similar catalogs have been constructed for the southern celestial hemisphere. Together, these catalogs are the basis for maps of the nearby universe; the use of automatic plate scanners to construct catalogs that reach deeper into the universe is an important recent advance.

The next step in mapping the galaxy distribution is obtaining a measure of the radial distances to the galaxies. This makes it possible to go from the two-dimensional map of positions on the sky to a three-dimensional map. Until the 1930s, extragalactic astronomers were limited to the ''flatland'' of maps of galaxy positions on the sky. In 1929, E. Hubble's discovery of the universal expansion opened the way

for exploration of the three-dimensional structure of the universe. The distance to a galaxy is proportional to its apparent radial velocity or redshift. Sources of departure from this simple relation that complicate the interpretation of maps in redshift space were discussed in the first section of this article.

Since the mid-1970s, several groups of astronomers have been measuring large numbers of redshifts of galaxies. Made possible by improvements in both optical and radio detectors and driven by broad scientific goals, this work has produced an increase in the number of galaxies with measured redshifts from less than 1500 in 1975 to nearly 40,000 in 1990. Both optical surveys and radio 21-cm-line surveys have been carried out. These efforts have profoundly changed the present view of the universe. In the 1920s and 1930s, Hubble recognized the existence of rich clusters of galaxies but pronounced the universe sensibly uniform on scales larger than a few million light-years. In the 1930s, the role of clusters of galaxies in the definition large-scale structure was further emphasized, and the existence of superclusters—clusters of clusters of galaxies—was suggested. However, as late as 1970, most cosmologists believed that 80–90% of all galaxies were randomly distributed in space.

In the 1970s and 1980s, three-dimensional redshift surveys dramatically changed this picture. The Local Supercluster became part of the generally accepted picture. The chain in Pisces-Perseus, already apparent in two-dimensional maps of the galaxy distribution, was mapped and shown to be a part of a three-dimensional structure. A void of 4×10^{24} cubic light-years, a region containing very few bright gal-

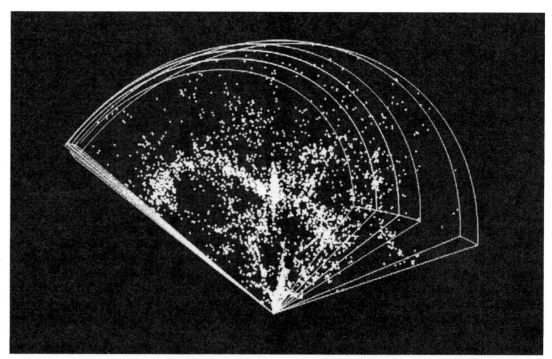

Fig. 4. Four slices of the redshift survey of the Harvard-Smithsonian Center for Astrophysics in the north Galactic cap. The Milky Way is at the apex; each slice extends to a redshift of 15,000 km/s or 9000 mi/s (approximately 500×10^6 light-years radius) and covers a strip of the sky 6° wide by approximately 100° long. Each of the 4000 points represents a galaxy.

axies, was discovered in the constellation Boötes. Shallow but large-area redshift surveys showed that clustering is the rule rather than the exception; there are few, if any, isolated galaxies.

In 1985, the first slice of the extended redshift survey of the Harvard-Smithsonian Center for Astrophysics, which included about 1100 galaxies, was completed. The entire survey will eventually include 15,000 bright galaxies in the northern celestial hemisphere. That high-spatial-resolution galaxy density map revealed remarkable coherent structures that contain most of the galaxies. These structures are surfaces that surround or nearly surround the low-density voids. The structure may be bubble- or spongelike. Clusters of galaxies are embedded in the surfaces.

Great Wall. The distribution of galaxies in eight additional slices has now been mapped, including four in the north Galactic cap. **Figure 4** shows the distribution of galaxies in all four of these slices. This map contains 4000 galaxies and confirms the original suggestion that galaxies are located primarily in thin surfaces. The most remarkable of these is the Great Wall, which extends nearly horizontally across the map. The wall is about one-half to two-thirds of the way between the Earth and the outer (curved) boundary of the survey. This Great Wall is the largest known structure, and it probably extends outside the boundaries of the survey. It contains thousands of galaxies like the Milky Way. Current estimates are that it extends for about 500×10^6 by 200×10^6 light-years but is less than 15×10^6 light-years thick. The Great Wall is probably composed of the surfaces surrounding several adjacent bubbles. Recent very deep redshift surveys suggest that great walls and large voids are common features of the distribution of galaxies.

Future directions. The pattern observed in the distribution of galaxies has profound implications. One of the simplest and perhaps most disconcerting is that the largest voids that can be seen are the largest that can be detected within the limits set by the dimensions of the surveys. There are no direct limits on even larger structures. These enormous dark but not necessarily empty regions are common. It is clear that if the universe could be sampled in volumes 100×10^6 light-years on a side, the distribution of galaxies would differ substantially from one volume to the next. Some volumes (patches) might be devoid of bright galaxies; others might contain many galaxies in one or several thin surfaces. This represents a profound change in the picture of the universe.

For observational cosmologists, one challenge is to see how big a region must be surveyed before one region looks essentially the same as another. The diameters of the largest voids seen so far are about 1% of the radius of the visible universe—still small. If similarly striking voids with much larger diameters are found, it might be necessary to abandon the fundamental idea in cosmology that the universe is homogeneous on some sufficiently large scale. Deeper surveys are necessary to answer this question. Very much deeper surveys of galaxies at large redshifts can

also reveal directly how the maps of the universe looked billions of years ago.

For background information SEE COSMOLOGY; GALAXY, EXTERNAL; HUBBLE CONSTANT; UNIVERSE in the McGraw-Hill Encyclopedia of Science & Technology.

Bibliography. M. Aaronson et al., The velocity field in the Local Supercluster, *Astrophys. J.*, 258:64–76, 1982; J. Cornell (ed.), *Bubbles, Voids, and Bumps in Time: The New Cosmology*, 1988; A. Dressler and S. M. Faber, New measurements of distance to spirals in the Great Attractor: Further confirmation of the large-scale flow, *Astrophys. J. (Lett.)*, 354:L45–L48, 1990; M. Geller and J. Huchra, Mapping the universe, *Science*, 246:897–903, 1989; J. Huchra et al., A survey of galaxy redshifts, IV. The data, *Astrophys. J. Suppl.*, 52:89–119, 1983; D. Lynden-Bell et al., The spectrometry and photometry of elliptical galaxies, V. Galaxy streaming toward the new supergalactic center, 326:19–49, 1988; V. Rubin and G. Coyne (eds.), *Large Scale Motions in the Universe*, 1988; A. Sandage and G. Tammann, *The Revised Shapely-Ames Catalogue of Bright Galaxies*, 1981; M. Strauss et al., A redshift survey of IRAS galaxies. I. Sample selection, *Astrophys. J.*, 361:49–62, 1990.

Gastropoda

Studies focusing on the evolution of plant or animal life-history characteristics are essentially analyses of the economics of energy allocation. An organism obtains a finite amount of energy that must be apportioned among competing demands, such as growth, reproduction, and maintenance. For instance, energy devoted to growth is unavailable to support the production of new offspring. Particular strategies of energy allocation have therefore evolved to maximize an organism's fitness (how well it will be represented in future generations) in a particular environment. In addition, energy must be allocated at a particular time during the ontogeny of an organism. A life history, then, can be defined as the age-specific schedules of growth, reproduction, and survival. Studies of the evolution of life-history characteristics seek to explain patterns of variation in growth rates, age at maturity, number and size of offspring, reproductive effort, and mortality. The goal is to be able to make predictions about which traits might evolve under any particular environmental setting.

A knowledge of life-history variation is important for understanding population dynamics, biotic interactions, and the role of particular species in structuring communities. More importantly, though, the fitness of an organism is intimately tied to its life-history characteristics in that only through reproduction and survival can an organism increase in abundance and persist in an environment. With the environment being altered by humans both locally and globally, it is important to understand how or if organisms will be able to compensate for modification or destruction of their natural habitats. The variation of life histories along environmental gradients provides important insights into the compensatory capabilities and mechanisms of individual species.

A typical life history comprises a suite of characters molded by natural selection to deal with a particular environment. Organisms are often dichotomized with respect to the suite of traits they exhibit into those with rapid development, early reproduction, large reproductive effort, small body size, many small offspring, and short life-spans versus those with slow growth, delayed reproduction, small reproductive effort, large body size, few large offspring, and longevity. It is important to keep in mind that these are relative terms and organisms fall on a continuum.

Two classes of models have been proposed to explain under what conditions each suite of characters will evolve. Stochastic models incorporate variability in the fecundity and mortality schedules induced by environmental fluctuations, whereas deterministic models have fixed fecundity and mortality schedules. The deterministic models predict that high levels of density-independent mortality, wide fluctuations in population density, or repeated episodes of colonization (referred to as r-selected) should select for the former suite of characters, whereas density-dependent mortality or constant population density (K-selected) will favor the latter suite of traits. Similar combinations of life-history characteristics are predicted by the stochastic models but for different reasons. Fluctuating environments that impact juvenile mortality should favor the latter suite of characters, while those affecting adult mortality should favor the former suite of traits.

Wave action as environmental gradient. Environmental gradients represent natural experiments on the evolution of life-history characteristics because the factors that potentially influence these traits typically vary across gradients. For example, wave energies can have a profound and powerful affect on the biology of intertidal and shallow subtidal organisms. The effects can be measured at the individual, population, or community level and may be direct or indirect.

Direct effects are typically the most obvious and include the actual impact of waves; impact of debris, such as rocks and logs, carried by waves; and the continual acceleration and deceleration of the surrounding water medium. In fact, water velocities have been measured that impose drag on organisms that would be comparable to what trees in a forest would experience if they were exposed to a 1000-m/h (1600-km/h) wind. Indirect effects tend to be more subtle and occur through the modification of a number of environmental components. For instance, wave energies are known to alter the efficiencies of predators, the type and abundance of food, competitive abilities, and the physiological stress that organisms experience. Organisms with distributions that span several exposure regimes are likely to experience changes in predation intensity, foraging time, competition, reproductive success, and mortality rates, all of

which may impose important constraints on the suite of life history characteristics that evolve under local environmental conditions.

To examine how this variation might influence life histories, several populations of a common intertidal snail *Nucella lapillus* arrayed along a wave exposure gradient on the coast of Massachusetts were studied. *Nucella lapillus* is an extremely abundant predatory snail with a broad geographic distribution that includes both Europe and North America. Vertically it ranges from near high-water neap tides to just below low-water spring tides and feeds primarily on barnacles and mussels. Mature females deposit egg capsules on the undersides of rocks and in crevices, from which metamorphosed juveniles emerge. The lack of a dispersal stage during the life cycle is crucial in understanding patterns of variation in *N. lapillus* because it implies that gene flow among populations is limited and thus populations can be adapted to local selective pressures.

Patterns of variation. Life history characteristics varied considerably among populations experiencing different levels of wave action. Snails from wave-swept shores grew more slowly and stopped growing at a smaller size than did those from more sheltered shores. The pattern of growth on the exposed shores had two important ramifications for other life histories. First, because the snails grew more slowly, they became sexually mature at a much smaller size than did their counterparts on protected coasts. However, all populations matured in their second year, so no difference existed in the timing of maturation. Second, populations from exposed shores were dominated by small individuals, which is important because fecundity is often correlated with adult size. Thus, the population on an exposed shore as a whole might be less fecund because of its smaller-size adults, but this turned out to be false for *N. lapillus*. Snails on the exposed shore suffered much higher mortality relative to those from more protected shores. The difference in mortality among populations reflected the higher mortality of large snails on exposed shores. The reproductive biology of *N. lapillus* also differed dramatically among exposure regimes. Snails from exposed shores produced about twice as many egg capsules, each of which contained about twice as many offspring relative to snails from more protected shores. The hatchlings emerging from the exposed-shore egg capsules were about 20% smaller, but the exposed-shore snails, because they produced nearly four times as many egg capsules, expended considerably higher reproductive effort.

Similar patterns of life history variation have been observed in several other intertidal snails. For example, *Littorina rudis* from wave-swept shores grew more slowly, matured at a smaller size, and produced more and smaller offspring than did individuals from more protected shores. The same pattern was also found in the fresh-water snail *Lymnaea peregra* from shores differentially exposed to wave action. The congruence of life-history characteristics along gradients of wave energies suggest that these ''strategies'' represent successful solutions to the demographic, physiological, and developmental conditions imposed by wave action. What remains to be addressed is how much of this variation is under genetic control and what environmental factors are responsible for the observed variation.

Environmental and genetic factors. The extent to which this life-history variation is genetic has rarely been considered. *Nucella lapillus* was reciprocally transplanted between exposed and protected shores, and individuals were reared from both exposure regimes under uniform conditions in the laboratory to separate the genetic and phenotypic components of variation in life histories. The effect of the treatment on growth rates indicated that there was virtually no difference in growth rates in laboratory-reared snails, suggesting that differences in growth were not genetic. Moreover, snails grew faster on the protected shore regardless of their origin, confirming the laboratory results and indicating that growth rates are suppressed on wave-swept shores. The results were identical for a similar experiment conducted with three species on the California coast (the limpets *Collisella digitalis* and *C. scabra* and the whelk *Nucella emarginata*). Thus, growth rates, at least, appear to be highly plastic. The suppression of growth on exposed shores may be related to a reduction in foraging time and efficiency. The hydrodynamic forces that develop when waves break on shore limit the foraging time because they force snails to remain in crevices to avoid being dislodged. In addition, wave action reduces feeding efficiency by continually jostling the predator.

Because the size at maturity and the size structure of a population are to a large extent determined by the growth rate, these traits are also likely to be plastic. Surprisingly little is known about the plasticity of reproductive features, although in the fresh-water snail *Physella virgata virgata* they are known to shift in response to the presence of a predator. The higher reproductive effort at the exposed shores is probably favored by natural selection to offset the higher mortality rates. As the theoretical models predict, where mortality is high, especially for adults, energy allocated to reproduction should be high. The difference in energy investment per offspring among snails from different exposure regimes is also likely to be genetic. Snails on protected shores suffer greater levels of predation and physiological stress relative to exposed shores. The greater investment per offspring results in larger hatchlings, which is advantageous under these conditions because larger hatchlings can tolerate physiological stresses better, can survive longer periods of starvation, can choose from a wider array of prey, and are less vulnerable to predation.

For background information *SEE GASTROPODA; POPULATION DYNAMICS* in the McGraw-Hill Encyclopedia of Science & Technology.

Ron J. Etter

Bibliography. K. M. Brown and J. F. Quinn, The effect of wave disturbance on growth in three species of intertidal gastropods, *Oecologia*, 75:420–425, 1988; T. A. Crowl and A. P. Covich, Predator-induced life-history shifts in a freshwater snail, *Science*, 247(4945):949–951, 1990; R. J. Etter, Life history variation in the intertidal snail *Nucella lapillus* across a wave-exposure gradient, *Ecology*, 70:1857–1876, 1989; S. C. Stearns, Life history tactics: A review of the ideas, *Quart. Rev. Biol.*, 51:3–47, 1976.

Genetic engineering

Recent advances in genetic engineering include investigations of heme proteins by using recombinant technology, *Agrobacterium*-mediated plant transformation, and the regeneration of genetically transformed plants.

HEME PROTEINS

Proteins incorporating a heme prosthetic group are central to many biological processes such as electron transfer and oxygen transport. Because of their relative importance and informative and easily measured light-absorbing properties, heme proteins have been extensively studied and characterized. Recombinant deoxyribonucleic acid (DNA) technology has proven to be a powerful tool in these investigations by making possible the exchange of one amino acid in the peptide backbone for another of similar or different characteristics. The changed properties of a variant protein provide clues about the role of the mutated amino acid in the function of the protein as a whole.

Site-directed mutagenesis. Specific amino-acid site-directed mutagenesis is made possible by the ability to synthesize, cut, and connect pieces of DNA. This so-called cut-and-paste processing of genetic material is at the heart of all recombinant manipulations. Genes coding for proteins of interest can be obtained either by molecular cloning techniques or by chemical synthesis. The synthetic approach offers the opportunity to optimize the factors controlling expression in the bacterial host and to engineer unique restriction sites that allow for convenient manipulations of DNA (**Fig. 1**). In the synthetic system, a series of overlapping and complementary oligonucleotides are chemically synthesized and annealed in the organism. The product is ligated enzymatically and cloned into an autonomous plasmid suitable for the host expression cell.

Once the protein gene is in hand, numerous methods exist for altering a specific part of the DNA code. Cassette mutagenesis involves cutting both strands of DNA around the planned mutation site, removing the cut-out DNA, and replacing the strand with new DNA of the desired sequence. Other techniques that do not require convenient restriction endonuclease sites involve annealing a mismatched mutagenic primer to the native DNA, replicating the DNA in laboratory ware, and then selecting for replicated products with the altered sequence.

DNA encoding either the native or mutagenic protein is of use only if the gene can be expressed,

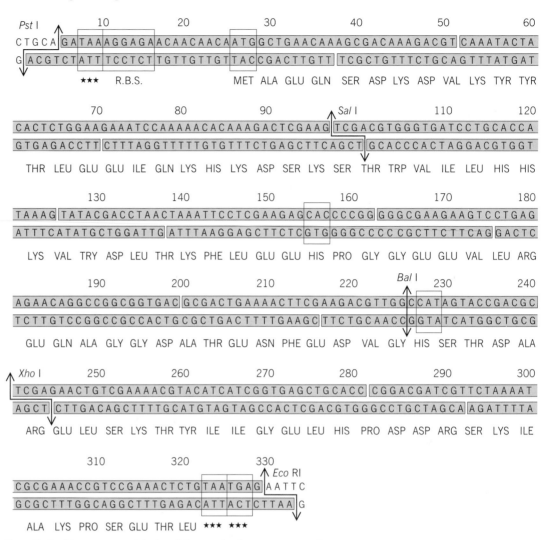

Fig. 1. Nucleotide sequence and design of the gene for the core segment of rat liver microsomal cytochrome b_5. The boxed nucleotide sequences depict the 17 individual, separately synthesized oligodeoxyribonucleotides. The bracketed regions of the sequence indicate, respectively, a stop codon (***) downstream from the lac promoter region of the pUC13 plasmid to prevent the synthesis of a fusion protein, a ribosome binding site (RBS) optimized for the expression of proteins in *Escherichia coli*, the initiator codon (MET), the codons for the axial histidine ligands (HIS), and two terminator codons prior to the *Eco* RI site to halt the synthesis of the soluble form of the cytochrome (***). The arrows show how the unique restriction enzymes *Pst* I, *Sal* I, *Bal* I, *Xho* I, and *Eco* RI cut the gene, allowing for cassette mutagenesis. (*After S. Beck von Bodman et al., Synthesis, bacterial expression, and mutagenesis of the gene coding for mammalian cytochrome b_5, Proc. Nat. Acad. Sci., 83:9443–9447, 1986*)

thus producing the desired enzyme either in the test tube or preferably in a host organism. More and more cell types can be made to carry and express exogenous DNA. At present, numerous bacterial (that is, *Escherichia coli* and *Bacillus subtilis*), yeast, mammalian, and plant cells can be made to produce protein and transfer altered genes to progeny. The overexpression of heterologous proteins in *E. coli* requires conserved features in the 5′-untranslated DNA sequence, including a strong promoter and an efficient ribosome binding site. Whereas cytochrome b_5, sperm whale myoglobin, and cytochrome P450$_{cam}$ (a camphor-metabolizing cytochrome) may be produced in *E. coli* in large amounts as complete enzymes, other proteins such as cytochrome *c* peroxidase are found to be expressed without the heme group, or not at all, like human hemoglobin or myoglobin. A sophisticated expression system for human hemoglo-

bin has been developed based on the production of a fusion protein. In these constructions, the globin sequence is placed after a highly expressed *E. coli* protein. The hybrid protein produced from this construction contains a protease cleavage site just prior to the globin sequence. The hybrid protein is expressed in *E. coli*, and the globin is liberated by subsequent cleavage using trypsin. The apoglobins produced by this method can be reconstituted with heme; they are identical, by a variety of criteria, to native human myoglobin or hemoglobin. By using these techniques, numerous variant proteins have been designed to answer fundamental questions about the structure and function relationships in heme proteins as well as proteins in general.

Heme environment. In all heme proteins, the fifth coordination site of the heme is complexed with a specific amino acid residue of the polypeptide chain.

The sixth site is either bound to an amino acid residue, as in most cytochromes, or reversibly bound by a small molecule such as oxygen. In the five-coordinate position, there are a variety of metal ligands, including cysteine (cytochrome P450 and chloroperoxidase), histidine (myoglobin, hemoglobin, horseradish peroxidase, and others), and tyrosine (catalase). An excellent example of mutagenic perturbation of a sixth heme ligand is the observation that the introduction of a methionine as one of two native histidine ligands in cytochrome b_5, an electron transfer protein, causes the variant to exhibit catalytic oxidative demethylase activity.

The ability to alter active-site amino acid residues that are thought to participate directly in catalysis or ligand binding may be one of the most exciting possibilities of site-directed mutagenesis. In both hemoglobin and myoglobin, the oxygen-binding sites are formed primarily by the heme and the so-called distal histidine. Site-directed mutagenesis has been used to investigate thoroughly the effects of this histidine on oxygen autoxidation, affinity, orientation, and discrimination. In binding of carbon monoxide to sperm whale myoglobin, a surprising finding has resulted from the replacement of the distal histidine with glycine. The observed angle of the heme plane to carbon of approximately 70° for this mutant is in contrast to the current dogma, which predicts an angle of nearly 90° when steric bulk in the form of the distal histidine has been removed.

The control of substrate orientation and mobility by hydrophobic contacts and hydrogen bonding through residues of active sites has been studied extensively in bacterial cytochrome $P450_{cam}$. One such investigation involved the removal of the hydrogen bond between the camphor's carbonyl and a protein tyrosine by mutating the unique active-site tyrosine to a phenylalanine. The predictable result of this mutation is that loosening the contacts between camphor and protein caused deterioration of both camphor's ability to exclude water from the active site and the strict regiospecificity of camphor hydroxylation. This mutant delineated the importance of this specific interaction not only for the regiospecificity of hydroxylation but also for the spin state of the iron center, the linkage of cation and substrate binding, and the interaction of the main protein domains. These results imply that with the judicious choice of engineering strategies, it is feasible to construct cytochromes P450 that afford new products or metabolize novel organic compounds.

Interprotein recognition and binding. Protein–protein recognition and formation of complexes play an essential role in the establishment of biological specificity and optimal enzymatic activity. Site-directed mutagenesis was used in several systems to investigate the role of charged surface residues implicated in stabilizing complex formation. Site-directed mutants where the surface charges of cytochrome $P450_{cam}$ were neutralized and reversed resulted in impairment of the putidaredoxin–cytochrome $P450_{cam}$ association. In contrast, electron transfer rates between cytochrome c and cytochrome c peroxidase are not optimized in the electrostatically stabilized complexes.

A series of mutations of surface charge with the rat liver cytochrome b_5 system has yielded the important experimental observation that individual salt bridges do not necessarily yield the same contribution to complex stability. The significance of this finding was realized through the use of a high-pressure technique, which gives information about volume changes associated with complex dissociation. The observed dissociation volumes agree well with those predicted for the number of salt bridges proposed in the native and mutant complexes. Furthermore, the linear relationship between volume and free-energy changes for these complexes has provided the first direct experimental evidence for the suggested entropic drive resulting from loss of solvent structure at the complex interface.

Complex formations were also studied by using site-directed mutagenesis as a secondary tool to place a functional chemical group at a specific locus on the protein surface. Engineering a cysteine on the surface of cytochrome b_5 made possible the specific labeling of the protein with environmentally sensitive fluorescent probes, allowing for the detection of association without sterically interfering with the process. The utility of this approach is well represented in the characterization of the association reactions of cytochrome b_5 with cytochrome $P450_{cam}$ and metmyoglobin.

A special case of protein–protein interactions studied by site-directed mutagenesis is the allosteric interaction between hemoglobin subunits. Beta chains of human hemoglobin were mutated in order to provide more insight into the mechanisms of the Bohr effect, where a decrease in oxygen affinity is observed at lower pH's.

Electron transfer. The role-specific amino acids in interprotein electron transfer have also lent themselves to study by mutagenic techniques. Specifically, aromatic amino acids have been proposed by virtue of either their aromatic character or orientating properties to facilitate electron transfer. In the cytochrome $P450_{cam}$ system, a series of proteins was made where the aromatic C-terminal tryptophan of the reductant partner of cytochrome $P450_{cam}$ was either deleted or replaced by an aromatic or aliphatic residue. Efficient electron transfer was maintained only with aromatic substitutions.

Stephen G. Sligar; Djordje Filipovic; Matthew D. Davies

AGROBACTERIUM-MEDIATED PLANT TRANSFORMATION

A naturally occurring system exists for transferring deoxyribonucleic acid (DNA) from *Agrobacterium tumefaciens* to plants. Manipulation of this system by recombinant DNA techniques, in conjunction with tissue culture methods, allows transgenic plants to be produced that contain and express genes from widely varying organisms. These introduced genes are passed to future plant generations in the same manner as genes that occur naturally within the species.

Wild-type Agrobacterium tumefaciens. This free-living, gram-negative soil bacterium causes crown gall disease in many plants. It contains a large tumor-inducing (Ti) plasmid that carries the two major regions of extrachromosomal DNA necessary for gall production. These are the transfer DNA (tDNA), which is inserted into the plant nuclear genome, and the virulence (*vir*) region, which encodes the products that mediate tDNA transfer. Phenolic compounds with low molecular weights are released from a plant on wounding, and these act as specific inducers of *vir* gene expression. This leads to single-stranded cleavage within tDNA border sequences and to the production of single-stranded copies of the tDNA. After integration into the plant DNA, the genes encoded in the tDNA are expressed, producing both an uncontrolled proliferation of plant tissue (the gall) and food for the bacterium. This food is in the form of opines, the exact type (for example, nopaline, agropine, or octopine) depending on the *Agrobacterium* strain.

Plant transformation. The tDNA is bordered by simple DNA sequences that are 25-base-pair direct repeats, and any DNA placed between these borders becomes tDNA. Therefore, by using recombinant techniques, the natural tDNA can be deleted and replaced with a different tDNA in the plant genome. Within the *Agrobacterium, vir* products act at a distance; this new tDNA can be on the Ti plasmid, producing a cointegrate vector, or it can remain on a totally separate plasmid, producing a binary vector.

The structural genes that are placed within the tDNA theoretically can be from any organism. However, they must be associated with the DNA control elements that allow them to be expressed in plants, the minimum being a promoter and a terminator. For example, a chimeric gene consisting of a 35S promoter (from cauliflower mosaic virus), a structural element coding for the enzyme neomycin phosphotransferase II (from Tn5 transposon), and a nopaline synthase terminator (from *Agrobacterium* itself) will be expressed in many plant cells. This confers resistance in the cells against the antibiotic kanamycin, which is broken down by the neomycin phosphotransferase II.

Individual plant cells or small numbers of cells can be made to regenerate into whole plants when grown on special media in tissue culture. Transformation of such material at an early stage, followed by regeneration, can result in entire plants wherein each cell contains the introduced DNA. These transgenic plants are capable of passing the introduced genes to their offspring. Such transgenic plants have been produced by cocultivating modified *Agrobacterium* strains with material that is regenerating through somatic embryogenesis, through organogenesis, or from protoplasts, followed by selection for the expression of the introduced genes. In some cases, even cocultivating *Agrobacterium* with germinating seeds has resulted in the production of transgenic plants. One of the most successful transformation methods for solanaceous plants is the leaf disk cocultivation method outlined in **Fig. 2.**

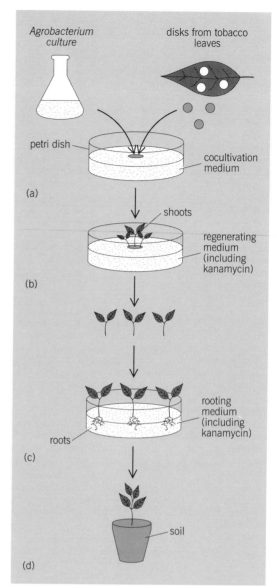

Fig. 2. *Agrobacterium*-mediated production of transgenic tobacco plants via regenerating leaf disks. (*a*) *Agrobacterium* containing tDNA with the gene coding for neomycin phosphotransferase II is mixed with surface-sterilized tobacco leaf disks. (*b*) Cells around the cut edge of the disks are transformed, are hence resistant to kanamycin, and produce shoots on regenerating medium containing the antibiotic. (*c*) Individual shoots are then detached and form roots on rooting medium containing the antibiotic. (*d*) Rooted plants are transferred to soil and further tested to confirm that they are transformed.

Transformation is confirmed by the expression of the introduced genes, the presence of their RNA transcript, Southern blots of plant genomic DNA showing integration of the tDNA, and mendelian inheritance of the introduced trait.

Marker and reporter genes. Transformation is a comparatively rare event; therefore, transformed material must be selected from that which is untransformed. Several selectable marker genes are available that work well in plant systems. Some examples code for neomycin phosphotransferase II, producing resistance to kanamycin; for hygromycin phosphotransferase, producing resistance to hygromycin; and for dihydrofolate reductase, producing resistance to methotrexate.

A reporter gene is one whose product is not normally found in untransformed material; it can be detected easily when the gene is expressed in transformed material. Reporter genes are often placed within the same tDNA as the selectable markers. These code for compounds that are also novel to plant material and can be scored easily. One of the most useful reporter genes is *uidA* from *E. coli*, which codes for β-glucuronidase. With one particular substrate this enzyme will produce a bright-blue precipitate, thus allowing visual identification of those plant tissues that are transformed. With another substrate it produces a fluorescent product that can be quantified, thus allowing determination of the degree of expression of the introduced gene. Perhaps the most spectacular of the reporter genes are those coding for luciferase. Plants that contain one of these genes will emit light when watered with the substrate luciferin.

Intertransformant variability. An introduced gene can show considerably different expression among transformants, with some transgenic plants exhibiting 200 times the expression level of others produced from the same experiment. There is no obvious explanation for this intertransformant variability. Although more than one copy of the tDNA can be inserted into an individual plant cell, the expression of the gene has not been found to increase with the number of insertions. The tDNA is inserted randomly within the plant genome, and it has been suggested that the activity of the introduced DNA is affected by the adjacent host DNA. However, this position effect has yet to be proven. During the transformation process, the tDNA can be truncated or deleted internally, and inverted repeats can be inserted at one or more loci. For example, the right border of the tDNA is more important to the transfer process than the left; therefore, truncations often involve losing the latter end. Again, however, intertransformant variability cannot be linked generally to these effects.

Species transformed by Agrobacterium. *Agrobacterium* is more infectious on dicotyledonous than monocotyledonous plants; transgenic plants can thus be produced from a wide range of dicotyledonous species. Solanaceous plants, including tobacco, tomato, potato, and petunia, are among the easiest to transform. The growing list of other transformable dicotyledonous species includes important crop plants such as alfalfa, cabbage, celery, cotton, flax, lettuce, rapeseed, soybean, and sugarbeet, as well as some trees (poplar and walnut). The transformation of one particular weedy species, *Arabidopsis thaliana*, has also been achieved. This is important, as the very small genome of this species makes it ideal for studies in plant biochemistry and genetics. While monocotyledons include the most important crop plants, the cereals, *Agrobacterium*-mediated transgenic plants have been successfully produced in this group only for asparagus.

Uses of plant transformation. Marker and reporter genes allow only the identification of successful transformation events, but single genes can be introduced that can have a major economic impact on plant growth, quality, and value. Structural genes coding for a desirable trait can be isolated from any organism. Adding the appropriate DNA-regulating elements (promoters, terminators, transit peptide coding sequences, enhancers, and so forth) to these genes before transformation effects the spatial and temporal control of their expression in plants.

Structural genes under the control of constitutive promoters have been used to confer protection from herbicides, viruses, and pests under field conditions. Some nonselective herbicides are highly efficient, rapidly degradable in the soil, and nontoxic to animals. Introduction of genes coding for products that confer resistance to these broad-spectrum herbicides, for example, by breaking down the active ingredient, introduces the necessary immunity to the transformed crop. Introduced genes coding for the coat proteins of viruses confer cross protection to virus infection. For example, coat protein from tobacco mosaic virus expressed in tomato produces partial resistance to infection and retardation of symptom development by both tomato and tobacco mosaic viruses. Expression in plants of the genes coding for a toxin found in *Bacillus thuringiensis*, which attacks the guts of lepidopteran insects, deters feeding and hence reduces damage from such pests.

There are many other examples of useful products in transgenic plants. Genes coding for antibodies and pharmaceutically active peptides have been expressed in plants, making them sources of new natural products with commercial potential. Nutritional quality can also be manipulated, for example, by placing a gene coding for a storage protein with comparatively high methionine into legumes, which are nutritionally deficient in this amino acid. Genes can also be introduced that modify existing biochemical pathways; for example, new flower colors can be produced in petunia. In addition, the expression of undesirable genes that are naturally present in a plant can be eliminated or reduced by introducing DNA that codes for antisense RNA for that gene.

Furthermore, an effective *Agrobacterium*-mediated transformation system provides a means not only for crop improvement but also for the basic study of plant physiology, biochemistry, and molecular genetics. *Agrobacterium*-mediated transformation can be used as a mutation system, inserting tDNA into plant genes and eliminating or altering their expression. It can also be used to introduce reporter genes under the control of different promoters, to determine the timing and location of activity of these controlling elements in plants. In addition, biochemical pathways can be examined by introducing genes to increase or decrease enzyme concentrations.   *Shaun Hobbs*

REGENERATION OF GENETICALLY TRANSFORMED PLANTS

Transformation is the introduction and integration of foreign DNA into the nuclear or organellar chromosomes of an organism. Regeneration in plants is the process by which a fully functional organism is produced from either a single cell, some tissue, an organ, or an isolated embryo. Laboratories are using a variety of techniques to transfer genes of interest into plant cells. Scientists have isolated genes, such as those conferring resistance to environmental stresses, insects, or disease, and transferred them into the cells of model or agronomic species. Then transformed cells and tissues are identified, isolated, and regenerated into mature, fertile plants that express the foreign gene.

Methodologies for plant transformation. There are essentially two methodologies for the transformation of plants: direct DNA transfer and *Agrobacterium*-mediated DNA transfer. Methods for transformation, the cells and tissues most commonly involved, and a representative species that has been successfully transformed and regenerated are listed in the **table.**

Electroporation and liposome fusion are methods for direct DNA transfer that require the use of protoplasts, that is, cells from which the plant cell wall has been removed. Protoplasts may be derived from the digestion of any appropriate source of meristematic, or potentially meristematic, tissue. As soon as the cell wall is adequately removed by using experimentally determined concentrations of cellulases, hemicellulases, and pectinases, the protoplasts are washed, suspended in an appropriate buffer, and transformed.

Electroporation employs two electrodes, the protoplasts, and a DNA solution in a convenient sterile chamber. An electrical discharge from a capacitor temporarily disrupts the protoplast membrane and permits DNA delivery into the cytoplasm. Proteins expressed from the foreign DNA may usually be detected within 24 h. This transient expression of the DNA (that is, transcription and translation of the introduced gene in its extrachromosomal or nonintegrated form) often is used to evaluate protoplast quality and the possibilities for success of the experiment. In various species, calcium and polyethylene glycol treatments have been substituted for electroporation, and they operate in a comparable manner.

Liposome fusion is essentially a directed endocytosis and has become an accepted and well-established technique for delivery of various biological macromolecules [DNA, ribonucleic acid (RNA), protein, antibiotics and other drugs] in both plant and animal systems. Liposomes are artificial vesicles that are compatible with the cell membrane and contain the macromolecules of interest. They fuse with the membrane of the plant protoplast and deliver DNA (or other molecules).

Whereas protoplasts are presently necessary for the successful use of electroporation and liposome technologies, the advantage of the remaining methods for transformation—microinjection, particle bombardment, and *Agrobacterium*-mediated transformation—is that intact cells and tissues can be used. In theory, the intact cell or higher organizational unit is more likely to integrate and express the foreign DNA and to be able to regenerate a fully functional plant.

The advantage of microinjection is exact delivery of a very small volume of fluid (approximately 2 picoliters) containing a defined amount of DNA. Fluorescent dyes are used to evaluate the success of microinjecting a single cell, nucleus, chloroplast, or mitochondrion. The technique is limited by the frequency for stable integration (and subsequent expression) of the foreign DNA into the chromosomes of the recipient cell. Estimates of the frequency of integration vary from 1.0% to 0.0001%. Microinjection was used successfully on embryogenic cells to produce transformed rapeseed.

Particle bombardment, also known as microprojectile technology, has made it possible to insert DNA into cells of nearly any plant tissue, including embryogenic cell cultures, apical meristems, and embryos. The basic apparatus resembles a gun with a shortened barrel that discharges into the plant material tungsten (or gold) particles, 1–5 μm in diameter, coated with DNA molecules. Currently, several different methods are used to accelerate the microprojectiles, from gunpowder to electrical discharge. Providing that the size of the particles is fairly uniform, it is still necessary to optimize the velocity and force with which the particles are accelerated in order to maximize the number of cells of the target tissue that receive the DNA and survive being bombarded.

Agrobacterium-mediated transformation requires the exposure of plant cells or tissues to a disarmed *A. tumefaciens* (or *A. rhizogenes*). This method is discussed in detail in the preceding section.

Identification and recovery of transformants. The efficient identification and recovery of transformed cells and tissues are aided by the use of genes, commonly called selectable markers, which confer an altered phenotype to transformed tissues. In plant systems, the most commonly used marker genes are isolated from bacteria and the altered phenotype is antibiotic resistance. After the transformation procedure is carried out, the plant tissue is cultured on growth media containing lethal concentrations of the antibiotic. Antibiotic selection is a more or less reliable process depending on the plant species and its capability to exclude or alter a particular antibiotic. In plant transformation, the gene for neomycin phosphotransferase (NPT II) is commonly used to select transformants; it confers resistance to the aminoglycoside kanamycin.

Methods of plant transformation

Method	Tissues employed	Transformed species
Direct DNA transfer		
Electroporation	Protoplasts	Tobacco, corn
Liposome fusion	Protoplasts	Tobacco, rice
Microinjection	Protoplasts, embryogenic cells	Tobacco, rapeseed
Particle bombardment	Cell, tissue, or organ	Corn, cotton, soybean
Agrobacterium-mediated DNA transfer	Cell, tissue, or organ	Tobacco, flax, alfalfa, cucumber

Genes for enzymes that deactivate or detoxify herbicides are also being used as selective agents. Some common examples are acetolactate synthase (ALS), which prevents amitosis (the inability of a cell to divide) caused by the herbicide chlorsulfuron; 5-enolpyruvylshikimate-3-phosphate synthase (EPSPS), which confers resistance to the herbicide glyphosate; and phosphinotricin acetyltransferase (PAT), which allows survival in the presence of the herbicide glufosinate. Use of these genes as selectable markers may mean that the regenerated transformant has at least two desirable traits: an agronomically useful or biologically interesting gene and some degree of herbicide resistance as provided by the selectable marker.

Another way to identify transformants is to use genes that will permit visual selection. This can be achieved via formation of a visible or luminescent compound when a substrate is added or through formation of naturally occurring plant pigments. In plants, one of the first widely used markers was the gene for β-glucuronidase (GUS). Transcription and translation of this gene produces the enzyme that reacts with the substrate X-gluc^R (5-bromo-4-chloro-3-indolyl-β-glucuronic acid). Transient or stable GUS expression results in the development of an intense, easily visible blue stain. Unfortunately, this histochemical reaction is lethal to the cells and tissues in which it occurs, and is useful only when the material can be subsampled or sacrificed. In contrast, the firefly or click beetle genes for luciferase elicit bioluminescence in the presence of the substrate luciferin in living cells. This reaction can be detected by using photographic or x-ray film and is easily quantified by using a scintillation counter or luminometer. **Figure 3** compares the transient activity of the firefly and click beetle luciferases in corn protoplasts 24 h after electroporation.

In cells or tissues that lack anthocyanins, regulatory genes such as the R and Bz from corn can activate the production of vacuolar pigments. Transient or stable expression of these marker genes is visible at the cellular level. These genes should prove to be particularly useful in comparing transient expression with stable expression in the same living tissue over time.

Regeneration of transformed plants. In recent years, the process of multiplying or cloning plants has moved into the laboratory. By using sterile techniques and tissue culture methods, very small pieces of the transformed parent plant can be used to generate literally thousands of copies.

Because of the capacity to regenerate, tobacco cells and tissues have been used as a model system in the development and testing of many plant biotechnologies. Tobacco may be regenerated from electroporated or microinjected protoplasts or from bombarded or *Agrobacterium*-infected leaves, leaf explants, or meristems. After transformation with the gene NPT II, tobacco cells or tissues are moved to a medium that suppresses growth of cells that are not expressing the introduced DNA. The medium allows or stimulates transformed cells to regenerate into a plant.

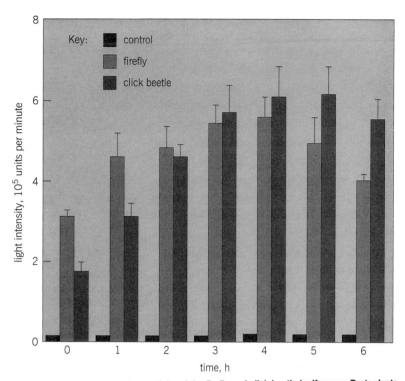

Fig. 3. Comparison of transient activity of the firefly and click beetle luciferases. Protoplasts from Black Mexican Sweet Corn suspension culture were electroporated with plasmids containing either the firefly or the click beetle luciferase gene. Transient activity of the luciferases was assayed 24 h later by adding 100 nanomoles of luciferin to the protoplasts and counting the light units produced per minute by using a scintillation counter. This histogram shows mean and standard deviation from three replicates of 10^6 cells. Control cells were electroporated with DNA lacking the luciferase gene.

During the first 3 weeks of culture, callus forms at cut edges of the explant and shoots emerge from the callus. Once shoots are produced, they may be sterilely excised and placed on rooting medium. Two weeks later, transformed plantlets will have healthy green leaves and rapidly growing roots. Plantlets with low expressing or scrambled NPT II genes will bleach (lose their chlorophyll) and produce roots that grow only across the surface of the medium. By the time the plantlets are 3–4 in. (8–10 cm) tall, they can be placed in soil, hardened off (slowly exposed to drier, cooler air), and sent to the greenhouse.

Agronomic monocots such as corn, rice, and wheat are more difficult to regenerate than dicots. This is partly due to their pattern of determinate growth, in which the apical meristem is used in the production of flowers, fruits, and seeds. To date, regeneration of transformed corn has required the use of protoplasts or embryogenic cell cultures. Embryogenic suspension cultures of A188 hybrid cells have been regenerated into plants following particle bombardment. Transformed cells carrying the gene bar that encodes phosphinotricin acetyltransferase (PAT) were selected 7–14 days after bombardment. Calli were regenerated, and shoots were rooted on hormone-free media.

Prospects. Further optimization of particle bombardment technology and use of visible marker genes should permit transformation and regeneration of other suspension cell cultures, meristems, and embryos. Transformation of tissues that are already organized could shorten significantly the many months presently required to regenerate transformed corn plants. Scientists in plant biotechnology should continue to improve plant transformation systems; methods for gene identification, isolation, and expression; and protocols for regeneration of mature, fertile plants. Advances are expected in the agricultural, food, and pharmaceutical areas. Advances may include (1) supplementation or replacement of native seed proteins by proteins useful as medicines or pharmaceuticals, (2) changing the nutritional qualities of vegetables, fruits, or seeds by altering carbohydrate, amino acid, or lipid composition, and (3) extending stress tolerance in a particular crop species beyond the present genetic limits.

For background information SEE AMINO ACIDS; BREEDING (PLANT); CYTOCHROME; DEOXYRIBONUCLEIC ACID (DNA); ENZYME; GENETIC ENGINEERING; NUCLEIC ACID; OLIGONUCLEOTIDE; PLASMID; PROTEIN; RECOMBINATION (GENETICS); REPRODUCTION (PLANT); RIBONUCLEIC ACID (RNA); TISSUE CULTURE; TRANSPOSONS in the McGraw-Hill Encyclopedia of Science & Technology.

Lynn E. Murry

Bibliography. R. Deblaere et al., Vectors for cloning in plant cells, *Meth. Enzymol.*, 153:277–292, 1987; R. T. Fraley, N. M. Frey, and J. Schell, *Genetic Improvements of Agriculturally Important Crops*, Cold Spring Harbor Laboratory, 1988; W. J. Gordon-Kamm et al., Transformation of maize cells and regeneration of fertile transgenic plants, *Plant Cell*, 2:603–618, 1990; K. Nagai and H. C. Thogersen, Generation of β-globin by sequence-specific proteolysis of a hybrid protein produced in *Escherichia coli*, *Nature*, 309:810–812, 1984; I. Potrykus (ed.), Gene Transfer to Plants—A Critical Assessment: Proceedings of an EMBO Workshop, *Physiologia Plantarum*, 79:1–220, 1990; W. Reznikoff and L. Gould, *Maximizing Gene Proteins*, 1986; C. A. Rhodes et al., Genetically transformed maize plants from protoplasts, *Science*, 240:204–207, 1988; K. K. Rodgers, T. C. Pochapsky, and S. G. Sligar, Probing of macromolecular recognition: The cytochrome b_5–cytochrome c complex, *Science*, 240:1657–1659, 1988; H. Sigel and A. Sigel (eds.), *Metals in Biological Systems*, vol. 25, 1989; S. G. Sligar, D. Filipovic, and P. Stayton, Site-directed mutagenesis of cytochromes P-450$_{cam}$ and b_5, *Meth. Enzymol.*, vol. 206, 1991; P. S. Stayton, M. T. Fisher, and S. G. Sligar, Determination of cytochrome b_5-association reactions, *J. Biol. Chem.*, 263:13544–13548, 1988; R. Walden, *Genetic Transformation in Plants*, 1989; P. Zambryski, J. Tempe, and J. Schell, Transfer and function of T-DNA genes from *Agrobacterium* Ti and Ri plasmids in plants, *Cell*, 56:193–201, 1989.

Geochemistry

Rocks are usually thought of as changing slowly, making rather obvious responses to the influence of their environment. However, they can sometimes have an internal dynamic that allows for their mineral content to become organized in a way that is not a simple response to an imposed pattern. Processes of chemical reaction and of mass and energy transport can interact to give a variety of patterns of mineralization. These patterns take the form of bands, spots, fingers, branching-tree structures, and other more complex spatial patterns, as well as of complex temporal behavior such as sustained periodic or chaotic oscillation of variables in time (**Fig. 1**). This patterning has both scientific and industrial importance.

Geochemical self-organization. Since about 1960, research in far-from-equilibrium systems has demonstrated the possibility for the spontaneous development of spatiotemporal patterning, through positive feedback in reaction-transport networks. The application of the resulting concepts to rocks can provide a unified framework for understanding many phenomena such as repetitive mineral banding and other types of patterning. The diversity and yet the universality of these phenomena make the study of geochemical self-organization a challenging new area of science.

Interest in the physicochemical basis of compositional patterning in rocks dates back to the beginning of the twentieth century with the work of R. Liesegang. Liesegang banding, a term referring to some types of layering in rocks, properly refers to the oscillatory spatial distribution of precipitate content that occurs when coprecipitates are interdiffused. Liesegang's major contribution was to show experimentally that such banding can occur without the mediation of periodically varying boundary or initial conditions for a wide range of coprecipitates. Because banding in this context arises spontaneously, it is an example of self-organization.

Perhaps the most readily understood example of self-organization is the genesis of Bénard convection cells. When a horizontal layer of fluid is maintained at constant temperature along the top and at higher, constant temperature along the bottom, an array of up- and downdrafts spontaneously emerges when the temperature difference between the top and bottom exceeds a critical value. The pattern follows no imposed template, such as a regular array of hot spots along the bottom. The distance between adjacent updrafts depends only on system parameters, in this case, the applied temperature difference, layer thickness, and fluid properties. The dynamics of the system mediated by the unstable density profile breaks the symmetry in the horizontal direction by amplifying omnipresent fluctuations in density, temperature, and velocity into a Bénard pattern of up- and downdrafts.

Transformation processes. Processes that transform rocks include diffusion, dispersion, pore fluid flow, elastic and nonreversible deformation, mineral and pore fluid chemical reactions, and thermal transport. These processes can couple in a number of ways that can lead to reaction-transport feedback, a necessary condition for self-organization. An example of such a feedback (flow self-focusing) is shown in **Fig. 2**. Flow self-focusing can destabilize a planar reaction front. If

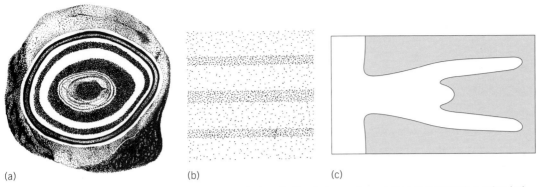

(a)   (b)   (c)

Fig. 1. Schematic view of some of the possible patterns of mineralization that can arise spontaneously through geochemical self-organization. (a) Concentric ring. (b) Banded. (c) Branching-tree.

the permeability in the altered zone is higher than in the unaltered zone downstream from the front, then a bump in the front focuses flow to its tip. But as the front advancement rate increases with the influx rate of the reactive fluid causing the front, the bumps may elongate, completing a positive feedback loop.

Work on chemical reaction-transport systems has shown that there is a wealth of patterning phenomena that can be sustained when such a system is driven out of equilibrium. These include propagating fronts, pulses, periodic wave trains, rotating spiral waves, static patterns, and periodic or chaotic states of temporal oscillation. The theoretical basis of banded or spotted self-organization mediated by reaction-transport systems has been the focus of a great deal of research since 1960. It was shown that the interplay of reaction and diffusional transport can lead to self-organization in systems maintained sufficiently far from thermodynamic equilibrium. *See Chaos*.

Far-from-equilibrium systems. There are various ways in which a geochemical system may be maintained far from thermodynamic equilibrium. Changes in the thermodynamic stability of minerals accompanying changes in pressure and temperature conditions, concomitant with subsidence and upheaval of the Earth's crust, lead to driving forces for reaction. A geological system may also be maintained out of equilibrium by the application of a nonhydrostatic stress or a temperature gradient across it.

Two additional scenarios by which a region of the Earth's crust may be maintained far from equilibrium are analogs of continuously stirred tank and batch reactors, commonly used in chemical engineering. A continuously stirred tank reactor consists of an inlet for reactants, a reaction chamber, and an outlet. Water–rock reaction fronts are similar to such a reactor; an example is an aquifer into which an inlet fluid is continuously being injected. If that fluid is undersaturated with one or more minerals in the aquifer, a zone of reaction advances continuously downstream. In the frame of reference moving with the reaction zone, reactants and products enter from both the up- and downstream sides, a situation similar to that in a continuously stirred tank reactor.

A batch reactor is a chamber into which reactants

are thrust. One natural example is a magma chamber. Both the exchange of mass between growing crystals and remaining molten rock that occurs as the chamber loses heat to its surroundings and the exchange of mass between melt and the walls of the chamber in response to thermal or chemical gradients produce nonequilibrium conditions under which self-organization can commence.

Significance. Geochemical self-organization is of interest from a number of perspectives. From the viewpoint of physics and chemistry, it is both the study of symmetry breaking in systems in which first-order phase transitions are coupled to transport and the analysis of the underlying complex dynamics. Geologically, geochemical self-organization provides alternative explanations for patterned phenomena that, from a more classical viewpoint, appear to have developed by an externally imposed template. The term classical geochemistry is generally restricted to systems that are at or close to equilibrium. From the standpoint of industry, developing an understanding of self-organization can have important implications for mineral and petroleum exploration and engineering, and waste disposal management. From the viewpoints of both the fundamental and the applied sci-

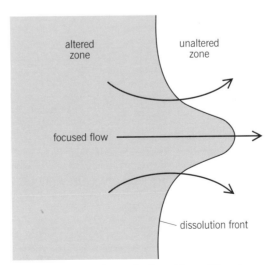

Fig. 2. Diagram showing mechanism of flow self-focusing.

ences, self-organization is important, because it allows a system to develop patterns of mineralization, petroleum, or other entities that are not a direct reflection of the pattern of influences imposed on the system by its environment.

Pattern genesis. It is important to demonstrate that a given patterning phenomenon is an example of self-organization and not simply a manifestation of similar patterns in initial or boundary conditions. To accomplish this, it is necessary to rule out the possibility that any known repetitive process (seasonal deposition rate, sea-level change, astronomical cycle, glacial epoch, biological structure, tectonic event, and so forth) could directly account for the pattern. On the other hand, theoretical or experimental models may be developed to account for the observed patterns. The demonstration of pattern genesis computationally by means of numerical simulation (**Fig. 3**) can yield a powerful, quantitative constraint on the determination of the precise processes thought responsible for a given phenomenon. Figure 3 shows self-organization of bands of enhanced quartz content as simulated by a reaction-transport mechanical model of differentiated layering. Omnipresent small deviations from uniformity are amplified into an intense pattern of alternating quartz-rich and -poor bands through a process of grain growth, migration of molecular species within pores, and reprecipitation elsewhere due to a mechanochemical feedback.

Geological processes often operate over time scales unattainable in the laboratory. Typically, geological studies involve only the end product of a coupling of a number of physical and chemical processes. Therefore, the delineation of the dynamics leading to a given phenomenon is often difficult. Thus, it is often problematic whether or not a given spatial pattern of

mineralization arose through geochemical self-organization.

There are a number of likely examples of patternings that have formed in a broad range of environments from sedimentary to igneous. Taken as a whole, the examples seem to indicate the widespread occurrence of these phenomena. As individual examples, many require extensive future geological and physicochemical research to show that they arose through self-organization. The following list contains some possible examples of geochemical self-organization; it begins with micrometer-scale intracrystal zoning patterns and concludes with kilometer-scale compartmentalization of sedimentary basins.

Oscillatory intracrystalline zoning of plagioclase feldspar, pyroxene, calcite, garnet, and sphalerite
Branching-tree or fractal crystal morphology patterns
Agate patterns
Banded skarns
Banded Mississippi Valley–type ore bodies
Banded redox fronts
Layered igneous rocks
Scalloped, fingered, or branching-tree-shaped reaction fronts arising during weathering diagenesis or metasomatism
Small-scale (microstylolite) diagenetic laminations in chalk
Roughly evenly spaced arrays of stylolites or other dissolution seams in carbonate rocks, sandstones, or coal
Diagenetic bedding in marl/limestone systems
Bands of authigenic phyllosilicate minerals in sandstones
Pressure seals of meter-scale (or greater) thickness composed of banded cements or arrays of stylolites
Layers of enhanced chemically induced compaction alternating with domains of cementation in sandstones
Metamorphic differentiated layering
Concentric shells of alternating light and dark minerals in igneous rocks (orbicules)
Kilometer-scale compartmentation of a sedimentary basin into hydrologically isolated domains with relatively permeable interiors

For background information *see* CHEMICAL EQUILIBRIUM; CHEMICAL REACTOR; GEOCHEMISTRY; OSCILLATORY REACTION; PRECIPITATION (CHEMISTRY); ROCK in the McGraw-Hill Encyclopedia of Science & Technology.

Peter Ortoleva

Bibliography. R. J. Field and M. Burger, *Oscillations and Traveling Waves in Chemical Systems*, 1985; G. Nicolis and I. Prigogine, *Self-Organization in Non-Equilibrium Systems: From Dissipa-*

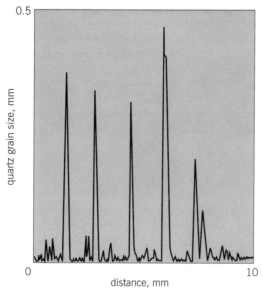

Fig. 3. Graph showing a simulation of pattern genesis. 1 mm = 0.04 in. (*After T. A. Dewers and P. Ortoleva, The self-organization of mineralization patterns in metamorphic rocks through mechano-chemical coupling, J. Phys. Chem., 93:2842–2848, 1989*)

tive *Structures to Order Through Fluctuations*, 1977; P. Ortoleva et al., Geochemical self-organization, I. Feedback mechanisms and modeling approach, *Amer. J. Sci.*, 287:979–1007, 1987; P. Ortoleva et al. (eds.), *Proceedings of the Workshop on Self-Organization in Geological Systems, Santa Barbara, 1988*, Earth Science Reviews, 1990.

Hazardous waste disposal

In situ vitrification (ISV) is a mobile hazardous waste treatment technology that utilizes electricity to melt contaminated solid material (such as soil, sludge, mine tailings, sediments, and asbestos) in place. The technology was conceived in 1980, and since that time, advances have been made in research and development.

In the early stages, the primary interest in the technology was in its application to radioactive waste. As work progressed, it became apparent that in situ vitrification also had applications to hazardous chemical wastes. Therefore, research and development of treatment techniques for both hazardous and radioactive waste have been conducted.

Process. In situ vitrification uses electrodes placed in the contaminated material, usually in a square array. Since waste materials are generally not electrically conductive, a starter path consisting of flaked graphite and glass frit is used (**Fig. 1***a*). When power is applied to the electrodes, the starter path is heated, and the soil attains a temperature of over 1600°C (2900°F); the adjacent contaminated material in contact with the hot starter path begins to melt. Since molten soil is electrically conductive, once the conductive path has been established the molten mass grows down and outward from the electrodes. The melt continues to grow as long as power is applied to the electrodes or until the melt reaches a size that corresponds to the capacity of the vitrification equipment.

The process of vitrification results in a volume reduction of the waste (Fig. 1*b*). This volume reduction results primarily from (1) elimination of pore space in the waste (typically 20–40% in soil), (2) elimination of all volume previously occupied by organic material, and (3) mineral decomposition. Many minerals can participate in decomposition. For example, calcium carbonate ($CaCO_3$; calcite) is decomposed into carbon monoxide (CO), which is incorporated into the melt, and carbon dioxide (CO_2), which is released to the off-gas system. Another example is the dehydration and melting of clays, which results in a density increase of roughly 20%. As a result, the surface of the molten mass at the end of the process is at a lower elevation than the original ground-surface grade.

Once the desired melt mass has been achieved, the electricity is turned off, and clean soil is used to fill the resulting depression. The mass is left in place to cool, resulting in a vitrified (glass) monolith with chemical and physical characteristics similar to those of natural

obsidian (Fig. 1*c*). The process is repeated as a batch sequence until all of the contaminated material has been vitrified.

The process of in situ vitrification utilizes an equipment system that is shown in simplified form in **Fig. 2**. Power is supplied to the unit by a 12.5-kVA or 13.8-kVA three-phase utility distribution system or portable generators. A mobile multiple voltage tap transformer converts the supplied power to two-phase voltage, which, in turn, supplies power to the electrodes. The processing area is covered by an off-gas collection hood in which the flow of air is controlled to maintain a negative pressure (0.5–1.0 in. or 1.25–2.5 cm of water).

A supply of air provides excess oxygen for combustion of pyrolysis products (primarily carbon and hydrogen) and organic vapors, if any, that evolve to the surface during the process. An induced draft system draws these gases from under the hood into the off-gas treatment system, which cleans the gases with (1) cooling, (2) pH controlled scrubbing, (3) dewatering (mist elimination), (4) heating (temperature and dew-point control), (5) particulate filtration, and (6) adsorption by activated carbon. The overall in situ vitrification process is monitored and controlled by a distributed microprocessor system. The entire system is transportable on four standard-size tractor/trailers for over-the-road transport.

Destruction of organic materials. The vitrification process destroys both organic and inorganic compounds by pyrolysis (thermal decomposition of the waste in the absence of oxygen into elemental components) in a strongly reducing environment. Most radionuclides are incorporated into the melt and retained in the solidified mass. Several processes contribute to the destruction of high percentages (commonly 99.995% before off-gas treatment and more than 99.99999% destruction and removal following off-gas treatment) of organic materials contained in the contaminated mass. These processes include molecular diffusion, carrier gas transport, and capillary forces.

Molecular diffusion. When organic materials are vaporized, they behave like any other gas with respect to

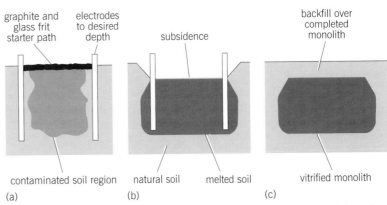

graphite and glass frit starter path

electrodes to desired depth

subsidence

backfill over completed monolith

contaminated soil region

natural soil

melted soil

vitrified monolith

(a)   (b)   (c)

Fig. 1. Application stages of the in situ vitrification process. (*a*) Configuration of electrodes and starter path relative to the contaminated soil region. (*b*) Subsidence and melted soil. (*c*) Vitrified monolith under backfill.

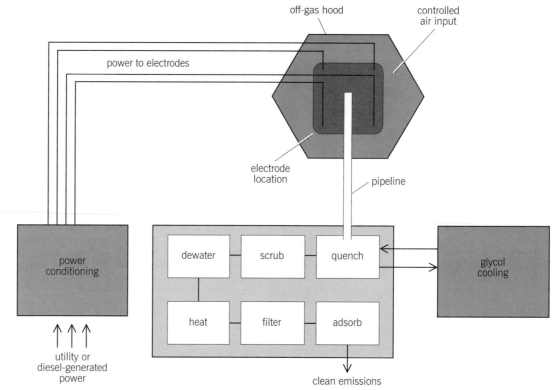

Fig. 2. Diagram of the equipment system for in situ vitrification. A backup off-gas treatment system containing similar equipment on a smaller scale is provided to maintain the process in the event of the failure of the main system.

the random motion of individual molecules. They are free to move in a zone between two temperature boundaries characterized by the vapor-point temperature and pyrolysis temperature. This zone (the dry zone) is located between the melt and the cool soil (**Fig. 3**). When an organic vapor molecule moves

away from the melt, it moves only as far as its vapor-point temperature boundary, because at this point it will condense to a liquid and become virtually immobile. When an organic vapor moves toward the melt and encounters its pyrolysis temperature boundary, it will decompose to its elemental constituents (primarily carbon and hydrogen).

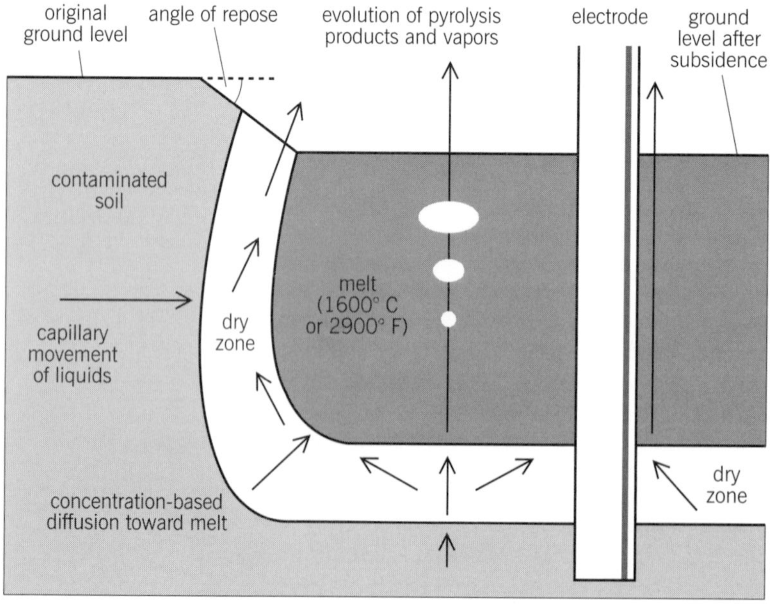

The continual decomposition of organic vapors at the pyrolysis temperature boundary results in a reduction of the concentration of that vapor adjacent to the melt. Gases will diffuse down a concentration gradient (Fick's law of gaseous diffusion). Consequently, the in situ vitrification process results in a net migration of organic vapors toward the melt to their destruction.

Research involving a similar process has also shown that volatile organic compounds will move from outside the heated soil region into the heated region, and from the heated region to the ground surface, there to be captured by the off-gas treatment system. These observations are consistent with the behavior of liquid and vapor organic compounds under exposure to the in situ vitrification process.

Carrier gas transport. Some organic contaminants will be carried by water vapor in a process similar to steam stripping. Water vapor has been shown to move through and alongside the molten zone on its way to the surface. Previous research demonstrated that all of the water in a given treatment volume is recovered by the off-gas treatment system. Most of this water vapor

Fig. 3. Diagram of the response of the contaminant to in situ vitrification.

Results from recent treatability tests

Contaminant	Concentration	Results
Polychlorinated biphenyls (PCBs)	500 ppm	>99.9999% destruction and removal
	3000 ppm	>99.99999% destruction and removal
	19,000 ppm	>99.99999% destruction and removal
Dioxin	100 ppb	>99.99999% destruction and removal
Iron/steel	16 wt %	Successful
Arsenic	27,000 ppm	Passed leach test
Radium-226/radon-222	20,000 pCi/g	Passed leach test; passed radium flux test
Pesticide/mercury mix	300 ppm (pesticide), mercury to 60 ppm	>99.99% destruction and removal; passed leach test
Arsenic	Up to 5% arsenic	Passed leach test
Mercury	Up to 0.5% mercury	97% removal
Radon-226	1352 pCi/g	99.9994% retention*
Uranium (natural)	2.9 µg/g	Passed leach test
Selenium	2260 µg/g	Passed leach test
Lead-210	7269 pCi/g	96.1% retention

*Retention refers to the amount of material retained in the vitrified product and does not include removal efficiency of the off-gas treatment system.

is subjected to temperatures that cause pyrolysis of associated organic materials.

Capillary forces. Capillary forces will move water and organic liquids toward areas of thinner films and lower film tension (that is, toward drier zones). In the in situ vitrification process, continual drying of contaminated material occurs at the vapor-point temperature boundary for the various liquids present (such as at 100°C or 212°F with respect to water). This continual drying effect draws liquids outside the boundary into the vitrification process. Therefore a net migration of liquids, including water and contaminants, toward the melt occurs on a continual basis during the vitrification process.

The movement of vapors and gaseous pyrolysis products in the subsurface during the vitrification process occurs along the path (or paths) of least resistance. Vapors that may move away from the melt will move only as far as their vapor-point temperature boundary, where they condense to the liquid phase. Therefore, contaminated material at the vapor-point temperature boundary has a much lower permeability than the dry material next to the melt. As a result, the vapors have only two paths to the surface: bubbling up through the melt, and transport through the dry zone.

Treatment of inorganic materials. Many rocks (especially igneous and metamorphic rocks) consist of an assemblage of silicates and oxides that are not in chemical equilibrium. As a result, rocks weather and break down into soil.

Even though the minerals once present in rock (and now in soil) have changed into different phases or into different minerals, most of the original rock-forming constituents are still present. If the chemistry of a soil is known, its molten characteristics and the minerals that may crystallize from the melt can be predicted. Since in situ vitrification usually results in rapid cooling, and since rapid cooling inhibits formation of crystals, glass is usually formed in preference to a crystalline mass.

Silicate glass can contain high concentrations of metals and radionuclides while maintaining high chemical durability. Because heavy metals and radionuclides are generally soluble in silicate melts, vitrification is used worldwide for the stabilization of high-level radioactive wastes. Solubility of these materials in glass and the high chemical durability of glass facilitate application of the in situ vitrification process to highly contaminated sites.

It has been demonstrated that glass formed by in situ vitrification is unaffected by extremes of temperature, is not biotoxic, should remain relatively stable for 1 million years, and will pass government tests that measure how fast the contaminants will leach out over time.

Advantages. Numerous tests have been performed on various types of hazardous organic, inorganic, and radioactive wastes. These tests demonstrate the applicability of in situ vitrification to the various forms and types of waste. The **table** presents a summary of recent test results.

In situ vitrification provides several advantages for the treatment of hazardous and radioactive waste. This process permits treatment of mixtures of various types of waste, including organic, inorganic, and radioactive. It is not necessary to excavate the waste prior to treatment, as excavation along with transportation increases health risks. The vitrification process results in the destruction of organic compounds; since they no longer exist, they do not pose any future danger. Toxic metals and radioactive waste are immobilized in a chemically durable glass for geologic time. Rubble in waste (such as rocks and pieces of metal or concrete) does not need to be segregated prior to treatment; these materials will melt along with other waste materials and be dissolved in the resultant glass.

The in situ vitrification process was first applied commercially in full-scale setup in 1990. Efforts to improve the technology and the operational economics are continuing. The technology is currently being used in the United States on National Priority List (NPL) sites identified by the Comprehensive Environmental Response Compensation and Liability Act (CERCLA), more commonly known as Superfund sites.

For background information SEE DIFFUSION IN GASES AND LIQUIDS; HAZARDOUS WASTE DISPOSAL; PYROLYSIS in the McGraw-Hill Encyclopedia of Science & Technology.

Dale M. Timmons

Bibliography. J. L. Buelt et al., *In Situ Vitrification of Transuranic Waste: An Updated Systems Evaluation and Applications Assessment*, Battelle Pacific Northwest Lab. Doc. PNL-4800, Suppl. 1, 1987; J. L. Buelt and W. F. Bonner, *In Situ Vitrification: Test Results for a Contaminated Soil Melting Process*, Battelle Pacific Northwest Lab. Doc. PNL-SA-16584, 1989; H. Dev et al., Field test of the radio frequency in situ soil decontamination process, *Proceedings of the Superfund '88; HMCRI 9th National Conference*, Washington, D.C., 1988; J. C. Green et al., Comparison of toxicity results obtained from eluates prepared from non-stabilized and stabilized waste site soils, *Proceedings of the 5th National Conference on Hazardous Wastes and Hazardous Materials*, Hazardous Materials Control Research Institute, 1988.

Herbal medicine

About 250 years ago, nearly all medicines were natural, and most natural medicines were herbal. Mineral and animal products were also used as medicines, as well as human urine (which contains many important medicinal compounds). Today, with life expectancies almost doubled, an estimated 25% of prescription drugs contain at least one compound derived from, or patterned after a natural product derived from, higher plants. A herbal medicine contains one or more herbs, which are defined as culinary or medicinal plants. Sometimes the definition of herbs is restricted to plants that produce no wood, but many medicinal plants do produce wood, for example, curare and quinine. **Table 1** lists some plants that are important sources of drugs.

Anticancer drugs. In the United States the most valuable crude medicinal export is a quasimedicinal plant, the ginseng (*Panax quinquefolius*); it is exported mostly to the Orient, where the mystical properties of the plant are held in high esteem. Perhaps the second most important economically is the rosy periwinkle, which contains alkaloids that are useful for alleviating several types of leukemias and cancer. Also known as Madagascar periwinkle, it is in widespread use as an ornamental. However, most of the periwinkle used for drugs is grown in Texas.

The second most important anticancer plant is the mayapple, source of the semisynthetic drug etoposide, which is prescribed for small-cell lung cancer and cancer of the testicles. Most of the drug seems to be derived from the Asian mayapple (*Podophyllum hexandrum*), which has apparently become endangered because of this use.

The third most important anticancer plant is the yew. The western yew (*Taxus brevifolia*) is a potentially important source of taxol, the most promising new drug for ovarian cancer; it may also find applications in other types of cancer, such as lung cancer and melanoma. However, several thousand trees are required to produce enough taxol for clinical trials, so in 1990 the National Cancer Institute (NCI) held a symposium in an effort to overcome the supply problem without endangering the species. Plans to preharvest yew trees from western timber tracts scheduled to be clearcut had to be postponed when the spotted owl was placed on the list of endangered species.

Other anticancer drugs that have been developed by the National Cancer Institute over the last 40 years include maytensine from the African tree *Maytenus*; camptothecin from the Chinese "happy tree," *Camptotheca*; cephalotaxine, harringtonine, and homoharringtonine from various plum yews of the genus *Cephalotaxus*; triptolides from the Chinese "yellow vine," *Tripterygium*; and with anolides from various species of husk tomatoes of the genus *Physalis*. Many compounds that initially showed great anticancer activity were found to be highly toxic in clinical trials and were discarded. Taxol itself has several undesirable side effects.

Chemopreventives. In 1990, reflecting a new attitude in the United States toward preventive medicine, the National Cancer Institute launched the Designer Food Program, attempting to formulate cancer-preventive foods (chemopreventives) and test them in clinical trials. In addition to studying the well-known chemopreventive vitamins and antioxidants, the National Cancer Institute will sponsor research involving chemopreventives occurring in such plants as burdock, cabbage, carrot, chili, citrus, clover, cress, flax, garlic, garlic-mustard, licorice, rosemary, and soybean. Isoflavones, protease inhibitors, phytic acid, phytosterols, and saponins, all found naturally in the

Table 1. Some important plant sources of drugs		
Drug	Plant genus	Common name
Xanthotoxin, visnagin	*Ammi*	Ammi
Tetrahydrocannabinol	*Cannabis*	Marijuana
Capsaicin	*Capsicum*	Chili
Chymopapain, papain	*Carica*	Papaya
Vinblastine, vincristine	*Catharanthus*	Periwinkle
Emetine, ipecac	*Cephaelis*	Ipecac
Quinidine, quinine	*Cinchona*	Quinine
Colchicine	*Colchicum*	Autumn crocus
Atropine, scopolamine	*Datura*	Stramonium
Digitalis glycosides	*Digitalis*	Foxglove
Steroids	*Dioscorea*	Wild yam
Ephedrine	*Ephedra*	Ma huang
Cocaine	*Erythroxylum*	Coca
Sterols, isoflavones, protease inhibitors	*Glycine*	Soybean
Codeine, morphine, thebaine	*Papaver*	Opium poppy
Physostigmine	*Physostigma*	Ordeal bean
Pilocarpine	*Pilocarpus*	Jaborandi
Podophyllin	*Podophyllum*	Mayapple
Reserpine, yohimbine	*Rauwolfia*	Indian snakeroot
Taxol	*Taxus*	Yew

soybean, have been discussed as cancer chemopreventives. It is possible that prevention of disease will be of increasing importance in public health not just for cancer but for many of the major causes of death in the United States.

Dietary interventions. Of the more than 2 million deaths that occurred in the United States in 1986, more than 65% were due to ailments that can be prevented by dietary interventions. Heart disease was the most prominent, causing 36.4% of deaths; digitalis [from leaves of the foxglove (*Digitalis purpurea*)] is still a major drug for curing heart ailments, and aspirin (a semisynthetic pharmaceutical, patterned after compounds found in willows and poplars) is being recommended for preventive management. Garlic and onion are recommended to reduce the tendency for blood clots, which often presage heart attacks. The second most serious disease is cancer, causing 22.3% of deaths; and cerebrovascular disease is third, causing 7.1% of deaths. Incidence of both cancer and cerebrovascular disease can be reduced by increasing consumption of vegetables, fruits, and fiber- or vitamin-rich herbs; increasing exercise; decreasing dietary fat; and ceasing smoking. Recent data have shown that xanthines such as theophylline might be useful in treating chronic obstructive pulmonary disease. Pneumonia and influenza are sometimes subject to herbal or phytomedicinal intervention. Adult-onset diabetes mellitus, which is subject to dietary intervention, can be ameliorated with legumes in the diet, and many spices have proven to be insulin-sparing, that is, they may lower the requirements for insulin, thus permitting lower doses. Milk thistle, artichoke, and colchicine are showing promise in cirrhosis and hepatosis; other herbs, such as dandelion and blessed thistle, are being widely recommended as preventives. Atherosclerosis can sometimes be alleviated by dietary and herbal (onion, garlic, ginger) intervention.

Problems of development. While there are still many natural drugs available, and while there seems to be an increasing demand for natural antioxidants, colorants, extenders, medicines and pesticides, most drugs are still synthetic or semisynthetic substances. It takes a very long time and a large investment to develop a new drug and prove it safe and efficacious, and it is not always easy to get patent protection on well-known natural products. Hence it is easier to patent a semisynthetic derivative of a natural product than the natural product itself. New-use patents can be obtained; for example, they have been issued for the use of a couple of old compounds, momordin [from bitter melon (*Momordica charantia*)] and trichosanthin [from Chinese cucumber root (*Trichosanthes kirilowii*)], that seem promising for the treatment of acquired immune deficiency syndrome (AIDS).

Disadvantages and precautions. All medicines, synthetic or natural, can be toxic in large enough doses. Probably all have side effects, positive and negative. In some cases, antidotes to a poisonous drug may be found in the same species that provides the drug. The opium poppy, used as an important medicinal plant for around 5000 years, contains codeine,

Table 2. Some phytochemical antidotes to phytomedicinal poisons

Antidote	Poison
Atropine	Foxglove, larkspur, monkshood, mountain laurel, yellow jessamine, veratrum
Epinephrine	Dieffenbachia
Quinidine	Foxglove, lily of the valley, oleander
Pilocarpine	Atropine, jimsonweed
Milk thistle	Mushroom

which in the United States is probably the second most important analgesic (after aspirin); it also is a very important antitussive compound. The more addictive morphine, with similar properties, can be converted to the illicit and very addictive drug heroin. Overdoses and withdrawal symptoms of heroin are treated with derivatives of yet a third opium alkaloid, thebaine. Another opium alkaloid, papaverine, has been injected as an aphrodisiac; and sanguinarine, also found in opium poppy, is an ingredient in some antiplaque toothpastes. More frequently, the antidote to a phytochemical will be found in an unrelated plant species (**Table 2**). An old and accepted use of a plant-source substance, ipecac (*Cephaelis ipecacuanha*), is to induce vomiting in case of ingestion of a toxin.

Neither self-diagnosis nor unsupervised self-treatment with herbal medicine is advisable. Without proper diagnosis, one might be treating for the wrong ailment. Unproved folk medicines, sometimes harmful and sometimes harmless, may prevent the individual from seeking more reliable and effective treatments. Many safe and salubrious herbs have dangerous look-alikes; for example, even experts sometimes misidentify mushrooms and carrot relatives. Often it is not possible to estimate the quantity of one biologically active compound in a medicinal herb, let alone the dozens of such compounds in it. One plant that looks exactly like another standing next to it may have 10 times as much of the active compound. For example, the therapeutic dose for digitalis is about half the lethal dose; hence, a tenfold variation in the active compound between plants could have fatal consequences.

Prospects. Many drugs will continue to be extracted from plants, purified, and quantified. Some people believe that such natural drugs are safer than synthetic drugs to which human genes and immune systems have not been exposed. However, the immune system is quick to recognize a new compound, be it natural or synthetic. While herbs and herbal medicines can be characterized as natural, it has not been determined whether "natural" is better. Currently, the high price of proving a new drug safe and effective, and the insecurity of patent protection for natural compounds that have been used for a long time, lead to the preference for the semisynthetic or synthetic drug.

For background information SEE PHARMACEUTICALS TESTING; PHARMACOGNOSY; PHARMACOLOGY in the McGraw-Hill Encyclopedia of Science & Technology.

James A. Duke

Bibliography. Council of Scientific and Industrial Research, *The Wealth of India*, 11 vols., 1948–1976; I. N. Dobelis (ed.), *Magic and Medicine of Plants*, 1986; J. A. Duke, *CRC Handbook of Medicinal Plants*, 1985; W. Lewis and M. Elvin-Lewis, *Medical Botany: Plants Affecting Man's Health*, 1977.

Human genetics

The field of human genetics is advancing rapidly. The Human Genome Project, a loose international scientific collaboration formed in 1988, emerged from advances in several technologies that made molecular biology applicable to human genetics on a large scale. New techniques such as the polymerase chain reaction (PCR) have opened up new areas of study. In particular, PCR is being used to study ancient deoxyribonucleic acid (DNA) for archeological applications.

Human Genome Project. The goal of the Human Genome Project is to develop tools to study genetic information by looking at the entire genomes, or significant portions of genomes, of large and complex organisms. The project proposes to map systematically the overall structure of DNA in the chromosomes of humans and other organisms of interest, so that the information can be used to address pressing questions in biology. The genome project is intended to be a means of significantly accelerating biomedical research.

The genome project will produce technologies and organized information. The technologies will provide ways to analyze DNA faster, less expensively, and more precisely than is currently possible. The information will be in the form of genetic linkage maps, physical maps, and DNA sequencing maps. In addition, there are four process goals: to establish databases; to develop new methods and instruments for analysis of DNA structure; to promote medical and other applications of research results; and to anticipate the social, ethical, and legal implications of those applications. These process objectives are not, strictly speaking, science or technology, but rather ways to manage information. The process objectives will thus not be discussed further in this article.

The underlying premise of the genome project is that genetics will be an increasingly powerful way to study human disease and other aspects of biology. The central justification of the genome project is that systematic mapping efforts will, in the long run, reduce costs and expedite research by building a research infrastructure for future investigations. Human geneticists have identified over 5000 genes, usually by studying genetic diseases inherited in families. The genes causing most genetic diseases, however, are unknown. The number of genes in humans is uncertain, but the best estimates are 50,000 to 150,000. More than 90% of human genes are thus still undiscovered. Of the 5000 identified characteristics, genes for fewer than 2000 have been mapped to an approximate location on the chromosomes. Of the

mapped genes, a few hundred have been isolated, cloned, and studied in some detail. The highest-resolution map of a gene is its DNA sequence. However, the DNA sequence of less than 1% of a complete set of human chromosomes (approximately 3 billion DNA base pairs) has been deposited in public databases.

The central scientific problem is to find out how a mutant gene causes disease. Human genetics has, until recently, started from protein and worked back to gene, reversing strategies used for other organisms. The few hundred aberrant proteins associated with genetic diseases were generally discovered by knowing enough about a disease process to guess what protein might be miscoded by a mutant gene. The Mendelian recessive inheritance pattern of sickle cell disease was first established in 1949, for example, and a protein anomaly in the beta-globin protein was independently discovered that same year. It was another 8 years before the amino acid substitution resulting from the sickle mutation was revealed. The fine structure of the human gene itself could be studied only in 1978, after the development of recombinant DNA techniques. Most genetic diseases have, to date, eluded the protein-based gene search because of the difficulty of accurately guessing what protein might be defective in a given genetic disease.

In yeast, bacteria, and other organisms with a rich tradition of genetics, understanding a gene generally starts from studying mutants with a particular phenotype. The path to the gene product begins by mapping the location of a mutation by genetic linkage, then isolating the DNA, identifying some sequences for the gene, and finally finding what the gene produces. The genome project enables researchers to apply the same strategy to find human genes. Human geneticists may now begin to understand the thousands of genetic diseases that have resisted exploration, by using the products of the genome project—a genetic linkage map, a physical map, and abundant DNA sequence information about the human genome as a whole.

Genetic linkage map. The first tool applied in the above strategy is a genetic linkage map. This starts with a set of markers blanketing the human chromosomes that can be used to correlate the inheritance of diseases or traits with the physical location of genes. The map is a collection of such markers ordered relative to one another on the chromosomes. As a disease or genetic trait is passed through members of a family, a genetic linkage map is used to trace the parts of particular chromosomes that are inherited in the same pattern as the disease or trait. The result is a statistical method of linking chromosome regions to the inheritance of genes. The markers themselves may be genes known from previous study whose location on the chromosomes is already known. More often, however, the markers are simply stretches of DNA that differ among individuals.

There are, on average, several million genetic differences between any two people. A few cause differences in appearance or other characters that are intuitively associated with inheritance. Most differences are merely variations that have accumulated over past generations and that have no clinical impact or known function. Geneticists take advantage of such variations as landmarks on the chromosomes. Once these landmarks are mapped to chromosomal locations, associations between the inheritance of such markers can be statistically correlated with how often a disease is inherited. Markers located near the genetic locus of a disease will almost always be inherited with the disease, whereas more distant markers will not.

The first fairly complete genetic linkage map was published in 1987. Genetic linkage mapping in the 1980s greatly helped to narrow the search for genes causing Huntington disease, Duchenne muscular dystrophy, cystic fibrosis, and dozens of other diseases. The 1987 map, however, needed to be made denser, so the process of genetic linkage could be made faster and more informative. The first objective of the human genome project is to refine this map, so that genes can be located with greater precision.

Physical maps. Once a gene is located on a chromosome, the next step is to study the DNA from that region, in search of the precise mutation associated with the disease. This requires a different kind of map, known as a physical map. Unlike linkage maps, distance on a physical map is measured by actual number of DNA base pairs, rather than how often a gene is separated from a DNA marker during meiosis. The order of genes on both kinds of maps is the same, but locating a gene by genetic linkage provides only a rough indicator of how close or far away the gene is from the marker on the chromosome.

Physical mapping of eukaryotic organisms began systematically in the early 1980s, with work on bakers' yeast, *Saccharomyces cerevisiae*, and the nematode *Caenorhabditis elegans*. The idea behind genomic physical maps is to establish sets of cloned DNA fragments containing all the DNA in an organism's genome. These clones can then be analyzed to reconstruct their order, so that the origin of each clone's DNA can be cataloged. With such a physical map, once a gene's approximate location is determined by genetic linkage, obtaining the DNA from that region would simply be a matter of consulting the catalog and selecting clones from the appropriate region. Once the DNA is in hand, search for the gene can commence.

DNA sequencing. Almost all strategies to search for a gene involve extensive DNA sequencing, that is, determining the order of nucleotide bases by using methods developed in the 1970s. Automation of these methods was successfully applied in the early to mid 1980s.

The DNA sequencing aspect of the genome project is the most technically daunting; the proper strategy to accomplish it is fraught with uncertainty. This sequencing is too difficult and too expensive with existing technology, although current instruments pushed to their limits are immensely useful in many sequencing applications. DNA sequencing of entire large genomes, however, awaits further technical developments. Systematic sequencing should begin

once it is an efficient way to catalog chromosomal structure; eventually databases will contain the complete sequence of large regions of the genome.

Use of genetic maps. Genetic factors are important in almost all diseases, particularly the chronic diseases that constitute the major disabling and fatal conditions in the developed world, that is, cancer, cardiovascular disease, Alzheimer's disease, arthritis, diabetes, and mental illnesses. These diseases are much more complex than single gene defects such as sickle cell disease or cystic fibrosis, but it is clear that genetic factors predispose to disease. For example, some people have malfunctioning receptor molecules to clear cholesterol from the bloodstream and are prone to heart disease. Some react against their own insulin-producing cells in response to infection and develop diabetes. Cancer stems from alterations of the DNA in some cells and is in this sense a genetic disease. In some families, Alzheimer's disease appears to be linked to a gene on chromosome 21, in other families to some other gene, and in other cases it may be more significantly influenced by environmental factors. Complete genetic maps are necessary to dissect the various genetic factors involved in these complex disorders.

As the Human Genome Project proceeds, and long after the maps near completion and the project winds down, interpreting the functional meaning of genes will require a much broader range of investigations. There must be parallel work along several fronts to make genetic maps useful. Clinical descriptions must be refined and physiological tests must be correlated to gene function before causal pathways are fully understood.

R. M. Cook-Deegan; J. D. Watson

DNA from ancient tissues. Recently, the technique of polymerase chain reaction (PCR) has been used to extract DNA from human mummies and skeletons. Examining ancient DNA will enable scientists to follow the molecular evolution of human genes through time. Genetic information will also enhance the study of burial patterns, familial relationships, marriage customs, and political organization in archeological populations.

Initial studies. DNA was first isolated from ancient tissues in 1984. The tissue used was muscle attached to skin of a quagga, an extinct zebralike animal. A phylogenetic tree was constructed by comparing DNA sequences taken from the quagga sample with DNA samples taken from a modern horse and two species of zebra. In 1985, DNA was extracted from the soft tissue of an Egyptian mummy about 2400 years old. The DNA was shown to be human by hydrogen-bonding it with an Alu sequence, which is a genetic sequence common in humans but not found in other animals. Since then, researchers have isolated DNA from a wide variety of sources, including fossil plants, woolly mammoths, preserved human brain tissue, and human skeletons.

PCR techniques. PCR generates thousands of copies of genetic sequences from very small samples. This is critical for archeological samples where only small amounts of DNA may be preserved, and in forensic cases where evidence at a crime scene may consist of as little as a single human hair.

PCR is similar to molecular cloning in that it produces the same product, that is, large quantities of specific genetic sequences. However, PCR is a much more rapid technique. While molecular cloning is dependent upon microorganisms growing the DNA sequences, which is a time-consuming practice, PCR uses an automated enzymatic process that exponentially increases the amount of target DNA with each cycle.

PCR is a three-step cycle that may be repeated 20–50 times. First, the DNA is denatured into single strands. Second, primer sequences (short DNA sequences that tell the copying enzyme, or polymerase, where to begin) attach to the single strands. Third, Taq polymerase, a heat-stable enzyme, builds a new strand by using the original DNA as a template. After the cycle is complete and new strands are formed, the process starts over again, with twice as many strands as the previous cycle.

Damage in ancient DNA. The PCR makes the study of ancient DNA feasible because it speeds up the laboratory preparation process and reduces the initial amount of DNA needed. There are some problems unique to ancient DNA that must be solved, however, before these procedures can become widely used.

Ancient DNA is usually damaged during the decay processes that occur immediately after death. The speed with which these processes occur appears to affect the DNA more than the time spent buried in the ground. For this reason, a skeleton thousands of years old may contain better-quality DNA than a skeleton only a few hundred years old. The speed and degree of decay are affected by environmental factors such as climate, soil type, exposure to oxygen, and cultural practices such as artificial mummification.

Current evidence indicates that ancient DNA is heavily oxidized at the pyrimidine nucleotides. The DNA also forms intermolecular links either with itself or with nearby proteins. Researchers have devised several strategies that minimize these problems and allow amplification, but the retrieved DNA still tends to be very small in size. The average length is 100–200 base pairs. It has been suggested that larger fragments may be obtained from bone, but further research is necessary.

Most current research has focused on isolating mitochondrial DNA (mtDNA) from these ancient tissues. Since each of the hundreds of mitochondria in every cell contains its own genetic material, there are hundreds of times as many copies of mtDNA sequences present in ancient DNA as there are copies of nuclear genomic DNA. It is therefore more likely that at least one of the mtDNA sequences will be undamaged by oxidation and crosslinking. Mitochondrial DNA is also useful for study because the entire genome has been sequenced and found to vary considerably between individuals.

Prospects. The potential information to be gained from the study of ancient genes is tremendous. Evolutionary biology will be revolutionized if it becomes possible to routinely isolate DNA from incompletely mineralized fossils. Ancient DNA obtained from more recent species will benefit evolutionary studies by allowing researchers for the first time to study mutation rates in large animals, such as humans, which reproduce much more slowly than small organisms such as yeast and bacteria.

Ancient genes will also provide important information for medical research. For example, the evolution of human parasites that cause infectious diseases such as tuberculosis or syphilis could be traced by isolating parasite DNA from infected mummified human tissues. Also, the relationship between disease and genetics could be better understood by studying gene frequency changes following catastrophic events such as black plague epidemics.

The benefits of this type of analysis to archeologists will be significant, particularly now that it is possible to extract DNA from bone. For instance, genetic information could be used to determine the sex of skeletons more accurately. This is especially helpful when analyzing the skeletons of infants and children because it is currently impossible to identify their sex with the traditional methods based on skeletal changes that occur during puberty. Most importantly, however, this technique will provide the opportunity to study genetic relationships at both the individual and population level. Previously, archeologists attempted to reconstruct genetic relationships by using cranial measurements and other markers which, while partially genetic, were also governed by developmental and environmental factors. Clearly, studying the genes themselves would provide a much more accurate indication of genetic relationships.

When a skeleton is excavated, information is gained from both the skeleton itself and from the context in which the body was buried. Articles, such as jewelry or weapons, buried with the individual, as well as grave type and location often indicate what kind of position that individual held in society. When DNA analysis of a group of individuals is combined with the information derived from the context of their burial, much can be said concerning the social and political organization of the group being studied.

Democracy is a fairly recent development. In the past, rulers, priests, artisans, and warriors often owed much of their social position and political influence to kinship. Even in less complex societies, kinship was important and villages were often organized by lineages, or extended family groups. These relationships can be reconstructed at an archeological site by studying the DNA of the skeletons. Other important information such as marriage alliances between groups can be determined by examining genetic variability in skeletons of the same sex from different sites. Migration patterns can be traced by comparing the DNA of skeletons excavated from numerous archeological sites within a region.

Now that these molecular genetic techniques are available, they have vast potential applications in the field of archeology.

For background information SEE ARCHEOLOGY; GENE AMPLIFICATION; GENETIC MAPPING; HUMAN GENETICS; MITOCHONDRIA; MOLECULAR BIOLOGY in the McGraw-Hill Encyclopedia of Science & Technology.

Sloan R. Williams

Bibliography. N. Arnheim, T. White, and W. E. Rainey, Application of PCR: Organismal and population biology, *BioScience*, 40(3):174–182, 1990; R. M. Cook-Deegan, The Human Genome Project: The formation of federal policies in the United States, 1986–1990, *Biomedical Politics*, 1991; R. Higuchi et al., DNA sequences from the quagga, an extinct member of the horse family, *Nature*, 312:282–284, 1984; S. Pääbo, Molecular cloning of ancient Egyptian mummy DNA, *Nature*, 314:644–645, 1985; M. Pines, *Mapping the Human Genome*, Occas. Pap. 1, Howard Hughes Medical Institute, Bethesda, December 1987; U.S. Department of Health and Human Services and U.S. Department of Energy, *Understanding Our Genetic Inheritance: The First Five Years, FY 1991–1995*, DOE/ER-0452P, April 1990; J. D. Watson, The Human Genome Project: Past, present, and future, *Science*, 248:44–49, 1990.

Hydrogen

Under normal conditions, hydrogen forms a gas consisting of diatomic molecules (H_2) with strong covalent bonds. At low temperature or at high pressure, hydrogen may be solidified to form an insulating (nonconducting) molecular solid. Under very high pressures above 100 gigapascals (1 megabar), the material is predicted theoretically to transform first to a conducting molecular solid (molecular metal) and then to a monatomic metal. The formation of a metallic modification of hydrogen was first predicted over a half century ago. More recent theoretical work has indicated that the material may possess very unusual properties, primarily as a result of its light mass. It may be a very-high-temperature superconductor and may have a liquid ground state (that is, at lowest possible temperatures). Finally, metallic hydrogen is likely to be the major component of the large gaseous planets, Jupiter and Saturn, so it is believed to be the most abundant metal in the solar system.

Development of the diamond-anvil cell during the past several years has permitted considerable progress in the experimental study of hydrogen at very high pressures. In 1985 the first successful optical observations and Raman scattering measurements of hydrogen at calibrated pressures above 100 GPa were carried out. This study documented that the molecular solid is stable at room temperature to at least 150 GPa. Work in 1988 at lower temperatures (~77 K or −321°F) showed that samples could be contained to higher pressures with cooling of the diamond-anvil cell. In these investigations the molecular solid was found to be stable to pressures above 200 GPa, where the density ρ is approximately 10 times that of the low-pressure solid ($\rho/\rho_0 \sim 10$). Further, a phase transition was observed spectroscopically in the molecular solid at about 150 GPa and 77 K (−321°F). Recent optical measurements have shown that the transition from the insulating to the metallic molecular form occurs in this pressure range, and that the low-temperature transition exhibits unusual critical phenomena.

Solid hydrogen. Hydrogen crystallizes as an insulating diatomic molecular solid at high pressure (5.4 GPa at room temperature) or at low temperature (14 K or −434°F at room pressure). The solid consists of fully rotating molecules with the covalent bonds intact. With increasing pressure the molecules are brought closer together, and interactions between molecules are increased. At some critical density, however, the molecular bonds are expected to no longer be stable, and the system will attain a lower free energy and a higher density if the molecules dissociate. In the very high-pressure state, therefore, the material is believed to be monatomic. With one electron per atom, there would be a half-filled band, which would make the material a metal. In this state, hydrogen may have properties similar to the alkali metals, which are just below hydrogen in the periodic table. The possibility of forming metallic hydrogen under pressure was first examined theoretically in 1935 by E. Wigner and H. B. Huntington. Calculations carried out in the late 1970s suggested that the insulator–metal transition may occur within the molecular solid below 200 GPa and that molecular dissociation occurs at higher pressures.

Demonstration of the formation of metallic hydrogen in the laboratory has been hampered by both the difficulty of generating sufficient pressure and the inadequacy of techniques used to characterize the state of the hydrogen, including accurate measurements of the pressure and stress conditions on the sample. Development of techniques based on the megabar diamond-anvil cell has permitted this problem to be examined in new ways. A variety of spectroscopic techniques have been developed for measurement of optical, vibrational, and structural properties of materials at very high pressures with this device. X-ray diffraction measurements have demonstrated that pressures in excess of 300 GPa can be achieved, and a calibrated pressure scale has been established. X-ray diffraction measurements on hydrogen itself have indicated that the solid crystallizes in the hexagonal close-packed (hcp) structure and that the solid retains this structure to at least 30 GPa ($\rho/\rho_0 \sim 5$).

Raman spectroscopy. Vibrational Raman spectroscopy, an inelastic light-scattering technique, has been a useful probe of the state of bonding in dense solid hydrogen since work on the high-pressure properties of the material up to 60 GPa was carried out in 1980. These studies indicated that, although the solid remains molecular over this pressure range, measurements of the H-H stretching vibration (vibron) indicate that a significant weakening of the molecular bond begins at pressures above 30 GPa. Recent work indicates that this trend continues up to pressures above 200 GPa. Measurements carried out at 77 K

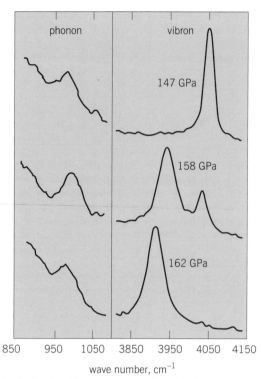

Fig. 1. Vibrational Raman spectrum of solid hydrogen at 77 K (−321°F) and at various pressures, showing the discontinuous change in the frequency of the H-H stretching vibration (vibron) at ~150 GPa.

(−321°F) showed that there is discontinuous shift of the vibron frequency at 150 GPa (**Fig. 1**). This observation is evidence for a phase transition in the molecular solid. Subsequent measurements carried out on deuterium indicated that the sharpness of the transition and the isotope effect on the transition pressure were consistent with an electronic change. Measurements of the temperature dependence of the spectrum have revealed very unusual behavior. The magnitude of the vibron discontinuity at the transition decreases significantly with increasing temperature and eventually disappears. This observation provides evidence for the disappearance of the phase transition at a critical point, as discussed below. The low-frequency vibrational spectra reveal information on the crystal structure and rotational states of the molecules. A change in the rotational transitions with increasing pressure suggests restricted rotational motion. A low-frequency lattice vibration (called an optical phonon), which is characteristic of the hexagonal close-packed structure, has also been measured (Fig. 1). The results suggest that this structure (or a closely related structure) is preserved in the high-pressure phase.

Infrared reflectivity. Measurement of infrared reflectivity is a direct test of metallic behavior. The reason is that the overlap of the valence and conduction bands will result in free carriers that give rise to intraband absorption and reflectance extending into the long-wavelength (or low-energy) region of the spectrum. In 1990, a series of infrared measurements carried out on hydrogen at room temperature showed

evidence for metallic behavior at pressures above 150 GPa (**Fig. 2**). In each experimental run a systematic increase in reflectivity in the infrared region below 1 eV was clearly evident, with very little change in the visible range (1.5–2.5 eV). The high-pressure spectra can be fitted with a simple free-electron model. An analysis of the pressure dependence of the spectra gives a transition pressure of 149 ±10 GPa at room temperature. The results are consistent with metallization by band-gap closure in the molecular solid. The solid has the unusual property that it remains essentially transparent as it becomes metallic. Under these conditions, the material initially has a relatively low charge-carrier density (conductivity) and should be considered a semimetal like graphite, bismuth, or antimony.

Phase diagram of solid hydrogen. The evidence for metallization at 295 K (71°F) and a critical point in the transition at low temperature has important consequences for the assignment of the transition. The very important constraint in this regard is the evidence from the lattice-mode spectra for the lack of any major crystallographic structural changes accompanying the transition. N. F. Mott has shown that band overlap may be accompanied by a discontinuity in volume and conductivity. Such discontinuities are expected to terminate in a critical point at finite temperature. The appearance of a discontinuity, the lack of crystallographic transition, and the evidence for critical behavior, together with the measurement of infrared reflectivity at higher temperature, thus fit an insulator–metal

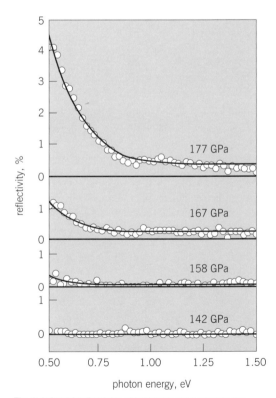

Fig. 2. Infrared reflectivity of hydrogen measured in the diamond-anvil cell at 295 K (71°F) and at a series of pressures. The curves are fits to the spectra with a simple free-electron model.

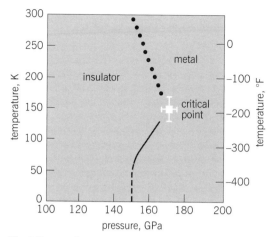

Fig. 3. Proposed pressure–temperature phase diagram for hydrogen in the vicinity of the insulator–metal transition.

transition. It is possible that a rotational ordering transition, in which the molecules become aligned, may occur at the same range of pressures as the insulator–metal transition. At high temperatures (295 K or 71°F) the insulating phase continuously transforms to the metallic phase (**Fig. 3**). The critical point constitutes the lower temperature bound on the continuous transformation, and at temperatures below that of the critical point T_c the transition becomes discontinuous. The behavior is analogous to a critical point in a single-component liquid–vapor transition. There is a close similarity between the insulator–metal transition for hydrogen and that of $(V,Cr)_2O_3$ alloys, which have been studied extensively since the late 1960s.

Studies above 200 GPa. The optical properties of hydrogen change dramatically at pressures above 200 GPa. Specifically, the samples develop a gradual absorption at visible wavelengths in this pressure range such that ~1-micrometer-thick samples are essentially opaque at pressures estimated to be in excess of 250 GPa. This behavior appears to be due to direct interband and intraband transitions in the molecular metallic solid. Raman-scattering measurements have failed to show evidence for molecular bonding in the highest-pressure samples (estimated to be 250–300 GPa). While this may be indicative of molecular dissociation, and hence the formation of a monatomic metallic phase, direct evidence for such a transition is not yet available. Future work will likely involve direct electrical and magnetic measurements of the solid at very high pressures. These measurements will be useful for tests for superconductivity in metallic hydrogen.

For background information SEE CRITICAL PHENOMENA; FREE-ELECTRON THEORY OF METALS; HIGH-PRESSURE PHYSICS; HYDROGEN; PHASE TRANSITIONS; PLANETARY PHYSICS; RAMAN EFFECT; REFLECTION OF ELECTROMAGNETIC RADIATION; SUPERCONDUCTIVITY in the McGraw-Hill Encyclopedia of Science & Technology.

Russell J. Hemley; Ho-kwang Mao

Bibliography. R. J. Hemley and H. K. Mao, Critical behavior in the hydrogen insulator-metal transition, *Science*, 249:391–393, 1990; H. K. Mao et al., Infrared reflectance measurements of the insulator-metal transition in solid hydrogen, *Phys. Rev. Lett.*, 65:484–487, 1990; H. K. Mao and R. J. Hemley, Optical studies of hydrogen above 200 gigapascals: Evidence for metallization by band overlap, *Science*, 244:1462–1465, 1989; E. Wigner and H. B. Huntington, On the possibility of a metallic modification of hydrogen, *J. Chem. Phys.*, 3:764–770, 1935.

Insect physiology

Insect metamorphosis, the transformation of larvae to pupae and then to adults, has provided model systems for understanding fundamental molecular aspects of developmental and physiological processes. Insect larvae feed and grow until they become limited by their exoskeleton. For further growth they must molt, that is, synthesize a new and larger exoskeleton and shed the old one. Molting is initiated by a family of steroid hormones, the ecdysteroids (see **illus.**), synthesized by the prothoracic glands. The activity of these glands is modulated by prothoracicotropic hormones (PTTHs), which are neurohormones produced by the insect brain. The release of PTTHs from their storage site is controlled by one or more environmental signals, such as photoperiod. The actual stage of the molt, that is, larval, pupal, or adult, is controlled by the juvenile hormones (see illus.): when the level of juvenile hormone is relatively high and the ecdysteroid reaches a critical level, the insect molts from a larva to another larva. When juvenile hormone is absent, the pupa undergoes metamorphosis to the adult.

Hormone synthesis. In the case of *Bombyx mori*, the commercial silkworm, the genes that encode two forms (big and small) of PTTH have been cloned. The small PTTH gene specifies a peptide (bombyxin) with an amino acid sequence bearing a strong resemblance to that of mammalian insulin. This structural similarity suggests that the *Bombyx* small PTTH acts upon target cells (in this case, the prothoracic gland) via a molecular mechanism analogous to insulin's mode of action. It appears that the result of the action of PTTH on the prothoracic gland cells is the phosphorylation of a protein that appears to be important in the translational control of protein synthesis. Its phosphorylation may elicit a series of biosynthetic steps in the prothoracic gland leading to the synthesis of ecdysteroids. The primary ecdysteroid products of the prothoracic gland of *Manduca* and *Bombyx*, as well as of higher Diptera, are ecdysone and 3-dehydroecdysone (see illus.), which are converted in peripheral tissues to the active molting hormone, 20-hydroxyecdysone. While ecdysone has been considered to be a prohormone, the emerging consensus is that it and perhaps other ecdysteroids play vital and distinct roles in the molting process.

Another important aspect of ecdysteroid biosynthesis involves its role in a "hibernation" strategy called diapause that is utilized by many insects to survive

Structure of some important juvenile hormones (JH) and ecdysteroids in insects. Juvenile hormone bis-epoxide (JH B₃) is a recently detected form of juvenile hormone.

winter. The diapause state is elicited by environmental cues such as decreasing or short day length and low temperatures that portend winter's onset. Typically, the length of day and temperature experienced in an early phase of the life cycle "programs" the insect to develop continuously through metamorphosis or, alternatively, to enter a diapause state at the appropriate time. For each species, the photoperiodic cues that induce diapause and the stage at which diapause occurs are specific, although examples of diapause exist for all stages of the insect life cycle, including embryonic and adult stages. Larval diapause in the blowfly, *Calliphora*, and pupal diapause in *Sarcophaga* and *Manduca* are associated with a failure of the prothoracic gland to respond to PTTH that consequently results in the absence of a peak of ecdysteroids to induce molting. Ultimately, the onset of metamorphosis is halted until warm temperatures and longer day lengths return and stimulate the resumption of development. By a combination of photoperiod and low temperature, adult diapause has been elicited in *Drosophila melanogaster* and is typified by an adult in which yolk deposition in the developing oocyte is not initiated. These insects remain in the diapause state for weeks. At the appropriate time, they are stimulated to continue postemergence maturation upon the synthesis and secretion of juvenile hormone that presumably acts to stimulate the oocyte, resulting in yolk deposition.

Hormone response. The action of ecdysteroids in responsive cells has been studied extensively. Conveniently, the chromosomes of the larval salivary gland of *D. melanogaster* become polytene; that is, they replicate several times without a cell division, allowing for easy examination under the light microscope. When these glands are placed in culture and exposed to 20-hydroxyecdysone, specific sites on the chromosomes, representing specific genes, begin to display puffs. These puffs are actually ribonucleic acid (RNA) and indicate transcriptional activity that is presumably elicited by the action of ecdysteroids. For many years, it has been postulated that 20-hydroxyecdysone, like steroid hormones generally, interacts with an intracellular receptor protein that, in turn, somehow enhances transcription of the early puff genes.

The genes encoding several mammalian steroid hormone receptors have been utilized as probes to identify similar sequences in the *Drosophila* genome in an attempt to recover one or more ecdysteroid receptors. At least 11 *Drosophila* genes that encode a protein with a molecular structure typical of steroid hormone receptors have been recovered. However, none have so far proven to specify a gene product that interacts with ecdysteroids to bring about changes in the expression of genes responsive to ecdysteroid, the indisputable criterion necessary to establish any candidate as an ecdysteroid receptor. Presumably, some of these receptors will eventually prove to be receptors for the two major hormones, 20-hydroxyecdysone and juvenile hormone.

Future directions. Almost every aspect of the insect's life, in addition to molting and metamorphosis, is controlled by hormones, including caste differentiation in social insects such as bees and termites, the production of pheromones such as the sex attractants of adult moths and butterflies, and regulation of homeostasis by peptide hormones. Indeed, one of the major advances in the field of insect physiology relating to hormonal control of developmental and physiological processes has been the identification and elucidation of the structure of numerous peptide hormones, most if not all originating in the nervous system. In terms of evolution, it is perhaps even more cogent that many of the peptide hormones already characterized in vertebrates have been identified, at least immunocytochemically, in insects. It is of interest that insects, which evolved a half billion years before the first vertebrate, possess molecules that have persisted so long during evolution and play major roles in vertebrate physiology and biochemistry. The same can be said of the mammalian steroid hormones, since androgens, estrogens, and adrenal cortical steroids have been identified in insect tissues, but again, their roles in the insect are unknown.

The small size of the insect had proven to be a difficult obstacle in studying the molecular basis of metamorphosis until the advent of recombinant deoxyribonucleic acid (DNA) technology. With the use of this modern approach, the prospect has emerged that new methods of biological insect control may become apparent, thus eliminating the need for ecologically harmful pesticides.

For example, recent molecular studies on insect baculoviruses in which their pathogenecity to insects has been increased by recombinant DNA technology has set the stage for biological manipulations. The baculoviruses are specific to arthropods and particularly Lepidoptera (moths and butterflies). Thus, insects such as the gypsy moth can be controlled by viruses that are totally innocuous to higher animals. The baculovirus–insect system is already being utilized for the production of specific proteins. A gene for the protein in question is inserted into the baculovirus genome, and the virus is injected into lepidopteran larvae or applied to insect cell cultures. The foreign gene then directs the synthetic machinery of the insect cells to produce copious amounts of the protein in question. The possibility exists that pharmaceuticals as well as protein and peptide hormones can be produced in this manner and can then be utilized to interfere with the insect's endocrine machinery.

For background information *SEE ECDYSONE; INSECT PHYSIOLOGY; INSECTA* in the McGraw-Hill Encyclopedia of Science & Technology.

Lawrence I. Gilbert; Vincent C. Henrich

Bibliography. L. I. Gilbert et al., Peptides, second messengers and insect molting, *BioEssays*, 8:153–157, 1988; J. Koolman (ed.), *Ecdysone*, 1989; S. R. Palli et al., Juvenile hormone receptors in insect larval epidermis: Identification by photoaffinity labeling, *Proc. Nat. Acad. Sci. USA*, 87:796–800, 1990; D. S. Saunders, V. C. Henrich, and L. I. Gilbert, Induction of diapause in *Drosophila melanogaster*: Photoperiodic regulation and the impact of arrhythmic mutations, *Proc. Nat. Acad. Sci. USA*, 86:3748–3752, 1989.

Insectivorous plants

Although motion responses are common in plants, they are usually too slow to be observed easily. However, several plant species are capable of very rapid movement. One example is the closure of the trap, a modified leaf, of the Venus' flytrap plant (*Dionaea muscipula*). This trap closure, although not the fastest plant movement known, is notable because of the speed and complexity of morphological change. Such a movement in an organism that lacks nerves and muscle tissue is intriguing, especially since the plant must also detect the presence of prey in the right position within the trap, and must close quickly in a way that prevents fast insects from escaping. Movement in the Venus' flytrap represents a sophisticated alternative to typical animal movement systems.

Charles Darwin initiated detailed studies of carnivorous plants, including the Venus' flytrap, and developed the first models attempting to explain how traps close. He found that trap closure was usually initiated when one or more trigger hairs of the inner trap surface were successively touched within several minutes. In more recent studies it was discovered that the

base of the trigger hair was the apparent site for mechanoreception, although most of the inner trap surface is sensitive to strong stimuli. As the trigger hair is deflected, distortion of cells at its base opens stretch-activated membrane channels, leading to rapid movement of ions across the plasma membrane. This causes a change in the electrochemical balance and a reversal of the electric charge between the normally negative inside and positive outside of the basal cell, thus depolarizing the cell. The depolarization wave travels out in all directions from the base of the trigger hair, perhaps traveling from cell to cell via plasmodesmata, which are small tubelike connections that extend through the cells and connect the plasma membranes of adjacent cells. Membrane depolarization is associated with the movement of calcium and potassium ions into the cell and movement of hydrogen ions out of the cell. Calcium ions are important because they participate in propagating the depolarization wave and, as secondary messengers, initiate a number of metabolic events. Movement of ions into the cell also induces osmotic uptake of water, thereby increasing turgor pressure and creating the force for cell enlargement. It is the differential enlargement of specific motor tissues within the trap that powers closure.

Most early models of trap closure were overly simplistic, viewing it as a single event where lobes simply snapped together. These models failed to account for many of the different types of movement observed during closure. More detailed studies and models of trap closure were made possible with the discovery that traps could be anesthetized by blocking the movement of calcium ions into the cell and neutralizing hydrogen ions released from cells. This meant that anatomical comparisons could be made between fully opened traps and traps that were closed for various lengths of time. Based on results from studies of traps stimulated continuously for 5 s, three major stages of trap closure were defined (see **illus.**). The complete closure of an artificially stimulated trap may take up to an hour or more to complete. Struggling prey cause the trap to close more rapidly, but all three stages of closure still occur.

Stages of closure. The first stage, capture (illus. *a*), is the only truly rapid part of trap closure. During this stage the marginal tynes of the trap bend sharply inward, forming an intermeshed network that prevents large prey from escaping. Tyne intermeshing occurs within 1 s after stimulation followed by a slight bowing of the trap lobes. The duration of this stage is approximately 5–15 min after stimulation.

The second stage of closure, appression (illus. *b*), is characterized by contact of the two opposing trap lobes. This occurs approximately 20–30 min after stimulation and is significant because pressure from opposing lobes affects further trap closure.

Approximately 1 h after stimulation, the last stage, sealing (illus. *c*), is complete. During this stage the trap margins become tightly appressed and flattened against each other, forming a tight seal; and the lower part of the trap increases its outward bulge. This

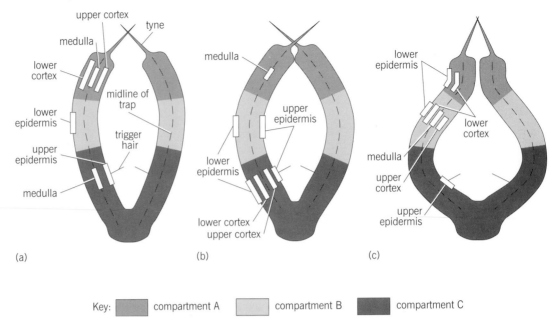

Key: ▨ compartment A ▢ compartment B ■ compartment C

Cross section of a trap during the three stages of closure, showing the major motor tissues responsible for movement: (a) capture; **(b)** appression; **(c)** sealing. A, B, and C represent the three major compartments from which movement originates during trap closure. The relative size of each box is an indication of the enlargement of that particular tissue. The midline of the trap indicates the separation of the inner and outer tissues. Tissues are symmetrical on both sides, so only one side is shown for simplicity.

results in the formation of a watertight digestive pocket.

The complex morphological changes that occur during closure appear to originate from enlargement of cells in three different compartments of the trap. Within each compartment, five potential motor tissues can participate in bending the lobes during closure. Selective enlargement of cells of one or several of these tissues bends the trap inward.

Based on data from studies of tissue dynamics during trap closure, the following events take place. During capture, a stimulus from the trigger hairs travels to the A and C compartments, where the outer cortical and medullary tissues enlarge, causing the tynes to bend inward. At the same time but to a lesser extent, the inner cortical tissue enlarges, tending to bend the trap outward. Thus, the enlargement of both cortical tissues creates opposing growth forces within the lobe. The net effect is to cause the trap to curve inward, rather than forming a simple, flat bend as would occur if only the outer cortical tissue had enlarged. The medullary and inner epidermal tissues of compartment C also enlarge slightly during this stage, forcing the bottom of the trap to bulge slightly outward.

During the appression stage, movement originates mainly from compartment C, where enlargement of outer cortical and epidermal tissues force the trap inward, thereby swinging the lobe shut. Opposing forces caused by enlargement of both inner and outer cortical and epidermal tissues increase the curvature of the C compartment.

One hour after stimulation, the sealing stage is complete and margins are tightly appressed. Enlargement of the outer cortex, medulla, and the inner cortex

in compartment B creates the main motor force during this stage. Trap morphology at the sealing stage is dependent upon the following: (1) the force exerted on compartment A by tissue enlargement in compartment B; (2) enlargement of inner epidermal tissues in compartment A; (3) pressure from the opposing trap lobes on compartment A; and (4) the force acting on tissues in compartment C as a result of tissue enlargement in compartment B combined with growth of the inner epidermis in compartment C causes compartment C to further bulge outward, thereby enlarging the digestive pocket.

Unanswered questions. Although the complex process of trap closure is beginning to be unraveled, a number of major questions remain. For example, it is not known how specific cells in a tissue (such as the outer cortex) are induced to enlarge while other cells in that same tissue in other compartments do not enlarge. Also unanswered is how some cells can delay response to signals from trigger hairs up to 30 min. The role of acidified cell wall loosening (the result of hydrogen ions pumped out of depolarized cells) in the closure process also remains to be elucidated.

At this point it is not known if the Venus' flytrap closure movement is unique to a small group of carnivorous plants or if it represents an evolutionary modification of a more widespread type of movement found in other plants. In any case, the movement responsible for trap closure represents a complex, highly coordinated, and finely regulated process of plant tissue dynamics.

For background information SEE INSECTIVOROUS PLANTS; NEPENTHALES; VENUS' FLYTRAP in the McGraw-Hill Encyclopedia of Science & Technology.

Wayne R. Fagerberg

Bibliography. C. Darwin, *Insectivorous Plants*, 1875; W. R. Fagerberg and D. Allain, A quantitative study of tissue dynamics during closure in the traps of Venus's flytrap *Dionaea muscipula* Ellis, *Amer. J. Bot.*, 78(5):647–657, 1992; B. E. Juniper, R. J. Robins, and D. M. Joel, *The Carnivorous Plants*, 1989.

Invertebrate physiology

Since 1980 the topic of lipid assimilation by marine invertebrates has received considerable study, mainly because of its relevance to global organic geochemistry. It is generally accepted that most marine lipids originate in phytoplankton growing in surface waters. One mechanism linking the lipids in these plant cells with those in sediments is grazing by herbivorous zooplankton. As part of the grazing process, unassimilated material is egested by the animals as fecal pellets. These pellets could be an important source of the lipids in marine sediments.

Lipids assimilated. Such considerations have stimulated a series of controlled laboratory feeding experiments in which comparisons have been made between the lipid composition of an invertebrate's diet and its fecal pellets. The two main objects of such experiments are to establish how dietary lipids are "edited" by passage through the guts of the animals, and how the lipid composition of the egested material relates to that of surface sediments.

Although much of the work has been done by using zooplanktonic crustaceans, notably the pelagic copepod *Calanus helgolandicus*, certain other species have also been used, including those with an estuarine habitat.

Fatty acids. Herbivorous feeding by *Calanus* leads to a substantial uptake of certain polyunsaturated fatty acids from the diet. This preferential removal of polyunsaturated fatty acids reflects the animal's need for such compounds in its phospholipids and wax esters. Such editing of dietary fatty acids is not confined to copepods but has also been observed with the krill *Euphausia superba*, the barnacle *Elminius modestus*, the shrimplike crustacean *Neomysis integer*, the crab *Pleuroncodes planipes*, the benthic bivalve *Scrobicularia plana*, and the annelid *Hediste (Nereis) diversicolor*. Nor is it characteristic only of herbivorous feeding; for example, "repackaged" fecal pellets released by *Calanus* feeding on fecal pellets egested by other zooplanktonic species also contain much lower amounts of polyunsaturated fatty acids. The substantial removal of dietary polyunsaturated fatty acids by zooplanktonic crustaceans could explain the very low amounts of these compounds detected in particulate material collected at shallow depths in the sea.

Although polyunsaturated fatty acids are rapidly assimilated, certain other fatty acids show levels in fecal material greater than those in the diet. Such compounds probably originate from bacteria present in the gut. Further evidence of enteric microbial

contributions to feces comes from experiments with *Neomysis, Scrobicularia*, and *Hediste*.

Sterols. Zooplanktonic crustaceans feeding herbivorously on diets essentially free from cholesterol release fecal pellets containing large amounts of it. As crustaceans cannot synthesize cholesterol de novo, its appearance in fecal pellets could represent formation of the compound from dietary plant sterols in excess of nutritional need. In addition, cholesterol could be contributed to fecal pellets from gut-lining epithelial tissue released during feeding. Cholesterol, like certain saturated and monounsaturated fatty acids, is therefore a feature of the lipid "signature" imposed by animals on their fecal pellets.

A plant sterol of particular interest is dinosterol. This compound is a possible tracer for dinoflagellate inputs to sediments. The fate of this sterol in *Calanus* was studied by using an artificial diet of starch granules containing dinosterol labeled with radioactive carbon (^{14}C), and a similar diet of ^{14}C-labeled cholesterol for comparison. Cholesterol was rapidly assimilated, with over 60% of the labeled cholesterol entering the animal's tissues. Dinosterol was not assimilated. Similar results have been obtained in parallel studies with *Neomysis*, where more than 15% cholesterol but less than 1% dinosterol ingested was retained in the tissues. In further studies with fecal pellets from copepods, no microbial alteration of fecal dinosterol was observed over 20 days, a period typical of the sedimentation rates for large particles in open ocean areas. Such findings uphold the view that dinosterol is a biological marker for dinoflagellate inputs to sediments, and that an important source of dinosterol in such sediments could be fecal pellets egested by zooplankton grazing on dinoflagellates.

Fatty alcohols. In most of the species used in these studies, fatty alcohols accounted for only a small fraction of total lipid. The exception was *Calanus*, where substantial amounts of fatty alcohols occur as components of wax esters. As with certain fatty acids and sterols, fatty alcohols form part of the animal signature imposed on the lipids in fecal pellets. Also present, however, are numerous fatty alcohols occurring neither in the animals nor in their plant diets. The presence of these fatty alcohols again indicates that there is an active biotransformation of lipids in the copepod gut.

The fatty alcohol fraction of phytoplanktonic lipids consists almost entirely of phytol, the ester-linked side chain of chlorophylls *a* and *b*. This alcohol is rapidly assimilated from diets by *Calanus* and used as the precursor of the hydrocarbon pristane. In addition to the small amounts of phytol released in fecal pellets, trace quantities of its reduced product dihydrophytol, a compound absent in both the animals and their plant diets, are released. It seems likely that in areas where calanoid copepods can dominate in the zooplankton, for example, in the Gulf of Maine, their fecal pellets could be the main source of dihydrophytol in the sediments.

Hydrocarbons. Although aliphatic hydrocarbons normally present in phytoplankton, such as *n*-alkanes and *n*-alkenes, are completely removed during passage through the copepod gut, the same does not hold for compounds such as polycyclic aromatic compounds and polychlorinated biphenyls (PCBs). Fecal pellets from the animals have been indicated as an important source of these pollutants in sediments.

Novel long-chain lipids. The ubiquitous golden-brown algae *Emiliania huxleyi*, as well as several other species of the class Prymnesiophyceae, contain certain long-chain alkenes, in addition to unsaturated methyl and ethyl ketones. These compounds have been detected in certain deep-sea sediments, such as the Cariaco Trench, suggesting that organisms such as *E. huxleyi* could be the origin of these novel lipids. None of the compounds has been detected in *Calanus*. However, when feeding on *E. huxleyi*, the animal released large quantities of them in its fecal pellets, in which form they could eventually be transported to the sediments.

Assimilation efficiencies. So far there has been only one quantitative study of the lipid assimilation efficiencies of zooplankton grazing on a range of food concentrations. The animal studied was *Calanus*. It was fed a diet of the alga *Scrippsiella trochoidea* at concentrations ranging from open ocean (highly diluted) to bloom conditions (highly concentrated).

Polyunsaturated fatty acids and phytol were almost completely assimilated at all concentrations. Over 90% of saturated and monounsaturated fatty acids was assimilated at the highest food concentration, but only 60% at the lowest. The reduced assimilation was probably due to the longer gut residence times at lower concentrations; the longer time results in the animal's egesting more of its own fatty acids in the fecal pellets. The total dietary sterols assimilated was 21% at the lowest food concentration and 11% at the highest. Clearly, sterols are less readily assimilated than either fatty acids or phytol.

Prospects. In the future, emphasis will probably shift from land-based studies to on-site studies at sea. Experiments will involve those species that are dominant in the zooplankton population, feeding on their natural diet. Egested lipids may then be compared with those in sediment-trap material in order to assess the importance of animal feeding in lipid deposition at a particular area and at a particular time of year.

For background information *SEE* LIPID in the McGraw-Hill Encyclopedia of Science & Technology.

Eric D. S. Corner

Bibliography. S. A. Bradshaw et al., Assimilation of dietary sterols and faecal contribution of lipids by the marine invertebrates *Neomysis integer*, *Scrobicularia plana* and *Nereis diversicolor*, *J. Mar. Biol. Ass. U.K.*, 69:891–911, 1989; S. A. Bradshaw et al., Changes in lipids during simulated herbivorous feeding by the marine crustacean *Neomysis integer*, *J. Mar. Biol. Ass. U.K.*, 70:225–243, 1990; E. D. S. Corner and S. C. M. O'Hara (eds.), Copepod faecal pellets and the vertical flux of biolipids, *The Biological Chemistry of Marine Copepods*, 1986; H. R. Harvey et al., Biotransformation and assimilation of dietary lipids by *Calanus* feeding on a dinoflagellate, *Geochim. Cosmochim. Acta*, 51:3030–3041, 1987; H. R. Harvey et al., The comparative fate of dinosterol and cholesterol in copepod feeding: Implications for a conservative molecular biomarker in the water column, *Org. Geochem.*, 14:635–641, 1989.

Irrigation (agriculture)

Rapid population growth since midcentury has created severe shortages of water for agricultural, industrial, and municipal uses in many arid and semiarid regions. Reuse of treated municipal sewage effluent for crop and landscape irrigation has become common in several countries. Since 1970, improvements in the design and operation of sewage treatment systems have greatly reduced the potential for harm to the public health and environment from irrigation with treated effluent. However, several issues remain unresolved, including assessment of potential health effects of low but detectable levels of viral pathogens and trace organic substances, determination of the best strategy for prevention of groundwater contamination with nitrogen in effluent, and identification of reliable wastewater treatment techniques suited to developing countries.

Development of sewage farming. Application of raw sewage to cropland was practiced as early as 1650 in Edinburgh, Scotland, but did not become widespread until the installation of sewer systems in England in the midnineteenth century. Even though the germ theory of disease had not gained acceptance, it was widely appreciated that the pollution of surface waters with sewage was linked to epidemics of cholera and other diseases. The fertilizer value of sewage was also recognized. However, three developments conspired to decrease the practice of land application in England and Europe in the late 1800s. Towns grew outward, reaching the sewage farms, which by then were often overloaded. Modern sanitary engineering techniques were developed, including the unit processes of chemical precipitation, biological treatment, filtration, and disinfection. Increased availability of animal manures and manufactured fertilizers lowered the value of sewage as fertilizer.

In the United States, a similar evolution occurred. Sewage farming was the most common method of treatment and disposal during the early twentieth century, but within a few decades it was replaced by the more land-efficient, less polluting methods of modern sanitary engineering. Newly built treatment plants removed wastewater solids by primary treatment (settling or sedimentation) followed by secondary treatment (biological processing and sludge removal). Sludge solids removed from the wastewater were dumped into the ocean, incinerated, or put in a landfill. It became a common practice to discharge the treated wastewater into the oceans and rivers.

During the 1950s in the United States, increased public concern with eutrophication of lakes and

Table 1. Wastewater treatment and quality criteria for irrigation

Treatment level	Coliform limits (median MPN*)	Type of use
Primary		Surface irrigation of orchards and vineyards Fodder, fiber, and seed crops
Oxidation and disinfection	≤23/100 ml	Pasture for milking animals Landscape impoundments Landscape irrigation (golf courses, cemeteries, and so on)
	≤2.2/100 ml	Surface irrigation of food crops (no contact between water and edible portion of crop)
Oxidation, coagulation, clarification, filtration,† and disinfection	≤2.2/100 ml, max. = 23/100 ml	Spray irrigation of food crops Landscape irrigation (parks, playgrounds, and so on)

*Median MPN = median most probable number of coliform bacteria as determined by standard water analysis procedure. 1 ml = 0.06 in.³
†The filtered effluent cannot exceed an average of 2 turbidity units during any 24-h period.

streams (caused in part by discharge of nutrient-laden wastewater) led to renewed interest in land application, which since has usually been referred to as land treatment. Public Law 92-500, the Federal Water Pollution Control Act Amendments of 1972, emphasized land application of wastewater and provided for incentives for reclamation and reuse of wastewater.

Land application of wastewater. Irrigation with reclaimed municipal wastewater is a common practice in Israel, Australia, Africa, the United States, and the regions in Europe where water is relatively scarce. In Israel, about 45% of the total wastewater is recycled for irrigation, a volume of 89,000 acre-ft (1.1×10^8 m³) per year. Since 1985, a large project in northern Israel has produced cotton on 35,000 acres (14,000 hectares) by using reclaimed wastewater. In Australia, reclaimed wastewater is used to irrigate more than 80 golf courses and several silvicultural plantations of eucalyptus, poplar, and casuarina trees.

By 1979, over 3000 land treatment systems, excluding septic tanks, were in use in the United States, with wastewater irrigation being most widely practiced in California. Yet, with less than 15% of all treated wastewater intentionally reused in California, the potential for increasing reuse is great. Annual reuse in California has increased by 71% since 1977 to 315,000 acre-ft (3.9×10^8 m³), with 63% of the reuse being for agricultural irrigation, 13% for landscape irrigation and impoundments, and the remainder for groundwater recharge, wildlife habitat, recreational impoundment, and industrial uses. More than 30 different food and nonfood crop species, including Christmas trees, flower seeds, and celery, are irrigated; livestock feed, fiber, and seed crops use more than one-third of the reclaimed effluent. One unusual use of wastewater is for leaching salts from agricultural fields. For new and expanding treatment facilities, urban landscape vegetation is most often the recipient of the treated wastewater. By 1987, 63 golf courses were irrigated with wastewater in California, and more were being planned. In the upper midwestern regions of the United States, wastewater is commonly used for production of forages and for poplar, larch, spruce, and other species of trees grown for wood, fiber, and fuel. A large project in Florida (25,000 acres or 10,000 hectares) involves irrigation

of citrus trees with treated wastewater from the city of Orlando. Some of this irrigation is solely for the purpose of frost protection. In Arizona, several projects have operated successfully for nearly two decades, with no degradation of soil.

Benefits and limitations of wastewater irrigation. Two benefits of irrigation with municipal wastewater are the provision of cost-effective pollution control and economical use of existing disposal facilities. An additional benefit is that supplies of potable water are conserved. This benefit is obviously less important or even absent in regions that are rich in water. On the other hand, even in the most water-short regions, reclaimed wastewater irrigation does not expand the regional water supply except where water that otherwise would be lost to unproductive evaporation or would be discharged to saltwater is reused. In California in 1987, of all the wastewater reused, 84% replaced fresh water that would have been used, thus conserving it for higher uses.

The potential benefits of wastewater include its fertilizer value, which in some cases is high enough to eliminate the need for purchased fertilizers, and its value as a dependable supply of water.

These benefits must be weighed against costs or problems associated with wastewater irrigation. First, institutional or legal constraints are sometimes present, for example, when the farmer is obligated by contract to accept all effluent, even when it is raining.

Second, in some regions, the level of salinity of reclaimed water is high enough to require planting salt-tolerant varieties, blending effluent with fresh water, or avoiding use of sprinklers during the daytime. When effluent is used, drip irrigation systems may be more prone to clog with mineral precipitates, sediment, or algal and bacterial slimes. Where effluent is valued as a water supply, communities may seek to protect its quality. For example, the Regional Water Quality Control Board in San Diego, California, has urged adoption of restrictions on self-regenerating water softeners. This type of water softener discharges large volumes of brine directly into the sewer. The resulting increase in sodium and salt levels in the wastewater degrades its value for irrigation.

A third factor that can lower the attractiveness of effluent to the user is that it may be costly or inconvenient to modify the design and operation of conventional irrigation systems as may be required by the government in order to protect public health. Health standards vary among states and countries. Wastewater reclamation criteria established in California are shown in **Table 1**. The most stringent requirements are for uses that are likely to result in the greatest public exposure, such as spray irrigation of food crops and playground landscape irrigation. No standard is set for virus levels, because studies have shown that advanced treatment, as specified for spray irrigation of food crops, produces an essentially virus-free effluent. In addition to standards for water quality, requirements are imposed on irrigation system design and site management. Examples include color-coded pipes to prevent accidental cross-connections between fresh and reclaimed water; back flow prevention or air-gap devices on fresh-water plumbing on the site; key-operated valves and no installation of hose bibcocks; avoidance of any runoff or spray drift; buffer zones between the effluent-irrigated vegetation and areas used by the public; nighttime irrigation; and algae control measures for storage ponds.

Current public health issues. It has been argued that there should be standards for wastewater reuse that are better suited to the needs of developing countries than the standards in Table 1. Tentative guidelines recommended by a group meeting under the auspices of the World Health Organization, the World Bank, and the United Nations Environment Programme are shown in **Table 2**. Helminth (nematode) disease transmission is a potentially serious result of raw wastewater use. The standard also reflects the group's belief that too stringent a standard for coliform bacteria and virus

particles raises the cost of sewage treatment so as to discourage reclamation and reuse.

A related issue is the selection of the most appropriate means for reducing exposure to virus particles. Because secondary effluent is not entirely free of virus particles and bacteria, a conservative approach is prudent. The health significance of differences in the efficiency of virus removal when virus levels are already very low is not known. Recent studies show that it is possible to produce essentially virus-free effluent by using direct filtration and a low level of chlorination—a less expensive method than conventional advanced treatment processes. However, for this less expensive treatment to be effective, a very high quality of secondary effluent with low levels of suspended solids and turbidity is required.

Another unresolved issue is the degree of risk to the public posed by the presence of refractory organic substances in effluent. Chlorination of effluent can produce trihalomethane or other halogenated compounds. Alternatives to chlorination, such as lime stabilization, are expensive and not widely used. It has been difficult to identify and quantify all the trace organic substances in treated effluent, and one of the concerns is that an unknown number of them will be mutagenic or carcinogenic. In California, total organic carbon standards are being considered where treated effluent is to be used for groundwater recharge. No studies have been done to establish the level of risk associated with the consumption of wastewater-irrigated produce or with skin contact that is posed by trace organics, but presently it appears that the risk is very small.

In Europe and the United States, public concern over nonpoint sources of nitrogen entering groundwater supplies has increased since 1985. Treated municipal wastewater often contains enough nitrogen for crop requirements to be exceeded, especially where a disposal objective leads to excessive rates of application. The most common solution to this problem is to select plant species that are particularly effective in stripping nutrients from the water but that still have some agronomic value, such as *Panicum maximum* (guineagrass) or *Medicago sativa* (alfalfa, lucerne). It is also possible to design and operate the wastewater treatment plant so as to produce effluent with a nitrogen content appropriate for the end use.

For background information SEE IRRIGATION (AGRICULTURE); SEWAGE; SEWAGE TREATMENT in the McGraw-Hill Encyclopedia of Science & Technology.

G. Stuart Pettygrove

Bibliography. C. R. Bartone, S. Arlosoroff, and H. I. Shuval, Development of health guidelines for water reuse in agriculture: Management and institutional aspects, in *Implementing Water Reuse: Proceedings of Water Reuse Symposium IV*, American Water Works Association Research Foundation, 1988; H. H. Hahn and R. Klute (eds.), *Pretreatment in Chemical Water and Wastewater Treatment*, 1988; W. J. Jewell and B. L. Seabrook, *A History of Land Application as a Treatment Alternative*, U.S. Environmental Protection Agency, EPA

Table 2. Tentative microbiological quality guidelines for treated wastewater reuse in agricultural irrigation[a]

Reuse process	Intestinal nematodes,[b] eggs per liter	Fecal coliforms, number per 100 ml[c]
Restricted irrigation[d]: of trees, industrial crops, fodder crops, fruit trees[e], and pasture[f]	Less than 1	Not applicable[c]
Unrestricted irrigation: of edible crops, sports fields, and public parks[g]	Less than 1	Less than 1000[h]

[a]In specific cases, local epidemiological, sociocultural, and hydrogeological factors should be taken into account, and these guidelines modified accordingly.
[b]*Ascaris, Trichuris*, and hookworms. 1 liter = 0.26 gal.
[c]1 milliliter = 0.06 in.^{3}
[d]A minimum degree of treatment equivalent to at least a 1-day anaerobic pond followed by a 5-day facultative pond or its equivalent is required in all cases.
[e]Irrigation should cease 2 weeks before fruit is picked, and no fruit should be picked off the ground.
[f]Irrigation should cease 2 weeks before animals are allowed to graze.
[g]Local epidemiological factors may require a more stringent standard for public lawns, especially hotel lawns in tourist areas.
[h]When edible crops are always consumed well cooked, this recommendation may be less stringent.
Source: *The Engelberg Report: Health Aspects of Wastewater and Excreta Use of Agriculture and Aquaculture, Report of Review Meeting of Environmental Specialists and Epidemiologists*, World Health Organization, 1985.

430/9-79-012, 1979; G. S. Pettygrove and T. Asano (eds.), *Irrigation with Reclaimed Municipal Wastewater: A Guidance Manual*, 1985.

Laser

Free-electron lasers (FELs) make up a class of potentially efficient devices capable of generating high-quality coherent radiation, continuously tunable from the microwave region to the x-ray region, at high average or high peak power. Most of the essential physics of free-electron lasers has been demonstrated, and the device is now becoming a research tool rather than a pure research topic.

Free-electron laser physics. The free-electron laser is characterized by a relativistic electron beam copropagating through an undulator field (or wiggler field) with an input radiation field. The undulator field, which gives electrons a periodic transverse motion, is typically a static periodic magnetic field. If the electron beam is of good quality and has a sufficiently high current density, input radiation at the appropriate frequency will grow, as in the case of a free-electron-laser amplifier. When the gain per pass through the undulator is low, many passes are required for the radiation to reach saturation, as in a free-electron-laser oscillator. A schematic of this oscillator is shown in the **illustration**. Here, the end mirrors reflect the radiation between the exit and the entrance of the undulator.

The free-electron laser is very different from conventional lasers. Its active medium is not atoms or molecules but free electrons. Its radiation frequency is continuously tunable and is not limited to discrete allowed quantum transitions. The free-electron-laser mechanism is a stimulated process, differing from the incoherent, low-power spontaneous radiation due to electron acceleration in the undulator field. In the free-electron laser, the electron kinetic energy is converted into radiation. Some electrons may lose kinetic energy, while others may gain kinetic energy. Under the appropriate resonance conditions, coherent radiation is generated as electrons on the average lose kinetic energy. Since each electron can gain or lose many photons, the physics can be understood in terms of classical mechanics.

Electron dynamics. The detailed electron trajectory through the undulator is complicated. The net effect of the electron dynamics, however, can be summarized by an interaction between the electrons and a wave traveling at a phase velocity slightly less than the electrons. This wave is called the ponderomotive potential wave, which is generated by the beating of the radiation field (frequency ω, wavelength $\lambda = 2\pi/k$, and wave number $k = \omega/c$) and the undulator fields (wavelength λ_u and wave number $k_u = 2\pi/\lambda_u$), where c is the speed of light. The beat wave has the same frequency as the radiation, but its wave number is $k + k_u$. The phase velocity of the beat wave, $v_{ph} = \omega/(k + k_u)$, is slower than the speed of light. When the axial electron-beam velocity v_{z0} is approximately equal to the phase velocity of the ponderomotive potential wave, the electrons are subject to an approximately constant axial field. Hence, the ponderomotive beat wave may interact strongly with the electron beam. Since some electrons are subject to an accelerating field while others experience a decelerating field, the initially uniformly distributed electrons become bunched at the ponderomotive wavelength. If the initial beam velocity is slightly larger than the phase velocity of the beat wave, the beam, on average, will lose energy during the bunching process. The stimulated radiation thus emitted from the periodic beam bunches is coherent and is many orders of magnitude higher in power than the spontaneous radiation.

Tunability. The wavelength of the radiation λ is tunable because it is a function of the energy of the electron beam and the magnetic field of the undulator, as shown in the equation below. Here the wavelengths

$$\lambda \simeq \frac{1 + (5.6 \times 10^{-5} B_u \lambda_u)^2}{2\gamma^2} \lambda_u$$

λ and λ_u are expressed in centimeters, B_u is the magnetic field (in gauss) of the undulator on axis (for planar undulator fields as shown in the illustration), $\gamma = 1 + 1.96\,E$, and E is the energy of the electron beam in megaelectronvolts. This radiation–wavelength relation is a consequence of requiring $v_{z0} \simeq v_{ph}$. Since the period of the undulator is difficult to change, the wavelength of the radiation can be decreased or increased most easily by increasing or decreasing the energy of the electron beam, or by opening or closing the gap separating the top and bottom arrays of undulator magnets, leading to a decrease or increase of the magnetic field of the undulator on-axis.

Efficiency. For a free-electron laser to be a practical device, the percentage of the electron energy converted to radiation must be large; that is, high efficiency is required. The intrinsic efficiency of the free-electron laser drops as the wavelength of the radiation decreases. At any particular wavelength, efficiency can be increased by a gradual decrease in

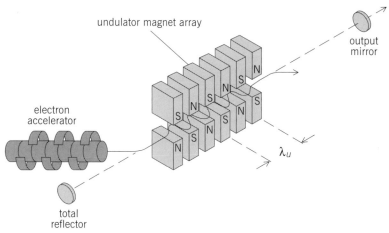

Free-electron-laser oscillator.

the period or the amplitude of the magnetic field of the undulator, among other schemes. Decreasing the period of the undulator decreases the phase velocity of the ponderomotive wave, forcing the trapped electrons to decelerate axially and to give up kinetic energy associated with their axial motion. Decreasing the amplitude of the magnetic field of the undulator decreases the kinetic energy associated with the transverse wiggling motion. The additional energy lost by the electrons as the result of such tapering of the undulator is converted to radiation energy, leading to an increase of the efficiency. Experiments have demonstrated an efficiency of 35% at millimeter wavelengths with a tapered undulator, whereas a uniform undulator in the same system gives only 6% efficiency.

The free-electron laser can provide very high peak or average power because high-power electron beams can be generated with high efficiency. In addition, the electron beam differs from a conventional laser medium in that it can support high-power radiation without being damaged.

Limits on velocity spread. The implementation of the free-electron laser is critically dependent on having an electron beam with a small effective axial velocity spread. Sources of effective axial velocity spread are the energy spread and emittance (a measure of the transverse velocity spread and beam size) of the electron beam and the undulator field errors (the differences between the actual undulator field and a pure sinusoid). The condition for effective axial velocity spread of the beam not to degrade the performance of the free-electron laser is that electrons with different axial velocities do not significantly separate over some relevant interaction length. This separation distance must be less than some fraction of the ponderomotive potential wavelength, $2\pi/(k + k_u)$, which is approximately the radiation wavelength for high-energy beams. The relevant interaction lengths are the undulator length in the low-gain regime; the e-folding length (a measure of the rate of growth of the laser amplitude with distance) in the high-gain, low-current (no space-charge effect) regime; and the plasma wavelength in the high-gain, high-current (space-charge-dominated) regime. The limits on the effective axial velocity spread of the electron beam are more stringent for shorter radiation wavelengths. A large effective axial velocity spread will significantly reduce electron bunching in the ponderomotive potential wave, leading to significant degradation of the growth rate, gain, and efficiency of the laser. Thus, free-electron lasers require electron beams of good quality, that is, with small energy spread and small transverse emittance.

Experiments. Electron-beam sources vary widely in energy, beam quality, current, pulse length, and repetition rate. Accelerators driven by radio-frequency power, such as radio-frequency (rf) linear accelerators (linacs), microtrons, and storage rings, have peak currents from a few amperes to hundreds of amperes, electron energies up to hundreds of megaelectronvolts, and electron micropulse lengths much shorter than the radio-frequency wavelength. Induction accelerators can have many kiloamperes of current, and presently have energies up to tens of megaelectronvolts and pulse lengths of a few tens of nanoseconds to a few microseconds. Electrostatic accelerators usually have low current and low energy but good beam quality. Lumped-parameter modulators usually can provide relatively inexpensive electron beams in the low-energy range with pulse durations up to several

Table 1. Free-electron-laser user facilities driven by rf linacs

Institution	Duke University	Stanford University*	Stanford University*	Vanderbilt University
Accelerator	MK III	Superconducting	Superconducting	MK III clone
Year of operation	1990	1990	1991	1991
Electron energy	27–44 MeV	20–200 MeV	20–200 MeV	20–45 MeV
Current	30 A	5 A	35 A	35 A
Wavelength	2–9 μm	0.5–5 μm	0.2–15 μm	1–8 μm
Lasing pulse duration				
Micropulse	1–4 ps	3 ps	2.2 ps	0.5–3 ps
Macropulse	1–8 μs	10 ms	20 ms	1–8 μs
Repetition rate				
Micropulse	2.856 GHz	12 MHz	12 MHz	2.856 GHz
Macropulse	30 Hz	10 Hz	20 Hz	60 Hz
Energy per pulse				
Extracted	5–10 μJ	3 μJ	15 μJ	4 μJ
Q-switched	500 μJ	—	500 μJ	—
Power				
Peak, extracted	1.25–10 MW	1 MW	7 MW	2 MW
Peak, Q-switched	125–500 MW	—	230 MW	—
Average, extracted	6 W	3.6 W	36 W	12 W
Bandwidth	Transform-limited	Transform-limited	Transform-limited	Transform-limited
Radiation mode				
Linearly polarized	Yes	Yes	Yes	Yes
TEM$_{00}$	Yes	Yes	Yes	Yes
Diffraction-limited	Yes	Yes	Yes	Yes

*Energy and power are based on radiation at 4 μm.

microseconds and hundreds of amperes of current. Electron beams from pulsed power sources can have very high currents, but they have low energies and pulse durations of tens of nanoseconds. Each of these electron-beam sources has been applied to the generation of free-electron-laser radiation. The output radiation characteristics are diverse as a result of the diversity of electron-beam sources.

Although free-electron laser concepts have been known since they were proposed by H. Motz in 1951 and the successful experiments by R. M. Phillips in the millimeter-wavelength regime in 1960, it is the work by J. M. J. Madey in the infrared wavelength regime in the 1970s that has made the free-electron laser a serious candidate for a new powerful radiation source.

Free-electron laser experiments are conducted in many countries. Although the conceptual configuration of the free-electron laser is simple, the implementation was arduous because existing electron-beam sources were not designed with free-electron lasers in mind. Most of the basic theory of the free-electron laser has been verified. All the experiments have demonstrated the importance of high-quality electron beams. Research efforts are under way to develop free-electron lasers in the direction of user-oriented and practical devices.

Applications. Many applications have been envisioned for free-electron lasers. Due to the current complexity and the cost of the device, applications are in the area of research rather than consumer products, and several user facilities are being developed. **Tables 1** and **2** list the output light properties at the currently funded facilities. Many additional user facilities are being proposed. Proposed accelerators are being designed for specific user needs. Important criteria for free-electron laser user facilities are electron-beam stability, reliability, ease of frequency tuning, and ease of operation.

In the microwave regime, the high-power and high-efficiency nature of the free-electron laser dominates applications. Free-electron lasers have been constructed for plasma heating in tokamaks and as radio-frequency sources for conventional accelerators.

In the infrared and visible wavelength regimes, there exist a wide range of applications in spectroscopy, chemistry, biology, medicine, isotope separation, condensed-matter studies, material sciences, and national defense. In addition to the tunability and high peak or average power, free-electron lasers have other desirable characteristics. Free-electron lasers driven by radio-frequency accelerators in this range can deliver laser pulses of picosecond durations. This short pulse duration, and the corresponding wide frequency range in the infrared, provides new spectroscopic capabilities. The micropulse nature also facilitates pump-probe-type experiments, which can provide quantitative studies of many surface phenomena.

Applications in the ultraviolet and x-ray ranges are exciting, but free-electron lasers become more difficult and more expensive as the wavelength decreases.

Table 2. Free-electron-laser user facilities driven by other than rf linacs

Institution	Duke University[a]	University of California, Santa Barbara
Accelerator	Storage ring	Electrostatic Van de Graaff
Year of operation	1993[b]	1990
Electron energy	0.5–1 GeV	2–6 MeV
Current		
Peak	300 A	2 A
Average	150 mA	
Wavelength	50–400 nm	63–400 μm
		193 μm–1.34 mm
		30.6–100 μm[c]
Pulse duration	30 ps	1–3 μs[d]
		3–20 μs[e]
Repetition Rate	2.7624 MHz	1 Hz
Energy per pulse		
Extracted	3 μJ	1.4–4.2 mJ
Q-switched	300 μJ	3–280
Power		
Peak, extracted	100 kW	1–1.4 kW
Peak, Q-switched	10 MW	~6–8 kW[f]
Average, extracted	0.037 W	0.0014–0.014 W
Bandwidth, Δλ/λ	Transform-limited	$(10^{-7})^d$
		0.05%[g]
Radiation mode		
Linearly polarized	Yes	Yes
Coupled-out	TEM$_{00}$	Mirror
Diffraction-limited	Yes	Yes

[a]Single bunch. Energy and power are based on radiation at 100 nm.
[b]Assuming funding is provided as requested.
[c]Third-harmonic operation, with beam energy 3–6 MeV, to become operational in 1991.
[d]Single-pulse-mode operation.
[e]Multiple-longitudinal-mode operation.
[f]40-ns pulse.
[g]Averaged over many pulses.

The energy of the ultraviolet radiation is high enough to cause ionization and bond breaking in deoxyribonucleic acid (DNA) and proteins. Radiation can also be absorbed and redistributed to vibrational states. Radiation within the water window between 2.4 and 4.4 nanometers will allow the study of living organisms in their natural aqueous environment. Coherent x-ray sources can be used for holography. Three-dimensional imaging of atomic structures may be feasible in the future. The criteria limiting the ultimate wavelength for the free-electron laser are electron-beam quality and laser optics.

For background information SEE HOLOGRAPHY; LASER; PARTICLE ACCELERATOR; PHASE VELOCITY; SYNCHROTRON RADIATION in the McGraw-Hill Encyclopedia of Science & Technology.

Cha-Mei Tang

Bibliography. C. Brau, *Free-Electron Lasers*, 1990; H. P. Freund and R. K. Parker, Free-electron lasers, *Sci. Amer.*, 260(4):84–89, April 1989; K.-J. Kim and A. Sessler, Free-electron lasers: Present status and future prospects, *Science*, 250: 88–93, October 5, 1990; T. Marshall, *Free-Electron Lasers*, 1985; J. M. Ortega, Free-electron laser and harmonic generation, *Synchro. Radiat. News*, 2(2):18–23, 1989, 2(4):21–24, 1989, 3(1):26–27, 1990; C. W. Roberson and P. Sprangle, A review of free-electron lasers, *Phys. Fluids*, B1:3–42, 1989.

Leaf

Plants were once thought to have animallike circulatory systems containing veins and arteries. Although this has long since been disproved, the term vein has persisted for the vascular bundles in a leaf. Veins support the leaf, distribute water and dissolved minerals, and collect photosynthates and other complex molecules and transport them to the stem for distribution. How the structure of veins enables them to function in different vascular plant groups is far from settled, but progress is evident. Research in the 1980s focused attention on the veins of dicotyledons, and they will be emphasized in this article.

Vein configurations vary considerably among dicots, but in general there is a conspicuous pinnately arranged set of major veins interspersed with an intricate reticulum of minor veins, of which the ultimate elements are simple or branched veinlets that end blindly. Most monocotyledon leaves are traversed instead by several parallel longitudinal major veins that gradually approach each other and merge near the leaf tip, and that are interconnected all along by minor veins arranged transversely like rungs of a ladder. Blind vein endings are uncommon. Nonflowering plant features such as open dichotomous venation, endodermis with Casparian strips, and transfusion tissue are virtually unknown in veins of dicots and monocots.

Since all green leaves have similar physiological tasks, the structural diversity among veins is indirect evidence that there are differences among plant groups as to how these tasks are performed.

Dicot venation. Typically, six somewhat intergrading but progressively smaller vein orders can be recognized, from the largest in the midrib down to the vein endings (**Fig. 1**). Orders 1–3 are called major veins; they resemble the vascular bundles of stems both in size and in xylem and phloem composition, and, like stem bundles, they probably serve almost exclusively as conduits for long-distance transport. Major veins typically have a fibrous or parenchymatous vertical flange extending to one or both epidermises, an architectural feature that may provide a pathway for water to reach the epidermis.

Vein order 4 often appears intermediate in size and thus is sometimes considered to be major venation and sometimes minor venation. Orders 5 and 6 are always minor venation (minor here refers only to vein diameter). Numerically, minor veins are far more significant than major veins; in several species they account for 86–99% of total vein length. Minor veins delimit most of the ultimate meshes of the vascular reticulum and form the blind vein endings within them.

Minor veins are ensheathed by a close-fitting layer of thin-walled cells, the bundle sheath, and they lack vertical extensions to the epidermis, so they appear to be suspended in the middle of the mesophyll. The bundle sheath is usually considered to be part of the vein, but it has also been interpreted as being an intermediate layer between vascular tissue and mesophyll. In many species, for example, some or all of the

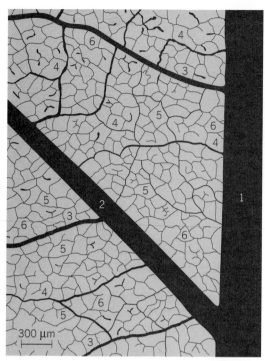

Fig. 1. Magnified view of a small area of soybean (dicot) leaflet showing vein orders and venation pattern: orders 1–3 (major), 4 (intermediate), 5–6 (minor). (*From A. D. Bán et al., Arquitetura foliar da soja Glycine max (L. Merrill), Agron. Sulriogrand., 17:25–50, 1981*)

lateral bundle sheath cells have two or three horizontal protrusions that interdigitate with mesophyll parenchyma cells in the same plane. In some plants, notably many of the legumes, lateral bundle sheath cells and horizontally extended mesophyll cells in the same plane combine with minor veins to form a coherent middle layer for storage and lateral transport of photosynthates throughout the leaf or leaflet. Even in plants that lack an anatomically distinct middle layer, the broad sheet of mesophyll cells in this plane directs photosynthates horizontally, if diffusely, toward the minor veins.

Xylem. Xylem in minor veins usually consists of only one or two files of water-conducting tracheary elements that virtually always have only annular or helical lignified wall thickenings. Xylem in minor veins is therefore equivalent to the protoxylem in stem bundles. In many species, however, the tip of most or all vein endings expands by proliferation or enlargement of the terminal tracheary elements into a conspicuous xylary knob (**Fig. 2**), which has long been speculated to store water. Recent experiments with dyes show that water accumulates in these vein endings before diffusing preferentially upward and outward into the sheath cells located directly and obliquely above the xylem. Dye patterns also indicate that the bundle sheath can reabsorb and recycle ions from the water before it passes through the cell wall and into the mesophyll air spaces as water vapor. In grasses, which lack vein endings, water departs preferentially from the tiny transverse veins that connect the longitudinal veins. Here, too, water moves directly and obliquely upward, but suberized lamellae in

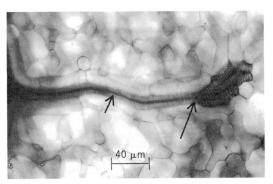

Fig. 2. Vein ending from a cleared leaf of coneflower (dicot) showing the terminal xylary knob at the far right. Focus is on the phloem to show a sieve tube (left arrow) that ends (right arrow) just before the xylary knob. (*From N. R. Lersten, Sieve tubes in foliar vein endings: Review and quantitative survey of Rudbeckia laciniata (Asteraceae), Amer. J. Bot., 77:1132–1141, 1990*)

walls of the conspicuous bundle sheath cells force water to pass through narrow gaps between the lamellae.

Phloem. Phloem in many minor veins consists of one to three filamentous sieve tubes (each a file of living but enucleate cells called sieve-tube members that transport sugars linearly) and associated blocky companion cells (nucleated cells that load sugars into, and remove sugars from, sieve tubes) along with two to four phloem parenchyma cells. The sieve tubes are extremely slender, and the companion cells extremely large (**Fig. 3**), which is just the reverse of their counterparts in larger leaf veins and stem bundles. This size reversal is the anatomical response to the physiological necessity of having to pump sugars into sieve tubes against a sugar gradient.

Companion cells, and sometimes adjacent parenchyma cells as well, may have extensive peglike or ridgelike cell wall invaginations, which greatly increase the surface area of the cell membrane (Fig. 3). These wall ingrowths serve two purposes: they allow the expanded cell membrane to more effectively "capture" sugars moving extracellularly from the mesophyll, and they allow reabsorption of ions from water leaving the xylem. These types of companion cells, known as transfer cells, thus pump sugars into sieve tubes and redistribute xylem-borne ions to prevent the excessive accumulation of ions in the leaf. Transfer cells in minor veins are more common in herbaceous species than among woody plants, a distribution pattern that may relate to a need for more rapid recycling of organic and inorganic solutes in nonwoody plants.

The pathway for solute movement into sieve tubes involves extracellular diffusion of solutes and active transport across cell membranes against a sugar gradient. Electron microscope studies show that frequency of plasmodesmata, the threadlike cytoplasmic connections between cells, decreases in the minor veins of some species. This could isolate the sieve-tube–companion-cell complex and thereby mandate active membrane transport. Other studies, however, have shown abundant plasmodesmatal connections to the phloem in minor veins of some species. Some

investigators maintain that plasmodesmata are the primary pathways for photosynthates to move into sieve tubes. However, it has not yet been shown that movement against a gradient can occur through plasmodesmata; these strands appear to be passive portals with no control over the direction of flow of molecules. Although this problem is still unresolved, accumulated evidence favors the extracellular pathway.

Variability in sieve tubes at vein endings. Recent anatomical studies show that blind vein endings in dicots, which may total a third or more of the length of all veins and thus constitute a significant proportion of leaf vasculature, are extremely variable among species with respect to how far they are penetrated by sieve tubes. Sieve tubes may parallel xylem to the tip of all vein endings, may vary in both occurrence and degree of penetration, or may be absent from all vein endings. In one tropical tree species, sieve tubes are absent not only from all vein endings but also from many of the surrounding minor veins. In contrast, there are several families, of which the best known are the Solanaceae (potato, tomato, and so on) and Cucurbitaceae (cucumber, pumpkin, and so on), in which a phloem strand occurs both above and below the xylem, with one or both of the strands always extending to the tip of the vein. A recent quantitative survey of 385 vein endings from a fairly typical dicot leaf revealed that about 30% lacked sieve tubes, 60% had them to some intermediate point (Fig. 2), and only 10% had sieve tubes extending to the vein tip.

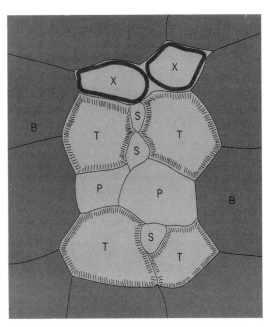

Fig. 3. Minor vein from pea leaflet in cross-sectional view. Two xylem vessels (X) are subtended by the phloem, which consists of three slender sieve-tube members (S) flanked by large companion cells with extensive transfer-cell (T) wall ingrowths; two phloem parenchyma cells (P) lacking wall ingrowths also occur. The minor vein is surrounded by bundle sheath cells (B). (*After B. E. S. Gunning et al., Specialized "transfer cells" in minor veins of leaves and their possible significance in phloem translocation, J. Cell Biol., 37:C7–C12, 1968*)

The finding that sieve tubes can be absent from a substantial proportion of minor veins contradicts the opinion held by most experimenters that all veins have both phloem and xylem, and raises some new questions about vein function. Minor veins, especially vein endings, certainly differ from larger vascular bundles occurring elsewhere, which indicates that they have some unique functions. Sophisticated experiments with dye markers to track water and sugar movement, coupled with quantitative comparative anatomical studies using both light and electron microscopy, hold the promise of linking structure with function for a better understanding of leaf venation.

For background information SEE LEAF; PHLOEM; PLANT TRANSLOCATION OF SOLUTES; PRIMARY VASCULAR SYSTEM (PLANT); XYLEM in the McGraw-Hill Encyclopedia of Science & Technology.

N. R. Lersten

Bibliography. D. A. Baker and J. A. Milburn (eds.), *Transport of Photosynthates*, 1989; M. J. Canny, What becomes of the transpiration stream?, *New Phytol.*, 114:341–368, 1990; D. G. Fisher, Leaf structure of *Cananga odorata* (Annonaceae) in relation to the collection of photosynthate and phloem loading: Morphology and anatomy, *Can. J. Bot.*, 68:354–363, 1990; N. R. Lersten, Sieve tubes in foliar vein endings: Review and quantitative survey of *Rudbeckia laciniata* (Asteraceae), *Amer. J. Bot.*, 77:1132–1141, 1990.

Magma

Chemically zoned magmatic bodies such as ash-flow sheets and lava flows are relatively common. Recent research has elucidated the role of eruption rates in the zonation of these bodies.

Magmatic bodies. Chemically zoned dikes and conduits generally consist of a central zone of silicic rocks with borders of more mafic rocks. Occasionally, some of these dikes and conduits can be seen to pass upward into chemically zoned lava flows (**Fig. 1**). These lava flows consist of silicic rocks overlying mafic rocks. Nearly all chemically zoned lava flows consist of silicic rocks overlying mafic rocks, and nearly all chemically zoned dikes and conduits consist of mafic margins and silicic cores.

In contrast to lava flows, ash-flow sheets commonly contain mafic rocks overlying silicic rocks. Chemically zoned ash-flow sheets are very common. Many of these magmatic bodies (lava flows, conduits, dikes, ash-flow sheets) contain evidence of two or more magma types with some magma mixing and commingling between the types.

Ash-flow sheets are a type of pyroclastic eruption that commonly precedes lava-flow eruptions. The eruption rate for most pyroclastics is at least two orders of magnitude greater (about 1000 m^3 or 35,000 ft^3 per second) than lava flows (about 5 m^3 or 177 ft^3 per second). Because many magmatic bodies degas as they approach the surface, the eruption rate changes from very high (pyroclastic eruptions) to low (lava-flow eruptions) after the magma body is degassed. This difference in eruption rates is very important in controlling the nature of the chemical zoning in lava flows compared with ash-flow sheets.

Eruption history. From the study of these zoned magmatic bodies, it has been concluded that they have erupted from chemically stratified magma reservoirs, with the less dense (more silicic) magma overlying the more dense (more mafic) magma. Because of its higher silica content, the silicic magma is more viscous than the mafic magma. During eruption of a stratified magma body, either through a conduit or a dike, the more viscous (silicic) upper magma flows more slowly than the underlying less viscous (mafic) lower magma (**Fig. 2**), and the lower magma is drawn up into the silicic upper magma (Fig. 2a). Depending on the length of the conduit or dike and the eruption rate, the less viscous mafic magma may get ahead of the more viscous silicic magma. High eruption rates and long conduit or dike length would favor the more mafic magma flowing ahead of the silicic magma. As the eruption rate changes, the depth at which the lower magma is drawn up changes. At high rates, there is a large draw-up depth (Fig. 2); conversely, at low rates, draw-up depths are greatly reduced. In Fig. 2, all of the variables except the eruption rate are held constant. In Fig. 2a the magma body is shown as one-half (the left side) of a box, with the conduit in the upper right-hand side of the diagram. (The right-hand side of the box is not shown.) Two draw-up depths are shown, d_1 for high eruption rates and d_2 for low eruption rates. Magma draw-up depths are dependent on conduit length, eruption rates, and viscosity of the magmas. The draw-up depths are maximum depths from which magma can be withdrawn from the magma chamber. The boundary between the high-viscosity silicic magma and lower-viscosity mafic magma is also shown. The arcs t_1 and t_2 represent different times during the high-eruption-rate stage. From t_1 to t_2 the eruption is entirely confined to silicic magma. At t_2 the interface between the silicic and mafic magma is breached, and the mafic magma would be at the conduit (or dike) entrance. At t_3 the maximum draw-up depth is reached (d_1). At this eruption rate and conduit length,

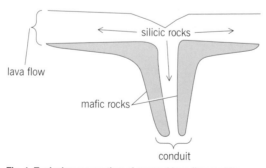

Fig. 1. Typical cross section of a zoned lava flow and its associated feeder dike or conduit. Zoned ash-flow sheets would have the opposite configuration of the flow: mafic rocks would be in the upper portion of the flow.

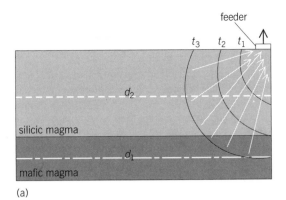

(a)

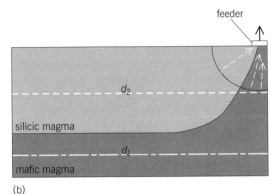

(b)

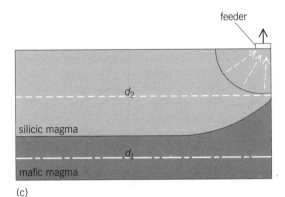

(c)

Fig. 2. Diagrammatic cross sections of a magma body (only one-half of the magma body is shown). Draw-up depths are labeled d_1 (high eruption rate) and d_2 (low eruption rate). The arcs near the feeder indicate the erupted volume after a period of time. Arrows show flow of magma to conduit. (a) Eruption during high rates (pyroclastic stage), with t_1, t_2, and t_3 indicating different times as the eruption proceeds. (b) Position of the silicic–mafic magma interface after t_3 during low eruption rates (lava-flow stage). Maximum draw-up depth is now d_2. (c) Continued eruption with a low rate. All of the mafic magma immediately below the feeder has been erupted. Only silicic magma is available for eruption because the maximum draw-up depth is d_2.

this would be the maximum draw-up for these conditions. At these high eruption rates, until t_2 only silicic magma is erupted to the surface; the arc t_2 in Fig. 2a does not extend into the mafic magma.

Ash-flow sheets. Zoned ash-flow sheets (zoned from silicic rocks at the base to mafic at the top) can be produced during the pyroclastic stage. During the initial stages of eruption, only the silicic top of the magma body is erupted until time t_2 (Fig. 2a), but the lower (mafic) layer is drawn toward the feeder. As

the high eruption rates continue, the lower layer is drawn into the feeder (after time t_2) and erupted to the surface. Figure 2b represents the configuration of the magma in the chamber after time t_3; both the silicic and mafic magmas are erupted simultaneously. At this point in time, the feeder consists of mingled mafic and silicic magma. The resulting ash-flow sheet consists of silicic rocks in the lower portions, which were erupted prior to t_2, and more mafic rocks in the upper portions, erupted after t_2. However, this more mafic upper portion of the ash-flow sheet is a mixture of the mafic magma and silicic magma. The pumice fragments that occur in the upper portion of the ash-flow sheet are from both the silicic and mafic layer in the chamber. Often the pumice fragments contain both magma types as discrete streaks. Occasionally, complete mixing of the two magma types occurs.

Dikes and conduits. Zoned dikes and conduits (zoned from mafic rocks at the margins to silicic in the cores) and lava flows (mafic in lower portion, silicic in upper portion) can be produced after the pyroclastic stage during the lava-eruption stage. At this stage the eruption rates are too low to draw up the mafic layer, but the mafic magma that has already been drawn up into the upper part of the chamber and conduit (dike) is still available for eruption (Fig. 2b). Now the draw-up depth is d_2. As the eruption continues at lower rate, the mafic magma that is near and in the feeder forms the lower part of the lava flows and the margins of the conduit or dike. However, because the supply of mafic magma is limited to the mafic layer that was drawn up into the upper part of the chamber and feeder during the pyroclastic stage, the supply of mafic magma is soon exhausted and only the upper silicic layer is erupted (Fig. 2c). During this stage of low eruption rate, which builds the lava flow, the mafic and silicic magmas initially erupt simultaneously; then, after the mafic magma is removed from the upper part of the magma chamber, only silicic magma is erupted. This leaves a configuration of silicic rocks over mafic rocks for the lava flow, and mafic margins and silicic core for the feeder conduit or dike.

Characteristics. Ash-flow sheets and lava flows are commonly zoned in an opposite fashion. Lava flows have silicic rocks overlying mafic rocks, whereas ash-flow sheets usually have mafic rock overlying silicic rocks. This is generally due to the variation in eruption rates as a stratified magma reservoir erupts to the surface. High eruption rates are associated with the eruption of ash-flow sheets, which commonly precede the eruption of lava flows. During these high eruption rates the draw-up depth continuously increases until a maximum depth is reached (d_1 in Fig. 2). This forms a zoned ash-flow sheet from silicic rocks in the lower portion followed by mafic rocks in the upper portion. As the eruption degasses, the eruption rate abruptly decreases, which results in a decrease in draw-up depth (d_2 in Fig. 2). During the building of the lava flow, magma is erupted from the top of the magma chamber and first consists of a limited supply of mafic magma (Fig. 2b). As the eruption continues, this magma, which is near the feeder, is used up, and only silicic magma is erupted (Fig. 2c). This results in a lava flow with mafic rocks in the lower portion, silicic rocks in the upper portion, and a feeder system zoned from mafic rocks at the margin to silicic rocks in the core.

For background information *see* Lava; Magma; Volcano in the McGraw-Hill Encyclopedia of Science & Technology.

Thomas A. Vogel

Bibliography. S. Blake and I. H. Campbell, The dynamics of magma-mixing during flow in volcanic conduits, *Contrib. Mineral. Petrol.*, 94:72–81, 1986; S. Blake and G. N. Ivey, Magma-mixing and the dynamics of withdrawal from stratified reservoirs, *J. Volcanol. Geotherm. Res.*, 27:153–178, 1986; B. C. Schuraytz, T. A. Vogel, and L. W. Younker, Evidence for dynamic withdrawal from a layered magma body: The Topopah Spring Tuff, southwestern Nevada, *J. Geophys. Res.*, 94:5925–5942, 1989.

Manufacturing engineering

Recent advances in computer-aided manufacturing include the development of computer-aided process planning and computer control of flexible manufacturing systems.

COMPUTER-AIDED PROCESS PLANNING

Process planning in manufacturing determines, in detail, the production steps necessary to convert a drawing of a part to the final manufactured product. Process planning provides the critical link between design and manufacturing. Often, decisions made during this phase have direct and indirect consequences on product cost, quality, and the ability of the manufacturing personnel to utilize the production facilities in an efficient manner. The process planning functions in an industry involve some or all of the following activities: selecting machining operations, sequencing machining operations, selecting machine tools, selecting cutting tools, planning setup requirements, selecting fixtures or work-holding devices and reference surfaces, determining production tolerances, calculating cutting parameters, determining tool paths, establishing numerical control (NC) programs for generation of parts, and estimating time and cost.

The degree of detail incorporated into a typical plan may vary from industry to industry; it depends, to a large extent, on the type of parts, production methods, and documentation needs. The process planning activity has traditionally been experience-based and has been performed manually. The manual approach scrutinizes engineering part drawings to develop plans and manufacturing instructions to develop economical and feasible plans. Here, the experience and skill of a process planner plays an important role in the quality of plans generated. Although it is a laborious, tedious,

and time-consuming approach, manual planning does, in fact, represent a sound approach when few plans have to be generated. A problem facing modern industry is the lack of skilled process planners. Manual process planning has other problems. Variability among planners' judgment and experience often leads to varying perceptions of what constitutes the best plan, thus leading to different plans for the same part; this ultimately results in inconsistent planning. Addition of new machinery and modifications in plant layout often make the old plans obsolete; often replanning is not performed because of lack of time.

The use of computers in process planning began in the late 1960s and has been influenced to a great extent by subsequent developments in numerical control and computer-aided design (CAD). Early attempts to create computer-aided process planning (CAPP) systems consisted of computer-assisted systems for report generation, editing, storage, and retrieval of plans. When used effectively, these systems saved up to 40% of a process planner's time. Such a system can by no means eliminate the process planning task, but it helps to reduce the clerical work of the process planner. Computer-aided process planning can eliminate many of the decisions required during planning; it has the following advantages: reduced demand on the skill of the planner, reduced planning time, more consistent and more accurate plans, and a potential for use in a computer-integrated manufacturing (CIM) environment. The basic approaches to computer-aided process planning can be classified as variant, semi-generative, or generative (automatic).

Variant process planning. The variant approach is based on the concept that similar parts will have

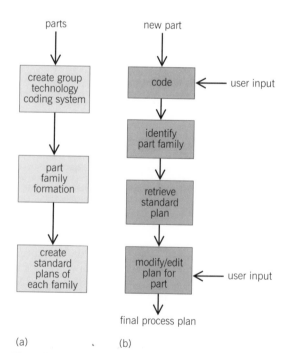

Fig. 1. Flow charts for a variant process planning system. (a) Initial setup. (b) In use.

similar plans. The computer can be used to assist in the identification of similar plans, to retrieve them, and to edit them to suit the requirements for specific parts.

In such systems for computer-aided process planning, schemes based on group technology are used to group parts into families by similar design or manufacturing attributes. A standard plan consisting of a process plan to manufacture the entire family is created and stored for each part family. Before the system can be of any use, coding, classification, family formation, and standard plan formation must be completed. The effectiveness and performance of the variant planning system depend to a large extent on the effort expended to create the coding scheme, family formation, and development of standard plans. This can be a very time-consuming process, often requiring several worker-years.

To use the variant planning system for new parts, the incoming parts are first coded by using the group technology coding and classification system. The code is then used to determine the part family to which the new part belongs. The standard plan for that family is retrieved from the database, and it is edited to suit the requirements for the specific part.

Figure 1 shows the various steps in the building and use of variant process planning systems. Variant systems are easy to build and use; however, there are several problems associated with them. Planning is limited to components that belong to previously defined families; detailed plans cannot be generated; and experienced process planners are still needed to modify the standard plans. Despite these problems, the variant approach is still an effective method, especially in industries where similar components are produced repetitively, and it is the most popular approach in industry today.

Semigenerative process planning. The semigenerative approach is basically an advanced variant system, incorporating some generative features that make the task of process planning easier. The human is still retained in the planning loop and interacts with the computer system during the generation of process plans. The generation of plans is more advanced in that standard plans are not directly stored but are generated for the individual component based on the component attributes captured in a group technology code. The decision logic for creating skeletal or rough plans for the parts is captured in the form of a decision tree or expert system, where the decisions to be made are governed by the attributes identified in the group technology code. The manufacturing knowledge and rules, factory conditions, and manufacturing experience are contained in the decision logic and provide the information to create the rough plan. Refinements to the rough plan and addition of key details are often performed manually via interaction with the computer. **Figure 2** shows the various steps in building and using semigenerative process planning systems.

This approach is limited by the extent of information captured in the group technology code. All the

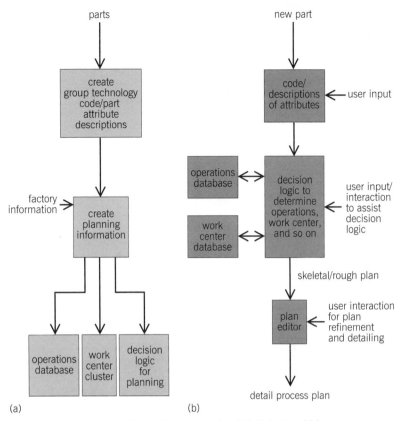

Fig. 2. Flow charts for a semigenerative process plan. (*a*) Initial setup. (*b*) In use.

information required for developing an extensive process plan cannot simply be captured in a group technology code without adding significantly to the complexity and detail of the code. The incompleteness due to the limitations of a group technology code and the associated lack of planning knowledge are overcome by retaining the human in the planning loop. This is the most popular approach for industrial implementations of computer-aided process planning systems.

Generative process planning. The generative (automated) process planning systems are designed to synthesize automatically all the information required to create a process plan without any human intervention. In this approach, the computer is used exclusively; it interprets the design data, generates required operations and sequence, selects tools and equipment, computes processing parameters, and generates the final cutter path [when integrated with a computer-aided design/computer-aided manufacturing (CAD/CAM) system]. A typical system can be viewed as consisting of several modules, each with a well-defined function, and input/output information, with interfaces to other modules and databases (**Fig. 3**). The system contains all the necessary logic, specialized formulas, and algorithms for tasks such as process optimization, tool selection, and cost calculation; it also contains the knowledge of manufacturing needed to check the geometric and processing require-

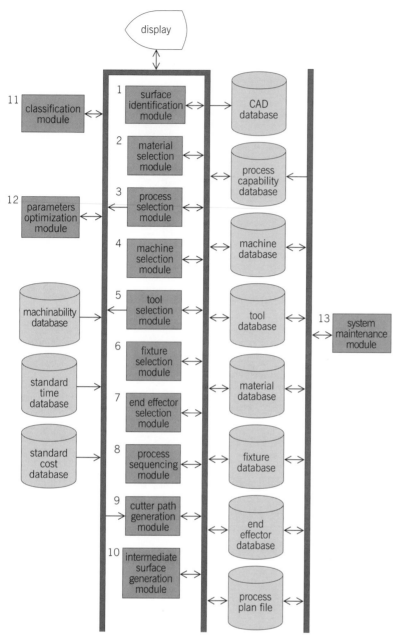

Fig. 3. Diagram showing modules and databases for generative process planning. (*After T. C. Chang and R. A. Wysk, An Introduction to Automated Process Planning Systems, Prentice-Hall, 1985*)

ments of the parts against existing in-house machines and tool capabilities.

The generative process planning approach has several advantages. It can generate consistent plans rapidly, and new components can be planned as easily as the existing ones. In addition, generative process planning has the potential of integrating with an automated manufacturing facility to provide detailed control information for production planning and scheduling activities. Successful implementation of this approach requires the following: development of a part representation scheme, acquisition of process planning knowledge, representation of process planning knowledge, and development of an inference procedure to enable use of the knowledge. Current

research activities in generative process planning have provided limited success in a very few specific and limited areas. The complexity of the problem, lack of adequate tools, and lack of understanding of the effect of interactions between the process planning functions have been significant factors in the slow development of completely automated systems.

Research in generative process planning is focused on several aspects of representation of parts and use of artificial intelligence techniques to represent and manipulate the manufacturing knowledge. The need for integration with CAD systems has led to the development of feature recognition techniques that would automate the interpretation of CAD models. Another emerging approach that is gaining popularity is the application of feature-based design to enable the creation of CAD part models using manufacturing features. This allows manufacturing knowledge to be associated directly with the part features. Planning is then performed individually for each feature. Shortcomings exist with both approaches, and the solution procedure remains an elusive problem, except for small well-defined domains of application. Another area that needs more development is automation of the complex reasoning required for all aspects of process planning. The techniques of artificial intelligence and expert systems have not yet evolved to the degree of sophistication possessed by the minds of human planners.

COMPUTER CONTROL OF FLEXIBLE MANUFACTURING CELLS

Flexible manufacturing cells (FMCs) are an integral part of computer-aided manufacturing and form the basis for all computer-integrated manufacturing activities. A typical flexible manufacturing cell comprises several numerical control machines, material-handling devices (such as robots, automated guided vehicles, and conveyors), fixtures, and tools capable of operation under computer control. Flexibility in a flexible manufacturing cell is provided by the ability to manufacture a range of parts (typically a group-technology–based family of parts), with varying requirements in processing, quantity, operation sequences, and flow through the cell. This flexibility is typically achieved through the use of reprogrammable hardware, with computers and software systems replacing the hard automation systems.

Architecture for control. The control architecture describes the control components (hardware and software) in terms of their configuration, tasks, and interactions, and forms the basis for all physical implementation.

The most commonly used control architecture is the hierarchical. In this architecture, various levels are created to reduce the complexity of the control problems. Each level in the hierarchy is provided with limited responsibility and authority, and it coordinates and plans its activities with the levels above and below it. The time horizon for planning activities decreases from top to bottom in the hierarchy. **Figure 4** shows a typical hierarchy used for shop floor control activities. The distinct levels in this hierarchy are shop, cell, workstation, and equipment.

The functions to be performed at each level are limited by the planning horizons. The planning horizon is the period of time over which the module is responsible for planning and maintaining local goals. The goals must be consistent with the goals of the supervisor (immediate higher level), and they are used to create a unified and coordinated course of action to be followed by subordinates (lower levels). This is achieved by each controller's decomposing the input commands from its supervisor into procedures that must be executed at that level, generating commands to be issued to the subordinate modules, and sending status back to the supervisor module. The decomposition is carried out to the lowest level, where the physical actions are generated by the equipment (**Fig. 5**). The term controller is often used to describe the computer and associated software that performs the desired functions at each level.

The equipment level corresponds to the physical devices such as NC machines, robots, measuring and inspection machines, material-handling and storage devices, and programmable fixtures. The equipment is usually controlled by a device that determines the various capabilities of the equipment. A particular equipment selection may perform a single operation or a variety of operations on a part. Equipment controllers are typically connected to commercial equipment, and they are necessary for running the individual pieces of hardware. These controllers usually execute actions by means of NC part programs or by specialized programming languages. The planning horizon ranges from several milliseconds to several minutes.

The workstation level corresponds to several integrated pieces of equipment. At the simplest level, a workstation could be a robot tending a single machine, along with the requisite fixtures, buffers, sensors, and so forth. A more complex workstation could be a single robot tending several NC machines. The workstation controller directs and coordinates the physical action for the group of equipment. Typical planning horizons range from several minutes to a few hours. The tasks include sequencing subsystems at the equipment level, allocating jobs to machines, coordinating the synchronized activities between machines and material-handling equipment, managing local buffers, monitoring equipment states, and executing part and information flow based on the states of individual machines.

A cell is viewed as several integrated workstations, coupled by material transport workstations. The primary activity of the cell is to provide organizational control of the workstations. Cells are typically organized around group-technology part families, and each cell is intended for particular part families. The functions at this level are responsible for sequencing jobs through workstations, and controlling the inter-workstation material handling. The planning horizon is from several hours to several weeks. Typical activities include balancing of workloads between workstations, creating subbatches, allocating tasks to workstations, assigning due dates to workstations, and

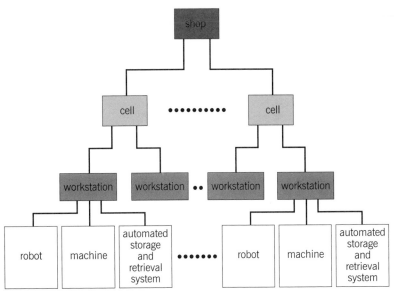

Fig. 4. Diagram showing hierarchical control architecture.

interfacing with the shop and workstation levels to maintain and update status.

The shop is the complete collection of cells; it is conceptually similar to the cell, except that the activities performed relate to cells instead of workstations. This level is responsible for coordinating the production and support activities that are carried out by the cell controllers. The shop control system has a planning horizon ranging from several weeks to several months. The activities include planning capacity, grouping orders into batches, assigning and releasing batches to cells, allocating resources to cells, and keeping track of individual orders. For maintaining flexibility, the architecture allows creation of virtual cells; these are dynamically created to allow workstations to be allocated to cells as desired. This concept allows software reconfiguration of cells without affecting the physical layout.

Other architectures that can be used for cell control activities include distributed control (peer-to-peer control). Distributed control permits coordination and synchronization of activities between various controllers without the intervention of a higher level or other control devices. However, the overall control prob-

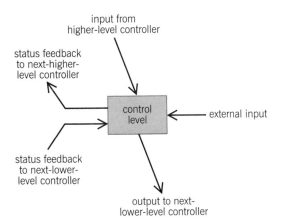

Fig. 5. Diagram showing control actions for individual controllers.

lems are more complex using distributed control, since each controller communicates directly with the other controllers in the system.

Operational modes of computer control. Computers in the control of manufacturing systems can be used in several ways, ranging from data input and processing devices to complete control of untended manufacturing cells. The complexity of software and intelligence required increases in the progression toward computer control of untended cells.

The majority of controllers in current installations perform monitoring, not control. The controller is distinguished from a monitor by its ability to perform active control, which is the ability to direct or regulate the process or actions without the intervention of an operator or higher-level device. When computers are used for monitoring, the controller can be shut down without affecting the process being controlled, the response times are not typically required in real time, and the software used is comparatively simple. The actions to be taken are typically decided by the human operator interacting with the controller.

Controllers for untended manufacturing cells require the ability to make decisions without the intervention of human operators, and all actions and decisions need to be coordinated between the various computers in the control system. Such a system must also have the capability to recover automatically from both hardware and software errors. Therefore, creating such a system with complete flexibility is very complex, and several control problems still need to be resolved.

Another distinction that can be made is between centralized and decentralized control. The difference is in the level of detail in the commands sent to the controllers. In a highly centralized system, the controllers receive explicit instructions from higher-level controllers, and the lower-level controllers merely execute them. The decision-making capability is limited. In the decentralized system, the controllers receive broad goals from which each controller creates its own plan to achieve the goal, and generates goals for lower-level controllers. Intelligence and decision-making capability of each controller is high. Centralized control is typically easier to implement but requires more communications between the levels, thus placing a greater burden on the communication network used to integrate the various controllers.

As the sophistication of flexible manufacturing cells increases, there will be increased needs in software capabilities for control requirements imposed by the rapidly changing environments. Control software will become the bottleneck in the implementation of untended flexible systems. Software systems and tools will be needed to allow plan floor operators and factory engineers to make required changes in the system without frequent intervention by computer professionals. Flexibility, reconfigurability, and software transportability will become key factors in development of control software. Current research efforts are under way in developing and using computer-aided software engineering (CASE) tools to assist in reducing in the control software development time and in increasing reliability and accuracy of the software.

For background information SEE ARTIFICIAL INTELLIGENCE; COMPUTER-AIDED DESIGN AND MANUFACTURING; COMPUTER-INTEGRATED MANUFACTURING; COMPUTER SYSTEMS ARCHITECTURE; CONTROL SYSTEMS; EXPERT SYSTEMS; MANUFACTURING ENGINEERING; ROBOTICS in the McGraw-Hill Encyclopedia of Science & Technology.

Sanjay Joshi

Bibliography. L. Alting and H.-C. Zhang, Computer aided process planning: The state-of-the-art survey, *Int. J. Prod. Res.*, 27(4):553–585, 1989; T. C. Chang and R. A. Wysk, *An Introduction to Automated Process Planning Systems*, 1985; National Electrical Manufacturers Association, *Industrial Cell Controller Classification Concepts and Selection Guide*, NEMA Stand. Publ. 1A-1, 1990; A. T. Jones and C. R. McLean, A proposed hierarchical control model for automated manufacturing systems, *J. Manuf. Sys.*, 5(1):15–25, 1985; S. Joshi and T. C. Chang, Feature extraction and feature based design approaches in the development of design interface for process planning, *J. Intellig. Manuf.*, 1:1–15, 1990; S. Joshi, R. A. Wysk, and A. T. Jones, A scaleable architecture for CIM shop floor control, *Proc. CIM-CON '90*, NIST Spec. Publ. 785, May 1990.

Manufacturing processes

The cost of manufacture of a particular mechanical component is a direct function of production quantity. Conventional automated manufacturing technologies are well suited for large production numbers but ill suited for low production quantities. As a result, manufacture of mechanical components in low production quantities is done mainly by manual methods, which lead to substantially longer production time and higher unit costs.

Currently, several manufacturing technologies that can be used effectively for low production are under development. These technologies, known as solid free-form fabrication or desktop manufacturing, produce free-form solid objects directly from a graphical

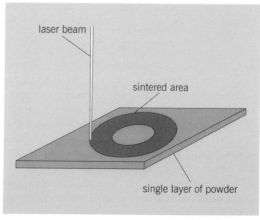

Fig. 1. Diagram of a single layer of the sintering process.

computer description of the object [provided by computer-aided design (CAD) software] without part-specific knowledge or human intervention. One such technology is selective laser sintering. This process begins by depositing a thin layer of powder into a container and heating it to just below its melting point. The first cross section of the object is traced on the powder layer with a laser (**Fig. 1**). The laser raises the temperature of the powder it contacts to the point of sintering. Traditionally, sintering implies high temperature and pressure and substantial time to allow solid-state diffusion. In selective laser sintering, a local melt is created that is more akin to liquid-phase sintering. Laser-beam intensity is modulated to sinter powder in areas to be occupied by the part. In areas not sintered, powder remains loose and may be removed once the part is complete. Successive layers of powder are then deposited and sintered until the entire part is complete. Each layer is sintered deeply enough to fuse it to an underlying layer (approximately 125 micrometers or 0.005 in.). Three-dimensional objects are produced as a result. Two applications are formation of complex metal parts in low production quantities, and rapid prototyping.

PRODUCTION OF METAL PARTS

The energy transfer characteristics of a laser to a metal surface have been studied in great detail since the advent of laser technology. However, there have been fewer studies involving laser-powder interactions under a scanning beam. One process was developed in the mid-1970s to melt a metallic powder layer onto a bulk-metal-part surface. The initial purpose was to take advantage of high cooling rates associated with rapid laser heating in the production of advanced material coatings for applications in the aerospace industry. It was determined that bulk parts could be made by using a continuous powder feed coupled to a reversing/translating or rotating mandrel under the laser. A near-net-shape, nickel superalloy turbine disk 5.2 in. (13.2 cm) in diameter was manufactured with this technique. However, as-lased parts exhibited poor tolerance and surface finish and required significant machining.

There are several differences associated with multiple-layer lasing of powder compared to surface lasing of bulk metals. First, the irregular powder surface allows multiple reflections of the beam and thereby affords better energy transfer to the material.

Fig. 3. Copper–solder metal parts made with selective laser sintering.

Second, if powder surrounds the region scanned by the laser, the cooling rate is lower, which may affect the microstructure of the part. Powder technology offers great flexibility in alloying and blending together different types of metals. Last, melting and resolidification on a bulk surface results in epitaxial growth, which may be deleterious to part quality. In powder it is possible to mitigate this by mixing together powders with different melting points.

Selective laser sintering of metallic parts can be done directly by using metallic powder or indirectly by using polymeric or ceramic materials and post-processing. **Figure 2** shows a brass casting made from a laser-sintered pattern.

Direct method. The short interaction time between the scanning laser and any point on the powder surface requires rapid local binding of discrete powder particles. This has been accomplished by melting one metal in a binary powder blend. **Figure 3** shows thin parts made directly by using a copper–solder powder blend. The solder melted during lasing and wet the solid copper; this is similar to liquid-phase sintering. Control of oxide formation in the melt is critical to good wetting and has been accomplished with inert or reducing atmospheres and with solid fluxes mixed with the powder. One technical area where this approach is promising is in production of tools and dies made of various carbides held together by a matrix of a metal with a lower melting point, such as cobalt.

Numerous issues must be considered when designing a powder blend for direct metal selective laser sintering. Besides oxide control, residual stress arising from the thermal gradients intrinsic to processing may cause layer distortion, particularly in the initial few layers. Raising the ambient temperature of the process chamber reduces the thermal gradient, with concomitant decrease in residual stress distortion. Another consideration is residual porosity in a part. The molten phase during sintering may not completely fill the space between remaining solid particles. Traditional powder-metallurgy techniques can be used to reduce porosity through postprocessing heat treatment. Here, part dimensions must be slightly larger than the desired final dimensions to accommodate shrinkage.

Indirect methods. Two indirect methods for making metal parts are generation of lost-wax patterns from wax

Fig. 2. Brass casting from a laser-sintered pattern.

powder and a ceramic "lost-lost-wax" process. Polymer parts are easier to sinter than metal due to their low melting point and softening characteristics. A wax pattern of the part is made, and then well-established lost-wax techniques are used to create a ceramic mold into which any metal can be cast. Another polymeric indirect technique involves coating metallic powder particles with a thin layer of polymer like polymethyl methacrylate. During selective laser sintering, the polymer surface material sinters together. The part is subsequently given a postprocessing heat treatment to volatilize the polymer and to sinter the remaining metal in the solid state to generate structural integrity.

The lost-lost-wax technique consists of sintering a ceramic mold directly from ceramic powder blended with a glass or ceramic of low melting point, thereby eliminating the need for a wax pattern. A significant manufacturing advantage is immediately realized, because not only can a mold cavity be created directly, but also the requisite mold features (sprues, gates, risers, and so on) can be generated simultaneously.

David Bourell; Joseph Beaman

RAPID PROTOTYPING

Manufacturers are seeking ways to expedite the process of new product design, development, and manufacturing. The product design segment of the overall manufacturing process is particularly time-intensive, with design reviews driving successive cycles of product modeling and prototyping that are, in turn, acquired through conventional machine and modeling operations, often external to the manufacturer's operations.

There is a trend toward flexible manufacturing as industries of all types respond to rapidly changing market requirements; customer service issues; design for manufacturing; and requirement for a greater variety of complex products, in smaller quantities, with shorter life cycles. Modelmaking and prototyping capabilities are critical components of a flexible manufacturing process and are now addressed by new technologies variously known as desktop manufacturing systems, solid free-form fabrication systems, or rapid prototyping systems. These technologies, although different in implementation, all address the concept of providing the transfer of the design geometry from computer-resident software to three-dimensional, solid fabrication in the real world.

The selective laser sintering process is one implementation of desktop manufacturing or rapid prototyping systems. Design models that are created in CAD, computer-assisted manufacturing (CAM), and computer-assisted engineering software are provided to rapid prototyping systems through a compatible interface known as the .STL file. This file format is currently the industrial standard; it is a method used to convert CAD databases into a form that is usable by rapid prototyping systems.

The .STL file is a faceted representation of the part designed in CAD. The .STL file is then transferred to the selective laser sintering machine for computer-based processing. The modulated laser beam selectively describes the part geometry based on the coordinate points. The design information is thus transferred from a computer representation to an actual physical object through a unified, completely digital process (**Fig. 4**).

Advantages. There are several advantages to users of selective laser sintering that are tied directly to the inherent capabilities of the technology. These include the ability to use a variety of heat-fusible powdered materials, the material-additive nature of the process, and its relative speed.

Multiple materials. Selective laser sintering is capable of sintering a wide variety of powdered materials. Actually, any material that softens and has decreased viscosity upon heating can potentially be used as an input material. Specific materials include acrylonitrile butadiene styrene (ABS) plastic, polyvinyl chloride (PVC), polycarbonate, and investment casting wax. Cast metal parts can be created rapidly through the sintering of investment casting wax by selective laser sintering and then using lost-wax casting as a secondary process. Ceramic parts can be fabricated through the process from polymer-coated ceramic powder, followed by oven-curing the object as a secondary process.

A traditional bottleneck in the design cycle process

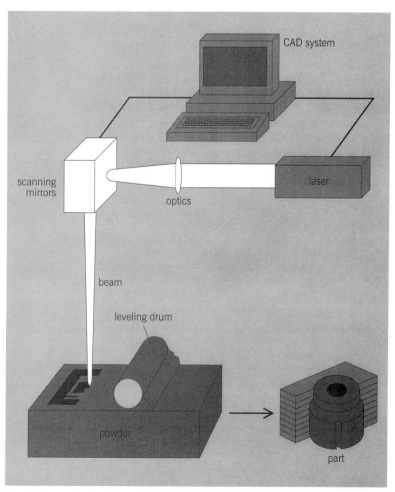

Fig. 4. Diagram showing the components of the selective laser sintering process.

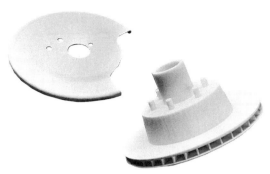

Fig. 5. Prototype of a disk brake assembly made from polycarbonate material.

has been the cost and inefficiencies associated with producing models and prototypes. The speed advantage gained through the use of selective laser sintering has a significant impact on the length of product design cycles, actually allowing a greater number of design cycles in an overall shorter period of time, resulting in a better final product.

Another related issue is that of producing functional, "testable" prototypes, in contrast to the simple, visual representations of plastic models or the time and cost associated with prototypes machined from metal or other materials. The ability of design engineers to test their designs for function, form, and fit in materials as similar to the final product as possible is critical in developing a superior product. Selective laser sintering provides this capability through the use of a variety of materials in the fabrication of prototypes and models. These materials may not be those employed in the final product, but they have a set of properties sufficient to provide correlation of data that would project final properties of the product. **Figure 5** shows an example of a prototype for a disk brake assembly made from polycarbonate material by using selective laser sintering.

The natural evolution of the technology is toward a process capable of sintering various metals. This capability would have widespread application throughout manufacturing industries, as over 60% of all design prototypes are currently formed of various metals by using conventional processes.

An additive process. Selective laser sintering does not require traditional dies or any type of part-specific tooling to produce parts. Therefore the process is particularly applicable where fewer, more complex designs are being prototyped or modeled. Large production runs warrant part-specific tooling for products with extended life expectancies or high volume requirements. Also, since selective laser sintering is a material-additive process, the creation of very complex parts is possible without the need for external support structures or clamping and repositioning of specific portions of the design geometry. The fabricated part is built one thin two-dimensional layer at a time from the bottom up. This allows the creation of very complex parts such as objects with internal cavities, overhangs, notches within notches, and other

geometries difficult, if not impossible, to achieve through conventional machining processes. The unsintered powder surrounding the fabricated part provides a simultaneous customized support structure easily removed upon part completion.

Relative speed. Selective laser sintering can produce a model or prototype at the rate of 0.5–1 in. (1.25–2.5 cm) per hour, dependent upon the relative complexity of the part design geometry. With the exception of ceramics, the process does not require any postproduction curing of the fabricated part. Current implementations of analogous technology all require some degree of postproduction curing to completely harden the liquid photopolymer resins. In addition, these technologies currently require time-consuming crosshatching scans of various types of lasers, interspersed with reapplication of the photopolymer resin, before removing the part for final curing. The software for selective laser sintering is designed to allow concurrent slicing of the part geometry files with fabrication processing.

Primary applications. Primary applications of selective laser sintering will be in three major areas: visual representations of product designs through modelmaking, functional prototypes, and production capabilities including both on-demand batch manufacturing and the production of molds and patterns required for large-scale production runs. These aspects are of interest in the automotive, aerospace, computer, consumer goods, foundry, and medical industries.

For background information SEE COMPUTER-AIDED ENGINEERING; COMPUTER-AIDED MANUFACTURING AND DESIGN; POWDER METALLURGY; SINTERING in the McGraw-Hill Encyclopedia of Science & Technology.

Carl Deckard

Bibliography. D. P. Colley, Instant prototypes, *Mech. Eng.*, pp. 68–70, July 1988; M. Fallon, Desktop manufacturing takes you from art to part, *Plast. Tech.*, February 1989; P. H. Lewis, Advances: Device quickly builds models of a computer's designs, *N. Y. Times*, March 16, 1988; H. L. Marcus et al., From computer to component in 15 minutes: The integrated manufacture of three-dimensional objects, *JOM*, 42(4):8–10, 1990.

Materials science and engineering

The need for development of materials for the enhanced performance of aerospace systems of the twenty-first century is critical. The materials of choice will be drawn from both monolithic and composite metals, polymers, and ceramics. It has been suggested that achievement of the desired material properties will require a transition from a so-called basic materials age to an era of engineered materials.

Advanced materials. Aerospace systems will be required to operate at higher elevations and to fly faster and farther, thus requiring engineered materials that are stronger, stiffer, hotter, and lighter than those in current use. The new materials will often be synthesized from the atomic level. Therefore, they will require a total engineering approach involving the traditional relationships of chemistry processing, structure, properties, and applications; considerations such as life prediction, producibility, durability, reliability, inspectability, and maintainability; as well as the important factor of cost.

As an example of increased demands involving performance, fighter aircraft must operate over broader combat envelopes with supersonic persistence, at greater turn rates, higher load factors, and maximum Mach numbers. The data shown in **Fig. 1** indicate the major role that improved materials play in structural weight reduction, greatly exceeding the contribution made by other advanced technologies such as structural concepts, automated design and active load control, and flutter suppression.

The requirement for improved materials is even more critical to achievement of higher performance in propulsion systems. The trend is toward mechanically simpler engines, with fewer components, operating under increased temperatures, made of lower-density materials, and bearing higher load; these lead to an increased ratio of thrust to weight.

Because of the ease of working with polymer matrix materials (which have the characteristics of low melting point and low reactivity), advances in engineered materials have occurred more rapidly with organic materials than with metals. However, this does not mean that carbon-based materials will take over from metals. Rather, the best material will be selected for any particular application, with cost always a concern.

In the past, the alchemist did not talk to the blacksmith. However, now there is interaction between those who study the fine detail of materials and those who fabricate components. Synthesis and processing of materials are seen as playing a key role in the production of a cost-effective product with enhanced physical and mechanical characteristics. A vast variety of manufactured materials, including synthetic-polymer fibers and plastics, metal alloys,

high-strength ceramics, and composite materials, have been developed.

Airframe. A major application of advanced materials is for airframes. Advanced materials for airframes are discussed under the categories of low-temperature (to about 500°C or 900°F) and high-temperature (above 500°C or 900°F). Generally, the organic polymer matrix composites, aluminum, and magnesium are classified as low-temperature; titanium, superalloys, intermetallic compounds, refractory metals, ceramics, and carbon/carbon composites are classified as high-temperature, although there is some overlap.

Low-temperature. The low-temperature materials for airframes are mainly either the advanced organic matrix composites or aluminum. By the beginning of the twenty-first century the higher-temperature thermoset polyimide organic matrix composites will be well established, as should thermoplastics. Following close behind will be hybrid materials and molecular composites. It has been forecast that in the United States production of composites for aircraft will be as high as 3.8×10^6 lb (1.7×10^6 kg) per year by 1995. The tough solvent-resistant thermoplastics will be available in unidirectional and cloth prepreg (fiber reinforcement coated with matrix resin) as well as commingled tows, fibers coated with resin from powder slurries, or extruded matrix coatings; this will allow great design flexibility and manufacturing options. Cost reduction will come from an integrated factory-of-the-future approach. A major step in this direction is represented by the all-composite experimental aircraft, the *Voyager* and the *Starship*, made from carbon fiber/epoxy laminate over an aramid core.

Advances are also being made in metals, especially aluminum, for example, the development of low-density aluminum-lithium (Al-Li) ingot for elevated temperatures, and high-strength alloys and metal matrix and hybrid lamellar materials. Development of low-cost processes and early maturity (involving a reproducible, adequate database) will play a major role in determining the material to be used in any particular application. Other advances for the future include corrosion-resistant magnesium alloys that are produced by either ingot metallurgy or rapid-solidification methods, even-lower-density alloys of aluminum-lithium-beryllium (Al-Li-Be), and graphite-aluminum and graphite-magnesium, particularly for space-environment use.

High-temperature. For future airframe construction, advanced aluminum alloys (possibly mechanically alloyed or intermetallic compounds such as modified Al_3Ti), advanced titanium alloys (such as rapidly solidified terminal alpha alloys, which have a low-temperature allotriomorphic structure at the Ti-rich end of the phase diagram), ordered intermetallic compounds based on Ti_xAl or Ni_xAl (where x generally = 1 or 3; Ni = nickel), advanced superalloys, carbon/carbon composites, ceramic composites, and refractory metals with environmental resistance are potential candidates, as are newer composites. A number of these proposed materials are quite complex

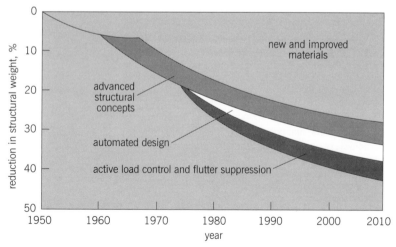

Fig. 1. Graph showing trends in structural weight reduction for fighter aircraft. The major contribution of new and improved materials is evident.

and therefore potentially expensive, such as reusable carbon/carbon composites. For the latter materials integrated oxidation protection systems have been proposed; these use combinations of outer surface coatings, inner layer sealing, inhibited matrices (carbon/carbon materials in which the matrix materials have chemical constituents that reduce or retard oxidation), and carbon-fiber coating to achieve improved resistance to cyclic oxidation (a condition in which surface regions can crack during temperature cycling).

Engine. To date, the two materials preferred for gas turbine engines have been metallic: titanium and nickel-base superalloys (hotter sections). Engines having increased thrust-to-weight ratios have been developed to a large extent through reduced density (titanium replacement of nickel-base superalloys in the compressor) and increased strength.

The large gains to be made by increased thrust-to-weight ratio render the expensive exotic materials more acceptable in engines than in airframes, if they lead to the desired enhancement of performance (**Fig. 2**). Since engines are generally hotter than airframes, low-temperature materials are defined as those useful up to 700°C (~1300°F), and high-temperature materials are defined as those that are candidates for components likely to have to function at temperatures above 700°C (~1300°F). As with the airframes, there is some overlap.

Low-temperature. The strong competition between the organic polymer composites and both monolithic and composite metals will occur here also. Substantial savings in weight and cost could result from replacing conventional metal components with composites useful to 400°C (750°F). Inner ducts, which are subjected to relatively high temperatures, will require the high-temperature organic materials such as bismaleimide, polyimide, or advanced thermoplastics.

In the field of metals, the processing methods of rapid solidification, mechanical alloying, composite fabrication, and vapor deposition have extended the useful temperature range of aluminum alloys to 300°C (570°F) and beyond. Lower-density Al-Li alloys have a useful temperature range up to 175°C (350°F), making them highly competitive with organic composites. Dispersion-strengthened rapidly solidified magnesium alloys with enhanced corrosion behavior could be used extensively. Titanium alloys with higher strength (to 1700 megapascals or 245,000 lb/in.²) and lower density (approaching that of aluminum) could also see increased use. Additionally, advanced high-temperature titanium ingot alloys (to 600°C or 1100°F), dispersion-strengthened, rapidly solidified alloys reinforced with silicon carbide (SiC; to 700°C or 1290°F), will see use in advanced engines. Titanium castings are also likely to see increased use, particularly if thin sections can be successfully cast from advanced alloys.

High-temperature. By far the greatest potential as well as the greatest challenge is present in the higher-temperature regime of the advanced turbine engine. However, only very limited replacement of superalloys is likely by the year 2000. Thus, it is worthwhile

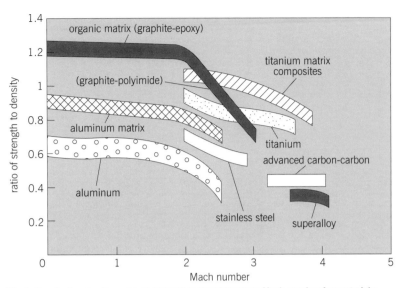

Fig. 2. Graph showing the ratio of strength to density versus Mach number for materials under consideration for advanced gas turbine engines. The organic matrix is graphite-epoxy at lower temperatures and graphite-polyimide at higher temperatures.

to consider briefly the advances that may occur with superalloys, although it is anticipated that an exotic spectrum of materials will become available beyond the year 2000. Control of grain size and shape, intricate components containing cooling passages, improved coatings, and mechanical alloying are likely to see increased use. Additionally, a major object of research for the superalloys in the 1990s will be to build in quality with improved melting practices.

Beyond superalloys, materials with capabilities of service at higher temperatures offer the potential for the elimination of the requirement of cooling air and also significant enhancement of efficiency. Among the materials receiving consideration for use at temperatures exceeding the capabilities of the superalloys are the intermetallic refractory metals, ceramics, and reusable carbon/carbon and metal matrix composites. The negative aspect of these families of materials lies in their high density and poor oxidation behavior (intermetallic refractory metals), brittleness (ceramics), and processing difficulties (carbon/carbon and metal matrix composites).

For the refractory metals, judicious alloying using the techniques of rapid solidification may allow the necessary protective oxide to form. Coatings offer a much less desirable option. Research in developing new composite materials is also being pursued, for example, tungsten fibers in a matrix of niobium.

The ceramics, including reinforced types, are attractive candidates, especially silicon nitride (Si_3N_4) and silicon carbide reinforced with fibers such as graphite and silicon carbide. The composite concepts allow some forgiveness to be designed into the material. Preliminary stress/design methodology, understanding of the behavior of these materials, and manufacturing processes have been developed. In terms of mechanical properties, the ceramic matrix composites offer a density that is 70% lower than that of the superalloys, a maximum use temperature of

1650°C (3000°F; compared to 1150°C or 2100°F for uncooled superalloys), and a strength of 170 MPa (25,000 lb/in.^{2}) at 1300°C or 2370°F, a temperature at which superalloys have almost zero strength. However, the challenge is to produce ceramic matrix composites with reliable and predictable behavior.

Development of reusable carbon/carbon composites is an equal or greater challenge. Presently, such composites are used in rocket nozzles, where the ablative characteristics provide the short life necessary, and in aircraft brakes, where temperature exposure is short. The long and complex manufacturing cycles necessary to produce matrix inhibition and coatings are very expensive, and as yet they are far from optimized. At present, service life is too short for commercial engines, but for the future carbon/carbon composites seem likely to have application for hot-section components such as tip shrouds, turbine vanes, and exit nozzles.

Another class of materials that appears promising are the metal matrix composites, and it has been proposed that by 2010 as much as 80% of the engine could be made of a mixture of composites, metal and ceramic matrix. The excellent specific strength of metal matrix composites is demonstrated by the estimated value for unidirectionally reinforced titanium aluminides with silicon carbide fibers, which is nearly twice that of the alloy Inconel 718 (density-corrected), over twice that of the alloy Ti-6Al-4V at 540°C (1005°F; V = vanadium), and twice that of unreinforced titanium aluminide at 760°C (1400°F).

The intermetallic compounds such as the titanium (Ti_xAl) and nickel aluminides (Ni_xAl, where x generally equals 1 or 3) have lower densities than conventional superalloys. Potential applications of these materials include replacement of single-crystal alloys in airfoil blades with Ni_3Al. Beyond these intermetallics are materials such as Nb-Al intermetallics, $(Ta,Ti)Al_3$, and molybdenum disilicide ($MoSi_2$), where the challenge of building in useful levels of forgiveness is even greater.

Rapid solidification of superalloys and perhaps other materials will probably receive further attention in the next decade in applications where there seems to be a potential for circumventing the alloying constraints indicated by the equilibrium phase diagram. Applications may be found for layered structures, produced by techniques such as electron-beam vapor deposition, although thermal stability will be a concern. Mechanically alloyed materials strengthened by oxide dispersion have not yet been developed to their full potential; it is expected that advances here could come in conjunction with fiber reinforcement, giving a potential use temperature that is 200°C (360°F) or more beyond current superalloys. Tailoring of surface regions by using techniques such as chemical vapor deposition, ion implantation, and reaction magnetron sputtering could lead to enhancement of resistance to wear, fatigue, and corrosion. Another form of tailoring, which is already being evaluated, is the tailoring of disk hub regions to enhance fatigue performance and disk blade regions to improve creep behavior.

General constraints. In all of the concepts discussed above, it is imperative that the materials behave in a reliable, predictable manner, so that the useful life of the component can be defined. However, these materials are unlike those in present use. New "lifing" concepts are needed so that it will be possible to make the most effective use of these new materials in a relatively short time, with the goal being improved performance and reduced cost of ownership. Processing in a so-called factory of the future will be needed to build in reliability and to maintain cost at acceptable levels.

For background information SEE AIRFRAME; ALLOY; CERAMICS; COMPOSITE MATERIAL; ENGINE; HIGH-TEMPERATURE MATERIALS; INTERMETALLIC COMPOUNDS; MACH NUMBER; METAL MATRIX COMPOSITE; POLYMERIC COMPOSITE in the McGraw-Hill Encyclopedia of Science & Technology.

F. H. Froes

Bibliography. W. Betz et al. (eds.), *High Temperature Alloys for Gas Turbine and Other Applications*, vol. 1, pt. I, 1986; F. H. Froes, Aerospace materials for the twenty-first century, *Swiss Mater.*, 2(2):23, 1990; R. N. Hadcock, *AIAA Aerospace Conference and Technical Show Proceedings*, Los Angeles, February 12–15, 1985; Materials for economic growth (entire issue), *Sci. Amer.*, no. 4, October 1986; Office of Technology Assessment, *New Structural Materials Technologies*, 1986; J. Williamson et al. (eds.), *Evolution of Aircraft/Aerospace Structures and Materials*, AIAA, Dayton, Ohio, April 1985.

Medical chemical engineering

Since about 1970, the application of heat to treat cancer has been the subject of extensive developmental and clinical investigation throughout the world. Hyperthermia, which has been used to treat various human ailments since the time of Hippocrates, currently involves raising the temperature of a tumor-bearing body region by using immersion methods (hot-water bath, hot-air suit), extracorporeal heating of blood, ultrasonic heating, or electromagnetic heating. The therapeutic temperature range for an effective treatment is 42–45°C (108–113°F). When applied as an adjuvant procedure with chemotherapy and radiation therapy, regional hyperthermia has been successful in inducing a complete destruction of the tumor in many cases. Because normal tissue is as susceptible to the deleterious effects of heat as cancerous tissue, heat transfer in the hyperthermic region must be well understood and predictable. The study of bioheat transfer involves applying the basic principles of conservation of mass and energy to predict heat transfer and temperatures inside vascularized tissue under hyperthermic conditions. Recent advances in bioheat transfer include both theoretical and experimental methods by which therapeutic hyperthermia can be analyzed.

Use of theoretical models. The application of a valid heat-transfer model is necessary in order to quantify the thermal effects of hyperthermia for several reasons. In the clinical setting, the physician cannot measure the temperatures everywhere within a treated region. Typically, one or two temperature probes are used to monitor the heated tissue; however, a treatment will often be unsuccessful unless the entire tumor is heated to the therapeutic level. In addition, a subtherapeutic heat dose can induce thermotolerance in the cancer cells, reducing the effectiveness of subsequent hyperthermia treatments.

The lack of available temperature measurements underscores the importance of bioheat transfer modeling in planning a hyperthermia treatment. A mathematical model of heat transfer in human tissue that is exposed to a device that induces hyperthermia can be used to predict the temperatures within this body region. This type of model must include not only the relevant physical phenomena but also the physiological mechanisms that control body temperature. With such a model, the existence of potential cold and hot spots within the treated region may be known in advance. This type of heat-transfer model is invaluable to the development of a safe and efficacious hyperthermia protocol.

Whole-body modeling. Whole-body thermal models have been developed to quantify the extent of the systemic heating caused by regional hyperthermia treatments, especially those in the abdomen and the thoracic core. **Figure 1** shows a schematic view of a whole-body thermal model of a human and the associated circulatory system.

Models of whole-body heat transfer subdivide the human into numerous body segments. These are subdivided into several compartments of tissue and blood, usually separating the arterial, venous, and capillary circulations within a given body segment. The energy-balance equations for each of the tissue segments must include terms for conduction, convection, radiation, and heat sources within tissues. The important heat-source terms are the metabolic heating of tissue, evaporative heat loss, muscular work, and the presence of the hyperthermic heat source that has been applied externally during a therapeutic hyperthermia treatment. The terms for convective heat transfer involve heat transfer between the skin and the ambient environment as well as convection in the microcirculation between the vascularized tissue and the arterial blood and venous blood that perfuse the tissue. A heat-transfer coefficient between the blood compartment and surrounding tissue is often utilized to model the skin–environment effect, while several different methods have been utilized to model convective heat transfer in the microcirculation.

Variations in the temperatures of blood and tissue as well as the rates of blood flow throughout the body during a local hyperthermia treatment can be computed by using the principles of conservation of mass and energy. The calculations of the changes in local body temperatures and blood flows during a hyperthermia treatment are based on experimental models of human thermoregulation and the appropriate principles of heat transfer.

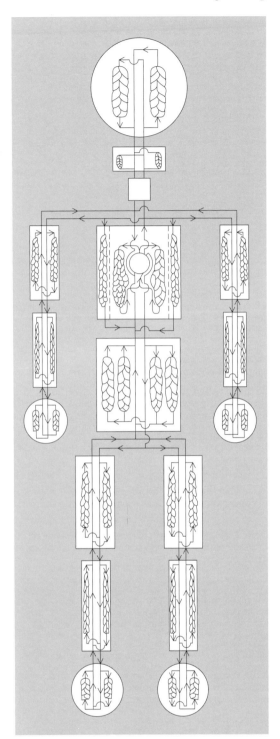

Fig. 1. Schematic diagram of a segmental whole-body thermal model of a human with a circulatory system.

Thermoregulation. Human thermoregulation can be modeled mathematically by using a set-point control system. The degree of vasodilation and vasoconstriction of blood vessels in the skin layers has been shown empirically to depend on an integrated whole-body skin temperature signal, the hypothalamic temperature, and the differences between local skin temperatures and their respective set-point temperatures. In addition, regional blood flow depends on both the

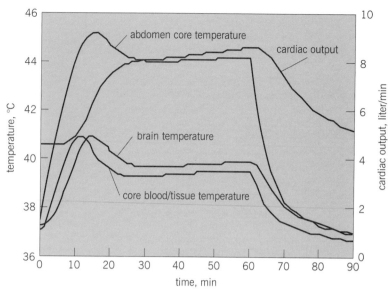

Fig. 2. Graphs showing simulated whole-body response to regional hyperthermia (1000 W) in the abdomen. The hyperthermia heating source is turned on at time zero and off after 60 min. °F = (°C × 1.8) + 32°, 1 liter = 1.06 qt.

local temperature and the central thermoregulatory command. Typically, basal (resting) skin blood flow is doubled for every 5°C (9°F) increase in local temperature above the set-point temperature of that region, and halved for every 5°C (9°F) decrease in the local temperature below the same set-point temperature.

Experiments have also revealed that the change in skin blood flow is not only a function of the skin and hypothalamic temperatures but also a function of time. A first-order time-constant type of delay in the human cardiovascular response has been observed experimentally, and this behavior can be easily incorporated

into mathematical models of human thermoregulation.

The blood flow to muscle during a hyperthermia treatment varies with the local muscle temperature, but not in the same manner as that in the skin. Measurements made in human subjects have revealed that muscle blood flow remains at a basal level until the local muscle temperature reaches a threshold temperature between 41 and 43°C (106 and 109°F). Once above this temperature, the blood flow to the muscle can be modeled to vary with the local muscle temperature in a linear manner, while simultaneously subjected to a first-order time-constant behavior similar to that described above for the thermoregulatory mechanism of skin blood flow. The rates of blood flow to the core and fat layers of the various body segments are usually kept constant at the basal level under all physiological conditions. Blood volumes in the individual segments, however, may vary during a hyperthermia treatment, due to the shunting of blood between body regions.

Evaporative heat loss. Heat is transferred by evaporation from the skin to the ambient air according to the thermoregulatory control equation for total sweating. The partial pressures of water in the skin and air, which are functions of temperature and relative humidity, are significant in determining the extent to which heat may be shed from the body via evaporation. Heat is also lost by evaporation of water from the lungs to the ambient air. The evaporative heat lost via this route depends on the ventilation volume and the partial pressure of water in the ambient air.

Simulation results. **Figure 2** illustrates the predictions of changes in temperature and blood flow during a simulated 1-h hyperthermia treatment in the abdomen using a whole-body thermal model of a human. Systemic heating effects are indicated by the significant increases in cardiac output and temperatures of body core (core blood and abdomen core). Comparisons of these modeled treatments with experimental data obtained from clinical procedures have shown that such mathematical models are capable of qualitatively predicting the whole-body response to local hyperthermia.

Microcirculatory heat transfer. Heat transfer in vascularized tissue depends significantly on the amount of blood flow with which the tissue is perfused as well as on the temperature of this blood. This effect was first modeled in 1948 by H. Pennes, who described the thermal effect of blood flow as a temperature-dependent heat source or sink term proportional to the volumetric blood perfusion rate in the tissue and the temperature difference between the arterial blood entering the tissue and the exiting venous blood. The Pennes model, also known as the bioheat equation, assumes that in a manner analogous to the mass transport of oxygen from blood to tissue, all of the effective heat transfer takes place in the capillaries, and there is no precapillary or postcapillary heat gain or loss. Consideration of microvascular geometry in tissue has led to the formulation of alternatives to the bioheat equation.

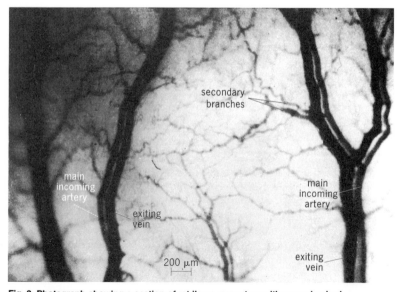

Fig. 3. Photograph showing a section of rat ileum mesentery with several paired arterio-venous networks. (*From S. Baez, Skeletal muscle and gastrointestinal microvascular morphology, in G. Kaley and B. M. Altura, eds., Microcirculation, vol. 1, University Park Press, 1977*)

Normal human muscle and mesentery tissues are embedded with a branching tree structure of both arteries and veins, whose diameters range from 50 to 500 micrometers. **Figure 3** illustrates the paired, countercurrent flow arrangement of arteries and veins in the mesentery of rat ileum, which is typical of that observed in skeletal muscle as well. The temperature of the arterial blood that perfuses the tissue is highly dependent upon the degree of heating or cooling to which the flowing blood has been exposed before it reaches the capillary level. The thermal behavior of this flow network has been characterized in a manner similar to a countercurrent heat exchanger. A particularly useful method computes tissue temperatures by using an effective thermal conductivity that accounts for the size and spacing of the blood vessels in the countercurrent arterio-venous network. This effective thermal conductivity depends on the velocity of blood in the countercurrent vessels, the size of the vessels, as well as the inherent thermal conductivity of the tissue. As the blood velocity decreases, the thermal significance of the countercurrent pair decreases; consequently the arterial and venous temperatures approach the surrounding tissue temperature, which is an indication of thermal equilibration.

The relative merits of the Pennes and effective thermal conductivity bioheat transfer models are currently being investigated theoretically and experimentally. It is apparent, however, that both vascular architecture and metabolic state, which affects the velocity of blood in the vessels, play a significant role in the heat-transfer characteristics of perfused tissue.

Implications. Bioheat-transfer models are a valuable method by which to plan and evaluate a hyperthermia treatment. A subject of much interest in hyperthermia research today involves the experimental verification of these proposed models via the measurement of both temperature and blood flow in a region undergoing hyperthermia. With the recent advances in noninvasive techniques of simultaneous measurement of temperature and perfusion, more data of this type are being collected. As these data become available, bioheat transfer will yield an increased understanding of the physiological and physical principles that govern the thermal response of tissues to heat.

For background information SEE CANCER (MEDICINE); HEAT TRANSFER; MEDICAL CHEMICAL ENGINEERING; THERMOREGULATION; THERMOTHERAPY in the McGraw-Hill Encyclopedia of Science & Technology.

Caleb K. Charny

Bibliography. A. Shitzer and R. C. Eberhart (eds.), *Heat Transfer in Medicine and Biology*, 1985; D. H. K. Lee (ed.), *Handbook of Physiology: Reactions to Environmental Agents*, American Physiological Society, 1977; K. M. Sekins et al., Local muscle blood flow and temperature responses to 915 MHz diathermy as simultaneously measured and numerically predicted, *Arch. Med. Rehab.*, 65:1–7, 1984; S. Weinbaum and L. M. Jiji, A new simplified bioheat equation for the effect of blood flow on local average tissue temperature, *Trans. ASME, J. Biomech. Eng.*, 107:131–139, 1985.

Medical imaging

The development in the early 1980s of magnetic resonance imaging (MRI) for clinical diagnosis has prompted the development of a new class of contrast agents for use in living organisms. The pharmaceuticals developed for contrast enhancement in magnetic resonance imaging differ from the classical iodinated agents used for x-ray and computerized tomography applications in that they generally contain a paramagnetic metal ion encapsulated in an organic chelate or ligand.

Magnetic resonance imaging. As a diagnostic tool, magnetic resonance imaging delivers superior anatomical resolution (important for staging cancer) and sensitivity (important for diagnosing the presence and course of disease) as well as higher inherent tissue contrast (important for both diagnosis and staging) without the use of ionizing radiation. Other advantages include its multiplanar capabilities and lack of imaging artifacts from bone and teeth. Currently, magnetic resonance imaging is used extensively for studying brain and spine tumors, and its use in the musculoskeletal system, heart, abdomen, and excretory organs is rapidly growing.

Magnetopharmaceuticals. The recognition of the importance of paramagnetic materials and their effect on the relaxation times of resonating protons occurred almost simultaneously with the discovery of the magnetic resonance process itself in 1946, when it was demonstrated that paramagnetic ferric ions in solution could reduce the longitudinal (spin-lattice) nuclear relaxation rate of water protons. Research in the 1950s extended this work to a variety of paramagnetic transition metals and produced a theoretical description of the effect.

Throughout the 1960s, these results provided the framework for the development of paramagnetic shift reagents and relaxation probes that greatly assisted the development of nuclear magnetic resonance (NMR) spectroscopy. The development of the technique of magnetic resonance imaging led to the development of contrast-enhancing media. The first use of paramagnetic ions and chelate complexes studied in dogs to alter relaxation times in tissues was described in 1978.

Chelate effect. Although the paramagnetic metal ions present in enhancement agents for magnetic resonance imaging are the active component, these ions are generally highly toxic in living organisms, forming insoluble hydroxide particulates that aggregate in the liver. However, metal ions have the capability of binding to a variety of donor atoms, resulting in metal complexes that are produced through the formation of coordinate bonds between a metal cation and an electron-donating ligand. Metal chelate complexes are produced from the formation of a complex between a single ligand that forms more than one coordinate bond to a metal ion. This multiple bonding of a single ligand is known as the chelate effect and results in a significant increase in the stability of the metal complex formed.

The chelation of naked metal ions will markedly affect their proton-relaxation–enhancing properties as well as the toxicity of the metal ions themselves. Solvent molecules have a greatly restricted access to the metal ion nucleus (the inner coordination sphere) in chelate complexes; thus the transfer of unpaired-electron-spin information (magnetization) is less effective than in free-metal-ion systems. Chelation can also impact the spin state, the net molecular magnetic moment, and the electron-spin relaxation characteristics of metal ions. On the other hand, upon chelation the solubility and toxicity in living organisms of the free metal ions are also radically altered, generally imparting characteristics to the substances that render them useful as contrast media by lowering toxicity and modification of biodistribution, plasma residues, and elimination routes.

Ideally, a specific localization of the paramagnetic chelate is desired to maximize the diagnostic value. For contrast enhancement agents in magnetic resonance imaging, it is sufficient that the relaxation rates of the target issue be enhanced relative to other tissues. The chemical structure of the metal chelate effectively determines the biodistribution and excretion routes; complete clearance of the agent is desired to reduce acute as well as chronic toxicological effects.

The stability and toxicity of the metal chelate in the living organism are intimately related. Toxic effects from metal chelates can arise from (1) free metal ion, via chelate dissociation and possible transmetalation; (2) free ligand, via chelate dissociation; (3) intact metal chelate; and (4) metabolites of these three processes. Both metal ions and ligands tend to be more toxic than the metal chelate. Metal chelates have less avidity than metal ions for binding to proteins, enzymes, and membranes; and in the living organism the ability of metal ions to sequester metal chelates is much less than that of free ligands. These two factors generally reduce the toxicity of metal chelates relative to the corresponding metal ion and ligand. Thus the balance between proton relaxation enhancement and low toxicity drives the process of designing new contrast media for magnetic resonance imaging.

Contrast agents. Work on the development of the first commercial contrast enhancement agent, gadolinium diethylenetriaminepentaacetate bis-N-methylglucamine [Gd(DTPA)(NMG)$_2$], commenced around 1981, and its first reported use in humans occurred in 1984. During the early years of the development of contrast media for magnetic resonance imaging, the need for such media was strongly debated. The primary reasons then presented for the adoption of magnetic resonance imaging over the alternate imaging modalities were its noninvasive nature, coupled with the exquisite quality of unenhanced images (with respect to anatomical resolution and inherent contrast).

However, it was soon recognized that enhancement of image contrast between normal and diseased tissue could increase the specificity of diagnosis. Unenhanced magnetic resonance imaging often poorly delineates, or misses completely, a variety of pathological tissue conditions, including some meningiomas and small

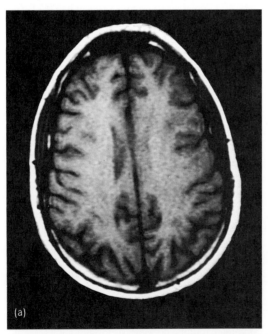 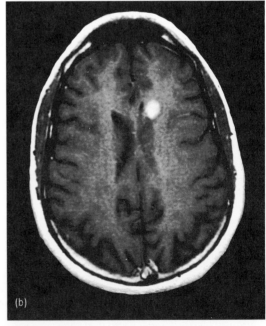

Fig. 1. Brain imaging for detection of a lesion. (*a*) Precontrast image obtained by using a spin-echo pulse sequence; no lesion can be detected. (*b*) Image obtained following intravenous administration of gadodiamide, 0.1 mmole per kilogram body weight; the spin-echo pulse sequence reveals a homogeneous enhancing anaplastic astrocytoma.

metastatic lesions with little associated edema. Magnetic resonance imaging enhanced by contrast also offers the promise of identifying smaller lesions earlier in the progression of disease, increasing the edge definition of tumor mass from edema, abscess, and nonmalignant mass, as well as differentiating a recurrent tumor from fibrous tissue resulting from surgery or radiation. Theoretically, the use of contrast media could decrease the time required per scan, compensating for the time required for contrast agent administration protocols and providing speedier procedures for patients; this has not yet been realized. Taken together, with the significant incidence of cancer, these reasons provided the necessary impetus for further research and development of contrast media.

Conceptually, the desire for organ-specific contrast media capable of determining the status of organ function and associated tissue perfusion remains. The extent to which contrast media has been accepted by the clinical magnetic resonance imaging community at the present time can be gauged by both the rapidity and the extent to which $Gd(DTPA)(NMG)_2$ has come to be utilized. This agent received approval from the U.S. Food and Drug Administration for indications in the head and spine in June 1988, and over 1.2 million contrast-enhanced scans were completed in the United States alone by mid-1990, with an ever-increasing range of applications becoming evident.

$Gd(DTPA)(NMG)_2$ is an ionic (that is, charged), nonspecific extravascular agent that clears very rapidly from the body, with an elimination half-life from blood of approximately 20 min. Recently, nonionic (uncharged) extracellular agents, such as Gd-(DTPA-BMA) [(DTPA-BMA) = diethylenetriamine-pentaacetic acid bis(methylamide)] and Gd(HP-DO3A) [(HP-DO3A) = 4,7-tris(carboxymethyl)-10-(2′-hydroxypropyl)-1,4,7,10-tetrazacyclododecane], have been developed to address the need for low-osmolal contrast media for extracellular imaging applications. **Figure 1** demonstrates the use of gadodiamide injection for the detection of a blood–brain barrier lesion, anaplastic astrocytoma.

The first tissue-specific contrast enhancement agent for magnetic resonance imaging expected to be approved for liver imaging is manganese dipyridoxyl diphosphate [Mn(DPDP)]; this substance and its metabolites are taken up by the hepatocytes of the liver and excreted through the biliary system. This biodistribution allows imaging of focal liver disease such as primary and metastatic cancer in addition to diffuse liver disease such as hepatitis and cirrhosis. An example of use of Mn(DPDP) in detection of liver cancer, hepatocellular carcinoma, is shown in **Fig. 2.**

Future developments in contrast enhancement media for magnetic resonance imaging are likely to focus on the development of site-specific magnetopharmaceuticals, contrast media for ultralow-field systems, and especially for fast-scanning (millisecond) techniques that may allow for the detection and quantification of blood perfusion for stroke and myocardial infarct applications.

For background information SEE CHELATION; COM-

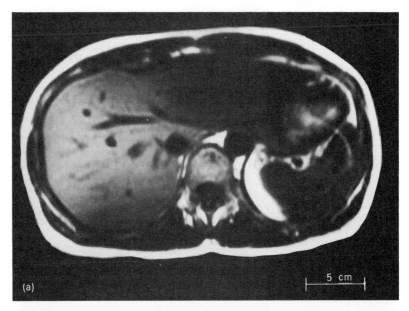

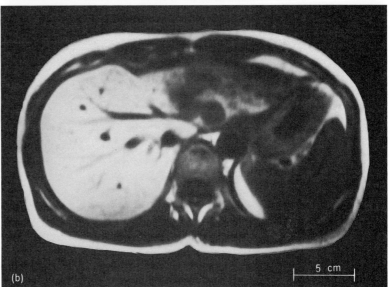

Fig. 2. Liver imaging for detection of cancer. (a) Precontrast image obtained by using a spin-echo pulse sequence; a diffuse low-signal-intensity abnormality is seen involving the left hepatic lobe. (b) Image obtained 30 min after intravenous infusion of Mn(DPDP), 0.01 mmole per kilogram body weight; the spin-echo pulse sequence now reveals multiple nodular abnormalities in the left hepatic lobe. These discrete spherical lesions are distinguished from edema and surrounding, residual noncancerous liver tissue.

PLEX COMPOUNDS; COORDINATION CHEMISTRY; MEDICAL IMAGING in the McGraw-Hill Encyclopedia of Science & Technology.

Scott M. Rocklage

Bibliography. W. P. Cacheris et al., The relationship between thermodynamics and the toxicity of gadolinium complexes, *Magn. Reson. Imag.*, 8:467–481, 1990; M. Laniado et al., First use of GdDTPA/dimeglumine in man, *Phys. Chem. Phys. Med. NMR*, 16:157–165, 1984; M. H. Mendonca-Dias et al., Contrast agents for nuclear magnetic resonance imaging, *Biol. Trace Elem. Res.*, 13:229–239, 1987; V. M. Runge (ed.), *Enhanced Magnetic Resonance Imaging*, 1989.

Metal cluster compounds

Small metal clusters are so-called molecular surfaces that lie in the transition between a molecule and a bulk material. They may be a totally new class of materials with unique chemical, electronic, and optical properties. This is a very active field; significant changes and new insights are anticipated over the next few years.

The chemical and physical properties of metal clusters depend on the number of constituent atoms, the nature of those atoms, and the overall charge on the cluster. A rich diversity of dramatic size-selective effects has been observed in the rate of chemical reactions, electron-binding energies, evolution of electronic band structure, number of molecules chemisorbed, and dissociative reactions of molecules on the cluster surface when the number of constituent atoms varied from three to about a hundred. Rapid advances in experimental techniques, not only for synthesizing metal clusters one atom at a time but also for characterizing their composition as well as electronic structure, binding energy, and chemical reactivity with a wide range of reagents, have significantly increased the information available concerning these fascinating systems.

Chemical reactions have been an extremely sensitive probe of the electronic and structural dynamics of these clusters, especially since direct structural probes have yet to be demonstrated. In this size range, cluster properties are not simply delineated by molecular, solid-state, and liquid-state descriptions; rather a transitioning among them is observed. Significant advances in theoretical tools are also contributing to an overarching theoretical picture about both the bonding and structure of such systems. So far, only limited ground rules that govern the behavior of these interesting systems have emerged. The very question of exactly what constitutes a metal cluster is open for discussion, since the metal–insulator transition cannot be described in the simple terms that had previously been thought to apply. The discoveries of dramatic size-selective behavior for gas-phase clusters have stimulated new work involving preparation of materials made of monosized clusters, with the anticipation that new and unique materials will emerge.

Experimental advances. Size-selected ''naked'' metal clusters in the gas phase are produced in the subfemtomole to picomole range, from a few hundred to a few billion clusters. To date, there are no techniques that can consistently produce a large sample of naked clusters of a specific number of constituent atoms that are visible and subject to traditional structural and chemical characterizations. Organometallic synthesis can produce ligated metal clusters of some defined number of constituent atoms, but these ligated clusters are limited by known synthetic pathways and bonding rules. Stripping the ligated clusters to produce naked clusters has met with only very limited success. As a result, this area of materials science has been researched by using highly sophisticated ultrasensitive in place techniques to probe samples of a few hundred to a few billion clusters, a sample size not usually studied by inorganic, organometallic, and catalytic chemists and materials scientists.

Rapid advances in the development and adaptation of experimental tools based on gas-phase chemical physics have played a critical role in creating the field of metal cluster chemistry. One critical development has been the combination of pulsed supersonic molecular-beam technology with laser-photoionization mass spectroscopy and photoelectron spectroscopy. Similar to the production of artificial snow by the rapid expansion of water under high-pressure air, metal clusters are produced by the rapid condensation of metal vapor in a high-pressure atmosphere of helium or one of the other inert gases. Metal vapor is produced in the laser vaporization plasma of metal targets in the throat of a supersonic nozzle. The use of state-of-the-art tunable laser photoionization and other forms of mass spectroscopy have made it possible to study the properties of these clusters as a function of cluster size. **Figure 1** shows such an experimental arrangement. The chemical properties of the clusters can be studied by injecting them into a high-pressure flow reactor before forming the molecular beam for mass analysis, or specific-size cluster ions can be injected selectively into a flow reactor or a magnetic bottle for study. Actually, only a few experiments require that the clusters be size-selected. The field has been able to advance rapidly because of the rapid development of measurement techniques that do not require the collection of a physical sample of mass-selected clusters.

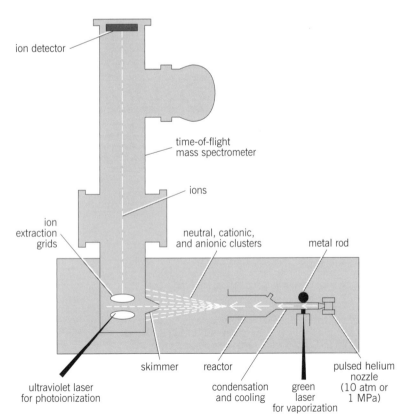

ion detector

time-of-flight
mass spectrometer

ions

ion
extraction
grids

neutral, cationic,
and anionic clusters

metal rod

skimmer   reactor

pulsed helium
nozzle
(10 atm or
1 MPa)

ultraviolet laser
for photoionization

condensation
and cooling

green
laser
for vaporization

Fig. 1. Diagram of the apparatus used to prepare and study charged and neutral gas-phase clusters.

Chemical reactions. Size-selective chemisorption studies have been reported for activation of dihydrogen (H—H), dinitrogen (N—N), dioxygen (O—O), carbon monoxide (CO), carbon dioxide (CO_2), hydrogen sulfide (H_2S), methane (CH_4), RO—H, R—H, and RS—H (R = an organic group) bonds on clusters of many metals, as well as for several oxides, carbides, nitrides, and sulfides. This discussion emphasizes the chemisorption of dihydrogen, methane, and benzene on a variety of metal clusters, both neutral and charged.

Chemical reactions have been used to postulate assumptions about cluster structure. One of the most elegant experiments in this area has been done at Argonne National Laboratory, where scientists have measured how many molecules of ammonia can bind to cobalt clusters of different sizes. They discovered that cobalt-55 binds 12 ammonia molecules, while clusters that contain more or fewer cobalt atoms will bind more than 12 ammonia molecules. Based on this information, it can be postulated that the structure of cobalt-55 is that of an icosahedron. The logic is as follows. On well-defined crystal surfaces, ammonia preferentially sticks to single atoms, and an icosahedral structure has 12 corners; thus those are the ammonia binding sites. Once they are used up, no more ammonia molecules will bind. Clusters that are smaller or larger have a rougher surface that provides for more single-atom binding sites, hence more ammonia molecules will bind. Further support for this model comes from the observation that molecules that do not require single-atom binding sites, for example, hydrogen, can coexist with ammonia on the cluster. An icosahedral structure is especially stable; for reactions where the surface has to reconstruct to accept the chemisorbed species, it may be relatively unreactive.

Hydrogen chemisorption, the first step. The oxidative addition of hydrogen to metal (M_n) clusters, as in the reaction below, exhibits dramatic size dependence for

$$H_2 + nM \rightarrow M_nH_2$$

vanadium, niobium, tantalum, iron, cobalt, nickel, palladium, and aluminum clusters, both neutral and charged. Positively charged gold clusters react with hydrogen, while neutral and negatively charged clusters do not. Rhodium and platinum clusters do not show size dependence for hydrogen chemisorption, but they generally chemisorb hydrogen readily. Copper does not react under the conditions studied for any cluster size.

Figure 2 shows plots of the chemisorption rate constant versus cluster size for neutral and positively charged iron clusters with dihydrogen. The rate-controlling reaction mechanism for both the neutral and positively charged clusters is similar, the difference being primarily due to the charge-induced dipole for the hydrogen molecule approaching the metal cluster or cluster ion. Aside from this electrostatic effect, which is monotonic with cluster size, there are two distinct regions of reactivity that show cluster size-specific behavior; one region comprises clusters containing fewer than eight atoms, the other more than

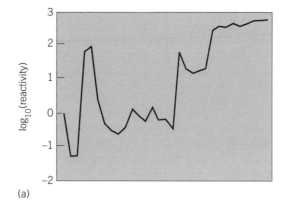

(a)

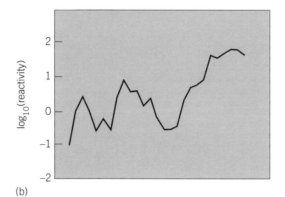

(b)

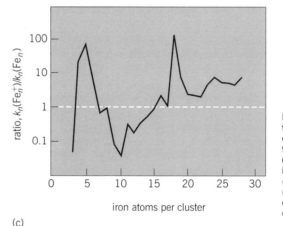

(c)

Fig. 2. Graphs showing the chemisorption rate constant (k_n) as a function of cluster size. (a) Positively charged iron atom (Fe_n^+). (b) Neutral iron atom (Fe_n). (c) Ratio of rate constant of Fe_n^+ to that of Fe_n).

eight. The rate-limiting step in the oxidative coupling reaction involves electron transfer to overcome the activation barrier leading to the weakening of the H—H bond and the formation of the metal hydride bond. In clusters of fewer than eight atoms, the electron transfer primarily moves an electron from the hydrogen molecule to the metal cluster; for those with more than eight atoms, the motion is from the metal cluster toward the hydrogen molecule. Thus the individual electron binding states, or the Fermi level of the cluster, becomes a critical determinant; these are specific for each cluster size and structure. This model is useful to explain the relative reactivity of hydrogen, alkanes, and nitrogen with clusters of niobium, vanadium, tantalum, palladium, rhodium, aluminum, and gold. The model falls short of explaining why certain cluster sizes are completely unreactive while those all around them are very reactive, and why certain clusters have isomers that are not reactive.

Two or more cluster isomers are species that have the identical mass, that is, the same number of constituent atoms, but for some reason display different chemical behavior. Mostly this is detected by the fact that some clusters of a specific size react at a substantially different rate. Examples are niobium [Nb(9), Nb(11), Nb(12), Nb(12)$^+$, Nb(19)$^+$] and vanadium [V(9)], where the numbers refer to metal atoms in the cluster. When reacted with hydrogen gas, some clusters of each specific size above reacted readily, while others either did not react at all or reacted at a substantially different rate. This has suggested that there are two different structures for these clusters, one which is quite reactive and the other not. The unreactive cluster may be very tightly bound, having a highly symmetric structure, and is not able to rearrange, that is, reconstruct, to accommodate the hydrogen adducts. Clusters that do not react can be activated, and evidence for cluster isomers disappears as the temperature of the clusters is increased. At even slightly higher temperature, clusters are expected to be more fluxional and reconstruct readily.

Calculations of electronic structure continue to identify a large number of possible ways in which the atoms in the cluster can be assembled with relatively little energy difference between many different structural configurations. In general, clusters tend to be more reactive at higher temperatures, though there is tough competition among constructive chemistry and decomposition of the clusters as a result of the breaking of metal-metal bonds that leads to the destruction of the specific cluster. Many metal-metal bonds are of the same order of magnitude as M—H, M—O, and M—C bonds. This ability to reconstruct and break metal-metal bonds has general significance for cluster reactivity as the question of how many molecules can be chemisorbed on a metal cluster is explored.

Saturation hydrogen coverage. Experiments have used the isotope of hydrogen known as deuterium. **Figure 3** shows the deuterium (D) uptake normalized to the number of metal (M) atoms in the cluster for a number of metals. These are the general trends for nickel, platinum, and rhodium. The ratio of deuterium to metal (D/M) increases as the cluster size decreases. For larger-size clusters, the ratio D/M tends toward unity. This picture is very different for clusters of iron, where the D/M ratio tends to stay near unity regardless of cluster size, as long as surface states dominate. For clusters of niobium, vanadium, and tantalum, small clusters are observed to have the ratio D/M > 1, though the numbers are much more modest, ranging from 1.4 to 1.6, depending on the metal. Aluminum tends to pick up only a single molecule of hydrogen, though it is quite reactive with atomic hydrogen from other sources. Small clusters of gold ions are quite active with the ratio D/M = 3 for the trimer, declining for larger cluster ions, becoming inactive by the 15-atom cluster ion.

The very active noble metals suggest that significant incorporation of hydrogen is occurring and that the formation of metal hydride clusters is occurring readily. The capacity of the cluster to restructure is critical. This is facilitated by the fact that the M—H and M—M bonds in these systems are quite close; and the formation of three center bonds, such M—H—M, is favored. This is not the case for iron and aluminum.

Methane and light hydrocarbon chemisorption. Clusters of platinum, palladium, positive platinum ions, and positive gold ions chemisorb methane dissociatively, while clusters of iron, niobium, rhodium, and aluminum do not. The role of the charge is to change the probability of the overall reaction, but it also influences the size-selective, nonmonotonic behavior, which seems to be more intimately tied to the oxidative coupling capability of the cluster and its ability to reconstruct to accommodate the reaction products. It should be noted that the dissociative chemisorption of the first small hydrocarbons in the alkane series appears to block further adsorption sites, and thus surface coverage is rapidly saturated. There appears to be a correlation between the reactivity of the cluster and its ability to form compact structures with three or fewer coordinated surface atoms in a close-packed structure.

Benzene. Chemisorption of benzene on niobium and vanadium clusters shows no size dependence, but the subsequent decomposition chemistry of benzene on the cluster surface shows strong size dependence. The correlation appears to follow the same rules that hold for C—H and H—H bond activation that involves an oxidative coupling mechanism. Once the initial chemisorption and dissociation have occurred, subsequent chemisorbed molecules appear to be inactive with respect to further decomposition; that is, the system is poisoned.

For background information *SEE CHEMICAL STRUCTURES; FLUXIONAL COMPOUNDS; LASER SPECTROSCOPY; MASS SPECTROSCOPY; METAL CLUSTER COMPOUNDS; ORGANOMETALLIC COMPOUND* in the McGraw-Hill Encyclopedia of Science & Technology.

Andrew Kaldor

Bibliography. D. M. Cox et al., Abnormally large deuterium uptake on small transition metal clusters,

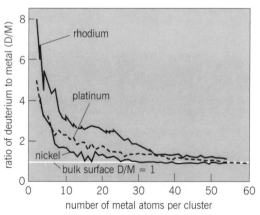

Fig. 3. Graphs showing deuterium uptake relative to the number of metal atoms in the cluster.

Catalysis Lett., 4:271–278, 1990; T. Klots et al., Magic numbers through chemistry: Evidence for icosahedral structure of hydrogenated cobalt clusters, *J. Chem. Phys.*, 92:2110, 1990; R. Pool, Clusters: Strange morsels of matter, *Science*, 248:1186–1188, June 8, 1990; I. Prigogine and S. A. Rice (eds.), *Evolution of Size Effects in Chemical Dynamics*, pt. 2, vol. 70 in *Advances in Chemical Physics*, 1988.

Metal hydrides

There are still surprises in store in the study of the bonding and reactivity of the hydride ion (H^-), the simplest ligand in coordination chemistry. New bonding modes and dynamic processes have been uncovered recently in transition-metal complexes. Also emerging is a new understanding of the dual nature of the hydride ion, sometimes H^- and sometimes H^+. These discoveries impinge on the important role of metal hydrides in industrial catalysis and hydrogen storage.

Binary metal hydride compounds. The term hydride implies the presence of an anionic hydrogen (H^-) bonding via electrostatic attractions to a metal ion (M^+), as in M^+H^-. This applies only to saltlike hydrides such as sodium hydride (NaH) or calcium hydride (CaH_2), where hydrogen with electronegativity (χ) of 2.2 is combined with a metal (M) with $\chi < 1$. Here the H^- has a large ionic radius of up to 0.15 nanometer and reacts (often violently) as a strong base with compounds as weakly acidic as water, alcohols, and amines, as in reaction (1). Hydrides are

$$MH + HOH \rightarrow M^+OH^- + H_2 \qquad (1)$$

also reductants; for example, they react with oxygen and reduce ketones to alkoxides. These are the typical chemical indicators of the so-called hydridic character.

However, the polarization of the hydride can be completely changed as the electronegativity of the metal is altered, to the point where the hydride shrinks to a covalent radius of 0.04 nm and actually acts as a strong acid or oxidizing agent. For example, the soluble complex hydridotetracarbonylcobalt [$CoH(CO)_4$] is still called a metal hydride, but it is acidic, not hydridic. It produces acetic acid when it reacts with the acetate ion as in reaction (2). The

$$CoH(CO)_4 + CH_3COO^- \rightleftharpoons$$
$$\text{Acetate ion}$$
$$Co(CO)_4^- + CH_3COOH \qquad (2)$$
$$\text{Acetic acid}$$

electron-withdrawing carbonyl ligands change the effective electronegativity of Co from 2 to about 3.

Most metal-hydride bonds have a covalent component that increases as the electronegativity of the metal increases. Metals with electronegativities in the range 1–2 form binary compounds that are hydridic, ill-defined solids with some covalent and some metallic character. Examples are magnesium hydride (MgH_2), titanium hydride ($TiH_{1.7}$), and lanthanum hydride

($LaH_{2.87}$). These metal hydrides or hydrides of related alloys like lanthanum-nickel ($LaNi_5$) and titanium-iron (TiFe) have the property of releasing gaseous hydrogen [$H_2(g)$] when warmed; often the metal will reabsorb $H_2(g)$ with release of energy [reaction (3)].

$$TiFeH_{1.95} \underset{\text{cool}}{\overset{\text{heat}}{\rightleftharpoons}} TiFe + H_2(g) \qquad (3)$$

Thus large volumes of hydrogen can be stored as a hydride and released on demand. Since H_2 is a nonpolluting combustible fuel, these metal hydrides will play an important role in storing energy for a future society that heats it homes and propels its vehicles with H_2. Hydrogen is no more dangerous than natural gas; but the hydridic storage material must be kept free of water (H_2O) and dioxygen (O_2). Successful demonstration buses and cars and solar heating and cooling systems have already been constructed based on metal hydride technology.

Metals located around rhodium (Rh; $\chi = 2.2$) in the periodic table (the platinum metals) form completely covalent metal-hydride bonds. Hydrogen gas adds rapidly and reversibly to the surface of platinum-group metals to give reactive hydrogen atoms (via homolytic cleavage of the H—H bond) that bond to one metal or that bridge between two or three metal atoms (**Fig. 1**). The ease of formation of these surface hydrides explains why finely divided ruthenium (Ru), Rh, iridium (Ir), nickel (Ni), palladium (Pd), or platinum (Pt) are such active heterogeneous catalysts for the hydrogenation of olefins and other unsaturated organics. The substrate that is to be hydrogenated must also be able to coordinate to the metal surface for the transfer of hydrogen atoms to occur. This process can be reversed at higher temperatures, particularly on Pt. Thus Pt metal catalysts are used to reform petroleum feedstocks [restructure the

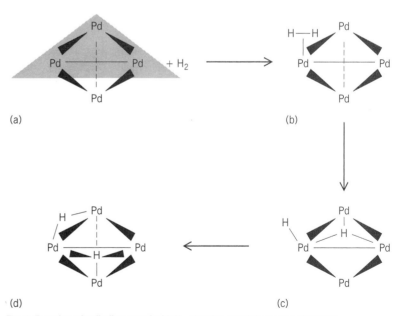

Fig. 1. Reaction of palladium metal with H_2, showing possible modes of hydride coordination. (*a*) Surface atoms (triangle) and one interior atom. (*b*) η^2-Dihydrogen on Pd. (*c*) A terminal hydride and a triply bridging hydride after H_2 splits homolytically. (*d*) A doubly bridging hydride and a hydride in a tetrahedral, interstitial site.

carbon (C) backbone] by coordinating C—H bonds to the surface and then breaking them into transitory Pt-H and Pt-hydrocarbon species.

Palladium is the only platinum metal that reacts with H_2 [and its isotopic form, deuterium (D_2)] to give a binary hydride, $PdH_{0.98}$. There is considerable controversy over the claim that sufficient quantities of D atoms can be made by electrochemical means to "dissolve" into the interstitial cavities of the metal lattice of a Pd electrode so as to favor a cold-fusion nuclear reaction. Figure 1d shows an interstitial hydride in a tetrahedral site.

An important heterogeneous catalyst for the production of methanol from synthesis gas provides a good example of how hydrogen gas can be split into a metal hydride and a proton. In reaction (4), where the

$$2H_2 + CO \xrightarrow{\text{ZnO/Cu}} CH_3OH \qquad (4)$$

heterogeneous catalyst is zinc oxide/copper metal (ZnO/Cu), H_2 is thought to coordinate to the Zn^{2+}. It then splits into hydride, H^- coordinated to Zn (ZnH^+), and a proton (H^+), which adds to the oxide; this is the heterolytic splitting of hydrogen, and there is now evidence that this process involves an intermediate like $Zn(H_2)$ with H_2 side-on bonded to the metal [reaction (5)]. It is the hydride H^- on electropositive

$$ZnO + H_2 \rightarrow Zn(H_2)(O) \rightarrow [ZnH]^+(OH)^- \qquad (5)$$

Zn^{2+} that is thought to attack CO coordinated to a nearby Cu atom and thus initiate the reduction.

Transition-metal hydride coordination complexes. Considering that platinum metal surfaces form such reactive hydrides, it seemed possible that soluble metal complexes could be made where every metal atom, instead of just the surface atoms, could form a catalytically active, hydride site. Work in the 1950s and 1960s demonstrated that soluble metal hydride catalysts could be stabilized by using triphenylphosphine (PPh_3; P = phosphorus, Ph = C_6H_5) and related ligands. Wilkinson's catalyst [$Rh(Cl)(PPh_3)_3$; Cl = chlorine] is used extensively by organic chemists for the selective hydrogenation of C≡C and C≡N (N = nitrogen) bonds and many other transformations that involve the transfer of hydrogen atoms to or from the Rh. This homogeneous (soluble) catalyst reacts with $H_2(g)$ and ethylene at 1 atm (10^5 pascals) and 20°C (68°F) to give the intermediate dihydride $Rh(Cl)(H)_2(C_2H_4)(PPh_3)_2$ [I], which forms

(I)

just before the transfer of H atoms to the carbons to give ethane and regenerate the catalyst. A more recent elaboration of this catalyst involves the use of chiral

phosphine ligands to impart to the Rh an enzymelike specificity in the production of one optical isomer of hydrogenated product from a non-optically active starting olefin complex. Amino acid derivatives possessing high optical purity, particularly the precursor in the synthesis of L-dopa, are being prepared industrially by using a chiral ditertiaryphosphine ligand in a rhodium catalyst. A related ruthenium complex has been used to hydrogenate geraniol to one enantiomer of citronellol with 99% selectivity.

Polyhydride complexes. Also discovered in the 1960s were metal polyhydride complexes of the type $MH_x(PR_3)_y$, where $x = 2$–7 and $y = 5$–1, depending on the transition metal (the chromium, manganese, iron, and cobalt groups), and R = phenyl, ethyl, methyl, or other functional groups. These soluble complexes give nuclear magnetic resonance (NMR) spectra for ^1H and ^{31}P that change with temperature, since the small hydrogen atoms can travel over the surface of the metal atom but are frozen into directional covalent bonds at low temperature (<200 K or $-99°F$); usually the PR_3 ligands also move in these so-called fluxional processes.

Neutron diffraction on single crystals is the method of choice for accurately determining the position of hydrides in these complexes. There are only a few nuclear reactors in the world where this can done, so the number of polyhydride structure determinations that have been made is limited. So far the structures have revealed that the ligands coordinate at vertices of regular polyhedra with the large PR_3 groups usually staying as far away from each other as possible; **Fig. 2** illustrates these features for the core of $OsH_6(PPhPr_2)_2$ (Pr = *iso*-propyl; Os = osmium). There is renewed interest in these structures since there is NMR evidence that some polyhydrides, like the catalytically active complex $RuH_4(PPh_3)_3$, contain H_2 coordinated with the H—H bond intact.

Soluble clusters. Soluble clusters with properties of Pt metal surface hydrides have been characterized recently. The complex $[CuH(PPh_3)]_6$ has an octahedron of Cu atoms with hydrides bridging six of the eight triangular faces. A larger "raft-shaped" cluster (II) with hydrides around the edges is formed when

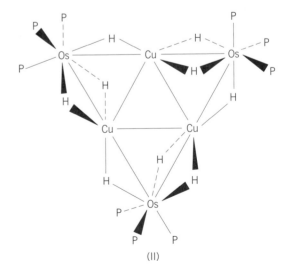

(II)

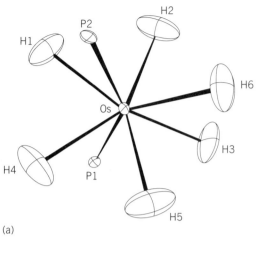

(a)

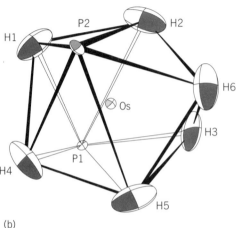

(b)

Fig. 2. Perspective views of (a) OsH_6P_2 core with thermal ellipsoids and (b) the coordination polyhedron of the complex $OsH_6(PPhPr_2)_2$ as determined by single-crystal neutron diffraction. *(After J. A. K. Howard et al., Crystal and molecular structure of bis(diisopropylphenyl-phosphinehexahydroosmium [OsH_6(PC_{12}H_{19})_2]: Single-crystal neutron diffraction study at 20 K, Inorg. Chem., 26:2930–2933, 1987)*

this Cu complex is reacted with a polyhydride [reaction (6); Me = methyl]. Clusters like [OsAu₂-

$$\frac{1}{2}[CuH(PPh_3)]_6 + 3[OsH_4(PMe_2Ph)_3] \rightarrow$$

$$[CuOsH_3(PMe_2Ph)_3]_3 + 3PPh_3 + 3H_2 \qquad (6)$$

$(H)_3(PPh_3)_5]^+$ (Au = gold) can be built up in a similar fashion by reaction with $AuPPh_3^+$. The compound $[Pt_3(H)_2(PPh_2CH_2CH_2PPh_2)_3]^+$ has hydrides bridging over and under a triangle of Pt atoms. Work is now focusing on whether such complexes display the reactivity of Pt metal catalysts.

Transition-metal carbonyl clusters. Ligands other than phosphines are known to stabilize transition-metal hydrides. A huge variety of transition-metal clusters containing carbonyl ligands and hydrides such as triangular $Os_3H_2(CO)_{10}$, tetrahedral $K[Ru_4H_3(CO)_{12}]$, and polyhedral $K_2[Rh_{13}H_3(CO)_{24}]$ are known (K = potassium). The hydride ligands in the first two complexes bridge two metal atoms (on the edge of the polyhedron) like Pt metal surface hydrides, whereas the last complex has interstitial hydrides like PdH_2.

The hydrides in all these complexes are acidic ($pK_a \sim 11$ in CH_3CN solution) because of the carbonyl ligands; the interstitial hydride reacts much slower with bases than the bridging ones. An exception is $K[Ru_3H(CO)_{11}]$, which has a hydridic ligand. This complex is an active catalyst for the water-gas shift reaction used in the production of synthesis gas.

Dihydrogen complexes. A discovery in 1984 by G. Kubas and coworkers at Los Alamos National Laboratory was proof of the existence of complexes that are stable at 20°C (68°F), where molecular hydrogen coordinates to a transition metal and the H—H bond remains intact. Octahedral complexes $M(CO)_3$-$(\eta^2\text{-}H_2)(PPr_3)_2$, where M = molybdenum (Mo; III) or

(III)

tungsten (W), were shown by neutron diffraction to have an H-H distance of 0.082 nm at low temperature in the crystal. This is only slightly longer than the H-H distance in free H_2 (0.074 nm). The η^2-dihydrogen nomenclature indicates that the H_2 ligand is side-on bonded to the metal with metal-hydrogen distances similar to those observed for hydrides (0.16–0.17 nm). Nuclear magnetic resonance measurement of the large H-D coupling constant of the complexes $M(CO)_3(\eta^2\text{-}HD)(P[C_6H_{11}]_3)_2$ in solution showed that the H—D or H—H bond is retained in solution. The H_2 reversibly dissociates and re-adds to the metal in these complexes at 20°C (68°F). Since then, many new η^2-dihydrogen complexes, both stable and unstable at room temperature, have been characterized. The dihydrogen in the complex $[Fe(H)(\eta^2\text{-}H_2)$-$(PPh_2CH_2CH_2PPh_2)]BF_4$ [(IV); B = boron, F =

(IV)

fluorine] has an H-H distance of 0.082 nm (neutron study) and is trans to an ordinary iron hydride group in the octahedral metal complex; this compound in solution can be warmed to 50°C (122°F) in a vacuum without loss of H_2. Nuclear magnetic resonance stud-

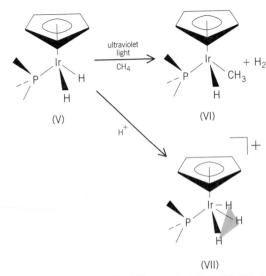

Fig. 3. Reactions of the dihydride complex Ir(C₅R₅)(H)₂(PMe₃) [V]. Its photolysis in cyclohexane solution in the presence of methane gives Ir(C₅Me₅)(CH₃)(H)(PMe₃) [VI]. Its reaction with HBF₄ in ether gives a trihydride Ir(C₅H₅)(H)₃(PMe₃)⁺ [VII] with unusual properties.

ies have revealed that the η^2-H₂ ligand is involved in many dynamic processes, including rapid spinning like a propeller on the metal-H₂ axis, exchange of its H atoms with other hydrides in the molecule, and breaking of the H—H bond to give a tautomeric dihydride species in reversible equilibrium with the η^2-H₂ form. There is evidence that the trihydrogen ligand, η^3-H₃, might also exist.

Homolytic cleavage of the η^2-H₂ ligand is inhibited when the complex is not too reducing; in this case the complex usually has electron-withdrawing ligands like CO or has a positive charge. This also renders the H₂ acidic. The dihydrogen in the complex [Ru(C₅H₅)(CO)₂ (η^2-H₂)]⁺ has a pK_a value less than zero. Thus dihydrogen, when coordinated to a transition-metal ion, can become exceedingly acidic, and this explains the heterolytic cleavage of dihydrogen. The heterolysis of H₂ is thought to occur at a Ni(III) center in the enzyme hydrogenase. In summary, the η^2-H₂ ligand can probably be found in any reaction of H₂ with a metal surface or metal complex.

Other hydride complexes. More reducing metal hydride complexes that are neutral and contain electron-donating groups like C₅R₅ (R = H, Me) or PMe₃ are being investigated as reagents and possibly catalysts for the functionalization of otherwise inert C—H bonds. The hydrides in the complexes Ir(C₅Me₅)(H)₂(L) [L = CO, PMe₃ (V)] resist forming an H—H bond and do so only when irradiated with ultraviolet light (**Fig. 3**). Dihydrogen is lost upon photolysis, and a reactive species Ir(C₅Me₅)(L)* is generated that is capable of coordinating CH₄ and breaking the C—H bond to give the stable methyl hydride (VI). This reaction and related ones are currently being studied in the hope of developing a catalyst capable of converting CH₄ into a more useful hydrocarbon. Homogeneous catalysts are also being sought for the reduction of carbon monoxide (CO).

The unusual hydrides RhH(octaethylporphyrin) and TaH₂(OSiBu₃)₃ [Ta = tantalum, Si = silicon, Bu = butyl] both display promising reactivity toward CO at 20°C (68°F).

A recent nuclear magnetic resonance study of unusual complexes of the type [Ir(C₅H₅)(PR₃)(H)₃]⁺ (VII) that were prepared by the reaction in Fig. 3 has provided new insights into motion of hydrides on the surface of the metal atom. The structure as determined by single-crystal neutron diffraction shows that the hydrides (with Ir-H distances of 0.16 nm) are arranged on the vertices of an isosceles triangle with sides 0.168 × 0.168 × 0.21 nm. It was found that the hydrides can exchange positions by undergoing quantum-mechanical tunneling along the 0.168-nm sides of the triangle, even at 20°C (68°F).

For background information *SEE CATALYSIS; CHEMICAL BONDING; COMPLEX COMPOUNDS; COORDINATION CHEMISTRY; ELECTRONEGATIVITY; METAL HYDRIDES; NEUTRON DIFFRACTION* in the McGraw-Hill Encyclopedia of Science & Technology.

Robert H. Morris

Bibliography. R. H. Crabtree et al., H—H, C—H, and related sigma-bonded groups as ligands, *Adv. Organomet. Chem.*, 28:299–338, 1988; D. M. Heinekey et al., Structural and spectroscopic characterization of iridium trihydride complexes, *J. Amer. Chem. Soc.*, 112:909–919, 1990; R. Kirchheim et al., Metal-hydrogen systems: Fundamentals and applications, *Z. Physik. Chem. Neue Folge*, issues 163 P1 and 164 P2, 1989; R. T. Weberg and J. R. Norton, Kinetic and thermodynamic acidity of hydrido transition-metal complexes, 6. Interstitial hydrides, *J. Amer. Chem. Soc.*, 112:1105–1108, 1990.

Metallography

Fractography has traditionally been an observational (qualitative) technique for identifying and describing the features in a fracture surface, with the objective of relating the topography of the fracture surface to the causes and basic mechanisms of fracture. The goal of quantitative fractography, on the other hand, is to express the features and important characteristics of a fracture surface quantitatively in terms of the true surface areas, lengths, sizes, numbers, shapes, orientations, and locations, as well as the distributions of these quantities.

Fractals. The early preoccupation of mathematicians, physicists, geometers, and others with the irregular, chaotic processes of nature led to the development of the subject of fractals. This work has been extended to include the study of fracture surfaces. Materials specialists are attempting to relate the fractal properties of a wide range of fractured materials to fracture mechanisms, surface configurations, or microscopical features. As is usual with materials investigations, the approach is largely experimental.

Fractals are a measure of the roughness of surfaces and curves and are a natural choice for characterizing fracture surfaces. The fractal dimension is nonintegral and has fractional values between the euclidean dimensions of 0, 1, 2, and 3. The set (or family of shapes) exhibits scaling, which involves scale-dependent behavior, and self-similarity, which implies that the degree of irregularity is identical at all scales. That is, the apparent roughness of a curve or surface has the same visual appearance when viewed at different magnifications. SEE FRACTALS; FRACTON.

Profiles. Because rough, irregular surfaces are so difficult to evaluate experimentally, most investigations deal with the trace (or profile) generated by a section plane intersecting the surface. The trace characteristics are linked geometrically to the attributes of the fracture surface. In many cases, the sectioning planes can be described as vertical or horizontal to the effective fracture plane.

The fundamental fractal equation for irregular planar curves is the Richardson-Mandelbrot relationship (1),

$$L(\eta) = L_0 \eta^{-(D_M - 1)} \qquad (1)$$

where $L(\eta)$ is the apparent length of the profile as a function of the size of the measuring unit (or measuring step) η ; D_M, the (Mandelbrot) fractal dimension, is a dimensionless constant that characterizes the degree of roughness of the fractal curve; and L_0 is a constant with the dimensions of length. $L(\eta)$ should increase without limit as $\eta \to 0$.

The linear form of Eq. (1) is given by Eq. (2). The

$$\log L(\eta) = \log L_0 - (D_M - 1) \log \eta \qquad (2)$$

fractal plot consists of $\log L(\eta)$ versus $\log \eta$ with the absolute value of the slope equal to $|D_M - 1|$. Equation (2) specifies a straight line over the entire range of η values, and this will be obeyed experimentally provided the profile is fractal. This requirement has been ignored by several investigators, who arbitrarily linearize local segments of curved fractal plots and then apply Eq. (2) to each segment. Another problem arises because of the indeterminancy in the choice of a meaningful, fixed profile length from the infinite fractal plot. Moreover, it is apparent that Eq. (1) is not dimensionally consistent. One device that rectifies this inexactitude is the replacement of η by the dimensionless fraction η/η_0, where η_0 is an arbitrary reference length. In this way, the slope, and hence the value of D_M, is not changed when plotting Eq. (2).

A good example of strict adherence of fractal behavior to Eqs. (1) and (2) is provided by the so-called pathological Koch figures. As the measuring unit gets smaller, the perimeter progressively becomes more complex, and its length increases without limit as the subdivision process proceeds to infinitesimally small segments.

Self-similarity is also exhibited, but in a more statistical sense, in the irregular planar curves of nature. **Figure 1** illustrates the idea, wherein the apparent roughness has the same visual shape and configuration at higher magnifications. The profile length thus appears to increase continually with magnification. Equivalently to this kind of apparent length change, the profile length $L(\eta)$ also appears to become greater when measured

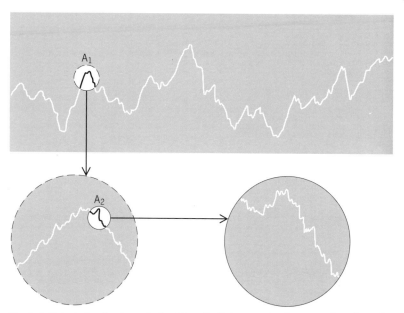

Fig. 1. Self-similarity of an irregular fractal profile. Enlargements of curve portions A₁ and A₂ are shown. (*After H. C. Ward, Profile description, Surface Topography in Engineering: Short Course at Teeside Polytechnic, Middlesbrough, Cleveland, England, 1976***)**

with a smaller measuring unit η. Visual inspection of rough fracture profiles is not enough to establish self-similitude. The latter can be demonstrated only by a fractal plot that yields a straight line extending indefinitely at very small and very large values of η.

It is important to establish whether the linear fractal plot is due to self-similarity or to some other effect. Irregular, nonfractal planar curves that have a completely fixed, unchanging configuration (that is, no self-similitude) can also yield linear curves on the fractal plot. **Figure 2** illustrates a nonfractal curve of fixed configuration measured with different sizes of measuring units. It is obvious that the length appears to increase with smaller measuring units. This is nothing but a simple geometrical effect and does not invoke fractal concepts. As a result, several reported studies of fractal behavior are of doubtful validity.

Slit island analysis. A vertical section through a fracture surface generally results in a long, irregular planar curve (the profile). Horizontal sections, on the other hand, nip off peaks in the fracture surface, forming flat sections of irregular outline known as perimeters. The areas and perimeters of these closed loops are related by the fractal equation due to Hack-Mandelbrot [Eq. (3)], where D_{SI} is the fractal

$$P^{1/D_{SI}} = KA^{1/2} \qquad (3)$$

dimension obtained according to this method, the slit island analysis (SIA); P is the perimeter length; A is the area of the island; and K is a constant dependent on the island shape. Measurements of perimeter length and area are made while holding the measurement unit η constant.

The linear form of Eq. (3) is given by Eq. (4),

$$\log P = D_{SI} \log K + (D_{SI}/2) \log A \qquad (4)$$

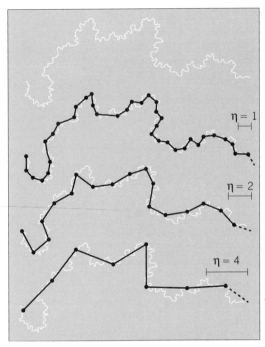

Fig. 2. Variation of the apparent length of a nonfractal curve as a function of size of measuring unit. (*After D. Paumgartner, G. Losa, and E. R. Weibel, Resolution effect on the stereological estimation of surface and volume and its interpretation in terms of fractal dimensions, J. Microsc., 121(1):51–63, 1981*)

where the value of D_{SI} is obtained from the slope of the log P versus log A fractal plot. Typical plots obtained by slit island analysis are shown in **Fig. 3**. If the plot is not linear, the structure is not self-similar and D_{SI} will not be constant. Moreover, for Eq. (4) to be valid, the island shapes must be the same and must remain constant from section to

section. It is highly unlikely that the shapes of the islands from an irregular fracture surface conform to these conditions.

Reversed sigmoidal curves. One of the most important findings to emerge from all known fractal plots of fracture profiles is that the plots are not linear. The more extensive fractal plots show well-defined asymptotic behavior at the extremities of the fractal curve and are known as reversed sigmoidal curves. Studies with more limited range in η show only curved segments of the full reversed sigmoidal curves. Four examples of nonlinear behavior are shown in **Fig. 4** for different metallic alloys. Note that $R_L(\eta)$, the (linear) profile roughness parameter, is used instead of $L(\eta)$. $R_L(\eta)$ equals $L(\eta)/L'$, where L' is the projected length of the profile. The fractal curves for real materials are not self-similar, because they are ultimately limited by the size of the atom. Accordingly, the fractal plot for real materials cannot proceed to infinity as it can for mathematical models or special geometrical constructions.

The fractal dimension D_M is not constant when the fractal plot is curved, but constant values have been obtained in two ways. One method replaces the curved fractal plot with a series of linear segments. This provides several values of D_M, one for each segment, and each segment has its own size range in η. The problem is that these local values of D_M cannot relate unambiguously to microscopical or fractographic features that have the same corresponding size range. This follows because each segment of the fractal plot consists of contributions from the entire length of all profiles within that η range.

The other method linearizes the entire reversed sigmoidal curve by a suitable function, $\exp(-\alpha\eta^{\beta})$, where α and β are constants. The curve of this function is the reverse of the standard sigmoidal curve used in kinetic processes and expresses the amount untransformed. The fraction of the apparent length change between the two asymptotic limits is given by Eq. (5), where $R_L(\eta)$ is the variable, $R_L(\infty) \rightarrow 1$ as

$$\frac{R_L(\eta) - R_L(\infty)}{R_L(0) - R_L(\infty)} = e^{-\alpha\eta^{\beta}} \qquad (5)$$

η → ∞, and $R_L(0)$ is an adjustable constant whose value is uniquely determined by the experimental data. The constant $R_L(0)$ is related to the "true" profile length, since its value is determined for η → 0. The linear form of Eq. (5) is obtained by taking double logarithms of both sides. Excellent straight lines are obtained from the experimental plot. The slope β is then equated to that of the modified fractal equation (6), giving $D_L = \beta + 1$. The term

$$R_L(\eta) = C\eta^{-(D_L-1)} \qquad (6)$$

D_L is the modified fractal dimension for profiles and has a constant value.

A similar treatment is developed for fracture sur-

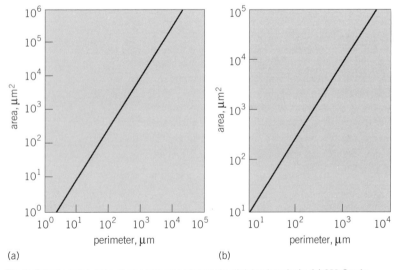

(a)   (b)

Fig. 3. Experimental plots of area versus perimeter in slit island analysis. (a) 300-Grade Maraging Steel; tempering time 1.5 h; tempering temperatures are indicated (*after B. B. Mandelbrot, D. E. Passoja, and A. J. Paulley, Fractal character of fracture surfaces of metals, Nature (London), 308(5961):721–722, 1984*). (b) Dual-phase steel (*after Z. G. Wang et al., Relationship between fractal dimension and fatigue threshold value in dual-phase steels, Scripta Metallurg., 22:827–832, 1987*).

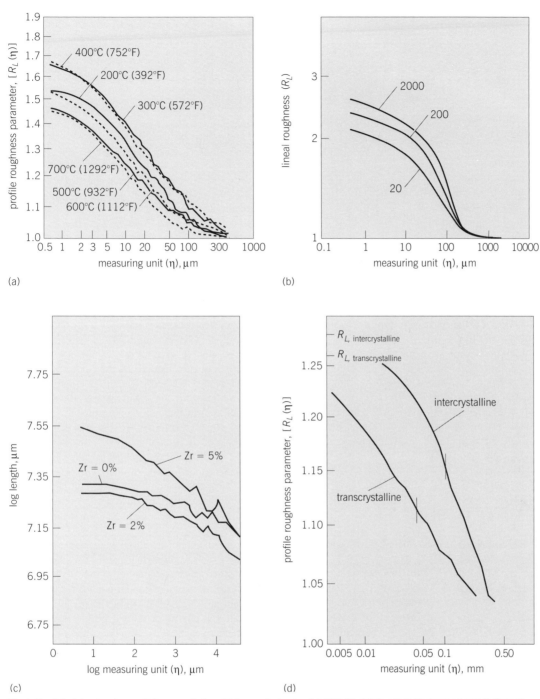

Fig. 4. Fractal plots showing partial reversed sigmoidal curve behavior. (a) 4340 Steel (*after E. E. Underwood and K. Banerji, Fractals in fractography, Mater. Sci. Eng., 80:1–14, 1986*). (b) 31% manganese steel; the magnification at which the measurements were made is indicated for each curve (*after R. H. Dauskardt et al., On the interpretation of the fractal character of fracture surfaces, Acta Metallurg. Mater., 38(2):143–159, 1990*). (c) Titanium alloys Ti-6A1-2V with 0–5% zirconium (*after C. S. Pande et al., Fractal characterization of fractured surfaces, Acta Metallurg., 35:1633–1637, 1987*). (d) Ferritic-pearlitic steel; R_L terms indicate asymptotic limits at very small values of η (*after K. Wright and B. Karlsson, Fractal analysis and stereological evaluation of microstructures, J. Microsc., 129(pt. 2):185–200, 1983*).

faces that show reversed sigmoidal curve behavior in their fractal plots. Instead of using the surface area directly, the surface roughness parameter $R_S(\eta^2)$ is employed. It is expressed in terms of an areal measuring unit η^2 and equals $S(\eta^2)/A'$, where $S(\eta^2)$ is the surface area and A' is its projected area. The fraction of the apparent area change between the two asymp-

totic limits is given by Eq. (7), where $R_S(\eta^2)$ is the

$$\frac{R_S(\eta^2) - R_S(\infty)}{R_S(0) - R_S(\infty)} = \exp[-\alpha(\eta^2)^\gamma] \qquad (7)$$

variable, $R_S(\infty) \to 1$ when $\eta^2 \to \infty$, and $R_S(0)$ represents the "true" surface area when $\eta^2 \to 0$. The linear form of Eq. (7) yields straight lines of slope γ. When γ is equated to the slope of the fractal equation for irregular surfaces [Eq. (8)], the term $D_S =$

$$R_S(\eta^2) = C(\eta^2)^{-(D_S-2)/2} \qquad (8)$$

$2\gamma + 2$ is obtained, where D_S is the constant fractal dimension for surfaces. The experimental confirmation of this analysis has been excellent. Furthermore, it has been demonstrated that $D_S - D_L = 1$, and that $D_L > 1$ and $D_S > 2$, as would be expected from fractal theory.

Fracture surfaces. Before the fractal characteristics of fracture surfaces can be investigated, it is necessary to obtain values of either the surface area or the surface roughness parameter. This is done readily by means of the important parametric equation (9) that relates the experimentally available profile roughness parameter R_L to the surface roughness parameter.

$$R_S = 4/\pi \, (R_L - 1) + 1 \qquad (9)$$

Knowing R_S, the fractal plot can be constructed and D_S obtained by using Eqs. (7) and (8).

Published accounts of fractals and fracture-related properties include such topics as embrittlement, corrosion fatigue, toughness, fatigue thresholds, impact energy, and microcracking. The scope and volume of this fractal-related research should continue to increase. Meanwhile, experimental techniques must be refined and standardized, and fractal theory developed specifically for materials.

For background information SEE CORROSION; EMBRITTLEMENT; FRACTALS; METAL, MECHANICAL PROPERTIES OF; METALLOGRAPHY in the McGraw-Hill Encyclopedia of Science & Technology.

Ervin E. Underwood

Bibliography. American Society for Metals, *Metals Handbook*, vol. 12: *Fractography*, 9th ed., 1987; B. B. Mandelbrot, *The Fractal Geometry of Nature*, 1982; B. B. Mandelbrot, D. E. Passoja, and A. J. Paulley, Fractal character of fracture surfaces of metals, *Nature (London)*, 308(5961):721–722, 1984; E. E. Underwood and K. Banerji, Fractals in fractography, *Mater. Sci. Eng.*, 80:1–14, 1986.

Meteorological instrumentation

At present, many scientific challenges are facing society, among them the possibility of climate change. Airborne observation with its versatility is an important tool for monitoring atmospheric changes as well as land-use changes that can affect the atmosphere.

Significant advances have been made since the early 1950s in mounting sophisticated instruments on aircraft in order to obtain measurements of atmospheric compositions. These airborne observations have enhanced understanding of the environment not only in atmospheric sciences but also in disciplines such as oceanography and other Earth sciences.

Prior to World War II, few atmospheric observations had been made from aircraft. Extensive series of

observations from an aircraft were first made in 1946 and 1947 in Ohio and in Florida during the Thunderstorm Project. These observations contributed to understanding of the transformation of cumulus clouds into thunderstorms or cumulonimbus clouds. Airborne observations gradually expanded as a result of the availability of improved aircraft sensors and instruments.

The advantages of airborne observations lie in maneuverability and mobility. An aircraft, for example, can probe in depth the troposphere and stratosphere both by vertical profiling and by sampling on a horizontal plane. Both capabilities are important for three-dimensional observations of highly localized or transient phenomena. Another advantage of airborne observations is that measurements can be obtained in remote areas, such as the Antarctic and Arctic, and over vast regions of the oceans.

However, there are disadvantages associated with the high mobility of the aircraft. Airborne measurements require fast response in instruments that have to perform in a vibrating, turbulent environment. Corrections may have to be made, taking into account compressibility, adiabatic and frictional heating, and distortion of flow induced by the aircraft.

Observation methods. The most common use of research aircraft is for in-situ measurements. For this method, sensitive monitoring instruments are attached to the aircraft platform and flown directly into the atmospheric location of interest. These measurements permit fundamental discoveries by generating a basis for understanding cloud physics, atmospheric chemistry, airflow dynamics, radiation, and air pollution. These measurements are often used to calibrate either ground-based or satellite-borne remote-sensing devices.

Research aircraft are also used to release deployable sensors, such as dropwindsondes (devices that measure speed and direction of wind in addition to temperature, pressure, and humidity) and oceanographic probes. A dropwindsonde released from an aircraft can provide a vertical profile (sounding) of temperature, pressure, humidity, and winds as it descends. Such observations are important around storms and in remote areas where no other means for observation are available. In oceanography, aircraft-launched drifting buoys and expendable bathythermographs (for temperature profile measurement in the upper several hundred meters of the ocean) provide data from the surface and subsurface of the ocean.

Aircraft can also be used as platforms for remote sensing, carrying sophisticated instruments such as radar or lidar to map large areas of interest. Such devices can provide detailed cross sections of storm structure, boundary-layer conditions, and distribution of trace gases in the atmosphere. As part of the development of remote sensing, many prototype instruments for orbiting satellites are often first installed and tested on aircraft.

Aircraft platforms. In the United States, a wide variety of aircraft platforms are operated by the federal government, universities, and even by the private sector to obtain desired airborne measurements. The *Airborne Geoscience Newsletter* lists about 30 aircraft platforms, ranging from single- and twin-engine aircraft with limited range and payload capability to long-range four-engine turboprop and turbojet aircraft. Some of these aircraft are available only as a platform for specialized equipment, while others are equipped with fully instrumented meteorological measurement systems.

Several of these platforms are modified to allow easy interphasing of specialized instruments with aircraft such as those shown in the **illustration.** These aircraft (a King Air, a Sabreliner, and an Electra) support a variety of atmospheric science research projects sponsored by the National Science Foundation (NSF); they are operated by the Research Aviation Facility (RAF) of the National Center for Atmospheric Research (NCAR).

The selection of a particular research aircraft for a given field program is determined by the range, altitude, and payload capability required to carry the necessary scientific equipment, and the instrumentation that must be available. In large field experiments, it is not unusual to have several research aircraft collect measurements simultaneously.

Instrumentation. A fully instrumented research aircraft is like a flying laboratory, equipped with a variety of equipment for making measurements. These include instrumentation for atmospheric state parameters (temperature, pressure, humidity), gust-probe instrumentation (for measuring both the mean horizontal wind and turbulent fluctuations in all three components of the wind velocity) for turbulent flux measurements (rate of transport of atmospheric properties), cloud physics instrumentation, radiometers (short-wave, long-wave, and ultraviolet), sensors for electric field strength, samplers for atmospheric trace gases, dropwindsonde (dispensing and data acquisition), oceanographic dropsonde (dispensing and data acquisition), sensors for remote radiometric surface temperature measurement, lidar (for vertical profiles, aerosols, and clouds), airborne radar (for cloud and precipitation studies), video cameras, accurate positioning systems, and data collection systems with real-time measurement displays.

Measurements. Without instrumented research aircraft it would be impossible to measure certain atmospheric processes and chemical compositions directly. There are many scientific challenges facing researchers in studies of atmospheric phenomena. For example, more accurate prediction of climate change rests partly on better understanding of radiative mechanisms for heat transfer in and between air and oceans.

Measurements of the concentrations and fluxes of atmospheric trace gases—for example, carbon dioxide, ozone, chlorofluorocarbons, methane, and nitrous oxide—provide an important input into models of global atmospheric chemistry that are designed to estimate the contributing causes as well as the magnitude of the greenhouse effect from these gases. Aircraft-assisted studies provide investigators with measurements made over large geographic areas, with

Typical aircraft used in atmospheric research: Electra (top), Sabreliner (left), and King Air (right). (*National Center for Atmospheric Research/National Science Foundation*)

vertical profiles that extend from the surface of the sea well into the stratosphere.

Aircraft-borne instruments measure chemical composition of the atmosphere in three dimensions. Aircraft measurements are also used to validate satellite measurements. For example, the satellite-observed "ozone hole" over Antarctica in 1987 was verified by the high-altitude ER-2 aircraft of the National Aeronautics and Space Administration (NASA) and provided the first direct measurements of the chlorine chemistry in this region.

Airborne observations play an important role in understanding atmospheric dynamics, through studies of boundary-layer processes, clear-air mesoscale circulations, and a variety of weather systems such as fronts and large convective systems (thunderstorms, hurricanes). Many convective weather systems have regions within them that are unsafe for aircraft penetration. However, these areas could be investigated with remote-sensing devices from an aircraft, including Doppler radar, lidar, dropwindsondes, or dual-channel microwave radiometers. At higher altitudes the stratospheric–tropospheric air exchange is of global importance. These exchange events are highly localized and transient in nature, and they have been identified as precursors to development of extreme meteorological events. Measurements from aircraft can provide necessary observations of these atmospheric structures.

Understanding of clouds and precipitation physics has been enhanced by direct observations from aircraft. Cloud interiors can be characterized by in-situ measurements such as liquid water content and particle size, shape, and concentration. Such measurements are usually limited to smaller convective systems. Knowledge of the relationships involving cloud microphysics, precipitation development, and dynamic processes within large convective systems is rather limited, primarily because of the danger to aircraft from turbulence, hail, and lightning. Understanding of these complex convective structures would benefit from further study by a well-instrumented aircraft designed to penetrate storms safely.

Airborne observations are used to investigate cloud radiative processes. Information about cloud radiative feedback mechanisms is also important in estimating the magnitude of possible climate responses to human activity. Clouds such as cirrus, stratus, and growing cumulus clouds can affect the Earth's radiation budget. Measurements of the reflectance and absorption of incoming solar radiation are taken during many of the programs for atmospheric measurements.

During the past several decades, aircraft observations have become an important and necessary resource for researching air–ocean interactions and for studying other oceanographic phenomena. The observations are made by means of in-situ measurements, air-deployable oceanographic sensors, and remote-sensing devices. For these measurements a turboprop aircraft is the preferred type of platform. The turboprop aircraft has the range and endurance to allow study of the large-scale phenomena, and, once on station, it enables the study of mesoscale phenomena as they evolve. In many instances, during joint atmospheric oceanographical experiments the data set that is obtained from an aircraft is complemented with the measurements obtained from a research vessel.

For background information SEE APPLICATIONS SATELLITES; GREENHOUSE EFFECT; METEOROLOGICAL INSTRUMENTATION; METEOROLOGICAL RADAR; METEOROLOGICAL SATELLITES; REMOTE SENSING in the McGraw-Hill Encyclopedia of Science & Technology.

Paul Spyers-Duran

Bibliography. *Airborne Geoscience Newsletter*, NASA, quarterly; H. R. Byers and R. R. Braham, Jr., *The Thunderstorm*, 1949.

Methods engineering

A fundamental aspect of design and manufacture is the transformation of the functional requirements of the product into geometric dimensions and tolerances of features of individual components; this is followed by economic production of these features within the imposed bounds of dimensions and tolerances. There is a consensus of opinion that lack of a computer-integrated facility for analysis and management of tolerances represents a major road block to integration of design and manufacturing functions. An integrated approach to tolerancing embraces all design activities concerned with selection of engineering materials and precise determination of the effects of small errors or variations on overall function. It is also concerned with the assessment of corresponding difficulties and cost of production, as well as the inspection of components. Its ultimate aim is to identify optimum or near-optimum tolerances for each component so that product performance represents the best compromise

between specified design requirements and the costs or difficulties of production, inspection, and assembly. An important class of tolerancing problems that occurs in design and manufacturing requires solution of algebraic equations involving stochastic variables. The availability of computer-based methods for solution of these problems represents an important contribution and a step toward the concept of computer-aided tolerancing.

Design and production. At the design stage, one result of design activities is the specification of the precise form of the individual components that are assembled to make a product. The intended function of the product will impose requirements on some critical elements. In **Fig. 1** the distance between the left-hand bushing and the left shaft shoulder (X_Σ) is an example of a critical element. Variability of this distance should be controlled to prevent interference as well as axial displacement of the gears. Geometric analysis of similar assemblies usually leads designers to express critical requirements as algebraic functions of dimensions associated with individual components. Since it is known that parts cannot be made to exactly the same size, designers look for means to account for the stochastic nature of component dimensions. As the likelihood of extreme component dimensions combining during random assembly is small, the potential benefit from consideration of the stochastic nature of components is the possibility of assigning larger tolerances to the dimensions of individual components, reducing the cost of manufacture. Thus, important problems at the design stage are the expression of critical dimensions as functions of relevant dimensions of individual components and the application of suitable methods to anticipate statistically their combination.

At the production stage, raw materials are transformed into components that satisfy design specifica-

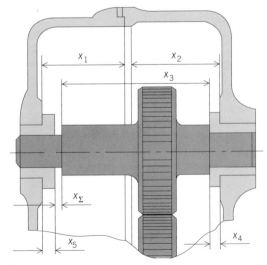

Fig. 1. Example of a critical requirement X_Σ in a gear train assembly. The value of X is given by the summation $X_\Sigma = X_1 + X_2 - X_3 - X_4 - X_5$. (*After O. Bjorke, Computer Aided Tolerancing, ASME Press, 2d ed., 1989*)

tions. Some key activities involved at this stage are outlined in **Fig. 2***a*. Production usually starts by developing a tentative process plan (process sequence) that will allow transformation of raw material shape into final component shape. The process plan is an outline of operations to be performed, their sequence, machine tool types and tooling involved, and a description of each operation procedure. Feasibility of a tentative process plan is then checked with a tolerance chart analysis (tolerance chart control). A tolerance chart is a graphical representation of each operation contained in the process sequence (Fig. 2*b*). The result of the analysis is a set of equations of the form

$$d_i = \sum_{j \in S_i} X_j,$$

showing the influence of process operation (X_j's) on each part design specification d_i. The purpose of the tolerance chart is to show how individual cuts combine to produce each desired drawing dimension. This usually results in a set of linear algebraic expressions that shows the relationship between each desired drawing dimension and individual machining cuts that contribute to it. The dimension produced by a specific machining cut will exhibit variability, the extent of which depends on machine tool capability, work material, tool material, and so forth. The production department is then faced with a tolerance problem similar to that of the design stage; it will also look for solution methods suitable for easy determination of the results of stochastic variable combinations. The need therefore exists for suitable computer-based modules that can be used easily by both design and production personnel to solve these types of problems. Several methods for solution have become available. These methods can be broadly classified into those relying on approximate analytical techniques and those based on computer simulation.

Analytical methods. A variety of approximate analytical methods have been developed to deal with tolerance analysis. Most of the early methods looked for ways to anticipate the distribution of linear combinations of stochastic dimensions through simple computations making use of the central-limit theorem. This theorem enables the sum of a large number of independent random variables to be approximated by a normal distribution. While useful under some conditions, normal distribution approximations present some limitations. Indeed, design and manufacturing problems often lead to tolerancing chains involving at most two or three stochastic dimensions, as is typical of tolerance chart problems; these dimensions may also be mutually correlated and exhibit truncated nonnormal distributions. Application of the central-limit theorem under these conditions is not possible. In addition, the normal distribution approximation cannot adequately model asymmetric cases, and its theoretically infinite bounds may be unrealistic for modeling finitely bounded part dimensions.

A system known as TOLTECH (a commercial

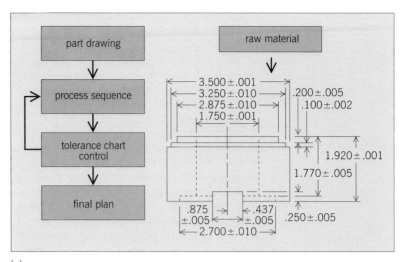

(a)

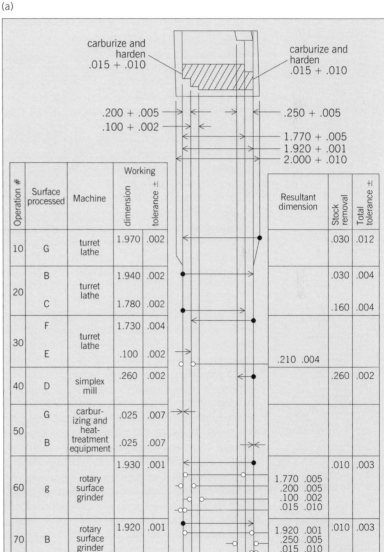

(b)

Fig. 2. Production stage operations. (*a***) Key steps in formulating design specifications. (***b***) Tolerance chart showing the sequence and interaction of operations. Dimensions are given in inches (1 in. = 2.54 cm). Solid circles represent the position of the datum surface being used. Open circles represent dimensions that result from the combination of two or more operations (***after D. F. Eary and G. E. Johnson, Process Engineering for Manufacturing, Prentice-Hall, 1962***).**

software package) overcomes some of the difficulties associated with the normal distribution approximation. The system is based on a classification of geometric elements or assemblies frequently found in designs; the system provides formulas to compute statistical parameters of each element. Design or manufacturing tolerance chains are then built up from a series of appropriate geometric elements. The TOLTECH system uses a beta distribution to approximate the resultant of a combination of stochastic variables. The beta distribution has a finite range and can approximate symmetric as well as asymmetric distributions. Furthermore, it can deal with tolerance chains having as few as two dimensions. The method is elegant, is computationally simple, and can be applied to tolerance analysis as well as synthesis problems. However, a comprehensive evaluation of the precision and performance of beta distribution approximation for tolerance computations remains to be carried out and compared to other approximate methods.

There is a third and major method for solving tolerancing problems in design and manufacturing. It is a method that is essentially an adaptation of techniques developed for reliability computations in structural design. With this procedure, a typical design stage problem can be cast in the following manner. Given design dimensions $X_1, X_2, \ldots, X_n$, which are stochastic variables with known distributions, and m design requirements $R_1, R_2, \ldots, R_m$, which are functions of the X_i design parameters, it is necessary to determine the probability that all requirements are simultaneously satisfied. Each functional requirement R_j defines a safe region in an n-dimensional X-space. This method maps the X-space into another n-dimensional space with new variables $Y_1, Y_2, \ldots, Y_n$, which are independent and have each a standard normal distribution. The design point, corresponding to the mean in X-space, will be mapped to the origin of the Y-space. A reliability index expressing the minimum distance from the origin to the hypersurface delineating the failure region is then used as a measure of the probability content of the failure region. This method can also be used to distribute tolerances on different dimensions and enable maximum relaxation of these tolerances while still satisfying assembly constraints. In theory, this method can accommodate mutually correlated random variables having any distribution. Multiple nonlinear functional requirements are also permitted. Computations, however, appear to be much easier if linear functional requirements with uncorrelated variables are present. This method thus appears to embody a number of desirable aspects commonly found in tolerancing problems. In its general form, however, it appears to be computationally difficult.

Simulation. In addition to approximate analytical methods, Monte Carlo simulation of tolerances remains a powerful tool for the study of combinations of stochastic dimensions. This method enables study of an artificial stochastic model of a particular process. Once a system or process is modeled with mathemat-

ical expressions, it can be evaluated for varying parameter configurations. Monte Carlo simulation examines a system through repeated evaluation of the mathematical model as its parameters are randomly varied to mimic their true behavior. High-speed computers essentially use random number generators to simulate a system by evaluating its mathematical model for numerous sets of random samples drawn from appropriate distributions. Measurements on selected criteria are compiled, and resultant distributions are examined to evaluate the system performance. Parameter distributions are modified and the simulation process repeated until a satisfactory result is obtained. Several software packages are available for simulation of tolerances. A typical package is a module that runs on a personal computer and allows design and manufacturing engineers to generate automatic simulations of tolerancing problems. A user first describes the mathematical equations of the problem to the system. A menu is then used to select and specify appropriate distributions for each variable of the model. The system automatically generates and runs a Monte Carlo simulation of the model. After execution, a user can view histograms of resultant dimensions and interrogate the result interactively for estimates of probabilities. This process can be repeated at will until a satisfactory solution is reached.

The methods described above constitute an important contribution to the goal of computer-aided tolerancing. Much remains to be done, however, before this goal is attained, particularly in the areas of automatic dimensioning and tolerancing, automatic geometric analysis of products, functional tolerancing, three-dimensional tolerance chart analysis, and inspection of tolerances.

For background information SEE ANALYSIS OF VARIANCE; DISTRIBUTION (PROBABILITY); ESTIMATION THEORY; METHODS OF ENGINEERING; MONTE CARLO METHOD; STATISTICS in the McGraw-Hill Encyclopedia of Science & Technology.

E. A. Lehtihet

Bibliography. O. Bjorke, *Computer Aided Tolerancing*, 2d ed., 1989; E. A. Lehtihet and B. A. Dindelli, TOLCON: Microcomputer-based module for simulation of tolerances, *Manuf. Rev.*, 2(3):179–188, September 1989; D. B. Parkinson, Assessment and optimization of dimensional tolerances, *Comput. Aided Des.*, 17(4):191–199, May 1985.

Microwave solid-state devices

A general feature of the development of semiconductor devices is the ever-increasing ability to define sharp interfaces within a device structure; this is clearly the chief factor in the miniaturization of transistors for integrated circuits. Since 1987, a new generation of solid-state microwave devices has emerged as a direct result of the ability to engineer the doping level and the chemical composition of each layer in a multilayer semiconductor structure. It is now possible to focus on the physics of device operation with a view to designing each atomic layer in the semiconductor for its role in the operation of the final device. Although such design might focus on one key aspect of device performance (for example, efficiency), sound physical principles lead to a simultaneous improvement in a whole range of other device attributes, such as lower noise, reduced sensitivity to operating temperature, greater reliability, and higher manufacturing yield. This article describes a source and a detector that exemplify this new generation of microwave devices.

Heterojunction Gunn diode. The gallium arsenide Gunn diode was invented in the 1960s. As an oscillator, it relies on a negative differential resistance associated with the transfer of hot electrons from the central low-mass Γ valley to higher-mass satellite L and X valleys in the gallium arsenide conduction band. The transferred electrons have a low mobility because of both the increased mass and the increased scattering rates that follow from the increased density of states. In practice, a large electric field applied to a moderately doped layer of gallium arsenide heats the electrons. The current rises with increasing bias, reaches a maximum, and then decreases as an increasing fraction of the electrons are transferred to the low-mobility satellite valleys. Over the years, more has been demanded of the Gunn diode. Fundamental frequency operation is useful up to about 60 GHz, where the efficiency is a few percent. Beyond this frequency, the physics of heating the electrons, the intervalley transfer itself, and the complex pattern of electron density instabilities that propagate through the remainder of the semiconductor layer severely degrade the efficiency. It is possible to work at higher frequencies by using the second- or higher-harmonic content of the instability, but again sacrificing efficiency of conversion from dc to microwave and millimeter-wave power. One percent efficiency at ~94 GHz represented the most advanced Gunn diode performance in 1988.

It had been clear for some time that an improved cathode structure for launching the hot electrons would improve the performance of the Gunn diode. The primary reason is as follows: As electrons are being heated by a strong electric field, they can lose their excess energy by the emission of optic phonons (of energy ~0.04 eV). This energy loss process is very effective (and temperature-dependent, as discussed below) and degrades the device efficiency. For high-speed devices, the length of material taken up by this heating process is a considerable fraction of the total active length of the device, and can be considered dead space as far as device performance is concerned. If there is some way of providing the ~0.3 eV excess energy over a very short distance, electrons may not have the opportunity to lose energy by phonon emission. Such a method of launching the hot electrons (**Fig. 1**) consists of a layer of the alloy $Al_xGa_{1-x}As$, with x increasing linearly from 0 to ~30% over 50 nanometers, followed by a heterojunction (a sharp interface where the aluminum content returns to zero) and a very thin layer (~10 nm) of heavily doped

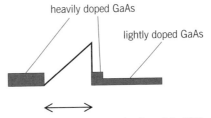

~50-nm layer of AlGaAs with Al increasing from 0 to 30%

(a)

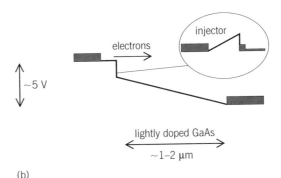

(b)

Fig. 1. Heterojunction Gunn diode. (a) Layer structure of a hot-electron injector for the diode, showing the conduction-band and valence-band profiles. (b) The same profile, and that of the entire diode, under forward bias.

gallium arsenide, before an otherwise conventional drift length for a Gunn diode. The graded aluminum gallium arsenide alloy layer represents a quasi-electric field that is reduced to zero by an applied bias. Depletion in the thin heavily doped layer reduces the electric field after the alloy layer to that which is optimal for intervalley transfer and establishing the electron density instabilities.

This launching pad is able to more than double the efficiency of a gallium arsenide Gunn diode operating in the second-harmonic mode at 94 GHz to over 2.5%, leading to output powers approaching 100 mW at room temperature. Comparable efficiency improvements are obtained in the fundamental mode of operation at lower frequencies. Most of the other systems-critical figures of merit are also improved. For example, conventional devices exhibit a great variation of output power with temperature (a factor of 3 lower at the upper end of the military specification range at 176°F or 80°C compared with that at −40°F or −40°C), largely because the energy loss rate due to optic phonons increases with temperature. Since electrons in the new device remain cold until they are able to transfer, this temperature variability is reduced by a factor of 4. Again, one of the chief sources of noise in the Gunn diode results from the random nature of the position down the device where electrons actually transfer, given the many optic phonon losses sustained en route. In the new device, the position for intervalley transfer is more closely tied to the position of injection. The new devices have a single-sideband noise of typically −86 dBc/Hz at 100 kHz off-carrier, compared with −80 dBc/Hz in conventional devices. (The abbreviation dBc stands for decibels with respect to the carrier power level.) This tying of the length of the active device is also responsible for a halving of the drift in frequency with temperature in the new devices compared with the old. The new cathode structure also increases the device impedance somewhat, which greatly eases the circuit matching. Finally, the performance of these devices in accelerated life tests is very encouraging. All high fields in the new devices occur in single-crystal semiconductor layers, in contrast to appearing at metal-semiconductor junctions which have been proposed as alternative injector structures, where the microcrystalline structure at the metal-semiconductor interface places a major limitation on reliability. Thus a careful engineering of the cathode structure in a Gunn diode improves nearly all the device performance attributes of relevance to systems application.

Planar-doped barrier diode. The rectifying metal-semiconductor contact, the Schottky diode, is an important component in microwave circuits: the non-linearity of the current-voltage characteristics is the basis for mixing and detecting microwave signals. The limitations of Schottky diodes include the difficulty of their reliable and reproducible manufacture, the strong temperature dependence of the current-voltage characteristics, the fact that the devices must be biased to the point of maximum curvature of the current-voltage characteristic (typically $\sim$0.5 V) for optimum efficiency, and their high component of close-to-carrier noise. Here again, a carefully designed multilayer structure can improve upon nearly all the desirable features of a Schottky diode. A planar-doped barrier diode, consisting of an n^+-i-p^+-i-n^+ doping sequence, all in gallium arsenide (**Fig. 2**), where the intrinsic (denoted by i) layers are of unequal length, has been used to provide the asymmetry in the current-voltage characteristics. The mode of operation is simple: The p^+ layer is thin (less than 10 nm thick) and highly doped ($\sim$$10^{18}$/cm^3), and it is fully depleted. The resulting negative charge centers in this layer provide a triangular electrostatic barrier to the motion of other electrons. The barrier height is determined by the doping-thickess product in the p^+ layer, and the barrier shape by the relative lengths of the two intrinsic arms. In practice, a useful model for designing the current-voltage characteristics requires the consideration of other factors, such as the spillover of electrons from the n^+ layers into the intrinsic regions, and the finite thickness of the p^+ layer, but these can be handled to the extent that quantitative models require no parameters other than the doping levels and layer thicknesses. There is considerable freedom, not available with a metal-semiconductor contact, to prepare low-barrier mixers or to move the point of maximum curvature of the current-voltage characteristics toward zero bias.

The microwave performance of gallium arsenide planar-doped barriers with 50-nm and 500-nm intrinsic arms has matched in all aspects the device and circuit performance of Schottky diodes (whether in silicon or gallium arsenide), and in many cases superior performance has been achieved in key aspects.

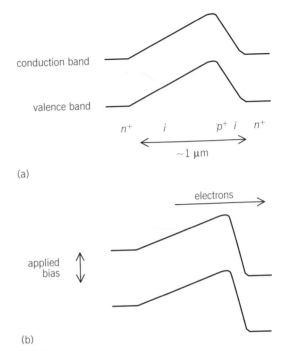

(a)

(b)

Fig. 2. Planar-doped barrier diode. (*a*) Layer structure, showing the conduction-band profile. (*b*) The same profile under forward bias.

First, reliability and reproducibility follow from the degree of control over the growth of semiconductor multilayers. Second, whereas the height of a Schottky barrier decreases with temperature and so accentuates the strong temperature dependence of thermionic emission, the height of the planar-doped barrier increases with temperature and cancels some of this temperature dependence. Third, the planar-doped barrier is a single-crystal semiconductor device, and the absence of the microcrystalline morphology of a practical metal-semiconductor interface results in planar-doped barriers being able to withstand pulsed incident microwave powers up to 500 times as intense as with a Schottky diode. Furthermore, the noise associated with carriers passing through the metal-semiconductor contact is considerable but not well understood; the planar-doped barrier diodes have exceptionally low close-to-carrier noise, which is important in Doppler-shift radar applications, for example. Planar-doped barriers achieve a tenfold increase in dynamic range of detecting ability, and they can be driven with a fivefold reduction in local oscillator power. The only potential weakness so far discovered, namely that the devices are greater than 0.5 micrometer thick, becomes important above 100 GHz, where it places a transit-time and RC (resistance-capacitance) limitation on the upper frequency of operation.

Other devices. The heterojunction Gunn diode and the planar-doped barrier diode are two examples of new-generation microwave devices. Research on others is well advanced. Nearly all existing components can be subject to such a treatment. New tunnel diodes and IMPATT diodes (impact avalanche and transit time diodes) can be expected, and microwave and millimeter-wave transistors are starting to feature heterojunction design elements. Higher-power and lower-noise microwave systems, and new millimeter-wave systems operating above 100 GHz are emerging.

For background information *SEE* B*AND THEORY OF SOLIDS*; M*ICROWAVE SOLID-STATE DEVICES*; S*EMICONDUCTOR HETEROSTRUCTURES*; S*OLID-STATE PHYSICS* in the McGraw-Hill Encyclopedia of Science & Technology.

M. J. Kelly; N. R. Couch; M. J. Kearney

Bibliography. N. R. Couch et al., High performance graded gap AlGaAs injector Gunn diodes at 94 GHz, *IEEE Electr. Devices Lett.*, 10:288–290, 1989; M. J. Kearney et al., Asymmetric planar doped barriers for mixer and detector applications, *Electr. Lett.*, 25:1454–1456, 1989.

Microwave tube

The orbitron microwave maser combines features of conventional microwave tubes and standard masers and lasers. It uses synthetic atoms composed of free electrons orbiting positively charged metal nuclei. The metal nuclei consist of long thin wires inside a metal structure, and the free electrons must be supplied by an external source. These synthetic atoms replace the atoms or molecules used in conventional masers. The advantage of these synthetic atoms is that their orbit frequencies can be changed by altering the applied potential or the size of the nuclei (wire radius). In particular, the orbitron maser can emit in the submillimeter range between conventional microwave tubes and masers and lasers. It is a simple, low-voltage device with no applied magnetic field.

The orbitron microwave maser was designed specifically as a source of submillimeter microwave radiation, where there now exists a technological gap between conventional microwave tubes and masers and lasers. The problem with conventional tubes is that the frequency is determined by a mechanical structure that is difficult to fabricate on the small scale required for producing high frequencies. The difficulty with masers and lasers is that the resonant structure is a molecule or atom, which in general will not operate in the submillimeter range (0.1–1.0 mm).

Competing advanced microwave devices for this frequency range include the gyrotron and the free-electron laser. Although both these devices produce intense submillimeter radiation, they require heroic measures. The gyrotrons require magnetic fields of 40 teslas and semirelativistic ($\approx$0.5 MeV) electron-beam velocities to reach wavelengths of $\lambda = 0.3$ mm. The free-electron laser requires multimegavolt electron-beam velocities to provide strong relativistic frequency upshifts. In contrast, the orbitron reaches this same wavelength range without the high electron-beam velocities, and with no magnetic field. In fact, a weak external magnetic field generally quenches microwave emission. The orbitron maser overcomes the problems of this frequency range by using maser techniques with synthetic atoms, which have resonant frequencies that can be adjusted into the submillimeter range. An example of such a device is shown in **Fig. 1**, in which a positive central wire is placed inside a negative container. *SEE* L*ASER*.

Orbitron operation. In this device, free electrons can be generated by several processes. In the original device a gas discharge was used. In later devices a hot filament produced the free electrons. In the orbitron these electrons are attracted to the central wire. However, since the electrons generally are produced with some random, transverse energy, many miss the central wire and go into orbit around it. Negative end electrodes prevent the electrons from escaping at the ends of the wire. The net result is that the electrons are permanently trapped inside the system until they escape, either by some collision process or by an instability. The orbiting electrons can be made visible by adding gas (**Fig. 2**).

The cathode structure forms a microwave cavity resonator. Microwave radiation from the orbiting electron cloud around the wire is trapped inside the cavity. This trapped microwave radiation perturbs the orbiting electron cloud, enhancing the radiation emission (stimulated emission) and producing radiation at distinct frequencies. These frequencies are determined by modes of the cavity, but also must be in the range of frequencies permitted by the orbiting electrons (**Fig. 3**). In the preliminary version of the orbitron, this radiation merely leaked out of the cavity along the central positive wire. In later models, waveguide outputs have been used.

Emission frequencies. The significant feature of the orbitron maser is its ability to generate very high frequency radiation without the use of high voltages

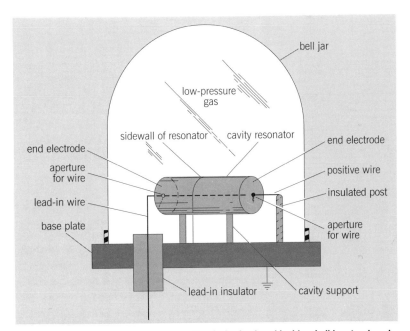

Fig. 1. Diagram of the first orbitron maser. The device is placed inside a bell jar at reduced gas pressure ($\approx$5 μm Hg or 0.7 Pa). A high-voltage positive pulse on the central positive wire initiates a glow discharge and causes intense microwave emission. End electrodes confine electrons. (*After USM Patent #4,459,511, Alexeff*)

Fig. 2. Orbiting electrons around an etched positive wire. The thicker segments correspond to positive nuclei, and electrons orbit about them. A trace of gas produces the light emission by collisions between electrons and gas atoms.

or strong magnetic fields. The basic frequency ν (in hertz) of electrons at an orbit radius r around a cylindrical wire of radius r_0 inside a cylindrical can of radius r_1 is given by Eq. (1). Here e is the electron

$$\nu = \frac{1}{2\pi r}\left[\frac{eV}{m_e \ln (r_1/r_0)}\right]^{1/2} \quad (1)$$

charge (1.60×10^{-19} coulomb), V is the applied voltage (in volts), m_e is the electron mass (0.91×10^{-30} kilogram), r, r_0, and r_1 are expressed in meters, and ln is the natural logarithm (base $e = 2.71828$). For electrons grazing the surface of the wire, the highest frequency is produced. For example, if r_0 and r are 1/40 mm (1 mil), r_1 is 1 cm (0.4 in.), and V is 10,000 volts, the resultant frequency is 1.09×10^{11} Hz or 2.75 mm in wavelength. A comparable magnetic field–induced frequency in a magnetic cyclotron maser is given by Eq. (2), where B is

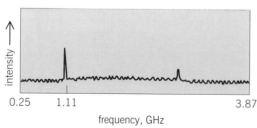

Fig. 3. Microwave spectrum of steady-state orbitron emission using a glow discharge as an electron source. Vertical scale (intensity) is logarithmic.

$$\nu = \frac{1}{2\pi}\frac{eB}{m_e} \quad (2)$$

the magnetic field in teslas, and all other variables are as defined previously. For $B = 0.4$ (the limit for permanent magnets), this equation gives $\nu = 1.112 \times 10^{10}$ Hz, or $\lambda = 2.68$ cm. Thus, the electrostatic device produces a wavelength about 10 times shorter than the magnetic device. Raising the voltage and reducing the wire size both lead to higher-frequency operation. It has been found that the upper frequency limit of these devices generally follows the scaling law of Eq. (1).

By going to much thinner wires, with radii of 9×10^{-3} mm, and etching the wires to provide thicker sections for the electrons to orbit around (Fig. 2), microwave radiation has been produced at a frequency of 1 THz (wavelength of 0.3 mm). The radiation was detected by an indium antimonide bolometer crystal in liquid helium at a temperature of 4.2 K ($-452°F$). The frequency was determined by filtering the radiation through a succession of wire-mesh, high-pass filters. About 1 W of terahertz radiation was produced in a repetitive, 1-microsecond pulse. The upper frequency limit was determined by the band limit of the receiver, not the orbitron maser.

If the electrons are in noncircular orbits, the frequency of electron motion is given approximately by Eq. (1), with r interpreted as the maximum excursion that the electron makes from the wire. In addition, these noncircular orbits can emit at harmonics of their periodic motion, producing even higher-frequency emission.

Improvements and advances. Electron confinement in the orbitron maser is improved when the central wire is electrochemically etched to form a series of alternately thick and thin segments (Fig. 2). The electrons are confined around the thick wire segments, which simulate individual positive nuclei in atoms. This is demonstrated by the trapping of electrons around the thick segments, as shown by the gas-induced glow discharge in Fig. 2 and by computer simulations. This trapping prevents the electrons in small, high-frequency orbits from spiraling off the ends of the wire and enhances very high-frequency microwave emission.

Further advances in the orbitron maser development include the observation of stimulated emission; operation in a quasioptical mode between parallel mirrors; operation in a vacuum with electrons produced by a hot, oxide-coated tungsten filament; operation in the steady state with a gas-discharge electron source; operation with a pulsed-line high-energy source (2.2 kW in a pulse at 10 cm); and operation with a split-cathode system to produce narrow lines. The device is being developed commercially, for example, as a source of radiation for millimeter radar.

Use as receiver. The orbitron maser can also be used as a radio-frequency receiver. If a microwave signal at a wavelength of 30 cm is injected into an operating hot-cathode orbitron at a frequency near that of an emitted fundamental line, the power emitted by that line can be modified, and the electron current changes. Changes in the emission current up to a factor of 2.5 have been observed. This observation suggests that the orbitron maser could be used as a simple, robust detector for submillimeter radiation.

Efficiency. The efficiency of an orbitron maser operating as a broad-band source appears to be high. By using a Q-spoiled device, it was observed that the anode wire became incandescent when the microwave emission was terminated by Q-spoiling, but became dark when microwave emission was permitted. This observation suggests that most of the applied power is emitted as radio-frequency radiation, although over many spectral lines. In a given spectral line the emission at best so far corresponds to a few percent efficiency.

Computational simulation predicted about 15% efficiency for an orbitron operating on an end-injected electron beam. This efficiency is comparable to that obtained with conventional traveling-wave electron tubes if the electronic structure is not tapered. In a tapered tube the electromagnetic wave slows down along the tube length to stay at the same velocity as the electron beam (in resonance) as the electron beam loses energy and slows down. Tapering in an orbitron maser should be easy to do by means of, for example, a tapered central positive wire.

Multianode orbitrons. High-power operation of a single-anode orbitron maser has been difficult because of the heat load imposed on the central wire anode. Consequently, multianode-wire orbitrons have been constructed to share the heat load. The orbiting electrons clouds around each wire appear to phase-lock to the shared cavity field, producing narrow-line emission at much higher power. Up to 11 anode wires have been used successfully.

For background information SEE GYROTRON; LASER; MASER; MICROWAVE TUBE in the McGraw-Hill Encyclopedia of Science & Technology.

Igor Alexeff; Fred Dyer; Mark Rader

Bibliography. I. Alexeff and F. Dyer, Millimeter microwave emission from a maser by use of plasma-produced electrons orbiting a positively-charged wire, *Phys. Rev. Lett.*, 45:351–354, 1980; J. M. Burke, W. M. Manheimer, and E. Ott, Theory of the orbitron maser, *Phys. Rev. Lett.*, 56:2625–2628, 1986; D. Chernin and Y. Y. Lau, Stability of laminar electron lasers, *Phys. Fluids*, 27:2319–2331, 1984; *Conference Digest of the 13th International Conference on Infrared and Millimeter Waves, December 5–9, 1988, Honolulu*, SPIE vol. 1039, 1989; A. K. Ganguly, H. P. Freund, and S. Ahn, Nonlinear theory of the orbitron maser in three dimensions, *Phys. Rev. A*, 36:2199–2209, 1987; V. L. Granatstein and I. Alexeff (eds.), *High-Power Microwave Sources*, 1987.

Mid-Oceanic Ridge

The axis of the Mid-Oceanic Ridge undulates up and down hundreds of meters like a gentle roller coaster over distances of 30–300 km (18–180 mi; **Fig. 1**). The rise axis is not strictly a continuous, uninterrupted structure along its length; the deep points are located at ridge-axis discontinuities. Some of these discontinuities are transform faults and propagating rifts; however, most of the discontinuities are nonrigid structures, such as overlapping spreading centers. These ridge-axis discontinuities define a fundamental segmentation of mid-oceanic ridges on length scales of 10–1000 km (6–600 mi). Many of the smaller discontinuities, such as overlapping spreading centers, migrate along the ridge at various rates and in varying directions, so that the intervening ridge segments may lengthen or shorten with time and may move north or south along the ridge. These segments appear to behave independently, so that the amount of magma available for eruption and the intensity of tectonic activity often varies between neighboring segments. Magmatic and tectonic activities also vary systematically along the length of each individual segment, with magmatism and volcanism least active near discontinuities and most active near shallow areas between the discontinuities.

Magma supply model. High-resolution bathymetric images of the East Pacific Rise show that the rise crest is generally broadest along the shallow portions of each segment and most narrow (often a razor-backed ridge) along the deeper portions where transform faults and overlapping spreading centers occur. These observations led to the hypothesis that the Mid-Oceanic Ridge's supply of melt from the upper mantle is enhanced beneath the shallow, swollen areas of each segment, where melt is fed to shallow crustal magma reservoirs, and is depleted at the ends of each segment near overlapping spread-ing centers and other discontinuities. Magma might flow beneath the ridge away from centers of upwelling.

Thus, a model of magma supply for ridge segmentation has been proposed. As the plates spread apart, episodes of decompression partial melting of mantle rocks are triggered at depths of 30–60 km (18–36 mi). The buoyant melt segregates from residual solid rock and ascends, filling shallow magma chambers within the crust along the ridge axis. These melts swell the crustal magma reservoirs locally, and the buoyant forces associated with the melt and surrounding halo of hot, low-density rock create a local shoaling of the ridge crest. Continued injection of melt leads to local eruptions, as well as to migration of magma away from the locus of upwelling, expanding the axial magma chamber along the strike. The laterally migrating magma loses hydraulic head with increasing distance from the center of magma replenishment, creating a graded increase in the depth of the ridge axis along the strike. As magma migrates at depth along the ridge, the overlying brittle carapace of frozen lava stretches and fractures; these fractures then serve as conduits of the flow of magma to the sea floor, and volcanic eruptions follow in the wake of an advancing, cracking front. The process outlined above occurs repeatedly as plate separation continues. In this magmatic model for the spreading center, ridge-axis discontinuities occur at the distal ends of magmatic

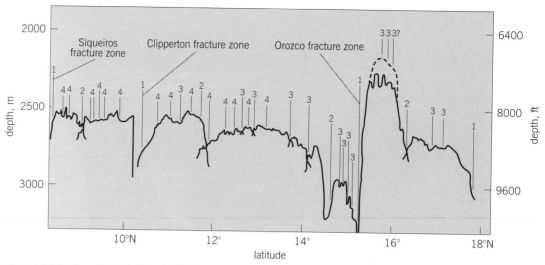

Fig. 1. Axial depth profile of the East Pacific Rise from 8 to 18°N latitude, extracted from *Sea Beam* records. Numbers indicate the orders of the discontinuities identified. Long-wavelength undulations in the axial depth profile are punctuated by first- and sometimes second-order discontinuities, whereas short-wavelength undulations are marked by third-order discontinuities. Fourth-order discontinuities have little or no bathymetric signature.

pulses and at deep points along the ridge axis, and they define the ends of ridge segments.

Hierarchy of segmentation. There appears to be a hierarchy in segmentation based on the lengths and longevities of segments and on the sizes and durations of the discontinuities that define them. **Figure 2** shows a proposed hierarchy of ridge-axis discontinuities of orders 1 through 4 for slow- and fast-spreading centers. The most common first-order discontinuity, regardless of the spreading rate, is the transform fault, where rigid plates slide past each other. The transform faults offset the ridge by at least 20 km (12 mi), persist for millions to tens of millions of years, and partition the ridge on a length scale of 200–500 km (120–300 mi; total range 100–1000 km or 60–600 mi). At fast-spreading centers, where the ridge is usually marked by an axial high (profile A-A′ in Fig. 2a), the most common type of second-order discontinuity is a large overlapping spreading center, where the ridge is offset at least 2 km (1.2 mi). Such centers partition the ridge on a length scale of 50–200 km (30–120 mi) and persist for less than several million years. Third-order discontinuities are small overlapping spreading centers that offset the ridge only 0.5–2 km (0.3–1.2 mi). They segment the ridge on a length scale of 30–100 km (18–60 mi) and appear to last on the order of 10,000 years. On an even shorter length scale of 10–50 km (6–30 mi), fourth-order discontinuities correspond with little or no offset of the ridge and are characterized by slight deviations in axial linearity (devals; Fig. 2a). They persist less than 10,000 years and may be as short-lived as major eruptive events (years to hundreds of years). On slow-spreading ridges, the length scales of second-, third-, and fourth-order discontinuities are a bit shorter, while the time scales are likely to be a bit longer than for discontinuities of the same order at fast-spreading rates. The structure and morphology of higher-order discontinuities are also different from their fast-spreading counterparts because of the presence of a deep rift valley

(profile B-B′ in Fig. 2b) and thicker lithosphere along the axis, and because of less frequent volcanic eruptions. For example, second-order discontinuities are bends in the axial rift valley, and the spreading centers do not overlap significantly. Third-order discontinuities are gaps between chains of volcanoes within the rift valley floor, while fourth-order discontinuities may be small gaps within single chains of volcanoes. For all spreading rates, there is a continuum between discontinuities of orders 1 through 4, and a given discontinuity can evolve to a higher or lower order.

Geophysical evidence. To "see" beneath the ridge, a host of seismic measurements and measurements of the Earth's gravity field have been made on the East Pacific Rise between 9 and 13°N latitude. Multichannel seismic measurements show a very bright reflector of sound energy approximately 1.2–2.5 km (0.89–1.6 mi) beneath the sea floor, along the shallow portions of each ridge segment. This reflector often deepens and then disappears near overlapping spreading centers and transform faults. This conspicuous reflector has been interpreted to be the roof of an axial magma chamber that serves as the principal magma reservoir for creation of new oceanic crust. All that can be said with certainty is that the reflector is the top of a region of low sound velocity, but most workers in this field agree that the most reasonable cause of the reflector is a long, narrow, shallow body of magma and hot rock beneath the ridge. Measurements made across the ridge by using two ships (one to shoot and one to listen) and seismometers placed on the ocean floor show that the mostly molten (>50% melt) chamber is not large. It is only 2–4 km (1.2–2.4 mi) wide and less than 1 km (0.6 mi) thick in most places. However, the magma chamber is surrounded by a wider reservoir of very hot (perhaps slightly molten) rock that has only recently solidified and may be 6–10 km (3.6–6 mi) wide and 3–6 km (1.8–3.6 mi) thick. This region of hot rock extends at least to the base of the oceanic crust and probably a few kilometers into the upper mantle. Precise measurements of the Earth's gravitational field in the same area support the interpretation that the region of hot and slightly molten rock is sufficiently wide and buoyant to support the axial ridge, which is 4–10 km (2.4–6 mi) wide and 200–400 m (660–1320 ft) high, and that the chamber of true magma (>50% partial melt) is very narrow. A thin pool of nearly 100% melt approximately 4 km (2.4 mi) wide caps the top of this broader reservoir of magma and hot rock and provides the large acoustic impedance contrast needed to explain the very bright reflector observed 1.2–3.5 km (0.74–2.2 mi) beneath the sea floor. In summary, the axial magma chamber may resemble a mushroom in cross section with a narrow stalk of partial melt feeding a 4-km-wide (2.4 mi), but very thin, lens of pure melt (Fig. 2). This chamber shrinks and disappears near overlapping spreading centers and transform faults. The broader reservoir of hot rock bends to stay beneath the ridge at the overlapping spreading centers, and it is offset at transform faults.

History of segments and discontinuities. The history of the behavior of ridge-axis discontinuities and the ridge segments they define can be investigated by carefully mapping the disturbed sea floor. Several swath-mapping expeditions have extended the high-resolution coverage out on the flanks of the East Pacific Rise to crustal ages of greater than 2×10^6 years. The bathymetric and tectonic maps show that large overlapping spreading centers (offsets >3 km or 1.8 mi) leave wakes of deformed oceanic crust up to

80 km (48 mi) wide as they migrate up and down the ridge axis. Within the wake of disturbed oceanic crust there are curved fossil ridge tips 10–20 km (6–12 mi) long that have been cut off at overlapping spreading centers as predicted from laboratory-based studies of overlapping spreading centers and crack propagation. The sea floor within these disturbed regions (discordant zones) is 100–300 m (330–990 ft) deeper than surrounding sea floor, just as the overlapping spreading centers are 100–300 m (330–990 ft) deeper than the shallow, magmatically robust portions of the ridge segments. The crust generated at overlapping spreading centers may be thinner and deeper by at least several hundred meters than oceanic crust created elsewhere along each segment, because overlapping spreading centers are distal to sources of magma supply and tend to be starved of magma. Detailed seismic and gravity measurements need to be made in these areas to test this idea.

Measurements of the Earth's magnetic field over overlapping spreading centers support the idea that these centers occur where the magma supply is low. Lavas that erupt at overlapping spreading centers are often much more strongly magnetic than lavas erupted elsewhere along the ridge. It turns out that lavas that erupt from small magma chambers that alternately freeze and become replenished tend to have more iron-rich minerals in a highly magnetized state than lavas that erupt from magma chambers that are large enough not to solidify between episodes of replenishment of magma. Thus, the wakes of discontinuities

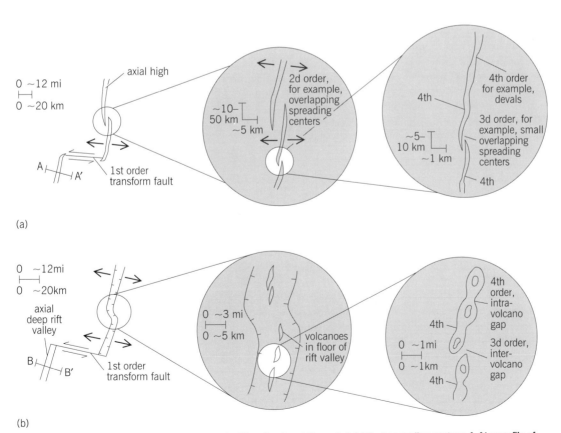

(a)

(b)

Fig. 2. A proposed hierarchy of ridge-axis discontinuities of orders 1 through 4. (a) Fast-spreading centers; A-A′ = profile of axial high. (b) Slow-spreading centers; B-B′ = profile of deep rift valley.

are detected as large disturbances in the measured magnetic field, because the high magnetization of crust created at discontinuities is preserved for several million years.

Based on the age of the crust into which the discordant zones extend and the patterns of these off-axis wakes, it can be demonstrated that second-order segments persist as discrete entities for up to several million years. The discontinuities may oscillate slowly in position by 10–20 km (6–12 mi) on the ridge, or they may migrate along the ridge many tens of kilometers at rates of 20–100 mm (0.8–4 in.) per year, the same magnitude as the spreading rates. This migration tends to occur in jerks; a ridge segment may lengthen by propagation of its tip at rates of several hundred millimeters per year, but it may then retreat and shorten for a period of time before making the next surge forward. In this way, the ridge tips at a second-order discontinuity appear to be dueling as they surge back and forth along the ridge, generally making slow progress in one direction or the other.

In contrast, the ridge segments defined by third- and fourth-order discontinuities with offsets less than 2–3 km (1.2–1.8 m) leave no detectable off-axis wakes. Thus, they appear to be short-lived, less than 10,000–100,000 years in most cases. Third-order discontinuities, such as overlapping spreading centers with small offsets (<3 km or 1.8 mi), have a distinctive shape on sea-floor maps, and the two cases that have been studied seismically show clear breaks in the axial magma chamber reflector. On the other hand, fourth-order discontinuities may have little or no tectonic signature, at most a subtle bend or tiny offset (<500 m or 1640 ft) of the spreading axis. Given that the present state-of-the-art ability to pinpoint the axis is only slightly better than 500 m (1640 ft) on fast-spreading centers and usually somewhat worse than 500 m (1640 ft) on slow-spreading ridges, fourth-order discontinuities are very difficult to map. In some cases the axial magma chamber deepens slightly and, more rarely, exhibits an apparent break, but in most cases it is fairly continuous beneath fourth-order discontinuities.

Several kinds of second-order discontinuities have been investigated at slow-spreading rates. However, the existence and behavior of third- and fourth-order discontinuities at slow spreading rates are still poorly understood.

For background information SEE LITHOSPHERE; MAGMA; PLATE TECTONICS; TECTONOPHYSICS; TRANSFORM FAULT in the McGraw-Hill Encyclopedia of Science & Technology.

Ken C. Macdonald

Bibliography. R. S. Detrick et al., Multi-channel seismic imaging of a crustal magma chamber along the East Pacific Rise, *Nature*, 326:35–41, 1987; R. M. Haymon and K. C. Macdonald, The geology of deep-sea hot springs, *Amer. Sci.*, 73:441–449, 1985; C. H. Langmuir, J. F. Bender, and R. Batiza, Petrological and tectonic segmentation of the East Pacific Rise, 5°30′–14°30′N, *Nature*, 322:422–429, 1986; K. C. Macdonald et al., A new view of the mid-ocean ridge from the behaviour of ridge-axis discontinuities, *Nature*, 335:217–225, 1988; K. C. Macdonald and P. J. Fox, The mid-ocean ridge, *Sci. Amer.*, 262:72–79, 1990; H. Schouten, K. D. Klitgord, and J. A. Whitehead, Segmentation of mid-ocean ridges, *Nature*, 317:225–229, 1985.

Military aircraft

Aircraft have been used extensively in almost every large-scale military action since 1940. The defending forces have, in most cases, countered hostile aircraft with as wide a range of defensive weapons as possible. The problem of shooting down an aircraft or missile with these air defense weapons is dependent not just on the quality of the weapons but also, to a great extent, on the vulnerability of the intruding aircraft. Vulnerability of an aircraft begins with its detection. Detectability of the intruder depends on its radar, infrared, visual, and acoustic signatures or observables.

Radar detection. Radar is the most reliable method for detecting aircraft. Reducing an aircraft's vulnerability to detection by radar is perhaps the most important application of stealth technology. As a solid body, an aircraft reflects radar energy according to the aspect or view presented to the radar receiving antenna. The reflected energy pattern about the aircraft has a large number of peaks and troughs with values differing by several factors of 10. The reflection or signature characteristics depend primarily on the size and arrangement of the aircraft, its surface shape, and the nature of the materials used in its construction. Metallic surfaces reflect better than electrically nonconductive ones.

Measurements of radar reflectivity are made by using either the aircraft itself or models. For design purposes, or for the simulation of a hostile aircraft, a

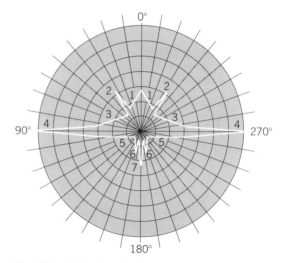

Fig. 1. Calculated radar reflection plot for the F-18 fighter. The numbers indicate features of the fighter: 1 = aircraft nose and forebody; 2 = air intaker; 3 = leading edge of wing; 4 = overall lateral reflection; 5 = trailing edge of wing; 6 = vertical stabilizer; 7 = rear fuselage.

Table 1. Estimated radar signatures

Radar target	Nose-on, average, X-band radar signature, m²*	
	Without stealth	With stealth
Aircraft carrier	10,000–20,000	Not applicable
Destroyer	1000–1500	Not applicable
Bomber aircraft	10–200	0.1–1.0
Fighter aircraft	1–20	0.01–0.1
Cruise missile	0.1–1.0	0.001–0.01
Bird	0.0001–0.1	Not applicable

*1 m² = 10.8 ft².

sufficiently accurate reflection plot can also be produced by computation. This is done by simulating the aircraft configuration with geometrically simple shapes for which formulas are available to calculate the reflectivity. Examples of such shapes are the cone, cylinder, sphere, plane mirror, and two-cornered reflectors. The fictional aircraft constituted by these shapes can then be considered from any desired direction. **Figure 1** shows a computer-generated reflection diagram for the F-18 fighter with 360° horizontal illumination. Integration of reflection diagrams for the horizontal and vertical illumination gives the average radar cross section, which in this case is slightly more than 1 m² (10.8 ft²).

Radar signature estimates for various types of targets with and without stealth design are shown in **Table 1**. Signatures for ships and birds are included as reference values.

Various design techniques and materials have been developed to reduce the radar signature of an aircraft. Smaller aircraft have lower radar and visual signatures. Components are arranged to hide or shield high-signature features of the configuration, bury engines, shield engine inlets, and carry ordnance internally. Surfaces having three-dimensional curvature are used, which minimize the reflected energy (signature) from incident radar beams. Finally, selective applications are made of radar-absorbing materials and coatings, and materials having switchable conductivity.

Radar antennas aboard the aircraft are also effective radar reflectors. They normally reflect radar energy in the frequency band in which the antenna was designed to operate, and this could increase a stealth aircraft's chances of being detected by hostile radars. One type of antenna that has a high radar signature is the slotted

planar array, the type most fitted to modern fighter aircraft. However, conformal phased-array antennas installed on the B-1B and B-2 bombers are considered stealth radars. The radomes that enclose these radar antennas act as filters. Electromagnetic elements within the walls of the radome itself hide the antenna from certain radar frequencies. These special radome radio-frequency filters allow the internal radar to transmit and receive on one frequency while reflecting unwanted frequencies.

Infrared detection. Aircraft are sources of heat and can, therefore, be located and tracked with infrared sensors. These sensors detect a target by making use of the temperature difference between the heat emitted by the target and the cooler surroundings. The main source of radiation at supersonic speeds is the aircraft itself, which is heated by boundary-layer friction. At subsonic speeds, the primary heat sources are the hot engine parts and the exhaust plume. Unlike visual detection, infrared sensing can detect targets at night and in many inclement weather situations.

Engine infrared emissions must be eliminated or masked for a stealth aircraft to be successful. Methods for achieving this include shielding, active or passive cooling, and application of special materials and coatings that absorb or reflect energy.

The cool air from the bypass fan section of a turbofan engine, for example, can be mixed with the hot core exhaust gases to reduce infrared emissions from the exhaust plume. Baffles can also be fitted to exhaust nozzles to reduce infrared emissions. The baffles separate exhaust flow, allowing the exhaust gases to cool faster, and also redirect the exhaust gas flow so that the infrared emissions are masked and hidden from the hostile weapons.

Visual detection. Detection of an aircraft by an observer on the ground with the unaided eye depends on weather conditions. For Central Europe the range at which an aircraft can be identified is, on the average, less than 2.2 mi (3.5 km) and hardly ever exceeds 3.1 mi (5 km), even with good visibility. These low visual identification ranges ensure relatively good protection for fighters, since they travel this distance in about 10 s. Visual detection of aircraft during combat is usually easy at close range. The objective of stealth technology is to reduce the visual signature so that the mission is not jeopardized as the aircraft gets close to its target.

Camouflage is the most widely used method of

Table 2. Stealth aircraft

Aircraft	First flight	Maximum takeoff gross weight, lb (kg)	Maximum speed, Mach number or knots (km/h)	Status
SR-71	April 1965	172,000 (78,000)	M 3.35	Retired
TR-1A	August 1981	40,000 (18,160)	375+ (695+)	Operational
F-117A	June 1981	52,500 (23,835)	High subsonic	Operational
B-1B	October 1984	477,000 (216,558)	M 1.25+	Operational
B-2A	July 1989	350,000 (158,000)	435 (806)	Development
YF-23A(ATF)	August 1990	50,000 (22,700)	M 2.0+	Development
YF-22A(ATF)	September 1990	50,000 (22,700)	M 2.0+	Development

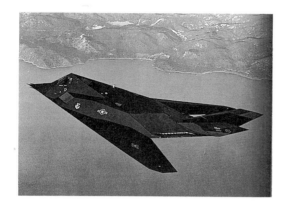

Fig. 2. F-117A stealth fighter. (*Lockheed Aeronautical Systems Co., Lockheed Corp.*)

reducing an aircraft's visual signature. Correct application of camouflage colors, appropriate to the terrain in which the aircraft will be operating, allows the aircraft to blend into the background. One material that has been successful is a paint, developed by the U.S. Army, that reduces an aircraft's infrared and visual signatures. In addition to reflecting infrared radiation, the roughly textured paint combined with special pigments diffuses sunlight, which aids in blending the aircraft into the background.

Another specific visual signature that must be controlled is glint, or light reflected off aircraft canopies. Canopy glint can be reduced considerably with the right combination of tint and laminate. This technique also improves pilot vision from the cockpit, especially under hazy conditions. The latest versions of the F-15 and F-16 fighters have polarized laminates applied to their canopies.

Contrails and smoke from the engines represent still

another visual signature and could mark the path of a stealth aircraft. Engine technology has largely eliminated the smoke problem by burning fuel more efficiently, but condensation trails (contrails) are more difficult to eliminate. Aerodynamic contrails form if the air flowing past the aircraft contains enough moisture and its temperature drops below the dew point of the ambient air. Engine exhaust contrails form when moisture-laden exhaust is expelled into cold air and condenses immediately. Exhaust contrails usually occur above 30,000 ft (9000 m). Fuel additives are sometimes used to minimize them.

Acoustic detection. Aircraft engines are noisy, and so are propeller and helicopter blades. If a stealth aircraft cannot be detected by radar or infrared tracking systems or by sighting, at some point it will probably be audible to hostile forces.

Noise is most pronounced in turbine and turbofan engines. Acoustic absorber materials can be selectively applied to absorb high-frequency harmonic vibrations produced by the high-speed airflow exhausted from the primary fan nozzles. Absorbers can also be used to dampen noise generated by the rotating stages of the engine.

Other methods include the use of screech liners in afterburners and sandwich composite skins with pyramidal structures pointing inward to absorb engine noise. Procedures that reduce acoustic signatures also contribute to minimizing infrared signatures. Laminated coatings on the exhaust nozzle, for example, reduce both infrared and acoustic signatures. For piston engines, modified mufflers are used to reduce engine noise.

Integrated design. The design integration of low-observable technologies leads directly to major trade-offs between signature reduction, flight performance, and program cost. The reductions that can be achieved are very dependent on the aircraft configuration and the threat sensors that must be countered. Computer programs have been developed that allow the design optimization of conflicting signature reduction features.

Characteristics of various stealth aircraft are listed in **Table 2**. All of these aircraft have stealth technologies incorporated in their design. The designs of the F-117A and the B-2A were motivated primarily by stealth requirements.

The F-117A arrangement (**Fig. 2**) has inlets on top of the wing which feed the airflow to engines buried in the fuselage. Engine exhaust is discharged through low-signature, slotted horizontal nozzles on the wing-fuselage afterbody. Surfaces are angled sharply from the vertical so that primary reflections are up or down, away from the radar receiver. Unavoidable reflections in the plane of the radar receiver are concentrated along a few narrow paths (spiky returns).

The B-2A (**Fig. 3**) also has inlets on top of the wing set back from the leading edge, buried engines, and a notched wing trailing edge. Unlike the F-117A fuselage, which consists of flat surfaces oriented at specific angles, the B-2A is a blended wing-body arrangement which has complex curved exterior

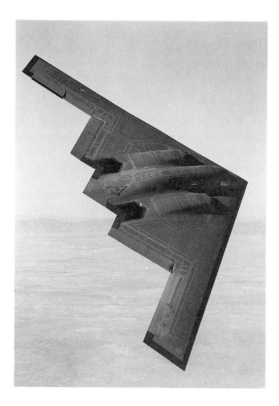

Fig. 3. B-2A stealth bomber. (*Northrop Aircraft Div., Northrop Corp.*)

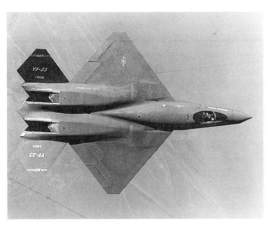

Fig. 4. YF-23A(ATF) stealth fighter. (*Northrop Aircraft Div., Northrop Corp.*)

surfaces throughout the configuration. These blended three-dimensional surfaces are also used on the YF-23A(ATF) [**Fig. 4**]. The YF-23A has buried engines, nose-to-wing chines, and widely separated tail surfaces canted outward at 45°.

For background information SEE AIRCRAFT NOISE; INFRARED IMAGING DEVICES; MILITARY AIRCRAFT; RADAR; RADAR-ABSORBING MATERIALS in the McGraw-Hill Encyclopedia of Science & Technology.

Robert J. Strohl

Bibliography. M. A. Dornheim, Display of F-117A reveals new details of stealth aircraft, *Aviat. Week Space Technol.*, 132(18):30–31, April 30, 1990; B. Gunston, *Stealth Warplanes*, 1988; J. Jones, *Stealth Technology*, 1989; W. B. Scott, YF-23A previews design features of future fighters, *Aviat. Week Space Technol.*, 133(1):16–21, July 2, 1990.

Mining

Recent advances in mining have involved computer-assisted ore-body modeling and simulation in design of new mines, computer-aided mineral exploration, and the use of portable spectrum analyzers for on-site element analysis.

ORE-BODY MODELING AND SIMULATION

Computers of all sizes are used in all phases of mining activity, from exploration through detailed mine scheduling and production reporting. The most commonly used computers are personal computers, which are relatively inexpensive and easy to obtain, with a wide selection of general and mining-specific software available. Advances in the development of hardware for personal computers provide sufficient memory for running programs, hard-disk space, and the speed required to handle the information needed to model and plan all but the very largest deposits. For large deposits, the mine model is developed on a minicomputer-based workstation or mainframe computer, and the engineers retrieve subsets of the model to a personal computer for more detailed work on small areas. Another attraction for using a personal

computer as an engineer's tool is the dedication of the computer to interactive design; the engineer has complete control locally of programs, techniques, and necessary interpretation based on engineering judgment.

Computer environment. Personal computers can be used for simple-to-complex applications by technicians, engineers, and geologists not familiar with computers, as well as by those who have extensive computer knowledge or programming experience. Personal computers are used in small companies as well as large companies for a wide variety of applications. Applications range from simple spreadsheet and mapping to use of extensive integrated software for data handling, mine modeling, mine planning, and scheduling. Available software includes programs for economics, geologic, environmental, metallurgical, and geographic information systems, and many other applications.

Geologists and mining engineers are becoming more sophisticated computer users. As user knowledge increases, the need for flexible and appropriate software and more powerful computer hardware also increases. Over 800 programs are available for use on personal computers and workstations for specific mining applications, and thousands of more general packages are available. Many general packages can be adapted to the needs of mining companies for data handling, mathematical calculations, surveying, mapping, and many other applications.

Workstations consisting of very powerful personal computers or minicomputers are beginning to appear in mines and mine offices. Such installations also often include digitizers, plotters, scanners, and printers that provide complete systems for collecting and analyzing data and producing maps, geologic models, and mine models. An important feature of workstations is high-quality graphics displayed on a video terminal. The visual displays help the geologist and engineer model and plan mines very quickly. A deposit model can be shaded and rotated quickly, providing a tool for visual evaluation of deposits that was not feasible previously. The interpretive part of model development can be done on screen before final plots are produced as hard copy.

Personal computers and workstations are stand-alone computers, but they often contain databases, deposit models, and other information that must be shared with other users. One way to share information is to transfer it to diskettes or data cartridges and carry the physical media to other computers by hand. This so-called sneaker-net system is satisfactory in small companies where the data transfer needs are only occasional. In situations where there are more than a few computers, a local-area network (LAN) is a popular solution for connecting computer hardware and sharing peripheral equipment such as printers and plotters. Such technology is needed most in locations where several different departments frequently need access to the same data. For example, assay results produced by a laboratory are used by geologists for refining the mine model and at the same time may be used by the mine planning group for ore control. Another example is an operating surface mine where timing is crucial in evaluating the grade of an area to be blasted. Within 2 days, the blast hole samples must be assayed, data transferred to mine planning, a new model of the blast generated, new maps drawn, and the grade variations marked in the pit before the material can be moved to the proper locations (waste, low-grade, high-grade).

Ore-body modeling. Models of ore bodies are generated from a combination of drill hole information, surface or underground samples, and geology. Information compiled from all sources is plotted on maps either by hand or by using a computer. A geologist adds interpretations to the base computer maps by hand. Interpretations are then transferred to a digital format that can be used in computer programs to constrain a model. The most common use of geological interpretation is for development of a rock model, which is used to limit the areas affected by interpolation of the data points. That is, data points outside a geologic zone will not be used to influence values at locations inside that zone.

Mining companies depend on computer techniques to help produce the best model and mine design possible. The model must be built with information that is often sketchy. Hand methods of determining ore reserves and mine plans are still used in very small operations, but are not feasible for modern mining operations in the fast-paced industrial climate.

Commercial mine modeling software offers a variety of modeling techniques described below.

Regular block model. A block model is commonly expressed as a three-dimensional set of blocks of the same rectangular size. The three-dimensional block model is most applicable for disseminated and massive deposits with relatively linear, gradational changes in mineralization. Constraints can be imposed on the three-dimensional block model such as ground surface and geological contacts. The software that builds and evaluates the three-dimensional model also can determine values for each block such as rock type, grade, or other attributes.

Gridded block model. A gridded block model contains rectangular blocks in two dimensions and has a variable third dimension. This model is often used for bedded or tabular deposits. Block values are determined by using techniques similar to those used in the 3D model. Sample influence applies only within the bed or layer and does not extend to adjacent layers.

Irregular block model. This type of model consists of rectangular blocks, and the blocks have different x,y dimensions within the same model. Usually, the third dimension is constant. The model is initiated with same-size blocks, and as geological boundary definition is applied, subblocks are created along the boundaries that provide a closer approximation of the detailed geology requirements than fixed-size rectangles. The boundary is more closely approximated by making smaller blocks where the boundary intersects the larger blocks. The technique keeps the model smaller, and it avoids the problems associated with a very large

area that must be defined with very small blocks to accommodate detail. These models are used to define deposits that are regular in part but possess a variability not readily definable with a regular block model.

Cross-section model. Cross-section models are built by digitizing geological and other boundaries on cross sections. The process includes displaying a cross section showing drill holes and downhole information on the screen, and defining geological outlines interactively with a screen cursor or digitizer. Another approach is to plot the drill hole cross sections on paper, define the geological boundaries manually, then digitize the manually drawn boundaries.

Cross-section models can be interpreted in two ways. One is a computer technique that reproduces the standard manual method. The manual method calculates volumes by projecting the geologic or ore outlines between sections and calculating an average grade within the outlines. Computer programs can reproduce the manual technique of calculating the area of polygons (outlines) and projecting that area halfway to the adjacent sections. Additional capabilities offered by computer programs include connecting critical points between sections, constructing three-dimensional shapes to fit the volume, and calculating the volume of the three-dimensional shape.

Another method of evaluating cross-section models is to transfer the geological outlines to a two- or three-dimensional block model. The outlines can be used to define control of attribute assignment within the three-dimensional block model and to provide a limit for volume calculations within defined zones. The geological boundaries defined on sections can be used to define a rock model in the framework of a three-dimensional block model. The rock model defines the type of rock in each block; it is used to limit extrapolation of values to those areas within selected rock types.

String model. A string consists of a group of x,y,z coordinates. The assumption is that the coordinates are points located on a polygon or an open line. String models are usually handled as a series of cross sections, and volumes are calculated between sections. The volume calculation may be done by the computerized manual technique of projecting the area halfway between the section and adjoining sections. More definition may be attained by generating additional sections between existing sections. An advantage of string models is that a model can be changed by simply changing the strings. Some programs offer interactive capabilities for editing strings.

Surface model. A surface model can be visualized as the process of laying a cover over a series of data points. The cover is calculated by building triangles between data points, polynomial equations, a least-squares fit, spline interpolation, or other methods. The model can also be generated by overlaying a grid and estimating the value at each grid node. A digital terrain model is a type of surface model that is usually constructed by generating triangles between data points. Some computer programs show a surface model in two or three dimensions as contours, triangles, or grid blocks. Programs are also available that can transfer a triangulated surface to a grid model.

Solid model. A solid model (also known as wire frame model) is usually generated by digitizing geology, mine workings, or other characteristics on sections. The sections may be oriented along the main axes of a block model, or they can have any orientation irrespective of a model. The information from the sections is then compiled into a three-dimensional model. Outlines from the sections are connected to form the three-dimensional shape. The three-dimensional boundary may appear as straight lines or as a network of triangles. The solid model is used to define geological boundaries within a block model. Programs are also available that can shade and fill in the three-dimensional surfaces with color, so that the result looks very realistic.

Three-dimensional component modeling is a variation of solid modeling. Quantities of material and grades are calculated based on the true volume within the model boundaries, rather than transferring the solid model to a block model. Attributes such as grade, rock type, rock stability, and other characteristics are assigned to zones within the model.

The methods used most for mineral deposit evaluation are the three-dimensional block models and the two-dimensional gridded and cross-section models. Interest in three-dimensional models is increasing as computer hardware becomes capable of displaying the three-dimensional characteristics of the models to best advantage.

A new technique for defining subsurface features, geotomography, is in research and development stages. Waves generated by a transmitter in one location (borehole) are received by receptors at other locations; the data collected at the receptors indicate the wave-diffusion properties of material between the locations. Experience with the properties of various rock types provides a framework for evaluating the data and determining the meanings of the information collected. The data are processed by using experimental neural networks and produce an image of the conductivity of the rock in the test area. The method can be used to determine relative porosities of rock, presence of changes in rock types, and cavities. Geotomography has potential for providing more detail on geology of an area than borehole interpolation. *See Geophysical exploration*.

Simulation. In the mining industry, simulation techniques are used in a few selected applications such as ventilation, production modeling, and some aspects of mine design. For surface mine design, several optimization techniques are available in computer programs. The most commonly used optimization technique is known as the floating cone method; it is used to determine ultimate open-pit outlines. Others that are gaining in popularity are a Lerchs-Grossman directed graph technique for pit optimization and approaches using linear programming.

Simulation of underground room-and-pillar mine development is available in commercial software as well as from public sources such as universities and government agencies. The techniques used are extensions of computer-aided design (CAD). A computer-aided system allows a user to define a library of standard shapes such as sections of underground openings, pillars, or other geometric components. A mine design is created by identifying the starting points for selected shapes and defining the area or volume where the shape is to be copied. For example, a drift can be laid out by selecting a cross section and indicating where the drift will start and end. The computer-aided-design program then draws the outline of the drift, and the user can proceed to add more shapes as needed to define the mine outlines. A program that simulates a mine design works in much the same way, but is more automated and does more of the detail work for the user.

Much of the commercial software available provides tools for the mining engineer to design mines by using the computer for calculation, but requiring the engineer to provide some interpretation. These products offer a semiautomated approach to mine design that uses the computer for the tedious and time-consuming part of the process, freeing the engineer to explore alternatives.   *Betty L. Gibbs*

COMPUTER-AIDED MINERAL EXPLORATION

Since the start of the computer era in the early 1950s, the continuous improvement in computer hardware has yielded an increasing number of computer-aided methodologies in mineral exploration. Mineral exploration can be defined, with regard to the discovery of new mineral deposits, as all the activities and evaluations necessary before an intelligent decision can be made in terms of establishing size, initial flowsheet, and annual output of new extractive operations. Thus, in a broad sense, all findings arising from the conduct, management, and economics of mineral exploration call for a decision as to action to be taken. There is a progression of exploration stages, each stage requiring a key decision.

Mineral exploration comprises four broad areas of application: information storage and retrieval, data reduction and analysis, reserve estimation and analysis, and economic evaluation. Data reduction and analysis may be performed at a remote site on a mini- or microcomputer with graphics capabilities; the other three application areas usually required a central mainframe that was connected to field offices by telephone lines and accessed through terminals and graphic display devices. The introduction of more powerful mini- and microcomputers led to changes in these requirements. In fact, microcomputer-based technology now allows rapid communication between field parties and regional exploration offices. The 1990s will see a continuing trend toward the use of microcomputers, assuming that considerable computing power will be available on the desktops of exploration geologists.

Information storage and retrieval. Computer-aided geologic information storage and retrieval begins during detailed reconnaissance of favorable areas and continues through surface appraisal of target areas and detailed three-dimensional sampling and preliminary evaluation. Computers permit the systematic organization, storage, analysis, and display of exploration data. Many types of geologic information, such as geochemical analysis, stratigraphic sections, and drill hole assays, are collected, stored, and displayed for the purpose of mineral resource assessment and development. To facilitate automated processing of this information, geologic databases composed of files, records, and data elements must be developed within a database system. Thus the capability of the computer to store large amounts of data permits the convenient retrieval and analysis of all geologic data collected during the stages of reconnaissance and target investigation.

It is anticipated that in the 1990s the use of hand-held devices (based on microtechnology) for entering field data will become widespread. These devices coupled with battery-run laptop computers will allow field personnel to perform data reduction and analysis in remote areas where minimal support facilities are available.

Data reduction and analysis. The utility of the computer in data reduction and analysis of geochemical and geophysical survey data is to reduce field data to meaningful anomalies (potential areas of interest), plot them on maps, and further analyze the significance of these anomalies in terms of exploration targets. Recent computer-aided techniques allow for effective comparison of the observed field data with target-specific profiles obtained from controlled field or laboratory experiments. For instance, during the reconnaissance stage, different kinds of indirect observations need to be considered, such as those obtained from airborne geophysical methods and regional geochemical analysis. Of great importance is the feasibility of treating the data mathematically by filtering, that is, removal of noise in the data; upward continuation, that is, continuity of trend in an upward direction; and calculations of dip, depth, and thickness of exploration targets. In this way, the computer's role is extended to the interpretation phase, and the interpretation capabilities of the computer will play an increasingly important role in integrating exploration field data.

Perhaps the most recent development in computer-aided techniques for reconnaissance is in the area of artificial intelligence (AI) through the use of expert systems (ES). An expert system is an intelligent computer program that uses knowledge and inference procedures to solve a kind of problem whose nature requires significant human expertise for solution. One of the earliest expert systems developed was an exploration application known as PROSPECTOR. This application deals with the probabilistic interpretation of data on the soil and geology of the site in a manner comparable to the performance of expert geologists. It is projected that in the 1990s models developed by expert systems will be used to guide mineral exploration.

Reserve estimation and analysis. Subsequent to detailed surface appraisal, the task of target investigation is the estimation of grades and tonnage of the target deposit that will be used for eventual economic evaluation. Estimates are made by using the analysis of drill-hole core samples. Implicit in this procedure are sample preparation (including chemical analysis) and mathematical modeling.

While reserve estimation was perhaps one of the first computer applications, the most significant strides were made with the introduction of geostatistics. The ability to express the reliability of the estimates, heretofore not possible with any other method, represented the most significant advance made in this application area. Computer-aided methods of ore reserve estimates using geostatistics are now well established; and it may be specified that all new discoveries are systematically estimated with the application of geostatistical techniques known as kriging.

When a particular estimate is classified as inferred, indicated, or measured, the method used to estimate the reserves should be clearly described, and the key assumptions that were made should be indicated. If a computer method was used, it is important to mention which programs were used and which parameters were chosen. Geostatistical methods are extremely varied and should be described in detail. Of utmost importance in making a geostatistical ore reserve estimation is to ensure that the geostatistical parameters, including the variograms, are compatible with the geological interpretation.

Economic evaluation. The early incorporation of knowledge gained through financial analysis provides a means for resolving many common exploration and development problems economically. Such applications range from exploration planning, reconnaissance, land acquisitions, and exploration drilling to project evaluation.

The purpose of exploration planning is to define the commodities being sought and to outline the general regions within which active exploration will be conducted. The planning decisions are affected by economic considerations; timely financial analysis provides pertinent information that is useful in the allocation of exploration funds, time, and personnel. For example, two areas may have a similar geological potential for mineralization, but favorable taxation aspects may allow lower-grade material to be classified as ore in one of the localities. Minimum requirements for ore grade depend on the geographic locations and depth of the target deposit; when equivalent physical conditions are assumed, substantial economic differences exist between countries and between political subdivisions within a country. Therefore computer-aided planning models based on financial analysis are used in making allocation of substantial exploration resources.

Reconnaissance is the initial stage of field exploration. Direct and indirect geological evidence of mineralization is analyzed and compared to minimum expectations. The target deposit must survive this preliminary screening; the quality of the screen depends in part on the exploration geologist's understanding of the economic aspects of prospecting. Computer-aided financial analysis appropriate to the reconnaissance area will indicate, quantitatively, the minimum tonnage and grade requirements for an economic venture.

Land acquisition is often necessary after completion of a successful reconnaissance. When land is acquired through negotiation with present landowners, the economic framework within which the future mine must operate is established. Computer-aided financial analyses indicate the effects of the alternatives of royalty or land purchase on the target hypothesis.

Exploration drilling yields approximations of grade and tonnage reserve. Due to the natural variation in mineralization, a cut-off grade must be established, and an average grade at that cut-off is calculated. Computer-aided financial analyses provide information that affects decisions impinging on the discovery of the target deposit or termination of interest. One method consists of introducing several grade cut-offs, calculating the associated tonnages and average grades, and then determining the minimum economic tonnage requirement for each average grade. By analyzing the results of such calculations, an optimum cut-off may be selected at which the reserves of target deposit should be assessed. Arbitrary cut-offs, either too high or too low, could indicate that the target deposit was not sufficiently profitable to be of further interest. This method is particularly appropriate when the target deposit is marginal by the company's criterion, since it prevents undue emphasis on so-called standard cut-off grades.

Project evaluation, including preliminary economic analysis, uses the computer to make a first estimate of minable reserves by simulating either an open-pit or underground mining method. Following such a procedure, the computer analysis also aids in the development of a preliminary mine plan in which the total minable reserve is divided into sequential planning periods such as quarterly, annual, or 5-year intervals. These planning periods then become the input to a preliminary financial analysis. Sensitivity analysis, risk analysis, and the ability to ask "what if?" type of questions are all part of this application.

The major contribution of the computer in such applications is that it imparts to the exploration team the ability to build an economic model combining geological estimates with parameters based on economic development estimates to produce criteria of expected profitability and risk in order to support the investment decision. Advances in the capabilities of microcomputers coupled with commercially available software make it possible to perform an economic evaluation without the assistance of a support staff of programmers and systems analysts.   *Alfred Weiss*

ELEMENT ANALYSIS IN THE FIELD

Recent technological advances have led to the successful development of instruments capable of providing qualitative and quantitative analysis of several elements and materials in the field, in their naturally occurring place and condition, without the need for special handling or sample preparation.

Numerous applications exist for remote or in-place tests and measurements of heavy metals in mining, exploration, characterization of hazardous waste sites, testing for lead content in paint, and space exploration, among others.

In-place analyzer. This is a tool for valuation, and it provides a means for improved mining decisions. Tests and studies have quantified the problems associated with current conventional sampling techniques, and they have compared the results obtained by conventional methods with those obtained through in-place analysis. Field tests and usage have demonstrated the improved precision in valuation resulting from the in-place method. Because of the many mining decisions that depend on the estimated grade of ground to be mined, the economic benefits of using an in-place analyzer are potentially very substantial.

Greater precision in estimating the mineral content of unmined ground becomes increasingly important as mines strive for optimum exploitation of their deposits. The variable distribution of minerals within a deposit, along with the difficulty and the cost of high-density sampling, limits the precision that is currently possible with traditional sampling methodology.

An inherent problem exists in the conventional method of sampling. The erratic values of ore grades necessitate the collection of large numbers of samples for statistically accurate valuations and estimation of ore reserves. Owing to the cost of personnel and laboratory assays and to time constraints, current sampling techniques do not allow for the collection of sufficiently large numbers of samples.

The most common method of sampling in current practice is known as chip sampling. Its purpose is to obtain an estimate of the value of blocks of ground prior to mining. This information is used to plan the rate and sequence of mining; yet the limits on the amount of sampling that is possible in practice cause these block estimations to possess considerable uncertainty. This inevitably leads to incorrect mining decisions, which may be very costly.

These factors, combined with recent technological advances in computer science and electrical engineering, have stimulated the development of hand-held instruments that provide information for real-time decisions, enable rapid and denser sampling, display estimated grades immediately, cost less than current chip sampling methods, and lead to more accurate valuation than can be achieved currently with chip sampling.

A typical in-place analyzer is a lightweight instrument specifically designed for field and mine operations. The design of this instrument is based on the principles of energy-dispersive x-ray fluorescence (EDXRF) for determining mineral concentrations. The use of this type of instrument in the field is expected to alleviate many of the problems of conventional sampling, and the device improves the quality of mining decisions.

The analyzer contains a small radioactive source that emits low-energy (<1 MeV) gamma rays. During an analysis the sample is exposed to the radioactive source. The electrons of the atoms in the sample absorb some of the gamma rays and then reemit secondary gamma rays and x-rays. The electrons of each element emit x-rays at unique so-called fingerprint energies. A hand-held sensor or downhole probe sensor detects and measures each x-ray; the measurement is then sent to a micro field-hardened (durable) multichannel analyzer. The resulting data produce a spectrum of energies and their frequencies. An algorithm is utilized to process this spectrum and derive a measurement of the concentration of each element.

These detector (sensor) systems are linked to a data storage and display subsystem, the control console of the instrument. The console communicates with the sensors, processes the data, displays results to the operator, and stores the results of the measurements. Data can then be downloaded to a computer or printer.

The instrument console is a small low-power (2-W) computer system. The system is fitted with rechargeable batteries that are sufficient to power the console and the sensors for up to 12 h and a memory large enough to store measurements and spectrums.

The typical hand-held scanner is a device that can be used to analyze rock walls, chip samples, drill cuttings, sludges, solutions, walls in houses, or any accessible location. The downhole probe will operate in surface or underground drill holes. The scanner and probe system contains a sealed radioactive source, a detector, a power supply, an amplifier system, and an analog-to-digital converter.

The algorithm that is used with these instruments is based on statistical correlations found between areas of the spectrum and laboratory assay values of the samples; it compensates for interfering elements and changes in sample density.

Uses in mining. The key to being able to exploit any ore body optimally lies in predicting the grade of the ore body accurately and in sufficient detail for selecting the most cost-effective mining strategy.

The type of errors that arise in estimating the value of an unmined block of ground depend on the sampling configuration. In simple terms, these errors have two major components. The first errors occur when the grades of samples from the exposed face are used to estimate the mean grade of the face. The second error arises when estimated face values are used to determine the grade of the block.

The estimated face, and hence block values, may vary greatly. The grade of a face is taken as the average value of a suite of chip samples taken from the face. However, the mean value of different suites of samples taken from the same face would vary, owing to the considerable variations in grades that occur along the face. Clearly, this variability would also depend on the precision of the measuring technique.

Face valuation. This is a function of the accuracy of measurement and the density of the sampling. Because of variations in mineral concentration along the face, the accuracy of face valuation becomes a question of the specific sampling density that is feasible in

practice as well as the accuracy of the grade determination for individual samples.

In chip sampling, a measurement error arises when the pulverized chip sample is subsampled. Data collected from various mines have shown that the error in assaying chip samples is approximately 20% of the grade of the sample. In contrast, the error of measurement with the in-place analyzer is independent of the grade; instead, it is determined by the time taken to make the measurement.

In conventional sampling, the time and personnel required to take the samples to the assay office to have them processed place a limit on the frequency of sampling that is practicable. Sampling intervals can be reduced to as little as 0.3 m (1 ft) when using the in-place analyzer to make spot measurements without an increase in staff. Furthermore, when the analyzer is used in the scanning mode, it may even be practical to scan the entire face without requiring additional personnel or incurring additional costs for assays.

In two studies, data derived from contiguous chip samples for two faces have been used to calculate the error in estimating the grade of the faces for different sampling intervals. The errors in estimating the grade of the face were calculated by using chip sample data taken at various sampling intervals. A similar exercise was done for the case of the in-place analyzer. The measuring errors associated

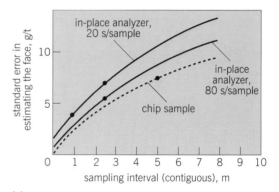

(a)

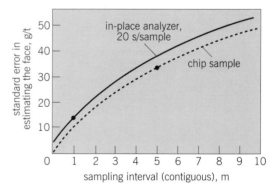

(b)

Fig. 1. Graph showing errors in estimating the average grade of a face, using different sampling intervals and measuring techniques. 1 m = 3.28 ft. (a) Average grade 3.14 g/metric ton (0.1 oz/ton) (b) Average grade 14.6 g/metric ton (0.47 oz/ton).

with using the analyzer were taken into account, but the errors associated with chip sampling and assaying have not been considered. The results of these calculations are given in **Fig. 1.**

Figure 1a illustrates quantitatively how measurement methodology and sampling interval affect the error in estimating the face. From a practical point of view the figure shows that relatively inaccurate in-place analyzer measurements that are 20 s in duration spaced 2.5 m (8.2 ft) apart lead to a more precise valuation of the face than is possible by using chip samples spaced 5 m (16.4 ft) apart. The figure also demonstrates the benefit of making longer analyzer measurements. However, the important conclusion to emerge from a study of Fig. 1a is that 20-s analyzer measurements spaced 1 m (3.3 ft) apart will yield a much more precise estimation of the face than chip samples spaced either 5 or 3 m (16.4 or 9.8 ft) apart. Bearing in mind that a 1-m (3.3 ft) sampling interval with the in-place analyzer is entirely practical, the extension of this conclusion is that use of the analyzer would significantly improve the reliability of data used to make mining decisions. Reference to Fig. 1b shows that this conclusion applies even more to the higher-grade example. This is expected, because the in-place analyzer error, as a fraction of the grade, diminishes as the grade increases.

While the above analysis has been based on using the in-place analyzer to make spot measurements, even greater benefits are possible when the analyzer is used to scan the face.

For the two cases considered, the in-place analyzer generates more reliable estimates of the average grade of the exposed faces. This is due to the increased sampling frequency (sample size) that is practical with the in-place analyzer.

Block valuation. With reference to the total variance, and having dealt with the sample-to-face variance, attention is now given to the face-to-block variance. In order to establish the magnitude of the face-to-block variance, data from bulk and contiguous chip sampling have been analyzed. The data are for adjacent blocks with the dimensions 4 m × 4 m × 1 m (13 ft × 13 ft × 3.3 ft) that were contiguously chip-sampled on one face. The true grade of the face was taken as the average of the grade of the contiguous chip samples, on the basis that the ±20% error in the individual assay values would yield a negligible error when the average grade of the face was calculated. The grade of each of these blocks was determined by bulk sampling. The measured grades of the faces and blocks were then used to calculate the error in estimating the grade of the blocks from the grade of the exposed face.

The magnitude of the error in estimating the grade of the blocks increases with increasing grade. The physical interpretation of this is that high-grade areas have a more variable distribution of mineralization than do low-grade areas.

Using the in-place analyzer to obtain substantial improvements in the sample-to-face variance would

therefore significantly reduce the error of estimating the grade of the next 4 m (13 ft) to be mined.

Effect on mining decisions. The important financial implications of achieving a more precise valuation become apparent when a decision of whether or not to mine a block of ground has to be made. Imprecise block valuation is the cause of two costly mistakes.

The first occurs when potentially profitable (payable) blocks are not mined because of erroneously low valuations; the second occurs when unprofitable (unpayable) blocks are mined at a loss because of erroneously high valuations. This is illustrated qualitatively in **Fig. 2**, which shows the benefit of using more precise valuation techniques.

In Fig. 2 the blocks that are plotted in segment A are those that are left behind erroneously; the blocks plotted in segment B are those that are mined at a loss. Thus, a more precise valuation technique would decrease these errors and improve the profitability of mining. The potential use of the in-place analyzer to achieve this benefit has been determined quantitatively in field tests. The value of using the analyzer in place of conventional chip sampling has yielded a benefit of millions of dollars.

Scale of valuation. A knowledge of the spatial distribution of the target metal can be of considerable value when making certain mining decisions. In particular, such information could significantly influence the selection of the mining method to be adopted, for example, longwall versus scattered mining or, depending on the direction of payshoots (mineralized zones of ore), updip versus breast mining. It could also be important in the siting of strips of ground to be left as stabilizing pillars, differentiating between payable and unpayable faces, and in planning the sequence and rate of mining various faces for the purpose of grade control.

One of the disadvantages of conventional chip sampling in practice is that the sampling density is too low for reliable identification of patterns of metal distribution on a small scale. The much higher sam-

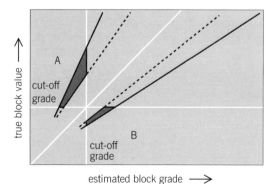

Fig. 2. Schematic representation of the boundaries of the scatter diagram obtained when true block values are plotted against estimated block grades. Segment A: blocks that are erroneously left behind. Segment B: blocks mined at a loss. Solid lines indicate current precision; broken lines, improved precision. Shaded areas indicate reduced errors, improved profits.

pling density possible with the in-place analyzer improves this position.

These examples indicate that the analyzer could play a valuable role in providing more precise and detailed grade estimations for making important mining decisions.

Time scale of valuation. When mining is done in areas that approach unprofitability, and where the decision of whether or not to continue mining is taken each time the face is sampled, immediate availability of the grade estimate can provide considerable benefit. In most mines the grade estimates become available from 1 day to weeks after the face has been sampled, depending on the location of and workload in the assay office. During this time a considerable amount of unprofitable mining may take place. The in-place analyzer can provide grade information immediately, considerably reducing such loss.

Applications. While the potential application appears to be wide, perhaps the least beneficial use of the in-place analyzer would be as a direct (equivalent precision) replacement for chip sampling. Even then the benefits would be impressive and would include a reduction in personnel, a smaller assay facility, and immediate availability of grade information. It should be possible for the cost of the instrument to be offset against reductions in personnel alone in less than 1 year in most cases.

However, more substantial benefits lie in training and using available personnel to enhance precision and resolution in valuation, and hence to improve mining decisions. There are many mining decisions that depend on the estimated grade of the ore to be mined, and it is therefore difficult to assess the financial benefits that each improved mining decision might yield. However, on an industry basis, the benefits are potentially very substantial.

For background information SEE ARTIFICIAL INTELLI-GENCE; COMPUTER; EXPERT SYSTEMS; LOCAL-AREA NET-WORKS; MINING; SIMULATION; SURFACE MINING; UNDER-GROUND MINING, X-RAY FLUORESCENCE ANALYSIS in the McGraw-Hill Encyclopedia of Science & Technology.

Lawrence T. Lynott

Bibliography. B. L. Gibbs, Mining software trends and applications, *Min. Eng.*, 42:974–988, 1990; R. L. Grayson, Y. J. Wand, and R. L. Sanford (eds.), *Use of Computers in the Coal Industry*, 1990; D. G. Krige (ed.), *Proceedings of the 20th International Symposium on Application of Computers and Operations Research in the Mineral Industry*, South African Institute of Mining and Metallurgy, 1987; R. V. Ramani (ed.), *Proceedings of the 19th International Symposium on Application of Computers and Operations Research in the Mineral Industry*, 1986; J. M. Stewart and M. Nami, *The Gold Analyzer: Gold Mining Technology*, South African Institute of Mining and Metallurgy, 1986; A. Weiss (ed.), *Computer Methods for the 80s in the Mineral Industry*, Society of Mining Engineers of the American Institute of Mining, Metallurgical and Petroleum Engineers, Inc., 1979; A. Weiss, *21st International Symposium on Application of Computers and Operations Research in the Mineral Industry*, Society of Mining Engineers, 1989.

Mobile radio

The two unique features of mobile cellular telecommunications systems are the concept of frequency reuse for increasing the spectrum efficiency, and the scheme for handing off communication from one frequency to another when a mobile unit enters a new cell. An analog mobile cellular telecommunications system called Advanced Mobile Phone Service (AMPS) was developed in the 1970s. The first installation of a cellular system was in Tokyo in 1979, using a minor modification of AMPS. The first AMPS cellular system installed in the United States was in Chicago in 1983. While analog cellular systems are in use in most of the world, standards for digital cellular systems have been developing to increase radio capacity. Besides the time division multiple access (TDMA) being considered, another particularly promising technology is code-division multiple access (CDMA).

Analog AMPS. In the analog AMPS system, mobile units are compatible with all the cellular systems operating in the United States, Canada, and Mexico. A spectrum of 50 MHz (824–849 MHz for the mobile transmit and 869–896 MHz for the base transmit) is shared by two cellular providers in each market (city). Each one operates over a bandwidth of 25 MHz in a duplex fashion, that is, 12.5 MHz in each direction between cell sites and mobile units. There are 416 channels, comprising 21 setup channels and 395 voice channels. The channel bandwidth is 30 kHz.

Digital AMPS. In 1987 the capacity of the AMPS cellular system started to show its limitations. The growth rate of cellular subscribers far exceeded expectations. In 1987 the Cellular Telecommunication Industry Association (CTIA) formed a subcommittee for Advanced Radio Technology to study the use of a digital cellular system to increase capacity. At that time the Federal Communications Commission (FCC) had clearly stated that no additional spectrum would be allocated to cellular telecommunications in the foreseeable future. Therefore, the analog and digital systems would share the same frequency band. In December 1989 a group formed by the Telecommunication Industry Association completed a draft of a digital cellular standard.

The digital AMPS, which must share the same spectrum with the analog AMPS, is a duplex time-division multiple-access (TDMA) system. The channel bandwidth is 30 kHz. There are 50 frames per second in each channel. Three time slots per frame can serve three calls at the same time in one channel. The speech coding rate is 8 kilobits per second. An equalizer is needed in the receiver to reduce the intersymbol interference that is due to the spread in time delay caused by the time arrival of multipath waves.

European GSM system. At present, not all mobile telephone systems in Europe are compatible. In 1983, in response to the need for compatibility, a special task force, the Special Mobile Group (GSM), developed a digital cellular system. The operating principles of the GSM system resemble those of the digital AMPS, but the system parameters are different (see **table**).

Design parameters. The frequency reuse concept makes possible the high capacity of cellular systems. According to this concept, two cells, separated by a distance D, can use the same frequency without interference between them. These two cells are called cochannel cells, and if the separation is less than D, the interference between them is called cochannel interference. The cochannel interference reduction factor (CIRF), q, is defined by Eq. (1), where R is the

$$q = \frac{D}{R} \qquad (1)$$

cell radius. The quantity q is a constant depending on the particular cellular system. In AMPS, q equals 4.6. For a given value of q, the larger the size of the cell, the greater the separation required between two co-channel cells.

From the value of q and a geographical hexagon diagram, the number of frequency reuse cells K can be shown to be given by Eq. (2), which indicates how

$$K = \frac{q^2}{3} \qquad (2)$$

many cells, forming a cluster, would share the total allocated number of frequency channels in cellular systems. For example, in an AMPS system, $q = 4.6$, so that $K = 7$ (**Fig. 1**). The total of 416 channels divided by 7 becomes 59 channels per set. Each set is assigned to one of the seven cells in a cluster. Then those cells that have the same sets but are in different clusters, called cochannel cells, are separated by 4.6 R.

There is a close relationship, Eq. (3), between

$$K = \sqrt{\frac{2}{3}(C/I)_s} \qquad (3)$$

the minimum required carrier-to-interference ratio, $(C/I)_s$, in the signal picked up by the receiver and the number of frequency reuse cells, K. This relationship is based on a worst case in which six cochannel cells

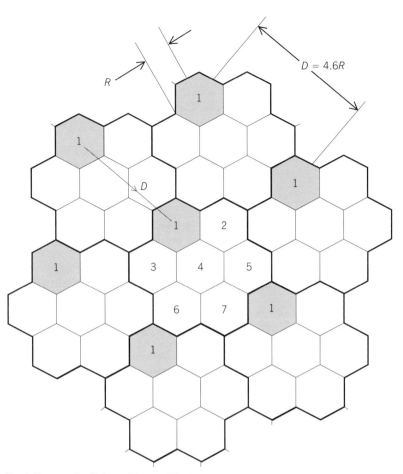

Fig. 1. Hexagonal cells in an Advanced Mobile Phone Service (AMPS) system. R = radius of cell, D = minimum separation of cochannel cells, $q = D/R$ = 4.6, K = number of cells in a cluster = 7. Clusters are indicated, and the six cells that effectively interfere with cell 1 in one cluster are numbered 2 through 7. The shaded cells are cochannel cells.

would simultaneously serve their mobile units in different clusters but interfere with the target cell, as shown in Fig. 1. The propagation loss in a real mobile radio environment is around 40 dB/dec; this value is also used to derive Eq. (3). Here, 40 dB/dec means that a 40-decibel loss can be observed from 1 to 10 mi or from 2 to 20 mi (decade). In other words, an inverse-fourth-power law is obeyed over the distance r; the power is proportional to r^{-4}.

Capacity. In cellular systems, an increase in capacity can be achieved by reducing the required cochannel-cell separation D. There are two ways of accomplishing this: either by letting q be fixed, and reducing the cell radius R; or letting R be fixed, and reducing q.

Enhancement in microcells and in-building systems. The former approach leads to small cells or microcells in which the cell radius is about 0.6 mi (1 km). The signal coverage pattern for a cell would be very difficult to control for a cell radius less than 0.6 mi (1 km) with present technology. In-building communication can also be considered microcell approach. Every building can be treated as a cell. The building walls have a signal penetration loss of about 20 dB and up, and form a natural shutter for reducing the interference coming from outside the building.

Parameters of digital cellular telecommunications systems		
Parameter	Digital AMPS	GSM
Mobile transmit band	824–849 MHz	890–915 MHz
Base transmit band	869–896 MHz	935–960 MHz
Total bandwidth	50 MHz	50 MHz
Number of channels	416	124
Channel bandwidth	30 kHz	200 kHz
Number of frames per second in each channel	50	217
Number of slots per frame	3	8
Speech coding rate	8 kilobits/s	13 kilobits/s

Enhancement in digital systems. The approach of reducing q in order to increase capacity leads to digital cellular systems. The parameter q is related to the minimum required carrier-to-noise ratio, $(C/I)_s$. In digital systems, $(C/I)_s$ can be smaller than in analog systems because a digital waveform is less susceptible to degradation from noise than an analog waveform. From Eqs. (1)–(3), it follows that if $(C/I)_s$ is smaller the value of q also becomes smaller.

Calculation. In every system, the minimum required value of C/I, $(C/I)_s$, must be derived from the given channel bandwidth B and the desired voice quality. The evaluation of voice quality should be based on a subjective test. The voice quality has five rating levels; "excellent" and "good" are the top levels. The standard criterion for voice quality states that a $(C/I)_s$ level should be chosen based on 75% of listeners saying that the voice quality is excellent or good at that level. With known values of $(C/I)_s$ and B, the radio capacity m can be expressed as Eq. (4),

$$m = \frac{M}{K} \qquad (4)$$

where m is the number of channels per cell; and M is the number of frequency channels in frequency-division multiple access (FDMA), the number of time-slot channels in time-division multiple access, or the number of traffic channels in a code-division multiple-access radio channel.

A variety of multiple-access schemes is used in digital systems (**Fig. 2**), and the calculation of the radio capacity of each scheme is different. In a frequency-division multiple-access system (Fig. 2*a*) a frequency channel is assigned to each call for accessing the system, while in a time-division multiple-access system (Fig. 2*b*) each call is assigned a time slot. The equivalent-channel bandwidth is based on the time-slot channel. If a carrier channel of 30 kHz has three time slots, the equivalent-channel bandwidth is 10 kHz. Furthermore, the $(C/I)_s$ per time slot is the same as $(C/I)_s$ per channel in time-division multiple-access systems.

In a code-division multiple-access system (Fig. 2*c*) a code sequence is assigned to each call. All the code sequences use the same radio channels. The optimum code-division multiple-access system would reuse one wide-band (spread-spectrum) radio channel in every cell in the entire system. The differences between different cells are the different assigned sets of M code sequences. The code sequences can be considered the traffic channels. In this case, the value of q is 2, and K becomes 1.33, which is the smallest possible value.

Cellular CDMA. Cellular code-division multiple access has been developed and has demonstrated a unique power to increase capacity. The minimum number of frequency-reuse cells, K, which is 1.33, is the smallest value that can be implemented in cellular systems. The other main advantage of code-division multiple access is in the nature of human conversation. The average human-voice-activity cycle is about 35–40%, meaning that the average person talks only 35–40% of the time. When users assigned to a traffic channel are not talking,

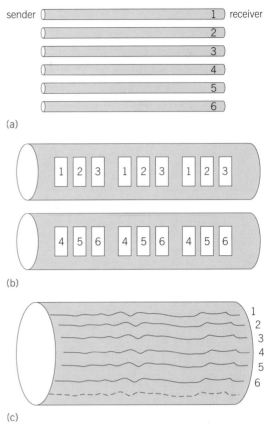

(a)

(b)

(c)

Fig. 2. Different multiple-access schemes for the case in which a set of six channels is used in one cell.
(*a*) Frequency channels in frequency-division multiple access (FDMA). (*b*) Time slots in time-division multiple access (TDMA). (*c*) Code sequences in code-division multiple access (CDMA).

all others on the channel benefit. Thus, the voice-activity cycle reduces mutual interference by 60–65%, increasing the true capacity of the channel. Code-division multiple access is the only technology that takes advantage of this phenomenon.

Advantages. Cellular code-division multiple access has numerous advantages. There is no hard handoff, since all cells use a common radio channel; the system can take advantage of the human-voice-activity cycle; there is only one radio channel per cell site; the system can use cell sectorization (by implementing directional antennas) to increase capacity (for example, in theory, a three-sector cell can increase capacity three times); there is less multipath fading in a wide-band radio channel; the transition from analog to digital telecommunications is relatively easy; there is no need for frequency management; and there is no need for an equalizer, which must be used in time-division multiple access. The system has soft capacity, meaning that a radio channel always can accommodate more traffic channels by sacrificing voice quality. Finally, since a portion of the radio spectrum is designated for code-division multiple access, it would interfere less with analog telecommunications than other digital systems.

Concerns. An important concern of cellular code-division multiple access is its reliance on power control. A reverse-link (mobile-to-base) power control is used primarily to avoid near-to-far interference. This type of interference is introduced when a strong signal from a near-end mobile unit interferes with the relatively weak signal from a near-end mobile unit to the far-end mobile unit received at the cell site. A forward-link (base-to-mobile) power control is used to increase the capacity.

Another concern is that a slight increase in the required value of $(C/I)_s$ can result in a large decrease in radio capacity.

For background information SEE ELECTRICAL COMMUNICATIONS; MOBILE RADIO; MULTIPLEXING in the McGraw-Hill Encyclopedia of Science & Technology.

William C. Y. Lee

Bibliography. Electronic Industries Association, *Cellular System Mobile Station–Land Station Compatibility Specification*, Interim Stand. 3D, March 1987; Electronic Industries Association, *Dual-Mode Subscriber Equipment: Network Equipment Compatibility Specification*, Interim Stand. 54, Proj. 2215, December 1989; GSM/Permanet Nucleus, *GSM System*, July 1989; W. C. Y. Lee, *Mobile Cellular Telecommunications Systems*, 1989; W. C. Y. Lee (ed.), Special issue: Overview of CDMA, *IEEE Trans. Vehic. Technol.*, vol. 40, no. 2, May 1991.

Molecular biology

Through its well-developed genetics, the yeast *Saccharomyces cerevisiae* is amenable to very sophisticated manipulations. It has the capacity to carry out a number of co- or posttranslational modifications and fold proteins correctly; also, it can be grown to high cell densities in fermentors. These features make it a good host for the production of heterologous proteins with biological activity. Some problems, such as a machinery for glycosylation (the addition of carbohydrates on the protein backbone) that differs from that of mammalian cells, also must be taken into consideration and may make yeast an inappropriate host in certain circumstances.

Yeast-derived hepatitis B vaccines are currently in use, and several other recombinant proteins are undergoing clinical trials; more products and developments are expected to follow.

Vectors. The basic tool of recombinant deoxyribonucleic acid (rDNA) technology is the vector used to introduce heterologous genetic information (foreign DNA) into the host. In yeast, these are essentially of two generic types: plasmid vectors and integrative vectors.

Plasmid vectors. Typically, plasmid vectors used for heterologous gene expression are high-copy, autonomously replicating genetic elements. They are based on an endogenous yeast plasmid (called the 2-micrometer circle plasmid) found in most laboratory strains of *S. cerevisiae* and other yeasts (see **illus.**). Their maintenance in yeast requires the use of selec-

tive markers; these usually consist of a gene coding for a metabolic enzyme whose chromosomal copy is inactivated by a mutation or mutations in the recipient strain. Maintenance of the extrachromosomal plasmid is therefore required for the yeast to grow in a medium lacking the metabolite chosen for selection. Some markers, because they are poorly expressed from their plasmid copy, drive the selection of cells carrying very high plasmid copy numbers.

Integrative vectors. The utilization of integrative vectors is an alternative that offers several advantages. For example, integrative vectors can be stably integrated in the chromosomes, thus allowing for growth of the transformed cells in the absence of selective pressure (such as in a complex, undefined medium). In addition, such integrated vectors will be homogeneously distributed in the cell population, whereas plasmids usually have a segregation bias. This allows the synthesis of a gene product at a well-defined level and with the same efficiency in each cell of the population. This is of importance if one is interested in the expression, in the same cell, of several proteins in well-defined amounts. Integration into the chromosomal DNA is realized by homologous recombination between a chromosomal gene and its cloned counterpart on the vector. If the chosen segment is present in multiple copies on the chromosomes, then cells with several copies of the vector integrated can be selected.

All yeast vectors also contain bacterial DNA, therefore allowing them to be replicated and selected in the bacteria *Escherichia coli*. This organism is used for amplification of all intermediate recombinant plasmids that are constructed during the multistep proce-

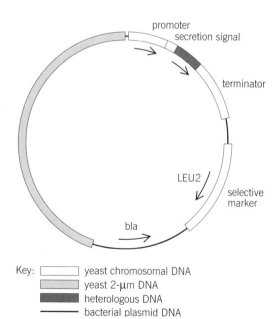

Structure of a typical extrachromosomal expression vector for *Saccharomyces cerevisiae*. In the example shown, a complete copy of the yeast 2-micrometer plasmid was cloned in the bacterial vector pBR327 (thin line). The gene *bla* codes for beta-lactamase, the enzyme that renders the bacteria resistant to beta-lactam antibiotics. The arrows indicate the direction of transcription of the genes.

Key:
- yeast chromosomal DNA
- yeast 2-μm DNA
- heterologous DNA
- bacterial plasmid DNA

promoter
secretion signal
terminator
selective marker
LEU2
bla

dure leading to the final expression vector. The vectors are called shuttle vectors because of their capacity to be moved back and forth between bacterial and yeast hosts.

Expression of heterologous protein. High-level production of heterologous protein requires the use of a strong promoter to drive transcription of the foreign coding sequence. Proper processing and stability of the chimeric transcript also require transcription termination signals. Typically these elements are arranged in an expression cassette (see illus.). The expression cassette contains both yeast-derived and heterologous DNA in the following sequence: promoter, signal sequence or other precursor sequences, coding sequence of the heterologous protein, yeast transcription termination signal. The selective marker for maintenance of the recombinant plasmid is, in the illustrated example, a copy of the LEU2 gene, coding for β-isopropylmalate dehydrogenase, an enzyme from the leucine biosynthetic pathway. Strains transformed with this plasmid are either deleted or mutated at the chromosomal LEU2 locus.

Promoters. A number of yeast promoters have been used for heterologous gene expression. These can be either constitutive (expressed in any growth conditions) or regulated. A constitutive promoter is generally chosen from the genes encoding enzymes of the glycolytic pathway because these are expressed at very high levels. In favorable cases, heterologous proteins expressed with these promoters on high-copy-number plasmids constitute 5–10% of proteins in the cells. In rare cases, where the protein is especially stable (for example, the human enzyme copper-zinc superoxide dismutase), levels above 30% have been observed. Constitutive expression of proteins that are toxic to the cell results in poor growth and recombinant plasmid instability. One way to overcome these problems is to use regulated promoters. These can be activated at a late stage in the culture, when biomass has already accumulated. Examples of such promoters include those from the acid phosphatase gene, which is regulated 200-fold by phosphate depletion, and from the galactokinase gene, which is induced 2000-fold by galactose. In attempts to combine inducibility and strength, hybrids between regulated and constitutive promoters are also used.

Secretion. Another way of circumventing toxicity problems is to direct the heterologous protein to the outside of the cell. This has two other advantages: Since the level of endogenous proteins that are secreted is quite low, purification is greatly simplified. In addition, disulfide bonds are more likely to be properly formed in proteins passing through the secretory pathway than they are in proteins expressed in the reducing environment of the cytoplasm. The major effector of protein secretion is the presence at their amino terminal of a sequence with well-defined characteristics called the signal sequence. This signal is cleaved from the precursor protein during its transit through the secretory pathway. A number of different signal sequences have been employed to direct the secretion of foreign proteins from *S. cerevisiae*. In a

few cases, expression of the naturally occurring form of the precursor has resulted in the secretion of the correctly processed, mature proteins. More commonly, fusions to leader sequences (signal sequences) from yeast proteins such as invertase have been used. The most successful development comes from the use of the gene that encodes the α-factor of the mating pheromone of *S. cerevisiae*. This small protein is secreted in the culture medium through complex processing of a large precursor. Fusion of heterologous sequences at the precursor-mature junction (the junction point between the part of the protein that is cleaved and the "mature" final protein that is secreted) results in accurate and efficient processing. Small polypeptides such as epidermal growth factor, β-endorphin, interferon, or leech hirudin are effectively secreted into the culture medium. It should be noted that these polypeptides, like the α-factor, are not naturally glycosylated. SEE PROTEIN.

Problems in recombinant proteins. When intended for human therapeutic use, recombinant proteins should be as close in structure as possible to their naturally occurring counterparts in order to minimize negative effects that may be generated by unexpected biological activities or immunogenicity. Such undesirable properties may result from contaminants present in the preparation, as well as from features of the recombinant protein itself, for example, imperfect glycosylation, heterogeneous or "incorrect" amino terminal, or improper folding.

Glycosylation. As in other eukaryotes, addition of carbohydrate chains onto proteins occurs when proteins are transported to the cell surface. These additions are either nitrogen-linked at asparagine residues or oxygen-linked to serine or threonine residues. Although the structure of the core oligosaccharide transferred to asparagine residues of yeast glycoproteins is identical to that found in mammalian cells, subsequent modification is quite different, leading to long outer chains of mannose residues. The structure of oxygen-linked carbohydrates is also different from that of mammalian glycoproteins. As a result of these differences, heterologous glycoproteins secreted from yeast may be inactive or antigenically different from the natural proteins. Efforts have been made to eliminate or minimize this problem through the use of yeast mutants that are defective in the ability to add outer-chain oligosaccharides. Nevertheless, this problem is probably the single major limitation of yeast as a host for the production of foreign proteins.

Amino terminal modifications. Initiation of translation of a polypeptide from its messenger ribonucleic acid (mRNA) requires the first codon to code for a methionine residue. If not targeted to secretion via a signal sequence or another type of precursor, the protein will remain in the cytoplasm and will not necessarily be processed at its amino terminal. However, the natural counterparts of a number of recombinant proteins intended for pharmaceutical use do not have methionine at the amino terminal. Recent progress in the understanding of protein degradation inside cells has allowed the development of an elegant vector system based on the ubiquitin hydrolase system of yeast. Fusion of heterologous sequences to the carboxyl terminal of ubiquitin results in correct and efficient processing at the fusion point. This allows production of proteins with any amino terminal residue in the cytoplasm. Other types of amino terminal modifications have been found to occur in *S. cerevisiae*: copper-zinc superoxide dismutase is amino-terminal-acetylated as it is when synthesized in human erythrocytes, and a number of viral proteins have been found to be fatty-acid-acylated with myristic acid as they are when synthesized in mammalian cells.

Other modification. Fatty acid acylation with palmitic acid as well as phosphorylation of heterologous proteins have also been described in yeast.

Other yeasts. For reasons summarized above, *S. cerevisiae* stands out as the favorite yeast of both fundamental biologists and biotechnologists. However, other yeast genera are receiving increasing attention and have considerable potential for heterologous gene expression. Initially developed as sources of single-cell proteins, the methylotrophic yeasts *Pichia pastoris* and *Hansenula polymorpha* have been shown in several instances to be extremely efficient producers of heterologous proteins. Other yeasts may have specific advantages. In particular, glycosylation patterns that are more related to higher eukaryotes appear to exist in some species.

For background information *SEE* GENE AMPLIFICATION; GENETIC ENGINEERING; MOLECULAR BIOLOGY; PLASMID; PROTEIN in the McGraw-Hill Encyclopedia of Science & Technology.

                                                        Michel J. De Wilde

Bibliography. P. J. Barr, A. J. Brake, and P. Valenzuela (eds.), *Yeast Genetic Engineering*, 1989; D. V. Goeddel (ed.), *Methods in Enzymology*, vol. 185: *Gene Expression Technology*, 1990.

Moon

Efforts to expand the human presence in the solar system will require major facilities in Earth orbit or possibly on the Moon. Such facilities will need large amounts of materials for structural modules, life support, propulsion, and shielding. The heavy transportation costs for lifting materials from the Earth's surface make a terrestrially based space system inherently inefficient. A more cost-effective approach would be to provide as much of the required materials as possible from other bodies in space.

Mining. Mining anywhere involves exploration—for resources or site selection; excavation—actually making the hole in the ground for mining or construction; and materials handling and processing—transport, beneficiation, or resource extraction.

Exploration. The Moon has been explored to a much greater extent than other extraterrestrial bodies. Most of this exploration, however, has been directed toward planetary science rather than toward resource recovery and base construction; thus it is of insufficient resolution for planning surface operations. Previous observations of the Moon's surface by remote sensing have been carried out both from lunar orbit and from Earth. Orbital studies have included photography, radar, infrared radiometry, ultraviolet and gamma-ray spectrometry, and x-ray fluorescence measurements. Earth-based studies have included photography, radar, infrared eclipse temperature measurements, and imaging spectrometry. Preliminary results of the lunar orbital studies have shown that criteria related to site suitability and minability can be detected from orbit by photography, radar, and infrared radiometry.

The Surveyor and Apollo missions collected considerable data on lunar soil conditions at multiple surface sites, with some subsurface insight provided by electromagnetic and seismic probing. Inferences have been made from Apollo experiments involving soil mechanics as to potential physical properties, and Earth-based tests on returned samples also provided some insight into the behavior of the regolith and the strength of the constitutive material. Questions of sample disturbance and changed environment, however, have not allowed great confidence in the predictions for conditions on the Moon, particularly at depth. There is reasonable confidence in particle size distributions and in mineral contents for the regolith, if only because of relative consistency between sites. Estimates of the strength properties of the regolith, however, vary over a large range and make firm conclusions difficult. Friction angles were proposed from as low as 13° to nearly 90°, and cohesion values varied from 0 to several kilopascals. The impact of these extremes on the design of an excavating system would be significant. Difficulties in drilling and trenching on the Moon during the Apollo missions seem to suggest that the higher values are probably more representative of conditions.

Excavation. The process of excavating material generally can be divided into breaking the material, loading, and transporting. Mining might also include beneficiation of the excavated material, but the processing of the ore into usable product is considered a separate process. For mining or excavating on the Moon, the regolith or rock will also have to be broken, loaded, and transported, although not as a direct analog of terrestrial processes. The constraints of an extraterrestrial environment and the unique properties of the lunar material suggest that new techniques, or significant variations of existing systems, be developed for accomplishing the unit operations of excavating in the lunar environment. The **illustration** shows a proposed lunar mining facility for the production of liquid oxygen from ilmenite, a fairly common oxygen-rich component of lunar soil.

The breaking, fragmenting, or loosening of the regolith material is accomplished by putting energy into the material in excess of its binding energy. On Earth, this is almost universally accomplished by blasting or mechanical means (cutters, teeth, buckets), with other electrical, thermal, and mechanical methods used in special cases. Generally the alternate methods to blasting or mechanical cutting suffer from being energy-intensive. In most cases, blasting is the

Proposed lunar mining facility for producing liquid oxygen from ilmenite, a fairly common oxygen-rich component of lunar soil. (NASA)

most efficient method of breaking rock, and it has even been demonstrated that explosives can be used to loosen the lunar regolith to allow loading by buckets or scrapers. However, concerns over delivering and handling explosives and the probable difficulty in drilling emplacement holes do influence consideration of explosive fragmentation. Also, blast design is highly site-specific, requiring subjective evaluation; thus it demands significant human intervention.

Mechanical fragmentation offers safer, more continuous operation and is better suited to automation. However, the cutting edges or tools used to excavate the lunar material will experience wear, possibly severe, and will have to be replaced periodically. Also, cutting forces (machine mass) and energy (power) for excavating a unit volume of material on the Moon will probably be greater than required for excavating terrestrial soils. Experiments have demonstrated that neither a scaled hydraulic excavator nor a scaled dragline bucket was able to excavate in a lunar simulant that had been compacted to 100% relative density (excavation was possible after blasting, however). The probable difficulty in excavating highly densified lunar regolith must be considered when evaluating equipment and establishing the depth of excavation.

Excavation methods can be divided into two groups: those that are essentially continuous and combine the operations of breaking, loading, and transport; and those that are cyclic and unitize the operations of breaking, loading, and transport. An example of a continuous system is a bucketwheel excavator or drum miner loading onto a conveyor belt; while a cyclic operation may consist of explosive loosening of the material, digging and loading of the material by bucket, and transport by separate vehicle.

A continuous system is generally a simpler operation, but can require more complex equipment. Cyclic methods can take advantage of simpler and multiuse

equipment, but have more pieces of equipment involved in more complex operations. Launch mass requirements for both approaches could probably be mitigated through use of local materials as ballast. Power per unit volume excavated will be similar; but the cyclic operations, by their nature, will require higher peak power. In terms of automation, the simpler operations of the continuous system provide a major advantage over the number and complexity of operations for the cyclic approach. Individual reliability will probably go to the simpler equipment possible for unit operations, while overall reliability, or more appropriately availability, will depend on redundancy and interchangeability. The continuous approach offers the possibility of including the beneficiation operation onboard the excavator and dramatically reducing the volumes of material it is necessary to transport around the lunar surface. Equipment for both approaches can be designed to minimize exposed seal or bearing surfaces, with the exception of cable-driven systems such as draglines or drum slushers, where the rubbing contact between cable and sheave or drum can be a problem.

Materials handling and processing. Whether for basing operations or resource recovery, the containment and transport of material can be divided into two levels: local—at the excavation itself; and interlocal—between the excavation point and the use, processing, or storage point. Local handling involves containment of the material at the cutting or breakage point of a particular excavating system and the transfer to the interlocal level if applicable. Interlocal materials handling involves containment and transport over distance, including any required transfer or storage points. Little or no atmosphere and low gravity will probably dictate a closed materials-handling system where the containment and transport system will physically move the material rather than use any fluid or gas transporting medium.

The low grades of target minerals in lunar soil and rock suggest that excavated materials be beneficiated locally to the greatest extent possible to minimize the interlocal transport of large quantities of materials. The processing of lunar materials for extraction of useful products has been studied by numerous investigators. The common requirement for suggested processes involves a large energy input or highly reactive imported reagents. This again suggests that the excavated material be beneficiated to the highest grade possible prior to processing. Even the preparation of construction materials from lunar regolith (soil) for road beds, aggregate, or shielding will require significant beneficiation. The lunar regolith has a median particle size of 0.04 mm (0.002 in.) and an average size of 0.07 mm (0.003 in.), being mostly dust.

Future missions. Upcoming missions to explore and map the Moon further will be critical to the siting and planning of bases and resource recovery operations. In addition, it is deemed essential that future crewless missions to the lunar surface be planned. These surface missions should involve a rover vehicle to explore key sites in more detail, and they should include appropriate tools or capabilities to resolve unanswered questions related to physical properties and excavatability.

Remote sensing for future missions must include the capability to resolve the physical characteristics of the lunar surface that affect site selection and surface operations. The apparent homogenization of the surface material by millennia of meteoric impact may dictate a choice only between mare and highlands in terms of resource concentrations. However, other local conditions such as topography, regolith properties and depth, and crater density will affect specific site selection and the design of surface operations.

Future missions should include lunar surface explorations as well. In addition to confirming remote-sensing observations at selected sites, experiments must be included to answer the unresolved questions on the excavatability of the lunar regolith. These experiments should be designed to simulate excavation processes and to record the relevant parameters of tool forces and wear. It is felt the best way to accomplish this would be to place on a lunar rover appropriate tools for digging and grading at various locations. The simplest, yet sufficient, set of tools need be only a small, instrumented, backhoe-type arm that could trench to a significant depth, and a blade that could be lowered to test the movement of surficial materials and gradability.

Additional studies. Earth-based studies in support of future missions, feasibility determinations, and preliminary design need to be undertaken as soon as possible because of the long lead times in mission definition and procurement. Efforts are needed in the areas of materials properties and behavior, excavation systems, beneficiation and extraction, and operational control.

The mechanical properties of the lunar regolith are probably as well understood as is possible with the current samples and soil mechanics tests. These properties are considered conservatively in terms of foundation design and use as construction materials. Uncertainties related to the excavatability of the regolith have already indicated the need for additional lunar surface tests for future missions, and also suggest several simulant-based terrestrial tests that should be performed.

Potential excavating systems (continuous miners, buckets, or blades) should also be tested in lunar simulants, but of necessity at reduced scale. These tests might also include beneficiation techniques (for example, electrostatics or electromagnetics) to improve the grade or processability of the excavated material. Testing of beneficiation techniques should eventually proceed to actual lunar samples.

The proposed processing technologies to extract useful products should be tested—first on simulants, then on lunar samples. New techniques, such as in-place processing, could be investigated.

Finally, operational control strategies need to be developed and tested for the excavating and mining systems under consideration. Systems will need to be evaluated for their potential for autonomous operation or their need for human intervention or teleoperation. One promising way to accomplish this would be through scale-model testing of a selected system. These tests could be integrated with the excavation system tests suggested earlier. Having such a scaled excavator would also be of benefit for testing materials, components, lubricants, or control systems.

For background information SEE MINING; MOON; REMOTE SENSING; SPACE FLIGHT in the McGraw-Hill Encyclopedia of Science & Technology.

Russell J. Miller

Bibliography. L. E. Bernold and S. Sundareswaren, Laboratory research on lunar excavation, *Engineering, Construction, and Operations in Space II*, ASCE, Space 90, Albuquerque, April 1990; E. R. Podnieks and W. W. Roepke, Mining for lunar base support, in W. W. Mendell (ed.), *Lunar Bases and Space Activities of the 21st Century*, Lunar and Planetary Institute, Houston, 1985.

Mycotoxin

In the transformation of planted crops to food or feed, losses occur that are attributable to crop consumption or damage by animals, insects, and microorganisms. Microorganisms, particularly molds, can leave behind secondary products of their metabolic activity. Some of these products are innocuous; some, such as penicillin, may be useful; and some have the potential to cause harm if consumed. Under some circumstances, toxic residues from mold growth (mycotoxins) in food or feed have been deleterious to human or animal health (**Table 1**).

Prior to the late 1960s, except for limits on ergot sclerotia in wheat and rye, there was no attempt by any country to regulate limits on mycotoxins in food or feed. At that time, the discovery of the aflatoxins, the observation of their distribution and carcinogenicity, and the availability of new chemical technologies for isolating and characterizing small quantities of complex organic compounds combined to cause a marked increase in mycotoxin research. Over 100 previously unidentified toxic compounds produced by molds have since been added to the list of toxic mold metabolites. At least half of these compounds show some degree of carcinogenic potential, but only a few have been identified as natural contaminants of food or feed (**Table 2**).

The only serious attempt to link exposure to a specific mycotoxin with a specific cancer in humans has been in regard to aflatoxin with liver cancer. There has been speculation that exposure to one or more of the trichothecenes may be related to esophageal cancer, and that ochratoxin A may be a factor in the endemic nephritis observed in the Balkans with the concurrent increased risk of kidney tumors. In no case has causation been established. The record so far is of acute toxicoses, mostly in domestic animals.

Regulation. The United States and Canada established maximum tolerated levels of aflatoxins in peanuts in 1968. Since then, regulations for the

Table 1.   Diseases that have been caused by moldy food or feed and the molds and mycotoxins implicated

Disease	Humans affected	Livestock affected	Mold	Mycotoxin
Aflatoxicosis	+	+	Aspergillus flavus, A. parasiticus	Aflatoxins
Ergotism	+	+	Claviceps purpurea, C. paspali, C. fusiformis	Ergot alkaloids
Alimentary toxic aleukia	+	+	Fusarium sporotrichioides, F. poae	(T-2 toxin)*
Urov or Kashin-Beck	+		Fusarium spp.	
Drunken bread toxicosis (scabby grain toxicosis)	+	+	Fusarium spp.	
Yellow rice disease (cardiac beriberi)	+		Penicillium citreo-viride, P. citrinum, P. islandicum	(Luteoskyrin, islanditoxin, cyclochlorotine, citrinin, citreoviridin)*
Stachybotryotoxicosis	+	+	Stachyobotrys atra	(Various macrocyclic trichothecenes)*
Nephropathy of swine		+	Penicillium viridicatum, Aspergillus ochraceus	Ochratoxin A, citrinin
Facial eczema (sheep)		+	Pythomyces chartarum	Sporidesmin
Slobbering (cattle)		+	Rhizoctonia leguminicola	Slaframine
Hyperestrogenism (swine)		+	Fusarium graminearum	Zearalenone
Refusal and emesis (swine)		+	Fusarium graminearum	Deoxynivalenol
Dendrochiotoxicosis		+	Myrothecium roridum	(Roridin A, verrucarin A)*
Leukoencephalomalacia (equine)		+	Fusarium moniliforme	Fumonisins
Moldy feed syndrome†		+	Penicillium cyclopium, P. rubrum, P. purpurogenum, P. viridicatum	(Cyclopiazonic acid, penitrem A, rubratoxins A&B, viomellien, xanthomegnin)*

*Produced by molds isolated from implicated commodities; no reported detection in implicated food or feed.
†Exclusive of specific diseases listed above.

control of aflatoxins in commodities susceptible to aflatoxin contamination have been enacted by 52 countries. The original rationale for the limit on aflatoxin, on the basis of much speculation and little knowledge, was to keep to the lowest practical level an unavoidable contaminant that was a possible human carcinogen. Most of the countries that subsequently established limits for aflatoxin invoked this simple rationale, even after more knowledge of the risk factors had been developed. Fifteen of the countries with regulatory limits for aflatoxins also have limits for one or more of the following mycotoxins:

Table 2.   Mycotoxins that have been identified as natural contaminants of foods or feeds

Mycotoxin	Contaminated food or feed (most common occurrence)	Translatable* toxicology
Aflatoxins	Corn, cottonseed, peanuts, tree nuts, copra	+
Altenuene	Tomato	
Citrinin	Barley, oats	+
Cyclopiazonic acid	Peanuts	
Deoxynivalenol	Corn, wheat, barley	+
Ergot alkaloids	Rye, wheat	+
Fumonisins	Corn	+
Nivalenol	Corn, wheat	+
Ochratoxin A	Barley, corn, green coffee, oats, rice, sorghum, wheat	+
Patulin	Apples	+
Penicillic acid	Corn, dried beans	+
Sterigmatocystin	Cheese, green coffee	+
T-2 toxin	Barley, corn, sorghum	+
Zearalenone	Corn, wheat	+

*Convertible to human or livestock injury.

patulin, ochratoxin A, deoxynivalenol, zearalenone, T-2 toxin, chetomin, stachybotryotoxin, and phomopsin. Only one country has provided a rationale for that action (Canada for deoxynivalenol). All regulation for patulin, according to a nongovernment source, is for quality-control purposes only. In most countries there is little or no scientific basis for mycotoxin regulation.

Control. Molds can attack crops growing in the field and during harvest; most of the molds that have been involved in mycotoxicoses are those that invade crops in the field. The specific invading mold and the success of its attack are determined by numerous factors, such as inoculum density, plant species and variety, prevailing temperature, tissue damage caused by weather, insect activity, cultivation and harvesting practices, and moisture and fertilization as they affect plant vigor. Some of these factors can be controlled to some extent.

Harvested crops can be protected from mold invasion by various devices. Grains, legumes, nuts, and some fruits are dried to moisture levels that are lower than those needed to support mold growth, and are stored under conditions to maintain that moisture status—for example, in storage areas that are ventilated to remove moisture of respiration and to prevent moisture condensation. Most fruits are stored in controlled atmospheres at reduced temperatures. When the established controls are breached, massive invasion by storage molds can occur. Most of these molds constitute a different group from the field molds, but some toxigenic species can be found in both environments, for example, *Aspergillus flavus* (aflatoxins), *Penicillium expansum* (patulin), and *Aspergillus ochraceus* (ochratoxin). All of the reported mycotoxicoses (Table 1) have involved situations in which either harvested crops were unprotected, permitting extended growth of field molds, or the protective devices had not been properly employed, allowing growth of storage molds.

Mold growth is usually confined to individual kernels, seeds, or fruit and results in observable damage. Therefore, the incidence of mold damage, and thus of any associated mycotoxin, can be reduced by removal of the damaged units. The practicality of this type of sorting depends on the size of the unit, the visibility of the damage, and the value of the crop. The effectiveness of a sorting process for reducing the level of any mycotoxin for which there is a legal limit should be checked by analysis for that toxin, using a representative lot sample. This last stipulation is usually difficult to achieve because of the nonrandom nature of the typical contamination.

All the mycotoxins that have been identified as natural contaminants have chemical structures with components that should make them susceptible to oxidative degradation or cause them to adsorb to polar surfaces. Oxidation by exposure of foods to ultraviolet light, or to heat and air, has been studied for some mycotoxins—for example, aflatoxin, ochratoxin, and zearalenone—and have been found effective. The most extensively studied decontamination treatment is the use of ammonia for destruction of aflatoxin in peanut or cottonseed meals, cottonseed, and corn, all to be used for animal feed. Variants of ammonia treatment are being used extensively in a number of African countries for reduction of aflatoxin levels in peanut meal. Other variants of ammoniation are being used for treatment of cottonseed, cottonseed meal, and corn in a number of states in the United States. The Food and Drug Administration (FDA) has not yet approved ammoniation for feedstuffs in interstate commerce. An aluminosilicate powder, approved by the FDA as an ingredient to prevent caking of feeds, was found to adsorb aflatoxin and prevent its absorption by the consuming animal, but primary use of the aluminosilicate to absorb aflatoxin has not yet been approved.

Economic aspects. The cost of mycotoxins to the economy goes beyond the adverse impact of the mold growth itself on crop yield and quality. In general, the lowest-quality portion of a crop, that portion most likely to contain mycotoxins, is used for animal feed; and feedstuffs are likely to have received the least care in storage, enhancing the possibility of mycotoxins from storage molds. When toxin levels are sufficiently high, the result can be quantifiable animal losses and quantifiable destruction of animal feed. However, the more likely scenario is subacute toxicosis observed as reduced feed efficiency, reproductive failures, and reduced resistance to disease. Poor animal performance may be difficult to detect against a normal variable background. It is only when animal losses are high that the effort to determine whether a mycotoxin could be the cause may be warranted. There have been no reports of human death or illness from mycotoxins in foods in developed countries during normal peri-

ods. Such events appear to be confined to people in underdeveloped areas with limited dietary selection and inadequate crop-handling facilities, or to areas under wartime stress.

For background information SEE AFLATOXIN; MYCOTOXIN in the McGraw-Hill Encyclopedia of Science & Technology.

Leonard Stoloff

Bibliography. Council for Agricultural Science and Technology, *Mycotoxins: Economic and Health Risks*, Task Force Rep. 116, November 1989; P. Krogh (ed.), *Mycotoxins in Foods*, 1987; S. Natori, K. Hashimoto, and Y. Ueno (eds.), *Mycotoxins and Phycotoxins '88*, 1989; P. S. Steyn and R. Vleggaar (eds.), *Mycotoxins and Phycotoxins*, 1986.

Natural gas

Practically all of the proven reserves of natural gas that have been tabulated in the world today reside in conventional sandstone or carbonate reservoirs. Such reservoirs generally have porosities greater than 10% and permeabilities exceeding 1 millidarcy.

Proven reserves are the quantities of the resource that, with reasonable certainty, are known to exist and can be produced under current economic and operating conditions. The proven reserves of natural gas in the world in 1990 were estimated to be about 4000×10^{12} ft^3 (113×10^{12} m^3). There appear to be ample supplies of natural gas from conventional resources for the world's needs in the foreseeable future. However, when unconventional sources of natural gas are considered, there is, for all practical purposes, an unlimited amount.

By the early 1990s the world was producing approximately 75×10^{12} ft^3 (2.1×10^{12} m^3) per year, which would imply there are only about 50 years of gas left of proven reserves. However, the amount of proven reserves that is tabulated has increased each year. In reality, there are vast resources of methane beyond the officially tabulated reserves from conventional reservoirs.

Most resources can be depicted as residing in a triangle where the least expensive and easiest to obtain are extracted from the top of the triangle. But there are limited amounts of this resource. To tap the larger amount of resource lower in the triangle usually entails greater expense and advanced technology. The top of the resource triangle in the **illustration** represents practically all the gas produced to date as well as remaining recoverable resources from conventional rock. Beneath that is an enormous amount of natural gas from unconventional sources. Production from many sources of unconventional gas is beginning to occur around the world at the present time.

Tight gas formations. When sands or carbonates have porosities and permeabilities that are low enough to inhibit gas from flowing at commercial rates, they are known as tight formations of all the unconventional sources of natural gas. The most emphasis to

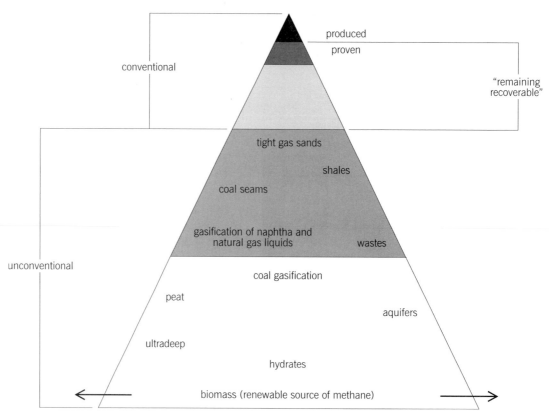

Resource triangle for recoverable natural gas.

date has been on tight gas sands, which exist in great quantities in almost every gas province of the world. Such sands may have less than 10% porosity and severe restrictions to flow due to small openings within and between the pore spaces.

Under such conditions, flow of gas into a well bore would be insufficient for commercial production. To induce greater flow, techniques such as massive hydraulic fracturing are employed. With this procedure, high pressures are applied down the well bore until the rock splits. These cracks are then extended deep into the formation by pumping fluids laden with a proppant (a support to maintain the opening) such as sand. In deep wells (over 8000 ft or 2400 m), higher-strength proppants such as sintered bauxite or special ceramic beads are required, because sand tends to crush under high stress. Drilling horizontal wells perpendicular to natural fractures is another technique to enhance flow from tight sands.

Significant tight gas accumulations have been identified in the United States and in many other parts of the world, including South America, Australia, Spain, France, and Germany.

Shales. When large volumes of organic mud, silt, and sand are deposited in marine environments and buried to great depths, the combined effect of heat, pressure, and chemical reactions can transform the organic shale to source rocks for oil or gas. The hydrocarbon may migrate to nearby porous formations and, with appropriate trapping conditions, become conventional reservoirs. In some cases, however, the source rock itself may become potential reservoirs.

This condition occurs in the Appalachian Basin in the eastern United States. When the shales in this region are naturally fractured, there is the potential for commercial production. Although the wells generally have low productivity [typically less than 200,000 ft^3 (57,000 m^3) per day], they are capable of producing for many decades.

Experiments have also been carried out on the direct gasification of shales in the eastern United States as well as the western oil shales. Although this technology is not economical at this time, the potential from gasification of shales is many times the existing free gas within the shales.

Coal. In addition to its applications for supplying direct heat or generating electricity, coal has significant potential as a source of methane. In fact, natural gas may be produced from coal by several different processes, such as coal seam demethanization, in-place gasification, or mining and processing through surface plants.

Coal seam demethanization. Even shallow coal seams have a significant amount of methane. In the past few years, there have been a large number of wells deliberately drilled to coal seams to extract methane on a commercial scale.

The methane in a coal seam may be trapped within natural cleats (fractures) or within thin sand lenses embedded in or immediately adjacent to the coal. However, most of the gas within a coal seam is adsorbed on the coal structure. Often, the cleat system within the coal contains water, and arrangements must be made to pump off this water so the gas can desorb.

As pressure decreases, gas is released from the coal surface. After desorption, the gas is transmitted through the natural fracture system. The rate of this transmission depends upon the frequency of natural fractures and how permeable they are to fluid flow.

In-place coal gasification. This technique is considered when coal seams are not suitable for mining, because they are too deep, too steep, or too thick. In this process, burning is initiated in one borehole, and the product gas flows to another hole for recovery. If air is used to supply the burn front, low-Btu gas results. If pure oxygen is used, medium-Btu gas is produced, which can then be upgraded to pipeline quality. While there have been a number of pilot projects, many technical questions must be answered before there can be wide-scale application.

Surface coal gasification. A significant amount of synthetic gas (analogous to natural gas) could be produced from coal through surface gasification plants. Several coal gasification plants have been proposed around the world, especially in Europe and the United States.

Naphtha and natural gas liquids. Natural gas may also be produced from processing light hydrocarbon liquids such as naphtha or natural gas liquids (NGL), which include propane and butane. There are presently 13 such facilities in the United States; they are used primarily for production of additional gas during heavy-demand periods. In special situations, such liquids are used as the sole source of natural gas; for instance, in Hawaii gas is produced from butane and pentane.

Peat gasification. Peat forms when vegetation is deposited in bogs isolated from local drainage systems under conditions of moderate-to-high rainfall. With little or no overburden, peat is considered the earliest stage of coal formation. Recently, there has been considerable interest in using peat as a raw material for synthetic fuels. Peat can be converted to natural gas by either thermal or biological methods. Experiments have also been conducted on degasification of circulated peat bog waters.

Research has shown that peat is easier to gasify than coal, and in fact yields up to 3½ times more methane than lignite.

Geopressured aquifers. Beneath Texas and Louisiana and the adjacent offshore regions in the Gulf of Mexico, large amounts of gas dissolved in hot high-pressure brine have been identified. Estimates of the amount of gas vary over wide ranges, indicating a large uncertainty of technical and economic viability. Although most earth scientists agree that large quantities of natural gas are probably dissolved in brines, much of it is in shale or sand formations that have too low a permeability for reasonable production. Furthermore, as the brine is produced, the pressure rapidly decreases, so that less than 5% of the accumulation can be recovered before flow ceases.

Hydropressured aquifers. Water-producing formations under normal pressure gradients (0.45 lb/in.^{2} per gradient foot or 10.2 kilopascals per gradient meter) that have associated gas are known as hydropressured aquifers. Generally, such aquifers occur at shallower depths and may contain only one-half to one-third as much gas per unit volume as geopressured waters. These formations do not flow naturally; therefore, pumps are required to lift sufficient volumes of water to the surface. The amount of gas production per well may be very small; very few fields of this type are presently in production.

Because of the large resource base around the world in hydropressured aquifers and the relatively simple technology of gas extraction, this is a potentially important source of energy. The economics of development will vary by country depending upon the cost of labor, equipment, environmental constraints, and (most importantly) the gas-to-water ratio of the particular aquifer.

Ultradeep gas. Sedimentary deposits are believed to exist to a depth of 50,000 ft (15,400 m) in some basins of the world. Methane remains stable at temperatures up to 1022°F (550°C); before such temperatures are reached, oil is converted to methane. Therefore, methane will probably be the only hydrocarbon found in the deepest and oldest geological formations.

At depths beyond 25,000 ft (7700 m; ultradeep), limitations to drilling equipment, logging tools, tubular goods, and safety devices are quite severe. The high temperatures and the probability of encountering hydrogen sulfide, carbon dioxide, and corrosive salt-saturated fluids compound the problems of production.

Notably, there is a more favorable success ratio for holes beyond 15,000 ft (4600 m) than is generally found at shallower depths. While the United States has more wells and a higher density of drilling than any other country in the world, it is estimated that less than 3% of the deep sediments have been explored. Yet, huge reserves have already been identified. Although a number of them are generally located between 15,000 and 25,000 ft (4600 and 7700 m; less than the ultradeep classification), the implication is that there are very large sources of deep gas.

Further interest in deep potential has been sparked by ideas on the possibility of abiogenic sources of gas. Finding rocks that can act as reservoirs that will allow the high rates needed for commercial production is the key to deep and ultradeep drilling success.

Hydrates. Enormous amounts of natural gas exist in a hydrate form in arctic regions and beneath the oceans. Estimates have been made that as much as 35×10^{18} ft^3 (1×10^{18} m^3) of natural gas may exist as hydrates.

Gas hydrates are icelike crystalline compounds of water with bound gas. At room temperature and pressure, hydrates will simply be reduced to water while releasing more than 100 times their volume in gas.

Production of gas from subsurface hydrates will be difficult. For any reasonable rates of production, it will probably be necessary to inject some auxiliary fluids. This could be an alcohol such as methanol, which is commonly used to prevent hydrate formation in natural gas pipelines. Another possibility is to inject steam into the formation.

Economic extraction of gas from hydrates is much farther into the future than most of the other unconventional sources. However, recognition of this incredible volume of energy has stimulated serious investigations and experiments.

Biomass and wastes. There is potential for generation of methane as long as the Sun continues to shine. Many plant types are efficient storehouses of solar energy. Considerable research on extraction of methane from biomass has been done.

For example, the giant salt-water sea kelp, which can grow at a rate of 2 ft (0.6 m) per day and reach 200 ft (60 m) in the adult state, has been tested as a source of methane. It is estimated that 60,000 mi^2 (100,000 km^2) of kelp farms offshore could provide the raw material for producing 22×10^{12} ft^3 (0.62×10^{12} m^3) of methane per year. This would be enough to satisfy the current needs of Canada and the United States. Once a kelp farm has been established, it is expected that it would sustain itself indefinitely. The harvested kelp is converted to methane with the help of microorganisms in large airtight containers, known as biodigesters. Initial experiments on containment of the kelp in ocean farms have shown that enormous problems remain to be solved before a commercial process is feasible.

Fresh-water plants such as water hyacinth are also rich sources of methane. This plant, common in temperate zones, often chokes fresh-water lakes or canals. In an experiment in Florida, the growth of hyacinth plants is enhanced by wastewater from a conventional primary sewage treatment plant. In the process, the plants also further treat the wastewater. Then they are harvested along with the waste sludge and biologically gasified to methane.

Many experiments are being carried out on municipal, industrial, and agricultural wastes. Recovery of gas from wastes not only adds to the energy supply but also improves the quality of life for the surrounding communities by reducing environmental pollution as well as generating additional income. Biogas recovered from decomposition of garbage at two landfill sites in the Los Angeles area is sufficient to heat 12,000 homes. Other landfill sites across the United States and Canada are currently producing or being prepared for degasification.

For background information SEE AQUIFER; COAL GASIFICATION; HYDRATE; METHANE; MINERAL FUEL AREAS; NATURAL GAS; SOLAR ENERGY in the McGraw Hill Encyclopedia of Science & Technology.

R. E. Wyman

Bibliography. American Gas Association, *The Gas Energy Supply Outlook 1989–2010*, September 1989; National Petroleum Council, *Unconventional Gas Sources*, June 1980.

Neural network

Since about 1970, interest in neural and cognitive phenomena has extended beyond the boundaries of neurobiology and psychology and into the fields of mathematics, physics, and computer science. Two theoretical frameworks have emerged that are grounded in fundamentally different assumptions: One approach, based on symbol processing, explicitly assumes that the principles underlying cognition are best expressed in the language of formal computational theory and are independent of biological considerations. The other approach, based on processing in networks of neuronlike elements, integrates ideas from neuroscience as basic assumptions. The latter framework was simultaneously formulated by a number of independent research groups and individual researchers; hence the field of neural networks has been given several different names, including connectionism, parallel distributed processing (PDP), artificial neural systems (ANS), and neurocomputing. In the 1980s, it virtually developed into a scientific discipline in its own right, overlapping artificial intelligence, neurobiology, and cognitive psychology.

Biological neuron. A neuron is a living cell, similar in most respects to other cells in the body, with some important exceptions. For example, neurons do not divide after the nervous system is developed (this development is complete in humans shortly after birth), and they have specialized apparatus for transmitting and receiving electrochemical signals. The parts of a neuron most relevant to its information-processing capability are the cell body (soma), the axon, the dendrites, and the synapses (**Fig. 1a**). Generally, a cell transmits a signal through its axon, which divides into several branches and subbranches to form a tree, or arborization; swellings (or boutons) at the "leaves" of the tree tend to lie very close (within about 20 nanometers) to a dendrite or the soma of another neuron and contain a neurotransmitter. The signal induces the release of the neurotransmitter into the gap, or synapse, between the cells, which induces a local disruption in the membrane of the second (target) cell; the voltage signals propagate to the soma, where they are combined; if the integrated voltage exceeds a value known as the cell's firing threshold, then the neuron transmits a pulse, known as an action potential. The frequency of generation of action potentials increases with increasing integrated voltage.

Formal neuron. Neural network models consist of interconnected nodes, or units, which include some of the functional features of neurons. The response of each unit is usually computed by the following three-step process (Fig. 1b): each input component s_i is multiplied by a corresponding weight factor w_i; the products are summed to give a net activation x; and the response r is a function of x.

A linear unit is defined as a unit whose response function r is a linear combination of its inputs, that is, $r = x$. The mathematics of linear units has been developed to a much higher level than that of nonlinear units. However, the linear unit is problematic on two counts: it is inconsistent with laboratory data; and it is computationally limited, that is, the class of functions computable by a network consisting of only linear units is severely restricted. Hence, nonlinear units are much more powerful. A typical compromise is the semilinear unit, defined by a response function that is a nonlinear function of a linear combination of the input components; that is, $r = f(x)$. A common choice for $f(x)$ is the threshold function, defined as 1 if x exceeds a threshold θ; otherwise $f(x) = 0$. The popularity of the threshold function stems from two sources: it is analogous to the firing threshold observed in neurons, and the binary nature of the output is appealing for logical applications.

Pattern recognition. Neural networks are particularly well suited to applications in pattern recognition. Each neuron in the network performs a pattern recognition function; in general, the magnitude of a neuron's response depends on the overlap, or similarity of the components of the stimulus s_i with the corresponding weight values w_i. For example, consider a retina consisting of an array of receptors activated by the image formed by the optics of the eye. If these receptors are connected to a unit with positive and negative weights, then the unit will be maximally responsive if the image excites all of the units with positive weights and none of the units with negative weights. Pattern recognition units can be arranged by layers into networks that compute features of increasing complexity (**Fig. 2**). Interestingly, neurophysiological studies of neurons in the visual system show a similar progression of complexity from the retina all the way to the cortex.

Satisfaction of simultaneous constraints. The responses of pattern recognition units, or feature detectors, can be interpreted as representing the probability that a particular hypothesis about the world is true. By interconnecting feature detectors such that the weights between units reflect the correlations of the corresponding events, the resulting networks can yield reasonable interpretations of data that are incomplete or inconsistent. Thus, the weights reflect constraints (both positive and negative) on these hypotheses that are imposed by regularities in the world. This approach has been used to account for many cognitive phenomena, including stereovision, problem solving, and recall of spatial information.

An example is the constraint satisfaction network designed by J. L. McClelland and D. E. Rumelhart to simulate the results of response time experiments that have revealed that letters are identified more quickly in words than in nonword letter groups. The network consists of three levels (**Fig. 3***a*). The first level detects oriented lines at particular locations in the visual field; units at the second level indicate the likelihood of particular letters at any of four positions; and units at the third level indicate combinations of four letters that form words. The weights are determined according to a few rules; for example, letter units that are inconsistent with other letter units have negative connections (no two letters may occupy the same position), word units have positive connections with the units representing the constituent letters, and so on. The activities of selected units at the letter and word level for a particular input are shown in Fig. 3*b* and *c*.

It is a common misconception that neural network simulations of phenomena at a high cognitive level,

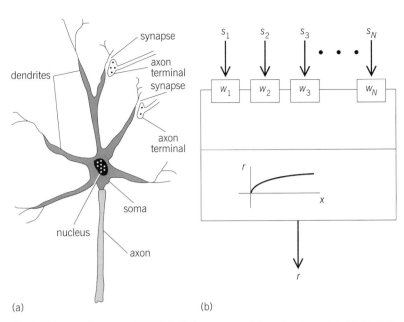

Fig. 1. Diagrams of neurons. (*a*) A biological neuron consisting of a soma, an axon, several dendrites, and several synapses. (*b*) A formal neuron, which includes many of the functionally important parts of its biological counterpart. s_i = input component; w_i = weight factor; $x = \sum_i w_i s_i$ = net activation; r = response.

such as language, are meant to have implications at a neurobiological level. To the contrary, although these models are grounded in processing assumptions inspired by biology, it must be strongly emphasized that they are not meant to be interpreted as evidence that there should be neurons that respond specifically to letters or words.

Learning. Since the response properties of a neural network are determined by the weights between the units, learning processes must involve a change in the weights. Thus, learning procedures for neural networks can be expressed mathematically as differential equations in the weights. Considerations of biology

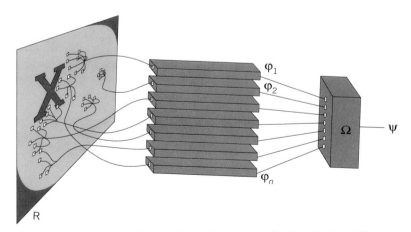

Fig. 2. Pattern recognition unit. A unit that receives external stimuli as direct input (the units labeled $\varphi_1 \ldots \varphi_n$) can compute only simple functions over the components of its receptive field. By computing a simple function Ψ over the responses of the φ units, a higher-level unit (Ω) can compute relatively complex functions of the pattern of activity in the external world. For example, the φ units may detect line segments, and the Ω unit may specifically respond to the letter X. (*After M. L. Minsky and S. A. Papert, Perceptrons, MIT Press, 1969, paperback text ed., 1987*)

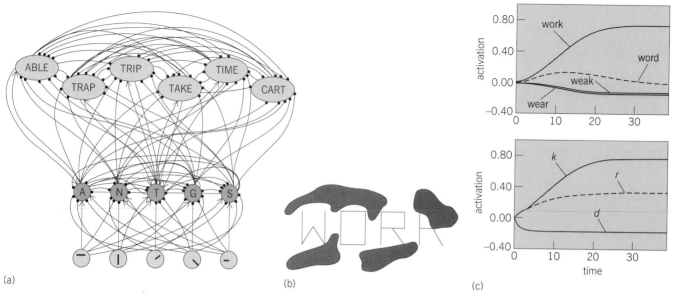

Fig. 3. McClelland and Rumelhart letter recognition model. (*a*) A small set of units from the network is shown with the connections. The network architecture includes three levels: the feature level (bottom) consists of units that respond selectively to line segments; the letter level (middle) consists of 104 units, one for each of the 26 letters in any of four positions; the word level (top) consists of units corresponding to 1179 four-letter words. (*b*) The visible segments corresponding to the word WORK with segments from the K missing are presented to the feature level of the network. (*c*) The activities of the units corresponding to various words (top) and letters (bottom) are shown over time. The axes are labeled in arbitrary units, with the time units corresponding to iterations of the computer simulation. Letter perception is thus based on bottom-up influence from the feature level, as well as top-down influence from the word level. This is in agreement with data from experiments in letter recognition. (*After J. L. McClelland and D. E. Rumelhart, An interactive activation model of context effects in letter perception, part 1: An account of basic findings, Psychol. Rev., 88:375–407, 1981*)

and physics suggest that weight changes depend on local variables. In particular, a weight change is thought to depend on the activities of the two units linked by the synapse between them.

Most learning procedures are based on the neurophysiological postulate proposed by D. O. Hebb in 1949: "When an axon of cell A is near enough to excite a cell B and repeatedly or consistently takes part in firing it, some growth process or metabolic change takes place in one or both cells such that A's efficiency, as one of the cells firing B, is increased." Several learning procedures of this form have been investigated.

Units in neural networks can be explicitly trained to give desired responses to a number of patterns through an iterative process generally referred to as error-correction learning. Given a training set **T** of stimulus patterns and the corresponding desired responses, the following six-step procedure is usually used for error-correction learning: (1) initialize the weights to random values; (2) randomly select a stimulus–response pair from **T**; (3) present the stimulus to the unit, thereby generating a response; (4) calculate an error value δ by subtracting the generated response from the actual response; (5) modify the weights by an amount proportional to the product of the error and the activity of the input, that is, $\Delta w_i = \delta s_i$; (6) stop if the learning is sufficient (the criterion varies from case to case) or else return to step 2.

Back propagation of error. The version of error-correction learning described above is valid for training a network only if an error value can be calculated for every unit. Since the desired response of the intermediate, or hidden, units is not generally known, this presents a serious dilemma for training networks of several layers. A solution to the multilayer training problem is to compute an effective error value for the hidden units. The error attributed to a particular unit is calculated as a function of the errors in the units to which it connects and the weights of those connections. Thus, once the responses of the output units have been computed and compared with the corresponding desired values, effective errors can be computed for the hidden units by a process of backward propagation of error. Once the effective errors of all the hidden units have been computed, the same error-activity product (step 5 in the learning sequence above) can be applied to modify the weights.

The back propagation rule for weight modification has become the leading technique for training neural networks. It has been used to generate several models of human performance in cognitive tasks, including visual perception, language, and memory. An impressive example is a 1987 demonstration by T. Sejnowski and C. Rosenberg, in which a network was trained to pronounce text; that is, the input units encoded a sequence of textual characters, and the output units represented phonemes. The output was used to control a voice synthesizer, which produced actual sounds. At the beginning of the learning procedure, the output sounded like static, but after many repetitions through a text of 400 words, it was intelligible. Also, like many other networks trained with back propagation, the learning generalized well; that is, after training on the original text, the network was able to successfully pronounce new input text.

For background information SEE NERVOUS SYSTEM (VERTEBRATE); NEUROBIOLOGY; NEURON in the McGraw-Hill Encyclopedia of Science & Technology.

Paul Munro

Bibliography. J. A. Anderson and E. Rosenfeld (eds.), *Neurocomputing*, 1988; D. O. Hebb, *Organization of Behavior*, 1949; J. L. McClelland and D. E. Rumelhart, An interactive activation model of context effects in letter perception, part 1: An account of basic findings, *Psychol. Rev.*, 88:375–407, 1981; T. J. Sejnowski and C. R. Rosenberg, Parallel networks that learn to pronounce English text, *Complex Sys.*, 1:145–168, 1987.

Nitrogen fixation

Nitrogen is a major ingredient of proteins, nucleic acids, and other organic molecules that are essential for life. However, plants and animals cannot convert atmospheric nitrogen into a biologically useful form. The only organisms that can utilize atmospheric nitrogen are prokaryotes (bacteria and cyanobacteria). Nitrogen-fixing bacteria and cyanobacteria are collectively referred to as diazotrophs.

Nitrogen cycle. Nitrogen fixation is one of the critical processes in the global nitrogen cycle. Diazotrophs convert atmospheric nitrogen into ammonia, which plants can use directly. Alternatively, the ammonia is released into the soil, where other microbes convert it to nitrate. The released nitrate is then assimilated by plants into organic compounds such as amino acids and proteins. Animals can obtain nitrogen only by eating plants or other animals. During the breakdown of plant and animal bodies, soil bacteria degrade the organic nitrogen compounds into nitrogen gas, which then returns to the atmosphere.

Fixation of nitrogen. Diazotrophs fix nitrogen with a complex enzyme system called nitrogenase. Nitrogenase consists of two distinct proteins which can only function when bonded to one another in a specific way. Nitrogenase reduces molecular nitrogen (N_2) by adding hydrogen ions and electrons to form ammonia (NH_3), as indicated by the equation below.

$$N_2 + 6H^+ + 6e^- \xrightarrow{\text{nitrogenase}} 2NH_3$$

Nitrogen fixation is energetically expensive because it requires 12 molecules of adenosinetriphosphate (ATP) to make each molecule of ammonia.

In aquatic habitats, nitrogen fixation is carried out by free-living cyanobacteria (also known as blue-green algae). In terrestrial habitats, nitrogen is fixed by many kinds of free-living bacteria and by bacteria that form symbiotic associations with a number of plant species. Cyanobacteria form symbioses with almost all groups of plants, including mosses, liverworts, ferns, cycads, and one genus of angiosperms, *Gunnera*. Two groups of symbiotic relationships, however, are of particular interest to scientists, in terms of the amount of nitrogen that these groups fix and of their impact on agriculture, horticulture, and forestry. In the first group, plants of the Leguminosae family form nitrogen-fixing root- and stem-nodulating symbioses with the soil bacterium *Rhizobium*. Many of the plants in this group are important agricultural crops. A second diverse group of plants, which are known collectively as actinorhizal plants and which play an important role in managed ecosystems, horticulture, and forestry, form nitrogen-fixing root-nodule associations with the filamentous soil bacterium *Frankia*.

Legume–rhizobial symbioses. Plants of the legume family include such diverse species as soybeans, peas, peanuts, lupines, and redbud trees. These plants have localized swellings of root tissue called root nodules. These nodules are made of plant cells inside which the symbiotic nitrogen-fixing bacteria live. The development of this intimate relationship is dependent on the infection process, a complex sequence of interactions between the plant and the microbe. Rhizobia in the soil are attracted to the root by chemical or electrical signals, whereupon they colonize the root surface. The bacteria then enter the plant by invading root hairs, entering naturally occurring cracks or wounds on the root surface, or penetrating between the epidermal cells of the root surface.

Many of the agriculturally important legumes are infected through their root hairs. This infection mech-

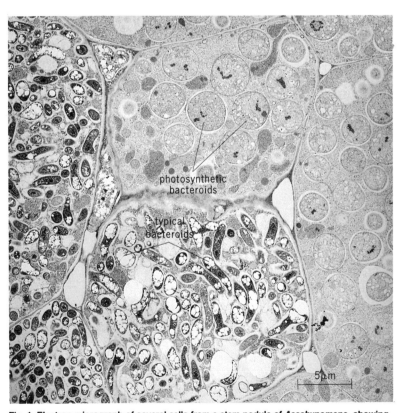

Fig. 1. Electron micrograph of several cells from a stem nodule of *Aeschynomene*, showing the two types of rhizobial bacteroids that can coexist inside a nodule. The typical bacteroid forms are enlarged rods, branched rods, or globular structures. The bacteroids of the photosynthetic *Rhizobium* are large spherical structures that resemble free-living photosynthetic bacteria, including the ability to synthesize bacteriochlorophyll A.

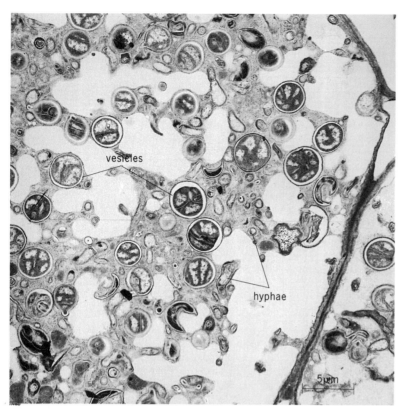

Fig. 2. Electron micrograph showing part of two cells from *Elaeagnus angustifolia*, the Russian olive. A dense network of hyphae forms in the host cell. At the tips of the hyphae, enlarged spherical vesicles form, and are the site of nitrogen fixation.

anism, which is the typical way in which legumes are infected, begins with the bacteria inducing curling of the root hair by softening the hair cell wall. The bacteria then break through the wall. In response to this invasion, the plant cell forms a tube of cell wall, called the infection thread, around the bacteria. As the bacteria penetrate the cell, the tube is continually deposited around them. The bacteria then pass from the root hair into cells of the root cortex. The infection thread then branches into other cells of the root cortex. Bacteria are released from the tips of these infection thread branches into the interior of the infected cells. During the release, the bacteria become enclosed either singly or in groups by host-plant cell membrane. At this stage in infection, the bacteria differentiate into bacteroids, which are enlarged rods, branched rods, or globular forms of the microorganism (**Fig. 1**). Bacteroids are structurally and metabolically different from the free-living *Rhizobium*.

It is in the bacteroid condition that the nitrogen fixation genes (*nif*) are switched on and the fixation of atmospheric nitrogen by the symbiont begins. Although nitrogenase readily fixes nitrogen, it is also poisoned and deactivated by oxygen. To prevent this, legumes produce leghemoglobin, an oxygen-absorbing protein similar to hemoglobin. The interior of leguminous nodules is typically pinkish red due to the presence of leghemoglobin. After the bacteroids reduce nitrogen to ammonia, the ammonia is converted by the host plant to amino acids and ureides (derivatives of urea) such as glutamine, citrulline, asparagine, and allantoic acid. These substances are then transferred to the plant's vascular system and exported to the entire plant.

Stem nodules occur in two genera of leguminous plants, *Sesbania* and *Aeschynomene*. The nodules form by the infection of adventitious roots developing from the stem and require wet or flooded conditions for infection by *Rhizobium* to occur. Because these nodules are in photosynthetic tissue, it is easy for the host plant to provide the bacteria with photosynthetic products used to power nitrogen fixation.

Recently, some interesting observations have been made in stem nodules of *Aeschynomene*. Some nodules contain two forms of rhizobial bacteroid (Fig. 1), one a more typical bacteroid and another that is large and spherical and contains internal membranes reminiscent of photosynthetic bacteria. The spherical form of bacteria has been isolated and, when used to infect the host plant, produces nodules containing only the spherical bacteroid. The spherical form, a *Rhizobium* species, produces bacteriochlorophyll A, a photosynthetic pigment found usually only in photosynthetic bacteria. This discovery could be significant for improving the efficiency of leguminous crops. The host plant must provide the bacteria with 12 lb (or kg) of photosynthate for every pound (or kilogram) of nitrogen fixed. If the bacteria were to provide a significant part of that photosynthate requirement themselves, the host plant's productivity would rise.

Actinorhizal plants. Actinorhizal plants are a taxonomically diverse group of woody shrubs and trees belonging to eight families. Nitrogen-fixing nodules form on the roots of actinorhizal plants in response to infection by the filamentous soil actinomycete *Frankia*. In certain host plants such as *Alnus* (alder) and *Myrica* (bayberry), the root becomes infected as it does in legumes, that is, by entry of the microorganism into a root hair. From there the filament grows through cell walls and invades certain cells in the root cortex, forming a prenodule. Hyphal filaments branch from there through the root tissue to find and invade developing lateral roots. These lateral roots become nodules. Certain actinorhizal plants are infected by an alternative mechanism. In members of the family Elaeagnaceae such as *Elaeagnus* (for example, *E. angustifolia*, Russian olive), hyphal filaments of *Frankia* penetrate between the epidermal cells of the root. The filamentous bacteria then grow into the intercellular spaces of the root cortex. The cells of the root cortex secrete substances that nourish the bacteria. Instead of a prenodule forming, the hyphae invade a developing lateral root, thereby altering the root's development and transforming it into a root nodule. Members of two other large actinorhizal families, Rhamnaceae and Rosaceae, are infected by this direct type of infection. Thus, more than 60% of actinorhizal genera may be infected by this mechanism. Given that this is such a simple mechanism in comparison to root hair infection, it is a prime candidate for extending the nitrogen-fixing symbioses to other presently nonnodulating species. Scientists are currently studying the possibility of extending the symbiosis within the fruit crops of family Rosaceae (apple, pear, plum, cherry, raspberry, and so on), hoping to develop nitrogen-fixing fruit trees.

Invasion of the host plant by bacteria profoundly changes the structure and metabolism of the symbiotic bacteria. After entering the cell, the hyphae branch repeatedly and form a fairly dense cluster (**Fig. 2**). The ends of the hyphae then swell into large round or pear-shaped, compartmentalized structures called vesicles. These vesicles, which develop thick walls to block oxygen diffusion, are where nitrogen fixation occurs in actinorhizal plants.

For background information SEE LEGUME; NITROGEN CYCLE; NITROGEN FIXATION in the McGraw-Hill Encyclopedia of Science & Technology.

Iain M. Miller

Bibliography. J. R. Gallon and A. E. Chaplin, *An Introduction to Nitrogen Fixation*, 1987; C. R. Schwintzer and J. D. Tjepkma (eds.), *The Biology of Frankia and Actinorhizal Plants*, 1990; J. I. Sprent and P. Sprent, *Nitrogen Fixing Organisms: Pure and Applied Aspects*, 1990.

Nobel prizes

The Nobel prizes for 1990 included the following awards for scientific disciplines.

Physiology or medicine. Joseph Murray of Boston's Brigham and Women's Hospital and E. Donnall Thomas of the Fred Hutchinson Cancer Research Center in Seattle were awarded the prize for their pioneering work in using transplanted organs and tissues to treat human patients. Murray performed the first successful transplant of a human organ, a kidney, in 1954. Thomas, in 1956, was the first to successfully transfer bone marrow from one individual to another.

Despite repeated efforts since about 1900, almost all efforts at transplantation ended in organ rejection. Working first with dogs, then with identical twins, and finally with unrelated patients, Murray and Thomas helped define and then overcome the poorly understood immunological mechanisms behind organ rejection.

After successfully transplanting a kidney from one identical twin to the other in 1954, Murray sought a way to overcome organ rejection between individuals who are not genetically identical. Working with George Hitchings and Gertrude Elion, Murray developed a drug regimen based on 6-mercaptopurine that suppressed the immune system. By using this regimen, in 1962 Murray performed the first successful transplant from an unrelated donor.

Thomas investigated certain forms of leukemia in which the bone marrow stops functioning. He reasoned that if the patient's diseased marrow could be replaced with healthy marrow from a donor, the patient's condition would improve. Like Murray, he faced problems of tissue rejection. By using a technique combining irradiation of the patient's marrow

and administration of the drug methotrexate that diminishes graft-versus-host disease (a kind of autoimmune reaction), Thomas and his team of researchers at the University of Washington paved the way for bone marrow transplants that are becoming increasingly successful. The team also began typing tissues based on the major histocompatibility complex (MHC), thereby greatly improving the likelihood of finding suitable donors.

Physics. Jerome I. Friedman and Henry W. Kendall of the Massachusetts Institute of Technology and Richard E. Taylor of Stanford University were awarded the prize for demonstrating that protons, neutrons, and similar elementary particles are made up of quarks, which are even more fundamental. Their work was in some respects analogous to the experiment in 1911 in which Ernest Rutherford showed that atoms consisted of a heavy nucleus and orbiting electrons.

Quarks were first proposed in 1963 by M. Gell-Mann and G. Zweig, chiefly as a means of classifying the numerous particles that were then being discovered, but initially quarks were regarded as merely mathematical constructs. In experiments from 1967 to 1973 at the Stanford Linear Accelerator Center, a large team of scientists led by Friedman, Kendall, and Taylor directed a beam of electrons, raised to an energy of 20 GeV in the 2-mi-long (3.2-km) particle accelerator, at targets of stationary protons and neutrons contained in vessels of liquid hydrogen or deuterium. Two large magnetic spectrometers were designed to measure accurately the energies and angular deflections of the scattered electrons.

In the study of deep inelastic scattering reactions, in which the electrons shatter the protons and neutrons, a large number of electrons were found to be deflected through large angles. This indicated the presence of hard, pointlike structures inside the particles (much as the large-angle scattering of alpha particles from atoms in Rutherford's experiment indicated the presence of a small, dense nucleus at the center of each atom), and these constituents were quickly identified with quarks. Further analysis established that the quarks were immersed in a sea of electrically neutral gluons that held them together.

Chemistry. Elias J. Corey of Harvard University received the prize in chemistry for the development of theories and methods of organic chemical synthesis. His landmark work has made possible the production of a wide variety of complex biologically active substances and has stimulated research involving pharmaceuticals and other useful chemicals.

Corey and his research group developed syntheses for approximately 100 compounds whose structures duplicated or closely resembled those of naturally occurring compounds. This achievement was facilitated by their development of a method of planning organic chemical synthesis that came to be known as retrosynthetic analysis. Unlike previous syntheses that were accomplished largely as a result of intuition and trial-and-error, those based on retrosynthetic analysis proceed from a careful classification of all the significant features desired in the structure of the end product, such as chains, rings, and functional groups. Using this information, the researcher works backward, one step at a time, seeking precursors that will produce these features. In each step, precursors are sought for those precursors from the previous step that have been judged desirable. The result of the retrosynthetic analysis is the construction of pathways (trees) of possible synthetic routes that will finally yield the desired end product.

The rules of retrosynthetic analysis can be expressed as algorithms, so that a computer program can be devised for generating the trees of the precursors that will lead to the desired end product. Corey's computer program is known as LHASA (logic and heuristics applied to synthetic analysis).

Corey's most significant work is considered to be the synthesis of a group of 20 prostaglandins known as octylcyclopentylheptanoic acids. These are produced in very small amounts in every cell in the human body. In general, synthesis of these prostaglandins and related compounds has led to the development of valuable pharmaceuticals for treating a variety of ailments, including heart disease and rheumatoid arthritis.

Noise measurement

A major advance in the knowledge of human hearing is the rediscovery that exposure to intense sound produces lasting hearing threshold deficits. As early as 1700, it was recognized that tinsmiths suffered hearing loss as a consequence of their incessant hammering. The phenomenon was rediscovered in the nineteenth century as boiler-maker's deafness. With the latest rediscovery, the need to protect workers from intense long-term noise exposure has been codified in occupational safety regulations throughout the world. Measurement of industrial noise serves one of two objectives: protection of worker hearing or noise abatement.

Health effects of noise exposure. Exposure to high-level industrial noise may cause injuries of two types: trauma and deafness. Trauma is evidenced by rupture of the tympanic membrane or disarticulation of the ossicular chain. This type of injury is produced by explosive pressure gradients and, in the United States, is reportable to the Occupational Safety and Health Administration (OSHA) as a traumatic injury. Hearing loss may be temporary, temporary threshold shift (TTS), or permanent. In the industrial setting, it may become apparent to a worker who notices that the car radio is too loud in the morning and insufficiently loud after work. Noise-induced permanent hearing threshold shift (NIPTS) develops upon extended exposure to high-level noise. NIPTS has evolved empirically from extensive dose–response data collected in many countries since about 1960.

These data form the basis of regulations that protect workers' hearing. Before the effect of industrial noise exposure (or avocational noise exposure) can be assessed, the hearing thresholds must be corrected for the effect of aging. Median hearing threshold shifts due to age are shown in **Fig. 1a** for a non-noise-exposed population. There is an accelerated growth of hearing loss at the higher frequencies in the fifth and sixth decades of life.

Age-corrected hearing levels according to the model of D. W. Robinson are shown in Fig. 1b. Curves A and B are examples of hearing thresholds for various populations; curve A shows the threshold at 4 kHz for 10% of the population, and curve B the

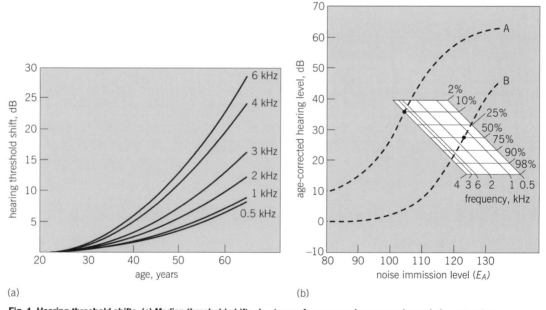

Fig. 1. Hearing threshold shifts. (a) Median threshold shifts due to age for a non-noise-exposed population, at various frequencies. (b) Threshold shifts from noise exposure for various populations. The text explains how to find the hearing threshold shift at any frequency, and percentage of the population for a given noise immission level. (*After D. W. Robinson, ed., Occupational Hearing Loss, Academic Press, 1971*)

threshold at 1.5 kHz for 50% of the population, for different noise immission levels. Noise immission level E_A is the cumulative energy to which the individual is exposed over time. In each case, the shape of the curve is the same and represents the underlying distribution of hearing loss for a specific percentage of the population from noise exposure, excluding the component due to age. The hearing threshold shift at any frequency and percentage of the population for a given noise immission level can be found by choosing the frequency and percentile of interest on the inset axes, and, using a transparent overlay, tracing curve A and moving it to the point chosen, maintaining it parallel to curve A. The hearing decrement for a given immission level on the abscissa can then be read on the ordinate. Immission level is calculated from Eq. (1), where L_A is the

$$E_A = L_A + 10 \log \left(\frac{T}{T_0} \right) \qquad (1)$$

average noise level to which the person has been exposed over T, the time period in years, and the reference time period T_0 is 1 year.

Implicit in this model is the assumption that the hazard to hearing is directly proportional to the acoustical energy to which the individual has been exposed. This hypothesis has been codified by most countries into regulations that limit the exposure of workers. As discussed below, in the United States an additional factor is applied to correct for the effects of intermittent exposure such that the calculated hazard is not proportional to energy but depends more heavily on the duration of the exposure than on the level of the exposure.

Descriptors of noise exposure. Protection of hearing requires knowledge of the level, duration, spectrum, and type of noise exposure.

Level. Noise amplitude is specified in decibels, so that the scale of sound pressure level is logarithmically compressed. Industrial noise exposures can be categorized as continuous or impulsive. Continuous noise characteristically exhibits small amplitude variations over long time periods. As such, it may exhibit statistical variability but small kurtosis. Impulsive noise may be characterized by high variability and kurtosis in excess of 4.0.

Duration. Duration for a continuous noise exposure is simply the time between onset of the sound or entry of the worker into the noise field and termination of the sound or exit of the worker from the field. Most commonly the measurement is made with a stop watch. Precise determination of onset and decay times is unnecessary because the duration is usually on the order of minutes or hours.

Spectrum. Regulatory limits to noise exposure are specified in terms of A-weighted sound levels. This frequency weighting function deemphasizes the low-frequency components of the sound, so that the response of the measuring device is effectively converted so as to mimic that of the middle-ear transmission pathway. In this way, the measured sound levels

more closely represent the energy transmitted to the inner ear, where this energy acts on the hair cells. For purposes of determining effective hearing protection, the most accurate method requires that the noise be measured in octave bands.

Characteristics of impulsive noise. Examples of two types of impulsive noise are shown in **Fig. 2**. The Friedlander profile (Fig. 2*a*) is typical of gunfire in a nonreverberant environment and of other explosive sources. Waveforms of this type are infrequently encountered except in firing ranges situated on open, flat terrain. The profile of Fig. 2*b* is typical of impulsive noise encountered in a wide range of industrial processes such as metalforming, materials handling involving dropped material, and impact fastening such as nailing and stapling.

Impulsive noise is characterized by the maximum (peak) level and the duration of the impulse. In the Friedlander wave, the A duration is measured between the first positive pressure excursion and the return to zero pressure. In the typical industrial impact, the B duration is characterized as the time between the first positive excursion and the time at which the envelope of the signal decays 20 dB.

Damage-risk criteria for impulsive noise exposures are not as well established as those for continuous noise, because there is uncertainty about the mechanism of damage and a lack of data for dose–response relationships for those exposures. One hazard rating that has met with some acceptance has been developed by the Committee for Hearing, Biomechanics and Bioacoustics (CHABA) of the National Academy of Sciences. Recognizing the differences in spectrum

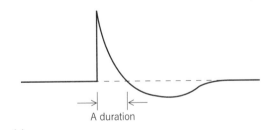

(a)

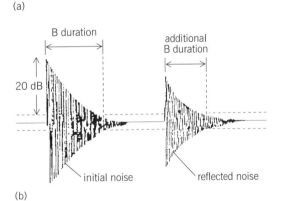

(b)

Fig. 2. Two types of impulsive noise waveforms.
(*a*) Friedlander profile, typical of explosive sources in a nonreverberant environment. (*b*) Waveform that is typical of industrial processes. (*After R. R. A. Coles and C. G. Rice, Hazards from impulse noise, Ann. Occup. Hyg., 10:381–388, 1967*)

between A and B impulses, the CHABA criteria trade impulse duration with level (**Fig. 3**). The peak level allowable is also reduced in proportion to the number of impulses in an 8-h day by 10 dB for every 10-fold increase in the number of impulses.

Time-weighted average level. If the purpose of measuring industrial noise is to assess hazard, then the principal tenet is that the average effective noise exposure must be measured over some time period. In countries that use the average energy L_{eq} as a predictor of hazard, this is straightforward. The equivalent continuous sound level is given by Eq. (2), where

$$\overline{L}_{eq} = 10 \log \left\{ \frac{1}{T} \int_{t_1}^{t_2} \left[\frac{p^2(t)}{p_0^2} \right] dt \right\} \qquad (2)$$

$T = t_2 - t_1$ is the time of the exposure, $p(t)$ is the instantaneous sound pressure in pascals, and $p_0 = 20$ micropascals, the reference pressure. In the United States, measurement of noise exposure is a more complex undertaking that requires that the dose first be calculated.

Under the rules adopted in the United States by the Occupational Safety and Health Administration, the daily noise dose is calculated from measurements as follows.

For each noise level to which the worker is exposed during the shift, the allowable exposure time T_n, in hours, is calculated from Eq. (3), where T_n is the

$$T_n = \frac{8}{2^{(L_n - 90)/5}} \qquad (3)$$

allowable exposure time at that level for 100% dose and L_n is the exposure level.

For a single exposure, the dose is related to the exposure time and the allowable exposure time by Eq. (4), where D_n is the dose and C_n is the actual exposure

$$D_n = 100 \left(\frac{C_n}{T_n} \right) \qquad (4)$$

time at that level. For composite daily exposures, the total dose is calculated from Eq. (5).

$$D = 100 \left(\frac{C_1}{T_1} + \frac{C_2}{T_2} + \cdots + \frac{C_n}{T_n} \right) \qquad (5)$$

The time-weighted average (TWA) level is calculated from the dose by using Eq. (6), where D is the

$$\text{TWA} = 16.61 \log \left(\frac{D}{100} \right) + 90 \qquad (6)$$

calculated composite dose in percent.

In countries that use an equal-energy trading function, in which a doubling of duration is the same as a doubling of energy, the equations for calculation of dose are Eq. (7), where the parameters are as defined

$$T_n = \frac{8}{2^{(L_n - 90)/3}} \qquad (7)$$

above, and Eqs. (4) and (5).

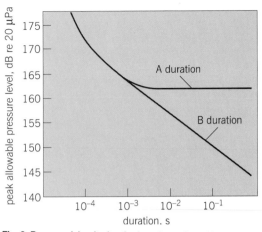

Fig. 3. Damage risk criterion for impulse noise. (*After Committee on Hearing, Biomechanics and Bioacoustics (CHABA), National Academy of Sciences, Hazardous Exposure to Intermittent and Steady-State Noise, 1965***)**

Equipment. Measurement of noise in the industrial setting is performed with either a sound-level meter or a noise dosimeter.

Sound-level meter. The sound-level meter measures the root-mean-square sound pressure level averaged over a short duration, either 125 milliseconds for a FAST response setting or 1 s for a SLOW response setting. An integrating sound-level meter is capable of calculating the average sound energy over a selectable time period. Sound-level meters are disadvantageous for industrial monitoring because they sample employee exposure in time and approximate the spatial (and therefore the actual level) distribution of exposure sound levels. Sound-level meters are too large to be worn by an employee for an adequate time to obtain an accurate exposure measurement. For industrial noise measurement other than for noise control, sound-level meters may be used principally for initial surveys and for area monitoring in which the acoustic environment is sampled, worker movements through the space are determined by time-and-motion studies, and the overall exposure is calculated by summation of dosage with the equations above.

Dosimeter. Noise dosimeters are alternative devices with which to measure noise exposure. Dosimeters are typically pocket-sized and are worn by the individual being monitored. A microphone typically mounted at the shoulder is attached to the device with a cable. This has the advantage that the actual environment in which the person is exposed throughout the shift is monitored from a point close to the ear.

Similar to the integrating sound-level meter, the dosimeter integrates the sound pressure incident on the microphone; however, it provides an optional integration constant that depends on the exchange rate used for the measurement. Operation of the dosimeter is described by Eq. (8), where T_c is the 8-h criterion

$$D = \left(\frac{100}{T_c} \right) \int_0^T 10^{(L - Lc)/q} \, dt \qquad (8)$$

exposure time; T is the exposure duration; L is the

A-weighted SLOW sound level; L_c is the criterion
sound level (in the United States, 90 dB for exposure compliance and 85 dB for hearing conservation
measurements); and q is the exchange rate parameter, which is equal to 10 for 3-dB exchange, as in
Eq. (7), and 5/(log 2) for 5-dB exchange, as in Eq.
(3). Thus, the dosimeter can produce a measurement based on an exchange rate of 3 dB or 5 dB per
doubling of time as required by the appropriate
regulatory agency. The output of this device may be
in terms of total dose or time-weighted average
sound level. This is the minimum information
required for determining compliance with various
governmentally mandated worker-protection programs.

Alternatively, a dosimeter may be equipped with
digital memory in which to store periodic averages
of the incident sound level. Such a time-history
dosimeter may be programmed to store average
sound levels in intervals ranging from several milliseconds to hours. Noise exposure time histories
are advantageous in situations where the worker is
exposed to different sources throughout the shift.
By correlating the noise levels with tasks, it is
possible to determine which sources contribute the
principal exposure and to target those for mitiga-
tion.

Assessment. The measurements of employee
noise exposure may be used to predict the hazard to
a population from the exposure. This procedure is
based on comparison of hearing thresholds in noise-
exposed and non-noise-exposed groups. The International Organization for Standardization has developed a procedure for estimating the hearing
impairment in a population, given the group exposure level, duration of the exposure, and age. This
standard is based on the equivalent continuous noise
level (3-dB trading rule) measured with an integrating sound-level meter or a dosimeter. Procedures
are specified for calculation of the hearing threshold
distribution of the noise-exposed population, which
is to be compared with data for a non-noise-exposed
population. By subtracting the threshold for the
nonexposed population from that of the exposed
population, the long-term risk of hearing loss can be
calculated.

For background information *SEE ACOUSTIC NOISE;
HEARING (HUMAN); LOUDNESS; NOISE MEASUREMENT* in
the McGraw-Hill Encyclopedia of Science & Technology.

John Erdreich

Bibliography. R. R. A. Coles and C. G. Rice,
Hazards from impulse noise, *Ann. Occup. Hyg.*,
10:381–388, 1967; Committee on Hearing, Biomechanics and Bioacoustics (CHABA), National Academy of Sciences, *Hazardous Exposure to Intermittent and Steady-State Noise*, 1965; International
Organization for Standardization, *Acoustics: Determination of Occupational Noise Exposure and
Estimation of Noise-Induced Hearing Impairment*, ISO1999, 1990; D. W. Robinson (ed.), *Occupational Hearing Loss*, 1971.

Nondestructive testing

Recent advances in nondestructive testing include the
development of computer vision inspection and
thermal-wave imaging.

COMPUTER VISION INSPECTION

Sophisticated manufacturing systems require automated inspection and test methods to guarantee quality. Although devices for monitoring quality do not
usually influence the manufacturing processes directly
as do feedback control systems, they are used to
determine the acceptance or rejection of parts and to
monitor the calibration of fixtures and the condition of
cutting tools. Without automatic quality-monitoring
systems, the effectiveness of an automated manufacturing system would be questionable. Bottlenecks
could appear in each inspection station because of the
relatively low speed of inspection and measurements
performed by humans as compared with modern
production rates. Other practical reasons for implementing automatic quality-monitoring devices include
freeing human inspectors from routine tasks, performing inspection in unfavorable environments, reducing
the costs of human labor, and analyzing statistics on
test information.

There are many automatic inspection and measurement systems currently available in industry, such as
automatic gaging and sorting systems, computer-
supported coordinate measurement machines, and
computer vision systems. Among them, computer
vision systems are the most versatile of the noncontact
inspection systems. Their applications are less restricted by the geometrical configuration of the parts to
be inspected. Many particular features of objects, such
as angles, radii, and nonlinear curves, cannot be
measured by contact systems because of difficulties in
making precise reference models. Computer vision
systems also avoid scratches and distortion on critical
dimensions, materials, or finishes that are too sensitive for mechanical contacts. Today, computer vision
systems can be applied in all manufacturing processes:
incoming receiving, forming, assembly, warehous-
ing, and shipping.

Basic concepts. While there are many variations in
the equipment and image-processing techniques being
used for computer vision inspection, the inspection
process takes basically the following steps. The part to
be inspected is first brought to the scanning station.
The part is then illuminated by a light source, and the
deflected light is focused and directed through a series
of lenses to form an image on the image sensor. Once
the image is in electrical form, the voltage level is
converted into a digital number representing the illumination. An image that is obtained electronically
consists of an array of picture elements (pixels). In a
simple black-and-white image, each pixel assumes a
value of 0 or 1. In more sophisticated systems, pixels
may assume codes corresponding to the gray levels or
colors. After a digital picture is obtained, a feature-
extraction technique isolates the features relative to the
inspection problem. The necessary features are then

measured, and a mathematical analysis is performed to classify the part as confirming or nonconfirming.

A diagram of a typical installation of a vision-aided inspection station is shown in **Fig. 1**. The system includes a conveyor, an in-view detector, a video camera, light sources, a computer, a vision control interface unit, and a reject mechanism. The camera may use a vidicon or a solid-state sensor to scan a scene.

Vidicon cameras record an entire image as a charge-density pattern on a photoconductive surface for subsequent scanning by an electron beam. The beam essentially reads the voltage necessary to bring the image pattern back to a refresh level. Vidicons were popularly used in early vision systems. However, they are fragile, being susceptible to damage from shock and vibration, and it is difficult to eliminate geometric distortions.

Solid-state cameras use an integrated circuit fabricated on a silicon chip with an array of extremely small and accurately spaced photodetector elements in which charge packets of electrons are generated by incident light. The chip also includes a complementary array of storage elements, on which the charges are accumulated and stored in a one-to-one correspondence with the sensors, and a scanning circuit for the sequential readout of the stored charges. Solid-state cameras are relatively stable and shock-resistant, and they are not subject to significant image distortion. Their applications in vision systems are increasing.

If the digital representation of an image is given by a discrete function, $f(i,j)$, $i = 1, \ldots, M$ and $j = 1, \ldots, N$, then the resolution of the digital image is given by the values of M and N. Currently, most computer vision systems process up to 512×512

pixels per array. In applications requiring greater definition, the object must be scanned with a linear array scan camera, in sections with multiple cameras, or through a step-and-repeat process. Processing speed is proportional to both number of pixels and object positioning time. Looking to the future of solid-state cameras, there is a need for higher resolution and precision and for improving quality of pixels (that is, higher and more uniform sensitivity) and color discrimination.

Most of the machine vision systems today analyze only the two-dimensional projected image of a scene. However, there is technology to obtain three-dimensional information of an object. One method is to use stereo or multiple cameras to obtain pseudo three-dimensional images. Another method, known as the multiple illumination technique, is based on the fact that objects produce shadows depending on their height and the direction of the illumination. By illuminating an object successively from different sides, the differences in successive images can be used to obtain the location and height of the object.

The speed of image processing and decision making is limited by the computer technology. Today, there are available commercially several parallel pipeline image-processing architectures where each processor can process certain tasks independently in a set of input data. The results of a parallel image-processing system are high speed and reliability.

Industrial applications. A variety of applications of automated inspection by computer vision have been developed.

Consumable goods inspection. The perception of consumable goods quality, such as for fruits, vegetables, or meat, is based largely on product appearance. Competition has motivated many major manufacturers to make zero defects a goal. Today, this goal can be achieved by using optical feature extraction technology. Commercial vision systems are available that work through a process known as Fourier feature extraction. This process analyzes an image and arrives at a compact representation known as the Fourier signature. Fourier signatures are then used to compare the overall appearances of objects in real time.

Dimensional inspection of complex parts. Metal stampings are subject to cracks, breaks, and tears in certain areas because of the stresses produced by the drawing operations. Many stampings are quite complex, making operator inspection time-consuming and complex. A three-dimensional vision system has been developed that is used for the inspection of engine castings. A laser-beam light source and a triangularization algorithm determine distance to the surface at any given point. A set of four fixed sensors scans portions of the surface until the entire surface is done.

Printed-circuit-board inspection. Printed circuit boards are composed of relatively few classes of similar components that can be modeled with generic prototypes at several levels of specificity. The inspection algorithm for each class of components can be implemented with a small, modular procedure. The two-dimensional structure of printed circuit boards implies

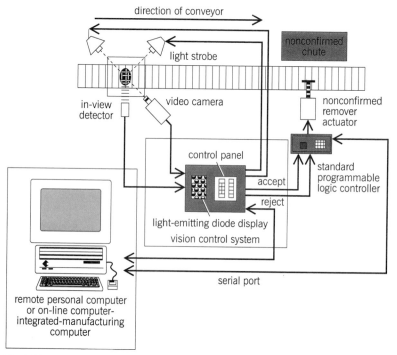

Fig. 1. Diagram of a typical installation of a vision-aided inspection station.

that the shape of components is unaffected by position and orientation, even though the components may appear in different locations. The boards are organized with their components on a rectilinear grid, and they need to be inspected only in two directions, x and y. Therefore, efficient one-dimensional signal-processing techniques can be used to analyze intensity profiles from scan lines.

Alphanumeric character inspection. Computer vision systems are being utilized to automate inspection and verification functions that previously required a human capability. Because of increased efficiency and 100% inspection, quality is improved and inventory control is more accurate. Industrial applications include hot-stamped characters on aircraft wires, embossed black-on-black serial numbers on sidewalls of rubber tires, and expiration dates and commodity codes on pharmaceutical labels.

Container and label inspection. Containers may be inspected for missing, crooked, or incorrect labels, bottle-fill height, and missing or damaged caps.

It is expected that within this decade 90% of all visual industrial inspection activities will be done with computer vision systems. Immediate, practical applications for industrial inspection devices in the United States are estimated in the hundred of thousands.

Jose A. Ventura

THERMAL-WAVE IMAGING

One of the promising technologies in the field of nondestructive testing is photothermal (thermal-wave) imaging. Complementary to conventional methods, the thermal-wave technique for nondestructive testing came to maturity in the late 1980s, and has only recently been introduced for commercial use, notably in instruments designed for testing integrated circuits and silicon and semiconductor devices.

Mode of action. The imaging is achieved by focusing an intensity-varying heat source onto a sample, and detecting the change in the thermal field as the focused source is scanned across the sample surface. The heating beam in most thermal-wave imaging systems consists of laser or particle beams, which are used in either a pulsed or sinusoidally modulated mode; the devices used for detecting the induced thermal field include microphones, ultrasonic transducers, infrared devices, and laser probes. These techniques depend on the fact that the modulated heat source creates a thermal wave that propagates through the sample, in the same way that an acoustic wave travels through a material. The interaction of the thermal wave with cracks and other defects in the sample leads to partial reflection of the thermal wavefront back to the surface, where it is then detected and analyzed to provide information about the feature (**Fig. 2**). Back-surface detection is also possible.

The primary difference between thermal-wave nondestructive testing and other methods is that by nature thermal waves are heavily damped, dying out at depths the order of a thermal wavelength, typically a few micrometers to a few millimeters. At first sight, this may seem to be a disadvantage. However, this characteristic means that the contrast of thermal-wave images is dominated by features that lie within a thermal wavelength of the source. Thus, by varying the thermal wavelength, the sample volume that contributes to the image can be varied. A second consequence is that thermal-wave imaging is well suited for the testing of defects at shallow depths; important applications include the characterization of thin films and coatings on substrates (**Fig. 3**). In contrast, alternative techniques of nondestructive testing have poorer resolution or contrast, or they are adversely affected by defects located near the surface. The variation of the thermal wavelength is, furthermore, easily accomplished by simply changing the frequency at which the laser intensity is modulated.

There do exist other techniques competitive with photothermal nondestructive testing in characterizing near-surface defects. Eddy-current techniques, for example, are also based on a critical wave phenomenon; however, these methods are restricted to flaw detection in conductive, metallic samples. Spectroscopic ellipsometry may also be used for depth profiling of layered sources; however, thermal-wave methods are better suited for optically opaque samples, as is usually the case.

Detection schemes. Thermal-wave imaging systems are based on many different detection schemes, each of which has its own advantages and disadvantages, depending on the nature of the source and the detector used.

Gas cell–microphone detection. Historically, this was the first photothermal detection method, reported by Alexander Graham Bell in the 1880s, and called the photoacoustic technique. In the modern version, the sample is enclosed in a gas-tight cell containing a microphone, and the exciting laser beam is admitted through an optical window. The periodic heating of the sample leads to a cyclic expansion and contraction

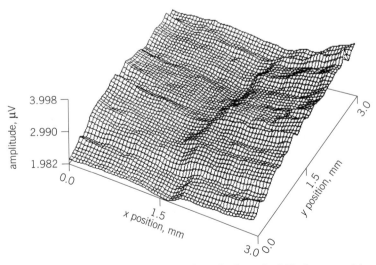

Fig. 2. Thermal-wave image of a microcrack in an aluminum sheet. The image was taken by using a pyroelectric film–pin combination to detect the thermal waves transmitted through the sample. 1 mm = 0.04 in. (*After M. Munidasa and A. Mandelis, Photothermal imaging and microscopy, in A. Mandelis, ed., Progress in Photothermal and Photoacoustic Science and Technology, vol. 1, Elsevier, 1991*)

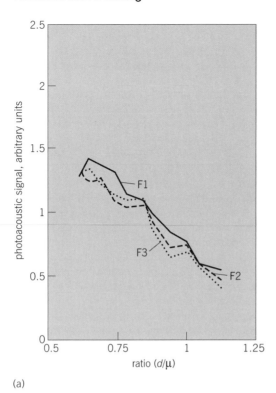

(a)

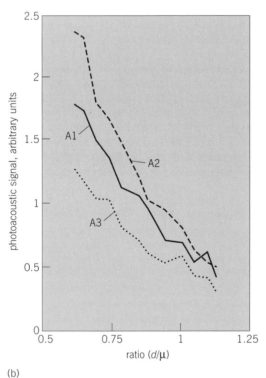

(b)

Fig. 3. Thermal-wave depth profiles for magnetic tape samples showing (a) a homogeneous distribution of metal oxide (sample F) and (b) an inhomogeneous distribution (sample A). F1–F3 and A1–A3 denote different locations on the samples. d = sample thickness; μ = a quantity proportional to the thermal wavelength; at $d/\mu = 1$, the thermal wavelength is approximately equal to the tape thickness. (*After S. B. Peralta and A. W. Williams, Characterization of γ-Fe_2O_3 homogeneity in magnetic recording media by photoacoustics, Appl. Phys. Lett., 54:2405–2407, 1989*)

of the gas in the cell, detectable as a periodic pressure gradient by the microphone. This detection scheme is easily analyzable because of the planar symmetry; this symmetry, unfortunately, precludes the detection of closed vertical cracks, an important defect class. A further disadvantage is that microphones have a limited frequency range—less than about 10 kHz—which limits the obtainable thermal wavelengths and thus the system resolution. A more serious disadvantage is that the sample must be physically small and enclosed in a cell.

Piezoelectric detection. In this technique the sample is attached to a piezoelectric material such as lead zirconium titanate. The exciting laser beam leads to the propagation of thermoelastic waves through the sample. This vibration is picked up by the piezoelectric detector and converted into a measurable voltage. Some difficulties with this scheme are the need for physically bonding the detector to the sample, as well as the presence of many possible modes of thermoelastic vibration in the solid, which complicates theoretical analyses.

Mirage effect. One technique that avoids the difficulties outlined above utilizes a method known as optical-beam deflection or the mirage effect. The heat induced by the pump beam is conducted to the thin gas layer adjacent to the sample surface. Consequently, the density of this layer of gas will be different from the ambient gas density, and the gas layer effectively acts as a lens. A weak probe beam skimming the sample surface, passing through the so-called thermal lens, is periodically deflected, with the deflection a function of the amplitude of the induced thermal field. The deflection can be quite large: in a methanol atmosphere, an optical path a few meters long leads to a deflection on the order of centimeters. In addition to being contactless, this technique can operate up to a bandwidth of a few hundred kilohertz.

Photothermal deflection. The heating beam also induces the sample surface to buckle and form submicroscopic bumps on the order of 0.01 nanometer in height. A weak laser-beam probe is focused on, and reflected from, the focus. As the bump forms, the probe beam undergoes a deflection proportional to the temperature field at that point. This technique is especially suited to the testing of concave surfaces, where the methods discussed above are not easy to apply. Photothermal deflection is complicated however by the fact that in general the deflection of the probe beam is also affected by the mirage effect. Furthermore, this detection technique is strongly dependent on the roughness and reflectivity of the surface.

Interferometric techniques. The material displacement (stress or strain) of the sample due to the heating beam may also be detected by optical interferometry. In this technique, the sample forms one mirror in a Michelson interferometer, or a similar device. The thermal expansion or contraction of the sample is then detected as a shift in the interference pattern formed in the instrument.

Photothermal radiometry. In this technique, the local induced thermal field is detected directly through the infrared emission of the heated point with a focused infrared detector. In contrast to the methods discussed above, radiometry does not depend on heat flow to the air, and thus an analysis is not complicated by the phase delays due to the presence of air. However, variations in the surface emissivity of the sample and the transmittance through the air can complicate the thermal-wave image.

Photothermal reflectance. The surface reflectance of a sample is also dependent on the induced temperature field. In the photothermal reflectance method, a probe beam is incident on the heated area and the reflected intensity is detected by a photodiode. A map of the reflectance change across the sample surface may then be processed to yield the thermal-wave image.

Limitations and advantages. In the techniques discussed above, the source of thermoelastic waves is the laser-beam focus on the sample surface. The resolution is therefore limited by the beam spot size and its closest approach to subsurface scatterers or defects. Although the intrinsic resolution may be better than the thermal or acoustic wavelength, this resolution deteriorates with increasing scatter depth. Consequently, only features less than 1 micrometer below the surface can be investigated, if submicrometer resolution is required. This restriction may be circumvented if a subsurface, localized source of energy is employed. In practice, this is achieved by replacing the laser beam with an electron or ion beam, whose penetration depth is dependent on the particle energy. Furthermore, since the energy is primarily deposited near the end of that depth, the energy source may be considered to be localized within a region determined by the so-called bloom of the beam. By adjusting the range to the depth of features of interest, better resolution is achieved than with use of a surface source. Another advantage is that since resolution depends on the proximity of the source, scatterers at other depths go out of focus, enhancing the contrast for the defects at the chosen depth.

Prospects. In general, the principles of thermal-wave nondestructive testing have been established theoretically and experimentally. For many detection schemes, it is possible to carry out calculations of images for different subsurface defects, and quantitative agreement with experiments has been obtained. These techniques produce high-resolution images, and they have been proven in the characterization of semiconductor devices and in other applications. It is notable that, compared with optical and acoustic imaging, thermal-wave imaging is a young technique, the first images having been obtained only a few years ago. It is anticipated that further development of instrumentation for thermal-wave imaging will establish it as an important method complementary to other techniques of nondestructive testing.

For background information *see* Computer vision; Infrared imaging devices; Nondestructive testing in the McGraw-Hill Encyclopedia of Science & Technology.

Samuel B. Peralta

Bibliography. A. Mandelis (ed.), *Photoacoustic and Thermal Wave Phenomena in Semiconductors*, 1987; S. B. Peralta, H. H. Al-Khafaji, and A. Williams, Thermal wave imaging using harmonic detection of the photoacoustic signal, *Nondestructive Testing and Evaluation*, 6:17–23, 1991; G. Schaffer, Machine vision: A sense for CIM, Spec. Rep. 767, *Amer. Machin.*, pp. 101–120, June 1984; J. A. Sell (ed.), *Photothermal Investigation of Solids and Fluids*, 1989; D. O. Thompson and D. E. Chimenti (eds.), *Review of Progress in Quantitative Nondestructive Evaluation*, vol. 8, 1989; S. Yeralan and J. A. Ventura, Computerized roundness inspection, *Int. J. Oper. Res.*, 26(12):1921–1935, 1988; N. Zuech and R. K. Miller, *Machine Vision*, 1987.

Nuclear reactor

Design phases for several advanced nuclear reactors are under way. Among these are the PIUS light-water reactor and the modular high-temperature gas-cooled reactor (MHTGR).

PIUS Reactor

PIUS (originally an acronym for process-inherent ultimately safe) is a passive and simplified, reconfigured 600-MW-electric pressurized-water reactor (PWR), based on well-established light-water reactor technology, designed with the aim of creating a supersafe reactor. In light-water reactors, a severe accident can occur only when the reactor core is overheated. The primary goal of light-water reactor safety, therefore, is to protect the core and its fuel. In PIUS, such protection is accomplished by submerging the core in a large, ever-present pool of water and ensuring that the cooling capability of this water is greater than the power of the core. Ever-present functions that are natural parts of the process, utilizing thermal-hydraulic characteristics, in combination with inherent and passive features, further enhance this approach. Core protection is thus ensured without reliance on detection or actuation equipment, without the functioning of any active components, and without operator action.

Passive shutdown principle. The operating principles of PIUS can be described with the aid of a simple model. A vertical pipe with openings at its top and bottom is placed in a large pool of water (**Fig. 1a**). A nuclear heat source inside the pipe, near its bottom, heats the surrounding water, making it less dense than the water outside the pipe. Thus, the heated water is buoyant and rises. A natural circulation through the pool results as water at the bottom of the pipe is replaced by the cooler pool water.

If a parallel leg with a pump is connected to the vertical pipe (Fig. 1b), then the pump can drive water through this outer leg to maintain a flow around the resulting loop, while stopping the flow to and from the pool. A definite boundary between hot and cold water is formed, known as a thermal barrier. In this configuration, the temperature of the water continually increases, resulting in decreasing density. If the thermal barrier is to remain intact, the pump speed must be continually adjusted to compensate.

The addition of a steam generator to the pumped segment of the loop is required to make it into a power-production loop. When the heat inputs and outputs are balanced, a stable circulating loop can be maintained, requiring only small variations in pump

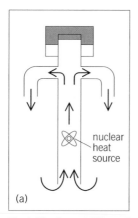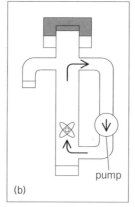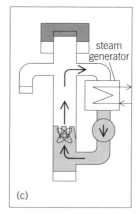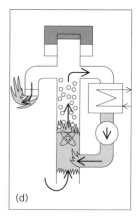

(a)   (b)   (c)   (d)

Fig. 1. Simple model illustrating operating principles of PIUS nuclear reactor. (a) Vertical, open pipe in large pool of water with heat source inside the pipe. (b) Configuration resulting from addition of a parallel leg with a pump. (c) Configuration resulting from addition of a steam generator to the pumped segment. (d) Result of conditions that could lead to overheating and damage to the reactor core. (ABB Nuclear Reactors Inc.)

speed. Thermal barriers (called density locks) are established at the openings at the top and bottom of the pipe.

In this configuration, any set of conditions that could result in overheating and damage to the reactor core (for example, reactor overpower, loss of a coolant pump, loss of coolant, or steam-generator undercooling) will also upset the hydraulic balance that keeps the loop and pool waters apart (Fig. 1*d*). Such conditions will cause the water in the pipe to become hotter and thus decrease in density. Pool water will then rush in through the lower density lock, reestablishing the natural circulation illustrated in Fig. 1*a*. The pool water contains boron, a poison to a nuclear chain reaction, and thus shuts down the reactor as well as cooling it.

Arrangement of components. The basic arrangement is shown in **Fig. 2**, and major plant data are listed in the **table**. A core of the type found in existing pressurized-water reactors is located near the bottom of a reactor pool. Reactivity is controlled by regulating the boron concentration and temperature of the coolant. Control rods are not used in PIUS for shutdown or for power shaping. Burnable absorber is used extensively, and boron concentration is kept low throughout the operating cycle. Hence, the temperature reactivity coefficient of the moderator is strongly negative under all operating conditions.

The reactor pool, a highly borated water mass, is enclosed in a prestressed concrete reactor vessel. The pool water is cooled by two systems: one with forced circulation and one with entirely passive function, utilizing naturally circulating water and heat dissipation in dry, natural-draft cooling towers. The passive system ensures pool cooling in case of an accident or station blackout and prevents boiling of the pool water. Even if all pool cooling fails, the in-place water inventory ensures core cooling for a protracted period of time (several days); no other light-water reactor can claim such safety.

Heated coolant from the core passes up through a riser pipe to the upper plenum, leaves the reactor vessel, and passes to the once-through steam genera-

tors, which are mounted on two sides of the concrete vessel. The main coolant pumps are located below and structurally integrated with the steam generators.

The cold-leg pipes enter the vessel at the same level as the hot-leg nozzles, and the return flow is directed downward to the reactor core via a downcomer. A siphon breaker arrangement in the upper part of the downcomer serves to prevent siphoning off of the reactor pool water inventory in a hypothetical cold-leg pipe rupture event. During normal operation, it does not affect the water circulation. At the bottom of the downcomer, the return flow enters the reactor core inlet plenum.

A 3-ft-diameter (1-m) pipe below the core inlet plenum provides a bottom opening to the enclosing reactor pool. A tube bundle arrangement is provided there to ensure stable layering of hot reactor-loop

PIUS nuclear reactor plant data	
Characteristic	Value
Core thermal power	2000 MW
Net electric power	640 MW
Cooling water temperature	15°C (60°F)
Number of fuel assemblies	213
Core height (active)	2.50 m (8.2 ft)
Core equivalent diameter	3.76 m (12.3 ft)
Average linear power density	11.9 kW/m (3.9 kW/ft)
Average power density	72 kW/liter (2.0 MW/ft^3)
Core inlet temperature	260°C (500°F)
Core outlet temperature (mixed mean)	290°C (554°F)
Operating pressure (pressurizer)	9 MPa (1280 lb/in.^{2})
Core mass flow	13,000 kg/s (28,630 lb/s)
Average burnup	45,500 megawatt-days per metric ton of uranium (41,300 megawatt-days per ton)
Equilibrium ingoing enrichment (12-month cycle)	3.5%
Concrete-vessel cavity diameter	12.2 m (39.5 ft)
Volume	3300 m^3 (870,000 gal)
Concrete-vessel total height	44 m (145 ft)
Concrete-vessel thickness	7–10 m (23–33 ft)
Number of steam generators and coolant pumps	4

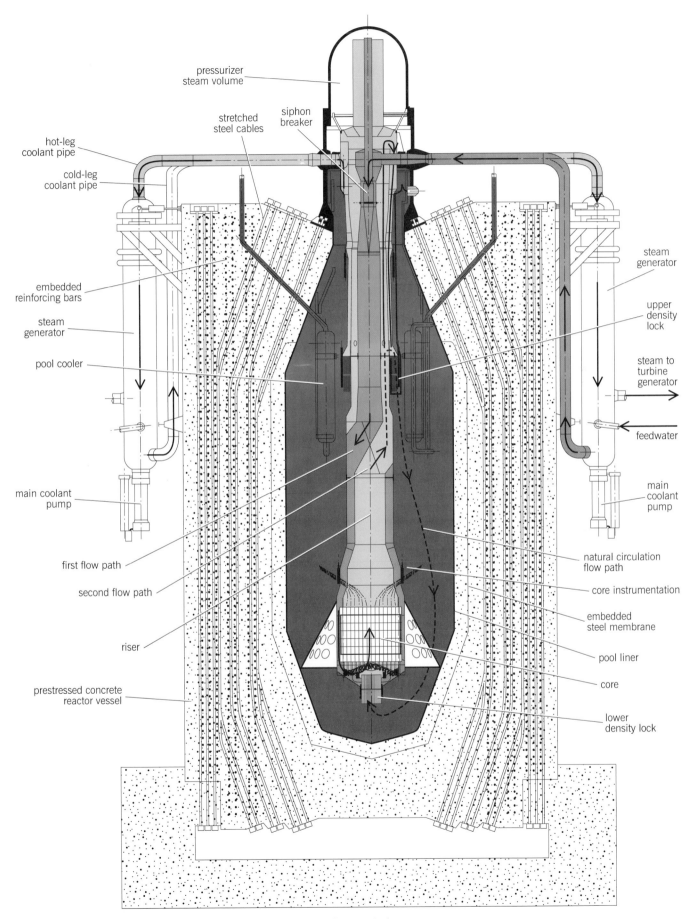

Fig. 2. Arrangement of components of PIUS nuclear reactor. (*ABB Nuclear Reactors Inc.*)

water on top of colder pool water. This pipe, with the bundle arrangement and stratified water, is called the lower density lock, one of the unique components of the PIUS design philosophy. The speed of the main coolant pumps is controlled by the position of the interface between hot and cold water, as determined by temperature measurements. As described above, there is also an upper density lock.

As in other light-water reactors, the nuclear steam supply system is enclosed in a containment structure, in turn partly enclosed in a reactor building. Pools for spent fuel and for reactor internals during refuelings are arranged on top of the containment inside the reactor building.

Compared with existing light-water reactor designs, a number of safety-grade systems have been deleted in PIUS, thereby allowing major simplification. Such deletions result in a plant that is simpler to operate and maintain than existing light-water reactor plants while concurrently being supersafe.

Safety performance. The primary emphasis during the development work has been to prevent core degradation accidents in an entirely passive way, while ensuring that plant economy and operability are not sacrificed.

Accident analyses performed by means of RIGEL, a computer code developed to analyze PIUS-type systems, and by other codes confirm that the safety goals are fulfilled; no accident sequence leading to core degradation has been identified. Intervention of active systems is not needed to ensure safety but only to maintain the reactor in operation and to prevent it from reverting to a state of shutdown and natural-circulation core cooling.

The design is based on established light-water reactor technology, and the existing regulatory framework can be utilized for licensing. The concrete vessel and the absence of control rods represent important changes from current technology, but they are, together with totally passive safety systems, key elements for the favorable safety performance. The absence of control rods is actually an advantage since mechanical devices and electrical interacting detector and insertion systems are eliminated, as well as the risk of serious reactivity excursions due to control-rod malfunctions.

The response to a hypothetical double-ended cold-leg pipe rupture illustrates the favorable safety performance. Primary-loop water flows out through both ends of the rupture, borated water enters from the pool, and the reactor shuts down.

The hot-leg outflow stops when the water level drops below the nozzle level. The pressure inside and outside the reactor is equal, and the large outflow from the reactor stops due to the siphon breaker. The core is cooled by naturally circulating pool water, and decay heat is absorbed in the pool. The reactor pool is cooled by the passive system, having four groups each with its cooling tower. The pool-water temperature will remain below the boiling point at atmospheric pressure, even if one group fails.

The accident therefore does not cause fuel damage,

and the reactor pool water will not start boiling. Hence, radioactive releases to the containment will be very small, and the increased iodine concentration due to spiking remains in the pool water. The whole-body dose at the plant fence has been calculated to about 1 millirem (10^{-5} sievert), several orders of magnitude less than for other light-water reactors. A significant reduction of emergency preparedness requirements is a design goal for PIUS.

Design verification. Essentially, PIUS is a pressurized-water reactor utilizing existing light-water reactor technology, with conservatively chosen major design parameters: lowered core power density, lowered linear heat rating, negative power coefficient throughout the operating cycle, and lower reactor pressure and temperature than existing pressurized-water reactors.

The self-protective thermal hydraulics, which ensure protection against core degradation accidents, have been successfully demonstrated in the ATLE test rig, a model of a complete PIUS-type system with an electrically heated, simulated fuel assembly. Several transients were performed to verify the safety behavior, and the very close correspondence between predictions by the RIGEL code and observed results proves the reliability of that code.

Remaining departures from current reactor technology, except the absence of control rods, have been verified through testing or have a sound basis in technology outside reactor technology. Adequate evidence of the feasibility and practical operability of the new features has been verified by tests and experiments. Further tests are planned for optimization and verification of detailed design and engineered performance.

Construction and operation. Cost estimates for turnkey plants, based on Swedish conditions, have yielded a cost advantage over a boiling-water reactor plant of similar size.

Plant construction procedures have been reviewed, based on experience, policy, and practice from construction of other light-water reactor plants, adjusted to the characteristics of PIUS. This resulted in a schedule of 36 months from first pouring of structural concrete to fuel loading, and 42 months to full-power operation, for one of a series of plants. Modularization and premanufacturing will be used extensively to shorten this site construction period.

Compared with other light-water reactor plants, the PIUS secondary-side steam has lower pressure and temperature, and turbine thermal efficiency is somewhat lower. The associated penalty in fuel-cycle costs is offset by reduced operation and maintenance costs.

Potential. The prime potential for PIUS is for medium-size electric power generation needs on a worldwide basis. The simplification in plant design and the anticipated simplicity in operation and maintenance of the plant, as well as the increased error tolerance, should make it particularly suitable also for deployment in industrially less advanced countries.

Another potential use for PIUS is as a cogeneration plant, where its safety features can make siting closer to densely populated areas possible, in contrast to present reactor designs. A special version of PIUS, designated SECURE, is designed for urban siting and heat supply to district heating systems, with a thermal power rating of 200–400 MW.

Tor Pedersen; Reimar F. Duerr

HIGH-TEMPERATURE GAS-COOLED REACTOR

Gas-cooled reactors have had a long and varied history, which dates back to the very early stages of nuclear energy development. Most of the early development centered on low-temperature systems using a graphite moderator, metal-clad fuel, and carbon dioxide coolant. Commercial deployment of such systems started in the mid-1950s, primarily in the United Kingdom and France, with the natural uranium-fueled MAGNOX stations. These were followed by the higher-temperature, low-enriched-uranium-fueled advanced gas-cooled reactor (AGR) stations, deployed solely in the United Kingdom, starting in the mid-1970s. Although these two pioneering programs have concluded (some of the early MAGNOX stations are being decommissioned, and the final advanced gas-cooled reactor stations at Heysham 2 and Torness are now complete and have been recently put into service), experience from the over 1000 reactor-years of operation comprises a very valuable database for the ongoing development and design programs on higher-temperature reactors.

From the beginning, it was recognized that higher gas temperatures would lead to greater benefits from gas cooling, in particular, the ability to attain modern fossil-fired steam conditions, thereby permitting more efficient electricity production. It was this goal, coupled with the vision that such higher gas temperatures might also lead to broader applications of nuclear energy, such as providing industrial process heat, that motivated the development of the high-temperature gas-cooled reactor (HTGR). The distinguishing characteristics of this reactor type are its core consisting of graphite moderator and ceramic fuel and its use of inert helium gas as the coolant. At present, development work in the United States is focused on the modular version of the HTGR, the MHTGR.

Design concept. The heart of the MHTGR concept is coated-particle fuel. Microspheres of uranium oxycarbide fissile fuel and thorium oxide fertile fuel are coated with multiple layers of refractory materials: pyrolytic carbon and silicon carbide. These layers enclose each fuel particle and the fission products generated during power operation, hence acting as a tiny thick-walled containment vessel. The fuel particles are pressed into fuel rod compacts that are contained in prismatic graphite fuel elements (**Fig. 3**). The safety characteristics of the MHTGR are governed by the inherent capability of the coated-particle fuel to withstand elevated temperatures without significant failure, and the capability of the design concept to passively limit the temperature rise in the fuel during transients associated with postulated severe events.

Core design. In the MHTGR core design, the graphite fuel elements are configured in an annular geometry with both inner and outer graphite reflectors. The annular core geometry was selected to enhance the surface-to-volume ratio of the core, allowing an output in the range of 350 MW of thermal power, while retaining the capability to dissipate decay heat by passive means.

Reactor module. Each reactor module (**Fig. 4**) consists of the annular reactor core contained within a steel vessel, connected via a concentric cross-duct vessel to a single, helically coiled, once-through steam generator within a vessel located to the side and below the elevation of the reactor vessel. A variable-speed, electric-motor-driven main circulator is located on top of the steam generator vessel. For decay heat removal during maintenance when the main heat transport system is unavailable, the plant concept incorporates a small shutdown helium-to-water heat exchanger and electric-motor-driven circulator located at the bottom of the reactor vessel. Control-rod drives and their associated control rods and reserve shutdown material hoppers are installed through penetrations on top of the reactor vessel. (The reserve shutdown material hoppers are cannisters containing pellets that can be dropped into the core to serve as a shutdown material if the control rods are unable to go down into the core.) Off-line refueling of the reactor is also accomplished through these same penetrations.

In normal operation, the helium coolant flows down through the coolant channels in the graphite fuel elements, is collected and mixed in an annular chamber below the core, and then transports the reactor heat through the center duct of the concentric cross-duct vessel to the steam generator. After flowing down through the helical bundle of the steam generator, the helium flows up in the annulus between the steam generator shroud and the vessel to the circulator. The compressed helium is then routed back to the reactor vessel via the outer annulus of the concentric cross-

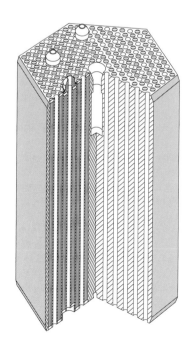

Fig. 3. Prismatic fuel element of the modular high-temperature gas-cooled reactor (MHTGR) being developed in the United States. (*General Atomics*)

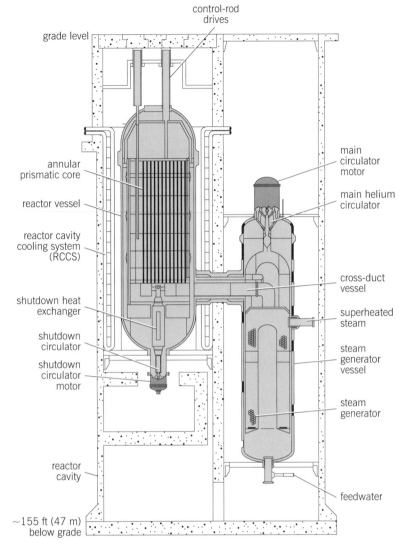

control-rod
drives

grade level

annular
prismatic core

reactor vessel

reactor cavity
cooling system
(RCCS)

shutdown heat
exchanger

shutdown
circulator

shutdown
circulator
motor

reactor
cavity

~155 ft (47 m)
below grade

main
circulator
motor

main helium
circulator

cross-duct
vessel

superheated
steam

steam
generator
vessel

steam
generator

feedwater

Fig. 4. Cross-sectional view of reactor module of modular high-temperature gas-cooled reactor. (*General Atomics*)

duct vessel and to the top of the core via flow channels in the annulus between the core barrel and the reactor vessel. Thus, cooler helium continuously bathes the walls of all three steel vessels. Feedwater enters at the bottom of the steam generator vessel, and superheated steam at a temperature of 1000°F (538°C) and pressure of 2500 lb/in.^{2} (17.2 megapascals) exits through a side-mounted nozzle on the vessel.

Plant design. The overall plant concept comprises multiple modules located side by side in below-grade-level enclosures in a slide-along configuration. The number of modules and the schedule for their deployment would be selected to match load growth or financing constraints. Each module transmits energy in the form of steam to the adjacent energy conversion area. With an energy conversion efficiency of approximately 38%, each module can provide a net output of approximately 135 MW of electrical power. The reference MHTGR plant configuration consists of four reactor modules connected to two turbine generators with a net output of about 540 MW of electrical power.

Passive safety concept. Three alternatives are available for the removal of decay heat. The first is the non-safety-related main heat-transport system, which would transfer the decay heat via the steam generator and a turbine bypass to the condenser. In the event that the main heat-transport system is not available due to planned maintenance or component failure, decay heat would be rejected through the shutdown cooling system. The shutdown cooling system is also a non-safety-related system that rejects decay heat via a closed cooling-water loop and individual air-blast heat exchangers.

In the event that neither of the above two active cooling systems is available, decay heat would be rejected through the third alternative, the reactor cavity cooling system (RCCS), a continuously operating, safety-related, passive heat-removal system. In this system, ambient air is directed via an intake–exhaust structure and concentric ducts to air-cooling panels located within the reactor enclosure. The air is heated within the panels by decay heat conducted and radiated from the reactor and is returned to the environment via the inner portion of the concentric ducts and the intake–exhaust structure. Passive decay-heat removal will occur without exceeding fuel design limits or incurring plant damage, even with loss of forced helium flow or depressurization of the primary system.

This capability, coupled with the always-negative temperature coefficient of reactivity of the reactor, assures a power shutdown to decay levels as the reactor temperature increases. Further, while the probability of coincidently losing all three heat-transport systems is diminishingly small, an evaluation of even that "beyond design basis event" was made. The results indicate that while some investment-related plant damage may occur, decay heat would be passively rejected to the ground, and the resulting fuel temperatures would remain within design limits and not be appreciably above those if the reactor cavity cooling system were available. Moreover, several other accidents well beyond those that would be considered as a basis for licensing a plant were evaluated, including total control-rod withdrawal with delayed scram, unrestricted air ingress, unrestricted water ingress, and the simultaneous failure of all heat-removal systems coincident with massive failure of the cross-duct vessel.

For all such events, radiological dose levels requiring protective action would not be exceeded even at the exclusion area boundary, using a site with a radius of 1400 ft (425 m). A technical basis is thus provided for the elimination of early notification (sirens and so forth) and emergency evacuation and sheltering drills as required elements of an off-site emergency plan. In addition, the above results are achieved with no reliance on ac-powered systems or operator actions.

Economic feasibility. Any consideration of returning to the deployment of smaller nuclear plants faces the issue of competitive economics. During the 1960s and early 1970s, competitive pressures led to the commercial offering of ever-larger nuclear plants in pursuit of the perceived economies of scale associated with such capital-intensive ventures. Experience has shown, however, that the commitment to such a massive capital investment over a long time frame entails inordinate risks. These risks have been realized in some cases with disastrous economic consequences and, even if not realized, remain as factors to be considered in any evaluation.

Achieving competitive economics with the MHTGR is heavily dependent on the simplicity provided by the passive safety concept. This concept eliminates the need for many expensive safety systems and enables the necessary safety and investment-protection features of the reactor modules to be both physically separated and functionally decoupled from the turbine plant. As a result, only the nuclear island, consisting of the reactor modules, steam generators, and all the necessary nuclear service systems, need be constructed to nuclear standards, whereas the turbine plant can be constructed to conventional standards.

In addition, modularity not only in the context of multiple, smaller-power-output reactor modules but also with respect to the modular design of components, systems, piping, instrumentation, controls, and so forth, throughout the plant permits a large fraction of the plant to be factory-fabricated and preassembled. Factory fabrication facilitates quality control and, from experience, is less expensive than field fabrication and construction. An additional major benefit of large-scale plant modularization and factory fabrication is the reduced risk of on-site construction delays and the accumulation of interest charges.

Economic analyses indicate that the reference four-module 540-MW-electric MHTGR plant will be competitive with modern coal-fired plants for many regions in the United States. However, cost-reduction and design-optimization efforts are under way to improve overall competitiveness.

Prospects. The focus of the MHTGR program in the United States is on the development of the basis for a commitment to a lead project. Such a project is targeted for operation around the year 2000, and would provide the overall demonstration of licensing, performance, and infrastructure required for commercialization. In the longer term, the MHTGR offers the possibility of direct-cycle gas-turbine applications and process-heat applications beyond the reach of other nuclear concepts.

For background information SEE NUCLEAR FUELS; NUCLEAR POWER; NUCLEAR REACTOR in the McGraw-Hill Encyclopedia of Science & Technology.

L. D. Mears

Bibliography. D. Babala, U. Bredolt, and J. Kemppainen, A study of the dynamics of SECURE reactors: Comparison of experiments and computation, *Nucl. Eng. Des.*, 122:387–399, 1990; Gas-cooled Reactor Associates, *A Utility/User Summary Assessment of the Modular High Temperature Gas-Cooled Reactor Conceptual Design*, GCRA 87-011, 1987; K. Hannerz et al., The PIUS pressurized water reactor: Aspects of plant operation and availability, *Nucl. Technol.*, 91(1): 81–88, July 1990; T. Pedersen, PIUS: Status and perspectives, *IAEA Bull.*, 31(3): 25–29; U.S. Department of Energy, *Conceptual Design Summary Report Modular HTGR Plant*, DOE-HTGR-87-092, 1987.

Operations research

The application of operations research has benefited in recent years from advances in complementary areas such as artificial intelligence. Artificial neural systems are generating a great deal of interest in this role. In addition to applications in speech processing and image recognition, artificial neural systems offer an alternative approach to decision problems involving pattern recognition and combinatorial optimization. Recent research has established neural networks as a significant force in complex pattern recognition and has devised improved algorithms. In particular, the powerful back-propagation model was introduced, and further study has yielded improved implementations. Also, combinations or hybrid neural networks have yielded powerful systems. Finally, the role of neural networks in difficult optimization problems has enlarged.

Artificial neural structure. **Figure 1** illustrates a typical artificial neural structure, consisting of nodes (circles) and links connecting them. Information

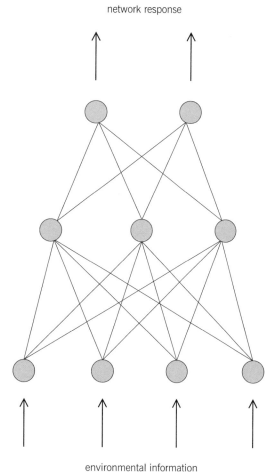

network response

environmental information

Fig. 1. Artificial neural structure; circles represent nodes.

from the outside, in the form of a pattern of numbers, causes the system to respond. First-layer or input nodes typically only receive and transmit the pattern through connections to nodes at the layer above. Connections have associated weights that magnify or reduce the signal passing through them. A unit, j, has an associated weight vector $\mathbf{w}_j$ that is composed of the weights on the connections from all nodes at the previous layer. Other processing units receive information, which is a weighted version of the transmissions of the previous-layer nodes, and compute a response (activation) via a specified activation function $f(\cdot)$. In general, for a processing unit j, activation y_j is computed from Eq. (1), where $\text{net}_j = \text{net}$

$$y_j = f(\text{net}_j) = f\left(\Sigma_i x_i w_{ij} - \Theta_j\right) \qquad (1)$$

input to node $j = \Sigma_i x_i w_{ij} - \Theta_j$, x_i is the ith element of the input pattern $\mathbf{x}$, w_{ij} is the ith element of the weight vector for unit j, and Θ_j is the bias of node j. The function f may be linear or nonlinear, and it usually implements a kind of thresholding (**Fig. 2**). The activation of each unit at a layer then fans out via the weighted connections to the next layer, thus serving as the input vector to a successive layer. At the topmost layer, the node activations supply the overall system output. The activations of nodes are computed from locally available information. That is, all computations performed at a node require only information carried by connections from adjacent-layer nodes. Such a structure allows all nodes at a layer to process information simultaneously or in parallel. Its characteristics coarsely model the biological neuron by obeying this locality constraint.

The network stores information as weights $\mathbf{w}$ on its connections. These parameters should ultimately enable the network to classify patterns submitted to it into appropriate classes. The network is not explicitly provided with these weights; rather, it learns them via exposure to examples from the classes during its training stage. The choice of training patterns affects the capability of the network to distinguish properly. *SEE NEURAL NETWORK.*

Supervised learning. In learning to store information that will allow it to determine class memberships of patterns, the network follows a specified algorithm for changing its weights. In a one-layer (of connections) model, the Widrow-Hoff delta rule learning

procedure is often used. In this case, output nodes have linear activation functions. Although not new, such learning serves as a foundation for more complex algorithms and as an important participant in hybrid, or combination, systems. The delta rule changes the weight vector associated with a particular output unit in the direction of the objective of the system: to minimize a function of the error between actual and desired output. Because the desired output is explicitly used to change weights, the learning is termed supervised. While such networks perform important functions, they do not have the ability to handle nonlinear problems.

The powerful back-propagation algorithm, which allows multiple-layer networks to learn nonlinear decision regions, generalizes the delta rule procedure by using nonlinear, differentiable activation functions for all nodes. Back propagation has already seen numerous applications, including financial forecasting and manufacturing control. However, early experimenters found that the algorithm could be slow to converge and could also perform suboptimally. Recent research has attempted to apply nonlinear optimization theory while obeying locality constraints and implementing techniques on neural structures. Direct results have included improved choices for learning rates, parameters that profoundly affect network speed and performance.

Unsupervised and hybrid learning. The function of unsupervised networks, which more closely emulate biological neural systems, lies in finding prototype vectors to represent meaningful categories. Essentially, patterns are grouped by similarity. The network cannot be trained to learn, nor does it require, desired class associations; hence the label unsupervised. However, the groups that result can facilitate classification on several different levels. In addition, the network computations are fast and simple.

Unsupervised learning networks are useful as stand-alone systems, but for technical applications their role in a hybrid system greatly enhances their function. In hybrid systems, a supervised stage is combined with an unsupervised stage to expedite learning. Hybrid systems exploit the fast processing of unsupervised learning, which results in a reduction in dimensionality of the data. The succeeding supervised stage operates on much simpler, and quite possibly linearly separable, patterns, so that it may employ a fast, supervised algorithm such as the delta rule.

Optimization neural networks. In a completely different role, neural networks have provided an approach to combinatorial optimization problems. In contrast to pattern-recognizing networks, these systems have fixed, rather than adaptable, weights. The system state comprises the individual node states, which are usually initialized at random values. These change in accordance with an update rule. For instance, the activation state of node i, v_i, is given by the threshold function (2), where $\text{net}_i \equiv$ input to node i,

$$v_i(\text{net}_i) = +1 \text{ if } \text{net}_i > 0, \ -1 \text{ if } \text{net}_i \leq 0 \qquad (2)$$

and the input to node i is again a weighted sum of the

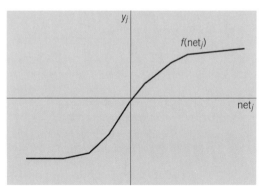

Fig. 2. Example of a nonlinear activation function.

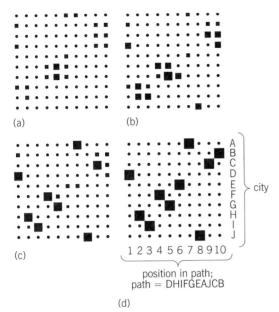

(a)   (b)

(c)

A
B
C
D
E
F
G
H
I
J
} city

1 2 3 4 5 6 7 8 9 10

position in path;
path = DHIFGEAJCB

(d)

Fig. 3. Evolution of a solution to a traveling salesman problem, the convergence of the 10-city analog circuit of a tour. The linear dimension of each black square is proportional to the activation state of the individual node. (*a–c*) Intermediate times. (*d*) Final state, showing how it is decoded into a tour (the solution). (*After J. Hopfield and D. Tank, "Neural" computation of decisions in optimization problems, Biol. Cyber., 52:141–152, 1985*)

activations of all nodes connected to i. An energy function must be derived whose minima (low energy points) correspond to good solutions to the problem. In addition, the network's weights must be chosen so that a stable state corresponds to a minimum point of the function. Operation proceeds with the network changing state by updating nodes sequentially or in parallel until a stable (unchanging) point is reached.

The system first noted for solution of combinatorial optimization problems was the Hopfield network. This network received a great deal of attention for solving the classical traveling salesman problem: find the shortest closed route through n cities (or points) such that each city is visited exactly once. The traveling salesman problem is a combinatorial optimization problem that is both difficult to solve to optimality and genuinely important in a large number of applications. The neural network for solving this problem consists of nodes that now do not require feedforward organization. Rather, the nodes correspond to cities visited in a particular order. For the traveling salesman problem, each node connects to every other. Weights on connections must be specified a priori, again to ensure that stable states of the network correspond to minima of the energy function. The network can then be initialized and the states of individual nodes updated in either a sequential (one at a time) or a parallel (many at a time) mode. **Figure 3** shows the evolution of a solution to a particular traveling salesman problem by using Hopfield networks. *See Optimization*.

Unfortunately, the above described approach has not led to success for reasonably sized problems (100

cities or more). Most limiting is the possibility that the system will not yield a feasible solution, a feat that is easily accomplished with standard heuristics. In addition, determination of the parameters of the energy function, and its structure, poses a difficult problem. Recent research has identified new energy functions, and relationships among terms, that promise to yield more consistently successful implementations. The energy function chosen is analogous to a penalty function in nonlinear optimization. Other researchers have successfully applied neural networks with learning to the traveling salesman problem. These models tend to implement geometric heuristics. In addition to the widely publicized application to the traveling salesman problem, neural networks have seen some success in solving other hard combinatorial optimization problems. In some cases, feasibility but not optimality can be guaranteed; in others, feasibility itself remains difficult to ensure.

Neural networks offer a promising alternative to expert systems, statistical methods, and optimization algorithms for appropriate problems. For pattern recognition problems, they provide a viable alternative to conventional procedures. However, if a large amount of historical data is unavailable or there exist simple rules relating data to classifications or diagnoses, it is prudent to devise a heuristic or expert system. The role of neural networks as optimizers of combinatorial problems also holds promise. Although experimental, neural network approaches to optimization have shown enough potential to warrant continued research.

For background information *see* Algorithm; Control systems; Cybernetics; Decision theory; Nonlinear programming; Operations research; Optimization in the McGraw-Hill Encyclopedia of Science & Technology.

Laura I. Burke

Bibliography. R. Lippmann, An introduction to computing with neural networks, *IEEE ASSP Mag.*, 4(2):4–21, 1987; D. Rumelhart and J. McClelland, *Parallel Distributed Processing I*, 1986; T. Wassermann, *Introduction to Neural Computing*, 1988.

Optical fibers

There has been a great increase in the demand for telecommunications services, largely brought about by such factors as the increasingly international nature of business and banking, the creation of far-flung computer networks, and the greatly increased use of facsimile services. Of all the transmission media, such as coaxial cable, microwave links, and satellites, only optical fibers have the bandwidth to meet that demand economically. (The low-loss transmission window of silica-glass fibers covers a wavelength span in the near-infrared extending from about 1500 to 1600 nanometers, or a bandwidth of about 12,000 GHz.) This article describes an emerging technology that enables use of a much greater fraction of the potentially vast transmission capacity implied by that bandwidth than is achieved at present.

Limitations of electronic regeneration. Optical fibers are naturally suited for essentially error-free digital transmission, with the presence or absence of short bursts or pulses of light representing the bits, that is, the 1's and 0's, respectively, of such transmission. In the presently used technology, periodically along the fiber path (typically once every 50 mi or 80 km), the light pulses, weakened by their trip through the fiber, are detected (converted into electrical pulses), amplified and reshaped electronically, and then converted back into light pulses for the next leg of the journey. Unfortunately, however, this process of electronic regeneration tends to limit the rate at which the information can be transmitted. In the first place, it mitigates against the use of multiple channels, or wavelength-division multiplexing (WDM), since at each point of regeneration the channels must first be optically separated, or demultiplexed, then each processed by its own regenerator, and finally all multi-

plexed back together again. Furthermore, the cost of each regenerator tends to increase very rapidly as the single-channel speed, or bit rate, is increased. Thus, at present, wavelength-division multiplexing is seldom used, and the maximum terrestrial bit rate is 1.7 gigabits/s, while the rate per fiber pair on the recently laid transatlantic undersea cable, TAT-8, is only about one-seventh that value. Such rates obviously represent but a tiny fraction of the potential capacity of optical fibers.

Use of optical amplifiers. The above limitations are largely removed, however, if the signal remains strictly optical over its entire path. In this all-optical approach to overcome the fiber loss, the regenerators must be replaced by optical amplifiers. Recently, a very practical form of amplifier has been developed for this purpose. It consists of several meters (1 m = 3.3 ft) of fiber, quite similar to the transmission fiber, whose core has been doped with the rare-earth element erbium. When optically excited (pumped) with light of shorter wavelength (bands centered about 1480 or 980 nm), such a fiber can provide the necessary optical power gain to compensate for the loss in many tens of miles of ordinary fiber, for signals whose wavelengths lie in the broad range from about 1530 to about 1560 nm. The erbium-fiber amplifiers have many technical advantages: they can be spliced directly into the transmission fiber with negligible loss or reflection; they produce no sensible distortion of the signal pulses; and they require only modest pump power, about 10–25 mW, which is easily supplied by an electrically pumped, semiconductor laser diode. (The pump light is efficiently coupled into the main fiber, without disturbance of the signals, through use of a wavelength-division coupler.) Erbium-fiber amplifiers are clearly very economical replacements for high-speed regenerators.

Dispersive effects. In the all-optical approach, however, it is also necessary to have a way to overcome the various dispersive effects of the transmission fiber. One such effect is chromatic dispersion, that is, the tendency for light of different wavelengths, or frequencies, to travel at different speeds down the fiber. Of necessity, a pulse contains a finite band of frequencies, whose width is inversely proportional to the pulse's temporal width (**Fig. 1a**). In the abovementioned low-loss window (centered about approximately 1550 nm) of silica-glass fibers, chromatic dispersion causes the higher-frequency components of a pulse to run ahead and the lower-frequency components to lag behind, thus causing the pulse gradually to broaden in time (Fig. 1b). Left unchecked, such broadening will eventually cause adjacent pulses (bits) to merge with one another, and the information will be lost.

The other major form of dispersion stems from the fact that in fibers the speed of light depends slightly on the light's intensity. The effect of such index nonlinearity is to cause the higher-intensity center of a pulse to lag behind its weaker wings. Although this form of dispersion does not sensibly alter the shape and width

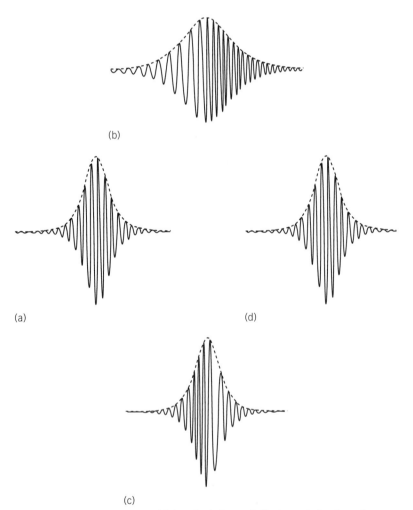

(b)

(a)   (d)

(c)

Fig. 1. Instantaneous waveforms of light pulse before and after transmission through an optical fiber. (a) Initial pulse. (b) The pulse after it has been subject to the effects of chromatic dispersion only. Separation of higher from lower frequencies has resulted in broadening of the pulse. (c) The pulse after it has been subject to the effects of index nonlinearity only. New frequencies have been added. (d) The pulse after it has been subject to both dispersion and nonlinearity to form a soliton, so that it is identical to the initial pulse. The soliton's hyperbolic-secant envelope shape is accurately represented here.

Fig. 2. Running on a mattress as a simple analog to the optical soliton: the moving valley pulls along slower runners and retards the faster ones. In like manner, the light pulse creates a moving valley of higher index material, thereby keeping the various frequency components of the pulse together.

of the pulse in time, it does tend to raise frequencies in the trailing half of the pulse and to lower them in the leading half (Fig. 1c). [The effect is a bit like that of pulling back on the center of a coiled spring.] The net result is that the spread of frequencies representing the pulse is increased.

Solitons. Now it can be shown from the equation governing the propagation of light in the fiber, and has been verified experimentally, that there exists a pulse of the right shape (the hyperbolic-secant function; Fig. 1d) and the right intensity such that the two dispersive effects just described exactly cancel one another. That pulse is known as the soliton. The soliton is a strictly nondispersive pulse; that is, neither its shape and width in time nor its frequency spectrum changes as it propagates down the fiber. For further understanding, the simple analog shown in **Fig. 2** is helpful.

Strictly speaking, the above description of the soliton applies only in a perfectly lossless fiber. But it has been shown both theoretically and experimentally that solitons can also be transmitted through a chain of amplifiers and ordinary fiber. The only requirement is that the scale length for significant pulse broadening by chromatic dispersion be at least several times longer than the spacing between amplifiers. Since that scale length can easily be made at least several hundreds of kilometers long, while the amplifier spacing would typically be no more than about 20 mi (33 km), the condition is easily met. Then, in order to have solitons, it is merely required that the intensity, as averaged over the path between amplifiers, be equal to the soliton intensity in a lossless fiber.

Dispersion-shifted fibers. The chromatic dispersion of silica-glass fibers passes through zero at a certain wavelength, known as λ_0. In principal, at least, so-called dispersion-shifted fibers can be manufactured with λ_0 at any desired place within the low-loss transmission window. This fact has led to the suggestion of transmitting pulses at λ_0 (where there is presumably no dispersive broadening) as an alternative to the use of solitons. Unfortunately, there are two practical difficulties with that scheme. First, manufac-

turing tolerances make it nearly impossible to maintain a perfectly constant value of λ_0 over long fiber spans. Second, with transmission at λ_0, on average there is no chromatic dispersion to cancel the nonlinearity, and the frequency spectrum of the pulses is broadened by a large factor. Computer simulations have shown that the combination of these two factors leads to significant pulse broadening and other distortions, which increase rapidly with increasing bit rate. No such limitation exists with solitons. Soliton transmission also works best with dispersion-shifted fiber, but with λ_0 being a few tens of nanometers shorter than the signal wavelength, for the achievement of a small but nonzero chromatic dispersion. It has been shown both theoretically and experimentally that soliton transmission is stable against modest fluctuations in the fiber's dispersion. Furthermore, the path-average value of the fiber's dispersion is not at all critical, as it is for transmission at λ_0.

Effects of noise. With solitons, then, the limits to transmission speed and distance are set only by the effects of noise. In the all-optical system, the only significant noise corresponds to photons, emitted at random by the amplifiers, known as spontaneous emission. The spontaneous emission is amplified along with the signal pulses, so it grows as the transmission distance increases. If the spontaneous emission becomes too strong (in relation to the signal pulse energy), it can cause errors in either of two ways. First, a noise burst can cause a 0 to be mistaken for a 1, or conversely, it can subtract sufficient energy from a 1 to make it look like a 0. Second, by way of the intensity-dependent speed of light, the spontaneous emission can modulate the velocities of the pulses, resulting in a random jitter in pulse arrival times; this jitter is known as the Gordon-Haus effect. If the jitter is large enough, some pulses will arrive displaced entirely out of their proper time slots, and again an error will result. Nevertheless, it has been predicted theoretically that, with the proper choice of fiber and pulse parameters, essentially error-free transmission (error rate less than 10^{-12}) should be possible over 5600 mi (9000 km, the undersea distance across the Pacific Ocean) at single-channel rates up to about 5 gigabits/s.

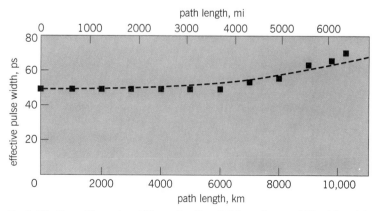

Fig. 3. Effective soliton pulse width as a function of distance traversed. The data points are experimentally measured values. The curve is the theoretically expected behavior, where the broadening corresponds to jitter in pulse arrival times only.

Experimental test. The above ideas have been given a partial experimental test in a closed loop consisting of three 15.6-mi (25-km) lengths of fiber interspersed with erbium-fiber amplifiers. A train of soliton pulses was made to circulate a number of times around the loop until the desired length was reached. The measured effective pulse width as a function of distance was in close accord with that predicted to result from the Gordon-Haus jitter (**Fig. 3**), implying that the individual pulses have suffered little or no detectable broadening, even over 6200 mi (10,000 km).

Wavelength-division multiplexing. As already indicated, one of the principal advantages of the all-optical approach is that it greatly facilitates wavelength-division multiplexing. In wavelength-division multiplexing with solitons, pulses in a shorter-wavelength channel have higher velocity than those in a longer-wavelength channel, so the former slowly overtake and pass through the latter. As a consequence of the fiber's nonlinearity, the solitons attract each other during such a collision. In a lossless fiber, however, the resultant acceleration and deceleration is perfectly symmetrical, so that there is no net change in velocity of either pulse following the collision. Furthermore, the solitons emerge from the collision with initial shapes and energies intact. Nevertheless, the variation in pulse energy obtaining in a chain of ordinary fiber and amplifiers threatens to upset the required symmetry of the collision. But it has been shown that as long as the collision length (the distance that the solitons travel down the fiber while passing through each other) is two or more times greater than the spacing between amplifiers, for all practical purposes the effects of the collision are just the same as in lossless fiber. This idea has also recently been confirmed experimentally.

The requirement on collision length sets an upper bound on the maximum allowable frequency or wavelength spacing between channels. Nevertheless, considerable wavelength-division multiplexing could be achieved. For example, with respect to the 5600-mi (9000-km) transmission cited above, it is at least theoretically possible to have as many as five wavelength-division multiplex channels of 4 or 5 gigabits/s each, for a total transmission rate of 20–25 gigabits/s on a single fiber. All this would be accomplished in a total wavelength spread no greater than about 1 nm. Furthermore, transmission in an all-optical system has the potential to be bidirectional (counterpropagating signals do not interact), as opposed to the unidirectional transmission in a system with regenerators. Thus, the total transmission capacity of a single fiber might well be at least 20 times greater than with conventional technology. Although much yet remains to be proven experimentally, it is already clear that the all-optical approach using solitons has tremendous technological and economic potential.

For background information SEE MULTIPLEXING; NONLINEAR OPTICS; OPTICAL COMMUNICATIONS; OPTICAL FIBERS; SOLITON in the McGraw-Hill Encyclopedia of Science & Technology.

Linn F. Mollenauer

Bibliography. J. P. Gordon and L. F. Mollenauer, Effects of fiber nonlinearities and amplifier spacing on ultra long distance transmission, *J. Lightwave Technol.*, 9:170–173, 1991; R. J. Mears et al., Low noise erbium-doped fibre amplifier operating at 1.54 μm, *Electr. Lett.*, 23:1026–1027, 1987; L. F. Mollenauer et al., Demonstration of soliton transmission at 2.4 Gbits/s over 12,000 km, *Electr. Lett.*, 27:178–179, January 17, 1991; L. F. Mollenauer et al., Experimental study of soliton transmission over more than 10,000 km in dispersion shifted fiber, *Opt. Lett.*, 15:1203–1205, 1990; L. F. Mollenauer, S. G. Evangelides, and H. A. Haus, Long distance soliton propagation using lumped amplifiers, *J. Lightwave Technol.*, 9:194–197, 1991.

Optimization

In the traveling salesman problem, a salesman starting at home base must visit customers in other cities and then return home. The objective is to minimize travel time. The traveling salesman problem is an optimization problem because there is an objective to be maximized or minimized; it also is a combinatorial problem because there is a finite number of alternatives. For example, if the starting city is Washington, D.C., and the customers are located at each of the 48 continental state capitals, then there are 48 factorial ($48! = 48 \times 47 \times \cdots \times 2$) alternative solutions corresponding to each of the permutations of the integers 1 to 48. The travel time of any one permutation can be found easily by summing the times between adjacent cities in the permutation. The difficulty is the number of permutations. The number 48! is so huge that even with a computer that can evaluate a billion permutations per second it would take billions of centuries to compute the travel times of all 48! permutations. However, with methods developed in the past few years, a 48-city problem can be solved in seconds on a microcomputer, and problems with as many as 2000 cities have been solved on mainframe computers.

Applications. A remarkably rich variety of problems can be represented by discrete optimization models. An important and widespread area of application concerns the management and efficient use of scarce resources to increase productivity. These applications include operational problems such as the distribution of goods, airline crew scheduling, and production scheduling. They also include planning problems such as capital budgeting, facility location, and portfolio analysis, and design problems such as communication and transportation network design, very large scale integrated (VLSI) circuit design, and automated production system design.

A version of the production scheduling problem is mathematically identical to the traveling salesman problem. Suppose a number of jobs have to be processed on a machine, and if job j is processed immediately after job i, there is a setup time of $c(i,j)$. Thus, the problem of minimizing the total setup time is a traveling salesman problem in which each job

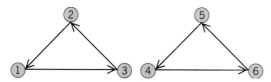

Fig. 1. Two subtours in a six-city problem.

corresponds to a city that must be visited by the salesman, $c(i,j)$ is the travel time between cities i and j, and there is an artificial job or city 0 corresponding to the home base with $c(0,j) = c(j,0)$ for all j.

The purpose of this analogy is to illustrate that the problems cited above, although they come from diverse areas, have similar mathematical structures and can be attacked by a common methodology. The traveling salesman problem has emerged as a generic combinatorial optimization problem, and many of the recent algorithmic advances were first developed and tested on it. The traveling salesman problem can be used to describe the methodology for solving these problems. SEE OPERATIONS RESEARCH.

Integer linear programming formulation. An important step in solving a combinatorial optimization problem is to formulate it as an integer linear program. The objective of such a program is to maximize or minimize a linear function of several variables subject to linear inequality and equality constraints, with the additional requirement that some or all of the variables be integers.

To formulate the traveling salesman problem as an integer linear program, the variables shown in Eq. (1) are introduced, with $x(i,j) = 1$ if city j is the imme-

$$x(i,j) = 0 \text{ or } 1 \qquad \text{for all } i \text{ and } j \qquad (1)$$

diate follower of city i and $x(i,j) = 0$ otherwise. The requirements that each city be entered and left exactly once are stated as Eqs. (2)–(3). The constraints of

$$\sum_j x(i,j) = 1 \qquad \text{for all } i \qquad (2)$$

$$\sum_i x(i,j) = 1 \qquad \text{for all } j \qquad (3)$$

Eqs. (2) and (3) are not enough to define a feasible solution (tour) for the traveling salesman, since these constraints are also satisfied by subtours. For example, in a six-city problem, Eq. (4) satisfies Eqs.

$$x(1,2) = x(2,3) = x(3,1) = x(4,5)$$
$$= x(5,6) = x(6,4) = 1, x(i,j) = 0 \qquad (4)$$

(1)–(3), but it does not correspond to a tour (see **Fig. 1**). Note that a tour must satisfy Eq. (5). In

$$x(1,4) + x(1,5) + x(1,6) + x(2,4) + x(2,5)$$
$$+ x(2,6) + x(3,4) + x(3,5) + x(3,6) \geq 1 \quad (5)$$

general, subtours can be eliminated by adding the constraints given in Eq. (6) for all proper subsets S of

$$\sum_{i \in S, j \notin S} x(i,j) \geq 1 \qquad (6)$$

the cities of size at least 2. The number of constraints of the type shown in Eq. (6) is huge, but they can be handled successfully by algorithms that generate constraints only if and when they are needed.

If $c(i,j)$ is the time, distance, or cost of going directly from city i to city j, the traveling salesman problem can be formulated as the integer linear program shown in expression (7), where $x(i,j)$ satisfies Eqs. (1)–(3) and (6).

$$\min \left\{ \sum_i \sum_j c(i,j)\, x(i,j) \right\} \qquad (7)$$

Heuristics. Heuristic or approximate algorithms are designed to find good, but not necessarily optimal, solutions quickly. Two basic ideas are applicable to a wide variety of combinatorial optimization problems.

The first is the idea of a so-called greedy heuristic in which there is an attempt to build up a solution by taking the best possible local step. For the traveling salesman problem, there are many greedy tour-building heuristics. Perhaps the simplest one is nearest neighbor, in which the tour starts at some arbitrary city and then the closest city not yet visited is selected as its immediate successor.

The second idea is that of local search. A heuristic of this type tries to improve a given solution by considering only limited changes. When no such changes yield an improvement, the heuristic stops. The simplest such heuristic for the traveling salesman problem deletes k links from a tour and forms another tour by replacing these links by k others, if such a change yields a tour of shorter distance. The case of $k = 2$ with $c(i,j) = c(j,i)$ for all $i \neq j$ is illustrated in **Fig. 2**. Since local-search heuristics stop when a locally optimal solution is found, they are typically applied many times with randomly chosen starting points.

A different, and usually more reliable, approach for

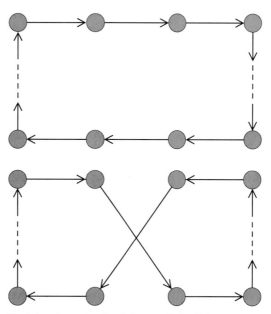

Fig. 2. Local search: a local change with two links.

trying to find a global optimum by using local search is to allow with positive probability local changes that make the solution worse. The probability of accepting an inferior solution decreases as a function of the increase in the objective function and the number of iterations. Random algorithms of this type are known as simulated annealing. Under suitable assumptions, they can be shown to converge to a global optimum. Furthermore, empirical evidence indicates that they perform well in practice.

There are some simple combinatorial optimization problems for which simple greedy or local-search algorithms are guaranteed to find optimal solutions or solutions that are within a prespecified error bound of being optimal. But for most combinatorial optimization problems that arise in practical applications, more elaborate approaches are needed to guarantee optimality.

Branch-and-bound. The conventional way to solve combinatorial optimization problems is by a controlled enumeration procedure known as branch-and-bound. In an integer linear program in which all of the variables are required to equal 0 or 1, all of the potential solutions could be enumerated on a tree as shown in **Fig. 3**. However, for an n variable problem, total enumeration would be out of the question since there are 2^n potential solutions.

To control the enumeration, the idea of a relaxed problem is introduced. A relaxation of combinatorial optimization problem (P), denoted by (RP), is a problem that has the following relation to (P): (1) every feasible solution to (P) is feasible to (RP); and (2) if (P) is a minimization problem, and x is a feasible solution to (P), then the objective value of x in (RP) is equal to or less than the objective function value of x in (P). From these two properties it follows that (a) if (RP) has no feasible solution, then (P) has no feasible solution; (b) the optimal value of the objective function in (RP) is equal to or less than the optimal value of the objective function in (P); (c) if an optimal solution to (RP) is feasible to (P), then that solution is optimal to (P).

The idea of branch-and-bound is to solve (RP) in place of (P). If either condition (a) or (c) is obtained,

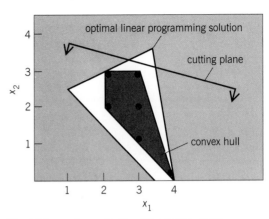

Fig. 4. Diagram for a set of feasible solutions in two variables for an integer linear program. The white region represents the points that satisfy the linear constraints. The integer points (circles) within the darkly shaded region are the feasible solutions.

(P) is solved. Otherwise the problem is branched by consideration of two subproblems, one with some variable $x(j) = 0$ and the other with $x(j) = 1$. The process is continued by solving the relaxation of each of the subproblems. The enumeration is controlled by not having to branch further from any subproblem whose relaxation satisfies either condition (a) or (c) or whose optimal objective value is equal to or greater than the value of the best feasible solution known to the original problem (P). This latter condition is referred to as pruning the tree by bounds.

For the bounding process to be effective and efficient, two things are required: heuristics for finding good feasible solutions to (P), and relaxations that both are easy to solve and provide tight bounds.

Problem-specific relaxations are available for the traveling salesman problem and other combinatorial optimization problems. However, the only algorithms that are robust, in the sense of using a relaxation that is applicable to essentially all combinatorial optimization problems, use linear programming relaxations.

Linear programming relaxations. **Figure 4** shows the set of feasible solutions to an integer programming problem in two variables. The white region represents the points that satisfy the linear constraints. The integer points (circles) within the darkly shaded region are the feasible solutions to an integer linear program. The problem of optimizing over the white region is called the linear programming relaxation of the integer linear program. It satisfies the relaxation conditions given above; that is, if the integer linear program is problem (P), then linear programming relaxation can serve as the relaxed problem (RP). Moreover, a linear programming relaxation can be solved efficiently. When the solution to a linear programming relaxation is fractional, that is, one or more variables required to be integral have fractional values, then one of these variables can be chosen for branching. Alternatively, the linear programming relaxation can be strengthened.

As shown in Fig. 4, linear programming relaxations

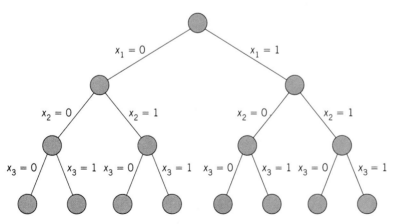

Fig. 3. An enumeration tree for a branch-and-bound procedure.

can be improved by adding linear constraints (cutting planes) that are satisfied by all of the feasible integer solutions but are not satisfied by a fractional optimal linear programming solution. The strongest such cutting planes, which are called facets, define the convex hull of feasible integer solutions, the darkly shaded region in Fig. 4. When all of the facets are available, then it is guaranteed that the integer linear program can be solved by solving its linear programming relaxation.

Recent research on many combinatorial optimization problems has provided methodology for iteratively improving linear programming relaxations by efficiently finding facets of the convex hull that are not satisfied by a fractional optimal solution to a linear programming relaxation. This approach, used in combination with fast heuristics and embedded in a branch-and-bound algorithm, has made it possible to find and verify optimal solutions to very large combinatorial optimization problems. In particular, it is now possible to solve to optimality traveling salesman problems with more than 1000 cities (or a million variables), while conventional branch-and-bound algorithms would be unlikely to cope with more than 100 cities.

For background information *SEE* ALGORITHM; COMBINATORIAL THEORY; LINEAR PROGRAMMING; OPERATIONS RESEARCH; OPTIMIZATION in the McGraw-Hill Encyclopedia of Science & Technology.

G. L. Nemhauser

Bibliography. G. L. Nemhauser, A. H. G. Rinnooy Kan, and M. J. Todd (eds.), *Optimization*, Handbooks in Operations Research and Management Science, vol. 1, 1989; G. L. Nemhauser and L. A. Wolsey, *Integer and Combinatorial Optimization*, 1988; R. G. Parker and R. L. Rardin, *Discrete Optimization*, 1988.

Organic chemical synthesis

Organometallic compounds have found increasing utility in organic synthesis in recent years. Not only does a substantial portion of the most useful new synthetic methodology employ organometallics, but also the synthesis of complex natural products and many important industrial processes rely heavily on them as intermediates. The emphasis in organometallic chemistry in recent years has been on the development of new, milder organometallic reagents that are more selective for reaction with the desired functional group in the presence of other functional groups (chemoselectivity) or more selective for the formation of one particular stereoisomer (enantioselectivity and diastereoselectivity). This latter area of investigation is particularly important, since most natural products and other biologically active substrates, including most pharmaceuticals, are chiral.

Synthesis of alkanes, alkenes, and alkynes. The cross-coupling of organic halides and organometallic reagents is one of the most important methods for the formation of carbon-carbon bonds [reaction (1), where

$$R'-X + R-M \longrightarrow R-R' + M-X \qquad (1)$$

X = halogen, M = metal, and R and R' = organic groups]. Virtually all types of organic groups can now be efficiently cross-coupled, including alkyl, aryl (C_6H_5), allylic ($H_2C=CHCH_2$), benzylic ($ArCH_2$), vinylic ($RHC=CH$), allenic ($RHC=C=CH$), alkynyl ($RC\equiv C$), and propargylic ($RC\equiv CCH_2$) groups. Since the highly reactive organometallics RLi (Li = lithium) and RMgX (Mg = magnesium) accommodate few important organic functional groups, they are rapidly being replaced by less reactive or more selective organometallics.

Copper (Cu) salts catalyze the cross-coupling of organic halides and RMgX or stabilized carbanions like $^-CH(CO_2R)_2$, or equivalent amounts of the copper-containing reagents RCu or R_2CuLi can be employed. With new routes to organocopper reagents bearing important functional groups like esters (CO_2R) or cyano (CN) groups, and the introduction of so-called higher-order organocopper reagents such as RCu(CN)M [M = Li or MgX], $R_2Cu(CN)Li_2$, and $(R'C\equiv CCuR)Li$, these reagents are finding significant utility for such cross-coupling processes.

By using nickel (Ni) or palladium (Pd) catalysts, organometallic compounds of Li, Mg, boron (B), aluminum (Al), silicon (Si), tin (Sn), zinc (Zn), and zirconium (Zr) can also be employed. The combination of an organoboron or an organotin compound (particularly those with alkyl, vinylic, and aryl groups) and aryl or vinylic halides in the presence of a palladium catalyst is particularly versatile. When the catalyst employed possesses a chiral ligand, the reaction forms one enantiomer predominantly, often in high optical purity. These reagents also allow the cross-coupling of organic tosylates ($R'O_3SC_6H_4CH_3$), triflates ($R'O_3SCF_3$), ethers ($R'OR$), sulfides ($R'SR$), and sulfones ($R'SO_2R$).

The inter- and intramolecular, palladium-catalyzed reactions of acyclic or cyclic alkenes and aryl, alkynyl, or vinylic halides or triflates continue to provide a useful route to substituted alkenes [reaction (2)].

$$R-X + H_2C=CHR' \xrightarrow{Pd\ catalyst} RHC=CHR' \qquad (2)$$

Organocopper and organopalladium reactions are among the most useful for double-bond transposition [reaction (3), where X = halogen, O_2CR, O_2CNHR,

$$R_2CuLi + H_2C=\overset{\overset{X}{|}}{C}HCHR' \longrightarrow$$
$$RCH_2CH=CHR' \qquad (3)$$

OR, SO_2R, or Me_3N^+, and reaction (4), where

$$(RO_2C)_2CHLi + H_2C=\overset{\overset{X}{|}}{C}HCHR' \xrightarrow{Pd\ catalyst}$$
$$(RO_2C)_2CHCH_2CH=CHR' \qquad (4)$$

X = halogen, O_2CR, NO_2, O_2COR, OAr, OH, or Me_3N^+; Me = CH_3, the methyl group]. New leaving groups (X) have been reported for these reactions, and the palladium process proceeds as well with nucleophiles like amines (HNR_2) that contain heteroatoms. Good asymmetric induction by chiral ligands is now possible in the palladium reaction.

Electrophilic, free-radical, or palladium-catalyzed coupling of allylic silanes and stannanes provides another useful route to alkenes [reaction (5), where

$$H_2C=CHCH_2MR'_3 \xrightarrow[\text{or } RX \text{ (plus Pd)}]{E^+ \text{ or } R\cdot} (R)ECH_2CH=CH_2 \quad (5)$$

M = Si or Sn and E = electrophile].

Chromium reagents readily convert carbonyl compounds to alkenes [reaction (6), where X = bromine

$$RCHO + HCX_2Y \xrightarrow{CrCl_2} RCH=CHY \quad (6)$$

(Br) or iodine (I) and Y = any halogen, $SiMe_3$, or SPh; Ph = C_6H_5, the phenyl group].

Acetylenes can be conveniently alkylated by copper and palladium intermediates [reactions (7) and (8),

$$RC\equiv CH \xrightarrow[\text{2. } E^+]{\text{1. } R'Cu \cdot MgX_2} \underset{R}{\overset{R'}{>}}C=C\underset{H}{\overset{E}{<}} \quad (7)$$

$$RC\equiv CH + R'X \xrightarrow[\substack{\text{catalyst: CuI} \\ \text{base}}]{\text{catalyst: } PdCl_2L_2} RC\equiv CR' \quad (8)$$

where R' = aryl or vinyl in reaction (8)]. The numbers at the arrows indicate the order in which the reagents are added.

Synthesis of alcohols. Alcohols can be obtained from epoxides by employing organometallic reagents of copper, such as R_2CuLi, $RCu(CN)Li$, or $R_2Cu(CN)Li_2$, in the presence or absence of boron trifluoride etherate [$BF_3 \cdot OEt_2$; reaction (9)].

$$R_2CuLi + H_2C\overset{O}{-}CHR' \xrightarrow{(BF_3 \cdot OEt_2)} RCH_2\overset{OH}{\underset{|}{C}}HR' \quad (9)$$

Versatile new methods for the addition of organic halides, esters, or alcohols to carbonyl compounds have recently been developed [reaction (10), where

$$RX + R'_2C=O \xrightarrow{M} \xrightarrow{H_2O} RC\overset{OH}{\underset{|}{R}}'_2 \quad (10)$$

R = allylic, CH_2CN, CH_2CO_2R, vinylic, or alkynyl; X = halogen, triflate, O_2CR, or OH; and M = Al, indium (In), Sn, lead (Pb), Ni, chromium (Cr), manganese (Mn), ruthenium (Ru), or samarium (Sm)].

Organometallics of Sn, titanium (Ti), Cu, Zn, Cr, and cerium (Ce), often accommodating important organic functional groups, have become important for the addition of organic groups R to aldehydes (RCHO) or ketones (R_2CO) to form alcohols. Organometallics of Ti and Mn will selectively add to aldehydes in the presence of ketones, or vice versa. The enantio- and diastereoselectivity of the addition of organozinc reagents (R_2Zn) in the presence of chiral amino alcohols and the addition of chiral allylic boranes ($H_2C=CHCH_2BR^*_2$, where R* is chiral) afford alcohols of very high optical purity.

Synthesis of carbonyl compounds. The reaction of organometallics and acid halides remains one of the most important routes to ketones [reaction (11)].

$$RM + ClCR' \overset{O}{\underset{\|}{}} \longrightarrow RCR' \overset{O}{\underset{\|}{}} \quad (11)$$

While organocopper reagents are quite effective here, recent work has focused on the use of organometallics of Al, Sn, Pb, vanadium (V), cadmium (Cd), mercury (Hg), Zn, Mn, and Ce, which will cleanly afford the carbonyl product while often tolerating other important organic functional groups. Palladium catalysis is required to effect the reactions involving Al, Sn, Pb, Hg, and Zn.

The direct coupling of organic halides and acid halides in the presence of Zn-Pd, Zn-Cu-Pd, or Ni provides a useful new route to ketones that seems likely to gain widespread acceptance.

Organocopper reagents continue to be the most important organometallics for conjugate addition to unsaturated aldehydes, ketones, esters, and amides [reaction (12), where X = H, R, OR, or NR_2].

$$RM + \overset{}{>}C=C\overset{}{-}\overset{O}{\underset{\|}{C}}-X \xrightarrow{E^+} R-\overset{}{\underset{|}{C}}-\overset{E}{\underset{|}{C}}-\overset{O}{\underset{\|}{C}}-X \quad (12)$$

Functionally substituted organocopper reagents can now be utilized and the addition of trimethylsilyl chloride (Me_3SiCl) promotes such processes. High asymmetric induction (enantioselectivity) can be achieved when chiral groups are appended to either the carbonyl substrate or the organocopper reagent, or when chiral ligands are employed. Organometallics of Si, Sn, Zn, Ni, and Pd have also found synthetic utility in such reactions.

Transition-metal-catalyzed carbonylation affords another important route to aldehydes [reaction (13a), where R = alkyl, aryl, or vinylic, X = halogen or triflate, $HSnBu_3$ = tri-n-butyltin hydride, L = ligand, Pt = platinum] or ketones [reaction (13b),

$$RX \xrightarrow[\substack{\text{catalyst:} \\ PdL_4 \text{ or } PtL_2}]{\substack{CO \\ H_2 \text{ or } HSnBu_3}} RCH \overset{O}{\underset{\|}{}} \quad (13a)$$

$$RX \xrightarrow[\text{Pd catalyst}]{\substack{CO \\ R'M}} RCR' \overset{O}{\underset{\|}{}} \quad (13b)$$

where R = alkyl, aryl, or benzylic; X = halogen, diazonium salt, or triflate; M = boron or tin; and R' = aryl, alkyl, allylic, or vinylic]. The industrially important hydroformylation reaction that combines alkenes ($RHC=CH_2$), carbon monoxide, and hydrogen to give either straight-chain (RCH_2CH_2CHO) or branched-chain ($RCH(CH_3)CHO$) aldehydes using a transition-metal catalyst remains an active area of investigation.

Carboxylic acids and derivatives can be prepared via transition-metal [cobalt (Co), rhodium (Rh), Ni, Pt, or Pd] catalyzed carbonylation of organic halides and triflates [reaction (14), where R = allenic, prop-

$$R-X + CO + HOR'(HNR'_2) \xrightarrow{\text{catalyst}} RCOR'(NR'_2) \overset{O}{\underset{\|}{}} \quad (14)$$

argylic, vinylic, aryl, or benzylic; X = halogen or triflate]. Running these reactions in the presence of two liquid phases (water and an immiscible organic solvent) and intramolecular versions of this process have recently received attention. Under appropriate conditions, α-keto derivatives $RCOCO_2R'$ (R = alkyl, aryl, benzylic; catalyst = Pd, Co) can also be obtained.

A number of important new routes to optically active carboxylic acids and derivatives have been developed recently by using chiral acyl iron (Fe) complexes or metal enolates [reaction (15), where

$$\text{RCHCX} \xrightarrow[\text{2. } E^+]{\text{1. base}} \text{RCHCX} \qquad (15)$$

M = Li and X = $CpFe(CO)(PPh_3)$ or M = B, Sn, Ti, Zr, and X = NR_2; E^+ = alkyl halide, epoxide, aldehyde, or ketone; $CpFe(CO)(PPh_3)$ = cyclopentadienyl iron carbonyl triphenylphosphine].

Synthesis of carbocycles and heterocycles. Numerous new syntheses of carbocycles promoted by transition metals have appeared in recent years [reactions (16–18); Ac = acetyl, Me = methyl], and several

$$\xrightarrow[\text{Pd catalyst}]{Me_3SiCH_2CCH_2OAc} \qquad (16)$$

$$H_2C{=}CH(CH_2)_3HC{=}CH_2 \xrightarrow[Cp_2Zr]{CO} \qquad (17)$$

$$RC{\equiv}C(CH_2)_nC{\equiv}CR \xrightarrow[\text{2. } H^+]{\text{1. Ti}} \qquad (18)$$

natural products have been prepared by intramolecular versions of the Pauson-Khand reaction (19).

$$HC{\equiv}C(CH_2)_3HC{=}CH_2 \xrightarrow[Co_2(CO)_8]{CO} \qquad (19)$$

Arene chromium complexes are versatile intermediates in organic synthesis because of their ability to promote anion addition to the aromatic ring; they also activate the arene toward both anion and cation formation and direct reactions to the opposite face of the ring [reactions (20) and (21)].

$$\xrightarrow[\text{2. } E^+ \ \text{3. [O]}]{\text{1. base}} \xleftarrow[]{\text{1. RM} \ \text{2. [O]}} \qquad (20)$$

$$\xrightarrow[\text{2. } E^+ \ \text{3. [O]}]{\text{1. base}} \xleftarrow[\text{(X = H)}]{\text{1. } H^+ \ \text{2. } Nuc^- \ \text{3. [O]}} \qquad (21)$$

Aryl-, alkoxy-, and amino-substituted chromium carbene complexes react with alkenes, allenes, acetylenes, imines, and isonitriles to form a variety of naphthalenes, indenes, cyclobutenones, β-lactams, and other cyclic derivatives [reaction (22)].

$$(CO)_5Cr{=}C{\Big\langle}^{OCH_3}_{C_6H_5} \xrightarrow[]{RC{\equiv}CH} \qquad (22)$$

Metal carbenes of Ti, tungsten (W), molybdenum (Mo), rhenium (Re), Ni, and Rh are involved in industrially important olefin and alkyne metathesis and polymerization processes.

Numerous useful cyclizations catalyzed by Rh that involve insertion of C-H, O-H and N-H have been reported in the last few years [reaction (23)], and

$$R(CH_2)_3CCCOR \xrightarrow[]{Rh_2(OAc)_4} \qquad (23)$$

applications of this unique methodology in the synthesis of natural products as diverse as prostaglandins and β-lactams have been developed.

Transition metals also efficiently promote skeleton rearrangements, olefin isomerization, and the oligomerization or polymerization of alkenes, allenes, dienes, and alkynes. The inter- or intramolecular cyclotrimerization of alkynes to arenes is catalyzed by Mo, W, niobium (Nb), and Rh. Molecules as complicated as steroids have been efficiently prepared by related intramolecular processes [reaction (24)]. In the

$$\xrightarrow[]{CpCo(CO)_2} \qquad (24)$$

presence of nitriles, cobalt and rhodium catalysts cyclotrimerize acetylenes to pyridines.

It is anticipated that organometallic chemistry will continue to play a leading role in the development of new organic synthetic methodology, to find increasing utility in the synthesis of natural products and pharmaceuticals, and to gain even greater importance in industrial processes.

For background information SEE CATALYSIS; ORGANIC CHEMICAL SYNTHESIS; ORGANOMETALLIC COMPOUND; REACTIVE INTERMEDIATES; STEREOCHEMISTRY in the McGraw-Hill Encyclopedia of Science & Technology.

Richard C. Larock

Bibliography. J. P. Collman et al., *Principles and Applications of Organotransition Metal Chemistry*, 1987; F. R. Hartley (ed.), *The Chemistry of the Metal-Carbon Bond (The Chemistry of Functional Groups)*, vol. 4: *The Use of Organometal-*

lic Compounds in Organic Synthesis, 1987; F. R. Hartley and S. Patai (eds.), *The Chemistry of the Metal-Carbon Bond*, vol. 2: *The Nature and Cleavage of Metal Carbon Bonds*, 1985; H. Werner and G. Erker (eds.), *Organometallics in Organic Synthesis 2*, 1989.

Organometallic compound

Organic transformations mediated by transition metals often proceed with high chemo-, regio-, and stereoselectivities. In addition, catalysts based on transition-metal compounds are used extensively in the petrochemical industry. The organic chemist's rather large repertoire of metal-assisted reactions is largely dominated by reagents containing middle and, especially, late transition metals (particularly those in the nickel group), perhaps in part because an understanding of the fundamental chemistry of these metals was achieved somewhat earlier than that of the early transition metals in the scandium through vanadium groups. In turn, the relatively slow exploitation of early organo-transition-metal chemistry may be partially attributed to the extreme reactivity of many of these complexes with air, water, and heteroatomic functionalities in organic substrates, and to the large number of facile intramolecular decomposition pathways available to them. Nevertheless, the 1970s saw the development of at least two important reagents based on early transition metals: Schwartz's reagent (see below) and Tebbe's reagent, a titanium-based methylene-transfer reagent developed by F. Tebbe that has proven to be a useful alternative to traditional phosphorane-based reagents.

As understanding of the organometallic chemistry of early transition metals has advanced in the intervening years, increasingly sophisticated methods for controlling their reactivity have emerged. During the late 1980s, several interesting systems based on elements from the scandium, titanium, and vanadium groups were developed that perform useful organic transformations in both stoichiometric and catalytic reactions. The following survey is confined to only those systems that exploit the reactivity of a metal-carbon bond.

Zirconium-mediated transformations. Chlorobis(cyclopentadienyl)hydridozirconium, referred to as Schwartz's reagent (I) after J. Schwartz who developed much of its chemistry, has been an important reagent since well before the 1980s for the hydrozirconation of olefins and acetylenes. Hydrozirconation may offer some advantages over the traditional, and in many cases closely analogous, hydroboration alternative. Insertion of a carbon-carbon multiple bond into the zirconium-hydride bond yields a zirconium alkyl or alkenyl species. Further treatment of these species leads to the regio- and stereospecific transformation of the original unsaturated hydrocarbon to halides, alcohols, aldehydes, acids, esters, sulfonates, and specifically deuterated hydrocarbons. This process is now widely used, and Schwartz's reagent is commercially available.

Recent years have seen a remarkable broadening of the scope of zirconium-induced organic transformations. Alkyne and benzyne complexes of bis(cyclopentadienyl)zirconium (zirconocene) serve as convenient synthetic equivalents of otherwise unstable or inaccessible reaction intermediates, and they often allow for their manipulation in highly specific ways. Zirconocene alkyne complexes (II) are readily available from Schwartz's reagent and an alkyne, as shown in reaction (1), where R_1, and R_2 are alkyl, aryl, or

$$\text{(1)}$$

silyl groups. The numbering of the reagents in reaction (1) indicates sequential treatment of the starting materials with the indicated reagents.

Treatment of the zirconocene alkyne complexes (II), or zirconacyclopropenes, with a second equivalent of alkyne leads to the zirconacycles (III) that may be hydrolyzed smoothly to 1,3-dienes, as in reactions (2), where H_3O^+ = hydronium ion and I_2 = iodine.

$$\text{(2)}$$

Regioselective placement of the substituents R_1–R_4 (where R_3 and R_4 are groups such as R_1 and R_2, but from the second alkyne in the sequence) is typically quite good and can be predicted. Conventional diene syntheses require isomerically pure vinyl halides as starting materials and thus may be less convenient. If the zirconacycles (III) are treated with iodine (I), 1,4-diiodo-1,3-dienes are obtained. These compounds are not readily accessible by other means, and they are potentially useful as coupling partners (reagents) in conventional vinyl halide–based olefin syntheses.

Treatment of the similar zirconocene benzyne complexes (IV) with nitriles (RCN) yields *meta*-acylated aromatic compounds [reaction (3)]. The most com-

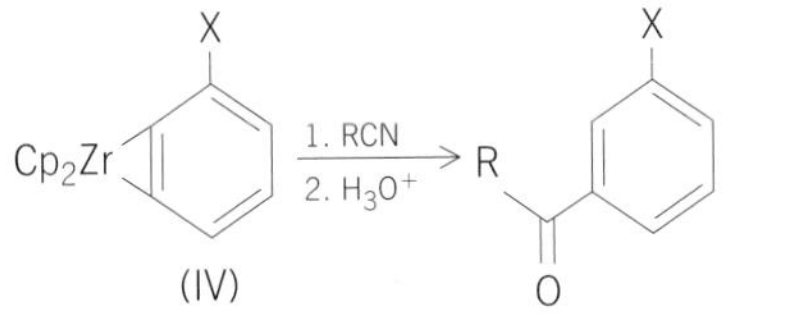

(3)

(IV)

mon organic method for the acylation of aromatic rings, the Friedel-Crafts reaction, typically yields ortho- and para-substituted products. The substituent X may be an alkyl or alkoxy group.

Zirconacycles (III) are also versatile precursors for the formation of a variety of heterocyclic compounds containing main-group atoms. The zirconium moiety may be replaced with an atom from groups III, IV, V, or VI upon treatment with an appropriate electrophile, for example, gallium chloride ($GaCl_3$), germanium chloride ($GeCl_4$), dichlorophenylphosphine [$(C_6H_5)PCl_2$], or disulfurdichloride (S_2Cl_2) [see **Fig. 1**].

Intramolecular couplings have also been performed with zirconocene reagents, by somewhat different methods. The bicyclic zirconium compounds (V) are obtained when zirconocene is generated in solution and treated with enynes (VI), compounds containing a terminal double and a terminal triple bond. Exocyclic hydrocarbons or halides may be obtained by treatment with acid or iodine as above; or the zirconium fragment (Cp_2Zr) may be extruded by treatment with carbon monoxide (CO), yielding cyclopentadieneones (VII), as shown in reaction (4).

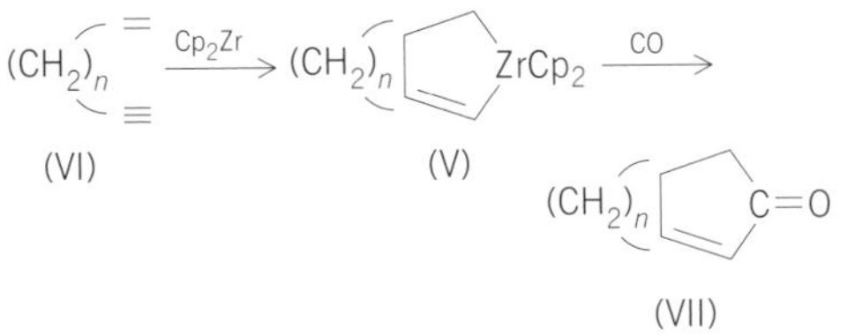

(4)

Insertion into zirconocene imine complexes, or zirconaaziridines, provides a synthetic route to amines and substituted pyrroles. The imine complexes are

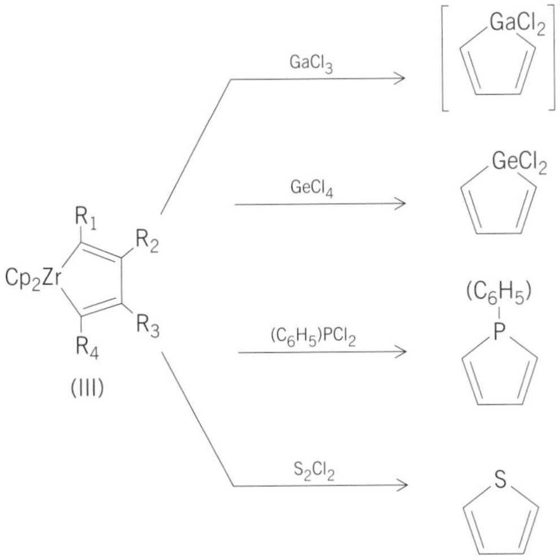

Fig. 1. Examples of main-group heterocycle preparation from zirconacyclic intermediates.

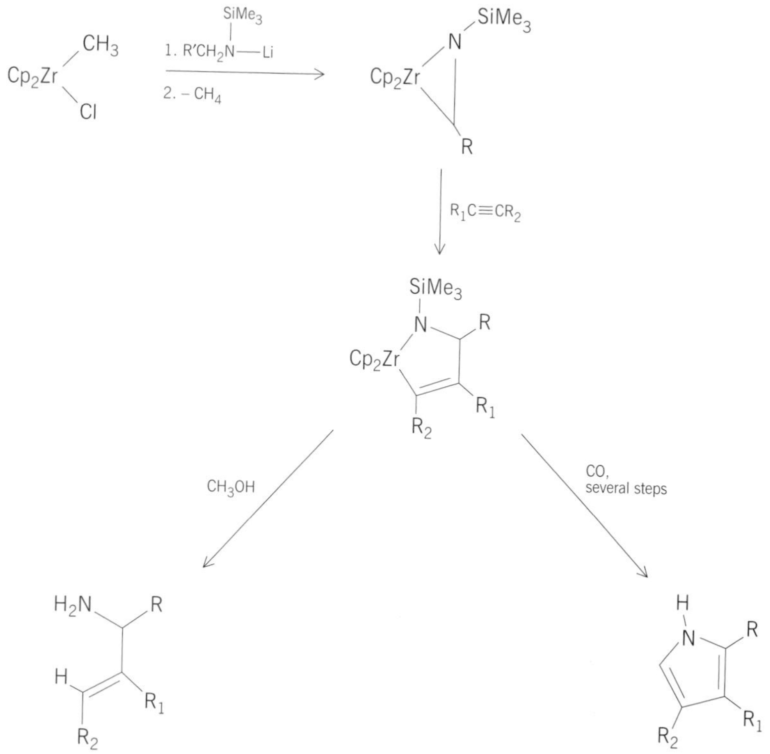

Fig. 2. Preparation of amines and pyrroles from zirconacyclic intermediates. Me = CH₃.

readily obtained from the lithium salt of the appropriate amine and a variant of Schwartz's reagent. As **Fig. 2** illustrates, alkynes ($R_1C\equiv CR_2$) [and alkenes] selectively insert into the zirconium-carbon bond rather than the zirconium-nitrogen bond. The intermediates are five-membered zirconacycles, which may be hydrolyzed or treated with CO to obtain open-chain or cyclic products. For reasons that remain unclear, treatment with CO leads to the deoxygenated pyrrole products. The pyrrole synthesis is particularly remarkable for the simplicity of the starting materials (an amine, an alkyne, and carbon monoxide) from which the pyrrole is assembled.

Niobium-mediated transformations. Niobium-based reagents have been developed over the last several years that conveniently effect a number of useful organic transformations. The niobium(III) chloride $NbCl_3$(DME), where DME represents a coordinated dimethoxyethane molecule, was first prepared in 1983 and has already become available commercially. Treatment of $NbCl_3$(DME) with alkynes in tetrahydrofuran (THF) solvent leads to the formation of niobium alkyne complexes (VIII). Several examples have been isolated and characterized. Niobium undergoes a formal two-electron oxidation in this process, and the alkyne complexes, like the zirconium complexes above, are best thought of as niobacyclopropenes. These intermediates have been shown to undergo a remarkably versatile condensation with 1,2-arylaldehydes, yielding upon workup substituted naphthols, as shown in reaction (5). Large substituents

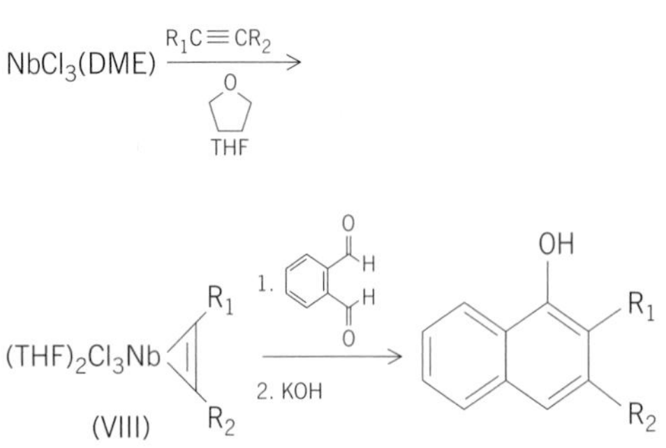

NbCl$_3$(DME) $\xrightarrow[\text{THF}]{R_1C\equiv CR_2}$

(5)

(THF)$_2$Cl$_3$Nb (VIII) $\xrightarrow[\text{2. KOH}]{\text{1.}}$

are specifically placed meta to the remaining hydroxyl group, that is, with one intervening open position between them on the ring.

As with the zirconium-based reagents, imines as well as alkynes may be used as coupling partners. Treatment of organic imines with NbCl$_3$(DME) presumably leads to formation of imine complexes [(IX) in reaction (6)], though these intermediates have not

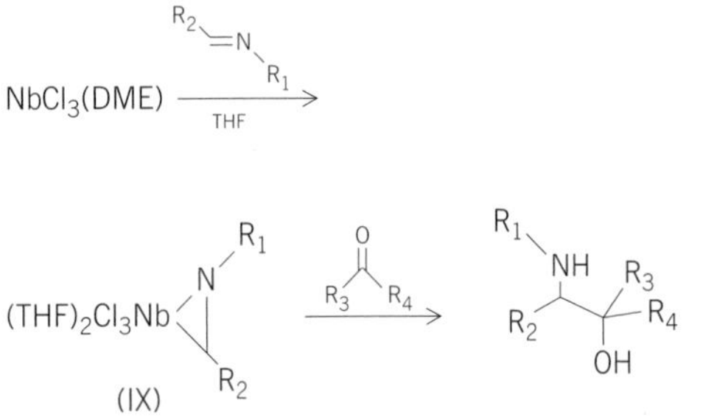

NbCl$_3$(DME) $\xrightarrow[\text{THF}]{}$

(6)

(THF)$_2$Cl$_3$Nb (IX) $\xrightarrow{}$

been isolated. Coupling of the imine complexes with organic carbonyl compounds such as aldehydes, ketones, or keto-esters leads to amino alcohols with the formation of two new chiral centers: the adjacent carbons bearing the nitrogen and oxygen atoms. Diastereomeric mixtures are obtained, with modest selectivity.

Scandium group and lanthanides. Although none of the reactions discussed so far have been catalytic, the relative ease of preparation or commercial availability of these titanium and vanadium group reagents makes them attractive for use by organic chemists. The organometallic compounds of the scandium group transition metals and lanthanides are generally much less readily available and are often more difficult to prepare and handle (that is, more sensitive to air and water). Nevertheless, some interesting catalytic processes for the dimerization, cyclization, and cyclic hydroamination of olefins have recently emerged from the chemistry of these highly electrophilic elements.

The chemistry of early transition metals with olefins has been extensively studied, largely in the context of efforts to understand the mechanism of the Ziegler-Natta polymerization of ethylene and α-olefins. The fundamental reaction steps are direct insertion of the unsaturated bond into a metal-hydride or metal-alkyl bond, and the reverse reaction, transfer of an alkyl group or, more commonly, hydrogen from the second carbon in the chain (the β-carbon) to the metal, releasing olefin. The scandium hydride com-

plex (X) shown in **Fig. 3a**, though structurally related to a family of known α-olefin polymerization catalysts, reacts with olefins to yield not polymers but dimers. The catalytic reaction proceeds at room temperature or above with a modest turnover rate of approximately 45 per hour. The products are formed by two sequential insertion steps, first into a scandium-hydrogen (Sc—H) bond and then into the newly formed Sc—C bond. β-Hydrogen transfer then occurs, liberating the product and regenerating the scandium hydride species. Insertion of a third olefin is strongly disfavored because of the steric congestion near the reactive scandium center. Figure 3a illustrates the stepwise process. This dimerization may be applied to the cyclization of α,ω-diolefins, as shown in Fig. 3b. Some modest functionality (thioethers, tertiary amines, silyl groups) is tolerated in the diolefin substrates.

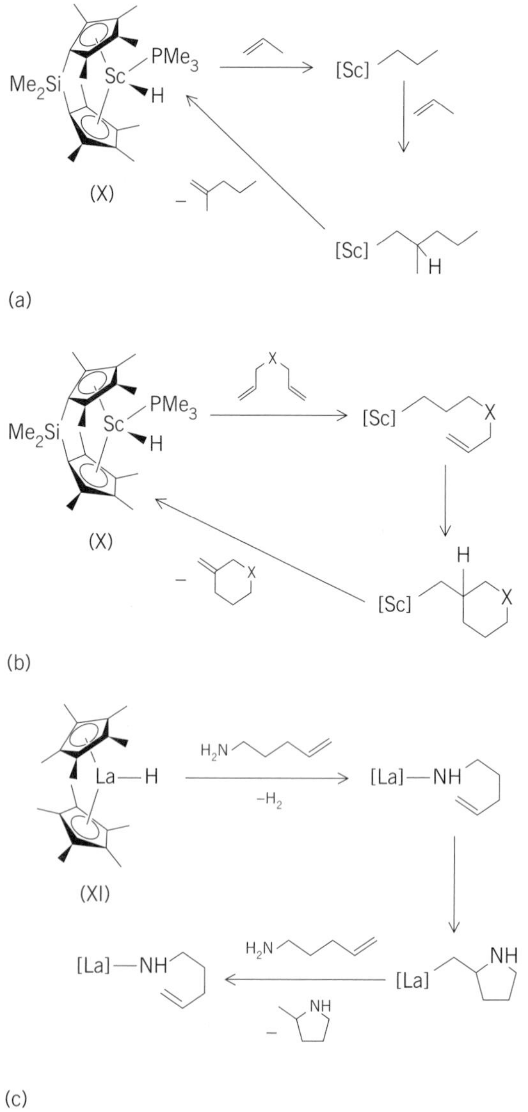

(a)

(b)

(c)

Fig. 3. Examples of catalytic dimerization and cyclization reactions involving scandium and lanthanum hydrides. (a) Dimerization by a stepwise process. Me = CH$_3$. (b) Cyclization of α,ω-diolefins. X = CH$_2$, S, SiMe$_2$, or NMe. (c) Catalytic cyclization of α,ω-amino olefins.

Bis(pentamethylcyclopentadienyl)lanthanum hydride (XI), shown in Fig. 3c, catalyzes the cyclization of α,ω-amino olefins. In the first step of this reaction, a nitrogen-hydrogen (N—H) and a lanthanum-hydrogen (La—H) bond react to form a lanthanum amide species and dihydrogen (H_2). Subsequent insertion of the pendant olefin moiety into the lanthanum-nitrogen bond follows, forming the new carbon-nitrogen bond and closing the heterocyclic ring. A second molecule of amine then reacts with the new lanthanum-carbon bond, liberating the product and closing the catalytic cycle. In the example shown, the carbon bearing the methyl group (CH_3) is a new chiral center. Bis(pentamethylcyclopentadienyl)lanthanum hydride is an achiral molecule that does not impart enantioselectivity to the reaction; racemic mixtures are obtained. Chiral analogs have been recently prepared wherein a methyl group in one of the pentamethylcyclopentadienyl ligands is replaced by a bulky chiral substituent. The reaction then proceeds with good enantioselectivity.

For background information SEE CATALYSIS; FRIEDEL-CRAFTS REACTION; ORGANOMETALLIC COMPOUND; STEREOSPECIFIC CATALYST; TRANSITION ELEMENTS in the McGraw-Hill Encyclopedia of Science & Technology.

W. Donald Cotter; John E. Bercaw

Bibliography. J. P. Collman et al., *Principles and Applications of Organotransition Metal Chemistry*, 1987; B. M. Trost and I. Fleming (eds.), *Comprehensive Organic Synthesis*, 1990.

Paleolimnology

Human activity has a measurable effect on natural ecosystems. For want of historical data, however, it is often difficult to document and understand the long-term impact of human disturbance of environments. If extended and continuous environmental histories can be recovered, analyses of those records can provide insights into the development and function of ecosystems and the ecological impact of human behavior. Paleolimnology is the study of the sedimentary deposits of lakes and of the processes by which these deposits are formed. Because lake sediments are made up of materials that originate largely from adjacent terrain, studies of lacustrine histories can implicate human activities in ecosystem change. Such studies can provide bases for reconstructing the histories of lakes, for documenting events of human occupation in lake basins, and for determining the impact of human populations on the structure and function of ecosystems.

Ecosystems and human disturbance. Ecology is the study of extant, living ecosystems. These natural environments are molded by a variety of processes that occur at different rates and over long periods of time. Understanding the mechanisms of ecosystem development and maintenance, therefore, requires a historical perspective. That perspective is provided by techniques of paleoecological analysis.

Large ecosystems, such as contiguous forest or desert, are usually long-lived, and changes in an entire system usually occur slowly. On a global scale, geological, meteorological, and topographical characteristics establish the growth potentials for the biological components of any ecosystem. In turn, broad processes of climatic change and paleomagnetic, volcanic, or tectonic events directly influence the long-term stability of biological components.

During the latter part of the Quaternary geological period, the character of natural ecosystems has also been influenced increasingly by *Homo sapiens*. As human populations have come to dominate the Earth's surface, their agriculture, urbanization, waste disposal, and other cultural activities have become as much a part of developmental processes of ecosystems as the physical decomposition of litter or nitrification by microbial communities.

Ecologists have only recently begun to study the long-term factors that govern ecosystem structure. Unfortunately, in most cases the historical information required for rigorous definition of past environmental conditions is lacking or inadequate. Therefore, such studies often must rely on comparative evaluations of contemporaneous systems of different ages, where the different ages are thought to be representative of different stages of ecosystem development, with research focusing on particular characteristics or processes that are assumed dynamic from one stage (age) to the next. Paleolimnological investigations provide environmental records that can be correlated with social histories to determine the points of articulation between human behavior and ecosystem structure and function.

Study of paleolimnology. The unit of study for paleolimnology is the lake and its associated drainage basin (the paralimnion). The latter is the discrete terrestrial territory that is hydrologically specific to a given lake. Beyond the physical limits of the drainage basin, water and transported materials flow to other catchments and lakes.

Lakes and their terrestrial surroundings often are considered as distinct ecosystems, but adjacent aquatic and terrestrial systems are linked by processes that transfer energy, materials, and nutrients from one component to the other (**Fig. 1**). Lakes serve as traps or catchments of accumulating sedimentary products of these processes, as they take place in and around their basins. Materials come into a lake through atmospheric fallout, geological weathering, and biological transfer. In turn, they can be lost from lake waters by volatilization, outflows, emerging organisms, or sequestering in the lake's sediments.

The physical, chemical, and biological characteristics of lacustrine sediments are determined by environmental conditions within both the lake and its drainage basin, but particularly the latter. Because lakes are the downhill recipients in the relationship, their physical, chemical, and biological processes are affected profoundly by the types and magnitudes of these transfers. In mature terrestrial ecosystems, loss to the aquatic sector is minimized and stabilized by the

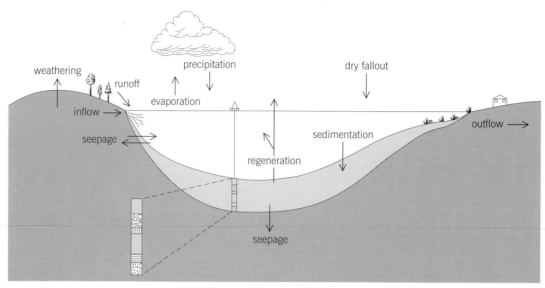

Fig. 1. Diagrammatic representation of a lake and its drainage, showing the major pathways by which water and other substances enter and leave the lake. Investigation of historical conditions in the lake and watershed is accomplished by stratigraphic analysis of accruing lake sediments. (*After M. Binford et al., Ecosystems, paleoecology and human disturbance in subtropical and tropical America, Quatern. Sci. Rev., 6:115–128, 1987*)

presence of standing vegetation. When the biological component of the terrestrial system is disturbed, however, transfers to the lake are altered and accelerated.

Applications of paleolimnology. Paleolimnological study can document past events in a lake's drainage because changes in the watershed are preserved as a record of alterations within the sediments on the lake bottom. The sediments accumulate in an ordered manner through time; therefore, they contain a continuous record of processes and events of the past. Because they are continuous, stable, and interpretable, lake sediments can provide historical data for studies of long-term processes in individual ecosystems.

The paleolimnological record, in conjunction with archeological data or historical information, or both, can be used to establish the impact of human activity on a basin. Particularly in closed lakes, which have no stream outflow, evidence of natural and human-induced changes in the larger ecosystem is preserved relatively undisturbed. Such evidence includes pollen, carbonized fragments of terrestrial vegetation, mineral and chemical inputs from terrestrial runoff and rainfall, and fossils of animals and plants that lived in lake waters.

Paleolimnology offers the possibility of defining predisruptive, disruptive, and postdisruptive conditions in ecosystems; therefore, paleolimnological studies have played a role in the resolution of numerous modern environmental problems, including the identification of the effects of road construction, industrial enrichment of heavy metals in sediments, acid deposition (acid rain), and the cultural eutrophication of lakes. One specific study that has focused on both prehistoric and modern issues of societal influence and environmental management involves investigation of the impact of ancient Maya occupation and subsistence activity on the tropical forest landscape of

the Yucatán peninsula during almost two millennia (about 1000 B.C.–A.D. 1000) of Maya history.

Central Petén historical ecology project. The pre-Hispanic Maya occupied the tropical forest region of modern Belize, Guatemala, Honduras, and Mexico from approximately 1000 B.C. to A.D. 1525. Their civilization flourished in much of the area during the Classic period of their history (A.D. 300–900).

Maya civilization. Over much of the region, Maya civilization fell into seemingly rapid decline in the eighth and ninth centuries, a collapse that entailed dissolution of native political and economic structures, cessation of elite arts, and dramatic population loss. The causes of the collapse have long been debated, but one set of hypotheses is of more than academic interest, given modern threats to the world's reserves of tropical forests. These hypotheses concern the long-term interactions between the Maya and their fragile lowland forest environment. Perhaps Maya architectural and agricultural activities so altered the landscape as to cause structural and functional changes in their ecosystem, ultimately undermining its productivity and support capability for human populations.

Initiation and execution of project. In 1972 a historical ecology project was initiated to investigate the effect of Maya occupation on the tropical forest ecosystem. The study focused on a chain of lakes that formed at the close of the Pleistocene in the modern Department of Petén, Guatemala, where previous research had indicated a long, continuous history of Maya occupation. Six lakes having variable sizes, locations, topographies, and settlement histories were chosen for study (**Fig. 2**).

Archeological surveys and excavations were carried out within sampling transects in each lake basin to determine the size and density of Maya populations through time. Biological and limnological studies of the modern, largely uninhabited terrestrial and aquatic systems were undertaken to provide a baseline of data from which to understand past states of ecosystem history. Lacustrine sediments were then sampled for study by drilling into the floor of the lakes to remove columns or cores of sequentially deposited settlement layers. Study of the constituent pollen, chemistries, aquatic fossils, and organic materials suitable for radiocarbon dating provided quantitative data on changes in the sequence of biological, chemical, and sedimentary inputs into the lakes.

Results of project. While there was variation in the natural and social histories of the lake basins, the regional picture was one of dramatic landscape modification and disturbance keyed to relative amounts of Maya activity in the basins (**Fig. 3**). The size of Maya populations can be shown to have increased exponen-

tially through time at a rate of approximately 0.17% per year, for a doubling time of 408 years, from 25 persons per square kilometer by the end of the Middle Preclassic period (about 300 B.C.) to as many as 250 persons per square kilometer in the Classic period. The ninth century witnessed rapid demographic decline to below Middle Preclassic levels, low densities that were relatively unchanging until Maya abandonment of all basins after the Spanish conquest (A.D. 1697 in this region of Guatemala).

The tropical forest, stable since the early Holocene, was progressively removed by expanding Maya populations. This deforestation would have diminished natural habitats and affected terrestrial resources. As landscape was put into settlement and subsistence production for long periods of time, protracted manipulation and exposure contributed to structural degen-

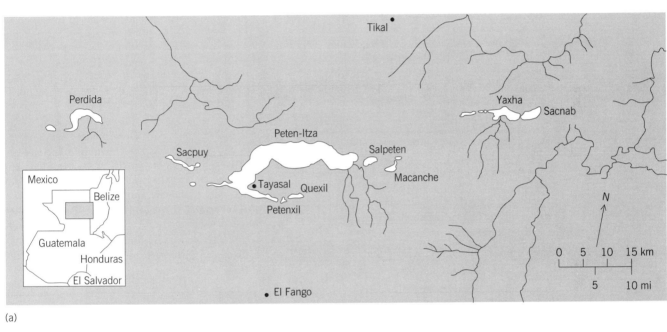

(a)

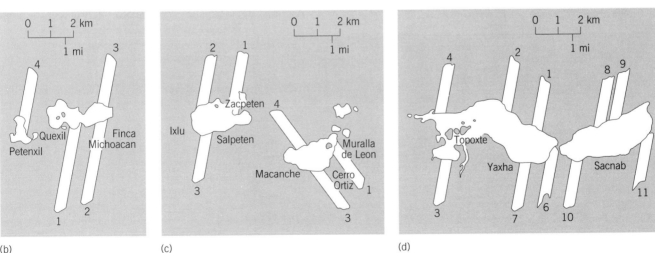

(b)   (c)   (d)

Fig. 2. Map of the Central Petén lakes region, Guatemala, showing the locations of the six lakes studied by the historical ecology project, and the location of the archeological survey transects within each of the six lake basins. (a) General view, with larger region shown in inset. (b) Petenxil and Quexil. (c) Salpetén and Macanche. (d) Yaxha and Sacnab. (After D. S. Rice, The Petén Postclassic: A settlement perspective, in J. A. Sabloff and W. V. Andrews, eds., Late Lowland Maya Civilization: Classic to Postclassic, University of New Mexico Press, 1986)

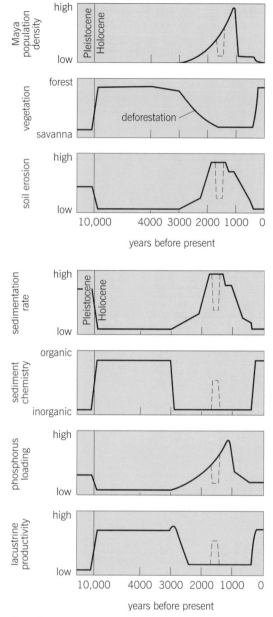

Fig. 3. Summary diagram showing the impact of long-term Maya settlement on the terrestrial and aquatic systems of the Central Petén lakes, Guatemala. *(After M. Binford et al., Ecosystems, paleoecology and human disturbance in subtropical and tropical America, Quatern. Sci. Rev., 6:115–128, 1987)*

eration of the soil and altered local water flow, the products of which ultimately found their way to the lake sediments. When organic surface soils were depleted, sediments became increasingly inorganic and became dominated by siliceous silts and clays.

Increased population densities not only exaggerated sedimentation rates and the deposition of clay-rich muds but also accelerated phosphorus loading in the lake. The movement of phosphorus at levels above normal minimums results primarily from human deflection of the element to surface soils through deforestation, sewage and waste disposal, and human interments. Phosphorus inputs to lake sediments ap-

peared to proceed at a rate of 0.5 kg of phosphorus per person per year, which is roughly equal to the annual physiological output for humans. Most of the phosphorus reaching the lake was probably locked in insoluble compounds by limestone-derived soils and removed from the terrestrial environment through erosion, then buried in lake sediments. There it was made permanently unavailable in any form to support forest growth, crops, or human populations.

In sum, archeological and paleolimnological analyses demonstrated that the Maya had to cope with declining access to natural resources, progressive deterioration of accessible soils, destabilization of water flows and storages, and diminution of essential nutrients within the ecosystem. These stresses undoubtedly contributed to the Maya collapse. Environmental degradation and the very slow recovery of the natural ecosystemic structure after human abandonment of the region provide a sobering model for future economic development of the world's tropical forests.

For background information *SEE ECOLOGY; ECOLOGY, APPLIED; ECOSYSTEM; LIMNOLOGY* in the McGraw-Hill Encyclopedia of Science & Technology.

Don S. Rice

Bibliography. M. Binford, E. S. Deevey, and T. L. Crisman, Paleolimnology: An historical perspective on lacustrine ecosystems, *Annu. Rev. Ecol. Syst.*, 14:255–286, 1983; E. S. Deevey et al., Maya urbanism: Impact on tropical karst environment, *Science*, 206:298–306, 1979; D. S. Rice, P. M. Rice, and E. S. Deevey, Paradise lost: Classic Maya impact on a lacustrine environment, in M. Pohl (ed.), *Prehistoric Lowland Maya Environment and Subsistence*, Pap. Peabody Mus. Archaeol. Ethnol., vol. 77, pp. 91–106, 1985; R. G. Wetzel, *Limnology*, 1975.

Paleontology

Recent research in the field of paleontology has included work on evolutionary relationships as indirectly recorded in biological molecules such as nucleic acids and proteins, and studies on the earliest environment of vertebrates.

THE EVOLUTIONARY CLOCK

The dimension of time, as an objective component of large-scale evolutionary studies, has traditionally been the domain of paleontology. Fossils derived from independently dated rocks serve as calibration points for phylogenies and allow for the determination of rates of evolution. However, modern techniques of determining biochemical structure demonstrate conclusively that evolutionary time is also indirectly recorded in the sequence of constituent moieties in long-chain biological molecules, such as nucleic acids and proteins. These molecules are reflective of the genetic code, which is evolutionarily derived by modification of the nucleotide sequence of a species' ancestors. The evolving structure of these molecules can be resolved into the branching pattern of cladogenesis. The time dimension derives from the sequence of systematic bifurcations that chart the phylogenetic history of clades. *SEE PEPTIDE; PROTEIN*.

There are a number of techniques for evaluating the relative temporal significance preserved in the sequence of base pairs in nucleic acids, or in the sequence of amino acids in proteins. For these techniques to be of predictive value in estimating divergence dates between clades, assumptions are usually made as to the selective neutrality of the differences involved such that substitutions and inversions of base pairs, or amino acids coded by them, occur at an average rate over time. Rate constancy can be statistically tested, and based on those assumptions, it results in a measurable, metronomic, clocklike change in the genetic code and its fundamental product, protein. The fossil record currently provides the only independent tests to this "molecular clock." However, the recovery of biologically significant molecular fossils is an exciting new development at the interface between paleontology and molecular biology. *SEE HUMAN GENETICS*.

Fossils and geologic time. The broad observation that the remains of ancient life, that is, fossils, are generally different in succeeding geological strata and that the sequence of fossils is largely consistent from place to place led to the concepts of faunal and floral succession. These concepts, in turn, led to the measurement of relative geologic time and contributed to the formulation of evolutionary theory. Improvements in chronology, related to but largely independent of fossils, have greatly clarified the absolute time scale of Earth history. Most numerical values of geologic time are directly or indirectly derived from radiometric age determinations of high-temperature, naturally occurring, radioactive minerals. The exceptions are derived from radiocarbon dating and certain other minor techniques applicable on the scale of 100,000 years or less, but not applicable throughout most of geologic time. Combining radiometric dates with the rock record of the Earth's magnetic field allows the precise formulation of a detailed numerical global time scale. Dates derived from all these methods provide the fundamental framework of absolute geologic time. This framework is the independent time scale of evolution of life on Earth.

Molecular clock. Evolution is basically the change in organisms through time. This can be detected in the nucleic acids. These long-chain molecules are made up of species-specific sequences of purines and pyrimidines, which code for a specific sequence of amino acids and thus a particular protein. Differences in segments of the genetic code between species, as reflected in differences in the nucleotide sequences, can be determined precisely. Considering the enormous size of nucleic acid molecules, assuming that selectively neutral changes are a major component of the differences between species, and following the central limit theorem of statistics, there should be an overall average rate of change for nucleic acid molecules. Thus, greater differences in nucleic acids should be seen in pairs of species that diverged from a

common ancestor in more ancient times than in pairs that diverged in more recent times. This is the basis for the concept of a molecular clock.

Techniques to study molecular differences. The extent of molecular differences between two species can be evaluated by using a number of different techniques. The most precise, because it is the most fundamental, is the direct comparison of nucleotide sequences of specific portions of the genome. Other techniques include deoxyribonucleic acid (DNA) hybridization, which relies on similarity of structure to determine the strength of bonding between DNA strands of two different species. The degree of relatedness of a pair of species is reflected by the temperature at which hybridized strands of DNA separate. Various immunological techniques, which rely on protein differences to elicit immune responses commensurate with the degree of structural difference in the protein of species under comparison, are also utilized. Immunological techniques and DNA hybridization provide measures of the degree of divergence between pairs of species, but they do not address specific differences in the genome.

All of these techniques measure differences of temporal significance, but it is by no means certain that molecular changes occur at a constant rate between or within lineages; in fact, there is clear evidence of cases that do not evolve at the same or at constant rates. Moreover, there is evidence that natural selection operates at the level of proteins. Nevertheless, there may be portions of the genome, or segments of proteins, that remain relatively unaffected by natural selection. Regardless of the constancy of the molecular clock, independent calibration of molecular evolution comes from the fossil and rock records by the geological dating of divergence points in phylogeny. Divergence points are those points in time when two species diverge from a common ancestor.

Cladistics and divergence dates. The most extensively utilized method of phylogenetic analysis is that of cladistics, whereby the evolutionary relationships of taxonomic units are resolved into a branching pattern of sister groups based on the possession of shared, derived characters. Cladistic analysis of features seen in preservable parts, such as bones and teeth, can be applied to homologous characters of both fossil and living organisms; but currently, cladistic analyses of DNA sequences are limited almost exclusively to living taxa because the relevant molecules are usually not found in fossils. The applicability of cladistics to molecular data has, in itself, elucidated the history of life as never before possible. For example, a realistic evolutionary tree, based on nucleic acid, has been reconstructed for the relationships of major metazoan groups, for which an adequate fossil record of origins does not exist. On the other hand, molecular studies have made almost no contribution to understanding relationships among the vast number of extinct species known only from the fossil record. Thus, the fossil and molecular approaches to evolutionary history are mostly complementary, but in part overlap in their applicability. It is in cases of overlap where the fossil record provides chronological tests to the molecular clock.

The oldest record of a representative of a given clade in the fossil record, based directly or indirectly on absolute radiometric age determinations, represents the minimum age limit of the divergence point between it and its sister group. In this way, the fossil record provides temporal calibration to a cladogram, and by extension may provide for the calculation of rates of molecular evolution, or tests of rate constancy.

The determination of the geological age of divergence points is imprecise for a number of reasons. Clearly, if a member of a clade is unambiguously present as a fossil, the branch point must predate the age of the fossil. A major weakness of this method stems from the inadequacies of the fossil record, the difficulty in evaluating the systematic position of often fragmentary fossils, and the uncertainties in identifying ancestors. This weakness can be overcome by a discriminating use of those fossils that are most appropriate for constructing reasonable hypotheses of divergence dates. These hypotheses of divergence times could be tested against each other, because complex lineages must yield a concordant pattern of decreasing fossil-based divergence dates in going from higher to lower systematic categories within a cladogram. Greater density of fossil-based divergence dates provides for greater precision in determining rates of evolution, whether molecular or morphological.

Confidence intervals can be placed on the first occurrence of a taxon in the fossil record (that is, on the lower bound of its stratigraphic range) by using the equation below, where α is the size of the confidence

$$\alpha = (1 - C_1)^{-1/(H-1)} - 1$$

interval as a proportion of the range, C_1 is the confidence level, and H is the number of horizons at which the fossil occurs within the range.

The size of the confidence interval decreases with increasing number of fossiliferous horizons within the range. The most important underlying assumptions are that fossilization events are independent, the fossiliferous horizons are randomly distributed throughout the range, and collection has been uniform across the range. The lower boundary of the range should not be truncated by faulting or other geological phenomena that would preclude the occurrence of fossils.

Examples of preserved molecular fossils. Significant tests of the molecular clock would come from comparison of homologous segments of nucleic acid or protein that have been extracted from both fossils and living organisms. Such long-chain molecules as these, however, appear to be preserved only in special circumstances and usually only in very young fossils. Nevertheless, there are significant examples of preservation. Amino acid chains from collagen can sometimes be extracted from fossil bone and radiocarbon-dated if the fossils are no more than a few tens of thousands of years old. DNA has been extracted,

characterized, cloned, and amplified from the desiccated soft tissue of the marsupial wolf (*Thylacinus cynocephalus*), extinct since the last century, from a 13,000-year-old ground sloth (*Mylodon*) from a cave in South America, and from human mummies. An exciting development is the recovery and sequencing of DNA from one of the chloroplast genes belonging to a *Magnolia* specimen 17–20 million years (m.y.) old. This opens the possibility of more detailed and integrated studies of molecular and morphological evolution, utilizing both fossil and living organisms, for both cladistic and time-related purposes.

Louis L. Jacobs

EARLIEST VERTEBRATE ENVIRONMENT

The Harding Sandstone of Colorado (about 460 m.y. old) contains one of the earliest accumulations of fossil jawless fishes, which are the earliest group of vertebrates. Many studies have stated that the Harding Sandstone provides evidence that the earliest vertebrates inhabited the sea. However, a recent analysis of depositional environments has revealed that vertebrate fossils at a new locality of Harding Sandstone were preserved in riverine and related deposits. This suggests that, contrary to current opinion, the earliest vertebrates lived in fresh water.

Determining the original vertebrate environment. Historically, the study of the original vertebrate environment was approached by two methodologies: comparative physiology of living vertebrates, and paleontological associations. The former method assumes that most living vertebrates retain a primitive osmoregulatory (water regulation) adaptation. The latter method assumes that it is possible to make paleoenvironmental interpretations of an entire geological formation based on the occurrence of marine invertebrate fossils. Both methods provide inconclusive results. It is difficult to assess whether adaptations of modern forms are indeed ancestral retentions or are the products of recent evolution. Also, the environment of deposition of an entire formation cannot be determined on the basis of the restricted occurrence of marine invertebrate fossils, because formations are usually composed of deposits from many depositional environments, and sometimes fresh-water and marine sediments can lie very close to one another. Nonetheless, two schools of thought exist favoring either a fresh-water or marine origin of the vertebrates.

Fresh-water school. Those favoring a fresh-water origin of the vertebrates base their conclusions primarily on data from vertebrate physiology. Most vertebrate kidneys contain glomeruli, which are primarily water-filtering structures. As blood is sent through the kidney via tiny arterioles, a certain amount of the plasma (blood fluid) is filtered through glomeruli into collecting tubules that pass the fluid eventually to the bladder. Millions of glomeruli are contained in a typical vertebrate kidney, and their size is an ultimate constraint on the amount of plasma that can be filtered. The degree of filtration depends on the environment. For instance, most marine vertebrates live in an environment that has greater ionic concentration (more salts) than their body fluids, so water tends to move out of their bodies by osmosis through exposed epithelia. Fresh-water vertebrates, in contrast, have body fluids of higher ionic concentration than their surroundings and tend to gain water by osmosis through their exposed epithelia. The result is that fresh-water vertebrates are generally in danger of overdilution of body fluids and marine vertebrates are generally in danger of overconcentration of salts in their body fluids. Osmoregulation is necessary because any type of ionic imbalance will hinder nerve and muscle function.

In the most evolutionarily derived state, the glomeruli of marine vertebrates are reduced or absent. This is probably because renal water filtration is not generally advantageous in the sea since water must be conserved. Fresh-water vertebrates contain large glomeruli. Since water is constantly moving into the body by osmosis, the glomeruli must filter and excrete large amounts of it.

Virtually all the major groups of vertebrates contain large glomeruli, including the most primitive forms, which are the hagfishes and lamprey. Over 50 years ago, the argument was that the ancestral vertebrate must have filtered large quantities of water because all of its living descendants have glomerular kidneys or are derived from ancestors with glomeruli. Only fresh-water vertebrates need to filter large quantities of water; thus, in general, vertebrate osmoregulatory physiology suggests that the ancestor of all the vertebrates was adapted to fresh water.

Marine school. However, it is possible that the glomerulus evolved in the sea to regulate certain ions in the body fluids, and not as a water filter. Those who favor a marine origin of the vertebrates state that after the glomerulus evolved in the sea, the vertebrate kidney became preadapted for life in fresh water. The marine scenario explains why hagfishes, which are the most primitive living vertebrates, have body fluids that are isosmotic (having the same salt concentration) with seawater: the earliest vertebrates inhabited the sea, and later ones made a transition into fresh waters, where their body fluids became more dilute over time. Proponents of this scenario state further that the ancestors of extant marine vertebrates did not reevolve the adaptation of isosmicity. Rather, there was, among other strategies, an evolutionary loss of glomeruli in marine lineages that resulted in water conservation, and this is why dilute body fluids are now found in marine and fresh-water vertebrates alike.

Current opinions. Most recent textbooks that discuss the early vertebrate environment state that the group arose in the sea. This conclusion is based partially on the fact that the hagfish has body fluids with the same ionic concentration as seawater, as do most marine invertebrates. However, this could be viewed either as an ancestral retention or as an adaptation to the sea after ancestors moved there from fresh waters. Modern textbooks also cite the association of marine invertebrates with the earliest vertebrate fossils. Such associations have been reported from South America, Australia, and North America, in-

cluding the Harding Sandstone. However, it is very difficult to assess the nature of the fossil associations because of the general lack of sedimentological data in most reports of the earliest vertebrate localities.

Depositional environment. Although suggestive of fresh-water ancestry, osmoregulatory evidence, by itself, is inconclusive. It is thus necessary to approach the problem of the original vertebrate habitat by determining the type of depositional environments that contain the oldest bone fragments. This can be achieved through facies analysis, whereby one can make physically constrained paleoenvironmental interpretations based primarily on sedimentary structures. This method, developed since the mid-1960s, is rooted in studies of modern sedimentary processes. For instance, by analyzing the sedimentary structures produced by modern fluvial (riverine) systems, it is possible to recognize ancient fluvial systems because similar types of sedimentary structures were produced in the past. Such structures define depositional facies. Different environments can thus be distinguished on the basis of different facies associations, and a sequence of strata can be viewed as superimposed past environments. One can then determine in which environments certain fossils are preserved, and whether associated fossils in a formation are indeed contained within the same depositional facies.

A new locality of Harding Sandstone. In the northern Sangre de Cristo mountains of central Colorado, there is an uninterrupted sequence of Ordovician (500–430 m.y. old) sedimentary rocks. In its middle portion, there is a new locality of fossiliferous Harding Sandstone that crops out near the Bushnell Lakes of San Isabel National Forest. A facies analysis has revealed that the Harding Sandstone was the result of a terrestrial depositional system that prograded over a low-gradient carbonate shelf. It is composed primarily of nearshore marine deposits, both intertidal and subtidal, but also contains some fluvial deposits composed of granular pebble conglomerate. Within these latter deposits, bone fragments constitute up to 45% of the total rock volume. Vertebrate fossils occur only rarely in the intertidal and never in the subtidal deposits (see **illus.**). The fluvial units contain no trace

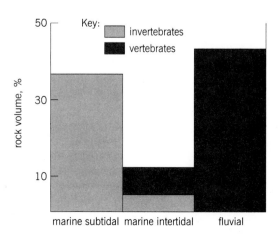

Histogram of fossil types (invertebrates and vertebrates) and distributions in the sediments at Bushnell Lakes. The depositional environments were determined by facies analysis.

of invertebrates, although they are preserved in marine facies at Bushnell Lakes.

The distribution of fossils suggests that vertebrates lived in fresh-water streams and became incorporated into the stream sediments after death. It seems, however, that some bone fragments were occasionally washed into the nearshore marine environment. If one were to infer that the vertebrates were marine animals, it would be difficult to explain why there is such a paucity of their remains in the marine sediments and why the fluvial deposits are so rich in bone fragments. Generally, the presence of fossils in the nearshore is not conclusive evidence that the organisms lived there. Indeed, the nearshore zone receives sediment from both incoming marine currents and seaward-flowing river currents. Thus, remains from both marine and fresh-water species often accumulate in the intertidal zone. On the other hand, fossil accumulations in fluvial deposits can indicate a fresh-water life habitat. Because river and stream currents flow toward the sea, it is unlikely that remains of marine organisms would get incorporated into stream-bed sediments. Even if some individuals of a marine species did somehow get transported upstream into fluvial environments after death, one would still expect to find most of the conspecifics preserved in the sea. The fact remains that very few vertebrate fossils at Bushnell Lakes are contained in marine deposits; thus the most conservative assumption is that they lived in fresh water. SEE TAPHONOMY.

Early on, the fresh-water school was bolstered by negative paleontological evidence. There is an unusual scarcity of bone in marine sediments of pre-Devonian age (345–395 m.y. old), yet marine invertebrates are abundant in such units. It was assumed in the 1930s that vertebrates were preserved in yet-undiscovered fresh-water deposits.

The facies analysis at Bushnell Lakes provides the first positive evidence that the earliest vertebrates were associated with fresh-water deposits. Other Ordovician localities have been reported to contain vertebrate fossils in marine units, but it remains to be proven that they were not washed into the sea from fluvial environments. It is entirely possible that during the Ordovician there were both marine and fresh-water vertebrates, but facies analyses at other Ordovician localities are necessary in order to determine if this is true. Eventually a coherent picture of the earliest environments of vertebrates will result from integration of physiological evidence and facies analysis.

For background information SEE ANIMAL EVOLUTION; DATING METHODS; GENETIC CODE; GEOLOGICAL TIME SCALE; NUCLEIC ACID; ORDOVICIAN; ORGANIC EVOLUTION; OSMOREGULATORY MECHANISMS; PALEONTOLOGY; PHYLOGENY; PROTEINS, EVALUATION OF; STATISTICS in the McGraw-Hill Encyclopedia of Science & Technology.

Greg Graffin

Bibliography. R. H. Denison, A review of the habitat of the earliest vertebrates, *Fieldiana, Geol.*, 11:357–457, 1956; E. M. Golenberg et al., Chloroplast DNA sequence from a Miocene *Magnolia* species, *Nature*, 334:656–658, 1990; R. W. Griffith, Freshwater or marine origin of the vertebrates?, *Comp. Biochem. Physiol.*, 87A:523–531, 1987; C. R. Marshall, The fossil record and estimating divergence times between lineages: Maximum divergence times and the importance of reliable phylogenies, *J. Mol. Evol.*, 30:400–408, 1990; A. D. Miall, *Principles of Sedimentary Basin Analysis*, 1984; S. Pääbo, Ancient DNA: Extraction, characterization, molecular cloning, and enzymatic amplification, *Proc. Nat. Acad. Sci. USA*, 86:1939–1943, 1989; A. S. Romer and B. Grove, Environment of the early vertebrates, *Amer. Midl. Nat.*, 16:805–856, 1935.

Parkinson's disease

Parkinson's disease is a chronic neurodegenerative disorder for which there is no known cause. The initial symptoms are usually mild, but the disease typically follows a slowly progressive course over years. Symptoms commonly include involuntary shaking at rest, severe slowness of voluntary movement, an increased and abnormally distributed tension in muscles with an associated ratchetlike resistance to passive movement at a joint by an examiner, and a balance disorder causing frequent falls because of the difficulty in correcting for minor changes in the center of gravity. Parkinson's disease has a characteristic gait disorder characterized by a stooped posture and slow, shuffling steps with hesitancy at the initiation of motion followed by an involuntary acceleration in the velocity of movement, often culminating in falls. The disease is rare before age 40 and increasingly common thereafter.

In the early 1960s, loss of dopaminergic neurons in the substantia nigra area of the midbrain was recognized as underlying the clinical expression of Parkinson's disease. The administration of the lipophilic dopamine precursor, L-dopa, was shortly thereafter recognized as a breakthrough in the treatment of Parkinson's disease. Treatments with L-dopa and subsequently developed dopaminergic agents have caused a significant decrease in Parkinson's disease mortality. Before the L-dopa era, the mortality rate was three times greater for an individual with Parkinson's disease than for an individual of the same age without the disease; by 1979 that rate had fallen to between 1 and 1.9. Although treatment with dopaminergic drugs has ameliorated the major signs of the disease, particularly early in the course of the illness, Parkinson's disease continues to pose a major health and economic burden. It is in this context that the search for agents that interfere with the underlying disease process is conducted.

Clues from a parkinsonism-causing toxicant. Clues to the cause of Parkinson's disease remained elusive until, in the early 1980s, California neurologists identified a series of unusually young patients with rapidly developing, severe parkinsonism. The clinical pattern was so distinct from that usually encountered that the neurologists were compelled to seek new causes for the parkinsonism. All patients were intravenous drug abusers, and their parkinsonism was found to be associated with the injection of the compound MPTP (1-methyl-4-phenyl-1,2,3,6-tetrahydropyridine).

A vigorous period of investigation showed that MPTP caused parkinsonian signs in primates and many other animals, as well as pathologic changes in the brains of animals that were similar to changes seen in Parkinson's disease. In one human autopsy, the pathologic changes following MPTP poisoning were remarkable in their specific parallel to the primary pathologic changes in Parkinson's disease. These striking similarities led to the hypothesis that Parkinson's disease may be caused by a mechanism similar to that of MPTP toxicity, and that exposure to an MPTP-like agent might cause typical Parkinson's disease.

Understanding of the toxic mechanism of MPTP is not yet complete, but much was learned in the late 1980s. MPTP itself is not neurotoxic, but must be converted to 1-methyl-4-phenylpyridinium ion (MPP+). This conversion is mediated by the enzyme monoamine oxidase B (MAO B). Following this, MPP+ is taken into the dopaminergic neuron and, via mechanisms that are not yet fully delineated, causes cell death. Work in animals with the MPTP model of parkinsonism has provided the foundation for the evaluation of possible protective agents in human Parkinson's disease.

Monoamine oxidase connection. Monoamine oxidase B is the enzyme that converts MPTP to its toxic metabolite MPP+. Animal studies have shown that the monoamine oxidase B inhibitor, selegiline (also known as deprenyl), can protect against all aspects of MPTP neurotoxicity. This evidence, as well as evidence that monoamine oxidase B inhibitors can protect against the toxic effects of oxidation in dopaminergic neurons, led to the initiation of several clinical trials of selegiline as a protective agent in Parkinson's disease.

Clinical trials of selegiline. The first reports that selegiline may alter the progression of Parkinson's disease came from Austrian physicians, who compared their clinical impressions of two groups of patients—those treated with L-dopa alone and those treated with L-dopa in combination with selegiline. The patients treated with selegiline combined with L-dopa appeared to outlive those receiving only L-dopa by about 15 months. Since these observations were essentially anecdotal, their validity remained uncertain.

The first carefully controlled trial was a prospective, double-blind, randomized evaluation of the effects of selegiline in patients with early, untreated Parkinson's disease. Importantly, none of the patients in this study were taking any other antiparkinsonian medication, so that the specific effect of selegiline therapy could be determined. A second, larger trial of selegiline and another putative protective agent, the antioxidant tocopherol, was also begun.

Both clinical trials used the statistical method of survival analysis to determine the differences between treated and untreated subjects, calling the time elapsed from starting the study to the time that a second, antiparkinsonian drug was needed the "survival time." In each study, patients were at an early stage in the course of their illness and had only mild symptoms at study entry. Importantly, they were permitted no drugs to alleviate the symptoms of Parkinson's disease. Patients with atypical disease were excluded.

The results of both studies were similar and dramatic. The use of selegiline in otherwise untreated Parkinson's disease patients caused a significantly longer survival time. In the first study, selegiline-treated patients "survived" nearly twice as long without additional treatment as did those receiving placebo. In the second, larger study, results were so striking that the study's monitoring committee recommended that the group consider halting the study on ethical grounds because the benefit in the selegiline-treated group was dramatically apparent. After 1 year, twice as many patients in the placebo group had required additional antiparkinsonian treatment, as compared to the selegiline-treated group. As a consequence, in 1989, nearly 2 years before the expected conclusion of the trial, all patients were placed on selegiline therapy.

Protection or symptomatic treatment. The biggest concern in evaluating the results of selegiline therapy in Parkinson's disease is differentiating whether the drug truly protects against progression of the disease or merely alleviates symptoms. If selegiline merely allows the postponement of other treatments because it relieves the symptoms of Parkinson's disease, the results of the studies, while not without some benefit, are of much less import. In the second, larger study, mild symptomatic improvement was demonstrated by statistical tests, but not appreciated by either the patients or their doctors during the first 3 months of the study. One other small double-blind study has shown similar mild improvement of symptoms. Several small, open (without placebo control) studies have reported mild improvement of symptoms in a few subjects. The lack of an obvious symptomatic effect in most patients suggests that even if there is a minor symptomatic improvement this is not the major mechanism of selegiline's effect.

To evaluate further the question of symptomatic effects, both of the major controlled trials of selegiline have included a washout observation period, when the patient is examined 1 or 2 months after stopping selegiline. If the effect is primarily one of symptomatic relief, patients would be expected to get worse as the drug effect wanes. Although final results for all subjects are not available, evidence to date suggests that washout effects are minimal, arguing in favor of selegiline being a truly protective mechanism.

Prospects. The final determination of a protective versus symptomatic action for selegiline will require laboratory studies as well as additional clinical observations. Different doses of selegiline and of other agents with similar mechanisms but different pharmacologic profiles may provide greater protection from disease progression, or combinations of agents may be more effective than single agents. Certainly, many additional studies evaluating the role of protective therapies in Parkinson's disease will be necessary before the best intervention is determined, but the studies done to date appear to have opened the door to a new therapeutic strategy for Parkinson's disease.

For background information SEE MONOAMINE OXIDASE; PARKINSON'S DISEASE in the McGraw-Hill Encyclopedia of Science & Technology.

Caroline M. Tanner; J. William Langston

Bibliography. W. Birkmayer et al., Increased life expectancy resulting from addition of L-deprenyl to Madopar treatment in Parkinson's disease: A longterm study, *J. Neural Transm.*, 64:113–127, 1985; Parkinson Study Group, DATATOP: A multicenter controlled clinical trial in early Parkinson's disease, *Arch. Neurol.*, 46:1052–1060, 1989; Parkinson Study Group, Deprenyl delays disability in early Parkinson's disease: A controlled clinical trial, *N. Engl. J. Med.*, 321:1364–1371, 1989; J. W. Tetrud and J. W. Langston, The effect of deprenyl (selegiline) on the natural history of Parkinson's disease, *Science*, 245:519–522, 1989.

Peptide

The number of chemically identified regulatory peptides has been increasing rapidly since the mid-1980s, and about 60 peptides have now been identified in various mammalian species (that is, humans, rats, oxen, and pigs). That estimate must be very conservative, however, since new mammalian peptides continue to be discovered, and many of the novel sequences being found in invertebrates will probably have mammalian analogs as well. Therefore, the total number of peptides in any mammal (and by analogy, in any species) may well be an order of magnitude larger than the current estimate. Furthermore, as the pool of known sequences has increased, the tendency of peptides to occur in families characterized by substantial similarities of structure has become more apparent.

The history of most peptide families begins when a tissue extract with activity in a bioassay is purified and a novel peptide is sequenced. Several invertebrate assay preparations, mostly isolated visceral and somatic muscles from mollusks and insects, have been especially productive of new peptides. Once a sequence is in hand, the techniques of radioimmunoassay or molecular genetics can be applied to search for the rest of the family. Such searches reveal entirely new families with surprising frequency, especially in distantly related species. For example, probes of tissue extracts from vertebrates, echinoderms, and cnidarians, with a radioimmunoassay for FMRFamide, a molluscan tetrapeptide, have yielded unrelated, novel families in each of these groups.

Nearly 30 peptide families have now emerged, about 12 of them in the vertebrates. For more than a decade, attempts have been made to define, characterize, and explain the significance of these families.

But peptide families are heterogeneous, and thus resist firm definition and characterization. Distinguishing between peptides that are similar because they share a common ancestry, and peptides that are similar due to convergence, and discerning the functional roles of the association are problems.

Origin and diversity. All secretory peptides are encoded by genes as segments of larger precursor molecules. Upon translation, the precursors are packaged in secretory vesicles and processed by proteolytic enzymes to produce their active peptides. Peptide families seem to arise by gene duplication; either the entire gene is copied, or a copy is made of only a segment including the region encoding a particular peptide and its processing signals. In either case, subsequent point mutations in one of the resulting copies would yield a novel peptide and, perforce, a peptide family.

Within any species, most families for which genetic details are known comprise the products of similar but distinct precursors. But those precursors may produce only one peptide each, as in the pancreatic peptide–related peptides of mammals and the adipokinetic hormones of locusts; multiple peptides, as in the FMRFamide-related peptides of mollusks; or a combination of the two possibilities, as in the enkephalin-containing peptides. In very few families (and these may be ones for which the complete genealogy is not yet known) is the entire membership composed of only two or more segments processed out of the same precursor [as in the family of small cardioactive peptides (SCPs) in gastropod mollusks].

Sequence similarities can also occur by chance, and chance is to be suspected when two peptides have a relatively modest sequence similarity and are borne on notably dissimilar precursors. Yet even where the primary sequences are not very similar, as in the family of insulin and insulinlike growth factors in vertebrates, the tertiary structures of two peptides and the organization of their precursors may be sufficiently alike that divergence from a common genetic origin seems likely. Information about genetic organization, as well as about amino acid sequence, is necessary for an understanding of peptide family relationships.

Phyletic distribution. Congeners of peptide families discovered in one species are routinely identified in other closely related species. The set of all such congeners constitutes an extended peptide family. An important question concerns the extent to which peptide families are actually restricted in their phyletic distributions. Since extended peptide families are usually thought of as comprising congeners that have evolved from a common ancestral precursor encoded by an ancestral gene, this question bears on peptide evolution; for example, widely ranging peptide families must have evolved earlier than restricted ones in metazoan phylogeny. Since the ancestral molecules are no longer available, the evidence for homology must rest, again, on the structural similarities between the genes, precursors, and peptides constituting the extant family.

The phyletic limits of peptide families range from broad to narrow. At one extreme, insulinlike molecules of rather similar structure occur throughout the vertebrates and share structural similarities with the prothoracicotrophic hormone of insects and with a growth-regulating peptide produced by neurosecretory cells in the aquatic snail *Lymnaea stagnalis*. Since this superfamily includes peptides in vertebrates, arthropods, and mollusks, it must have diverged from an ancestral gene present in the earliest metazoans; therefore, insulinlike peptides are probably ubiquitous in metazoans.

The family of FMRFamide-related peptides is also polyphyletic, but more narrowly distributed than the insulinlike peptides. In particular, peptides from four protostomian phyla (Mollusca, Annelida, Arthropoda, and Nematoda) have sequence similarities so conserved that they must all be homologs. In contrast, although certain cnidarian, echinoderm, and vertebrate peptides are modestly similar to FMRFamide, and are even recognized by FMRFamide antisera, these resemblances are probably fortuitous; that is, they are examples of convergence.

If the examples mentioned above are set aside and only known sequences are accepted as evidence, then most vertebrate and invertebrate peptide families would appear to be restricted to a single phylum. A few discoveries, however, run counter to that conclusion: the molluscan peptide eledoisin is a close relative of the mammalian tachykinins; the opioid peptides of mammals seem to have been sequenced in mollusks; and a congener of the mammalian peptide neurotensin has been completely characterized in lobsters. Recent studies have also tended to broaden ranges; for example, homologs of the pigment-dispersing hormones of crustaceans occur in insects as well, and the large family of peptides related to the adipokinetic hormones of insects and the pigment-concentrating hormones of crustaceans has recently been extended to mollusks.

In summary, most peptide families seem to have restricted ranges, but the restriction may only reflect the technical difficulties, which are rapidly disappearing, of identifying peptides in unusual places.

Functional significance. Most peptides have a repertoire of fundamental actions on a variety of cells and tissues. Each of these actions is initiated by binding to one of a set of receptor proteins. Thereafter, an ion channel may be opened directly, or an effect may be mediated by one of several common intracellular mechanisms. A single peptide, such as cholecystokinin, may have complementary actions at different sites in an organ system, such as the digestive tract, and may thereby contribute to the integration of the system as a whole.

The appearance and coevolution of peptide families and their receptors would seem to be an effective and efficient mechanism for augmenting the number and variety of regulatory agents in a species. First, when sibling peptides are encoded by separate genes, their precursors can be expressed in different tissues. Even when two or more sibling peptides are encoded by the same gene, alternative modes of transcription and processing can provide tissue-specific expression. Indeed, peptide families commonly include analogs that are secreted by nonneural cells in such tissues as gut (and its derivative glands), heart, and skin.

The specificity of action of sibling peptides is conferred in the first instance by their complementary receptors. However, sibling peptides with different sites of production and distinct targets can still act, in a pharmacological experiment, at most of the complementary family of receptors. In mammals, for example, pancreatic polypeptide and its two homologs, peptide YY (produced in the colon) and neuropeptide Y (produced in the brain), can inhibit pancreatic secretion, inhibit colonic motility, and constrict blood vessels. Physiological specificity in such instances is ensured by the physical separation of the targets and of the neurons transporting the peptides to them.

Related control functions in an organism are frequently allocated among the peptides in a family. For example, pairs of sibling peptides—cholecystokinin and gastrin, peptide YY and pancreatic polypeptide, and substance P and substance K—collaborate in regulating different functions of the vertebrate gut. Other instances of complementarity are instructive. First, neurosecretory products of the egg-laying hormone gene, and homologous peptides secreted as pheromones by the atrial gland, affect reproductive behavior in the sea slug *Aplysia*. Second, the tetrapeptide and heptapeptide analogs of FMRFamide have distinctive effects on muscle tone, heart function, and posture in pulmonate snails. Third, SCP-containing neurons in *Aplysia* produce and release two similar homologs, SCP_A and SCP_B, and the actions of these peptides have so far been indistinguishable. Therefore, separate receptors for SCP_A and SCP_B have either not yet evolved or have not yet been found in this mollusk.

Finally, the actions of effector organs are usually modulated by neuropeptides from several different families as well as by classical transmitters. This is pointedly illustrated by studies that show that a single effector, the accessory radula closer muscle of *Aplysia*, is regulated by more than 10 endogenous agents. Therefore, the effect of a single peptide on the cells of a particular tissue can certainly be designated as its action, but its function or physiological role in that tissue is necessarily fractional or participatory. From this perspective, the effector emerges as a kind of olfactory organ, its physiological responses dependent upon the particular combination and the relative concentrations of regulatory agents delivered to its vicinity by neurons or by the circulation.

For background information SEE BIOASSAY; HORMONE; PEPTIDE in the McGraw-Hill Encyclopedia of Science & Technology.

M. J. Greenberg

Bibliography. S. Falkmer et al. (eds.), *Evolution and Tumor Pathology of the Neuroendocrine System*, 1984; D. G. Fautin (ed.), *Biomedical Importance of Marine Organisms*, Mem. Calif. Acad. Sci., 1988; M. J. Greenberg and M. C. Thorndyke (eds.), Consistency and variability in peptide families, *Biol. Bull.*, 177:[1]67–229, 1989; D. T. Krieger et al. (eds.), *Brain Peptides*, 1983.

Perissodactyla

The perissodactyls, or odd-toed hoofed mammals, include the horses, rhinoceroses, tapirs, and their extinct relatives. Recent discoveries have radically changed the notions about perissodactyl evolution. New fossils from China show that perissodactyls arose in Asia about 55 million years ago (m.y.a.), along with their close relatives, the elephants and their kin. Anatomical evidence suggests that the living hyraxes, also called conies, long thought to be related to elephants, are probably related to perissodactyls. The remaining nonhyracoid perissodactyls are now divided into three groups: the Hippomorpha (horses and their extinct relatives), the Titanotheriomorpha (the brontotheres, extinct rhino-sized animals with paired blunt horns), and the Moropomorpha (tapirs, rhinoceroses, and their extinct relatives, including the clawed chalicotheres).

Before about 34 m.y.a., the brontotheres, archaic rhinos, and primitive tapirs were the dominant large mammals around the world. After these groups became extinct, horses and rhinoceroses were the most successful perissodactyls, although they have been ecologically replaced by even more successful cattle, antelopes, deer, and their relatives. Today, most nondomesticated perissodactyls are endangered in the wild. Rhinoceroses, for example, are being killed by poachers at an alarming rate.

Origins. Perissodactyls were once thought to have originated in Central America from the phenacodonts, an archaic hoofed mammal group. In 1989, a specimen from Chinese deposits about 55 m.y. old was described and named *Radinskya*. This specimen showed that perissodactyls must have originated in Eurasia some time before 55 m.y.a. (see **illus.**). Its similarities to the most primitive relatives of the elephants suggested that elephants and perissodactyls have a close common ancestry in Eurasia about this time. This close relationship is also suggested by the hyraxes, a group of woodchucklike hoofed mammals found today in Africa and the Near East. Although they were long thought to be related to elephants, new anatomical evidence suggests that hyraxes are probably perissodactyls. Hyraxes and other perissodactyls have very striking similarities in their internal-ear anatomy, in the structure of their hooves, in the muscles of their limbs, and even in the irises of their eyes. If elephants, hyraxes, and other perissodactyls are all closely related, then they must have diverged in Eurasia-Africa about 60 m.y.a., shortly after the extinction of the dinosaurs.

Horses. Once perissodactyls spread from Asia to North America and Europe, they began to diverge into three main groups: the Hippomorpha (horses and their relatives), Titanotheriomorpha (brontotheres), and Moropomorpha (tapirs, rhinos, chalicotheres, and their relatives). Among the hippomorphs, the earliest horses occurred in North America about 53 m.y.a., and their close relatives, the palaeotheres, occurred in the European archipelago about the same time. (High sea levels isolated Europe into a number of small islands at this time.) The name of the earliest true horse in North America was long thought to be *Eohippus* or *Hyracotherium*, but the correct name is probably *Protorohippus*. From this four-toed, dog-sized ancestor, horses underwent a well-known evolutionary history, becoming larger and longer limbed with reduced side toes. By about 30 m.y.a., horses had become very common in North America.

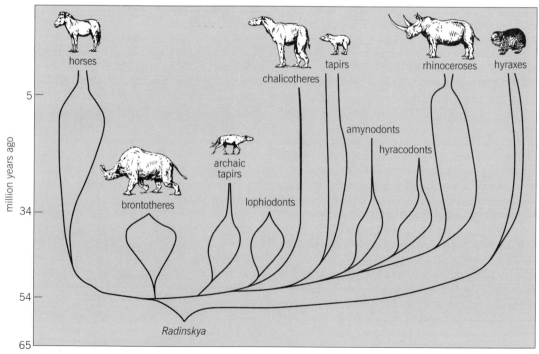

Evolutionary relationships of the major group of perissodactyls.

The horses *Mesohippus* and *Miohippus*, long thought to be sequential segments of the main trunk of the horse family tree, are in fact bushy side branches that overlap by millions of years. From *Miohippus*-like ancestors, some horses became specialized for eating gritty grasses and developed high-crowned, ever-growing teeth. Others remained leaf-eating browsers with low-crowned teeth. By 15 m.y.a., horses were so diverse that as many as 12 species occurred in one locality; they were ecologically equivalent to the diversity of antelopes on the modern African savanna. About 5 m.y.a., most of this diversity of horses (particularly the primitive browsing types and the three-toed hipparions) had waned, leaving only the living genus *Equus*. Although the primary area of horse evolution was always North America (with occasional migrations to Eurasia), horses disappeared from this continent at the end of the last ice age, only to be reintroduced by Columbus in 1493.

Brontotheres (titanotheres). Brontotheres started out as small, pig-sized animals and evolved into larger, cowlike animals during the span of about from 50 to 40 m.y.a. At the end of their evolution, they reached elephantine proportions, with large paired blunt horns on the tip of the nose. Brontotheres were the largest land mammals in North America and Eurasia until about 34 m.y.a. when they became extinct. Their extinction, long a mystery, now appears to be related to massive climatic changes. These climatic changes, triggered by the first Antarctic glaciation, caused changes in the vegetation that led to extinction of many archaic browsing mammals, such as the brontotheres and the primitive tapirs.

Tapirs, rhinoceroses, and chalicotheres. The third major group of perissodactyls is the Moropomorpha, which includes the tapirs, rhinos, and their extinct relatives. They originated about 50 m.y.a. as dog-sized animals that were virtually indistinguishable from the earliest horses. Until about 34 m.y.a., archaic tapirs were very diverse in both Asia and North America. Although they already had leaf-cutting teeth like those of modern tapirs, they did not develop the tapir's trunk or proboscis until later in their evolution. Like the brontotheres, most archaic tapirs became extinct with the climatic changes 34 m.y.a. that nearly wiped out their forest habitat.

One descendant of the archaic tapirs, the lophiodonts, lived in the European archipelago 40–50 m.y.a. The lophiodonts independently developed a proboscis and huge leaf-cutting teeth, so that they looked very much like modern tapirs. Recent evidence has shown that the lophiodonts are closely related to another extinct group, the chalicotheres, whose relationship to other perissodactyls has long been a mystery. Chalicotheres looked like heavy-bodied horses, with claws on their feet instead of hooves. Some of them could knuckle-walk on their long gorillalike forelimbs. Their claws were used for hauling down branches for feeding. Chalicotheres were rare but persistent in the Northern Hemisphere until the beginning of the last ice age, when they disappeared from Asia and Africa.

The most diverse and successful group of perissodactyls was the rhinoceroses. From their origin in tapirlike forms about 50 m.y.a, rhinoceroses diverged into three groups: the amynodonts (aquatic hippolike and tapirlike forms), the hyracodonts (long-legged runners), and true rhinoceroses. Amynodonts and hyracodonts both suffered from the extinction event of about 34 m.y.a., although both managed to survive until about 20 m.y.a. One group of hyracodonts became gigantic. Of these, a hornless long-necked hyracodont called *Paraceratherium* (formerly known as *Baluchitherium* or *Indricotherium*) was the largest land mammal that ever lived. It reached 20 ft (6 m) at the shoulder and weighed over 44,000 lb (20,000 kg, or about three times the weight of the largest elephant). *Paraceratherium* probably browsed on the tops of trees in Asia about 35 m.y.a.

The oldest known true rhinoceros was described in 1989 from deposits in Oregon about 40 m.y. old. The earliest true rhinoceroses were hornless, and remained small in body size until the extinction of brontotheres about 34 m.y.a. From about 34 to 20 m.y.a., rhinoceroses were the largest herbivorous mammals in North America or Eurasia. Then mastodonts migrated out of their isolation in Africa and took over the megaherbivore niche. True rhinoceroses have occupied many different habitats. Some were generalized forest browsers, and others grazed tough grasses with their ever-growing teeth. One group was short-legged and fat like the hippopotamuses, and must have had an aquatic habitat much like that of modern hippos. Most extinct rhinoceroses were hornless, but some groups evolved a single horn on the nose or forehead, or tandem horns on both nose and forehead. Two groups of rhinos independently evolved paired horns on the tip of the nose. Rhinoceroses suffered severe extinctions during the major climatic changes about 5 m.y.a. They disappeared completely from North America, and only five living species managed to survive in Africa and Eurasia.

Conservation. Perissodactyls have been declining for the last 10 m.y., driven to near extinction by competition with cattle, antelope, deer, and other hoofed mammals with a more efficient, cud-chewing digestion. Perissodactyls suffered even more severe extinctions during the climatic crisis of 5 m.y.a., and during the great extinction of large mammals at the end of the last ice age. Most of the living species are very rare or endangered in the wild. Among the horse family, the domestic horse is very widespread, but its ancestor, Przewalski's horse, is nearly extinct in central Asia. Although the plains zebra is common, the Grevy's and mountain zebras are endangered, and the quagga became extinct in the nineteenth century. Wild asses and onagers are now extremely rare. Tapirs are endangered in the jungles of both South America and southeast Asia. All of the five living species of rhinoceros are now endangered in the wild, since rhinoceros horn is worth its weight in gold for use in folk medicines and Arab dagger handles. There are only a few thousand white rhinos, a few hundred Indian rhinos, and a few dozen Sumatran or Javan rhinos surviving. Once abundant, the black rhinoceros is now being decimated by poachers. Of 65,000 in 1970, only a few hundred remain today, and it may be extinct in the wild by the turn of the century. Despite over 50 m.y. of evolutionary success, the fate of wild perissodactyls is uncertain as long as human populations continue to expand and destroy wild habitat.

For background information SEE EXTINCTION (BIOLOGY); HYRACOIDEA; MAMMALIA; PERISSODACTYLA; RHINOCEROS; TAPIR in the McGraw-Hill Encyclopedia of Science & Technology.

Donald R. Prothero

Bibliography. R. L. Carroll, *Vertebrate Paleontology and Evolution*, 1988; D. R. Prothero, The rise and fall of the American rhino, *Nat. Hist.*, 96:26–33, 1987; D. R. Prothero and R. M. Schoch (eds.), *The Evolution of Perissodactyls*, 1989; R. J. G. Savage and M. R. Long, *Mammal Evolution: An Illustrated Guide*, 1986.

Pesticide

The introduction of synthetic pesticides, following World War II, ensured a stable food supply in the developed countries and led to the control of many diseases, including yellow fever and malaria. While plant growth regulators, herbicides, defoliants, fungicides, and insecticides have proved to be useful pesticides, not all have had a benign effect on the environment. A number of environmentalists and health specialists have voiced a growing concern about pesticide use; consequently, agrochemicals have come under close scrutiny. During the 1990s, it is anticipated that many pesticides will be withdrawn from the marketplace. Unfortunately, it is virtually impossible to grow crops in many parts of the world without pesticides, especially in humid subtropical areas such as the southeastern United States. In juxtaposition, the world population keeps expanding. Therefore, it would appear that a collision course has been set: increased population and uninsured harvests. However, a concerted effort to isolate natural products from diverse sources for serious consideration as substitutes for hard pesticides is being fueled by countries that have had some serious environmental problems, such as Japan. These natural products, or soft pesticides, have certain common features. They are active at low concentrations; they tend to be target-specific; and they are biodegradable. Soft pesticides offer chemical templates for biotransformational or synthetic alteration, which would thereby alter their target specificity. Their environmental origins are broad; they have been isolated from plants, microorganisms, and marine sources. Their uses may be broad as well; a compound gathered from one source may have application in an apparently unrelated system.

Plant substances that affect plants. It had been postulated for over 30 years that Irish potato (*Solanum tuberosum*) leaves produce a natural product that induces tuber formation. This was confirmed by

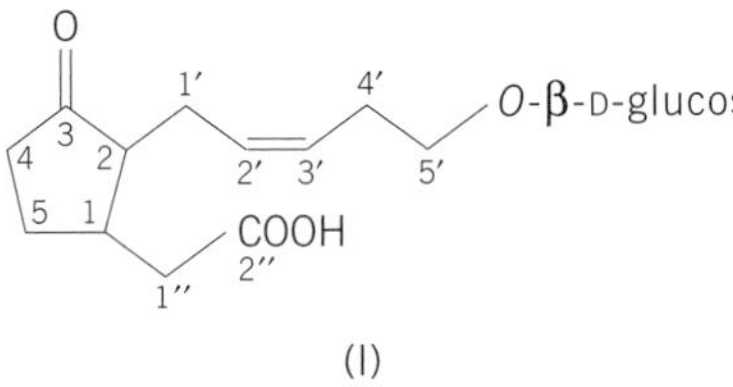

(I)

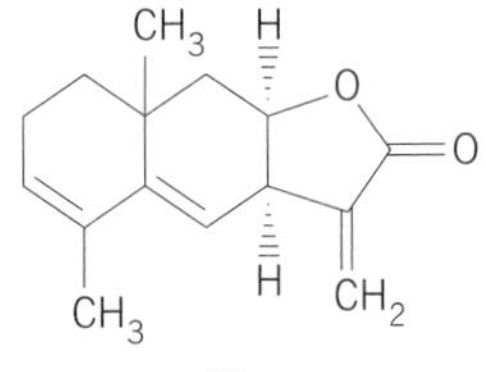

(II)

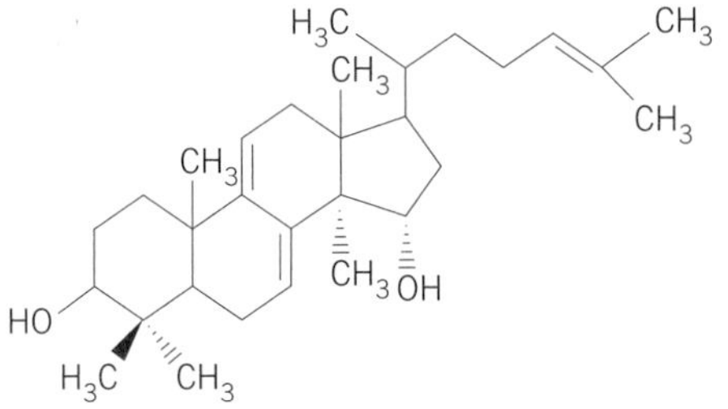

(III)

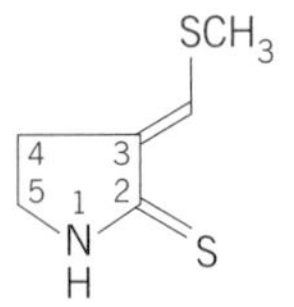

(IV)

grafting experiments that demonstrated translocation to stolon tips. Recently, the active compound has been isolated from potato leaves and identified as 3-oxo-2-(5'-β-D-glucopyranosyloxy-2'-Z-pentenyl)-cyclopentane-1-acetic acid (I). While the material is slightly unstable in the pure state, it nevertheless induced tuber formation in single-node stem segment cultures. Such an elicitor would particularly aid in potato production by reducing the growing time and by reducing the amount of fungicides used annually, which is currently about 18 applications per crop.

Another potentially useful plant growth regulator, obtained from powdered roots of *Inula racemosa*, is isoalantodiene (II), a potent root initiator in cuttings of *Vigna radiata*. To date, only one natural metabolite, indole-3-acetic acid, has been shown to initiate rooting. A synthetic congener, indole-3-butyric acid, is widely used in plant cuttings and air layering to promote rooting.

In the search for antitumor pharmaceuticals, the bark of *Melodorum fruticosum* has been extracted to yield a number of biologically active products. Of these, polycarpol (III) has some signal properties, among which is the ability to stimulate the growth of *Lemna*, an aqueous surface plant.

Potential herbicides are being sought from higher plants. From *Raphanus sativus*, radish, comes the compound 3-(E)-(methylthio)methylene-2-pyrrolidinethione (IV), which inhibited experimental etiolated cress seedlings. Ironically, the compound had previously been synthesized but was not reported as a plant growth inhibitor or as a natural product.

Fungal substances that affect plants. There are formidable numbers of fungal metabolites that have been isolated for potential use on plants, and the number is growing. These fall into two groups: those that stimulate plant growth and increase yields, for example, the gibberellins, from *Gibberella fujikuori*; and those that have selective herbicidal properties, for example, moniliformin, from *Fusarium moniliforme*.

A novel discovery is acremoauxin A (V) from *Acremonium roseum*, an indole analog, (2S)-2-(3-indolyl)propionyl mannitol. In bioassays with Chinese cabbage seedlings, effects were similar to those obtained with indole-3-acetic acid. Also, in rice plants, the compound exhibited responses similar to those obtained with indole-3-acetic acid or brassinosteroids. The overall activity was assessed at about one-tenth that of the synthetic broad-leaved herbicide 2,4-dichlorophenoxyacetic acid.

Structurally curious metabolites, ampullicin (VI) and isoampullicin (VII), have been discovered in an *Ampulliferina*-like fungus. These compounds greatly accelerated root growth in lettuce seedings. Rapid root growth is an important step in the establishment of seedlings.

Spiciferone (VIII), isolated from *Cochliobolus*

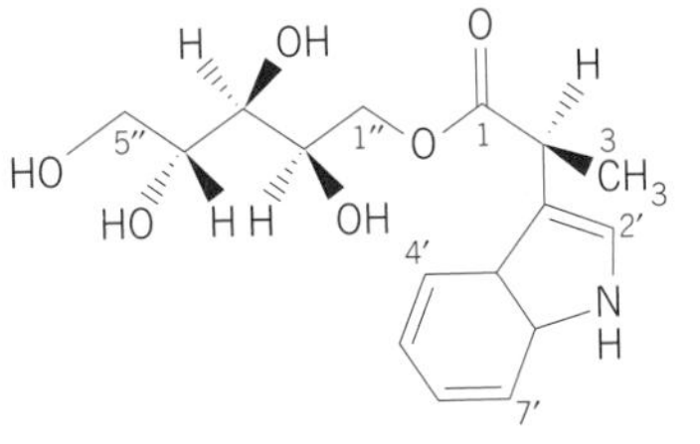

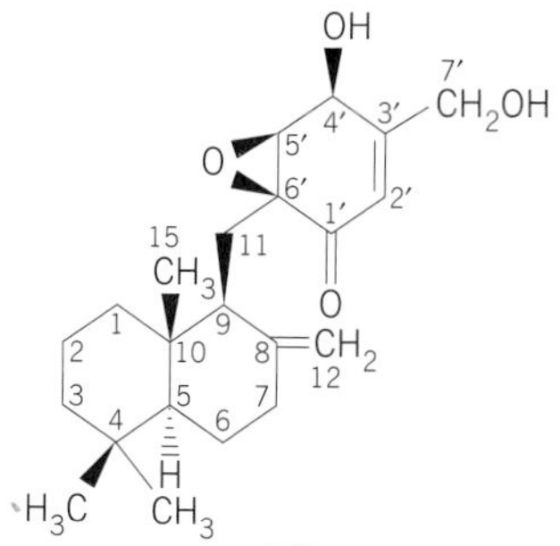

(IX)

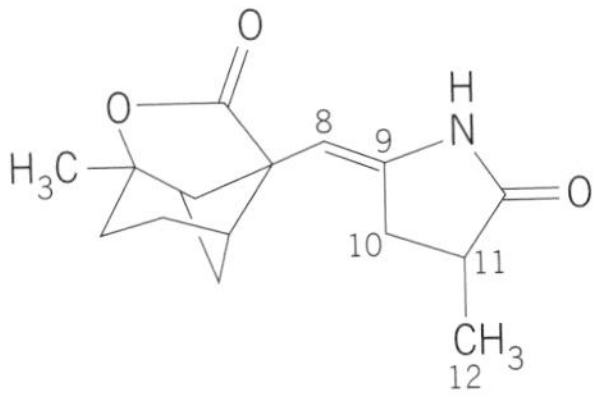

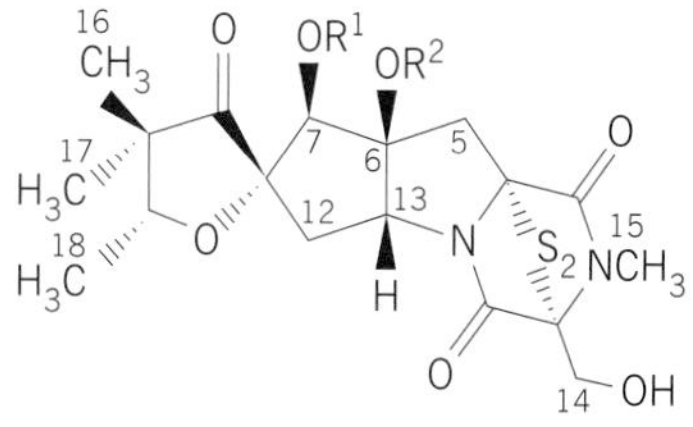

(X)

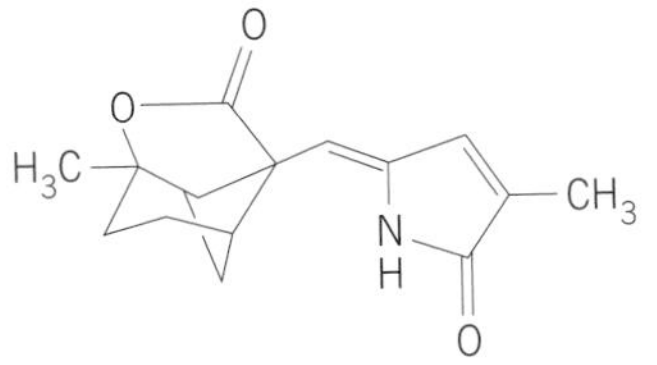

spicifer, which causes a leaf spot disease in the Gramineae, rapidly promoted growth in rice seedlings and caused second leaf sheaths to reach 138% of untreated plants.

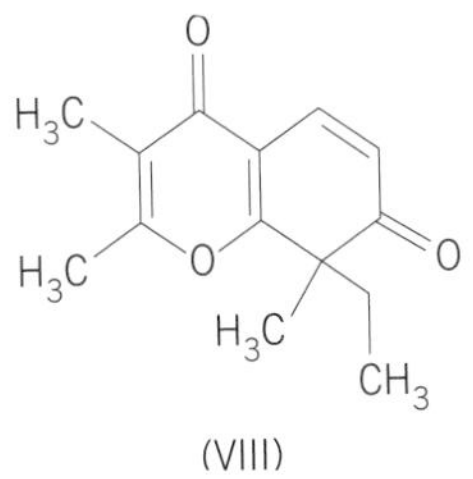

A pathogen that has devastated entire communities of elm trees throughout the world is Dutch-elm disease, caused by *Ceratocystis ulmi*. Much research has been conducted to save the few trees that remain. A number of natural products have been isolated from *Penicillium brevi-compactum*, and of these a particularly promising agent, 11-(5'-epoxy-4'-hydroxy-3'-hydroxymethylcyclo-2'-hexenone)-$\Delta^{8(12)}$-drimene (IX), has been as active as the standard treatment, hyalodendrin, in controlling *C. ulmi* in cell cultures. The molecule features a number of functional groups that may be derivatized to increase biological activity further.

A rapid method to produce resistant lines of economically important crops is to challenge tissue cultures with pure toxins isolated from their phytopathogens. One endeavor has been with *Leptosphaeria maculans*, a pathogen that causes blackleg in rapeseed, *Brassica napus* and *B. campestris*; some of the phytotoxins isolated have been sirodesmin PL, where R^1 = acetate and R^2 = H; deacetylsirodesmin PL, where R^1 and R^2 = H; and 6-monoacetyl-sirodesmin (X), where R^1 and R^2 = acetate. Each produced large lesions (20–30 mm) when inoculated into rapeseed, mustard, and cabbage. Resistant cells are then grown into whole plants.

Plant substances that affect insects and nematodes. The literature abounds with examples of plant products that affect insects and nematodes. Compounds extracted from alfalfa and marigolds provide two examples that illustrate the structural diversity of these metabolites. They may confer resistance on plants, and the genes responsible for producing them may eventually be transferred to the appropriate species. Another possibility is that the substances may be applied exogenously.

(V)

(VI)

(VII)

(VIII)

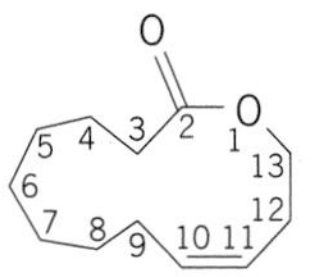

(XI)

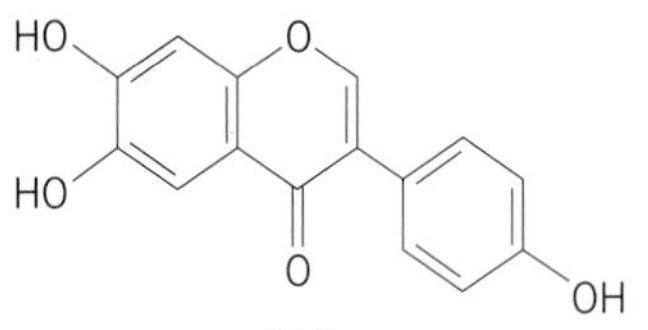

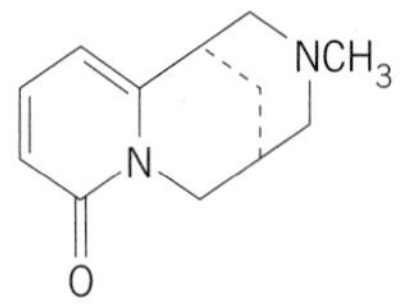

(XII)

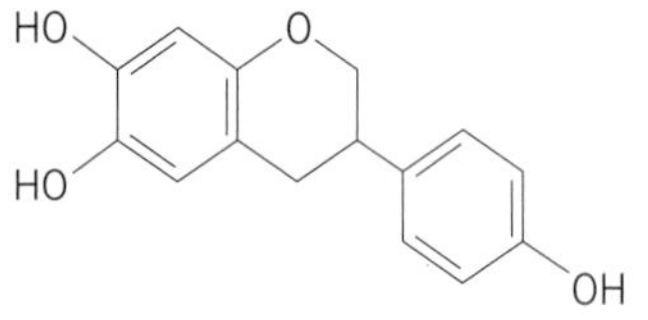

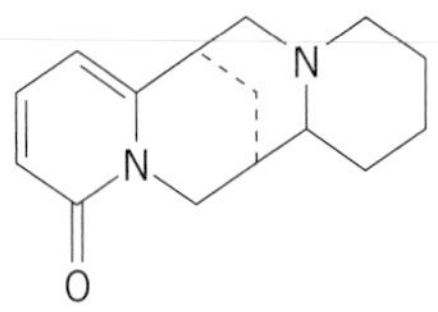

(XIII)

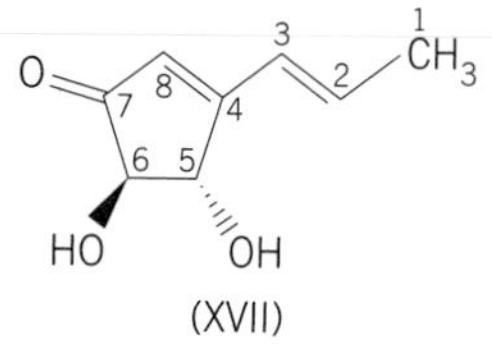

(XIV)

(XV)

(XVI)

(XVII)

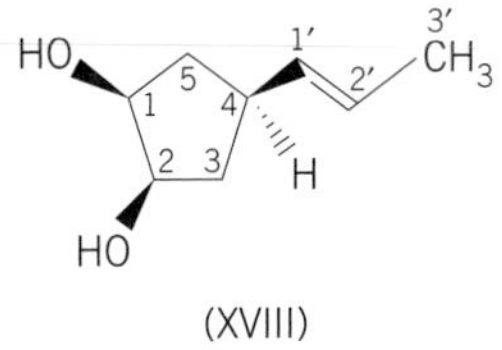

(XVIII)

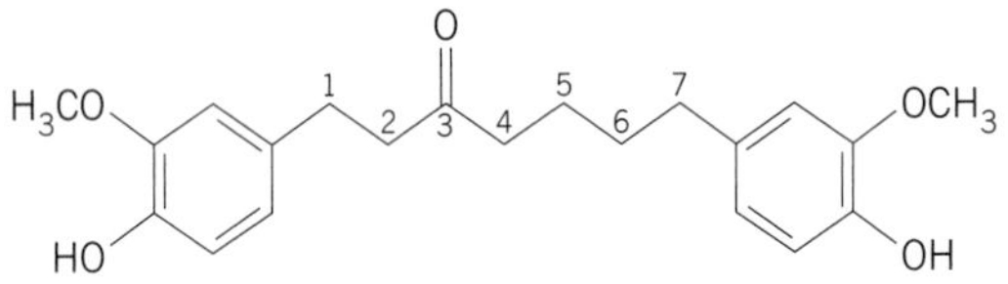

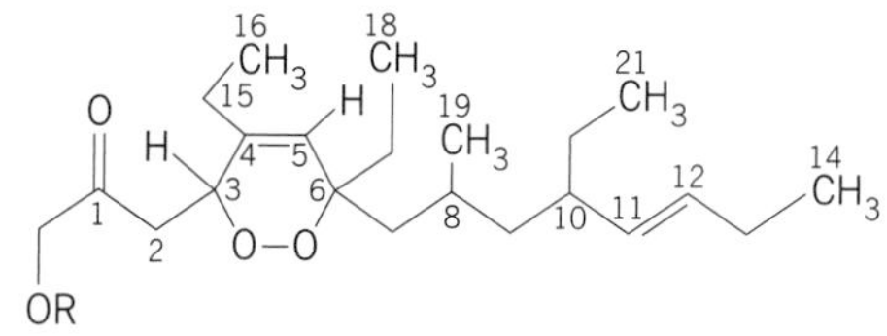

(XIX)

The alfalfa species *Medicago rugosa* is particularly resistant to adult weevil (*Hypera postica*) feeding. Steam distillate fractions from leaves and stems gave six compounds in small amounts. Of these, *cis*-(*Z*)-oxacyclotridec-10-en-2-one (XI) inhibited alfalfa feeding on phagostimulant-treated membrane filters.

Nematodes pose a constant problem in crop production. While the damage that they may inflict is seldom seen as total devastation, their effects on yields can be significant. Certain plants inhibit the spread and growth of nematodes: among these are marigolds (*Tagetes* spp.) that are used in domestic vegetable gardens. Recently, extracts of dried roots of *Sophora flavescens*, used in Chinese medicine as an anti-inflammatory, antipyretic, and anthelmintic, were

also found to have nematicidal properties. Methanolic extracts gave two natural products, (−)-*N*-methylcytisine (XII) and (−)-anagyrine (XIII), which were active against the pinewood nematode, *Bursaphelenchus xylophilus*; (−)-*N*-methylcytisine was approximately twice as active as (−)-anagyrine. Evaluations are under way to determine the structure–activity responses.

Plant substances that affect fungi. An urgent need exists for safe fungicides in both field and storage applications. Those presently on the market may be withdrawn by 1996. Thus, an intense search is under way for natural product fungicides.

The condiment ginger, *Zingiber officinale*, is an Asian drug with many medicinal uses. The gingerenones A, B, and C have been isolated from its

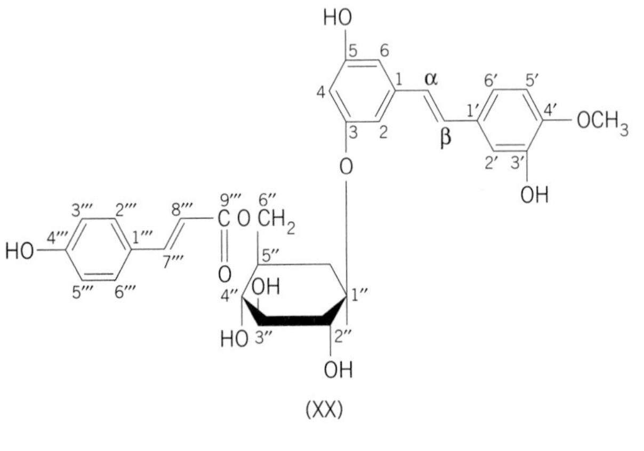

(XX)

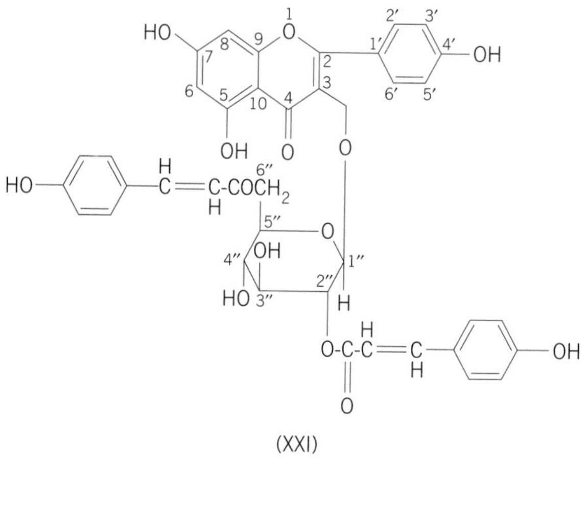

(XXI)

rhizomes, and these new structures have been evaluated in biological tests. Gingerenone A (XIV) was moderately active as an anticoccidium agent, but more importantly, it completely inhibited elongation of invading hyphae of *Pyricularia oryzae*, a particularly pernicious rice pathogen.

Certain isoflavanoids (XV) and isoflavones (XVI) from members of the Leguminosae have been shown to be active against storage fungi, especially *Aspergillus flavus*, the organism that attacks stored peanuts, corn, and cottonseed, and produces the aflatoxins that are potent carcinogens.

Fungal substances that affect fungi. Medical literature is replete with examples of fungal metabolites that control fungi, and there are several examples in the mycological literature.

Two metabolites, terrein (XVII) and *E*-4-(1-propenyl-1-yl)-cyclopenta-1,2-diol (XVIII), have been isolated from *Aspergillus terreus*. These compounds have been used to control so-called take-all fungus (*Gaeumannomyces graminis* var. *tritici*), which causes root rot in wheat and rye grass. The compounds were antagonistic to the fungus, but the activity lasted only 5 days.

Marine substances that affect fungi. Marine natural products, which have been accessed for medicinal use, are now being evaluated as sources of biodegradable agrochemicals. The sponge *Plakortis angulospiculatus* has yielded two cyclic-peroxide acids (XIX), cyclic-peroxide 1 where R = H and cyclic-peroxide 3, where R = H and where there is an 11,12 dihydro function. These cyclic-peroxide acids are active against *Candida albicans* and *Aspergillus nidulans*.

Plant substances that affect marine life. An intensive search is under way in Japan for antifouling substances specifically to control the blue mussel, *Mytilus edulis*. The plant product rhaponticin (XX) from *Eucalyptus rubida*, when mixed with cupric sulfate, has a high level of repellant activity. Another plant product with repellant activity is kaempferol 3-*O*-(2″,6″-di-*O*-(*E*)-*p*-coumaroyl-β-D-glucopyranoside (XXI) from the oak, *Quercus dentata*. In addition, a series of isothiocyanates [CH_3NCS; $CH_3(CH_2)_{1-11}NCS$; and $CH_3(CH_2)_{13\ or\ 15}$] from wasabi, *Wasabia japonica*, and horseradish, *Cochlelia armoracia*, have also given excellent control of *M. edulis*.

For background information SEE AGRICULTURAL CHEMISTRY; AGRICULTURAL SCIENCE (PLANT); CHEMICAL ECOLOGY; FUNGISTAT AND FUNGICIDE; HERBICIDE; INSECTICIDE; PESTICIDE in the McGraw-Hill Encyclopedia of Science & Technology.

Horace G. Cutler

Bibliography. W. A. Ayer, I. Van Altena, and L. M. Brown, Three piperazinediones and a drimane diterpenoid from *Penicillium brevi-compactum*, *Phytochemistry*, 29:1661–1665, 1990; H. G. Cutler (ed.), *Biologically Active Natural Products Potential Use in Agriculture*, ACS Symp. Ser. 380, 1988; R. P. Doss et al., (*Z*)-oxacyclotridec-10-en-2-one, an alfalfa weevil feeding deterrent from *Medicago rugosa*, *Phytochemistry*, 28:3311–3315, 1989; K. Endo, E. Kanno, and Y. Oshima, Structures of antifungal diarylheptenones, gingerenones A, B, C and isogingerenone B, isolated from the rhizomes of *Zingiber officinale*, *Phytochemistry*, 29:797–799, 1990; E. L. Ghisalberti, M. J. Narbey, and C. Y. Rowland, Metabolites of *Aspergillus terreus* antagonistic towards the take-all fungus, *J. Nat. Prod.*, 53:520–522, 1990; S. P. Gunasekera et al., Two new bioactive cyclic peroxides from the marine sponge *Plakortis angulospiculatus*, *J. Nat. Prod.*, 53:669–674, 1990; M. Sakado, T. Hase, and K. Hasegawa, A growth inhibitor, 3-(*E*)-(methylthio)-methylene-2-pyrrolidinethione from light-grown radish seedlings, *Phytochemistry*, 29:1031–1032, 1990; N. Yamashita et al., An acylated kaempferol glucoside isolated from *Quercus dentata* as a repellent against the blue mussel *Mytilus edulis.*, *Agr. Biol. Chem.*, 53:1383–1385, 1989.

Petroleum engineering

Recent advances have involved development of database access systems for petroleum exploration, that is, computer mapping, and application of geophysical techniques to the production and exploitation of petroleum reservoirs.

COMPUTER MAPPING

In petroleum exploration, relational databases and advanced computer graphics have radically altered the work environment of computers from processing applications to interpretive usages. Extended network databases and local-area networks have been used with great success in production applications; it is anticipated that within the next few years much more attention will be given to object-oriented systems. In addition to processing applications, there is a heavy emphasis on facile gathering of data and extraction of selected items to provide effective displays and interpretations.

The analysis of subsurface petroleum information poses a dual problem: there is either too much or insufficient data for any given application. In exploration, the volumes of data from seismic processing and interpretation can strain the limits of most mass-storage devices, and sophisticated processing routines can require supercomputers. On the other hand, data from wells can be minimal to nonexistent and may require extreme extrapolation and model-based solutions. Production applications can also be computer-intensive. As an example, reservoir oil simulation is feasible only in advanced modules on supercomputers. In most cases, there is no happy medium. Because of this dichotomy, computing environments in the petroleum industry run the full range from laptop field applications to main-frame installations. The computing requirements have become visually oriented, requiring a large dependency on interactive, graphic-display and visualization routines.

In general, petroleum computing can be viewed on

three levels: geological computing, geophysical computing, and engineering applications. Recent geological computing trends have focused on database and spatial system configurations, with specialty applications such as cross-section balancing or geochemical modeling. Geophysical computing tends to be computer-intensive; interpretive installations are, like all interactive workstation environments, driven by graphics. Engineering applications are also computer-intensive; they are generally classified as either simulation or process types.

Geological computing. The recent proliferation of high-quality workstation environments has given an individual geoscientist the tools to create maps and perform analyses on a large scale and to incorporate data to a much larger extent than was previously possible. The demand for data has led to the development of major products for exploration applications. Databases, language structures, and hardware that facilitate access and management of data are becoming a necessary part of computing environments in companies of all sizes. These databases are commonly built around efficient languages. A few commercial products such as ORACLE and INGRES currently dominate this market. These programming languages are designed to perform as a central module in a developed package; they are not designed for use by exploration professionals as a daily front-end product.

In addition to database software, geographic information systems (GIS), initially developed for resource planning and cultural analysis, are being adapted to geological computing. The internal structure of geographic information systems differs from conventional computer-aided design (CAD) technology in the referencing of digital elements in a geographic (spatial) relation to all other elements in the database, not to layers (two-dimensional drawings) as symbols independent of their locations.

Because of this structure, geographic information systems offer significant advantages over CAD systems in the manipulation, storage, analysis, and querying of large spatial databases. Unfortunately, because the technology was developed for land-use planning applications, it has yet to be demonstrated that geographic information systems have the capability to handle common subsurface data easily. The main problem arises where there is a multiplicity of values and interpretations at a single point such as a shotpoint or well. While the spectacular graphic and analytical environment of geographic information systems can facilitate advanced surface modeling and interpretation, these systems have not yet achieved the seemingly elusive goal of providing a database for all exploration data in a common system.

Geophysical computing. The most spectacular advances in geophysical computing revolve around the interpretive workstation. These products generally are integrated minicomputers, dedicated to tasks of seismic interpretation. The display, manipulation, and interpretation of both two- and three-dimensional seismic data can now be done interactively. Both stand-alone workstations and integrated main-frame systems have been developed to perform this difficult task. Seismic lines and traces are displayed in color, and the system permits interactive identification of horizons from line to line. Most of these programs have automated facilities for enhancement of attributes as well as for correction of errors in interpretation between seismic lines (mis-ties). The computer-aided interpretation frees the explorationist or production geophysicist from the extremely large volume of paper-record sections; at the same time it allows for consistent and realistic mapping data.

Seismic interpretive workstations include integrated base mapping and well-log analysis programs. The results of interactive seismic interpretation, whether two- or three-dimensional, is meaningful only in the context of a map. The values from the interpretation of seismic data can be automatically posted to appropriate locations and easily mapped.

The ability of these workstations to handle three-dimensional interpretation becomes important in describing the volumetric geometry of potential reservoirs. By examining slices of the geological features, interpreters are able to document potential reserves as well as to assess the risks. Many of the drafting and display problems of the past can be handled easily in the elegant base-mapping routines available commercially. The production of a final map for presentation evolves rapidly from the work product.

Engineering computing. The demands of engineering computing in development of petroleum resources are intense, and hardly any area remains untouched. The giant reservoir simulation routines known as black-oil simulators have given way to even more sophisticated segmented-reservoir simulators. These modeling packages generally require supercomputing speeds to adequately produce reasonable representations of reservoirs. The addition of geostatistics (kriging) and fractal geometries in the simulation procedure hold promise for the analysis of reservoir heterogeneities. Almost any other quantitative task, from on-site well-log analysis to remotely telemetered production data, falls in the realm of the computer. One active area is the development of highly portable components for evaluation of production.

Hardware considerations. While significant development of computing platforms will continue to influence the expensive, central-processing-unit-intensive, and sophisticated applications, the development of efficient communications links and networks as well as mass storage devices and high-resolution graphics devices will drive the technology in the petroleum industry. As prices and sizes of workstations shrink to the desktop level, more and more computing power will be at the individual's control.

One of the main weak links in the development of this technology is the data-entry stage. As in all computing, the design of devices for the input of raw data will be strongly influenced by the ability of the system to operate on this input data. Seismic workstations operate with standardized tape input, but input of geological and engineering data still requires painstaking digitization. Currently, optical scanners can be used for only a fraction of the data input required for most practical applications.

The trend is for distribution and archiving of data on optical devices such as laser optical disks and compact-disk read-only-memory (CD-ROM) data storage devices. These optical disks have the ability to store not only digital information but image data as well. Expert systems can be used for classification of computer graphics images of fossils. The storage and retrieval of these images on optical disk constitutes the next step.

Future directions. In terms of implementation of products, both geological and geophysical computing are at a crossroads. Most, if not all, commercially developed products have been designed not for inter-communication between products but rather for efficient internal storage and performance for a given platform and process application. Many vendors hold internal data structures confidential and do not release conversion programs; others insist on reading data only in the given internal format. Unfortunately, most companies have a multitude of products, each selected for a specific optimal application. The ensuing babel of communication can delay or prohibit the integration of investigations from one application to another. Development of external standards for product communications and implementations for exploration computing has been initiated only recently.

The direction of product development will be influenced by the resolution of the problems involving the establishment of communication standards for the industry. The American Petroleum Institute (API), the American Association of Petroleum Geologists (AAPG), the Society of Exploration Geophysicists (SEG), the Society of Petroleum Engineers (SPE), and other national groups such as the Petroleum Industry Data Dictionary (PIDD) are leading these efforts. In addition to generic standards, product-specific standards are being developed, for example, by the Petrophysical Open Systems Corporation (POSC) and Openworks. It is anticipated that a manageable system-communications environment will result from these combined efforts.

On-location microcomputers are being used for geophysical applications as well as traditional geological and engineering fieldwork, creating on-site data verification and review. The portability of micro systems permits checking the quality of acquired data and completing preliminary interpretations during the acquisition phase, leading to efficient and robust collection of data.

The single most important product of exploration and development is still the map—a two-dimensional portrayal of the relationship of multiple variables. The speed, accuracy, and efficiency of data analysis required to create a useful map have been radically changed by digital processes. As the search continues for resources that are becoming increasingly scarce, computational coordination as well as sophistication will determine the success of all of these ventures.   *Brian Robert Shaw*

PRODUCTION GEOPHYSICS

Production geophysics is the application of geophysical techniques to problems related to the production and exploitation of petroleum reservoirs. It has become more apparent in the last few years that the internal heterogeneity of a petroleum reservoir has a profound influence on its production performance. The internal structure of a reservoir is complex and generally poorly understood. Structural deformations, fractures, lithologic variations, and diagenetic alterations all contribute to the creation or destruction of conduits and barriers to fluid flow. In addition, it may be important to monitor changes in fluid flow or composition during the producing life of the field. Since production geophysics is the only method available that can image the reservoirs under in-place conditions, it is rapidly becoming an active field of applied research aimed at improved descriptions and understanding of reservoirs and their fluid-flow behaviors.

Seismic methods. One of the most significant developments in production geophysics has been the use of seismic waves to image detailed rock and reservoir properties. Other geophysical methods are also used, including magnetic measurements, low-frequency electromagnetic measurements and detailed gravity measurements, both at the surface as well as in the wellbores. These methods generally have limited resolution, but they are relatively inexpensive as compared to seismic methods. They have met with different degrees of success and are continuing to be developed.

The use of seismic techniques for subsurface studies is not new; it has been successfully applied by earthquake seismologists and petroleum explorationists for many decades. However, application of seismic techniques to surface studies has been retarded by the limited resolution. As the elastic seismic waves propagate through the Earth, they are attenuated because of spherical divergence. Their higher-frequency components are even more attenuated due to scattering and absorption, resulting in a drastic decrease in resolution. For this reason the majority of successful cases in production geophysics reported to date have been limited to closed spacing imaging (short spacing between source and receiver) such as is found in shallow reservoirs or reservoirs very close to the wellbore.

An important development in seismic technology in the 1980s was the use of one or more boreholes for seismic imaging. By placing acoustic sources at appropriate distances from the receivers in one or more boreholes, it was demonstrated that higher-frequency seismic signals can be propagated farther in the subsurface, resulting in a greater improvement in resolution. Although this development in borehole geophysics is still in its infancy, it has already shown much promise, especially in terms of reservoir characterization and management, mapping of reservoir heterogeneity, monitoring of secondary or tertiary recovery, and as indicator of proximity to possible oil and gas accumulations.

Seismic technology. The formulation of the elastic-wave equation that governs the behavior of seismic-wave propagation through the Earth has concerned earthquake seismologists for many decades. By studying body waves such as compressional waves and shear waves propagating through the Earth's interior, seismologists have learned more about the constituent and elastic properties of its solid and liquid core, solid mantle, and thin crust. The major differences between earthquake seismology and petroleum exploration seismology are scales and knowledge of the location of seismic disturbances. Earthquake seismic waves have periods in minutes and resolution in kilometers, whereas in exploration seismics the period is tenths of a second and the resolution is tens of meters. In production seismology, there is a need for even higher-frequency seismic surveying and better resolution. Resolution on the order of meters is often required.

Seismic technology has undergone a quiet revolution since 1980. For the first time since the seismic tool was introduced, geophysicists can exploit it closer to its capacity. The availability of larger, faster, and less costly computers permits resolution of some of the theoretical complexities of elastic-wave propagation, so that deeper insight into the wave field phenomena can be obtained. The advances in computer technology and in microelectronics have produced many new tools for production geophysics.

Three-dimensional seismic survey. This is a method of acquiring surface seismic data by placing acoustic sources and receivers in an areal pattern (**Fig. 1**). One example of a simple three-dimensional layout is to place the receivers along one line and shoot into these receivers along a perpendicular line. These shots are produced by devices such as vibrators or other impulsive sources. The common midpoints of a three-dimensional survey also form an areal surface. When all these traces or time series are displayed, they form a closely spaced data volume that represents an acoustic image of a three-dimensional subsurface. Because of the large data set involved, processing of a

three-dimensional seismic survey can be very demanding on computing power. It is almost impossible to interpret three-dimensional data without the use of powerful workstations with sophisticated computer graphics. Because of the high costs this entails, the use of three-dimensional seismic surveys is often restricted to detailing specific fields rather than regional exploration.

Three-dimensional seismic surveying provides a more accurate and detailed image of the subsurface. It offers significantly higher signal quality than the data in two dimensions that are commonly acquired. It also improves both spatial and temporal resolutions. Because of its ability to provide geologic details, three-dimensional seismic surveying is being applied to production problems at rapidly increasing rates. A measure of this increase is that three-dimensional surveys now account for half of the seismic activity in the Gulf of Mexico and North Sea; the first commercial three-dimensional surveys were conducted in these areas in 1975.

Three-component seismic survey. This refers to a seismic survey in which three-component receivers, three-component sources, or both are used.

Seismologists have identified two modes of body-wave propagation in the subsurface. These waves are labeled primary (P; compressional) and secondary (S; shear). During the passage of the P wave, the rock changes volume but not shape in response to alternating compressive and tensional stresses. The S wave, on the other hand, changes shape but not volume in response to the passage of shear stresses. The shear waves move perpendicular to the direction of propagation. The speed at which each body wave travels is a function of the elasticity of the rock.

Conventional seismic surveys typically use velocity phones in the vertical direction only, which is adequate for P-wave exploration. To record S waves in three dimensions, however, three-component geophones are required. It is common practice to place the three-component phones orthogonally in the vertical direction and in the directions parallel and perpendicular to the direction of shooting.

If only P-wave sources are used in the survey, only converted shear waves are recorded by using the three-component phones. These are shear waves generated through conversion of P-waves at boundaries of acoustic discontinuities. Some surveys involve both P-wave and S-wave sources in three orthogonal directions, producing a total of nine seismic sections, one for each source–receiver direction.

A wealth of information related to rock and reservoir properties in these multicomponent data sets has become available only very recently. It will be some time before petroleum geophysicists have developed methods of interpretation; more research is needed. Some of the potential applications that have been reported are very encouraging; they include locating directions of fractured beds, estimating density of fractures, and identifying lithology and fluid contents.

Vertical seismic profile survey. This is a measurement procedure in which a seismic signal generated at the

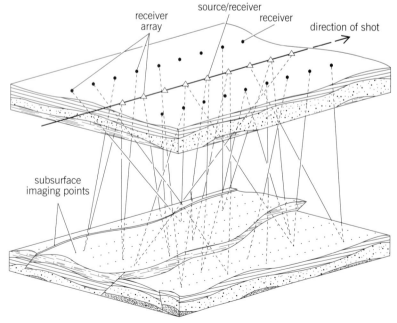

Fig. 1. Diagram showing a typical layout for a three-dimensional seismic survey.

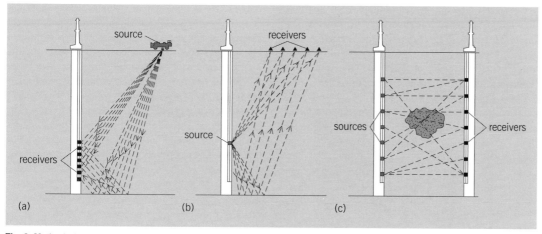

Fig. 2. Methods for determining borehole geometry. (a) Offset vertical seismic profile survey. (b) Reverse vertical seismic profile survey. (c) Cross-well tomography.

surface of the Earth is recorded by geophones secured at various depths to the wall of a drilled well. When the source is directly at the top of the well, it is known as zero-offset vertical seismic profiling; otherwise it is known as offset vertical profiling.

A primary result of the vertical seismic profile survey is the time-to-depth relationship from which interval velocities as a function of depth can be determined. A zero-offset vertical seismic profile survey provides a link between the acoustic log and surface seismic information in two important ways. First, it calibrates the integrated acoustic log times to the vertical seismic profile times, which are directly related to the seismic times. Synthetic seismograms generated by using calibrated acoustic logs match the surface seismic data better. Second, the zero-offset vertical seismic profile links seismic reflection character in time to the well information in depth. It helps to correlate the seismic response to the lithologic interfaces. The vertical seismic profile also helps to explain some of the ambiguities and uncertainties associated with processing and interpretation of surface seismic data.

An offset vertical seismic profile can also be used as a predictive tool for exploitation. It can image the reservoirs ahead of a drill bit, investigate reservoir conditions several hundred meters around the wellbore, or monitor the propagation of a steam front in a thermal enhanced recovery process.

Reverse vertical seismic profiling uses a reverse configuration in which the source is placed at a given depth in a borehole and the receivers are deployed in different patterns at the surface. It is a more time-efficient and cost-effective method of downhole imaging.

Cross-well tomography. Also known as cross-well seismology, this is a method of seismic imaging in which both sources and receivers are placed in separate boreholes (**Fig. 2**). By placing both sources and receivers at depths below the highly attenuating near-surface materials such as the weathered layers, seismic waves having frequencies up to several kilohertz can be propagated, resulting in a great

improvement of resolving power. The limiting factor is the amount of acoustic energy with which downhole sources can be actuated without risking the safety of the wellbore. This in turn limits the distance between the wells at which surveys can be safely run.

In a typical cross-well seismic survey between two wells, travel-time and amplitude measurements for each source–receiver location across the wells are made. A seismic wave that is transmitted through a medium rather than being reflected is known as a transmissive seismic wave. As each ray of such a wave passes through the reservoir under investigation, it is being modified according to the elastic properties of the reservoir. Tomography is an inverse processing technique that is used to reconstruct a two-dimensional image of these reservoir properties. Tomographic reconstruction techniques were developed originally for use in medicine. Conventional tomography assumes straight ray paths. When the effect of diffraction and scattering is considered in the mathematical model, it is known as diffraction tomography.

Cross-well seismic tomography has been used successfully to estimate velocities between wells, mostly for the purposes of monitoring enhanced oil recovery. Active research is being conducted in imaging reservoir heterogeneities such as faults, pinch-outs, or facies changes; initial results are encouraging.

Real-time seismic monitoring. This is a seismic technique used to monitor the change in the conditions of a reservoir as it is producing through time. It has been demonstrated to work most effectively in heavy-oil thermal recovery projects where steam front progressions have been successfully monitored by using repeated three-dimensional seismic surveys.

For background information SEE ACOUSTIC TOMOGRAPHY; DATA COMMUNICATIONS; DATABASE MANAGEMENT SYSTEMS; GEOPHYSICAL EXPLORATION; PETROLEUM ENGINEERING; PETROLEUM PROSPECTING; PETROLEUM RESERVOIR ENGINEERING; SEISMOLOGY; WAVE MOTION in the McGraw-Hill Encyclopedia of Science & Technology.

Tai P. Ng

Bibliography. American Association of Petroleum Geologists, *Geobyte*, bimonthly; A. R. Brown, *Interpretation of Three-Dimensional Seismic Data*, AAPG Mem. 42, 1986; *Computer and Geosciences*, 8 times per year; S. H. Danbom and S. N. Domenico (eds.), *Shear-Wave Exploration*, Geophysical Development Series, vol. 1, 1986; D. H. Johnston, Recent advances in exploitation geophysics (two articles), *Lead. Edge*, 8(9):22–28, 1989; A. Nur, Four-dimensional seismology and (true) direct detection of hydrocarbons: The petrophysical basis, *Lead. Edge*, 8(9):30–36, 1989; Special issue on reservoir geophysics, *Lead. Edge*, vol. 8, no. 2, 1989; Society of Petroleum Engineers, *J. Petrol. Tech.*, monthly.

Pharmacognosy

In the fall of 1989, three cases of a very rare disease, subsequently named eosinophilia-myalgia syndrome (EMS), were seen by three physicians in New Mexico. All three patients had been taking L-tryptophan food supplements, and this was thought to be the cause of their illness. Subsequent investigations showed that there had been an epidemic of eosinophilia-myalgia syndrome in the United States in the summer and fall of 1989, affecting thousands of people in all 50 states. Epidemiologic studies demonstrate that the syndrome was apparently caused by an impurity (or contaminant) in tablets and capsules containing L-tryptophan that was produced by a single Japanese manufacturer. Dozens of impurities can be found in small amounts in the implicated products containing L-tryptophan, and studies are still under way to identify the specific causative agent. This disease outbreak has important implications for the regulation of the food supplement industry and the safety of products made by similar manufacturing processes. A full understanding of the pathogenesis of eosinophilia-myalgia syndrome may also provide insight into the development and prevention of related diseases.

L-Tryptophan is an essential amino acid present in a wide variety of foods. Although it was used by many people for insomnia, depression, premenstrual syndrome, and other less frequent indications, such as weight loss and drug detoxification programs, it was not classified as a drug. Like vitamins, minerals, and other amino acids, it was classified as a food supplement, and therefore treated as such by the Food and Drug Administration.

Symptoms of EMS. Eosinophilia-myalgia syndrome is marked by a striking elevation of eosinophils and the presence of severe, incapacitating myalgia, although in some patients the myalgia may be less pronounced or inapparent. A variety of other tissues or organs may be involved, and neurologic, pulmonary, dermatologic, and vascular symptoms are frequently seen. The clinical course of the disease may be prolonged, even after the cessation of L-tryptophan use. Medication (usually glucocorticoid steroids) is of only modest benefit in many patients. The disease can be quite severe; of the cases that were reported to the Centers for Disease Control, over 30% of the patients had already been hospitalized. Twenty-seven deaths in patients with eosinophilia-myalgia syndrome (1.75% of reported cases) had occurred as of August 1990.

Epidemiology of EMS. The initial case reports from New Mexico were followed by two epidemiologic studies confirming that there was a strong relation between use of products containing L-tryptophan and the onset of eosinophilia-myalgia syndrome. It was found that all individuals who developed eosinophilia-myalgia syndrome had taken L-tryptophan before the onset of the illness.

Triggered by news reports of the patients in New Mexico, cases were soon being reported from other parts of the country. The Centers for Disease Control, working with all state and territorial health departments, quickly established a national reporting and surveillance system for eosinophilia-myalgia syndrome. By August 24, 1990, 1536 cases had been reported to the Centers for Disease Control. Since reporting of cases to state health departments by physicians is voluntary, it is likely that this is an underestimate of the total number of cases that occurred. Furthermore, as physicians became more familiar with the syndrome, the clinical spectrum of the disease was seen to be broader than initially appreciated, suggesting that there was also underreporting of atypical or very mild cases. It is probable, therefore, that as many as 5000 or more cases of eosinophilia-myalgia syndrome may have occurred in this epidemic.

The national surveillance showed that the epidemic had begun in early summer of 1989 and had reached its peak in the fall. In November 1989, the Food and Drug Administration responded to the rapidly increasing number of cases by instituting a recall of products containing L-tryptophan from retail outlets. This led to a rapid termination of the epidemic.

Ages of patients varied from 4 to 85 years, with a median of 48 years. Ninety-four percent of patients were non-Hispanic white, and 83% were female. The geographic distribution of cases suggests higher rates in western states. This demographic pattern is consistent with known usage patterns for food supplements in general, and also for specialty food supplements such as amino acids. Recent epidemiologic studies tend to corroborate that the distribution of the syndrome reflects the degree to which L-tryptophan was taken by different population groups rather than a unique susceptibility of groups such as females.

Worldwide, eosinophilia-myalgia syndrome has been reported from approximately a dozen other countries, but the incidence rates are much lower than in the United States. Again, this probably reflects a much higher use of such products in the United States. In addition, unlike the United States, many other industrialized countries treat L-tryptophan as a drug and require pharmaceutical-grade L-tryptophan (a higher grade, or more purified form) to be prescribed by physicians.

Eosinophilia-myalgia syndrome is a unique disease. Prior to the epidemic in 1989, sporadic cases of disease consistent with the diagnosis of eosinophilia-myalgia syndrome have been seen, but in the entire world literature for such cases the total number that have been reported is only a small fraction of the number of cases that occurred in the L-tryptophan-related outbreak. Most of the earlier sporadic cases have been diagnosed by descriptive terms such as eosinophilic fasciitis or eosinophilic perimyositis. The clinical appearance of eosinophilia-myalgia syndrome, however, is similar to that seen in an epidemic that occurred in Spain in 1981. The latter disease became known as toxic oil syndrome when it was found to be caused by an adulterated rapeseed oil that was sold for food use rather than its intended industrial use. Over 20,000 cases of toxic oil syndrome, including 315 deaths, were seen in the Spanish epidemic. Although toxic oil syndrome and eosinophilia-myalgia syndrome are not identical, they are so closely related that it is likely that a common pathogenetic mechanism is involved.

Causative agent of EMS. Since it was suspected that a contaminant of L-tryptophan might be the causative agent of eosinophilia-myalgia syndrome, much of the epidemiologic and laboratory investigation has focused on the search for such a contaminant. Three state health departments (Oregon, Minnesota, and New York) have reported results of case-control studies in which they compared the specific brands of L-tryptophan used by patients with the syndrome to brands consumed by users of L-tryptophan who did not get sick. The studies indicated that certain brands posed a much higher risk than others. Subsequently, L-tryptophan products used by individuals in the studies were traced back through the distribution channel (retailer, distributor, bottler, importer, manufacturer), and it became apparent that virtually 100% of patients with eosinophilia-myalgia syndrome used L-tryptophan produced by a single Japanese manufacturer. Preliminary results from a selective national tracing of L-tryptophan used by those with the syndrome again showed that virtually all had used L-tryptophan manufactured by the same company.

With the source of the contaminated L-tryptophan established, subsequent investigations have focused on the laboratory analysis of specific contaminants in the L-tryptophan, and the production process at the implicated manufacturing plant. Initial laboratory investigations demonstrated that bacteria, viruses, endotoxins, inorganic chemicals, and radioactive substances were not significant. A variety of organic chemical contaminants have been identified in L-tryptophan samples from the implicated manufacturer. These include complexes of L-tryptophan, a variety of indoles, beta carbolines, and bacitracin. At this point, the contaminant that most closely fits the epidemiologic data, and is therefore most actively pursued as the potential culprit, is the di-L-tryptophan aminal of acetaldehyde. Other contaminants are still being identified and investigated; however, any candidate causative agent will have to be confirmed in cell culture or animal studies.

Implicated L-tryptophan was produced by a bacterial fermentation process, using a bacterial organism that was genetically engineered to produce large amounts of L-tryptophan. At the end of the fermentation production process, L-tryptophan is separated and purified. To analyze how a particular contaminant might have appeared, many factors need to be considered, including the bacterial strain that was used, the starting materials in the fermentation process, the operating conditions for the fermentation process, and the purification steps. It is now clear that very potent bioactive compounds can be produced by this type of production process. Investigators are currently reviewing the process in detail for further clues as to how specific contaminants may have been produced.

For background information *SEE GENETIC ENGINEERING; PHARMACOGNOSY; TRYPTOPHAN* in the McGraw-Hill Encyclopedia of Science & Technology.

Henry Falk

Bibliography. Analysis of L-tryptophan for the etiology of eosinophilia-myalgia syndrome, *Morbid. Mortal. Week Rep.*, 39:589–591, 1990; E. A. Belongia et al., An investigation of the cause of the eosinophilia-myalgia syndrome associated with tryptophan use, *N. Engl. J. Med.*, 323:357–365, 1990; L. Slutsker et al., Eosinophilia-myalgia syndrome associated with exposure to tryptophan from a single manufacturer, *JAMA*, 264:213–217, 1990; L. A. Swygert et al., Eosinophilia-myalgia syndrome: Results of national surveillance, *JAMA*, 264:1698–1703, 1990.

Plant hormones

The plant hormones (auxin, cytokinins, gibberellins, abscisic acid, and ethylene) play a key regulatory role in plant development. Plant cells produce these hormones at extremely low concentrations. The activities of neighboring or distant cells are affected by the hormones that are transported to them. In addition, it is now recognized that secondary messengers, released into the cytoplasm in response to cellular stimulation by one of these primary messengers, play an important role in cellular responses. The known secondary messengers are calcium, diacyl glycerol, cyclic 3'-5'-adenosine monophosphate (cAMP), and cyclic 3'-5'-guanosine monophosphate (cGMP).

Hormone receptors. Two general types of hormone receptors are found in animals. Those related to the steroid hormone receptors bind the hormone at the cell membrane. The hormone–receptor complex is then translocated to the nucleus and acts there to stimulate transcription of specific genes. The more common type of receptor is an integral cell membrane protein that transduces the hormonal signal via release of a secondary messenger into the cytoplasm. The following evidence suggests that the latter type of receptor mediates effects of certain plant hormones.

When exposed to certain stimuli, such as light or wounding, plant cells produce rapid changes in the electrical charge at the cell membrane. The speed of this response would not be possible unless these charge variations were transduced by biochemical pathways that involved posttranslational rather than transcriptional regulation of proteins. Certain effects of auxin, ethylene, and abscisic acid occur within a few minutes, which is also too short a time to be caused by de novo protein synthesis. Reversible protein phosphorylation is the only universal mechanism known in eukaryotes for such rapid cellular responses to environmental stimuli.

Secondary messenger system. Only one common type of signaling system has been found to exist among eukaryotes. It involves protein phosphorylation that is dependent on secondary messengers. In this system, shown in the **illustration**, an extracellular hormone is produced by one or more cells in response to an environmental stimulus or internal developmental cue. A single hormone molecule binds to the extracellular portion of a receptor located on the surface of responsive cells; the cytoplasmic portion of the receptor then activates guanine nucleotide–binding proteins (G proteins). In the activated state [in which guanosine triphosphate (GTP) rather than guanosine diphosphate is bound], G proteins either inhibit or stimulate enzyme effectors, resulting in an increase in the concentration of a secondary messenger. For example, adenylate cyclase, an enzyme located within the cell membrane, is stimulated by a specific G protein to produce cAMP. Phospholipase C, an enzyme that hydrolytically breaks down phospholipids in the cell membrane, is the effector that increases the concentration of diacyl glycerol. Inositol triphosphate, another product of phospholipid breakdown, acts to trigger release of calcium from intracellular stores (such as the endoplasmic reticulum in animals) into the cytoplasm. Thus phospholipase C activation increases, both directly and indirectly, cytoplasmic concentrations of two secondary messengers.

Secondary messenger molecules act by binding to, and thus activating, protein kinases, which are enzymes that posttranslationally add phosphate to spe-

cific amino acids of various cellular proteins; phosphorylation modifies the activities of these proteins, resulting in altered cell activity or morphology. Protein phosphatases are enzymes that remove the phosphate from proteins modified by kinases, thus eventually attenuating the signal. Protein kinase activation has more prolonged effects on cells through phosphorylation of those proteins that affect transcription of specific genes. Altered transcription then causes cellular responses via protein synthesis.

A small initial input signal can be amplified at different steps in the pathway and multiplied to effect many different types of target molecules by this cascade of biochemical events. The amplification and multiplication provided by this type of signaling system could account for the diverse effects of extremely low concentrations of plant hormones. Individual components of this type of signaling system have recently been discovered in higher plants, supporting the view that its operation is universal in eukaryotes.

Plant signal transduction components. These components include proteins with the conserved structural features of G proteins and protein kinases that are activated by secondary messengers. Other proteins, structurally similar to DNA binding proteins that are responsive to secondary messengers, regulate gene transcription in plants. Plant protein phosphatases have been found to be similar in activity and specificity to those active in animal signaling systems.

The best-characterized plant hormone receptor is that for auxin. The auxin-binding protein is a peptide that is associated with the endoplasmic reticulum. The site of auxin action is, however, known to be extracellular. The small size of auxin-binding protein and the fact that it is not an integral protein of the cell membrane are inconsistent with an identity as a hormone receptor that is coupled with a secondary messenger. One possible mode of action is for an auxin-binding protein–auxin complex to bind to a receptor that is coupled with a secondary messenger. *See* Auxin.

Some genetic components in plants that are responsible for high levels of gene transcription are identical to elements that are responsive to secondary messengers in other organisms, although it is not known yet which hormonal signals modulate the activity. Other genetic elements have been found that confer gene transcription in response to abscisic acid, but it is not yet known precisely how the hormone effect is transduced.

Calcium as a plant secondary messenger. Intracellular calcium levels have a well-documented and important regulatory function in plants. Phenomena such as stomatal movement, mobilization of seed protein reserves at germination, touch-induced morphological changes, and gravitropic and phototropic curvature are mediated by changes in intracellular calcium. *See* Plant movements.

Levels of inositol triphosphate are increased in plants in response to certain environmental stimuli. Since inositol triphosphate induces calcium release

Schematic drawing of cellular signaling system involving protein phosphorylation that is dependent on a secondary messenger.

from the plant vacuole to the cytoplasm, the vacuole may be the principal source of signal-induced cytoplasmic calcium fluxes. Plants also contain calmodulin, a highly conserved calcium-binding regulatory protein that transduces some calcium effects in animals and microorganisms. Protein phosphorylations that are dependent on calcium and calmodulin occur in plant cells responding to various environmental stimuli, indicating that calcium functions in transduction of these responses. Since plant hormones are known to be involved in the response to these stimuli, it is possible that calcium serves as a secondary messenger in these responses. Antagonists of calcium or calmodulin activity interfere with many plant responses, thus supporting this view. SEE ROOTS.

The role of cAMP as a secondary messenger in plants has been the subject of a great deal of controversy, largely because the levels measured are frequently substantially lower than in other eukaryotes. However, protein kinase activity that is dependent on cAMP has been demonstrated in several different plants. Additionally, mating responses of a unicellular plant, *Chlamydomonas*, utilize cAMP as a secondary messenger to activate protein kinase in the transduction cascade.

Hormone effects on secondary messengers. Although a complete transduction scheme is not yet known for any plant hormone, recent evidence suggests that calcium acts as a secondary messenger that mediates effects of auxin and abscisic acid. Auxin application stimulates formation of inositol triphosphate in plant cell cultures within minutes. Stomatal closure that is induced by abscisic acid is preceded by an increase in calcium within the guard cells. Uptake of exogenous diacyl glycerol mimics some of the effects of abscisic acid on guard cells, implying that the transduction pathway for abscisic acid involves formation of both this second messenger and calcium through inositol phospholipid breakdown by phospholipase C. Calcium antagonists interfere with responses to both auxin and abscisic acid. Major challenges in plant hormone research will be to link steps in transduction pathways that are dependent on secondary messengers and to isolate the plant hormone receptors.

For background information SEE PLANT GROWTH; PLANT HORMONES; PLANT MORPHOGENESIS in the McGraw-Hill Encyclopedia of Science & Technology.

Brenda J. Biermann

Bibliography. J. Braam and R. W. Davis, Rain-, wind-, and touch-induced expression of calmodulin and calmodulin-related genes in *Arabidopsis, Cell*, 60:357–364, 1990; C. Ettlinger and L. Lehle, Auxin induces rapid changes in phosphatidylinositol metabolites, *Nature*, 331:176–178, 1988; H. Ma, M. F. Yanofsky, and E. M. Meyerowitz, Molecular cloning and characterization of GPA1, a G protein α subunit gene from *Arabidopsis thaliana, Proc. Nat. Acad. Sci. USA*, 87:3821–3825, 1990; M. R. McAinsh, Abscisic acid-induced elevation of guard cell cytosolic calcium precedes stomatal closure, *Nature*, 343:186–188, 1990.

Plant movements

Gravitropism, or geotropism as it was once known, is the vectorial growth response of a plant organ to the force of gravity. A similar response can be obtained with the experimental application of centripetal force. Germinating seeds must be able to direct the growth of the embryonic shoot and root in order to establish independence from the seed. Gravity is the environmental factor that directs the growth of the seedling germinating below the surface of the soil. Thus, the response of plant organs to gravity is essential to the survival of plants.

Classification. At a very basic level, there are two growth responses of a plant to the force of gravity. Plant growth toward the direction of the force of gravity is known as positive gravitropism, and growth away from the force of gravity is negative gravitropism. An example of positive gravitropism is the growth of the primary root downward into the soil. An example of negative gravitropism is the growth of the seedling shoot upward from the soil.

Another method of classification places plant growth responses to gravity into one of four categories. Orthogravitropic (orthogeotropic) response is growth of an organ vertical or parallel to the force of gravity, for example, the growth of monocotyledonous or dicotyledonous seedling. Plagiogravitropic (plagiogeotropic) response is growth of an organ at an angle to the force of gravity, for example, the growth of lateral branches and lateral roots. Diagravitropic (diageotropic) response is growth of an organ perpendicular to the force of gravity, such as the growth of rhizomes or stolons parallel to the soil surface. Agravitropic (ageotropic) response is the lack of directional growth of an organ in response to gravity, for example, the growth of some mutant plants (agravitropic mutants), some roots grown in water or liquid culture, and higher-order branch roots and stems. Since orthogravitropic responses are the most intensely studied of these four categories, this response will be emphasized in the discussion.

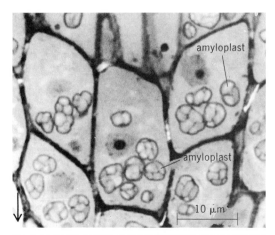

Fig. 1. Light micrograph of statocytes in the cap of a primary root of rice (*Oryza sativa*). Statoliths (amyloplasts) are observed in the lower portion of the cells. The direction of the force of gravity is indicated by the arrow at the lower left.

Perception of gravity stimulus. There are three components of the gravitropic response. (1) The plant organ perceives the gravity stimulus. (2) The stimulus is transmitted or transduced to the region of the organ where the growth response will occur. (3) A growth response occurs.

The perception of the gravity stimulus by higher plants is thought to occur by a similar mechanism. Exposure to gravity must exceed a critical period or presentation time before a response occurs. This

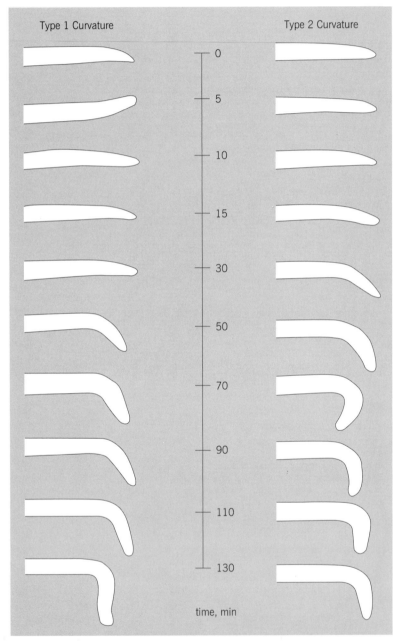

suggests that perception of gravity is through the intracellular movement of cellular components. Cellular components that move in response to changes in the gravitational field are known as statoliths and occur in specialized cells known as statocytes. The location of statocytes varies with the organ. In roots the statocytes are located primarily in a discrete region, the columella, of the root cap (**Fig. 1**). Statocytes are found in pulvini of mature grass and cereal plants. In dicot stems and monocot coleoptiles the sites of gravity perception are scattered along the length of the organ. Thus, much of the work involving the sensory mechanism has been performed by using roots and pulvini.

Evidence suggests that the statoliths in the statocytes of many plants are starch-containing, membrane-bound plastids known as amyloplasts (Fig. 1). The kinetics of gravity perception closely correlates to the gravity-induced movement of these starch-containing amyloplasts. Further evidence that amyloplasts are responsible for gravity perception is that starch-depleted organs lose their sensitivity to gravity and require a much longer presentation time before a response can be detected. In addition, removal of the root cap, and with it the amyloplast-containing statocytes, results in the loss of sensitivity to gravity in most species.

The exact physiological or biochemical changes during the perception phase of gravitropism have not been elucidated. Evidence in at least one plant species suggests that differential pressure exerted by the amyloplasts on the endoplasmic reticulum or plasma membrane is sensed by the membranes. The change in the membranes would be central to the perception of gravity. Other investigators have reported high concentrations of calcium associated with the amyloplast membranes. Calcium has been shown to alter membrane polarity and transport and thus could account for the membrane change. It has also been suggested that starch statoliths interact with the intracellular network of microtubules and microfilaments known as the cytoskeleton. Elements of the cytoskeleton are known to have direct associations with membranes; thus the membranes could indirectly sense the movement of statoliths via cytoskeletal connections. The nature of the membrane change has not yet been determined, but could involve a change in ion transport or permeability.

Transduction of stimulus and response. Although the perception of gravity is basically similar in shoots and roots, the transduction of the stimulus and the growth response by these organs is different due to the spatial association of the site of perception and the site of response. Therefore, the transduction of the gravity stimulus and the differential growth response in these organs are best discussed separately.

In the coleoptiles of monocots, negative gravitropic response occurs. This is due to an increase in cellular elongation growth on the lower side of the horizontally oriented organ compared to the upper side. This differential growth response can be explained by an asymmetrical distribution of auxin, a substance that

Type 1 Curvature   Type 2 Curvature

0
5
10
15
30
50
70
90
110
130

time, min

Fig. 2. Kinetics of two types of curvature of intact roots of maize (*Zea mays*). The scale indicates the time in minutes that has elapsed since the root was placed in a horizontal position. Type 1 curvature is observed in about 30% of the roots examined and is characterized by an initial negative response followed by a significant positive response after 50 min. Type 2 curvature is observed in about 70% of the roots examined and is characterized by significant positive curvature within 30 min and an overshoot, curving beyond the vertical at 70 min. This is followed by an oscillatory readjustment of the angle of curvature toward the vertical. (*Dr. Timothy J. Mulkey, Department of Life Sciences, Indiana State University*) controls plant growth. The greater concentration of auxin in the lower side of the coleoptile causes the cells there to elongate faster than those cells on the upper side, resulting in upward growth. The source of the auxin is the tip of the coleoptile, so that the auxin must be transferred some distance, to the site of elongation behind the tip.

Shoots of dicots also undergo negative gravitropism, but much less is known about the transduced signal that results in the growth response. Evidence indicates that auxin asymmetry occurs in these stems as in coleoptiles, but to a smaller degree. Another growth-regulating substance, gibberellin, has been reported to be asymmetrically distributed in these stems with the greater concentration located on the lower side. However, there is no evidence for the lateral distribution of gibberellin, and the asymmetry was measured well after the gravity response had occurred. Evidence also points to an asymmetric distribution of cations, potassium in the lower and calcium in the upper portion of the stem. It is uncertain what, if any, role the distribution of potassium and calcium ion levels has in the gravitropic responses of these stems, but this distribution could be a result of the asymmetric distribution of auxin there.

In an attempt to address the shortage of substantial evidence to support a theory of gravity response in shoots, many investigators have proposed that the outer cell layers of the stem or the epidermis are responsible for the regulation of growth in stems (and coleoptiles) and therefore control the gravitropic response. This suggests that the lower outer cell layers grow more rapidly than the upper outer cell layers and thus control stem growth. Stripping of the epidermis and nearby outer cell layers on the lower surface reduces gravitropic response. The response can be restored by the addition of exogenous auxin to the lower surface. Thus, auxin controls this growth and apparently increases in the lower cell layers.

Cereal grass shoots have at the base of their leaves distinct regions known as pulvini, which have the potential for growth only if placed horizontally. Gravity is perceived and the growth response is produced by the statocytes or cells in the immediate vicinity. Auxin is asymmetrically distributed to the lower portion of the organ, possibly as a result of the synthesis, release, or activation of a precursor molecule upon gravistimulation. The presence of auxin in the lower cells of the pulvinus produces elongation growth of these cells, whereas cells in the upper portion elongate very little, producing a negative gravitropic response.

Traditionally, root gravitropism has been explained by a mechanism that is similar to that of gravitropism in coleoptiles. Gravity is perceived by the root cap, which results in the asymmetrical distribution of auxin in the lower region of the root. Since root cell elongation is inhibited by the increase of auxin concentration above ambient levels, the asymmetry produced by gravistimulation results in the inhibition of root cell elongation on the lower side and increased elongation of cells on the upper side; this results in a positive gravitropic response (**Fig. 2**). In some cases, a change in the kinetics of curvature results in an initial negative gravitropic response followed by the usual positive response; this is known as type 1 curvature. This may be a result of differential sensitivities or transport of auxin in these tissues. The purely positive response is known as type 2 curvature.

Considerable controversy surrounds the role of auxin in root gravitropism. Some investigators maintain that auxin has no role in root gravitropism and that growth inhibitors from the root cap, abscisic acid (ABA), or unknown growth inhibitors are asymmetrically distributed and responsible for the gravitropic response. Other investigators speculate that an early event of the gravitropic response results in the accumulation of auxin or increases the sensitivity of cells in the lower portion of the root to the growth-inhibitory effects of auxin.

Most investigators agree that the sensory mechanism of gravitropism involves the movement of some cellular component; however, the physiological or biochemical result of the perception and the transduction of the stimulus have yet to be defined.

For background information *see* A*uxin;* P*lant growth;* P*lant movements* in the McGraw-Hill Encyclopedia of Science & Technology.

Martin A. Vaughan

Bibliography. M. L. Evans et al., How roots respond to gravity, *Sci. Amer.*, 255:112–119, 1986; P. B. Kaufman et al., How cereal shoots perceive and respond to gravity, *Amer. J. Bot.*, 74:1446–1457, 1987; R. Moore and M. L. Evans, How roots perceive and respond to gravity, *Amer. J. Bot.*, 73:574–587, 1986; B. G. Pickard, Early events in geotropism of seedling shoots, *Annu. Rev. Plant Physiol.*, 36:55–75, 1985.

Plant pathology

When a plant is invaded thoroughly by a virus, it is usually protected against the effects of a second infection by a more severe strain of the same virus. Although many terms have been used to describe this, the effect is commonly referred to as cross protection. Cross protection was used formerly to determine the relatedness of virus strains, but more reliable tests are now used for this purpose. Because the first virus, the protectant virus, often does little or no damage to the plant, it was suggested that cross protection might be used to control plant virus diseases in the field. Difficulties in implementing cross protection as an agricultural practice, however, discouraged early attempts to use it as a control. Recently these difficulties were overcome, and practical control of several virus diseases has been successful. Developments in biotechnology have shown that the basic mechanism of cross protection can be put to practical use without the drawbacks involved with using mild strains of active viruses as protectants.

Drawbacks. There are at least four drawbacks to commercial use of cross protection. (1) Protection is not always obtained or may be incomplete. Only that tissue invaded by the protectant virus is protected from subsequent infection, and experimental inoculation with high doses of a severe strain may overcome protection. (2) The protectant virus might spread to other kinds of susceptible plants, in which it could cause a severe disease. It was thought possible that the protectant virus might mutate to a more severe form and be destructive. (3) Instances are not uncommon where infection of a diseased plant with a second, unrelated virus results in a disease more severe than would be caused by either of the two viruses separately. Thus, severe disease may develop if an unrelated virus were to infect plants purposely infected with a mild virus to protect them against a different disease. (4) Much work and some expense may be required to infect each of thousands or hundreds of thousands of small annual seedlings with a protectant virus. This would be less of a problem with a perennial crop, which would not need to be infected each year, such as fruit trees.

Commercial use. Despite these drawbacks, cross protection recently has been used successfully against several devastating plant virus diseases. Situations in which cross protection should be considered as a control were described a number of years ago for the swollen shoot disease of cacao in West Africa. Diseases appearing to be candidates for such control should be endemic—that is, always present throughout a large area—and impossible to eradicate or control in other ways. A candidate disease should be one that spreads rapidly and results in losses that are so great that some reduction in yield, which might be caused by a mild strain over a long period, would not be considered an objectionable drawback. Effective protection by a mild strain of the causal virus is, of course, a prerequisite. Successful application of cross protection to a number of serious diseases has illustrated the soundness of these principles.

Tristeza. Protection against the virus causing the devastating tristeza (quick decline) of citrus was achieved by selecting mild strains of the virus in naturally infected orange trees in Brazil and Florida. Careful testing of strains showed that those selected on the basis of mildness in one kind of citrus were not necessarily mild in others. Field tests of 45 Brazilian strains indicated that 3 were satisfactory for orange, 2 for Galego lime, and 1 for Ruby Red grapefruit. Mild strains differed in their ability to protect; the protection was overcome when numerous aphid vectors carrying a severe strain of virus were placed on a protected tree. Such breakdown of protection, however, did not occur naturally. These tests illustrated that it is important to test the protected plants under conditions to which the crop is normally exposed rather than under unrealistic experimental conditions that are so severe as to overcome a useful degree of protection that would remain under field conditions. From experience in Brazil, a number of criteria were recommended for obtaining protectant virus strains: (1) Select from a large number of isolates those that protect best. (2) Use mild field strains from the variety of citrus to be protected. (3) Collect mild strains from orchards where most trees show severe symptoms. Such strains have probably already protected against the severe strains. (4) Evaluate mild strains by vector transmission as well as by graft infection.

The application of cross protection to a perennial crop such as citrus can be done by propagating many new trees as clones from a mother tree infected with a single mild strain. This is normal propagation procedure and does not involve additional time or labor to produce trees containing the protectant virus. The capacity of mild strains of tristeza virus to protect against severe strains did not change after transfer to other trees.

Severe symptoms of tristeza occurred in some protected trees. The reason for this is not known, but it might have resulted from heavy exposure to severe virus strains. The success of cross protection as a control, however, should be measured in terms of increased income from orchards of protected trees rather than complete protection of each tree.

Tomato mosaic. Another example in which cross protection has been effective is tomato mosaic, caused by a highly infectious virus that spreads rapidly by contact during cultural operations. Attempts to prevent the disease by sanitary and cultural practices were not effective, so control efforts turned to cross protection. The first mild strains were produced by incubating infected tissue at temperatures above 93°F (34°C). The strains thus obtained caused some reduction in yield and were eventually supplanted by a mild strain produced by mutation of tomato mosaic virus induced by nitrous acid treatment.

Because tomato is an annual crop, it was necessary to infect large numbers of small seedlings with the protectant virus. This was done efficiently by using high-pressure spraying with plant sap containing the protectant virus plus finely divided carborundum, an abrasive. A commercial paint sprayer operated by a compressor was used to spray seedlings when they were in compact groups in propagating flats.

As is usual with cross protection, some protected plants developed severe symptoms, but these occurred late in the season and did not greatly affect the yield of fruit. Cross protection has been widely used in Europe, Great Britain, and Japan and has resulted in increased yields. This control, however, has been largely replaced by the use of recently developed tomato varieties that are genetically resistant to tomato mosaic virus. Resistant varieties are the ultimate control for plant disease because they require no additional work or expense by the grower.

Papaya ringspot. Another disease for which control by cross protection has been successful is papaya ringspot. This important disease is caused by an aphid-transmitted virus that infects only a few other kinds of plants. Resistant varieties of commercial quality are not available, and other attempts at control have not been successful.

In Hawaii, papaya ringspot is confined to the island of Oahu, where it is endemic. It does not occur on the island of Hawaii in the Puna district, the state's major production area. Thus, cross protection trials on Oahu seemed sufficiently isolated to offer no threat to the major production area in case the protectant strain mutated to a severe form. Besides papaya, only plants in the cucurbit family are infected by papaya ringspot virus. The protectant strain of the Hawaiian virus was symptomless in cucurbits and protected them from severe strains. Thus, there seemed to be no danger of it causing destruction in other crops.

Attempts to find mild strains of papaya ringspot virus occurring naturally, as had been done with citrus tristeza virus, were not successful in either Hawaii or Taiwan. Attempts were then made to induce mild mutant strains of severe Hawaiian virus by the nitrous acid treatment. Eventually a strain was obtained that did not cause symptoms in papaya but did protect plants from the effects of severe strains encountered in the field.

Although papaya is a perennial in the tropics, plants are started from seed. Closely spaced propagation of papaya seedlings provided an opportunity for infecting them with the protectant virus by high-pressure spray inoculation, as had been done with tomato mosaic virus. One person could inoculate 10,000 seedlings in 2 h and achieve 100% infection.

In Hawaii, papaya subjected to natural infection by aphids was well protected by the symptomless strain. In Taiwan, extensive exposure to severe strains occurred when protected and unprotected plants were mixed in the same experimental plot, and severe symptoms developed in protected plants. Although symptoms were delayed, there was no economic benefit because symptoms occurred before fruit was set. When protected and unprotected papaya were established in separate blocks, however, not adjacent to diseased papaya orchards, and when severely diseased plants were removed every 10 days, the protected papaya gave a 111% increase in monetary return. This was due both to more good-quality fruit and to increased yield.

Because some protected papaya plants in Taiwan eventually developed severe symptoms, it was evident that protection was less efficient in Taiwan than in Hawaii. The protectant strain that had been produced from a severe Hawaiian strain protected better against severe Hawaiian strains than against severe Taiwan strains. It thus may be preferable to produce or select protectant strains from the locality where they will be used.

Molecular basis. The molecular basis for cross protection has been the subject of speculation and controversy. For many years when cross-protection tests were used as evidence of relatedness of viruses, precautions were necessary in interpreting such tests. Viruses were considered related if one protected against the other. If there was no protection, however, no conclusions could be drawn regarding relatedness. Also, depending on how the tests were made, reactions resembling cross protection might occur between obviously unrelated viruses.

Although the mechanisms within the plant that result in cross protection have not been clearly de-

fined, there is evidence that only part of the virus is involved. Most plant viruses consist of a nucleic acid core (the genetic part) surrounded by a coat of protein that is different for each virus. When a virus infects, the protein coat must first be removed in order for the nucleic acid to interact with components of the plant cell to produce new virus, including the coat protein. For most plant viruses, the protein coat is determined by a single virus gene. Evidence suggests that this virus protein has a major role in cross protection.

Genetic engineering has been used to aid in providing cross protection. Virus genes, including the gene that controls the production of virus coat protein, can be incorporated into the genetic makeup of crop plants. Tobacco plants have been produced that contain the coat protein of tobacco mosaic virus, but do not contain infective virus. Tobacco was used in these experiments because it is susceptible to tobacco mosaic virus and is easy to manipulate experimentally. Plants containing coat protein are resistant to the severe effects of infection by tobacco mosaic virus. As is the case with plants protected by a mild strain of virus, the modified plants are not completely immune; resistance can usually be overcome with increased concentrations of virus for inoculation. Virus movement throughout the plant, however, is usually restricted; therefore, symptoms of virus infection may appear later in plants containing coat protein than in normal plants, if the symptoms appear at all. Since 1986, the development of several modified plants that are resistant to virus infection has been reported, including alfalfa resistant to alfalfa mosaic virus, tomato resistant to tobacco mosaic and tomato mosaic viruses, and potato resistant to potato viruses X and Y.

The virus gene responsible for coat protein production is inherited by offspring of genetically modified plants in the same manner as a natural plant gene. This provides a source of resistance to virus that can be used in commercial agriculture in the same way as resistant varieties developed by conventional genetics are used. Furthermore, when the mechanisms responsible for resistance mediated by coat protein are found, and when the mechanisms for cross protection are fully identified, there may be a wider use of cross protection to control plant virus diseases.

For background information SEE PLANT PATHOLOGY; PLANT VIRUSES AND VIROIDS; VIROIDS in the McGraw-Hill Encyclopedia of Science & Technology.

Robert W. Fulton

Bibliography. R. N. Beachy, L. S. Loesch-Fries, and N. Tumer, Coat protein-mediated resistance against virus infection, *Annu. Rev. Phytopathol.*, 28:451–474, 1990; A. S. Costa and G. W. Muller, Tristeza control by cross protection: A U.S.-Brazilian cooperative success, *Plant Dis.*, 64:538–541, 1980; R. W. Fulton, Practices and precautions in the use of cross protection for plant virus disease control, *Annu. Rev. Phytopathol.*, 24:67–81, 1986; D. Gonsalves and S. M. Garnsey, Cross-protection techniques for control of plant virus diseases in the tropics, *Plant Dis.*, 73:592–597, 1989; A. F. Posnette and J. M. Todd, Virus diseases of cacao in West Africa, IX.

Strain variation and interference in virus 1A, *Ann. Appl. Biol.*, 43:433–453, 1955; Shyi-Dong Yeh et al., Control of papaya ringspot virus by cross protection, *Plant Dis.*, 72:375–380, 1988.

Plant physiology

Plants have evolved several strategies to successfully grow in environments where stressful conditions frequently occur. Stress escapers are those plants that escape seasonal stresses, such as drought and extreme temperatures, by completing their life cycles during the favorable part of the year. Stress avoiders are those plants with mechanisms that enable them to avoid the development of stressful conditions inside their tissues. For example, a plant may maintain a high water status even when there is soil drought. Stress-tolerant plants are those that can survive the stressful conditions in their tissues. Plants that are unable to avoid or tolerate stress and are irreversibly damaged are known as stress-sensitive.

Plasma membrane in stress physiology. The plasma membrane constitutes a dynamic interface between the cell and its environment. It is the primary sensor and transducer of environmental stresses. A normally functioning plasma membrane has a liquid crystalline fluidity that appears to vary directly with the degree of desaturation of its phospholipid component and is affected by temperature and cell water content. The fluidity of the membrane decreases as the temperature decreases; at a critical temperature, fluidity of the membrane changes from liquid crystalline to more rigid, solid crystalline. This change impairs cellular function. As the temperature increases above normal, fluidity increases until finally the plasma membrane's functional integrity is lost. Thus, increased desatura-tion at low temperatures and decreased desaturation at high temperatures may help maintain normal membrane fluidity and cellular integrity. With a severe dehydration of the cell, the organization of the plasma membrane changes from a lamellar to a hexagonal configuration. On rehydration, the inability of the membrane to resume a lamellar configuration results in membrane dysfunction.

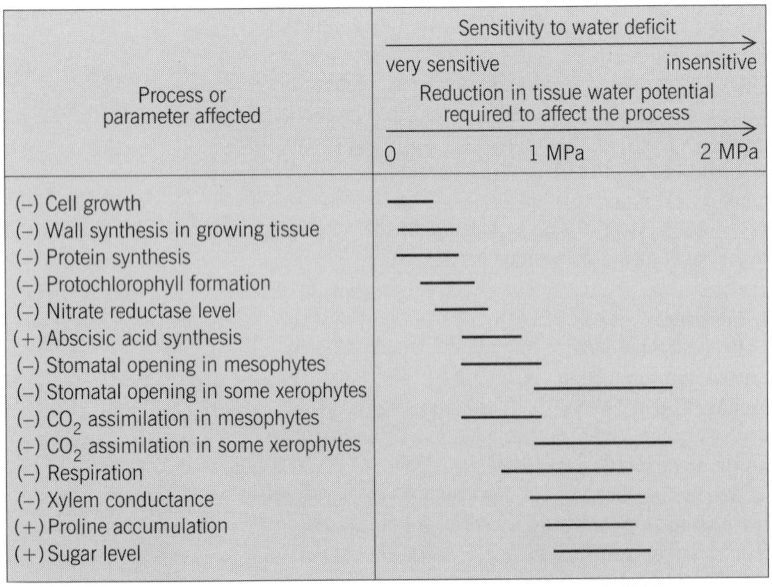

Water deficit (drought) stress. The uptake of carbon dioxide (CO_2) and the loss of water occur through the stomatal opening and are, therefore, unavoidably linked. In an actively photosynthesizing plant, a continuous stream of water moves from the soil, through the plant, and into the atmosphere. As soil dries out, water absorption by the plant lags behind its transpirational demand and a water deficit develops in the plant's tissues. Because of water deficit, water potential of the tissue declines (becomes more negative) and turgor pressure is reduced (becomes less positive).

Some of the plant processes affected by water deficit are shown in **Fig. 1**. Since turgor acts as a driving force for cell growth, even a small reduction in turgor inhibits growth directly and almost instantaneously. Therefore, inhibition of growth by water deficit is considered primarily a biophysical phenomenon. Decrease in turgor also alters membrane properties such as tension, thickness, and charge density.

As water deficit intensifies, the hormone abscisic acid starts accumulating and causes stomatal closure through regulation of ion fluxes and turgor in guard cells. The rate of photosynthesis starts declining, partly because of stomatal closure, thereby restricting CO_2 supply, and partly because of the disruptive effect of water deficit on thylakoid membranes in the chloroplast where photosystem components are anchored. With water deficit, photosynthesis becomes increasingly vulnerable to photoinhibition, and chloroplasts may be damaged by photooxidation. A prolonged, severe water deficit causes the breakdown of chlorophyll and chloroplast structure and hastens leaf senescence.

Plant species vary widely in their habitat with respect to soil moisture and to their responses to water deficit. Hydrophytes grow in water, mesophytes grow in soil with moderate water supply, and xerophytes grow in arid soils. Desert ephemerals are actually mesophytes that escape drought by completing their life cycles during the short rainy season. Since soil moisture and atmospheric humidity and temperature fluctuate daily and seasonally, almost all land plants experience water deficit some time during their life cycles. Most higher plants and all crop plants have drought-avoiding mechanisms that enable them to survive and function under a mild or moderate degree of water deficit. Through these mechanisms, plants maintain nearly normal water status in their tissues by limiting transpiration through stomatal closure or by maintaining water absorption.

Many lower plants, such as algae, lichens, and mosses, can withstand complete desiccation. Such protoplasmic drought tolerance is rare in higher plants

Process or parameter affected	Sensitivity to water deficit		
	very sensitive → insensitive		
	Reduction in tissue water potential required to affect the process →		
	0	1 MPa	2 MPa
(−) Cell growth			
(−) Wall synthesis in growing tissue			
(−) Protein synthesis			
(−) Protochlorophyll formation			
(−) Nitrate reductase level			
(+) Abscisic acid synthesis			
(−) Stomatal opening in mesophytes			
(−) Stomatal opening in some xerophytes			
(−) CO_2 assimilation in mesophytes			
(−) CO_2 assimilation in some xerophytes			
(−) Respiration			
(−) Xylem conductance			
(+) Proline accumulation			
(+) Sugar level			

Fig. 1. Relative sensitivities of some plant processes or parameters to water deficit (reduction in tissue water potential). The plus sign signifies an increase in the process or parameter, and the minus sign a decrease. (*After T. C. Hsiao, Phil. Trans. Roy. Soc. Lond., B273:479, 1976*)

and is absent in crop species. True drought tolerance is associated with an extremely slow rate of growth, small cells, small and few vacuoles, and thick cell walls. These plants have sacrificed high productivity in order to achieve survival.

Flooding stress. When soil is inundated, air spaces between soil particles become filled with water, and air is expelled. Thus, roots are subjected to anaerobiosis or anoxia, which is the primary cause of injury from flooding.

Root growth rapidly ceases during anoxia, and the plant wilts. After a day or two, stomata close. Photosynthesis and translocation of the assimilates are also inhibited. As expected, anoxia interrupts aerobic respiration and lowers production of adenosinetriphosphate (ATP).

Cessation of root growth is soon followed by cell death. Generally, cytoplasmic pH is kept slightly alkaline by the pumping of protons from the cytoplasm into the cell vacuole. Energy from ATP hydrolysis maintains the pH gradient. In hypoxic roots, the free energy available from ATP hydrolysis is considerably lower and is insufficient to maintain the pH gradient. Thus, protons leak from the cell vacuole into the cytoplasm, ultimately resulting in cellular decompartmentation and death.

Plants can be hardened against anoxia by preexposing them to hypoxic conditions (for example, 4% oxygen). When subsequently subjected to anoxia, the hardened plants survive longer. Many plants are resistant to transient or permanent flooding. In almost all cases, the mechanisms underlying this resistance involve avoidance of anoxia. Many plants develop aerenchyma, which is a tissue that contains a network of intercellular spaces. The hormone ethylene, which increases during flooding, plays a major role in the development of aerenchyma. Exogenous application of ethylene causes the development of aerenchyma in several plants. Aerenchyma is continuous throughout the stem and roots. Thus, the root tissues remain in contact with air although the soil is waterlogged. Another important adaptation to flooding is the development of adventitious roots that absorb water and mineral ions.

Anoxia suppresses the synthesis of normal proteins but induces synthesis of about 20 anaerobic proteins. A major anaerobic protein has been identified as an alcohol dehydrogenase isozyme. Several other anaerobic proteins have been identified as isozymes of various glycolytic enzymes. Thus, new forms of ordinary enzymes are produced under anoxia.

High-temperature stress. Temperatures above the optimum cause an acceleration of the normal developmental sequence. Stressful high temperatures cause direct injury to membranes and inactivate enzymes.

In most higher plants, the optimum temperature is between 68 and 86°F (20 and 30°C). Plant parts that lack transpirational cooling and have impervious outside layers, such as fruits, are more vulnerable to heat injury. Effects of heat stress depend upon the tissue type, physiological process, and amount and rate of increase in temperature. Visible effects or symptoms of heat stress are common and include sun scald or bark burn on the sun-exposed sides of tree trunks and in fruits such as tomato and cantaloupe. Seedlings are particularly susceptible to heat since they come in direct contact with the hot soil surface. Flower abscission, pollen sterility, reduced fruit set, and reduction in yield are major agronomic effects of heat stress. Among physiological processes, photosynthesis is relatively more sensitive to heat stress than is respiration; therefore, net assimilation decreases at high temperatures. The site of sensitivity has been identified as the photosystem II reaction center.

Plants cope with high temperature through either avoidance or tolerance of heat stress. An avoidance mechanism is typical of leaves and involves transpirational cooling. This enables the plant, in some cases, to maintain leaf temperature of up to 18–22°F (10–12°C) below that of the air. However, a sudden increase in temperature or heat shock affects the pattern of protein synthesis profoundly. Within a short time, the synthesis of normal proteins is shut off, but their messenger ribonucleic acids (mRNAs) are generally conserved. The synthesis of several other proteins, known as heat shock proteins, is initiated. Some of these proteins are highly conserved. So far, there have been only limited studies on plant heat shock proteins. A correlation between heat tolerance and abundance of some heat shock proteins in wheat has been demonstrated. A similar correlation between heat shock proteins and heat tolerance has been reported in barley, sorghum, and soybean.

Low-temperature stress. Low-temperature stress includes chilling stress and freezing stress.

Chilling stress. Temperatures below the optimum inhibit growth and development by slowing down biochemical reactions. The activity of many enzymes is sensitive to chilling. However, temperatures below a certain critical temperature injure the plant. For most of the chilling-sensitive plants, this critical temperature is between 50 and 59°F (10 and 15°C). This temperature is dependent on the phospholipid component of the plasma membrane and coincides with the temperature at which membrane fluidity changes from liquid crystalline to solid crystalline.

A common visible symptom of chilling injury is the presence of water-soaked areas on the surface of the damaged plant part. Chilling causes a rapid inhibition of root and shoot growth. Interestingly, the inhibition of root growth does not involve changes in turgor, but is due to decreased extensibility of the cell wall. At the cellular level, vacuolization, stoppage of protoplasmic streaming, disruption of cellular membranes, and the resulting decompartmentation are common consequences of chilling stress. Photosynthesis sharply declines; photosystem II is particularly sensitive to chilling. In addition, translocation of assimilates is rapidly inhibited, as is the aerobic phase of respiration and electron transport.

Plants can be hardened against chilling by preexposing them to temperatures slightly above the injurious one. During such hardening, abscisic acid increases, and changes in the phospholipid and protein components of the plasma membrane are observed.

Synthesis of new proteins during chilling has been shown in several plants. Since products of anaerobic respiration, such as ethanol, accumulate in sensitive plants during chilling, it is possible that some chilling-induced proteins are similar to those induced by flooding, for example, alcohol dehydrogenase.

Freezing stress. In temperate, alpine, and arctic regions, perennial plants and seeds of freezing-sensitive annuals can withstand freezing temperatures. Also, some cultivars of such forage crops as alfalfa, clover, and lotus are able to overwinter. Freezing tolerance is acquired seasonally and is an inducible response. Gradually declining temperatures and decreasing day length induce freezing tolerance. In the laboratory, freezing tolerance can be induced by exposing the plant to low but above-freezing temperatures. During the acquisition of freezing tolerance, endogenous abscisic acid level increases. Exogenous application of abscisic acid can substitute for cold acclimation completely in cell suspensions but only partly in entire seedlings (see **Fig. 2**).

Damage due to freezing is characterized by total membrane disruption and structural collapse. The mechanism by which freezing injury occurs is still controversial. Cell death occurs because of intracellular ice formation. Whether the intracellular ice originates inside the cell or represents propagation of the extracellular ice is not clear. Spontaneous ice formation occurs at about −40°F (−40°C). In the acclimated plant, the freezing point of cell sap is lowered because of solute accumulation and the rapid loss of water that moves out of the cells in response to the much lower water potential of the extracellular ice. Thus, dehydration has been shown to induce freezing tolerance in some plants. Deep supercooling (cooling to very low temperatures without ice formation) occurs in many trees and shrubs.

Recently, synthesis of new proteins and activation of new genes due to cold acclimation or abscisic acid treatment have been shown in several plants. Some of the proteins show a partial homology with drought-induced proteins. The functional nature of these genes is being actively investigated.

Salinity stress. Under conditions of soil salinity characterized by excessively high concentrations of sodium chloride or sodium sulfate, plant growth suffers and yield is reduced. Plants show a multiplicity of responses to salinity. The presence of excess salt in the soil decreases the water potential of the soil and, therefore, imposes an osmotic or dehydration stress on the plant. Thus, the effects of salinity are intertwined with those of drought. A plant responds to conditions of salinity either by avoiding the absorption of salt or by taking up salt. In either case, the plant either may be damaged or may adapt to the situation. Plants that are able to exclude salt uptake resist salinity stress by osmotic adjustment. When plants are unable to avoid salt absorption, they may employ several strategies to resist salinity stress. A plant may sequester the salt into a vacuole and thus keep its cytoplasmic machinery relatively unaffected. In succulent plants, salt is distributed throughout the plant. This results in dilution of salt concentration. Some plants have evolved special glands through which salt may be secreted to the outside. Plants devoid of such mechanisms sustain injury because of either salinity-induced drought (when salt is not absorbed) or the toxic effects of salt (when salt absorption cannot be avoided).

Mechanical stress. Plants are subjected to mechanical stresses such as wind, pressure of water spray or raindrops, and wounding by insects. The effects of the physical pressure of wind and rain can be duplicated by manually rubbing the plant stem. Major effects of such mechanical stimulation are a decrease in length and an increase in diameter of the stimulated part. Interestingly, mechanically stimulated plants become more resistant to several types of environmental stress, such as drought, heat, and chilling. Recently, it has been shown that mechanical stimulation rapidly activates genes for proteins related to the calcium-binding protein calmodulin. It has been suggested that mechanical stimulation activates some pressure-sensitive ion channels, resulting in a sudden increase in the cytoplasmic calcium. This increase in calcium is highly injurious to the cells. It is thought that calcium-binding proteins are synthesized to sequester the excess calcium. Application of exogenous gibberellic acid reverses the inhibitory effects of mechanical stimulation on growth.

A common mechanical injury to plants is that resulting from wounding by insects. Some plants respond to insect bites by the defense mechanism of rapidly synthesizing proteinase inhibitors of proteolytic enzymes in the insect gut. Since insects cannot digest the proteins, they become sick. Intriguingly, these proteinase inhibitors do not inhibit the plant's own proteolytic enzymes. The wounding-induced production of proteinase inhibitors is not localized to the wounded tissue but is systemic. Thus, wounding must produce a translocatable signal.

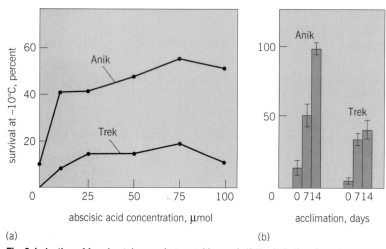

Fig. 2. Induction of freezing tolerance in two cultivars of alfalfa, Anik (freezing-tolerant) and Trek (freezing-sensitive), (a) by abscisic acid and (b) by cold acclimation. Symbols at the top of the bars in b represent standard error of the mean. −10°C = 14°F. (From S. S. Mohapatra et al., Abscisic acid-regulated gene expression in relation to freezing tolerance in alfalfa, Plant Physiol., 87:468–473, 1988)

For background information SEE ABSCISIC ACID; ETH-YLENE; PHOTOSYNTHESIS; PLANT GROWTH; PLANT HOR-MONES; PLANT METABOLISM; PLANT PHYSIOLOGY; PLANT RESPIRATION; PLANT-WATER RELATIONS in the McGraw-Hill Encyclopedia of Science & Technology.

Rajinder S. Dhindsa

Bibliography. A. Blum, *Plant Breeding for Stress Environments*, 1988; J. L. Key and T. Kosuge (eds.), *Cellular and Molecular Biology of Plant Stress*, 1985; J. Levitt, *Responses of Plants to Environmental Stresses*, vols. 1 and 2, 1980; H. Mussell and R. C. Staples (eds.), *Stress Physiology in Crop Plants*, 1979; P. Steponkus, Role of the plasma membrane in freezing injury and cold acclimation, *Annu. Rev. Plant Physiol.*, 35:543–585, 1984.

Plant viruses and viroids

The ability of a plant virus to spread from cell to cell is essential for establishing a viral infection. However, unlike animal cells, plant cells have a cell wall barrier that viruses must cross to move from cell to cell. Virus–host interactions required for spread of plant viruses are just beginning to be understood. Most of the recent advances have come from studies of virus-encoded proteins that have been shown, or are presumed, to facilitate virus spread.

Virus–plant interaction. Following the initial entry and replication of a virus, the infection takes one of two directions. If the virus fails to move to neighboring cells, the plant remains healthy with only the initially inoculated cells containing progeny virus. This leads to an aborted infection, and the plant

appears to be resistant to the virus. If the virus is capable of moving from cell to cell, however, a more productive, widespread infection is established in the plant. While the specific interactions between the virus and host ultimately define the nature of the viral disease, virus movement is an important component in determining pathogenicity and virulence as well as host range.

Types of virus movement. Virus movement in plants can be divided into two types: slow, short-distance movement from an infected cell to adjacent healthy cells; and long-distance movement from infected tissue, such as leaves, to other tissues via the vascular system.

Short-distance (local) movement. Plant viruses move from cell to cell through plasmodesmata (see **illus.**), which are channels that span the cell wall and provide cytoplasmic continuity between adjacent cells. These intercellular connections play an important role in cell-to-cell communication and provide the route through which water and metabolites move from cell to cell. However, based on the architecture and the molecular size exclusion limits of plasmodesmata, as well as the size of plant viruses or their genomes, it is generally assumed that the structure must be modified during virus infection for progeny virus to move from cell to cell. Indeed, structural changes in plasmodesmata have been observed in many viral infections, and in the case of tobamoviruses a virus-encoded protein is known to modify the function of plasmodesmata.

In many virus–host interactions, viral progeny spread only locally, for example, only in the inoculated leaf. These virus–host combinations can be divided into two groups. In the first group, the virus moves from cell to cell throughout the infected leaf,

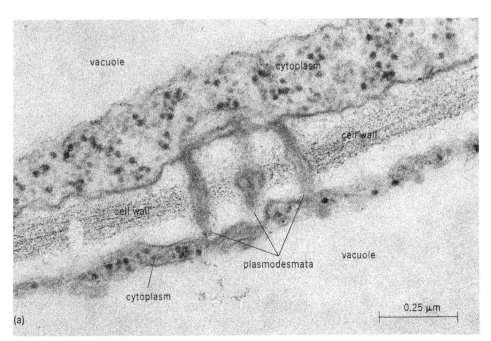
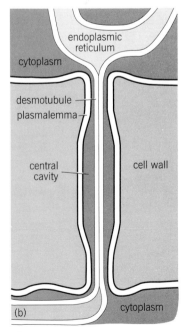

Plasmodesmata. (*a*) An electron micrograph of plasmodesmata connecting adjacent cells in a tobacco leaf (*courtesy of P. Moore*). (*b*) Diagram illustrating the components and basic structure of a simple plasmodesma (*from A. W. Robards and W. J. Lucas, Plasmodesmata, Annu. Rev. Plant Physiol. Plant Mol. Biol., 41:369–419, 1990*). The desmotubule is a tubular extension of the endoplasmic reticulum that runs through the center of the plasmodesma.

but movement to uninoculated tissues, such as other leaves, is blocked. In the second group, the virus moves from cell to cell but is confined to a small area around the site of infection. This is usually manifested by the development of a local necrotic lesion and is defined as a hypersensitive response. Both types of localized infection limit the severity of the viral disease.

Long-distance (systemic) movement. Depending on the virus–host combination, some viruses are capable of entering the vascular system and are transported to uninoculated tissues of the plant. Unlike short-distance movement, which is slow, long-distance spread is rapid and, aside from a few exceptions, occurs through the phloem tissue. Viruses that are capable of moving into and replicating in many cell types have a broad tissue range and spread systemically. Therefore, viruses that move systemically are generally more pathogenic than viruses that are capable of only short-distance movement.

Genes involved in virus movement. While a number of plant viruses encode a protein tentatively identified as required for virus movement, to date only the movement proteins of two tobamoviruses, tobacco mosaic virus, and tobacco mild green mosaic virus, have been positively identified. Tobamoviruses have ribonucleic acid (RNA) genomes that encode four proteins; two proteins are postulated to be subunits of the viral replicase, a third is the coat protein that protects the viral RNA in the mature virus particle, and the fourth is the movement protein. During infection the movement protein is required for either type of movement of the virus progeny.

Tobacco mosaic virus. Genetically engineered and naturally occurring mutants of tobacco mosaic virus that contain a nonfunctional movement protein gene are incapable of moving from initially infected cells to neighboring healthy cells, although replication and encapsidation occur unabated. Therefore, a key to understanding how tobamoviruses as well as many other plant viruses move is to determine how the movement protein interacts with the host cell and viral genome to facilitate movement of viral progeny. Transgenic tobacco plants have been developed that express a gene encoding the tobacco mosaic virus movement protein. The gene is stably integrated into the plant's genome. In these transgenic plants the plasmodesmata are altered in function. Plasmodesmata in the transgenic plants allow passage of molecules with at least 10-fold greater molecular mass than do plasmodesmata in plants that do not express the tobacco mosaic virus movement protein gene. This finding is supported by immunocytochemical and biochemical experiments to localize the movement protein to plasmodesmata and the cell wall, respectively, in both transgenic and tobacco mosaic virus–infected tobacco leaf tissue. Routine electron microscopy techniques show no detectable differences in ultrastructure between plasmodesmata of transgenic and nontransgenic plants. Since the mechanism by which the tobacco mosaic virus movement protein modifies plasmodesmata has yet to be determined, the

localization of the movement protein to plasmodesmata suggests that the movement protein interacts directly with and modifies the structure of the channels. Of course, the ability of the tobacco mosaic virus movement protein to alter the function of plasmodesmata does not preclude the possibility that the protein influences virus movement in additional ways.

Other plant viruses. Criteria for assigning putative movement functions to virus-encoded proteins have predominantly depended on the molecular size of the proteins and the immunocytochemical localization of these proteins in infected tissue. Plant viruses that invade multiple tissues presumably encode a movement function. These putative movement proteins are similar in size to movement proteins of tobacco mosaic virus and tobacco mild green mosaic virus. Indeed, many if not all plant viruses encode proteins of 30–40 kilodaltons. Immunocytochemical studies on infected tissue have localized three such putative movement proteins. The 38-kDa protein of cauliflower mosaic virus, a deoxyribonucleic acid (DNA) virus, is associated with plasmodesmata in infected cells. Interestingly, the plasmodesmata in cauliflower mosaic virus–infected tissue have an enlarged diameter and a wide tubule extending from the plasmodesmata into the cytoplasm. Virus particles have been observed in these tubular structures. The mechanism used by cauliflower mosaic virus and related viruses to modify plasmodesmata is apparently different than that of tobacco mosaic virus.

Localization of the putative movement proteins of two tripartite RNA viruses offers conflicting results. A 32-kDa protein of alfalfa mosaic virus has been detected in the middle lamella of the cell wall in infected tissue but has not been localized to plasmodesmata, whereas a 35-kDa protein of cucumber mosaic virus was localized to the nucleoli of infected cells. These proteins are considered to be analogs. If, indeed, they provide movement function to their respective viruses, then they must function by very different mechanisms. It is possible that the protein of cucumber mosaic virus modifies the function of plasmodesmata indirectly by regulating the production of host proteins. Although general conclusions about viral movement proteins cannot be made at the present time, it is likely that a number of mechanisms are involved.

Viruses without movement function. Some viruses may not encode a movement protein. For example, the luteoviruses and some geminiviruses are limited to phloem tissue. These viruses may not be capable of moving into nonvascular tissue. It is interesting in this respect that transgenic tobacco plants, which express the tobacco mosaic virus movement protein gene, allow potato leafroll virus, a luteovirus, to move into nonvascular leaf cells, where the virus replicates. In the case of potato leafroll virus, the failure of the virus to move into nonvascular tissue is due not to an inability to replicate but to the absence of a gene product that encodes movement function. Likewise, potato leafroll virus moves into nonvascular tissue in plants that are also infected with the potato virus Y, a

---

potyvirus. Presumably the potato virus Y provides a movement protein that is utilized by potato leafroll virus again indicating that the latter does not encode a movement protein.

Host genes affecting movement. Unfortunately, very little is known about plant proteins involved in virus movement. The fact that viral movement proteins are required for virus spread, and in the case of tobacco mosaic virus, alter plasmodesmata function, indicates that host proteins must be modified or regulated for virus movement to occur. To date, the only direct evidence that a viral movement protein interacts with a host protein comes from studies of tobacco mosaic virus in tomato. Tomato lines homozygous for the *Tm-2* gene are resistant to infection by most strains of tobacco mosaic virus. While the *Tm-2* gene product has not been identified, mutations have been identified in the movement protein gene that allow tobacco mosaic virus to overcome the resistance. Therefore, the mutant movement protein can interact with, or fails to be inactivated by, a host protein and facilitates virus movement in a plant in which the wild-type movement protein cannot function.

In conclusion, a better understanding of how plant viruses move will eventually allow the modification and manipulation of the virus and host-encoded proteins involved in virus movement so as to produce plants that are resistant to viral diseases. In addition, very little is known about plasmodesmata and the role they play in cell-to-cell communication. Since the tobacco mosaic virus movement protein and probably the movement proteins of other plant viruses interact with and modify these structures, the viral movement proteins will be important tools in studying the structure and function of plasmodesmata.

For background information *SEE PLANT PATHOLOGY; PLANT VIRUSES AND VIROIDS; VIRUS* in the McGraw-Hill Encyclopedia of Science & Technology.

Carl M. Deom; Roger N. Beachy

Bibliography. C. M. Deom, M. J. Oliver, and R. N. Beachy, The 30-kilodalton gene product of tobacco mosaic virus potentiates virus movement, *Science*, 237:389–394, 1987; R. Hull, The movement of viruses in plants, *Annu. Rev. Phytopathol.*, 29:213–240, 1989; A. W. Robards et al. (eds.), *Parallels in Cell-to-Cell Junctions in Plants and Animals*, 1990; S. Wolf et al., Movement protein of tobacco mosaic virus modifies plasmodesmatal size exclusion limit, *Science*, 246:377–379, 1989.

Plasma physics

In the latter 1950s, several researchers working in the area of plasma physics suggested the construction of a plasma device that, although entirely unsuitable for the production of hot, so-called thermonuclear plasmas, would allow basic plasma physics experiments to be performed with relative ease. Several approaches were tried, one of them resulting in the appearance of a device called a Q-machine. The letter Q, for

quiescent, was to emphasize the expectation that this type of plasma device would be relatively free of the low-frequency instabilities that had plagued most laboratory plasmas in use in the 1950s.

In a Q-machine, alkali-metal (cesium, potassium, and so forth) plasmas are generated by a method of ion production discovered in the 1920s by I. Langmuir and coworkers. They found that, if a stream of cesium atoms is allowed to impinge on a hot tungsten surface at a temperature of at least about 1000°C (1800°F), almost all cesium atoms emerge as ions. Each atom loses one electron to the tungsten surface by surface ionization. The phenomenon occurs because the ionization potential of cesium (3.87 V) is smaller than the work function of tungsten (4.52 V). It occurs also with other combinations of metals and alkali atoms that have the same relation between the work function of the metal and the ionization potential of the atom, for instance, with a rhenium surface (work function ~5 V) and potassium atoms (ionization potential 4.3 V).

In a typical Q-machine (see **illus.**), a beam of neutral cesium atoms emerges from an atomic-beam oven and strikes a tungsten plate kept, by electron bombardment, at a temperature of about 2000°C (3600°F). The cesium atoms are ionized by surface ionization, while the tungsten plate is hot enough to emit copious amounts of thermionic electrons. Thus, the two constituents of a plasma (positive ions and electrons) are provided by the hot tungsten plate. A magnetic field **B** of several thousand gauss (several tenths of a tesla) confines the plasma in a column about 1 m (3 ft) in length and 3 cm (1.2 in.) in diameter (the diameter of the tungsten plate). At the end opposite the generating plate, the column is terminated either by a second hot tungsten plate that reflects most of the plasma ions (double-ended Q-machine) or by a cold metal plate that acts as a sink for the ions (single-ended Q-machine).

In general, the base pressure maintained in the

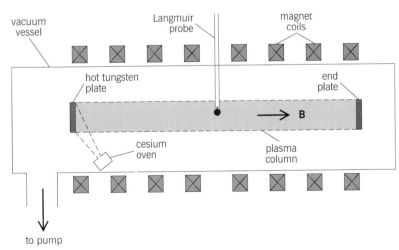

Diagram of a Q-machine. The plasma column is produced by surface ionization of alkali metals on the hot tungsten plate, which emits thermionic electrons. The magnetic field produced by the magnet coils confines the plasma in the radial direction. Typical dimensions of the plasma column are a length of about 1 m (3 ft) and diameter of about 3 cm (1.2 in.).

vacuum vessel of the device is on the order of 10^{-6} torr (10^{-4} pascal), so that ion-atom and electron-atom collisions are very rare, with collision mean-free paths much larger than the machine length. Some Q-machines, however, have been designed to operate at gas pressures (argon, helium, neon, and so forth) as large as about 10^{-3} torr (10^{-1} Pa), for the purpose of investigating the effect of ion-atom and electron-atom collisions on various plasma phenomena.

The ion and electron densities n are generally in the range 10^8 cm$^{-3} \leq n \leq 10^{12}$ cm^{-3}, the largest densities being attainable in double-ended operation. The ion and electron temperatures are $T_i \approx T_e \approx 0.2$ eV, that is, close to the temperature of the hot tungsten plate. At a plasma density $n \approx 10^9$cm^{-3}, the collision mean-free path for ion-ion (or ion-electron) collisions is about 1 m. Thus, for densities of approximately this value or less, collisionless plasma phenomena may be investigated, whereas at much larger densities ion-ion and ion-electron collisions may be important.

The plasma diagnostics tool most frequently used is the Langmuir probe. Other methods that have been used are microwave beams, gridded ion-energy analyzers, and laser beams.

Past experiments. Many interesting plasma phenomena have been investigated in Q-machines.

Plasma confinement. Experiments on plasma confinement, and diffusion across magnetic field lines, have been performed in uniform magnetic fields, in magnetic field geometries in which the field lines could have a variable curvature (U-bends), in magnetic-mirror geometries, and in stellarator configurations. One of the primary aims of this type of investigation was to distinguish between plasma losses across magnetic field lines arising from collisions among charged particles (collisional diffusion) and losses due to a somewhat mysterious process that apparently operated in early "thermonuclear" plasmas, and was generally referred to as Bohm diffusion. It was presumably associated with plasma instabilities.

Transport properties. A typical example of an experiment on the transport properties of a fully ionized plasma is the determination of the plasma resistivity obtained by analyzing the current-voltage characteristic of a Q-machine plasma column.

Waves and instabilities. By far the vast majority of experiments performed in Q-machines falls under this heading. Some of the wave and instability types investigated are ion-acoustic waves, electrostatic ion cyclotron (EIC) waves, drift waves, and the Kelvin-Helmholtz instability.

Ion-acoustic waves are similar to normal sound waves, with the important difference that the restoring force for the oscillation arises from a self-consistent electric field produced by a small charge separation of the ion component relative to the electrons. The inertia in the oscillation is provided almost entirely by the massive ions. An important phenomenon of wave-particle interaction, referred to as Landau damping, was demonstrated by an analysis of the (collisionless) damping of these waves.

Electrostatic ion cyclotron waves are waves propagating at a very large angle to the magnetic field **B** with a frequency somewhat in excess of the ion cyclotron frequency, $\omega_{ci} = eB/m_i$, where e and m_i are the positive ion charge and mass, respectively. The quantity ω_{ci} is the angular frequency of the gyromotion of a single ion spiraling around a magnetic field line. The motions of the plasma ions and of the electrons combine to produce a longitudinal (or compressional) wave in which the wave vector **K** is parallel to the electric field **E** of the wave. By varying the magnetic field strength B, the wave frequency is observed to change according to the relation $\omega \approx \omega_{ci} = eB/m_i$.

Drift waves are low-frequency waves found, in nonhomogeneous plasmas, to propagate in a direction nearly normal to the **B** field and to the density gradients. In a Q-machine plasma column the density gradient is directed everywhere radially inward, while the **B** field is in the axial direction. Thus, wave propagation occurs predominantly in the azimuthal, or θ, direction, all around the column. The waves have $K_\parallel \ll K_\theta$, where $K_\parallel$ is the component of the wave vector **K** along **B** and K_θ is the azimuthal component.

The Kelvin-Helmholtz instability may arise in a stratified fluid if different layers are in relative motion. A shear may be present in the plasma velocity along **B** or across **B** (or in both). If the velocity shear is sufficiently large, the interface between two adjacent plasma layers becomes unstable and is eventually disrupted, in much the same way as a wind blowing horizontally over a lake produces small ripples of the water surface.

The above is only a very partial discussion, and not even representative of many additional types of investigations that have been carried out.

Recent and current work. In recent years, Q-machines have been utilized to conduct laboratory experiments that should help to understand plasma phenomena occurring in the near-Earth space environment, for example, in the ionosphere and the magnetosphere. There are definite advantages in this type of laboratory work as compared to on-site space observations. Quite apart from the fact that laboratory experiments are much less expensive to conduct than, for example, collecting data from spacecraft, a laboratory experimenter can generally set up the conditions of the experiment in such a way as to be able to examine special aspects of the phenomenon under investigation. The experimental conditions can also be varied and adjusted for detailed tests of theories of various space phenomena. Space-related researches recently (or currently) conducted in Q-machines include the investigation of electrostatic waves in the lower ionosphere (E region) excited by relative drifts of electrons and ions; so-called electric double layers and associated particle-beam formation; and effects due to large concentrations of negative ions on various types of low-frequency plasma waves, and related phenomena occurring in so-called dusty plasmas, such as those of planetary environments, comet tails, and interstellar plasmas.

For background information SEE IONOSPHERE; MAG-NETOSPHERE; PLASMA DIAGNOSTICS; PLASMA PHYSICS; WAVES AND INSTABILITIES IN PLASMAS in the McGraw-Hill Encyclopedia of Science & Technology.

N. D'Angelo

Bibliography. N. D'Angelo, *Adv. Plasma Phys.*, 2:1 1969; R. W. Motley, *Q-Machines*, 1975; N. Rynn and N. D'Angelo, Device for generating a low-temperature, highly-ionized cesium plasma, *Rev. Sci. Instrum.*, 31:1326–1333, 1960; R. W. Schrittwieser (ed.), *Current-Driven Electrostatic Ion-Cyclotron Instability*, 1988.

Plasma processing

Methods and technologies that utilize a plasma to treat or manufacture materials have become important for many advanced materials-processing applications. In the 1960s, plasma chemical techniques were limited to scientific investigations in research laboratories. Since then, plasma technology has evolved into numerous commercially important applications, and many different types of plasma processing machines and methods have been invented.

Processing plasmas. Each small volume of plasma gas is composed of at least three interpenetrating subgases: electrons, ions, and neutral gases. The neutral gas can usually be divided into a number of subspecies such as atomic, molecular, and radical species. The plasma state of matter differs from the lower-energy ordinary gases by the presence of high densities of charged species and the high energies, that is, high temperatures, of each subgas. Thus, each plasma particle is moving at a high velocity and has many collisions with other particles as it travels through the gas. Although the electrons and ions arrange themselves so that each small plasma volume with dimensions larger than the Debye length is electrically neutral, electromagnetic forces can be applied and electric energy can be absorbed by the plasma through the charged species.

The electron gas is heated directly by the applied electric fields, and, in turn, the heated electron gas transfers its energy to the neutral and ion gases by elastic electron-neutral and electron-ion collisions and

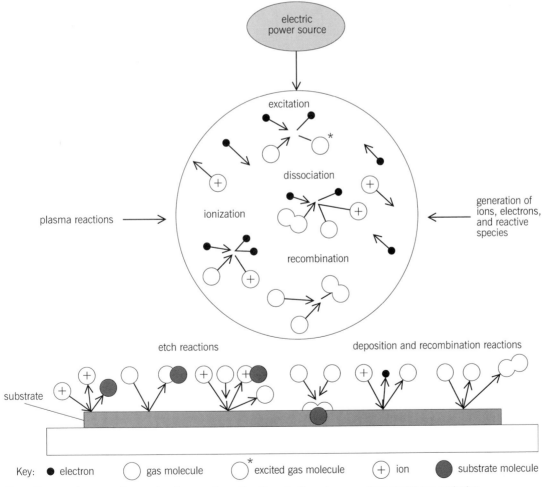

Fig. 1. Plasma volume reactions and surface reactions. Equations for the volume reactions shown are: excitation, $M + e^- \rightarrow M^* + e^-$, where M denotes a molecule, and M^* the molecule in an excited state; dissociation, $M_2 + e^- \rightarrow 2M + e^-$; ionization, $M + e^- \rightarrow M^+ + 2e^-$; recombination, $M + M \rightarrow M_2$.

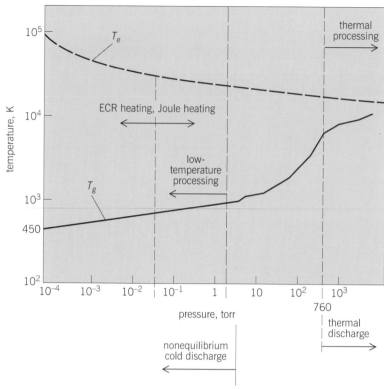

Fig. 2. Typical electron and gas temperatures in electric discharges. °F = (K × 1.8) − 459.67. 1 torr = 133 Pa.

Electron and gas temperatures. Depending on the input power and gas pressure, electric discharges can create plasmas with different properties and species concentrations. One important difference between discharges is how the plasma gas energy is distributed between the light electron gas and the heavier neutral, atomic, and molecular gases. **Figure 2** displays the electron temperature T_e and heavy-particle temperatures T_g as functions of pressure for a typical electrical discharge. The electrons always have the highest temperature, since they are heated directly by the electric field and transfer only a small amount of energy per collision to heavier particles. The heavy-particle subgases are in good collisional contact with each other, since they usually have similar masses. Thus, in Fig. 2 the heavier gases have been combined into one gas with a temperature of T_g.

At high pressures (greater than 100 torr or 13 kilopascals), the electron temperature decreases and the gas temperature increases. At the very high pressure of 1 atm (100 kPa), the electrons are only slightly hotter than the heavier particles. The plasma then becomes a very hot (over 2000 K or 3000°F), thermal, arc discharge, which can be used in thermal plasma processing applications where high-temperature chemistry is important.

At low pressures, the heavier gases have the lower temperatures of 400–500 K (260–440°F), while the electrons become very energetic with temperatures in excess of 50,000 K (80,000°F). These discharges are often called cold plasmas or nonequilibrium plasmas since the gas temperature is low and thus will only moderately heat the walls or substrate surfaces. However, the energetic electrons create through inelastic collisions a high-density mix of atoms and radicals and hence produce a very chemically active medium.

Plasma reactors. **Figure 3** displays examples of several commercially important plasma reactors that are utilized in integrated-circuit processing. In each, the processing chamber is evacuated by vacuum pumps, and then the process gas is introduced in a controlled manner so that the desired gas mix, gas flow, and chamber pressure are achieved. Electrical power is applied, and the plasma is generated. Processing pressures are application-dependent, usually varying from 10^{-5} to 1 torr (10^{-3} to 10^2 pascals). Thus, as shown in Fig. 2, the discharge is characterized as a cold plasma, where the neutral and ion gas temperatures are low and the electron temperature is high. The activated gases from the plasma react with the process substrate, that is, the silicon wafers shown in Fig. 3, and residual gas is evacuated. When processing is complete, the process chamber is returned to 1 atm (100 kPa) by applying dry nitrogen (N_2) gas or an inert gas, and the material is removed from the chamber, usually into another environment for additional processing.

Barrel etch reactor. In this type (Fig. 3a), the wafers are placed in a quartz support stand, called a boat, and the boat is inserted into the plasma chamber. Radio-frequency electricity, usually at 13.56 MHz, is applied to the process chamber through a tuning network

by inelastic electron–heavy particle collisions. The inelastic collisions are the special chemical reactions that give a plasma its unique chemistry (**Fig. 1**). Some, called volume reactions, occur within the plasma volume, while others occur on the walls or boundaries of the plasma. Some important plasma volume chemical reactions are ionization, dissociation, excitation, recombination of atoms or radicals into molecules, and deexcitation of excited species.

Surface reactions. In many plasma processing applications the surface of a material object, often called a substrate, must be treated. The substrate is located inside or at the boundary of the plasma, and thus the surface reactions produce the desired effect on the substrate surface. The plasma surface reactions (Fig. 1) include recombination of gases; removal of substrate material, called etching reactions; and deposition reactions, which may involve plasma species or plasma and substrate molecules. Thus, depending on the specific plasma chemistry, the substrate surface can be removed molecular layer by layer, or thin films can be deposited or grown on the substrate surface.

Figure 1 displays the three mechanisms responsible for plasma etching. In sputtering, a high-energy ion hits the substrate surface and mechanically ejects substrate material. In the second process, a neutral radical chemically combines with substrate material, forming a volatile product. In the third etching mechanism, called ion-enhanced etching, the energetic ion alters a thin layer on the substrate, which then allows a radical molecule to react and volatilize the substrate surface.

to either capacitive electrodes, as shown in Fig. 3a, or inductive coils, and an electrical discharge is initiated. The electrons, ions, atoms, radicals, and molecules of the plasma then diffuse in between and surround and contact the wafers to perform the desired etching. If a perforated metal screen, called a tunnel, is placed around the wafers, the plasma excitation is limited to the region shown in Fig. 3a. Thus, the etch tunnel prevents the energetic ions and electrons from contacting the wafers, and etching is performed only by neutral molecules and radicals diffusing through the screen. This reactor has the inherent problem of lack of temperature control of the wafers, and etching uniformity is difficult to achieve. However, these reactors are still often used for noncritical processes, such as stripping and cleaning, because of their large capacity and low cost.

Parallel-plate reactor. This is the most common plasma reactor now in use (Fig. 3b), and has many applications as both a plasma etcher and a plasma-enhanced chemical vapor deposition reactor. It consists of two closely spaced parallel plane electrodes. Process gases either flow in from the side, as shown in Fig. 3b, or flow out from the center. When radio-frequency (rf) excitation is applied to the electrodes, a plasma is formed in the space between them. The substrates can be placed on either electrode, and, in contrast to the barrel etcher, the substrates can be temperature-controlled by either heating or cooling the electrode. Substrate surfaces are directly exposed to the charged species as well as to the excited neutral radicals produced in the plasma.

ECR reactor. A relatively new plasma reactor technology makes use of the resonant coupling of microwave energy into an electron gas at electron cyclotron resonance (ECR). This resonance occurs when the electron cyclotron frequency equals the excitation frequency. In an actual discharge, this condition can be satisfied in a volume or surface layer within the discharge where the static magnetic field strength is adjusted to resonance, and where a component of electric field is perpendicular to the static magnetic field. The electrons are accelerated in this volume and, in turn, ionize and excite the neutral gas. The result is low-pressure, almost collisionless, plasmas that can be varied from a weakly to a highly ionized state by changing discharge pressure, gas flow rates, and input microwave power.

Several electron cyclotron resonance plasma reactors are under development, one of which is shown in Fig. 3c. It makes use of rare-earth magnets and a cylindrical microwave cavity formed from an aluminum cylinder, a sliding short, and a water-cooled bottom plate. Microwave power is coupled into the cavity through a coaxial input port. A dome-shaped quartz tube confines the process gas to a region where the 2.45-GHz microwave fields produce a 10-in.-diameter (25-cm) cylindrical discharge. The input gas is introduced into the discharge region through 12 pinhole openings in an annular ring and gas feed tube (not shown).

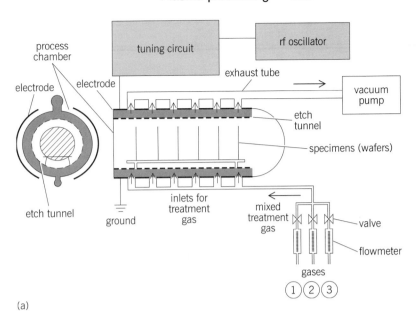

(a)

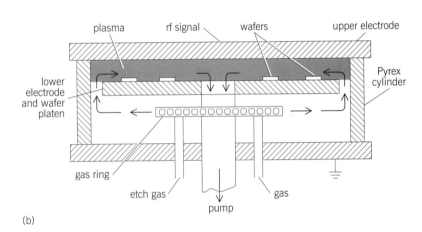

(b)

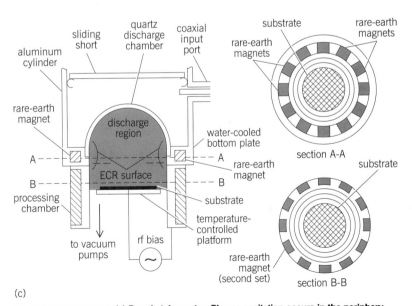

(c)

Fig. 3. Plasma reactors. (a) Barrel etch reactor. Plasma excitation occurs in the periphery (shaded area) of the process chamber. (b) Parallel-plate reactor, shown for plasma etching application. (It is also used for thin-film deposition.) (c) Microwave electron cyclotron resonance (ECR) reactor.

Twelve rare-earth magnets are equally spaced in a circle around and adjacent to the gas feed ring and the quartz discharge chamber. They are arranged with alternate poles forming a multipolar static magnetic field that is zero at the center and increases in the radial direction. An undulating three-dimensional surface, at which the magnetic field is equal to the value of 875 gauss (0.0875 T) required for electron cyclotron resonance, is located inside the discharge zone. The electromagnetic fields inside the cavity are focused on the quartz tube, and particularly on the electron cyclotron resonance layers, for efficient coupling of the electromagnetic energy into the electron gas.

The charge species diffusion can be controlled with the addition of a processing chamber surrounded by a multipolar static magnetic field produced by a second set of magnets. The substrate is placed on a temperature-controlled platform in the chamber, a short distance downstream from the active electron cyclotron resonance applicator. The control of species densities and independent adjustment of potential over the substrate surface with an rf bias provide the control required for etching submicrometer linewidths on integrated-circuit wafers.

Applications. Plasma processing machines and methods are now used routinely in many industrial applications. Plasma-assisted chemical vapor deposition (PACVD) of thin films and plasma etching are commonly employed in fabricating integrated circuits since processing can take place at very low wafer temperatures. Plasma etching is required to produce the circuit dimensions of 0.5–1.0 micrometer required for 4- and 16-megabit dynamic random access memories.

Other important applications include insulating films, optical coating of lenses, coatings against corrosion, and hard coatings. PACVD technology is being developed for manufacturing photovoltaic (solar cell) modules. Recently, diamondlike and diamond films and superconducting thin films have been produced by various PACVD techniques. Hard coatings extend the life of cutting tools by increasing their hardness and wear resistance. Tungsten carbide (WC) tools are routinely coated with titanium carbide (TiC), titanium nitride (TiN), or TiC_xN_{1-x}. Hard coatings have also found application in erosion protection in gas-turbine engines and rocket nozzles.

For background information SEE ARC DISCHARGE; ELECTRICAL CONDUCTION IN GASES; INTEGRATED CIRCUITS; PLASMA PHYSICS in the McGraw-Hill Encyclopedia of Science & Technology.

Jes Asmussen, Jr.

Bibliography. J. Asmussen, Electron cyclotron resonance microwave discharges for etching and thin-film deposition, *J. Vac. Sci. Technol.*, 47:883–893, 1989; D. M. Manos and D. L. Flamm, *Plasma Etching*, 1989; S. M. Rossnagel, J. J. Cuomo, and W. D. Westwood, *Handbook of Plasma Processing Technology*, 1990; T. Sugano, *Applications of Plasma Processing to VLSI Technology*, 1980.

Polymer

Concern for the environment as well as appreciation for the limited nature of the resources of this planet have aroused public consciousness. The need for greater utilization of resources is being recognized. With the limited number of options for the safe disposal of waste materials, a situation exacerbated by the closure of several inground disposal sites and by mounting legislative pressures, several solutions are being sought. The U.S. Environmental Protection Agency has set a goal of 25% material recovery from solid waste, from a current practice of about 10%. While plastics are a small fraction (about 8% by weight) of the total solid waste, their larger bulk volume (about 20%) and their general nondegradability on land or in water have prompted aggressive efforts for their recovery and recycling.

Plastics recycling. Of the over 6×10^{10} lb (2.37×10^{10} kg) of plastics produced worldwide annually, currently less than 1×10^9 lb (4.5×10^8 kg) are recycled. While a significant fraction of industrial plastic waste is being recycled, these plastics are not associated with consumer waste, which arises predominantly from packaging articles such as containers and films and automotive parts such as bumpers and fenders. A limited amount of successful work has been carried out on recycling postconsumer articles such as automobile tires and textiles, two major applications for synthetic polymers. The current work involves mostly postconsumer plastics from food packaging items such as soft drink bottles [polyethylene terephthalate (PET)], milk containers [high-density polyethylene (HDPE)], and other household articles. Some polyvinyl chloride (PVC) and polycarbonate (PC) are also being recycled. Work is being carried out in recycling polystyrene (PS) [for example, foamed polystyrene cups, trays, and lightweight packaging foam].

In 1989 approximately 3.4×10^8 lb (1.54×10^8 kg) of plastics were reported to have been recycled in the United States versus 3×10^8 lb (1.35×10^8 kg) in 1988: about 1.6×10^8 lb (7.26×10^7 kg) of polyethylene terephthalate from soft drink bottles, about 1.2×10^7 lb (5.5×10^6 kg) of polyethylene terephthalate x-ray films, and about 1×10^8 lb (4.54×10^7 kg) of high-density polyethylene. The total increase, while significant, is a relatively small percentage of the virgin polymers produced.

The quality of these recycled polymers varies substantially. Apart from impurities such as paper, adhesive residues, and metals (aluminum particles), the polymers may differ significantly in molecular weight, branching, and additives. Residual impurities from the purification processes, for example, a flotation process using detergents or residues from an alkaline wash used to remove aluminum, are not unlikely. Besides mechanical means, studies involving selective extraction with appropriate solvents are being explored in order to obtain polymer components of high purity from the mixed polymer waste. Some studies using supercritical fluids, for example, carbon dioxide under pressure, as the solvent have also been reported.

A critical aspect of polymer recycling is collection and sortation into individual polymer components. Density-based separation of the ground mixture, by suspension in water, separates these polymers into two major groups: a lighter fraction of polyolefins and a heavier fraction that may be polyesters, polyvinyl chloride, polycarbonate, and others. Careful initial separation minimizes the cross contamination extensively. Careful separation of polyvinyl chloride is necessary to avoid degradation of the other polymers during subsequent melt processing operations. Hydrocyclone technologies and automatic separation systems using detection via spectroscopic techniques (such as infrared or ultraviolet) are also subjects of active research.

Polyethylene terephthalate. The progress in polymer recycling is limited by the lack of substantial markets for the recovered products. The opportunities for large-volume applications requiring food contact, for which these polymers were originally used, are not available for these recycled polymers because of the potential presence of unsafe impurities as well as the lack of confidence in guaranteeing their purity for continuing compliance with the strict food-contact regulations. Polyethylene terephthalate, currently the largest-volume recycled plastic, can be purified by depolymerizing to lower-molecular-weight oligomers, followed by filtration, and then subsequently repolymerized to uniform high-molecular-weight polymers, suitable for the manufacture of fibers for carpets, fabrics, geotextiles and so forth. Lower-molecular-weight polyols can also be prepared by depolymerization in the presence of glycols. Such polyols often are used as ingredients for making polyurethane foams and as components for the polyesters for the fiberglass industry. The purified polyethylene terephthalate, as blends or as glass-fiber-reinforced composites, are also used as materials for molding and extrusion purposes. Clean recycled polyethylene terephthalate can be blow-molded into containers (for example, oil cans or containers for liquid detergents) for several industrial and household applications. Thermoformed articles such as cartons for packaging eggs are being made from sheets extruded from recycled polyethylene terephthalate.

High-density polyethylene. Recycled polyethylene, primarily from milk jugs and detergent packaging, is recovered by washing and purifying; it is often used to blow-mold containers. Containers for some household, non-food-contact applications (for example, detergent bottles or motor oil cans) are being made from such high-density polyethylene. An obstacle to the recovery and reuse of high-density polyethylene is its ability to absorb several low-molecular-weight impurities, such as the fatty acid by-products (for example, butyric acid) resulting from the biological degradation of dairy products. Large-scale injection-molding applications of these polymers are limited because of their high molecular weight and resulting high melt viscosities. High-density polyethylene obtained from

purification of the base cups of soft drink bottles has low melt viscosity and can be used for some applications involving injection molding.

Flexible films and foams. Used flexible films and low-density rigid foams offer a greater challenge for the recycling of the individual polymers therein. However, some industrial activity is being initiated for their collection, cleaning, and reuse. Polystyrene, primarily from foam products, is the subject of such recycling efforts by several organizations. The low density of these foamed articles makes their transportation expensive; hence their recycling is often limited to areas of high population density. Recycling of polyvinyl chloride is also in its infancy. While the recycling of polyvinyl chloride insulation from the wire and cable industries has been established, recycling from household and industrial postconsumer usage is just emerging. Such recycled polyvinyl chloride products are often used for applications such as pipes or sidings for houses.

Mixed materials. Recycling of multilayer articles, such as containers for oxygen-sensitive foods (for example, tomato ketchup), while perceived as difficult, often can be addressed technologically. Careful modification of the mixed plastic material obtained by such recycling with polymeric compatibilizers (polymeric interfacial agents that permit production of uniform blends having desirable end properties from normally immiscible polymers) and additives can yield products of value. An example is the demonstration of production of an automotive fascia made from such recycled multilayered bottles, in an effort by a consortium of several industries involved with such polymers. Similar experiments are being pursued for the recycling and reuse of the polyolefin and polypropylene components of personal hygiene products such as diapers.

Inseparable mixtures of polymers and residues from plastics recycling processes, generally classified as commingled plastics, can be used for melt-fabrication into so-called plastic lumber, which can be used to make articles such as benches for parks or planks and boards for docks and marinas. These plastic articles have significantly better water resistance than the wood products. The U.S. Navy is involved in an experimental program to accumulate and return to shore all plastic articles after their use on ships for fabrication of such articles.

Automotive plastics. The recycling of plastics from discarded automobiles and other durable goods is in its infancy. Major manufacturers of plastics for automobiles are beginning to address the recycling issue. While the components made from thermoplastic materials (for example, bumpers or fenders made from thermoplastic polyesters, polycarbonates, and blends) are the objects of the initial focus, efforts are also under way to address the issues of thermoset materials such as the styrene-based sheet molding compound (SMC). The approaches to recycling of thermosets are generally limited to their regrinding and reuse as fillers, or in production of monomers and oils for making new polymers, or for using as fuels.

Prospects. An interesting approach to the recycling of polymers for several limited service applications has been developed by a company based in Switzerland. This involves designing polymers from acrylic and styrenic monomers, and other monomers with acidic or basic functional groups, such that the polymers become soluble on demand in water at controlled pH. The polymer can be reclaimed by changing the pH again. Several companies are exploring this technology as a potentially viable option for several applications involving disposable plastics and adhesives.

Polymer recycling is an emerging field of technology. Substantial growth in this area can be expected in the 1990s. Molecular design of friendlier polymers and additives (such as flame-retarding agents, and colors and pigments), avoiding heavy metals (such as lead, cadmium, and mercury), and engineering design of fabricated parts with a special focus on their recyclability after use can be expected. Engineering advances in the technologies for sortation, identification, and purification can be foreseen. Importantly, the infrastructure required for collection and accumulation of plastic waste will be established. Legislative pressures and responsible social habits should contribute substantially to this vital issue.

For background information *SEE PLASTICS PROCESSING; POLYMER* in the McGraw-Hill Encyclopedia of Science & Technology.

P. M. Subramanian

Bibliography. J. Maczko, An alternative to landfills for mixed plastic waste, *Plastics Eng.*, 46:51–53, April 1990; *Mod. Plastics*, Waste Solutions suppl., April 1990; *Plastics World*, spec. issue, September 1989; J. H. Schut, A barrage of news from the recycling front, *Plastics Technol.*, 36:109–119, July 1990; L. Wielgolinski, *Preprints*, Soc. Plastics Eng. RETEC, October 30–31, 1989.

Porphyrin

Nature employs porphyrins to bring the rich chemistry of metal ions to such diverse biological functions as oxygen transport, electron transfer, energy conversion, and catalysis. These flat molecules (**Fig. 1**) bind metal ions tightly without blocking access to the metal center. In most biological roles, the porphyrin molecule is bound to a protein that fine-tunes the chemical properties of the metal atom and controls the environment of the metalloporphyrin complex. Nature regulates the function of the porphyrin by specifying the composition and structure of the protein to which it is bound. Recent research has revealed a new way in which some of the utility and versatility of these porphyrin-containing proteins can be mimicked with more robust synthetic materials. Tetraphenylporphyrin molecules form beautiful, intensely colored crystals that are honeycombed with microscopic tunnels; an example is zinc tetraphenylporphyrin (Fig. 1c), a synthetic species that has been often used to model the natural systems. The tunnels in these so-called porphyrin sponges provide restricted access to the metal atoms in much the same way that protein molecules do. Just as the properties of the natural porphyrins are programmed by the protein, the structural regularity of the crystalline sponges provides a simple and rational means whereby they can be programmed for a variety of applications.

The utility of these materials results from the fact that tetraphenylporphyrin molecules cannot pack efficiently in a pure crystalline solid. Crystallization from most solutions results in the incorporation of guest molecules into the tunnels of the porphyrin host lattice. Materials of this type are known as lattice clathrates. **Figure 2**a and b shows the molecular packing for the two most common structures exhibited by porphyrin sponges.

The term porphyrin sponge alludes not only to the microporous nature of these materials but also to the softness associated with a clathrate lattice in which there are no strong bonds between neighboring host molecules. The tunnels in these crystalline materials distort to accommodate guest species of diverse size and shape. Over 100 porphyrin sponges have been structurally characterized. They range from the species known as dry sponges with no guests to those in which the guests constitute nearly 50% of the crystal volume. The fact that these materials adopt very similar structures demonstrates the degree to which the molecular packing is controlled by the large and relatively rigid porphyrin. It also suggests a way in which guest incorporation can be made selective.

Programming a sponge. Typically, the tunnels in these clathrate materials contain two inversion-related guest molecules per porphyrin molecule; that is, there are alternating left-handed and right-handed guest sites in each tunnel created by the centrosymmetric packing of host molecules. If a host lattice is constructed from porphyrin complexes in which a single ligand molecule has been attached to the metal atom, this ligand will occupy one of the guest sites. Because of the symmetry of the porphyrin packing, such a sponge will preferentially incorporate into the second guest site a molecule of the same size and shape as the ligand, but one of opposite handedness. Because replacement of one ligand by another is a simple chemical process, this represents a powerful means of

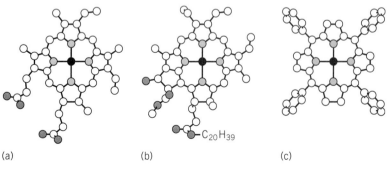

(a) (b) —C₂₀H₃₉ (c)

Key: ○ carbon ○ nitrogen ○ oxygen ● metal

Fig. 1. Porphyrin molecules. (a) Heme. (b) Chlorophyll. (c) Zinc tetraphenylporphyrin.

programming a sponge for incorporation of guests of a specific size, shape, and handedness.

Dynamics. If the metal ions in porphyrin sponges are to participate in chemical transformations, reactant and product molecules must be transported through the sponge. Desorption and absorption studies of unreactive guests have been used to examine both the structural changes involved in these processes and the kinetics of these changes.

X-ray diffraction measurements have revealed that when clathrates of either of the types shown in Fig. 2*a* and *b* are heated to drive out the guest, the desolvated sponge adopts the structure shown in Fig. 2*c*. The same or different guests can be reintroduced into the solid from the vapor phase. In most cases, the structure of the resulting clathrate is identical to that observed when the host is crystallized from a solution containing the guest.

Gas chromatography and solid-state nuclear magnetic resonance measurements have shown that the new guest molecules are initially adsorbed on the surface of the desolvated sponge with a significant reduction in vapor pressure. Then, typically on a time scale of hours, the clathrate structure is reformed.

Polymeric sponges. In some applications, the flexibility of the sponge structure may be undesirable. For reasons of increased guest mobility or reduced solubility, a more rigid structure might be required. Zeolites, the most economically important lattice clathrates, have found a wide range of technological applications that take advantage of their rigid aluminosilicate frameworks. Zeolites exist in numerous structural forms, the most suitable of which can be chosen for a particular application. These materials are not, however, very amenable to rational synthetic modification. By forging links between the host molecules of a porphyrin sponge, it is possible to impart the structural rigidity of a zeolite while retaining the desirable chemical properties of the porphyrin. One such polymeric sponge is shown in **Fig. 3.** In this material, hydrogen bonding between hydroxyl substituents on the phenyl groups links the porphyrin molecules into a three-dimensional network. Materials of this type can be expected to exhibit clathrate properties quite different from those of the original sponges.

Applications. There are numerous potential scientific and technological applications of these materials. The feasibility of several classes of applications has already been demonstrated; others remain to be explored. These applications take advantage of the fact that clathrates based on porphyrin molecules crystallize with quite predictable structures, and the fact that these structures can be manipulated in a controlled way through alterations of both the host and the guests. While chemists have developed a large complement of tools that make it possible to design and construct complex molecules, success in the engineering of crystal lattices has been limited. It should be possible to apply the concepts demonstrated in the design and application of porphyrin sponges to the engineering of a wide range of molecular crystals.

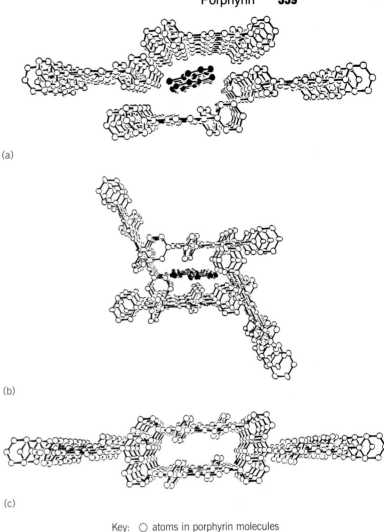

(a)

(b)

(c)

Key: ○ atoms in porphyrin molecules
● atoms in guest molecules

Fig. 2. Molecular packing in porphyrin sponges. (*a, b*) Most common structure types. (*c*) Desolvated structure, after the guest molecule has been driven out by heating.

Crystallization. A simple application of porphyrin sponges is the conversion of liquids to solids. Desolvated sponges will absorb a wide range of guest species from the liquid or vapor phase, or single crystals can be prepared by crystallization of the host from solutions containing the guest. Structural characterization of the guest is greatly facilitated if the guest is in a crystalline form. This approach has been used to determine the structures of several molecules that are ordinarily liquids at room temperature.

Stabilization, storage, and delivery. Stabilization of reactive species can be achieved through the isolation and dilution provided by a porphyrin host. Incorporation lowers the vapor pressure, simplifying storage of volatile species. Sponges might provide a convenient way to control delivery of a reagent in various chemical, technological, or medical procedures.

Separations. There are many ways in which porphyrin sponges can be used to carry out chemical separations, since sponges can be programmed to preferentially incorporate molecules of a specific size and shape. Such a separation could take the form of a

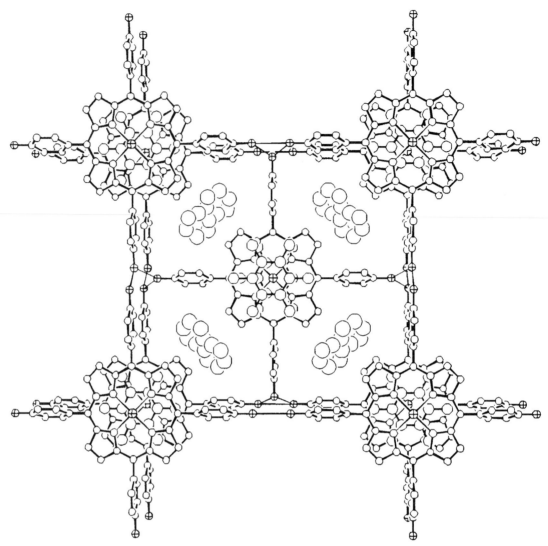

Fig. 3. Polymeric sponge.

simple crystallization. The use of polymeric sponges with fixed pore sizes may be useful in chromatographic separations. It might even be possible to construct membranes based on these materials that have a size-selective permeability.

Catalysis. There is an enormous chemical literature related to porphyrin catalysis. The ubiquity of porphyrin-based catalysis in nature has prompted a multitude of investigations both in the laboratory and in living organisms. Incorporation of the catalytic species into a sponge lattice provides much better control over the catalyst environment and the interactions between the porphyrin and potential substrates than can be achieved in fluid solution. Of particular interest in this regard are porphyrin-catalyzed oxidations. Investigations of guest transport properties and the construction of polymeric sponges are directed toward the development of such applications.

Artificial photosystems. Porphyrins, in the form of chlorophyll, harvest light energy and act as the initial electron donors in photosynthesis. The properties of porphyrin sponges suggest that they might be used to construct artificial photosynthetic systems. The hosts are good light absorbers, and electron donors and acceptors can be introduced as guest species. The kinetics of guest transport are again important in any practical energy conversion scheme.

Nonlinear optics. There is considerable interest in the use of lattice clathrates in the construction of nonlinear optical devices that can convert light of one color to another. The absorption of visible light by porphyrins precludes the use of conventional porphyrin sponges in many such applications, but this absorption can be eliminated by the use of reduced porphyrin derivatives. Candidates for such applications are currently under investigation.

For background information SEE CLATHRATE COMPOUNDS; PORPHYRIN; ZEOLITE in the McGraw-Hill Encyclopedia of Science & Technology.

Charles E. Strouse

Bibliography. M. P. Byrn et al., Tetraarylporphyrin sponges: Composition, structural systematics, and applications of a large class of programmable lattice clathrates, *J. Amer. Chem. Soc.*, 112:1865–1874,

1990; G. D. Stucky and J. E. MacDougall, Quantum confinement and host/guest chemistry: Probing a new dimension, *Science*, 247:669–678, 1990.

Precipitation (meteorology)

Although nearly every country regularly records rainfall amounts and snow depths, a complete and reliable picture of the global distribution of precipitation is presently unavailable. This results partly from absence of data over the world's oceans and spatially uneven terrestrial rain-gage distributions. Another reason is the underestimation by gages of the actual amount of precipitation that reaches the Earth's surface; gage designs vary widely between different countries.

Scientists have long recognized these problems; but until recently, most maps of global precipitation (including those in many atlases) were drawn from a limited number of observations and often were highly generalized. With the advent of high-speed computers and the concern for potential global climate changes (resulting from, for example, increases in atmospheric carbon dioxide), it has become both necessary and possible to develop more accurate and comprehensive global precipitation climatologies.

Spatial distributions. Most rain gages tend to be located in industrialized countries and particularly near cities. Consequently, the densities of rain-gage placements are usually lower in mountainous, polar, and desert regions. It is difficult to locate precipitation gages in nearly inaccessible locations, such as over rugged terrain and in thick jungle growth, although precipitation variability often may be large and of great importance in these areas. A typical rain-gage network will usually consist of fewer than 400 gages.

Estimation of precipitation over the ocean is even more problematic than over land. Since water covers nearly 73% of the Earth's surface, knowledge of precipitation over these areas is extremely important. However, rain-gage estimates of oceanic precipitation are not very reliable. Capture of spray from surf and a ship's constant motion reduce accuracy of ship-based rain gages. Furthermore, a fair-weather bias exists, because ships are frequently navigated to avoid storms. Oceanic estimates, therefore, must be approximated by using extrapolations from land-based observations.

The most reliable long-term estimates of oceanic precipitation are based on empirical relationships between observed rainfall and weather reports at coastal stations. Oceanic estimates then are obtained by applying this relationship to ship-based weather reports. Results are consistent with rainfall over low atolls as well as measurements from satellites, radars, and fixed-position ships.

Rain-gage biases. As wind speed increases, the amount of precipitation recorded by a rain gage decreases due to deformation of the wind field by the rain gage. This effect is greater during snowfall than during rain. To reduce this problem, gage shields are sometimes used to reduce the wind speed near the orifice. It has been estimated that global precipitation is underestimated by an average of 8% due solely to this effect.

Moisture adhering to the internal walls of the gages also leads to a decrease in global precipitation estimates by an average of about 2%. This source of error varies significantly among gage designs. Evaporation from a gage during the time between the end of rainfall and its measurement also can cause underestimates, particularly in warm climates. A decrease in global estimates of precipitation by about 1% can be attributed to evaporation. Other less serious sources of error exist, including splashing into and out of the gage and random errors such as leakage, incorrect

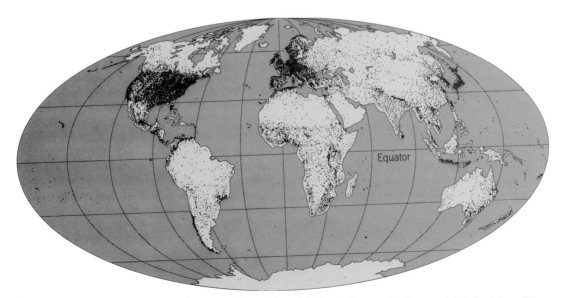

Fig. 1. Map showing distribution (dots) of the 24,635 terrestrial precipitation stations used in the new global climatology. This represents a station network that is far denser than that used in most earlier climatologies. (*After D. R. Legates, A high-resolution climatology of gage-corrected, global precipitation, in B. Sevruk, ed., Precipitation Measurement, WMO/IAHS/ETH International Workshop on Precipitation Measurement, St. Moritz, Switzerland, pp. 519–526, 1989*)

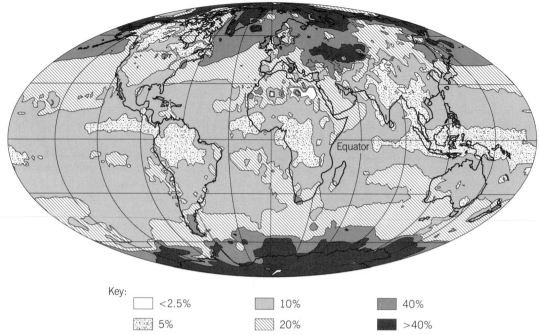

Key:
  ☐ <2.5%   ▨ 10%   ▨ 40%

  ▨ 5%   ▨ 20%   ■ >40%

Fig. 2. Map showing percent error (undercatch) in gage-measured annual precipitation.

placement of the gage, spilling of the catch, and inaccurate measuring or recording techniques.

Improved global precipitation climatology. A new global precipitation climatology has been developed that includes long-term monthly precipitation averages for 24,635 land-based stations (**Fig. 1**), the highest station density ever utilized. Oceanic estimates of

mean monthly precipitation were obtained by using more than 1 million ship-based weather reports from 1950 through 1972 and procedures noted above.

The most important contribution of this new climatology lies in its removal of gage-induced biases resulting from the effect of wind, wetting and evaporative losses, and even peculiarities associated with

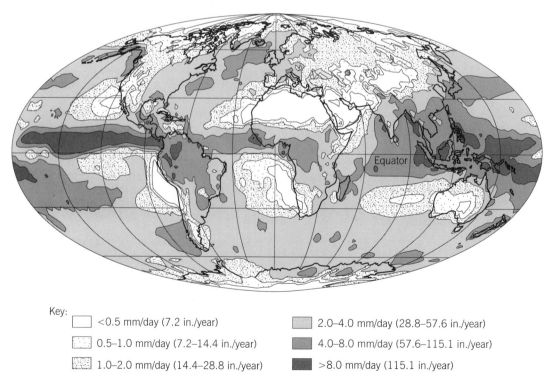

Key:
  ☐ <0.5 mm/day (7.2 in./year)   ▨ 2.0–4.0 mm/day (28.8–57.6 in./year)

  ▨ 0.5–1.0 mm/day (7.2–14.4 in./year)   ▨ 4.0–8.0 mm/day (57.6–115.1 in./year)

  ▨ 1.0–2.0 mm/day (14.4–28.8 in./year)   ■ >8.0 mm/day (115.1 in./year)

Fig. 3. Map showing annual precipitation. Errors induced by the gage measurement process have been estimated and removed.

the Soviet Tretyakov gage during blowing snow. Differences in rain-gage designs, siting characteristics, and local effects near each rain gage were included to estimate and remove the biases.

Error in annual gage-measured precipitation, expressed as a percent of the corrected precipitation, is mapped (**Fig. 2**). A strong latitudinal influence on gage errors exists. Near the Equator, errors in measured precipitation are very low, usually 2.5–5%. In midlatitudes (between 30° and 60° North and South), errors are larger and approximately 10–20% of the corrected precipitation. Gage errors increase near the poles, often exceeding 40%. This poleward increase reflects the largest source of gage error—the effect of wind on snowfall.

From this climatology, average annual global precipitation is 1123 mm (44.2 in.), while terrestrial and oceanic precipitation are 820 mm (32.3 in.) and 1251 mm (49.3 in.), respectively. These figures are 11% higher than estimates suggested previously. However, they are commensurate with estimates of global evapotranspiration, which must equal precipitation in the long term.

Mean monthly estimates of precipitation were interpolated to a 0.5° grid in both latitude and longitude. The resulting annual mean field is shown for a 2° by 2° subset (**Fig. 3**). Precipitation is greatest along the Equator where intense surface heating exists and the tropical easterlies converge. By contrast, surface divergence produces the major deserts of the world—the Sahara, Gobi, Kalahari, Atacama, Sonoran, and the Australian Outback—over land areas between 20° and 30° North and South latitude. These desert areas often extend over adjacent ocean waters kept relatively cool by coastal upwelling. Precipitation is enhanced over and near relatively warm waters in midlatitude.

Other influences on precipitation patterns can be seen. In the Northern Hemisphere in particular, precipitation is highly variable over mountains (such as the Alps and the Rockies), while relatively smooth patterns exist over level terrain. Precipitation is higher on the windward side of the mountains than on the leeward side. Dramatic variations in the longitudinal distribution of precipitation in the Southern Hemisphere are associated with Africa, South America, and Australia.

Future advances. While this precipitation climatology has resulted in a more accurate picture of global precipitation, further refinements are forthcoming. Data exchange agreements are presently being discussed between climatologists from the United States and other countries, particularly the Soviet Union. Studies are addressing temporal variability of precipitation as well as its long-term average. Research also is increasing to produce better depictions and understanding of precipitation patterns in underdeveloped countries (for example, in sub-Saharan Africa). In addition, scientists in the Soviet Union, Western Europe, and North America recently have begun a collaborative effort toward better estimation of gage measurement errors.

The largest steps to increase accuracy in global precipitation estimates, however, probably will be made through remote-sensing procedures. Both satellites and radars can be used to obtain estimates of precipitation. These procedures provide for a much finer resolution than can be obtained through traditional rain gages, and can be used to provide coverage of both remote land areas and the oceans. Remote-sensing procedures have not been used to produce global climatologies because of their short observation periods and problems in estimating stratiform precipitation from satellites.

For background information SEE EVAPOTRANSPIRATION; PRECIPITATION (METEOROLOGY); PRECIPITATION GAGES in the McGraw-Hill Encyclopedia of Science & Technology.

David R. Legates

Bibliography. C. E. Dorman and R. H. Bourke, A temperature correction for Tucker's ocean rainfall estimates, *Quart. J. Roy. Meteorol. Soc.*, 104: 765–773, 1978; D. R. Legates, A climatology of global precipitation, *Publ. Climatol.*, vol. 40, no. 1, 1987; D. R. Legates and C. J. Willmott, Mean seasonal and spatial variability in gauge-corrected, global precipitation, *Int. J. Climatol.*, 10:111–127, 1990.

Primates

For many years, the earliest known anthropoids that could be confidently identified in the fossil record were early Oligocene [32–34 million years ago (m.y.a.)] in age and all had come from one geological formation in Egypt. In recent years, however, early anthropoids have also been discovered at other African and Arabian localities. More significantly, even older and more primitive anthropoids have now been found in late Eocene deposits (35–40 m.y.a.) in Egypt. Detailed studies of these and of the growing sample of early Oligocene fossils have enhanced the understanding of anthropoid origins, adaptations, and evolution.

Early anthropoid fossils. The earliest known anthropoid primates appear in Africa just before the Eocene–Oligocene boundary (37 m.y.a.). Before this time, the only primates found in the fossil record are prosimians, which date back to the earliest Eocene (55 m.y.a.) in North America and Europe.

There is no clear fossil evidence placing anthropoids in South America or Asia earlier than in Africa. The age of the earliest New World monkey (*Branisella boliviana*) is only 20–25 m.y., or about 15 m.y. younger than the earliest African anthropoids. A small sample of fragmentary teeth and jaws found in the late Eocene of Burma may represent two genera of presumed anthropoids (*Amphipithecus* and *Pondaungia*). However, the fossils are very incomplete, and their anthropoid status is uncertain; they may be adapid prosimians. Thus, the fossil record of Eocene and Oligocene anthropoids is currently restricted to Africa, and most finds there come from one site, the Fayum Province of Egypt.

Fayum primates. The Jebel Qatrani Formation, source of the Fayum fossils, can be divided into two distinct stratigraphic levels, a lower, older level (probably 35–40 m.y.a.) that lies near the Eocene–Oligocene boundary, and an upper, younger level that is early Oligocene (32–34 m.y.a.) in age. Anthropoids from the upper level include the well-known parapithecids (*Apidium* and *Parapithecus*) and the so-called dawn apes or propliopithecids (*Aegyptopithecus* and *Propliopithecus*). In recent years, intensive paleontological efforts have been focused on the Fayum's lower levels, where primates are very rare, in order to discover older and more primitive anthropoids.

Until the 1980s, only one anthropoid lower jaw, the type specimen of *Oligopithecus savagei*, had been found in the lower strata of the Fayum. Persistent fieldwork has now produced several new anthropoid fossils from the same site, including isolated upper and lower teeth of *Oligopithecus*. In addition, lower jaws of an entirely different anthropoid, *Qatrania wingi*, have been found there.

Oligopithecus is related to the dawn apes found higher in the Fayum section, and is classified with them in the family Propliopithecidae. *Oligopithecus* weighed about 1.98–2.2 lb (900–1000 g), about the same as that of a modern squirrel monkey (*Saimiri*). The chewing surfaces of the molars bear sharp crests and edges, suggesting a partially insectivorous diet rather than the highly frugivorous diet of the dawn apes. *Oligopithecus* and the dawn apes have only two premolars on each side of the upper and lower jaws, an exclusive resemblance to all later Old World monkeys, apes, and humans (Catarrhini, or Old World higher primates). For this reason, *Oligopithecus* and the dawn apes are believed to lie near the ancestry of later catarrhines.

Qatrania is smaller than *Oligopithecus*, and it retains three premolars in each jaw. Its molar cusps are more bulbous than those of *Oligopithecus*, suggesting a more frugivorous diet. *Qatrania* is classified with *Apidium* and *Parapithecus* in the family Parapithecidae, which apparently did not give rise to later descendants. Until 1989, *Qatrania* and *Oligopithecus* were the world's earliest known anthropoids. However, these genera have now been superseded by new discoveries at even lower stratigraphic levels in the Fayum.

World's earliest anthropoids. Two species of anthropoids that are probably of late Eocene age (36–40 m.y.a.) have been found at the recently discovered quarry L-41. One of the new species is *Catopithecus browni* (**Fig. 1**), known from upper and lower teeth and jaws, and a crushed skull. It is closely related to, and may be ancestral to, *Oligopithecus*. The other species, *Proteopithecus sylviae*, is currently represented by a single upper jaw similar in size to that of modern tamarins. It is unclear from this one specimen if *Proteopithecus* is a close relative of *Catopithecus* or is an early parapithecid.

The anthropoid status of *Catopithecus* is confirmed by its postorbital septum, by the fusion of the midline suture of the frontal (forehead) bone, and by the placement of a ringlike ectotympanic bone (the loop of bone holding the eardrum) at the bullar

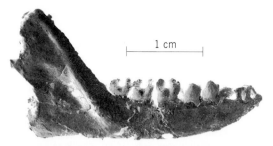

Fig. 1. Lower right jaw of the oldest known anthropoid primate, *Catopithecus browni*, found in late Eocene sediments of Egypt.

aperture (the bony lateral opening of the ear). Like catarrhines, it has only two premolars per jaw. However, in molar structure and in the shape of the mandible, *Catopithecus* closely resembles prosimians. The lower molars have several features generally found only in prosimians, for example, a raised anterior portion of the tooth and a basinlike posterior portion; and the upper molars are each dominated by only three major cusps with just a slight, crestlike fourth cusp, unlike the four evenly balanced cusps of most later anthropoids.

The size of the crushed braincase of *Catopithecus* suggests a small brain by anthropoid standards, perhaps not exceeding the average brain size of a modern lemur. Even the younger Fayum anthropoid, *Aegyptopithecus zeuxis*, had a brain proportional in size only to those of some prosimians. *Catopithecus* had small orbits relative to its skull size, indicating that it was diurnal. Also the orbits are relatively widely spaced, suggesting that the olfactory sense may have been well developed in comparison with that of most modern anthropoids.

Thus, *Catopithecus* is a mosaic of specialized catarrhine features, generalized anthropoid features, and primitive prosimian features. On the one hand, the cranial characters reveal its phylogenetic relationship to anthropoids; on the other, the dental features may offer phylogenetic clues as to its prosimian ancestry. This complex mosaic of primitive and derived characters in one fossil species suggests that the prosimian–anthropoid transition and the early evolutionary radiation of the major anthropoid taxa may have occurred at a very primitive, prosimianlike stage, which was recognizably anthropoid only on the basis of a relatively few structural details, such as the postorbital septum. In addition, this mosaic demonstrates that many traits common to most modern anthropoids, such as those of the teeth, jaws, and skull, must have evolved convergently in platyrrhines (New World higher primates) and catarrhines.

Discoveries outside Egypt. Early anthropoid fossils have now turned up at sites in Algeria, Angola, and on the Arabian Peninsula (which was part of Africa, not Asia, during the Eocene and early Oligocene). An anthropoid canine, *Propliopithecus* sp., was described from Malembe, Angola, in 1986; a lower molar of *Biretia piveteaui* was reported from Nementcha, Algeria, in 1988; and several isolated molars from *Oligopithecus savagei* and *Propliopithecus* sp. were found at Thaytiniti and at Taqah, Oman, in 1988 (see **Fig. 2**). The teeth from these sites are very similar to those from the Fayum, suggesting that all five sites are of similar age. Paleomagnetic dates from Oman indicate an age of 35.8 m.y., or late Eocene. The sparse and incomplete fossils so far obtained outside the Fayum provide little additional information concerning anthropoid origins, except for proving that by the early Oligocene, anthropoids were widely distributed on the Afro-Arabian continent.

Prosimian ancestors of anthropoids. During the Eocene (34–55 m.y.a.), prosimians flourished in great abundance and diversity. The many dozens of Eocene species may be classified into one of two families, Adapidae and Omomyidae. Adapids are often characterized as lemurlike, while omomyids are considered to be tarsierlike, but it is not always clear what the precise relationships are between Eocene taxa and modern lemurs and tarsiers. At the broadest level, the question of anthropoid origins revolves around whether the lemurlike adapids or the tarsierlike omomyids were ancestral to anthropoids. Both of the Eocene families may be divided into several adaptively divergent subfamilies, which further complicates attempts at phylogenetic reconstruction.

The prosimianlike attributes of the earliest anthropoids provide some clues concerning which of the Eocene groups may have been their closest prosimian relatives. The early anthropoids were not tarsierlike in cranial or dental morphology. The Fayum anthropoids have ringlike ectotympanic bones, broad interorbital distances, small orbits, lower jaws fused together at the midline, spatulate incisors, and large sexually dimorphic canines. In contrast, tarsiers have tubular ectotympanics, compressed interorbital regions, giant orbits, separate lower jaws, stout pointed incisors, and small monomorphic canines. Cranial features that previously hinted at a tarsioid ancestry of anthropoids (including an anterior entry of the internal carotid artery into the base of the skull, and orbits that are separated by only a thin septum of bone in the midline) have now been shown to be absent from the early anthropoids. This suggests evolutionary convergence between tarsiers and some later anthropoids, rather than exclusive common ancestry.

The prosimians with the greatest resemblance to the Fayum anthropoids are some of the small, lemurlike members of Adapidae, especially members of the tribe Protoadapini. However, the resemblances between protoadapins and anthropoids may have arisen by convergent evolution. Thus, one cannot exclude the unlikely possibility that anthropoids may have come from an ancient tarsiiform primate that was not as specialized as the modern tarsiers. In particular, the extinct subfamily Omomyinae contains several genera, for example, *Chumashius, Rooneyia,* and *Hemiacodon,* that resemble adapids and anthropoids in the morphology of their incisors, molars, and locomotor adaptations. However, even these relatively anthropoidlike omomyids are too specialized in the dentition and ear region to have given rise to

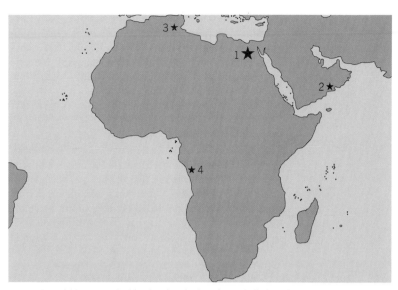

Fig. 2. Map of Africa and Arabia showing the locations of all sites that have yielded fossilized remains of Eocene and Oligocene anthropoid primates. (1) The Fayum depression of Egypt, the site of most of the fossils. (2) Thaytiniti and Taqah, Oman. (3) The Nementcha Mountains of Algeria. (4) Malembe in Cabinda, Angola.

anthropoids. Furthermore, the attributes evident in the earliest anthropoids more closely resemble those of prosimians of the tribe Protoadapini than those of Omomyinae.

Prosimian–anthropoid transition. The origin of anthropoids from a prosimian ancestor is an important evolutionary transition from the point of view of human ancestry, but it also bears significance to broader evolutionary questions. This prosimian–anthropoid transition now constitutes one of the best examples in the fossil record of evolution across a subordinal taxonomic boundary. Many attributes associated with extant anthropoids are missing in the Fayum taxa, particularly those from the lower strata, whereas other anthropoid characters, such as a fused lower jaw and details of the middle ear, are already present in some Eocene prosimians. Classification of late Eocene primates into either Prosimii or Anthropoidea must rely heavily on a relatively small handful of taxonomically diagnostic characters, such as the postorbital septum. This demonstrates that the complete package of anthropoid attributes observed among modern taxa did not arise suddenly as part of a relatively sudden evolutionary shift, but rather that anthropoid attributes were accumulated piecemeal over millions of years. This favors the theory that evolutionary transitions between the major taxonomic subdivisions, which have been defined on the basis of extant species, may actually represent a long process of microevolutionary changes that can be examined only in fossil species.

For background information *SEE MONKEY; PRIMATES* in the McGraw-Hill Encyclopedia of Science & Technology.

Tab Rasmussen

Bibliography. G. C. Conroy, *Primate Evolution,* 1990; J. G. Fleagle, *Primate Adaptation and Evolution,* 1988; D. T. Rasmussen and E. L. Simons,

New specimens of *Oligopithecus savagei*, early Oligocene primate from the Fayum, Egypt, *Folia Primatol.*, 51:182–208, 1988; E. L. Simons, Discovery of the oldest known anthropoidean skull from the Paleogene of Egypt, *Science*, 247:1567–1569, 1990.

Production planning

As manufacturing processes have become more complex, frequently being designed for automation and flexible manufacturing, scheduling has become the focus of a great deal of research. This article describes recent advances in scheduling of small-lot assembly systems, dynamic scheduling systems for production, and the use of conservation laws and polymatroid structure in dynamic scheduling.

SMALL-LOT ASSEMBLY SYSTEMS

Recent research has yielded advances in scheduling small-lot assembly systems, initiating a science base for the time management of material flow. Mathematical models that describe the nature of kitting operations in a dynamic, stochastic environment are used to set due dates for vendor deliveries and to schedule flow throughout a multiechelon system such as that shown in **Fig. 1**. Schedules are prescribed to minimize the total cost (TC) of time management, including job tardiness penalties and holding costs of inventories for both in-process and finished goods. Additionally, related models reschedule to compensate for random disruptions, assuring cost-effective due-date performance while promoting shop efficiency.

System performance. Before an assembly operation can begin, required components are collected in a kit, so the flow of material must be managed to ensure that kits are composed according to schedule. Historically, the kitting process has been poorly understood and, consequently, poorly managed. Kitting time ($\mathbf{KT}_J$) is the maximum of the ready times (R_j) of constituent components $j = 1, \ldots, J$ [Eq. (1)].

$$\mathbf{KT}_J = \max\{R_1, R_2, \ldots, R_J\} \qquad (1)$$

Since ready times are random variables, Eq. (2)

$$E[\mathbf{KT}_J] > E[R_j] \text{ for } j = 1, \ldots, J \qquad (2)$$

applies, where E is the expected value operator. Most often, managers assume that $E[\mathbf{KT}_J] = \max_j E[R_j]$; and they create schedules that cannot be achieved, since they neglect the possibility of late deliveries, which increase the expected cost of time management ($E[\mathbf{KT}_J]$). The discrepancy can be exacerbated if J or the variances of ready times are large.

Even if inventories are well managed, some components may be out of stock when kitting is scheduled. Guarantee of component availability can be achieved by traditional inventory management methods only at a prohibitive cost. Consequently, various devices are used to compensate, including dedicating available components to a priority kit while waiting for delivery of missing components, or releasing the partial kit into production. These alternatives increase costs; for example, dedicated components might be used to complete other kits, and late deliveries require special handling.

In multiechelon systems which produce subassemblies to be used in subsequent operations, flows must be coordinated to ensure timely kitting at all echelons. Typically, each echelon is managed by a different organization, making it difficult to achieve adequate coordination.

Material requirements planning (MRP), which assumes that components are available when required and that lead times are deterministic, cannot resolve these problems. In fact, planned lead times are typically determined in an ad hoc manner, rather than by dealing with the underlying phenomenon scientifically. Furthermore, the material requirements planning approach does not necessarily lead to minimum costs.

Recently, just-in-time (JIT) or pull systems have been applied to manage material flow. Originally designed for the high-volume, level-production setting, the just-in-time approach is also limited in the small-lot environment. The small-lot manufacturer may not have sufficient influence on suppliers to make just-in-time deliveries possible. Requiring one order to be pulled by the independent demand of another does not give effective control over due-date performance.

The electronics industry exemplifies these problems. Circuit cards require numerous components, many with highly variable lead times. Automated assembly processes require complete kits, and the products (for example, computers) entail multiechelon assembly.

Mathematical models lend insight into the relationships that influence system performance. Descriptive models highlight underlying functional relationships, while prescriptive models specify schedules to minimize the total cost of time management. Solutions prescribed by enriched models may simply require adaptation by managers before implementation.

Stochastic models. These are models that describe the kitting process and prescribe optimal safety lead times. Additionally, models of transient material flow are described as a method for scheduling.

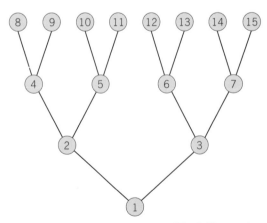

Fig. 1. Four-echelon assembly system. (*After L. Wang and W. E. Wilhelm, A New Paradigm for Scheduling Assembly Operations, Working Paper, Department of Industrial Engineering, Texas A&M University, 1990*)

Kitting. For an assembly whose components are special-ordered, the kitting process may be modeled by dealing with random variables pairwise, for example, define $\mathbf{KT}_1 \equiv R_1$, then $\mathbf{KT}_j = \max[\mathbf{KT}_{j-1}, R_j]$ for $j = 2, \ldots, J$. This approach has been used with the assumption that component ready time $\{R_j | j = 1, \ldots, J\}$ has the multivariate normal distribution. Given the mean and variance of each ready time (R_j) and the correlation between each pair of ready times, well-known equations can be used to approximate the moments of $\mathbf{KT}_j$, and its distribution may be approximated by the normal. This approximation introduces some error, since in general the distribution of $\mathbf{KT}_j$ is not normal. Nevertheless, extensive empirical tests have shown that the approximation gives a good, practical description of kitting time, and they were subsequently used to identify various inherent relationships. For example, as J increases, $E[\mathbf{KT}_j]$ increases at a decreasing rate, showing quantitatively the difficulty involved in kitting a large number of components (**Fig. 2**). In addition, increasing correlation between ready times decreases $E[\mathbf{KT}_j]$ (**Fig. 3**), facilitating the management of kitting and suggesting an advantage in not using independent vendors. This model can be used to prescribe due dates for deliveries, setting safety lead times that minimize $E[\mathbf{TC}]$. This is a scientific approach that offers significant advantages in comparison with traditional methods of setting safety lead times.

Transient flows. The approach has been extended to approximate the time-dependent performance of production and assembly systems including flow shops, assembly networks, and cellular assembly configurations. The starting time of an operation is determined by the ready times of all required resources; an example is Eq. (3),

$$S_{ij} = \max[H_{ij}, Q_{ij}, R_{ij}] \qquad (3)$$

in which S_{ij} = starting time of job j on station i, H_{ij} = ready time of station i to start job j, Q_{ij} = ready time of job j for this operation, and R_{ij} = ready time of a robot to move job j to station i. Operation finish time F_{ij} is given by Eq. (4),

$$F_{ij} = S_{ij} + P_{ij} \qquad (4)$$

where P_{ij} is the associated processing time. By assuming that processing times are independent and normally distributed, and approximating the distribution of starting times with the multivariate normal, the moments and distributions of start and finish times can be estimated, providing a schedule of time-dependent operations. The pairwise correlations between finish times that define S_{ij} are estimated by a specialized algorithm and used to evaluate Eq. (3).

The significance of recursion models is that they permit a rapid (substantially faster than equivalent simulation models) and accurate evaluation of management policies that affect material flow, so that schedules may be devised to satisfy due dates that have been promised to customers. Recursion models have evolved to a rather mature state and can incorporate numerous features encountered in industrial

settings; these include parallel machines at a station, job shop routing, and material handling within and between cells.

Enhanced recursion models. Recently, the mixture of Erlang distributions has been employed as the assumed distribution for processing times as well as the approximation distribution for start and finish times. This distribution is known to provide a good fit for the distribution of any nonnegative random variable; thus it offers the potential for generalizing recursion models and for improving their accuracy. The model describes accurate schedules for multiechelon assembly systems in which independent finish times deter-

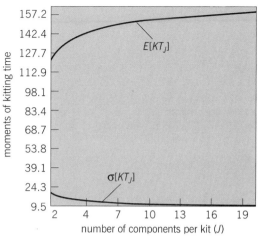

Fig. 2. Relationship of the moments of kitting time to the number of components per kit (*J*). *E* [KT$_J$] = expected kitting time. [σKT$_J$] = standard deviation of kitting time. Pairwise correlation between component ready times = 0.0. C_j (coefficient of variation or component ready time) = 0.02. K_j (safety lead-time parameter for component due date) = 1. (*After W. E. Wilhelm and L. Wang, Management of component accumulation in small-lot assembly systems, J. Manuf. Sys., 5(1):27–39, 1986*)

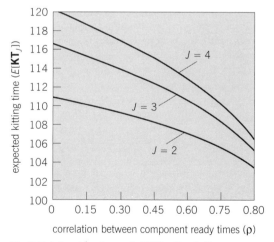

Fig. 3. Relationship of expected kitting time to the correlation between component ready times (ρ). C_j (coefficient of variation of component ready time) = 0.2. *J* = number of components in the kit. K_j (safety lead-time parameter for component due date) = 1.0. (*After W. E. Wilhelm and L. Wang, Management of component accumulation in small-lot assembly systems, J. Manuf. Sys., 5(1):27–39, 1986*)

mine the start time of each operation (excluding resource sharing among operations and repetitive production).

Applicable to job shop production, this model incorporates analytical expressions that define expected earliness and tardiness, provides measures of kitting timeliness, and permits rescheduling to compensate for random disruptions. Employed within an optimization procedure, the model rapidly prescribes schedules that minimize the $E[TC]$. Significantly, it has been shown that cost-optimal flow control can be achieved by managing vendor deliveries rather than by delaying material in process.

Deterministic models. A complementary approach has been developed to prescribe the allocation of components to kits and to schedule kitting operations in multiechelon assembly systems so as to minimize the total cost of time management. Even if inventories are well managed, it is unlikely that all required components will be available as scheduled. Thus, components on hand and those promised for future delivery must be allocated effectively to achieve a balance between satisfying due dates for end products and promoting efficiency in the assembly shop. This problem was formulated as an integer program that can be applied daily to prescribe kitting decisions that compensate for random disruptions. The model is resolved by using a specialized solution approach involving Lagrange relaxation, dynamic programming, and dominance properties. The approach performed significantly better than a commercial optimization package on a set of randomly generated test problems.

Implications for materials managers. These research results describe the nature of the kitting process and its influence on material flow. Models of time-dependent flow can be applied to develop schedules that accommodate random delivery times and coordinate flows throughout multiechelon systems. Stochastic models suggest flow management tactics that give the most cost-effective means of ensuring responsiveness to customer requirements. Deterministic models prescribe short-term compensation for random perturbations, ensuring cost-effective allocation of components to kits and balancing the costs of inventory and schedule performance while promoting shop efficiency. These interrelated models initiate a science base for the time management of material flow in small-lot assembly systems.   *Wilbert E. Wilhelm*

DYNAMIC SCHEDULING SYSTEMS

Dynamic scheduling systems for production control the flow of jobs through machines in the manufacturing process. Jobs enter the system, are processed consecutively by a number of machines, and leave on completion. The scheduling system may set the due date of each job, release the job into the system, and pick the route that it will follow; at each machine it may set the processing speed and determine the sequence in which waiting jobs will be performed. Objectives include satisfying the demand for completed jobs, maximizing throughput and utilization of the machines, minimizing waiting times and times for work in process, and reducing variability.

With increasing complexity of modern manufacturing processes, rising cost of capital outlay, and greater automation, scheduling systems play an increasingly crucial role. The use of flexible manufacturing methods introduces a much wider variety of different job types into the process. At the same time, these methods allow greater flexibility in routing the jobs through the machines. Greater automation necessitates centralized control, but also massive databases with rapid access to current position information that can be tapped by the scheduling system. Such demands outpace the ability of the shop floor manager to control the process effectively by old methods; to aid in the process, large and complex decision-support systems are being installed in many plants.

There are four distinct approaches to production scheduling: deterministic scheduling, stochastic scheduling, queueing networks, and expert systems.

Deterministic scheduling. This approach deals with the scheduling of a batch of jobs as a static, finite, combinatorial optimization problem with known deterministic data. A systematic study has identified 4536 distinct problems classified by machines, jobs, and objectives. Machine configurations include single machine, parallel machines with identical or differing speeds, flow shops of machines in series, and job shops built up of a network of machines. Jobs are characterized by release times, due dates, and processing requirements, as well as by possible precedence relations among jobs, setup times, and switching costs.

Some of the scheduling problems of single machines have straightforward, easily implemented solutions. The flow time is defined as the average waiting time per job. It is minimized by sequencing the jobs according to shortest processing time first. Weighted flow time is minimized by Smith's rule, that is, by scheduling the jobs in decreasing order of the ratio of the weight to the processing time. Maximal lateness is minimized by Jackson's rule, that is, scheduling by the earliest due date first. Problems that involve more than one machine are usually more complex. Shortest processing time first minimizes flow time on parallel machines. On a two-machine flow shop the time until completion of the last job (the makespan) is minimized by Johnson's rule, that is, starting to schedule jobs that will require more processing on the second machine according to shortest processing time first on the first machine, then scheduling jobs that will require more processing on the first machine according to the longest processing time first on the second machine. Other, seemingly simple problems are not as easily solved. Minimization of makespan on two or more parallel machines is equivalent to partitioning the jobs as equally as possible between the machines; however, this partitioning problem is known to be NP-hard. At the present state of the art, NP-hard problems are assumed to require an amount of computations that grows exponentially with the size of the problem; this makes it impractical to try to solve such problems completely.

In all, efficient algorithms, which require an amount of calculation polynomial in the problem size, have been devised for 416 deterministic scheduling problems. In addition, 3817 problems were shown to be NP-hard, which places them in a large class of combinatorial optimization problems for which it is unlikely that polynomial time algorithms exist. For such problems, polynomial time heuristics, which achieve good but suboptimal solutions, have to be used. A simple heuristic for minimization of makespan on m parallel machines is longest processing time first, which has worst-case performance, measured by the ratio of the heuristic and the optimal solutions, of

$$\frac{4}{3} - \frac{1}{3m}$$

Moreover, the longest-processing-time-first heuristic is asymptotically optimal, in average case analysis, under the assumption of a probabilistic population of problems; fully polynomial approximation schemes are also available for this problem.

For the more demanding problem of minimizing makespan in a job shop, where each job has its own route through the machines, even very small problems may require excessive amounts of computation. For example, a problem that involves 10 machines and 10 jobs has occupied researchers for several years; it finally required 5 h of computation on a large computer to solve. The so-called sliding-bottleneck heuristic iteratively identifies a sequence of flow-shop problems that are a bottleneck for the current solution, and solves them approximately, to provide a reasonable suboptimal solution for the job shop.

Stochastic scheduling. This approach considers scheduling with incomplete information about future processing times, but the probability distributions of the processing times are assumed to be known. Here expected values are used, and the optimization of the expected value of the objective function is sought. Far fewer problems have been analyzed under incomplete information, though some general patterns can be discerned. For a single machine, shortest expected processing time first and Smith's rule remain optimal. On parallel machines, shortest expected processing time first does not always minimize expected flow time; a sufficient condition is that the processing times of the jobs be stochastically ordered. However, the use of shortest expected processing time or of Smith's rule as a heuristic is, under some reasonably mild assumptions, asymptotically optimal when the number of jobs becomes large. This is in contrast to the worst-case performance of Smith's rule in the deterministic case, which is not asymptotically optimal.

The expected flow time can be reduced by allowing preemption of jobs. Preemption means that a job is interrupted and taken off the machine before it is completed, and its processing is resumed at a later time. On a single machine, a dynamic scheduling rule, which dynamically calculates a priority index for the currently processed job, can be used to determine the optimal time to preempt the job and replace it by the waiting job that possesses the highest index. The priority index is given by the ratio of the probability of job completion to the required time, maxi-

mized over all positive stopping times (a Gittins index).

Several stochastic scheduling problems can be solved when the processing times are assumed to be exponentially distributed. Examples are a generalization of Jackson's rule and of Johnson's rule as well as threshold policies for dispatching jobs to parallel machines of unequal speeds.

Queueing networks. Many applications require models for scheduling problems in which jobs arrive in an infinite random stream rather than as a single batch with given release times. Scheduling a single machine that processes a stream of jobs of several types is modeled by a single queue with nonhomogeneous customers. The simplest model for stochastic arrival times of nonhomogeneous jobs is to assume that jobs arrive in a Poisson stream. This means that arrivals occur singly, in an unpredictable but uniform rate; furthermore, each arrival has fixed probabilities of belonging to any of the different types. Under Poisson arrivals, the flow time is minimized by the so-called $c\mu$ rule, which gives priority to jobs with the highest product of the cost rate c and the processing rate μ. The processing rate is defined as the reciprocal of the expected processing time, and hence the $c\mu$ rule is actually identical to Smith's rule discussed above. A more complex single-machine scheduling problem, known as Klimov's problem, was introduced in 1973. Here jobs may require several stages of processing and may incur variable holding costs at different stages; the scheduler has to choose, at any time at which the machine completes a processing stage, which job should be processed next. The choice is between continuing the current job, choosing one of the partially completed waiting jobs, or starting a newly arrived job. A priority index that is the ratio of holding cost reduction per unit processing time, maximized over stopping times, determines an optimal order for the jobs; it is also a Gittins index.

Scheduling a stream of jobs through a flow shop is modeled by a system of queues in tandem. An important problem is the optimal order of the machines, and the division of the total processing time of the jobs into processing stages performed by the different machines. Some special cases in which the order of the machines is unimportant include deterministic and exponential processing times. In those cases, it is best to balance equally the average workload of the machines. In more general situations, approximation methods, such as renewal approximation and diffusion approximation, can be used to determine an efficient order of the machines and a division of processing into stages. If there is only finite buffer space between the machines, then blocking can occur. In that case it is sometimes optimal to arrange the machines in a bowl-shaped order, with more processing allocated to the first and last machines of the flow shop, and less work to the machines in the middle.

Job shops are modeled by networks of queues. While exact analysis of such systems is intractable, they can be approximated under heavy traffic conditions by fluid or diffusion approximations. Various optimization problems (such as setting due dates, control of the release of jobs into the system, routing through the system, and scheduling at the machines) can be reformulated as control problems for the diffusion approximation and solved numerically. The resulting optimal control of the diffusion approximation is then reinterpreted to yield a control policy for the original queueing network. Such techniques have been successfully implemented for networks of two to four queues, and they are potentially very promising. *See Queueing theory.*

Expert systems. Recent developments involve expert systems that are used to combine the various tools of deterministic and stochastic scheduling and queueing networks into software that can be used in a production facility. Dynamic production scheduling systems for a whole plant include several modules implemented hierarchically. The different modules operate at various time horizons, time scales, and degrees of centralization. A master program operates centrally, over a long time horizon, by receiving as input the overall production goals, and by using an overall status description of the plant. Those goals are disaggregated into subproblems for several parts of the plant, each with a limited number of machines and jobs, and a short time horizon. The subproblems are solved locally, subject to a more detailed and more frequently updated local status description. They are re-solved at relatively short time intervals to provide dynamic control of the actual production. It is at this local level that most of the techniques described above are implemented.

The downward communication from the master program to the local modules is supplied by a pricing mechanism, which sets due dates and holding costs for the various jobs. This part of the system incorporates the general goals and philosophy of the plant management, and is implemented as a large mathematical program. Upward communication is provided by transfer of the local modules' status to a central database and aggregation of these data for the master program. The installation and efficient operation of on-line status reports, information flow, and database management are perhaps the costliest part of the system.

Artificial intelligence techniques are employed to incorporate knowledge-based rules into the system. Front-end software that enables the managers and engineers at the local as well as the central level to monitor the operation, and to intervene in the scheduling process or override it, is also an essential part of most expert systems.   *Gideon Weiss*

CONSERVATION LAWS AND POLYMATROID STRUCTURE

Despite their complexity, many modern, human-made systems, such as multiuser computer systems and computerized manufacturing systems, are often operated under simple priority rules. Apart from their obvious advantage of being simple and easy to implement, is there any scientific justification for using such rules? Can a class of systems as general as possible be characterized in which such rules are proven optimal in some precise sense? The answers lie in the concept of conservation laws and the related polymatroid structure. While the latter is a purely mathematical entity, the former is easily stated in verbal terms, is intuitively verifiable, and is present in a wide range of systems.

System characteristics. In many complex human-made systems, the resource, or server, is shared among different types of jobs that generate service requests at random time epochs. Each job requires a certain amount of service (processing) time, which is typically random and perhaps not known until the job's arrival (that is, when the service request is generated). An arrived job often has to wait (before receiving service) if the resource is not immediately available (for instance, if it is fully occupied by other jobs), giving rise to queueing and delay, which are typical phenomena in such systems. Some simple examples are a computerized machine tool that processes different types of parts that are fed to it from an upstream machining or assembly stage, and the central processing unit of a computer system that processes jobs generated by multiple users. A basic problem in operating such systems is to design a control scheme that schedules the server (that is, decides which job to serve) on a real-time basis. Such a scheme is known as a control or scheduling policy. Since jobs follow different arrival mechanisms, have different service requirements, incur different delay costs, and yield different revenue (for being processed), a viable control policy must strike a sensible trade-off among all these factors. The setting in which such a policy will be carried out is typically both dynamic and random.

Apparently the problem is overwhelmingly complex. However, in practice it is often observed that most systems are operated under some simple-minded priority rules. That is, jobs are assigned a priority index: $1, 2, \ldots, n$ (for example, in decreasing order of priority), and the server always serves a job that has the highest priority index. Jobs of a lower priority will be served only if no job of higher priority is present in the system. A natural question is whether this simple rule can be optimal in some ways. A better way to address the same question is perhaps to characterize a class of systems, as general as possible, in which simple priority rules are optimal.

To begin with, it is necessary to specify the class of admissible policies, of which the priority rule is a member, and against which the priority rule is to be compared. The first requirement is that any admissible policy be nonanticipative; that is, it can make use only of past history and the current state of the system, while all future information is deemed inaccessible. Second, it is also required that admissible policies be nonidling; that is, the server is not allowed to be idle as long as there are jobs to be served. Third, no admissible policy will affect the arrival mechanism, the service requirement, and the cost and revenue structure of the jobs; nor will any additional cost be incurred for switching among different job types. However, admissible policies can be either preemptive or nonpreemptive. A preemptive policy allows the server to switch to another job before the current service is completed (and upon switching back, service is resumed from the point at which it was previously interrupted). A nonpreemptive policy insists that once the server starts working on a job it must complete the job before switching to another job. (Obviously, allowing preemption makes the class of policies more general.)

Even with all the restrictions, the class of admissible policies is still rich enough to capture all reasonable ways in which a wide range of systems are operated. Among these policies, a set of priority rules stands out for its simplicity, which greatly facilitates implementation. In particular, priority rules are static policies, in the sense that the priority indices are determined off line, that is, making no use of the feedback information of the running system. Obviously, the optimality of priority rules must be a very strong statement, since it says that these simple-minded policies are just as good as any sophisticated policies that fully exploit on-line information.

Thus, it becomes necessary to specify the system performance measures and related objective functions that priority rules will optimize. A performance measure typically refers to the expectation or average of, say, the delay of jobs, the workload or congestion level in the system, or the throughput rate (that is, the number of job completions per time unit). The main conclusion below is that the optimality of priority rules is rigorously justified for a class of systems in which the performance measures satisfy conservation laws and the objective is a linear function.

Conservation laws and the polymatroid structure. Let $E = \{1, 2, \ldots, n\}$ denote the set of all job types. Verbally, conservation laws can be summarized into the following statements: (1) the total performance (that is, the sum) over all job types in E is invariant under any admissible policy; (2) the total performance over any given subset, $A \subset E$, of job types is minimized (or maximized) by offering priority to this subset over all other types.

A simple example is a system comprising two job types, type 1 and type 2. Each job (of either type) brings into the system a certain amount of work (service requirement). If the server serves (that is, depletes work) at unit rate, then (1) the total amount of work, summing over all jobs of both types that are present in the system, is independent of the actual policy that schedules the server, as long as it is nonidling; and (2) if type 1 jobs are given preemptive priority over type 2 jobs, then the amount of work in system summing over type 1 jobs only is minimized, that is, it cannot be reduced further by any other admissible policy.

For the formal definition, $x(i)$ denotes the performance measure under study of type i jobs, $i \in E$; $\mathbf{x} = (x(i))_{i=1}^n$ is the performance vector. For any $A \subseteq E$, $|A|$ denotes the cardinality of A, and

$$x(A) = \sum_{i \in A} x(i);$$

in particular, $x(\emptyset) = 0$. $\mathcal{A}$ denotes the space of admissible policies, and $\mathbf{x}^u$ the performance vector under an admissible policy $u \in \mathcal{A}$; π denotes a permutation of the integers $1, 2, \ldots, n$. The expression $\pi = (\pi(1), \ldots, \pi(n))$ denotes a priority rule, in

which type $\pi(1)$ jobs are assigned the highest priority, and type $\pi(n)$ jobs the lowest priority.

For a definition of conservation laws, the performance vector $\mathbf{x}$ is said to satisfy conservation laws if there exists a set function $b: 2^E \mapsto \mathfrak{R}_+$, satisfying Eq. (5a), or respectively function $f: 2^E \mapsto \mathfrak{R}_+$, satisfying Eq. (5b),

$$b(A) = x^\pi(A),$$
$$\forall \pi: \{\pi(1), \ldots, \pi(|A|)\} = A, \quad (5a)$$
$$\forall A \subseteq E;$$

$$f(A) = x^\pi(A),$$
$$\forall \pi: \{\pi(1), \ldots, \pi(|A|)\} = A, \quad (5b)$$
$$\forall A \subseteq E;$$

such that for all $u \in \mathcal{A}$, Eqs. (6a) and (6b)

$$x^u(A) \geq b(A),$$
$$\forall A \subset E; \quad (6a)$$
$$x^u(E) = b(E);$$

$$x^u(A) \leq f(A),$$
$$\forall A \subset E; \quad (6b)$$
$$x^u(E) = f(E).$$

are satisfied.

The functions $b(A)$ and $f(A)$ in the above definition correspond respectively to the minimal and maximal total performance summing over the job types in the subset A. It is important to note that this minimal or maximal performance is required to be independent of the priority assignment among the types within the subset A (as long as any type in A has priority over any type in $E - A$). This requirement is reflected in the qualification imposed on π in Eqs. (5a) and (5b).

The performance space as characterized in Eqs. (6a) and (6b) is a polytope, that is, the intersection of half spaces. In general, it is not any easy matter to characterize this polytope. For instance, it has $n!$ vertices, and there is no easy way to identify these vertices. However, with the above formal definition of conservation laws, it is possible to identify a very strong and special structure of this polytope; namely, it is a polymatroidal polytope, of which the vertices correspond to the priority rules (the correspondence is one to one).

Polymatroid is a special discrete mathematical structure that is often useful in solving some combinatorial optimization problems. For instance, when the performance polytope has a polymatroid structure, its vertices can be simply read off as follows. Denote, for instance, the polytope in Eq. (6a) as $\mathcal{B}(b)$, and denote its vertex set as $\mathcal{B}_v(b)$. When $\mathcal{B}(b)$ has a polymatroid structure, it is known that given any permutation π there exists a vertex $\mathbf{v}(\pi) \in \mathcal{B}_v(b)$ with the coordinates shown in Eq. (7).

$$v(\pi(1)) = b(\pi(1)),$$
$$v(\pi(i)) = b(\pi(1), \ldots, \pi(i)) - b(\pi(1), \ldots, \pi(i-1)),$$
$$i = 2, \ldots, n \quad (7)$$

Indeed, any vertex $\mathbf{v}(\pi) \in \mathscr{B}_v(b)$ that corresponds to a permutation π can be expressed as in Eq. (7), that is, $\mathscr{B}_v(b) = \{\mathbf{v}(\pi), \pi \in \text{II}\}$, where II is the set of all $n!$ permutations of the n members of E.

Optimality of priority rules. The above result implies that when the performance vector under study satisfies conservation laws, the system performance under any admissible control constitutes exactly (that is, no more, no less) the polytope $\mathscr{B}(b)$; and furthermore, any vector in the performance space can be realized by a control that is a randomization of at most n priority rules. If $\mathbf{x}$ is such a performance vector of interest, then $x(i)$ is the work of type i jobs that are present in the system, and $c(i)$ is the unit cost incurred by the work of type i. If it is desired to find an admissible control $u \in \mathscr{A}$ in order to minimize the total cost

$$\sum_{i=1}^{n} c(i)x^u(i)$$

then the optimal control problem can now be transformed into the linear programming problem shown in expression (8).

$$\min \sum_{i=1}^{n} c(i)x(i) \text{ such that } \mathbf{x} \in \mathscr{B}(b) \qquad (8)$$

Because the state space is a polymatroidal polytope, the above linear programming problem is known to have a "greedy" type of solution. Specifically, if π is a permutation such that expression (9)

$$c(\pi(1)) \geq c(\pi(2)) \geq \ldots \geq c(\pi(n)) \qquad (9)$$

holds, then the corresponding vertex, $\mathbf{v}(\pi) \in \mathscr{B}(b)$, solves the problem shown in expression (8), and hence the optimality of the priority rule π. In other words, the priority rule that assigns priority to job types according to the order of the cost coefficients $c(i)$ [highest cost having highest priority] is optimal among all admissible policies in $\mathscr{A}$.

Objective functions more general than linear functions are also possible, for instance, functions whose partial derivatives (that is, marginal costs) can be ordered among job types. A wide range of systems and performance measures have now been verified that satisfy the conservation laws. Indeed, since these laws are intuitive and easily stated in verbal terms, they can often be verified based on first principles. They provide engineering intuition as well as scientific insight to the class of complex systems in which priority rules make sense.

For background information SEE ALGORITHM; COMPUTER-INTEGRATED MANUFACTURING; CONTROL SYSTEMS; CRYPTOGRAPHY; EXPERT SYSTEMS; INVENTORY CONTROL; MATERIAL REQUIREMENTS PLANNING; OPERATIONS RESEARCH; OPTIMIZATION; PRODUCTION PLANNING in the McGraw-Hill Encyclopedia of Science & Technology.

David D. Yao

Bibliography. J. F. Chen and W. E. Wilhelm, *Kitting in Multi-echelon Assembly Systems*, Working Paper, Department of Industrial Engineering, Texas A&M University, 1990; J. C. Gittins, *Bandit Processes and Dynamic Allocation Indices*, 1989; D. P. Heyman and M. J. Sobel, *Stochastic Models in Operations Research*, 1982; E. L. Lawler et al., *Sequencing and Scheduling: Algorithms and Complexity*, Center for Mathematics and Computer Science, Amsterdam, Rep. BS-R8909, June 1989; K. W. Ross and D. D. Yao, Optimal dynamic scheduling in Jackson networks, *IEEE Trans. Automat. Control*, 34:47–53, 1989; R. O. Roundy et al., *A Price Directed Approach to Real Time Scheduling of Production Operations*, School of Operations and Industrial Engineering, Cornell University, Rep. 812, July 1988; J. G. Shanthikumar and D. D. Yao, Multiclass queueing systems: Polymatroidal structure and optimal scheduling control, *Oper. Res.*, 1991; L. Wang and W. E. Wilhelm, *An Algorithm for Applying Recursion Models to Cellular Production/Assembly Systems*, Working Paper, Department of Industrial Engineering, Texas A&M University, 1989; L. M. Wein, Optimal control of a two station Brownian network, *Math. Oper. Res.*, 15:215–242, May 1990; D. J. A. Welsh, *Matroid Theory*, 1976; W. E. Wilhelm and L. Wang, Management of component accumulation in small-lot assembly systems, *J. Manuf. Sys.*, 5(1):27–39, 1986; W. E. Wilhelm and B. Carroll, A PERT-based paradigm for optimally scheduling and rescheduling time-managed operations in certain assembly systems, Working Paper, Department of Industrial Engineering, Texas A&M University, 1990; R. Wolff, *Stochastic Modeling and the Theory of Queues*, 1988.

Protein

This article discusses how newly synthesized proteins are transported across cellular membranes and how computers are used to analyze protein and nucleic acid sequences.

PROTEIN TRANSPORT ACROSS CELLULAR MEMBRANES

Eukaryotic cells and even the simpler prokaryotic cells contain cellular membranes and membrane-bounded compartments with distinct protein compositions. Specific proteins are transported to these distinct membranes and compartments (or to the extracellular space) via a sort of zip code system. A given zip code, termed a topogenic sequence, is represented by a short sequence within each protein that is to be transported. The various topogenic sequences are recognized by specific cellular machines. These cellular machines then effect localization.

Topogenic sequences. Three types of topogenic sequences have been distinguished. (1) Signal sequences specify translocation of particular proteins across specific translocation-competent membranes (not all cellular membranes are competent to translocate proteins). (2) Stop-transfer sequences interrupt the protein conductance across the membrane, thereby leading to integration of proteins into the lipid bilayer of membranes. (3) Sorting sequences specify transport following translocation (or integration), thereby either retaining a protein in a compartment (or membrane) or shipping it to other, translocation-incompetent membranes (or compartments) by vesicular carriers.

Signal sequences. The structure of a given signal sequence is membrane-specific. For example, each of the many structurally and functionally diverse proteins to be translocated across (or integrated into) the endoplasmic reticulum contains a signal sequence that is targeted for the endoplasmic reticulum, and each of the many proteins to be translocated across (or integrated into) the peroxisome membrane contains a signal sequence that is targeted for the peroxisomal membrane. Although the primary structure of the known examples of signal sequences that are targeted for the endoplasmic reticulum varies considerably, the signal sequences all contain common secondary structural features. Therefore, only one set of identical machines is needed for recognition of multiple versions of signal sequences that are targeted for the endoplasmic reticulum. Likewise, only one set of identical machines is needed for recognition of signal sequences that are targeted for the peroxisomal membrane.

Signal sequences either remain uncleaved from their proteins following translocation or are cleaved by specific proteases, termed signal peptidases. Signal sequences may be located anywhere within the polypeptide chain. The primary structure of various signal sequences that are membrane-specific has been elucidated. Proteins that, through genetic engineering, have had their signal sequence mutated or removed can no longer be translocated. Likewise, proteins that are not normally translocated can be translocated across virtually any of the cellular translocation-competent membranes after addition of a proper membrane-specific signal sequence by genetic engineering techniques.

Stop-transfer sequences. Unlike signal sequences, stop-transfer sequences do not appear to be membrane-specific. These sequences are represented by a stretch of about 20 apolar amino acid residues flanked on both sides by charged amino acids.

Sorting sequences. So far, there are only two clearly elucidated examples of sorting sequences. One sequence consists of a stretch of four amino acids at the carboxy terminal end of all proteins that are retained in the lumen of the endoplasmic reticulum. These proteins thus contain two topogenic sequences: a signal sequence targeted for the endoplasmic reticulum that gets them into the lumen of the endoplasmic reticulum, and a sorting sequence that retains them in this compartment. Another clearly elucidated example for a sorting sequence is for lysosomal enzymes of yeast. Again, these proteins contain two topogenic sequences: an endoplasmic reticulum–targeted signal sequence that permits translocation into the lumen of the endoplasmic reticulum, and a sorting sequence that permits separation from other proteins that are segregated into the lumen of the endoplasmic reticulum.

Decoding machines. The machines decoding the various topogenic sequences and effecting localization are just beginning to be identified and characterized. The term translocon has been proposed to comprise all components that are involved in the signal-sequence-mediated translocation process. The most detailed data so far have been accumulated for the translocon of the endoplasmic reticulum.

The signal sequence that is targeted for the endoplasmic reticulum is recognized by a cytosolic signal recognition factor. This factor is a ribonucleoprotein particle, known as signal recognition particle. Signal recognition particle consists of one molecule of 7S ribonucleic acid (RNA) and six proteins. One of the signal recognition particle proteins has been shown to bind the signal sequence and to function as a protein that binds guanosine triphosphate. Signal recognition particle binds with low affinity to ribosomes but with high affinity to ribosomes that contain nascent polypeptide chains with exposed signal sequences that are targeted for the endoplasmic reticulum. This high-affinity binding leads to a transient arrest in the elongation of the nascent polypeptide. The signal recognition particle–ribosome–nascent chain complex is then targeted to a homing receptor that is exclusively localized in the endoplasmic reticulum. This receptor binds the complex via signal recognition particle and is therefore also known as signal recognition particle receptor. The signal recognition particle receptor consists of two subunits, a and b. Both are proteins that bind guanosine triphosphate. The interaction of the complex with the signal recognition particle receptor leads to dissociation of the signal sequence from signal recognition particle and to the release of the elongation arrest. The signal sequence is then free to interact with a subunit of a protein-conducting channel in the endoplasmic reticulum. This interaction is presumed to open the channel for protein translocation across the membrane. The existence of such a protein-conducting channel has been shown by electrophysiological methods, but the components representing it remain to be conclusively identified. Among the candidates are signal peptidase, comprising a complex of five different proteins.

In addition, heat shock proteins, also termed chaperones, located on the cis side (the cytosol) or the trans side (the lumen) of the endoplasmic reticulum membrane may interact with portions of the protein to be translocated, thereby facilitating the translocation process.

It is thought that the protein-conducting channel decodes the stop-transfer sequence. By interacting with a subunit of the channel during its passage through the channel, the stop-transfer sequence is thought to cause a lateral opening of the protein-conducting channel to the lipid bilayer, followed by displacement of the stop-transfer segment of the chain from the channel to the bilayer and subsequent closure of the channel. Thus, unlike ion-conducting channels, which open and close in only one dimension (across the bilayer), the protein-conducting channel in the endoplasmic reticulum would open and close in two dimensions, that is, across the bilayer and in the plane of the bilayer.

It is likely that other translocons consist of similar components and work in a similar fashion to the endoplasmic reticulum translocon, although variations of the general themes are likely. For example, in cases where translocation occurs through two membranes (such as in chloroplasts and mitochondria), a protein-conducting channel in the outer membrane of these organelles may be physically linked to a protein-conducting channel in the inner membrane. Or, in the case of protein translocation into the nucleus that is mediated by a signal sequence, the components of the protein-conducting channel are likely to be part of a large complex of proteins (that is, the nuclear pore complex) and not directly embedded in the lipid bilayer of the nuclear envelope membranes. This channel would therefore not be able to decode stop-transfer sequences; that is, it could not function in the integration of proteins into the nuclear envelope membrane. Moreover, unlike the protein-conducting channels of the other translocons, that of the nuclear pore complex does not require that the protein be retained in a relatively unfolded state during translocation. Thus, its mechanism of opening and closing is likely to be very different from that of the other protein-conducting channels.

The components of the decoding machinery for sorting sequences have not yet been identified except for a receptor for the above-mentioned carboxy terminal sorting sequence of proteins to be retained in the lumen of the endoplasmic reticulum. It is thought that all proteins translocated into the lumen of the endoplasmic reticulum are then transported by vesicular carriers (independent of signal sequences) to the next downstream compartment, the Golgi complex. Proteins to be retained in the lumen of the endoplasmic reticulum are therefore subject to escape, and it is thought that the above-mentioned receptor is required for retrieval of these escaped proteins.   *Günter Blobel*

COMPUTER ANALYSIS OF PROTEIN AND NUCLEIC ACID SEQUENCES

During the 1950s, chemical methods were devised for determining the actual sequence of amino acids in a protein. Proteins are chains of amino acids, ranging in length from 40 to 4000 units, the average being about 350. The order in which the amino acids occur, called simply the sequence or primary structure, is encoded by the order of purine and pyrimidine bases in corresponding segments of deoxyribonucleic acid (DNA) or ribonucleic acid (RNA). By the middle 1960s, enough protein sequences had been collected that computers were introduced to store and analyze the data. The advent of recombinant DNA procedures in the late 1970s greatly accelerated the acquisition of new sequence data, and at that point computer data management of the vast numbers of sequences reported became the only reasonable course.

Computer analysis of protein and nucleic acid sequences stored in large sequence data banks is providing an unparalleled view of how life on Earth has emerged. There has been a continuing expansion of genes and gene products as a result of the natural tendencies of RNA and DNA to hyperduplicate portions of their lengths during normal replication. These duplicated regions diverge in sequence as the result of base substitutions, deletions, and insertions during subsequent replicative events. Similarly, the sequences of the same genes in diverging organisms are also subject to change. Computers can organize all these sequences according to their similarities and reconstruct the history of events.

Sequence comparisons. The same protein from different organisms can have a different sequence of amino acids, the degree of the difference being inversely proportional to the relatedness of the organisms (**Fig. 1a**). Thus, two closely related animals might have an enzyme, the sequences of which are 90% identical, but the same enzyme from a bacterium might be only 40% identical with the animal proteins. Computer programs were devised to align the sequences and cluster them according to their similarity. In general, the phylogenetic trees that emerged from such analyses mirrored traditional biological relationships (see Fig. 1 and **Fig. 2**).

It also became clear that the number of protein types was not unlimited and that even within a single organism there were families of proteins, often with similar functions but with varying degrees of sequence resemblance. These protein families are the result of past gene duplications, the descendants of which have been modified by random base substitutions in the DNA and reflected in different amino acid sequences. The same computer programs used to establish organismal relationships are used to trace the history of a family of proteins. The extent of this genetic expansion was greater than originally anticipated, and sequence comparisons soon revealed that many proteins with apparently distinctive functions were derived from a common stock. As an example, the protein angiotensinogen, which is the precursor of a hormone involved in the regulation of blood pressure and water balance, was found to belong to a well-known family of protease inhibitors. Similarly, the receptor protein for adrenaline was found to be related to the visual pigments of the eye. These unexpected resemblances were uncovered by systematic computer searching.

```
                       *                                        *
human       V L S P A D K T N V K A A W G K V G A H A G E Y G A E A L E . .
rabbit      V L S P A D K T N I K T A W E K I G S H G G E Y G A E A V E . .
kangaroo    V L S A A D K G H V K A I W G K V G G H A G E Y A A E G L E . .
chicken     V L S N A D K N N V K G I F T K I A G H A E E Y G A E T L E . .
frog        S L S A S E K A A V L S I V G K I G S Q G S A L G S E A L T . .
lungfish    M R F S Q D D E V L I K E A W G L L   H Q I P N A G G E A L A . .
(a)
```

```
                         *   *           *         * * *
alpha       V L S P A D K T N V K A A W G K V G A H A G E Y G A E A L E . .
beta        V H L T P E E K S A V T A L W G K V   N V D E V G G E A L G . .
gamma       G H F T E E D K A T I T S L W G K V   N V E D A G G E T L G . .
delta       V H L T P E E K T A V N A L W G K V   N V D A V G G E A L G . .
zeta        S L T K T E R T I I V S M W A K I S T Q A D T I G T E T L E . .
myoglobin   G L S D G E W Q L V L N V W G K V E A D I P G H G Q E V L I . .
(b)
```

Fig. 1. Computer alignment of amino acids (denoted by single capital letters) in hemoglobin chains. The asterisks denote positions where all six sequences have the same amino acid. (a) The first 30 amino acids in hemoglobin alpha chains from various vertebrate animals that have diverged from each other over the course of the last 400 million years. (b) The first 30 amino acids of six different hemoglobin chains found in all humans. The different genes for these sequences are the result of a series of gene duplications over the past 450 million years.

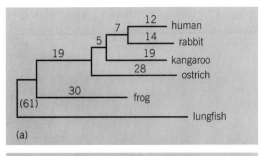

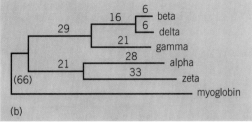

(a)

(b)

Fig. 2. Phylogenetic trees constructed from the full sequences of the hemoglobins aligned in Fig. 1. The numbers indicate the relative distances between branch points and sequences. (a) Vertebrate hemoglobin alpha chains. (b) Human hemoglobin chains.

Such searches must be able to identify similar sequences, not just identical ones. Indeed, the most sensitive algorithms are sometimes able to pick out protein sequences that are only 10–15% identical to the sequence being queried. Thus, when a new amino acid (or nucleic acid) sequence is determined, it is now routine to compare it with all known sequences, which are stored in large sequence data banks. Some of these banks store nucleic acid sequences (which can be translated into the corresponding protein sequences by the computer); other banks store only the amino acid sequences of proteins. SEE PEPTIDE.

Once candidate sequences are found, the computer can verify the statistical likelihood of the match. These operations usually involve an empirical evaluation of the measured similarity of the sequences against what might be expected from a comparison of random sequences of the same lengths and compositions. The computer can also be used to make multiple alignments of all the sequences in a given family and to construct a phylogenetic tree based on their relative similarities (see Figs. 1 and 2). The main problem in aligning sequences has to do with the observation that nucleic acid (and therefore protein) sequences not only change by base substitution but also suffer from genetic deletions and insertions of bases. Accordingly, allowance must be made for these naturally occurring events by allowing gaps in the alignment. Obviously, any pair of sequences can be made to look similar if unlimited gaps are allowed; therefore scoring systems have been devised that give positive points for similar amino acids (or bases) and negative points for gaps.

Prediction of three-dimensional structures. Considerable effort has been invested in predicting the three-dimensional structures of proteins on the basis of their amino acid sequences alone. It is a part of the dogma of protein chemistry that the sequence contains all the information for how the protein folds into three dimensions. As a consequence, if all the forces exerted by the individual amino acids on each other and the surrounding solvent were known, it should be possible to predict the final structure based solely on the primary structure. So far, all such efforts have failed. In contrast, comparison studies of various kinds have been quite successful. For example, if the three-dimensional structure of one protein is known as the result of x-ray crystallographic analysis, then it is possible to model the structure of a related protein. Computer-graphic regimens have been developed that make this a relatively straightforward procedure.

It is also possible to predict certain localized elements of protein structure by consideration of overlapping segments of the amino acid sequence. These segments, typically ranging from 5 to 20 amino acids in length, can be evaluated with regard either to some property such as polarity or, in what is termed a knowledge-based approach, to a table of functions based on sequences of known secondary structures (alpha-helices, turns, and so on). Aided by these programs, experienced analysts are often able to predict membrane-spanning regions, alpha-helical segments, and other characteristic features of proteins.

For background information SEE AMINO ACIDS; CELL MEMBRANES; ENDOPLASMIC RETICULUM; GENETIC CODE; PROTEIN; PROTEINS, EVOLUTION OF in the McGraw-Hill Encyclopedia of Science & Technology.

R. F. Doolittle

Bibliography. G. Blobel, Intracellular protein topogenesis, *Proc. Nat. Acad. Sci. USA*, 77:1496–1500, 1980; R. C. Das and P. W. Robbins (eds.), *Protein Transfer and Organelle Biogenesis*, 1988; M. O. Dayhoff (ed.), *Atlas of Protein Sequence and Structure*, vol. 5, National Biomedical Research Foundation, 1978; R. F. Doolittle, *Of URFs and ORFs: A Primer on How to Analyze Derived Amino Acid Sequences*, 1987; W. M. Fitch and E. Margoliash, Construction of phylogenetic trees, *Science*, 155:279–284, 1967; S. B. Needleman and C. D. Wunsch, A general method applicable to the search for similarities in the amino acid sequence of two proteins, *J. Mol. Evol.*, 48:433–453, 1971.

Psychology, physiological and experimental

Eyeblinks are a universal activity among humans, other mammals, and most vertebrates. A blink occurs when the two eyelids appear to touch and the pupil is momentarily hidden from view. A blink has three phases—closing, closed, and reopening. The closing phase takes about half as long as the reopening phase, and in most blinks the lids do not actually touch. Together, the three phases last about a quarter of a second, with considerable variability between individuals. In nonblink eye closures, the closing phase is slower, the closed phase lasts longer, and the reopening phase occurs more quickly. Together, the three phases can last from a half second to several seconds.

Types of blinks. There are three types of blinks, one of which is voluntary and two involuntary. Voluntary blinks are infrequent and occur when a conscious decision is made to close the eyes briefly. Involuntary blinks may occur as a defensive or protective reflex against potentially harmful events, such as a strong puff of air or a sudden burst of bright light. These blinks rarely occur. Most common are the spontaneous blinks that do not occur in direct response to any identifiable event. They occur about 15,000 times a day.

Measurement techniques. There are five methods for recording blinks: the electrooculogram (EOG), the electromyogram (EMG), infrared reflectance, photography, and direct observation. An EOG is an electrically recorded graph of movements associated with the eye (see **illus.**). The three primary characteristics of a blink are amplitude, duration, and frequency. As shown in the illustration, the vertical excursion of the EOG wave defines blink amplitude, which identifies how far the upper lid travels before the lids come together. The width of the wave at 50% of its amplitude yields information on the duration of the lid closure. The number of times such a wave occurs yields blink frequency, which is usually expressed as blink rate (blinks per minute, or bpm). In cases when blinks are made following voluntary attention to an event, blink latency is measured. This is the time it takes for a blink to occur after the event, such as after reading a sentence. The EMG method measures blinks by recording the muscle activity of the eyelids as the eye closes and reopens. In the infrared reflectance technique, an infrared beam is transmitted to the eyeball, which reflects the signal when the lids are open and which does not reflect the signal when the lids are closed.

The photographic method for measuring blinks uses either high-speed film or videotape. The direct-observation method permits the assessment of blink rate. An observer counts the number of apparent lid closures and uses a stopwatch to estimate the time elapsed during the count. If 110 blinks are counted in 5 min 30 s, dividing 110 by 5.5 yields a rate of 20 bpm. A continuous 5-min sample is recom-

mended for a reliable count, although in naturalistic settings several shorter time epochs can be accumulated. A modification of this method is to measure the time between two successive blinks. For example, a 1.5-s interblink interval yields a rate of 40 bpm (60 s divided by 1.5 s). The use of filming or videotaping increases the reliability of the observational method by permitting more than one count of the blinks.

Biological factors. Blinking prevents potentially harmful stimuli from entering the orbit of the eye. Once a foreign particle, such as a speck of dirt, is on the eye, a blink sweeps it aside by the wiper action of the lids. The larger upper lid carries the debris downward, and at the time of apparent closure the lower lid, with a lateral movement, propels the matter toward the corner of the inner eye, where it can be removed by the eye's drainage system or a swipe of the fingertip. A blink also keeps the cornea moist and healthy by the re-formation of a tear film layer. By a sweeping action, the lids spread out the fluid secreted by the lacrimal (tear) glands and distribute mucin, the natural detergent of tears. A blink once every 15–30 s is enough for this wetting and cleansing function. Blinking can also prevent visual blurring during head movements and changes in eye fixations by readjusting the lens of the eyeball. Tension and fatigue in the eye muscles can be reduced by blinking, although there is some disagreement on this point. Ambient temperature and humidity have no significant effect on blink frequency.

The average human rate is approximately 15–20 bpm for a relaxed quiet state and varies widely, depending upon thoughts, feelings, and actions. Since normal adults need only 2–4 bpm to keep the eyeball moist, most blinks are physiologically unnecessary. Furthermore, since blind individuals have the same blink rate as sighted individuals, the significance of blinks goes beyond visual functions.

Cognitive activity. Blinking and thinking are related. Activities requiring complex thinking, such as mental arithmetic or memorization of numbers, tend to increase blink frequency. Doing two tasks at once, such as listening to tones and memorizing letters, increases blinking. An important aspect of these tasks is the inward direction of attention to cognitive functions. Vocalization also increases blinking. Conversation and responding to interview questions result in a further increase in blink frequency. Even a simple task such as reciting the alphabet aloud tends to increase blink frequency. This effect of vocalization suggests that blinking can be an indicator of generalized muscle tension and an expression of released motor tension.

When a cognitive task requires close attention to outside visual events, blink frequency decreases, perhaps to optimize input of information. Rates can drop to 3 bpm during reading unless the time on the task is protracted. In that case, blink frequency increases, probably due to fatigue. During reading, blinks appear at strategic points. For example, blinks tend to appear when turning a page, at the end of a line, or at the end of a sentence, particularly a lengthy one. These blinks

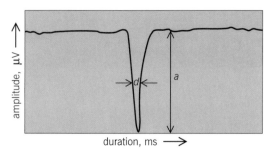

Electrooculographic (EOG) tracing of an eyeblink, which represents an electrical change (potential difference) between the cornea and the fundus (back of the eye). Electrodes were placed 3 cm above and 2 cm below the right eye. Amplitude is represented by line *a* (380 microvolts), and duration is shown by line *d* (120 milliseconds).

may indicate a pause in the processing of information and may be a signal for the brain to rest between information inputs. During reading, a weak blink of low amplitude and short duration can occur, resulting in only a partial closure of the eye, perhaps indicative of an attempt to avoid missing visual information. Blinks also occur just before or after difficult parts of a task, possibly facilitating an erasure function by eliminating remnants of older information and preparing the brain for newer information.

Mood states. Blinking is related to a variety of mood states. Increased blink frequency generally reflects negative mood states, such as nervousness, stress, boredom, and fatigue. For instance, in courtroom proceedings, witnesses show increased blinking during cross-examination. Blinking increases during heightened time pressure or when addressing a large audience. A good example of the blink–stress association (so-called Nixon effect) comes from the August 8, 1974, resignation speech of President Richard Nixon, who made over 50 bpm when discussing being forced out of office. He also showed rapid bursts of blinks, called eyeblink storms (3 blinks per second or more). Eyeblink storms reflect underlying nervousness and fear and often accompany stuttering, which is also an apparent indicator of apprehension. Blinking increases when errors are made in memorizing digits, just before a decision is made to quit a difficult problem-solving task, and during poor performance brought on by fatigue, which are all situations conducive to negative mood states.

Slower blink rates are observed during positive mood states. For instance, blinking slows down after relaxation has been hypnotically induced. Blinking also decreases during successful problem solving and during simple repetitive thoughts (such as counting silently), both of which should produce positive mood states. These relationships between feelings and blinking support the hedonia-blink hypothesis, which states that increased blinking accompanies unpleasant feelings and decreased blinking accompanies pleasant feelings. According to this view, blink frequency can be used to detect lying only if accompanied by negative feelings. Thus the psychopathic liar would show decreased blinking if pleased with the deception.

The information on cognition and mood leads to a two-factor theory of blinking: (1) Blink frequency is increased during unpleasant mood states and is decreased during pleasant mood states (hedonia hypothesis). (2) Blink frequency is increased when attention is directed inward and is decreased when attention is directed outward (attention hypothesis).

Clinical groups. Blinking differs in clinical groups, depending on brain functioning. Parkinsonian patients have lower blink rates, presumably because of damage to the basal ganglia (located in the forebrain) or the substantia nigra (located in the midbrain), which is a structure that produces the neurotransmitter dopamine. Schizophrenics, who are presumed to have increased dopamine functioning, show increased blink rates. When they are treated with dopamine antago-

nists, their blink rate drops. This relationship between dopamine and blinking is called the dopamine-blink hypothesis. Drug treatment of schizophrenics is also associated with a reduction in negative mood states.

Neuroanatomy. Since blinking is affected by thought and mood changes that depend on appraisal of events, the cortex (outer covering of the brain that controls higher mental functions) is involved in blink regulation. Important subcortical areas that appear to facilitate blinking are the basal ganglia and lateral geniculate body (located in the forebrain), substantia nigra, and the pontine reticular formation (located in the hindbrain). The substantia nigra and basal ganglia (forebrain area) are degenerated in Parkinsonian patients and appear to be necessary for blinking. The cerebellum seems to inhibit blinking, since damage to it in animals results in increased blinking.

Animals. There are marked species differences in blink frequency. Invertebrates do not blink. Fishes and other animals that live underwater have no eyelids. At one end of the phylogenetic scale, blink rate is relatively fast, such as in primates (for example, sudanese monkey, 43 bpm). At the other end of the phylogenetic scale are species that blink infrequently, such as rodents (for example, mice and rats, < 2 bpm), reptiles (for example, lizards, < 4 bpm), and amphibians (for example, frogs, < 2 bpm). Birds vary considerably (for example, parrots, 26 bpm, and ostriches, 1 bpm). Nocturnal animals blink less frequently than daytime animals. Blink rate is faster for herbivores (such as cows, 22 bpm) than for carnivorous mammals (such as lions, < 1 bpm). Blinking in animals appears to serve the same physiological functions as in humans.

For background information SEE BRAIN; EMOTION; EYE (INVERTEBRATE); EYE (VERTEBRATE); PSYCHOLOGY, PHYSIOLOGICAL AND EXPERIMENTAL; REFLEX; STRESS (PSYCHOLOGY) in the McGraw-Hill Encyclopedia of Science & Technology.

J. J. Tecce

Bibliography. A. Hall, The origin and purposes of blinking, *Brit. J. Ophthalmol.*, 29:445–467, 1945; C. N. Karson, Physiology of normal and abnormal blinking, *Adv. Neurol.*, 49:25–37, 1988; E. Ponder and W. P. Kennedy, On the act of blinking, *Quart. J. Exp. Physiol.*, 18:89–110, 1927; J. A. Stern et al., The endogenous eyeblink, *Psychophysiology*, 21:22–33, 1984.

Queueing theory

Recent advances in queueing theory include the characterization of humans in queueing systems, the development of generalized models of queueing networks, and the development of cyclic service queueing systems.

QUEUES

At supermarkets, banks, fast-food restaurants, movie theaters, almost everywhere that services and goods are purchased, there seem to be people waiting in lines. In Great Britain and in much of western Europe, waiting lines are called queues. The scientific study of queues started in 1915 when the Danish telephone engineer A. K. Erlang developed the first mathematical models of a queue to help the Copenhagen Telephone Company size its first telephone switching systems. Today there is considerable scientific, technological, and entrepreneurial activity focused on queues and in particular on improving the waiting-time experience of customers "standing in line."

Queueing system. **Figure 1** is a diagram of a simple queueing system. The input is the stream of arriving customers desiring service. Once at the service facility, a customer may enter service immediately, or if all servers are busy, the customer joins a queue of other waiting customers. Eventually service is provided to the customer, after which the customer departs from the system.

In practice, a queue may not be easy to identify. Drivers in rush-hour traffic are captive in a moving queue of motor vehicles. Applicants for public housing usually enter their names on lists that constitute queues of waiting families; recently in Boston the wait in such a queue was projected to be 13 years! Queueing is not always simply standing in line.

Until recently those who studied queues focused on easily identified and measured performance criteria, such as the average duration of the wait, the percentage of people who waited more than X minutes, or the percentage of time during which servers were busy with customers. Within the past several years this technocratic focus has been broadened to include more subtle human-oriented concerns.

Order of service and perceived fairness. Surprisingly, in some studies in fast-food restaurants, interviewed customers said that they would prefer joining a single-line system with average wait $2X$ to a multiline system with average wait X. The single-line system guarantees the "socially just" first-come first-served queue discipline; the multiline system allows a customer who entered the restaurant after another customer to receive service before the latter, simply by the good luck of having chosen a faster line. Many customers become infuriated by such violations of fairness, even when the total wait in line is quite short for all customers. As a result, single-line serpentine queues have been implemented throughout many United States service organizations.

First-come first-served is not always the fair way to select people from the queue. In a hospital emergency medical room, a waiting patient with a minor injury

would generally agree that a recently arrived patient with a life-threatening medical problem should receive medical attention first. In supermarkets, those waiting with overflowing shopping carts do not mind latecomers with stipulated small amounts such as "12 or fewer items" who quickly pass through the express lane. Thus, some queues have priority structures, implying that fairness may mean other than first-come first-served. However, in many queues in day-to-day experience, fairness usually implies first-come first-served.

Replacing empty waiting time. Waiting in a queue is considered by some to be a subtle form of imprisonment. Queue dwellers, in the absence of anything else to do, focus on the second-by-second passage of time, thereby lengthening the perceived wait and heightening frustration. Sometimes a simple distraction can remove the frustration, or even change a negative experience into a positive one.

In the 1950s and 1960s, when high-rise office buildings and hotels sprung up across the United States, building occupants complained loudly about the long waits for elevators. The ingenious solution was to install floor-to-ceiling mirrors adjacent to the elevators, resulting in a plummeting of complaints about delay, although the statistics of delay were unchanged. Other stratagems have included live piano music and other entertainment in bank lobbies; and installation of a so-called participatory musical sculpture in a subway station, where people waiting for a train can play music.

Recently companies have sprung up that offer services to inform and entertain line dwellers. An early company offered closed-loop television showing videotapes to those standing in line at amusement parks. A more recent entry is a satellite service from Los Angeles that provides a moving-word display of news, sports, features, and advertising. This service has been enthusiastically greeted by those standing in lines for tellers and for automatic teller machines at banks. A live satellite TV service has been inaugurated that broadcasts news to those in lines at supermarket check-out counters. Another TV service focuses on those waiting in doctors' offices. Perhaps the most successful new business development focusing on changing waiting time into productive time is mobile cellular telephones, allowing people stuck in rush-hour traffic to carry out telemarketing and other professional work.

Customer-provided service. Human servers are expensive. Demographic projections for the 1990s suggest that new entrants into the labor force will be decreasing in numbers, thus making the hiring of human servers more difficult. As a result, a trend to have customers provide their own service is growing, often with the aid of a machine. Perhaps the first major move toward having customers provide their own service occurred in the 1950s with the introduction of supermarkets as replacements for small grocery stores, putting an end to the type of shopping in which the groceries were gathered for the customer by a store

Fig. 1. Diagram of a simple queueing system.

employee. The supermarket self-service concept is now implemented widely in national discount stores, department stores, home supplies stores, and drug stores.

Many gasoline service stations have converted to total or partial self-service, with the customer providing virtually all the service; the employee simply collects money. With such a service policy, queue delays previously caused by motorists waiting for an attendant to fill the tank have vanished.

Those holding appropriate bank cards use the automatic teller machines in the vast majority of banking transactions, thus greatly reducing the need for human tellers. Banks report that well-utilized networks of automatic teller machines cost only about a third as much per transaction as human tellers; in addition, customers using them can virtually avoid banking queues by performing transactions during off-peak hours. Machines that resemble automatic teller machines are now proliferating throughout other industries; there are automatic ticketing machines at airports, food-ordering machines at some fast-food restaurants, automated "parking lot attendants," and tourist-information video machines. Most of these automated services are available 24 h a day, allowing many customers to plan use so as to avoid queue delays.

An operations research study performed by a fast-food company revealed that a large cost in providing human service to customers is in the preparation of soft drinks. As a result, in many of its large restaurants, customers asking for soft drinks are now sold empty cups and are asked to fill them at drink machines.

Service time options. Waiting customers respond better when they are informed of the approximate duration of the anticipated delay. Customer satisfaction with the queue experience, as with most service experiences, depends directly on the difference between expectation and reality. It is when reality falls short of expectation that dissatisfaction occurs. Announcing the anticipated delay can help the service provider manage the customer's expectations.

As examples, at most of the major attractions in a large theme park there are signs stationed along the queue channel advising patrons of the typical wait from that point. Restaurateurs provide similar information to dinner groups who arrive without reservations. Airline pilots advise passengers of anticipated delays, often caused by air traffic control, prior to takeoff. Even police 911 operators, when recording the details of a request for police service, are now often instructed to advise the caller of the anticipated wait prior to the arrival at the scene of a dispatched police car.

Some firms are offering service time guarantees; if the wait exceeds some prespecified threshold, the customer receives a bonus. In addition, customers are increasingly finding options for avoiding some queues. An example is the express check-in and check-out at hotels and car rental companies.

Innovations. Probably most of the queue innovations are the products of research. That research combines mathematical modeling methods with customer interviews and in-the-field testing of emerging technologies. New services and products now being offered to customers standing in line are being buttressed with extensive videotaped evaluations and sophisticated statistical procedures for estimating queue performance. One new procedure, the queue inference engine (QIE), allows an evaluator to discern the statistics of the queue without ever observing it, requiring only the service start and stop time for each customer served. The ultimate goal of most of this work is to reduce the time spent waiting in line, either removing the delays or replacing empty time with fulfilled time. *Richard C. Larson*

QUEUEING NETWORKS

A mathematical model of a system of service facilities is known as a queueing network. The service facilities are called queues, and the items receiving service are called customers or jobs. As depicted in **Fig. 2**, the customers from outside the network arrive at one of the queues, wait for their turn to receive service, receive service, and then either leave the network or are routed to another queue, where they again wait, receive service, and so on. To understand the models, it is helpful to think of systems that might be modeled. For example, the system might be the local motor vehicle bureau, with several different waiting lines, or a factory with many different machines making many different products.

The theory of queueing networks is part of queueing theory, being concerned with queueing models containing more than one queue. Queueing theory, in turn, is a major subject of operations research and a branch of probability theory. Queueing models are intended to represent the uncertain ways in which customers arrive to request service and service is performed.

A queueing network model. A typical queueing network model contains several queues and several classes of customers. As part of the model definition,

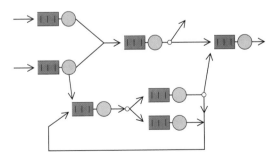

Fig. 2. Network of queues. Lines and arrows represent flows of customers, large circles represent the servers, rectangles represent extra waiting space, and small circles indicate that a choice is being made.

the number of servers and the number of waiting spaces at each queue are specified. Also specified are the routing, service requirements, and arrival pattern for each customer class. The routing might be a definite itinerary, that is, a prescribed sequence of queues that each customer will visit, perhaps with some queues being visited several times and others not at all; or the routing might be random, specified by a probability that each queue is the next queue after each service completion from a given queue. Moreover, the manner in which waiting customers are selected for service at each queue is specified. This might be first-come first-served, last-come first-served, at random, or according to some priority scheme. Finally, for each customer class the network may be open or closed. For an open class, customers from that class arrive from outside the network at some given rate. For a closed class, the class population is fixed, so that a new customer enters the network the instant any customer leaves the network.

System performance. Given that a queueing network model has been constructed, the object of further analysis is to describe system performance. The principal performance measures are server utilizations (the proportion of time each server is busy), throughputs (the flow rate of customers of one class or several classes out of individual queues or the entire network), queue lengths (the number of customers either waiting or being served at each queue), and sojourn times or response times (the time each customer spends at individual queues or in the entire network). For open classes, the network throughput is simply the external arrival rate, and the throughputs at individual queues are just the arrival rates there, which include arrivals from other queues as well as external arrivals. Thus, for open classes, the queue throughputs can be found easily by solving a system of linear equations called the traffic-rate equations. For closed classes, finding the throughputs requires more work. The queue lengths and sojourn times are random quantities, so that they are described by their probability distributions or associated summary measures such as long-run averages.

The performance measures of interest may be calculated, either exactly or approximately, by using mathematical analysis; or they may be estimated by using computer simulation. With simulation, a typical realization of the queueing network is generated on the computer by using a random-number generator to produce realizations of random quantities with prescribed probability distributions. Then the performance measures are estimated by sample averages. The quality of the performance measure estimates can also be estimated by using statistical methods. This is typically done by giving confidence intervals for the quantities being estimated. *See Simulation.*

Different performance measures exhibit very different kinds of behavior. For example, for an open class, the server utilizations and the throughput are directly proportional to the arrival rate as long as the arrival rate is below its upper limit for stability, whereas the average queue length and the average sojourn time at a queue with unlimited waiting space grow exponentially without bound as the arrival rate approaches this critical level (**Fig. 3**). Consequently, systems that can be modeled as queueing networks usually must operate with significantly less than the maximum possible throughput. A major purpose of the performance analysis is to find a balance between the desire to have high resource utilization and the desire to decrease congestion experienced by customers.

The analysis of queueing networks is largely descriptive, but the performance measure descriptions are usually obtained in order to improve system design. For example, it may be possible to determine a good number of servers at a queue, a good number of customers in a closed class, or a good configuration for the entire network. In some cases, optimal (best by some criterion) values can be determined by mathematical optimization (for example, calculus or mathematical programming), but usually improvement must be obtained by trial and error. An exciting new possibility is systematic and efficient optimization via computer simulation, but this approach is still in the exploratory stages.

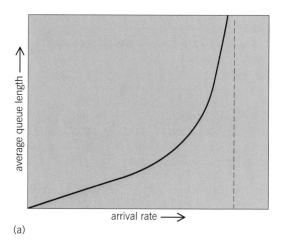
(a)

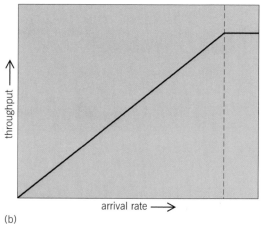
(b)

Fig. 3. Graphs showing the behavior of (*a*) average queue lengths and (*b*) throughputs in response to increasing arrival rate. The broken line represents the critical level for the arrival rate.

Applications. Queueing networks have received considerable attention since about 1970, primarily because they have proven to be valuable tools for analyzing the performance of computer systems. With the rapid evolution of computers, there is frequent need for performance analysis of computer systems. A classical application of queueing networks is the study of multiprogrammed operating systems, which allow some programs to use input/output (I/O) devices such as disks while other programs use the central processing unit (CPU). This sharing of computer resources among many programs creates greater efficiency, but it also creates competition for the resources.

A simple queueing network model of multiprogramming is the central-server model (**Fig. 4**). There are three kinds of queues: a single queue containing all the terminals, the central processing unit, and a queue for each disk. The central processing unit and the disks are represented as single-server queues. The queue containing the terminals has infinitely many servers. The model has a single closed customer class; that is, there is always a fixed number of customers in the network, with each customer corresponding to a program. Then service at the central processing unit and the disks corresponds to tasks associated with executing the programs.

A program begins its execution when a customer moves from the terminals to the central processing unit. After a random processing (service) time at the central processing unit, the program requires input/output from one of the disks, with the particular disk being chosen randomly according to appropriate probabilities. After input/output service, either execution is complete, in which case the customer returns to the terminal queue; or further processing is required at the central processing unit, and the customer moves back to it and the previous scenario is repeated. When the customer eventually returns to the terminal queue, the program execution is complete. After a random "think" period (service time at the terminal queue), a new customer moves from the terminal queue to the central processing unit queue, initiating the execution of another program.

There are other important applications of queueing networks in addition to computer systems. The next most frequent application is probably to the performance analysis of data networks, which are closely related to computer systems. In data networks, messages are broken up into smaller pieces called packets, which are sent from source to destination through a backbone network.

Many other systems have been analyzed by queueing networks. Manufacturing facilities motivated early work on queueing networks, and they still provide many applications. Queueing networks have also been applied to analyze airports and hospitals.

Theory. A major reason for the applied success of queueing networks is a well-developed mathematical theory for a large class of these models. The nicely behaved queueing networks have product-form steady-state distributions; that is, in equilibrium the

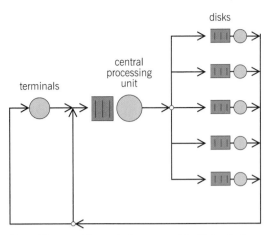

Fig. 4. Diagram of the central server model with terminals. Lines and arrows represent flows of customers, large circles represent the servers, rectangles represent extra waiting space, and small circles indicate that a choice is being made.

network behaves the same as independent single queues (with appropriate arrival rates). This property holds because the process representing the vector of queue lengths is Markov (the future depends on the past only through the present) with a certain reversibility property (the process viewed backwards in time looks just like the original process). More complicated non-product-form queueing networks are also being analyzed, often by bounds and approximations. Among the more successful approximations are diffusion-process approximations and other many-customer approximations (obtained by considering the limiting behavior as the number of customers gets large in some sense), which are related to statistical mechanics. *Ward Whitt*

Cyclic Service Queueing Systems

Cyclic service queueing systems appear as performance analysis models for a variety of manufacturing processes, traffic signaling controls, and computer communication networks. Cyclic service systems are also known as polling models, patrolling-repairman-type queueing systems, and alternating priority queues. Beginning in 1957 with the work of C. Mack, addressing the analysis of patrolling repairman systems, interest in the study of cyclic service systems has expanded rapidly. Impetus for elementary cyclic server models arises principally from the popularity of certain demand-access computer communication systems (polling systems).

Basic model. The basic cyclic service model consists of N queues attended by a single server (**Fig. 5**). The server patrols the N queues in a cyclic order. While the server is working at queue i, customers queued at i are served according to a prespecified discipline (examples of such queueing disciplines will be discussed below); the service discipline need not be the same for each queue. According to the service discipline, the server will at some point abandon

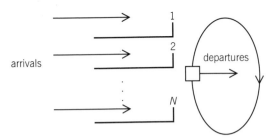

Fig. 5. Diagram of a cyclic service queueing system.

remaining customers in queue i and begin a walk, of random length, ending at queue $1 + i$ (modulo N). Upon completing the walk (that is, arriving at the next queue) the server will immediately begin serving waiting customers; if no customers are waiting, the server begins a walk toward the next queue in the cycle.

Customers arrive at each queue according to a stochastic process (the arrival process need not be the same at each queue). Usually arrivals are modeled as N independent Poisson streams having rates λ_i, where $i = 1, 2, \ldots, N$. The service times of customers in queue i are drawn from the general distribution F_i with mean s_i. The server walk times between i and $1 + i$ (modulo N) are drawn from the general distribution G_i with mean w_i. Henceforth, it is assumed that the (Poisson) arrival streams, service times, and walk times are mutually independent.

When $\lambda_i = \lambda$, $F_i = F$, and $G_i = G$ for $i = 1, 2, \ldots, N$, and all N queues have the same service discipline, the system is called symmetric; otherwise, the system is called asymmetric. The analysis of cyclic service systems is often simplified when the system is symmetric. In such systems, all queues are statistically identical, and the system behavior can be inferred by studying a particular queue in isolation (as a vacationing server queue).

Single-buffer systems (patrolling repairman problem). A simple special case of cyclic service queueing systems is obtained when the capacity of each of the N queues is fixed at one. When an arriving customer finds the queue occupied, the customer is not allowed to join the queue and is lost forever. Thus, in any particular queue there can be at most one customer awaiting or receiving service. Such systems are often referred to as single-buffer polling systems.

Single-buffer polling systems were first studied within the context of patrolling repairman systems. When queue capacities are limited to one, the cyclic service system is equivalent to a collection of N machines maintained by a single repairman. Here, the repairman cycles among the machines, stopping to repair those that are failed. Following repair, a machine is restored to a good-as-new status. The lifetime of machine i is exponentially distributed with parameter λ_i, its repair time is distributed F_i, and the walk time from i to $1 + i$ (modulo N) is distributed G_i.

Down time of machines and cycle time of the repairman are random variables that are often investigated when characterizing the operation of patrolling

repairman systems; mean values of these variables are usually taken as the principal system performance measures. For a symmetric system, where C is the cycle time of the repairman, D is the time between a machine's failure and the arrival of the repairman, and R is the number of machines repaired in a single cycle, it has been shown that the expected cycle time $E[C]$ is given by Eq. (1) and the expected down time $E[D]$ by Eq. (2). It follows from Eqs. (1) and (2),

$$E[C] = N \cdot w + s \cdot E[R] \tag{1}$$

$$E[D] = (N - 1) \cdot s - \frac{1}{\lambda} + \frac{N^2 \cdot w}{E[R]} \tag{2}$$

where w = mean walk time and s = mean service time, that finding the expected cycle time of the repairman $E[C]$ and the expected down time of a machine $E[D] + s$ is straightforward once the expected number of repairs per cycle $E[R]$ is computed. When repair times and walk times are constant, a simple expression for $E[R]$ can be given. However, there appears to be no simple expression for $E[R]$ when F and G are taken as general distributions.

When the assumption of symmetry is relaxed, the analysis of patrolling repairman systems becomes considerably more complicated. Recently, it has been shown that the behavior of asymmetric systems can be characterized by a semiregenerative stochastic process defined on a finite-state space. That is, each time that the repairman completes an activity (service or walk), the future of the system becomes a probabilistic replica of its future after time zero, given that the state of the system at zero and at the activity completion are the same. When F_i and G_i are replaced with particular distribution functions, the ergodic distribution of this process can be computed and then used to determine expected cycle time and expected machine down times. While this approach accommodates general asymmetric systems, the computational burden grows explosively as N becomes large.

Infinite-buffer systems (polling models). The recent popularity of demand-based multiple-access computer/communication networks has led to many useful developments in the theory of cyclic service systems. In contrast to patrolling-repairman-type systems, demand-based multiple-access networks have very large buffer capacities. Such networks are often modeled as infinite-buffer cyclic service systems. Infinite-buffer systems are such that each of the N queues has unlimited waiting room; hence, all arriving customers are allowed to join their respective queue where they wait to begin service. The class of models employed in the study of demand-based multiple-access networks are usually referred to as polling models.

Since each of the N queues accommodates an unlimited number of customers, each queue must be assigned a service discipline that determines the manner in which waiting customers are served. The service discipline need not be the same for each queue. There are three basic disciplines that have been most extensively investigated in the polling model literature: exhaustive service, gated service, and limited service. With each of these disciplines, customers within any queue are served in a first-come first-served fashion.

Under exhaustive service, the server will continue working at a queue until all waiting customers have been served; arrivals that appear while the server is working at the head of the line will be served before the server departs for the next queue in the cycle. Under gated service, only those customers already queued at the time of the server's arrival will be served during the present cycle; customers arriving during the server's present visit are held for service during the next cycle. For a queue operating under limited service, at most a fixed number of customers are served in any cycle. Most often, limited service at a queue is such that at most one customer is served per cycle.

Customer waiting times and server cycle time stand as the most important measures of performance in polling systems; some of the basic results often employed in studying these performance measures are shown in Eqs. (3)–(5). For polling models operating with any mixture of the three basic service disciplines described above, the load on the system is given by Eq. (3), where $\rho_i = \lambda_i \cdot s_i$. Mean cycle time of the

$$\rho = \sum_{i=1}^{N} \rho_i \qquad (3)$$

server in polling models is determined by Eq. (4).

$$E[C] = \frac{1}{1-\rho} \sum_{i=1}^{N} W_i \qquad (4)$$

Waiting time W_i of an arbitrary customer in the i-th queue (that is, the time that an arbitrary customer waits in steady state to begin service) can be examined through a pseudoconservation law due to O. J. Boxma and W. P. Groenendijk. This pseudoconservation law does not generally yield simple expressions for the expectation of W_i; however, when the polling system is symmetric, closed-form expressions become readily available. Further, for symmetric systems, Eq. (5) follows from the pseudoconservation law. The order-

$$E[W]_{\text{exhaustive}} \leq E[W]_{\text{gated}} \leq E[W]_{\text{limited}} \qquad (5)$$

ing relationship of Eq. (5) is intuitive by reasoning that customers should wait longer under disciplines where the server is more likely to depart the queue, leaving work behind.

Computation of mean customer waiting times in asymmetric polling systems can often be accomplished by using numerical techniques. This approach requires expressions obtainable from the generating function of the joint probability distribution for the number of customers in each queue; such generating functions are easily found for most basic polling systems.

As is the case in many models of collections of queues, an understanding of the mathematical structures of these systems is available. Explicit moment calculations for some of these models are indicated above. Investigations of other properties require a

study of the numerical problems inherent in these models.

For background information SEE COMPUTER; DATA COMMUNICATIONS; NONLINEAR PROGRAMMING; OPERATIONS RESEARCH; OPTIMIZATION; PROBABILITY; QUEUEING THEORY; SIMULATION; STATISTICAL MECHANICS; STATISTICS; STOCHASTIC PROCESS in the McGraw-Hill Encyclopedia of Science & Technology.

M. A. Wortman; R. L. Disney

Bibliography. O. J. Boxma, Workloads and waiting times in single-server systems with multiple customer classes, *Queueing Sys.*, 5:185–214, 1989; O. J. Boxma and W. P. Groenendijk, Pseudo-conservation laws in cyclic-service systems, *J. Appl. Prob.*, 24:949–964, 1987; D. Bertsekas and R. Gallager, *Data Networks*, 1981; R. L. Disney and P. C. Kiessler, *Traffic Processes in Queueing Networks*, 1987; R. W. Hall, *Queueing Methods for Services and Manufacturing*, 1991; K. L. Katz, B. M. Larson, and R. C. Larson, Prescription for the waiting-in-line blues: Entertain, enlighten, and engage, *Sloan Manag. Rev.*, Winter 1991; F. P. Kelly, *Reversibility and Stochastic Networks*, 1979; R. C. Larson, Perspectives on queues: Social justice and the psychology of queueing, *Oper. Res.*, 35:895–905, 1987; R. C. Larson, The queue inference engine: Deducing queue statistics from transactional data, *Manag. Sci.*, 36:586–691, 1990; E. D. Lazowska et al., *Quantitative System Performance: Computer System Analysis Using Queueing Network Models*, 1984; C. Mack, The efficiency of N machines uni-directionally patrolled by one operative when walking time is constant and repair times are variable, *J. Roy. Stat. Soc.*, series B, 19(1):173–178, 1957; D. H. Maister et al. (eds.), *The Service Encounter*, 1985; H. Takagi, *Analysis of Polling Systems*, 1986; H. Takagi (ed.), *Stochastic Analysis of Computer and Communication Systems*, 1990; J. Walrand, *An Introduction to Queueing Networks*, 1988.

Radiative transfer

Early engineering studies in heat transfer were motivated by the need to manipulate heating and cooling rates in basic industrial and residential systems. The temperatures involved in these systems were relatively moderate. Technological advances, however, have continuously pushed many engineering systems to higher or lower temperature levels, making radiative transfer an important mode due to its characteristic temperature dependence.

Unlike heat transfer by conduction and convection, which depends on the temperature difference of the elements or surfaces to approximately the first power, radiative transfer (also known as radiation heat transfer or radiation transport) depends on the difference of the individual body temperatures raised to the power of 4–5. Radiative transfer, therefore, becomes more dominant at higher temperatures, such as in combustion systems. At very low temperatures, for example, in the cryogenic range, the radiant energy involved is much smaller than that at room or high temperature. However, radiation may also become a dominant mode of heat transfer in many low-temperature applications, such as cryogenic insulations and cryoelectronics, due to lack of conduction and convection under vacuum conditions. Radiation is also the major (sometimes the sole) mode for thermal control of space vehicles.

Interaction of radiative transfer with convection and chemical reaction occurs mostly in high-temperature systems, such as flames, fires, industrial furnaces, and combustion chambers. Mathematical formulation of the combined modes is much more complicated than that for radiation or convection alone. A nonlinear integrodifferential equation generally results for the energy equation of such problems. Analytic solutions are usually not obtainable; therefore, results are mostly presented numerically. Due to this complexity, most studies have tended to treat either the pure radiation aspect or the pure convection problem. Significant progress has been made in recent years on radiative transfer in fires, furnaces, and fluidized- and packed-bed combustors. Some of the major findings and applications involving radiative transfer in recent years are briefly described below.

Thermal insulation. Thermal insulation has been a subject of great importance to engineers, and was one of the major concerns in the early development of heat-transfer technology. Recent advances in such technologies as aerospace, nuclear, and cryogenic engineering have extended considerably the ordinary temperature range of operation and have presented a great number of formidable engineering problems at extreme temperature limits. There has been a surge of interest in thermal insulation for high-temperature and cryogenic applications since about 1970. Recently, thermal insulation for the intermediate- and room-temperature range has also received considerable attention as a result of the worldwide worsening energy problem. It is realized that considerable energy conservation can be achieved through widespread and effective application of thermal insulation in various energy systems, such as buildings, industrial pipings, and the storage and transport of liquefied natural gas.

It has been increasingly recognized that radiative transfer plays an important role in the thermal perfor-

mance of all types of thermal insulation, including powder, fibrous materials, and evacuated multilayers (often called superinsulation). Radiative transfer is also important in all new insulation concepts and techniques. Among many new concepts advanced for thermal insulation in recent years, two appear to have great potential for applications.

One concept concerns the reduction of gas conduction through porous insulation by the use of ultrafine powders (of diameter 7 nanometers). Since gas conduction is characterized by the molecular mean free path, it can be reduced greatly if the pore size between solid particles is less than the mean free path. The resulting thermal conductivity, mostly due to radiation through the void space, can be lower than that of still air, and high-vacuum-like insulation performance can be achieved without evacuating the system. The loosely packed powder structure has a porosity of 97–98%, but the particles can be further compressed to yield denser structures having porosity of about 90% in order to reduce further the contributions of gas conduction and particulate radiation.

The other promising concept deals with thermal insulation in flow systems. In high-temperature systems such as heat exchangers, exhaust chimneys, furnaces, and combustion chambers, the recovery of residual energy from exhaust gases is often a major concern. Many methods have been devised to recover as much as possible of the energy that otherwise is lost to the ambient environment. One of the most effective methods for reducing and controlling these energy losses is by the conversion of the exhaust-gas enthalpy into thermal radiation in the upstream direction. This can be implemented easily by installing a radiation-shield-like porous medium normal to the gas-flow direction (see **illus.**). A porous medium, consisting of sintered fine particles or fibers, or a few layers of screens, exhibits excellent characteristics for the effective conversion between fluid enthalpy and thermal radiation since it combines the high convective heat-transfer rates between the solid matrix (porous medium) and the gas with the high-radiative-emission power of solid bodies. Due to the solid matrix, part of the incoming fluid enthalpy (H_1) is used to raise the temperature (thermal energy) of the solid matrix, which in turn emits and scatters radiation in the upstream direction (H_3) and downstream direction (H_4). This results in a reduction of the outgoing fluid enthalpy (H_2). The thermal performance of the porous medium in a flowing gas system depends on the conversion of gas enthalpy into thermal energy of the solid matrix and on the directional characteristics of the radiation emitted by the porous medium. The porous medium, for example, a few layers of screens, can be designed such that the flow pressure drop is minimal.

Flame and combustion systems. Flame radiation has become increasingly recognized as a critical factor in many combustion phenomena such as in large-scale fires and in industrial furnaces and combustors. For instance, flame radiation may exert a strong influence on fire detection, ignition, and spread; indeed, in large-scale fires with long path lengths, flame radia-

Schematic of energy recovery by thermal insulation in flow systems. For notation, see text.

tion constitutes over 80% of the energy required for fuel gasification. In industrial furnaces and combustors, flame radiation is essential to predicting thermal performance and consequently improving the fuel economy.

Thermal radiation from flames may generally be classified as either luminous or nonluminous. Nonluminous radiation arises mainly from combustion gases such as carbon dioxide and water vapor whose radiation is concentrated in spectral bands. The combined contributions from gases and soot particles then constitute the radiation from luminous flames. Accurate prediction of the radiation from flames requires detailed information on the radiative properties of the combustion gases and soot particulates. The appropriate quantity to describe flame radiation is the total emissivity. Extensive experiments and analyses for the determination of flame emissivity exist in the literature. Despite the many efforts to treat this problem, the computational procedure is still quite involved, particularly when temperature and gas partial pressures vary along the path. Therefore, it is desirable to develop simpler yet accurate methods to facilitate calculations of general flame radiation.

Some interesting ideas have been advanced recently on improving the thermal efficiency of conventional furnaces and combustors, which use flame radiation as the dominant heat-transfer mechanism. In order to attain high emission of radiation from the flame, heating chambers of volume much larger than the object to be heated are usually required. This often gives rise to large temperature variations on the heated object. Improved industrial furnaces of smaller size are possible by simply adding porous (flow-permeable) walls surrounding the heating chamber. In an enclosure surrounded with porous walls, higher and more uniform temperatures can be obtained due to the strong radiation field that results from the conversion of the gas enthalpy into thermal radiation emitted back into the load. In addition, the thermal insulation capabilities of the porous medium make possible a significant reduction in the cost of thermal insulation exterior to the furnace. In combustor systems, heat transfer and combustion augmentation by radiation can be achieved through the use of porous media. High and uniform temperatures provided by radiation heating may be used to improve the efficiency of combustion processes. Higher combustion temperatures may also reduce the emission of pollutants and the excess-air ratio requirements. The latter is especially important in situations where nonoxidating atmospheres are required.

Direct-absorption solar receiver. Thermal radiation plays an essential role in solar-energy collection, particularly in solar receivers. In conventional solar receivers, the working fluid is contained in tubes, and the incident solar flux impinging on the tube outer surface is absorbed and transferred to the fluid by conduction and convection. Upper limits on the incident solar flux, and hence on the temperature of the fluid and on the efficiency of the solar receiver, are imposed in order to maintain the temperature of the tubes within permissible limits. An attractive alternative design is the concept of a direct-absorption solar receiver, where the solar flux is incident on an absorbing and emitting falling-liquid film. Radiant energy is directly absorbed by the film, for example, a flowing molten salt, as the radiant flux propagates through the fluid. Lower temperatures of the receiver materials and high fluid temperatures are obtained, compared to those achieved in conventional solar receivers, since the radiant energy is absorbed directly by the working fluid. Higher fluid temperatures are possible when the solar absorption properties of the fluid film are enhanced by the addition of very small, highly absorbing particles.

Direct-absorption solar receivers have several important advantages over conventional receivers. First, the cost of the receiver may be significantly reduced since the heat-transfer surfaces (tubes) are eliminated. Second, higher fluid temperatures are possible, increasing the efficiency of the receiver. Finally, solar receivers with higher irradiation flux levels, and hence of smaller size for a given heat load, are possible since the temperature of the tubes is no longer of concern.

High-temperature heat exchangers. Improvements in the efficiency of high-temperature heat exchangers are related to the enhancement of heat-transfer rates, particularly through the use of radiative transfer. Heat transfer by radiation can be enhanced by several means, such as by inserting radiation-emitting plates in the gas flow and by adding solid absorbing particles in nonradiating gas. More innovative techniques, based on the conversion between the fluid enthalpy and thermal radiation, may also be used in advanced high-temperature heat exchangers.

Significant augmentation of radiative transfer can be obtained by the use of radiation-emitting plates. Radiant energy absorbed by these plates is transferred to the surrounding fluid by convection. This simple idea may be applied easily to shell-type heat exchangers to enhance the overall inner or outer heat-transfer coefficients. Enhanced heat-transfer coefficients outside the tube, as is usually required in gas-gas heat exchangers, can be achieved by the use of radiation plates of high emissivity and tubes covered with high-emissivity paint. Higher inner heat-transfer coefficients, as required in fluidized coal combustors for preheating combustion air, can be obtained by inserting loosely twisted tapes in the tubes that provide large radiative shape factors. Heat-transfer augmentation of 10–70% has been reported for the inner surface of a tube.

For background information SEE ENTHALPY; HEAT TRANSFER; RADIATIVE TRANSFER in the McGraw-Hill Encyclopedia of Science & Technology.

Chang-Lin Tien

Bibliography. G. J. Hwang (ed.), *Transport Phenomena in Thermal Control*, 1989; W. M. Rohsenow et al. (eds.), *Handbook of Heat Transfer Fundamentals*, 1985; R. Siegel and J. R. Howell, *Thermal Radiation Heat Transfer*, 2d ed., 1981; R. Viskanta and M. P. Mengüc, Radiation heat transfer in combustion systems, *Prog. Energy Combust. Sci.*, 13:97–160, 1987.

Railroad engineering

The need to transfer increasing numbers of people within urban areas by more effective and environmentally sensitive means while reducing the ever-increasing congestion on highways has led to much development in recent years in urban railroad transportation. Extensive research, applying new technologies to the development of urban railroads, has paid off in terms of practical applications.

There has been great importance placed on saving energy; thus, there has been a strong demand for lighter rail cars and the more efficient use of electrical power in rail systems. The need to reduce construction, maintenance, and operations costs led to the introduction of the driverless, automatic guideway transit (AGT) system.

Technological innovations in power electronics have provided the gate-turnoff thyristor (GTO); the four-quadrant chopper; variable-voltage, variable-frequency (VVVF) inverter control; and rail cars driven by three-phase alternating-current (ac) rotary and linear motors. Cars using such equipment can achieve a reduction of up to 40% in power consumption as compared to conventional rheostat-controlled vehicles.

Car bodies. Aluminum alloy car bodies have been introduced to reduce weight and save energy. The most recent manufacturing technologies use an alloy of aluminum, magnesium, and silicon, which has excellent extrusion workability. The construction uses long, hollow extrusions of a large sectional size, which extend along the entire length of the car body. This reduces the number of transverse beams needed, eliminating a considerable number of parts and nonautomatic welds. This, in turn, reduces fabrication time and increases the welding reliability, thus reducing capital and maintenance costs.

Development of the stainless-steel body, now in common use, began with the stainless-steel skin, which was designed to prevent corrosion of the outer plate and to eliminate coatings that required maintenance. Later, semistainless-steel and all-stainless-steel cars were developed to further reduce weight and eliminate the need for maintenance. In addition, structural design employing the finite element method, large computer systems, adoption of high-strength stainless steel (SUS310), and advanced fabrication methods enabled the production of stainless-steel cars that are now as light as those made with less dense metals.

Wheels and truck assemblies. Transit systems place emphasis on reduction in travel times between stations that can be as little as 1500–3000 ft (500–1000 m) apart. It is, therefore, the acceleration and braking rates that become important and not necessarily high speed. For this reason, automatic guideway transit systems using rubber tires have been developed, since these tires provide twice as much adhesion as steel wheels and enable high acceleration and retardation rates to be achieved. This, in turn, has produced a radical redesign of truck assemblies. Apart from passenger comfort and the reduction in noise and vibration levels, vehicles with rubber tires have the ability to climb grades of up to 10% compared with a typical 6% grade limit for steel-wheeled vehicles (**Fig. 1**).

In the more traditional steel-wheel-on-rail system, simple, lightweight, bolsterless, and steerable trucks have been introduced, which contribute not only to saving energy but also to reducing track wear, damage, and noise.

Propulsion-drive systems. Urban transit vehicles are normally supplied with direct-current (dc) power

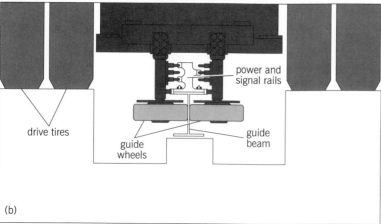

Fig. 1. Rubber-tire automated guideway transit (AGT) systems. (a) Cars in Orlando, Florida, system (*AEG Westinghouse Transportation Systems, Inc.*). (b) Guideway system. The guide wheels lock the vehicle to the steel guide beam that runs down the center of the guideway (*after Automated People Mover Systems, AEG Westinghouse Transportation Systems, Inc., 1990*).

derived from a standard utility source. Typically, traction-supply line-side substations provide nominal voltage in the 600–1000-V range. However, the on-board vehicle voltage to the traction motor must be variable in order to control the motor speed and, in turn, the speed of the train.

Until the early 1970s, only schemes using direct current with on-board train resistors (rheostats) were used for voltage control. These resistors, which handled high currents, were cumbersome and power-consuming. The use of power semiconductors (thyristors) produced the chopper circuits, which are a far more efficient means of varying voltage and hence speed of the direct-current motors.

Frequency can also be varied by thyristor circuits, which are then used as a method for controlling speed in alternating-current motor drives. The ability to control the speed of alternating-current motors was further enhanced by the introduction of gate-turnoff thyristors. These are switched on by the application of positive gate pulses and switched off by negative gate pulses. Accordingly, the forced commutation circuits of conventional thyristors are not required; therefore, switching circuits which use fewer semiconductors but which have high capacities, up to 3000 A and 4500 V, have been designed.

The development of large-capacity gate-turnoff thyristors and the application of computer control technology led to the manufacture of variable-voltage, variable-frequency inverters. The main benefit of the variable-voltage, variable-frequency inverter control and gate-turnoff thyristors lies in their ability to control the speed of large alternating-current motors, thus allowing these motors to be used for traction purposes, with the following advantages:

1. Alternating-current drive motors have no armature windings or brushes, and so less maintenance is required and high reliability can be expected.

2. Since a number of contactors and switches can be eliminated, the equipment size, weight, and maintenance can be reduced.

3. A high-adhesion performance, by accurate power control, can be achieved.

4. In comparison to a direct-current motor, an alternating-current motor of equivalent physical size is more powerful, so that a motor with a larger power capacity can be installed in the same space.

Braking systems. In general, there are two modes of braking, friction and dynamic (electrical) braking. When a train is utilizing dynamic braking, the traction motors are operated as generators, converting the kinetic energy of the moving vehicle into electrical energy.

Early electrical systems with direct-current motors dissipated this electrical energy as heat in resistors (rheostatic braking). A more efficient method of dissipating the electrical energy is to reuse it by returning power to the utility system (regenerative power). In practice, most utility companies prefer not to purchase this regenerated power; hence transit systems using this approach return power to the overall train system network.

Given this regenerative power, good synchroniza-

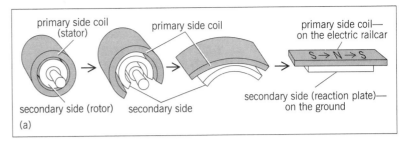

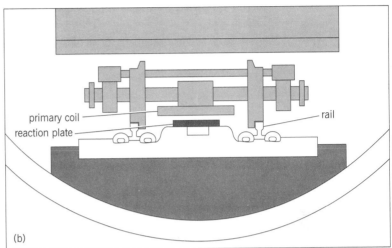

Fig. 2. Linear-motor drives. (a) Development from rotary motor to linear motor (*after Z. Umeda, ed., Urban Transportation Systems in Japan, Japan Rolling Stock Exporters Association, 1990*). (b) Structure of the drive (*after Linear Metro: Linear Motor-Driven Subway System, Japan Subway Association, 1990*). (c) Sky-Train in Vancouver, British Columbia, Canada. Reaction plate is visible between the tracks.

tion and control of all trains throughout the system are required in order for the power to be used effectively. Sending too much power back to the line, beyond what is called the limit of receptivity, would cause waste by dumping the power once more as heat through specifically designed on-board resistors. It is, therefore, much easier and more cost-effective to control this power by using automatic guideway transit operations, which can be fully programmed to take advantage of this arrangement.

Use of linear motors. A linear motor produces linear rather than rotational motion by "unwrapping" and laying flat the structure of a conventional electric motor. The principle is that, as in a rotary motor, a varying magnetic field, generated by an alternating current flowing through the primary coil, interacts with another varying magnetic field simultaneously generated by the secondary coil; the resultant repulsive and attractive forces create the propulsion. In transit applications, the linear motor on the vehicle corresponds to the primary coil, the reaction plate fixed between the rails is the secondary coil, and the magnetic fields generated by the currents in the two coils create the driving force (**Fig. 2**).

The main advantages from the application of the linear motor to trains are reduced car heights and increased maximum gradients. The floor height can be reduced because of the flat shape of the linear motor. Therefore, the car height can be reduced without sacrificing space and riding comfort, thus reducing the overall vehicle weight and dimensions. This, in turn, permits a reduction of tunnel cross sections and reduces construction costs. In the case of elevated tracks, the kinematic and structural clearance gage can be reduced, so that a more slender structure can be designed with costs reduced accordingly.

A conventional steel-wheel train is somewhat restricted in climbing capability (maximum gradient of 6%). In a linear-motor application, the nonadhesive drive is not dependent on traction between wheel and rails and is capable of gradients up to 10%. This allows much more flexibility in route selection, and thus further savings in costs are achieved.

Automated guideway transit systems. These systems use all of the developments discussed above. Their basic advantages are lower capital investment costs (because of smaller civil and structural construction); lower operating costs (because of driverless operation); and features that make them potentially more attractive to the user, including higher service frequency and longer daily operating schedules—for the same equipment investment, automatic operation makes it possible to adjust to complex and fluctuating demand quickly, particularly in terms of peak and off-peak train frequency, without a corresponding increase in personnel.

Transit car safety is maintained at a central control point. At this center, train operation control is executed by central traffic control (CTC) equipment, with automatic train operations executed by the automatic train operation (ATO) equipment. The function of the operational control center, therefore, can be divided, by using this equipment, into (1) operation planning,

Fig. 3. Light rail vehicle in Sacramento, California.

including train scheduling; (2) operation control, including control of departure, stopping, speed, door opening and closing, stop-position adjustment control, and route control; and (3) operation monitoring, including monitoring of train tracking, and control of operation disturbances and equipment status.

At the operation control center, information received from the central traffic control equipment is displayed on an operational display panel, and centralized control for the trains is executed by interactive action between the operators and the computers. Thus the automatic, crewless operation of the train is executed safely by utilizing the expertise of the skilled operators in central control working with the data output and input from the various computer banks. All aspects of safety are thus monitored, controlled, and shown on visually displayed boards and, more recently, on rear-projection television.

One of the largest urban automated guideway transit systems in operation, which also uses the linear motor for propulsion, is the Sky-Train in Vancouver, British Columbia, Canada (Fig. 2c). Lower operating costs and higher quality of service have also led to the adoption of the VAL rubber-tire automated-guideway transit system in Lille, France. Both systems are planning further expansion together with a rubber-tire system in Miami, Florida (Fig. 1a). Automated guideway transit systems have recently been commissioned in Japan, for example, in Kobe and Yokoh. Typically the trains operate between 0 and 55 mi/h (25 m/s) with low headways (time between trains) of under 2 min.

Other systems. Light rail transit (LRT) systems, unlike automated guideway transit schemes, do not require a dedicated right of way; for this reason they are increasingly being introduced worldwide, using many of the developments outlined above (**Fig. 3**).

Systems utilizing magnetic levitation with linear-motor drives are at various levels of study and design. Small-scale schemes using these machines are currently operating and being built in England, Germany, Japan, and the United States.

For background information SEE RAILROAD CONTROL SYSTEMS; RAILROAD ENGINEERING; SUBWAY ENGINEERING in the McGraw-Hill Encyclopedia of Science & Technology.

Anthony Daniels

Bibliography. *Automated People Mover Systems*, AEG Westinghouse Transportation Systems, Inc., Pittsburgh, 1990; E. E. Riches, *The Birmingham, England–Airport Maglev Transit Link*, GEC Transportation Projects, Ltd., Manchester, England, 1988; R. Royaux, *The VAL, a French Automated Guideway Transit System: Achievements and Future Prospects*, Alsthom Rev. 10, GEC Alsthom, 1988; Z. Uneda (ed.), *Urban Transportation Systems in Japan*, Japan Rolling Stock Exporters' Association, Tokyo, 1990.

Rape

In the 1960s, Canadian plant breeders developed lines of rapeseed (*Brassica napus* and *B. campestris*) that produced oils containing low levels of erucic acid, a 22-carbon, monounsaturated fatty acid, and seeds with low levels of glucosinolates. The term Canola was adopted to help market these improved varieties and is now used to refer to all edible *Brassica* varieties in North America. Reducing the level of erucic acid in Canola oil produced a fatty acid composition that has been recognized by nutritionists as beneficial in human diets. Glucosinolates are toxic substances produced by many plants of the Cruciferae family to protect the seeds and foliage from herbivore damage. Low levels of glucosinolates in the meal residue, the by-product of seed oil extraction, allow it to be used as a high-protein animal feed supplement. The improved oil and meal characteristics of Canola quality varieties have increased the production and utilization of this crop throughout the world.

Nutritional characteristics. Nutritional studies indicate that oils with no cholesterol, low levels of saturated fatty acids, moderate levels of polyunsaturated fatty acids, and high levels of monounsaturated fatty acids may significantly reduce serum cholesterol and thus the risk of atherosclerosis and coronary heart disease. Canola oil has been recognized as having an almost ideal fatty acid composition in this respect; it has the lowest level of saturated fatty acids of any major source of edible oil (**Fig. 1**). The relatively high level of monounsaturated fatty acids and moderate levels of polyunsaturated fatty acids found in Canola oil also promote the increase of high-density-lipoprotein (HDL) cholesterol, which is believed to

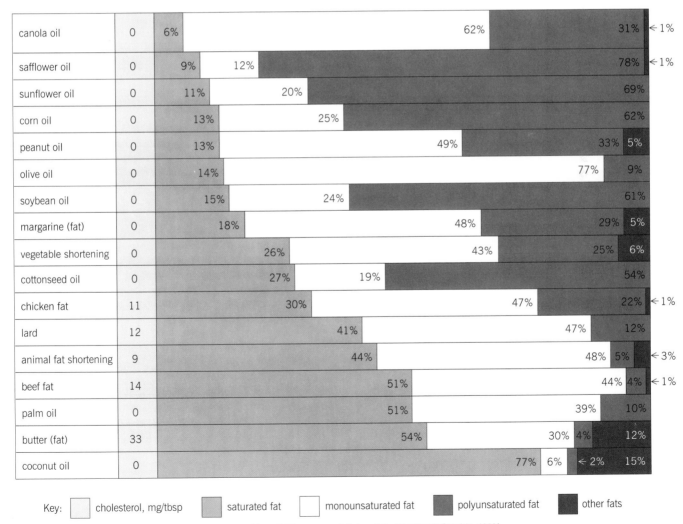

Fig. 1. Comparison of cholesterol and fatty acid composition of 17 sources of dietary fats. (*Procter & Gamble, 1989*)

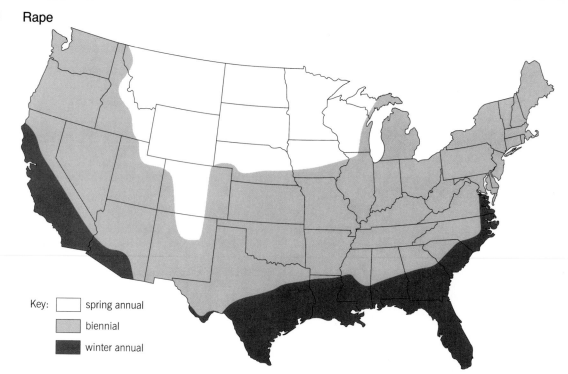

Fig. 2. Areas of Canola adaptation across the continental United States. (*From D. L. Auld, D. R. Brady, and K. A. Mahler, Rapeseed and Canola breeding in North America: The U.S. potential, Proceedings of the 5th Crucifer Genetics Workshop, University of California, April 7–9, 1989*)

protect against atherosclerosis and coronary heart disease. The relatively high concentration of linolenic acid, an omega-3 fatty acid, in Canola oil may reduce the formation of blood clots responsible for both coronary attacks and strokes. Linolenic acid is also a direct precursor to longer-chain polyunsaturated fatty acids of the omega-3 series, whose fatty acids may help counteract obesity and inflammations.

Canola production. Rapeseed and Canola is grown on 4.2×10^7 acres (1.7×10^7 hectares) in Canada, Europe, and Asia and produces an estimated 2.4×10^7 tons (2.2×10^7 metric tons) of harvested seed. The oil extracted from this seed provides nearly 15% of the world's edible vegetable oil. Prior to 1985,

when the U.S. Food and Drug Administration first allowed Canola oil to be used in food products, there were less than 50,000 acres (20,000 hectares) of rapeseed and Canola produced annually in the United States. Current demand for Canola oil in the United States has increased interest in growing this oilseed crop.

Canola can be grown on most agricultural lands that have well-drained soils with a pH of 5.8–6.2. Because of the wide range of growth habits found in the oilseed *Brassica* species, most agricultural production zones in the United States can grow Canola as a commercial crop (**Fig. 2**). In the northern tier of the United States, Canola can be grown as a spring annual since biennial varieties cannot survive the winters of this region. Across much of the central United States, biennial varieties of Canola similar to those currently used in northern Europe can be grown. Areas of the southern United States with relatively mild winters can produce Canola as a winter annual using varieties with a range of vernalization requirements and cold tolerance. Several American, Canadian, and European organizations have active programs to develop Canola varieties that are adapted to the diverse production environments found across the United States.

Canola is seeded and harvested by using the same grain drills and combines currently used to grow cereal crops such as wheat and barley. Biennial and winter annual Canola varieties are usually seeded approximately 6 weeks before the first anticipated hard frost. Spring-planted Canola is usually seeded when soil temperatures rise above 50°F (10°C) to reduce potential insect damage and optimize seed yield potential. Canola crops require the application of nitrogen, phosphorus, potassium, sulfur, and boron

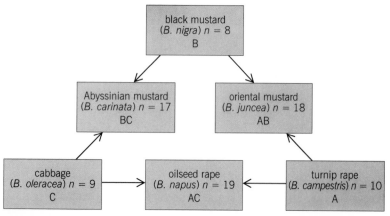

Fig. 3. Haploid chromosome number (*n*) and genomic relationship of six domesticated species of *Brassica* grown as oilseed crops. The three species with higher chromosome numbers, *B. napus*, *B. juncea*, and *B. carinata*, are the result of interspecific hybrids between two diploid species. (*After G. Robbelen, R. K. Downey, and A. Ashri, eds., Oil Crops of the World, McGraw-Hill, 1989*)

fertilizers at levels similar to those needed to produce winter wheat. Weeds, insects, and diseases of Canola are controlled by using a combination of cultural practices and judiciously applying pesticides currently registered in both Canada and Europe.

Genetic enhancement. Canola's wide area of adaptation, high seed yield potential, and high concentrations of oil and protein make this crop a model system for genetic manipulation. The six domesticated species of *Brassica* have three sets of genomes (**Fig. 3**): the A genome represents the 10 chromosomes derived from turnip; the B genome represents the 8 chromosomes from black mustard; and the C genome represents the 9 chromosomes from cabbage. Canadian researchers recently used crosses between *B. nigra* (B genome) and *B. campestris* (A genome) to transfer the genes for Canola quality oil and meal into Oriental mustard (*B. juncea*), which contains both the A and B genomes. Oriental mustard plants can tolerate higher temperatures and drier soil conditions than either *B. napus* or *B. campestris*. Canola quality varieties of Oriental mustard will allow this improved oilseed crop to be grown in more arid regions in North America and Asia. Additional interspecific crosses will continue to expand the genetic base and improve the yield of Canola.

Researchers have also used advanced techniques of cell biology, mutagenesis, and genetic engineering to improve Canola. These techniques have allowed the development of rapeseed lines with pest resistance, selective herbicide resistance, and unique chemical compositions. Other researchers are developing pollination control systems that would allow commercial production of hybrid varieties that could improve yields by over 20%. These developments should reduce the cost of producing Canola while improving the utility of the oil and meal derived from new varieties.

For background information SEE ARTERIOSCLEROSIS; CHOLESTEROL; FAT AND OIL (FOOD); RAPE in the McGraw-Hill Encyclopedia of Science & Technology.

Dick L. Auld

Bibliography. T. H. Applewhite (ed.), *Proceedings of World Conference on Biotechnology for the Fats and Oils Industry*, 1988; G. Robbelen, R. K. Downey, and A. Ashri (eds.), *Oil Crops of the World*, 1989; M. Vaisey-Genser and N. A. Michael Eskin, *Canola Oil: Properties and Performance*, 1987.

Reactive intermediates

Most chemical processes involve reactive intermediates. That is, starting materials (reactants) are not converted directly to products as in reactants → products but rather are converted to products through one or more transient intermediates as in reactants → intermediates → products. A multitude of distinct intermediates have been detected by one or more experimental techniques in an even larger number of reactions of organic, biological, or inorganic

reactants. (Reactants are often called substrates if the reaction is a catalyzed process.) Although the number of intermediates documented in the scientific literature with precisely defined atomic compositions and reasonably well-defined structures is formidable, it is nevertheless possible to classify these reactive species as belonging to a few genetic types. The principal classes of intermediates in organic and biological reactions (with their distinguishing properties in parentheses) are as follows: carbocations (positive charge on carbon), radicals (neutral charge but with an unpaired electron), carbanions (negative charge on carbon), carbenes (neutral carbon center with only six valence electrons), radical cations or cation radicals (positively charged with an unpaired electron), and radical anions or anion radicals (negatively charged with an unpaired electron). There are intermediates analogous to these classes derived from organometallic and inorganic compounds. Their properties, although well studied in many cases, are not as well understood as those derived from organic compounds. Atoms or molecules in their excited electronic states (excited states for short) also constitute reactive intermediates.

The detailed energetic course of a chemical reaction (the mechanism) as well as the products of the reaction are often dictated by the type and nature of reactive intermediates. As such, knowledge about these intermediates is usually of practical as well as intellectual significance.

Polyoxometalates. Knowledge about some aspects of reactive intermediates has recently been increased through the use of polyoxometalates. Some of the polyoxometalates can be converted to potentially useful high-valent oxometal species that function as reactive intermediates in key oxidation processes. In addition, many polyoxometalates in the presence of light can produce carbocations, radicals, or carbanions from organic or biological molecules.

The polyoxometalates are a large class of inorganic clusterlike complexes composed principally of transition-metal (TM) ions and oxide ions. The transition-metal ions are usually in their d^0 electronic configurations, that is, they formally contain no d electrons. As a consequence, polyoxometalates in their resting or most commonly occurring states do not contain unpaired electrons, and thus they are colorless or lightly colored (yellow or orange) and diamagnetic. The principal transition-metal ions that are found in polyoxometalates are molybdenum (Mo^{VI}), tungsten (W^{VI}), vanadium (V^V), niobium (Nb^V), and tantalum (Ta^V), in descending order of the number of distinct structural families of polyoxometalates associated with each. There are two large subclasses of polyoxometalates: the isopolyoxometalates (also known as isopoly compounds or isopoly anions), which contain only the transition-metal and oxide ions; and the heteropolyoxometalates (also known as heteropoly compounds or heteropoly anions), which contain one or more heteroatoms in addition to the transition-metal and oxide ions. The heteroatoms can be either main-group or transition-metal ions. Structures of one representative isopoly-oxometalate and one heteropolyoxometalate are illustrated in **Fig. 1.**

High-valent polyoxometalate-based oxometal intermediates. Some compounds of transition metals can be oxidized to high-valent forms where the metal contains a terminal oxygen atom. Many of these high-valent oxometal species (TM=O) are reactive intermediates in the transfer of oxygen from donors (DO) to organic or biological substrates (S) catalyzed by the transition-metal-containing species (**Fig. 2**). This catalyzed oxygenation process is believed to be operable in a host of biological and industrial oxidation reactions based on some transition metals, such as iron (Fe) and manganese (Mn). Among the biological systems that function through the intermediacy of high-valent TM=O species is the enzyme cytochrome P-450. This enzyme is the most potent broad-spectrum oxidant or oxidizing agent of organic/biological molecules found in the biosphere and one of the most studied biological systems in the 1980s.

A limiting problem with a number of synthetic transition-metal complexes designed by chemists to catalyze oxygen transfer processes, as well as with cytochrome P-450 itself, is that in all these cases the environment of the TM=O intermediate is constituted principally by organic groups. Unfortunately, all organic groups are ultimately (thermodynamically) unstable with respect to oxidation to carbon dioxide (CO_2) and water (H_2O) by most oxidants, including TM=O and even air [21% dioxygen (O_2)]. Thus the rate of oxidative degradation in all these systems (k_{deg} in Fig. 2) is finite, and the catalyst sooner or later is destroyed. In contrast, if the transition-metal catalyst is a transition-metal-substituted polyoxometalate compound such as $PW_{11}(TM)O_{39}^{5-}$ [P = phosphorus], which can be prepared by substituting one of the d^0 W^{VI} ions in the oxidatively resistant parent Keggin structure (Fig. 1b) with the appropriate transition-metal ion such as Mn^{II}, it acts as an oxidatively resistant catalyst for oxygen transfer. Since k_{deg} is effectively 0 for these completely inorganic complexes, the desired oxygen transfer pro-

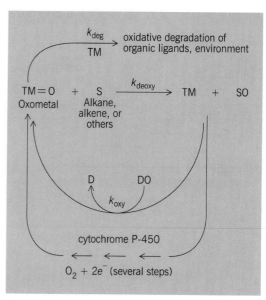

Fig. 2. General features of the oxygenation of organic substrates (S), including alkanes, mediated by transition-metal (TM) complexes. The oxygen donor (DO) can, in principle, be a range of reduced oxygen species (for example, C_6H_5IO, $ROOH$, H_2O_2, R_3NO, OCl^-) or possibly O_2. This scheme does not imply mechanism (detailed path of reaction) beyond the involvement of the oxometal intermediate, TM=O. k_{deg} = rate of oxidative degradation. k_{oxy}= rate constant for oxygenation of the TM center. k_{deoxy}= rate constant for transfer of oxygen from TM=O to substrate.

cesses ($k_{deoxy} + k_{oxy}$; Fig. 2) continue for many catalytic cycles.

As the properties of the transition metal and hence the reactive intermediate TM=O derived therefrom can be altered synthetically in a number of ways, these systems open up avenues of investigation into the nature of oxygen transfer from the transition metal to organic substrates, k_{deoxy}. One particular subset of this research area involves the oxidation of hydrocarbons by such species, including the active form of cytochrome P-450. Hydrocarbon (C—H) bonds may react with TM=O intermediates exclusively by abstraction of a hydrogen atom, forming a hydroxy-TM-organic radical (R) pair, [TM—OH · R], as is indicated by the data from most systems examined in depth to date; there may be an initial electron transfer from hydrocarbon to TM=O to generate a cation radical; or there may be some other pathway that is operable in some systems. Also a process of great significance that is even less well understood than the reaction of TM=O with C—H bonds is the reaction of TM=O with alkenes (TM=O + alkene substrate → TM + epoxide product). A number of mechanisms proposing an equal number of thermally unstable reactive intermediates have been proposed for these processes. *See* ASYMMETRIC SYNTHESIS.

Polyoxometalate photocatalysis. Upon absorbing near-ultraviolet or blue light, some structural families of polyoxometalates (P_{ox}), including decatungstate (Fig. 1a), form excited states (P_{ox}*) that can be reduced (that is, they take up electrons, e^-) by oxidizing a range of organic substrates, including alkanes (RH), without alteration of the basic structure

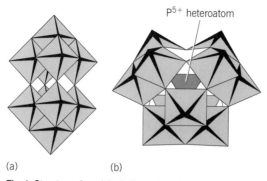

(a)   (b)

Fig. 1. Structures in polyhedral notation of representative polyoxometalates. (a) The isopolytungstate, decatungstate ($W_{10}O_{32}^{4-}$). (b) The Keggin heteropolytungstate, α-$PM_{12}O_{40}^{3-}$, M = Mo or W. In polyhedral notation, each octahedron represents one metal atom surrounded by six oxygen atoms (or a tetrahedron in the case of the 4-coordinate central atom in b). The metal atom is not seen directly in polyhedral notation.

of the complex [reactions (1) and (2)]. The resulting

$$P_{ox} + h\nu \rightarrow P_{ox}* \qquad (1)$$

$$P_{ox}* + RH \rightarrow P_{red} + RH \text{ oxidation products} \quad (2)$$

reduced forms of the polyoxometalates (P_{red}) can be subsequently reoxidized to restore the initial oxidized (electron-poor) form of the polyoxometalate (P_{ox}) by reduction of either dioxygen to form water [reaction (3)] or protons to form dihydrogen [H_2; reaction (4)],

$$2P_{red} + 2H^+ + \tfrac{1}{2}O_2 \rightarrow 2P_{ox} + H_2O \qquad (3)$$

$$2P_{red} + 2H^+ \rightarrow 2P_{ox} + H_2 \qquad (4)$$

completing cycles that are catalytic in P_{ox}. That is, P_{ox} facilitates a net process, but it is not destroyed itself in the course of the process. Reactions (1)–(4) are written for n (the number of electrons) = 1, although some families of polyoxometalates can be reduced or photoreduced by more than one electron.

In reaction (2), a number of reactive intermediates can be and usually are generated. The broad spectrum and very high reactivity of $P_{ox}*$ with respect to organic substrates facilitate production of reactive intermediates even from classes of organic compounds that usually show little reactivity. One such class, the alkanes (RH), comprises the most abundant class of organic compounds and includes the principal fossil fuels. Alkanes show little reactivity with most conventional chemical reagents. The only commercially significant oxidation reactions that alkanes undergo are halogenation and autoxidation (or combustion); both processes are minimally selective and hard to control. The photooxidation of RH by decatungstate and other photochemically active polyoxometalates [reaction (2)] leads to a radical pair [PH·R] or an ion pair [PH⁻ ⁺R], intermediates whose decompositions to useful products can be controlled to some extent. Alkanes can be dehydrogenated by such systems to the corresponding alkenes (R[—H]); and they can produce preferentially either the most substituted or least substituted forms (isomers) of the alkenes, depending on the polyoxometalate and the reaction conditions. Production of the least substituted R[—H], a useful and very unusual reaction, is achieved by minimizing the oxidation of radical intermediates. Photooxidation of RH by some systems can actually produce carbanion intermediates by reduction of previously generated radical intermediates.

Trapping the carbanions with electrophiles (species with positive charges or positive-charge character) can lead to the net replacement of unactivated alkane C—H bonds with carbon-to-carbon (C—C) bonds, an unprecedented type of process. A specific example is shown in reaction (5). This complex reaction involves

(I)   (II)

the generation and trapping of three types of reactive intermediates, the excited state of the polyoxometalate, $W_{10}O_{32}^{4-}*$; the radical, R·; and the carbanion, R⁻, [where RH is the caged polycyclic hydrocarbon reactant (I)]. First, $P_{ox}*$ abstracts the hydrogen atom from reactant (I) to form R·, R· is subsequently reduced by the P_{red} ($W_{10}O_{32}^{6-}$) that accumulates during the reaction to form R⁻, then R⁻ reacts with the nitrile carbon of the acetonitrile solvent (CH_3CN). Hydrolysis (reaction with H_2O) of the resulting imine then generates the product (II).

Thus, the intrinsic oxidation-reduction properties of polyoxometalates in both their ground and excited states facilitate not only the generation of reactive intermediates from some very unreactive compounds, including alkanes, but also the controlled and sequential generation of more than one type of reactive intermediate. The fact that the generation of these intermediates can be rendered catalytic in the polyoxometalate and that the intrinsic oxidation-reduction properties of the polyoxometalate complexes themselves can be extensively altered establishes polyoxometalate-based photocatalytic routes as promising vehicles for further defining the oxidation-reduction properties and uses of some classes of reactive intermediates.

For background information SEE ALKANE; COMPLEX COMPOUNDS; ELECTRON CONFIGURATION; REACTIVE INTERMEDIATES; TRANSITION ELEMENTS in the McGraw-Hill Encyclopedia of Science & Technology.

Craig L. Hill

Bibliography. C. L. Hill (ed.), *Activation and Functionalization of Alkanes*, 1989; P. R. Ortiz de Montellano (ed.), *Cytochrome P-450*, 1986; M. T. Pope, *Heteropoly and Isopoly Oxometalates*, 1983; C. M. Prosser-McCartha and C. L. Hill, Direct selective acylation of an unactivated C—H bond in a caged hydrocarbon: Approach to systems for C—H bond functionalization that proceed catalytically at high substrate conversion, *J. Amer. Chem. Soc.*, 112:3671–3673, 1990; R. F. Renneke, M. Pasquali, and C. L. Hill, Polyoxometalate systems for the catalytic selective production of nonthermodynamic alkenes from alkanes: Nature of excited state deactivation processes and control of subsequent thermal processes in polyoxometalate photoredox chemistry, *J. Amer. Chem. Soc.*, 112:6585–6594, 1990.

Remote sensing

Remote sensing of the Earth's surface and atmosphere using microwaves has gained impetus in recent years due to the advances in microwave radar technology and data-processing techniques. Significant advances in synthetic-aperture radar (SAR) technology have enabled high-resolution microwave remote sensing from space-borne platforms such as satellites and space shuttles; and in the near future, the space station to be launched by the National Aeronautics and Space Administration (NASA) will provide an additional facility.

Remote sensing in general implies sensing or detecting, with the help of devices known as sensors, the electromagnetic radiation either reflected or emitted by an object. The remote sensing of the Earth's surface (land and ocean) and atmosphere, in order to gather useful data, is of particular interest. Both photography and infrared imaging have been used extensively for this purpose. The visible electromagnetic radiation reflected from the Earth's surface is utilized in photography from airborne and space-borne platforms. In the infrared imaging technique, the radiation emitted by the Earth's surface in the infrared region of the electromagnetic spectrum (wavelengths from 0.7 to 1400 micrometers) is sensed. The disadvantage of both visible photography and infrared imaging is that the Earth's surface cannot be observed when obscured by clouds and in darkness. This is overcome in microwave remote sensing (wavelengths from 0.1 to 30 cm).

Advantages of microwave remote sensing. The most important reason for using microwaves is their ability to penetrate clouds and rain. **Figure 1** shows the transmission of microwaves between space and Earth. Ice clouds that obscure the Earth and hence preclude aerial photography are transparent to microwave propagation. Water clouds have a severe effect only for wavelengths less than 1 cm. Rain has a significant effect only below a wavelength of 2 cm.

Microwaves also are able to penetrate deeper into vegetation than optical waves. The penetration depth depends upon the moisture content and density of the vegetation as well as the microwave frequency. Lower-frequency microwaves penetrate deeper than higher-frequency waves. Therefore, lower-frequency waves can provide information about the bottom

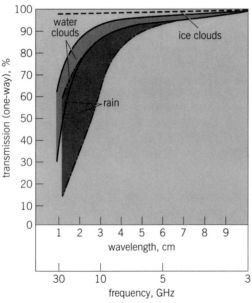

Fig. 1. Effect of clouds and rain on microwave transmission from space to Earth. (*After F. T. Ulaby, R. K. Moore, and A. K. Fung, Microwave Remote Sensing, Active and Passive, vol. 1: Microwave Remote Sensing Fundamentals and Radiometry, Addison-Wesley, 1981*)

layers of the vegetation and the ground, and the higher frequencies can provide data about the top layers. Further, microwaves can significantly penetrate the soil. The penetration at lower frequencies is significant for dry soil. The depth of penetration is greater than that obtainable with visible and infrared radiation.

The sensors used in microwave remote sensing can be classified into two categories depending on their operational modes: active sensors and passive sensors.

Active microwave remote sensing. In this technique, active sensors are used. These sensors contain a transmitter and a receiver. The transmitters generate microwave signals, and the receivers receive the signals reflected from the scene and process them to obtain data. Active microwave sensors include radar imagers, scatterometers, and altimeters. In fact, all these sensors are essentially microwave radar systems.

The term radar stands for radio detection and ranging. The signal from the transmitter is coupled into space through a transmitting antenna. The signal travels to the region under observation and is reflected from there to a receiving antenna that couples the reflected wave into the receiver circuits. The receiver then processes the signal and sends it to an indicator or recorder.

Radar imagers. The commonly used imaging radars for remote sensing are side-looking airborne radars (SLARs). The antenna has a beam that is narrow horizontally and wide vertically, and it points to the side of the aircraft on which the radar is mounted. A short pulse of microwave energy is transmitted; after it strikes a target, the pulse returns to the radar. The time delay associated with the signal is a measure of the distance between the radar and the target. The returned signal is used to modulate the intensity of the beam on an oscilloscope. The oscilloscope display is transferred to a film that moves in synchronism with the aircraft. As a result, a two-dimensional picture of the surface is obtained.

The picture, however, is distorted because the across-track dimension of the image, which is determined by the time measurement associated with the slant range R from the radar to the point on the ground, does not accurately correspond to the horizontal distance on the ground.

In the across-track dimension, the pixel size r_p is given by Eq. (1), where c is the velocity of light, τ is

$$r_p = \frac{c\tau}{2 \sin \theta} \qquad (1)$$

the pulse duration, and θ is the angle to the vertical. The pixel size is dependent on the distance because $\sin \theta$ is small near the vertical and approaches unity near grazing incidence. Thus, the pixel size changes in the across-track direction, depending on the range.

Further, the along-track resolution of the radar is determined by the along-track width of the antenna beam, β_h, which is inversely proportional to the along-track length of the antenna, l, in terms of the wavelength of operation, λ. The along-track resolution r_a can be shown to be given by Eq. (2).

$$r_a = \beta_h R \simeq \frac{\lambda R}{l} \qquad (2)$$

Therefore, it is a function of the slant range and is different in various parts of the image.

The foregoing drawbacks of side-looking airborne radars are overcome in synthetic-aperture radar systems. Synthetic-aperture radars are capable of producing high-resolution images. They employ a technique known as aperture synthesis, wherein, effectively, signals from an array of antennas are coherently measured. The along-track resolution of a synthetic-aperture radar system is independent of the distance to the observed objects. Therefore, such a system is useful on both aircraft and spacecraft. The pixel dimension in the along-track direction is given by Eq. (3), where l is the along-track length of the antenna.

$$r_a = \frac{l}{2} \qquad (3)$$

Thus, for fine resolution the synthetic-aperture radar system requires only a small antenna, whereas the side-looking airborne radars, which use the real aperture, need a very large antenna. (The lower bound for a synthetic-aperture radar antenna is dictated by antenna directivity considerations. In general, the directivity of an antenna is directly proportional to the length of the antenna in terms of the wavelength of operation.) The pixel dimension in the across-track direction for a synthetic-aperture radar system is given by Eq. (4), where B is the bandwidth. Since synthetic-

$$r_p = \frac{c}{2B \sin \theta} \qquad (4)$$

aperture radars use a frequency-modulated pulse (chirp), the bandwidth B can be made very large, resulting in very small pixel sizes. In a side-looking airborne radar, on the other hand, short pulses are required to achieve the same resolution. A very short pulse requires a high peak power, whereas in a chirp radar used in synthetic-aperture radars, the bandwidth can be made large without excessively high peak-power requirements. The advantage of synthetic-aperture radar for obtaining a high-resolution beam is illustrated in **Fig. 2.** An example of a *Seasat* synthetic-aperture radar image is shown in **Fig. 3.**

Synthetic-aperture radar technology is continuously advancing. In the 1980s, several spacecraft carrying synthetic-aperture radars were launched, and several launchings are planned for the 1990s. The short-term synthetic-aperture radar missions in the future will use shuttle platforms, and long-term missions will use satellites. Currently, NASA is planning for a synthetic-aperture radar facility aboard its future space station. The Earth Observing System (EOS) program, which is a major international effort to understand the interaction between the incoming energy, water, atmosphere, and life on Earth, will begin in the mid-1990s. This program will incorporate many microwave remote sensing devices, including synthetic-aperture radar, electronically scanned thin-array radiometers, and other

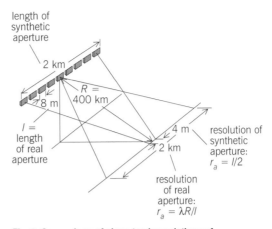

Fig. 2. Comparison of along-track resolutions of synthetic-aperture radar (SAR) and side-looking airborne radar (SLAR; with real aperture), showing advantage of SAR in the case of a spacecraft radar with wavelength $\lambda = 4$ cm. Symbols explained in text. 1 m = 3.3 ft. 1 km = 0.6 mi. (*After F. T. Ulaby, R. K. Moore, and A. K. Fung, Microwave Remote Sensing, Active and Passive, vol. 1: Microwave Remote Sensing Fundamentals and Radiometry, Addison-Wesley, 1981*)

types of advanced microwave sensing equipment, on space platforms. SEE BIOSPHERE.

Scatterometers and altimeters. The remote sensing radars that are nonimaging are either scatterometers or altimeters. Scatterometers have been used for obtaining data on winds at sea as well as measuring characteristics of returns from targets for radar design purposes. The *Skylab* in 1973 and *Seasat* in 1978 carried scatterometers. Various airborne scatterometers have also been used. Some scatterometers have a relatively narrow beam and use either pulse modulation or continuous waves.

Radar altimeters have been used to measure characteristics of waves and tides. The altimeters have been placed on airborne as well as space-borne platforms.

Passive microwave remote sensing. This method uses a passive sensor consisting of an antenna and a receiver. Passive microwave sensors are known as microwave radiometers. The energy received by a radiometer comes from the target as a result of both emission and reflection. The microwave power emitted by an object is proportional to its absolute temperature. The received signal in a radiometer is phase-incoherent and covers a wide frequency range, unlike that in a conventional radar. Further, the ratio of the signal power received to the noise power at the input of the receiver is much smaller than in a conventional radar. Therefore, radiometers are designed to be highly sensitive receivers.

The EOS mission of the 1990s will use an advanced scanning microwave radiometer (ASMR) in which the beam is switched electronically. This instrument will also have multifrequency measurement and multiple-beam scanning capabilities. The primary objectives of the ASMR are to measure sea-surface temperature, wind speed, atmospheric water vapor, cloud and precipitating liquid water, sea ice concentration, and terrain snow coverage.

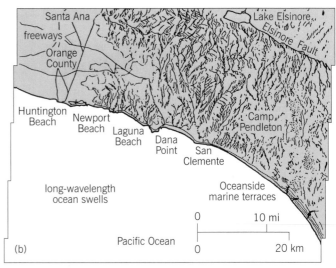

Fig. 3. Digitally processed *Seasat* synthetic-aperture radar (SAR) image of part of the southern California coast, with a resolution of 25 m (82 ft). (*a*) Image. (*b*) Map identifying principal features. (*NASA; Jet Propulsion Laboratory*)

Applications. Present and potential application of airborne and space-borne imaging radars in remote sensing include the following areas: (1) geology—observing structure and lithology; (2) hydrology—measuring soil moisture, mapping watersheds, mapping floods, mapping areas of surface water such as ponds, lakes, and rivers, and mapping snow; (3) agriculture—mapping crops, monitoring agriculture practices, identifying field boundaries, monitoring growth and harvest progress, identifying stress areas, monitoring rangelands, and monitoring water problems, as in hydrology; (4) forestry—monitoring cutting practices, mapping fire damage, identifying stress areas, and, in the future, estimating timber volume; (5) cartography—mapping topography in remote cloudy areas, mapping land use, and monitoring land-use changes, urban developments, and so forth; (6) polar region study—monitoring and mapping sea ice, mapping continental ice sheets, monitoring iceberg formation and movement, and monitoring glacial changes; and (7) oceanography—monitoring wave patterns, ice spills, ship traffic, and fishing fleets.

Areas for the application of airborne and space-borne microwave radiometers in remote sensing include the following: (1) hydrology—measuring soil moisture distribution for river-stage and flood forecasts, observing watershed surface drainage characteristics, mapping floods, identifying water surfaces, and measuring snow cover extent, snow water equivalent, and snow wetness; (2) agriculture—observing soil moisture distribution for crop-yield estimation and for irrigation scheduling, and delineating freeze–thaw boundaries; (3) polar regions study—monitoring and mapping sea ice types, and mapping continental ice sheets; (4) oceanography—monitoring surface wind speeds, temperature, salinity, and oil spills; (5) severe storm study—monitoring tropical cyclones, obtaining rain maps, obtaining temperature and humidity profiles, and measuring sea temperature and wind; monitoring severe local storms, obtaining temperature and humidity profiles, and observing soil moisture and rain as feasible; (6) meteorology and climatology (primarily over oceans)—obtaining temperature profiles, measuring integrated water vapor, obtaining water-vapor profiles, observing rain, and measuring ocean temperature and surface wind speed; (7) study of the stratosphere, mesosphere, and lower thermosphere—obtaining atmospheric temperature profiles and magnetic-field profiles, and measuring the abundance of atmospheric gases.

For background information *SEE APPLICATIONS SATELLITES; MICROWAVE; PASSIVE RADAR; RADAR* in the McGraw-Hill Encyclopedia of Science & Technology.

S. N. Prasad

Bibliography. D. M. Lee Vine et al., A multifrequency microwave radiometer of the future, *IEEE Trans. Geosci. Remote Sens.*, 27(2):193–198, 1989; K. Tachi, K. Arai, and Y. Sato, Advanced microwave scanning radiometer (ASMR) requirements and preliminary design study, *IEEE Trans. Geosci. Remote Sens.*, 27(2):177–182, 1989; F. T. Ulaby, R. K. Moore, and A. K. Fung, *Microwave Remote Sensing, Active and Passive*, vol. 1: *Microwave Remote Sensing Fundamentals and Radiometry*, 1981, and vol. 3: *From Theory to Applications*, 1986.

Reproduction (plant)

During sexual reproduction in plants, sperm cells fuse with egg cells. A fertilized egg develops into an embryo that grows from the seed to become a new plant. Sperm cells occur in pairs linked in the pollen tube. In the laboratory they may be separated from the pollen and purified.

Sperm development. In flowering plants, sperm cells form in pollen grains or pollen tubes (**Fig. 1a**). Following meiosis of a pollen mother cell, the haploid nucleus divides asymmetrically to form the vegetative and generative cells. The vegetative cell makes up most of the pollen grain and pollen tube cytoplasm. The generative cell is embedded in the vegetative cell and separated from it by two membranes: the cell membrane of the generative cell and the inner cell membrane of the vegetative cell. The generative cell divides to form the two sperm cells. If the generative cell divides within the pollen grain before germination of the pollen tube, the resulting mature pollen grain is tricellular, with two sperm cells plus the vegetative cell (Fig. 1a). If the generative cell divides later, that is, during germination of the pollen tube, the pollen grain is bicellular at maturity and contains one vegetative cell and one generative cell.

Plant sperm structure. Because sperm cells are embedded in the vegetative cell of pollen grains or tubes, they are surrounded by two membranes: the sperm cell membrane and the inner cell membrane of the vegetative cell. In freeze-fractured pollen grains, the membranes surrounding the sperm cells show no distinctive particle domains or patterns. Furthermore, although the two membranes are closely appressed, no membranous connections or bridges link the sperm and vegetative cell membranes.

Sperm cells are elongate and reduced in volume compared to most plant cells (**Fig. 2**). Sperm cells do not swell proportionately when the pollen grain swells at germination because sperm cytoplasm is much less concentrated than vegetative cell cytoplasm. The volume of sperm cells is kept low by osmosis as water moves out of sperm into the more concentrated cytoplasm of the vegetative cell.

The sperm nucleus is compact, with tightly packed deoxyribonucleic acid (DNA), and is surrounded by a nuclear envelope that has few pores. This condensed chromatin correlates with low transcription rates. Sperm nucleoli are small or absent, indicating that there is little synthesis of ribosomes.

Cytoplasmic organelles, including mitochondria, endoplasmic reticulum, dictyosomes, ribosomes, and, in some species, plastids, suggest that sperm cells are metabolically active. A cytoskeleton is formed of microtubule bundles arranged obliquely along the cell axis. The sperm of flowering plants lack flagella, and no intrinsic movement has been observed.

Sperm remain closely associated with the vegetative nucleus in the pollen tube. Pairs of sperm cells attach to each other and to the vegetative nucleus, thereby forming a structure called the male germ unit (**Fig. 3**). The sperm, loosely joined at one end, are encased in the inner vegetative cell membrane. One sperm extends into interstices of the vegetative nucleus and attaches to the vegetative cell membrane by unknown means. In some species, serial ultrathin sections of sperm in the pollen grain (prior to germination of the pollen tube) show that the sperm are separate. However, after the pollen tube forms, sperm cells are connected as a male germ unit, indicating that the male germ unit may form by a mechanism unrelated to cell division. The male germ unit serves to deliver both sperm to the vicinity of the egg and central cell for synchronous double fertilization.

The size, shape, and contents of the two sperm cells may differ. Serial sectioning of pairs of sperm cells in

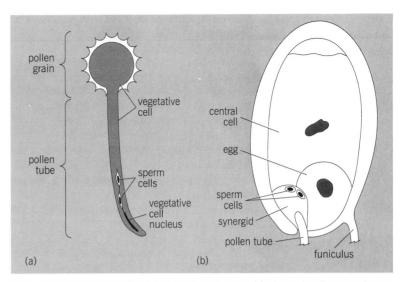

Fig. 1. Location of sperm cells (a) in the pollen tube and (b) in the ovule adjacent to the egg and central cell. As the pollen tube grows from the pollen grain, the male germ unit moves down the tube. The pollen tube grows into the ovule and deposits two sperm cells near the egg and central cell for double fertilization.

pollen grains has shown that they are often dimorphic, being of unequal size and, in some plants, having unequal numbers of mitochondria and plastids. This dimorphism may relate to their differing fates in double fertilization.

Double fertilization. The pollen tube grows through the stigma and style into the ovary, where it enters an ovule, bursts, and deposits the two sperm cells near the egg and the adjacent central cell (Fig. 1b). Two fertilization events occur: one sperm fuses with the egg to form the embryo of the seed, and the other

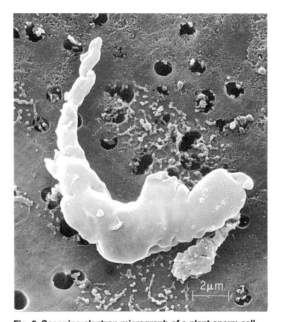

Fig. 2. Scanning electron micrograph of a plant sperm cell isolated on a filter. The nucleus is located in the thicker part of the cell. Pores in the filter are 1 μm in diameter. (*From D. Southworth and R. B. Knox, Flowering plant sperm cells: Isolation from pollen of Gerbera jamesonii (Asteraceae), Plant Sci., 60:273–277, 1989*)

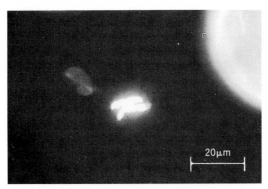

Fig. 3. Fluorescence micrograph of the male germ unit of *Gerbera jamesonii* released from a pollen grain and stained with the DNA-specific fluorochrome, 4',6-diamidino-2-phenylindole (DAPI). The nuclei of the sperm cells fluoresce brightly, indicating a dense packing of chromatin; the vegetative nucleus has more diffuse chromatin. (*From D. Southworth and R. B. Knox, Flowering plant sperm cells: Isolation from pollen of Gerbera jamesonii (Asteraceae), Plant Sci., 60:273–277, 1989*)

sperm fuses with the central cell to form the endosperm that nourishes the embryo. Because there are two types of sperm and two fertilization events, preferential fertilization may occur. In *Plumbago*, the sperm cell that usually fuses with the egg is smaller and contains more plastids and fewer mitochondria than the sperm cell that fuses with the central cell. Plastids and mitochondria are probably not responsible for preferential fusion; rather, they reflect the dimorphism of sperm cells.

Characterization of pollen proteins by polyacrylamide gel electrophoresis shows that the male germ unit contains some unique proteins not found in pollen cytoplasm. Plant sperm cells have been injected into mice to induce antibody formation. Half the antibodies that form are unique to sperm and do not react with vegetative cell cytoplasm. The other half of the antibodies recognize both sperm and vegetative cell components. However, there is no evidence for two types of sperm cell membranes.

In many plants, the DNA from sperm plastids and mitochondria is not inherited. One explanation is that sperm cell cytoplasm does not enter the egg at fertilization. In barley, a cytoplasmic body containing mitochondria and plastids and recognizable as sperm cytoplasm forms outside the egg while the sperm nucleus is inside the egg. In *Plumbago*, however, sperm plastids enter the egg.

Isolation of sperm cells. Sperm cells separated from pollen cytoplasm provide useful material for the basic study of reproduction in plants. In tricellular pollen, sperm can be isolated directly from pollen. However, in bicellular pollen, pollen tubes must first germinate to allow division of the generative cell into two sperm. Because the vegetative cell membrane encases sperm cells but is not attached to them, sperm can be released easily from ruptured pollen by gentle grinding or by osmotic shock. A combination of filtration to remove pollen walls and centrifugation to separate vegetative cell organelles produces a fraction that is rich in sperm cells. Isolated sperm cells gradually change from an elongated shape to a spherical one, indicating a change in the cytoskeleton. The sperm cells swell and shrink in response to osmotic changes, giving evidence that their cell membranes are functional.

Isolated sperm cells have been fused with isolated egg cells in an electric field. It is possible that sperm cells will prove to be useful for artificial hybridizations or for delivery of selected genes.

Gymnosperm sperm structure. The four extant divisions of nonflowering seed plants differ from one another and from flowering plants in sperm structure. In the cycads and in *Ginkgo*, sperm are flagellated and swim to the egg cell. Cycad sperm are large, up to 400 micrometers in diameter, with about 40,000 flagella arranged spirally over the anterior hemisphere. *Ginkgo* sperm are smaller (75 μm in diameter) and similar in structure. The sperm nuclei of both cycads and *Ginkgo* are relatively large, and the chromatin is dispersed. Only the primary fertilization event of sperm and egg fusion has been reported in the cycads and *Ginkgo*.

The sperm of conifers are not flagellated. Two sperm are released from the pollen tube in the vicinity of the egg. The larger sperm fuses with the egg cell and the smaller one degenerates. In *Ephedra* each pollen tube produces one nonflagellated binucleate sperm. The egg also is binucleate, containing the egg nucleus and the ventral canal nucleus. One sperm nucleus fuses with the egg nucleus to form a zygote and eventually one or more embryos. After this primary fertilization event, the second sperm nucleus fuses with the ventral canal nucleus. The fate of this fusion product is unknown, although the second fusion suggests a parallel with double fertilization in flowering plants.

For background information *SEE FLOWER; POLLEN; POLLINATION; REPRODUCTION (PLANT)* in the McGraw-Hill Encyclopedia of Science & Technology.

Darlene Southworth

Bibliography. H. L. Mogensen, Exclusion of male mitochondria and plastids during syngamy in barley as a basis for maternal inheritance, *Proc. Nat. Acad. Sci. USA*, 85:2594–2597, 1988; B. A. Palevitz and M. Cresti, Microtubule organization in the sperm of *Tradescantia virginiana*, *Protoplasma*, 146:28–34, 1988; S. D. Russell, Isolation and characterization of sperm cells in flowering plants, *Annu. Rev. Plant Physiol. Plant Mol. Biol.*, 42:189–204, 1991; D. Southworth and R. B. Knox, Flowering plant sperm cells: Isolation from pollen of *Gerbera jamesonii* (Asteraceae), *Plant Sci.*, 60:273–277, 1989.

Ribonucleic acid (RNA)

Gene expression is the process by which information stored in the deoxyribonucleic acid (DNA) of chromosomes in cassettes called genes is decoded and put to use by living cells. It is mediated by the faithful transcription of ribonucleic acid (RNA) from DNA and by the modification, transport, function, and, ultimately, breakdown of that RNA. Genes of all types can contain one or more internal elements (intervening sequences) that are transcribed and subsequently removed during RNA maturation. The process by which intervening sequences are removed, termed RNA splicing, is a frequent and ubiquitous phenomenon; it has been observed in all eukaryotes examined, including fungi, plants, insects, amphibians, and mammals, and recently also in archaebacteria and certain bacteriophages.

RNA splicing may be categorized according to the origin of the mature RNA segments (exons): in cis splicing they originate from a single RNA transcript, and in trans splicing from discontinuous transcripts such as occurs in *Trypanosoma* and *Caenorhabditis*. Cis splicing can be subdivided on the basis of the presence and nature of cis- and trans-acting factors and the biochemical mechanism by which splicing occurs into at least four categories: group I self-splicing, group II self-splicing, precursor–messenger RNA (pre-mRNA) splicing, and precursor–transfer RNA (pre-tRNA) splicing. Several terms will be useful in the following discussion of these splicing categories. The basic building block of RNA is the nucleoside, which consists of a ribose moiety joined to either a purine base (adenine, A, or guanine, G) or a pyrimidine base (cytidine, C, or uracil, U). The nucleotides are linked in linear array by phosphodiester bonds between 5'- and 3'-carbons of the ribose moieties. Directionality of an RNA sequence is indicated by reference to the orientation of the 5'- and 3'-carbon positions in the linked nucleotide chain. Hydrogen bonds may form between specific base pairs (A and U or C and G) called complementary bases and confer a rigid secondary structure on RNA. RNA transcripts are categorized by function; for example, mRNAs encode proteins, ribosomal RNAs (rRNAs) are components of the protein synthetic machinery, and tRNAs function in protein synthesis to deliver amino acids.

Group I self-splicing. In group I self-splicing, each transcript itself provides most or all of the elements necessary to effect the splicing reaction. Group I RNAs are typified by the nuclear 26s rRNA gene of *Tetrahymena*, which is interrupted by one intervening sequence (see **Fig. 1**a); other examples include several nuclear rRNAs, fungal mitochondrial mRNAs and rRNAs, plant chloroplast tRNAs, and some bacteriophage mRNAs. Splicing shows a requirement for the native structure, and activity is lost when that structure is perturbed by high temperature or chemical destabilization. In group I transcripts, the 5' splice site is preceded by a conserved polypyrimidine sequence, which shows complementarity to an intervening sequence segment called the internal guide sequence. The internal guide sequence also shows complementarity to the 5'-end of the 3' exon and may thus align exon splice sites. Group I intervening sequences also contain three conserved pairs of complementary elements whose secondary structure is thought to form a helical core necessary for productive splicing. Evidence for such a tertiary structure comes from comparative sequence analysis of group I intervening

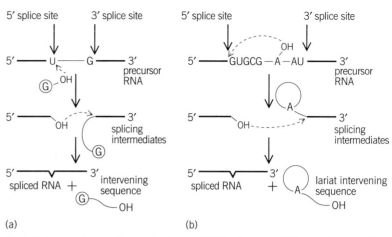

Fig. 1. Splicing pathways of group I and group II RNAs. Precursor RNAs are shown as 5' and 3' exons (heavy lines) interrupted by one intervening sequence (light line). Splice sites and conserved splice site nucleotides are indicated. For simplicity, conserved sequence elements within the intervening sequence are not shown. (a) Group I splicing; guanosine cofactor is shown as a circled G, 3'-hydroxyl groups as —OH. (b) Group II splicing; branch-point adenosine with 2'-hydroxyl group is shown as A—OH.

sequences as well as accessibility of intervening-sequence nucleotides to chemical and enzymatic probes.

Also required for splicing are a guanosine cofactor and magnesium or manganese. Splicing proceeds as two sequential transesterification reactions. In the first, the 3'-hydroxyl of the guanosine cofactor attacks the 5' splice site, liberating the 5' exon. The remaining intervening sequence-3' exon molecule acquires a 5'-guanosine. In the second transesterification, the 3'-hydroxyl of the 5' exon attacks the 3' splice site, releasing the intervening sequence and covalently joining 5' and 3' exons (Fig. 1a). For many group I transcripts, splicing has been observed in solution as well as in the whole cell. For those transcripts that cannot be spliced in solution, it is postulated that accessory factors may be necessary to stabilize the active conformation.

Group II self-splicing. Splicing of group II intervening sequences, defined by a small group of mitochondrial pre-mRNAs, shares certain features with group I reactions. Group II intervening sequences are also thought to provide the elements necessary to effect splicing. These include a conserved 38-nucleotide stem-loop region near the 3'-end and the consensus sequences GUGCG and AU at the 5' and 3' boundaries, respectively. Magnesium is also necessary; however, the function of the guanosine cofactor is replaced by an adenosine residue located in the intervening sequence. The first transesterification again yields the free 5' exon; but, in this instance, the remaining intervening sequence-exon molecule forms a branched structure called a lariat (Fig. 1b). Subsequent exon joining proceeds as in the second transesterification of group I RNAs, releasing ligated exons and free intervening sequence. As for group I reactions, both genetic and biochemical evidence suggest that accessory factors may play a noncatalytic role.

Pre-mRNA splicing. Similar to group II splicing, pre-mRNA splicing proceeds as two sequential transesterifications. In the first, cleavage occurs at the 5′ splice site with formation of an intervening sequence-3′ exon lariat intermediate. In the second, cleavage occurs at the 3′ splice site with concomitant ligation of 5′ and 3′ exons. Also, all cleavages produce 3′-hydroxyl and 5′-phosphate termini. Pre-mRNA splicing is distinguished by the unique nature of the conserved cis- and trans-acting factors and the requirement for energy in the form of adenosinetriphosphate (ATP).

Three cis-acting elements have been identified. The first is a 5′ splice site nine-nucleotide consensus sequence of which the six nucleotides at the 5′ terminal of the intervening sequence are sufficient for activity. The second element is a consensus sequence at the 3′ splice site consisting of a pyrimidine-rich region of variable length, a nonconserved position, another pyrimidine, and an AG dinucleotide (**Fig. 2**).

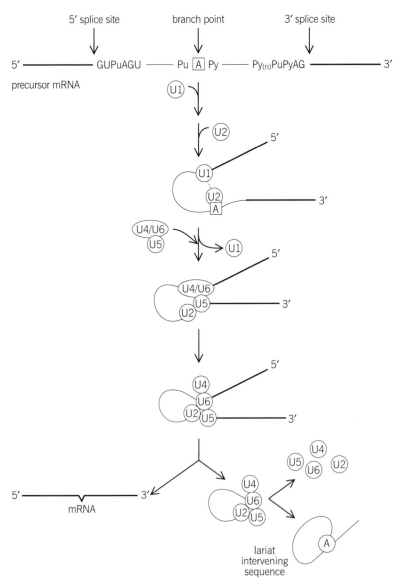

Fig. 2. Pre-mRNA splicing pathway. Precursor RNA is shown as 5′ and 3′ exons (heavy lines) interrupted by an intervening sequence (light line); conserved splice site and branch site nucleotides are shown. Branch-point adenosine (A) is shown in a square; Pu and Py indicate purine and pyrimidine, respectively; circles represent small nuclear ribonucleoprotein particles (snRNPs).

The third cis-acting element is the branch point, which in yeast is the highly conserved UACUAAC sequence of which the sixth-position adenosine provides the 2′-hydroxyl active in 5′ splice site cleavage. Stringency of conservation varies between species; for example, the polypyrimidine tract of the 3′ consensus sequence appears to be necessary in mammals for efficient 5′ splice site cleavage but unnecessary in yeast. Higher eukaryotes show weak conservation of branch points with some preference for the branched adenosine to be preceded by a purine and followed by a pyrimidine; however, strong position constraint is observed. The vast majority of vertebrate branch points thus far identified are located a distance of 18–37 nucleotides upstream of the 3′ splice site regardless of the length of the intervening sequence. Indeed, correct splicing of an artificially inserted pre-mRNA intervening sequence from a tRNA via the pre-mRNA mechanism suggests that the minimal sequences required for splicing are contained within the intervening sequence.

Essential to pre-mRNA splicing are the trans-acting factors constituting a group of small nuclear ribonucleoprotein (snRNP) particles that associate with pre-mRNA in active complexes known as spliceosomes. Each snRNP contains one or two RNA species identified as U1, U2, U4, U5, U6, or U4/U6, together with five to eight polypeptides common among all snRNPs. Some snRNPs have unique polypeptides, which may contribute to their specificity of interaction. Binding of certain snRNPs and their substrates occurs at least in part due to sequence complementarity of RNA components. The proposed dynamics of the pre-mRNA splicing reaction are diagrammed in Fig. 2. It is thought that proper pre-mRNA folding is effected by association with snRNPs. Furthermore, it is suggested that energy required in pre-mRNA splicing may be used in spliceosome formation and rearrangement and not necessarily in the excision–ligation reactions directly. Trans splicing in *Caenorhabditis* and *Trypanosoma* proceeds by the pre-mRNA splicing mechanism and is thought to involve many of the same components.

Pre-tRNA splicing. Nuclear pre-tRNA splicing proceeds via a unique pathway. The base composition, sequence, and length of pre-tRNA intervening sequences are not conserved, although the mature tRNA secondary and tertiary structures are highly conserved and present in the pre-tRNA. Length of the intervening sequence (9–60 nucleotides) is generally shorter than that in other splicing classes. In eukaryotic nuclear pre-tRNAs, intervening sequences are uniformly located one nucleotide 3′ to the anticodon (the nucleotide triplet that directly decodes mRNA sequence information). Only two cis-acting factors are recognized for pre-tRNA splicing: the conserved tertiary structure of the mature segment of the tRNA and the single-stranded nature of the 3′ splice site. Actual splicing of pre-tRNAs is executed by the trans-acting factors, tRNA endonuclease and ligase (**Fig. 3**). Both enzymes have been purified from *Saccharomyces* and have been shown to be capable of acting indepen-

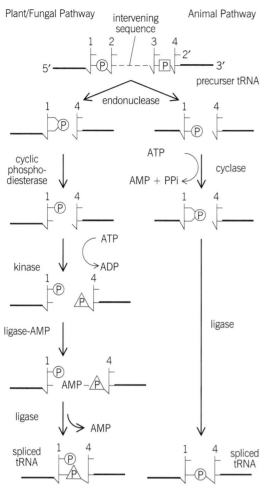

Fig. 3. Pre-tRNA splicing pathways of plants and fungi versus animals. Precursor tRNA is shown as 5′ and 3′ exons (heavy lines) interrupted by an intervening sequence (broken lines). Vertical numbered lines refer to ribose moieties of the ribonucleotides at the splice sites. Lines protruding right and left from the vertical indicate 2′, 3′, and 5′-hydroxyl or -phosphate linkage, top to bottom respectively. Species differences in phosphate linkage are emphasized. Symbols around phosphates (P) are used to indicate site of origin in either 5′ or 3′ exon.

kinase, adenylylate synthase, and RNA ligase. There is a requirement for ATP, and the ligation product contains a 2′-phosphate at the splice junction, which is subsequently removed by a separate phosphatase activity. Again, tRNA ligase enzymes from different species may vary with respect to substrate and ATP usage, as shown in Fig. 3.

Conclusion. The differences in tRNA splicing pathways appear to reflect a systematic lineage (that is, plants and fungi versus animals), which might suggest an evolutionary origin for tRNA splicing subsequent to divergence of animals from plants and fungi. The apparently unique reliance of pre-tRNA splicing on protein enzymes and cis-acting elements within the tRNA exons suggests an evolutionary origin distant from (or perhaps unique) that for group I, group II, and pre-mRNA splicing.

The ubiquitous nature of RNA splicing as well as the mechanistic similarities of group I, group II, and nuclear pre-mRNA splicing suggest a rather ancient origin for splicing and primal catalytic nature for RNA. Possible selective advantages of split genes include exon shuffling by recombination within intervening sequences; regulating gene expression by kinetic regulation of the splicing mechanism, differential splice-site selection, and sequestration of information within intervening sequences; and association of splicing with other activities such as anticodon base modification in tRNA splicing.

For background information *SEE* E*XON;* G*ENE;* G*ENE ACTION;* G*ENETIC CODE;* I*NTRON;* R*IBONUCLEIC ACID (RNA)* in the McGraw-Hill Encyclopedia of Science & Technology.

Heather G. Belford; Chris L. Greer

Bibliography. T. R. Cech, Conserved sequences and structures of group I introns: Building an active site for RNA catalysis—A review, *Gene*, 73:259–271, 1988; M. R. Culbertson and M. Winey, Split tRNA genes and their products: A paradigm for the study of cell function and evolution, *Yeast*, 5:405–427, 1989; F. Michel et al., Comparative and functional anatomy of group II catalytic introns—A review, *Gene*, 82:5–30, 1989; P. A. Sharp, RNA splicing and genes, *JAMA*, 260:3035–3041, 1988.

Ribozyme

Rapid progress during 1990 led to a detailed structural model and kinetic description of a ribozyme, which is a catalyst made of ribonucleic acid (RNA). In the near future, methods for the evolution of RNA molecules in the laboratory should lead to the development of new ribozymes. Recent insights have suggested the primacy of ribozymes in the origin and evolution of life.

Structure. The precise positioning of the component bases of a ribozyme determines how a substrate will be bound and activated. Thus, understanding how a ribozyme accelerates reactions requires at least a rudimentary knowledge of that ribozyme's structure. The most intensely studied ribozyme, an RNA from

dently, although the cleavage and joining reactions are thought to be concerted in the whole cell.

Differences in endonuclease activity and products occur among species. Yeast endonuclease cleaves the pre-tRNA to yield 5′-hydroxyl and 2′,3′-cyclic phosphate termini, and is an integral membrane protein multimeric in nature, consisting of up to three unique polypeptides. *Xenopus* tRNA endonuclease appears to be soluble, and cleavage products do not include a 2′,3′-cyclic phosphate. However, in both species, evidence suggests that endonuclease recognizes elements of mature tRNA domain and determines cleavage sites specifically by position relative to the anticodon stem (a characteristic base-paired sequence adjacent the anticodon area of the tRNA).

Ligase has been purified and its gene cloned from *Saccharomyces*. The 95,400-dalton polypeptide is soluble and contains all the activities of joining, that is, 2′,3′-cyclic phosphodiesterase, 5′-polynucleotide the single-celled ciliate *Tetrahymena thermophila*, is able to remove itself from longer strings of RNA by a process called self-splicing. The first step in this excision process is cleavage of the RNA at a specific position by a guanosine molecule. The ribozyme directs the guanosine to the exact bond to be broken by holding it in a network of hydrogen bonds known as a base triple. A base triple is the association of a single guanosine base with an existing double, or Watson-Crick, base pair (**Fig. 1***a*). Although this base triple was originally identified by structural modeling, it has been experimentally verified by engineering a mutant version of the ribozyme that can form base triples with molecules other than guanosine (Fig. 1*b*).

Other base triples in the *Tetrahymena* ribozyme are part of a structural hierarchy that folds linear information (the sequence of the RNA) into the three-dimensional architecture responsible for catalysis. Initially, about half of the *Tetrahymena* sequence folds into base-paired helices. Some of the remaining single-stranded segments then associate with these preformed stretches of helix to make base triples that pull together otherwise remote regions. Since each nucleotide has a different combination of hydrogen bond donors and acceptors, only certain combinations of nucleotides can work in a base triple. The base triples postulated for the *Tetrahymena* ribozyme were probed by introducing sequence changes singly or in combination into each member of the triplet; in many cases, a self-splicing ribozyme was recovered only when the three bases were coordinately changed so as to restore the original hydrogen bonding network.

In a theoretical tour de force, F. Michel and E. Westhof have synthesized this limited information on ribozyme structure with other data garnered from the evolutionary record to produce a model of the overall structure of the *Tetrahymena* ribozyme. By observing that evolutionary relatives of the *Tetrahymena* ribozyme often accumulated similar multiple sequence changes, they were able to deduce that the covarying bases were in physical contact with each other (just as experimentally changing triplets of bases in concert provided evidence for base triples). In addition, they were guided by experiments that used a chemical probe that cut or modified sequences on the surface of the ribozyme while leaving the ''inside'' or core intact. These data allowed the linear information present in the RNA sequence to be roughly apportioned in space.

Ribozyme structures are also beginning to be determined by direct physical methods. Hammerhead ribozymes are very small RNAs sometimes found as part of plant viruses; one portion of the hammerhead sequence binds complementary RNAs, and another portion then cleaves the bound sequence. Nuclear magnetic resonance spectroscopy is being used to discern distances between bases within the structure; this information, in turn, should allow a three-dimensional representation of the ribozyme to be constructed.

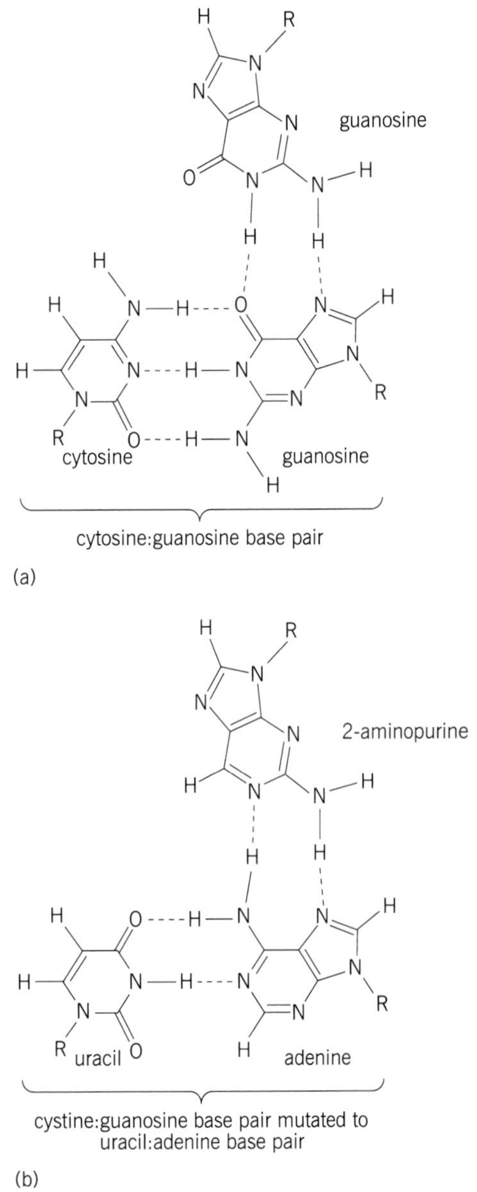

Kinetics. Individual steps in a ribozyme-catalyzed reaction can be followed by changing substrates or by altering reaction conditions. Analysis of such kinetic experiments helps to elucidate the steps that make up the ribozyme reaction. The kinetic profile of a modified *Tetrahymena* ribozyme that can perform only the initial cleavage reaction in self-splicing has recently been worked out. The reaction begins with a slow step in which a short helix containing the precise sequence destined for cleavage is formed (**Fig. 2**); guanosine then binds to the helix, which is quickly cleaved before it can unwind again; finally, the products of this cleavage are slowly released, and the ribozyme is freed (released) for a new reaction. This series of events, and their relative speeds, has important consequences for the function of the ribozyme in its cellular environment: since the ribozyme holds the helix very tightly, it can avoid releasing the cleaved pieces prior to the completion of the self-splicing reaction. Kinetic studies have also yielded the surpris-

cytosine:guanosine base pair

(a)

cystine:guanosine base pair mutated to uracil:adenine base pair

(b)

Fig. 1. Guanosine binding site of the *Tetrahymena* ribozyme. (*a*) The nucleoside guanosine is positioned by the *Tetrahymena* ribozyme via a base triple. The network of hydrogen bonds that holds guanosine in place is represented by broken lines. (*b*) When the cytosine:guanosine base pair is changed to a uracil:adenine base pair, the ribozyme can bind 2-aminopurine in place of guanosine. Some of the hydrogen bond donors (NH groups) and acceptors (N and O groups) switch places, but the overall pattern of interactions is maintained.

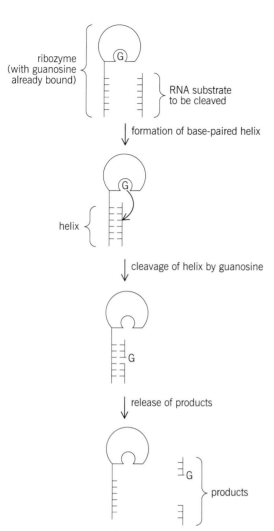

ribozyme (with guanosine already bound)

RNA substrate to be cleaved

formation of base-paired helix

helix

cleavage of helix by guanosine

G

release of products

G

products

Fig. 2. Kinetics of RNA cleavage by the *Tetrahymena* ribozyme. The *Tetrahymena* ribozyme can use a bound guanosine molecule to cleave RNA. In this diagram, the guanosine (G) is shown already bound to the ribozyme. The RNA substrate to be cleaved binds by base-pairing to a specific sequence on the ribozyme. This positions the RNA substrate near the bound guanosine. It should be noted that the RNA substrate could bind to the ribozyme before the guanosine binds to the ribozyme.

ing result that deoxyribonucleic acid (DNA) sequences can be cleaved by the *Tetrahymena* ribozyme, although at a slower rate than RNA sequences are cleaved.

Test-tube evolution. Since ribozymes can both function as catalysts and replicate as nucleic acids, schemas can be devised for the test-tube selection of ribozymes with new and unusual properties. For example, a set of sequence variants of the *Tetrahymena* ribozyme was selected to carry out a task that the ribozyme normally performs poorly: ligating deoxyribonucleic acid (DNA) to RNA. Those variants that succeeded in this ligating were preferentially amplified via reverse transcription and the polymerase chain reaction. New RNA was made from the amplified DNA, and the cycle was repeated; in each cycle, variants that ligated DNA were preferentially amplified, resulting in the isolation of a single type of RNA from the original diverse pool.

This technique has also been used to analyze the structure of the *Tetrahymena* ribozyme. A region of the ribozyme was randomly mutated, and functional variants were selected for. Only certain combinations of bases in the mutated region worked, and models of these combinations suggested a new structural interaction within the ribozyme. These selection methods should be generally useful in engineering ribozymes that can function at high temperatures or with different substrates.

As a first step in creating ribozymes that can catalyze an even wider range of reactions, RNAs have been selected that bind to a range of compounds. A population of random strings of RNA was passed over affinity columns containing organic dyes. Molecules that stuck to the columns were amplified, and the cycle was repeated. After several cycles, the mix of random information had been drastically winnowed, and only those few molecules (around 1 in 10^{10}) that could bind the dyes remained. Since the majority of protein enzymes catalyze reactions with the help of cofactors that are roughly the same size and complexity as these organic dyes, this process has important implications for the creation of ribozymes that can catalyze reactions by simply juxtaposing the substrate and cofactor.

Origin of life. A major problem in envisioning the origin of life has become known as the chicken-and-egg paradox: which came first, the information or the machinery necessary to reproduce the information? This paradox is handily resolved by the dual nature of ribozymes, which can be replicated and can perhaps act as machinery for their own replication. While no RNA replicase (that is, a ribozyme that can reproduce itself) yet exists, progress has been made toward engineering such a molecule. The *Tetrahymena* ribozyme has been modified so that it can ligate short RNA oligonucleotides that have annealed to a complementary template. This feat is similar to the tasks carried out by protein polymerases, which string together nucleoside triphosphates onto a complementary template. Moreover, the *Tetrahymena* ribozyme has been shown to be separable into three pieces; such a tripartite ribozyme is inherently easier to replicate than the intact molecule. Thus, if given the correct oligonucleotides and the correct template (RNAs encoding each of the three pieces), this ribozyme should soon be able to replicate itself in the test tube. Such an achievement could be heralded as the first example of artificially created life, and emphasizes the probable role of ribozymes in the genesis of modern organisms.

While the presence of ribozymes at life's inception is becoming more certain, the lineage of modern ribozymes is hotly debated. At one extreme, it is argued that modern ribozymes, such as the *Tetrahymena* ribozyme, are direct descendants of some of the earliest molecules. The discovery of ribozymes that are virtually identical in both cyanobacteria and chloroplasts has bolstered this view, since these ribozymes must date back at least to the origin of the plant kingdom. On the other hand, it has long been suspected that some ribozymes are transposable elements, able to leap into RNA (and perhaps DNA) molecules and to be transmitted between species. Such biological flexibility obscures their origins and suggests that they might have arisen recently as selfish genetic elements. The *Tetrahymena* ribozyme has recently been shown to be able to splice itself into, as well as out of, RNA. In addition, other ribozymes have been found to use cellular proteins to aid them in catalysis; this adaptation appears to be of recent origin, implying that the acquisition of the ribozymes themselves was a recent event.

Regardless of whether modern ribozymes are the direct descendants of early catalytic RNAs, clues to the existence of ancient ribozymes have been detected in modern metabolism. Ribosomal RNA, an essential component of protein translation machinery, has long been thought to have preceded proteins. This conjecture has recently been supported by finding that ribosomes stripped of most of their accessory proteins can still catalyze some simple reactions. The existence of other primordial ribozymes has been inferred from metabolic fossils, that is, modern biochemistry that appears to have been derived from an earlier metabolism based on RNA and ribozymes. For example, the use of transfer RNA in the synthesis of chlorophyll is not chemically necessary; alternate pathways that do not involve transfer RNA also exist. The simplest explanation for the use of transfer RNA in making this essential compound is that it is a historical accident, a remnant of a time when ribozymes (which would necessarily use RNA cofactors such as transfer RNA) were the principal catalysts.

For background information SEE EXON; GENETIC CODE; INTRON; LIFE, ORIGIN OF; MOLECULAR BIOLOGY; RIBONUCLEIC ACID (RNA); RIBOZYME; TRANSPOSONS in the McGraw-Hill Encyclopedia of Science & Technology.

Andrew Ellington

Bibliography. S. Benner, A. D. Ellington, and A. Tauer, Modern metabolism as a palimpsest of the RNA World, *Proc. Nat. Acad. Sci. USA*, 86:7054–7058, 1989; T. R. Cech, Self-splicing of group I introns, *Annu. Rev. Biochem.*, 59:543–568, 1990; D. Herschlag and T. R. Cech, Catalysis of RNA cleavage by the *Tetrahymena thermophila* ribozyme, 1. Kinetic description of the reaction of an RNA substrate complementary to the active site, *Biochemistry*, 29:10159–10180, 1990; F. Michel et al., The guanosine binding site of the *Tetrahymena* ribozyme, *Nature*, 342:391–395, 1989; D. L. Robertson and G. F. Joyce, Selection *in vitro* of an RNA enzyme that specifically cleaves single-stranded DNA, *Nature*, 344:467–468, 1990.

Rice

Wild rice (*Zizania palustris*) is an aquatic, annual, cross-pollinating grass species with a diploid chromosome number of 30. The plants are monoecious, with pistillate florets superior to staminate florets on a culm. Wild rice is wind-pollinated and protogynous;

the female florets emerge from the boot, and stigmas can be receptive to pollen while the male florets on the culm are still in the boot. Self-fertilization is possible under controlled pollination procedures. Wild rice seeds must be stored in water or in moist soil in order to remain viable for germination; dormancy is broken by the cold temperatures that occur over winter.

Wild rice has been gathered as a staple food crop by North American Indians for centuries and as a food and cash crop in the twentieth century. Some 50 years ago, significant attempts to cultivate the species in paddies were initiated, and domesticated production reached commercial importance in Minnesota in about 1970. Today wild rice is produced on approximately 20,000 acres (8100 hectares) of paddies in Minnesota and approximately 8000–10,000 acres (325–400 hectares) of paddies in California. California yields per unit of production are frequently two to three times larger than those in Minnesota.

Wild rice (**Fig. 1**) has grown for centuries in shallow lakes and slow-moving streams of north-central North America. The wild rice kernels disarticulate, or shatter, from the plant as each kernel matures. Indians gathered the rice using canoes to move through a rice bed; one person propelled the canoe with a push pole, and another person pulled stems over the canoe with a short stick and gently tapped the ripening panicles with another stick. The ripe kernels would fall to the bottom of the canoe, and the undamaged plants would be reharvested at 3- or 4-day intervals as kernels continued to mature. The gathered grain was parched in a heated container, the hulls were removed by treading, and the winnowed grain was stored for later use as food or for trade. In Minnesota, natural stands of wild rice are still harvested in this

Fig. 1. Wild rice (*Zizania palustris*).

manner, although modern processing plants are generally used to process the grain.

Domesticated wild rice production. Artificial paddies, leveled and diked to maintain water levels and to facilitate water removal prior to harvest, are used to grow wild rice commercially. Fertilizers, herbicides, and insecticides are used, as are improved varieties with nonshattering characteristics. Once a stand of rice is established in a paddy, the harvesting process results in a self-perpetuating stand. The seeds that are lost from the plants during harvest lie dormant until spring, when germination takes place in the flooded paddy. Large numbers of seeds can be lost to shattering, resulting in extremely dense stands in the succeeding growing season. Stands are thinned to a desirable number of plants (four per square foot of paddy) by using air boats with a set of trailing knives that cut some of the excess plants at the soil line under the water.

Prior to development of improved varieties, growers used a wide variety of mechanical devices (**Fig. 2**) to gather the maturing seed. Now modern rice combines (**Fig. 3**) are used to collect the ripened grain in a one-pass harvest. The grain is trucked to processing plants and spread on asphalt or concrete floors to cure. After curing, the rice is parched in gas-fired parchers until gelatinization of the starch occurs. The hulls are removed with mechanical hullers, and the grain is cleaned, sized, and packaged for sale. Most of the paddy-grown processed rice is sold to food processors for inclusion in mixes; the remainder is packaged as fancy-grade rice.

Paddy production of nonshattering wild rice in Minnesota is complicated by the cropping system, which works against the domestication process of the plant breeder and results in selection for increased shattering. The seeds that mature early and tend to shatter fall from the plant; nonshattering seeds are collected by the combine and processed for food. Thus, the new crop is based on seeds that tend to shatter easily. Methods of preventing volunteer reseeding are not practical; the primary means of eliminating an established variety from a paddy involve alternate cropping or removing the paddy from production for 2 or more years. However, the rice kernel can remain viable for up to 4 years.

Fig. 3. Commercial wild rice is harvested from nonshattering varieties by using modern combines on a track propulsion system to move over the soft paddy floor.

California has quickly developed a successful wild rice industry. In California, the cropping environment enhances the domestication process, with a longer, more favorable growing season, and with losses to diseases, shattering, and insect damage minimal relative to those in Minnesota. Because wild rice that shatters into the paddy in California does not usually germinate the following season, growers can reseed paddies every year and thereby have consistent stands for optimum production. The lack of germination of the seeds that were lost to shattering may be attributed to inadequate temperature conditions for breaking seed dormancy, or to desiccation of exposed seeds following harvest.

Development of nonshattering varieties. Initial attempts to grow rice in a paddy used seed harvested from natural stands, but discovery of a few plants that retained their ripe seed led to nonshattering wild rice varieties. Nonshattering types retain their seed after maturity of the kernels, while shattering types lose the kernels at onset of maturity. The degree of seed retention is a quantitatively inherited trait and can be improved through plant breeding. Resistance to shattering is controlled by two complementary genes in which shattering is dominant.

A major plant breeding program on wild rice was initiated at the University of Minnesota in 1972. Prior breeding efforts had resulted in development of the four open-pollinated, nonshattering varieties that became available to growers in 1969–1974. Three new varieties released by the Minnesota Agriculture Experiment Station in the period 1978–1985 have a reduced plant height and an earlier date of maturity than the older varieties. A private breeding program in California has developed varieties adapted to California conditions.

Crop loss. Harvest yield can be affected by a number of factors, including contamination by weed species, disease (leaf blight due to the fungus *Bipolaris* is the most common), insect infestation (the wild rice worm, which is the larval stage of *Apamea apamiformis*, can devastate a crop), and wildlife (crayfish and ducks damage stands at seedling stage, and flocks of foraging birds cause large losses of grain at harvest time). Environmental hazards can also

Fig. 2. Mechanical harvesters such as this collected rice kernels without destroying the plants.

cause serious loss of crop. An early frost can kill the plants or cause premature seed shattering, and huge losses of the mature seeds can result if high winds occur just prior to harvest.

Chemicals and wild rice production. The use of chemicals to control weeds, diseases, and insects is a cause for public concern, particularly in an aquatic environment. The chemicals can be retained in the paddy water that will eventually reach natural runoff systems and groundwater. Despite environmental studies that showed little, if any, effect on the environment, action by the Environmental Protection Agency and others has eliminated many of the pesticides previously used to produce paddy-grown wild rice. For example, the herbicides 4-chloro-2-methyphenoxy acetic acid (MCPA) and 2,4-dichlorophenoxy acetic acid (2,4-D) and the fungicide dithane M-45 are no longer approved for use on wild rice. The insecticide malathione is approved for control of rice worm, and a systemic fungicide known as tilt has temporary approval for leaf blight control.

For background information *SEE AGRICULTURAL SCIENCE (PLANT); AGROECOSYSTEM; BREEDING (PLANT); RICE; SEED GERMINATION* in the McGraw-Hill Encyclopedia of Science & Technology.

R. E. Stucker

Bibliography. W. R. Fehr and H. H. Hadley (eds.), *Hybridization of Crop Plants*, 1980; P. M. Hayes, R. E. Stucker, and G. G. Wandrey, The domestication of American wild rice (*Zizania palustris*, Poaceae), *Econ. Bot.*, 43:203–214, 1989; R. Sherman (ed.), *Wild Rice Production in Minnesota*, Agr. Extens. Serv., Minn. Agr. Extens. Bull. 464, 1982; T. Vennum, Jr., *Wild Rice and the Ojibway People*, 1988.

Root

Roots are the underground parts of plants. They have several important functions, including storing food (as in carrots and beets), absorbing water and minerals, and anchoring the plant. Roots accomplish these functions by having large surface areas, long lengths, and many branches. For example, a 4-month-old rye (*Secale cereale*) plant has 13 million roots with a combined length of more than 6700 mi (11,000 km).

Like shoots, roots respond to a variety of environmental stimuli. The response usually involves growth. A root grows from its tip and forces its way through the soil with the help of the root cap, a thimble-shaped organ that covers the tip of the root.

Cells in the columella (center of the root cap) contain 15–30 amyloplasts (starch-storing organelles). These organelles are the densest component of cells and therefore sediment in response to gravity. The sedimentation of these amyloplasts was once suspected to be the mechanism by which plants perceive gravity. However, roots of a mutant of *Arabidopsis* that lacks starch and amyloplasts are still strongly responsive to gravity, indicating that sedimentable amyloplasts are not necessary for root response to

gravity. However, starch may be necessary for full sensitivity to gravity by these roots. No differential growth of the upper and lower sides of the cap occurs before or during root curvature. This suggests that the regions for gravitropic sensing and curvature are spatially separate during all phases of gravitropism. *SEE PLANT MOVEMENTS.*

Root caps are critical to the various responses of roots to environmental signals such as gravity, moisture, and light. For example, roots grow downward in response to gravity, toward moisture, and away from light. Roots whose caps have been removed do not respond to these environmental signals, indicating that the root cap is an important environmental sensor for roots. Perception of these environmental signals involves the generation of electrical asymmetries in the root cap. However, although the perception of signals such as gravity occurs in the root cap, the actual response to the signal is differential growth that occurs in the elongating zone of the root, which is 0.08–0.24 in. (2–6 mm) behind the root cap. Thus, a growth-controlling signal must move from the site of perception (the root cap) to the site of the response (the elongating zone) in the root. For many years it was believed that responses to environmental signals such as gravity were controlled by abscisic acid, a plant growth regulator. However, the use of mutants again provided definitive experiments that demonstrated that abscisic acid is not necessary for a root's response to gravity. Mutant roots of corn (*Zea mays*) that produce no abscisic acid respond strongly to gravity. Similarly, roots treated with fluridone, an inhibitor of abscisic acid synthesis, are strongly responsive to gravity. These results indicate that the factor controlling root gravitropism is not abscisic acid. Moreover, abscisic acid does not cause horizontally oriented roots to become responsive to gravity.

Control of gravitropism by calcium and auxin. Most evidence now indicates that root gravitropism is controlled by calcium ions (Ca^{2+}) and auxin, another plant growth regulator. It includes the following observations: (1) Asymmetries of calcium across the root tip induce gravitropiclike curvature. (2) Chelating calcium with substances such as ethylenediamine-tetraacetic acid (EDTA) abolishes gravitropic curvature. (3) Horizontally oriented roots move calcium down across their tips. (4) Auxin-transport inhibitors such as triiodobenzoic acid (TIBA) abolish gravitropic curvature and movement of calcium across root tips. (5) Auxin moves down across the caps of horizontally oriented roots. (6) The asymmetric movement of auxin from the apex toward the base in intact roots does not occur in decapped roots. (7) In horizontally oriented roots, auxin-regulated ribonucleic acids (RNAs) become distributed asymmetrically within 20 min and the greatest asymmetry coincides with the onset of rapid bending. Taken together, these results clearly correlate gravitropic curvature and the expression of genes controlled by auxin, and are consistent with the suggestions that the root cap is not only the site of perception but also the location of the initial redistribution of calcium and auxin that ultimately leads to curvature. *SEE AUXIN; PLANT HORMONES.*

To understand how roots integrate all of these events to produce a response, consider a horizontally oriented primary root of corn (*Z. mays*). Perception of gravity in the root cap generates electrical asymmetries in the cap. These electrical asymmetries move calcium ions and auxin to the lower side of the root cap and subsequently to the elongating zone (in greater concentrations along the lower side of the root). There, they inhibit cellular elongation, thus slowing growth along the lower side of the root. As a result of this slower growth, the root curves downward.

Other responses such as hydrotropism (growth toward moisture) and phototropism (growth away from light) by roots occur via similar mechanisms involving differential growth in the elongating zone.

Orienting roots horizontally changes the electrical potential in cortical cells in the elongating zone of the root. In vertically oriented roots, the intracellular potentials of the outer cortical cells located 0.08 in. (2 mm) behind the root tip are approximately -115 mV. When roots are oriented horizontally, cortical cells on the upper side of the root hyperpolarize to -154 mV within 0.5 min, while cortical cells depolarize to about -62 mV. Such an electrical asymmetry does not occur in the mature region of the root located behind the elongating zone. The short lag-period (less than 0.5 min) suggests that these electrical changes may result from gravity perception in the elongating zone rather than from transmission of a signal from the root cap. Gravity-induced electrical changes in the elongating zone of roots may regulate growth by affecting hormonal transport, sensitivity, or distribution.

Movement of signal from root cap to elongating zone. The growth-controlling signal that moves from the root cap to the elongating zone moves at least partly outside the cells. Much evidence indicates that mucigel, a mucilaginous substance, is an important part of this pathway. Indeed, roots whose mucigel has been removed are less responsive to many environmental stimuli than are untreated roots. Also, mutants whose roots do not produce mucigel are nonresponsive to signals such as gravity. In caps of these roots, peripheral cells produce and package the mucilage in dictyosome-derived vesicles. However, these vesicles do not fuse with the plasmalemma, and therefore the mucilage is not secreted onto the surface of the root. However, adding artificial mucilage to the tips of these roots restores responsiveness to gravity, indicating that growth effectors moving from the root cap to the root move through mucigel.

Nonresponsive roots. Not all roots are equally responsive to environmental signals. For example, primary roots are the first roots produced by seedlings and are usually responsive to environmental signals such as gravity. Conversely, secondary (or branch) roots originate from primary roots, are smaller than primary roots, and are usually much less responsive to environmental stimuli than are primary roots. Caps of secondary roots produce mucigel. Furthermore, caps of these roots contain and are responsive to artificially

induced gradients of growth effectors such as calcium ions and auxin. The reduced responsiveness of these roots to environmental stimuli apparently results from their inability to produce a gradient of growth effectors. In the absence of such a gradient, growth along both sides of the root remains equal and the roots grow straight, showing no response to the stimulus. This explains why secondary roots of many plants grow in whatever direction they originate. Secondary roots that respond slowly to stimuli produce a weak gradient of effectors, thus accounting for their slow response. The size of the gradient of growth effectors depends largely on the size of the root cap—the larger the cap, the larger the gradient and the greater the response. Secondary roots of castor bean (*Ricinus communis*) less than 0.8 in. (2 cm) long are nonresponsive to gravity. When these roots become 1–1.6 in. (3–4 cm) long, their caps enlarge and the roots become responsive to gravity.

For background information *SEE AUXIN; BIOPOTENTIALS AND IONIC CURRENTS; PLANT HORMONES; PLANT MOVEMENTS; ROOT* in the McGraw-Hill Encyclopedia of Science & Technology.

Randy Moore

Bibliography. R. Brouwer et al., *Structure and Function of Plant Roots*, 1981; M. L. Evans, R. Moore, and K. Hasenstein, How roots respond to gravity, *Sci. Amer.*, 255:112–119, 1986; L. Feldman, Regulation of root development, *Annu. Rev. Plant Physiol.*, 35:223–242, 1984; P. J. Gregory, J. V. Lake, and D. A. Rose (eds.), *Root Development and Function*, 1987; W. Haupt and M. Feinlieb (eds.), *The Physiology of Movements*, 1979.

Satellite navigation systems

Two similar satellite navigation systems are now under development, the Global Positioning System (GPS) in the United States and GLONASS (Global Orbiting Navigation Satellite System) in the Soviet Union. When fully implemented, both systems will provide accurate three-dimensional position and velocity information and precise time to users anywhere in the world.

GLOBAL POSITIONING SYSTEM

The Global Positioning System has been under development since the early 1970s, and continuous navigation coverage is expected by the end of 1991. This availability plus the rapidly falling price of navigation equipment will trigger enormous growth in the number of users, and in a few years the Global Positioning System should become the world's most widely used electronic navigation system. Except for GLONASS, which is several years behind the Global Positioning System, no other system provides the same combination of accuracy, three-dimensional positioning, worldwide coverage, and all-weather performance. Not only is the Global Positioning System already being used for navigation on land, at sea, and in the air, but it is revolutionizing the art of surveying and may even be used to control farm tractors and earthmoving equipment and to aid in aircraft landings.

The past few years have seen dramatic changes in the Global Positioning System. The crash of *Challenger* affected its deployment, a new rocket was developed to launch Global Positioning System satellites (called *NAVSTAR*), the constellation plan has changed several times, operational deployment has begun, the price of user equipment has started to fall rapidly, civilian accuracy has been intentionally degraded by the Department of Defense, and correction signals have been developed to greatly improve the accuracy.

Orbit configuration. The original concept for the Global Positioning System was to have a 24-satellite constellation, with 8 satellites in each of three orbit planes. However, because of intense budget pressure, the number of planned satellites was later reduced to 18. In trying to optimize worldwide coverage with far fewer satellites, the configuration also was changed to 3 satellites in each of six orbit planes. Even with these changes, the marginal level of coverage with only 18 satellites, especially considering occasional satellite failures or down time, began to build support for better coverage. As a result, the plan is now firmly back to 24 satellites, but this time with 4 satellites in each of six orbit planes (**Fig. 1**). The objective is always to have at least 21 satellites always operational.

NAVSTAR satellites have a design life of 5–7 years. Some will last much longer, but others could fail early or be lost during an unsuccessful launch. Therefore, 3–5 replacement satellites will be needed each year just to maintain a full constellation of 24

satellites. Fortunately, enough satellites are now on order from two different companies to assure full service at least through the year 2000.

Satellite deployment. During the concept validation phase of the Global Positioning System development program, 10 prototype or block I *NAVSTAR* satellites were launched successfully by Atlas rockets. However, it was originally intended to launch all of the production or block II *NAVSTAR* satellites with the space shuttle, so the loss of *Challenger* and its crew on January 28, 1986, caused more than 2 years of delay in deploying an operational system. To compensate, the U.S. Air Force commissioned a redesign of the dependable Delta rocket. The result is the Delta II, which by the end of 1990 had successfully launched 10 block II satellites out of 10 attempts (**Fig. 2**). On average, about 5 *NAVSTAR* satellites are being launched each year to complete the constellation.

By the end of 1991, there should be 18 or more operational satellites, enabling any user with a clear view of the sky to receive signals from at least 3 satellites at any time anywhere on Earth. With an independent measure of altitude plus three satellite signals, a user can determine two-dimensional position. (Since time must be computed as part of the navigation solution, three satellite signals permit the user to determine the three parameters of time, latitude, and longitude.) Thus, at this level of deployment, continuous, worldwide navigation is available for all marine users and for aircraft or land vehicles with altimeter aiding.

With at least 21 operational satellites expected by the end of 1992, users will be able to receive four signals at all times. This will permit the determination of time, latitude, longitude, and altitude, or full three-dimensional navigation. The Global Positioning System is expected to be declared fully operational in 1993, with a complete constellation of 24 satellites.

Accuracy. There are four basic levels of navigation accuracy in the Global Positioning System. The first is called the precise positioning service (PPS), which is the best accuracy available without external aiding. The errors in the precise positioning service are due to satellite clock error, satellite orbit error, uncompensated ionospheric and tropospheric propagation effects, multipath signal reception, and deficiencies in the receiving equipment. These errors are magnified by dilution of precision (DOP) because of the geometry of the satellite positions relative to the user. Dilution of precision is defined as the ratio of probable navigation error to the expected error in each satellite signal measurement. With all effects included, the precise positioning service should yield a navigation error of 16 m (52 ft) spherical error probable (SEP), which is the radius of a sphere containing 50% of the three-dimensional navigation solutions, averaged over the Earth for 24 h. Another value for the precise positioning system is 28 m (92 ft) 2 drms, which is the radius of the circle containing 95% of the two-dimensional navigation solutions.

A current, quite controversial policy of the U.S.

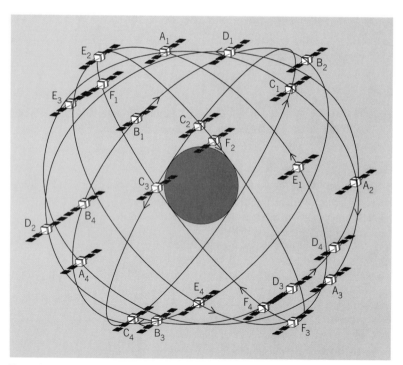

Fig. 1. Configuration of the 24 *NAVSTAR* satellites of the Global Positioning System, approximately 20,200 km (12,550 mi) above the Earth. A–F designate orbit planes, and 1–4 designate positions of satellites in an orbit.

Department of Defense is to deny the full accuracy of the Global Positioning System to any user without special authorization, which means most civilian users. The method of denying accuracy is to dither the clock and offset the orbit description in each satellite signal. The process is called selective availability (S/A), and only authorized users will have the cryptographic hardware, software, and keys to recover full precise positioning service accuracy. All others are limited to 100 m (328 ft) 2 drms. It is clear that the civil community will increasingly demand better accuracy. The originally announced selective availability error was to have been 500 m (1640 ft) 2 drms, but civilian concerns were effective in changing the policy to 100 m (328 ft). Further policy reviews can be expected.

The third level of accuracy is achieved by differential GPS. This is navigation relative to a reference station placed at a known location. The reference station uses the satellite signals to calibrate its own clock, and then it computes the range and range-rate errors of each satellite signal based on the clock solution and its known location. These errors are communicated to local users, in order to correct their own measurements before a navigation solution is computed. The result is a dramatic improvement in accuracy, generally to 5 m (16 ft) 2 drms or better. Because the satellites are so high above the Earth, these corrections are useful over a very large distance, such as 1000 km (621 mi), which is about the distance from Chicago to Washington, D.C. Also, the required communication data rate of 50–100 bits/s is quite modest.

Differential GPS already is being used extensively. For example, the North Sea and the Gulf of Mexico are blanketed by commercial correction signals that support the offshore oil exploration industry. Several of these correction signals are provided by communication satellites, and more are planned. The U.S. Coast Guard and many European countries are planning to transmit correction signals from existing marine radio-beacon transmitters. It seems likely that the majority of Global Positioning System users eventually will take advantage of differential corrections.

The fourth level of accuracy is also achieved with differential corrections, but the precision is greatly improved by using the phase of the carrier signals as a vernier. The problem with this technique is that carrier cycles are ambiguous; one is just like another. Special techniques must be used to resolve the ambiguities. The great advantage is that, once resolved, measurements of the 19-cm (7.5-in.) carrier wavelength can be made with millimeter precision.

Thousands of surveys already have been performed with Global Positioning System equipment using these techniques, which permit two or more widely separated points to be located relative to each other with an accuracy of a few millimeters. Especially as equipment cost rapidly drops, it is reasonable to assume that the Global Positioning System will thoroughly revolutionize the art of surveying. The technique is also being extended to moving vehicles,

Fig. 2. Delta II launch of a *NAVSTAR* satellite. (*McDonnell Douglas*)

which will permit such applications as precise aircraft landing aids and the control of earthmoving equipment.

GPS applications. The Global Positioning System soon will become the most widely used electronic navigation and positioning system in the world. On land, at sea, and in the air, it will be used for military, professional, commercial, and consumer applications. As the price of equipment drops, it will be used by hikers and bicyclists, in private automobiles with a digital map display, and in all forms of commercial vehicles. At sea, it will be used on all types of craft, from pleasure boats to supertankers. All large aircraft and a large percentage of small aircraft eventually will be equipped with Global Positioning System navigation, and it is possible that differential GPS will become the basis for precise landing aids. Surveying is being revolutionized, and there will be numerous engineering applications.

Integrity issues. The concern about the Global Positioning System that is voiced most often by the aviation community involves integrity. The concern is that a satellite might experience a failure that causes navigation error to grow at a critical time without this error being noticed by the navigator. A system with integrity may not have the correct answer, but it must warn the user if there is a problem.

There are two basic ways to solve the concern about the integrity of the Global Positioning System. The first is to monitor the satellite signals from a known point or points and quickly tell the user if a problem is detected. A differential GPS system inherently addresses the integrity problem while also improving navigation accuracy. The INMARSAT organization is proposing to include Global Positioning System and GLONASS integrity transponders on its future communication satellites. Thus, a user could combine signals of the Global Positioning System with the INMARSAT integrity channel and obtain rapid notification from the ground of the problem with the Global Positioning System, as well as the advantage of another satellite signal for navigation measurements. *See* COMMUNICATIONS SATELLITE.

The second way to improve integrity is to have an excess of satellite signals. Four signals are necessary for three-dimensional navigation, five signals permit the user to detect a problem, and six or more signals allow the user to isolate the bad signal and ignore it. More satellites would help. There is increasing interest in combining signals from the Global Positioning System with those of the GLONASS system (as discussed below). This may solve the integrity issue, improve overall performance by having twice as many satellites available, and further unite the world by cooperative use of signals from the two superpowers.

Thomas A. Stansell

GLONASS

GLONASS is a satellite-based navigation system currently being implemented by the Soviet Union for use in aviation, marine, and terrestrial applications. The system is designed to provide three-dimensional position and velocity data and precise time to equipped users. GLONASS has been in its development phase for many years, with the first GLONASS satellites launched in October 1982. The satellite deployment schedule calls for approximately 10–12 satellites in orbit by 1991 and a constellation of 24 operational satellites by 1995 (**Fig. 3**). In many respects, GLONASS resembles the Global Positioning System. When GLONASS is fully deployed in about 1995, its 24 operational satellites are to be in three uniformly spaced orbital planes at an altitude of 19,100 km (11,900 mi) above the surface of the Earth. This constellation of satellites will provide passive navigation capabilities (in that the users need not transmit) to users on and above the Earth's surface worldwide. The signals are continuously transmitted from the satellites and provide accurate, real-time, all-weather navigation and position-determination services.

Satellite launches and configuration. As of November 1990, there had been 46 spacecraft launched by the Soviet Union in the GLONASS program (see **table**), of which 8 were operational. The launches have been on a somewhat regular basis since their initiation in 1982, except for a 1-year period from May 31, 1989, to May 18, 1990.

Eight GLONASS satellites are planned in each of the three orbit planes, which are inclined 64.8° to the Equator, with an orbital period of 11 h 15 min. A four-stage Proton SL-12 launch vehicle boosts the spacecraft into orbit from the Baikonur cosmodrome at Tyuratam, western Russia. All satellites so far have been placed in only two of the three orbit planes.

Signal characteristics. The GLONASS satellites have Earth-coverage antennas that transmit circularly polarized signals. The two-frequency transmissions provide for the correction of ionospheric propagation delay in those GLONASS receivers that receive both frequencies. The principal frequency (F_1) is centered at 1607 MHz, and the second (F_2) at 1250 MHz (both in the L band); both have allocations extending 10 MHz to either side of the band centers. The GLONASS signals used for civil applications are transmitted from the satellites in the frequency range 1602–1616 MHz. Each satellite transmits on a different frequency channel separated by 562.5 kHz. The carrier frequency for the signal transmitted by the n-th satellite channel is given by the equation below, where

$$f_n = f_0 + n\Delta F_1$$

$f_0 = 1602.0$ MHz, $\Delta F_1 = 562.5$ kHz, and $n = 1, 2, \ldots, 24$. The second GLONASS frequency band (F_2), at 1240–1260 MHz, has its carrier frequencies separated by 437.5 kHz. The F_2 carrier frequencies (f'_n) can be determined by modifying the above equation with $f'_0 = 1245.0$ MHz and $\Delta F_2 = 437.5$ kHz.

The phase of each carrier frequency is modulated by using a binary digital code, which provides accurate ranging information to the user. The digital code sequence may appear to be random, or noiselike, but in fact its characteristics are known in detail. These codes are termed pseudorandom noise (PRN), or pseudonoise (PN), codes. They are normally generated by using feedback shift-register devices contained within the electronics of the GLONASS satellites and in the user equipment.

Fig. 3. GLONASS satellite, showing solar power panels and antenna system.

GLONASS spacecraft launch data and status*

International designation	Launch date	Cosmos launch number	GLONASS number	Frequency channel	Orbit plane
1982–100 A, D, E	Oct. 12, 1982	1413, 1414, 1415	1, 2, 3		1
1983–84 A, B, C	Aug. 10, 1983	1490, 1491, 1492	4, 5, 6		1
1983–127 A, B, C	Dec. 29, 1983	1519, 1520, 1521	7, 8, 9		3
1984–47 A, B, C	May 19, 1984	1554, 1555, 1556	10, 11, 12		3
1984–95 A, B, C	Sept. 4, 1984	1593, 1594, 1595	13, 14, 15		1
1985–37 A, B, C	May 17, 1985	1650, 1651, 1652	16, 17, 18		1
1985–111 A, B, C	Dec. 24, 1985	1710, 1711, 1712	19, 20, 21		3
1986–71 A, B, C	Sept. 16, 1986	1778, 1779, 1780	22, 23, 24		1
1987–36 A, B, C	Apr. 24, 1987	1438, 1439, 1440	25, 26, 27		3[†]
1987–79 A, B, C	Sept. 16, 1987	1483, 1484, 1485	28, 29, 30		3
1988–9 A, B, C	Feb. 19, 1988	1917, 1918, 1919	31, 32, 33		1[†]
1988–43 A, B, C	May 21, 1988	1946, 1947, 1948	34, 35, 36	12,[‡] 23, 24[‡]	1
1988–85 A, B, C	Sept. 16, 1988	1970, 1971, 1972	37, 38, 39	18, 7, 10[‡]	3
1989–1 A, B, C	Jan. 10, 1989	1987, 1988, 1989[§]	40, 41	9,[‡] 6,[‡] —	1
1989–39 A, B, C	May 31, 1989	2022, 2023, 2024[§]	42, 43	16, 17, —	3
1990–45 A, B, C	May 19, 1990	2079, 2080, 2081	44, 45, 46	15,[‡] 3,[‡] 21[‡]	3

*In part, after P. Daly, University of Leeds, United Kingdom.
[†]Launch failed to achieve final orbit. Probable cause was failure of Proton launch vehicle fourth stage.
[‡]Active channel as of Aug. 8, 1990.
[§]Cosmos 1987, 1988, 2022, and 2023 are GLONASS. Cosmos 1989 and 2024 are Soviet Etalon geodetic satellites.

GLONASS employs two pseudonoise digital codes, both of which are transmitted from all spacecraft. The pseudonoise sequence used for civil applications is generated by a nine-stage shift-register configuration resulting in a code sequence length of $2^9 - 1$, or 511 bits. This sequence is timed to repeat every millisecond, so the bits, or chips, occur at a rate of 511 kHz. This frequent repetition of the civil code allows rapid acquisition of the sequence by the GLONASS receivers. A second pseudonoise sequence transmitted by GLONASS is generated at a 10 times higher rate (5.11 MHz), and this code has a substantially longer period (several seconds). The low-rate pseudonoise code is used for civil applications and modulates the F_1 carrier frequency, while the high-rate code modulates both carrier frequencies.

The satellites also continuously transmit a 50-bits/s low-rate data channel, which includes satellite position and velocity predictions, satellite clock corrections, and related information. This information is superimposed on all signals in both frequency bands.

Ground control and monitoring. The synchronization of the GLONASS satellites and the formulation of the data messages transmitted by the satellites are accomplished by the ground control and monitoring stations. The estimated future positions for the satellites are periodically uploaded to the satellites from a ground control station. Geographically separated monitor stations receive the GLONASS signals, evaluate their characteristics, and use these data to correct system timing, formulate data messages, and correct for anomalies or disturbances. The ground control and monitor network is located entirely within the Soviet Union.

Receiver operation. The GLONASS user equipment measures the ranges and range rates (Doppler components) between the user antenna and the satellites in view. These measurements may be simultaneously obtained from several satellites, or sequentially obtained, one satellite at a time, depending upon the number of processing channels available in the equipment and other design factors. The receiver determines the range to a satellite by measuring the time difference between the coded signal received from the satellite and an internally generated identically coded signal. Accurate ranges to four or more satellites within view can be obtained by the GLONASS receiver and are used to determine the user position.

The ranges to the satellites are not initially known precisely because the synchronization between the GLONASS receiver and the satellites is not accurately established. The satellites have accurate and stable atomic standard oscillators, or clocks, which maintain excellent frequency stability, providing accurate GLONASS system time. The receivers normally use crystal oscillators that typically have timing errors and are therefore offset from system time. This receiver timing error, or clock bias, needs to be determined if the receiver is to accurately measure the signal propagation time interval (and therefore the range) between transmission by the satellite and reception by the receiver. Fortunately, the receiver clock bias can be determined from the measured ranges to 4 satellites. These range measurements are termed pseudoranges since they are initially incorrect because of the receiver's clock-bias error. Measurement of four (or more) ranges determines the four unknowns of latitude, longitude, altitude, and system time. Velocity is normally determined by measurement of the Doppler shift of the received carrier signals.

To accomplish the navigation solution, the precise positions in space of the satellites are required for the time of their measurement by the receiver. Since the satellites are orbiting the Earth, their positions in space can be expressed by their orbital elements with appropriate corrections, or for short periods of time by simplified trajectory information. The GLONASS system uses simplified cartesian coordinate trajectory information transmitted in the Greenwich geocentric coordinate system. Each satellite transmits position

Fig. 4. GLONASS Shkiper marine receiver.

and velocity information on its data channel during a 30-s interval and data on other satellite system conditions during the next 2.5 min. By using the satellite locations and the measured ranges, a position solution can be obtained by the computer in the receiver.

GLONASS operates with dynamic vehicles as well as with stationary or slowly moving users. Position accuracies are specified as 100 m (330 ft) in latitude and longitude and 150 m (500 ft) in altitude, although test results are significantly better. Velocity accuracy is specified at 15 cm/s (0.5 ft/s) in all three dimensions, and timing accuracy is to within 1 microsecond. These accuracies are specified as 95% confidence-level values.

User equipment. The Soviet Union has developed a unit of marine user equipment designated the Shkiper (**Fig. 4**). This unit determines the position and velocity of a ship as well as accomplishing a number of additional tasks. The Shkiper is automated in its operation, provides for the digital processing of signals by the use of microprocessors, and weighs 21.5 kg (47.3 lb). The data are either displayed on a screen or provided as a digital printed tabulation. Other equipment for GLONASS is under development in the Soviet Union, including a single-channel avionics set and a six-channel civil aviation receiver.

GPS-GLONASS hybrid receiver development. As mentioned above, there has been strong recent interest in combining the capabilities of the Global Positioning System with those of GLONASS. The objective is to develop common user equipment standards for the civil aviation use of the two systems. A joint program involving the U.S. Federal Aviation Administration and the civil aviation authorities of the Soviet Union has been initiated.

Both the Global Positioning System and GLO-NASS are spread spectrum systems with pseudonoise codes. The Global Positioning System separates its civil signals, which are transmitted from all satellites at a single frequency (1575.42 MHz), by the use of different 1023-bit pseudonoise coded transmissions from each satellite. GLONASS uses a single 511-bit

pseudonoise code with all satellites, but each satellite transmits at a different frequency. There are other differences between the systems; for example, the orbital period of the Global Positioning System is about 0.5 sidereal day (11 h 57.9 min), while the GLONASS period is 11 h 15 min. For a satellite in the Global Positioning System, the observation time at a given location advances about 2 min each orbit, and the ground track repeats every other orbit. GLONASS completes 17 orbits in 8 days and then repeats.

There are two principal areas of concern in combining the two system characteristics into a single receiver: the coordinate systems (WGS-84 with GPS and SGS-85 with GLONASS) and the time references. These and other matters relating to the feasibility of common standards are under discussion by the Radio Technical Commission for Aeronautics (RTCA) Special Committee 159, Eurocae Working Group 28, the Airlines Electronics Engineering Committee (AEEC) Satellite Subcommittee, and the two governments.

GLONASS applications. The GLONASS system, similar to the Global Positioning System, has a large number of applications, some of which are currently realized. GLONASS will clearly have many uses as a navigation system in determining position, course, and speed for aircraft and for marine and land vehicles. GLONASS will have strong applications in geodetic and cadastral surveying. The system is expected to provide guidance capabilities for aircraft, ships, and other vehicles in the future. Navigation satellite systems such as GLONASS have applications in the control of space vehicles as well as in the determination of their position, velocity, and acceleration during launch and reentry.

For background information SEE SATELLITE NAVIGATION SYSTEMS; SPREAD SPECTRUM COMMUNICATION in the McGraw-Hill Encyclopedia of Science & Technology.

Keith D. McDonald

Bibliography. P. Daly, Aspects of the Soviet Union's GLONASS satellite navigation system, *Navigation, J. Inst. Navig.*, 41(2):186–197, 1988; FANS/4 WP/70, ICAO, Montreal, May 6, 1988; B. Gallagher (ed.), *Never Beyond Reach*, 1989; *Global Satellite Navigation System GLONASS Interface Control Document*, Institute of Space Device Engineering and Glavkosmos USSR, 1990; *Proc. IEEE PLANS-88*, Orlando, Florida, November 28–December 2, 1988; T. A. Stansell, Jr., Civil GPS from a future perspective, *Proc. IEEE*, vol. 71, no. 10, October 1983; T. A. Stansell, Jr., *The Transit Navigation Satellite System: Status, Theory, Performance, and Applications*, Magnavox R-5933, 1978.

Sex determination

Some species of flowering plants have separated male and female functions into different flowers, which increases the rate of outcrossing. Most unisexual flowers arise by the suppression of stamens or pistils during floral development. The plant hormones auxin, cytokinin, and gibberellic acid have been implicated in this process.

The angiosperm flower consists of whorls of organs that are specialized for sexual reproduction: the sepals, petals, stamens, and pistils. Stamens produce the male gametophyte, pollen, which contains the male gametes, or sperm. Pistils produce the female gametophyte, or embryo sac, which includes the egg cell, the female gamete. The majority of angiosperm species have flowers with both stamens and pistils and thus are hermaphroditic. In approximately 8% of angiosperm species, the flowers have only functional stamens or pistils at maturity and thus are unisexual. Unisexual flowers develop from immature flowers that are perfect, that is, flowers in which both stamen and pistil primordia are present; at some time in development of these flowers, however, either the stamen or pistil primordia are arrested and eventually degenerate. A possible exception is hemp (*Cannabis sativa*), in which there is no morphological evidence of the development of the inappropriate organ. Although most of the tissues of the flower are sporophytic and only gametophytic tissue has a sex, flowers that have only pistils are considered female and flowers that have only stamens are considered male.

There are two main groups of plants with unisexual flowers; those species that have staminate and pistillate flowers on separate plants, such as the maidenhair tree (*Ginkgo biloba*), hemp, asparagus (*Asparagus officianalis*), and campion (*Melandrium alba*), are dioecious, and those species that have staminate and pistillate flowers on a single plant, such as maize (*Zea mays*), begonia (*Begonia* sp.), and cucumber (*Cucumis* sp.), are monoecious. In many dioecious species, there are distinguishable sex chromosomes. Sex determination usually is controlled in these plants by the presence of a Y chromosome; a flower is male if its cells have a Y chromosome, regardless of the number of X chromosomes, and a flower is female if its cells have only X chromosomes. In monoecious species the genetic complement of the cells in the staminate and pistillate flowers is identical. Thus, the different pattern of development seen in the two types of flowers in monoecious species must be controlled by differential gene expression in response to developmental and hormonal cues in different parts of the plant. Some monoecious plants may exhibit temporal dioecism. An example is jack-in-the-pulpit, a perennial herb that has male flowers in its first seasons when it is small, but develops female flowers once it has attained a certain size.

The desire to understand the genetic and molecular mechanisms that regulate development of unisexual flowers, coupled with the agricultural need for unisexual flowers in hybrid seed production, has stimulated research in the genetic, physiological, and molecular control of sex determination. There are other types of floral polymorphisms, such as andromonoecism, in which a single plant has both staminate and perfect flowers, and gynodioecism, in which a species has

individuals with either pistillate or perfect flowers. However, most of the current research focuses on species that are simply monoecious or dioecious.

Dioecious species. Sex is determined in asparagus by sex chromosomes. Plants in this species that are XY normally develop only staminate flowers, while XX individuals produce only pistillate flowers. In some growth conditions, however, XY individuals are andromoecious, producing male and perfect flowers. Genetic studies have shown that the Y chromosome controls male development with the functioning of two dominant genes, one that is a suppressor of pistil development and the other that is an activator of stamen development. In the absence of these two functions, pistillate flowers develop.

Sex determination in mercury (*Mercurialis annua*), a poisonous herb found in Europe, has been shown to be controlled by the action of three genes: A, B1, and B2. These genes, which are not located on a single sex chromosome, affect levels of the hormones auxin and cytokinin. The dominant alleles of A, B1, and B2 promote male development and vary in their strength such that A is stronger than B1, and B1 is stronger than B2. It is possible to cause a genetically male plant to develop female flowers by the application of cytokinin. Similarly, a genetic female will develop male flowers if treated with auxin. Biochemical analysis of female plants' shoot apices, which are the sites of developing flowers, show high levels of the cytokinin *trans*-zeatin and low levels of auxin. The shoot apices of male plants have no detectable *trans*-zeatin and high levels of auxin. In order to learn how different concentrations of these two hormones control sex differentiation at the molecular level, current work focuses on isolating genes that are expressed only in the developing flowers of male or female plants.

Monoecious species. Flowers that develop at the base of cucumber plants are male, followed by perfect flowers farther along the stem and female flowers at the tip. Addition of auxin promotes female development, and the addition of gibberellic acid masculinizes treated plants by causing female flowers to appear at later nodes on the plant. Mutations at the *st* locus also

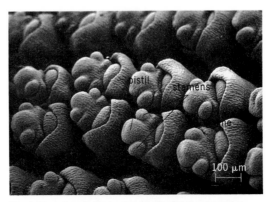

Fig. 1. Scanning electron micrograph of immature maize florets (on an ear) before any morphological evidence of sex differentiation. A glume is a leaflike structure surrounding grass florets.

feminize the plant by shortening the period of male flower development. Sex determination is completely disrupted in plants with mutations at the *m/m* locus; these plants develop all perfect flowers.

Monoecism has been much studied in maize (*Zea mays*) because of its highly evolved secondary sex characters and because of the large collection of mutations that affect sex determination. The inflorescence that bears the male florets (the tassel) develops from the shoot apical meristem. These inflorescences are thin, branched, and exposed to the wind. The inflorescences that bear pistillate florets (the ears), develop from axillary buds. These inflorescences are thick, unbranched, and enclosed by the husk leaves through the suppression of internode elongation below the ear. Like most monoecious plants, florets on the tassels and ears of maize plants are initiated as perfect flowers (**Fig. 1**). Later in development, pistils are suppressed in florets on the tassels (**Fig. 2**), and stamens are suppressed in florets on the ears (**Fig. 3**), resulting in staminate tassels and pistillate ears.

There are two main classes of mutations that alter this pattern of sex determination. The *dwarf* mutations cause a reduction in the stature of the plant and fail to suppress stamens on the ears, resulting in andromonoecism. Physiological studies have shown that most of the *dwarf* mutants are defective in the biosynthesis of the plant hormone gibberellic acid. This indicates that gibberellic acid is important in suppressing stamens in florets on the lateral inflorescences. The *tassel seed* mutations are characterized by the development of pistils instead of stamens in tassel florets. The biochemical basis of the phenotype of these mutations is still unknown. Genetic analysis has shown that the various *tassel seed* genes interact with one another and probably act in a single or a few genetic pathways. The approach of insertional mutagenesis and transposon tagging shows great promise for the molecular characterization of the *tassel seeds* and other morphological mutations in maize.

Control of floral organ identity. A more fundamental process in sex determination in plants is the differentiation of floral primordia into functional organs. Most

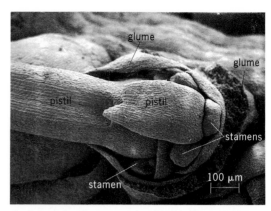

Fig. 3. Scanning electron micrograph of a pair of pistillate maize florets on an ear, showing developing pistils and stamens whose development has been arrested.

progress in understanding this process is taking place in the analysis of homoeotic mutations, which change the identity of an organ to that of a homologous organ. It is possible to identify homoeotic mutations in flowers because there is a strict sequence found in almost all flowering plants such that the successive whorls of organs are sepals, petals, stamens, and pistils. In the analysis of flowers with altered patterns of organs, it is assumed that it is simpler to change the pattern of differentiation of a primordium by mutation than it is to change the location of a primordium that is differentiating normally.

The homoeotic mutation *deficiens* of snapdragon (*Antirrhinum majus*) is one of the best-understood mutations affecting floral development. In plants expressing the *deficiens* mutation, stamens develop like carpels (a component of the pistil) and petals develop like sepals. Vegetative parts of the plant are not affected by this mutation. Thus, the wild-type gene is thought to provide some specific function that makes stamens different from carpels and petals different from sepals. The phenotype of the mutant also supports the idea that stamens and carpels are fundamentally more similar to each other than to sepals and petals, which are likewise more similar to each other. The *deficiens* gene has been isolated molecularly, and its nucleotide sequence has been determined. The sequence was found to be similar to that of two genes, the mammalian gene *SRF* and the yeast gene *GRM/PRTF*, both of which encode deoxyribonucleic acid (DNA) binding proteins. The sequence similarities suggest that the wild-type *deficiens* gene product is a regulatory protein involved in general cellular signals. Expression studies have shown that the wild-type *deficiens* messenger ribonucleic acid (mRNA) accumulates in all floral organs, and is not detectable in vegetative tissues, as would be predicted by the phenotype of the mutation.

Several homoeotic mutants of thale cress (*Arabidopsis thaliana*) affecting floral organ identity are currently being analyzed genetically and molecularly. *Apetala* 3 is a temperature-sensitive mutation that, like *deficiens* in snapdragon, converts petals to sepals and stamens to carpels. *Pistillata* 1 and 2 have similar

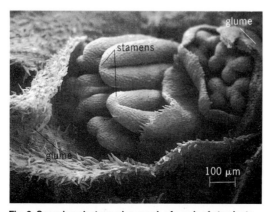

Fig. 2. Scanning electron micrograph of a pair of staminate maize florets on a tassel, showing well-developed stamens; the degenerating pistils are hidden in the middle of the stamens.

phenotypes. Flowers expressing the *agamous* pheno-type have the whorl sequence sepal-petal-petal, which is then repeated many times in a single flower. The *agamous* gene has been isolated, sequenced, and compared to other sequences of known function. *Agamous* also shows sequence similarities to mammalian *SRF* and yeast *GRM/PRTF*.

Another approach to understanding the control of differentiation of the various floral organs is the isolation of genes that are expressed specifically in one floral organ type. Although many such specific genes have been identified by this process, a clear picture of how these genes function has not emerged; much work remains to be done before the molecular basis of floral organogenesis is elucidated.

For background information SEE AUXIN; FLOWER; GENE ACTION; GIBBERELLIN; SEX DETERMINATION; SEXUAL DIMORPHISM in the McGraw-Hill Encyclopedia of Science & Technology.

Erin E. Irish

Bibliography. E. E. Irish and T. Nelson, Sex determination in monoecious and dioecious plants, *Plant Cell*, 1:737–744, 1989; J. P. Louis, Genes for the regulation of sex differentiation and male fertility in *Mercurialis annua* L., *J. Hered.*, 80:104–111, 1989; H. Sommer et al., *Deficiens*, a homeotic gene involved in the control of flower morphogenesis in *Antirrhinum majus*: The protein shows homology to transcription factors, *EMBO J.*, 9:605–613, 1990; M. Westergaard, The mechanism of sex determination in dioecious flowering plants, *Adv. Genet.*, 9:217–281, 1958.

Silicate systems

Rates of diffusion and chemical reactions involving solids are generally considered temperature-dependent and have not been studied very much as a function of pressure. Pressure itself inhibits diffusion in most solids, because of structural compression; similarly it decreases mobility and increases viscosity in most liquids. Such considerations are of importance in rheological behavior and reaction rates in the deep earth in response to temperature and pressure gradients. The revolutionary plate tectonic concept requires convection in the (solid) mantle of the Earth, and it emphasizes the importance of rheological behavior in silicates and oxides at depth. Recent work has shown that in a number of geologically important solids, high pressures may greatly enhance diffusion and reaction rates.

Aluminum-silicon interdiffusion in feldspars. The feldspars are aluminosilicates of sodium, calcium, and potassium; they constitute over 60% of the Earth's crust. Only the potassium and sodium end-member compositions, potassium aluminosilicate ($KAlSi_3O_8$) and sodium aluminosilicate ($NaAlSi_3O_8$), will be considered here. Aluminum (Al) and silicon (Si) are both tightly surrounded (coordinated) by four oxygen atoms in a continuous corner-sharing network, with the charge-balancing cations in cavities in the Al-Si

array. The Al-Si sites may be populated at random in a disordered array or in specific sites of an ordered configuration.

The Al and Si atoms are tightly bonded to the oxygen atoms in tetrahedral cages. Diffusional interchange is so difficult that the temperature regions of stability of the ordered and disordered structural forms of $KAlSi_3O_8$ remain undetermined, while the equilibrium thermal states of $NaAlSi_3O_8$ have been known only since 1985. The fully ordered low-temperature form of $KAlSi_3O_8$ (microcline) has not been produced from the high-temperature modification (sanidine) owing to the great barrier to thermal diffusion of the Al and Si, particularly at the low temperatures (certainly <500°C or 930°F) at which microcline is stable. The ordered form of albite ($NaAlSi_3O_8$) that is stable at low temperatures is known as albite (or low albite), and the disordered form that is stable at higher temperatures is known as high albite. The melting points of both alkali feldspars are not far from 1100°C (2010°F), yet microcline must be held at temperatures of around 1050°C (1920°F) at atmospheric pressure for days, weeks, or even longer before it disorders fully to sanidine. Even the liquids of the K and Na feldspars at the melting points are exceedingly viscous and extraordinarily difficult to crystallize.

After World War II, the development of apparatus capable of sustaining pressures of several kilobars of water pressure at temperatures of 900°C (1650°F) or more led to a revolution in experimental mineralogy and petrology. Hydrothermal experiments in the kilobar range could be carried out on silicates resistant to change or reaction under anhydrous conditions. The solubility of silicates in supercritical water is quite large, and increases with pressure. The dominant mechanism of reactions under hydrothermal conditions is one involving solution and reprecipitation, or recrystallization. Nevertheless, even this boost was inadequate to produce equilibrium reactions or configurations in a number of important mineral systems, including the structural-state relations of the feldspars. Furthermore, water at high pressures greatly reduces the melting temperatures of the substances, in albite from over 1100°C (2010°F) at 1 bar (10^5 pascals) to about 700°C (1290°F) at 12 kbar (1.2 gigapascals) water pressure. This places a distinct limitation on the pressure-temperature range of hydrothermal reactions.

Diffusional and reaction enhancement. In recent years, high-pressure devices (piston-cylinder presses) that use solids such as talc or common salt (NaCl) as pressure media have come into use, and can routinely sustain pressures up to 35 kbar (3.5 GPa) at temperatures of 1800°C (3270°F) or more. Hydrothermal experiments can also be carried out in them by sealing water with the sample in a noble-metal capsule. For some time, it had been observed that some reactions were enhanced at high pressures, even in systems without added water. These observations were not readily explained; pressure might be expected to slow them down. A few years ago, it was first observed that both the ordering and disordering processes involving Al-Si could be greatly accelerated at high pressures in

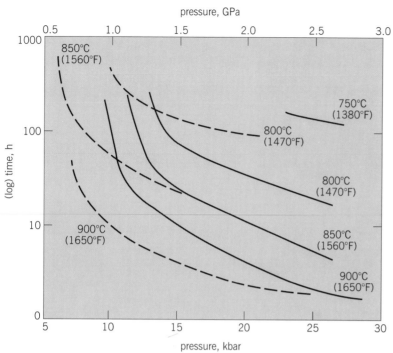

pressure, GPa

Curves for aluminum-silicon disorder rate in microcline (solid curves) and albite (broken-line curves) as a function of pressure. Each curve represents the time necessary to achieve the disordered configuration at the temperature indicated. *(After J. R. Goldsmith, Enhanced Al/Si diffusion in KAlSi₃O₈ at high pressures: The effect of hydrogen, J. Geol., 96:109–124, 1988)*

albite by using piston-cylinder devices. This was done without water in the sample capsule, and indeed, at pressures which, if any water were present, would melt the albite. The equilibrium states of ordered and disordered albite were established by approaching the equilibrium value from both higher and lower temperatures (reversing the equilibrium, as it were). Low albite (at about 17 kbar or 1.7 GPa) begins to disorder at about 650°C (1200°F) and does so continuously, with intermediate (stable) degrees of order, until essentially complete disorder to high albite is attained at about 900°C (1650°F).

It was also observed that pressure alone was not responsible for the enhanced diffusion, but that hydrogen, probably as the hydrogen ion (H^+), was required. The NaCl pressure medium generally used contains small amounts of moisture that cannot normally be eliminated. At the high pressures and temperatures of the experiments, dissociation of water in the NaCl (or talc) cell is greatly increased, providing the hydrogen. The samples are encapsulated by noble metals permeable to hydrogen, permitting it to penetrate them. The importance of hydrogen was established by decreasing its activity by two different techniques: using an essentially anhydrous pressure medium such as glass, or capturing hydrogen before it can become effective by surrounding the sample with a getter such as ferric oxide (Fe_2O_3). The hydrogen reacts with the Fe_2O_3 to produce ferrosoferric oxide (Fe_3O_4) and water (H_2O). Although the diffusing and activating species of hydrogen cannot be ascertained with certainty, diffusion rates in the silicates in this and other studies indicate that those containing large

atoms such as the oxygen in H_2O, hydroxide ion (OH^-), and hydronium ion (H_3O^+) are unlikely. Metallurgical studies have shown that the hydrogen probably diffuses through the platinum as protons (H), not as the dihydrogen molecule (H_2).

The Al-Si disordering rate as a function of pressure and at constant temperatures is shown in the **illustration.** For albite the curves represent the time as a function of pressure for low albite to reach the equilibrium state of disorder at 800, 850, and 900°C (1470, 1560, and 1650°F). The curves for microcline were similarly determined, and they indicate the times at 750, 800, 850, and 900°C (1380, 1470, 1560, and 1650°F) for disorder to sanidine to be complete. The pronounced steepening of the two sets of curves at the lower pressures is striking; the sluggishness of the Al-Si interdiffusion at lower pressures is apparent. In albite at 900°C (1650°F), there is a tenfold increase in reaction rate from 8 to 24 kbar (0.8 to 2.4 GPa), and a similar rate increase at 850°C (1560°F) from approximately 7 to 22 kbar (0.7 to 2.2 GPa). At lower pressures, data are not available because of the exceedingly slow reaction, but the rate falls off very much more rapidly. A drop from 8 to 7 kbar (0.8 to 0.7 GPa) decreases reaction time by a factor of 2 at 900°C (1650°F), and disordering cannot be accomplished by a solid-state diffusional mechanism in times available in the laboratory at pressures lower than 6 kbar (0.6 GPa) at 850–950°C (1560–1740°F). Microcline cannot be disordered in reasonable laboratory times at pressures much below 9–10 kbar (0.9–1.0 GPa). The effect of pressure is enormous.

A suggested mechanism. Enhanced reaction is dependent on a process that lowers the activation energy for diffusion by promoting the breaking of the strong silicon-oxygen (Si—O) and Al—O bonds in the structure. Hydrogen has been implicated, and it seems likely that hydroxyl (OH) groups are involved in the process. The large effect of pressure in increasing the dissociation of moisture in the pressure cell provides the hydrogen that diffuses through the metal capsule. In reacting with the oxygen of the silicate network to form OH groups, hydrogen must have been oxidized to H^+ within the sample, as unaccompanied ions cannot enter the capsule, and it is unlikely that cations also diffused so rapidly through it.

The hydrogen ions can produce OH groups, which have a smaller charge (−1) than the coordinating oxygen atoms (−2); and higher coordination (tighter packing) might thus be more readily induced in the surrounding oxygens because of the decreased net (repulsive) charge. This would be aided and abetted by high pressures, and transient coordination of 5 or even 6 around Al and Si that are normally surrounded by 4 oxygens could be attained. Thus protons may play a dual or associated role involving bond snipping and coordination change. Although Al is more susceptible to higher coordination than is Si, the transient increase in coordination, which could greatly enhance the interchange of the Si and Al atoms, could also involve Si. The formation of Al-OH-Si units reduces the formal bond strength, and transient OH groups may result from proton hopping. A similar mechanism of proton jumping has been suggested by others for the high electrical conductivity of ice. This bond-breaking ability of OH groups, amplified by pressure, can greatly accelerate diffusional processes; and relatively few migrating protons might produce significant interchange of Al and Si. The process takes place in single crystals without disrupting the structure, and interchange of Al and Si does not recrystallize the feldspar. The relative importance of the two roles of pressure, that of increasing dissociation and of promoting the transient coordination increase, is unknown. These bond-breaking factors also enhance the diffusion of oxygen in silicates. Oxygen isotope exchange reactions are being studied in which the oxygen of silicates is exchanged rapidly with that of carbonates and of carbon dioxide at high pressures and temperatures.

Implications for the Earth. Solid-state diffusion is a thermally activated process, normally inhibited by an increase in pressure. The enhancement of Al-Si diffusion in $NaAlSi_3O_8$ and $KAlSi_3O_8$ under anhydrous, high-pressure conditions in the presence of hydrogen is extraordinary and unexpected. Pressure can thus become a cooperative and effective partner to temperature.

The possible effect of ionic transport on the rheological properties of silicates merits consideration. Phenomena such as brittle versus ductile behavior of crustal material observed in earthquake studies are commonly thought of in terms of the effect of increased temperature as a function of depth. The current results suggest that in the presence of hydrogen the pressure effect may be at least equally important. Furthermore, at pressures corresponding to those at or greater than lower crustal depths where cracks, fissures, and voids are much less important than in the upper crust, recrystallization due to solution processes in the presence of water may be much less important than diffusional processes activated by hydrogen at high pressures. Very little is known about the availability of hydrogen at depth, but it diffuses through solids much more rapidly than any other substance. At the higher pressures, any water present would be highly dissociated, and a source of H^+ to activate the diffusive process.

For background information SEE COORDINATION NUMBER; CRYSTAL STRUCTURE; DIFFUSION IN SOLIDS; FELDSPAR; HIGH-PRESSURE MINERAL SYNTHESIS in the McGraw-Hill Encyclopedia of Science & Technology.

Julian R. Goldsmith

Bibliography. J. R. Goldsmith, Al/Si interdiffusion in albite: Effect of pressure and the role of hydrogen, *Contrib. Mineralogy Petrology*, 95:311–321, 1987; J. R. Goldsmith, Enhanced Al/Si diffusion in $KAlSi_3O_8$ at high pressures: The effect of hydrogen, *J. Geol.*, 96:109–124, 1988; J. R. Goldsmith and D. M. Jenkins, The high-low albite relations revealed by reversal of degree of order at high pressures, *Amer. Mineralog.*, 70:911–923, 1985.

Simulation

Recent advances in computer simulation have included combining the technologies of artificial intelligence, expert systems, and discrete simulation; and applying variance reduction techniques to simulation.

ARTIFICIAL INTELLIGENCE IN DISCRETE SIMULATION

The art and science of discrete simulation is rapidly changing. Recent research has been directed toward exploring the potential for combining artificial intelligence (AI), expert systems (ES), and discrete simulation technologies. Such systems will allow models to be quickly developed, validated, and run, with as much of the necessary expertise as possible built into the software. This is a natural development since the goals of artificial intelligence/expert systems and simulation are similar, that is, to provide computer-based modeling that facilitates decision making. Furthermore, the two technologies are highly complementary; the methodologies of simulation modeling are useful in analysis of problems (predictive) but limited in synthesis guidance (prescription), while the reverse is true of technologies based on artificial intelligence and expert systems.

The merging of artificial intelligence and simulation actually started years ago. One of the earliest simulation languages, SIMULA, introduced the concepts of object-oriented programming, which were later picked up by people working with artificial intelligence and incorporated into languages for artificial intelligence/expert systems such as LISP, Smalltalk, and C++.

Modeling. Current advances have been driven by several problems as well as by the opportunity presented by the confluence of developments in several areas of technology. Simulation modeling studies tend to have low productivity. The traditional modeling life cycle is very labor-intensive and time-consuming. Furthermore, since most of the so-called intelligent functions are performed by hand, a great deal of knowledge is discarded at the completion of each project. Most models are treated as throw-away items, as is the analysis required to solve the problem. Attempts are being made to conserve effort by incorporating knowledge into the software. With this approach, as much work as possible is being shifted from the human to the machine, thus freeing the modeler for handling decisions that truly require the attention of a human.

The current practice of simulation modeling is as much an art as a science. There is a severe shortage of trained and experienced personnel. Therefore, one of the goals for the development of techniques of simulation modeling based on artificial intelligence/expert systems is to capture the knowledge of the experts in the software, so that engineers, scientists, and managers can do modeling studies correctly and easily without excessively elaborate training.

Data management. Yet another opportunity has been presented by the progress in the field of knowledge representation and database management since

1980, especially the explosive emergence of relational and object-oriented database technology. Most serious applications of artificial intelligence/expert systems and simulation modeling deal with large amounts of data and the need to access those data efficiently. The development of excellent database programs for desktop computers and the evolution of sophisticated inquiry interfaces are natural adjuncts to data-hungry programs of artificial intelligence/expert systems and simulation modeling.

The progress being made in the technology for artificial intelligence/expert systems opens the door for a rethinking of the simulation-modeling process for design and decision support. The problem-solving paradigm as practiced in simulation modeling studies in the past is essentially a search. The goal of the search is to find the combination of parameter values that will optimize the response values and the controllable variables of the system. The burden of conducting that search and integrating the results of model behavior into a coherent solution currently rests with the human user. Much of the work in the field of artificial intelligence on automated reasoning has dealt with the inference and control mechanisms in the inference engine and the use of the facts and rules in the knowledge base so as to efficiently search for a solution. These developments are playing a vital part in the design of goal-seeking simulation modeling systems.

Software. The state of the art in bringing the technology of artificial intelligence/expert systems to bear upon simulation is very early in its development cycle. Two basically different approaches are being pursued. A number of developers of simulation software have taken the approach of developing intelligent, automatic programming interfaces to existing simulation systems. In these hybrid systems, an interactive interface allows the user to describe the system to be simulated in terms of placement or selection of graphical icons, menu choices, or answers to computer-controlled interrogations. Presently, such systems are by necessity limited to a specific domain such as computer networks, flexible manufacturing systems, materials handling, or electronics assembly. In such systems, the program automatically writes the model and experiment to be run in an existing simulation language such as SIMAN, GPSS, Simscript, or Simula. The user need not be familiar with these languages and may not even be aware of the language being used to write the model.

Other software developments are directed at intelligent back ends to existing simulation systems that will aid the user in analyzing the results and suggesting modifications. In these systems, a goal is set and the model is executed; and if the desired results are not achieved, the symptoms are analyzed by the software and suggested modifications are presented to the user. The last is usually accomplished by a program consisting of a set of rules of the form "IF this symptom or condition exists THEN suggest this action." This approach has also been used to develop expert diagnosis programs for debugging simulation models.

In addition, there are software programs under development that have as their ultimate aim a modeling system that follows a paradigm very different from any in existence today. In these goal-seeking modeling systems the user's knowledge about the study objectives, goals, model performance standards, and model component behavior is declared. The modeling software then assists the user to assemble a model that will generate the information necessary to evaluate performance criteria. The modeling software, after constructing the appropriate model, uses automated reasoning to guide its execution through to achievement of the desired goal. If it is impossible to achieve the declared goal under the constraints defined by the user, the system advises why the solution was not possible. One of the aims of such systems is to allow the user to define and simulate systems at different levels of abstraction, and to check the completeness and consistency of the model. The emphasis is upon developing a system of knowledge representation and a mechanism for a goal-directed search that is compatible with the paradigms of both artificial intelligence/expert systems and discrete simulation.

Prospects. Several commercial knowledge-based simulation systems that include the application of the concepts of artificial intelligence and logic programming have already become available. Future developments will continue to generate ever more powerful environments for discrete simulation modeling and will eventually lead to truly expert systems. The goal is to simplify and make accessible for the seminaive user the expertise of the most knowledgeable and experienced simulation experts and to build into the modeling system most of the decisions that are now made by the analyst.   *Robert E. Shannon*

Variance Reduction Techniques

When simulation is applied to a stochastic dynamical system, the simulation's output is itself random; it is only in averaging a large number of observed outputs that performance measures can be estimated accurately. Because of the central limit theorem of probability theory, the typical convergence rate of the averaged output of a simulation to the desired solution is $t^{-1/2}$ in the computation time t. More precisely, the accuracy of an averaged estimator obtained after t units of computer time have been expended can be expressed as in (1),

$$\left(\frac{\sigma^2 \eta}{t}\right)^{1/2} \tag{1}$$

where σ^2 = variance per unit of simulated time and η = rate at which computer time is expended per unit of simulated time. Formula (1) implies that in order to add one significant figure of accuracy to the current estimated solution (that is, increasing accuracy by a factor of 10), it is necessary to increase the computer time budgeted to calculating the solution by a factor of 100. Consequently, stochastic simulation is viewed as a numerical technique that is encumbered by a rather slow convergence rate.

Fortunately, a class of procedures, known as efficiency improvement techniques, can significantly improve the convergence characteristics of such simulations. The mathematical structure of the dynamical system is used to calculate more efficiently the performance measures of interest. This is done by reducing the size of the product $\sigma^2\eta$ that appears in formula (1). An improvement in efficiency can be obtained by reducing σ^2 or η, or both. Reducing η is often a matter of improving the efficiency of the computer code that generates the simulation, whereas reducing σ^2 (that is, variance reduction) involves using statistically based ideas that are commonly known as variance reduction techniques.

Common random numbers. If it is desired to compare the performance of two competing stochastic systems, as measured through the mean difference of their associated performance measures, it seems statistically fairer to subject them to as similar a sequence of driving inputs as possible, so that both systems are simultaneously either "lucky" or "unlucky." This variance reduction technique is known as the method of common random numbers. In the simulation of complex systems, typically an attempt is made to synchronize the simulation of the two systems by identifying separate random-number streams to drive the various logical subcomponents of the systems (for example, the various arrival processes to a network of queues). Each of the various input streams in the two systems is then appropriately synchronized (for example, by using the same stream of uniform random numbers to generate interarrival times in the two systems, the arrival time sequences in the two systems will be highly correlated with one another). The degree of variance reduction is strongly related to the extent to which the two systems (and their associated performance measures) respond similarly to their driving input sequences. SEE QUEUEING THEORY.

Control variates. In most simulations, a great deal of statistical information is known about characteristics of the driving input sequences. For example, in a manufacturing simulation the distribution used to generate machine failure times has known statistical characteristics. Control variates exploit this information. If $\overline{X}(t)$ is the averaged estimator for the system's performance measure and $\overline{Y}(t)$ is the average of all the machine failure times collected in the first t units of computer time, then $\overline{Y}(t)$ converges to the mean machine failure time μ, which is calculable from the distribution used to generate failure times. Then, for any scalar λ, the estimator given by (2)

$$\overline{X}(t) - \lambda[\overline{Y}(t) - \mu] \qquad (2)$$

will be, for large t, an appropriate estimator of system performance. The quantity $\overline{Y}(t) - \mu$ is known as the control variate. The parameter λ is then chosen so as to optimize the accuracy of the resulting estimator. The optimal choice of λ is related to the correlation between $\overline{X}(t)$ and $\overline{Y}(t)$, as well as to the variance of $\overline{Y}(t)$. Since these quantities are typically unknown, the optimal choice of the control coefficient must be estimated from the statistical data generated by the simulation. This estimation is not a difficult procedure, and consequently the method of control variates is widely applied to stochastic simulations. The extent of the variance reduction obtained with this technique is related to the degree of correlation between $\overline{X}(t)$ and $\overline{Y}(t)$.

Importance sampling. In many applications, the goal is calculating the probability of a rare event. For example, it may be desired to determine the probability that a complex system will fail in the first 100 h of operation. If the system's components are highly reliable, such an event may be quite improbable. In such a case, conventional simulation may yield a computed estimate that is relatively inaccurate, because it may take enormous amounts of computer time to observe even a small number of system failures.

A variance reduction technique that is frequently applicable to such simulations of rare events is the method known as importance sampling. In the context of the reliability example discussed above, the highly reliable components would be replaced by less reliable ones and the resulting system would be simulated. The less reliable system would fail more frequently. Thus, the important failure event is sampled more heavily. In order to obtain numerical results that are pertinent to the original highly reliable system, the outputs from the simulation are multiplied by an appropriately derived compensating factor. In many rare-event settings, such an importance sampling approach is the only practical means of obtaining accurate solutions within a realistic amount of computer time.

Conditional Monte Carlo. In some stochastic simulations, it is possible to integrate out some of the randomness that is present by analytically computing an appropriately chosen conditional expectation of the averaged estimator. This conditional expectation, while still random, has lower variance than the original estimator. The technique is applicable only to those systems in which such conditional expectations can be derived easily. Because of the analytical simplicity of the exponential distribution, such conditioning is often possible in systems in which a large number of the input distributions are exponential.

Stratification. In some systems, a fraction of the runs are randomly assigned a certain property (for example, a fraction of the runs simulate the company's operation in a weak demand scenario). In this variance reduction technique, the sample is stratified so that there is a deterministic preassignment of a fraction of the runs to the stratum in which all runs are required to have the property. By eliminating the sampling variability in the assignment of the given property to the various runs, variance is reduced.

A variety of other variance reduction techniques are also available. For example, the techniques known as splitting and Russian roulette have been applied successfully to many simulations of particle systems. In addition, infinitesimal-perturbation analysis and likelihood-ratio methods have been developed to calculate efficiently derivatives of performance measures with respect to various input parameters (for example, the derivative of steady-state queue length with respect to mean service time). Such derivatives are useful for both the optimization and sensitivity analysis of complex stochastic systems.

For background information SEE ARTIFICIAL INTELLIGENCE; DATABASE MANAGEMENT SYSTEMS; DECISION THEORY; DISTRIBUTION (PROBABILITY); EXPERT SYSTEMS; MONTE CARLO METHOD; PROGRAMMING LANGUAGES; QUEUEING THEORY; STATISTICS in the McGraw-Hill Encyclopedia of Science & Technology.

Peter Glynn

Bibliography. P. Bratley, B. L. Fox, and L. E. Schrage, *A Guide to Simulation*, 1987; P. W. Glynn, Optimization of stochastic systems, *Proceedings of the 1986 Winter Simulation Conference*, pp. 52–59, 1986; J. M. Hammersley and D. C. Handscomb, *Monte Carlo Methods*, 1964; R. Reddy, Epistemology of knowledge based simulation, *Simulation*, 48(4):162–166, 1987; R. E. Shannon, Knowledge based simulation techniques for manufacturing, *Int. J. Prod. Res.*, 26(5):953–973, 1988; L. E. Widman, K. E. Loparo, and N. R. Nielson (eds.), *Artificial Intelligence, Simulation and Modeling*, 1989.

Soil

Recent advances in soil science include elucidation of the turnover of carbon and nitrogen through the microbial biomass in soil, and development of the technique of soil solarization.

TURNOVER OF CARBON AND NITROGEN

Element turnover through the microbial biomass comprises uptake, transformation, and exudation processes by microbial cells. Since the vast majority of the microbial species in soil is heterotrophic, utilization of inorganic carbon compounds is needed for energy supply and biosynthesis of the microbial cells. Exudation of carbon from microbial cells is in either inorganic form, that is, mostly carbon dioxide (CO_2), or organic form. The release of microbial organic material is due to death and subsequent lysis of cells and by grazing activity of predatory organisms. Although the microbial biomass constitutes a relatively small part of the organic fraction in soil, usually 1–5%, it is also a relatively labile pool and therefore its importance in element cycling exceeds its size.

Rate of cycling. The turnover of other elements, such as nitrogen, phosphorus, and sulfur, is linked to the cycling of carbon through the microbial biomass, but the relationships vary. Nitrogen-mineralizing enzymes are rarely found outside microbial cells. Moreover, the species containing nitrogen that occur most frequently in the soil (ammonium, nitrate, amino acids, and proteins) are rather easily available for use by the microbial population; thus the turnover of nitrogen is very closely linked to the cycling of carbon. In contrast, phosphorus-mineralizing enzymes, that is, phosphatases, are often found outside microbial cells. Several phosphorus compounds are

strongly adsorbed to abiotic soil constituents and are relatively unavailable for microbes; therefore the phosphorus cycle is less closely linked to the carbon cycle than is the nitrogen cycle.

When assuming steady-state conditions, it may be possible to estimate microbial turnover rates from the input rate of organic matter and the size of the microbial biomass. On this basis, turnover rates of 1.25–2.5 years have been calculated. However, it is questionable whether this gives a proper indication of the cycling rate of carbon and nitrogen. The methodology is poorly developed for determining input rates of material and microbial biomass into soil. Only a fraction of the microbial population participates actively in transformation processes; the active fraction of the microbial population in soil is estimated at most 10% of the total microbial population under undisturbed conditions. Wide discrepancies exist between calculations on the level of the total population and calculations on the basis of specific cells or species. For example, the aforementioned values of turnover times of microbial biomass in soil implied a doubling time of the microbial population of 0.9–1.7 years, whereas doubling times of bacteria in soil of less than 20 h have been observed.

Inputs. The availability of organic matter for microbial metabolism is determined by input of material into soil, by the chemical constitution of the organic compounds, and by the physical accessibility of material. Organic materials in soil are derived mainly from plants. Total annual inputs in soil vary widely and depend on the type of plant cover and farming practice, for example, applications of manure or other waste material and tillage practices. An important source of continuous input of organic materials is root exudates and their derivatives; annual inputs in agricultural and forest soils are estimated to be 900–3000 kg carbon/ha (800–2700 lb/acre).

The chemical constitution of organic material determines the availability of the material as a substrate for microbes. Yet, the very wide variety of compounds and the great complexity of chemical bonds within compounds make it impossible to describe correctly the relationships between individual substrates and the microbes that utilize these substrates. In this instance also, inadequate methodology hampers the measurement of the quantity of substrate available in soil.

Bioavailability. Earlier techniques of fractionating the organic matter of the soil by means of different chemicals have been criticized for not allowing measurement of bioavailability. In more recently developed approaches, the availability of organic material as a substrate for microbes is determined on the basis of the association of organic materials with soil particles and aggregates. This fractionation method is based on the concept that the physical environment of the soil largely determines the availability of organic matter to microbes and, therefore, the turnover of carbon and nutrients through the microbial biomass.

Two types of physical impediments to substrate availability in the soil may be distinguished. Associations of organic compounds with soil particles, in particular clay minerals, have been shown to affect microbial transformations. Entrapment of organic matter in aggregates or other ways of surrounding organic compounds by inert soil particles may strongly influence the accessibility of this material to microbes. Finer-textured soils generally show lower turnover rates of carbon and nitrogen than sandy, coarser soils.

Disruption of soil structure either by natural processes, such as the remoistening of dry soils or the thawing of frozen soils, or by human activity, such as plowing, may drastically speed up the turnover of carbon and nitrogen. This may have dramatic consequences; an example is the increased turnover rate of organic matter in soils that occurred after cultivation of virgin prairie grassland in North America, resulting in losses of 60% or more of the original levels of carbon and nitrogen.

The main environmental variables that determine the microbial turnover of carbon and nitrogen are temperature and moisture. The combination of high temperature and high rainfall in the humid tropical conditions of Nigeria results in a decomposition rate of crop residues that is two times higher than of the same material in the hot but dry conditions in South Australia, and in a rate of decomposition that is four times higher than that in the temperate climate of England.

Nitrogen mineralization. Intracellular metabolism leads to the production of new cellular material and to waste products such as CO_2, fermentation products, and ammonium ions, which may be transformed into other inorganic or organic nitrogen products. The ratio of carbon to nitrogen (C:N) in the microbial biomass in soil is approximately 5 to 15 with an estimated average of 8. Assuming a carbon utilization efficiency of 40% (a figure often reported for soil microbiological transformations), 1 unit of nitrogen is needed for the utilization of 20 units of carbon. Thus, when organic materials with a C:N ratio of 20 or less are decomposed, net nitrogen mineralization may occur; when the C:N ratio of the organic matter is higher, theoretically the net immobilization of mineral nitrogen from soil will take place.

This classical rule of thumb has long been used by scientists and farmers to estimate the mineralization of nitrogen from organic materials applied to soil. However, deviations from this rule may easily occur; for example, it is possible for the efficiency of utilization of the organic material by the microbes to be less than 40%, which may be the case for recalcitrant complex materials.

The release of nitrogen and carbon from the microbial population is the result of cellular metabolism, cellular death, and trophic interactions of microbes and other soil organisms. Bacteria and fungi are grazed mainly by protozoa and nematodes, which in turn are a substrate for the micro- and mesofauna in soil. The flow of energy and material through the different trophic levels leads to the release of carbon and nitrogen from the cellular tissue. The importance of trophic interactions for the turnover of carbon and nitrogen in soil is still a matter of much debate. Protozoa grazing on bacteria have been shown to increase the availability of nitrogen from soil organic matter for plant uptake by approximately 30%. In arable soils the contribution of trophic interactions to the flux of nitrogen was estimated to be 40%. The relative importance of trophic interactions to the turnover of nitrogen seems to depend also on the agricultural management practice of the soil.

Prospects for manipulation of nitrogen turnover and, thus, for its optimization may lie in the manipulation of some of the aspects mentioned earlier, such as the management of soil structure, the use of organic amendments, and the manipulation of trophic interactions. However, nitrogen is not only the product of turnover processes but also a controlling factor in microbial turnover processes. Recent work has demonstrated that upon application of large amounts of mineral nitrogen to the soil, permitting unlimited plant growth, the net mineralization of native soil organic nitrogen was reduced, whereas under nitrogen-limiting conditions, relatively more (up to 30%) of the organic nitrogen in the soil was mineralized. As a possible mechanism, it has been proposed that when sufficient mineral nitrogen is available, microbes prefer utilization of root-derived materials that are low in nitrogen to the utilization of more recalcitrant native soil organic matter, whereas under nitrogen-limiting conditions this substrate preference does not occur and native soil organic matter, relatively high in nitrogen (C:N ratio of 10 to 12), is utilized as substrate as well. Consequently, in the latter case, net mineralization occurs.

Nitrogen, which is essential for crop growth, is taken up by plant roots from the soil solution mainly as inorganic nitrogen (ammonium or nitrate ions). Inorganic nitrogen present in the soil solution originates from mineralized soil organic matter, inorganic and organic fertilizers, and wet and dry deposition from the atmosphere. The annual contribution of atmospheric deposition and mineralized soil nitrogen to the nitrogen requirement of crops is generally less than 25 and 100 kg nitrogen/ha (22 and 90 lb/acre) respectively, and since there are inevitable losses, nitrogen fertilizers are applied to meet the total nitrogen requirement. This requirement varies from about 200 kg/ha (180 lb/acre) for a crop such as oats to about 400 kg/ha (360 lb/acre) for intensively managed grassland.

Predicting nitrogen requirements. Most nitrogen fertilizer recommendations may take into account the amount of inorganic nitrogen already present in the soil at the end of winter, but may not consider the amount to be mineralized during the summer. During the growing season, mineralization rates and yield levels of soil nitrogen may vary considerably from field to field. Therefore, two methods are being developed that explicitly take into account nitrogen mineralization and expected yield level. With the balance-sheet method, a balance sheet is drawn up in which the nitrogen requirement of the crop is given on one side and the contributions in the form of fertilizer nitrogen, inorganic nitrogen already present in the soil, and the nitrogen mineralized during the growing period on the other side. The other, even more refined method makes use of a simulation model. Simulation models calculate on a daily basis the growth and the nitrogen uptake of a crop and the nitrogen supply to a crop by using weather data and field and crop parameters as inputs. A simulation model can thus be used as a tool to indicate the fertilizer nitrogen requirement of a crop at any time during the growth period. *See Simulation.*

Recommendations based on the balance-sheet method and on simulation models rely heavily on a correct prediction of the amount of nitrogen mineralized during the growing period. A basic understanding of the mechanisms of nitrogen mineralization–immobilization is necessary to find simple ways to determine mineralization levels on a field scale. As yet, the present knowledge of the microbial turnover processes in soil are not sufficient to make precise predictions. Instead, most predictions are based on empirical data obtained from chemical extractions over short time periods or by extrapolation of data obtained from 2–4-week mineralization studies conducted under ideal environmental conditions.

J. A. van Veen; J. J. Neeteson

SOLARIZATION

Soil solarization is a hydrothermal process involving solar radiation that disinfests soil of plant pathogens and weed pests and also effects other changes in the physical and chemical properties of soil that lead to improved growth and development of plants.

Solarization of soil is accomplished by placing a sheet of plastic on moist soil for 4–6 weeks during the warm summer months. The plastic sheeting must be laid close to the surface of the soil to reduce the insulating effect of an air layer under the plastic; the edges of the plastic are buried or anchored in the soil. Solar radiation then heats the soil, and the resulting heat and the moisture in the soil are conserved by the plastic cover. Over time, the heat and moisture bring about changes in soil microbial populations, release of soil nutrients, and changes in the soil structure that are beneficial to plants subsequently established there.

Technology. The effects of soil solarization were reported in Israel in 1976; it is a mulching process, and it may seem to resemble the practice of using plastic mulches to warm soils to increase seed germination and seedling growth in early spring months. However, soil solarization differs from such use of plastic in that killing soil temperatures (approximately 97°F or 37°C or higher) are attained for many soil-borne organisms, seeds, and young plants or seedlings.

Soil solarization has been especially useful in regions with high solar radiation where daily temperatures in summer months often reach 90°F (32°C) or more; it is a nonchemical, easily applied method that is of little or no danger to workers; it embodies the objectives of integrated pest management and reduces the need for toxic chemicals for plant disease and pest control.

Soil preparation. The area or field to be solarized should be disked, rototilled, or worked by hand to provide a smooth surface so that the plastic sheeting will lie flat. The presence of large clods or plant debris will raise the plastic above the soil surface and create air pockets that are effective insulators and prevent the high rise in soil temperature needed for solarization. Laying the plastic sheeting across deep furrows should be avoided. Soil can be prepared for solarization in several ways, the most common being level soil or raised beds. For crops grown on-the-flat or for solarization of areas such as lawns, the edges of the plastic sheeting are anchored or buried about 8 in. (20 cm) deep (**Fig. 1**). For bedded-up soil, the plastic sheeting is laid over the length of the bed and the edges buried in the furrows (**Fig. 2**). Plastic sheeting can be obtained in widths up to 13 ft (4 m) and in various thicknesses. For large-scale use of solarization on flat or level soil, after the first sheet is laid and anchored in the soil, additional sheets are glued together on one side and anchored in the soil on the other side. The ends of the plastic sheeting are also anchored in the soil.

Fig. 1. Transparent polyethylene plastic (1.5 mils thick) for soil solarization. (*a*) Plastic sheets laid on flat or level soil. (*b*) The subsequent sheets being attached by gluing the edge on one side to the sheet already laid and burying the other edge in soil.

Fig. 2. Transparent polyethylene plastic being laid on raised soil beds for solarization; the edges are buried in the furrows to hold the plastic in place.

Soil moisture. Soil solarization is best accomplished when the soil is moist (at least 70% of field capacity), a condition that increases the thermal sensitivity of soil-borne pathogens and pests and also enhances heat conduction within the soil. When conditions permit, plastic sheeting may be laid on dry soil, and then irrigation water can be run under the plastic in order to wet the soil to a depth of at least 24 in. (60 cm). If the field or area is level, shallow furrows made by tractor wheels during the laying of plastic sheeting provide convenient courses for the water. If the area or field is not level, the soil should be preirrigated and the plastic sheeting then laid within a few days, or as soon as the soil will support machinery without compaction. When soils are bedded up before solarization, the plastic is laid over the bed and anchored on each side in the furrows (Fig. 2). Irrigation water is then run in the furrows, and the moisture subs into the raised beds (moves by capillarity).

Experiments with perennial or woody plants have shown that soil under the canopy of these plants or trees can be effectively solarized without harming the plants. Plastic sheets up to 430 ft^2 (16 m^2) in area are cut to the center to bring the plastic around the base of the trunk. The cut edges and the plastic around the trunk are then sealed with a transparent plastic tape. The outside edges are anchored in the soil. In postplant soil solarization, it is sometimes more convenient to irrigate the soil before applying the plastic. After an initial preirrigation or irrigation under the plastic, no further irrigation is advisable since the soil will remain moist for several months. In semiarid regions, postplant solarization of soil around fruit trees greatly reduces the amount of water needed for irrigation.

Plastic film. Plastic sheeting used in soil solarization provides a barrier to losses of moisture and heat from soil due to evaporation and heat convection. Clear or transparent polyethylene plastic sheeting is more effective for heating soil than black plastic, since the transparent plastic allows greater transmission of solar energy; it gives maximum transmittancy of radiation from wavelengths of 0.4 to 36 micrometers. Other

plastics are also effective, and in some instances superior to polyethylene. Colored transparent plastics tend to reduce the formation of water droplets on the underside or soil side of the plastic sheeting that may act as reflectors of radiation. Polyvinyl chloride films also have been used effectively for soil solarization in greenhouses.

The thickness of the polyethylene sheeting also affects the heating of the soil; film 1 mil (25 μm) thick results in temperature increases of soil several degrees higher than thicker films that reflect more solar energy. Although 1-mil (25-μm) or 1.5-mil (37.5-μm) polyethylene plastic sheeting is durable and proportionally less expensive than thicker films and will last for the solarization period, the usual practices with soil solarization, at least in gardens or for solarization of individual trees or small areas, employ plastic sheeting 2–4 mils (50–100 μm) in thickness. Polyethylene sheeting used for soil solarization requires the presence of an ultraviolet inhibitor to prevent its deterioration by sunlight.

Soil temperature. Day length, air temperature, cloud cover, and the darkness of soils all have marked effects on the heating of soil during solarization. The ideal time is during the summer months, when the days are longest and air temperatures are highest. Maximum temperatures in the upper 6–12 in. (15–30 cm) of soil are reached 3–4 days after solarization begins. Depending on soil depth, maximum temperatures of solarized soil are commonly between 108 and 131°F (42 and 55°C) at 2 in. (5 cm) depth and range from 90 to 97°F (32 to 36°C) at 18 in. (46 cm) depth (**Fig. 3**). The white curve in Fig. 3*a* refers to the time at different constant temperatures that is required to kill 90% of the microsclerotia or population of propagules of *Verticillium dahliae* in soil; it reflects the great excess of heat in solarized soil relative to the amount needed to kill this pathogenic fungus. The data for this curve were obtained in other experiments. Dark soils reach higher temperatures during solarization than lighter-colored soils due to their greater absorption of solar radiation.

Modes of action. Soil solarization encompasses changes in the biological, chemical, and physical properties of soil, changes that are often interdependent, dynamic, and evident for two or more growing seasons. Reduction in plant diseases and pests, coupled with better soil tilth and increases of mineral nutrients in solarized soil, contributes to the improved growth response of plants and increases in the quantity and quality of crop yields.

Biological effects. Soil microorganisms whose growth and development are inhibited at temperatures of about 86°F (30°C) or higher include most plant pathogens and some pests. The death rates of these organisms in moist soil at temperatures of about 97°F (36°C) or higher increase linearly with temperature and time. In contrast, other microorganisms in soil are thermotolerant, and their populations decrease less during solarization. Thermotolerant organisms include Actinomycetes and *Bacillus* species and certain mycorrhizal fungi; especially in the absence of soil-borne

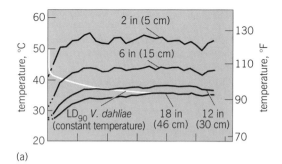

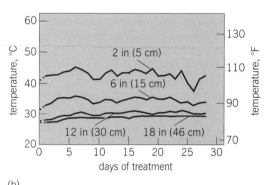

(a)

(b)

Fig. 3. Daily maximum soil temperatures (a) in moist soil during solarization with 25-μm clear polyethylene sheeting and (b) in dry nonsolarized soil at indicated depths. Each temperature value is an average of four experiments made during June and July at Davis and Shafter, California. (After G. S. Pullman, J. E. DeVay, and R. H. Garber, Soil solarization and thermal death: A logarithmic relationship between time and temperature for four soil-borne plant pathogens, Phytopathology, 71:959–964, 1981)

pathogens and pests, they are often highly beneficial to plant growth. In some soils, solarization causes biological changes that suppress the reestablishment of pathogens.

Chemical effects. Among the chemical changes that occur in soils during solarization are significant increases in soluble nutrients, such as nitrogen, calcium, and magnesium, which are associated with the increased growth response of plants grown in solarized soils.

Physical effects. The physical effects of soil solarization are largely unknown, but marked improvement in soil tilth occurs. During solarization soil moisture tends to rise to a cooler soil surface at night, whereas during the day the moisture tends to move toward a cooler temperature zone deeper in the soil. This diurnal cycling of soil moisture is believed to leach sodic layers and thus accounts for the reduction of salinity caused by solarization in some soils.

Applications. Soil solarization is a safe, nonchemical, and effective method for plant disease and pest control. When used in conjunction with agricultural chemicals and biological control agents, soil solarization can be extended to regions where environmental conditions are less suitable for it. Use of solarization is increasing in orchard, field, and truck crop farming as well as in greenhouse and nursery operations; it is also finding increased use for garden and landscape im-

provement. Considering the multiple benefits of soil solarization and its lasting effects, the costs are competitive with other methods for control of soil-borne diseases and pests.

For background information see AGRICULTURAL SOIL AND CROP PRACTICES; IRRIGATION (AGRICULTURE); MYCOR-RHIZAE; SOIL; SOIL ECOLOGY; SOIL MICROBIOLOGY; SOLAR RADIATION in the McGraw-Hill Encyclopedia of Science & Technology.

James E. DeVay

Bibliography. J. Katan, Solar heating (solarization) of soil for control of soil-borne pests, *Annu. Rev. Phytopathol.*, 19:211–236, 1981; J. Katan et al., Solar heating by polyethylene mulching for the control of diseases caused by soil-borne pathogens, *Phytopathology*, 76:683–688, 1976; J. J. Neeteson and J. A. van Veen, Mechanistic and practical modelling of nitrogen mineralization-immobilization in soils, *Advances in Nitrogen Cycling in Agricultural Ecosystems: Proceedings of the Symposium on Advances in Nitrogen Cycling in Agricultural Ecosystems*, Brisbane, 1988; E. A. Paul and J. N. Ladd (eds.), *Soil Biochemistry*, 1981; G. S. Pullman, J. E. DeVay, and R. H. Garber, Soil solarization and thermal death: A logarithmic relationship between time and temperature for four soil-borne plant pathogens, *Phytopathology*, 71:959–964, 1981; J. J. Stapleton and J. E. DeVay, Soil solarization: A non-chemical approach for management of plant pathogens and pests, *Crop Protect.*, 5:190–198, 1986; J. J. Stapleton and J. E. DeVay, Thermal components of soil solarization as related to changes in soil and root microflora and increased plant growth response, *Phytopathology*, 74:255–259, 1984; J. J. Stapleton, J. Quick, and J. E. DeVay, Soil solarization: Effect on soil properties, crop fertilization, and plant growth, *Soil Biol. Biochem.*, 17:369–373, 1985; J. A. van Veen and P. J. Kuikman, Soil structural aspects of decomposition of organic matter by micro-organisms, *Biogeochemistry*, 11:213–233, 1990; J. A. van Veen, R. Merckx, and S. C. van de Geijn, Plant- and soil-related controls of the flow of carbon from roots through the soil microbial biomass, *Plant and Soil*, 115:43–52, 1989.

Soil conservation

Erosion decreases the productivity potential of soil, except in rare locations where soils are unusually deep and fertile. Crop yields are decreased as topsoil is removed by erosion. Lower crop yields provide less income to sustain farm families and less food for the world. Recent research has identified and defined factors causing soil erosion and has developed management alternatives to combat these factors and conserve valuable soil resources. The most significant recent progress involves controlling erosion caused by irrigation on highly productive lands.

Furrow erosion processes. When water enters the furrow, erosive forces are created by wetting and water flow that exceed cohesive forces holding soil particles together and in place. The condition of the soil in the furrow when it is contacted by water largely determines if erosion will occur. When soils are dry, oxygen and nitrogen are adsorbed on the internal surfaces of aggregates. If these soils are wetted suddenly, water molecules rapidly displace the molecules of oxygen and nitrogen, and these gases join the air entrapped in the gaseous phase of the soil, causing pressure forces sufficient to break apart the soil aggregates. The bursting of small clods resembles tiny explosions. When water is applied early in the morning following a cool night, much less erosion results than when the water is applied during or following a hot dry afternoon.

Other eroding forces are shear forces caused by flowing water and transported materials. The furrow stream size and slope along the furrow are factors affecting water velocity that cause shear forces on the perimeter of the furrow. Under slow velocities, there may be practically no detachment of particles. Bonds between primary soil particles hold soil aggregates together. The strength of these bonds represents the soil cohesion or stability, which varies with clay content and type, organic matter content, compaction, adsorbed cations, time and water content since the last disruption, the wetting rate, and the chemical components in the water that is wetting the soil. The processes of formation and disruption of soil bonds are not yet well understood, but recent research has provided considerable enlightenment about factors controlling bond strength and cohesive forces. For example, bond strength increases with time since the last disruption, with increasing clay content, and with increasing amounts of gums and resins from decomposing plant residues present in the soil.

Factors affecting furrow erosion. The greater the slope along the irrigation furrow, the higher will be the water velocity and the resulting shear forces. Recent research has shown that erosion is approximately a two- to three-power function of furrow slope. Also, erosion is about a 1.5-power function of stream size. These two factors give rise to the eroding forces. Factors that impede furrow erosion are crop residues in the furrows and surface roughness of the furrows, which slow down the water velocity and thereby reduce the shear or erosion forces. The amount of residue on and in the soil surface can be controlled by cropping sequences and the kind and number of tillage operations between crops. Furrow surface roughness can also be controlled by tillage operations, the previous crop, and the water content of the soil when tilled. Analyses of this information show that furrow erosion can be controlled because a number of these factors can be controlled.

Erosion and sediment loss. Most of the furrow erosion takes place along the upper one-third of the furrow length and along about the last 15 m (50 ft) immediately above the tailwater ditch (the channel made to carry runoff water from the field) where a convex field end has developed. The cause of this development, which is a condition of increasing slope toward the ditch, is that past management has kept the

tailwater ditch 15 cm (6 in.) or more deeper than the furrow ends to facilitate drainage. This practice has caused accelerated erosion from the tailwater ditch, upslope along the furrows. The erosion along the upper one-third of the field exposes subsoils, which are generally less productive than topsoils. Usually, most of the eroded topsoil is subsequently deposited on the lower half of the field, but some of it may reach the tailwater ditch and be carried away in the drainage water.

These erosion processes are a natural consequence of the irrigation requirement, which is to supply enough water to meet crop needs over the entire length of run or furrow length. To meet this requirement, the size of the stream placed in the upper end of the furrow is often large enough to be erosive. As the stream progresses down the furrow, its size diminishes as a result of infiltration into the soil. Usually about one-third of the way toward the lower end of the field, the stream size has decreased sufficiently that it no longer has enough erosive energy to overcome cohesive forces; erosion ceases. A short distance farther, the stream becomes still smaller, and it no longer has enough energy to transport the soil that it eroded upslope. This soil is often known as sediment once it is suspended. At this point, sedimentation begins, and soil materials eroded from upslope areas are deposited in the furrows. The quantity of deposition depends upon the stream size and field slope, but generally most of the material is deposited before it reaches the lower end of the furrow. Hence, this process causes topsoil redistribution (**Fig. 1**), which seriously reduces crop yield on the eroded area and seldom increases it on the deposition area. This occurs because there is a critical requirement of topsoil depth for maximum crop productivity. When erosion decreases topsoil depths to less than this critical depth, crop yields are decreased. However, when topsoil is deposited onto areas that already have topsoil depths greater than the critical depth, crop yields are not changed.

Reducing erosion and sediment loss. Five years of intensive research has demonstrated that furrow erosion can be almost eliminated by changing cropping sequences and using conservation tillage practices on furrow-irrigated land. Studies have shown that cereals and corn can be grown without tillage following

Fig. 1. Aerial photograph illustrating the exposure of light-colored subsoils as a result of furrow erosion of the upslope one-third of the fields.

Fig. 2. Buried-pipe erosion and sediment-loss control system. (*a*) First irrigation after installation. (*b*) Basins filled with sediment after five irrigations.

alfalfa killed with herbicide, and that yields are maintained and requirements for nitrogen fertilizer are decreased. The same irrigation furrows used for alfalfa are cleaned and used to irrigate the cereal or corn. These furrows have uniform infiltration rates, and they erode very little, if any. Similarly, either cereal can be grown following corn or corn can be grown following cereal without tillage with the same effect of eliminating furrow erosion. When row crops such as dry beans or sugarbeets are to be grown, the smallest number of tillage operations possible to prepare a seedbed and plant the crop are used. This involves shallow tillage, generally only slightly deeper than the furrows from the previous crop, and it is done with equipment that leaves residue on and mixed into the soil surface. Studies have shown that furrow irrigation can be done successfully in the presence of crop residues on and mixed into the soil surface.

All of the tillage practices used in the newly developed management systems can be done without additional tillage implements. Some minor adaptations may be needed for seeding corn without previous tillage following alfalfa or cereal. Usually, mounting a cutting coulter (a serrated, sharp disk) or a small bull tongue shank (a vertical metal tool for cutting the soil) ahead of these seeders assures adequate seeding depth. Furthermore, results have shown that the number of tillage operations required over a 7-year cropping sequence can be reduced from near 40 to less than 10. Yields of the specific crops are not affected by their position in the sequence. The savings resulting from fewer tillage operations is reflected as a significant increase in net farmer income. Thus, farming to

prevent or reduce erosion on furrow-irrigated land benefits the farmer economically in the short term as well as in the long term by protecting the productivity of the soil resource.

The loss of eroded soil or sediment from convex field ends can be almost entirely eliminated by installing a buried-pipe erosion and sediment control system. The pipe is buried along the lower end of the field, and it has vertical inlets at intervals. The tops of these intervals are set at the level desired to correct the convex end when sediment is accumulated to the tops of these inlets. Small earthen dams are placed on the downslope side of these inlets to form small sediment basins along the lower end of the field. As sediment from the upper part of the convex end and further upslope settles and fills the basins, the convex end is corrected. The pipe remains as a means to drain runoff water from the field, eliminating the need for a tailwater ditch. **Figure 2***a* shows a newly installed system in operation, and Fig. 2*b* illustrates the same system after five irrigations with the convex end corrected.

Erosion control on sprinkler-irrigated land. The same erosion control principles apply to lands irrigated by sprinklers and by furrows. The differences are that sprinkler irrigation does not require furrows for water distribution and that it can be used on steeper lands than furrow irrigation. Greater amounts of residue can be tolerated under sprinklers.

The most serious erosion problem under sprinkler irrigation is on the steep slopes. In addition to conservation tillage systems to maintain residue on and in the soil surface, a new practice known as reservoir tillage has been developed and evaluated. This practice forms thousands of small water storage basins in the soil surface (**Fig. 3**). As water is applied by sprinkler systems, their reservoirs catch the water and store it temporarily until it infiltrates the soil. This prevents runoff from steep slopes, thus preventing erosion, and assures that adequate water infiltrates the soil to support the crop. These small reservoirs function best when they are depression-scooped (removing several shovelfuls of soil at a point) or pressed into the land surface rather than being formed by earthen dams in furrows. Erosion on sprinkler-irrigated land can be eliminated by using reservoir tillage and appropriately designed sprinkler systems.

Fig. 3. Reservoir tillage on sprinkler-irrigated land.

For background information SEE AGRICULTURAL SOIL
AND CROP PRACTICES; SOIL; SOIL CONSERVATION; SOIL
MECHANICS in the McGraw-Hill Encyclopedia of Science & Technology.

David L. Carter

Bibliography. D. L. Carter, R. D. Berg, and B. J. Sanders, The effect of furrow irrigation erosion on crop productivity, *Soil Sci. Soc. Amer. J.*, 49:207–211, 1985; W. D. Kemper et al., Furrow erosion and water and soil management, *Trans. ASAE*, 28:1564–1572, 1985; B. A. Stewart and D. R. Nielsen (ed.), *Irrigation of Agricultural Crops*, 1990.

Soil ecology

Recent advances in soil ecology include development of quantitative methods for studying belowground ecosystems in forests, the sensitivity of forests to air pollution, and the role of nematodes in deep-rooting systems in desert environments.

BELOWGROUND ECOLOGY OF FORESTS

The belowground portion of terrestrial ecosystems has historically received less study than the aboveground. This has been due in part to methodological difficulties, the inability to see and measure objects and processes belowground, and the perception that it is possible to understand the whole ecosystem while remaining ignorant of this hidden realm. Since 1970, it has become apparent that studying only the half of the ecosystem that is aboveground can badly mislead workers seeking to understand how the system functions and responds to disturbances such as acid rain or heavy-metal contamination. Today, scientific studies emphasize the flow of matter and energy between the various components, the processes that control these flows, and the role of plant root systems in regulating the structure and function of the whole forest.

Belowground components. In the simplest view, a belowground ecosystem contains only two components: plant roots and the growth medium (soil). Typically, belowground is defined as the zone that starts at the surface organic layer on top of the mineral soil and extends down to consolidated rock (bedrock), in which biological activity is absent and no organic matter accumulates. A more realistic view recognizes the belowground ecosystem as a complex and extremely heterogeneous mosaic of many interacting components (for example, roots, microorganisms, soil animals, organic matter, and mineral particles). Each component can be subdivided into many types and structures with differing functions in the belowground.

This definition of the belowground is valid for temperate zone forests but not for wet tropical forests. In tropical forests, root growth is not restricted to the soil; roots may grow into and on top of freshly fallen leaves, over exposed rocks, into streams, or up the trunks of other trees (apogeous roots; **Fig. 1**). Apogeous roots are commonly found climbing up the trunks of trees (such as palms) that collect nutrients in rainwater more efficiently because of their branching form. Roots are also produced adventitiously from branch tissues when organic material collects in branch crotches and cavities. This type of root growth in the tree canopy resembles the standard air-layering technique used in horticulture to propagate ornamentals. In addition, epiphytes such as bromeliads and orchids have root systems growing along the branches of trees, a role played by mosses in cool, damp northern forests.

Much of past soil research concentrated on studying soil (bulk soil) that had low biological activity and from which roots were removed. But the most critical area in any soil is the very narrow region, less than 0.06 in. (2 mm) wide, that surrounds fine roots. It is in this region, known as the rhizosphere, that biological activity is most pronounced, so that this soil has distinctive characteristics. Much of the rhizosphere may be discarded when roots are separated from soil; thus, traditional methods of studying soils underestimate the level of biological activity.

In most soils, mineral particles make up the bulk of the soil, ranging from microscopic, chemically active clay to larger, more inert sand and gravel. These serve as the primary source for nutrients such as calcium, potassium, phosphorus, and trace minerals, which are released through chemical dissolution (weathering).

Organic matter includes all material from recently dead leaves, branches, and roots to advanced decay products such as humic acids. Dead organic matter serves as an important storage reservoir for many nutrients, especially nitrogen, in all but the youngest soils. It also provides energy for saprophytic organisms (those that derive energy from breaking carbon bonds in dead organic matter).

Humic acids are large (molecular weight 5000–1,000,000+) organic complexes that are highly resistant to decomposition. They serve as a long-term storage pool of slowly released nutrients, especially nitrogen and phosphorus. They also provide a signif-

Fig. 1. Apogeous root of guaba (*Inga vera*) with large cluster of nitrogen-fixing nodules growing up the stem of a Sierra palm (*Prestoea montana*) in Puerto Rico.

icant amount of cation-exchange capacity, and they are important in soil development because of their ability to chelate iron and aluminum ions and transport them downward in the soil (a process known as podzolization).

Microorganisms include single-celled bacteria and actinomycetes, algae, and fungi (both symbiotic and free-living). They are responsible for many important functions, including decomposition and mineralization (the chemical breakdown of organic matter and the release of nutrients in plant-available form through the secretion of enzymes into the soil) and nitrogen (N_2) fixation (the conversion of atmospheric N_2 into biologically available ammonium, accomplished by certain specialized free-living or symbiotic bacteria and actinomycetes). In most forests, plant growth is limited by the rate of microbially mediated nutrient release.

Soil animals range from microarthropods (invisible to the naked eye), through ants, termites, and earthworms, to large ground-dwelling mammals and reptiles such as gophers and tortoises. Smaller animals (primarily invertebrates) are responsible for the initial physical breakdown of dead organic matter by eating and excreting it in a more readily decomposable form. A wide variety of animals excavate tunnels or burrows in the soil, with primarily beneficial effects. Soil aeration and drainage are improved by the creation of channels and macropores. Organic matter may be mixed downward into the soil, improving soil structure, fertility, and water-holding capacity. Conversely, large amounts of mineral soil may be moved from belowground to the surface, counteracting the downward leaching of nutrients and bringing fresh weatherable minerals into the primary rooting zone. This bioperturbation leads to a continual rejuvenation of the upper soil.

Root systems may be divided into several structural types that have an increasing impact on soil biological activity as the size of the root systems decreases. Coarse woody roots (>0.2 in. or 5 mm in diameter) are long-lived (typically the same age as the plant) and have annual increases in diameter and growth rings similar to aboveground stem and branches. Small woody roots 0.06–0.2 in. (1–5 mm) in diameter may extend 10–30 ft (3–10 m) from the base of the tree and serve as a pipeline for the transport of water and nutrients. Fine roots are generally short-lived, but they have a very high surface area; they provide most of the absorptive surface for the plant. In a 55-year-old hardwood forest in New Hampshire, for example, roots less than 0.024 in. (0.6 mm) in diameter contributed only 20% of the root system biomass but 85% of the surface area. Herbaceous plants primarily have fine roots.

Very importantly, the vast majority of plants have symbiotic associations known as mycorrhizae located on fine root tips where a fungus lives in intimate contact with root cortical cells (**Fig. 2**), which provide major absorptive additions to the regular root system. This mutually beneficial relationship occurs because the fungus is unable to use complex carbon sources as

Fig. 2. Dense growth of mycorrhizal roots (short, light-colored structures) on root system of Douglas-fir (*Pseudotsuga menzeseii*) seedlings.

food and is dependent on the plant for simple sugars. In return, the fungus greatly increases the ability of the root system to absorb water and nutrients by increasing the root surface area involved in the acquisition of these resources.

Belowground processes. The general nature of the processes that allow ecosystems to maintain themselves has been understood for decades, but recent studies indicate the need for major revisions in the traditional picture. Except for photosynthesis, all major ecosystem processes (such as weathering, decomposition, mineralization, N_2 fixation, and nutrient uptake) take place belowground. Losses of soluble nutrients by leaching from an ecosystem are minimized by biota because of the tight linkage of growth-death-decomposition-uptake cycles; released nutrients are taken up quickly by microorganisms and plants. This allows lush forests (for example, wet tropical forests) to grow in very nutrient-poor soils, since the pool of nutrients is continuously recirculated.

Roots were traditionally viewed as passive in these cycles, serving primarily to absorb water and nutrients. However, roots have a much more dynamic role in belowground processes in the newly emerging view. For example, they stimulate soil mineral weathering by releasing carbon dioxide (CO_2) and organic acids with low molecular weights (primarily Krebs-cycle acids such as oxalic, citric, and malic), and thus may actually increase the loss of nutrients from the rooting zone.

It has long been thought that the major source of energy for the belowground ecosystem is through the death of aboveground plant parts (leaves and stems). Since root systems usually contain much less standing biomass than the aboveground portion of plants, it was assumed that they contribute correspondingly less dead organic matter to the soil to be decayed by microorganisms. However, measurements since 1980 have shown that in many ecosystems fine roots and mycorrhizae may be the major source of energy supplied to the soil.

In subalpine fir forests of Washington state, fine roots (with less than 3% of the living biomass of the forest) contribute two-thirds of the dead organic matter to the soil because of their rapid turnover. This is an extreme example, but in all cases roots contribute energy to the soil far out of proportion to their biomass. These measurements do not include root contributions from other sources (such as exudation of soluble carbohydrates, sloughing of root cap cells during growth, and the growth and death of external mycorrhizal hyphae), which may be considerable but have yet to be measured in any forest.

Populations of microorganisms that are dependent on a direct supply of energy from the roots (including mycorrhizal fungi and many rhizosphere bacteria) may decline if this energy source is removed, as happens when a forest is harvested. In nutrient-poor, droughty, or otherwise harsh areas, the loss of these organisms may make reforestation difficult. The degradation of previously forested land caused by the severing of the interdependent plant–soil linkage is evident in many areas of the world.

Patterns of decomposition and nutrient cycling are also being reexamined. Plants have long been thought to be dependent on the action of free-living saprophytic microorganisms to release mineral nutrients in a form that plants can absorb. While this dependency is still recognized, plant roots may have a more important role in decompositional processes than was previously thought. Though roots do not release enzymes that break down long-chain organic molecules, they are able to produce enzymes that convert some organic substrates to mineral form. For example, many mycorrhizal fungi secrete phosphatases, which release inorganic phosphate from organic molecules. In addition, some mycorrhizal fungi have the ability to use small protein and peptide molecules as a source of nitrogen, thus possibly short-circuiting the decomposition-uptake pathway.

Mycorrhizal fungi may affect nutrient cycling patterns through their ability to connect the root systems of different plants by hyphal bridges. Though the tendency of woody roots to graft to each other is well documented, these connections are usually between individuals of the same species. Because a few mycorrhizal species can infect many different plant species, they may potentially connect the root systems of a wide variety of plants in a community. Unlike the aboveground, where individuals are basically distinct (except for clonal plants such as grasses and ferns), the belowground may consist of a highly interconnected

Fig. 3. Exposed roots of Western hemlock (*Tsuga heterophylla*) in coastal Washington, demonstrating extreme intermingling between roots of different trees. (*Photograph by R. Edmonds*)

network of root systems (**Fig. 3**), possessing the potential for exchanging nutrients and carbon between living plants, or from dying to living ones, without ever entering the death-decomposition-uptake pathway. These connections have been demonstrated in small-scale experiments, but their extent in natural ecosystems is currently unknown.

Current information on the belowground subsystems in natural forest ecosystems is very sparse. Much of this is due to the methodological problems of studying a system that cannot be visualized without disturbing it. Current technological innovations (for example, fiber optics and nuclear magnetic resonance) should contribute greatly to advancing knowledge of this hidden half.   *David Publicover; Kristiina Vogt*

LITTER DECOMPOSITION AND AIR POLLUTION

Until recently, the majority of studies on the sensitivity of forests to air pollutants focused on the direct effects of air pollutants on aboveground tree components. Recent data collected from some forested areas of North America and Europe, where air pollution is suspected to be contributing to forest decline, indicate that in many cases the major effects of air pollution on forests are mediated via the processes of plant–soil nutrient relations.

Scientists have designed experiments to study the effects of air pollution on temperate forest ecosystems by establishing experimental plots along concentration gradients of atmospheric deposition. Recognizing that efficient nutrient cycling is critical for sustained tree growth and site productivity, researchers have placed greater emphasis on studies of the effects of air pollution on litter decomposition and nutrient cycling. Results of recent studies suggest that exposure to low to moderate levels of acidic precipitation and sulfur pollutants does not disrupt litter decomposition. Decomposition of pine litter in southern California is not inhibited by high concentrations of ozone. Litter

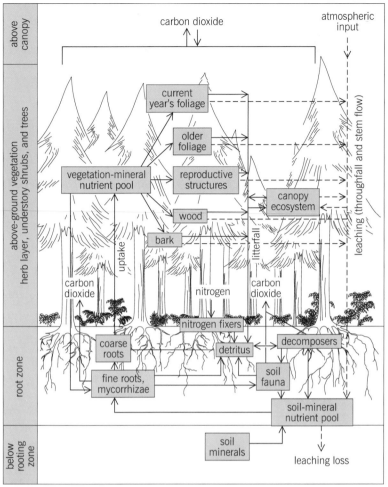

Fig. 4. Conceptual model of nutrient cycling in coniferous forests. Mycorrhizae are fungi that form a symbiotic relationship with roots and enhance root uptake of water and nutrients. Nitrogen fixers are microorganisms that convert atmospheric nitrogen (N_2 gas) to a form usable by plants. (*After D. W. Johnson et al., Nutrient cycling in forests of the Pacific Northwest, in R. L. Edmonds, ed., Analysis of Coniferous Forest Ecosystems in the Western United States, Institute of Ecology, 1982*)

and concomitant nutrient release controls the rate of forest production (**Fig. 5**). Sufficient interference with nutrient cycling will affect site productivity and ecosystem functioning. When exposure to air pollution affects the processes of nutrient cycling, forest productivity and ecosystem stability are likely to be reduced.

Air-pollution gradients. Air pollutants exist in aerosol, gaseous, or particulate forms; they can enter ecosystems via wet or dry deposition. Air pollutants usually occur in mixtures. Some of the more common pollutants of biological importance to which forests are exposed include photochemical oxidants such as ozone, oxides of sulfur and nitrogen, acidic compounds (mainly acids formed from water vapor and oxides of sulfur and nitrogen), and heavy metals (such as lead, nickel, cadmium, copper, and silver).

Pollutants emitted from a local, well-defined location are said to originate from a point source. A coal-burning electric power plant and a copper smelter are examples of point sources. Regional pollutants or pollutant precursors disperse over relatively large areas; or they originate from an extensive area containing many sources of pollution, such as motor vehicles and industrial sites. When many point sources are located within a given area, they may give rise to a regional pollution problem. Air-pollution gradients typically extend from a point source to approximately 2–25 mi (3–40 km) or more from the source. Regional pollution gradients may span a distance from approximately 150 mi (240 km) to many hundreds of miles or kilometers, often traversing several states, provinces, or even countries.

Studies of the impact of air pollution on ecosystems along air-pollution gradients have the advantage of measuring effects in natural field settings. Observations of different effects along a pollution gradient, however, does not prove causality. A major drawback of gradient studies is that environmental and site parameters other than air-pollution exposure also vary along air-pollution gradients, and this makes interpretations of results more difficult and less conclusive. The greatest amount of information, understanding, and confidence in results can be gained when results from studies using air-pollution gradients are combined with those from studies in which nontarget variables (such as temperature, precipitation, and soil and plant heterogeneity) are controlled. Such experiments are commonly performed by using plant growth chambers, fumigation chambers, tree-branch fumigation chambers, or open-air (no chamber enclosure) exposure facilities in which plants are exposed to known concentrations of air pollutants in their natural setting.

Factors determining decomposition rate. The principal factors controlling the rate of litter decomposition are soil fertility, nutrient content of vegetation and litter, vegetation type, age of plant material on entering the forest floor, chemical composition of the litter (litter quality), moisture content of litter, temperature, and solar radiation. Foliage of hardwoods produces nutrient-rich litter that decomposes rapidly, compared with foliage from conifers. Conifers retranslocate nutrients

decomposition is reported to be retarded in areas of Germany, where forests are exposed to highly acidic precipitation containing high levels of sulfur and nitrogen, as well as of heavy metals. Litter decomposition is also reduced in forests located close to pollution sources emitting highly inhibitory substances, such as heavy metals or sulfur pollutants.

Dead foliage, roots, and branches from trees, as well as dead understory plants, constitute the plant-litter component of a forest ecosystem. Nutrients lost from trees in litter fall are slowly released from litter by the processes of leaching and by biological decomposition carried out by the fauna and microflora of the forest floor. Most of the nutrients released from decomposing litter are once again available for use by plants or other microorganisms. This nutrient recycling is crucial for conservation of nutrients within an ecosystem (**Fig. 4**). In many ecosystems, either water or nutrient availability is the rate-limiting factor for plant growth. When the availability of one or more nutrients (primarily nitrogen) determines the growth rate of forest species, the rate of litter decomposition

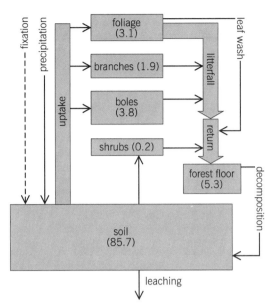

Fig. 5. Nitrogen cycle in a young Douglas-fir ecosystem. The distribution of nitrogen in the system components and the movement of nitrogen among them is indicated by the size of the boxes and by the arrows. The percentage of the nitrogen in the ecosystem contained in each component is given in parentheses. (*After R. F. Powers, Mineral cycling in temperate forest ecosystems, in R. J. Laacke, ed., California Forest Soils, University of California, 1979*)

internally to younger foliage from senescing foliage prior to needle abscission, and thus conifers produce nutrient-poor litter. Conifer litter is also high in lignin and tannins, materials that decompose very slowly.

The effects of air pollution on the rate of litter decomposition are superimposed on the existing environmental and biological factors that determine the rate of litter decomposition in an ecosystem. Also, the extent of the effects of air pollution on decomposition is dependent upon the concentration of pollutants, duration and timing of exposure, and sensitivity of plants and decomposers to the pollutants.

Air pollution can affect litter decomposition in various ways: by directly affecting decomposer organisms, by causing physiological changes in plants from which the litter originates, by altering soil chemistry and plant–soil nutrient relations, and by causing shifts in plant community composition (that is, plant species tolerant to pollution stress are favored over more sensitive species). When the physiology of plants is significantly affected by air pollution, or when atmospheric deposition alters plant–soil nutrient relations, litter quality, nutrient content, and decomposability are likely to be altered as well.

Effects of pollutants. The biological activity of plants and decomposers alike will be greatly reduced near point sources emitting high concentrations of heavy metals, sulfur dioxide, particulates, or combinations of such potentially deleterious pollutants. In cases of moderate pollutant exposure, microbial decomposer communities are often sufficiently diverse and adaptable to continue decomposer activity in spite of the pollutant stress. Microbial species that are

inhibited by pollutants may become less prevalent and be replaced to a degree by other species.

Some microbes have been found to grow faster when exposed to particular pollutants. Bacteria tend to be sensitive to high acidity, while many fungi grow well under acid conditions; thus acid precipitation is more likely to inhibit bacterial decomposition than fungal decomposition.

Nitrogen is often the limiting nutrient for plant growth and decomposer activity. Pollutants that contain nitrogen, and do not occur at levels that are inhibitory to plants or decomposers, may increase plant growth, plant and litter nitrogen content, and rate of litter decomposition. Stimulatory effects due to nitrogen deposition may not be long-lasting, as other factors may limit the growth of plants or decomposer organisms.

Heavy metals. The most dramatic and visible examples of the impacts of air pollutants on litter decomposition and nutrient cycling are along gradients downwind from point sources of heavy metals. Heavy-metal pollution almost always involves at least two metals. Close to a point source, high concentrations of metals can typically be found in the vegetation, litter layers, and soil. Litter has been found to accumulate in such situations because of the toxicity of heavy metals to the decomposer organisms. This is essentially due to an inhibitory effect on microbial enzymes. Over a period of years, more and more nutrients may become unavailable for plant use as they accumulate in the litter layer or are leached from the litter layer and through the soil profile. Productivity of sites located in the heavily polluted region of the gradient is greatly reduced, as the high concentration of metals in the ecosystem commonly reduces growth of many plant species.

Fresh pine needles contain about 20% lignin by dry weight. Lignin would account for an even greater percentage of the dry weight of partially decomposed litter needles. Several researchers have suggested that the content of phenolic compounds in litter, such as lignin, is an important regulator of nitrogen cycling. Lignin degradation is especially sensitive to heavy-metal toxicity, which may explain why the later stages of litter decomposition have been found to be more sensitive to heavy-metal toxicity than the earlier stages.

Sulfur pollution. Litter decomposition was studied at three pine forest sites located along a sulfur gradient downwind of a gas-processing plant in west-central Alberta, Canada. The forest had been exposed to sulfur dioxide and elemental sulfur dust. The pH of the litter layer was 3.5 at the highly polluted site compared with 4.7 at the less polluted sites. Populations of litter-colonizing bacteria and fungi were reduced at the highly polluted site. Litter decomposition was reduced, and litter accumulated on the forest floor at this site, even though litter fall rates did not differ significantly among the three sites. Results from this study suggest that the chemical characteristics of the needles were important in control of decay rates during the early stages of decomposition, while environmental factors and pollutant inputs became more important during the later stages.

Urban smog. Symptoms of damage attributed to ozone have been observed since the early 1960s on ponderosa and Jeffrey pine trees in the San Bernardino Mountains, east of Los Angeles, California. Fumigation experiments in chambers and other controlled studies have confirmed that ozone was causing the symptoms observed in ponderosa pine. An ozone concentration gradient spanning approximately 34 mi (55 km) has been characterized in the San Bernardino Mountains, with the highest ozone concentrations occurring on the western end of the gradient. Other pollutants such as sulfur compounds and, to a greater extent, nitrogen compounds also increase in a westerly direction across the air-pollution gradient.

Litter from ponderosa pine trees in the highly polluted sites in the San Bernardino Mountains sustains a more numerous and diverse population of decomposer fungi, is higher in nitrogen content, and decomposes faster than litter from pine trees growing in the low-pollution sites. The composition of the fungal community of ponderosa pine litter differs significantly in the high- and low-pollution sites. Until recently, it was thought that premature needle abscission in ozone-stressed ponderosa pine was the cause of younger, more nutrient-rich, and therefore more easily decomposable litter in the high-pollution plots. Subsequent studies revealed that foliage of sugar pine and incense cedar, which are relatively tolerant to ozone, was higher in nitrogen in the polluted sites and that litter of these species also decomposed faster in the high-pollution sites. Apparently, the presence of higher levels of soil nitrogen in the high-pollution sites results in higher levels of nitrogen in foliage and litter in both ozone-sensitive and ozone-tolerant conifer species.

Atmospheric deposition of nitrogen in the highly polluted areas of the San Bernardino Mountains may be stimulating litter-decomposing organisms and contributing to site fertility, thereby increasing the nitrogen content of foliage and litter in the high-pollution sites. The main effects of air pollution on litter decomposition seems to be mediated through increasing site fertility and production of litter that is higher in nitrogen and therefore decomposes at a faster rate.

Acid precipitation. Litter decomposition is usually retarded by acid precipitation only when it is extremely acidic (pH 2.0). Litter and soil organic matter have a high buffering capacity; thus, high acidic inputs are required to change the pH of litter. Evidence suggests that direct inhibition of litter decomposition occurs only in areas receiving very high loads of acidic deposition, or when acidic pollutants are accompanied by pollutants such as heavy metals or sulfur dioxide. Further studies are needed to better understand the long- and short-term effects of acid precipitation on litter decomposition due to the acidity itself, and the effects due to input of sulfur and nitrogen from anthropogenic sources.　　*Mark E. Fenn*

Nematode Herbivory in Desert Soils

Nematodes are abundant in all ecosystems, even deserts. Some nematodes directly affect plant production through herbivory by feeding on plant roots.

Studies of nematodes have been limited to surface roots and soil, usually less than 5 ft (1.5 m) deep. Plant-feeding nematodes do occur on *Prosopis glandulosa* (mesquite) roots to depths of 45 ft (15 m). In some ecosystems, such as deserts, the majority of herbivory may occur at soil depths that are rarely studied.

Measurement approaches. Nematodes are dominant soil invertebrates in most ecosystems, and they influence nutrient cycling by grazing on plant roots and on soil microfauna and microflora. Although the contribution of these microscopic fauna to total soil metabolism, as measured by evolution of CO_2, is estimated at less than 2%, they may have large effects on plant production through feeding on roots. The overall effect of plant-feeding nematodes is the physical and physiological disruption of roots, especially young roots, with subsequent reduction in root biomass and aboveground function.

It is difficult to quantify the effect of nematode herbivory on plants directly. A direct approach in field studies would measure plant production in control and nematicide-treated plots. This is impractical for all but very shallow-rooted ecosystems. Estimates of decreased plant production due to nematode herbivory that use this type of direct approach range from 6–13% in grassland ecosystems to 0.5–20% in agricultural systems. Most estimates of nematode herbivory are based on soil and rooting depths of less than 5 ft (1.5 m).

Consequently, indirect approaches for measurement of herbivory are necessary to assess the relative impact of the nematode on the plant under various environmental conditions. An appropriate approach for both deep- and shallow-rooted systems would be expressions of nematode numbers relative to root mass; or a herbivory index, that is, a calculation of an index weighted for nematode density and potential root damage by genera or species. The herbivory index approach would require knowledge of the type of life cycle, endoparasitic or ectoparasitic, of each nematode genus and the known effect on a plant species. Nematode genera that are endoparasites, such as *Meloidogyne* or *Meloidodera*, are considered to be more detrimental to roots via disruption of new cortical and stelar tissues; they might have a higher impact on plants than would other plant feeders. Endoparasites spend a majority of their life cycle in the root, protected somewhat from the soil abiotic environment, while ectoparasites rely on well-developed survival mechanisms, such as anhydrobiosis, to endure desert soil extremes.

Although nematodes are abundant in arid and semi-arid soils, the importance of nematode herbivory to plant production and nutrient cycling in the deep-rooting desert plants is just now being explored.

Deep-rooting desert plants. Plant roots occur to soil depths of ≥150 ft (50 m) in many ecosystems, especially deserts, grasslands, and savannas. These roots give access to deep soil moisture; they may increase nutrient uptake by the plant, particularly if infected by root symbionts such as rhizobia and mycorrhizal fungi. *Prosopis glandulosa*, a woody legume, is a unique desert plant, because its root system can adjust to varying soil conditions by penetrating to depths ≥15 ft (5 m), remaining close to the surface, or assuming distributions between these extremes. The plasticity of the rooting system can be used to study roots and soil nematode communities at depth.

A series of mesquite communities occurs in the northern Chihuahuan Desert, 19 mi (40 km) northeast of Las Cruces, New Mexico. In this desert, the 100-year annual precipitation is 8.3 in. (211 mm), with 60% of the annual precipitation occurring between July and October. Maximum temperatures in summer range from 94 to 104°F (36 to 40°C), and winter minimum temperatures regularly drop below 32°F (0°C). Four plant communities (grassland, dunes, arroyo, and playa) have variable rooting depths of greater than 4.5 ft (1.5 m). The grassland sites contain widely scattered mesquite and grasses (*Bouteloua, Aristida, Erioneuron*), and water is limited to incident rainfall. The dunes are dominated by low-growing mesquite that has accumulated wind-blown sand. The playa and arroyo receive accumulations of water into low-lying areas. The rooting depths at these sites from the shallowest to deepest are grassland < dunes < arroyo < playa.

Sampling of deep-rooting plants. Sampling to the depth of rooting in deserts requires special equipment to penetrate the different soil types, particularly the hard caliche, a calcium carbonate formation. A continuous sampling-tube-drilling system of a type frequently employed for soil environmental contamination studies was used to remove undisturbed soil cores from the rooting zones of plants at each of the sites. The corer took intact soil samples at increments of 4.5 ft (1.5 m) to a depth of 45 ft (15 m). To prevent contamination from soil that might contain nematodes from other depths, the corer was flame-sterilized prior to each new depth increment sampled. Seasonal sampling (winter, spring, and summer) accompanied the main growth events (phenology) of the plants.

To assess the abundance of nematodes associated with roots, both nematodes and roots from the soil samples were extracted with a semiautomatic elutriator. This is an automated system that washes the roots and nematodes from the soil samples onto fine sieves. Roots were separated from organic debris by hand-picking, and fresh root masses were measured. Plant-feeding nematodes were washed from the sieves, and counted and identified as to genus.

Distribution of nematodes with depth. Information on the four variable rooting depths of *P. glandulosa* showed that the distribution of the plant-feeding nematodes was generally related to root distribution (**Fig. 6**). Nematodes were recovered in all three seasons from the four sites. At all sites except the dunes and one depth at the playa, nematodes were found within the entire sampled root zone. Nematode density and root mass of *P. glandulosa* at all four sites decreased with depth. Nematodes were found to the maximum depth of recovered roots only at the playa, 33–36 ft

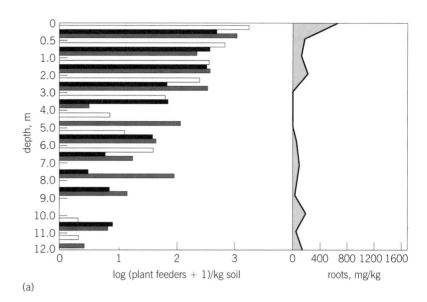

(a)

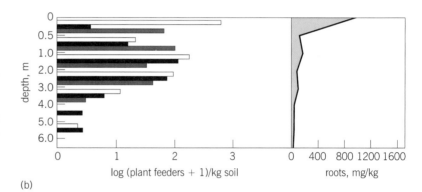

(b)

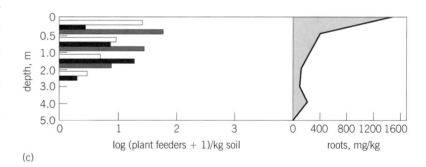

(c)

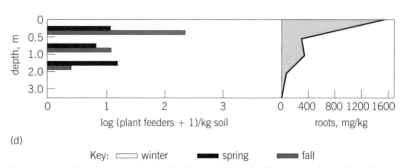

(d)

Key: ☐ winter   ■ spring   ■ fall

Fig. 6. Correlation of seasonal depth distribution of plant-feeding nematodes and root fresh mass for four Chihuahuan Desert sites: (*a*) playa, (*b*) arroyo, (*c*) dunes, and (*d*) grassland. The nematodes and roots decrease with depth. 1 m = 3.3 ft. 1 kg = 2.2 lb. 1 mg = 3.5 × 10⁻⁵ oz. (*After D. W. Freckman and R. A. Virginia, Plant-feeding nematodes in deep-rooting desert ecosystems, Ecology, 70:1665–1678, 1989*)

(11–12 m), and they occurred as deep as 15–18 ft (5–6 m) at the arroyo, 6–9 ft (2–3 m) at the dunes, and 3–6 ft (1–2 m) at the grassland. The highest numbers of nematodes were not always located with the greatest amount of roots. The dunes and grasslands had the greatest concentration of nematodes and roots in the surface 1.5 ft (0.5 m). At the playa, 75% of the plant-feeding nematodes and 90% of the roots were below 1.5 ft (0.5 m). The arroyo offered a major contrast, with 66% of the root mass recovered above 1.5 ft (0.5 m), but most of the nematodes were below that depth.

Diversity of nematode species. The diversity of genera varied among the four sites, although three genera dominated. These were *Meloidodera* sp., an endoparasite, and *Pratylenchus* and *Tylenchorhynchus*, both ectoparasites. The deep-rooting playa site had the greatest diversity, with seven genera, and it also had the greatest abundance of total nematodes ($38,600/\text{ft}^2$ or $415,000/\text{m}^2$) to 45 ft (15 m) depth and endoparasites (three genera for a total abundance of $113,000/\text{ft}^2$ or $1,220,000/\text{m}^2$) to 45 ft (15 m) depth.

Nematodes in deep-rooting systems. Nematodes were always positively associated with plant roots. The densities and diversity of species was greatest at the playa site. Although nematode density decreased with depth, both endo- and ectoparasitic nematodes occurred to depths of 45 ft (15 m). By using the indirect index of nematodes per gram of root or the herbivory index (based on nematode density weighted by an impact factor for each genus), the playa mesquite production was more affected by plant-feeding nematodes than was that of the other mesquite sites (see **table**). The index of nematodes per gram of root was highest at the playa, and the herbivory index was four times higher than at the other three sites. Depth of rooting did not determine the herbivory impact. The other three sites (grassland, arroyo, and dunes) also had nematodes at depths greater than 5 ft (1.5 m), but varied in herbivory index. The arroyo, with a rooting depth to 18 ft (6 m), had lower nematode densities than the shallower-rooted grassland, and dune mesquite had more ectoparasites and the lowest herbivory index.

It appears that in some ecosystems the majority of herbivory can occur at soil depths rarely studied. The occurrence in *Prosopis* sites of these nematode genera, which are known to cause economic losses in agricultural crops, suggests that nematode herbivory affects *Prosopis* production. However, plants have evolved mechanisms to cope with their environment, of which herbivorous nematodes are a part. It is possible that plants might tolerate nematode herbivory with little adverse effect on plant production. However, evidence that herbivory can actually benefit plant growth or fitness in natural ecosystems is still controversial.

For background information SEE ACID RAIN; AIR POLLUTION; FOREST ECOLOGY; HERBIVORY; HUMUS; MYCORRHIZAE; NEMATA; RHIZOSPHERE; ROOT; SOIL; SOIL ECOLOGY in the McGraw-Hill Encyclopedia of Science & Technology.

Diana W. Freckman

Bibliography. Biotic Interactions in Soil, *Agr. Ecosys. Environ.*, special issue, vol. 24, nos. 1–3, 1988; N. C. Brady, *The Nature and Properties of Soils*, 10th ed., 1990; M. E. Fenn and P. H. Dunn, Litter decomposition across an air-pollution gradient in the San Bernardino Mountains, *Soil Sci. Soc. Amer. J.*, 53:1560–1567, 1989; D. W. Freckman (ed.), *Nematodes in Soil Ecosystems*, 1982; D. W. Freckman and E. P. Caswell, The ecology of nematodes in agroecosystems, *Annu. Rev. Phytopath.*, 23:275–296, 1985; D. W. Freckman and R. A. Virginia, Plant-feeding nematodes in deep-rooting desert ecosystems, *Ecology*, 70:1665–1678, 1989; J. Harley and S. Smith, *Symbiosis*, 1983; R. P. Larkin and J. M. Kelly, A short-term microcosm evaluation of CO_2 evolution from litter and soil as influenced by SO_2 and SO_4 additions, *Water Air Soil Pollut.*, 37:273–280, 1988; D. A. Perry et al., Bootstrapping in ecosystems, *BioScience*, 39:230–236, 1989; C. E. Prescott and D. Parkinson, Effects of sulphur pollution on rates of litter decomposition in a pine forest, *Can. J. Bot.*, vol. 63, 1985; B. N. Richards, *The Microbiology of Terrestrial Ecosystems*, 1987; T. R. Seastedt, R. A. Ramundo, and D. C. Hayes, Maximization of densities of soil animals by foliage herbivory: Empirical evidence, graphical and conceptual models, *Oikos*, 51:243–248, 1988; W. H. Smith, *Air Pollution and Forests: Interaction Between Air Contaminants and Forest Ecosystems*, 2d ed., 1990; M. J. Swift, O. W. Heal, and J. M. Anderson, *Decomposition in Terrestrial Ecosystems*, 1979.

Expressions of nematode herbivory for four ecosystems in the northern Chihuahuan Desert, New Mexico*

Ecosystems	Nematodes, $\times 10^5$/g root	Herbivory index[†]
Playa	2.31	16.63
Arroyo	0.39	1.70
Dunes	0.17	3.67
Grassland	0.24	3.45

*After D. W. Freckman and R. A. Virginia, Plant-feeding nematodes in deep-rooting desert ecosystems, *Ecology*, 70:1665–1678, 1989.
[†]Weighted factor calculated as Σ(density/genera × impact factor for each genus), where impact factor is *Meloidogyne* = 1.0, *Meloidodera* = 1.0, *Pratylenchus* = 0.5, *Helicotylencus* = 0.4, *Paratylenchus* = 0.05, *Tylenchorhynchus* = 0.05, *Xiphinema* = 0.7. 1 g = 0.35 oz.

Soil microbiology

The ability of soil microorganisms (principally bacteria and fungi) to degrade pesticides is of interest for a variety of reasons. The pesticides that are applied must be biodegradable to ensure that they do not persist in soil long enough to percolate into surface or groundwater supplies. However, microorganisms have been known to degrade some pesticides so rapidly that these products are rendered useless in some locations. Also, some pesticides may be metabolized by soil microorganisms and transformed into compounds whose properties are greatly different from those of the parent compound; this includes the possibility that the pesticide may be converted into a more toxic chemical. Finally, the emerging environmental biotechnology industry is using microorganisms to clean up soils that are contaminated with high concentrations of pesticides because of accidental spills or long-term dumping. A great deal of ongoing research is directed at the complex ecological relationships of soil microorganisms in order to determine how best to take advantage of their properties. In addition, most of the information in this article could be applied equally well to biodegradation of other chemical contaminants in soils.

Soil microorganisms. These represent a largely uncharacterized kingdom of tremendous genetic and physiological diversity. A typical agricultural soil may contain from 10^6 to 10^9 bacterial cells per gram and an equal mass of fungi. Soil microorganisms play a number of vital roles in soil processes, including aggregation of soil particles, biogeochemical cycling of nitrogen, sulfur, phosphorus, and other important elements, and carbon cycling, including the degradation of organic matter. Despite decades of study, investigators estimate that fewer than 1% of all soil bacteria have been characterized and classified. Thus it is probable that the capacity of microorganisms to carry out various biochemical reactions may be much greater than is currently known.

Soil bacteria have been divided into two conceptual categories: autochthonous bacteria, which grow and metabolize at a constant, slow rate, and allochthonous bacteria, which normally lie dormant and undergo explosive growth when the appropriate conditions, such as addition of nutrients, arise. Although the dividing line between these categories is not always clear, they illustrate the boom-or-bust concept of microbial life in soils. Even though microorganisms that can degrade a chemical may be present in a particular soil, they may be inactive, leading to the persistence of the chemical.

Pesticide-degrading activity. It might be expected that microorganisms could degrade a particular pesticide if its molecular structure, or parts of the structure, are similar to chemicals commonly found in nature. Enzymes and the molecules they react with are often thought of as lock-and-key-type structures. If a part of a pesticide is shaped properly to fit into the enzymatic active site, the enzyme may be active. If, however, the pesticide does not resemble chemicals that the local microorganisms can degrade, the microorganisms may evolve the ability to degrade it over a period of months or years. Random mutations constantly alter the structures of enzymes as organisms grow, and if a new structure has greater affinity for a new source of carbon (the pesticide) the bacteria that contain it will have a competitive advantage over other bacteria. Because of the large number of microorganisms in soil and because their generation time can be as short as hours or even a few minutes, microorganisms can appear to evolve much faster than larger organisms.

An example is found in the sediments of the Hudson River, which has been contaminated with polychlorinated biphenyls (PCBs) for about 40 years. These compounds are very difficult to biodegrade, yet microorganisms in sediments downstream from the contaminated area are now able, albeit very slowly, to degrade polychlorinated biphenyls, while microorganisms in sediments upstream from the contaminated area cannot degrade these compounds.

In general, compounds with halogen substitutions and polyaromatic ring structures appear to be more foreign to most soil bacteria, and they take longer to degrade. It is often possible, however, to locate exotic environments where microorganisms capable of degrading a particular sort of compound may be found. For example, the marine tubeworm *Saccoglossus kowalewskii* naturally produces bromophenol as an allelochemical, a compound which inhibits other organisms. Bacteria from the vicinity of *S. kowalewskii* burrows can rapidly degrade some haloaromatic compounds.

Constraints on biodegradation in soil. Although bacterial degradation of a pesticide may be demonstrated in the laboratory, the pesticide often persists in the field, even though microorganisms capable of degrading it may be present. A number of factors can constrain microbial activity in soils, and these must be understood in order to promote decontamination of soils, as well as to use pesticides appropriately.

In soils, pesticide molecules are often adsorbed onto surfaces such as clay particles and humus. Such sorbed molecules are often unavailable to those microorganisms that may be able to take up chemicals only in solution. Many pesticides are very hydrophobic, and their sorption interactions are very strong. Some soil components are quite reactive chemically and may form covalent bonds with pesticide molecules.

Vital nutrients, such as nitrogen and phosphorus, that are required by microorganisms may be lacking in some soils. Although many pesticide-degrading microorganisms use the pesticide as a source of carbon and energy, not all can; thus, additional sources of carbon and energy may be required. In this case, degradation of the pesticide is referred to as cometabolism.

The oxidation-reduction potential of the soil and the availability of appropriate electron-accepting compounds may be crucial to degradation of pesticides. Many familiar microorganisms respire oxygen, so they will degrade a compound only if enough oxygen is present in the soil system. Other microorganisms substitute nitrate, iron(III), manganese(IV), sulfate, or carbon dioxide for oxygen, and will be active only in the absence of oxygen and in the presence of their particular electron acceptor. For example, long-chain hydrocarbons, such as oily waste, are degraded rapidly only in the presence of oxygen, while many highly chlorinated compounds are degraded only under anaerobic conditions.

Water content may be the most critical factor affecting pesticide degradation in many soils. Sufficient water must be present to sustain microbial life and to form a solution of nutrients and pesticide that the microorganisms can take up. The amount of water in a soil also determines how fast oxygen can diffuse through the soil, thus helping to control the oxidation-reduction potential.

The concentration of a pesticide in the soil is also an important factor. The rates of most biochemical reactions, including pesticide biodegradation, are dependent upon the concentration of the substrate because of the kinetic mechanisms of enzyme activity. If the concentration is great enough to be highly toxic, microbial activity can be inhibited. Also, if the concentration is very low, it may not generate a sufficiently strong chemical signal to induce (turn on) appropriate biochemical pathways in the microorganisms. Sometimes a contaminant chemical is not a good inducer, and a different chemical must be present to turn on biodegradation pathways.

Similarly, the presence of other pesticides and contaminants can interfere with biodegradation. They may be toxic to the microorganisms, or they may interfere with the metabolic pathway by either repressing its activation or short-circuiting active pathways.

The amount of time elapsed since the contamination event occurred may be important. The microbial population may have to become acclimated to the contaminant, resulting in a lag time before degradation occurs. This lag time may be due to several factors, including mutation of metabolic pathways, as described above; growth and multiplication of the degrading population to sufficiently high numbers; or prior degradation of compounds that interfere with degradation of the contaminant.

Bioremediation. There are two general methods in which microorganisms are used to clean up (bioremediate) contaminated soils. The most common is to stimulate the existing bacterial population in the soil by adjusting conditions to favor degradation of the contaminants. This usually consists of addition of nutrients, along with tillage to promote oxygenation; but it may also include adjustment of any of the other factors discussed above. The second general method is the addition of a laboratory-bred bacterial strain to the soil, along with adjustment of conditions to favor biodegradation. This method is more difficult, since many laboratory strains do not survive well under field conditions; also it may be difficult for exogenous bacteria to get into the interior of soil aggregates, where much of the contaminant may be sequestered and where they are protected from predation by protozoans.

Bioremediation of contaminated soils is an attractive alternative to more conventional decontamination methods for both economic and logistic reasons. Standard decontamination methods usually involve excavating the soil, placing it into storage drums, then hauling it to either a long-term storage facility or an incinerator. This process is costly and also raises the possibility of further contamination. Bioremediation, on the other hand, can be done on-site; it destroys the contaminant permanently and is usually much less expensive. The chief drawbacks of bioremediation are that it is often rather slow, it is subject to weather conditions, and the effects of the constraints must be determined for each site and each contaminant. Several large firms have recently been formed to provide bioremediation services, and research into more effective ways of promoting bioremediation is being carried out in industrial, university, and government laboratories.

For background information *SEE BIOGEOCHEMISTRY; ENZYME; NITROGEN CYCLE; POLYCHLORINATED BIPHENYLS; SOIL MICROBIOLOGY* in the McGraw-Hill Encyclopedia of Science & Technology.

Todd O. Stevens

Bibliography. D. Kamely, A. Chakrabarty, and G. S. Omenn (eds.), *Biotechnology and Biodegradation*, Advances in Applied Biotechnology Series, vol. 4, 1990; G. S. Omenn (ed.), *Environmental Biotechnology: Reducing Risks from Environmental Chemicals through Biotechnology*, Basic Life Sciences, vol. 45, 1988; E. A. Paul and F. E. Clark, *Soil Microbiology and Biochemistry*, 1989.

Solid-state chemistry

The structures of solids are often complex, even though the chemical formulas might suggest the opposite. Since there are a huge (and in principle infinite) number of possible structures for a given stoichiometry, involving a whole range of coordination numbers, it is not feasible to ask the basic question of why a structure is stable. Instead, the pertinent question is why one particular arrangement is more stable than an alternative that for one reason or another might be a viable candidate. The ready availability of fast and large computers has had a significant impact in the past few years on the way in which this question is answered, and understanding is sought concerning the structures of solids. Numerical methods not only are able to provide accurate results in areas where quantitative results have been completely absent, but also often permit quantification and support of ideas that previously could be phrased only in a qualitative fashion. Some structural problems from the area of main-group chemistry serve as useful illustrations of recent advances.

Bonding. There are several ideas developed for molecular chemistry that can be used profitably to study the structures of solids. Notations (1) and (2)

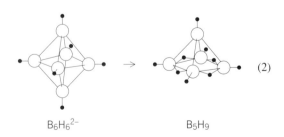

show how molecules adjust geometrically to the addition of extra electrons to a stable geometrical framework. A simple viewpoint can be used to describe how such molecules hold together. Modern theory uses the ideas of molecular orbital theory. The atomic orbitals located on the constituent atoms, which are used to describe the motion of the electrons in the free atoms, overlap to produce a set of molecular orbitals, used in turn to describe the motion of the electrons in the molecule. Molecular orbitals fall into three categories. Lying lowest in energy are bonding orbitals. Occupation of these energy levels by electrons leads to an attraction between the nuclei. Higher in energy are the nonbonding levels, often largely located on one atom; therefore they are often very much like atomic orbitals. Because of this, electrons residing in such orbitals do not contribute to the attractive or repulsive forces between pairs of nuclei. Highest in energy are the antibonding orbitals. Occupation of these orbitals by electrons leads to strong repulsive forces between the nuclei.

Thus, for a chemical bond to be formed between two atoms, there needs to be an excess of occupied bonding orbitals over occupied antibonding ones. In the phosphorus tetrahedron (P_4), there are a total of six bonds between the phosphorus atoms, along each edge of the tetrahedron. This is readily accounted for by placing six pairs of electrons in P—P bonding orbitals, keeping all of the antibonding orbitals empty, and placing four pairs of electrons into nonbonding or lone-pair orbitals. Thus the six bonds here result from the presence of six pairs of electrons in bonding orbitals. With reference to notation (1), addition of electrons to the stable P_4 tetrahedron would lead to population of P—P antibonding orbitals. Now the total bonding has been reduced, since the presence of each pair of electrons in an antibonding orbital nullifies the bonding effect of one pair of electrons in a bonding orbital. This results in the breaking of bonds; in fact, one bond should be broken per extra electron pair. This is just what is seen in the series of related molecules shown here. On moving from the left to the right, the number of electrons associated with the core of the molecule increases by two each time. In P_4 there is a total of 20 valence electrons, in B_4H_{10} there is a total of 22, and in P_4cy_4 (where cy = cyclohexyl, C_6H_{11}) there is a total of 24. (Each phosphorus atom contributes four electrons and each cyclohexyl shares one of its electrons with phosphorus when bonding to the core, leading to 24 electrons overall.) In PCl_3 and S_2F_2, there are a total of 26 valence electrons. Both have three bonds.

Of importance here is to see how the structure of the molecular orbital diagram changes during this bond-breaking process and what drives it energetically. As the atoms move apart during the breaking of this bond, the energy of these initially antibonding electrons drops. At the same time a pair of electrons initially located in the bonding region rises in energy (**Fig. 1**). Changes in energy associated with antibonding orbitals are invariably larger than those associated with their bonding counterparts, and so this process is energetically favored. The result is that two pairs of electrons, one from the antibonding region and one

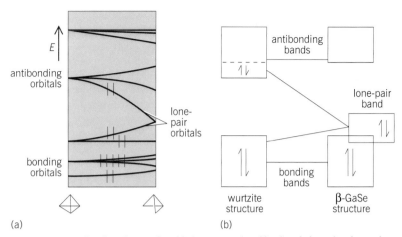

Fig. 1. Diagrams showing changes in orbital energy on breaking bonds in molecules and solids. (*a*) Changes in the molecular orbital levels on breaking a bond in the P_4 tetrahedron. Energies of orbitals denoted by short vertical lines remain unchanged. (*b*) Analogous changes associated with the energy bands of solids of the type shown in Fig. 2.

from the bonding region, are transformed into two pairs of lone-pair electrons, one at each end of the severed bond. The number of linkages decreases by one each time.

In notation (2) the response of the boron-hydrogen octahedron ($B_6H_6^{2-}$) with a total of 14 electrons holding the skeleton together to the addition of two electrons is to eject a BH group. In other words, the situation is similar to that in notation (1) except that all the bonds to one particular atom or group of atoms are broken. The hypothetical $B_5H_5^{4-}$ octahedral fragment that is produced is referred to as a nido octahedron, and it too has a total of 14 skeletal electrons. It is isoelectronic, in fact, with the known pentaborane (B_5H_9) molecule. If this species is written as $B_5H_9\square$, where $\square$ represents a vacancy, then the number of electrons per site is the same in both species if the vacancy is included. As before, the extra electrons, which would go into antibonding orbitals of the parent, are located in lone-pair orbitals that point into this vacant site after the BH unit has been ejected. Notations (1) and (2) thus show two different ways to accommodate these extra electrons.

Structures. **Figure 2** shows how the structures of a series of solids may be viewed in a fashion exactly analogous to that in notation (1). The diamond structure is a prototypical covalent solid-state structure. With four electrons per atom, four bonds are formed to its four symmetrically located neighbors. The zincblende structure is very strongly related to diamond. Every other site is occupied by metal (zinc) and sulfur and, with an average of four electrons per atom (two from zinc and six from sulfur), has the same electron count. There exists a very rare hexagonal diamond structure whose structure is very similar indeed to the more common cubic variety and an analogous zinc sulfide (ZnS) structure, that of wurtzite shown in Fig. 2*a*. This structure, just like those of diamond and zincblende, is stable for a total of eight electrons per AB unit, or four electrons per atom. The structures shown for β-gallium selenide (β-GaSe; Fig. 2*b*) and wolfsbergite (Fig. 2*c*)

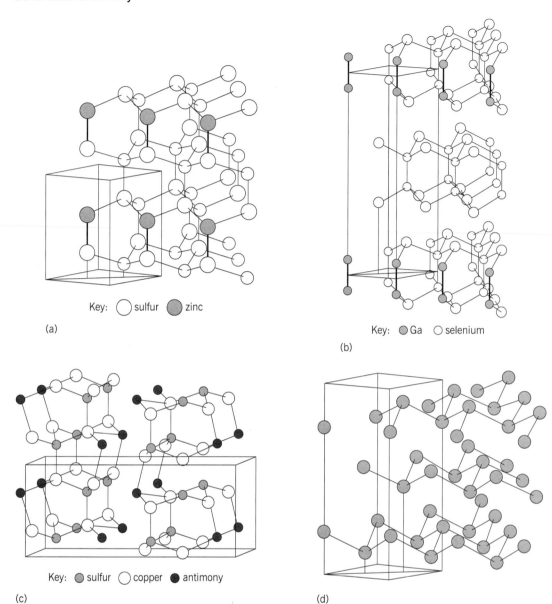

Key: ◯ sulfur ⬤ zinc

(a)

Key: ⬤ Ga ◯ selenium

(b)

Key: ⬤ sulfur ◯ copper ⬤ antimony

(c)

(d)

Fig. 2. Structures of some simple solids related by using the solid-state equivalent of notation (1). (a) Wurtzite structure found for zinc sulfide (ZnS). (b) β-Gallium selenide (β-GaSe). (c) Wolfsbergite (CuSbS₂). (d) Arsenic.

are found for solids with one extra electron per formula unit. By analogy with notation (1), it would be expected that some of the bonds between the atoms in wurtzite would be broken in these structures; and they are. Half of the atoms are four-coordinate and half are now three-coordinate. (There has been a rearrangement of

atoms in the β-GaSe structure compared to zincblende, leading to the formation of Ga-Ga contacts.) The two structures in Fig. 2b and c differ only in the way the bonds between pairs of layers of atoms have been severed. These are arranged in horizontal planes for β-GaSe and vertical planes for wolfsbergite. For arsenic, with ten electrons per atom pair, it would be expected that the fission of more linkages would occur. There are several choices. For example, a structure where every atom is three-coordinate or a structure containing four- and two-coordinate atoms in equal numbers are two possibilities. The structure of arsenic (Fig. 2d) is derived from the β-GaSe structure by breaking a second layer of horizontal bonds. Adding more electrons proceeds along the same lines. The structure of elemental selenium, for example, consists of spiral chains containing two-coordinate atoms that is simply derived from the arsenic structure.

Figure 3 shows the solid-state analog of notation (2). As noted, the diamond structure and that of zinc

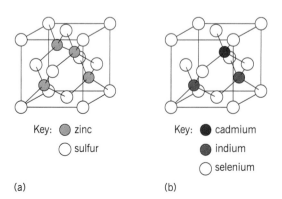

Key: ⬤ zinc
◯ sulfur

(a)

Key: ⬤ cadmium
⬤ indium
◯ selenium

(b)

Fig. 3. Two closely related sulfide structures. (a) Sphalerite structure found for ZnS. (b) Cadmium indium selenide (CdIn₂S₄).

sulfide are stable for a total of eight electrons per atom pair or four per atom. The total electron count in cadmium indium selenide ($CdIn_2S_4$) is 32 per formula unit or $32/7 = 4.6$ electrons per atom, an unusual electron count. In fact, the structure of $CdIn_2S_4$ is closely related to that of the zincblende form of ZnS, the only difference being an atomic vacancy in the cationic network. It could thus be written as $\square CdIn_2S_4$. Therefore, although the electron count per atom of the material is certainly higher than that of the ZnS parent, the electron count per site (including the vacancy) is now $32/8 = 4$, and it becomes the same in both structures. This result has been known for some years as the Grimm-Sommerfeld valence rule.

Figure 1 shows how the electronic situation in the molecular and solid-state cases are extremely similar. For molecules containing a finite number of atoms and thus chemically important orbitals, there are a finite number of molecular orbitals. In solids where there is an enormous number of atoms, there is a correspondingly large number of orbitals. These orbitals are collected into energy bands with properties actually very similar to those of molecular orbitals in molecules. Thus, whereas on bond breaking, individual antibonding orbitals are converted into nonbonding ones in molecules, a part of the antibonding band containing a part of the set of antibonding orbitals is converted into a band of nonbonding ones in solids.

Computations. **Figure 4** shows the results of some very accurate calculations for some possible structures of a typical coordinate II–VI system (for example, ZnS), as a function of the volume of the unit cell. Computations of this type are essential in order to identify the lowest-energy structure from a set of possibilities, especially if, as in the present case, the atomic coordination numbers are different. As can be seen, the numerical results correctly identify the cubic structure of sphalerite (a polymorph of zinc sulfide) as the lowest energy of the set. Not shown, for clarity, is that for the hexagonal wurtzite form. It lies just a little higher in energy. The cubic rock salt structure lies

higher in energy too, and it has a minimum corresponding to a shorter bond length. The diagram thus predicts that materials such as gallium arsenide (GaAs) and ZnS should adopt the rock salt structure under pressure, which in fact they do. In quantitative terms, the calculations are also able to provide a value for the pressure at which this transformation should occur. Since the change in energy with volume can be read off from Fig. 4, the pressure needed to change the volume sufficiently that the rock salt structure now lies lowest in energy may also be calculated. These are in very good agreement with experiment. Although it might be possible to predict the general form of the curves of Fig. 4, the numerical details, perhaps the most important part of the plots, require the use of highly accurate calculations. However, in contrast to the description of the processes taking place in Figs. 2 and 3, a good qualitative description of the electronic reasons behind the results of Fig. 4 is not yet possible.

For background information SEE COORDINATION NUMBER; CRYSTAL STRUCTURE; MOLECULAR ORBITAL THEORY; SOLID-STATE CHEMISTRY; VALENCE in the McGraw-Hill Encyclopedia of Science & Technology.

Jeremy K. Burdett

Bibliography. J. K. Burdett, Perspectives in structural chemistry, *Chem. Rev.*, 88:3–30, 1988; J. K. Burdett, Application of "molecular" theories to the structure of the crystalline state, *Nature*, 279:121–125, 1979; J. R. Chelikowsky and J. K. Burdett, Ionicity and the structural stability of solids, *Phys. Rev. Lett.*, 56:961–964, 1986.

Solution mining

Leach mining in place is a relatively new, lower-cost method for solution mining. In this technique, metal is extracted from an ore deposit by injecting and recovering a leach solution that dissolves the metal of interest selectively. The leach solution is injected into the ore through a series of wells, where it dissolves the target metal as it flows through the natural fractures and pore spaces of the ore (**Fig. 1**). The metal-bearing solution is then recovered through another series of wells and pumped to a processing plant on the surface, where the metal is removed from solution. In-place leach mining has been used to commercially recover uranium from sandstone deposits since 1963. Commercial production expanded rapidly in the 1970s, and in 1988 about one-third of all uranium mined in the United States was produced by in-place leach mining. So far, uranium is the only commodity that is recovered from undisturbed ore deposits by this method.

Recent research in technology for in-place leach mining is focused on recovering copper from undisturbed, fractured hard-rock deposits that have been oxidized. In the United States, geologic characterization, laboratory core leaching, and field site characterization studies by the Bureau of Mines are being used to evaluate the Santa Cruz copper oxide deposit near Casa Grande, Arizona, in preparation for an in-place copper mining test. In addition, the hydrol-

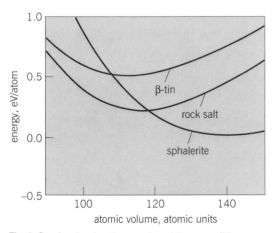

Fig. 4. Graphs showing the energies of three possible structures for a typical coordinate II–VI system as a function of atomic volume. (*After J. R. Chelikowsky and J. K. Burdett, Ionicity and the structural stability of solids, Phys. Rev. Lett., 56:961–964, 1986***)**

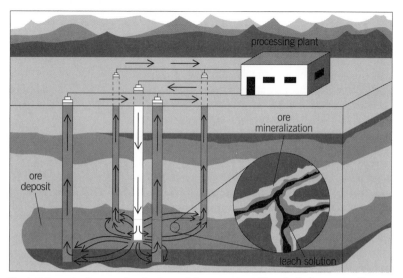

Fig. 1. Schematic diagram of an in-place mining operation with a single five-spot well pattern.

ogy of in-place leach mining of an unsaturated copper oxide deposit is being investigated at the Casa Grande Mine, also near Casa Grande, Arizona. Successful development of this type of leach mining technology for undisturbed, fractured hard-rock deposits is important, because the technique could be applied to a number of commodities, including copper, gold, silver, manganese, and other critical and strategic metals.

Geologic characterization. This is the study of ore to determine its chemical and textural makeup. These two characteristics are important in determining whether or not an ore deposit is amenable to in-place leach mining. The chemical makeup will determine if the particular ore minerals can be readily leached when contacted by a suitable leach solution (lixiviant), and if the gangue minerals will consume excessive amounts of lixiviant. The ore texture will determine if the natural fluid-flow channels (fractures and pore spaces) in the rock will allow lixiviant access to significant amounts of ore minerals. In the case of copper oxide deposits, the major copper minerals (such as chrysocolla, atacamite, azurite, and malachite) are all readily soluble in dilute solutions of sulfuric acid (1–5%). Many copper oxide deposits have occurrences of ore minerals primarily in the host rock fractures; this allows for easy access to lixiviant. Ore mineralization that is disseminated throughout the host rock matrix is much more difficult to contact with lixiviant.

Geologic characterization studies are conducted on polished thin sections of ore samples by using a polarizing microscope and an electron probe microanalyzer equipped with an x-ray analyzer. The polarizing microscope provides a direct visual image of the ore texture (**Fig. 2**), while the microanalyzer provides qualitative and quantitative chemical information, as well as x-ray mapping capabilities that are used to display elemental distributions within a sample. Geologic characterization studies should be conducted before and after core-leaching tests to help explain the reactions that occur during the leaching process.

Core leaching. Laboratory leaching tests are commonly used to evaluate the ability of a lixiviant to dissolve the metal of interest selectively from the ore host rock. In these, the ore is usually broken or crushed and leached in columns, or it is ground and leached in small containers. Evaluation of copper oxide deposits for suitability for in-place leach mining involves leaching intact drill-core samples that are mounted in reaction cells that limit the flow of lixiviant to the natural flow channels in the core. There are two major reasons for this. First, it is likely that the core-leaching experiments more closely simulate the nature of the solution–rock interface in place, because the lixiviant flows through the fractures where the copper mineralization is concentrated and is not exposed to the anomalous surface area of gangue mineralization that would result from crushing. Second, core-leaching experiments allow for an evaluation of permeability changes taking place during leaching. Permeability of the ore may increase as copper minerals are dissolved and fractures open up, or permeability may decrease if chemical precipitation or clay swelling blocks flow channels. Although the field permeability of a deposit will most likely be much higher than the permeability of the core leached in the laboratory, correlation of permeability changes during the experiments with data on the fluid chemistry and sample mineralogy will yield information that may be useful for determining the potential impact of these mechanisms during actual in-place leach mining in the field.

The core-leaching tests provide data on the rate of copper dissolution with time, overall copper recovery, sulfuric acid consumption, and the rate of gangue mineral dissolution with time. Two types of core-leaching tests are conducted: single-pass, in which fresh lixiviant is continuously injected into the sample, and recycle, in which the metal-bearing solution is stripped of copper, rejuvenated to its original lixiviant strength, and recycled into the sample. Both types of tests are conducted with constant flow through the sample. The recycle tests provide data on the buildup of other metals in solution and their effects on the leaching process. Pre- and postleach chemical analysis of the samples are conducted in addition to the

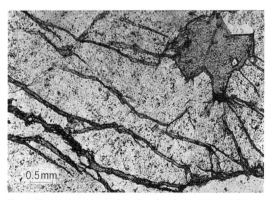

Fig. 2. Photomicrograph of fracture-hosted copper oxide mineralization (dark area at upper right).

geologic characterization studies. Typical test results show a 70–80% copper recovery in 100 days of leaching in copper oxide cores.

Field site characterization. An in-place copper mining field test is being conducted at the Santa Cruz site by the Bureau of Mines and the Santa Cruz Joint Venture. This site has had no previous mining activity, has reserves of 10^9 tons of copper that is soluble in 0.55% sulfuric acid solution, and is located 1200–2000 ft (365–610 m) below the surface; the deposit is naturally saturated with water. The purpose of this test is to provide technical, economic, and environmental data upon which the merits of in-place copper mining can be judged.

Current activities are concentrating on testing to characterize the structural geology and hydrology of the site in order to provide data for environmental permit applications. Drill-core fractures have been analyzed to provide data on the primary-fracture-system orientations. Since fractures in the rock will provide the major fluid-flow pathways, knowledge of the orientations of the primary fracture system will provide insight into the preferred fluid-flow directions and can provide a basis for planning well-field orientations.

Hydrologic characterization studies at the Santa Cruz site have revolved around the installation of one five-spot well pattern. Well spacing is 127 ft (39 m) from corner to corner. These five wells have been drilled to a depth of about 1850 ft (560 m), cased with 6-in.-inside-diameter (150-mm) fiberglass-reinforced plastic, and cemented in place with acid-resistant cement (**Fig. 3**). The casing and cement have been perforated between about 1600 and 1800 ft (490 and 550 m) with a shaped-charge perforator to provide fluid access to the well in the ore zone. Water injection tests have been conducted in each well to determine permeability and hydraulic conductivity in the ore zone. In addition, interference tests have been conducted by injecting water in one well and monitoring the water head level in the other four wells. Preliminary analysis of these test data indicate that sufficient quantities of water can be injected into the ore zone, and that sufficient communication between wells exists to continue with the test program.

The remaining hydrologic characterization test to be conducted will be a tracer test. A combination sodium chloride/sodium bromide tracer will be injected in the center well of the five-spot well pattern while pumping will be conducted in the four corner wells. Chemical analysis of the produced fluid will provide data on the speed with which the tracer fluid moves through the ore zone in each direction and the amount of tracer fluid that is recovered relative to the amount injected. With the completion of the tracer test, sufficient data should be available to complete the environmental permit applications for the in-place leach mining test.

Hydrology. The hydrology of in-place mining at the Casa Grande Mine is being studied. The in-place mining activities are carried out in the underground workings of the former block-cave mining operation

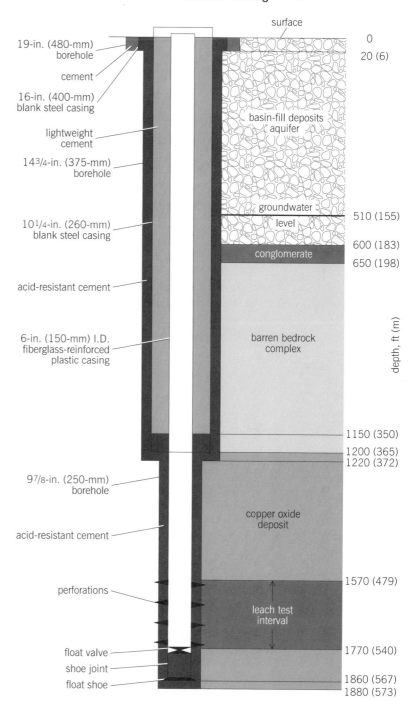

Fig. 3. Diagram showing the components of a single well in a five-spot well installation.

about 700–900 ft (200–275 m) below the surface. The copper oxide ore ranges from 0.8 to 1.2% and is initially unsaturated. The main hydrologic objectives of the operation are to be able to maintain adequate lixiviant injection and recovery flow rates and to distribute the lixiviant effectively throughout the ore zone.

A total of 58 wells have been constructed in a pattern of 15 parallel fans, located 12.5 ft (3.8 m) apart along the drift. Each fan contains three to five wells drilled at angles ranging from 6 to 89° from the horizontal. The wells are cased with 1-in. (25-mm) stainless-steel pipe that is cemented in place from the collar to the top of the ore zone. The pipe is perforated in the ore zone for fluid access to the ore. Each well is connected to a solution injection and return line, and can operate as either an injection or recovery well.

Acquisition of pressure and flow data from injection and recovery wells is an important part of the hydrologic investigation of the in-place leaching operation. Pressure and flow sensors have been installed in a number of individual wells and main flow lines. A hydrologic data acquisition system capable of continuous pressure and flow monitoring in up to 64 wells has been installed in the mine. A hydrologic model is used to analyze the data and estimate permeability versus time.

As a result of this ongoing research, it has been found that the characterization of permeability conditions through time is necessary for maintenance of optimum pressure conditions in injection and recovery wells during the in-place leach mining process; that individual well pressure and flow histories reflect the changes that occur in ore zone permeability; and that operational flexibility, which permits easy conversion of wells from injection to recovery mode, or vice versa, is an essential design requirement for accommodating changes in permeability conditions in in-place leach mining.

For background information *SEE* H*YDROMETALLURGY;* L*EACHING;* S*OLUTION MINING* in the McGraw-Hill Encyclopedia of Science & Technology.

Jon K. Ahlness

Bibliography. D. H. Davidson et al., *Draft Generic In Situ Copper Mine Design Manual*, U.S. Bur. Mines OFR4(2)-89, vol. 2 of *Final Report for U.S. Bureau of Mines Contract J0267001, 1988*, NTIS: PB 89-148225/AS, 1989; D. J. Millenacker and J. K. Ahlness, In situ copper leach mining: An alternate mining method, *Proceedings of the 1989 Multinational Conference on Mine Planning and Design*, University of Kentucky, May 23–26, 1989; R. D. Schmidt, K. C. Behnke, and M. J. Friedel, *Hydrologic Considerations of Underground In Situ Copper Leaching*, Preprint 90-179, SME-AIME, 1990.

Sonic boom

When an aircraft travels faster than sound, its disturbance of the air generates a pattern of shock waves that eventually reaches the ground as an audible sonic boom. Supersonic transport (SST) designs of the 1960s (Concorde, Tu-144, and the American SST) generate sonic booms that are unacceptable in populated areas. This was a leading reason for cancellation of the American SST, and is why the Concorde flies supersonically only over water. If sonic boom were not a problem, there could be a market for 500–1000 passenger-carrying supersonic aircraft. Major advances in aircraft technology over the past two decades offer the potential of an advanced supersonic transport with significantly lower sonic boom. There

have been no magic breakthroughs in reducing sonic boom, but rather increasing opportunity to apply the basic principles of sonic-boom minimization.

Nature of sonic boom. The basic principles of sonic boom generation and its propagation to the ground are illustrated in **Fig. 1.** The aircraft disturbs the air in a manner very closely related to aircraft geometry and aerodynamic loads, that is, lift and drag. The near-field pressure signature generally has positive pressure in the forward portion and negative pressure in the rear, and has fine structure that is readily identified with particular details of the aircraft. After it has propagated a short distance from the aircraft (typically one to two times the aircraft length), the disturbance is small compared to ambient air pressure, and the wave propagates very nearly as an acoustic (that is, linear) wave. Propagation is along conical wavefronts, which are inclined at the Mach angle μ, given by the equation below, where M = Mach number. In a

$$\mu = \sin^{-1} \frac{1}{M}$$

uniform atmosphere, the acoustic amplitude diminishes as $r^{-1/2}$, where r is the radial distance from the aircraft. In the real atmosphere, the amplitude is affected also by changes in air density and sound speed as the boom propagates toward the ground. In this region, the shape of an acoustic signature remains constant.

The pressure disturbance, although weak, is strong enough that it is not a true linear acoustic wave. Nonlinear effects cause it to become distorted as it propagates over long distances. High-pressure regions propagate a little faster and low-pressure regions a little slower. This distortion process, often referred to as aging, causes the signature to stretch out in duration or length. High-pressure regions tend to catch up to lower-pressure regions ahead of them, resulting in the formation of shock waves and the loss of detail in the signature. The mid-field signature (see Fig. 1) retains

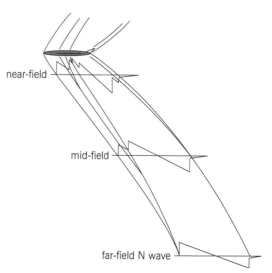

Fig. 1. Sonic-boom generation, propagation, and aging, showing near-field, mid-field, and far-field signatures.

only some of the original detail. Eventually, in the far-field, the forward positive and rear negative-pressure regions coalesce into front and rear shocks of nearly equal strength that are connected by a linear expansion region. This far-field signature, known as an N-wave, has lost all of the fine structure and has duration and shock strength related primarily to the size, weight, and altitude of the aircraft, and is only secondarily related to its detailed design features. An N wave continues to age and stretch, although at a diminishing rate, as distance increases. Nonlinear aging causes attenuation of the N wave to be greater than that of an acoustic wave. In a uniform atmosphere, the N-wave shock amplitude decreases as $r^{-3/4}$.

The duration of an N-wave sonic boom varies from about 100 milliseconds for small fighter aircraft at low altitudes to several hundred milliseconds for larger aircraft at high altitudes. Shock pressure varies significantly with altitude and aircraft size, but is typically 2 lb/ft^2 (96 pascals) for cruise condition of existing supersonic transports. Research in the 1960s established that an N-wave sonic boom having a shock pressure of 2 lb/ft^2 (96 Pa) was not acceptable to the exposed public, but this research did not establish what type of boom might be acceptable.

Early analyses of sonic boom considered that the far-field N-wave signature was inevitable, and concentrated on reduction of the shock amplitude of such booms. It is difficult to reduce substantially an N-wave sonic boom, since its overall characteristics are closely related to the aircraft size and weight. However, as a sonic boom propagates from thin air at high altitude to denser air at low altitude, the rate of aging diminishes rapidly, so that ground signatures for larger aircraft retain mid-field characteristics. This creates the possibility of reducing adverse impact by manipulating the signature shape rather than by just reducing amplitude. The benefit of this possibility is tempered by the open question of which aspects of a sonic boom signature are the most intrusive, and what shape would be better than an N wave.

Perception and impact. A sonic boom, when heard outdoors, generally sounds like two sharp cracks—one associated with each shock. A shock pressure of 2 lb/ft^2 (96 Pa) is well below the threshold of direct physical harm to people or animals. This shock pressure on rare occasions can break windows or trigger incipient damage in brittle structures such as plaster. The sound of the sonic boom is startling and intrusive, and can be very annoying. When heard indoors, the sound is modified by the vibroacoustic response of the building. The sonic boom itself is attenuated (and the sharpness of the cracks is reduced) inside a building, but it causes the building to vibrate and generates additional noise due to rattling of the building and small objects. Indoors, people complain of rattle more than anything else. It is not clear how annoying the rattle itself is, versus the rattle occurring at the same time that the attenuated boom is heard.

The human ear is most sensitive to sounds between 500 and 4000 Hz. If a sonic boom is passed through a filter similar to the sensitivity of the human ear, the result is a pair of spikes corresponding to the pair of shock waves; the shock waves are made up of sound in this sensitive range. The shape of the wave between the shocks, being composed of low-frequency sound, has very little effect on the magnitude of the spikes. The assumption that only shock strength matters leads to the concept of shaped extended-duration, flat-top, and minimum-shock sonic booms (see **Fig. 2**). The sonic booms are shaped to minimize the amplitude of the shocks. An extended-duration boom reduces shock amplitude in exchange for longer duration. A flat-top boom reduces duration by allowing plateaus following the bow shock and preceding the tail shock. A minimum-shock boom is an extension of this idea in that there can be a maximum pressure that exceeds that of the shocks. The shape of the signature near the maximum is gentle, however, so that low frequencies below hearing sensitivity are involved.

If two otherwise similar aircraft are designed so that one has an N-wave sonic boom and the other has a shaped minimized sonic boom, the minimized boom will have lower shock amplitude. This will be quieter outdoors. The minimized boom will, however, have increased energy at low frequencies. Depending on how this relates to the frequency response of build-

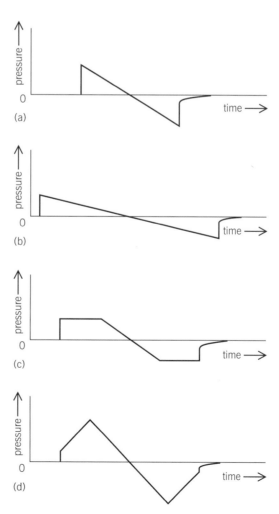

Fig. 2. Various reduced-noise sonic booms compared with a normal N wave. (a) Normal N wave. (b) Extended-duration sonic boom. (c) Flat-top sonic boom. (d) Minimum-shock sonic boom.

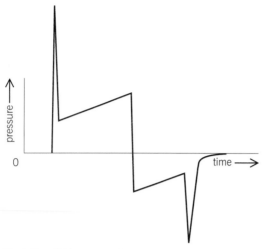

Fig. 3. Near-field pressure signature that will age into a minimum-shock sonic boom.

ings, the rattle and perceived vibration indoors could actually increase. Studies are currently under way at the National Aeronautics and Space Administration (NASA) Langley Research Center to establish the extent to which indoor noise level would decrease as a result of shaped minimum booms. These studies are part of a research program to establish the level of sonic boom that would be acceptable to the public.

Designing for low sonic boom. In the absence of a clear criterion for acceptable sonic-boom characteristics, aircraft design work is currently proceeding on the assumption that flat-top- and minimum-shock-shaped signatures will be beneficial. To achieve these types of boom, an aircraft must be shaped so as to generate a near-field signature with particular characteristics. **Figure 3** shows the type of near-field signature that is required. Most of the pressure disturbance is concentrated in a spike near the nose. This allows the rest of the signature to have lower amplitude. As the signature ages, the spike is absorbed into the bow shock, and is completely gone by the time it reaches the ground. The signature behind the spike can be flat, yielding a flat-top boom. It can have rising pressure with initial slope gentle enough not to age into a shock, yielding a minimum-shock boom. Aircraft designs that can achieve these goals have been theoretically predicted, and validated by wind-tunnel tests.

Key design features for a reduced-boom aircraft are concentration of aerodynamic load near the front, and smooth shape (including blending the wing and fuselage), which allows a plateau or gentle slope. Making the aircraft as long as possible increases the opportunity of achieving a desirable mid-field signature, and also allows for a smaller cross-sectional area, which will reduce the overall pressure. These details are not necessarily consistent with efficient aircraft design. Any aircraft design is based on a series of compromises, and details for boom reduction must be included in this trade-off. Supersonic transports designed in the 1960s did not have enough performance margin to allow for such compromises. Advances in aircraft structures, materials, propulsion systems, and

aerodynamics now provide a margin that can be compromised in favor of reduced sonic boom. To make the best use of this margin, sonic-boom details must be integrated into the design process. Current efforts by NASA and the aircraft industry are directed toward this integration. The extensive use of computers greatly improves the practicality of this effort, since an alternative concept may now be analyzed in hours rather than weeks or months. Over the next few years, the combination of design tools for sonic-boom reduction and research into the impact of reduced booms may very well show that a supersonic transport with acceptable sonic boom is possible.

For background information *SEE* LOUDNESS; SONIC BOOM; SUPERSONIC FLIGHT in the McGraw-Hill Encyclopedia of Science & Technology.

Kenneth J. Plotkin

Bibliography. C. M. Darden, *Study of the Limitations of Linear Theory Methods as Applied to Sonic Boom Calculations*, AIAA Pap. 90-0368, 28th Aerospace Sciences Meeting, January 8–11, 1990; D. J. Maglieri, H. W. Carlson, and H. H. Hubbard, Status of knowledge of sonic booms, *Noise Control Eng.*, 15:57–64, 1980; D. J. Maglieri and K. J. Plotkin, Sonic boom, in H. H. Hubbard (ed.), *Flight Vehicle, Acoustics, Theory and Practice*, vol. 1: *Noise Sources*, NASA Ref. Pub., WRDC-TR-90-3052, November 1990; R. Seebass and A. R. George, Sonic boom minimization, *J. Acous. Soc. Amer.*, 51:686–694, 1972.

Space flight

The year 1990 witnessed a number of important scientific achievements in space as well as major disappointments (**Table 1**). Despite some notable successes, the United States space program continued to be plagued by technical and managerial problems. Perhaps as a result of the economic and political turmoil in the Soviet Union, the Soviet space program, still the largest and most active in the world, showed further signs of stress. Overall, Soviet rocket launchings in 1990 numbered well below their historic highs. **Table 2** categorizes the 116 successful launches during 1990 according to country or organization.

The U.S. National Aeronautics and Space Administration (NASA) flew a total of six successful shuttle missions despite a stand-down of 5 months due to hydrogen leaks. The 1990 NASA launches included two that were devoted to Department of Defense payloads (February 28, STS 36; and November 15, STS 38), deployment of the *Syncom 4* communications satellite, and retrieval of the Long Duration Exposure Facility (January 9, STS 32). Deployment of the European Space Agency's *Ulysses* (October 6, STS 41) and the Hubble Space Telescope (April 24, STS 31), as well as the Astro 1 astronomy mission (December 2, STS 35), were achieved by shuttle launches.

The *Magellan* spacecraft began detailed radar mapping of the planet Venus in August, and the *Galileo* spacecraft made the first of two gravitational-

Table 1.　Some significant space launches in 1990

Payload or vehicle	Date	Country or organization	Purpose or outcome
Columbia, STS 32	Jan. 9	United States	Retrieval of Long Duration Exposure Facility after 6 years in orbit (intended for 18 months); great import to science
Muses-A	Jan. 24	Japan	First Japanese lunar probe
Spot 2	Jan. 31	France	Remote-sensing data for commercial use
MOS 1B	Feb. 7	Japan	Ocean surveillance and science spacecraft
Soyuz TM-9	Feb. 11	Soviet Union	179-day flight; science at space station Mir
Atlantis, STS 36	Feb. 28	United States	Major new reconnaissance spacecraft emplaced
Offeq 2	Apr. 3	Israel	Second science satellite
AsiaSat	Apr. 7	Asian consortium	Its first commercial communications satellite; launched by People's Republic of China
Palapa B-2R	Apr. 13	Indonesia	Communications satellite, previously retrieved (by STS 51A, Nov. 1984) after perigee motor failure; operating now; commercial launch
Discovery, STS 31	Apr. 24	United States	Emplaced Hubble Space Telescope
Cosmos 2079, 2080, 2081	May 19	Soviet Union	Major navigation satellites for maritime, air, and land use
Roentgen	May 29	United States and Germany	X-ray science satellite; cooperative effort; United States launching
Kristall	May 31	Soviet Union	Science laboratory attached to Mir; carried two manned maneuvering units (MMUs)
Soyuz TM-10	Aug. 1	Soviet Union	131-day flight; science at Mir
Marco Polo 2	Aug. 18	United Kingdom	Communications satellite for British customers; commercial launch
Feng Yun 1-2	Sept. 3	People's Republic of China	Sun-synchronous weather satellite
Discovery, STS 41	Oct. 6	United States; European Space Agency	Ulysses probe to Sun by way of Jupiter; will study solar poles
Galaxy	Oct. 12	Hughes Communications Co.	Communications satellite
SBS 6	Oct. 12	Hughes Communications Co.	Business communications satellite
INMARSAT 2	Oct. 30	INMARSAT	First fully owned INMARSAT satellite for maritime, air, and land use; commercial launch
Atlantis, STS 38	Nov. 15	United States	Classified payload, believed to be LaCrosse-type satellite with side-looking radar, placed in geosynchronous orbit
Satcom C-1	Nov. 20	GE American Communications Co.	Communications satellite
G-Star	Nov. 20	GTE	Communications satellite for GTE Spacenet
Gorizont 22	Nov. 23	Soviet Union	First communications satellite for Russian Soviet Federated Socialist Republic; there will be three
Columbia, STS 35	Dec. 2	United States	Astro 1 telescopes; studied distant galaxies and quasars
Soyuz TM-11	Dec. 2	Soviet Union	Carried first Japanese journalist on commercial flight; science at Mir

assist passes by Earth in its journey to the planet Jupiter. Both *Magellan* and *Galileo* were launched in 1989. The *Ulysses* spacecraft, launched by the space shuttle *Discovery* (October 6, STS 41), is on its way to study the poles of the Sun.

Table 2.　Successful launchings conducted in 1990*

Country or organization	Number of launches
Soviet Union	75
United States	20
European Space Agency	5
Japan	3
People's Republic of China	5
Israel	1
Commercial launches	7[†]
Total	116

*Achieved Earth orbit or beyond.
[†]These are not included in United States total since they were not launched by NASA or the Department of Defense.

Perhaps the greatest disappointment to NASA, and to the worldwide scientific community as well, followed successful deployment of the Hubble Space Telescope aboard the shuttle *Discovery*. Several weeks following deployment in April, the telescope's large mirror was found to have a spherical aberration that would prevent certain observations in the visible-light spectrum. Despite its flaws, the Hubble Space Telescope has undertaken unprecedented scientific work in spectroscopy, photometry, and ultraviolet imaging that is impossible from land-based instruments. Through the use of computer image processing to partially compensate for the flaw in the telescope's mirror, several visible-light observations were made of the Orion Nebula and, most remarkably, a giant storm on Saturn.

NASA continued to develop plans for achieving the Bush Administration's Space Exploration Initiative, including a return to the Moon. In May, President Bush announced the goal of a crewed mission to Mars no later than 2019.

In December, the Advisory Committee on the Future of the U.S. Space Program issued its final report recommending that NASA place primary emphasis on science and scientific missions. The report also urged a redesign of the space station *Freedom* and a reduction of the dependence on the space shuttle by developing a new crewless heavy-lift launch vehicle.

UNITED STATES SPACE ACTIVITY

NASA expanded the capabilities of its shuttle system during 1990 with two extended-duration missions flown by the shuttle *Columbia*, STS 32 in January and STS 35 in December. STS 32 was the longest shuttle mission ever flown, logging 261 h. The shuttle *Discovery* carried into orbit two payloads that scientists had long planned for: the Hubble Space Telescope in April (STS 31) and the *Ulysses* spacecraft in October (STS 41). *Atlantis* also made two flights during 1990 for the Department of Defense: STS 36 in February, which emplaced a major new reconnaissance spacecraft; and STS 38 in November, which placed into orbit a vehicle thought to be a LaCrosse-type spacecraft with a side-looking radar.

Fuel leaks. Shuttle launches were interrupted for 5 months after hydrogen leaks were detected on *Columbia* and *Atlantis* during external-tank loading operations. The problem on *Atlantis* was found to be related to the seals associated with the tank's 17-in. (43-cm) disconnect area. After the disconnect was replaced, *Atlantis* was returned to flight in November (STS 38).

The shuttle *Columbia* also suffered from a leak in the 17-in. (43-cm) disconnect area. The seal was replaced, but a separate leak in the main propulsion system remained difficult to pinpoint. A special investigatory team identified the problem as a defective seal in a prevalve of the main propulsion system. The seal had apparently been damaged during installation, following *Columbia*'s January 9 mission (STS 32). A test on October 30 verified that all problems with *Columbia* had been resolved, and the final shuttle mission of 1990 was flown by *Columbia* during December 2–10 (STS 35).

Hubble Space Telescope. After nearly a decade of anticipation, the Hubble Space Telescope was launched on April 24 aboard the space shuttle *Discovery* (STS 31; **Fig. 1**). It was expected that by orbiting outside the Earth's atmosphere, thus avoiding atmospheric distortion to astronomical observations, Hubble would offer greatly increased clarity and resolution. For example, Hubble may gather data that could help answer important questions related to the origins of the universe. In its first optical-engineering test, Hubble did, in fact, yield a valuable science observation, resolving the star cluster 30 Doradus three to four times better than the best ground-based observation.

The optimism over Hubble turned to disappointment with the discovery in June that the telescope's optical system is affected by a spherical aberration or error in the shape of the primary mirror that prevents the telescope from focusing light to a single, precise point. As a result, the telescope cannot observe very faint objects or distinguish faint objects in a crowded field.

A review board was appointed by the NASA Administrator to investigate and determine how the aberration had occurred and why it had not been detected before launch. After 5 months, the board concluded that a key device used to test the shape of the primary mirror had been assembled incorrectly. As a result, the mirror was improperly ground by 2–3 micrometers at its edge. The review board faulted both the contractor, Perkin-Elmer (now Hughes Danbury Optical Systems), and NASA's quality-control practices for the error and for failure to discover it in a timely manner.

Despite its major flaw, the Hubble Space Telescope is proving of considerable value to science. The instrument is still capable of observing objects in visible light much more clearly than ground-based telescopes, and it is able to make observations at ultraviolet wavelengths, which are largely blocked from the Earth's surface by the atmosphere.

The Wide Field/Planetary Camera (WF/PC), carried aloft with Hubble, observed with remarkable clarity a jet of material streaming away from the Orion Nebula, offering insights into this region, which is populated by young stars. The Faint Object Camera, built by the European Space Agency, returned the clearest image yet made of Pluto and its moon,

Fig. 1. Hubble Space Telescope during its deployment from the space shuttle *Discovery*. Most of the telescope can be seen as it is suspended in space by *Discovery*'s remote manipulator system, following the deployment of part of its solar panels and antennas. (*NASA*)

Charon. Most dramatically, the Wide Field/Planetary Camera took several hundred pictures as white spots, which have long been noted as an intermittent feature of Saturn's atmosphere, grew into an immense storm that spread around the planet's equator.

The discovery of the spherical aberration gave new emphasis to the previously planned Hubble Space Telescope servicing mission scheduled for 1993, when astronauts will replace the Wide Field/Planetary Camera with its backup unit, the WF/PC-2, which will be modified to compensate for the spherical aberration. Once the replacement is accomplished, NASA astronomers believe that Hubble will be able to accomplish most of its major scientific objectives.

Astro 1 mission. In December, astronomers used three ultraviolet telescopes and the broad-band x-ray telescope in the Astro 1 payload aboard the space shuttle *Columbia* (STS 35) to study the high-energy universe (**Fig. 2**). They made 394 observations of 135 objects. These included Jupiter and its satellite Io, one comet, and dozens of exploding stars, galaxies, and quasars.

Solar system observation. By the end of 1990, NASA's 1989 planetary missions, *Magellan* and *Galileo*, were returning interesting and significant data. On October 6, the space shuttle *Discovery* (STS 41) launched *Ulysses* on its mission to study the Sun's polar regions, areas that heretofore had not been studied closely.

The *Magellan* spacecraft returned radar images of Venus that showed geological features unlike any-

thing seen on Earth. One area featured what scientists have termed crater farms; another was covered by a checkered pattern of closely spaced fault lines running at right angles. Most intriguing were indications that Venus still may be geologically active, though much less so than Earth. Scientists hoped to map the entire surface of Venus and observe evidence of a volcanic eruption during *Magellan*'s mission, which continued into 1991.

Galileo flew by Venus in February, conducting the first infrared imagery and spectroscopy below the planet's dense cloud cover. NASA also expanded its commitment to study the Earth, receiving approval from Congress to undertake development of the Earth Observing System, a series of satellites and a sophisticated data management system that will use the perspective from space to observe the Earth as a global environmental system beginning in 1998.

Commercial space activities. Citing a failure by United States aerospace corporations to make strategic investments in commercial space ventures, the Department of Commerce issued a report warning that the industry's international competitive position may be in jeopardy. The Department also faulted the United States financial community for its apparent unwillingness to take the risks necessary to help create a viable commercial launch industry.

Nevertheless, there were some important achievements in 1990, including seven successful commercial launches. On June 23, a Martin Marietta Commercial Titan placed an *INTELSAT F6* spacecraft into orbit,

Fig. 2. Astro 1 payload being lowered into a transportation cannister on March 16, 1990, four days before its installation in the payload bay of the space shuttle *Columbia*. (*NASA*)

which has eased congested communications traffic on the International Telecommunications Satellite system; the satellite, which went into service in September, augmented the INTELSAT network by 120,000 voice and 3 television channels.

This successful launch followed an earlier effort (March 14) that resulted in failure. In that attempt the Titan second stage did not separate from the satellite on command from the ground, causing the satellite to enter an unusable 300-nautical-mile (550-km) orbit.

The United States commercial field gained a new entrant in 1990 as the construction engineering firm Brown & Root, known worldwide for its expertise in petroleum, mining, and electric power, announced formation of a space operations division. The firm, which plans to provide commercial launch facilities, proposes to use construction techniques that are proven for offshore drilling platforms and that could reduce launch-site and operations costs by up to 50%.

SOVIET SPACE ACTIVITY

The Soviet space program, while still the most active in the world in terms of launches, is beset with many problems, reflecting the turmoil throughout Soviet society. The program appears to be undergoing a major retrenchment, as budgets shrink under the weight of Soviet economic setbacks, and with largely disappointing results from efforts to market space-related goods and services.

Soviet shuttle. The Soviet shuttle program, which began spectacularly in 1988 with the maiden flight of *Buran*, the winged reusable space vehicle, has failed to live up to initial promises. Although the first flight was crewless, the Soviets predicted that the craft would soon be making up to four crewed flights each year. But no flights occurred in either 1989 or 1990. Although the Soviets insisted that *Buran* would dock with the space station *Mir* some time in 1991, the prospects for meeting this schedule remained uncertain.

The 40-ton (3.6-metric-ton) abandoned space station *Salyut 7–Cosmos 1686* was to have been visited by *Buran*, with parts returned to Earth for study. But the Soviets' apparent inability to pursue *Buran*'s intended flight schedule doomed *Salyut 7* to fall from orbit into extinction over Argentina in February 1991.

Space station Mir. The centerpiece of the Soviet space program, the large space station *Mir*, has also encountered major difficulties. According to plan, *Mir* should have been complete by 1990, with four large pressurized research modules mated to the space station's central docking unit and one to the aft port on *Mir*; as many as six scientist-cosmonauts were to have been stationed in *Mir* on a continual basis. But through 1990, only two Soviet cosmonauts routinely occupied the space station, and only three of the research modules were in place, including the *Kristall* science laboratory launched to *Mir* on May 31. On December 2, a Japanese television reporter was launched to *Mir* aboard *Soyuz TM-10* under terms of a commercial contract.

The Soviets planned a major reconfiguration of

Mir's electrical power system in order to keep the space station continually crewed into 1992. The work involves a series of extravehicular activities (EVAs) designed to transfer the solar panels from *Kristall* to the space station's *Kvant 1* module (astronomical observatory). When the panels are shifted in this way, they will be less subject to shadowing by the main module and will therefore produce more electric power.

Other changes to *Mir*'s electrical system involve new internal cabling to meet differing power requirements in various portions of the space station. A *Progress M-4* resupply spacecraft was launched on August 15 and docked at *Mir* on August 17, transporting electrical cables for the electrical system reconfiguration.

On July 17, Soviet cosmonauts who were engaged in extravehicular activity to repair damaged insulation blankets on *Mir* were forced to use emergency procedures to reenter the space station. The problem arose when the station's airlock hatch on the *Kvant 2* module malfunctioned and resulted in a precarious situation as the cosmonauts' extravehicular activity time (in excess of 7 h) approached safe limits. Reentry was effected only by dropping the pressure within the second section of the *Kvant 2* module. Repair of the hatch will require additional extravehicular activity, probably to be accomplished in tandem with the reconfiguration of *Mir*'s electrical system.

Military space program. Although major problems are evident with the Soviets' civilian space program, the military space program remains strong despite a reduction in military launch activity. Indeed, improvements in Soviet military spacecraft allow them to remain longer in orbit. Thus the reduced launch rate for military satellites (beginning in 1989, with an 18% drop over the total for the previous year) is a function of their greater longevity, rather than evidence of a declining military space effort.

EUROPEAN SPACE ACTIVITY

The first European launch of 1990 successfully injected a seven-satellite payload into Sun-synchronous orbit. The January 21 launch of Ariane V35 from the Guiana Space Center at Kourou carried the *Spot 2* Earth resources platform of the French national space agency, CNES, and six auxiliary *UoSAT/Microsat* satellites aloft. The launch marked the end of a nearly 3-month pause in the operations of the Ariane commercial booster, following a series of technical difficulties, first with the Japanese *Superbird B* (which had been originally slotted for the mission) and later with the launcher's Ferranti inertial reference system.

The *Spot 2*, a 4123-lb (1870-kg) satellite, is a replacement for the *Spot 1*, which provides Earth-imaging data to commercial customers. The *UoSAT* payloads carried instruments, including a radiation-environment experiment, experimental data-processing equipment, and experimental solar cells.

Service was interrupted on February 22, when an in-flight explosion resulted in loss of Ariane V36 and its payload. The accident occurred when one of the launcher's Viking 5 first-stage engines lost power during the early phase of climbout. Commercial service resumed on July 24, with the successful launch from Ariane V37 of two telecommunications satellites (the French *TDF-2* direct broadcast satellite and the German *DFS Kopernikus 2* telecommunications spacecraft).

A total of six Ariane commercial launches were carried out in 1990, including the October 12 Ariane V39 mission, which placed Hughes' *SBS 6* and *Galaxy 6* spacecraft into geostationary orbit. This marked the first time that Ariane had carried two spacecraft for the same customer.

Arianespace sought to broaden its commercial launch service capability in an effort to hold a major market share in the face of competition from the United States, the Soviet Union, and China. In particular, the French-based organization offered service for satellites in the 900–1800-lb (400–800-kg) range to meet the growing requirements for intermediate-sized spacecraft for telecommunications. In 1990 Arianespace introduced a commercial launch service for minisatellites (up to 110 lb or 50 kg) to address the growing trend toward use of these relatively inexpensive spacecraft. *SEE COMMUNICATIONS SATELLITE.*

The European Space Agency announced a 6-month delay in its formal decision to proceed with approval of full-scale development (phase 2) of the *Hermes* spaceplane. The delay is not expected to affect the *Hermes'* first flight, set for 1998. It was dictated by a need to complete the *Hermes'* baseline design as well as to evaluate design changes required by redefinition of NASA's international space station.

ASIAN SPACE ACTIVITIES

Japan and China continued their space activity in 1990.

Japan. With completion in 1990 of a new launch site for its H-2 booster, Japan marked a milestone in its efforts to become a major power in space during the twenty-first century. The facility, at the Tanegashima Space Center in southern Japan, rivals in size and capability the United States Titan 3 facilities at Cape Canaveral, Florida, and Europe's Ariane 4 facilities at Kourou, French Guiana.

When H-2 missions from Tanegashima commence later in the 1990s, they will provide Japan with a crewless heavy-booster and satellite launch capability rivaling that of the United States, the Soviet Union, and Europe. The H-2 is a product of Japanese development, unlike the Japanese Delta class N-1, N-2, and H-1 boosters, which have relied upon American technology and assistance. The H-2 is designed to resupply the *Freedom* space station as a backup to the United States shuttle, and is a key element in Japan's drive to develop an independent space-flight capability.

The Tanegashima Space Center includes a launch complex featuring a large vehicle assembly building (VAB), in which the H-2 booster will be placed in a vertical position on a mobile rail-mounted launcher platform. A seaside launch platform, a large H-2 pad

service structure, and a rocket engine test stand complete the ensemble.

With four times the capability of the H-1 booster, the H-2 is designed to launch 8800-lb (4000-kg) payloads into geosynchronous transfer orbit or 4400-lb (2000-kg) payloads into geosynchronous orbit, making it nearly identical in lifting power to the European Ariane 4.

On July 24, with launch of the Japanese Space Agency's *Muses-A Hiten* ("celestial maiden") spacecraft, Japan became the third country (after the United States and the Soviet Union) to place a spacecraft into orbit around the Moon. The payload, launched aboard an M-3S-2 booster, consisted of a primary 434-lb (197-kg) spacecraft, which carried a German cosmic dust counter, and a 27-lb (12.2-kg) lunar orbiter. A principal purpose of the mission was to demonstrate Japanese ability to manage complex spacecraft maneuvers, in preparation for the Japan-NASA Geotail experiment, scheduled for July 1992. The Geotail spacecraft, being built by the Japanese, will study the effects of the solar wind on Earth's magnetosphere.

Japan's first shuttle astronaut underwent intensive training in preparation for a Japanese-sponsored Spacelab mission, set for launch in late 1991 aboard the orbiter *Atlantis*. The planned 7-day flight, which will provide for extensive Japanese materials research, may be regarded as laying the foundation for future autonomous Japanese crewed space-flight capability.

The ambitious Japanese plans for future space activities include development of a space station, development of the *HOPE* crewless spaceplane, and design of Atlas-class rockets to launch lunar and planetary missions. A variety of specialized payloads are already under development in support of the program, including the *ERS 1* radar remote sensing satellite, for launch in 1992; the *Advanced Earth Observing Satellite (ADEOS)*, with United States and French participation, for launch in 1995; the *Astro-D* astronomy spacecraft and *Muses-B* very long range baseline interferometry (VLBI) spacecraft; the *Solar-A* solar imaging spacecraft; and lunar penetrators to be fired into the Moon's surface during a mission planned for launch in 1996.

China. China bolstered its position in the commercial launch market with the successful April 7 launch of *AsiaSat 1* from the Xichang Satellite Launch Center. The satellite was carried aboard a 203-ton (184-metric-ton) Long March 3 rocket.

AsiaSat 1 had been launched originally as *Westar 6*, then rescued from space by the United States space shuttle (mission 51A) in 1985. The satellite has 24 transponders and two beams, covering an area from Korea to the Middle East and from Mongolia to Malaysia. It is the first satellite ever to be rescued and returned to space.

The *AsiaSat 1* launch was regarded as politically important by the Chinese, who have been operating under Western criticism and United States legislative sanctions, following the Tiananmen Square massacre. President Bush exempted the United States–built satellite from sanctions and export controls in late 1989, allowing it to be launched from China.

On June 16, China launched its first Long March 2E rocket from the Xichang Satellite Launch Center. The Long March 2E is based on the Long March 2C and is distinguished by four strap-on liquid fuel boosters. The June launch, which placed a very small Pakistani research satellite and a full-scale model (dummy) of the *Aussat* into low-Earth orbit, precedes the launch of two Australian *Aussat B* telecommunications satellites, scheduled for 1991 and 1992.

In October, senior officials of the Chinese Ministry of Aerospace Industry unveiled plans to develop three new heavy-lift launch vehicles: the LM-X1, with a thrust of 3.7×10^6 lbf (1.7×10^7 newtons), which will be able to carry an 11-ton (10-metric-ton) payload to Mars or place at least 22 tons (20 metric tons) into low-Earth orbit; the LM-X2, designed to lift a 33-ton (30-metric-ton) payload to low-Earth orbit; and the LM-3B, with a thrust of 1.3×10^6 lbf (6×10^6 N), which will take a 5-ton (4.5-metric-ton) payload to geosynchronous transfer orbit. Observers believe that a version of the LM-X may be intended by the Chinese to serve as a launch vehicle for a possible space station, some time after the year 2000.

For background information SEE APPLICATIONS SATELLITES; COMMUNICATIONS SATELLITE; SATELLITE ASTRONOMY; SPACE FLIGHT; SPACE PROBE; SPACE SHUTTLE; SPACE STATION in the McGraw-Hill Encyclopedia of Science & Technology.

Robert J. Griffin, Jr.

Bibliography. Board of Investigations, NASA, *The Hubble Space Telescope Optical Systems Failure Report*, November 1990; Japan forging aggressive space development pace, *Aviat. Week Space Technol.*, 133(7):36–42, August 13, 1990; National Aeronautics and Space Administration, *NASA Activities*, 21(3):1–19, May/June 1990.

Space technology

The major changes in space transportation during the past few years reflect a number of trends for the future. The United States launch fleet now includes a number of vehicles besides the shuttle, and launch vehicles of other nations play an ever-increasing role in both global and domestic space activities. The need for lower launch cost, higher reliability, and greater flexibility in the range of payloads carried has brought consideration of a number of new launch system concepts. Perhaps the most important development, however, has been the realization that the United States needs a coherent national launch policy to meet both current and future space launch needs.

Space shuttle. The space shuttle continues to be the only United States vehicle able to carry humans to and from space. The shuttle also can carry greater mass to low Earth orbit than any other United States vehicle, about 45,000 lb (20,000 kg), reduced from its original rating of 55,000 lb (25,000 kg) after the *Challenger* failure in January 1986.

The National Aeronautics and Space Administration (NASA) has begun a major program to improve the shuttle's reliability and performance, since it is likely to remain the agency's principal means of access to space until at least the year 2000. Besides a complete redesign of the shuttle solid-rocket booster's O-ring seals (whose failure caused the loss of *Challenger*), NASA has begun development of a wholly new advanced solid-rocket motor (ASRM), which is expected to improve both reliability and overall shuttle payload-carrying capacity (by 8000–11,000 lb or 3600–5000 kg). Major redesigns have also been initiated on the shuttle's main liquid-propellant rocket engines, its brakes and landing gear, and other critical systems.

The shuttle's primary task in the 1990s will be to launch, deploy, assemble, and service the space station *Freedom*, expected to require over 20 flights during 1995–1999 and 2 or 3 flights per year thereafter. There is, however, considerable concern about the consequences to *Freedom* of losing another orbiter, as well as the excessive downtime that would result from even relatively minor launch problems such as those in 1990. Hence NASA is also evaluating two alternative shuttle-based configurations: use of liquid-propellant rocket boosters instead of the present solid-propellant rockets, and development of a crewless cargo version of the shuttle, called Shuttle-C.

Liquid boosters are inherently more reliable than solid ones because (1) they can be started up and checked for normal operation on the ground before releasing the shuttle for launch, and (2) they have engine-out capability; that is, each booster would employ four engines, any one of which, if it were to malfunction, could be shut down in flight without aborting the mission, thereby reducing the risk of catastrophic failure. The use of liquid boosters, whose hydrogen-oxygen propellants exhaust only water vapor, would also alleviate the growing concern about environmental effects of the chlorine-based solid-rocket propellants. These benefits must, however, be evaluated in terms of the development and operating costs of the liquid boosters, which are significantly higher than those of the advanced solid-rocket motor.

Shuttle-C would use the present shuttle boosters and big propellant tank but would replace the present piloted orbiter with an expendable cargo-carrier. Because it could then carry about three times the shuttle's payload to low orbit, it could cut the total number of shuttle flights needed to deploy *Freedom* by more than half, significantly reducing both the probability of failure and the consequent downtime during *Freedom*'s critical deployment phase. As with the liquid booster, however, this benefit must be traded off against the additional cost of developing Shuttle-C.

Expendable launch vehicles. Besides the shuttle, the United States launch fleet now includes a number of expendable launch vehicles (ELVs), most of which were developed from military missile launch rockets. The U.S. Air Force, because it could not depend on a single launch vehicle (the shuttle) for military preparedness needs, had begun development of a large-payload expendable launch vehicle even before the *Challenger* loss in 1986. That vehicle, now called Titan 4, currently provides the bulk of United States defense launch capability.

After the *Challenger* accident, a United States policy change precluded the use of the shuttle for launching commercial payloads such as communications satellites, and subsequently mandated NASA and the National Oceanic and Atmospheric Administration (NOAA) to employ commercial launch services for all payloads that do not require the shuttle's unique capabilities.

This policy change rejuvenated the nascent United States commercial launch industry, which before 1986 had been unable to compete with government-subsidized shuttle launch services. The three primary expendable launch vehicle families that had launched most United States spacecraft prior to the introduction of the shuttle were upgraded by their manufacturers, and they began to compete with the French commercial launch service company Arianespace, which had taken over more than half the commercial launches outside the Soviet Union during the post-*Challenger* hiatus. Three United States vehicle families—Delta and Delta 2; Atlas, Atlas 2, and Atlas 2S; and Titan 3—now share slightly more than half the commercial world market, and also are used by NASA, NOAA, and the U.S. Air Force for many United States government launches.

Smallsats. A recent development in space operations has been the emergence of military and civil requirements for very small satellites, well below the payload capabilities of the above-mentioned "big three." As a result, the commercial prospects for small expendable launch vehicles are growing rapidly. Such vehicles include the commercial Scout, a veteran NASA launcher; the Pegasus, a winged launcher carried aloft by a large airplane; the Taurus, derived from solid-propellant military missile stages; and several others in the early developmental stages. *See* COMMUNICATIONS SATELLITE.

Non–United States launchers. The 1990s will also see significant growth in space capabilities by many of the world's nations. The French company Arianespace, with its capable family of Ariane 4 launchers developed and funded by the European Space Agency (ESA), is currently the world's largest single purveyor of commercial launch services. The Ariane 5, scheduled for completion in the mid-1990s (again with ESA funding), will be the first commercially available heavy-lift launcher outside the Soviet Union. China also offers commercial launches on its CZ (Long March) family. Japan, India, and Israel have developed launchers to fly their own payloads but have not yet entered the commercial launch market. Japan, however, has formed a large consortium to market its new H-2 launcher, which would compete effectively with all three United States launch vehicle families when its development is completed in the mid-1990s. Pakistan, Brazil, and Iraq have expressed their intent to develop space launchers.

Soviet competition. By far the most significant threat to the growing United States commercial launch market, however, is the highly capable family of Soviet launch vehicles, which includes the Soyuz, the workhorse Proton, the enormous Energiya, the new Zenit (already adopted as the standard launcher by Australia's new commercial launch complex at Cape York), and even a shuttle, Buran, which is very similar to the United States shuttle. Up to 1990, security restrictions on United States–developed spacecraft technology kept the Soviet launchers out of the commercial world market, but the easing of such restrictions as a consequence of the new political developments in eastern Europe may open that market to the mass-produced (and therefore inexpensive) Soviet launchers.

Advanced launch system. The large-scale development of space, however, will require next-generation launch vehicles that can cut launch costs by a factor of 10 or more and at the same time can improve launch reliability significantly. (The demonstrated failure rate of current launch vehicles ranges from slightly more than 10% for Ariane 4 to somewhat less than 3% for the United States shuttle.) To do this, the United States developed a concept called the advanced launch system (ALS), originally to meet greatly expanded Department of Defense (DOD) launch needs. A reorientation of the Strategic Defense Initiative program eliminated much of that projected DOD demand, so that the development of the advanced launch system was reduced to a propulsion technology advancement program, with NASA conducting much of the design and testing. However, in 1990 the U.S. National Space Council determined that the advanced launch system should be reconfigured to meet not only DOD needs but all those of the United States national space program—defense, civil space activities of NASA and NOAA, and perhaps even commercial launch demand. In that context, the advanced launch system will be developed as a family of launchers suitable for all these operations in the twenty-first century. It will use a common propulsion system (the most costly and critical launch-vehicle component) and in various configurations will be able to carry payloads ranging from around 40,000 lb (18,000 kg) to perhaps as much as 250,000 lb (114,000 kg) to low Earth orbit.

Next-generation shuttle. The advanced launch system is being designed primarily as a cargo vehicle, but a reusable version could conceivably be developed as a dedicated personnel carrier to replace the shuttle early in the twenty-first century. Alternative personnel launchers could evolve from the present National Aerospace Plane (NASP) program, in which NASA and the Air Force are working with a team of five aerospace companies to develop an experimental craft, the X-30, which can take off like an airplane, accelerate to high speed in the atmosphere using air-breathing engines, and then use rockets to achieve orbit. Several such NASP-derived single-stage-to-orbit vehicle concepts are being considered as potential successors to the shuttle for use as dedicated personnel carriers, and perhaps also, if operating costs can be reduced to levels comparable to those of airline operations, for cargo.

Similar vehicles are under consideration in other countries: HOTOL (Horizontal Takeoff and Landing) in England; Saenger in Germany; HOPE (H-2 Orbiting Plane) in Japan; and Hermes, a shuttlelike craft launched by Ariane 5, being developed by the European Space Agency.

For background information SEE ROCKET ENGINE; SPACE FLIGHT; SPACE SHUTTLE in the McGraw-Hill Encyclopedia of Science & Technology.

Jerry Grey

Bibliography. J. M. Logsdon and R. A. Williamson, U.S. access to space, *Sci. Amer.*, 260(3):34–40, March 1989; Office of Technology Assessment, U.S. Congress: *Access to Space: The Future of U.S. Space Transportation Systems*, May 1990, *Affordable Spacecraft*, January 1990, *Round Trip to Orbit*, August 1989.

Speech technology

There have been recent advances in speech technology in the areas of text-to-speech synthesis and digital coding of speech.

TEXT-TO-SPEECH CONVERSION

Text-to-speech synthesis transforms linguistic information that is stored in a computer as data or text into an audible form that can be understood by humans. The speech is actually created by the computer, although the term voice synthesis is sometimes also used for digital recording of human voice. True text-to-speech synthesis can read anything, using an unlimited vocabulary. It is used in reading devices for blind people and as the voice of vocally handicapped people.

Use of text-to-speech technology has recently become a major adjunct to the rapidly growing digitized voice systems that provide telephone access to information. Text-to-speech converts electronic mail to voice mail for phone access. It permits field-service personnel to access large databases, such as parts inventories, by telephone. Use of text-to-speech technology is expanding rapidly because of improvements in delivery methods and speech quality.

Although a speech synthesizer is often thought of as a hardware device, usually a box or a board, text-to-speech synthesis originates as software. Fundamentally, it is a great deal of complex code that captures many sources of knowledge about language, and it is implemented on many different kinds of hardware platforms, including dedicated devices and standard microcomputers. A text-to-speech conversion process consists of two major elements. One, the vocal-tract module, is a sound-generating mechanism, analogous in function to the human vocal tract. The other, the text-to-parameter conversion module, takes as input the text to be spoken and produces a set of time-varying parameters that drives the sound-generating mechanism.

Vocal-tract module. When humans speak, air pressure is developed in the lungs and flows through the vocal folds in the larynx. The noise generated at the glottis or elsewhere in the vocal tract is modified as it passes through the oral and nasal cavities and radiates from the head as a speech waveform. The vocal tract can be modeled as a filter driven by an excitation source. The source is a periodic pulse waveform combined with a random noise waveform, each with a certain energy. The filter typically has three or more narrow-band resonances.

The text-to-speech vocal-tract module is a computer simulation of this source–filter model. The simulation requires a digital signal processor that can perform approximately 10 million operations per second. The inputs to this module are speech parameters. Correctly specifying the parameters of pitch, energy, and the resonance frequencies of the filter every 10 milliseconds makes possible the production of synthetic speech of high quality. This is a simulated waveform that is then fed to a digital-to-analog converter at 8000 samples per second for so-called telephone-quality speech.

Text-to-parameter module. A voice model requires from a few hundred to at most a few thousand computer instructions. However, the text-to-parameter module for a high-quality synthesizer requires from 100 to 1000 times as much memory as the voice model in order to capture the extensive linguistic knowledge that is required to generate correct parameters automatically. This module can be thought of as a model of how humans read aloud. Relatively little is known about how humans actually process language, but a great deal is known about the structure of language; the text-to-parameter module incorporates much of this knowledge.

Text normalization. Printed texts usually contain a great deal of symbolic material such as numbers, abbreviations, acronyms, and information that is signaled by graphic layout. The first step in converting text to parameters is text normalization: putting the information into a standard format by using complex rules. For example, the sequence "415" is read in three different ways depending on whether it is an area code, a part of an address, or a dollar amount. The pronunciation of abbreviations is also determined by context. Thus, the pronunciations of sequences of letters and punctuation marks are quite different depending upon where they occur, as illustrated by the following examples: "Dr. Jones lives on Jones Dr." "St. James St." "Jan. 22 is my wife's birthday; her name is Jan." Such phenomena can be handled reasonably successfully, but they require an extensive, nonalgorithmic computer program to execute a process that a human reader of English performs easily.

Pronunciation rules. The next task is pronouncing words correctly. English word pronunciation is the result of 2000 years of history, dozens of wars, and many population migrations. English therefore has by far the most complex relationship between its spelling and its word pronunciation of any alphabetic language. Most readers of English know the rule that a final "e" makes a preceding vowel tense—the rule that turns "dud" into "dude." A reasonably accurate text-to-speech system uses several thousand such rules to handle English.

It might be thought that pronunciation rules could be replaced with a large dictionary or table that would give the exact pronunciation of each English word. A text-to-speech system does use a built-in dictionary to pronounce some words, but a dictionary cannot store everything. The average United States junior high school student may encounter any of 500,000 different words. The 1970 United States Census listed over 2 million different last names. New words and names are introduced every day, which means that a dictionary will always be out of date. A text-to-speech dictionary can be used for words that are not regular enough to justify a rule, as well as for words whose pronunciation is dependent on syntax or context. Sophisticated rules must then pronounce everything else, including new words and names.

Prosodic rules. Specifying rhythm, intonation, and the emphasis or deemphasis of particular words requires the use of prosodic rules. In human speech, certain words sound more prominent than others; often these are the words that carry the information focus of the message. If the speech is part of a conversation, then the prosodic phenomena will be quite elaborate and may carry a significant portion of the meaning of the message. When speech is read from a text, especially by a person who is not interested in or knowledgable about the text, prosody will be more regular or monotonous. A text can be read with many different prosodic patterns, but some will sound wrong or unnatural to the listener.

Prosodies are generally the most difficult aspect of speech for a text-to-speech conversion system to model. A very good prosodic component would probably require that the computer actually "understand" the text. Human prosodies have been studied extensively, but since they are so complex, much of the information that is needed to generate prosody correctly from a text has not yet been explained by linguistic theory. However, an adequate reading may be achieved by employing simple rules that use the clues provided by punctuation marks in written texts (which indicate prosodic boundaries), and rules for emphasis or deemphasis of certain parts of speech. The resulting speech sounds somewhat mechanical, but simply making it more lively without a complex model of human prosody production makes it sound foolish.

Phonetic rules. After the text normalizer, the word pronunciation rules, and the prosody rules have been applied, the utterance to be spoken consists of a string of phonemes. Phonemes are the special symbols that are used by dictionaries to represent pronunciation. They are not actual speech sounds, but they represent the finest distinctions between sounds that most people who are not trained in phonetics normally notice. The concept of the phoneme covers a range of recognizable expressions. The distribution pattern for various phoneme realizations is part of what defines dialects. Phonemes are represented in ordinary speech in about the same way that the abstract ideal graphic letters of the alphabet are represented in messy handwriting.

Phonetic rules take phonemes as input and produce very detailed descriptions of the sounds of the utterance. For example, they assign a duration to each phoneme according to context. (The final vowel in a phrase may be lengthened by up to 50%.) Other rules remove parts of phonemes. (The aspiration of a stop such as /k/ is deleted in a number of different contexts.)

The major role of phonetic rules, however, is to describe the coarticulation between phonemes. Each phoneme very strongly influences the parameters in the adjacent phoneme. In some cases, this influence may extend over several adjacent phonemes. In American English, for example, the phoneme /t/ should sound quite different in the words "Tom," "cat," and "butter." Without phonetic rules, synthetic speech output is recognizable but very unnatural.

Voice-table module. A voice-table module converts the detailed phonetic description into numeric targets for use by the voice model. It is the phonetic rules in conjunction with the voice tables that are the primary determinants of synthetic speech intelligibility.

Intelligibility. Intelligibility is the likelihood that a human listener will be able to identify a particular spoken word. In one standard test, a modern, high-quality text-to-speech system typically scores approximately 95%, while low-end systems score from 60 to 75%.

Recent developments. Increased linguistic knowledge, improvements in software design, and advances in digital signal processing hardware have led to a number of important recent developments in text-to-speech synthesis. The technology has been implemented on many new kinds of hardware platforms for new kinds of applications. Some of these include pocket-sized talking dictionaries, with more than 100,000 entries, that synthesize words in real time by using a standard microprocessor chip; portable, battery-powered speech-output systems for blind users that read up to 700 words per minute; automotive text-to-speech conversion units, with built-in speakers, which speak text messages sent to trucks over satellites; and telephone systems that generate 16 text-to-speech processes simultaneously on one board, and can support 100 or more telephone lines with a single personal-computer-based unit containing multiple boards.

M. H. O'Malley

DIGITAL SPEECH CODING

Advances in coding algorithms and signal processing have led to sophisticated technologies for digital speech communication for a variety of applications, as well as greater flexibility in the design of terminals for the integrated communication of speech, images, and data. For traditional telephony, with a signal bandwidth of 3.2 kHz, the transmission rate for network-quality speech is now down to 16 kilobits per second (kbps). Robust communications quality, appropriate for cellular radio, has been realized at 8 kbps. Research attention is shifting toward coding at 4 kbps, with the focus on improving speaker identification and the naturalness of coded speech. For wide-band speech, with a signal bandwidth of 7 kHz, high-quality coding is now possible at 32 kbps, implying stereo conferencing or dual-language programming over a 64-kbps channel. Perceptually lossless coding of 20-kHz audio signal has been demonstrated at 128 kbps, with high-quality performance at rates as low as 64 kbps for some classes of signals.

Measurement of speech quality. Measurement of speech quality is a difficult and long-standing problem. A subjective rating scale of 1 to 5 is often used to quantify the level of digital speech quality. This is the so-called mean opinion score (MOS) scale used widely for evaluating coding algorithms for digital telephony. A score of 4.0 on the MOS scale signifies high-quality or near-transparent coding. Network quality implies high quality, but also further capabilities of the speech coder, as demanded by the communication network environment. An MOS score of 3.5 denotes communications quality. At this level, speech degradation is easily detectable, but it is not severe enough to impede natural communication. Finally, synthetic quality implies a signal characterized perhaps by high intelligibility but by an inadequate level of naturalness and speaker recognizability. These deficiencies are usually reflected by an MOS score not exceeding 3.0.

MOS measurements of speech quality are supplemented, especially in very low bit rate speech technology, by scores of the diagnostic rhyme test (DRT) and the diagnostic acceptability measure (DAM). The DRT is a word intelligibility measure, while the DAM reflects acceptability for speech communication in a broader multidimensional sense.

Coding of telephone speech. **Figure 1** depicts the current state of telephone speech coding in terms of standards activity, bit rate, typical applications, and decoded speech quality. The frequency range of the input signal is assumed to be 200–3400 Hz, and output speech quality is measured on the five-point

MOS scale. The **table** summarizes the DRT, DAM, and MOS performances of various speech coders, in the range from 2.4 to 64 kbps. Pulse-code modulation (PCM) is the simplest coding system, a memoryless quantizer. Adaptive differential PCM (ADPCM) is a type of waveform coding that uses redundancy-removing operations to present a signal of lower energy to the amplitude quantizer, resulting in a lower bit rate for a specified level of output speech quality. The linear predictive coder (LPC) or vocoder is an algorithm that produces intelligible but synthetic-sounding speech at very low transmission rates by means of a highly compact excitation-modulation model (**Fig. 2**a), where excitation refers to the energy source for the production of the speech signal. The synthetic quality in this system is accepted in applications where digital encryption and low transmission rate are of paramount importance. This could include commercial applications such as banking, but the principal customers by far are government and defense agencies. The code-excited linear predictive (CELP) coder is any one of a class of algorithms that combine the high-quality potential of waveform coding with the compression efficiency of a model-based vocoder. The idea is to use a time-varying excitation model much more sophisticated than that of a traditional vocoder, and to use waveform-coding principles to compute a distortion-minimizing excitation for every frame (say, 16 milliseconds) of input speech (Fig. 2b and c). Although vocoders and CELP coders are fairly complex, current signal processor technology permits implementation of both the encoder and the decoder on a single chip. Current goals in speech coding include the achievement of near-transparent or transparent quality at 8 kbps, and robust communications quality at 4 kbps and lower.

Digital telephony at 4 kbps. High-quality coding at 4 kbps is a primary focus in speech research. It is important for enhancing secure telephony in govern-

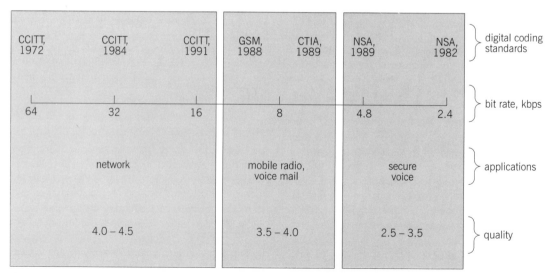

Fig. 1. Digital coding of telephone speech: standards, bit rates, typical applications, and ranges of speech quality. CCITT = International Telegraph and Telephone Consultative Committee; CTIA = Cellular Technology Industry Association (United States); GSM = Groupe Speciale Mobile (Europe); NSA = National Security Agency (United States). The frequency range of telephone speech is 200–3400 Hz, for a bandwidth of 3200 Hz.

DRT, DAM, and MOS scores for standard speech coders*

Coder[†]	DRT	DAM	MOS
64-kbps PCM	95	73	4.3
32-kbps ADPCM	94	68	4.1
16-kbps LD-CELP	94[e]	70[e]	4.0
8-kbps CELP	93[−]	68[e]	3.9
4.8-kbps CELP	93[−]	64	3.2[+]
2.4-kbps LPC	90	53	2.5[e]

*e indicates estimate, − indicates upper bounds, and + indicates lower bounds.
[†]PCM = pulse-code modulation; ADPCM = adaptive differential PCM; CELP = code-excited linear predictive coding; LD-CELP = low-delay CELP; LPC = linear predictive coder or vocoder.

ment and military applications, and for providing the capability of future band-efficient systems for digital radio, for example, cellular channels with a user bandwidth of 5 kHz. These are wireless access channels for moving vehicles in rural, suburban, or urban environments served by terrestrial base stations. Speech coding at bit rates on the order of 4 kbps is needed also in mobile satellite (MSAT) communication applications for providing wireless access to moving vehicles in remote areas; and in *INMARSAT* (*International Maritime Satellite*) applications with a 6.4-kbps target for total transmission rate (speech-coder bit rate plus overhead for channel error protection). Finally, speech coding at 4 kbps has an attractive implication for storage: 1 h of spoken material can be stored on a single 16-megabit memory chip. *SEE COMMUNICATIONS SATELLITE.*

Digital telephony at 8 kbps. Systems for first-generation digital cellular radio use bit rates of the order of 8 kbps for speech coding (Fig. 1). The proposed system in the United States has a user bandwidth of 10 kHz, and a total transmission rate of about 13 kbps for speech coding and channel error protection. The system will eventually replace the current use of analog frequency-modulated (FM) speech with a 30-kHz user bandwidth. The digital system provides greater robustness to channel noise and fading, as well as better reuse of individual carrier frequencies. As a result, the improvement in call capacity (number of users) will exceed the factor of 3 implied by the changeover from 30 to 10 kHz. The net gain is expected to be a factor of the order of 5 to 7.

In the United States, cellular telephony will be based on a CELP coding algorithm. Although speech quality at 8 kbps currently falls slightly below the high-quality threshold (MOS = 4.0), it provides improvements over analog FM telephony, especially at low levels of radio channel quality (say, a channel signal-to-noise ratio below 15 to 18 dB). The communication delay due to the speech coder-decoder (codec) is expected to be of the order of 40 ms, a value considered acceptable in the cellular application.

For digital cellular radio in Europe, the recommendation of the Groupe Speciale Mobile (GSM) is also for a hybrid coder (Fig. 2a), using a regular pulse excitation algorithm, with a bit rate of 13.2 kbps (out of a total transmission rate of 22.8 kbps) and a codec

delay of 40 ms. The coding technique is similar to that of the 9.6-kbps multipulse excitation coder used in the Skyphone system, which provides communications to airliners. *SEE MOBILE RADIO.*

Digital telephony at 16 kbps. Currently, the International Telegraph and Telephone Consultative Committee (CCITT) is considering defining a low-delay, network-quality speech standard at 16 kbps (Fig. 1). Possible applications include digital circuit multiplication equipment, voice mail, packetized speech, cordless telephones, and speech for videophone service.

An important challenge for the proposed algorithm is the combination of high quality with low processing delay. The delay requirement means that estimation of the time-varying model for the speech spectral envelope has to be performed in a backward-adaptive mode, by using a history of already quantized speech, rather than by buffering tens of milliseconds of input speech for forward spectral estimation. The challenge is to realize sufficient spectral estimation in spite of the presence of quantization noise in the past speech samples used for spectral analysis.

Coding of wide-band speech and audio. Figure 3 defines four commonly understood grades of audio bandwidth. If the audio signal is speech rather than music, gains in quality due to wider bandwidth are perhaps greatest in going from the telephone level to the commentary, or amplitude-modulation (AM) radio, level. The gains are in terms of increased intelligibility, naturalness, and speaker recognition. Low-frequency enhancement (50–200 Hz) contributes to increased naturalness and speaker presence, and high-frequency enhancement (3400–7000 Hz) provides greater intelligibility and fricative differentiation (for example, between ''s'' and ''f''). If the audio signal includes music, the highest frequency of interest extends to 15 kHz for FM-radio quality, and 20 kHz for compact-disk (CD) quality.

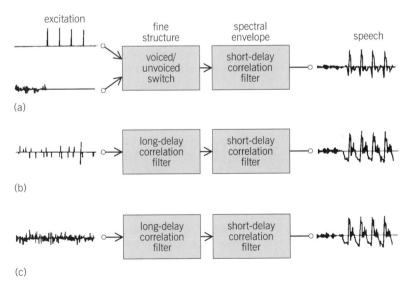

Fig. 2. Models of speech excitation in (a) linear predictive coder (LPC) or vocoder, and in the hybrid coders (b) multipulse LPC and (c) code-excited linear predictive (CELP) coder. Approximately the left one-third of each of the speech waveforms (on the right) represents unvoiced speech, while the right two-thirds represents voiced speech.

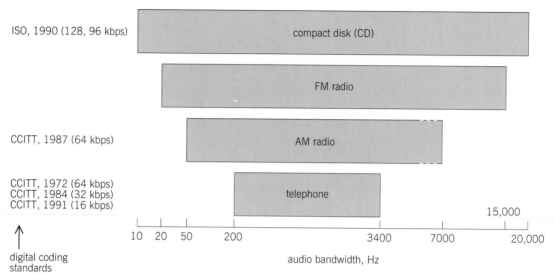

Fig. 3. Four grades of audio signal bandwidth and corresponding standards for digital coding. ISO = International Standards Organization; CCITT = International Telegraph and Telephone Consultative Committee. The quality ranges from 1 (unacceptable) to 5 (excellent).

Coding of 7-kHz audio. The naturalness of wide-band speech is a significant feature for extended telecommunication processes such as audio conferencing and program broadcasting. Integrated services digital networks (ISDNs) provide a natural framework for a 64-kbps algorithm to encode wide-band speech for such applications. The digital connectivity afforded by ISDNs has particularly made possible the inclusion of low frequencies down to 50 Hz in the transmitted audio band.

The CCITT standard for 7-kHz audio signals (G.721) is a 64-kbps algorithm developed primarily for ISDN teleconferencing and loudspeaker telephony. The G.721 coding algorithm is based on a two-band subband coder, with ADPCM coding of each of the subbands. The low- and high-frequency subbands are quantized by using 6 and 2 bits per sample, respectively.

Current work is addressing the problem of coding 7-kHz audio signals at 32 kbps with high quality. This will enable the transmission of two bilingual or stereo wide-band channels at 64 kbps. For stereo, the utilization of cross-channel correlations can provide a further increase of capability, for example, accommodation of a bandwidth greater than 7 kHz in the 64-kbps stereo system.

Coding of 20-kHz audio. The International Standards Organization (ISO) is committed to the standardization of a low-bit-rate coding algorithm for 20-kHz audio signals. Applications for low-bit-rate wide-band speech (and music) include electronic publishing, travel and guidance, teleteaching, multilocation games, multimedia memoranda, and database storage. Another major application for 20-kHz digital audio signals is in advanced television systems such as high-definition television (HDTV).

On a current compact disk, audio is stored at a rate of about 700 kbps per sound channel (16-bit PCM coding of signals sampled at 44.1 kHz). The emerging

ISO standard calls for a single-channel bit rate of 96 or 128 kbps. Production of high-quality audio at these very low rates calls for a new generation of coding algorithms in which the gains due to redundancy removal need to be augmented by the liberties offered by the human auditory process, as predicted by sophisticated models of just-noticeable distortion.

Prospects. Sophisticated algorithms for coding will lead to transmission techniques where speech quality is not limited by quantization noise, and the notion of enhancing speech quality by means of greater input bandwidth will become more pervasive. Coding systems in the 8–64-kbps range will thus provide graceful flexibility in terms of selected bandwidth and special features such as stereo separation in teleconferencing. Advances in coding will be supported by new technologies for wide-band transducers, noise-canceling systems for audio pick-up, and autodirective microphone arrays.

For background information *SEE DISK RECORDING; INFORMATION THEORY; MOBILE RADIO; PULSE MODULATION; VOICE RESPONSE* in the McGraw-Hill Encyclopedia of Science & Technology.

N. S. Jayant

Bibliography. B. S. Atal, High-quality speech at low bit rates: Multi-pulse and stochastically excited linear predictive coders, *Proc. ICASSP*, pp. 1681–1684, 1986; J. L. Flanagan, *Speech Analysis, Synthesis and Perception*, 2d ed., 1972; N. S. Jayant and P. Noll, *Digital Coding of Waveforms*, 1984; J. D. Johnston, Transform coding of audio signals using perceptual noise criteria, *IEEE J. Select. Areas Commun.*, 6(2):314–323, February 1988; M. H. O'Malley, Text-to-speech conversion technology, *Computer*, 23(8):17–23, August 1990; V. C. Welch, T. E. Tremain, and J. P. Campbell, A comparison of U.S. government standard voice coders, *Proceedings of ICASSP '89*, Glasgow, May 1989; I. H. Witten, *Principles of Computer Speech*, 1982.

Spinel

The spinel crystal structure takes its name from a mineral with the idealized chemical formula $MgAl_2O_4$. A famous example is the Black Prince's "ruby" in the Tower of London, whose rich red color is caused by a small fraction of chromium ions substituting for some of the aluminum in the chemical formula. As in most naturally occurring minerals, substitutions of this type are the rule rather than the exception; in fact, most minerals identified as spinels are chemically complex combinations of such ideal components. Some components important in naturally occurring spinels are listed in the **table**. In practice, the name given to any particular mineral specimen is that of its dominant component.

Spinels occur widely in nature, frequently as minerals of strategic significance in other fields of scientific inquiry. For example, species such as magnetite (Fe_3O_4) and titanomagnetite (Fe_3O_4–Fe_2TiO_4) mixtures are widespread accessory minerals in many types of rock and, being ferrimagnetic, dominate the magnetic properties of the rock. Since magnetite contains iron as both Fe^{2+} and Fe^{3+}, the magnetite component of a complex spinel may often be used to measure the oxidation state of its host rock at the time of the rock's formation. Oxidation state influences the chemistry of many geological processes, including the fundamental ones involved in the formation and early history of the Earth—or the other rocky planets for that matter. The most abundant mineral in the Earth's upper mantle, olivine ($Mg_{1.8}Fe_{0.2}SiO_4$), transforms to the spinel structure at pressures equivalent to 400 km (240 mi) depth in the Earth. This phase transformation may be responsible for many deep-seated earthquakes, and it also has substantial consequences for the mechanism of plate tectonics. Spinel minerals are economically important as the ores for many metals; for example, chromite ($FeCr_2O_4$) is virtually the only source of chromium. Technologically, synthetic spinels find applications as magnetic materials, refractories, and pigments.

Chemistry. In general, simple spinels have a chemical formula that may be written as AB_2X_4, where A and B are positively charged cations [for example, magnesium (Mg^{2+}) and aluminum (Al^{3+}) in spinel] and X is a negatively charged anion, such as oxide (O^{2-}). Although this discussion mainly treats the oxide spinels, it should be recognized that other anions form compounds with the spinel structure. Examples include thiospinels, with X = sulfide ion (S^{2-}) as in the mineral violarite ($FeNi_2S_4$); or spinels with X = fluoride (F^-), chloride (Cl^-), or even anion groups such as cyanide (CN^-). For the oxide spinels, the requirement of electrical neutrality in the formula unit means that, commonly, A is a divalent cation and B trivalent (a so-called 2-3 spinel, as in $MgAl_2O_4$), or A is tetravalent and B divalent (4-2), as in the mineral ulvospinel ($Ti^{4+}Fe_2^{2+}O_4$). Some of the cations commonly found forming oxide spinels are the following:

Charge 2^+:
Magnesium (Mg)   Nickel (Ni)
Manganese (Mn)   Copper (Cu)
Iron (Fe)   Zinc (Zn)
Cobalt (Co)   Cadmium (Cd)

Charge 3^+:
Aluminum (Al)   Cobalt (Co)
Vanadium (V)   Gallium (Ga)
Chromium (Cr)   Rhodium (Rh)
Manganese (Mn)   Indium (In)
Iron (Fe)

Charge 4^+:
Silicon (Si)   Manganese (Mn)
Titanium (Ti)   Germanium (Ge)
Vanadium (V)   Tin (Sn)

In addition, cations with other valence states may occur, providing charge balance is maintained, such as Li^+ in lithium ferrite ($Li_{0.5}Fe_{2.5}O_4$). Some spinels also contain cation vacancies, for instance maghemite (γ-Fe_2O_3), whose chemical formula could also be written as $\square\, Fe_{8/3}O_4$, where $\square$ denotes the vacancy.

Crystal structure. The spinel crystal structure is a relatively simple one. The basis of the structure is determined by the arrangement of the anions in a three-dimensional array known as cubic close packing. This is one of the most efficient ways in which spheres can fill space. There are two types of holes in this array: those surrounded by four anions, and those surrounded by six. A unit cell of this structure is defined so as to contain 32 anions, and will therefore also contain 64 holes of the first type and 32 of the second. In the spinel structure, one-eighth of the first and one-half of the second types of holes are filled by cations; because those in the first type are surrounded by four anions, forming the vertices of a regular tetrahedron, these cations are said to be in tetrahedral coordination, while those in the second type of hole are surrounded by six spheres forming the corners of an octahedron and are thus octahedrally coordinated (see **illus.**). With reference to the chemical formula unit of four anions, there are therefore two octahedral cation sites and one tetrahedral. The spinel structure is closely related to an even commoner and simpler crystal structure, the rock-salt structure (so named because it is the form adopted by table salt, NaCl), in which all the octahedral but none of the tetrahedral interstices in the cubic close-packed array of anions are occupied by cations.

Some commonly occurring natural spinels

Name	Formula	Cation distribution*
Spinel	$MgAl_2O_4$	N
Hercynite	$FeAl_2O_4$	N
Gahnite	$ZnAl_2O_4$	N
Chromite	$FeCr_2O_4$	N
Magnetite	$Fe^{2+}Fe_2^{3+}O_4$	I
Trevorite	$NiFe_2O_4$	I
Franklinite	$ZnFe_2O_4$	N
Jacobsite	$MnFe_2O_4$	N
Ulvöspinel	Fe_2TiO_4	I

*N = largely or completely normal; I = largely or completely inverse.

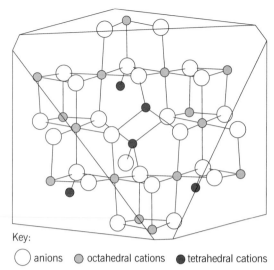

Key:

◯ anions   ⬤ octahedral cations   ⚫ tetrahedral cations

"Ball-and-spoke" model of the spinel crystal structure. The balls represent the positions of the ions in the structures, and the spokes show the shortest distances between nearest-neighbor ions. For clarity, only part of the cubic unit cell is shown.

Both cation sites lie on fixed or special positions in the unit cell. However, there is one further modification to this simple geometry allowed by the spinel space group symmetry. Keeping the unit cell size constant, the four anions surrounding a tetrahedrally coordinated cation may move either away from or in toward the cation by equal amounts, expanding or contracting the size of the tetrahedral coordination polyhedron. This will have a concomitant but opposite effect on the octahedral cation-to-anion distances, as well as causing a trigonal distortion of the octahedral coordination polyhedron. All six cation-anion bond distances remain the same, however. The anion array no longer has true cubic close-packing symmetry. The exact position of all anions in the unit cell may be described by just one additional geometric parameter, u, often called the oxygen parameter in oxide spinels. For perfect cubic close packing, u is ¼, and the tetrahedral cation-to-anion bond distance is 1.155 times that of the octahedral. Increasing u expands the tetrahedral and contracts the octahedral coordination polyhedra, while decreasing u has the opposite effect. In practice, u is found to vary from about 0.24 to about 0.275; the lower value occurs in silicate spinels that are stable only at high pressures, in which the small Si^{4+} cation occupies the tetrahedral site.

The tetrahedral (tet) and octahedral (oct) cation-to-anion bond distances (R) are given by

$$R_{tet} = a_0 \sqrt{3(u - 0.125)}$$

and

$$R_{oct} = a_0 \sqrt{3u^2 - 2u + 0.375}$$

where a_0 is the length of the unit cell. The individual bond distances thus completely determine both of the geometric variables in the structure.

Many modifications to the basic structure are possible, resulting in related structures with lower sym-

metry. One reason for such loss of symmetry is the presence of a nonspherically symmetrical cation, causing a Jahn-Teller type of distortion in the octahedral site. The manganese ion (Mn^{3+}) in the mineral hausmannite (Mn_3O_4) is an example. In other cases, the lower symmetry comes from further ordering of cations of different valences within either the octahedral or tetrahedral sites.

Cation distributions. Much of the interest in spinels stems from the ability of many cations to substitute simultaneously into both of the two geometrically very different cation sites. This ability may also lead to some degree of temperature-dependent disordering between the sites, which not only will be reflected in many of the spinel's physical properties but may also make an important contribution toward its thermodynamic stability.

A perfectly ordered spinel has what seems the obvious cation distribution, in which the single A cation in the formula unit occupies the tetrahedral site, with the two B cations on the octahedral. This is known as the normal cation distribution and may be written (A){B_2}O_4, where ( ) denotes the tetrahedral sites and { } the octahedral sites. The opposite extreme is the inverse distribution, in which one of the B cations occupies the tetrahedral site, while the octahedral sites are filled by a mixture of the remaining B cation with the A cation, that is, (B){AB}O_4. Any arrangement between these extremes is possible, at least conceptually—for example, a random distribution such as ($A_{1/3}B_{2/3}$){$A_{2/3}B_{4/3}$}O_4. The exact cation distribution may be described by the inversion parameter, x, defined as the fraction of B-type cations in the tetrahedral site. Thus x is 0 for a perfectly normal spinel, 1 for an inverse, and 0.667 for the random case. The random distribution makes a convenient point at which to divide spinels into those that are largely normal ($x < ⅔$) and those that are largely inverse ($x > ⅔$). This classification has been adopted in describing the spinels listed in the table.

The tendency for a cation to occupy a specific site in a spinel depends on its site preference energy. The actual cation distribution obtaining in a particular AB_2O_4 spinel then depends on the difference between the site preference energies of the A and the B cations, as well as thermodynamic variables such as temperature or pressure. The factors determining site preference energies are complex. The large preference for the octahedral site shown by certain transition-metal cations with incompletely filled d-electron orbitals (for example, Cr^{3+} and Rh^{3+}) may be explained by crystal field theory. For other cations, a tendency toward a more covalent type of bonding results in a preference for the tetrahedral site, for example Zn^{2+}. Finally, calculations of lattice energy show that, all other things being equal, the tetrahedral site prefers the larger cation in 2-3 spinels but the smaller cation in 4-2 spinels. The former result comes as a surprise to many mineralogists, who are accustomed to the opposite behavior in other minerals. For example, in $MgAl_2O_4$ spinel, the larger Mg^{2+} cation is indeed tetrahedrally coordinated, and Al^{3+} octahedrally, whereas in pyroxenes, amphiboles, and micas, Mg^{2+} is invariably octahedral, and much Al^{3+} is tetrahedral.

Lattice energy calculations also show that site preference energies are not constants independent of the spinel structure, but should vary somewhat with the degree of inversion; this has been shown to be true experimentally by careful measurements of cation distributions both in simple spinels as a function of equilibration temperature, and across binary spinel solid solutions at constant temperature. This variation in the site preference energies has important consequences for the prediction of thermodynamic activities of components in complex spinel solid solutions.

Cation distributions have mainly been experimentally measured by x-ray diffraction techniques. In Fe-containing spinels, the sites occupied by ^{57}Fe may be determined by gamma-ray resonance spectroscopy (the Mössbauer effect). Other methods, such as magnetic measurements and optical spectroscopy, have also been used where appropriate. Most of this kind of measurement has been made near room temperature on samples rapidly quenched from high temperature; and for those spinels that show increasing randomization of the cation distribution with increasing temperature, it is important to check whether such quenching preserves the high-temperature distribution. For spinels which contain cations in different valence states, for example, Fe_3O_4 (magnetite) with Fe^{2+} and Fe^{3+}, redistribution involves not the cations themselves but merely the transfer of an electron. This takes place extremely rapidly, even at subambient temperatures, so the high-temperature cation distribution can never be quenched in. A recent advance in the study of this kind of spinel is the use of thermopower and electrical conductivity measurements to elucidate Fe^{2+}–Fe^{3+} distributions at the temperature of interest.

Properties. Much of the technological interest in the spinel structure comes from the remarkable magnetic properties of many spinels, particularly the ferrites, that is, those spinels with Fe^{3+} as the B type of cation. A full description of the magnetic properties of spinels is beyond the scope of this article; suffice it to say that in some ferrites, below a certain temperature (the Curie temperature) the magnetic moments of paramagnetic cations become aligned on each site, the resulting moment of one site being antiparallel to the other. This leads to the phenomenon known as ferrimagnetism. Insofar as the magnetic properties of a ferrite depend on its chemical composition and its cation distribution, these can be tailored to suit a particular application—for example, by substituting different cations into the structure or by altering the thermal treatment.

For magnetite, the Curie temperature is 580°C (1076°F). This means that any hot rock containing magnetite (such as a volcanic lava) will acquire its magnetic properties from the Earth's magnetic field as it cools below this temperature. Such rocks then preserve a record of the Earth's magnetic field from this point in their history; thus they can be used as monitors of processes such as continental drift or seafloor spreading from eons past.

For background information *see* CHEMICAL STRUCTURES; CRYSTAL FIELD THEORY; CRYSTAL STRUCTURE; CURIE TEMPERATURE; JAHN-TELLER EFFECT; SPINEL in the McGraw-Hill Encyclopedia of Science & Technology.

Hugh O'Neill

Bibliography. R. J. Hill, J. R. Craig, and G. V. Gibbs, Systematics of the spinel structure type, *Phys. Chem. Minerals*, 4:317–340, 1979; H. St. C. O'Neill and A. Navrotsky, Cation distributions and thermodynamic properties of binary spinel solid solutions, *Amer. Mineral.*, 69:733–753, 1984; D. Rumble III (ed.), *Oxide Minerals*, 1976.

Sun

Because the Sun is the only star that is sufficiently close for its surface structure to be resolved, solar phenomena that result in local rather than global changes in temperature and luminosity have been studied in detail since the seventeenth century, when the 11-year sunspot cycle was discovered. The existence of cyclic variations in the Sun is generally regarded to be a consequence of the variation in the solar magnetic field, which exhibits a 22-year cycle. These variations are referred to as solar activity.

Origin and nature of solar activity. Cool stars (that is, stars with surface temperatures lower than the hydrogen ionization temperature of 10,000 K) have convection zones, which in the case of the Sun extends approximately one-quarter of the Sun's radius below the photosphere. In the convection zone, energy is transported by mass flow as well as by radiation. Standard stellar models are not yet able to predict the properties of convection zones. The convective process can be directly observed, however, in the pattern of bright cells separated by dark lanes in the photosphere. These bright granulation cells, typically 1000 km (600 mi) across, represent gas parcels that transport energy from below the surface, and therefore are hotter than their surroundings and hence brighter. The dark lanes represent cooled gas that is flowing back down below the surface. The granulation pattern is constantly changing (the mean cell lifetime is several minutes), demonstrating that solar convection is not a coherent process.

In addition to convection, the Sun exhibits two other characteristics that generate phenomena that are not predicted by conventional stellar models: the Sun has a magnetic field that varies with time, and it rotates not as a solid object but differentially, with a period of 25 days near its equator and more than 30 days near its poles. It is these characteristics that generate the phenomena associated with solar activity: sunspots, flares, an extended diffuse and expanding hot atmosphere (the corona and the solar wind), and the emission of x-rays and other short-wavelength radiation characteristic of temperatures from 50,000 to 50,000,000 K.

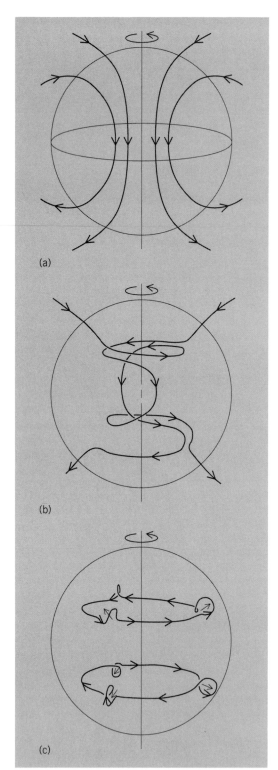

Fig. 1. Hypothesized mechanism of the solar magnetic cycle. In the dynamo believed to be responsible for the cycle, the interaction of convection and differential rotation stretches (a) an initially poloidal solar magnetic field to produce (b) a toroidal field. The twist of rising convective cells produces (c) a small poloidal field component from the toroidal field. The integrated effect of many such cells and the dominance of rising cells eventually reestablish the poloidal field, and the cycle continues. (After E. N. Parker, Cosmic Magnetic Fields: Their Origin and Activity, Clarendon Press, 1979)

Solar dynamo. The most widely accepted model for the operation of the solar activity cycle is the solar dynamo (**Fig. 1**). At the start of the cycle, the Sun's magnetic field has a classical dipole configuration. The magnetic flux lines are strongly anchored, or "frozen," in the highly conducting layers below the photosphere. The differential rotation of the Sun's convective layers causes the poloidal dipole field to be gradually stretched into a toroidal or azimuthal field, which increases the magnetic energy density at mid-latitudes initially. As the differential rotation continues to shear the magnetic flux lines, the increase in magnetic energy density progresses to lower latitudes. A sunspot results when locally the energy density in the convective layers becomes so great that a loop of dense magnetic flux breaks through the photosphere.

One alternative explanation for the solar cycle has generated considerable interest: possibly there is a source of a periodic magnetic fluctuation that originates deeper than the convection zone, such as a strongly magnetized radiative core, which generates a torsional magnetic oscillation with the 22-year magnetic period. One factor that supports this suggestion is that the period of the 22-year magnetic cycle appears to be more stable than that of the 11-year sunspot cycle.

Magnetic reconnection. Once created, the local enhancements of magnetic energy associated with sunspots begin to dissipate by generating currents or local electric fields by a process known as magnetic reconnection. This process can occur gradually or impulsively, heating the chromosphere and corona in the process, thus generating the high temperatures observed in the outer solar atmosphere. When dissipation occurs gradually, structures in which the hot coronal gas is contained by the magnetic field manifest themselves as bright loops of gas that emit x-rays (**Fig. 2**), or columnlike plumes of gas, concentrated at the Sun's magnetic poles (polar plumes). The gas contained in these structures ranges in temperature from 1,000,000 to 10,000,000 K. In contrast to the bright x-ray emission over regions of strong local magnetic field, regions where the magnetic field is weak are dark at x-ray wavelengths (Fig. 2), and are referred to as coronal holes.

Chromospheric supergranulation network. The chromosphere gas at temperatures between 30,000 and 500,000 K, which is present both in weak-field and strong-field regions, is concentrated into a network called the supergranulation network in weak-field regions. The lanes of this network are bright at far-ultraviolet wavelengths, while the cells within the bright lanes are dark. The magnetic field is concentrated in the bright lanes. The scale of the chromospheric supergranulation network is approximately 30,000 km (18,000 mi). Among the phenomena associated with the chromosphere network, which is visible in the mottled structures of Fig. 2, are surges of upward-flowing gas called spicules that may play a role in the transport of mass and energy from the high-density photosphere to the low-density chromosphere and corona.

Solar flares. When magnetic energy dissipation occurs impulsively, the mechanism involves the acceleration of particles to very high energies, from 100 to 1,000,000 times the mean thermal energy of the average particle in the coronal gas. These superthermal particles rapidly dissipate their energy by heating the chromospheric gas and by generating intense bursts of x-ray, extreme-ultraviolet, and gamma radiation. This phenomenon is known as a solar flare.

Fig. 2. Photograph of the solar atmosphere in the light of the ion Fe IX (that is, an iron atom with eight electrons removed) at a wavelength of 17.1 nanometers, which reflects the distribution of coronal plasma between 300,000 and 1,000,000 K. The image was obtained during a sounding rocket flight by A. B. C. Walker, Jr., T. W. Barbee, R. B. Hoover, and J. L. Lindblom. (*From A. B. C. Walker, Jr., et al., Soft x-ray images of the solar corona with a normal incidence Cassegrain multilayer telescope, Science, 241:1781–1787, 1988*)

Solar wind. The existence of the corona as a hot outer solar atmosphere results in the gradual loss, by gas outflow, of material from the Sun. This outflow, called the solar wind, creates a volume, extending beyond the orbit of Pluto, in which the particle density (about 5 particles per cubic centimeter at the Earth's orbit) is due to the Sun's influence, and excludes the local population of galactic interstellar gas. This volume of space dominated by the Sun's magnetic field and the solar wind is called the heliosphere. The solar wind flow exerts a braking effect on the Sun's rotation, and it is thought that over the 4.5×10^9 years of the Sun's life it has resulted in a considerable decrease in the Sun's rotation rate. The solar wind is not a steady uniform flow. There is enhanced flow from regions near the poles, such as the polar plumes shown in Fig. 2, and from coronal holes. Furthermore, large volumes of chromospheric gas are often expelled into the solar wind flow in connection with solar flares. These events are known as coronal mass ejections.

Terrestrial effects of solar activity. Although the phenomena associated with solar activity produce effects that are small compared to the total output of solar energy (the coronal x-ray luminosity is typically 10^{-6} of the total solar luminosity, and briefly a thousand times greater during a flare), active Sun phenomena such as the solar wind and x-ray and extreme ultraviolet emission can have important effects on the Earth. The solar short-wavelength radiation is responsible for the Earth's ionosphere, and energetic solar particles populate the Earth's radiation belts and cause auroral phenomena. Intense flashes of ultraviolet radiation and x-rays associated with solar flares cause transient changes in the Earth's ionosphere that affect radio propagation, and coronal mass ejections impacting the Earth's magnetosphere can cause fluctuations in the Earth's magnetic field (magnetic storms), which can affect electrical power distribution networks.

There has been considerable speculation that the solar activity cycle influences weather and climate cycles on the Earth. The demonstration of such effects and the investigation of the presence or absence of the solar cycle over a significant fraction of the 4.5×10^9 years of the Sun's lifetime require a record of solar activity that exceeds the reliable historical record of the sunspot cycle. Such indirect indicators of solar activity are referred to as proxy data. Proxy data are phenomena in the geological or biological record that demonstrate periodicity that can be linked more or less reliably to changes induced by solar activity. The best established link depends on the fact that the variation of the large-scale structure of the solar magnetic field modulates the penetration of high-energy particles from outside the solar system, the cosmic rays, into the heliosphere. The cosmic-ray bombardment of the Earth's upper atmosphere results in the formation of the radioactive isotope of carbon, ^{14}C. Precipitation of ^{14}C onto the Earth results in its incorporation into biological systems, specifically trees. Since trees undergo an annual growth cycle, the concentration of ^{14}C in each annual tree growth ring can in principle provide an index of the level of solar activity. Tree-ring structure is determined by many complex factors, as is the modulation of the cosmic-ray flux reaching the Earth's atmosphere. The most important factor affecting the latter is the strength of the Earth's magnetic field, which, however, can be independently determined from paleomagnetic data. When such effects are removed from the ^{14}C tree-ring variations, curve A in **Fig. 3** results. While there are some indications of a 22-year periodicity in ^{14}C concentrations, they are not convincing to all observers. The longer-period modulations of the intensity of ^{14}C concentration, approximately 300 years in duration (circled numbers 1–18 in Fig. 3), are much more compelling. Furthermore, they tend to correlate with other phenomena that indicate climatic conditions. The most compelling correlation is with the variations in the sunspot cycle recorded in the seventeenth century. This period of low sunspot counts is known as the Maunder minimum.

Correlations of proxy data with the 22-year magnetic cycle or the 11-year sunspot cycle are more controversial. However, one interesting proxy phenomenon, the changes in color in layers (called varves) in sedimentary rock found at Pichi Ricki Pass, South Australia, does display an interesting periodicity of 12.0 ± 1.75, extending over 1300 cycles.

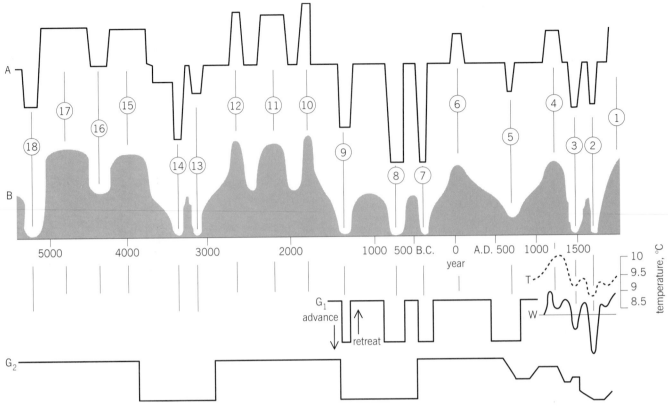

Fig. 3. Interpretation of radiocarbon deviations in terms of solar effects, with climate correlation. Curve A shows persistent radiocarbon deviations plotted schematically and normalized to feature 2 (the Maunder minimum); downward excursions indicate increased ^{14}C and imply decreased solar activity. Circled numbers identify postulated periods of abnormally high and abnormally low solar activity. Curve B is an interpretation of curve A as a long-term solar activity envelope (for example, of possible sunspot cycle variations). The remaining curves are four estimates of past climate: step curve G_1 shows times of advance and retreat of alpine glaciers; step curve G_2 shows times of advance and retreat for worldwide glacier fluctuations; curve T is an estimate of mean annual temperature in England (scale at right); and curve W is a winter severity index (colder downward) for the Paris-London area. °F = (°C × 1.8) + 32. (*Curve A data after J. A. Eddy, ed., The New Solar Physics, Westview Press, 1978*)

Periodicities of 23.7 and 314 varves are also present. This rock was laid down during the Precambrian Period, and has been interpreted to represent variations in temperature or some other climatic phenomena that are reflected in the nature of annual sedimentary deposits. If this interpretation is confirmed, it is of great significance because it demonstrates that the solar cycle has been present with a periodicity similar to its present periodicity for over 7×10^8 years.

Recent measurements have detected variations in the solar luminosity of the order of 0.1%, associated with the passage of large sunspot groups across the solar disk. However, there is as yet no evidence for any long-term luminosity variations associated with longer-term periodicities such as the 11-year and 22-year cycles, or the apparent 300-year cycle manifested by such phenomena as the Maunder minimum.

For background information SEE AURORA; COSMIC RAYS; IONOSPHERE; MAGNETOSPHERE; RADIOCARBON DATING; SOLAR CONSTANT; SOLAR WIND; STAR; SUN in the McGraw-Hill Encyclopedia of Science & Technology.

A. B. C. Walker, Jr.

Bibliography. R. N. Bracewell, Varves and solar physics, *Quart. J. Roy. Astron. Soc.*, vol. 29, 1988; J. A. Eddy (ed.), *The New Solar Physics*, 1978; G. Newkirk, Jr., and K. Frazier, The solar cycle, *Phys. Today*, pp. 25–34, April 1982; R. O. Pepin, J. A. Eddy, and R. B. Merrill (eds.), *The Ancient Sun*, 1979.

Syphilis

Syphilis is a systemic infectious disease caused by the bacterium *Treponema pallidum*. It is transmitted during sexual intercourse or other intimate contact, and it can also be transmitted from an infected woman to her fetus. Syphilis was a major public health problem before the antibiotic era; the Public Health Service estimated that about 2.5% of the United States population was infected with some stage of syphilis at the beginning of World War II. After the introduction of penicillin in the 1940s, the number of cases of primary and secondary syphilis in the United States declined by about 93%. During the second half of the 1980s, however, primary and secondary syphilis increased sharply. This increase has largely occurred in low-income minority heterosexual populations. An important contributor has been the exchange of sexual services for drugs, especially crack cocaine. Increases in heterosexual adult syphilis predict similar trends in congenital syphilis.

Clinical manifestations. Untreated syphilis is a chronic disease in which the microorganism spreads throughout the body through the bloodstream and can infect virtually every organ system. Sexual transmission requires contact with moist mucosal or cutaneous lesions. Although the disease becomes systemic within a matter of hours after infection, multiplication of *T. pallidum* at the site of entry produces the primary stage. Multiplication of the organism coupled with its increased dissemination to other tissues results in the secondary stage. Both stages are transient events, and approximately two-thirds of the infections will resolve without treatment due to the host's defense mechanisms. Untreated individuals enter the latent stage, in which there are no clinical signs or symptoms of the infection despite the fact that *T. pallidum* can still be demonstrated in some tissues and reactive serologic test results are still obtained. This latent period may last from 3 to 30 years with relapses to secondary syphilis occurring with decreasing frequency during the first 2 years of latency. Individuals can infect their sexual partners during the primary, secondary, and early latent stages, when skin lesions are present. Pregnant women are most likely to transmit the infection to the fetus during the secondary stage, when the organism is in their bloodstream. Although this form of syphilis, called congenital syphilis, is the result of prenatal infection rather than an inherited tendency, usage has established the term. A significant proportion of individuals with untreated syphilis will progress and develop tertiary syphilis, which may involve the skin, bones, central nervous system, heart, and arteries.

Epidemiology of primary and secondary syphilis. The incidence of primary and secondary syphilis in the United States increased from a low of 3.9 cases per 100,000 population in 1957 to 21.2 cases in 1990. Reported primary and secondary syphilis cases totaled 52,575 in 1990, an 18% increase over the 44,540 cases reported in 1989. This is the highest number of reported cases since 1949 and reflects a trend in increased numbers of primary and secondary syphilis cases that began in 1985 (**Fig. 1**). Although prostitution has long been associated with syphilis, it appears

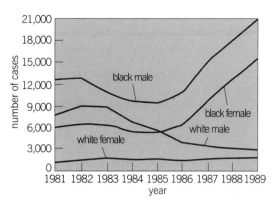

Fig. 2. Number of cases of primary and secondary syphilis by gender and race in the United States, 1981–1989.

to have been relatively unimportant in the overall epidemiology of syphilis in the United States during the 1960s and 1970s.

Before people were aware of the sexual practices that facilitated the spread of acquired immune deficiency syndrome (AIDS), syphilis was a common sexually transmitted disease among homosexual men. For example, of the 33,613 cases of primary and secondary syphilis reported in 1982, the sex ratio (males:females) among these cases was 2.9:1 with 42% of the infected men naming other men as sex partners.

Primary and secondary syphilis rates are higher in large urban areas, especially in low-income minority populations, than in less populated areas. Among white males, the number of reported cases of primary and secondary syphilis has decreased substantially since 1982. It has been presumed that this decline reflects changes in sexual behavior in response to the threat of AIDS. **Figure 2** shows that the increase in syphilis since 1985 has largely occurred in minority heterosexual populations. The causes for this increase have not yet been determined; however, two factors are probably important. From the 1970s onward, resources for the public health programs involved with the control of syphilis have progressively decreased. While this factor is not directly responsible for the increase, it almost certainly contributes to its magnitude. The second factor involves the dramatic increase in cocaine use in United States cities since 1985. Reports from individuals working in the area of syphilis control, reviews of partner referral interview records, and data from a case-control study all suggest that the exchange of sexual services for drugs, especially crack cocaine, is an important factor in the resurgence of syphilis.

Congenital syphilis. Trends in congenital syphilis reflect both the recent rise in the rate of heterosexual syphilis and the changing case definition of the condition. While steady declines in the incidence of congenital syphilis occurred from the 1950s through the early 1980s, substantial increases have been reported in recent years (Fig. 1). The rates of congenital syphilis are substantially higher for minorities than for whites. In 1988, the Centers for Disease Control issued new guidelines for classifying and reporting cases of congenital syphilis. These guidelines were

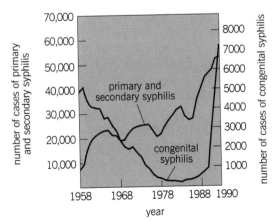

Fig. 1. Number of cases of primary, secondary, and congenital syphilis in the United States, 1958–1990.

revised in 1990 to broaden the case definition to include all liveborn and stillborn infants delivered by mothers with reactive serologic tests for syphilis and a history of untreated syphilis. This revision increased the sensitivity of the case definition for surveillance purposes. The lack of sensitivity of the old case definition makes it likely that the incidence of congenital syphilis had been underestimated in the past.

Several studies have suggested that the increased rates of congenital syphilis may also be related to either underutilization or inadequacy of prenatal care. Substance abuse may play a significant role in the lack of motivation to obtain care during pregnancy. In one recent study, 71% of the mothers of children with congenital syphilis admitted to substance abuse of crack cocaine, marijuana, or intravenous drugs. With escalating rates of heterosexual syphilis in many parts of the United States, it is essential to provide early prenatal care for women at high risk and to encourage serologic testing in both the first and third trimester.

For background information SEE SEXUALLY TRANSMITTED DISEASES; SYPHILIS in the McGraw-Hill Encyclopedia of Science & Technology.

Stephen A. Morse

Bibliography. Centers for Disease Control, Guidelines for the prevention and control of congenital syphilis, *Morbidity Mortality Week. Rep.*, 37(suppl. 1):1–13, 1988; Centers for Disease Control, Relationship of syphilis to drug use and prostitution—Connecticut and Philadelphia, Pennsylvania, *Morbidity Mortality Week. Rep.*, 37:755–764, 1988; J. M. Ricci et al., Congenital syphilis: The University of Miami/Jackson Memorial Medical Center experience, 1986–1988, *Obstet. Gynecol.*, 74:687–693, 1989; R. T. Rolfs et al., The perpetual lessons of syphilis, *Arch. Dermatol.*, 125:107–109, 1989.

Taphonomy

Taphonomy is the field of study within paleontology and archeology concerned with the paleoecological reasons for why ancient fossil records are biased in particular ways. It is a historical science concerned with reconstructing and understanding the processes that impact on the fossil record by altering the remains of organisms. Taphonomy has concentrated primarily on conditions affecting vertebrate fossils, but recent research has begun to explore factors that influence the successful preservation of invertebrates. A relatively new area in taphonomy is the study of plant remains and the agents that act upon them to bias their preservation in the fossil record.

An important element in taphonomy is uniformitarianism, which is James Hutton's theory stating that geologic deposits are reflections of past events that are identical to the natural processes occurring today. Taphonomists are often able to observe modern taphonomic processes in action or to initiate them through experiments that recreate a situation resembling the fossil record.

Taphonomic processes can have profound effects on the preservation of whole floras or faunas, as well as individuals or species. Differential preservation may be related to the kind of hard tissue present in a particular species. Age and size of the individual can also affect preservation.

Biotic and sedimentologic processes. Taphonomic processes can be divided into two major, often interrelated categories: biotic processes and sedimentologic processes. Biotic processes encompass such events as predation, scavenging, root and lichen etching, organic decomposition, and bioturbation. Sedimentologic processes involve fluvial movement, lake sediment accumulations, volcanic ash deposits, cave deposits, and various kinds of soil disturbance, such as frost heaving, solifluction, and subsidence. Other environmental factors, such as solar radiation and wind erosion, are also important elements in the biasing of the fossil record. The **table** summarizes major taphonomic processes.

While some taphonomic processes are responsible for the transport, surface alteration, breakage, or total destruction of the remains of organisms, others actually lead to their preservation. Although most attention is usually given to how taphonomic processes destroy or skew the fossil record, these mechanisms can also build deposits, as in the formation of coral reefs.

Death and transport. How the organism dies affects the likelihood that its remains will be preserved. Catastrophic death of a group of organisms by such events as avalanche or volcanic eruption can yield completely articulated skeletons of several individuals, whereas predation of a single wildebeest by a lion, followed by hyena scavenging, would leave few remains.

The chances of an organism's being preserved in the fossil record are also influenced by the location of its death or the location of deposition of its remains. Fish in lakes where sediments are rapidly depositing are more likely to be well preserved as fossils than are small arboreal mammals in rainforests. Rapid burial in fine sediments of proper soil pH and moisture content optimizes preservation, while prolonged exposure to sunlight, wind, and desiccation is not conducive to it.

However, the organism's remains may or may not be deposited within its natural habitat. An owl, for example, may carry its prey inside its digestive system for 12 h before regurgitating the undigestible parts in a pellet. In that time, the owl may fly many miles and eject its pellet near its roost rather than in the field where the prey lived. Similarly, seeds of plants can be carried long distances by animals that eat them. The means by which an organism reaches its resting place can be classified as autopod transport (transport by its own accord) or allopod transport (transport through the action of another organism). Primary agents are those transporters that are responsible for the death of the animal, while secondary agents, such as scavengers, transport the organism some time after death. Autochthonous fossils are those that have not been transported a significant distance from their place of death, while allochthonous fossils are those that have

Major taphonomic processes and their effects on vertebrate carcasses

Process	Causes	Effects
Predation	Raptors, carnivores	Transport, gnawing, breakage, digestion, species bias
Scavenging and bone ingestion	Invertebrates, rodents, carnivores, artiodactyls	Transport, gnawing, breakage, digestion, species bias
Bioturbation	Animals that trample or burrow	Movement, sorting, abrasion, breakage
Plant action	Roots, lichens	Movement, surface etching, breakage
Decomposition	Bacteria, soil pH	Deterioration, surface etching
Weathering	Exposure to air, ultraviolet radiation	Desiccation, split-line cracks, exfoliation
Eolian action	Wind-borne particle bombardment	Erosion, faceting
Hydraulic action	Fluvial action, flash floods, marine or lacustrine wave action	Transport, sorting, pitting, erosion
Pedoturbation	Solifluction, frost heaving, subsidence, cave roof fall, sediment loading	Transport, sorting, breakage, abrasion, plastic deformation

been transported some distance from the area where death occurred.

Postmortem destruction of the various parts of organisms depends heavily on the structures of those parts and their physical and mechanical properties. The size, shape, and composition of body parts, as well as the kinds of taphonomic processes involved, affect the probability that these elements will be preserved in the fossil record. Soft tissue, such as skin and organs, is the least likely to survive under normal postmortem conditions, although exceptions include plant and animal soft tissues recovered from peat bogs, permafrost, and extremely arid regions. Thin, delicate elements, such as fish or bat bones, are less likely to be preserved than the bones of robust animals. Similarly, seeds and nuts generally preserve better than flowers. The studies that have been conducted on the mechanical properties of bones have given paleontologists the information needed to predict the relative likelihood of preservation for individual elements of a particular species.

Invertebrates. The recent growth of taphonomic studies in invertebrate paleontology has also been impressive. For example, studies have shown that mollusk shells are winnowed by size during fluvial transport, affecting size distribution in the fossil record. In addition to the hydrodynamics of marine and fresh-water mollusk shells, studies have been conducted on predation, species bias, and potential errors in interpreting habitats of extinct species. For example, hermit crabs, which inhabit gastropod shells after the demise of the original occupant, can alter significantly the paleoecological record. The hermit crab selects a particular size of shell, modifies the shell it adopts, accelerates destruction of the shell, increases the quantity of organisms that attach to the shell (such as barnacles), and may move the shell to a habitat different from that of the living gastropod. If not recognized as taphonomic agents in gastropod assemblages, hermit crabs can skew the size ratios, frequencies, and habitat information of the gastropods represented in the assemblage.

Archeological assemblages. While paleontologists are concerned with how taphonomic processes may affect paleoecological reconstructions based on the surviving fossil evidence, archeologists also want to know how taphonomic processes interfere with their reconstructions of cultural events. As interpretations of taphonomic agents are made for an archeological assemblage, contrasts are made between cultural and natural-site formation processes. An example of a cultural process is the more frequent transport of whole carcasses of small species into a village site, as compared to the transport of partial skeletons of heavier animals. The breaking of bones to extract marrow with a hammerstone, butchering and burning of food refuse, and making of artifacts from bone, shell, and plant materials illustrate other examples of cultural formation processes. Human occupation sites often have their own specific set of taphonomic conditions, with domestic dogs and commensal rodents acting as scavengers, and trash heaps providing rapid burial of food refuse.

Recently, actualistic studies in taphonomy have been most widely developed in archeology. These involve the design and implementation of experimental simulations of assumed taphonomic processes. The archeological site, or a small portion of it, is reconstructed with proper soil conditions, topography, and so forth. The taphonomic agents, whether biotic or sedimentologic, are then invoked to act upon modern plant or animal remains. The resulting evidence is examined for surface alteration, breakage, destruction, transport, and other forms of modification. These results are compared to the archeological remains to see if the experimental conditions have succeeded in identifying the processes that affected the original ancient assemblage.

Botanical remains. Botanical taphonomy is one of the newest areas of research. In archeology, for example, paleobotanists are concerned with how accurate their identifications of botanical remains are, particularly when attempting to discern the earliest evidence for domesticated species. Carbonized botanical remains preserve far better than those that have not been burned, but the act of heating seeds and other macrobotanical remains to high temperatures may distort their morphology and complicate comparisons with modern wild and domestic species.

Reconstructing predator–prey relationships is an

important element in understanding the paleoecological impact of animals on plants. Increased work on the analysis of coprolites is helping to expand understanding of the dispersal of seeds by animals, and can help to explain why certain plant species become incorporated in particular fossil localities.

For background information SEE ARCHEOLOGY; FOSSIL; FOSSIL SEEDS AND FRUITS; PALEOBOTANY; PALEOECOLOGY; PALEONTOLOGY; TAPHONOMY in the McGraw-Hill Encyclopedia of Science & Technology.

Sandra L. Olsen

Bibliography. J. A. Efremov, Taphonomy: New branch of paleontology, *Pan-Amer. Geol.*, 74(2):81–93, 1940; S. M. Kidwell and A. K. Behrensmeyer, Overview: Ecological and evolutionary implications of taphonomic processes, *Palaeogeog. Palaeoclimatol. Palaeoecol.*, 63:1–13, 1988; M. Schiffer (ed.), *Advances in Archaeological Method and Theory*, 1981; S. E. Walker, Taphonomic significance of hermit crabs (Anomura: Paguridea): Epifaunal hermit crab—infaunal gastropod example, *Palaeogeog. Palaeoclimatol. Palaeoecol.*, 63:45–71, 1988.

Teleconferencing

Audio/video teleconferencing was introduced at the 1964 New York World's Fair. Corporate planning, however, at first resisted this promising approach because of its expense. Recently, rising travel expenses, routine travel delays and inconveniences, loss in productivity while away from the office, and advanced lower-priced full-duplex systems have led to increased interest in teleconferencing. The object of teleconferencing is to allow conferees in one office to see and hear conferees in another, similarly configured office at some remote location by using telecommunications, cameras, video monitors, and loudspeakers. Conferees sit around a conference table, which is usually trapezoidal or half-oval for good sight lines, so that the image in the video monitors gives the impression that the remote conferees are physically seated at the other end of the conference table. With good audio quality and adequate nonglare lighting to provide good depth of field, realistic and natural communication is possible.

While electronic advances have contributed to the popularity and ease of teleconferencing, acoustical advances and design guidelines have not kept pace, and in many instances the room is set up with inadequate planning. The importance of good electroacoustic design becomes apparent when outside interference raises background noise levels; sound quality and intelligibility in the transmitting and receiving rooms are poor; conversation within rooms is difficult; and interactions between room reflections, speakers, and microphones cause electroacoustic feedback. While some electronic techniques may alleviate the situation, acoustic problems should be addressed with acoustic solutions.

Design goals. The first design goal would be good noise isolation to provide a background noise level corresponding to Noise Criteria curve 20–30 (NC 20–30). Such a level of interfering noise is considered low enough to be appropriate and satisfactory. To accomplish this, floating walls, floors, and ceilings as well as sound-rated doors may be required. If this is not practical, then sound-rated doors should be used and walls should at least extend from floor to deck to minimize paths over wall partitions (flanking paths). A low-velocity heating, ventilation, and air-conditioning (HVAC) system with wrapped and lined ducts should be used, with air being allowed to fall into the room. High-velocity linear diffusors should be avoided. Once control of outside noise intruding into the teleconferencing room has been addressed, attention should be given to the control and distribution of sound generated within the room from loudspeakers and conversational speech.

Since internally generated sound from loudspeakers and conferees is not noise, the use of noise-control absorptive panels solely to manipulate interfering reflections is inappropriate. While increased sound absorption is an important part of reflection control, exclusive use leads to psychoacoustically dead spaces, where conversation in the transmitting room is difficult due to poor acoustical support of speech and the sound in the receiving room is perceived as dry and unrealistic. Interfering reflections should be controlled instead of simply being absorbed. **Figure 1** illustrates how sound can be attenuated by absorption, redirected by reflection, and distributed uniformly by diffusion. Absorbing and reflecting surfaces are generally well understood, but there is some confusion about diffusion. To scatter or diffuse sound, architectural acoustic design has often utilized effective ornamental surfaces, such as statuary, columns, parapets, coffered ceilings, and alcoves. Rising material and construction costs, however, have virtually eliminated much of this surface irregularity and have substituted flat reflective surfaces and acoustically absorbing suspended ceiling systems. These factors have led to a general degradation of interior room acoustics.

In 1983, a new approach to reflection control using a proprietary surface called an RPG, or reflection phase grating, was introduced. In contrast to absorbing surfaces, the RPG controls sound by diffusion in much the same way that frosted glass diffuses light instead of absorbing it. With the use of an RPG, sound from any direction can be uniformly distributed in many directions, thereby attenuating the sound in any particular direction. The RPG consists of a periodic array of (one-dimensional) wells or (two-dimensional) cells of equal width but different depths, separated by thin dividers (**Fig. 2**). When the depths are based on mathematical quadratic residue number theory sequences, a technique suggested by M. R. Schroeder, the surface becomes another proprietary surface known as a QRD.

Architects, acoustical consultants, and designers can utilize these surfaces and appropriate room geometry to design teleconferencing rooms that provide reflection control of incoming audio and conversational speech in a natural ambience, which promotes

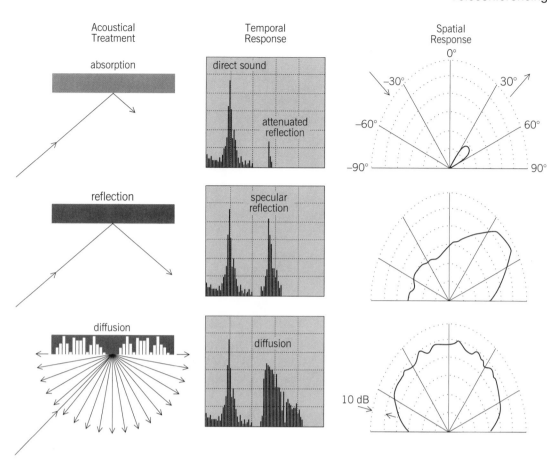

Acoustical Treatment

absorption

reflection

diffusion

Temporal Response

direct sound

attenuated reflection

specular reflection

diffusion

Spatial Response

0°

−30°   30°

−60°   60°

−90°   90°

10 dB

Fig. 1. Comparison of the spatiotemporal properties of acoustical treatments, illustrating how sound is attenuated by absorption, redirected by reflection, and distributed uniformly by diffusion. In the temporal response graphs, each horizontal increment represents a time of 10 milliseconds, and each vertical increment a sound intensity of 6 decibels.

high intelligibility, uniform speech coverage, excellent sound quality, and ease of conversation.

Design recommendations. Recommendations are given for such aspects as room dimensions; room layouts; front, side, and rear walls; and ceilings.

Room dimensions. Since teleconference rooms are acoustically small, low-frequency problems are often encountered. It is important to use appropriate dimensional ratios of the height, width, and length to provide a uniform distribution of low-frequency modes that minimizes boominess and provides an even sound decay with frequency. The ratios 1:1.4:1.9 have proven to be effective. To reduce low-frequency noise if room dimensions are fixed and low-frequency problems exist, Helmholtz slat absorbers (structures containing a series of Helmholtz resonators) or diaphragmatic panels with appropriate air cavities should be used. To minimize parallel surfaces and provide flutter echo control, walls and ceiling can be splayed outward about 1 in./ft (8.3 cm/m).

Room layout. The placement of speakers and video monitors and the design of speaker boundaries can draw on the design described here. A reflection-free zone is created over the conference area within which first-order room reflections from the front sides and ceiling are minimized with absorption and the predominant sound is from the loudspeakers. Reflections from the ceiling above the conference table and from side and rear walls are uniformly distributed by using

diffusor technology to bathe the conference area with useful diffuse reflections. The significance of this approach is that high-definition stereo audio and high speech intelligibility of reproduced sound are obtained over a wide area, simultaneously with good natural speech intelligibility, acoustic speech support, and a heightened sense of ensemble participation within the transmitting/receiving rooms. These design guidelines will ensure that these rooms will be acoustically ready in the future when high-fidelity broad-bandwidth audio transmissions are the norm. If an existing room is rectangular, it should be arranged so that the speakers radiate into the longer dimension, with windows, doors, and other reflective surfaces away from the speakers in the rear of the room, so that the surfaces immediately surrounding the speakers are absorptive.

Walls. To minimize speaker-boundary interference, speakers should be flush-mounted at the height of the video monitors in the front wall an equal distance on either side of the room's centerline. The front wall should be completely covered with an absorptive material, with openings for the video monitors, speakers, and cameras that are located above the video monitors.

Side walls are most effectively treated with a lower 2-ft (0.6-m) reflective area that provides noninterfering reflections and easy maintenance, a mid 4-ft-high (1.2-m) diffusive area consisting of

Fig. 2. A few components of the proprietary RPG Diffusor System, which can be used in teleconferencing room design. (a) QRD Diffusor, a broad-bandwidth wide-angle sound diffusor; (b) ABFFUSOR, broad-bandwidth sound absorber utilizing absorption and diffusion; (c) ABSORBOR, a fabric-upholstered graduated-density fiberglass sound absorber; (d) OMNIFFUSOR, a two-dimensional omnidirectional sound diffusor; (e) TERRACE, a two-dimensional omnidirectional sound diffusor without dividers; (f) FLUTTERFREE, a nonabsorbent flutter control acoustical moulding. (*RPG Diffusor Systems, Inc., Largo, Maryland*)

nonabsorptive flutter control panels, and an upper traditional-fabric-upholstered fiberglass section extending to the ceiling for reflection control above ear height. The side walls between the front speakers and the conference table should be absorptive to minimize reflections from these surfaces over the conference area.

The diffusive surface QRD can be used for the rear wall to diffuse reproduced sound from the speakers and internal conversation uniformly throughout the room. The rear corners can be used for dedicated low-frequency absorbers to minimize boom in the room.

Ceiling. One of the most important ingredients in teleconference design is an integrated suspended ceiling system that incorporates absorption between the front wall and conference table, diffusion located directly over the conference table and absorption panels, and lighting and HVAC distributed over the remaining area. Ceiling treatment with a diffusive surface improves intelligibility, speech coverage, and support. Additional low-frequency absorption can be obtained by adding fiberglass blankets in the ceiling space above the suspended grid.

For background information SEE ACOUSTIC NOISE; ARCHITECTURAL ACOUSTICS; TELECONFERENCING in the McGraw-Hill Encyclopedia of Science & Technology.

Peter D'Antonio

Bibliography. P. D'Antonio, *RPG Technical Brochure and Application Notes*, July 1987; F. A. Everest, *The Master Handbook of Acoustics*, 2d ed., 1989; M. R. Schroeder, *Number Theory in Science and Communication*, 1984.

Telerobotics

A teleoperator is a machine that extends a human operator's senses and motor skills to a remote location. The teleoperator typically has a mechanical arm and hand to perform useful manipulations on its environment or a propulsive means to move around in the environment, either or both controlled over a communication channel by a human operator. The teleoperator has video or other sensors and a communication channel for sensory feedback to the operator. In the most common type of teleoperator, the master-slave manipulator, which is used to handle radioactive or toxic materials, the operator positions a master hand-arm in multiple degrees of freedom, and servo-actuators drive corresponding motions of a slave hand-arm. In space and undersea applications, the human operator often positions one or more joysticks to command the rate of different teleoperator movements. In either case, the teleoperator's finger or arm joints may sense forces and transfer these back to the corresponding limb joints of the human operator.

The robot (which almost always means an industrial robot) and the teleoperator are related; in both, mechanical arms and hands manipulate objects to do useful work. But the robot and teleoperator have very different occupations. The robot acts on its own, autonomously, once it is programmed by a skilled technician and set in motion doing its assigned repetitive task, usually in a well-controlled factory production-line setting. The teleoperator is usually not found in a factory but rather in settings where unexpected contingencies demand the continuous control, perception, and intelligence of the human operator.

The telerobot is a new class of teleoperator, still controlled by a human but embodying features of the robot. The telerobot is initially programmed to allow for communication with the human operator in high-level language, a command language, by which the operator can assign subtask goals and criteria, and then to allow the telerobot to execute these subtasks while the operator monitors the procedure. This is called supervisory control. If task execution does not go as planned, the operator can revise the high-level instructions, or even switch to more primitive teleoperation control. Thus, the telerobot has the advantages of both the teleoperator (including the human capability to handle the unexpected, yet to do tasks remotely in a hazardous or inconvenient environment) and the robot (autonomy with speed and precision for well-defined parts of the task).

Communication. Supervisory control is rather like what occurs in a human organization wherein the supervisor communicates to a subordinate various goals, constraints, and other instructions to perform a specific task, and the subordinate then performs the task by using his or her own senses and intelligence to execute the instruction as well as conditions permit.

In the operations of a space teleoperator, and also in undersea operations where communication from the surface to a subsea teleoperator is by sound to avoid the encumbrance of an electrical or fiber-optical communication cable, there is a significant time delay (seconds) between the signal being sent by the human operator and the receipt of confirming feedback from the teleoperator. This delay causes instability, with control occurring in discrete and simple moves punctuated by waits for feedback. Thus, performance of a task can be stretched to many times its duration if the delay were not there.

In supervisory control, only the subgoal and task instructions are sent over the delayed communication channel, and the task execution is accomplished entirely at the site of the telerobot by means of control loops from the telerobot's own sensors through its computer and its actuators. In this case, the time delay is not destabilizing, and this is one reason telerobots are promising for space and undersea applications.

Applications. The long arm of the space shuttle, which grappled communication satellites from the shuttle's payload bay and pushed them into space, is a teleoperator. The Flight Telerobotic Servicer (**Fig. 1**), now being built for use in a variety of construction and servicing tasks on the space station of the National Aeronautics and Space Administration (NASA), is a telerobot. Other countries participating in the space station also have active space telerobot projects. The slightly larger-than-human-sized Flight Telerobotic Servicer has two arms for grappling and a third limb (shown at the bottom of Fig. 1). It has two video cameras, and a thruster-jet backpack for eventual free flight (although initially the backpack will be carried by the long manipulator arm). It also has a computer with a variety of automatic control modes to facilitate the task of the operator as well as to protect both equipment and astronauts from excessive forces and speeds.

In the deep ocean, Jason Junior, which was used to explore the wreck of the *Titanic*, was a teleoperated vehicle—without an arm, and remotely controlled by a human operator via an electrical cable from the surface. The new Jason vehicle being developed by Woods Hole Oceanographic Institution is a true telerobot, incorporating a manipulator arm and full computer.

In medical practice, flexible fiber-optic bundles are inserted into the gastrointestinal tract (from either end) or other body cavities for purposes of inspection. Some of these have cables for steering the end point or for manipulating tweezers or snares for simple cutting or biopsy (for example, removal of polyps). Such devices are also being used for arthroscopic surgery (repair of knee and shoulder joints). Other teleoperators are beginning to be used for repairing the retina of the eye. These devices are not yet telerobots, but various computer applications are being considered to assist in diagnosis and stabilize control of surgical procedures.

In many applications, it is important for the telerobot to reach around and avoid contact with some objects while grasping, cleaning, or making contact with others. This is an ideal task for computer aiding, allowing the human operator to direct the hand while a computer controls the elbow (or a more complex joint configuration) so that obstacles are avoided.

Telerobots are gradually being used for a number of other applications. In coal mining, the telerobot, guided from a safe distance by a human operator, automatically finds and follows the coal seam. In building construction, the telerobot is programmed to nail, staple, or perform other simple operations automatically while the operator guides it in a gross way. Telerobots for cleaning floors and windows use computer control to ensure that cleaning is effective (or else it must go back and do more) while the human sets the general task course and goals. Similarly, in

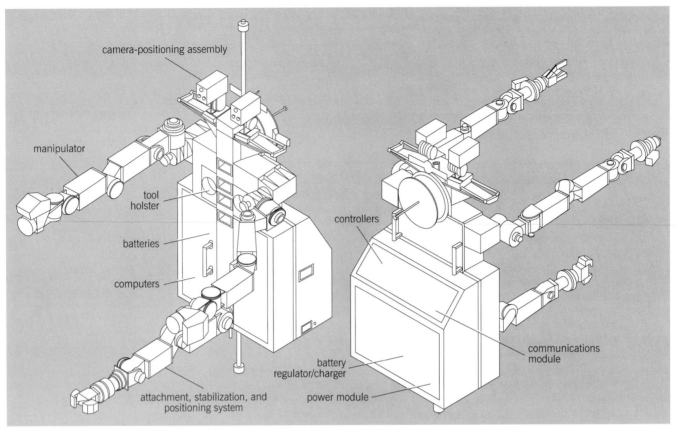

Fig. 1. Two views of the Flight Telerobotic Servicer.

agriculture, various soil preparation, planting, and harvesting operations are performed more or less automatically while the operator gives higher-level guidance.

A police sentry telerobot makes preassigned rounds, looks for certain kinds of visual or sound patterns that are indicative of trespassers, and reports to a human monitor if such evidence is found. Fire fighting is a challenging task for a telerobot, which can ascend or descend stairs, open doors, locate victims from visual and sound patterns, and possibly carry them out of burning buildings. There has probably been much interest in telerobots for military applications such as bomb disposal and espionage.

Not many telerobots are in use as yet, however, and those that are in use are more or less experimental compared to the older varieties of teleoperator. While the teleoperator is not expected to be replaced, the telerobot should find more uses as better computer control and artificial intelligence techniques become available.

Telepresence and virtual environments. Telepresence is a relation wherein the teleoperator (or telerobot) senses sufficient information about the remote task environment, and communicates this to the human operator in a sufficiently natural way that the operator feels physically present at the remote site. In recent years there have been serious research efforts to attain this ideal quality of sensory feedback. One approach is to present the video picture on a miniature

display attached inside a helmet that the user wears (**Fig. 2**). An electromagnetic sensor measures the movements of the user's head and transforms these to corresponding movements of a remote video camera. To complement the head-mounted display, technology is being developed by which the operator can send signals to move the teleoperator fingers and arm joints by simply moving his or her own fingers or arm in free space, without the intervention of a mechanical master or joystick device. This has been done by optical and electromagnetic instrumentation in a glove to measure the hand-arm configuration. Efforts are being made to incorporate skin vibrators in the instrumented glove to provide teletouch feedback from the remote environment.

If a head-mounted visual display gets its picture from a computer-graphic simulation of the environment, instead of from a remote video camera, the operator is said to experience a virtual environment. The virtual environment need not mock reality, but is limited only by the imagination of the programmer. Corresponding tactile sensations can also be provided by the computer environment, as can sound signals presented to the two ears so as to correspond correctly to the binaural locations of sounds in the real environment. Such a virtual environment can be used for training operators or for testing out telerobot procedures before they are tried in the real world.

Research problems. Telerobots are still very primitive, and there are many unsolved research problems.

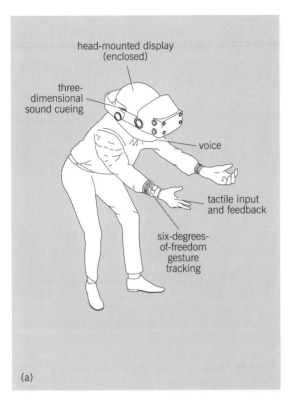

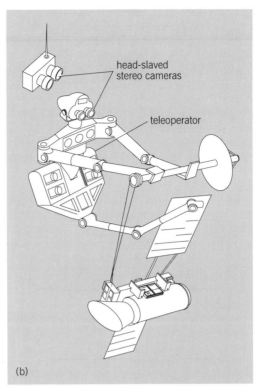

Fig. 2. Telepresence relation between (*a*) the human operator, with head-mounted display and instrumented glove, and (*b*) the teleoperator.

From a mechanical engineering perspective, telerobots need to be made lighter and less power-consuming for space, faster and stronger for Earth-based applications such as in construction, and smaller and more dexterous for entry into the human body. From the perspective of computer and control engineering, telerobots need combined speed and accuracy, with adjustable stiffness and damping characteristics for different tasks.

From the perspective of psychology and human-factors engineering, there are great challenges to devise better command languages. Numerous programming languages have been developed for industrial robots, but these still fall short of making it easy for both a telerobot and its human supervisor to confront a completely new task. One thing that has been learned from research on human–telerobot communication is that it is better to instruct the telerobot by pointing and measuring relative to other real objects in its environment than by presenting an abstract model of what is wanted. Just as in human instruction, the teacher and pupil have different viewpoints, different starting points, and different data available to them, and communication between them in abstract terms will never be perfect. There must be reliance on the specific data the pupil (in this case the telerobot) and the teacher already have in common.

For background information *SEE* R*EMOTE-CONTROL* *SYSTEM; R*EMOTE MANIPULATORS*; R*OBOTICS*; U*NDERSEA* *VEHICLES* in the McGraw-Hill Encyclopedia of Science & Technology.

Thomas B. Sheridan

Bibliography. T. B. Sheridan, Merging mind and machine, *Technol. Rev.*, pp. 32–40, October 1989; T. B. Sheridan, Telerobotics, *Automatica*, 25(4):487–507, 1989; J. Vertut and P. Coiffet, *Robot Technology*, vol. 3A: *Teleoperation and Robotics: Evolution and Development*, 1986.

Television

The concept of television as a broadcast-oriented remote eye was invented in the 1920s, immediately following the commercialization of radio as a broadcast-oriented remote ear. By the late 1930s, the relevant technology had matured sufficiently to allow the standardization of monochrome television systems for terrestrial broadcasting. Entertainment and news were envisioned as the principal applications. The television signal was analog, the picture completely occupied the screen, and the viewer was assumed to be passive with only an on-off control. Technology evolved further by the early 1950s so as to allow the standardization of color television systems, but the passive on-off nature remained. Since 1970, a number of technological changes have taken place that make the original concept of television obsolete, allowing more control by the viewer, and thereby making television interactive.

Technological advances. First, television is becoming digital. Algorithms for digitizing the television signal and compressing it to fewer bits have improved enormously. Current algorithms can compress raw digital television (200 megabits/s) to a few megabits per second. Advances in electronic circuits for processing and storage have reduced the cost of implementing these algorithms, making them practical.

Second, the media over which television is delivered to a viewer have become more varied: direct broadcast satellites (DBSs), transmitting to small antennas over individual homes; cable television (also called community-antenna television or CATV), using coaxial cable as a medium to support dozens of channels; fiber-optic networks proposed by telephone companies, offering enormous digital capacity; wideband local-area networks (LANs), laid out principally for computer data transmission; and finally, a variety of storage media including video cassette recorders (VCRs) and optical disks (including compact disks or CD-ROMs, for compact disk–read-only memory). While each medium has its own strengths and weaknesses, the combined result enables a viewer to access the equivalent of hundreds of television channels at any time.

Finally, the technology of picture creation has changed as well. The size, quality, and cost of a color television camera have improved by over three orders of magnitude since the 1950s, and the art of creating realistic visual images by a computer and merging them electronically with camera-generated images has become practical.

As a result of these developments, television, as a mode of communication, has become an important component of a broad variety of applications. Most of these new applications display a television signal that is more personalized, and allow the viewer to control its various parameters. Broadly, these applications may be grouped into two categories: interactive retrieval of pictorial information from a database or computing resource, and interactive communication.

Interactive picture retrieval. The interactive picture retrieval systems that have been proposed include the early teletext and videotex, interactive cable television systems, systems that use some type of compressed storage, and systems that use video windows.

Teletext and videotex. In the late 1970s, two forms of still-picture retrieval services, called teletext and videotex, were proposed and tried. In teletext, data signals are transmitted in place of video or other signals during some of the scan-line times of a television frame. A sequence of data represents one page of information, and a number of sequences are juxtaposed and repeatedly broadcast. A customer can request the acquisition and storage of any of the data sequences in the memory of a special receiver in the customer's television set when the receiver detects the transmission of the desired sequence. The special receiver then converts the stored data sequence into a displayed picture on the television set. Due to bandwidth restrictions and to avoid large access delay, only a small number of data sequences can be cycled.

In videotex, digital transmission on the switched telephone network (for example, 1200 bits/s using a modem) is used. A customer establishes a connection through the telephone network to a database and requests the transmission of a data sequence that describes a desired image. Again, a special receiver, attached to both a modem and a television set, stores the sequence and interprets and displays it as a picture on the television set.

The representation and coding of pictures in a compact form was standardized for both teletext and videotex. Despite many clever innovations, the quality of the pictures that can be accommodated in the narrow bandwidth remains low, giving the pictures a cartoon look. This is cited as one of the reasons for the lack of customer interest after several trials were conducted to evaluate this technology for a variety of applications such as home shopping, banking, and airline reservations. In the future, with wider bandwidths available from modern networks (such as the integrated services digital network or ISDN), the quality of pictures may be improved substantially by including photographs, animated graphics, and even motion video.

Interactive cable television. A number of systems that make cable television interactive have also been tried. In such systems, a viewer has a narrow-bandwidth channel (over either a two-way cable or a phone network connecting to the cable head-end) in the reverse direction (from viewer to the head-end) over which specific programs (single pictures or segments of picture sequences) can be retrieved from the database at the head-end. Thus the control information, which is digital and requires narrow bandwidth, travels in the reverse direction, whereas the high-bandwidth analog picture information travels in the usual direction. Although such services hold considerable promise, none of them has achieved permanent operation.

Compressed storage systems. Advances in the technology of compression, the low cost of CD-ROMs, and the reduced cost of decompression hardware are beginning to allow integration of audio, text, graphics, and digital video in workstations and networked computing systems. A variety of algorithms to jointly compress audio and video to playback rates below 1.5 megabits/s have been proposed (this rate is set by the transfer rate of a CD-ROM). At this transfer rate, a CD-ROM can store over a half hour of video. Unlike the compression algorithms used for point-to-point transmission (as in video-telephony), interactive applications require additional capabilities such as normal playback; random access, that is, access and reconstruction of an arbitrary video frame; reverse playback; fast search in both forward and reverse directions; reduced frame rates at higher spatial resolution; and zoom at high resolution. A coding algorithm that handles these requirements is being standardized by an international group, the Moving Pictures Coding Experts Group (MPEG). Such an algorithm exploits the spatial and temporal correlations present in the television signal and transmits only the perceptually important components with enough accuracy to be acceptable.

There are several competing digital representations that allow compressed storage and satisfy many of the above requirements. These representations come with not just the compression algorithms but also the software environments to interface to different computing systems and user commands to manipulate the pictures and audio. Digital Video Interactive (DVI), for example, represents picture sequences hierarchically in three dimensions (two spatial and one temporal), and can manipulate any of the hierarchies at high speed by using a programmable video processor attached to a computing system. It is hoped that advantages and disadvantages of each of these representations will become clear, leading to a consumer standard.

Video windows. The productivity of computer workstation users has been increased by opening windows (overlapped rectangular sections of the screen) that provide simultaneous transparent access to different computers located anywhere on a network. Many computer vendors have developed windowing systems in which each window acts as a character and graphics terminal. The natural next step is to perform video input and output, that is, to place video (single-picture or motion video) on part of the display or in a window; obtain digitized pixels from a frame of video for processing; and capture a portion of the screen (which may contain a computer-synthesized image) for some video recording. Unfortunately, most computers are not yet capable of handling raw video data rates (over 100 megabits/s), and therefore pictures thought of as arrays of pixels cannot be moved around arbitrarily. Hence, a high level of abstraction (for example, a frame) is frequently used, with specialized hardware doing the low-level computation-intensive operations. Such video windows will be useful for a variety of applications such as interactive simulation, where high-speed processing is done in a supercomputer but results are displayed on demand in the window of a workstation (called visualization); video postproduction; and communications.

Interactive communication. Another growing use of interactive video is as part of a multimedia desktop conferencing system. Wider-bandwidth local-area computer networks as well as integrated services digital networks offered by telephone companies are increasingly able to provide network support in terms of bandwidth and control to allow groups of people, using computers and telephones in their offices, to hold real-time discussions sharing voice, data, still images, and video. The conferees may talk to each other and may interactively retrieve, modify, and create identical video material on their computer screens, which may be at distant locations. The video windows, as described above, that are implemented in specialized hardware attached to the computer are an important component of such systems. The computer is used to initiate, conduct, and terminate the conferences and to work as an active agent to coordinate communications of the conferees. Unlike entertainment video, the video in desktop multimedia conferencing contains mostly a head-and-shoulders view of the conferees without frequent camera switching from one part of a scene to another. The results are a higher temporal correlation of the consecutive television frames and better compression efficiencies. Based on the integrated services digital network bit rates of multiples of 64 kilobits/s, standards for compression algorithms are evolving. At 64 kilobits/s, video is usable for several applications, although there are visible artifacts. At 384 kbits/s, video of much better quality is obtained.

For background information *see* Closed-circuit television; Computer graphics; Computer storage technology; Local-area networks; Optical recording; Teleconferencing; Television; Video telephone; Videotext and teletext in the McGraw-Hill Encyclopedia of Science & Technology.

A. N. Netravali

Bibliography. Institute of Electrical and Electronics Engineers, *GLOBECOM Conference Proceedings*, 1989; A. C. Luther, You are here... and in control, *IEEE Spectrum*, 25(9):45–50, September 1988; A. N. Netravali and B. G. Haskell, *Digital Pictures: Representation and Compression*, 1988; W. H. Ninke, Interactive picture information systems—where from?, where to?, *IEEE J. Selected Areas Commun.*, 1:237–244, February 1983; R. Scheffler, J. Gettys, and R. Newman, *X Window System*, 1988.

Thermoacoustic engines

A sound wave is usually thought of as consisting of coupled pressure and velocity oscillations, but adiabatic temperature oscillations always accompany the pressure oscillations. The combination of these temperature oscillations with the pressure and velocity oscillations produces a rich variety of thermoacoustic effects. Although these effects are too small to be noticed in everyday life, in extremely loud sound waves they can be harnessed to produce powerful heat engines: prime movers, heat pumps, and refrigerators. Whereas typical heat engines have crankshaft-coupled pistons or rotating turbines, thermoacoustic engines have no moving parts, or at most a single flexing moving part (such as a loudspeaker) with no sliding seals. These engines will be of use wherever simplicity, reliability, low cost, or small size are more important than high efficiency.

Examples. The Sondhauss tube (**Fig. 1a**) was the earliest thermoacoustic engine. Over 100 years ago, glassblowers noticed that when a small, hot glass bulb was being blown on a cool glass tubular stem, the stem tip sometimes radiated sound; early investigators understood that the sound was generated by oscillatory thermal expansion and contraction of the air in the tube, which in turn was due to acoustic motion of the air toward and away from the hot end of the tube. The Sondhauss tube was a prime mover, converting heat into mechanical work in the form of sound.

In the 1960s it was realized that the performance of such devices could be greatly enhanced by inclusion of proper heat-exchange structures. A heat-driven thermoacoustic refrigerator (Fig. 1b) illustrates these

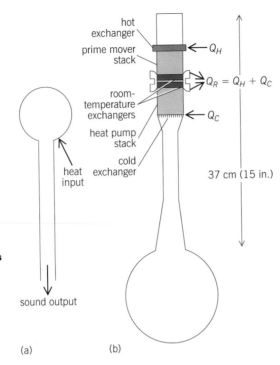

Fig. 1. Examples of thermoacoustic engines. (a) Sondhauss tube. (b) Heat-driven refrigerator. The rates of heat flow at the hot exchanger, cold exchanger, and room-temperature exchanger are, respectively, Q_H, Q_C, and Q_R.

structures, as well as the function of thermoacoustic engines as heat pumps and refrigerators. Its case consists of a 37-cm-long (15-in.) tube, closed at the top and opening at the bottom into a large spherical bulb, and containing 0.3-megapascal (44-lb/in.^{2}) helium gas. Near the top, there is a stack of thin, well-spaced stainless steel plates aligned parallel to the tube axis. Above the stack is the hot heat exchanger, consisting of spaced nickel strips; below, a similar set of copper strips forms a room-temperature heat exchanger, cooled by water circulating through a collar around the outside of the case. This is the prime-mover assembly, producing acoustic work from heat. When enough heat is supplied to the hot heat exchanger by an external heat source, the helium gas oscillates spontaneously at about 600 Hz, with peak-to-peak pressure oscillations as high as 40 kilopascals (6 lb/in.^{2}).

Just below the prime-mover stack, another stack and pair of heat exchangers function as a heat pump, driven by the acoustic work generated by the prime-mover stack. The heat-pump stack is identical to the prime-mover stack, and its upper, room-temperature heat exchanger and lower, cold heat exchanger are made of copper strips. When the hot-heat-exchanger temperature is high enough that the gas oscillates, the cold heat exchanger cools, as heat is pumped thermoacoustically from the cold heat exchanger to the room-temperature heat exchanger. Hence, the whole system functions as a refrigerator, with no moving parts, powered by heat delivered at high temperature. The refrigerator reaches −11°C (12°F) with no load, and can cool a load of $Q_C = 10$ W to −6°C (21°F).

Resonators. The case of a thermoacoustic engine forms an acoustic resonator, with the resonance serv-

ing as the flywheel to keep the engine operating. The resonator can be open, as in Fig. 1a where acoustic power is deliberately radiated away, or closed, as in Fig. 1b where the acoustic power is used within the resonator. In both cases, the acoustic oscillation is a pure tone, so that velocity, pressure, and temperature vary sinusoidally with time. The geometry of the resonator determines the frequency of the oscillation, just as the geometry (principally length) of an organ pipe determines its pitch.

Stored energy shifts back and forth between kinetic energy of moving gas and compressive energy of pressurized gas during each half cycle of the oscillation; at any given point in the engine, the pressure reaches its highest and lowest values at times when the velocity passes through zero, and vice versa. For example, in the Sondhauss tube at one time of the oscillation the air is motionless and the bulb is pressurized above ambient pressure. At this time, all the energy is stored in the compressibility of the pressurized air. A quarter cycle later, after the pressure has accelerated the air in the stem downward, the air velocity in the stem is at its largest (downward) value, and the pressure in the bulb is at ambient, so all the energy is stored kinetically in the moving air. Next, inertia of the moving air keeps it moving downward, pulling air

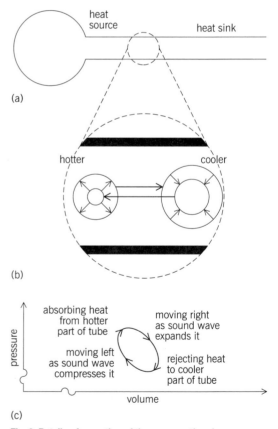

Fig. 2. Details of operation of thermoacoustic prime mover. (a) Sondhauss tube. (b) Magnified view of part of the stem, showing motion of a small parcel of air. (c) Thermodynamic cycle of the parcel.

out of the bulb and lowering the pressure there, so that a quarter cycle later the air comes to rest with the bulb below ambient pressure, shifting the energy back to compressibility. This low bulb pressure then pulls air back up the stem, so that after another quarter cycle the air is moving upward most rapidly and storing kinetic energy, and the bulb has been refilled to ambient pressure. Finally, inertia keeps the air moving upward, pressurizing the bulb, so that the system comes to rest at the starting condition after another quarter cycle.

Power production. The production of acoustic work in such a standing wave will be examined in the case of the Sondhauss tube. The motion of a typical, small parcel of air in this tube is shown in detail in **Fig. 2.** The standing wave carries the parcel back and forth as described above, compressing and expanding it, with phasing such that the parcel is at its most compressed state when at its farthest left position, and its most expanded state at its farthest right position. The presence of the tube wall with its steep temperature gradient adds an important new element to this simple acoustic oscillation: oscillatory heat transfer between the parcel and the tube wall. Here, the adiabatic temperature oscillations that accompany the pressure oscillations can be neglected. When the parcel is at its leftmost position, heat flows from the relatively hot tube wall into the parcel, expanding it; when the parcel is at its rightmost position, heat flows from it to the relatively cool tube wall, contracting the parcel. Since the expansion takes place at the high-pressure phase of the cycle and the contraction at the low-pressure phase, the parcel does net work on its surroundings. This net work, which is produced at the resonance frequency, generates the sound radiated from the Sondhauss tube and maintains the resonance itself.

The total power can be increased by increasing the cross-sectional area of the tube, but only by inserting heat exchange components. This is because parcels farther from the wall than a few times the thermal penetration depth δ, given in the equation below

$$\delta = \sqrt{\frac{K}{\pi f c_p}}$$

(where K is the thermal conductivity of the gas, c_p is its isobaric heat capacity per unit volume, and f is the oscillation frequency), do not have time to exchange much heat with the wall and hence cannot participate in the thermoacoustic process. Thus, to make all parcels useful in a large-diameter tube such as in the heat-driven refrigerator in Fig. 1*b*, the tube is filled with a stack of parallel plates spaced a few penetration depths apart, typically 1 mm (0.04 in.).

In the heat-driven refrigerator, the acoustic power produced as described above is then consumed by the lower stack, which uses it to pump heat up the temperature gradient from the cold heat exchanger to room temperature. There, as the gas oscillates along the stack, it experiences changes in temperature. Most of the temperature change comes from adiabatic com-

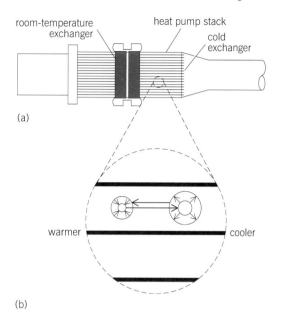

(a)

(b)

Fig. 3. Details of operation of thermoacoustic refrigerator. (a) Heat-pumping stack. (b) Magnified view showing motion of a small parcel of gas.

pression and expansion of the gas by the sound pressure, and the rest is a consequence of the local temperature of the stack itself. The resulting thermodynamic cycle is shown in **Fig. 3**, where, as before, one parcel of gas is followed as it oscillates. At the leftmost position of the parcel, it dumps heat to the stack, since its temperature was raised above the local stack temperature by adiabatic compression by the standing wave. Similarly, at its rightmost position, the parcel absorbs heat from the stack, since its temperature is below the local stack temperature. Thus, each parcel moves a little heat from right to left along the stack, up the temperature gradient, during each cycle of the sound wave. All the other parcels do the same thing, so that the overall effect, much as in a bucket brigade, is the net transport of heat from the cold heat exchanger to room temperature.

Current developments. Thermoacoustic engines will never reach the high efficiencies typical of diesel engines, steam turbines, or commercial refrigerators. Nevertheless, their lack of moving parts makes them more reliable and long-lived than conventional engines, and several thermoacoustic engines are under development. A loudspeaker-driven refrigerator is being studied for cooling satellite-borne sensors, and a heat-driven cryogenic refrigerator has been built for similar applications. A heat-driven underwater sound source and a heat-driven electric generator are also planned.

For background information SEE SOUND; STANDING WAVE; THERMODYNAMIC CYCLE; THERMODYNAMIC PRINCIPLES in the McGraw-Hill Encyclopedia of Science & Technology.

Gregory W. Swift

Bibliography. G. W. Swift, Thermoacoustic engines, *J. Acous. Soc. Amer.*, 84(4):1145–1180, 1988.

Tissue culture

Bioreactors are containers so designed that living cells inside manufacture some useful product. Somewhat like a farmer, the bioengineer raises individual cells in a nutrient solution within the specialized container and harvests the products of the thriving cells.

A bioreactor allows the bioengineer to target a specific biological product and maximize its synthesis, using a minimum of space, time, and cost. The products typically are complex chemicals that would be difficult or impossible to synthesize by other techniques. Currently, bioreactors are used primarily to make medically useful chemicals. However, researchers anticipate a wider variety of products, as more types of cells are exploited and bioreactor design is improved.

In the earliest bioreactors, bacteria and fungi were raised in batches in large mixing tanks, then harvested and processed for the chemicals they produced naturally. Then genetic engineering emerged: researchers learned how to alter the genetic material of bacteria in a controlled way. Genetic engineering made it possible to modify bacteria so that they could synthesize novel products (for example, insulin), extending their use in bioreactors.

At the same time, scientists were learning how plant and animal cells could be cultured after being totally isolated from the organism. For each type of cell from each type of organism, methods of isolation had to be refined, and nutrient and growth factor requirements had to be defined.

Finally, engineers began to design containers that enhanced cell survival, productivity, and product recovery, rather than just mixed the cells. Computerization led to improvements in culturing. Mathematical models of processes such as cell growth and productivity were created to study how the processes responded to specific changes in the culture conditions. As a result, variables in the culture conditions that could be improved were identified.

These are the types of advances, in both the biological and reactor components, that have made plant and animal cell bioreactors possible, and are also leading to further advances.

The use of plant and animal cells in bioreactors is a relatively new concept. Aside from bacteria, only algae, which may be considered simple plants, are currently being used commercially in bioreactors. However, plant and animal cells carry enormous promise in medical, agricultural, and industrial applications. Fulfillment of this promise will require development in the biological component and also the reactor component.

Biological component. Plant cells appear to be promising candidates for use in bioreactors for several reasons. First, much is already known about the culturing of isolated plant cells in solution. As a result, it should be technically simple to adapt them to bioreactors.

Second, plant cells have a wall that encloses the cell, like bacteria and fungi (but unlike animal cells).

This wall protects the cell from physical damage in the moving solution of the bioreactor, which is proving to be important in larger bioreactors.

Finally, and of greatest interest, plants produce a variety of complex and useful chemicals. In the plant, these chemicals serve special functions, such as producing unique odors and flavors, or serving as chemical attractants or repellants. Each chemical is typically produced by specific cells within the plant. When those cells are isolated and raised in a bioreactor, the chemical can be made in large amounts with fewer unwanted side products. The desired product then can be recovered for use outside the plant.

The diversity of natural chemicals that may potentially be produced this way is staggering. At present, only a small portion of the variety of chemicals that plants produce are known. These products can be grouped into five basic uses as follows.

Medicinal and pharmaceutical chemicals make up one group, and can be subdivided into alkaloids (atropine, cocaine, codeine, morphine, nicotine, quinine, reserpine, scopalamine, steroid alkaloids, tubocurarine, and vinchristine), glycosteroids (digitoxin and digoxin), and phenolics (aspirin). A second basic grouping is food and drink, which includes terpenoids (used for flavoring), such as camphor, limonene, menthol, and vanillin; and others, such as vegetable oils (linseed), perfumes (jasmine), herbs, spices, incense, and vitamins. Agrichemicals include fungicides and insecticides, such as pyrethrin. Industrial products, such as rubber and chicle, make up a fourth group. The final group contains specialty products, such as enzymes, peptides, textiles, and tobacco.

One interesting application for plant cells in bioreactors takes advantage of the plant cells' capacity for photosynthesis. In photosynthesis, the cells harness the energy of light to produce sugars, and absorb carbon dioxide and release oxygen. A plant cell–containing photosynthesizing bioreactor has a specific application aboard long space missions. Space travelers, by breathing, require oxygen and produce carbon dioxide, which is toxic to them. Rather than carrying bulky tanks of oxygen and cumbersome equipment to absorb the carbon dioxide, space travelers may eventually carry suitcase-size bioreactors to serve their gas exchange needs. The production of edible sugars by the cells may be a bonus, providing an additional food source for the trip.

Animal cells also are useful in bioreactors. Although in many ways biochemically similar to plant cells, animal cells make some unique products. These chemicals include growth factors, hormones, and many kinds of enzymes. Because of their complex structure, production of these "biologicals" is most economical in bioreactors.

A key point regarding the biological component of bioreactors is that each type of cell within a particular plant or animal is specialized in what it does and in what it produces. If the particular cells that produce a particular substance are isolated and then multiplied under ideal conditions in a bioreactor, the production of that desired chemical can be increased greatly.

A second key point regarding bioreactors is that their performance can be optimized. This leads to a discussion of the reactor part of the system.

Reactor component. Two parts of the reactor component of a bioreactor need to be designed for maximum effectiveness. The first is the solution that bathes the cells. The nutrients in the solution must be formulated not only to keep the cells alive but also to maintain them in a state of high productivity. In some cases, a product may be most rapidly synthesized when the cells are multiplying, although it is more common for complex syntheses to occur most efficiently in nondividing cells. The rate of division is readily controlled by solution composition.

The solution also must contain the appropriate ingredients for the efficient synthesis of the desired product. As in any synthesis, many variables can affect the efficiency of production.

The second part of the reactor component that must be considered is the hardware itself. It may be as simple as a tank with a stirring mechanism. If, however, the cells perform better when fixed in place rather than when floating, then sheets, beads, or tubes can be built into the tank, and the cells can be made to attach to them. The nutrient solution can be made to flow slowly over this solid support for the cells.

The construction of hardware is one of the most exciting areas of current research, particularly for animal cell culture. Animal cells are greatly influenced by the chemistry of their immediate surroundings. A simple change in the composition of plastic sheets or tubes within the bioreactor can significantly change the behavior and thus the productivity of certain types of animal cells.

Hardware can also be designed to maximize product recovery. For example, cells may secrete a diverse mix of products even though only one product is desired. By culturing the cells inside porous beads, larger molecules may be retained in the beads while smaller ones can escape and be recovered later.

Prospects for bioreactors. Bioreactors using plant and animal cells offer much promise. Current efforts are aimed at optimizing each cell type's normal capacity to synthesize certain products that the cell naturally produces. The emphasis of this effort centers on perfecting the various aspects of the reactor component. For example, new ways of encapsulating animal cells to protect them from physical damage and aid in product recovery are being developed.

The techniques of genetic engineering will continue to be a key element in the bioreactor field. With them, normal cell functions can be enhanced or inhibited, and new processes can be added.

In the future, cells may be selected not just for what they produce but also for what they do. Cells that produce specific enzymes may be used in bioreactors for direct use, say, in degrading or neutralizing toxic compounds or in removing waste materials. In this scenario, the bioreactor itself could be the marketed product. For example, the photosynthesizing plant cell bioreactor described earlier may be sold as an air-refreshing system for use in the home. Much larger versions might someday be needed for gas exchange on Earth, if the natural flora continues to be devastated.

More elaborate bioreactors may utilize multiple types of cells to complete complex syntheses. Such systems might mimic the animal organs in complexity. The products of these systems may be not only complex chemicals but also entire cells. These types of bioreactors are now being designed to produce such valuable products as synthetic blood and skin.

For background information SEE BIOCHEMICAL ENGINEERING; GENETIC ENGINEERING; INDUSTRIAL MICROBIOLOGY; PLANT MINERAL NUTRITION; TISSUE CULTURE in the McGraw-Hill Encyclopedia of Science & Technology.

Thomas G. Brock

Bibliography. M. Y. Chisti, *Air Lift Bioreactors*, 1989; F. Kargi and M. Z. Rosenberg, Plant cell bioreactors: Present status and future trends, *Biotech. Prog.*, 3:1–8, 1987; J. M. Lee and G. An, Industrial application and genetic engineering of plant cell cultures, *Enzyme Microb. Tech.*, 8:260–265, 1986; J. E. Prenosil and H. Pedersen, Immobilized plant cell reactors, *Enzyme Microb. Tech.*, 5:323–331, 1983; J. V. Twork and A. M. Yacynich (eds.), *Sensors in Bioprocess Control*, 1990.

Transmission lines

The high-voltage transmission lines that make up alternating-current (ac) electric power transmission networks are highly inductive. The distribution of power flow on an individual transmission line in a complex transmission network is determined by the interrelationship of the individual transmission line reactances that make up the network. The application of high-voltage series capacitors inserted into individual lines is a method being considered by power system engineers to modify the impedance structure of the network so as to adjust the power-flow distribution on individual lines and thus increase the power flow across such compensated lines. Although fixed series capacitors have been applied in specialized circumstances to improve the stability performance of generators connected into the power system via very long transmission lines, the application of variable, controlled series capacitors in complex network configurations as a means of power-flow control is a recent proposal that is receiving much study.

Power transfer equation. The power transfer across a transmission line is determined by the relative magnitudes and phase angles of the sending and receiving terminal voltages and the electrical characteristics of the network facilities connecting the sending and receiving terminals. An approximate representation of a transmission line is shown in **Fig. 1**. The equation below, called the power transfer equa-

$$P_{12} = \frac{E_1 E_2}{X_{12}} \sin (\delta_1 - \delta_2)$$

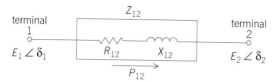

Fig. 1. Approximate representation of a transmission line, with parameters that appear in the power transfer equation.

tion, is a good approximation for an extrahigh-voltage transmission line where the inductive reactance component X_{12} is very much greater than the resistive component R_{12}. Here, P_{12} is the power transfer across the line, E_1 and E_2 are the magnitudes of the voltages at the terminals, and δ_1 and δ_2 are the phase angles of these voltages. For simplicity, the shunt capacitance (line-charging) characteristic of the transmission line is not shown in the figure, since it has little bearing on the power transfer across the line.

The power transfer equation provides insight into the techniques for controlling power flow on a transmission line operating as part of a transmission network. Although one means for control might be the terminal voltages (E_1 and E_2), as a practical matter this is not easily accomplished except by a very expensive voltage upgrade. The most attractive way of influencing the power transfer equation is to operate on the X_{12} term, the reactance of the transmission line in question.

Series compensation. While the inductive reactance of the transmission line could be modified by physical changes to the conductors, such as an increase in their size and capacity (reconductoring), this is generally an expensive undertaking and the reduction in line reactance is marginal. Furthermore, the size of the conductor that can be installed is most often limited by considerations of transmission tower strength.

A more attractive solution is to install an external capacitive reactance to compensate for the inherent inductive reactance of the transmission line. In this manner, the overall net reactance between terminals 1 and 2 is reduced, and the power transfer is increased proportionate to the decrease in X_{12}.

Series capacitors have been used to compensate transmission lines for many years. However, this application has been limited to only a few transmission lines in the United States, mainly very long distance lines, to increase the capacity of the line to carry heavy power flows. In recent times, as it has become increasingly difficult to construct new transmission lines, the concept of series compensation has become an attractive tool for enhancing the utilization of existing transmission lines when operated as part of a complex interconnected network. Furthermore, if such compensating devices can be made adjustable by means of some external control, then adjustable power-flow control can be achieved.

Application in a transmission interface. The maximum power flow that can be carried across a group of parallel transmission lines in a network (a transmission interface) is dependent on a number of factors,

including the capability of each transmission line and the relative reactance of each. Furthermore, the network must be capable of supporting the outage of the most heavily loaded transmission line without overloading the parallel transmission facilities, the flow on which would automatically increase as a result of the outage.

This is illustrated by **Fig. 2***a*, which represents a transmission interface between systems A and B. Line A-B represents a high-capacity, high-voltage transmission line having a loading of 2600 MW, whereas line A_1-B_1 represents a high-voltage line that carries about 600 MW. In this example, line A_1-B_1 is assumed to have a safe loading limit of 950 MW, due to system voltage and phase-angle limitations. The remaining transmission lines between systems A and B are represented in equivalent form by line A_2-B_2. The amount of power flow carried on each line is determined primarily by the relationship of the inductive reactances (neglecting the resistive component) of the individual lines. The power flow on each line is inversely proportional to the line's reactance. Thus, the line with lowest reactance carries the highest share of the power transfer between systems A and B, and the line with highest reactance carries the lowest share.

Figure 2*b* illustrates what happens when a contingency outage of line A-B takes place. The power loading that was carried by that line (in this example, 2600 MW) is automatically and instantaneously transferred to the remaining two lines between systems A and B, that is A_1-B_1 and A_2-B_2, imposing an incremental increase in power flow on the two remaining lines. The increased flows will be inversely proportional to the reactances of the two lines. Thus, the power flow on line A_1-B_1 increases to 950 MW (its safe loading limit), an increase of 350 MW over its precontingency flow value, whereas A_2-B_2 increases to 3050 MW, an increase of 2250 MW (which is the remaining portion of the interrupted 2600-MW flow

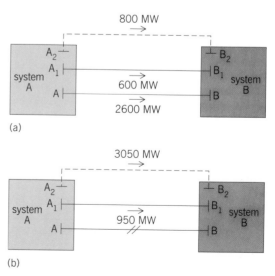

Fig. 2. Transmission interface between systems A and B, with no series compensation, (*a***) under normal conditions and (***b***) when outage of line A-B takes place.**

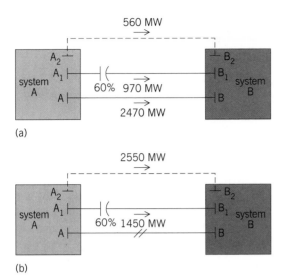

(a)

(b)

Fig. 3. Series-compensated transmission interface between systems A and B (*a*) under normal conditions and (*b*) when line A-B is taken out of service.

on line A-B). This increase on each line is due solely to the reactance relationship between the two lines. In this example, the thermal capability of line A_1-B_1 is 1525 MW; however, this capability cannot be used due to system constraints.

If a series-compensating capacitive device is installed in line A_1-B_1, this can be used advantageously to enhance the capability of the overall transmission interface between systems A and B to support higher levels of power transfers. If it is assumed that a series capacitor having a value of 60% of the inherent line reactance of line A_1-B_1 is installed, the resulting net line reactance for A_1-B_1 will be only 40% of its original value. This will result in a significant improvement in system transfer capability. First of all, the capability of line A_1-B_1 will be increased to its 1525-MW thermal limit due to the alleviation of the system voltage and phase-angle restrictions that constrained the line to a safe loadability of 950 MW.

The performance of the series-compensated transmission interface is illustrated in **Fig. 3*a***. A somewhat different power flow results, with line A_1-B_1 now carrying about 970 MW, whereas the flow on line A-B is slightly reduced to 2470 MW. Thus, the series-compensated line is more responsive to power transfers between systems A and B.

Figure 3*b* illustrates conditions when the critical line A-B is taken out of service. Line A_1-B_1 now carries 1450 MW, a substantial increase in power flow as compared to the condition of Fig. 2*b*. Thus, the series-compensated line is more responsive to the line outage condition, although it is operated within its line rating of 1525 MW.

In summary, the introduction of series capacitors in high-voltage transmission lines can have important system performance ramifications as follows: (1) the base power flow of the line will be increased; (2) system losses will change due to the change in power-flow pattern; (3) the compensated line will be more responsive to facility outages due to its lower effective inductive reactance with respect to the total interface; and (4) the loadability of the compensated line will increase due to the reduction in transmission angle resulting from the lower inductive reactance and the voltage support afforded by the capacitors.

Power-loss considerations. The impact of series compensation on system losses is a factor that needs to be carefully evaluated. As compensation is increased, resulting in increased power flows on the subject line, the power losses on the line increase. These increased losses may not be compensated by the lower losses of other transmission lines that might be unloaded. Therefore, in any application a careful analysis needs to be carried out to determine the appropriate level of compensation, taking into account the probability distribution of projected line loadings. A modular compensation device with the flexibility of being able to switch capacitor segments in increments would provide a means to optimize operation in order to minimize loss impact. Furthermore, this would provide power system operators with the flexibility to control power flows on critical transmission lines in accordance with actual system requirements, given the varying power-flow conditions that occur in day-to-day operation.

For background information SEE ALTERNATING-CURRENT CIRCUIT THEORY; TRANSMISSION LINES in the McGraw-Hill Encyclopedia of Science & Technology.

Raymond M. Maliszewski

Bibliography. J. L. Batho et al., Series capacitor installations in the B. C. Hydro 500 kV system, *IEEE PAS*, 96(6):1767–1776, 1977; T. J. Donahue and R. D. Rana, Analytical study of the loadability of AEP's Amos-Matt Funk 345 kV circuit, *Proc. Amer. Power Conf.*, 50:395, 1988; R. M. Maliszewski et al., *Power Flow Control in a Highly Integrated Transmission Network*, CIGRE Pap. 37-303, 1990; J. A. Maneatis et al., *500 kV Series Capacitor Installations in California*, IEEE Pap. 70 TP 580-PWR, 1970.

Transportation engineering

The past several years have seen a major emphasis on relieving vehicular congestion with relatively inexpensive initiatives that improve uses of the existing transportation infrastructure, especially in areas of high population densities.

Congestion problem. Traffic conditions in suburban and urban areas have worsened substantially in recent years. No longer only phenomena of large cities, congestion and the accompanying economic losses for both commuters and businesses have become commonplace. A Federal Highway Administration report, for example, estimated that the economic losses caused by vehicle congestion in 39 urban areas reached $44 billion in 1989.

The prevailing view had been that the suburban areas of large cities were places where people lived but did not work; they commuted from these suburbs to their workplaces in the cities. In recent years, this notion of suburbia has undergone a dramatic change. The development of shopping centers to serve suburban communities has been followed by unprecedented office-building and high-rise residential developments heretofore associated with the central cities. Not only have these developments, called suburban activity centers, brought masses of homeowners to the suburbs, but they have provided shopping centers and jobs to keep them there. Since 1950, for example, 86% of metropolitan population growth has been in the suburbs rather than the central cities.

The emergence of these suburban activity centers outside cities has contributed to a change in the fundamental character of commuting demand. Travel by public transit declined from 13% in 1960 to about 6% of all travel in 1980, while private vehicle travel rose from about 70% to over 85% of all travel in the same period. Meanwhile, the number of vehicles available to workers rose from 0.85 to 1.34 per worker. With severe funding constraints, environmental considerations, and limited space available for major new transportation construction, the primary emphasis has been on noncapital-intensive solutions to regional mobility issues.

Approaches to regional mobility problems. Previous solutions to suburban congestion problems have involved adding capacity to highway networks, either by new construction or by rebuilding older roadways. Funding had been typically from state and federal fuel-tax receipts and city or county property taxes. In many instances, this solution is inadequate since available funds are often unable to meet the growing demand for new highways and repair of older roads. Furthermore, added capacity may not alleviate the environmental and community impacts of congestion.

An integrated set of alternative strategies for addressing regional mobility problems includes (1) land-use and transportation planning, (2) organizational approaches, (3) demand management to reduce peak period traffic, (4) innovative suburban public transit services, (5) infrastructure design modifications, (6) innovative financing strategies, and (7) intelligent vehicle/highway systems. Each of these approaches is discussed below.

No single approach has been shown to be sufficient. Multiple solutions can be applied in an integrated and mutually supporting fashion. It is equally important that the selection and application of strategies be accomplished in a collaborative process involving private and public interests. Participants from the private sector include employers, developers, merchants, chambers of commerce, and private-sector transportation providers. Public-sector participation comes from county and city governments, transit agencies, metropolitan planning organizations, and state governments.

Land-use and transportation planning. These occur within both public- and private-sector frameworks, largely at the local level within a regional context. The private sector is involved in analyzing local and regional development markets, acquiring and holding land for future development, designing and marketing new development, constructing projects, and marketing specific areas and projects. All of this takes place within a public framework of planning, infrastructure development, and various land and building controls. There is a substantial variation in projects from one locale to another to meet differing development controls and procedures, market variations, and changing ordinances and codes.

Land-use plans are generally one of the major inputs to the transportation models and planning process that are used to produce long-range transportation plans. Ideally, short-range plans—transportation systems management elements (TSMs) and transportation improvement programs (TIPs)—are detailed elements of implementation of the long-range plan. However, local priorities, funding issues, and politics may dictate that other projects and improvements supplant implementation of long-range-plan elements.

Organizational approaches. Transportation management associations (TMAs) or organizations (TMOs) are nonprofit arrangements formed by local businesses, corporate employers, owners and developers of suburban and downtown properties, civic leaders, and public officials. They are designed to address community transportation problems that can be dealt with more effectively on a collective basis.

Association activities may be limited to transportation-system and -demand management activities or may also include other activity-center functions such as marketing and promotion. Over 50 transportation management associations or organizations are known to be operating in the United States, with more being developed. This has been the most frequent approach by suburban activity centers to addressing suburban mobility issues.

Transportation-demand management. Transportation-demand management (TDM) strategies involve programs or facilities that can change demands on the transportation system. Transportation-demand management involves changing travel behavior with respect to the use of cars. Success generally comes when employers play a vigorous promotional and organizing role.

Modified work schedules. Schedules such as flextime allow workers to arrive and depart at different times of the workday, spreading out and thus lessening the rush-hour congestion along many suburban corridors.

Parking management and pricing. These are used to improve the efficiency of existing parking facilities, reduce the need for new parking facilities, and plan for cost-effective parking expansion. Parking management strategies can work only when transportation alternatives are available to employees.

Ride sharing. This includes car pooling, van pooling, and bus pooling. Car pooling is encouraged by appointing a transportation coordinator, providing preferential parking, providing matching services, assisting in vehicle purchase or lease, allowing personal driver use of the vehicle, assisting in purchase of insurance, and cost sharing. The primary benefit to employers for providing a ride-share program comes from reduced requirements for parking. Employees benefit through reduced commuting costs.

Ordinances. These are enacted to reduce traffic congestion and improve air quality. Over 30 areas across the United States have experience in the development or implementation of transportation-demand management ordinances. For mandatory programs, transportation-demand management ordinances typically set a goal or standard that employers or developers must achieve, including the percentage of the employer's work force expected to commute by a mode other than single-occupant vehicle, the percentage reduction in vehicle trips, and peak-hour vehicle trip reduction.

Innovative transportation services. Transit service improvements can include changes to the transit system as well as employer-based programs to promote and subsidize transit usage in commuting. Approaches include suburban transportation centers, contracted transit services, employer-funded transit passes, transit service entrepreneurship programs, reverse-commute programs for inner-city workers, suburban-activity-center circulation buses, guaranteed rides home, and new technology opportunities.

Infrastructure design modifications. These are a series of design changes to existing facilities to reduce congestion and improve safety.

High-occupancy vehicle treatments. These are carried out in and around activity centers, and include provision of lanes for high-occupancy vehicles on major highways feeding the activity center, preferential parking for high-occupancy vehicles, curb bus lanes, and preferential toll facilities.

Mixed-use developments. These provide for complete employee needs, and include residences, retail businesses, offices, and transit access.

Bus shelters. These promote the use of buses by providing safe and dry places to wait.

Sidewalks. These and other pedestrian amenities can promote walking as an alternative mode of commuting or to complete a commuting trip.

Traffic engineering improvements. These can reduce congestion by using traffic-signal coordination and timing, bus and transit priority signal systems, one-way streets, reversible traffic lanes, intersection widening, left-right turn lanes, ramp-control metering and surveillance, and turning-movement and lane-use restrictions. Improved signing and street-name changes have also helped eliminate confusing discontinuities and inconsistencies on roadways.

Malls and auto-restricted zones. These prohibit automobiles from areas in which transit vehicles provide service. These areas are, by definition, not congested and very transit-accessible.

Innovative financing strategies. Such strategies by local communities in lieu of state and federal transportation funding often make possible needed projects that would otherwise be delayed or canceled. New approaches to financing of mobility projects have emerged. Benefit assessment districts are designated areas which have a special tax or special-benefit

assessment levied on properties to pay for certain transportation improvements within the district. Impact fees are charges assessed against new developments. Negotiations can be carried out between governmental bodies and individual developers to secure private participation in transportation financing. Toll financing involves the use of revenue (limited obligation) bonds based on anticipated tolls. Dedicated local taxes are a predictable source of funding for large capital investments.

Intelligent vehicle/highway systems. These are a series of emerging technologies used to improve traffic flow by unifying the vehicle and the highway.

Advanced traffic management technologies. These use real-time data in the synchronization of large-area traffic-signal systems. Advanced traffic management systems are the fundamental building blocks of the entire intelligent vehicle/highway system, with the primary goal of monitoring current network traffic.

Advanced driver information systems. These provide in-vehicle video displays and radio alerts of traffic information that enable the motorist to make intelligent route selection and diversion decisions.

Freight and fleet vehicle controls. These assist and promote commercial transportation. Some advanced technology is already in place, including systems such as automatic vehicle identification, automatic vehicle location, and vehicle-satellite data communications links.

Automated vehicle control. This includes automatic speed controls, and braking and proximity warnings. Automated vehicle control will provide considerably higher throughput of traffic for a given amount of right of way.

For background information SEE TRAFFIC ENGINEERING; TRANSPORTATION ENGINEERING in the McGraw-Hill Encyclopedia of Science & Technology.

James Costantino

Bibliography. A. E. Pisarski, *Commuting in America*, Eno Foundation for Transportation, Inc., Westport, Connecticut, 1987; Suburban mobility background paper, Houston Area Research Center, Houston, Texas, 1989; U.S. Department of Transportation, *Moving America: New Directions, New Opportunities (A Statement of National Transportation Policy)*, 1990; Urban Mass Transportation Administration, *The Status of the Nation's Local Mass Transportation: Performance and Conditions (Report to Congress)*, 1988.

Tumor

Cancer results from mutations that arise in two classes of interacting genes: those that facilitate cell growth and tumor formation, in which mutation or overexpression is oncogenic; and those that inhibit these processes, the tumor suppressor genes, whose loss is oncogenic.

The first indications that cancer involved the loss of genes came from two lines of evidence: (1) Cell fusions between pairs of cells, one normal and the other tumor-derived, gave the unexpected result that the hybrid cells, like the normal parent, were usually nontumorigenic. Thus, the tumor-forming ability of the tumor-derived cells was suppressed by some component of the normal cells. In further experiments, some hybrid cell populations regained tumor-forming ability following loss of specific chromosomes. This correlation led to the identification of particular genes or chromosomal regions encoding tumor suppressor functions in normal cells. (2) A strong inherited predisposition to particular forms of human cancer was correlated with identification of specific chromosome deletions by cytogenetic analysis. Retinoblastoma and Wilms' tumor, which are childhood cancers of the eye and kidney, respectively, are classic examples.

Both approaches have been very fruitful, leading to the association of many different tumors with specific chromosomal deletions. A variety of methods, based largely on recombinant deoxyribonucleic acid (DNA) technology, are now being used to identify the relevant tumor suppressor genes located within the chromosomal deletions, to clone and sequence these genes, and to establish their functions.

A pool of putative suppressor genes has come from the continuing search for chromosomal deletions associated with particular forms of cancer. For example, a deletion in chromosome 3p is characteristically found in small-cell lung cancer and in some renal cell carcinomas; a deletion in chromosome 5q is often found in colon polyposis and colorectal cancer; and localized deletions in chromosome 1, 11p, and 17q have been associated with breast cancer. In addition to these losses, which may be tissue-specific, a series of deleted or inactivated genes have been consistently found in many tumor types, including lung, colorectal, and breast carcinomas. Included here are loss of the retinoblastoma gene *RB* on chromosome 13q, loss of *p53* on 17p, and loss of a newly cloned gene called *DCC* on 18q. Specific genes associated with many of the deletions being found have not yet been identified.

These results demonstrate an important generalization. Some tumor suppressor genes appear to have a tissue-specific mode of action, whereas others have a common function shared by many or all epithelial tumors. Since the vast majority of cancers are epithelial in origin, identification of the common tumor suppressor genes will have wide application, whereas knowledge of tissue-specific suppressors may be of particular value in designing tumor-specific therapies.

Functional classification of tumor suppressor genes. Cancer arises as a clonal disease, starting within a single cell, which proliferates to form a clone and later a tumor. Thus, functional changes within cells precede the systemic effects of later disease and essentially determine its progress. Therefore, modern cancer research has focused on genetic changes identified and investigated at the cell and molecular levels. Some of these changes involve genes called oncogenes that encode proteins that accelerate proliferation and override normal cellular controls. Anticipated functions of tumor suppressor genes include (1) main-

tenance of chromosome stability to minimize the occurrence of mutations and chromosome aberrations, for example, genes encoding efficient DNA repair enzymes; (2) promotion of tissue differentiation, thereby removing cells from the proliferative pool; and especially (3) the regulation of cell proliferation itself.

The tumor suppressor genes so far studied affect cell proliferation, block differentiation, and interfere with the transfer of signals promoting cell proliferation from the cell surface to the nucleus. Additionally, a number of genes with postulated tumor suppressor activity have been or are being cloned. These include a gene on human chromosome 1 that regulates the expression of thrombospondin, a protein that inhibits the development of blood vessels. When tumors produce thrombospondin, they block vascularization, leading to tumor death for lack of nutrients and oxygen. Another gene, *nm23*, is associated with metastatic progression. In its presence, metastatic progression is decreased or inhibited.

By comparing genes expressed in tumor-derived cells and in the normal progenitors of the tumor, using a procedure called subtractive hybridization, it has been possible to isolate a collection of genes uniquely or mainly expressed in the normal cells. These genes, because of their differential expression, are candidate suppressor genes. In epithelial cells, these include genes encoding an extracellular matrix protein, fibronectin, and proteins of the intracellular intermediate filaments, the keratins. These findings suggest that structural proteins may also have an important regulatory role in normal cells and that this regulatory role is lost in tumor cells. There are a large number of genes now being investigated that promise to provide profound new insights into cellular regulation and tumor prevention.

Properties of three tumor suppressor genes. The *K-rev-1* gene was isolated from mouse fibroblasts transformed with the *K-ras* oncogene, on the basis of its ability to inhibit cellular properties associated with tumor-forming ability. The gene is structurally related to *ras* oncogenes but does not interfere with their expression. Rather, *K-rev-1* appears to function as a posttranslational competitive inhibitor, that is, by competition between the *K-ras* and *K-rev-1* encoded proteins. Since *K-ras* is involved in the signal transduction pathway that mediates proliferation signals from growth factor receptors on the cell surface, it is likely that *K-rev-1* inhibits this pathway. Thus, *K-rev-1* illustrates a mechanism of tumor suppression by competition at the protein level.

The *RB* gene, initially associated with tumors of the retina, is lost or inactivated in many different tumor types, including lung, colon, and breast, the three most common carcinomas in the West. The gene encodes a 110-kilodalton phosphoprotein that is located in the nucleus and that undergoes phosphorylation and dephosphorylation in each round of the cell cycle. Although the *RB* protein binds to DNA, it also interacts with other nuclear proteins, for example, kinases and phosphatases, involved in cell cycle control. Many of the mutations that inactivate the gene are presumed to interfere with the folded conformation and protein–protein interactions of the *RB* protein.

The gene called *p53* was identified initially by its encoded protein product of 53 kD, which binds to the SV40-virus-encoded T antigen; both proteins are nuclear. Mutant *p53* was studied in SV40-transformed rodent cells, where it was shown to be an oncogene. The steady-state level of wild-type *p53* protein in normal cells is barely detectable, whereas in tumor cells the mutant protein is long-lived and readily identified with suitable antibodies. Recent studies have established that the wild-type *p53* protein is a tumor suppressor gene that acts to regulate the cell cycle. Loss or mutation of the *p53* gene is the most common chromosomal change in carcinomas, and may be characteristic of all human cancers. In some of these tumors, the protein is missing. In others, a mutant protein is produced that may act as an oncogene. Thus, either loss or mutation of the *p53* gene may be oncogenic.

Senescence and control of cell proliferation. Normal human cells have a programmed proliferative potential. After a certain number of doublings, which vary with the cell type, the cells stop multiplying, although they continue to perform metabolic functions. Familiar examples are differentiated cells such as neurons and muscle cells, but most cell types, even fibroblasts, undergo an analogous process called senescence. Senescence is irreversible, as are other forms of terminal differentiation, and thus these cells are no longer available as tumor candidates. *See* Cell senescence.

Most cells in the body that are capable of proliferation, such as stem cells, are nevertheless nonproliferating much of the time in adults. They are quiescent, but in contrast to senescence, quiescence is reversible. The *RB* and *p53* genes, two of the common tumor suppressor genes whose expression is lost in many tumor types, are intimately involved in the regulation of cell proliferation. When either of these genes is lost or inactivated in presenescent cells, the cells become immortal; that is, they are no longer capable of senescence and can proliferate indefinitely. Thus, they become targets for tumorigenic transformation. Genetic studies by cell fusion between presenescent and immortal cells have demonstrated that senescence is determined by one or more dominant genes. Thus, immortalization can be inferred to involve loss of these genes that function by blocking proliferation. It should be noted that senescence probably involves several genes, since it is a very stable property of human fibroblasts and epithelial cells in culture. No instance of spontaneous immortalization of human cells has yet been reported.

Immortal cells are not tumor cells per se. Experiments using human cells in culture have demonstrated that the immortalization process does not induce tumor-forming ability but, by releasing the inhibition imposed by senescence, permits tumorigenic genetic changes to occur during subsequent proliferation.

A partial understanding of the functions of the *RB* and *p53* genes has been abetted by studies with DNA tumor viruses: SV40, adenovirus, and human papillomavirus (HPV). These viruses are each capable of inducing immortalization of human or rodent cells in culture. Each of the viruses encodes certain similar proteins with domains that recognize and form complexes with *RB* and *p53* proteins. It is likely that the viral proteins immortalize by binding *RB* and *p53* in such a way that these cellular proteins cannot function as normal regulators of proliferation. Thus, the viral proteins provide a model for the mechanism of immortalization, but in nonvirally infected tumor cells, other not-yet-identified proteins must play this role. Identifying the tumor or suppressor genes involved and restoring their normal function in tumor cells is a new approach to cancer therapy and prevention.

For background information *see* Cell senescence and death; Neoplasia; Oncogenes; Oncology; Tumor in the McGraw-Hill Encyclopedia of Science & Technology.

Ruth Sager

Bibliography. E. R. Fearon et al., Identification of a chromosome 18q gene that is altered in colorectal cancers, *Science*, 247:49–56, 1990; A. G. Knudson, Jr., Hereditary cancer, oncogenes and antioncogenes, *Cancer Res.*, 45:1437–1443, 1985; D. P. Lane and S. Benchimol, p53: Oncogene or anti-oncogene?, *Genes Develop.*, 4:1–8, 1990; R. Sager, Tumor suppressor genes: The puzzle and the promise, *Science*, 246:1406–1412, 1989; M. Schneffner et al., The E6 oncoprotein encoded by human papilloma virus types 16 and 18 promotes the degradation of p53, *Cell*, 63:1129–1136, 1990.

Turbellaria

The Turbellaria are a diverse group of flatworms generally regarded as being the closest of all modern Platyhelminthes to the ancestral flatworm stock. Turbellarians show many different degrees of elaboration of the anatomical and physiological features typical of the phylum; comparative studies of these features often indicate intermediate stages through which the members of the platyhelminth groups Monogenea, Digenea, and Cestoda may well have passed during their evolution. Such possible evolutionary stages are especially evident when nutritional adaptations to particular life styles are considered.

Life styles. The Turbellaria are predominantly aquatic, free-living animals. However, a significant number of species within the class live in permanent associations with a wide variety of invertebrate and vertebrate hosts. As a class, therefore, turbellarians show the beginnings of the trait of symbiosis (used here in its original sense of dissimilar organisms living together, not necessarily for mutual benefit). All other members of the phylum Platyhelminthes are obligate parasites. Further, each family of symbiotic Turbellaria is generally associated with only one or two types of hosts, just as are the Monogenea, Digenea, and Cestoda. Members of the family Temnocephalidae, for example, are associated mainly with certain fresh-water crustaceans, members of the family Umagillidae with certain echinoderms, and members of the family Grafillidae with certain mollusks.

Nutritional adaptations. As in most symbioses, those involving turbellarians have strong nutritional bases. Any habitat upon or within a given host type offers various potential foods, including free-living organisms occurring by chance in the host's immediate environment or in its respiratory or locomotor currents; cosymbiotes from other taxa; and the host's own food, secretions, and tissues. Most of the Monogenea and Digenea take up one or more of the last three types of food; the Cestoda take up digestive products and other soluble metabolites. In contrast, the symbiotic turbellarians show a continuous spectrum of feeding habits and diets, from predation supplemented by opportunistic commensalism, through full commensalism, to partial or total parasitism. As feeding habits and diets change along this spectrum, with concomitant changes in digestive physiology and food storage strategies, there is increasing metabolic dependence upon the host organism and thus an obvious foreshadowing of the nutritional physiologies of the Monogenea and Digenea. The spectrum culminates in those relatively few endosymbiotic turbellarians, living in the hemocoels of amphipod or isopod crustaceans, that have lost their functional endodermal alimentary systems. These endosymbionts take up all their nutrients in simple soluble form by absorption across their general body surfaces, in a cestodelike fashion.

Ectosymbionts. The evolution of parasitic habits in the Platyhelminthes conceivably began with the adoption of an ectosymbiotic life style in which turbellarianlike ancestors lived epizoically and used their hosts as feeding platforms from which they captured prey, some of which may have been cosymbiotic. This is seen in modern turbellarians such as temnocephalids, living on fresh-water crayfish, and in a few species of the order Tricladida, living on horseshoe crabs or fresh-water turtles. The diet of aquatic invertebrates, sometimes also living symbiotically on the host, is supplemented by opportunistic ingestion of fragments of the host's food. Significantly, these hosts, although widely separated taxonomically, all fragment their food before ingestion in such a way that particles drift from the mouth toward sites favored by the turbellarians. The ectosymbionts produce the full range of digestive enzymes and show few differences, other than in life style, from free-living flatworms. The Monogenea, typically ectoparasites that feed on blood and tissue of fishes, may well have evolved from ancestral forms living in this way on fish gills.

Endosymbionts. The least modified nutritional physiology of endosymbiotic species is seen in some umagillids that live in the intestine of sea urchins, feeding mainly on cosymbiotic bacteria and protozoa but occasionally ingesting wandering host cells and fragments of digesting food. A few species, though, show a shift in emphasis on dietary components and take mainly host tissues, demonstrating how the typical feeding pattern of gut-dwelling Digenea may have arisen. It is worth noting that a few digeneans living in amphibian hindguts often ingest cosymbiotic bacteria and protozoa as supplements to their diet of mucus, blood, and other host tissues.

Urchin-dwelling umagillids produce their own digestive enzymes. However, umagillids living in sea cucumbers and pterastericolids living in sea stars rely on enzymes within the ingested host tissues, and many no longer secrete their own enzymes. This is also the case in graffillids living in the intestine, stomach, and digestive glands of bivalve mollusks; they have become fully adapted to their hosts' specialized digestive physiologies, ingesting lysosomes released from the hosts' digestive glands and using these to continue digestion of the hosts' food within their own guts. Other graffillids, living in gastropod mollusks, rely on enzymes in ingested gut tissue.

In all these instances, the turbellarians retain functional endodermal alimentary systems even though the enzymes working within them may be of host origin. The logical conclusion to this trend of increasing metabolic dependence on the host would be to lose the alimentary system, leaving the hosts' enzymes to complete digestion, and then to take up soluble products across the body wall. This has not occurred so far as is known in gut-dwelling turbellarians, perhaps because some element of tissue feeding is essential. Evidence for supplementary epidermal uptake of solutes in gut dwellers is as yet equivocal. A fully cestodelike condition is seen, however, in fecampiid turbellarians that live in the hemocoels of amphipod and isopod crustaceans. Here, a functional gut is absent, and all nutrients are taken up across the epidermis, which surprisingly shows no major ultrastructural adaptation. A comparison could be made here with the larval stages of some cestodes, which live and grow in similar situations. However, in these larval cestodes the general tegument is much modified for nutrient uptake.

Parasitism. The most extreme nutritional adaptation occurring in symbiotic turbellarians is that seen in *Acholades asteris*, the only known member of the family Acholadidae. This species lives encysted within the connective tissue sheaths of the tube feet of the sea star *Coscinasterias calamaria* in Tasmanian waters. There is no trace of an endodermal alimentary system, but the epidermis and subepidermal glands have become modified into what is, in effect, a fully functional ectodermal gut. The turbellarian lies between the inner and outer collagenous layers of the connective tissue sheath; modified epidermal and subepidermal glands secrete digestive enzymes that attack the adjacent tissues, producing both soluble and finely particulate derivatives, which are then taken into the epidermal cells by pinocytosis. The pinocytotic vesicles merge into large phagosomes. Lysosomes produced within the cells fuse with the phagosomes, and digestion is completed within the resultant heterolysosomes. The host's regenerative capacity keeps pace with the damage, and as the turbellarian grows it becomes enclosed in a cyst formed entirely from host tissues. The tube foot remains fully functional even though up to three cysts may be present.

While it is very unlikely that the remarkable adaptation for tissue feeding seen in *A. asteris* can have had any specific parallel during the evolution of the Monogenea, Digenea, or Cestoda, its significance lies in its demonstration of the great plasticity and capacity for functional adaptation inherent in turbellarian stocks.

For background information *SEE* C*ESTODA;* D*IGENEA;* E*COLOGICAL INTERACTIONS;* M*ONOGENEA;* P*LATYHEL-* M*INTHES;* T*URBELLARIA* in the McGraw-Hill Encyclopedia of Science & Technology.

Joseph B. Jennings

Bibliography. P. Ax et al., *Free-living and Symbiotic Plathelminthes*, 1988; L. R. G. Cannon, *Turbellaria of the World*, 1987; J. B. Jennings, Epidermal uptake of nutrients in an unusual turbellarian parasitic in the starfish *Coscinasterias calamaria* in Tasmanian waters, *Biol. Bull.*, 176:327–336, 1989; J. B. Jennings, Nutrition and respiration in symbiotic Turbellaria, *Fortschr. Zool.*, 36:1–13, 1988.

Ultrafast molecular processes

The development of femtosecond pulse technology has led to continuing advances in the observation of the molecular motions involved in chemical reactions. Recently, sequences of femtosecond pulses have been used to control molecular motions, in addition to observing them.

Observation of molecular motions. Scientists have long searched for clues as to how chemical reactants proceed to products. Clearly, there must be some very fast rearrangement of atoms and molecules en route to the products, but the motions and configurations that this involves need to be determined. Although numerous techniques are available to analyze the structure and composition of stable compounds that exist before and after a reaction, the technology to probe the transient species in a reaction (otherwise known as transition states) was lacking until the mid-1980s. The experimental difficulties arise from the fact that the transition states often exist for times on the order of hundreds of femtoseconds (1 fs = 10^{-15} s). The problem is analogous to that of a camera in resolving a fast runner or a speeding car. In order to resolve the motion without the object being too blurred, a very fast shutter speed, on the order of one-thousandth of a second, is necessary. Producing a ''snapshot'' of a chemical reaction requires a device with the equivalent of a shutter speed at least a billion times faster than that of a camera.

Since 1980, developments in femtosecond laser technology have resulted in a pulse that is shorter in duration (6 fs is the current limit), higher in intensity, and more widely tunable in the visible spectral region than ever before. Direct time-resolved observation of the elementary motions of atoms and molecules— motions including crystal lattice vibrations, molecular vibrations, and intermolecular collisions in liquids— became possible. This permitted the observation of chemical reactions and other rearrangements that take place through such motions. However, the ability to observe ultrafast processes is not, by itself, sufficient to understand them. Because many different processes happen on this time scale, it is necessary to find some way to control which motions occur in order for the data to be interpretable. Recently, there have been some important developments in this area.

Complexity of molecular processes. An example of how complicated molecular processes can be is the reactive potential energy surface in **Fig. 1** for A + AB = AA + B. This is a typical exothermic reaction, where the reaction coordinate represents the path, in terms of nuclei spacing, that requires the least amount of energy for the reaction to occur. During the course of the reaction, the optical properties of the reacting species may change, theoretically allowing the intermediates to be observed. Optical properties include the absorption and emission of light, along with changes in the way the molecule scatters light. Complications arise because the molecules actually approach each other in an infinite number of orientations in three-dimensional space, which can result in no reaction at all or a totally different reaction with different products. Thus, there must be a way not only to observe ultrafast processes but to control which processes occur (by limiting the path of the approaching species, for example).

Observation of gas-phase dissociation. A recent example of femtosecond technology applied to following the progress of a chemical reaction was the observation of the gas-phase dissociation of cyanogen iodide (ICN), as in the reaction below. In this novel experi-

$$ICN \rightarrow [I\text{-}\text{-}CN] \rightarrow I + CN$$

ment, a femtosecond optical photolysis pulse is used to excite ICN to a higher-lying energy state. This state is dissociative in nature; that is, the reaction does not

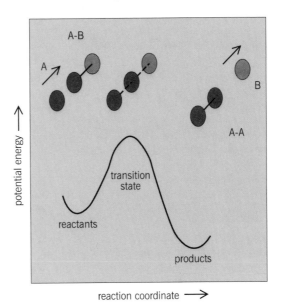

Fig. 1. Typical reactive potential energy surface for an exothermic reaction. A collision of the proper orientation and energy produces the transition state. Subsequent reaction produces the new chemical species A-A.

possess a barrier in its reactive potential energy surface as the reaction portrayed in Fig. 1 does. The end result is the production of I and CN fragments. The reaction coordinate in this case is the distance between the I and CN species. After the initial excitation, a method of probing the dissociation of ICN is necessary to follow the progress of the reaction. In this experiment, the CN species absorbs the wavelength of light used and subsequently emits light, whereas the I fragment does not. The CN emission is similar, though due to different physical processes, to the way the gas in a neon light is excited and then emits light as it relaxes to its unexcited (ground) state. A second, variably time-delayed pulse is used to excite the CN, and its emission is observed as a function of time. As more emission is observed, the more CN fragments there are, and a measure of the time it takes for dissociation of ICN is made. The reaction is found to be complete after roughly 600 fs. From this result, it is possible to infer, for example, the rotational excitation of the departing CN molecule. It is possible to analyze the results of the experiment because it is a relatively simple one, and the dissociation occurs essentially along one direction, the long axis of the molecule. The reaction also occurs in the gas phase, where it is free from interactions with neighboring molecules that occur in condensed phases.

Liquid-phase spectroscopy. Despite the difficulties of liquid-phase ultrafast spectroscopy, the importance of reactions in solution has spurred much effort in this area. The importance of the interaction between the reacting species and neighboring molecules is the center of focus. The progress of a chemical reaction such as that illustrated in Fig. 1 is governed not only by the shape of the curve but also by the way the local environment of the molecules involved in the reaction influences molecular motion in the transition-state region.

Friction is the general term used to describe the effect that neighboring molecules have on the motion of another molecule. It includes collisional friction, which is responsible for inhibiting molecular motion, and dielectric friction. Many reactions involve the transfer of charge, that is, an electron, in the course of the reaction. The lag of a solvent's ability to adjust to the sudden transfer of charge in the solute is referred to as dielectric friction. However, the solvent can also aid a reaction's progress because molecular collisions of a solvent molecule with a reacting solute molecule can provide the solute with enough energy to traverse the transition-state barrier and proceed to products. In other words, the reacting species must be given the energy to traverse the barrier to products, and collisions provide the transfer of energy that is frequently necessary. Solvent friction is necessary here also because, without it, the products could simply traverse backward on the potential energy surface and form the reactants again. Thus, solution chemistry is a fine balance between the solvent's ability to assist and impede a reaction's progress.

These molecular collisions occur on the ultrafast time scale and are therefore candidates for ultrafast

spectroscopy. A well-studied example of a solvent's influence on a chemical reaction is the photodissociation of the iodine molecule, I_2. Upon optical excitation, the iodine molecule should dissociate. Various experiments have implied that, upon dissociation, the paths of the iodine atoms may be impeded enough to force recombination of the two atoms. This has been termed the cage effect.

Control of molecular motions. The next frontier for ultrafast spectroscopy lies with the ability to optically control elementary molecular motions and chemical and structural rearrangements, in addition to simply observing them. Even a single femtosecond pulse can influence molecular and collective behavior in surprising ways. For example, a sufficiently short pulse can exert a sudden (impulse) force on vibrational modes of molecules or crystal lattices, resulting in vibrational oscillations that can be monitored in real time. This is analogous to a sudden tap on a mass attached to a spring. This technique is referred to as impulsive stimulated Raman scattering (ISRS). It is ideal for condensed-phase work, where vibrations are typically slow and easily excited through ISRS. Its limitation is that the femtosecond pulse is capable of exciting numerous oscillations that again hamper understanding of the data produced. Recently, however, the production of specified sequences of pulses whose relative timing and intensities are easily controlled and altered has been achieved, making possible the optical manipulation of molecular motion.

In an initial application of shaped femtosecond pulse sequences, a series of impulse forces was exerted on the molecules in an organic crystal lattice of perylene. Perylene is of interest in solid-state chemistry because, upon optical excitation, the molecules form excimers (short-lived dimers). Very little is known about the mechanism of the reaction, but it is believed to be assisted by crystal vibrations. In the experiment, each pulse in the sequence pushes neighboring molecules of perylene, a planar aromatic mol-

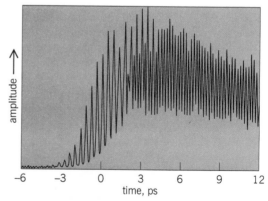

Fig. 2. Time-resolved observation of lattice vibrational oscillations in a perylene crystal driven by a femtosecond pulse sequence. The data are recorded with a variably delayed probe pulse; the time axis measures the delay between the probe pulse and the center of the excitation pulse sequence. (*After A. M. Weiner et al., Femtosecond pulse sequences used for optical manipulation of molecular motion, Science, 247:1317–1319, 1990*)

ecule, together such that they vibrate against each other. A sequence of evenly spaced pulses is timed to match the natural frequency of the intermolecular (lattice) vibrational mode. This is analogous to repetitively pushing a child on a swing, with exactly one push each oscillation cycle. Each pulse in the sequence amplifies the vibrational motion initiated by the previous pulses. When the timing of the pulses is altered and no longer matches the vibrational frequency, no response is observed.

Other lattice vibrational modes could also be driven and amplified by appropriately timed pulse sequences. **Figure 2** shows data recorded from the perylene crystal in which the excitation pulse sequence is timed to drive a lattice vibrational mode in which the molecules in each pair rotate together as a unit, undergoing librational oscillations about their initial positions. The pulse sequence, which can be observed in the data, consists of about 20 pulses spaced at 429-fs intervals. It starts with small pulses at about −3 picoseconds, has its highest-energy pulses at around the zero of time, and ends at about +3 ps. The amplitude of the vibrational response in the crystal is large enough to be observed after about the tenth pulse in the sequence, and continues to build until the end of the pulse sequence. The vibrational oscillations then continue and are gradually damped out after about 50 ps.

A single femtosecond pulse exerts a driving force on several lattice vibrational modes. With a sequence of pulses, one mode can be amplified selectively while the others are suppressed because they are not driven with the proper timing. More selective and more extensive optical control over molecular motion is offered through the use of femtosecond pulse sequences. It is hoped that sufficiently large molecular or lattice vibrational amplitudes can be driven such that new chemical reactions or crystalline phase transitions can be made to occur.

Prospects. The technology surrounding ultrafast spectroscopy is certain to grow throughout the 1990s. Attainment of shorter pulses will become easier, allowing new molecular processes to be studied. At the same time, pulse-shaping techniques will grow more advanced. While the 1980s produced a revolution in the ability to observe elementary molecular motions, the 1990s will produce a revolution in controlling these motions.

For background information *see* Chemical dynamics; Laser photochemistry; Lattice vibrations; Molecular structure and spectra; Optical pulses; Picosecond molecular processes; Raman effect in the McGraw-Hill Encyclopedia of Science & Technology.

Gary P. Wiederrecht

Bibliography. A. M. Weiner et al., Femtosecond pulse sequences used for optical manipulation of molecular motion, *Science*, 247:1317–1319, 1990; A. M. Weiner, J. P. Heritage, and E. M. Kirschner, High-resolution femtosecond pulse shaping, *J. Opt. Soc. Amer.*, B5:1563–1572, 1988; A. H. Zewail, Laser femtochemistry, *Science*, 242:1645–1653, 1988.

Vision

From the viewpoint of artificial intelligence, a vision system is a sophisticated computing system with one primary purpose: to construct useful descriptions of the environment from images. These descriptions typically focus on objects and their interrelationships: the three-dimensional shapes of objects, their motions relative to each other and to the viewer, the manner in which their surfaces absorb and scatter light, their arrangement in space, their textures, and their identities. The problems solved by a vision system, as it constructs these descriptions, are the reverse of those solved by a computer graphics system. Whereas a graphics system starts with a model of a three-dimensional world that it then must render as a two-dimensional image or sequence of images, a vision system starts with the two-dimensional images and must recover descriptions of the three-dimensional world and its objects.

Images. The inputs to a vision system are images, either a discrete sequence of images much like the discrete frames of a videotape, or a continuously varying family of images. In the discrete case, an image is a two-dimensional array of integers, say, $I(x,y)$. If the image is black and white, then each integer represents a shade of gray at a unique point in the image. For example, close inspection of a picture from a newspaper reveals that it is actually composed of a two-dimensional array of dots, varying from very dark arrays, consisting of larger dots, to very light arrays, consisting of smaller dots. The array of dots is integrated by the eye and perceived as a coherent image. The case of color images follows the same principle but is a bit more complex. Instead of a single integer for each point in the image, there are now three integers, indicating the intensities of red, green, and blue at each point. A common example of this idea is the image on a color television. In either case, whether the images are in color or in black and white, the problem to be solved by a vision system is quite difficult: Given a two-dimensional array of numbers (one row of which might be, for example, 231, 235, 223, 229,...), determine the corresponding state of the three-dimensional environment (for example, rec-

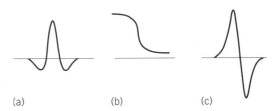

Fig. 1. Illustration of a process for finding edges in an image: the convolution of a with b to generate c. This is indicated symbolically as $a * b = c$. (a) The $\nabla^2 G$ function in profile; the full function is obtained by simply rotating this figure a half turn about a vertical line through its middle. (b) A section through an image containing an edge. (c) The result of convolving a with b; the point where the curve passes through the horizontal line is the zero crossing that indicates the existence and location of the edge.

ognize that the image is a picture of a child riding a bicycle).

Edges. Once a vision system has acquired an image, perhaps by means of a video camera, it must perform several tasks. One task is to locate and to classify edges and line segments in the image. Of the many approaches to this problem developed recently, perhaps the best known proceeds by convolving (filtering) the image with a two-dimensional gaussian $G(x,y)$ and then taking the laplacian of the result, where the laplacian is indicated by the equation below. These two steps can be combined into one

$$\nabla^2 = \frac{\partial^2}{\partial x^2} + \frac{\partial^2}{\partial y^2}$$

operation, namely convolution of the image with the function $\nabla^2 G$ ("del square g"). This function has the shape, roughly, of a Mexican hat. Once the image has been convolved with a $\nabla^2 G$, the result can be used to find edges and line segments; these correspond roughly to locations where the numerical values in the convolved image pass through zero, known as zero crossings. This process is illustrated in **Fig. 1**. When an image contains an edge, the brighter the image the higher the curve is (Fig. 1b). The region at which there is a transition from light to dark corresponds to an edge. By changing the variance (roughly, the width or dispersion) of the gaussian used in the function $\nabla^2 G$, edges can be sought at different scales of resolution: higher variance for coarser resolution, and lower variance for finer resolution. Once a vision system has obtained the edges of an image, it must interpret them, for example, by classifying them as arising from shading, object boundaries, surface markings (such as the grain of wood), or texture boundaries. This classification problem has not yet been solved in general, though progress has been made in special cases.

Three-dimensional structure. To facilitate navigation through the environment and to allow the manipulation and recognition of objects in that environment, a vision system must infer the shapes, locations, and motions of objects in three dimensions, and it must do this quickly and reliably. Among the cues to depth in images that have received attention are visual motion, stereo, shading, texture, surface contours, and occluding contours (roughly, the silhouettes of objects). In each case it has been found that the information in two-dimensional images by itself is insufficient to dictate a unique and reliable three-dimensional inter-

pretation. Therefore, a vision system must bring to bear background constraints or assumptions to guide the interpretation process. This has been a key insight.

Consider, for instance, the cosine surfaces depicted in **Fig. 2**. The pattern of curves is certainly planar, since it is printed on the page. However, it is nearly impossible to avoid perceiving (or, more precisely, misperceiving) the curves as lying on two nonplanar surfaces in three dimensions. This illustrates that human vision readily infers three dimensions from two, and that its inferential processes can at times reach incorrect conclusions. What is perhaps most remarkable is not simply that human viewers misperceive the figure but that they all report the same misperception. This suggests that, of the many three-dimensional interpretations that could be given to the figure, human vision employs consistently some assumption or constraint to pick out a unique interpretation. It has been proposed, for instance, that human vision interprets this figure by assuming that the curves lie on a surface for which the curves are in fact lines of curvature (in the differential geometric sense). This proposal may or may not be correct in detail, but it illustrates a strategy—namely, the search for reasonable assumptions or constraints on the set of possible interpretations—that has led repeatedly to progress in the study of machine vision and human vision.

Usually the constraints employed by human vision lead to a small number of consistent interpretations. Occasionally they fail to do so, with results that can be amusing, as in **Fig. 3**.

Motion. One powerful cue to depth is visual motion. From a sequence of images, such as the successive frames of a videotape, in which objects move slightly from frame to frame, it is often possible to perceive not simply two-dimensional motions but in fact three-dimensional motions and three-dimensional shapes. This ability is so well developed in human vision that it typically permits a person to drive a car safely even if one eye is not functioning. Recent theoretical work on this topic, usually called the perception of structure from motion, has led to several powerful theorems that state conditions under which a unique three-dimensional interpretation can be obtained, with a low probability of false interpretations. Here is one such theorem: If a vision system is given three distinct images of four or more feature points, points that are moving about in three dimensions, then (1) if the images allow the points to be interpreted as moving on a rigid object in three dimensions, generically they allow at most two such interpretations, and (2) the probability is zero that the imaged points can be interpreted as moving on a rigid object in three dimensions, given the fact that they were not so moving. This theorem and others of a similar nature provide a theoretical foundation upon which to build reliable machine vision systems for the interpretation of visual motion. These theorems also permit the principled development of algorithms that are insensitive to noise for the recovery of three-dimensional structure from image motion.

Manipulation and recognition. Motion, stereo, shading, and other cues allow a vision system to infer the three-dimensional structure of objects. But the

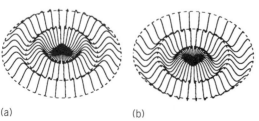

(a)   (b)

Fig. 2. Two planar figures that appear three-dimensional, with *b* being *a* turned upside down. The hills and valleys appear to exchange places between the two.

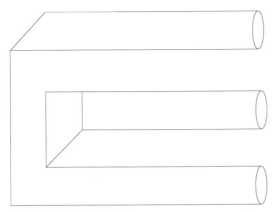

Fig. 3. The "devil's tuning fork." The assumptions used by human vision to analyze this figure lead to no globally consistent interpretation in three dimensions.

choice of how this three-dimensional structure should be represented, once it has been inferred, depends critically upon the tasks for which it is to be used. If, for example, the three-dimensional information is to be used to aid in grasping and manipulating the object, then it is preferable to use a viewer-centered representation, one that makes explicit the distance and orientation of the object relative to the viewer, for it is the three-dimensional shape and position of the object relative to the viewer that is critical for this task. If, on the other hand, the goal is recognition of the object, then it is preferable to use an object-centered representation, one that makes explicit the intrinsic geometry of the object independent of its position relative to the viewer, so that small changes in viewing position do not change fundamentally the description of the object and thereby impede the recognition process. Researchers in machine vision continue to refine both kinds of representation, and have successfully built systems that permit manipulation and recognition for certain restricted environments. The problem of devising general-purpose vision systems for recognition and manipulation is much more difficult and has not yet been solved.

Prospects. Recent advances in the understanding of vision are due in large part to increased multidisciplinary collaborations between experts in visual neurobiology, artificial intelligence, mathematics, and visual psychophysics. Continued research along these lines promises both technological and scientific advances. Technologically, the increasing speed of digital hardware, coupled with the new theories and algorithms being developed by vision researchers, is making possible a more versatile generation of machine-vision systems. As these systems increase in sophistication, they are likely to see broader application: in automated assembly, in the exploration of space, as the eyes of home and industrial robots, and as prosthetic devices for the visually impaired. Scientifically, continued progress can be expected in constructing mathematically rigorous theories and theorems for various aspects of visual information processing—from the detection of edges to the recovery and recognition of three-dimensional shapes.

Moreover, general principles should emerge, principles that when properly formalized and investigated may well transform the scientific study of vision.

For background information SEE ARTIFICIAL INTELLIGENCE; PERCEPTION; VISION in the McGraw-Hill Encyclopedia of Science & Technology.

Donald D. Hoffman

Bibliography. J. Aloimonos and D. Shulman, *Integration of Visual Modules*, 1989; D. Ballard and C. Brown, *Computer Vision*, 1982; B. Bennett, D. Hoffman, and C. Prakash, *Observer Mechanics*, 1989; B. Horn, *Robot Vision*, 1985; D. Marr, *Vision*, 1982; W. Richards, *Natural Computation*, 1988; J. Wolfe, *The Mind's Eye*, 1986.

Watermelon

The watermelon was popular 4000 years ago in Egypt and has been cultivated ever since for human and sometimes livestock consumption. Its popularity began to wane in the 1950s in the United States. Recent surveys have indicated that American consumers prefer a smaller watermelon, rather than a slice of a larger watermelon. The consumer also would prefer a watermelon with no seeds. Seedless watermelons were introduced in 1948 in Japan, but by 1987 they accounted for only 1% of the United States market; by 1990, however, more than 5% of the watermelon market was for seedless varieties. Similarly, among

Fig. 1. Diversity in watermelons grown in China. (*W. Ming, Northwestern Agricultural University, Yangling, Shaanxi, China*)

melons with seeds, small varieties are gaining in popularity. Per-capita watermelon consumption in the United States has also been increasing recently. Consumer demand for diversity in watermelon has been paralleled by the development of new breeding techniques that can provide that diversity.

Genetic diversity. The watermelon (*Citrullus lanatus*) belongs to the cucurbit family. It originated in central Africa but is now grown throughout the world in tropical and temperate climates. There are only a few species of the genus *Citrullus*, but a wealth of variability exists within the single species called watermelon (**Fig. 1**). This variability provides the genetic raw material to make new varieties that vary in size, shape, harvest date, flesh color, rind, and seed characteristics. Fruit size can vary from 1 to 50 lb (0.5 to 23 kg). Fruit is round, oval, blocky, or cylindrical and possesses a rind that is golden, light green, dark green, or various combinations of stripes. Fruit maturation time can vary from less than 70 days to more than 100 days. The rind can be so tough that only the sharpest knife will penetrate it or so fragile that a slight bump will crack the melon like an egg. Watermelon seeds can be as large as pumpkin seeds or as small as tomato seeds, and can be white, red, green, tan, brown, black, or bicolored. Watermelons can grow on small dwarf vines or vines extending 30 ft (9 m). Watermelon plants can be resistant to air pollution and to many types of diseases and insects. Watermelons can have 15% sugar or hardly any sugar at all, and the flesh can be soft or crisp. In short, the watermelon fruit has characteristics that can vary over a wide range.

Hybrid cultivars. Breeders have developed hybrids that are superior to the traditional open-pollinated cultivars. An open-pollinated cultivar is a pure line of watermelons produced from seeds year after year by pollination within a small, isolated population. A hybrid has two parental types. Each parent is

maintained in a manner similar to open-pollinated cultivars. Parents are chosen with specific hybrids in mind. For example, a cultivar with round fruit can be crossed with a cultivar with long fruit to produce a hybrid with fruit having a blocky cylindrical shape. However, producing a quantity of hybrid watermelon seeds sufficient for cultivation is an expensive and technical process, which is best understood in terms of watermelon flowers and breeding systems.

Breeding systems. Watermelon plants have two kinds of flowers, male and female (**Fig. 2**). For every female flower, six or seven male flowers are produced. In a field of plants, pollination is carried out by bees; pollen produced by male flowers is deposited on the stigma of female flowers. Pollen can be transferred between flowers on the same plant as well as between flowers on different plants; thus, many seeds are produced that are the offspring of two parents. To produce a large amount of pure hybrid seeds from two different parent lines, either self-pollinations must be eliminated or seedlings arising from such must be marked so that they can be identified and eliminated.

Self-pollinations can be eliminated by using male sterile genes that eliminate the production of pollen. Male sterility could be used in watermelons to produce hybrids, particularly a special kind of hybrid, the seedless watermelon. Seedless watermelons are produced from triploid seeds produced by transfer of pollen from diploid plants (having two sets of chromosomes) to female flowers of tetraploid plants (having four sets of chromosomes). Tetraploid pollen will not compete successfully with diploid pollen in fertilizing diploid flowers because diploid flowers produce much more pollen, and because tetraploid pollen tubes grow more slowly than diploid pollen tubes. Nevertheless, within the same fruit, when tetraploid plants are planted side by side with diploid plants, a number of tetraploid seeds from self-pollinations will arise, as well as triploid seeds from cross-pollinations. Currently, it is not possible to mechanically separate tetraploid and triploid seeds. Triploid fruit are mildly triangular and can be sorted out with additional labor at harvest. Male-sterile tetraploid watermelon maternal lines could be used to produce pure triploid hybrid seeds instead of a mix of tetraploid and triploid seeds. These lines would simplify the effort needed to generate diversity in types, such as fruit shape, color, nutrient content, maturity date, flavor, geographical adaptation, and plant size.

Tissue culture. Another special technique to improve watermelon cultivation is tissue culture. Immature seeds can be cultured on special media and cloned rapidly by repeated subculture of new shoots to produce thousands of copies of the same plant (**Fig. 3**). This technique may prove particularly useful in propagating triploid seeds. Currently, triploid seeds are expensive and germinate poorly under ordinary field conditions. Furthermore, the tetraploid parent requires considerably more breeding effort than the diploid parent to ensure the integrity of the triploid hybrid seeds. Consequently, fixing a desirable triploid by micropropagation offers a way to pre-

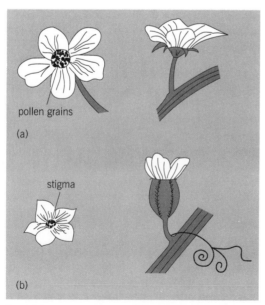

Fig. 2. Watermelon flowers. (*a*) Male, two views. (*b*) Female, two views.

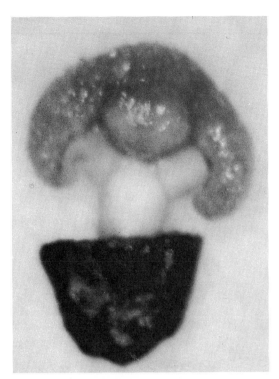

Fig. 3. Proliferation of tissue from a watermelon seed partially dissected 2 weeks after pollination and cultured. (*J. W. Adelberg, Department of Horticulture, Clemson University*)

serve a specific plant, and the value of the plant justifies asexual rather than sexual reproduction.

Small, seedless, and large-seed types. Scaling down the watermelon has been a goal of some watermelon breeders for a long time. Two of the older small watermelon varieties are Sugar Baby and New Hampshire Midget. Two recently developed small-fruited varieties are Minilee and Mickeylee. These taste better than the older small-fruited varieties and have greater disease resistance and storage life. They weigh 5–15 lb (2–7 kg). Other small-fruited varieties are also available. All of the small melons mentioned are open-pollinated types. Other small melons are hybrids derived from two parents as described above. Small melons frequently mature 10–25 days earlier than large melons. Small seedless types are desirable not only because the quality of the flesh is often better but also because such melons last longer in storage.

As noted before, seed size can be chosen in melons. Melons with a few small seeds would be almost as desirable as seedless types and would be considerably less difficult to produce since they could be produced through seeds. Also, even in seedless melons, unattractive white seed coats sometimes develop. Very small seed coats would be more desirable than large ones.

In some parts of the world, such as China, watermelon seeds are very popular. Varieties grown for seeds have a large number of small fruits with many very large seeds. Watermelon seeds have all the essential amino acids, a protein content of more than 25%, and a higher level of linoleic acid than soybean or sesame.

For background information *SEE POLLINATION; TISSUE CULTURE; WATERMELON* in the McGraw-Hill Encyclopedia of Science & Technology.

Billy B. Rhodes

Bibliography. C. F. Andrus, V. S. Sehadri, and P. C. Grimball, *Production of Seedless Watermelons*, USDA Tech. Bull. 1425, 1971; L. A. Risse et al., Storage characteristics of small watermelon cultivars, *J. Amer. Soc. Hort. Sci.*, 115(3):440–443, 1990; Taking a slice out of tradition, *Prog. Farmer*, p. 54, June 1986; J. Unrein, Popularity of seedless watermelons increases, *The Grower*, p. 52, 1990.

Welding and cutting of metals

Water-jet and abrasive jet-cutting are advanced industrial production methods that have been made possible by advances in computer-aided design (CAD) and computer-assisted control of tools. Computerized water-jet and abrasive-jet cutting methods offer the unique ability to machine very soft and very hard materials with a cool beamlike jet of water or a jet of water bearing very fine abrasive particles. This process is employed by many industries in a wide variety of applications. The technique is no longer merely a curiosity, and 24-h production lines have demonstrated that the process and hardware are very reliable and offer many technical and economic advantages over traditional methods. The water-jet and abrasive-jet methods feature an omnidirectional cutting tool that is ideally suited for automation and complementary robotic workcells.

Recent developments in manufacturing have been accompanied by use of new materials, higher production quotas, and intricate profiling manipulation. These represent only a few overwhelming challenges that are difficult to meet with conventional methods of cutting. Integrating two technologies such as a robot and a high-pressure intensifier system generally requires special attention to the basic theory of machine dynamics.

Intensifier. Water-jet cutting employs an intensifier system (**Fig. 1**) that concentrates and sustains a needlelike jet of water traveling at a supersonic speed (Mach 3). Typically, this jet is powered hydraulically by a 40–100-hp (29–75-kW) motor, and generates a laminar flow of water focused through a sapphire or diamond orifice, resulting in a powerful compressive-shear cutting action at close range.

The heart of the intensifier system is a double-acting reciprocating pump with a typical intensification ratio of 20:1. The hydraulic power is supplied from an electrically powered radial piston pump of variable displacement. As the pressure is increased, the intensifier shifts back and forth while the system maintains the pressure until an outlet is opened, such as the cutting nozzle being actuated. At this point the intensifier cycles at a rate proportional to the output flow rating of the outlet orifice, or nozzle. At full pressure, 55,000 lb/in.^{2} (378,950 kilopascals), the water is compressed by approximately 12% and is

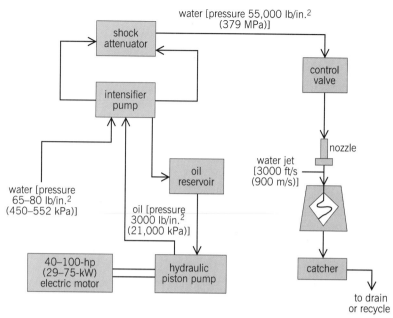

water [pressure 55,000 lb/in.²
(379 MPa)]

water [pressure
65–80 lb/in.²
(450–552 kPa)]

oil [pressure
3000 lb/in.²
(21,000 kPa)]

water jet
[3000 ft/s
(900 m/s)]

to drain
or recycle

Fig. 1. Schematic diagram of a high-pressure intensifier system.

relieved of pressure spikes as it travels through an attenuator, ensuring consistent nozzle pressure and suppressing any potential failure of the equipment due to fatigue.

Water. The incoming water fed to the intensifier is a site-specific issue that depends on the quality of the source water. The inlet water should have suspended particulate matter removed (filtered) to 0.5 micrometer absolute. Other methods such as reverse osmosis or mixed-bed deionization systems can increase the orifice life by at least a factor of 10, but the initial capital cost may dissuade some interested users. An analysis of the local water supply is recommended when orifice problems occur.

Plumbing. High-pressure stainless steel tubing should be rated at least at 60,000 lb/in.² (413.4 megapascals). Such tubing is available in sizes ranging from 0.25 to 9.16 in. (6.35 to 14.3 mm) diameter, and it should have a ratio of diameter to wall thickness of 3:1.

The 0.25-in.-diameter (6.35-mm) tubing is usually used at or near the cutting head of the robot; and it is usually plumbed in the least restrictive way, depending on the required movement of the robot. In one method, known as whip and coil, a free-moving length of high-pressure piping is connected from a solid base (usually through a flexible coiled pipe, and connected to a moving nozzle) or is moved by means of a robot; this exploits the flexibility of the tubing and allows movement of the robot, but this configuration usually takes up a large amount of air space and needs external recoil supports. Swivel joints are usually mounted on the robot arm in line with the joint rotation. The alignment of the center line of the swivel shaft rotation with the robot movement is critical, since side loads will be barely tolerated if installation is improper. The distance from the intensifier pump to the nozzle should be held below 450 ft (137 m).

The abrasive-jet system contains essentially all the standard water-jet hardware from the pump to the sapphire or diamond orifice, but has an abrasive mixing chamber attached and a wear-resistant mixing tube (**Fig. 2**).

The decision to use water-jet or abrasive-jet involves consideration of the costs, material hardness, desired surface finish, production rates, and so forth. The costs for abrasive-jet technology are considerably higher than for water-jet cutting, and they can escalate sharply with more abrasive use.

Cutting parameters. The cutting parameters depend on the type and style of the system required for a specific application. Although it is possible to cut through both hard and soft materials by using this process, discontinuities, such as large air gaps, will render a coherent jet into an inefficient spray. The speed of the water jet in abrasive-jet cutting accelerates the abrasive particles (garnet) down through the mixing tube and onto the target material. The process has now been transformed to erosive shear. The abrasive particles strike the workpiece at a rate over 100,000 times per second, simulating a micromachining process. In abrasive-jet cutting, the water is merely the vehicle and coolant. The abrasive jet is a very powerful tool for linear cutting and turning; it has recently been applied to milling and drilling. The abrasive jet leaves a polished or sand-blasted edge with a surface roughness or finish typically 65–110 μm.

The energy required for a specific system depends upon the number of nozzles being used and the amount of water needed to satisfy the established orifice size. The orifice size is 0.003 in. (0.07 mm) in diameter and upward in increments of 0.001 in. (0.025 mm).

Process parameters are chosen with a view of reducing inefficiencies during cutting. An intensifier

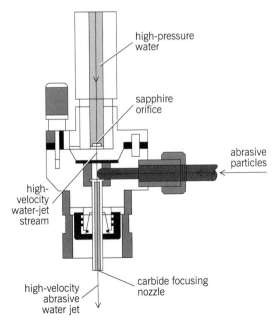

high-pressure
water

sapphire
orifice

abrasive
particles

high-
velocity
water-jet
stream

high-velocity
abrasive
water jet

carbide focusing
nozzle

Fig. 2. Schematic diagram of the abrasive-jet nozzle.

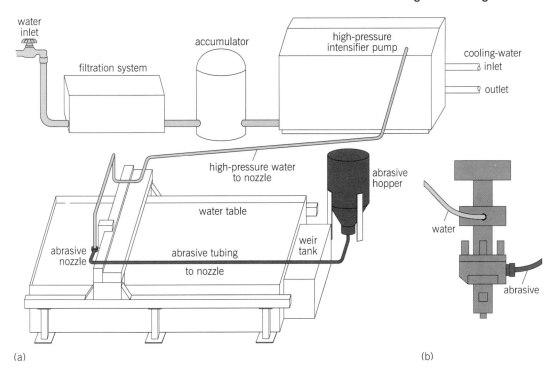

water inlet

filtration system

accumulator

high-pressure intensifier pump

cooling-water
inlet
outlet

high-pressure water to nozzle

abrasive hopper

water table

water

abrasive nozzle

abrasive tubing to nozzle

weir tank

abrasive

(a)

(b)

Fig. 3. Schematic diagrams of a gantry abrasive-jet system. (*a*) Typical layout. (*b*) Abrasive nozzle; the abrasive enters the nozzle and is combined with high-pressure water in the mixing chamber.

pump should not be operated at its maximum limit for too long a period of time. Considerable economies can be attained if the desired results can be achieved by reducing the pressure of the pump, the speed of the robot (to minimize joint fatigue), or the abrasive (if used). Cutting specifications include water-jet pressure, water-jet orifice size, stand-off distance, abrasive size and type, mixing tube length and diameter, jet angle (entry and exit), kerf angle, surface finish, speeds, tolerances, and costs.

Advantages and disadvantages. The outstanding benefits of water-jet and abrasive-jet cutting are absence of thermal effects or heat-affected zones, avoidance of chipping or cracking of brittle materials, low tooling costs, low material loss, elimination of most deburring, capability of single-pass cutting, minimal airborne dust, capability of omnidirectional shape and contour cutting, and an environmentally friendly impact.

Disadvantages of both water-jet and abrasive-jet cutting include an inability to carry out critical-tolerance cutting, unless intensive process monitoring is available. Some cuts may have a kerf angle, V-shaped, controlled by nozzle traverse speed, especially when cornering. Improperly pierced holes may leave a gouge mark. Some very slight work hardening may occur, but only in a few types of material.

In abrasive-jet cutting, garnet embedment may occur with some materials; the garnet can be removed with vibratory, polishing, or wash processes. Overall, the abrasive water jet is an excellent tool for alloys and composites that are difficult to machine.

Robotic applications. The typical water-jet nozzle weighs approximately (8 lb) (3.5 kg); considering the reaction of on/off back pressure and the weight of the end effector, a design that includes a robot specifies a payload of at least 22 lb (10 kg). An additional catcher would add another 50 lb (22 kg). An articulate robot design that is similar to an arm with 5 or 6 axes of movement (pedestal robots) with welding background or sealant dispensing is complementary to the continuous process of the water jet. Gantry-type robots are exceptional for their rigidity, and they are usually closer to machine tool specifications concerning accuracy and repeatability. The recommended programming method is to download a tool path that has been defined by computer-aided design (CAD). The point-to-point teach method involves recovery time if the program is lost or an engineering modification occurs.

Memory capacity is a significant consideration for instances where there are many programs or large programs that require large local memory or an interactive storage device. Another important consideration is whether the control can accept one program while executing another. It is necessary that the control processor be fast enough to ensure that the cutting head is tracking the path at the correct speed at all times.

Programs downloaded from CAD have a certain spacing of point definitions; this may conflict with the processing speed of the robots compared to the programmed cutting speed, resulting in oscillations of the motor drive and a rough, striated cut. This problem is distinct from those stemming from incorrect cutting pressures, speeds, abrasive flow, and so forth. The water jet or abrasive jet will reflect any mechanical vibrations or backlash, and this will be apparent on the cut edge. Pedestal robots may have more difficulty on sharp corners or circular interpolation moves. Designs for water-jet and abrasive-jet machines must avoid having uncovered bearings or ways. Way covers are essential, and some positive internal air pressure will greatly help prevent contamination. The controller should be well protected and ventilated. Ideally, the whole workcell should be located in an enclosed area, both for cleanliness and for noise suppression. The floor foundation should be of machine-tool quality to ensure that the robot and fixtures maintain their relationship. The floor grating can act as a catcher tank or the catcher can be located under the gantry, and the weir tank will allow sufficient drainage and dredging periodically. **Figure 3** shows the layout for a typical gantry abrasive-jet system.

It is not necessary that the fixtures be ruggedly strong, since the water jet creates minimal vertical or lateral forces. Fixture supports can be built with disposable sections and can use any cost-effective material suitable for deflecting a jet (tool steel, ball bearings, or wood). The accuracy of the final cut results from the repeatability of both the robot and the fixture. Safety devices include emergency stop buttons, light curtains, and software limits to robot joints.

For background information SEE COMPUTER-AIDED DESIGN AND MANUFACTURING; ROBOTICS; WELDING AND CUTTING OF METALS in the McGraw-Hill Encyclopedia of Science & Technology.

Duane E. Snider

Bibliography. G. Ayers, Principles of waterjet cutting, *Tappi J.*, 45:91–94, September 1987; M. Hashish, Advanced machining with waterjet cutting applications, *Automated Waterjet Cutting Processes*, pp. 54–63, 1990; D. E. Snider, Waterjet cutting in a production environment, *Proceedings of the 5th American Waterjet Conference*, pp. 231–236, 1989.

Wind shear

Microburst wind shear is one of the most serious hazards of aviation, particularly dangerous to heavily loaded jet transports flying low and slow on final approach, or during takeoff before a comfortable margin of airspeed has been established. For years the unrecognized cause of many accidents (often attributed to pilot error), this unique type of wind shear was first isolated and identified in 1975. A number of hasty solutions, involving both ground-based installations and airborne equipment, were proposed and advocated. Government and industry have now settled on a dual approach combining special pilot training and an airborne warning and recovery guidance system. This approach, now mandated by the Federal Aviation Administration (FAA), has brought the pilot and modern technology into a working relationship that promises a nearly perfect solution to the wind-shear hazard.

Nature of microburst. In June 1975, a Boeing 727 crashed in the approach lights at Kennedy International Airport in New York. T. T. Fujita investigated the accident and discovered and documented what is now known as microburst wind shear. A microburst is a vicious downblast of air that hits the ground and bursts outward in a vortex ring, creating violent wind

changes within the vortex curl. By far the most dangerous type of wind shear, a microburst differs not only in degree but in kind from ordinary wind shear. Pilots and engineers have learned to expect (and on-board computers have been programmed to deal with) the changes in horizontal winds normally encountered as an airplane descends. A microburst is different and requires a very different response.

Typically, a microburst is of very short duration. Appearing suddenly, it develops its dangerous potential and dissipates in a matter of minutes. It covers a very small space geographically; the core of a microburst may be as small as 1 nautical mile (1.8 km) or less in diameter. The downdraft in the core may reach velocities that can suddenly triple or quadruple the normal descent rate of an aircraft on approach, or drive a takeoff back into the ground.

Ground-based alerting system. Early attempts to protect against this phenomenon included the development of a low-level wind-shear alerting system (LLWAS), now installed at many major airports. This alerting system consists of wind direction and velocity sensors around the perimeter of the airport, plus a central sensor. Sensor data are computed and displayed in the control tower. Controllers then advise the pilots concerned. However, the information derived is seldom specific enough to justify aborting the takeoff or landing. Moreover, microbursts sometimes pass undetected between the sensors or occur in the approach or departure area outside the network. (At Dallas-Fort Worth Airport in August 1985, the LLWAS did not alert until after a major accident had occurred.) At best, the LLWAS tells the pilot that something may be present. But in a microburst encounter, the pilot needs unequivocal information that can be acted on immediately without further analysis.

Microburst scenario. In the typical microburst scenario, the pilot, on meeting the initial outflow headwind (point A in the **illustration**), finds the aircraft ballooning above the glide path with the airspeed suddenly higher than usual. The normal reaction is to reduce power and push the nose down. A few seconds later, the headwind outflow has disappeared, and a downdraft that can reach velocities of 1200–2400 ft/min (6–12 m/s) strikes from above (point B). Within another few seconds, the aircraft is in the outflow tailwind (point C) and falling rapidly below the glide path. The airspeed, so recently high, is suddenly 15–20 knots (8–10 m/s) below the so-called safe reference speed.

Two factors compound the problem. One is that the pilot does not realize quickly enough what is happening. (As few as 5 s can make the difference between safe escape and disaster.) The phenomenon may be encountered only once (or perhaps never) in a long career. By the time the pilot recognizes the problem, the inertial velocity downward may be irreversible.

Second, a lifetime of training makes it second nature to the pilot never to allow the airspeed to go below the so-called safe reference speed, and thus to automatically push the nose down to recover airspeed. In some microburst wind-shear encounters on record, this action reduced the angle of attack (which controls lift) to near zero. Loss of lift added to the force of the downdraft increased the descent rate (the inertial velocity downward) to the point where it could no longer be stopped in time.

Airborne warning system. Efforts were started very early to provide a wind-shear warning system aboard the aircraft to alert the pilot to the emergency well before warning could be provided by normal instrumentation. Popular at first was a concept in which the warning computer would observe only the horizontal winds of the microburst, comparing airspeed and ground speed and the rate at which these speeds were changing in relation to each other. This was a very simple way to see that a headwind was rapidly decreasing, or changing to a tailwind. This approach to measuring the threat was soon seen to be inadequate.

Some investigators, however, had recognized at the outset that an on-board system must deal with the energy loss to the aircraft caused by a combination of the horizontal and vertical winds of the microburst. A warning system was developed that sums the energy loss caused by the horizontal plus the vertical wind changes encountered when entering the outer edge of the microburst.

The on-board computer observes airspeed, pitch attitude, and angle of attack, and the rates at which these factors are changing, to determine what the airplane should be doing in normal airflow; that is, whether the aircraft should be climbing, descending, or holding steady altitude. Also, the computer determines whether the airspeed (and ground speed) should be increasing, decreasing, or holding steady.

Inertial sensors are also included in the wind-shear detection computer (if such sensors are not already available on the aircraft) to determine what the aircraft is actually doing. In a normal airflow situation, the inertial sensors confirm that the aircraft is doing what it should.

As the aircraft enters the outflow or headwind area of the microburst (point A in the illustration), however, the computer observes a rapidly increasing airspeed with no corresponding increase in the inertially derived ground speed. In the same way, the computer observes the force of the vertical downdraft as the aircraft enters the core of the downburst (point B). When the wind-shear computer determines that something outside the limits of normal turbulence is going on, the alarm is given over the cockpit loudspeaker.

Recovery guidance. With an effective warning device in place, it was soon evident that in the rapidly changing vertical winds of the microburst the pilot needed pitch guidance to optimize the aircraft's performance in the escape maneuver. To meet this need, the wind-shear computer, it was realized, could do more for the pilot than provide detection and warning.

Already constantly tracking the angle of attack, the wind-shear computer could very easily tell the pilot, through the pitch command bar of the flight director, what pitch attitude was required from moment to moment while flying through the danger area. Thus, recovery guidance was also included as an option.

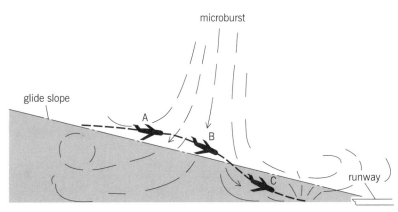

Winds in a microburst and their effect on a descending aircraft.

Regulations and installation. FAA has now mandated that a wind-shear warning system be installed and carried aboard certain categories of transport aircraft by certain dates. In addition, the FAA has mandated that all transport aircraft in certain categories also be equipped by certain dates with the recovery guidance option. A large number of airline and corporate jets are already equipped with an on-board wind-shear warning system. The stand-alone wind shear warning and recovery guidance systems now available have dimensions of the order of $4.5 \times 4.5 \times 9$ in. ($11.5 \times 11.5 \times 23$ cm), and weigh approximately 5 lb (2.3 kg). As an alternative to stand-alone equipment, the mandated wind-shear protection can be integrated in other equipment already on the aircraft, such as ground-proximity and stall-warning systems.

Training package. In parallel with the development of an on-board wind-shear warning computer, a training package has been developed by a consortium of airline people, airframe manufacturers, and FAA personnel to help pilots understand this phenomenon. Through classroom study and hands-on experience in a simulator, the pilot understands that during the critical seconds of a microburst encounter, airspeed is suddenly not the primary criterion, and learns that airspeed, pitch attitude, and angle of attack must be seen in a new relationship to prevent the downward velocity from going out of control.

Degree of protection. The presently mandated airborne warning system plus proper pilot training provides a very high degree of protection. The on-board warning systems are designed to have very high reliability. In present designs the possibility of a false warning is judged to be very near zero. The number of nuisance alerts (alerts on levels of shear that are not hazardous) are kept to a very low level, well below the level that would destroy pilot confidence in the system.

Because the threat is measured only as the aircraft enters the outer edge of the danger area, the presently mandated warning system is known as a reactive system. At first, many hypothesized that a reactive warning would come too late. However, a careful study of the eight accidents ascribed to wind shear by the National Transportation Safety Board between 1975 and 1985 gives clear evidence that in every case the warning from a reactive system could have provided the essential margin for safety. The FAA mandate endorses this conclusion.

There is continuing effort, nonetheless, to provide a look-ahead capability. Ground-based Doppler radar, an infrared sensor in the aircraft, and the enhancement of airborne radar with a Doppler capability are avenues currently being explored.

These efforts may eventually provide some additional protection. For the foreseeable future, the presently mandated system, backed up by pilot training, offers proven effectiveness, with relative ease of implementation and a high degree of reliability.

For background information SEE DOPPLER RADAR; INERTIAL GUIDANCE SYSTEM; STORM DETECTION; WIND in the McGraw-Hill Encyclopedia of Science & Technology.

Sam Saint

Bibliography. Federal Aviation Administration, Airborne low-altitude windshear equipment requirements, *Fed. Reg.*, 55(68):13,236–13,242, April 9, 1990; T. T. Fujita, Tornadoes and downbursts in the context of generalized planetary scales, *J. Atm. Sci.*, 38(8):1511–1534, 1981; T. T. Fujita and H. R. Byers, Spearhead echo and downburst in the crash of an airliner, *Month. Weath. Rev.*, 105(2):129–146, 1977; S. Saint, *The Missing Element in Wind Shear Protection*, SAE Tech. Pap. 830715, 1983.

Zooplankton

Recent research has included studies on the role of certain ciliates in the planktonic food web and studies on certain nudibranch larvae in an attempt to understand the evolution of nonfeeding planktonic larvae of marine invertebrates.

CHOREOTRICH AND OLIGOTRICH CILIATES

Choreotrich and oligotrich ciliates are now recognized as significant components of planktonic food webs in marine and fresh-water ecosystems. The recognition of their abundance and diversity has occurred since biologists began using more appropriate means of collecting, preserving, and staining these very delicate protists. Indeed, their production on occasion may well equal that of the most significant metazoan planktonic group, the copepods. Thus, the study of these ciliates is important in broadening overall understanding of the ecology of planktonic food webs.

Microbial elements in planktonic food webs. In recent years, planktonic food webs have been shown to be quite complex (**Fig. 1**). The microbial elements such as bacteria, flagellates, and ciliates are now considered important agents in the consumption and transfer of energy through the food web. The global consequences are significant since aquatic food webs probably fix more carbon each day (through photosynthesis by the phytoplankton and mixotrophic cili-

ates and flagellates) than all terrestrial ecosystems combined. Because of this, aquatic food webs are important regulators of atmospheric carbon dioxide, and could have an impact on stabilizing global climate. A thorough understanding of the roles of the microbial elements of the plankton will help to determine the global dynamics of carbon fixation and movement.

Two factors contributed to the historical underestimation of ciliates, and microbes in general, as significant components of planktonic food webs. First, there were too few scientists studying these organisms to have a significant impact on the scientific community. Second, the older methods of collecting and preserving, although successful for nonprotist zooplankton, were not appropriate for ciliates. Traditionally, plankton nets have been used to collect zooplankton. However, dragging a plankton net through the water will fragment the soft-bodied protists like ciliates, while bacteria and small flagellates will pass right through. Today, microbial components are studied by collecting water samples in bottle samplers and preserving the sample as soon as it comes on board ship. In this way, delicate organisms are visible on microscopic examination.

Major groups of planktonic ciliates. Even before bottle sampling became routine, one group of ciliates was generally recognized—the tintinnines. These ciliates build a solid lorica (a hard case) about themselves from cellular secretions and sometimes inorganic material in the water. Each lorica protects its delicate owner from fragmentation by planktonic nets. The lorica may also help tintinnines to avoid predation, not because predators find the lorica hard to ingest, but because the heavy lorica may allow these ciliates to sink out of the way of a predator that they sense to be approaching. An aloricate ciliate would not sink quickly enough to avoid capture.

The tintinnines belong to one of the two major groups of planktonic ciliates—the choreotrichid ciliates (**Fig. 2**). Choreotrichids are both loricate and aloricate, and are distinguished by having a set of oral ciliary structures that form a closed circle about the cell's anterior end. The cytostome or cell mouth is in the center of this circle. Some common genera of tintinnines are *Favella, Stenosemella, Tintinnopsis*, and *Helicostomella*, while common aloricate choreotrichids include species in the genera *Strombidinopsis, Strobilidium*, and *Lohmanniella*. The other major group—the oligotrichid ciliates (Fig. 2)—has oral ciliary structures that form an open circle: one set, called the collar, partly surrounds the anterior end; the other set, called the lapel, extends into an oral cavity at the bottom of which is the cytostome. Some common genera of oligotrichids are *Strombidium, Halteria, Laboea*, and *Tontonia*. The choreotrichid and oligotrichid ciliates collectively form the oligotrichs, one of the major subclasses of ciliates in the class Spirotrichea of the phylum Ciliophora.

Nutritional strategy. The oral cilia are used both to create a feeding current to bring prey to the oligotrich and to move the oligotrich through the water. The cilia on the body are very reduced and apparently not useful for movement. Oligotrichs are primarily heterotrophic; that is, they derive their energy from preformed organic compounds, usually packaged as other organisms. Their prey includes bacteria, phytoplankton, heterotrophic flagellates, and other ciliates. Tintinnines apparently prefer some species of dinoflagellate protists over others. It has been shown that the toxic dinoflagellates, which cause red tides, inhibit the growth rate and reduce the longevity of tintinnines. The abundance of heterotrophic oligotrichs is primarily determined by the abundance of their prey. In short-term studies, trophic coupling has been repeatedly observed: the prey of the ciliates peaks in abundance, and this is followed several days later by a peak in the abundance of the ciliates. Often, the abundant ciliates belong to one species; and, although this has not yet been proven, all may even be descendants from one cell.

Another major nutritional strategy of oligotrich ciliates, especially species in the genera *Strombidium* and *Laboea*, is mixotrophy. Mixotrophic ciliates combine two nutritional strategies, heterotrophy and autotrophy, and apparently they must do both. Like their purely heterotrophic relatives, they feed on bacteria, phytoplankton, heterotrophic flagellates, and other ciliates. However, when mixotrophic ciliates consume phytoplankton they are able, by an unknown means, to separate the plastids, which are the photosynthetic organelles, from the cytoplasm of their prey. By maintaining the plastids within their own cytoplasm, mixotrophic ciliates are able to fix carbon like the phytoplankton. It has been estimated that dense blooms of mixotrophic ciliates might fix as much carbon each day as the phytoplankton on which they feed.

In order to fully determine the role of oligotrichs in aquatic food webs, it is necessary to sample the water over periods of at least a year. When this kind of long-term sampling is done in different parts of the world's oceans, slightly different observations are made. In north temperate latitudes, there are relatively few ciliates throughout the colder months (**Fig. 3***a*). As spring approaches, the ciliates begin to increase in abundance, especially the larger ciliates (Fig. 3*b*), which are probably eating larger phytoplankton such as diatoms. During the summer months, smaller ciliates predominate, feeding on smaller prey, which are more abundant at this time of year. The ciliates may reach abundances of up to six per milliliter of seawater during the summer. In the tropics, the fluctuations in abundance are less pronounced, and the abundance on average is less than in temperate waters. *Strombidium* and *Strobilidium* species apparently predominate in both temperate and tropical waters.

Productivity. Abundance does not provide a good indication of the dynamic processes occurring in planktonic food webs. Productivity, the amount of biomass generated per unit time, is a much better measure for determining the role that any particular group plays in a food web. However, to calculate

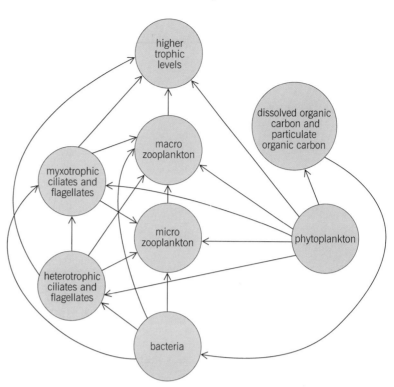

Fig. 1. Schematic model of some major components of the planktonic ecosystem. The arrows indicate direction of the movement of matter. Dissolved organic carbon and particulate organic carbon are transformed to living matter primarily by bacteria. Bacteria are eaten by larger organisms, as are the phytoplankton, which fix carbon dioxide by photosynthesis. (*From J. R. Beaver and T. L. Crisman, The role of ciliated protozoa in pelagic freshwater ecosystems, Microb. Ecol., 17:111–136, 1989*)

productivity rates, both the standing-crop biomass and the growth rate of the ciliates must be known. Both are very difficult to determine, and indeed, there are still no generally accepted and easy procedures devised. Generally, biomass is indirectly determined by estimating the cell volume. Then, by assuming a density for the cytoplasm, density is changed to carbon or energy equivalents based on conversion factors determined for other ciliates. Determination of growth rate is even more problematic. Ideally, growth rate of a population of ciliates would be measured by counting the individuals in the population at one time and returning a known time later and making a second count. However, this is almost impossible in a natural setting where ship schedules are tight and keeping track of parcels of water is complicated by water currents. Instead, water is collected, incubated in large containers, and sampled regularly to determine population growth rate. When this is not possible, growth rate is estimated based on the size of the ciliate: on average, smaller ciliates have higher growth rates than larger ones. By knowing the biomass and the growth rate, the productivity of the ciliate community can be calculated.

A food web with productivity estimates permits assessment of the relative importance of the different planktonic groups, since all of them are measured or expressed in the same units. By knowing the production of each compartment, an estimate can be made of the amount of energy that each compartment would need to maintain its production. Organisms are not 100% efficient at converting the food ingested to biomass: 10–30% conversion efficiencies are usual. Oligotrich ciliate production ranges from half of to equal to the production of copepods. On an annual basis, both these planktonic heterotrophic groups are far less productive than their prey, the bacteria and phytoplankton.

Remaining questions. The discovery that oligotrich ciliates, and all microbial components, are so productive has changed the ideas of the dynamics of planktonic food webs and stimulated active research on the microbial loop. A number of questions have been raised. For example, it is not known whether the microbial loop is a sink for carbon or a link for carbon.

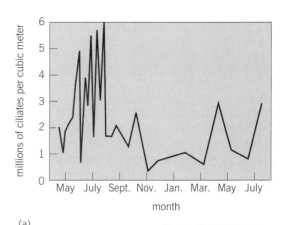

(a)

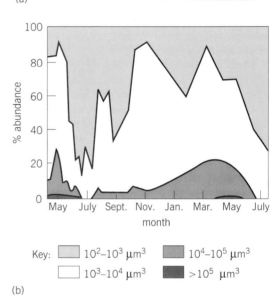

(b)

Key: ▢ 10^2–10^3 μm³   ▨ 10^4–10^5 μm³
▢ 10^3–10^4 μm³   ▪ >10^5 μm³

Fig. 3. Ciliates in waters surrounding the Isles of Shoals, Gulf of Maine. (*a*) The seasonal abundance of all ciliates. (*b*) The percentage abundance of different size classes of ciliates (*from D. J. S. Montagnes et al., The annual cycle of heterotrophic planktonic ciliates in the waters surrounding the Isles of Shoals, Gulf of Maine: An assessment of their trophic role, Mar. Biol. 99:21–30, 1988*).

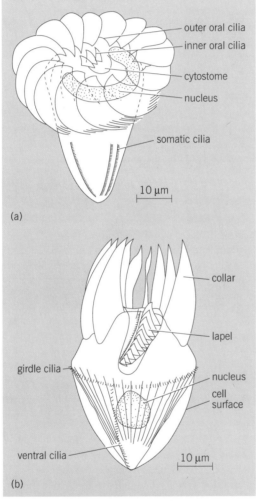

Fig. 2. Major groups of planktonic ciliates. (*a*) A choreotrichid ciliate whose oral ciliature forms a closed circle around the cell mouth or cytostome (*from D. H. Lynn and D. J. S. Montagnes, Taxonomic descriptions of some conspicuous species of strobilidiine ciliates (Ciliophora: Choreotrichida) from the Isles of Shoals, Gulf of Maine, J. Mar. Biol. Ass. U.K., 68:639–658, 1988*). (*b*) An oligotrichid ciliate whose oral ciliature forms a collar partially encircling the anterior end, and a lapel leading down into an oral cavity and the cytostome (*from D. H. Lynn et al., Taxonomic descriptions of some conspicuous species in the family Strombidiidae (Ciliophora: Oligotrichida) from the Isles of Shoals, Gulf of Maine, J. Mar. Biol. Ass. U.K., 68:259–276, 1988*).

In the former case the microbes just eat and die. In the latter case the microbes serve as prey for larger organisms in the food web, including economically important ones such as fish.

There are partial answers at this time. It appears that certain microbes are just sinks for carbon. In fact, that is how some major deposits of oil may have been produced. The White Cliffs of Dover (southeastern England) were certainly produced by the accretion of dead cell bodies of billions of protists. This process continues today. In contrast, a number of scientists have observed oligotrichs to be eaten by larger organisms, including copepods, jellyfish, mollusk larvae, and fish larvae. Thus, ciliates can be links, transferring the carbon they have ingested to other organisms. What is not yet known is how often this occurs and how significant it is in the life of a larval mollusk or fish.

Denis H. Lynn

EVOLUTION OF NONFEEDING PLANKTONIC LARVAL FORMS

While many species of benthic marine invertebrates reproduce by small eggs (~100 micrometers in diameter) that hatch as pelagic larvae, others may develop from larger eggs (several hundred micrometers in diameter) and hatch directly from the egg capsule as fully formed benthic juveniles.

Larval types. The latter group is said to display nonpelagic development. The mode of nutrition of the embryo and larva determines classification of the reproductive strategy as either lecithotrophic (all nutrition is provided by the parent within the egg) or planktotrophic (the larva obtains its nutrition independently from the plankton). As a generalization, planktotrophic larvae hatch at a small size and are poorly developed. During the obligatory pelagic phase the larva grows and increases in morphological complexity toward the state at which it is capable (competent) of settlement and metamorphosis. Once it achieves competence, the planktonic larva enters the facultative pelagic phase, during which it may delay metamorphosis until the appropriate physical or biological cues are encountered. Metamorphosis during the obligatory phase is simply not possible, even if the correct substratum or stimuli are encountered. Lecithotrophic species brood their offspring to varying levels of development: the young may be released at a postembryonic larval stage (as in many crustaceans) or as fully formed juveniles (as in some starfish). Lecithotrophic species include those that simply retain and protect the embryos externally, in addition to those that are viviparous and nurture internally the developing young.

Nudibranchs—which are mollusks that have no shell, lack a mantle cavity, and have a detorted visceral mass as adults—show clear recapitulation during their embryonic and larval development. Thus, the typical hatching larval form is a torted and shelled veliger, which undergoes metamorphosis to the adult form as a result of both loss of shell and operculum, and detorsion of the visceral mass—all in a matter of hours. The nudibranch larva belies the prosobranch ancestry of the opisthobranch lineage; such embryological conservatism can be used as an empirical basis to investigate the evolution of nonfeeding larvae among opisthobranch mollusks in particular and marine invertebrates in general. Species which display nonpelagic lecithotrophy undergo clearly vestigial "larval" stages within the capsule. For example, *Cadlina laevis* passes through a transitory veligerlike stage with a thin shell and rudimentary velum. Both of these structures regress during subsequent embryogenesis, while other larval features, such as the larval retractor muscle and operculum, never appear. Undoubtedly, these lecithotrophic larval forms have evolved from planktotrophic ancestors. Consequently, planktotrophy should be viewed as primitive or ancestral in this group (and probably in most other invertebrate higher taxa). This is not simply a matter of semantics, for the differing larval types displayed by even closely related species exert potent and far-reaching ecological and evolutionary consequences for individuals in terms of their ultimate fecundity, fitness, ability to colonize new habitats, and capacity to adapt genetically and physiologically to each local environment. Further, there are clear biogeographic trends in particular larval strategies that apply across the phyla and have yet to be satisfactorily explained. Generally, the larval strategies of benthic marine invertebrate species show marked latitudinal and depth gradients. There is a high proportional incidence of planktonic feeding larvae in shallow, low-latitude waters and an increase toward prevalence of nonplanktonic, nonfeeding embryos and larvae with increasing depth and latitude. For the reproducing individual (upon which natural selection acts), there are major ecological constraints, costs, and benefits attributable to each larval type.

Benefits of planktotrophy. These include minimal energetic investment by the parent in the egg itself and any protective capsule, both of which permit high numerical fecundity; rapid development to a hatching larva, thereby minimizing the embryo's exposure to benthic predators or infection; colonization of disjunct new habitats; escape from deteriorating local conditions; and minimal sibling-sibling or sibling-parent competition as benthic adults.

Costs of planktotrophy. These include lack of parental control over the microhabitat occupied by the offspring; enforced dispersal away from a suitable habitat (the latter implied by the reproductive success of the parent); high per-capita mortality of offspring in the plankton; and high larval mortality due to failure to locate an appropriate substratum.

Benefits of nonpelagic lecithotrophy. Offspring are not uncontrollably removed from the parental microhabitat; suitable local conditions may continue to be exploited; per-capita mortality of the embryos is low due to lack of exposure in plankton; there is potential for adult protection via brooding of developing embryos through to hatching.

Costs of nonpelagic lecithotrophy. The cost of parental investment in each egg and its protective capsule is high; there is maximal exposure to benthic mortality agents (in the absence of brood protection); fecundity is low; there is no escape from fluctuating or deteriorating local conditions; sibling-sibling or sibling-parent competition in adults is potentially high.

Pelagic lecithotrophy. Accordingly, it might appear that pelagic lecithotrophy presents the supreme compromise in that the hatching form is a free-swimming larva of an advanced state of development, which is pelagic for only a few hours or days and which does not require extrinsic planktonic nutrition to complete development. Such larvae (examples include the nudibranch mollusks *Adalaria proxima*, *Phestilla sibogae*, and *Tritonia hombergi*) develop to hatching moderately quickly, hatch from intermediate-sized eggs, and confer intermediate numerical fecundity on the parent. In consequence, there is an apparent minimization of the mortality risks in both the capsular (benthic) and planktonic (larval) periods of development. For *A. proxima*, fecundity (~1000 eggs) is intermediate between comparable planktotrophic (~100,000) and nonpelagic (~100) lecithotrophic species, and the obligatory phase is short, of perhaps 1–2 days duration. Nonetheless, the delay phase may last for at least several weeks. By means of this strategy, therefore, a varying degree of planktonic larval dispersal (perhaps up to ~60 mi or 100 km) might be predicted.

One critical evolutionary question is whether the pelagic lecithotrophic strategy is a stable specialization in its own right or merely an intermediate step between planktotrophy and nonpelagic lecithotrophy. In *P. sibogae* the larva differentiates a functional gut but does not need to feed during the brief free-swimming planktonic stage in order to complete development and metamorphosis to the adult form. Nevertheless, if *P. sibogae* larvae are held in the delay phase and fed in culture, they do accrete tissue mass over a period of up to several weeks. *Adalaria proxima* larvae also differentiate a functional gut, but somatic growth does not occur. In contrast, *T. hombergi* larvae do not differentiate a functional gut, and hatch as poorly swimming larvae that metamorphose to the carnivorous adult form soon after release from the capsule. Therefore, it is plausible that these species do illustrate a sequence of the order *Phestilla-Adalaria-Tritonia* in the evolution toward nonpelagic lecithotrophy.

Role of dispersal and genetic studies. Many authors have counterintuitively ascribed dispersal (and hence the colonization of new, disjunct habitats) as being the primary role or advantage of the pelagic strategy. The paradox is that evolution of larval forms among a wide range of taxa has generally been away from the primitive planktotrophy. Ultimate larval success depends entirely upon the encounter of the correct substratum upon which to metamorphose at the completion of development. Dispersal is, therefore, an unavoidable consequence of any planktotrophic strategy. Extant planktotrophic species have, however, clearly overcome the disadvantages, and display some level of consistent reproductive success.

Larval transport, and hence gene flow, prevents genetic differentiation of populations by either selection or random genetic drift. Knowing the duration of the planktonic phase for a given species, one can predict the dispersal potential conferred by that larva and the scale of population differentiation in the field. Accordingly, the genetic structure of sympatric populations of *A. proxima* and *Goniodoris nodosa* (an ecologically comparable species, but one with a planktotrophic larva of a 3-month obligatory pelagic phase) has been examined around the northern coastline of the British Isles. Differentiation of populations was predicted on scales of about 60 mi (100 km) and 600 mi (1000 km) for *A. proxima* and *G. nodosa* respectively. Genetic studies convincingly showed no differentiation of populations for *G. nodosa* over about 930 mi (1500 km) but demonstrated extraordinarily high levels of differentiation of populations of *A. proxima* on a scale as small as 2 mi (3 km). More importantly, however, the gene frequencies for particular *A. proxima* populations did not change significantly between generations. Because no adults survive beyond the annual spawning and because establishment of the succeeding generation arises from recruitment of these pelagic larvae, it is clear that the veligers essentially settle in the same microhabitat in which the parents laid the spawn masses. The behavior of *A. proxima* larvae appears to be such that planktonic dispersal is actually avoided or precluded—a finding that is totally at variance with the commonly perceived function of pelagic larvae.

Genetic studies have also shown *A. proxima* (pelagic lecithotrophic) to be closely related to, and to share a recent common evolutionary ancestry with, *Onchidoris muricata* (planktotrophic). The deduction is that *A. proxima* is the more advanced derivative. A question that remains, however, is whether pelagic lecithotrophy is merely transitional between planktotrophy and nonpelagic development or whether it is an evolutionarily stable adaptation. Other ultrastructural and biochemical studies suggest that the major obstacles to the transition from planktotrophy to obligate lecithotrophy lie at the subcellular level in the endocytotic-lysosomal system; necessarily there must be a shift from adaptation to particulate larval feeding to the mobilization of yolk reserves in the digestive gland diverticula. Many other features of the larval morphology and physiology, metamorphic behavior, and factors relating to the protection of developing embryos must accompany these biochemical shifts. The critical questions to be considered involve dispersal advantage. In order that they may be selectively favored during any evolutionary transition, the evolutionarily intermediate stages must confer increased fitness over their precursors. Given that, and accepting that nonpelagic lecithotrophy acts to minimize dispersal potential of the reproductive propagules, it is clear that larval dispersal cannot be the exclusive and predominant advantage of pelagic life and that many species have evolved away from this ancestral state. Indeed, it has been shown that the behavior of *A. proxima* larvae is geared to the avoidance of larval dispersal. That is not to say that larval dispersal is always disadvantageous. For those species whose fecundity offsets planktonic mortality, there may still be advantage in retaining the primitive mode despite the putative advantage of lecithotrophy.

For background information SEE BIOLOGICAL PRODUCTIVITY; CILIOPHORA; FOOD WEB; FRESH-WATER ECOSYSTEM; MARINE ECOSYSTEM; MOLLUSCA; TINTINNIDA; ZOOPLANKTON in the McGraw-Hill Encyclopedia of Science & Technology.

C. D. Todd

Bibliography. J. R. Beaver and T. L. Crisman, The role of ciliated protozoa in pelagic freshwater ecosystems, *Microb. Ecol.*, 17:111–136, 1989; J. N. Havenhand et al., Estimates of biochemical genetic diversity within and between the nudibranch molluscs *Adalaria proxima* (A & H) and *Onchidoris muricata* (Müller), *J. Exp. Mar. Biol. Ecol.*, 95:105–

111, 1986; S. C. Kempf and C. D. Todd, Feeding potential in the lecithotrophic larvae of *Adalaria proxima* and *Tritonia hombergi*: An evolutionary perspective, *J. Mar. Biol. Ass. U.K.*, 69:659–682, 1989; D. H. Lynn and D. J. S. Montagnes, Taxonomic descriptions of some conspicuous species of strobilidiine ciliates (Ciliophora: Choreotrichida) from the Isles of Shoals, Gulf of Maine, *J. Mar. Biol. Ass. U.K.*, 68:639–658, 1988; D. J. S. Montagnes et al., The annual cycle of heterotrophic planktonic ciliates in the waters surrounding the Isles of Shoals, Gulf of Maine: An assessment of their trophic role, *Mar. Biol.*, 99:21–30, 1988; P. G. Moore and R. Seed, *Ecology of Rocky Coasts*, 1985; P. C. Reid et al. (eds.), *Protozoa and Their Role in Marine Processes*, 1991; D. K. Stoecker et al., Abundance of autotrophic, mixotrophic and heterotrophic planktonic ciliates in shelf and slope waters, *Mar. Ecol. Prog. Ser.*, 50:241–254, 1989; C. D. Todd et al., Genetic differentiation, pelagic larval transport and gene flow between local populations of the intertidal marine mollusc *Adalaria proxima*, *Funct. Ecol.*, 2:441–451, 1988; P. Willmer, *Invertebrate Relationships: Patterns in Animal Evolution*, 1990.

Contributors

Contributors

The affiliation of each Yearbook contributor is given, followed by the title of his or her article. An article title with the notation "in part" indicates that the author independently prepared a section of an article; "coauthored" indicates that two or more authors jointly prepared an article or section.

A

Ahlness, Jon K. *Group Supervisor, In Situ Systems, U.S. Bureau of Mines, Minneapolis, Minnesota.* SOLUTION MINING.

Alexeff, Dr. Igor. *Electrical and Computer Engineering, University of Tennessee.* MICROWAVE TUBE—coauthored.

Armbruster, Dr. Harald. *Institut für Lebensmittelverfahrenstechnik, Universität Karlsruhe, Germany.* FOOD MANUFACTURING—coauthored.

Asmussen, Dr. Jes. *Department of Electrical Engineering, Michigan State University.* PLASMA PROCESSING.

Auld, Prof. Dick L. *Department of Plant, Soil and Entomological Sciences, University of Idaho.* RAPE.

B

Baars, Prof. Bernard J. *The Wright Institute, Berkeley, California.* CONSCIOUSNESS.

Babcock, Prof. Gerald T. *Chairman, Department of Chemistry, Michigan State University.* ENZYME.

Baer, Prof. Norbert S. *Conservation Center of the Institute of Fine Arts, New York University.* ART CONSERVATION CHEMISTRY.

Bangham, Dr. Charles R. M. *Molecular Immunology Group, Nuffield Department of Clinical Medicine, University of Oxford, England.* ANIMAL VIRUS—in part.

Barmatz, Dr. Martin. *Technical Group Leader, Jet Propulsion Laboratory, California Institute of Technology.* ACOUSTIC LEVITATION.

Barrett, Dr. J. Carl. *Chief, Laboratory of Molecular Carcinogenesis, Department of Health and Human Services, National Institutes of Health, Research Triangle Park, North Carolina.* CELL SENESCENCE.

Baulieu, Prof. Étienne-Émile. *Département de Chimie Biologique, Université Paris-Sud, France.* ANTIFERTILITY AGENT.

Beachy, Dr. Roger N. *Department of Biology, Washington University, St. Louis, Missouri.* PLANT VIRUSES AND VIROIDS—coauthored.

Beaman, Prof. Joseph. *Department of Mechanical Engineering, University of Texas, Austin.* MANUFACTURING PROCESSES—coauthored.

Belford, Dr. Heather G. *Department of Biological Chemistry, University of California, Irvine.* RIBONUCLEIC ACID (RNA)—coauthored.

Benson, Dr. Frank A. *Department of Electronic and Electrical Engineering, University of Sheffield, England.* CROSSTALK—coauthored.

Bercaw, Prof. John E. *Division of Chemistry and Chemical Engineering, California Institute of Technology.* ORGANOMETALLIC COMPOUND—coauthored.

Bergles, Prof. Arthur E. *Dean, School of Engineering, Rensselaer Polytechnic Institute.* COMPUTER.

Bezdek, Dr. James C. *Computer Science, University of West Florida, Pensacola.* FUZZY SETS AND SYSTEMS.

Biermann, Dr. Brenda J. *Department of Botany, University of California, Berkeley.* PLANT HORMONES.

Billinton, Dr. Roy. *Associate Dean of Graduate Studies, Research and Extension, University of Saskatchewan, Saskatoon, Canada.* ELECTRIC POWER SYSTEMS.

Bissell, Dr. Mina J. *Director, Cell and Molecular Biology Division, Lawrence Berkeley Laboratory, University of California, Berkeley.* DEVELOPMENTAL BIOLOGY—coauthored.

Blobel, Dr. Gunter. *Laboratory of Cell Biology, Rockefeller University.* PROTEIN—in part.

Borchers, Dr. Jeffrey G. *Department of Forest Science, Oregon State University.* FOREST AND FORESTRY—coauthored.

Bottjer, Dr. David J. *Department of Geological Sciences, University of Southern California.* ANIMAL EVOLUTION—coauthored.

Bourell, Prof. David. *Department of Mechanical Engineering, University of Texas, Austin.* MANUFACTURING PROCESSES—coauthored.

Brock, Dr. Thomas G. *Department of Human Genetics, University of Michigan Medical School.* TISSUE CULTURE.

Burdett, Dr. Jeremy K. *Searle Chemical Laboratory, University of Chicago.* SOLID-STATE CHEMISTRY.

Burke, Dr. Laura I. *Department of Industrial Engineering, Lehigh University.* OPERATIONS RESEARCH.

C

Camm, Dr. Edith. *Assistant Professor, Department of Botany, University of British Columbia, Vancouver, Canada.* CHLOROPHYLL—in part.

Carter, Dr. David L. *Supervisory Soil Scientist, Soil and Water Management Research, Agricultural Research Service, U.S. Department of Agriculture, Kimberly, Idaho.* SOIL CONSERVATION.

Carter, Dr. John V. *Department of Horticultural Science and Landscape Architecture, University of Minnesota.* COLD HARDINESS (PLANT).

Chaffin, Prof. Don B. *Director, Center for Ergonomics, University of Michigan.* ERGONOMICS.

Charny, Dr. Caleb K. *Albert Nerken School of Engineering, The Cooper Union, New York, New York.* MEDICAL CHEMICAL ENGINEERING.

Cheng, Dr. Edward S. *Goddard Space Flight Center, National Aeronautics and Space Administration, Greenbelt, Maryland.* COSMIC BACKGROUND RADIATION.

Cheryan, Prof. Munir. *Department of Food Science, University of Illinois, Urbana.* FOOD ENGINEERING—in part.

Chevray, Prof. Rene. *Chairman, Department of Mechanical Engineering, Columbia University.* FRACTALS.

Chin, Dr. Nai-Xun. *Department of Medicine, College of Physicians and Surgeons of Columbia University.* ANTIBIOTIC—coauthored.

Chu, Dr. Steven. *Department of Physics and Applied Physics, Stanford University.* ATOM.

Cook-Deegan, Dr. Robert Mullan. *Institute of Medicine,*

National Academy of Sciences, Washington, D.C. Human genetics—coauthored.

Corner, Dr. Eric D. S. *Retired; formerly, Plymouth Marine Laboratory, Plymouth, England.* Invertebrate physiology.

Costantino, Dr. James. *School of Information Technology and Engineering, George Mason University, Fairfax, Virginia.* Transportation engineering.

Cotter, Prof. W. Donald. *Division of Chemistry and Chemical Engineering, California Institute of Technology.* Organometallic compound—coauthored.

Couch, Dr. N. R. *GEC-Marconi Limited, Hirst Research Centre, Wembley, Middlesex, England.* Microwave solid-state devices—coauthored.

Courtens, Dr. Eric. *IBM Research Division, Zurich Research Laboratory, Ruschlikon, Switzerland.* Fractons—coauthored.

Crane, Dr. Kathleen. *Department of Geology and Geography, Hunter College, New York, New York.* Arctic Ocean—coauthored.

Crosby, Prof. E. Johansen. *Department of Chemical Engineering, University of Wisconsin, Madison.* Food engineering—in part.

Cudd, Dr. P. A. *Department of Electronic and Electrical Engineering, University of Sheffield, England.* Crosstalk—coauthored.

Cutler, Dr. Horace G. *Research Leader, Plant Physiology Research Unit, Agricultural Research Service, U.S. Department of Agriculture, Athens, Georgia.* Pesticide.

D

Dameron, Dr. Charles T. *Hematology-Oncology Division, University of Utah Medical Center.* Bioinorganic chemistry—coauthored.

D'Angelo, Prof. N. *Department of Physics and Astronomy, University of Iowa.* Plasma physics.

Daniels, Anthony. *Vice President, Transportation, Morrison Knudsen Corporation, San Francisco, California.* Railroad engineering.

D'Antonio, Dr. Peter. *RPG Diffusor Systems, Inc., Largo, Maryland.* Teleconferencing.

Davies, Dr. Matthew D. *Department of Biochemistry, University of Illinois, Urbana.* Genetic engineering—coauthored.

Deckard, Dr. Carl R. *DTM Corporation, Austin, Texas.* Manufacturing processes—in part.

Deom, Dr. Carl M. *Department of Biology, Washington University, St. Louis, Missouri.* Plant viruses and viroids—coauthored.

DeVay, Prof. James E. *Department of Plant Pathology, University of California, Davis.* Soil—in part.

De Wilde, Dr. Michel. *Director of Molecular and Cellular Biology Department, SmithKline Beecham Biologicals, Rixensart, Belgium.* Molecular biology.

Dhindsa, Dr. Rajinder S. *Department of Biology, McGill University, Montreal, Quebec, Canada.* Plant physiology.

Disney, Dr. Ralph L. *Department of Industrial Engineering, Texas A&M University.* Queueing theory—coauthored.

Doolittle, Dr. Russell F. *Center for Molecular Genetics, University of California, San Diego.* Protein—in part.

Dressler, Dr. Alan. *The Observatories of the Carnegie Institution of Washington, Pasadena, California.* Galaxy, external—in part.

DuBay, Dr. Denis T. *North Carolina Science and Math Alliance, Raleigh.* Acid rain.

Duerr, Reimar F. *United Engineers and Constructors, Inc., Philadelphia, Pennsylvania.* Nuclear reactor—coauthored.

Duke, Dr. James A. *Chief, Germplasm Resources Laboratory, U.S. Department of Agriculture, Beltsville, Maryland.* Herbal medicine.

Dyer, Dr. Fred. *Electrical and Computer Engineering, University of Tennessee.* Microwave tube—coauthored.

E

Ellington, Dr. Andrew. *Department of Molecular Biology, Massachusetts General Hospital, Boston.* Ribozyme.

Emerman, Dr. Michael. *Assistant Member, Division of Basic Sciences and Program in Molecular Medicine, Fred Hutchinson Cancer Research Center, Seattle, Washington.* Animal virus—in part.

Erdreich, Dr. John. *Ostergaard Acoustical Associates, West Orange, New Jersey.* Noise measurement.

Etter, Dr. Ron J. *Institute of Marine and Coastal Sciences, New Brunswick, New Jersey.* Gastropoda.

F

Fagerberg, Dr. Wayne R. *Department of Plant Biology, University of New Hampshire.* Insectivorous plants.

Falk, Dr. Henry. *Centers for Disease Control, Atlanta, Georgia.* Pharmacognosy.

Fenn, Dr. Mark E. *Research Plant Pathologist, Pacific Southwest Forest and Range Experiment Station, U.S. Department of Agriculture, Riverside, California.* Soil ecology—in part.

Filipovic, Dr. Djordje. *Senior Research Associate, Department of Biochemistry, University of Illinois, Urbana.* Genetic engineering—coauthored.

Fisk, Dr. Lennard A. *Associate Administrator, Office of Space Science and Applications, National Aeronautics and Space Administration, Washington, D.C.* Biosphere.

Franklin, Dr. Jerry F. *Bloedel Professor of Ecosystem Analysis, Forest Resources Management Division, University of Washington, Seattle.* Forest management and organization.

Freckman, Dr. Diana W. *Department of Nematology, University of California, Riverside.* Soil ecology—in part.

Froes, Prof. F. H. *Director, Institute for Materials and Advanced Processes, University of Idaho.* Materials science and engineering.

Fulton, Dr. Robert W. *Professor Emeritus, Department of Plant Pathology, University of Wisconsin, Madison.* Plant pathology.

G

Ghia, Prof. Kirti N. *Department of Aerospace Engineering and Engineering Mechanics, University of Cincinnati.* Fluid dynamics—coauthored.

Ghia, Prof. Urmila. *Department of Mechanical, Industrial and Nuclear Engineering, University of Cincinnati.* Fluid dynamics—coauthored.

Gibbs, Betty. *Gibbs Associates, Boulder, Colorado.* Mining—in part.

Gilbert, Prof. Lawrence I. *Chairman, Department of Biology, University of North Carolina.* Insect physiology—coauthored.

Glynn, Prof. Peter W. *Department of Operations Research, Stanford University.* Simulation—in part.

Goldsmith, Prof. Julian R. *Department of Geophysical Sciences, University of Chicago.* Silicate systems.

Graffin, Dr. Greg. *Department of Ecology and Systematics, Cornell University.* Paleontology—in part.

Gray, Dr. Henry F. *Naval Research Laboratory, Washington, D.C.* ELECTRONICS.

Greenberg, Dr. Michael J. *Director, The Whitney Laboratory, University of Florida.* PEPTIDE.

Greer, Dr. Chris L. *Department of Biological Chemistry, University of California, Irvine.* RIBONUCLEIC ACID (RNA)—coauthored.

Grey, Dr. Jerry. *Research and Engineering Consultant, New York, New York.* SPACE TECHNOLOGY.

Griffin, Robert J., Jr. *Department of Energy, Washington, D.C.* SPACE FLIGHT.

H

Hammond, Prof. P. *Department of Electrical Engineering, Southampton University, England.* ELECTROMAGNETISM.

Hartel, Dr. Richard W. *Assistant Professor of Food Engineering, Department of Food Science, University of Wisconsin, Madison.* FOOD MANUFACTURING—in part.

Hayes, William C. *Editor, Electrical World, McGraw-Hill, Inc., New York, New York.* ELECTRICAL UTILITY INDUSTRY.

Heiligenberg, Prof. Walter. *Neurobiology Unit, Scripps Institution of Oceanography, University of California, La Jolla.* ELECTROSENSORY SYSTEMS (BIOLOGY).

Hemley, Dr. Russell J. *Geophysical Laboratory, Carnegie Institution of Washington.* HYDROGEN—coauthored.

Henrich, Dr. Vincent C. *Department of Biology, University of North Carolina.* INSECT PHYSIOLOGY—coauthored.

Hessinger, Prof. David A. *Department of Physiology and Pharmacology, Loma Linda University, Loma Linda, California.* CHEMORECEPTION.

Hessler, Dr. Robert R. *Scripps Institution of Oceanography, La Jolla, California.* DEEP-SEA FAUNA—in part.

Hill, Prof. Craig L. *Department of Chemistry, Emory University.* REACTIVE INTERMEDIATES.

Hobbs, Dr. Shaun. *Plant Biotechnology Institute, National Research Council of Canada, Saskatoon, Saskatchewan, Canada.* GENETIC ENGINEERING—in part.

Hoffman, Dr. Donald D. *Department of Cognitive Sciences, University of California, Irvine.* VISION.

Holmes, Dr. David S. *Department of Biology, Clarkson University, Potsdam, New York.* BIOHYDROMETALLURGY.

Hose, Dr. Jo Ellen. *Biology Department, Occidental College, Los Angeles, California.* BLOOD—coauthored.

Howlett, Dr. Anthony R. *Cell and Molecular Biology Division, Lawrence Berkeley Laboratory, University of California, Berkeley.* DEVELOPMENTAL BIOLOGY—coauthored.

I

Irish, Dr. Erin E. *Department of Botany, University of Iowa.* SEX DETERMINATION.

J

Jablonski, Dr. David. *Department of Geophysical Sciences, University of Chicago.* ANIMAL EVOLUTION—coauthored.

Jacobs, Dr. Louis L. *Shuler Museum of Paleontology and Department of Geological Sciences, Southern Methodist University.* PALEONTOLOGY—in part.

Jacobsen, Dr. Eric N. *School of Chemical Sciences, University of Illinois, Urbana.* ASYMMETRIC SYNTHESIS.

Jayant, N. S. *Department Head, AT&T Bell Laboratories, Murray Hill, New Jersey.* SPEECH TECHNOLOGY—in part.

Jefferies, Dr. David J. *Department of Electronic and Electrical Engineering, University of Surrey, Guildford, England.* CIRCUIT (ELECTRICITY).

Jennings, Dr. J. B. *Department of Pure and Applied Biology, University of Leeds, England.* TURBELLARIA.

Jones, Dr. Alan M. *Department of Biology, University of North Carolina.* AUXIN—in part.

Jones, Dr. Scott W. *Sibley School of Mechanical and Aerospace Engineering, Cornell University.* CHAOS.

Joshi, Dr. Sanjay. *Department of Industrial and Management Systems Engineering, Pennsylvania State University.* MANUFACTURING ENGINEERING.

K

Kaldor, Dr. Andrew. *Corporate Research, Exxon Research and Engineering Company, Annandale, New Jersey.* METAL CLUSTER COMPOUNDS.

Kearney, Dr. M. J. *GEC-Marconi Limited, Hirst Research Centre, Wembley, Middlesex, England.* MICROWAVE SOLID-STATE DEVICES—coauthored.

Kelly, Dr. M. J. *GEC-Marconi Limited, Hirst Research Centre, Wembley, Middlesex, England.* MICROWAVE SOLID-STATE DEVICES—coauthored.

Kerner, Dr. Manfred. *Institut für Lebensmittelverfahrenstechnik, Universität Karlsruhe, Germany.* FOOD MANUFACTURING—coauthored.

Klock, Dr. Benny L. *Formerly, Advanced Weapons Technology Division, Defense Mapping Agency, Washington, D.C.* CARTOGRAPHY.

Kunov, Prof. Hans. *Director, Institute of Biomedical Engineering, University of Toronto, Ontario, Canada.* ACOUSTICS.

L

Langston, Dr. J. William. *Director, California Parkinson's Foundation, San Jose.* PARKINSON'S DISEASE—coauthored.

Larock, Prof. Richard C. *Department of Chemistry, Iowa State University of Science and Technology.* ORGANIC CHEMICAL SYNTHESIS.

Larson, Prof. Richard C. *Operations Research Center, Massachusetts Institute of Technology.* QUEUEING THEORY—in part.

Lee, Dr. William C. Y. *Vice President, Research and Technology, PacTel Cellular, Irvine, California.* MOBILE RADIO.

Legates, Dr. David R. *Department of Geography, University of Oklahoma.* PRECIPITATION (METEOROLOGY).

Lehtihet, Dr. E. A. *Department of Industrial and Management Systems Engineering, Pennsylvania State University.* METHODS ENGINEERING.

Lersten, Dr. Nels R. *Department of Botany, Iowa State University.* LEAF.

Lin, Dr. Fu-Kuen. *Head, Biomedical Science, Amgen Center, Thousand Oaks, California.* ANEMIA.

Lipps, Prof. Jere H. *Director, Museum of Paleontology, University of California, Berkeley.* EXTINCTION (BIOLOGY)—in part.

Lund, Prof. Daryl B. *Interim Executive Dean of Agriculture and Natural Resources, Rutgers University.* FOOD ENGINEERING—in part.

Lynn, Dr. Denis H. *Department of Zoology, University of Guelph, Ontario, Canada.* ZOOPLANKTON—in part.

Lynott, Dr. Lawrence T. *President, Scitec Corporation, Kennewick, Washington.* MINING—in part.

M

McDonald, Keith D. *Technical Director, Navtech Seminars, Inc., Arlington, Virginia.* SATELLITE NAVIGATION SYSTEMS—in part.

Macdonald, Dr. Ken C. *Department of Geological Sciences, University of California, Santa Barbara.* MID-OCEANIC RIDGE.

Malacinski, Prof. George M. *Department of Biology, Indiana University.* EMBRYONIC INDUCTION.

Maliszewski, Raymond M. *Vice President, System Planning, American Electric Power Service Corporation, Columbus, Ohio.* TRANSMISSION LINES.

Mao, Dr. Ho-kwang. *Geophysical Laboratory, Carnegie Institution of Washington.* HYDROGEN—coauthored.

Marder, Dr. Seth R. *Beckman Institute, California Institute of Technology.* COORDINATION CHEMISTRY.

Martin, Dr. Gary G. *Biology Department, Occidental College, Los Angeles, California.* BLOOD—coauthored.

Marzilli, Prof. Luigi G. *Department of Chemistry, Emory University.* BIOTECHNOLOGY.

Mate, Dr. C. Mathew. *IBM Research Division and IBM Magnetic Recording Institute, Almaden Research Center, San Jose, California.* ATOMIC FORCE MICROSCOPE.

Matz, Dr. Samuel A. *Consultant in Food Science, McAllen, Texas.* FOOD MANUFACTURING—in part.

Mauseth, Dr. James. *Department of Botany, University of Texas, Austin.* APICAL MERISTEM.

Mayer, Sandra M. *Senior Research Assistant, Division of Biology and Medicine, Brown University.* CHLOROPHYLL—in part.

Mazer, Dr. James J. *Argonne National Laboratories, Argonne, Illinois.* DATING METHODS—coauthored.

Mears, Dr. L. D. *General Manager, Gas-Cooled Reactor Associates, San Diego, California.* NUCLEAR REACTOR—in part.

Miller, Dr. Iain M. *Department of Biological Sciences, Wright State University, Dayton, Ohio.* NITROGEN FIXATION.

Miller, Prof. Russell J. *Director, Center for Space Mining, Colorado School of Mines, Golden.* MOON.

Mills, Dr. Claudia E. *Friday Harbor Laboratories, University of Washington, Friday Harbor.* DEEP-SEA FAUNA—in part.

Modlin, Dr. Robert L. *Division of Dermatology, University of California, Los Angeles.* CELLULAR IMMUNOLOGY—coauthored.

Mollenauer, Dr. Linn F. *AT&T Bell Laboratories, Holmdel, New Jersey.* OPTICAL FIBERS.

Moore, Prof. Randy. *Chair, Department of Biological Sciences, Wright State University, Dayton, Ohio.* ROOT.

Morse, Dr. Stephen A. *Director, Division of Sexually Transmitted Diseases Laboratory Research, Center for Infectious Diseases, Centers for Disease Control, Atlanta, Georgia.* SYPHILIS.

Morris, Prof. Robert H. *Department of Chemistry, University of Toronto,, Ontario, Canada.* METAL HYDRIDES.

Morrison, Prof. Sherie L. *Molecular Biology Institute, University of California, Los Angeles.* ANTIBODY.

Munro, Dr. Paul. *Department of Information Science, University of Pittsburgh.* NEURAL NETWORK.

Murry, Dr. Lynn E. *Zoecon Research Institute, Palo Alto, California.* GENETIC ENGINEERING—in part.

N

Neeteson, Dr. J. J. *Institute for Soil Fertility Research, Wageningen, The Netherlands.* SOIL—coauthored.

Nemhauser, Prof. George L. *School of Industrial and Systems Engineering, Georgia Institute of Technology.* OPTIMIZATION.

Netravali, Dr. Arun N. *AT&T Bell Laboratories, Murray Hill, New Jersey.* TELEVISION.

Ng, Dr. Tai P. *Manager, Geophysical Research, Canadian Hunter Exploration, Ltd., Calgary, Alberta, Canada.* PETROLEUM ENGINEERING—in part.

O

Odrey, Dr. Nicholas G. *Director, Institute for Robotics, Lehigh University.* CONTROL SYSTEMS.

O'Farrell, Dr. Patrick H. *Department of Biochemistry and Biophysics, University of California, San Francisco.* CELL CYCLE.

Ohlsson, Dr. Thomas. *SIK—The Swedish Institute for Food Research, Göteborg, Sweden.* FOOD ENGINEERING—in part.

Olsen, Dr. Sandra L. *Curator, Archaeology, Virginia Museum of Natural History, Martinsville.* TAPHONOMY.

O'Malley, Dr. M. H. *Berkeley Speech Technologies, Inc., Berkeley, California.* SPEECH TECHNOLOGY—in part.

O'Neill, Dr. Hugh. *Bayerisches Geoinstitut, Universität Bayreuth, Germany.* SPINEL.

Ortoleva, Prof. Peter J. *Department of Chemistry, Indiana University.* GEOCHEMISTRY.

P

Pardoll, Dr. Drew M. *Department of Medicine, Johns Hopkins University School of Medicine.* CELLULAR IMMUNOLOGY—coauthored.

Pedersen, Tor. *United Engineers and Constructors, Inc., Philadelphia, Pennsylvania.* NUCLEAR REACTOR—coauthored.

Peleg, Prof. Micha. *Department of Food Engineering, University of Massachusetts.* FOOD MANUFACTURING—in part.

Peralta, Dr. Samuel B. *Physicist, Science Section, Research Division, Ontario Hydro, Toronto, Ontario, Canada.* NONDESTRUCTIVE TESTING—in part.

Perry, Dr. David A. *Department of Forest Science, Oregon State University.* FOREST AND FORESTRY—coauthored.

Pettygrove, Dr. G. Stuart. *Extension Soils Specialist, Department of Land, Air and Water Resources, University of California, Davis.* IRRIGATION (AGRICULTURE).

Pickens, R. Andrew. *President, AvCom, Inc., Towson, Maryland.* COMMUNICATIONS SATELLITE—in part.

Plotkin, Dr. Kenneth J. *Principal Scientist, Wyle Laboratories, Arlington, Virginia.* SONIC BOOM.

Prasad, Dr. S. N. *Department of Electrical and Computer Engineering and Technology, Bradley University, Peoria, Illinois.* REMOTE SENSING.

Prothero, Dr. Donald R. *Department of Geology, Occidental College, Los Angeles, California.* PERISSODACTYLA.

Publicover, David A. *School of Forestry and Environmental Studies, Yale University.* SOIL ECOLOGY—coauthored.

Q

Queener, Dr. Stephen W. *Senior Research Scientist and Group Leader, Cell Culture Research and Development, Eli Lilly and Company, Indianapolis, Indiana.* ANTIBIOTIC—coauthored.

R

Rader, Dr. Mark. *Electrical and Computer Engineering, University of Tennessee.* Microwave tube—coauthored.

Rasmussen, Dr. Tab. *Department of Anthropology, University of California, Los Angeles.* Primates.

Rebek, Prof. Julius, Jr. *Department of Chemistry, Massachusetts Institute of Technology.* Deoxyribonucleic acid (DNA).

Reddy, Dr. A. S. N. *Laboratory of Plant Molecular Biology and Physiology, Department of Horticulture, Washington State University.* Auxin—in part.

Rhodes, Prof. Billy B. *Department of Horticulture, Clemson University.* Watermelon.

Rice, Prof. Don S. *Department of Anthropology, University of Virginia.* Paleolimnology.

Roach, Dr. W. T. *Retired; formerly, Meteorological Research Flight, Royal Aerospace Establishment, Farnborough, Hants, England.* Fog.

Rocklage, Dr. Scott M. *President, Salutar, Inc., Sunnyvale, California.* Medical imaging.

Rommens, Dr. Johanna. *Department of Genetics, The Hospital for Sick Children, Toronto, Ontario, Canada.* Cystic fibrosis—coauthored.

S

Sager, Prof. Ruth. *Chief, Division of Cancer Genetics, Dana-Farber Cancer Institute, Boston, Massachusetts.* Tumor.

Saint, Capt. Sam. *Aviation Consultant, Westerly, Rhode Island.* Wind shear.

Schowalter, Dr. Timothy D. *Department of Entomology, Oregon State University.* Forest ecology.

Schwartzberg, Prof. Henry G. *Department of Food Science, University of Massachusetts.* Food engineering—in part.

Scully, Dr. Brian E. *Department of Medicine, College of Physicians and Surgeons of Columbia University.* Antibiotic—coauthored.

Sebestyen, George S. *President, Defense Systems, Inc., McLean, Virginia.* Communications satellite—in part.

Segado, Prof. Roberto R. *Department of Food Science, University of Alberta, Edmonton, Canada.* Food engineering—in part.

Shannon, Prof. Robert E. *Department of Industrial Engineering, Texas A&M University.* Simulation—in part.

Shaw, Brian R. *BHP Petroleum (Americas) Inc., Houston, Texas.* Petroleum engineering—in part.

Sheridan, Prof. Thomas B. *Man-Machine Systems Laboratory, Department of Mechanical Engineering, Massachusetts Institute of Technology.* Telerobotics.

Sisler, Prof. Edward C. *Department of Biochemistry, North Carolina State University.* Ethylene.

Skatrud, Dr. Paul L. *Cell Culture Research and Development, Eli Lilly and Company, Indianapolis, Indiana.* Antibiotic—coauthored.

Sligar, Prof. Stephen G. *Department of Biochemistry, University of Illinois, Urbana.* Genetic engineering—coauthored.

Smithson, Dr. T. R. *Cambridge Regional College, Cambridge, England.* Carboniferous.

Snider, Duane E. *Water Jet Specialties, Inc., Burlington, Ontario, Canada.* Welding and cutting of metals.

Southworth, Prof. Darlene. *Department of Biology, Southern Oregon State College.* Reproduction (plant).

Spencer, Dr. Dorothy M. *PROIMI, San Miguel de Tucumán, Argentina.* Fungi—coauthored.

Spencer, Dr. J. F. T. *PROIMI, San Miguel de Tucumán, Argentina.* Fungi—coauthored.

Spies, Dr. Thomas A. *Research Forester, Forestry Sciences Laboratory, U.S. Department of Agriculture, Corvallis, Oregon.* Forest and forestry—in part.

Spyers-Duran, Paul. *Research Aviation Facility, National Center for Atmospheric Research, Boulder, Colorado.* Meteorological instrumentation.

Stansell, Thomas A., Jr. *Staff Vice President and Director of Advanced Programs, Magnavox Government and Industrial Electronics Company, Torrance, California.* Satellite navigation systems—in part.

Stevens, Dr. Todd O. *Terrestrial Sciences Section, Environmental Sciences Department, Battelle Pacific Northwest Laboratories, Richland, Washington.* Soil microbiology.

Stevenson, Dr. Christopher. *Diffusion Laboratories, Spring Mills, Pennsylvania.* Dating methods—coauthored.

Stoloff, Leonard. *Retired; formerly, Food and Drug Administration, Silver Spring, Maryland.* Mycotoxin.

Streuli, Dr. Charles H. *Cell and Molecular Biology Division, Lawrence Berkeley Laboratory, University of California, Berkeley.* Developmental biology—coauthored.

Strohl, Robert J. *Research Scientist, Battelle Memorial Institute, Columbus, Ohio.* Military aircraft.

Strouse, Prof. Charles E. *Department of Chemistry and Biochemistry, University of California, Los Angeles.* Porphyrin.

Stucker, Prof. R. E. *Department of Agronomy and Plant Genetics, University of Minnesota—Twin Cities, St. Paul.* Rice.

Subramanian, Dr. P. M. *Senior Research Fellow, Polymer Products Department, E. I. du Pont de Nemours and Company, Inc., Wilmington, Delaware.* Polymer.

Sues, Dr. Hans-Dieter. *Department of Paleobiology, National Museum of Natural History, Smithsonian Institution, Washington, D.C.* Extinction (biology)—in part.

Sundvor, Dr. Eirik. *Seismological Observatory, University of Bergen, Norway.* Arctic Ocean—coauthored.

Swift, Gregory W. *Physics Division, Los Alamos National Laboratory, Los Alamos, New Mexico.* Thermoacoustic engines.

Swift, Dr. James H. *Scientific Director, Scripps Institution of Oceanography, Oceanographic Data Facility, University of California, San Diego.* Arctic Ocean—in part.

Sykes, Frederick E. *Senior Design Engineer, Gennum Corporation, Burlington, Ontario, Canada.* Electronic power supply.

T

Tang, Dr. Cha-Mei. *Naval Research Laboratory, Washington, D.C.* Laser.

Tanner, Dr. Caroline M. *California Parkinson's Foundation, San Jose.* Parkinson's disease—coauthored.

Tecce, Dr. J. J. *Department of Psychology, Boston College.* Event-related potentials; Psychology, physiological and experimental.

Tien, Prof. Chang-Lin. *Department of Mechanical Engineering, University of California, Berkeley.* Radiative transfer.

Timmons, Dr. Dale M. *Senior Geologist/Corporate Marketing, Geosafe Corporation, Kirkland, Washington.* Hazardous waste disposal.

Todd, Dr. Christopher D. *Department of Biology and Preclinical Medicine, Gatty Marine Laboratory, University of St. Andrews, Scotland.* Zooplankton—in part.

Toledo, Dr. Romeo T. *Professor of Food Process Engineering, Department of Food Science and Technology, University of Georgia.* Food manufacturing—in part.

Tsui, Dr. Lap-Chee. *Senior Scientist, Department of Genetics, The Hospital for Sick Children, Toronto, Ontario, Canada.* Cystic fibrosis—coauthored.

U

Underwood, Dr. Ervin E. *Professor Emeritus, School of Materials Engineering, Georgia Institute of Technology.* METALLOGRAPHY.

V

Vacher, Prof. Rene. *Director of Research, Laboratoire de Science des Matériaux Vitreux, Université Montpellier, France.* FRACTONS—coauthored.

van Veen, Dr. J. A. *Institute for Soil Fertility Research, Wageningen, The Netherlands.* SOIL—coauthored.

Vaughan, Dr. Martin A. *Department of Biology, Rochester Institute of Technology.* PLANT MOVEMENTS.

Ventura, Dr. Jose A. *Department of Industrial and Management Systems Engineering, Pennsylvania State University.* NONDESTRUCTIVE TESTING—in part.

Verbeke, Dr. Judith A. *Department of Plant Sciences, University of Arizona.* FLOWER.

Vogel, Dr. Thomas A. *Department of Geological Sciences, Michigan State University.* MAGMA.

Vogt, Dr. Kristiina. *Director, Program in Belowground Ecology, School of Forestry and Environmental Studies, Yale University.* SOIL ECOLOGY—coauthored.

Vogt, Dr. Peter. *Naval Research Laboratory, Washington, D.C.* ARCTIC OCEAN—coauthored.

W

Walker, Prof. Arthur B. C., Jr. *Center for Space Science and Astrophysics, Stanford University.* SUN.

Watson, Dr. James Dewey. *Director, National Center for Human Genome Research, National Institutes of Medicine, Bethesda, Maryland.* HUMAN GENETICS—coauthored.

Weiss, Dr. Alfred. *President, Mineral Systems International, Stamford, Connecticut.* MINING—in part.

Weiss, Dr. Gideon. *School of Industrial and Systems Engineering, Georgia Institute of Technology.* PRODUCTION PLANNING—in part.

Whitt, Dr. Ward. *AT&T Bell Laboratories, Murray Hill, New Jersey.* QUEUEING THEORY—in part.

Wiederrecht, Dr. Gary P. *Department of Chemistry, Massachusetts Institute of Technology.* ULTRAFAST MOLECULAR PROCESSES.

Wilhelm, Prof. Wilbert E. *College of Engineering, Texas A&M University.* PRODUCTION PLANNING—in part.

Williams, Dr. Sloan R. *Los Alamos National Laboratory, Los Alamos, New Mexico.* HUMAN GENETICS—in part.

Winge, Prof. Dennis R. *Hematology-Oncology Division, University of Utah Medical Center.* BIOINORGANIC CHEMISTRY—coauthored.

Wortman, Dr. M. A. *Department of Industrial Engineering, Texas A&M University.* QUEUEING THEORY—coauthored.

Wyman, Dr. Richard E. *Director of Research, Canadian Hunter Exploration, Ltd., Calgary, Alberta, Canada.* NATURAL GAS.

Y

Yang, Prof. Wen-Jei. *Department of Mechanical Engineering and Applied Mechanics, University of Michigan.* FLUID FLOW.

Yao, Prof. David D. *Department of Industrial Engineering and Operations Research, Columbia University.* PRODUCTION PLANNING—in part.

Index

Index

Asterisks indicate page references to article titles.

P

Paintings: wall painting conservation 30–32
Paleobotany 465–466
Paleoecology: animal evolution 8–11*
   paleolimnology 317–320*
   taphonomy 464–466*
Paleolimnology 317–320*
   applications 318
   Central Petén historical ecology project 318–320
   ecosystems and human disturbance 317–318
   study of lake and its drainage basin 317–318
Paleontology 320–324*
   animal evolution 8–11*
   Carboniferous 54–55*
   cladistics and divergence dates 321–322
   current opinions 323
   determining original vertebrate environment 322
   earliest vertebrate environment 322–324
   evolutionary clock 320–322
   extinction (biology) 137–143*
   fossils and geologic time 321
   fresh-water school 322–323, 324
   marine school 323
   Perissodactyla 328–330*
   preserved molecular fossils 322
   primates 363–366*
   studies of Harding Sandstone 322–324
   taphonomy 464–466*
   techniques to study molecular differences 321
Panax quinquefolius 202
Pangaea 141, 142
Panicum maximum 216
Parapithecus 364
Paratylenchus 434
Parkinson's disease 324–326*
   clinical trials of selegiline 325
   clues from parkinsonism-causing toxicant 324–325
   monoamine oxidase connection 325
   prospects 325
   protection or symptomatic treatment 325
Pauling, Linus 106
Penicillin: molecular genetics of biosynthesis 16–18
Pennes, H. 236
Peptide 326–327*
   functional significance of peptide families 327
   origin and diversity of peptide families 326
   phyletic distribution of peptide families 326–327

Perissodactyla 328–330*
   brontotheres (titanotheres) 329
   conservation 329
   horses 328–329
   origins 328
   tapirs, rhinoceroses, and chalicotheres 329
Pesticide 330–333*
   fungal substances affecting fungi 333
   fungal substances affecting plants 331
   marine substances affecting fungi 333
   microbial degradation 434–436
   plant substances affecting fungi 332–333
   plant substances affecting insects and nematodes 331–332
   plant substances affecting marine life 333
   plant substances affecting plants 330–331
Petroleum engineering 333–338*
   computer mapping engineering computing 334
   future of computer mapping 335
   geological computing 334
   geophysical computing 334
   hardware considerations 334–335
   production geophysics seismic methods 335
   seismic technology 335–336
Pharmacognosy 338–339*
   causative agent of eosinophilia-myalgia syndrome 339
   epidemiology of eosinophilia-myalgia syndrome 338–339
   symptoms of eosinophilia-myalgia syndrome 338
Phillips, R. M. 219
Phonon: fractons 176–178*
Photosynthesis 171, 346–347, 360, 476–477
   chlorophyll 68–74*
Phestilla sibogae 499, 500
Physalis 202
Physella virgata virgata 189
Pistillata 414–415
PIUS reactor 297–301
   arrangement of components 298
   construction and operation 300
   design verification 300
   passive shutdown principle 297–298, 300
   potential cogeneration plant 300–301
   safety performance 300
Plakortis angulospiculatus 332
Planar-doped barrier diode 256

Plankton: zooplankton 496–501*
Plant growth: apical meristem 22–24*
   ethylene 133–135*
   use of plant transformation 193
Plant hormones 339–341*
   auxin 40–44*
   calcium as plant secondary messenger 340–341
   gibberellic acid and sex determination 413–414
   hormone effects on secondary messengers 341
   hormone receptors 339–340
   plant signal transduction components 340
   secondary messenger system 339, 340
Plant movements 341–343*
   classification of gravitropism 341
   insectivorous plants 211–213*
   perception of gravity stimulus 342
   root vs. shoot gravitropism 406–407
   transduction of stimulus and response 342–343
Plant pathology 343–346*
   commercial use of cross protection 343–345
   drawbacks to commercial cross protection 343–344
   molecular basis for cross protection 345
   see also Plant viruses and viroids
Plant physiology 346–349*
   chlorophyll 68–74*
   cold hardiness 76–78*
   ethylene 133–135*
   high-temperature stress 347
   low-temperature stress 347–348
   mechanical stress 348
   plasma membrane in stress physiology 346–349
   salinity stress 348
   water deficit (drought) stress 346–347
Plant transformation 191–196*
Plant viruses and viroids 349–351*
   genes involved in virus movement 349–350
   host genes affecting movement 351
   virus-plant interaction 349
   viruses without movement function 350–351
   see also Plant pathology
Plasma physics 351–353*
   past experiments in Q-machines 352
   recent and current work in Q-machines 352
   space-related researches 352

Plasma physics—*cont.*
   transport properties 352
   waves and instabilities 352
Plasma processing 353–356*
   applications 356
   barrel etch reactor 354–355
   electron and gas temperatures 354
   electron cyclotron resonance reactor 355–356
   parallel-plate reactor 355
   plasma reactors 354
   processing plasmas 353–354
   surface reactions 354
Plastics: polymer 356–358*
Platyhelminthes 483–484
Pleuroncodes planipes 213
Podophyllum hexandrum 202
Pollution: air *see* Air pollution
   emission standards for electric utility 113
   hazardous waste disposal 199–202*
   plastics recycling 356–358
   toxic metal wastes 46
Polymer 356–358*
   plastics recycling 356–357
   prospects for polymer recycling 358
Polymerase chain reaction: ancient DNA studies 205–206
Pondaungia 363
Porphyrin 358–361*
   applications 359–360
   dynamics 359
   polymeric sponges 359
   programming a porphyrin sponge 358–359
Precambrian: biotic transition 137–140
Precipitation (meteorology) 361–363*
   future advances 363
   improved global precipitation climatology 362–363
   rain-gage biases 361–362
   spatial distributions of rain gages 361
   see also Climatology; Fog
Pregnancy: antifertility agent 21–22*
Primates 363–366*
   discoveries outside Egypt 364–365
   early anthropoid fossils 363
   Fayum primates 363
   prosimian ancestors of anthropoids 365
   prosimian-anthropoid transition 365
   world's earliest anthropoids 364
Production geophysics 335–337*
   seismic methods 335
   seismic technology 335–336
Production planning 366–372*
   conservation laws and polymatroid structure 371–372